Enhanced WebAssign®

Your Instant Access to Practice, Problem Solving, and Mastering Concepts

Looking for more practice with problem solving? The Enhanced WebAssign online homework course that accompanies *Elementary Algebra* includes key elements from the textbook to help support you. Whether working in the textbook or online, you can access:

➤ The *five steps* as you practice solving a wide variety of application problems.

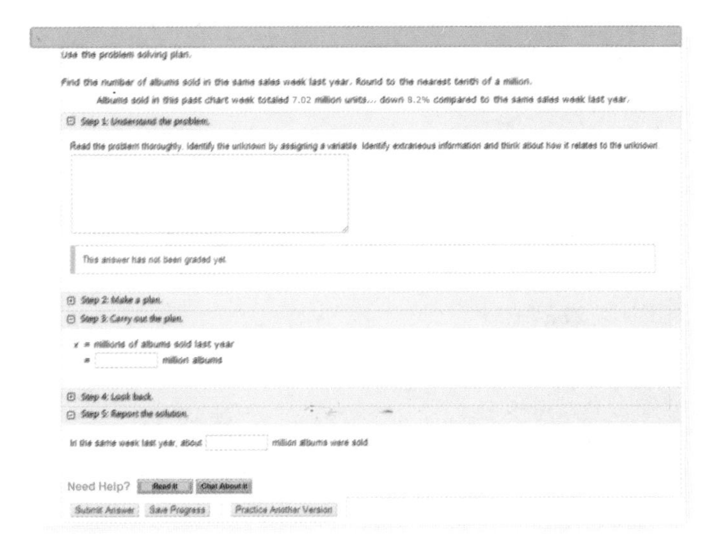

➤ *Find the Mistake* problems to help you think critically about your work and correct it as you analyze common errors.

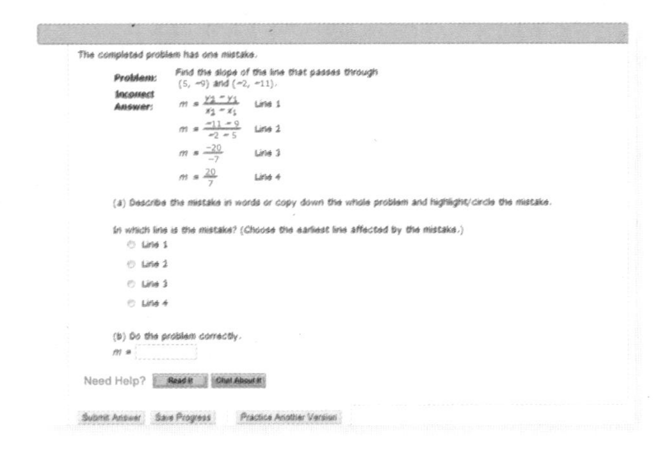

In addition, Enhanced WebAssign includes other key problem types chosen from the textbook to give you support in developing and refining problem-solving, critical thinking, and study skills. Tutorial videos and a customizable and interactive eBook also enhance your learning experience. Visit www.webassign.net to learn more.

ELEMENTARY ALGEBRA

LAURA J. BRACKEN
LEWIS-CLARK STATE COLLEGE

EDWARD S. MILLER
LEWIS-CLARK STATE COLLEGE

Australia • Brazil • Japan • Korea • Mexico • Singapore • Spain • United Kingdom • United States

Elementary Algebra
Laura J. Bracken, Edward S. Miller

Acquisitions Editor: Rita Lombard
Senior Developmental Editor: Erin Brown, Katherine Greig
Assistant Editor: Lauren Crosby
Senior Editorial Assistant: Jennifer Cordoba
Managing Media Editor: Heleny Wong
Associate Media Editor: Guanglei Zhang
Senior Brand Manager: Gordon Lee
Senior Market Development Manager: Danae April
Senior Content Project Manager: Cheryll Linthicum
Senior Art Director: Vernon Boes
Senior Manufacturing Planner: Becky Cross
Rights Acquisitions Specialist: Roberta Broyer
Production Service: Martha Emry
Photo Researcher: Terri Wright
Text Researcher: Terri Wright
Copy Editor: Barbara Willette
Art Editor: Leslie Lahr
Illustrator: Chris Ufer, Graphic World, Inc.
Text Designer: Diane Beasley
Cover Designer: Irene Morris
Cover Image: © iStockphoto.com/ Kyoungil Jeon
Compositor: Lachina Publishing Services

Library of Congress Control Number: 2012953159
ISBN-13: 978-0-618-95134-5
ISBN-10: 0-618-95134-2

Brooks/Cole
20 Davis Drive
Belmont, CA 94002-3098
USA

Cengage Learning is a leading provider of customized learning solutions with office locations around the globe, including Singapore, the United Kingdom, Australia, Mexico, Brazil, and Japan. Locate your local office at **www.cengage.com/global.**

Cengage Learning products are represented in Canada by Nelson Education, Ltd.

To learn more about Brooks/Cole, visit **www.cengage.com/brookscole.**

Purchase any of our products at your local college store or at our preferred online store **www.cengagebrain.com.**

Printed in the United States of America
1 2 3 4 5 6 7 17 16 15 14 13

To Tony and the rest of
the Hogan's family, who have
supported us in many ways as
we have written this book.

About the Authors

Courtesy Laura Bracken

LAURA BRACKEN, co-author of *Investigating Prealgebra and Investigating College Mathematics*, teaches developmental mathematics at Lewis-Clark State College. As developmental math coordinator, Laura led the process of developing objectives, standardizing assessments, and enforcing placement including a mastery skill quiz program. She has worked collaboratively with science faculty to make connections between developmental math and introductory science courses. Laura has presented at numerous national and regional conferences and currently serves as the regional representative for the AMATYC Placement and Assessment Committee. Her blog, *Dev Math Diary*, is at http://devmathdiary.wordpress.com/

Courtesy Laura Bracken

ED MILLER is a professor of mathematics at Lewis-Clark State College. He earned his PhD in general topology at Ohio University in 1989. He teaches a wide range of courses, including elementary and intermediate algebra. His terms as chair of the General Education Committee, the Curriculum Committee, and the Division of Natural Sciences and Mathematics have spurred him to look at courses as part of an integrated whole rather than discrete units. A regular presenter at national and regional meetings, Ed is exploring the use of multiple choice questions as a teaching tool as well as an assessment tool.

About the Cover

The cover image symbolizes the importance of self-empowerment or "stepping up" in order to achieve your goals. Success requires independence, persistence, responsibility, and a willingness to put your "best foot" forward—all key skills this developmental mathematics series seeks to foster.

Dear Colleagues,

Several years ago, we began writing materials to supplement our developmental mathematics textbooks and improve student success in our classes. Although we work at a four-year state college, it essentially offers open admission, serving as a community college and technical school for a large region. About 60% of our students must take developmental mathematics; many are enrolled in developmental English courses and are first-generation college students.

When we started to write *Elementary Algebra*, we identified the range of conceptual and procedural knowledge that students should learn to be prepared for their next mathematics course, whether that is intermediate algebra, introductory statistics, liberal arts mathematics, or quantitative literacy. We also considered how we could support students as they learn how to be successful in college. We used responses from faculty across the country as well as research on this student population to inform our work.

We concluded that as students learn the mechanics of algebra, they must also build a conceptual understanding of variable. They need to build connections between algebraic and graphical representations of relationships. They should improve their proficiency in identifying the quantitative relationships in application problems, and learn how to use algebra in solving these problems. While they are learning mathematics, they can also learn about the culture of college and develop strategies that will help them persist towards their certificate or degree.

To support these goals, we developed features and content that include:

- A five-step problem-solving plan based on the work of George Polya to organize student work on applications. Rather than memorize solutions to individual problems, the Five Steps help students develop strategies both for solving problems and for reflecting about the reasonability of their answers.
- Short readings called *Success in College Mathematics*, along with follow-up exercises, that address topics of relevance to students who are making the transition to college-level courses and help instructors identify issues outside of mathematics that may affect student success and retention.
- Examples with detailed, step-by-step explanation. Since students often rely heavily on examples when working on assignments and studying, each step of the detailed worked examples is accompanied by an annotation that details the mathematical operation performed. Color-coding and bold type are also used to help enhance the explanation of the work performed in each step of the solution.
- Multiple opportunities for review of arithmetic. Chapter R is an optional review chapter that includes a Pretest, Posttest, and instruction in arithmetic with whole numbers, integers, fractions, decimals, and percents. Just-in-time review or references to sections within Chapter R are included at key points within Chapters 1–9.
- End-of-chapter Study Plan and Review Exercises. These help guide students through test or exam preparation, step by step. The Study Plan includes study tables for each section in the chapter along with "Can I . . ." study questions that are linked to Review Exercises. References to Examples and Practice Problems from within the sections are also included for additional review, if needed.
- Algorithmic online homework capabilities through Enhanced WebAssign® that include unique question types from the textbook, as well as an interactive and customizable eBook (YouBook).
- Tools for instructors, such as Teaching Notes in the Annotated Instructor's Edition, along with Classroom Examples that mirror the examples provided in the student textbook. The PowerLecture CD-ROM is also available to instructors, and it contains ExamView® algorithmic computerized testing, Solution Builder (link to the complete textbook solutions), PowerPoint slides including selected worked examples from the textbook along with the corresponding Classroom Examples, and other resources for reference and instructional support.

We welcome your feedback as you use these materials. If you have any questions or comments please contact us at bracken@lcsc.edu and edmiller@lcsc.edu. We also invite you to follow our blog at http://devmathdiary.wordpress.com/

All the best to you and to your students.

Laura Bracken and Ed Miller

What Instructors and Students Are Saying About Bracken and Miller's Elementary Algebra

"This book has a smart approach to problem solving, helping students to develop good habits, and use applications throughout to bring meaning to the ideas being taught."

"[Students] agree that the book and the course helped them to develop their problem-solving skills and they feel more confident in their abilities."

—Barbara Goldner, North Seattle Community College

"Problem-solving plans help me organize data to complete problems."
"Very helpful. Gives great examples and explains problems very well."
"This book is easy to understand. It's very helpful and gives clear examples."
"I found breaking down word problems into sections was helpful."
"[Success in College Mathematics] helped assess personal goals."
[Find the Mistake] "If I could find the mistakes in other problems it would be easier to find the mistakes in my own work."

—Class Test Students, North Seattle Community College

"I absolutely LOVE these [Find the Mistake]. Error analysis is a huge tool that is often overlooked. It is also helpful to get them to write down their steps or "show their work."

—Jane West, Trident Technical College

"The fact that the applications include excerpts from current research and reports provides the real-life context that the authors desire as well as increasing student interest, and the number of problems with extraneous information will force students to read carefully for important information. These problems do not appear periodically; they are intentional and consistent."

—Shelly Hansen, Colorado Mesa University

"The five-step problem-solving plan is outstanding. The real-world applications are easier solved if this five-step strategy is used. This is one of the strongest parts of the textbook."

—Nicoleta Bila, Fayetteville State University

"I love these [Success in College Mathematics]. I think that many students take developmental math (and need to repeat it so much) because they're lacking in study skills and/or experiencing a transition-to-college culture shock. This feature is very useful in reminding the student that the burden of learning is on them, and it also helps the students be more aware of their own learning."

—Daniel Kleinfelter, College of the Desert

Brief Contents

CHAPTER R ARITHMETIC REVIEW 1

CHAPTER 1 THE REAL NUMBERS 35

CHAPTER 2 SOLVING EQUATIONS AND INEQUALITIES 109

CHAPTER 3 LINEAR EQUATIONS AND INEQUALITIES IN TWO VARIABLES 169

CHAPTER 4 SYSTEMS OF EQUATIONS AND INEQUALITIES 285

CHAPTER 5 EXPONENTS AND POLYNOMIALS 361

CHAPTER 6 FACTORING POLYNOMIALS 421

CHAPTER 7 RATIONAL EXPRESSIONS AND EQUATIONS 473

CHAPTER 8 RADICAL EXPRESSIONS AND EQUATIONS 547

CHAPTER 9 QUADRATIC EQUATIONS 589

APPENDIX 1 REASONABILITY AND PROBLEM SOLVING A-1

APPENDIX 2 DERIVING THE QUADRATIC FORMULA A-7

ANSWERS TO PRACTICE PROBLEMS PP-1

ANSWERS TO SELECTED EXERCISES ANS-1

INDEX I-1

Contents

CHAPTER R **Arithmetic Review** 1

Pretest 1
R.1 Evaluating Expressions and the Order of Operations 2
R.2 Signed Numbers 7
R.3 Fractions 14
R.4 Decimal Numbers, Percent, and Order 25
Study Plan for Review of Chapter R 32
Posttest 34

CHAPTER 1 **The Real Numbers** 35

SUCCESS IN COLLEGE MATHEMATICS
The Culture of College 35
1.1 The Real Numbers 36
1.2 Variables and Algebraic Expressions 46
1.3 Introduction to Problem Solving 54
1.4 Problem Solving with Percent 67
1.5 Units of Measurement and Formulas 78
1.6 Irrational Numbers and Roots 91
Study Plan for Review of Chapter 1 102
Chapter Test 107

CHAPTER 2 **Solving Equations and Inequalities** 109

SUCCESS IN COLLEGE MATHEMATICS
Time Management 109
2.1 The Properties of Equality 110
2.2 More Equations in One Variable 124
2.3 Variables on Both Sides 133
2.4 Solving an Equation for a Variable 145
2.5 Inequalities in One Variable 153
Study Plan for Review of Chapter 2 164
Chapter Test 168

CHAPTER 3 Linear Equations and Inequalities in Two Variables 169

SUCCESS IN COLLEGE MATHEMATICS
Learning Preferences 169
3.1 Ordered Pairs and Solutions of Linear Equations 170
3.2 Standard Form and Graphing with Intercepts 183
3.3 Average Rate of Change and Slope 196
3.4 Slope-Intercept Form 212
3.5 Point-Slope Form and Writing the Equation of a Line 230
3.6 Linear Inequalities in Two Variables 249
3.7 Relations and Functions 263
Study Plan for Review of Chapter 3 275
Chapter Test 282
Cumulative Review Chapters 1–3 283

CHAPTER 4 Systems of Equations and Inequalities 285

SUCCESS IN COLLEGE MATHEMATICS
Using Your Textbook 285
4.1 Systems of Linear Equations 286
4.2 The Substitution Method 298
4.3 The Elimination Method 313
4.4 More Applications 327
4.5 Systems of Linear Inequalities 344
Study Plan for Review of Chapter 4 357
Chapter Test 360

CHAPTER 5 Exponents and Polynomials 361

SUCCESS IN COLLEGE MATHEMATICS
Taking and Using Notes 361
5.1 Exponential Expressions and the Exponent Rules 362
5.2 Scientific Notation 375
5.3 Introduction to Polynomials 386
5.4 Multiplication of Polynomials 394
5.5 Division of Polynomials 404
Study Plan for Review of Chapter 5 415
Chapter Test 418

CHAPTER 6 Factoring Polynomials 421

SUCCESS IN COLLEGE MATHEMATICS
Learning from Tests and Quizzes 421
6.1 Introduction to Factoring 422
6.2 Factoring Trinomials 431
6.3 Patterns 442
6.4 Factoring Completely 449
6.5 The Zero Product Property 457
Study Plan for Review of Chapter 6 468
Chapter Test 471
Cumulative Review Chapters 4–6 471

CHAPTER 7 Rational Expressions and Equations 473

SUCCESS IN COLLEGE MATHEMATICS
Math Anxiety 473
7.1 Simplifying Rational Expressions 474
7.2 Multiplication and Division of Rational Expressions 481
7.3 Combining Rational Expressions with the Same Denominator 488
7.4 Combining Rational Expressions with Different Denominators 495
7.5 Complex Rational Expressions 507
7.6 Rational Equations and Proportions 518
7.7 Variation 532
Study Plan for Review of Chapter 7 541
Chapter Test 545

CHAPTER 8 Radical Expressions and Equations 547

SUCCESS IN COLLEGE MATHEMATICS
Learning from Tutors 547
8.1 Square Roots 548
8.2 Adding and Subtracting Square Roots 553
8.3 Multiplying Square Roots 557
8.4 Dividing Square Roots and Rationalizing Denominators 562
8.5 Radical Equations 568
8.6 Higher Index Radicals and Rational Exponents 577
Study Plan for Review of Chapter 8 585
Chapter Test 588

CHAPTER 9 **Quadratic Equations** 589

SUCCESS IN COLLEGE MATHEMATICS
Studying for a Final Exam 589
9.1 Quadratic Equations in One Variable 590
9.2 Completing the Square 598
9.3 The Quadratic Formula 606
9.4 Quadratic Equations with Nonreal Solutions 616
9.5 Quadratic Equations in the Form $y = ax^2 + bx + c$ 624
Study Plan for Review of Chapter 9 633
Chapter Test 636
Cumulative Review Chapters 7–9 637
Cumulative Review Chapters 1–9 639

APPENDIX
1 Reasonability and Problem Solving A-1
2 Deriving the Quadratic Formula A-7

ANSWERS TO PRACTICE PROBLEMS PP-1
ANSWERS TO SELECTED EXERCISES ANS-1
INDEX I-1

Index of Applications

AGRICULTURE, LANDSCAPING, AND GARDENING
Animal feed costs, 343
Apples, 385
Asparagus, 262
Bird feed mixture, 342, 355
Carrots, 198
Cattle feed mixture, 342, 355
Cheese, 198, 544
Chickens, 245, 585
Cleaning horse stalls, 529
Coffee, 258
Community garden, 104
Compost, 448
Dairy cows, 63, 284
Dairy products, 641
Farmer's market, **322–323**
Fertilizers, 122, 359
Fisheries, 343
Forest land, 260
Goats, 274, 360
Grapevine pruning, 529
Grass seed mixture, 311, 340
Hay prices, 144
Insecticide, 333
Irrigation, 229
Lawnmowers and lawn mowing, **116**, 531
Loans, 261
Mushrooms, **335–336**
Organic products/crops, 168, 208
Peaches, 280, 325
Pears, 325
Peas, 208
Potatoes, 374
Rain barrel, 226
Raised garden beds, 194
Sheep, 360
Strawberries, 466, 552
Tomatoes, 208, 247
Tree farm, 309
Weed killer, **329–330**, 537

ARTS, FILM, AND MUSIC
Album sales, 430
Costumes, 132
Drafting, 402
Grammy Awards, **71**
Gross receipts for Broadway shows, 448
Home theater speaker system, **72–73**
Latin music sales, 273
Movie box-office receipts, 262
Music, TV, and movie downloads, 63–64, 261, 323
Photography, 258, 342
Tickets sold and total receipts, 538
Walker Art Center endowment, 393
Website access to art, 167

BUSINESS AND BANKING
Advertising, 164
Annual sales, **117**
Bank accounts, 61, 65, 261, 272
Board of directors, 260
Break even, **140**, 142, 143, 152, 166, 168, 262, **290–291**, 296, **307–308**, 309, 311, 341, 357, 358, 360, 472, 487, 597, 640
Catering, 340, 341, 351, 358, 545
Cleaning services, 132, 144
Commissions, 413–414, 596
Company earnings, 63
Concession stand, 297, 360
Convention, 258
Cord blood banking service, 567
Corporate taxes, 506
Daycare, 64, 340, 360
Debit card purchases, 65
Employer medical benefits, 64, 65
Exchange rates, 623
Factory production, 225
Fatal work injuries, 164, 529
Foreclosures, 160, 623
401(k) plan, 529
403(b) plan, 261, 556
Grocery stores, 105, 529
Hotels and motels, 152, 343, 529, 538, 596
Investments, **83**, 89, **148–149**, 151, 152, 229
Job losses, 122
Law services, 144
Loans, 414
Losses, 104
Marketing, 120
Minimum wage, 210, 280, 403
Multimedia, 258
Natural gas production, 480
Newsletters, 529
Newspaper circulation, 284
Number of employees, 63, 121
Office supplies, 341, 472
Oil pipeline costs, 487
Online retailing, 182
Overdraft fees, 540
Paid sick days, 374
PayPal transactions, 640
Price per unit, 604
Profits, 76, 516
Publishing, 359
Revenues, 76, 208
Sales, **117**, 121, 297, **334**, 466
Service calls, 142
Shipping costs, 131–132, 152, 225–226
Short sales in real estate, **118–119**
Signage, 561, 576
Statistical service, 517
Stock value/price, 584
Telecommunications, 123
Ticket sales, 340
Tips, 144, 280, 641
Total salary, 64, 130, 131
Union representation, 70
U.S. gross domestic product (GDP), 236, **241–242**
Wages, 63, 76, 77, 162, 210, 280, 403, 506, 532
Warehouse storage, 325
Winemaker, 341, 355
Wireless data revenues, 64

BUSINESS RATIOS AND FORMULAS
Asset turnover ratio, 539
Commission, 166
Cost of insurance, 540
Current ratio, 539
Future value, **148–149**, 151, 152
Herfindahl-Hirschman Index (HHI), 99
Interest, **83**, 84, 89, 91, 106
P/E ratio, 90, 538
Percent markup, 77

COLLEGE
Binge drinking, 638
Budget cuts, 75
College foundation investments, 311
Community college, 66–67
Costs of delivering course, 342
Course schedule, 494
Courses with online requirements, 448
Debt after college, 640
Degrees awarded, 261, **266–267**
Enrollment, 76, 163, 208, 277, 467
Financial aid, 104, 105, 152, 261
Math tutoring center budget, **350**
Pell Grants, 66, 123, 168
Placement testing, 66
Prepaid tuition plan, 182
SAT scores, **255**
Stafford loans, 355
State funding, 63, 641
Student admissions, 105, 261
Student health insurance, 604
Student IDs, 265
Student living expenses, **214**, **350–351**, 640
Student loan repayment, 123, **377**
Student work hours, 62, 640
Tuition and fees, **58**, 75, 107, 182, 273, 342, 604

CONSTRUCTION AND HOME IMPROVEMENT
Boards and board feet, 61, 90, 151
Bolts, 342, 567
Bricklaying, 215
Building lot, 65, 77, **85**, 160

Page numbers in black indicate where the given entries appear within text or end-of-section exercises. Page numbers in **boldface** indicate where the given entries appear in Examples. Page numbers in blue indicate where the given entries appear in end-of-chapter exercises. Page numbers in red indicate where the given entries appear in Practice Problems.

Carpet, 168, 615
Communications cable trench, 226
Copper wire, 538
Doorway, 97
Doorway ramp, 596
Fencing, 506
Gable end of roof, **86**
Home building costs, 62
Home wind turbine, **594–595**
Lengthening of dock, **71–72**
Roofing, 91, **199**
Weekly salary, 63
Windmill tower water pump, 104
Windows, 258, 351, 538
Worker fatalities, 66, 529

CONSUMER AND FAMILY
Car score in *Consumer Reports*, 64
Cell phone ownership, 70, 165
Coffee cost, 62, 356, 359
Complaints, 262, 284
Cost of purchases, **73**, 76, 77, 105, 107
Credit card debt, 359
Discount and original price, **72–73**, 76, 77, **116**, 121, 164, 168, 273, 297, 311, 403, 487, 567, 640
Electricity bill and consumption, 62, 228
Engine size and gas savings, 62
Food costs, 62, 64, 281, 325, 341, 359, 545, 614
Foreclosures, 623
Health care expenditures, 64
Health insurance, 64, 65, 604
Home equity loans, 344
Inheritance, 65, 311
Interest earned on investment, **83**, 84, 89, 91, 106, 229, 283, 487
Investments, 261, 296, 311, 342, 355
Median household income, 167
Mortgages, **159**, 166, 343
Office supply costs, 343
Paper costs, 336
Paying off debt of adult children, 544
Pet care expenses, 75
Phone plan, 62, 64, 165, **197–198**, 208, 561
Popcorn consumption, 105
Property tax, 538–539
Reception budget, 261–262, 355
Rent, **56**, 62, 164, 226, 258
Retirement savings/plans, 122, 123, 261, 529
Sales tax, **73**, 76, 77, 105, 107
Social Security payments, 516–517
Soda serving size, 323
Square foot cost of condo, 64
Take-home pay remaining after house payments, **59–60**, 65
Takeout food, **321–322**, 325
Tax-free gifts, 350
Ticket prices, 245, 325
Toilet paper usage, 284, 516

COOKING, CRAFTS, AND HOBBIES
Cake/cupcakes, 488–489, 497, **525–526**, 641
Coffee, 276
Cookies, **61**, 66
Cork board, 66
Croissant, 88
Ingredients, 65
Knitting, 123
Meat grinder, 638
Model ship decking, 65
Olive oil, 104
Quilting, 90, 341, 402, 442, 529
Rag rugs, 104
Sausage, 66
Sewing, 65, 529

CRIMINAL JUSTICE
Death penalty appeals, 430
Prison vs. probation costs, 162
Prisoners, 456, 527
Robberies, 76
Violent crimes, 77, **197**, 279

DEMOGRAPHICS
Births in U.S., 198
China population, 63
Circular Area Profile, 574
Common surnames, 556
County population, 247
Deaths from chronic diseases, 530
Deaths without a will, 529
Haitian population living in poverty, 393
Hispanic population, 244
Millennial Generation in U.S., 194
Pennsylvania population in fiscally distressed municipalities, 122
Population in prison or on probation, 123, 527
Population of children in U.S., 343
Population without health insurance, 355
Teen births, 120
Total U.S. households, **117–118**
U.S. population increase, 78, 282, 284

EDUCATION
Budget, 194, 552
Electronic textbooks, 61
Enrollments, **235**, 374
Flesch-Kincaid Grade Level Readability Formula, 106
Homeless students, 132
Homework, 261
Library budget, 311
Online math test, 66
Preschool food budget, 360
Psychology, 105
SMOG reading level, 576
Teachers, 245, 297, 526
Test scores, **159**, 162, 311, 448
Time spent in school, 64

ENVIRONMENT AND POLLUTION
Carbon dioxide emissions, 59, 63, 228
Contaminated groundwater, 210
Energy conservation, 75
Greenhouse gases, 262
Landfill, 278
Pollution control costs, 64
Preserved acres, 64
Pulp used to make toilet paper, 516
Rain gauge, 283
Recycled crushed glass, **330–332**
Sewage spill, 466
Temperature, 65
Water waves, 100, 574
Well water delivered to reservoir, 385
Yard waste, 430

GEOGRAPHY
Balanced Rock in Arches National Park, 210
Elkhorn Slough, CA, 262
Temperature and, 90

GEOMETRY
Area of circle, **93**, 94, 106, 107, 182, 283, 466
Area of rectangle, **85**, 87, 89, 91, 107, 149, 168, 247, **395**, 398, **399**, 400, 402–403, 417–418, 419, 441, **462–463**, 463, 466, 470, 471, 536, 567, 597, **598–599**, **610–611**, 612, 614, 637
Area of triangle, **86**, 87, 88, 89, 100–101, 106, **399–400**, 400, 402, 418, 466, 530
Body surface area, 107
Circumference of circle, **93**, 94, 98, 106, 538, 576
Diameter of circle, **93**, 98, 101, 538, 567, 584
Heron's formula, 100–101
Perimeter of rectangle, **84**, 87, 88, 89, 91, 106, 149, 151, 247, 283, **336–337**, 339, 340, 341–342, 344, 359, 398, **399**, 400, 402–403, 417–418, 419, 472, 576, **610–611**, 612, 614, 637, 640
Perimeter of triangle, **84**, 87, 89, 151, **399**, 400, 402
Pythagorean theorem, **594–595**, 614, 632, 633, 637
Radius of circle, 99, 538, 574
Right triangle, 283, **594–595**, 596, 634, 636
Surface area of cylinder, 641
Trapezoid, 90, 151
Trapezoidal prism, 151
Volume of cone, 94, 615
Volume of cylinder, **93–94**, 94, 98–99, 101, 106, 356, 587
Volume of frustrum, 540
Volume of rectangular solid, **87**, 88, 106, 107, 283, **536**, 596
Volume of sphere, 94, **94**, 99, 106, 456, 641
Volume of water delivered by fire hose, 101

GOVERNMENT, LAW, AND POLITICS
Armed forces personnel, 75, 480
Army test scores, 280
Constitutional amendments, 66, 75
Cost of state care, 104
Cost overruns, 66
Electoral college votes, 267
FAA requirements, 261
False fire alarms, 59
Government payments, 121
Loan guarantees, 261
Municipal waste, 162, 245, 258, 430
National debt, **57**, 416

Pell Grant scholarships, 66, 123, 168
Political campaigns, **257–258**, 530
Prison vs. probation costs, 162
Property taxes, 121, 132, 538–539, 540
Public domain land acquisition, 384
Questionnaire response rate, 210
Social Security numbers, **264**
State employee pensions, 258, 261
State job cuts, 74
State revenue, 67
State sales tax rate, 76, 77
U.S. Treasury security holders, 383
Voting, 100, 162
Water service, **60**

HEALTH SCIENCE AND CAREERS

Alzheimer's disease, 544, 632
Antibiotics, 120
Basal metabolic rate, 89
Binge drinking, **159**, 530, 638
Body surface area, 637
Calories, 75, 123
Cancer, 122, 168, 530
Cerebral palsy, 494
Colonoscopy, 225
Cord blood storage, 567
Deaths from chronic diseases, 530
Deaths from pregnancy or delivery, 70
End-stage renal disease, 639
ER visits, **57**, 74, 545
Fatal work injuries, 164
Fire engine companies as first responders, 242
Flow rate through intravenous drip, 539–540
Flu shots, 122
Foodborne disease, 641
Fractional excretion of sodium, 540
Glomular filtration rate by kidney, 540
H1N1 influenza, 540
Health care costs, **69–70**, 104
Health care employees, 244
Heart attack or stroke, 529
Hepatitis A vaccine, 74
Herpes simplex virus, 375, **376–377**
HIV/AIDS, 162, 284, 456, 641
IV solution, **212–214**, 276
Lead poisoning, 604
Lidocaine solution, **119–120**
Measles, 162
Medicaid, 122
Medicare, 640
Medication dosage, 90, 419
Medications, 122, 466, **526–527**, 527, 530, 615
Meningitis vaccine, **160**
Mental health, 229
Nursing assistant, **524**
Nutrition and diet, 162, 225, **255**, **256**, 261, 327, **328–329**, 360, **533**, 537, 552
Obesity/overweight, 530
Periodontics, 374
Pre-eclampsia, 529
Rabies, 75
Rescue and recovery workers, 229
Sexually transmitted diseases, 529
Sickle cell anemia, 132
Smoking, 530
Tracheal tube, 283
Viruses, 383
Vitamin purchases, 76
Water fluoridation, 441
Weight loss, 65, 107
X-rays, 90, 403, 539

SCIENCE AND ENGINEERING

Acceleration, 151
Astronaut training, 132
Astronomy, **380**, 383, 419
Avogadro's number, 383
Biology, 273
Chemistry, 327, 333, 403, 614, 632
Density, 89, **147**, 151, 152, 576
Electronics, 516, 539, 543, 614
Escape velocity, 576
Frequency of light, 539
Fuel assemblies in nuclear power plant, 61
Geology, 210
Newton's constant of gravitation, 383
Parabolic water flow from fountain, 635
Physics, 516, 538, 539, 543, 545, 576, **593**, 596, 612, 614, **618–619**, 622
Planck's constant, 383
Radiography, 90, 403, 539
Shuttle orbit, **376**
Speed of light, 383

SOCIAL SCIENCES AND SOCIAL WORK

Foster care, 107
Poverty, 122, 393
Psychology, 164
Public assistance, 162, **308–309**
Senior citizen center food budget, 336

SPORTS AND RECREATION

Air shows and races, 261
Amusement parks, 59, 634
Baseball, 66, 263, **266**, **269**, 393, 596
Basketball, 530, 632, 641
Birthday party, 64
Bowling, 385
Camping, 88
Canoeing/boating, 342
Football, 597, 632
Golf, 167
Gymnastics, 162
Hiking, 341
Horses, 260
Lawn mower racing, 530
NASCAR, 76
Playground, 74, 152
Receipts donated to charity, 74
Recreational boat registrations, 244
Recreational vehicles, 66
Running time, 65
Saltwater fishing, 351
Soccer, 76
Swing set, 641
Team payroll, 66, 144

TECHNOLOGY, SOCIAL NETWORKING, AND COMPUTER SCIENCE

Apple products, 75
Blogging, 325
Calculators, 325
Color copier, 534
Computer and software budget, 350
Downloads, 63–64, 214, 261, 323
e-commerce revenue, 311
Internet, 272, 529
Microsoft Word 2010, 641
Mobile computing devices, 273
Range of VHF marine antenna, 101
Security cameras, 70
Smartphones, 122
Social media, 120, 284
Text messages, 65, 107
Wireless connections, 208

TRAVEL AND TRANSPORTATION

Air travel, 66, 100, 261, 284, 284, 430, 539
Antifreeze, 333, 340
Auto crashes/traffic fatalities, 64, 122, 208
Automotive formulas, 166, 539, 588
Buses, 261
Car engine, 588
Car ownership costs, 130, 131
Commuting, 132, 284
Cycling, **337–339**
Daily railroad passengers, 76
Deductible cost for driving, 534
Diesel fuel, **332–333**, 341
Drinking and driving, 74
Drivers asleep at wheel, 122
$d = rt$, **82–83**, 89, 91, 106, **148**, 151, 297, **337–339**, 339, 340, 341, 342, 344, 480, **535**, 539
Ethanol in fuel, 167
FAA requirements, 261
Freight charges, 247
Freight train, **610**
Gas mileage, 120, 122, 162, 225
Gas costs, **68–69**, 325, 340, 539, 540, 561
Gas usage, 277
High speed trains, 195
Hydroplaning, 100
Licensed drivers, 494
Light rail line, 104, 606, 614
Luggage restrictions, 261
Memorial Day travel, 66
Minivans, **117**
Parking, 74, 132, 160, 247, 272, 403
Plane with emergency supplies, 576
Red light camera programs, 343, 441
Road paving, 226
RV rentals, 182
Seatbelts, 74
Shipping containers, 556
Speed, 97, **97**, 166, **196**
Subways, 164
Taxis, 132, **141**, 143–144, 163
Tires, 166, 536, 539, 561, 576
Tourism, 121
Transit authority budget, 326
Trucking, **129–130**, **256**, 339, 340, 341, 480, 561
Tugboat, 340, 341
Turn signals, **264**
Velocity, **147**
Walking, 544

Preface

Our goal in writing Elementary Algebra is to share strategies and materials that we have developed to prepare our students for success in their next mathematics course, whether that is Intermediate Algebra or a college-level introductory course in statistics, liberal arts mathematics, or quantitative reasoning. We believe that successful students must develop conceptual understanding, procedural fluency, and confidence. They frequently need to develop new habits of self-sufficiency and personal responsibility.

Because many of our students struggled even with starting the process of solving an application, we developed a five-step problem-solving plan based on the work of George Polya. The Five Steps are introduced early in Chapter 1 in conjunction with basic application problems so that students can build the habits and confidence needed to solve more complex problems later in the course. The Five Steps provide structure as students practice translating their verbal or visual understandings of problem-solving situations into more abstract algebraic representations. A key component of this structure teaches students to look back and reflect on the reasonability of their answers, which promotes greater independence and self-sufficiency. We have included many opportunities for students to solve application problems throughout the textbook, using contexts that are familiar from daily life, pertain to various careers, and are relevant to their other academic courses. To encourage critical thinking rather than memorization, the *Problem Solving: Practice and Review* exercises provide continued practice in solving applications that are introduced earlier in the text.

Being successful in college requires more than learning academic content. Integrating student success information in courses like Elementary Algebra can help students feel part of a community, learn coping skills, and result in increased persistence and retention. We wrote the Success in College Mathematics readings and corresponding exercises to help students learn about the culture of college and to consider strategies for improving their performance.

When students pass an Elementary Algebra course with the skills, knowledge, and attitudes prerequisite for their next math class, they are empowered to move forward to a degree or certificate. Our experience class testing this textbook tells us that *Elementary Algebra* can be an important part of preparing students for that future.

Step Up to Success

Success in College Mathematics

To help students transition to college-level mathematics, Success in College Mathematics appear at the beginning of each chapter and address topics such as personal responsibility, study skills, and time management. Follow-up exercises appear at the end of each section to help students reflect on their own attitudes and habits and how they can improve their performance.

The Real Numbers

1

SUCCESS IN COLLEGE MATHEMATICS

The Culture of College

Going to college is exciting, challenging, and sometimes confusing. You are in a world with its own rules, traditions, and culture. To be successful, you need to learn about this world and what it expects from you. You may need to develop new habits and new ways of learning. You may need to seek help from ...er, your instructor, a college counselor, or other resources on campus. ...n, workshops, and syllabi are important sources of information about ...

...duction to each chapter in this book focuses on strategies for being ...in college. Each exercise set includes follow-up questions about using these ...n this first chapter, these questions assess some of your knowledge about ... Whether you are a commuter, live in a dorm, or take your classes on-line, ...learn how to be successful in college.

1.1 The Real Numbers
1.2 Variables and Algebraic Expressions
1.3 Introduction to Problem Solving
1.4 Problem Solving with Percent
1.5 Units of Measurement and Formulas
1.6 Irrational Numbers and Roots

35

SUCCESS IN COLLEGE MATHEMATICS

85. Describe your instructor's policy about late assignments.

86. Some instructors include attendance, class participation, or participation in an on-line discussion board as part of your grade. Explain how, if at all, attendance or participation affects your grade.

Examples with Step-by-Step Explanation

Each step of the worked examples is accompanied by an annotation that explains how the solution progresses, from the first line to the final answer. Color-coding and **bold type** are also used to help students easily identify the operation that occurs in each step, enabling the text to act as tutor when students are not in the classroom.

EXAMPLE 5 Solve: $-3(4a + 5) = 57$

SOLUTION ▸

$-3(4a + 5) = 57$	
$-3(4a) - 3(5) = 57$	Distributive property.
$-12a - 15 = 57$	Simplify.
$+15 \quad +15$	Addition property of equality.
$-12a + 0 = 72$	Simplify.
$\dfrac{-12a}{-12} = \dfrac{72}{-12}$	Division property of equality.
$1a = -6$	Simplify.
$a = -6$	Simplify.

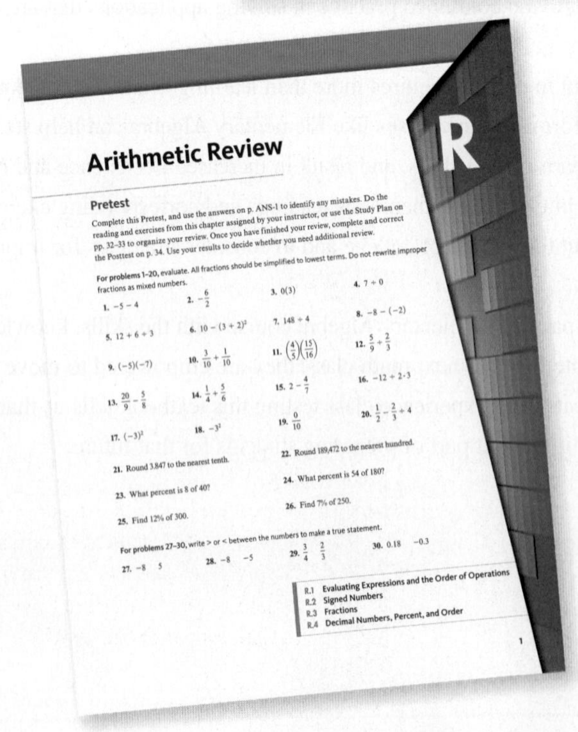

Arithmetic Review

R

Pretest

Complete this Pretest, and use the answers on p. ANS-1 to identify any mistakes. Do the reading and exercises from this chapter assigned by your instructor, or use the Study Plan on pp. 32–33 to organize your review. Once you have finished your review, complete and correct the Posttest on p. 34. Use your results to decide whether you need additional review.

For problems 1–20, evaluate. All fractions should be simplified to lowest terms. Do not rewrite improper fractions as mixed numbers.

1. $-5 - 4$ **2.** $-\frac{6}{2}$ **3.** $0(3)$ **4.** $7 \div 0$

5. $12 + 6 \div 3$ **6.** $10 - (3 + 2)^2$ **7.** $148 \div 4$ **8.** $-8 - (-2)$

9. $(-5)(-7)$ **10.** $\frac{3}{10} + \frac{1}{10}$ **11.** $\left(\frac{4}{5}\right)\left(\frac{15}{16}\right)$ **12.** $\frac{5}{9} + \frac{2}{3}$

13. $\frac{20}{9} - \frac{5}{8}$ **14.** $\frac{1}{4} + \frac{5}{6}$ **15.** $2 - \frac{4}{7}$ **16.** $-12 \div 2 \cdot 3$

17. $(-3)^2$ **18.** -3^2 **19.** $\frac{0}{10}$ **20.** $\frac{1}{2} - \frac{2}{3} + 4$

21. Round 3.847 to the nearest tenth. **22.** Round 189,472 to the nearest hundred.

23. What percent is 8 of 40? **24.** What percent is 54 of 180?

25. Find 12% of 300. **26.** Find 7% of 250.

For problems 27–30, write > or < between the numbers to make a true statement.

27. $-8 \quad 5$ **28.** $-8 \quad -5$ **29.** $\frac{3}{4} \quad \frac{2}{3}$ **30.** $0.18 \quad -0.3$

R.1 Evaluating Expressions and the Order of Operations
R.2 Signed Numbers
R.3 Fractions
R.4 Decimal Numbers, Percent, and Order

1

Chapter R Arithmetic Review

Chapter R is an optional review chapter that includes a Pretest, Posttest, and instruction in arithmetic with whole numbers, integers, fractions, decimals, and percents. Just-in-time review or references to sections within Chapter R are included at key points within Chapters 1–9.

Practice Problems

Following a set of worked examples in each section are a set of Practice Problems that mirror the examples. Students can use the Practice Problems to check their understanding of the concepts or skills presented.

Practice Problems

For problems 13–18,
(a) factor by grouping. Identify any prime polynomials.
(b) check.

13. $12x^2 + 8xy + 3xz + 2yz$ **14.** $10az - 15bz + 6ac - 9bc$
15. $35hk - 24 + 40k - 21h$ **16.** $7x^2 + 35xy^2 - 3xy - 15y^3$
17. $12xy - 4xz - 15y + 5z$ **18.** $p^2 + 6p + 6p + 36$

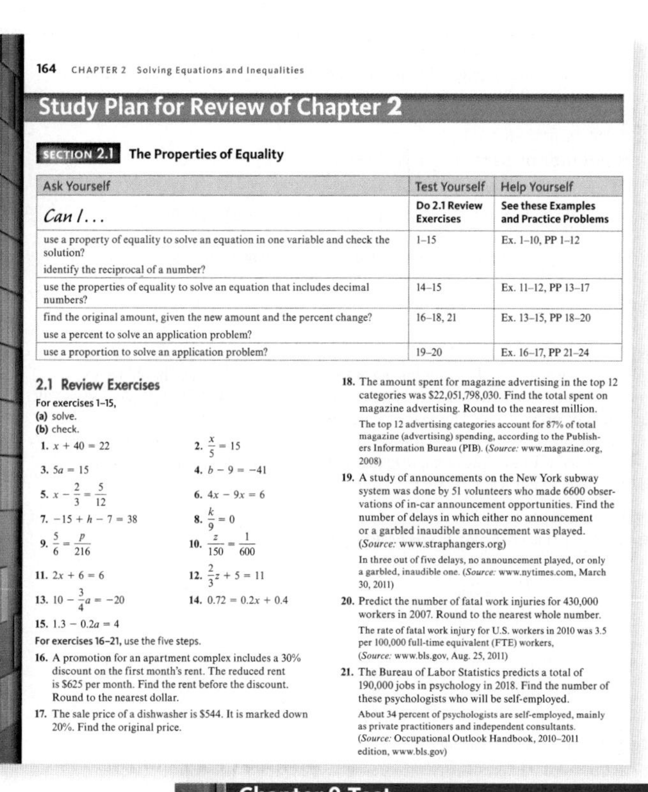

164 CHAPTER 2 Solving Equations and Inequalities

Study Plan for Review of Chapter 2

SECTION 2.1 The Properties of Equality

Ask Yourself	Test Yourself	Help Yourself
Can I . . .	**Do 2.1 Review Exercises**	**See these Examples and Practice Problems**
use a property of equality to solve an equation in one variable and check the solution? identify the reciprocal of a number?	1–15	Ex. 1–10, PP 1–12
use the properties of equality to solve an equation that includes decimal numbers?	14–15	Ex. 11–12, PP 13–17
find the original amount, given the new amount and the percent change? use a percent to solve an application problem?	16–18, 21	Ex. 13–15, PP 18–20
use a proportion to solve an application problem?	19–20	Ex. 16–17, PP 21–24

2.1 Review Exercises

For exercises 1–15,
(a) solve.
(b) check.

1. $x + 40 = 22$
2. $\frac{x}{5} = 15$
3. $5a = 15$
4. $b - 9 = -41$
5. $x - \frac{2}{3} = \frac{5}{12}$
6. $4x - 9x = 6$
7. $-15 + h - 7 = 38$
8. $\frac{k}{9} = 0$
9. $\frac{5}{6} = \frac{p}{216}$
10. $\frac{z}{150} = \frac{1}{600}$
11. $2x + 6 = 6$
12. $\frac{2}{3}z + 5 = 11$
13. $10 - \frac{3}{4}a = -20$
14. $0.72 = 0.2x + 0.4$
15. $1.3 - 0.2a = 4$

For exercises 16–21, use the five steps.

16. A promotion for an apartment complex includes a 30% discount on the first month's rent. The reduced rent is \$625 per month. Find the rent before the discount. Round to the nearest dollar.
17. The sale price of a dishwasher is \$544. It is marked down 20%. Find the original price.
18. The amount spent for magazine advertising in the top 12 categories was \$22,051,798,030. Find the total spent on magazine advertising. Round to the nearest million.
The top 12 advertising categories account for 87% of total magazine (advertising) spending, according to the Publishers Information Bureau (PIB). (*Source:* www.magazine.org, 2008)
19. A study of announcements on the New York subway system was done by 51 volunteers who made 6600 observations of in-car announcement opportunities. Find the number of delays in which either no announcement or a garbled inaudible announcement was played. (*Source:* www.straphangers.org)
In three out of five delays, no announcement played, or only a garbled, inaudible one. (*Source:* www.nytimes.com, March 30, 2011)
20. Predict the number of fatal work injuries for 430,000 workers in 2007. Round to the nearest whole number.
The rate of fatal work injury for U.S. workers in 2010 was 3.5 per 100,000 full-time equivalent (FTE) workers, (*Source:* www.bls.gov, Aug. 25, 2011)
21. The Bureau of Labor Statistics predicts a total of 190,000 jobs in psychology in 2018. Find the number of these psychologists who will be self-employed.
About 34 percent of psychologists are self-employed, mainly as private practitioners and independent consultants. (*Source:* Occupational Outlook Handbook, 2010–2011 edition, www.bls.gov)

End-of-Chapter Study Plan and Review Exercises

This tool offers students an effective and efficient way to prepare for quizzes or exams. The Study Plan, appearing at the end of each chapter, includes study tables for each section that will help students get organized as they prepare for a test or quiz. Each table contains a set of *Can I . . .* questions, based on the section objectives, for self-reflection. *Can I . . .* questions are tied to one or more Review Exercises that students can use to test themselves. Examples and/or Practice Problems are also linked to each question should students need reference for review.

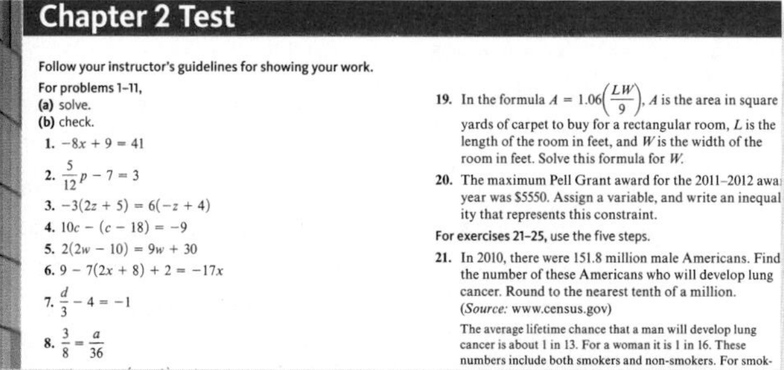

Chapter 2 Test

Follow your instructor's guidelines for showing your work.

For problems 1–11,
(a) solve.
(b) check.

1. $-8x + 9 = 41$
2. $\frac{5}{12}p - 7 = 3$
3. $-3(2z + 5) = 6(-z + 4)$
4. $10c - (c - 18) = -9$
5. $2(2w - 10) = 9w + 30$
6. $9 - 7(2x + 8) + 2 = -17x$
7. $\frac{d}{3} - 4 = -1$
8. $\frac{3}{8} = \frac{a}{36}$

19. In the formula $A = 1.06\left(\frac{LW}{9}\right)$, A is the area in square yards of carpet to buy for a rectangular room, L is the length of the room in feet, and W is the width of the room in feet. Solve this formula for W.
20. The maximum Pell Grant award for the 2011–2012 awa year was \$5550. Assign a variable, and write an inequal ity that represents this constraint.

For exercises 21–25, use the five steps.

21. In 2010, there were 151.8 million male Americans. Find the number of these Americans who will develop lung cancer. Round to the nearest tenth of a million. (*Source:* www.census.gov)
The average lifetime chance that a man will develop lung cancer is about 1 in 13. For a woman it is 1 in 16. These numbers include both smokers and non-smokers. For smok-

End-of-Chapter Test

Each chapter concludes with a sample test that students can take as preparation for an upcoming exam or to assess their understanding of chapter objectives.

Cumulative Review Chapters 1–3

Follow your instructor's directions for showing your work.

1. Use roster notation to describe the set of integers.
2. Evaluate: $-24 \div 3 \cdot 2 - (3 - 7)^2$
3. Simplify: $\frac{5}{6}(12x - 9) - \frac{1}{3}(15x + 8)$
4. Simplify: $-8(x^2 - 3x + 2) - (x - 4)$
5. Name the property illustrated by the equation $3 + 7 = 7 + 3$.
6. Write the distributive property.
7. Find the perimeter of a rectangle that is 6 in. long and 2.5 in. wide.
8. Find the simple interest earned on \$250 invested for 3 years at 3.2%.

26. Find the x-intercept and the y-intercept of the graph of $7x - 4y = 56$.

For exercises 27–30, graph.

27. $y = 3$
28. $x = -5$
29. $y \geq 4x - 7$
30. $5x - 3y > 20$
31. A linear equation is $-2x + 5y = 20$.
a. Find three solutions of the equation.
b. Graph the equation.
32. Use slope-intercept graphing to graph $y = -\frac{5}{6}x + 9$.
33. Write the equation of the graphed line.

End-of-Chapter Cumulative Review

To provide review during the course, Cumulative Reviews appear at the end of Chapters 3, 6, and 9. To assess readiness and provide practice for a final exam, a Cumulative Review for Chapters 1–9 appears after Chapter 9.

Problem Solving, Critical Thinking, and Reasoning

Five Steps for Problem Solving

Based on the work of George Polya, the Five Steps provide a framework for jump-starting and organizing student problem solving. Step-by-step worked examples throughout the textbook support students as they learn to solve a wide range of applications.

THE FIVE STEPS

Step 1 Understand the problem.
Step 2 Make a plan.
Step 3 Carry out the plan.
Step 4 Look back.
Step 5 Report the solution.

SOLUTION ▸ **Step 1 Understand the problem.**
The unknown is the original price of the lawnmower.

p = original price

Step 2 Make a plan.
The discount is the amount that the original price is reduced. Since number of parts = (percent)(total parts), the discount equals (percent discount)(original price) or $0.20p$. A word equation is original price − discount = sale price.

Step 3 Carry out the plan.

$p - 0.20p = \$399$	Original price − discount = sale price
$1p - 0.20p = \$399$	Rewrite p as $1p$.
$\mathbf{0.80p} = \$399$	Combine like terms; $1p - 0.20p = 0.80p$
$\frac{0.80p}{0.80} = \frac{\$399}{0.80}$	Division property of equality.
$\mathbf{1p} = \mathbf{\$498.75}$	Simplify.
$p = \$498.75$	Simplify.

Step 4 Look back.
Working backwards, if the answer for the original price is correct, we should be able to use it and the percent discount to find the sale price. Since $498.75 − (0.20)($498.75) equals the sale price of $399, the answer seems reasonable.
In this problem, we used a relationship that may apply to future problems: *original* + *change* = *new*. In this problem, *original* is the original price, change is the discount, and new is the sale price.

Step 5 Report the solution.
The original price of the lawnmower was $498.75.

Checking for Accuracy and Reasonability

To be successful in mathematics and other disciplines, an important skill for students to develop is checking their work for reasonability and accuracy. The explanations of reasonability in examples of the Five Steps and the checking of solutions in many of the worked examples provide a model for students learning to reflect on their own work. Being able to justify an answer helps students develop confidence and self-sufficiency.

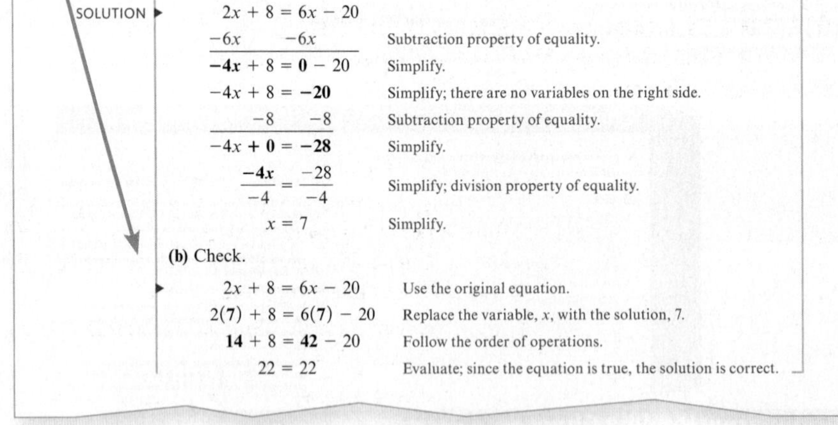

EXAMPLE 1 **(a)** Solve: $2x + 8 = 6x - 20$

SOLUTION ▸

$2x + 8 = 6x - 20$	
$-6x \quad\quad -6x$	Subtraction property of equality.
$\mathbf{-4x} + 8 = \mathbf{0} - 20$	Simplify.
$-4x + 8 = \mathbf{-20}$	Simplify; there are no variables on the right side.
$-8 \quad\quad -8$	Subtraction property of equality.
$-4x + \mathbf{0} = \mathbf{-28}$	Simplify.
$\frac{\mathbf{-4x}}{-4} = \frac{-28}{-4}$	Simplify; division property of equality.
$x = 7$	Simplify.

(b) Check.

$2x + 8 = 6x - 20$	Use the original equation.
$2(7) + 8 = 6(7) - 20$	Replace the variable, x, with the solution, 7.
$\mathbf{14} + 8 = \mathbf{42} - 20$	Follow the order of operations.
$22 = 22$	Evaluate; since the equation is true, the solution is correct.

Real Sources

Among the application problems appearing in this textbook are specially developed exercises and examples that reference information taken directly from news articles, research studies, and other fact-based sources. Many of these applications are set up in two parts, with a problem statement given first followed by an excerpt. Since the question appears first and then the information needed to answer the question follows, students experience problem solving in a more true-to-life way. Though similar in length to texts or tweets, the authentic excerpts often contain more information than is needed to solve the applications. Students will need to think critically to select the relevant information, and in doing so practice the skills needed to solve problems outside of the classroom.

83. According to an article on safety tips for blogging, two out of three teenagers provide their age on their blog, three out of five reveal their location, and one out of five reveals their full name. Find the number of 3675 teenagers who reveal their location on their blog. (*Source:* www.microsoft.com)

Problem Solving: Practice and Review

Beginning in Chapter 2, each section-ending exercise set includes a short set of applications-based exercises called *Problem Solving: Practice and Review.* Because these problems do not involve the concepts or skills taught in the section, students need to think critically about the information and relationships in the problem. Even in sections that have few or no applications, students can continue to practice their problem-solving skills by completing one or more of these exercises.

Problem Solving: Practice and Review

Follow your instructor's guidelines for using the five steps as outlined in Section 1.3, p. 55.

103. Find the percent decrease in the number of diagnoses of AIDS from 2008 to 2009. Round to the nearest tenth of a percent.

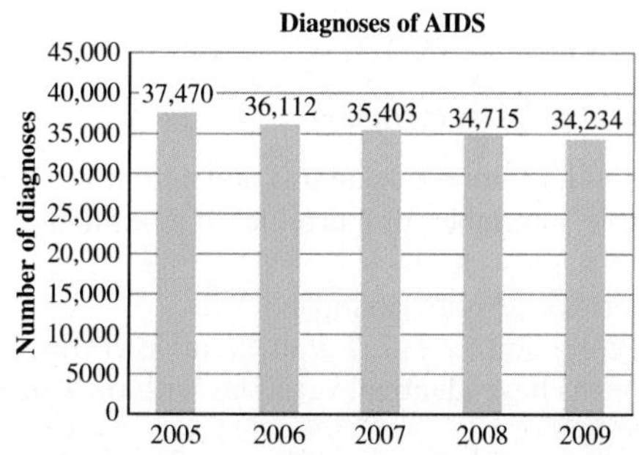

Source: gis.cdc.gov, Jan. 2012

104. A veteran drives three times a month to a VA Hospital that is 125 miles away from his home. Find the amount that the veteran will be reimbursed for his travel.

The current allowable reimbursement amount is 28.5 cents per mile with a $7.77 deductible for each one-way visit, or $15.54 for each round-trip visit. If your out-of-pocket costs exceed $46.62 in a given month, you can be reimbursed for the entire expense for any other authorized travel for that month. (*Source:* www.benefits.gov, Jan. 2012)

Find the Mistake

The section-ending exercise sets include *Find the Mistake* exercises. Each exercise is a problem and a step-by-step solution that includes one mistake. Students are asked to identify the error and then rework the problem correctly. These exercises help students learn how to find errors in their own work, which improves student persistence and self-sufficiency.

Find the Mistake

For exercises 89–92, the completed problem has one mistake.

(a) Describe the mistake in words, or copy down the whole problem and highlight or circle the mistake.

(b) Do the problem correctly.

89. Problem: Find the slope of the line that passes through $(7, 1)$ and $(9, 4)$.

Incorrect Answer: $m = \dfrac{y_2 - y_1}{x_2 - x_1}$

$m = \dfrac{9 - 7}{4 - 1}$

$m = \dfrac{2}{3}$

Concept Development

Learning the Language of Math

Vocabulary matching exercises that appear before the section-ending exercises help students improve their knowledge of vocabulary and notation.

1.2 VOCABULARY PRACTICE

Match the term with its description.

1. A symbol that represents an unknown number
2. A variable, a number, or a product of a number and a variable
3. $a(b + c) = ab + ac$
4. In the term $3x$, 3 is an example of this.
5. In the expression $5x + 7$, 7 is an example of this.
6. These terms have identical variables with the same exponents.
7. $a + b = b + a$
8. $a + (b + c) = (a + b) + c$
9. These numbers are elements of $\{\ldots, -3, -2, -1, 0, 1, 2, 3, \ldots\}$.
10. These numbers are elements of $\{0, 1, 2, 3, 4, \ldots\}$.

A. associative property of addition
B. coefficient
C. commutative property of addition
D. constant
E. distributive property
F. integers
G. like terms
H. term
I. variable
J. whole numbers

Writing Exercises

Integrated within the Practice Problems, end-of-section Exercises, and end-of-chapter Reviews, Tests, and Cumulative Reviews are exercises that require a written response. Students must reflect on concepts presented within the section in order to form a written response in their own words.

32. Explain why we cannot use the zero product property to solve $4y^2 - 15 = 10$ when it is written in this form.
33. Explain why the solutions of the equation $2(x + 5)(x - 1) = 0$ are the same as the solutions of the equation $(2x + 10)(x - 1) = 0$.
34. The equation $x(x - 9)(x - 9) = 0$ has three factors. However, it has only two solutions. Explain why.

41. Explain how to determine whether the boundary line on the graph of an inequality should be solid or dashed.
42. Explain how to determine whether an inequality is a strict inequality.
43. Explain why $(0, 0)$ is often a good choice for a test point when graphing an inequality.
44. Describe a situation in which $(0, 0)$ cannot be used as a test point when graphing an inequality.
45. Explain how to write the equation of a line in slope-intercept form given two points on the line.
46. Explain how to write the equation of a line in slope-intercept form given the slope and one point on the line.

Multiple Representations

Algebraic concepts are often presented in conjunction with a graph and/or a table. By presenting relationships in multiple ways, students are more likely to build connections between their concrete understandings and the abstract language of algebra.

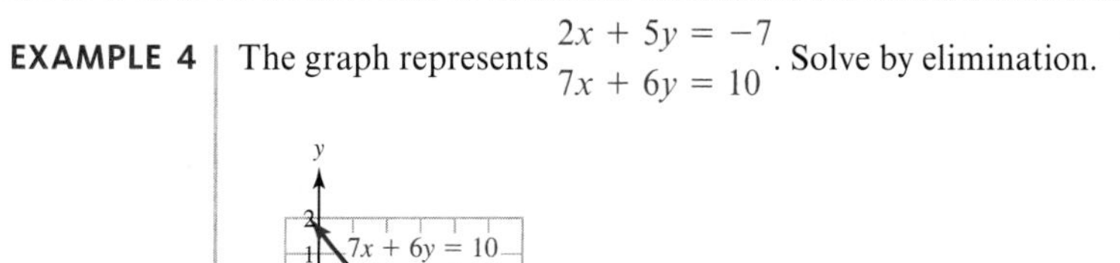

EXAMPLE 4 The graph represents $\begin{aligned} 2x + 5y &= -7 \\ 7x + 6y &= 10 \end{aligned}$. Solve by elimination.

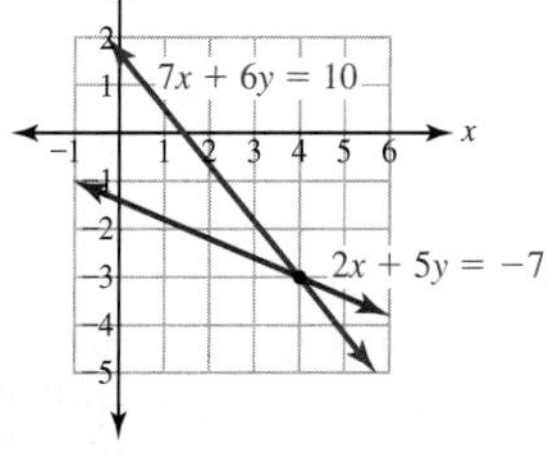

SOLUTION ▸ Choose either x or y to eliminate. In this example, we will eliminate x. To create x-terms that are opposites, multiply $2x + 5y = -7$ by 7, and multiply $7x + 6y = 10$ by -2.

Using Technology

For instructors who wish to integrate calculator technology, optional scientific and graphing calculator instruction is provided at the end of selected sections, where appropriate. The instruction that appears within the Using Technology boxes includes examples with keystrokes, screen shots, and Practice Problems. Follow-up *Technology* exercises appear at the end of section-ending exercises for continued practice.

Using Technology: Building a Table of Solutions

To build a table of solutions of a linear equation, use the TABLE SETUP screen and the TABLE screen.

EXAMPLE 10 Use a calculator to build a table of solutions of the equation $y = -\frac{3}{4}x + 6$ with a beginning x-value of 10 and an interval of 2 between x-values.

Press Y=. Type the equation. Go to the TBLSET screen; press 2nd WINDOW. Replace 0 with 10. This is the beginning x-value. Press ENTER.

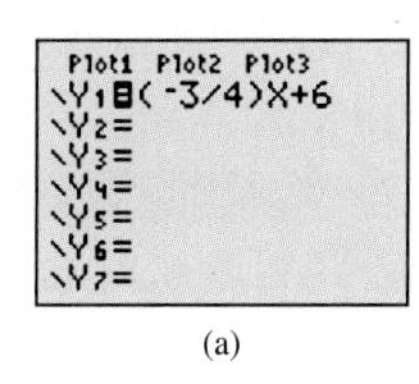

(a)

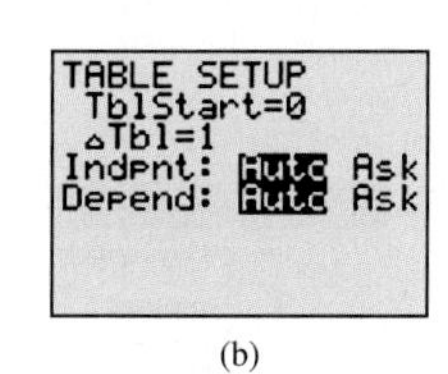

(b)

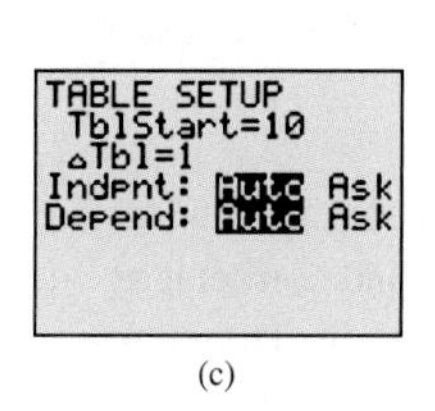

(c)

The symbol Δ means "change in." So ΔTbl means the change or interval between each x-value in the table. Replace 1 by typing 2. To see the table, press 2nd GRAPH to go to the TBL screen.

Enhanced WebAssign®

The perfect homework management tool to reinforce the material in this textbook.

Available exclusively from Cengage Learning, Enhanced WebAssign® (EWA) offers:

- An extensive online program that encourages the practice essential for concept mastery
- Intuitive tools that save you time on homework management
- Multimedia tutorial support and immediate feedback as students complete their assignments
- The Cengage YouBook, an interactive and customizable eBook that lets you tailor the textbook to fit your course.

Bracken/Miller and Enhanced WebAssign

The Enhanced WebAssign course developed for use with Laura Bracken and Ed Miller's *Elementary Algebra* mirrors the goal of their textbook: to empower students, providing them with tools to become better problem-solvers and critical thinkers. The EWA course includes algorithmically generated questions based on the text's end-of-section exercise sets and solution video tutorials, both of which promote the development of core algebraic skills. The course also provides the following problem types, derived from the textbook and carefully selected by the authors to support their approach.

1. The ***Five Steps*** for problem solving, based on the work of George Polya, is a framework the authors use to demonstrate how to solve a wide variety of application problems. By working through applications that involve the use of the five steps, students will learn how to follow an organized plan for solving problems, no matter their type.

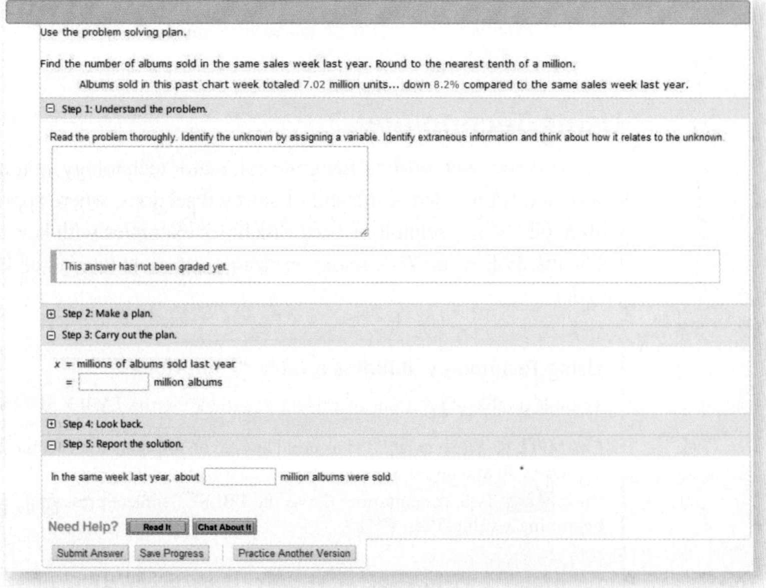

2. The ***Find the Mistake*** problems encourage students to think critically about their work and correct it as they analyze common errors and pitfalls. Drawn from section-ending exercises in the textbook, students are asked to identify the error and then rework the problem correctly. These exercises cultivate analytical skills and self-sufficiency.

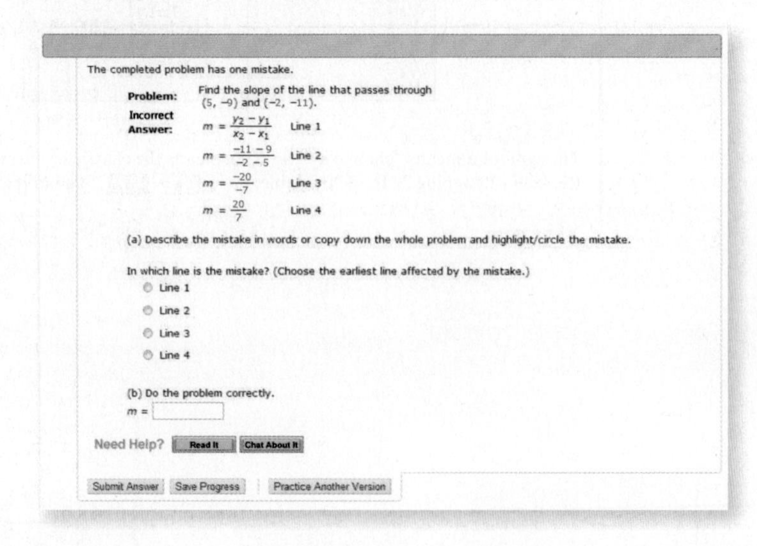

3. The ***Success in College Mathematics*** problems give students the chance to take charge of their learning. Based on the chapter opener narratives and corresponding end-of-section exercises from the textbook, these questions help students better understand college culture and address such topics as personal responsibility, study skills, and time management.

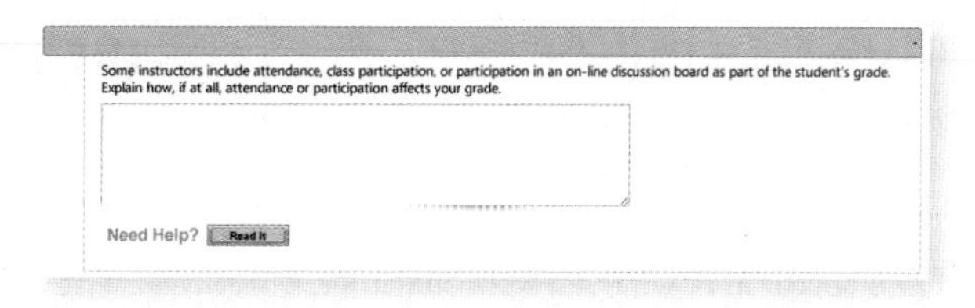

4. The ***Solving Equations Exercises with Checks*** encourage students to go through the analytical process of checking their solutions. These problems help promote self-sufficiency, an important skill students need at the college level.

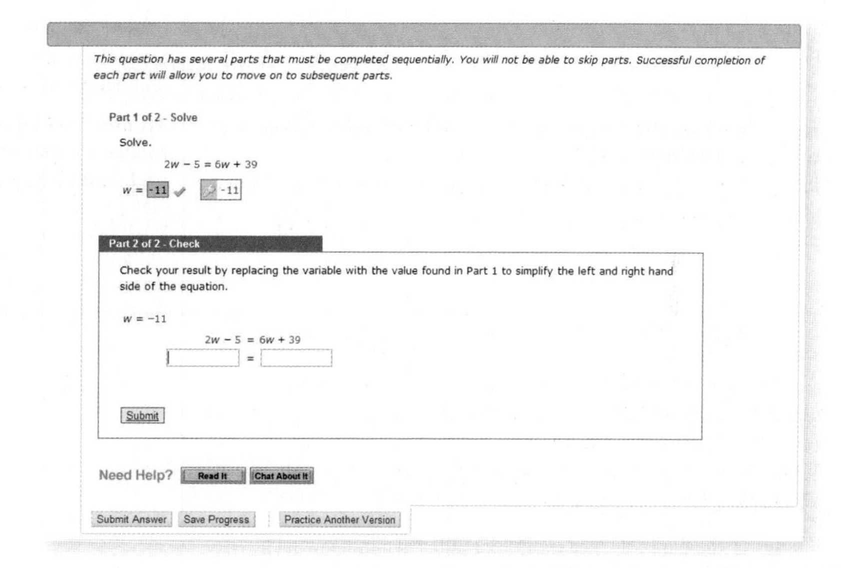

Other Enhanced WebAssign Resources

Also available for this course are interactive tools that address various learning styles helping students to study and improve their success.

Cengage YouBook

Engage your students with a customizable and interactive eBook that lets you embed web links, modify the textbook narrative as needed, reorder entire sections and chapters, and hide any content you don't teach to create an eBook that perfectly matches your syllabus.

Concept Mastery Videos

Extensive video resources—from Rena Petrello and Dana Mosely—reinforce concepts and help students prepare for exams. These are accessible through the Resources Tab in every Enhanced WebAssign course. Select videos are also embedded in the Cengage YouBook.

Resources

FOR THE INSTRUCTOR	FOR THE STUDENT
Annotated Instructor's Edition (ISBN: 978-0-618-95151-2) The Annotated Instructor's Edition features several teaching tools. Classroom Examples that are parallel to the worked examples in the student edition appear in the margin. These may be used to supplement lectures and also appear in PowerPoint form on the PowerLecture CD-ROM. Also included in the margin are Teaching Notes that offer instructors tips and suggestions for presenting the content. Answers to most exercises appear on page in close proximity. Answers not appearing on page are located in an appendix in the back of the book.	
Complete Solutions Manual by Alan Hain, Lewis-Clark State College (ISBN: 978-1-133-93532-2) The Complete Solutions Manual provides detailed solutions to all problems in the text.	**Student Solutions Manual by Alan Hain, Lewis-Clark State College** (ISBN: 978-1-133-94135-4) Go beyond the answers—see what it takes to get there and improve your grade! This manual provides step-by-step solutions to selected problems in the series, giving you the information you need to truly understand how these problems are solved.
Instructor's Resource Binder by Maria H. Andersen, Muskegon Community College (ISBN: 978-0-538-73675-6) This loose-leaf binder contains uniquely designed Teaching Guides that contain tips, examples, activities, worksheets, overheads, assessments, and solutions to all worksheets and activities.	**Student Workbook by Maria H. Andersen, formerly of Muskegon Community College and now Director of Learning and Research at Instructure** (ISBN: 978-1-285-06736-0) Get a head start with this hands-on resource! The Student Workbook contains all of the assessments, activities, and worksheets from the Instructor's Resource Binder for classroom discussions, in-class activities, and group work.
PowerLecture with ExamView (ISBN: 978-1-111-57491-8) This CD-ROM (or DVD) provides you with dynamic media tools for teaching. Create, deliver, and customize tests (both print and online) in minutes with ExamView® Computerized Testing Featuring Algorithmic Equations. Easily build solution sets for homework or exams using Solution Builder's online solutions manual. Microsoft® PowerPoint® lecture slides and figures from the book are also included on this CD-ROM (or DVD) along with the PowerPoint versions of the parallel Classroom Examples available in each Annotated Instructor's Edition. A graphing calculator appendix detailing specific calculator commands also appears along with additional support materials for use specifically with the textbook.	
Solution Builder This online instructor database offers complete worked solutions to all exercises in the text, allowing you to create customized, secure solutions printouts (in PDF format) matched exactly to the problems you assign in class. For more information, visit www.cengage.com/solutionbuilder.	
Enhanced WebAssign Printed Access Card (ISBN: 978-0-538-73810-1) Online Access Code (ISBN: 978-1-285-18181-3) Exclusively from Cengage Learning, Enhanced WebAssign® combines the exceptional Mathematics content that you know and love with the most powerful online homework solution, WebAssign. Enhanced WebAssign engages students with immediate feedback, rich tutorial content and interactive, fully customizable eBooks (YouBook) helping students to develop a deeper conceptual understanding of their subject matter. Online assignments can be built by selecting from thousands of text-specific problems or supplemented with problems from any Cengage Learning textbook. Consistent with the authors' approach, unique features and problem types from the text have been integrated into Enhanced WebAssign including the 5-Steps, Success in College Mathematics, Find the Mistake, and Solving Equation Exercises with Checks.	**Enhanced WebAssign** Printed Access Card (ISBN: 978-0-538-73810-1) Online Access Code (ISBN: 978-1-285-18181-3) Enhanced WebAssign (assigned by the instructor) provides you with instant feedback on homework assignments. This online homework system is easy to use and includes helpful links to textbook sections, video examples, and problem-specific tutorials.
Enhanced WebAssign: Start Smart Guide for Students Author: Brooks/Cole (ISBN: 978-0-495-38479-3) The Enhanced WebAssign: Start Smart Guide for Students helps students get up and running quickly with Enhanced WebAssign so that they can study smarter and improve their performance in class.	**Enhanced WebAssign: Start Smart Guide for Students** Author: Brooks/Cole (ISBN: 978-0-495-38479-3) If your instructor has chosen to package Enhanced WebAssign with your text, this manual will help you get up and running quickly with the Enhanced WebAssign system so that you can study smarter and improve your performance in class.

Ensuring Quality and Accuracy

To ensure that this textbook meets the highest standards for quality and accuracy, numerous quality assurance checks have been applied during development. During the writing and revision process, more than 100 instructors reviewed various aspects of the manuscript, including the overall approach, features, organization, depth of topic coverage, clarity of writing, and the quality, variety, and quantity of examples and exercises. Additionally, special advisors commented on specific issues related to content, art, and design throughout the process. Detailed responses from reviewers were carefully read and analyzed by the authors and publisher, followed by in-depth revisions. Feedback was also collected at focus groups that were held across the country at various stages of manuscript development. The insights of these focus group participants on an array of topics helped shape many of the features that are part of this textbook. The authors' colleagues at Lewis-Clark State College also played a vital role in the development of *Elementary Algebra*. They class-tested several iterations of the manuscript, with the authors continuously updating the manuscript based on comments from instructors as well as students.

As this textbook went into production, the process was designed to maintain these standards of quality and accuracy. A team of highly experienced art and text editors, as well as the authors, examined the manuscript prior to copyediting and art preparation. At each stage of production, the text and art samples were reviewed multiple times by proofreaders, copyeditors, the production editor, mathematicians working as accuracy checkers, and the authors. This review process was expanded to include the authors of both print and electronic supplemental materials, both to maintain accuracy and to create supplements that are compatible with the approach and content of the text.

At every step in development of this textbook, the emphasis was on creating valuable, worthwhile materials that meet the needs of developmental mathematics instructors and their students. An integral part of this emphasis is the focus throughout the process on quality assurance and accuracy. The immediate and extended family of mathematicians, editorial, art, and production professionals who worked on this text are among the best, committed to helping students learn mathematics and progress towards their goals.

Clockwise from top (keyboard): © ZTS/Shutterstock.com; © iStockphoto.com/Chris Schmidt; Courtesy of White Loop Ltd.; © iStockphoto/clu; © iStockphoto/Carmen Martinez Banus; (background) © istockphoto.com/Dougberry; © seanelliotphotography/Shutterstock.com; © Ian Leonard/Alamy

Acknowledgments

A textbook is the result of the efforts of a very large group of contributors. In writing these books, we incorporated the suggestions and insight of colleagues and students across the country. As classroom instructors, we recognize and are grateful for the time, commitment, and feedback of the following faculty and students.

Advisory Board Members

Kirby Bunas, *Santa Rosa Junior College*
Michael Kinter, *Cuesta College*
Marie St. James, *St. Clair Community College*

Reviewers and Class Testers

Dr. Laura Adkins, *Missouri Southern State University*
Darla Aguilar, *Pima Community College*
Kathleen D. Allen, *Hinds Community College–Rankin*
Jerry Allison, *Trident Technical College–Berkeley*
Jacob Amidon, *Finger Lakes Community College*
Sheila Anderson, *Housatonic Community College*
Jan Archibald, *Ventura College*
Dimos Arsenidis, *California State University–Long Beach*
Benjamin Aschenbrenner, *North Seattle Community College*
Michele Bach, *Kansas City Community College*
Brian Balman, *Johnson County Community College*
Leann Beaven, *University of Southern Indiana*
Alison Becker–Moses, *Mercer County Community College*
Rosanne Benn, *Prince George's Community College*
Rebecca Berthiaume, *Edison State College*
Armando Bezies-Kindling, *Pima Community College–Downtown*

Barbara Biggs, *Utah Valley University*
Dr. Nicoleta Bila, *Fayetteville State University*
Ina Kaye Black, *Bluegrass Community and Technical College*
Kathleen Boehler, *Central Community College–Grand Island*
Michael Bowen, *Ventura College*
Gail Brewer, *Amarillo College*
Shane Brewer, *College of Eastern Utah–San Juan*
Susan Caldiero, *Cosumnes River College*
Nancy Carpenter, *Johnson County Community College*
Jeremy Carr, *Pensacola Junior College*
Edie Carter, *Amarillo College*
Gerald Chrisman, *Gateway Technical College–Racine*
Suzanne Christian-Miller, *Diablo Valley College*
Dianna Cichocki, *Erie Community College–South*
John Close, *Salt Lake Community College*
Stacy Corle, *Pennsylvania State University–Altoona*
Kyle Costello, *Salt Lake Community College*
Debra Coventry, *Henderson State University*
Pam Cox, *East Mississippi Community College*
Edith Cranor-Buck, *Western State College of Colorado*
William D. Cross, *Palomar College*
Ken Culp, *Northern Michigan University*
Steven I. Davidson, *San Jacinto College–Central*
Susan Dimick, *Spokane Community College*
Dr. Julien Doucet, *Louisiana State University–Alexandria*
Karen Edwards, *Diablo Valley College*
Patricia Elko, *SUNY Morrisville*
Mike Everett, *Santa Ana College*
Rob Farinelli, *Community College of Allegheny County*
Stuart Farm, *University of North Dakota*
Julie Fisher, *Austin Community College*
Dr. Dorothy French, *Community College of Philadelphia*
Gada Dharmesh J., *Jackson Community College*
Scott Gentile II, *City College of San Francisco*
Jane Golden, *Hillsborough Community College*
Barbara Goldner, *North Seattle Community College*
John Greene, *Henderson State University*
Kathy Gross, *Cayuga Community College–Fulton*
Susan Hahn, *Kean University*
Shawna Haider, *Salt Lake Community College*
Dr. Haile Haile, *Minneapolis Community and Technical College*
Shelly Hansen, *Colorado Mesa University*
John T. Harris
Dr. Jennifer Hegeman, *Missouri Western State University*
Elaine Hodz, *Florida Community College at Jacksonville–Downtown*
Laura Hoye, *Trident Technical College*
Kimberly Johnson, *Mesa Community College*
Dr. Tina Johnson, *Midwestern State University*
Todd Kandarian, *Reedley College–Madera Center*
Dr. Fred Katiraie, *Montgomery College–Rockville*
Dr. John Kawai, *Los Angeles Valley College*
Dr. David Keller, *Kirkwood Community College*
Catherine Carroll Kiaie, *Cardinal Stritch University*
Barbara Kistler, *Lehigh Carbon Community College*
Daniel Kleinfelter, *College of the Desert*
Kandace Kling, *Portland Community College*
Alexander Kolesnik, *Ventura College*
Randa Kress, *Idaho State University*
Thang Le, *College of the Desert*
Kevin Leith, *Central New Mexico Community College*
Edith Lester, *Volunteer State Community College*
Mark Marino, *Erie Community College–North*
Amy Marolt, *Northeast Mississippi Community College*
William Martin, *Pima Community College*
Dr. Derek Martinez, *Central New Mexico Community College*
Carlea McAvoy, *South Puget Sound Community College*
Caroyln McCallum, *Yakima Valley Community College*
Carrie A. McCammon, *Ivy Tech State College*
Margaret Michener, *The University of Nebraska at Kearney*
Debbie Miner, *Utah Valley University*
Mary Mizell, *Northwest Florida State College*
Ben Moulton, *Utah Valley University*
Bethany R. Mueller, *Pensacola State College*
Dr. Ki-Bong Nam, *University of Wisconsin–Whitewater*
Sandi Nieto, *Santa Rosa Junior College–Santa Rosa*
Dr. Sam Obeid, *Richland College*
Jon Odell, *Richland Community College*
Michael Orr, *Grossmont College*
Becky L. Parrish, *Ohio University–Lancaster*
Dr. Shahrokh Parvini, *San Diego Mesa College*
John Pflughoeft, *Northwestern Michigan College*
Tom Pomykalski, *Madison Area Technical College*
Carol Ann Poore, *Hinds Community College–Rankin*
Michael Potter, *The University of Virginia's College at Wise*
Beth Powell, *MiraCosta College*
Brooke Quinlan, *Hillsborough Community College*
Genele Rhoads, *Solano Community College*
Tanya Rivers, *Western State College of Colorado*
Vicki Schell, *Pensacola State College*
Randy Scott, *Santiago Canyon College*
Jane Serbousek, *Northern Virginia Community College*
Sandra Silverberg, *Bergen Community College*
M. Terry Simon, *The University of Toledo*
Roy Simpson, *Cosumnes River College*
Zeph Smith, *Salt Lake Community College*
Donald Solomon, *University of Wisconsin–Milwaukee*
J. Sriskandarajah, *Madison College–Madison, WI*
Dr. Panyada Sullivan, *Yakima Valley Community College*
Mary Ann Teel, *University of North Texas*
Rosalie Tepper, *Shoreline Community College*
Rose Toering, *Kilian Community College*
Dr. Suzanne Tourville, *Columbia College*
Calandra Walker, *Tallahassee Community College*
Dr. Bingwu Wang, *Eastern Michigan University*
Dr. Jane West, *Trident Technical College*
Darren Wiberg, *Utah Valley University*
Dale Width, *Central Washington University*
Peter Willett, *Diablo Valley College*
Dr. Douglas Windham, *Tallahassee Community College*
Dr. Tzu-Yi Alan Yang, *Columbus State Community College*

Developmental Focus Group Participants

Dimos Aresenidis, *California State University–Long Beach*
Angelica Ascencio, student, *Cerritos College*
Mohammad Aslam, *Georgia Perimeter College–Clarkston*
Patricia J. Blus, *National Louis University*
Dona V. Boccio, *Queensborough Community College*
Dmitri Budharin, *Cerritos College*
Kirby Bunas, *Santa Rosa Junior College*
Ashraful Chowdhury, *Georgia Perimeter College–Clarkston*
Elisa Chung, *Riverside Community College*
Mariana Coanda, *Broward College*
Joseph S. de Guzman, M.S., *Norco College*
Cheryl Eichenseer, *St. Charles Community College*
Dr. Susan Fife, *Houston Community College*
Kathryn S. Fritz, *North Central Texas College*
Adrianne Guzman, student, *Cerritos College*
Dr. Kim Tsai Granger, *St. Louis Community College–Wildwood*
Edna G. Greenwood, *Tarrant County College–Northwest*
Larry D. Hardy, *Georgia Perimeter College–Newton*
Mahshid Hassani, *Hillsborough Community College*
Richard Hobbs, *Mission College*
Linda Hoppe, *Jefferson College*
Laurel Howard, *Utah Valley University*
Elizabeth Howell, *North Central Texas College*
James Johnson, *Modesto Junior College*
Janhavi Joshi, *De Anza College*
Susan Keith, *Georgia Perimeter College–Newton*
Michael Kinter, *Cuesta College*
Alex Kolesnik, *Ventura College*
Edith Lester, *Volunteer State Community College*
Bob Martin, *Tarrant County College*
Arda Melkonian, *Victor Valley College*
Ashod Minasian, *El Camino College*
Lydia Morales, *Ventura College*
Christina Morian, *Lincoln University*
Joyce Nemeth, *Broward College*
Marla Owens, *North Central Texas College*
Michael Papin, *Yuba College*
Svetlana Podkolzina, *Solano Community College*
Linda Retterath, *West Valley Mission College*
Marcus H. Rhymes, *Georgia Perimeter College–Clarkston*
Weldon Ritchie, *Le Cordon Bleu*
Kheck Segemeny, *Solano Community College*
Charlene Snow, *Solano Community College*
Pamela R. Sheehan, *Solano Community College and Los Medanos Community College*
Cara Smyczynski, *Trident Technical College*
Marie St. James, *St. Clair County Community College*
Deborah Strance, *Allan Hancock College*
Francesco Strazzullo, *Reinhardt University*
Chetra Talwinder, *Woodland Community College*
Jennie Thompson, *Leeward Community College*
J.B. Thoo, *Yuba College*
Binh Truong, *American River College*
Sally Vandenberg, *Barstow College*
Barbara Villatoro, *Diablo Valley College*
Dahlia N. Vu, *Santa Ana College*
Carol M. Walker, *Hinds Community College*
Tracy Welch, *North Central Missouri College*
Jason Wetzel, *Midlands Technical College*
Emily C. Whaley, *Georgia Perimeter College–Clarkston*
Rebecca Wong, *West Valley College*

We would also like to extend special thanks to other people who have helped us write. We are indebted to Charlie Van Wagner, our publisher, for his support, and to Lynn Cox, Richard Stratton, Mary Finch, Bill Hoffman, Angus McDonald, Maria Morelli, Peter Galuardi, Hazel McKenna, and Jay Campbell for their encouragement. The insight and experience of Erin Brown was our mainstay during writing. We cannot thank Cheryll Linthicum and Martha Emry enough for being in our corner during production. Rita Lombard, Katherine Greig, Gordon Lee, and Danae April helped us clarify the story behind this book and have developed a powerful marketing plan. We also are grateful for the efforts of Leslie Lahr, Barbara Willette, Marian Selig, Heleny Wong, Guanglei Zhang, Carrie Jones, Lauren Crosby, Jennifer Cordoba, Carrie Green, Roger Lipsett, Tammy Morgan, Linda Collie, and Susan Dimick. Our special thanks goes to mathematics instructors Scott Barnett and Alan Hain, whose experience, insight, and attention to detail has helped us maintain the highest standards of accuracy.

We have been very fortunate to have the support of colleagues and staff as we have class-tested this manuscript at Lewis-Clark State College. Our thanks to Alan Hain, Misti Dawn Henry, Burma Hutchinson, Matt Johnston, Masoud Kazemi, Jean Sawyer, Karen Schmidt, the staff in the print shop, and the people in the physical plant who delivered the endless boxes of copies.

Above all, we thank the students who gave their forthright feedback as they used preliminary and class test editions. Their efforts were invaluable as we wrote and accuracy checked this first edition.

Laura Bracken
Ed Miller

Dear Student,

In both our campus and on-line classes, we observe students developing the attitudes and habits that lead to success. Some of those observations are outlined below. As you begin the course, we encourage you to consider the tips provided. They may help you improve your performance as you work through the course.

TIP #1 Ask questions—discover what method works best for you.

Students who are most likely to succeed expect that some things will come more easily than others. When they are confused or discouraged, they seek help. Some students who are enrolled in a course that meets in a classroom may feel awkward about asking questions in front of peers or the instructor. These students often develop strategies for getting the help they need in a more private way, such as sending an email or text message to an instructor or campus tutor. Some students may be able to contact an instructor or tutor by phone. Since these students do not wait to ask questions or to get extra help, they are ready to learn what comes next. Asking questions and completing assignments on time is essential for being successful in mathematics classes. By doing the work and asking for help to complete the work when needed and as the work is assigned, these students seem to have a more rewarding experience.

Textbook Hint

Throughout the textbook, you will find many worked examples, and Practice Problems that correspond to the worked examples, that you can try on your own. Answers to most of the Practice Problems are found in the back of the textbook so that you can check to see how well you did. Answers to the odd-numbered section-ending exercises are also found in the back of the book so that you can check your work.

TIP #2 Make choices that promote long-term goals.

College is an environment where students are regarded as adults, responsible for their own learning. Tracking assignments and due dates for multiple classes, preparing for exams, and getting to know the policies of different instructors can overwhelm some students—especially those entering college for the first time. Students most likely to succeed often develop strategies for managing the requirements of their courses. They may create schedules to track assignment due dates or important exam dates. They may turn down opportunities to attend social events in order to allow themselves enough time to complete a project or to study. Successful students understand that attending college sometimes means making hard choices. A short-term sacrifice may be what it takes to achieve a more meaningful, long-term goal.

Textbook Hint

Success in College Mathematics, found at the beginning of each chapter, offers practical tips on making the transition to college-level courses and managing your time effectively so that you stay on top of your assignments. Refer to the Study Plan and Review at the end of each chapter, along with the Chapter Tests and Cumulative Reviews, to help with exam preparation.

© wavebreakmedia/Shutterstock.com

TIP #3 Learn to be self-sufficient.

Students most likely to succeed in math are students who have learned to be self-sufficient by checking their work for accuracy and reasonability. They often show steps of a solution, line by line, and look for any errors in arithmetic or algebra. These students also may have learned to determine whether the final answers they provide make sense—for example, they might verify that values reported as final answers are not too big or small in the context of the problems they are solving. Becoming a self-sufficient learner takes practice, but is a big asset to students in helping them improve their grades and achieve their academic goals.

Textbook Hint

Throughout the text, the worked examples will show you how to check your work for accuracy and reasonability and the section exercises will provide practice with these skills. You will also practice identifying errors in the Find the Mistake exercises.

TIP #4 Memorize less, understand more.

Students who are successful in math seem to rely on their understanding of basic principles and concepts rather than on memorizing how to do a specific problem. They look for connections and patterns and learn how to apply basic principles to new problems.

Textbook Hint

Section 1.3 outlines a series of Five Steps that can be applied to solve a wide variety of problems. You will also find that as you read the explanations next to each step of worked examples certain rules and concepts are used repeatedly—but to solve very different problems. Important properties, rules, and patterns are listed in boxes throughout the book. Be sure to complete the exercises in each section and in the Study Plan that ask you to explain your understanding of different concepts.

Arithmetic Review

R

Pretest

Complete this Pretest, and use the answers on p. ANS-1 to identify any mistakes. Do the reading and exercises from this chapter assigned by your instructor, or use the Study Plan on pp. 32–33 to organize your review. Once you have finished your review, complete and correct the Posttest on p. 34. Use your results to decide whether you need additional review.

For problems 1–20, evaluate. All fractions should be simplified to lowest terms. Do not rewrite improper fractions as mixed numbers.

1. $-5 - 4$ **2.** $-\frac{6}{2}$ **3.** $0(3)$ **4.** $7 \div 0$

5. $12 + 6 \div 3$ **6.** $10 - (3 + 2)^2$ **7.** $148 \div 4$ **8.** $-8 - (-2)$

9. $(-5)(-7)$ **10.** $\frac{3}{10} + \frac{1}{10}$ **11.** $\left(\frac{4}{5}\right)\left(\frac{15}{16}\right)$ **12.** $\frac{5}{9} \div \frac{2}{3}$

13. $\frac{20}{9} - \frac{5}{8}$ **14.** $\frac{1}{4} + \frac{5}{6}$ **15.** $2 - \frac{4}{7}$ **16.** $-12 \div 2 \cdot 3$

17. $(-3)^2$ **18.** -3^2 **19.** $\frac{0}{10}$ **20.** $\frac{1}{2} - \frac{2}{3} \div 4$

21. Round 3.847 to the nearest tenth. **22.** Round 189,472 to the nearest hundred.

23. What percent is 8 of 40? **24.** What percent is 54 of 180?

25. Find 12% of 300. **26.** Find 7% of 250.

For problems 27–30, write > or < between the numbers to make a true statement.

27. $-8 \quad 5$ **28.** $-8 \quad -5$ **29.** $\frac{3}{4} \quad \frac{2}{3}$ **30.** $0.18 \quad -0.3$

R.1 Evaluating Expressions and the Order of Operations
R.2 Signed Numbers
R.3 Fractions
R.4 Decimal Numbers, Percent, and Order

SECTION R.1 Evaluating Expressions and the Order of Operations

The Order of Operations

Addition, subtraction, multiplication, and division are **operations**. The result of addition is a **sum**, and the result of subtraction is a **difference**. The result of multiplication is a **product**, and the result of division is a **quotient**.

A **notation** is a way of representing an operation or expression. For example, the notations for multiplying 3 and 5 include $3 \cdot 5$, $3(5)$, $(3)(5)$, and 3×5. The notations for dividing 6 by 3 include $6 \div 3$, $\frac{6}{3}$, and $3\overline{)6}$.

To **evaluate** an expression, follow the **order of operations**. Write down the original expression. Since each of the lines in the solution is equivalent, begin each new line with an equals sign.

The Order of Operations

1. Working from left to right, evaluate expressions in parentheses or in other grouping symbols (whichever comes first).
2. Working from left to right, evaluate exponential expressions or roots (whichever comes first).
3. Working from left to right, multiply or divide (whichever comes first).
4. Working from left to right, add or subtract (whichever comes first).

EXAMPLE 1 Evaluate: $8 + 10 \div 2$

SOLUTION

$8 + 10 \div 2$

$= 8 \mathbf{+ 5}$ From left to right, multiply or divide.

$= 13$ From left to right, add or subtract.

A **base** raised to an **exponent** represents repeated multiplications. For example, 3^4 is an **exponential expression** that is equivalent to $3 \cdot 3 \cdot 3 \cdot 3$. The base in this expression is 3. The exponent is 4.

EXAMPLE 2 Evaluate: $5^3 - 4 \cdot 2 + 9$

SOLUTION

$5^3 - 4 \cdot 2 + 9$

$= \mathbf{125} - 4 \cdot 2 + 9$ From left to right, evaluate exponential expressions; $5^3 = (5)(5)(5)$

$= 125 \mathbf{- 8} + 9$ From left to right, multiply or divide.

$= \mathbf{117} + 9$ From left to right, add or subtract.

$= 126$ From left to right, add or subtract.

Grouping symbols such as **parentheses ( )** or **square brackets [ ]** show the parts of an expression that should be evaluated first.

EXAMPLE 3 Evaluate: $7 - (2 + 1)$

SOLUTION

$7 - (2 + 1)$

$= 7 - \mathbf{3}$ Inside parentheses, from left to right, add or subtract.

$= 4$ From left to right, add or subtract.

In following the order of operations, it is very important to work from left to right. If an expression contains both multiplication and division, do whichever operation comes first. In the next example, working from left to right, division comes first.

EXAMPLE 4 Evaluate: $18 \div 2 \cdot 3$

SOLUTION ▶

$18 \div 2 \cdot 3$

$= \mathbf{9} \cdot 3$ From left to right, multiply or divide.

$= 27$ From left to right, multiply or divide.

In the next example, working from left to right, multiplication comes first.

EXAMPLE 5 Evaluate: $18 \cdot 2 \div 3$

SOLUTION ▶

$18 \cdot 2 \div 3$

$= \mathbf{36} \div 3$ From left to right, multiply or divide.

$= 12$ From left to right, multiply or divide.

Work inside of grouping symbols first, following the complete order of operations. In the next example, the exponential expression inside the grouping symbols is evaluated first.

EXAMPLE 6 Evaluate: $5^2 - 2(3^2 - 8) + 7$

SOLUTION ▶

$5^2 - 2(3^2 - 8) + 7$

$= 5^2 - 2(\mathbf{9} - 8) + 7$ Inside parentheses, evaluate exponential expressions.

$= 5^2 - 2(\mathbf{1}) + 7$ Inside parentheses, from left to right, add or subtract.

$= \mathbf{25} - 2(1) + 7$ From left to right, evaluate exponential expressions.

$= 25 \mathbf{- 2} + 7$ From left to right, multiply or divide.

$= \mathbf{23} + 7$ From left to right, add or subtract.

$= 30$ From left to right, add or subtract.

After evaluating an expression inside of grouping symbols, choose the next operation by reading the expression from left to right.

EXAMPLE 7 Evaluate: $150 - 18 \div 6(3 + 1)^2$

SOLUTION ▶

$150 - 18 \div 6(3 + 1)^2$

$= 150 - 18 \div 6(\mathbf{4})^2$ Inside parentheses, from left to right, add or subtract.

$= 150 - 18 \div 6(\mathbf{16})$ From left to right, evaluate exponential expressions.

$= 150 \mathbf{- 3}(16)$ From left to right, multiply or divide.

$= 150 \mathbf{- 48}$ From left to right, multiply or divide.

$= 102$ From left to right, add or subtract.

When there is more than one pair of grouping symbols, work from the inside out.

EXAMPLE 8 Evaluate: $5 \cdot 8 - [9 - (12 \div 6 \cdot 2)]^2$

SOLUTION ▶

$5 \cdot 8 - [9 - (12 \div 6 \cdot 2)]^2$

$= 5 \cdot 8 - [9 - (\mathbf{2} \cdot 2)]^2$ Inside parentheses, from left to right, multiply or divide.

$= 5 \cdot 8 - [9 - \mathbf{4}]^2$ Inside parentheses, from left to right, multiply or divide.

$= 5 \cdot 8 - [\mathbf{5}]^2$ Inside brackets, from left to right, add or subtract.

$= 5 \cdot 8 - \mathbf{25}$ From left to right, evaluate exponential expressions.

$= \mathbf{40} - 25$ From left to right, multiply or divide.

$= 15$ From left to right, add or subtract.

Some expressions may include operations in the numerator (the top) and the denominator (the bottom). Evaluate the expression as if the numerator and denominator were each inside parentheses. The fraction bar acts as a grouping symbol. A fraction such as $\frac{50}{10}$ is equivalent to $50 \div 10$.

EXAMPLE 9 Evaluate: $\dfrac{90 - 5 \cdot 2}{50 \div 2 - 15}$

SOLUTION ▶

$\dfrac{90 - 5 \cdot 2}{50 \div 2 - 15}$

$= \dfrac{90 - \mathbf{10}}{50 \div 2 - 15}$ In the numerator, from left to right, multiply or divide.

$= \dfrac{\mathbf{80}}{50 \div 2 - 15}$ In the numerator, from left to right, add or subtract.

$= \dfrac{80}{\mathbf{25} - 15}$ In the denominator, from left to right, multiply or divide.

$= \dfrac{80}{\mathbf{10}}$ In the denominator, from left to right, add or subtract.

$= 8$ From left to right, multiply or divide.

Answers to all Practice Problems may be found in the back of the book.

Practice Problems

For problems 1–12, evaluate. Follow your instructor's guidelines for showing your work.

1. $4 + 12 \div 3$ **2.** $15 - 6 \cdot 2$ **3.** $30 - 4^2 + 1$

4. $13 - (7 - 3)$ **5.** $40 \div 2 \cdot 5$ **6.** $40 \cdot 2 \div 5$

7. $10 \div 2 \cdot 3 + 3^2 - 4$ **8.** $8 + 4 \div 2 + 5^2$

9. $6(9 - 2) - 4(3 + 1)$ **10.** $[(5 + 8 \div 2) - 3]^2$

11. $\dfrac{(12 - 4)^2}{2^2 \cdot 8}$ **12.** $3[(4^2 - 3^2)^2 - 36 \div 9 \cdot 2]$

Zero and One

When 0 is added to or subtracted from any number, the number does not change. When a number is multiplied or divided by 1, the number is not changed.

EXAMPLE 10 Evaluate the expression.

(a) $5 + 0$

SOLUTION ▶ $= 5$ The sum of a number and 0 is the number.

(b) $5 - 0$

▶ $= 5$ The difference of a number and 0 is the number.

(c) $5 \cdot 1$

▶ $= 5$ The product of a number and 1 is the number.

(d) $\dfrac{5}{1}$

▶ $= 5$ The quotient of a number and 1 is the number.

The product of 0 and a number is 0. The quotient of 0 divided by a number (except 0) is 0. However, *division by zero is undefined.*

EXAMPLE 11 Evaluate the expression.

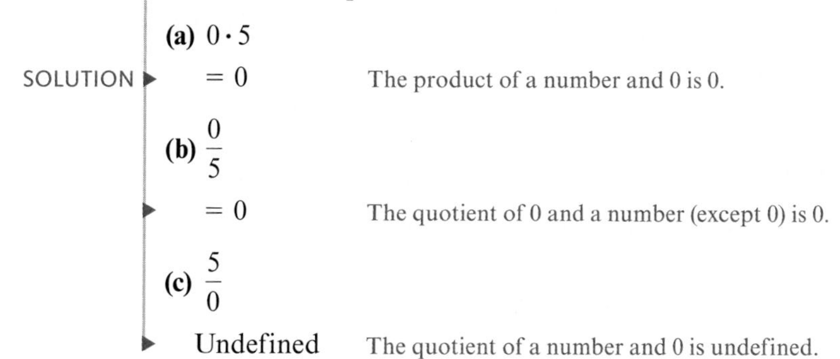

(a) $0 \cdot 5$

SOLUTION ▶ $= 0$ The product of a number and 0 is 0.

(b) $\frac{0}{5}$

▶ $= 0$ The quotient of 0 and a number (except 0) is 0.

(c) $\frac{5}{0}$

▶ Undefined The quotient of a number and 0 is undefined.

Practice Problems

For problems 13–23, evaluate.

13. $6 + 0$ **14.** $6 \cdot 0$ **15.** $6 \div 0$ **16.** $0 \div 6$

17. $6 \cdot 1$ **18.** $6 \div 1$ **19.** $6 - 0$ **20.** $0 + 6$

21. $\frac{5 - 5}{4^2}$ **22.** $\frac{3}{6 - 6}$ **23.** $1(15 - 3 \cdot 2)^2 \div 1$

Using Technology: Order of Operations on a Calculator

A calculator can speed up your work. However, having a calculator does not excuse you from being able to do and understand arithmetic.

Some calculators include the order of operations as part of their programming. Others do not. Some calculators have keys for parentheses. Others do not. Find out what kind of calculator you can use, if any, on homework and tests, and use only that calculator as you learn and practice.

Some scientific calculators have one entry line.

EXAMPLE 12 Use a scientific calculator with one entry line to evaluate $64 - 3 \cdot 5 + 4^2$.

This calculator follows the order of operations for arithmetic. Type the expression from left to right. As we press the keys, the display changes. Use the y^x key to type an exponent. **Do not press = until the end.**

The answer is 65.

If you press = before the end, the calculator no longer views this problem as a single expression. It "thinks" you are starting a new problem. It will apply order of operations only to the new entries.

Other scientific calculators have two entry lines. You can see what you have typed before you press =.

EXAMPLE 13 Use a scientific calculator with two entry lines to evaluate $64 - 3 \cdot 5 + 4^2$.

Type in the expression from left to right. Use the ^ key to enter an exponent.

6 4 - 3 × 5 + 4 ^ 2 = The answer is again 65.

In the next example, the expression includes parentheses. Most scientific calculators include keys for parentheses.

EXAMPLE 14 Use a scientific calculator with two entry lines to evaluate $64 - 3 \cdot (5 + 4^2)$.

Look for the (and) keys.

Press 6 4 − 3 × (5 + 4 ^ 2) = The answer is 1.

Graphing calculators are used in some elementary algebra classes. Instead of pressing =, press ENTER.

No matter what kind of calculator you use, it is extremely important that you always take the time to think about the answer. If you think, "That can't be right," it is time to redo the problem.

Practice Problems For problems 24–26, use a calculator to evaluate.

24. $60 - 12 \div 3$ **25.** $36 \div 2 \cdot 3 + 3^2 - 4$ **26.** $6(9 - 2) - 4(3 + 1)$

R.1 Exercises

Follow your instructor's guidelines for showing your work.

For exercises 1–80, evaluate.

1. 6^2 **2.** 7^2

3. 4^3 **4.** 5^3

5. $100 - 6^2$ **6.** $100 - 7^2$

7. $12 + 4^3$ **8.** $16 + 5^3$

9. $60 - 2 \cdot 6$ **10.** $90 - 3 \cdot 5$

11. $60 \div 2 \cdot 6$ **12.** $90 \div 3 \cdot 5$

13. $60 \cdot 2 \div 6$ **14.** $90 \cdot 3 \div 5$

15. $12(5 - 1)$ **16.** $15(6 - 1)$

17. $12 - (5 - 1)$ **18.** $15 - (6 - 1)$

19. $20 - 6 \div 2 + 1$ **20.** $18 - 12 \div 2 + 1$

21. $20 - 6 \div (2 + 1)$ **22.** $18 - 12 \div (2 + 1)$

23. $(20 - 6) \div 2 + 1$ **24.** $(18 - 12) \div 2 + 1$

25. $(9 - 1)^2$ **26.** $(10 - 1)^2$

27. $8 + 6(5 - 3)$ **28.** $9 + 5(6 - 2)$

29. $(8 + 6)(5 - 3)$ **30.** $(9 + 5)(6 - 2)$

31. $19 - (10 - 3)$ **32.** $17 - (12 - 2)$

33. $(19 - 10) - 3$ **34.** $(17 - 12) - 2$

35. $6 + 5 \cdot 0$ **36.** $9 + 2 \cdot 0$

37. $8 - 0 \div 3$ **38.** $5 - 0 \div 7$

39. $3 \div 0$ **40.** $7 \div 0$

41. $40 - 12 \div 2 \cdot 3$ **42.** $50 - 24 \div 3 \cdot 2$

43. $40 - 12 \cdot 2 \div 3$ **44.** $50 - 24 \cdot 3 \div 2$

45. $(40 - 12) \div 2 \cdot 3$ **46.** $(50 - 24) \cdot 3 \div 2$

47. $12 \div 3 \cdot 2 \cdot 4 - 9$ **48.** $15 \div 3 \cdot 2 \cdot 4 - 6$

49. $\dfrac{(12 - 2)^2}{29 - 3^2}$ **50.** $\dfrac{(14 - 4)^2}{29 - 2^2}$

51. $5 + 3(8 - 2) \div 2$ **52.** $3 + 5(11 - 5) \div 3$

53. $160 - 2 \cdot 5 \cdot 3^2$ **54.** $90 - 3 \cdot 5 \cdot 2^2$

55. $4^3 + 10 \div 2 \cdot 6$ **56.** $3^3 + 9 \div 3 \cdot 5$

57. $2 + 6^2 \div (3^2 - 5)$ **58.** $5 + 8^2 \div (20 - 2^2)$

59. $[(60 \cdot 2 \div 6) - (13 + 1)]^2$

60. $[(40 \cdot 3 \div 5) - (14 + 2)]^2$

61. $\dfrac{12(5 - 1)}{9 \cdot 2 - 2}$ **62.** $\dfrac{15(6 - 2)}{7 \cdot 2 - 2}$

63. $\dfrac{12(5 - 1)}{9(2 - 2)}$ **64.** $\dfrac{15(6 - 2)}{7(2 - 2)}$

65. $\dfrac{12 - (5 - 1)}{2^2 \cdot 2}$ **66.** $\dfrac{34 - (8 - 1)}{3^2 \cdot 3}$

67. $\dfrac{(9 - 1)^2 - 64}{2^3}$ **68.** $\dfrac{(10 - 1)^2 - 81}{2^3}$

69. $24 \div 6(5 - 3) + 18 \cdot 1$

70. $20 \div 5(8 - 2) + 17 \cdot 1$

71. $2[24 \div (8 - 6)(5 - 3)]$

72. $3[64 \div (9 - 5)(6 - 4)]$

73. $[5(19 - 10) - 3] - (8 - 4)^2$

74. $[9(17 - 12) - 2] - (9 - 4)^2$

75. $\left(\dfrac{6 + 5 \cdot 0}{6}\right)^2$ **76.** $\left(\dfrac{9 + 2 \cdot 0}{9}\right)^2$

77. $\dfrac{3 + 8^2 - 4^2}{3^2 - 9}$ **78.** $\dfrac{7 + 6^2 - 5^2}{4^2 - 16}$

79. $\dfrac{(8 - 2)^2}{13 - 4}$ **80.** $\dfrac{(10 - 4)^2}{14 - 5}$

81. a. Write your own example of an expression that includes three operations. Design the expression so that the evaluated expression equals a whole number.

b. Evaluate this expression.

82. a. Write your own example of an expression that includes four operations. Design the expression so that the evaluated expression equals a whole number.

b. Evaluate this expression.

83. A student was mixed up about the order of operations and always did multiplication first before she did any division. Write an expression in which doing this will result in an incorrect answer.

84. A student was mixed up about the order of operations and always did multiplication first before he did any division. Write an expression in which doing this will still result in a correct answer.

Technology

For exercises 85–88, use a calculator to evaluate. If the calculator has parentheses, use them.

85. $20 - 6 \div (2 + 1)$

86. $18 - 12 \div (2 + 1)$

87. $4 + (9 - 1)^2$

88. $3 + (8 - 1)^2$

SECTION R.2 Signed Numbers

Addition and Subtraction of Signed Numbers

The vertical lines on a number line graph are **tick marks**. The distance between each pair of adjacent tick marks is the same. The number below a tick mark shows its value. Together, these numbers are the **scale** of the number line graph. The x at the right end of a number line graph will be important in later work in algebra. To graph a number, draw a **point** on the number line graph.

EXAMPLE 1 Use a number line graph to represent 3.

SOLUTION ▶

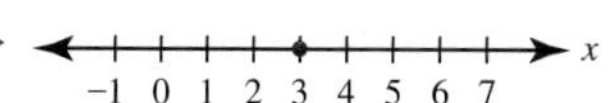

To show that the pattern of tick marks and numbers continue, each end of the number line graph in Example 1 has an arrow. To the left of 0 on the number line graph are the negative numbers.

EXAMPLE 2 Use a number line graph to represent -3.

SOLUTION ▶

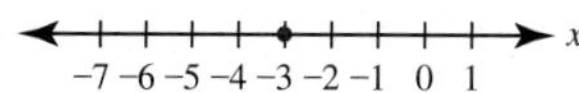

The numbers 3 and -3 are **opposites**. They are the same distance from 0 on the number line graph but are on opposite sides of 0. The opposite of 3 is negative 3; the opposite of negative 3 is 3. The same notation, a $-$ sign, represents a negative number and the opposite of a number. For example, -3 represents "negative 3," and it also represents "the opposite of 3."

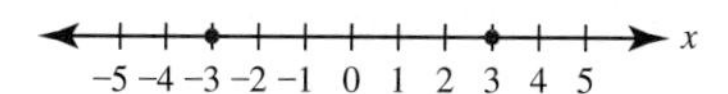

The **absolute value** of a number is its distance from 0 on the number line graph. The notation for absolute value is $|\ |$. Since the distance of 3 from 0 is 3, $|3| = 3$. Since the distance of -3 from 0 is also 3, $|-3| = 3$.

We can use a number line graph to visualize addition of signed numbers. Use a point to represent the first number in an addition problem. To add a positive number, move to the right.

EXAMPLE 3 Use a number line graph to evaluate $-9 + 5$.

SOLUTION ▶ Graph a point at -9. Move 5 tick marks to the right, ending at -4

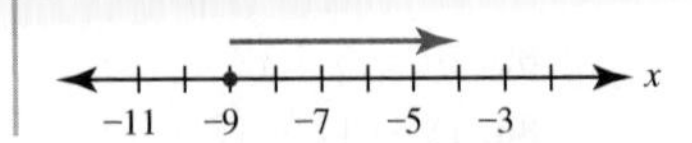

To add a negative number, move to the left. Notice that a negative number following the + sign is written in parentheses to separate the + and − signs.

EXAMPLE 4 Use a number line graph to evaluate $-3 + (-6)$.

SOLUTION ▶ Graph a point at -3. Move 6 tick marks to the left, ending at -9.

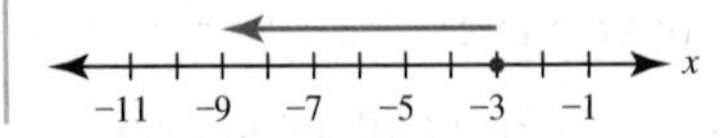

To add positive numbers on a number line graph, move to the right from the beginning point. To subtract, move to the left from the beginning point.

EXAMPLE 5 Use a number line graph to evaluate $8 - 2$.

SOLUTION ▶ Graph a point at 8. Move 2 tick marks to the left, ending at 6.

If the beginning point represents a negative number and we subtract a positive number, moving further to the left, the difference is a negative number.

EXAMPLE 6 Use a number line graph to evaluate $-3 - 6$.

SOLUTION ▶ Graph a point at -3. Move 6 tick marks to the left, ending at -9.

$$-3 - 6 = -9$$

Comparing Example 4, $-3 + (-6) = -9$, and Example 6, $-3 - 6 = -9$, we see that *subtraction of a positive number is equivalent to adding a negative number.*

EXAMPLE 7 Rewrite $-2 - 3$ as addition of a negative number, and evaluate.

SOLUTION ▶

$-2 - 3$

$= -2 + \mathbf{(-3)}$ Rewrite subtraction as addition of a negative number.

$= -5$ Follow the order of operations.

Since the opposite of a negative number is a positive number, subtraction of a negative number is equivalent to adding a positive number.

EXAMPLE 8 Evaluate: $2 - (-7)$

SOLUTION ▶

$2 - (-7)$

$= 2 + -(-7)$ Rewrite subtraction as addition of a negative number.

$= 2 + 7$ The opposite of -7 is 7; $-(-7) = 7$

$= 9$ Follow the order of operations.

Rewriting subtraction as addition of a negative number and rewriting the opposite of a negative number as a positive number are often done in one step.

EXAMPLE 9 Evaluate: $-9 - (-4)$

SOLUTION ▶

$-9 - (-4)$

$= -9 + \mathbf{4}$ Rewrite $-(-4)$ as $+4$.

$= -5$ Follow the order of operations.

Practice Problems

For problems 1–4, use a number line graph to represent the number(s).

1. 10 **2.** 8 **3.** 4 and its opposite **4.** -5 and $|-5|$

For problems 5–9, use a number line graph to evaluate.

5. $2 + (-6)$ **6.** $-2 + (-6)$ **7.** $-2 - 6$
8. $-2 + 6$ **9.** $2 - 6$

For problems 10–19, evaluate.

10. $8 - 3$ **11.** $3 - 8$ **12.** $-3 + 8$ **13.** $-3 - 8$
14. $-3 + (-8)$ **15.** $3 - (-8)$ **16.** $-3 - (-8)$
17. $-4 + 11$ **18.** $-4 - (-11)$ **19.** $-4 - 11$

Multiplication and Division of Signed Numbers

Repeated additions are equivalent to multiplication. Rewriting repeated additions of a positive number as multiplication shows that *the product of a positive number and a positive number is a positive number.*

$3 + 3 + 3 + 3$

$= 4(3)$ Rewrite repeated addition as multiplication.

$= 12$ Evaluate.

Rewriting repeated additions of a negative number as multiplication shows that *the product of a positive number and a negative number is a negative number.*

$-3 + (-3) + (-3) + (-3)$

$= 4(-3)$ Rewrite repeated addition as multiplication.

$= -12$ Evaluate.

Changing the order of the multiplication of two numbers does not change the product. Since the product of a positive number and a negative number is a negative number, then *the product of a negative number and a positive number is a negative number.* For example, $(-3)4 = -12$.

In Section 1.1, we will learn about the distributive property. Using this property, we can show that the product of a negative number and a negative number is a positive number.

$(-4)(-3)$

$= 12$

Sign Rules for Multiplication

(positive number)(positive number) = positive number

(positive number)(negative number) = negative number

(negative number)(positive number) = negative number

(negative number)(negative number) = positive number

In multiplying two nonzero numbers with the same signs, the product is a positive number. In multiplying two nonzero numbers with different signs, the product is a negative number.

EXAMPLE 10 Evaluate: $9(-4)$

SOLUTION ▶

$9(-4)$

$= -36$ (positive number)(negative number) = negative number

The operation of division "undoes" the operation of multiplication. For example, $2 \cdot 3$ equals a product of 6. If we divide 6 by 3, the quotient is 2. Similarly, since $2(-3) = -6$, $-6 \div (-3) = 2$. The rules for identifying the sign of a quotient are the same as those for identifying the sign of a product.

Sign Rules for Division

$$\frac{\text{positive number}}{\text{positive number}} = \text{positive number}$$

$$\frac{\text{positive number}}{\text{negative number}} = \text{negative number}$$

$$\frac{\text{negative number}}{\text{positive number}} = \text{negative number}$$

$$\frac{\text{negative number}}{\text{negative number}} = \text{positive number}$$

In dividing two nonzero numbers with the same signs, the product is a positive number. In dividing two nonzero numbers with different signs, the product is a negative number.

EXAMPLE 11 Evaluate: $\frac{-28}{-4}$

SOLUTION ▶

$\frac{-28}{-4}$

$= 7$ $\frac{\text{negative number}}{\text{negative number}} = \text{positive number}$

In the next example, the base is a negative number. Follow the order of operations, and multiply from left to right.

EXAMPLE 12 Evaluate: $(-4)^3$

SOLUTION ▶

$(-4)^3$	
$= (-4)(-4)(-4)$	Rewrite as repeated multiplications.
$= \mathbf{(16)}(-4)$	(negative number)(negative number) = positive number
$= -64$	(positive number)(negative number) = negative number

In Example 12, the base, -4, is inside parentheses, and $(-4)^3$ equals $(-4)(-4)(-4)$. In the next example, we evaluate -4^2. The base is not inside parentheses; -4^2 is the opposite of 4^2.

EXAMPLE 13 Evaluate: -4^2

SOLUTION ▶

-4^2	The opposite of 4^2.
$= \mathbf{-4 \cdot 4}$	Rewrite as repeated multiplications; the base is positive.
$= \mathbf{-16}$	(negative number)(positive number) = negative number

Practice Problems

For problems 20–30, evaluate.

20. $(12)(2)$ **21.** $(-12)(2)$ **22.** $(-12)(-2)$ **23.** $(12)(-2)$
24. $12 - (-2)$ **25.** $-12 + 2$ **26.** $-12 + (-2)$ **27.** $12 + (-2)$
28. 7^2 **29.** $(-7)^2$ **30.** -7^2

The Order of Operations and Signed Numbers

We can use the order of operations to evaluate expressions that include positive and negative numbers.

EXAMPLE 14 Evaluate: $12 \div (-4) - 3(6)$

SOLUTION ▶

$12 \div (-4) - 3(6)$	
$= \mathbf{-3} - 3(6)$	From left to right, multiply or divide.
$= -3 \mathbf{- 18}$	From left to right, multiply or divide.
$= -21$	From left to right, add or subtract.

Notice that parentheses can separate signs, as in $3 - (-4)$ or $3 + (-4)$. Parentheses can also be a notation for multiplication, as in $(-3)(-4)$ or $4(-3)$.

EXAMPLE 15 Evaluate: $-5^2 + 17 + 4(-3) + (-9)$

SOLUTION ▶

$-5^2 + 17 + 4(-3) + (-9)$	
$= \mathbf{-25} + 17 + 4(-3) + (-9)$	From left to right, evaluate exponential expressions.
$= -25 + 17 \mathbf{- 12} + (-9)$	From left to right, multiply or divide.
$= \mathbf{-8} - 12 + (-9)$	From left to right, add or subtract.
$= \mathbf{-20} + (-9)$	From left to right, add or subtract.
$= -29$	From left to right, add or subtract.

Some expressions may include operations in the numerator and the denominator. Evaluate the expression as if the numerator and denominator were each inside parentheses.

EXAMPLE 16 Evaluate: $\dfrac{-8 - 4^2 + 10}{-2(-1)}$

SOLUTION ▶

$\dfrac{-8 - 4^2 + 10}{-2(-1)}$

$= \dfrac{-8 - \mathbf{16} + 10}{-2(-1)}$ In numerator, from left to right, evaluate exponential expressions.

$= \dfrac{\mathbf{-24} + 10}{-2(-1)}$ In numerator, from left to right, add or subtract.

$= \dfrac{\mathbf{-14}}{-2(-1)}$ In numerator, from left to right, add or subtract.

$= \dfrac{-14}{\mathbf{2}}$ In denominator, from left to right, multiply or divide.

$= -7$ From left to right, multiply or divide.

Practice Problems

For problems 31–36, evaluate.

31. $30 - 24 \div (-2)(3)$

32. $-4^2 \cdot 3 - 12 \div (-2)^2$

33. $(-4)^2 + 3 - 12 \div (-2)^2$

34. $(-12)(-2) - (-12 - 2)$

35. $\dfrac{15 - 2 \cdot 3}{-3^2}$

36. $\dfrac{4 + 6(2 - 8)}{-3 + 19}$

Using Technology: Signed Numbers on a Calculator

On a scientific or graphing calculator, you will find a [(-)] key or a [+⊂-] key. These keys are used to type negative numbers. A negative number may be inside parentheses to separate signs. If this is the only purpose of the parentheses, we do not have to type them.

For the operation of subtraction, use a different key, the minus sign [-].

EXAMPLE 17 Evaluate $-6 - (-2)$ on a scientific calculator.

Press [(-)] [6] [-] [(-)] [2] [=] The answer is -4.

EXAMPLE 18 Evaluate $\dfrac{30}{-6}$ on a scientific calculator.

Press [3] [0] [÷] [(-)] [6] [=] The answer is -5.

EXAMPLE 19 Evaluate $\dfrac{-5}{0}$ on a scientific calculator.

Press [(-)] [5] [÷] [0] [=] The calculator shows *Error* or *Divide by 0 Error.*

Do not write "error" as your answer. The answer is "undefined."

Practice Problems For problems 37–40, use a calculator to evaluate.

37. $-18 - 20 \div (-4)$

38. $-3^2 - (4 - 6)$

39. $\dfrac{-18}{3(-2)}$

40. $(-10)(-2) - (-10 - 2)$

R.2 Exercises

Follow your instructor's guidelines for showing your work.

For exercises 1–6, use a number line graph to represent the number(s).

1. 4

2. 5

3. 6 and -6

4. 7 and -7

5. $|-2|$ and -2

6. $|-6|$ and -6

For exercises 7–14, use a number line graph to evaluate.

7. $-3 + 10$

8. $-2 + 7$

9. $-3 + (-10)$

10. $-2 + (-7)$

11. $3 - 10$

12. $2 - 7$

13. $-3 - 10$

14. $-2 - 7$

For exercises 15–100, evaluate.

15. $2 + 3$

16. $4 + 6$

17. $-2 + 3$

18. $-4 + 6$

19. $-2 + (-3)$

20. $-4 + (-6)$

21. $2 - 3$

22. $4 - 6$

23. $-2 - 3$

24. $-4 - 6$

25. $-2 - (-3)$

26. $-4 - (-6)$

27. $(2)(3)$

28. $(4)(6)$

29. $(-2)(3)$

30. $(-4)(6)$

31. $(2)(-3)$

32. $(4)(-6)$

33. $(-2)(-3)$

34. $(-4)(-6)$

35. $\frac{6}{2}$

36. $\frac{24}{4}$

37. $\frac{6}{-2}$

38. $\frac{24}{-4}$

39. $\frac{-6}{2}$

40. $\frac{-24}{4}$

41. $\frac{-6}{-2}$

42. $\frac{-24}{-4}$

43. $-6 \div 2$

44. $-24 \div 4$

45. $-6 \div (-2)$

46. $-24 \div (-4)$

47. 8^2

48. 5^2

49. $(-8)^2$

50. $(-5)^2$

51. -8^2

52. -5^2

53. $10 - 8^2$

54. $10 - 5^2$

55. $18 - 12 \div 2$

56. $16 - 12 \div 2$

57. $18 - 12 \div (-2)$

58. $16 - 12 \div (-2)$

59. $-18 - 12 \div (-2)$

60. $-16 - 12 \div (-2)$

61. $-3^2 + 8(-2)$

62. $-4^2 + 3(-2)$

63. $8(-3) - 1$

64. $7(-4) - 1$

65. $(1 - 8)^2 - 50$

66. $(1 - 10)^2 - 82$

67. $-2^2 - 15(-1)$

68. $-3^2 - 17(-1)$

69. $2 - 6 \div 2 + 1$

70. $3 - 12 \div 3 + 1$

71. $\frac{12 - 3}{-1 - 2}$

72. $\frac{11 - 5}{-1 - 2}$

73. $\frac{-12 - 6^2}{(3)(-2)}$

74. $\frac{-8 - 8^2}{(3)(-2)}$

75. $(4 - 9)^2 - 3(-1)$

76. $(5 - 8)^2 - 4(-1)$

77. $\frac{13}{2 - 2}$

78. $\frac{11}{4 - 4}$

79. $\frac{2 - 2}{9}$

80. $\frac{3 - 3}{14}$

81. $-6 - 3^2 - (-4)$

82. $-2 - 5^2 - (-1)$

83. $-16 + 18 \div (-2)$

84. $-8 + 20 \div (-4)$

85. $-16 - 18 \div (-2)$

86. $-8 - 20 \div (-4)$

87. $-16 - 18 \div 2$

88. $-8 - 20 \div 4$

89. $-6^2 \cdot 12 \div 3 - 4$

90. $-8^2 \cdot 10 \div 2 - 9$

91. $(-6)^2 \cdot 12 \div 3 - 4$

92. $(-8)^2 \cdot 10 \div 2 - 9$

93. $(-6)(-5) - (-4)$

94. $(-8)(-2) - (-5)$

95. $4 - 9 \div (-3) - (-5)^2$

96. $6 - 18 \div (-2) - (-4)^2$

97. $-2(6 - 9)^2 - 5$

98. $-3(8 - 13)^2 - 9$

99. $23 + (-9)^2 - 7 - (-18)$

100. $15 + (-7)^2 - 8 - (-22)$

Technology

For exercises 101–104, evaluate. Use a scientific or a graphing calculator.

101. $-80 \div (-2) + 4(-6)^2$

102. $-8^2 \cdot 2 - 9$

103. $54 - (-12) - 9 \div (-3)^2$

104. $-8^2 - 4 - 15 \div (-3)$

SECTION R.3 Fractions

Simplifying Fractions into Lowest Terms

The result of multiplication is a **product**. The multiplied numbers are **factors**. Since $1 \cdot 6 = 6$ and $2 \cdot 3 = 6$, the factors of 6 are 1, 2, 3, and 6. The **greatest common factor** of two numbers is the greatest factor that two numbers share.

EXAMPLE 1 Identify the greatest common factor of 24 and 36.

SOLUTION ▶ The factors of 24 are 1, 2, 3, 4, 6, 8, 12, and 24.

The factors of 36 are 1, 2, 3, 4, 6, 9, 12, 18, and 36.

The greatest common factor is 12.

In the fraction $\frac{3}{4}$, the **numerator** is 3 and the **denominator** is 4. The denominator tells us the *number of total parts* in one whole. The numerator tells us the *number of parts that we have*. If a fraction is equal to 0, the numerator is 0, and the denominator is any number except 0. If a fraction is equal to 1, the numerator and denominator are equal. If a fraction is in **lowest terms**, the numerator and denominator have no common factors except 1.

To simplify a fraction into lowest terms, find common factors of the numerator and denominator. If the same factor is in the numerator and denominator, it is a fraction that is equal to 1.

EXAMPLE 2 Simplify: $\frac{24}{36}$

SOLUTION ▶

$$\frac{24}{36}$$

$$= \frac{\mathbf{12} \cdot 2}{\mathbf{12} \cdot 3} \quad \text{The greatest common factor of 24 and 36 is 12.}$$

$$= \mathbf{1} \cdot \frac{2}{3} \quad \text{Rewrite; } \frac{12}{12} = 1$$

$$= \frac{2}{3} \quad \text{From left to right, multiply or divide.}$$

In Example 2, we simplified $\frac{24}{36}$ into an equivalent fraction written in lowest terms, $\frac{2}{3}$. Although these fractions represent the same amount, we simplify into lowest terms because it is often easier to do arithmetic with fractions in which the numerator and denominator are smaller numbers.

To save time when simplifying fractions, we often do not rewrite fractions such as $\frac{12}{12}$ as 1. Instead, we draw a line through the common factors in the numerator and denominator. We can draw these lines only because the fraction formed by the common factors is equal to 1 and because the product of 1 and any number is that number.

EXAMPLE 3 Simplify: $\frac{14}{21}$

SOLUTION ▶

$\frac{14}{21}$

$= \frac{2 \cdot \cancel{7}}{3 \cdot \cancel{7}}$ Find common factors.

$= \frac{2}{3}$ Simplify; this fraction is in lowest terms.

In the next example, before simplifying, rewrite the numerator as the product of the numerator and 1. The numerator of the simplified fraction is 1.

EXAMPLE 4 Simplify: $\frac{5}{30}$

SOLUTION ▶

$\frac{5}{30}$

$= \frac{\mathbf{1} \cdot 5}{30}$ Rewrite the numerator as $1 \cdot 5$.

$= \frac{1 \cdot \cancel{5}}{\cancel{5} \cdot 6}$ Find common factors.

$= \frac{1}{6}$ Simplify; this fraction is in lowest terms.

Simplifying a Fraction into Lowest Terms

1. Identify the greatest common factor in the numerator and denominator.
2. Simplify fractions that are equal to 1.

If the common factor is not the *greatest* common factor, the fraction will need to be simplified again.

EXAMPLE 5 Simplify: $\frac{36}{72}$

SOLUTION ▶

$\frac{36}{72}$

$= \frac{\cancel{9} \cdot 4}{\cancel{9} \cdot 8}$ A common factor of 36 and 72 is 9.

$= \frac{4}{8}$ Simplify; this fraction is not in lowest terms.

$= \frac{\cancel{4} \cdot 1}{\cancel{4} \cdot 2}$ A common factor of 4 and 8 is 4.

$= \frac{1}{2}$ Simplify; this fraction is in lowest terms.

Practice Problems

For problems 1–6, simplify. Some fractions may already be in lowest terms.

1. $\frac{4}{12}$ **2.** $\frac{15}{18}$ **3.** $\frac{4}{64}$ **4.** $\frac{53}{53}$ **5.** $\frac{42}{70}$ **6.** $\frac{21}{25}$

Improper Fractions and Mixed Numbers

If the numerator of a positive fraction is less than the denominator, it is a **proper fraction**. If the numerator of a positive fraction is greater than or equal to the denominator, it is an **improper fraction**.

EXAMPLE 6 Simplify: $\frac{50}{6}$

SOLUTION ▶ $\frac{50}{6}$

$= \frac{\cancel{2} \cdot 25}{\cancel{2} \cdot 3}$ The greatest common factor of 50 and 6 is 2.

$= \frac{25}{3}$ Simplify; this is an improper fraction.

A positive mixed number is the sum of a whole number and a proper fraction. To find the whole number part of a mixed number, use mental arithmetic or long division with remainders (do not add zeros after the decimal point). The whole number part of the quotient is the whole number part of the mixed number. The remainder is the numerator of the proper fraction; the divisor is the denominator of the proper fraction.

EXAMPLE 7 Rewrite $\frac{22}{5}$ as a mixed number.

SOLUTION ▶ $\frac{22}{5}$ $22 \div 5 = 4R(2)$

$$\begin{array}{r} 4 \\ 5\overline{)22} \\ \underline{-20} \\ 2 \end{array}$$

$= 4\frac{2}{5}$ Whole number + proper fraction

In algebra, *we do not usually rewrite improper fractions as mixed numbers.* In applications, mixed numbers are usually limited to situations that involve cooking, carpentry, or other work that uses measurements with fractions.

Practice Problems

For problems 7–9, simplify the fraction into lowest terms. Do *not* rewrite as a mixed number.

7. $\frac{40}{24}$ **8.** $\frac{54}{15}$ **9.** $\frac{120}{64}$

For problems 10–12, rewrite the improper fraction as a mixed number.

10. $\frac{31}{5}$ **11.** $\frac{24}{5}$ **12.** $\frac{80}{3}$

Multiplication

The product $\frac{1}{8} \cdot \frac{3}{4}$ is one-eighth of three-fourths. Multiply the numerators, multiply the denominators, and simplify the product into lowest terms.

EXAMPLE 8 Evaluate: $\frac{1}{8} \cdot \frac{3}{4}$

SOLUTION ▶

$\frac{1}{8} \cdot \frac{3}{4}$

$= \frac{3}{32}$ Multiply numerators; multiply denominators.

Since 3 and 32 have no common factors, the product is in lowest terms.

If there are common factors in the numerators and denominators of the fractions being multiplied, we can simplify before multiplying. Find common factors, and simplify fractions that are equal to 1. Then multiply the remaining factors in the numerator and multiply the remaining factors in the denominator.

EXAMPLE 9 Evaluate: $\left(-\frac{9}{20}\right)\left(-\frac{5}{12}\right)$

SOLUTION ▶

$\left(-\frac{9}{20}\right)\left(-\frac{5}{12}\right)$

$= \frac{\cancel{3} \cdot 3}{4 \cdot \cancel{5}} \cdot \frac{\cancel{5}}{\cancel{3} \cdot 4}$ Find common factors; (negative)(negative) = positive

$= \frac{3}{16}$ Simplify; multiply remaining factors; fraction is in lowest terms.

> **Two Methods for Multiplying Fractions**
>
> **Method 1**
> 1. Multiply numerators; multiply denominators.
> 2. Simplify the product.
>
> **Method 2**
> 1. Find common factors in the numerators and denominators.
> 2. Simplify fractions that are equal to 1.
> 3. Multiply any remaining factors in the numerator and denominator.

The **reciprocal** of $\frac{3}{4}$ is $\frac{4}{3}$. The reciprocal of $\frac{6}{1}$ is $\frac{1}{6}$. The number 0 has no reciprocal because division by 0 is undefined. The product of a number and its reciprocal is 1.

EXAMPLE 10 Multiply $\frac{3}{4}$ by its reciprocal.

SOLUTION ▶

$\frac{3}{4} \cdot \frac{4}{3}$ The reciprocal of $\frac{3}{4}$ is $\frac{4}{3}$.

$= \frac{\cancel{3}}{\cancel{4}} \cdot \frac{\cancel{4} \cdot 1}{\cancel{3} \cdot 1}$ Find common factors.

$= 1$ Simplify.

Practice Problems

For problems 13–16, evaluate.

13. $\frac{3}{4} \cdot \frac{7}{10}$ **14.** $\frac{5}{6} \cdot \frac{3}{10}$ **15.** $\frac{5}{6} \cdot \frac{6}{5}$ **16.** $\frac{9}{20} \cdot \frac{8}{27}$

Division

In $8 \div 4 = 2$, the **dividend** is 8, the **divisor** is 4, and the **quotient** is 2. To find the quotient of two fractions, rewrite the divisor as its reciprocal and change the operation from division to multiplication. The process of rewriting division as multiplication by the reciprocal of the divisor is sometimes referred to as "invert and multiply."

EXAMPLE 11 Evaluate: $\frac{1}{2} \div \frac{1}{8}$

SOLUTION ▶

$\frac{1}{2} \div \frac{1}{8}$ The divisor is $\frac{1}{8}$; the dividend is $\frac{1}{2}$.

$= \frac{1}{2} \cdot \frac{\mathbf{8}}{\mathbf{1}}$ Multiply by the reciprocal of the divisor.

$= \frac{1 \cdot \cancel{2} \cdot 4}{\cancel{2} \cdot 1}$ Find common factors.

$= 4$ Simplify; $\frac{4}{1} = 4$

Dividing Fractions

1. Rewrite the divisor as its reciprocal; change division to multiplication.
2. Multiply the fractions.
3. Simplify into lowest terms.

EXAMPLE 12 Evaluate: $-\frac{3}{10} \div \frac{9}{4}$

SOLUTION ▶

$-\frac{3}{10} \div \frac{9}{4}$

$= -\frac{3}{10} \cdot \frac{\mathbf{4}}{\mathbf{9}}$ Rewrite the divisor as its reciprocal; multiply.

$= -\frac{\cancel{3} \cdot \cancel{2} \cdot 2}{\cancel{2} \cdot 5 \cdot \cancel{3} \cdot 3}$ Find common factors; (negative)(positive) = negative

$= -\frac{2}{15}$ Simplify; the fraction is in lowest terms.

Practice Problems

For problems 17–20, evaluate.

17. $\frac{4}{9} \div \frac{2}{3}$ **18.** $-\frac{7}{8} \div \frac{1}{2}$ **19.** $\frac{3}{8} \div 4$ **20.** $6 \div \left(-\frac{2}{9}\right)$

Addition and Subtraction

To rewrite a fraction as an equivalent fraction with a new denominator, multiply by a fraction that is equal to 1. Multiplying by 1 does not change the value of the original fraction.

EXAMPLE 13 Rewrite $\frac{3}{5}$ as an equivalent fraction with a denominator of 20.

SOLUTION ▶ Since the new denominator is to be 20 and $5 \cdot 4 = 20$, multiply by a fraction that is equal to 1, $\frac{4}{4}$.

$\frac{3}{5}$ The new denominator will be 20.

$= \frac{3}{5} \cdot \frac{4}{4}$ Multiply by a fraction equal to 1; $\frac{4}{4} = 1$

$= \frac{12}{20}$ Multiply numerators; multiply denominators.

Rewriting a Fraction as an Equivalent Fraction with a Different Denominator

1. Multiply by a fraction that is equal to 1.
2. Multiply the numerators; multiply the denominators.

When we add or subtract fractions, we are adding or subtracting the number of parts. *The parts must be the same size.* Since the denominator describes the size of the parts, we can add or subtract only fractions with the same denominator. To add or subtract fractions with the same denominator, combine the numerators and do not change the denominator.

EXAMPLE 14 Evaluate: $\frac{3}{20} + \frac{7}{20}$

SOLUTION ▶

$\frac{3}{20} + \frac{7}{20}$ The denominators are the same, 20.

$= \frac{3 + 7}{20}$ Add the numerators; the denominator does not change.

$= \frac{\mathbf{10}}{20}$ $3 + 7 = 10$

$= \frac{\cancel{10} \cdot 1}{\cancel{10} \cdot 2}$ Find common factors.

$= \frac{1}{2}$ Simplify; this fraction is in lowest terms.

Adding or Subtracting Fractions with the Same Denominator

1. Add or subtract the numerators; do not change the denominator.
2. Simplify the sum or difference into lowest terms.

To add two fractions with different denominators, we need to first rewrite the fractions as equivalent fractions with the same (common) denominator. The **least common denominator (LCD)** is the smallest number that is a multiple of both of the denominators of the original fractions.

One way to find the least common denominator is to write a list of the multiples of each denominator. The smallest number that is in both lists is the least common denominator (LCD).

EXAMPLE 15 Find the least common denominator of $\frac{3}{4}$ and $\frac{5}{6}$.

SOLUTION ▶ The multiples of 4 are 4, 8, 12, 16, ...

The multiples of 6 are 6, 12, 18, 24, ...

The least common denominator (LCD) is 12.

In the next example, we rewrite the fractions as equivalent fractions with the least common denominator (LCD) of 12 and then add the fractions.

EXAMPLE 16 Evaluate: $\frac{3}{4} + \frac{5}{6}$

SOLUTION ▶

$\frac{3}{4} + \frac{5}{6}$ The least common denominator (LCD) is 12.

$= \frac{3}{4} \cdot \frac{3}{3} + \frac{5}{6} \cdot \frac{2}{2}$ Multiply by fractions equal to 1.

$= \frac{9}{12} + \frac{10}{12}$ Multiply numerators; multiply denominators.

$= \frac{19}{12}$ Add the numerators; the denominator does not change.

Adding or Subtracting Fractions with Unlike Denominators

1. Find the least common denominator (LCD).
2. Rewrite the fractions as equivalent fractions with the least common denominator.
3. Add or subtract the numerators; the denominator stays the same.
4. Simplify into lowest terms.

EXAMPLE 17 Evaluate: $\frac{16}{21} - \frac{3}{56}$

SOLUTION ▶ Find the least common denominator of 21 and 56.

The multiples of 21 are 21, 42, 63, 84, 105, 126, 147, 168, ...

The multiples of 56 are 56, 112, 168, ...

The least common denominator (LCD) of 21 and 56 is 168.

Rewrite each fraction with a denominator of 168, subtract, and simplify into lowest terms.

$\frac{16}{21} - \frac{3}{56}$ The least common denominator (LCD) is 168.

$= \frac{16}{21} \cdot \frac{8}{8} - \frac{3}{56} \cdot \frac{3}{3}$ Multiply by fractions equal to 1.

$$= \frac{128}{168} - \frac{9}{168}$$ Multiply numerators; multiply denominators.

$$= \frac{\mathbf{119}}{168}$$ Subtract the numerators.

$$= \frac{\not{7} \cdot 17}{\not{7} \cdot 24}$$ Find common factors.

$$= \frac{17}{24}$$ Simplify; this fraction is in lowest terms.

Fractions that are negative numbers are to the left of 0 on the number line. The negative sign can be written to the left of the fraction, in the numerator, or in the denominator. The fractions $-\frac{3}{20}$, $\frac{-3}{20}$, and $\frac{3}{-20}$ are equivalent. In adding or subtracting, the negative sign is often written in the numerator.

EXAMPLE 18 Evaluate: $-\frac{3}{20} - \left(-\frac{1}{6}\right)$

SOLUTION ▶ Find the least common denominator (LCD) of 6 and 20.

The multiples of 6 are 6, 12, 18, 24, 30, 36, 42, 48, 54, **60**, ...

The multiples of 20 are 20, 40, 60, ...

The least common denominator (LCD) of 6 and 20 is 60.

$$-\frac{3}{20} - \left(-\frac{1}{6}\right)$$

$$= -\frac{3}{20} \mathbf{+} \frac{1}{6}$$ Rewrite $-\left(-\frac{1}{6}\right)$ as $+\frac{1}{6}$.

$$= -\frac{3}{20} \cdot \frac{3}{3} + \frac{1}{6} \cdot \frac{10}{10}$$ Multiply by fractions equal to 1.

$$= -\frac{9}{60} + \frac{10}{60}$$ Multiply numerators; multiply denominators.

$$= \frac{\mathbf{-9}}{60} + \frac{10}{60}$$ Write the negative sign in the numerator.

$$= \frac{1}{60}$$ Add the numerators; this fraction is in lowest terms.

Practice Problems

For problems 21–27, evaluate.

21. $\frac{3}{16} + \frac{5}{16}$ **22.** $\frac{9}{10} - \frac{3}{4}$ **23.** $\frac{1}{9} + \frac{2}{15}$

24. $\frac{18}{25} - \frac{4}{15}$ **25.** $-\frac{9}{10} + \frac{3}{10}$ **26.** $-\frac{7}{8} - \left(-\frac{5}{12}\right)$

27. $-\frac{5}{2} - \frac{8}{3}$

Fractions and the Order of Operations

We can use the order of operations to evaluate expressions that include fractions.

EXAMPLE 19 Evaluate: $\frac{1}{10} - \frac{1}{10} \div \left(\frac{2}{3}\right)^2$

SOLUTION ▶

$\frac{1}{10} - \frac{1}{10} \div \left(\frac{2}{3}\right)^2$

$= \frac{1}{10} - \frac{1}{10} \div \frac{4}{9}$ From left to right, evaluate exponential expressions.

$= \frac{1}{10} - \frac{1}{10} \cdot \frac{9}{4}$ Rewrite division as multiplication by the reciprocal.

$= \frac{1}{10} - \frac{9}{40}$ From left to right, multiply or divide; the LCD is 40.

$= \frac{1}{10} \cdot \frac{4}{4} - \frac{9}{40}$ Multiply by a fraction equal to 1.

$= \frac{4}{40} - \frac{9}{40}$ Multiply numerators; multiply denominators.

$= -\frac{5}{40}$ Subtract the numerators.

$= -\frac{1 \cdot \cancel{5}}{8 \cdot \cancel{5}}$ Find common factors.

$= -\frac{1}{8}$ Simplify; this fraction is in lowest terms.

Practice Problems

For problems 28–30, evaluate.

28. $\frac{5}{9} \div \frac{1}{2} \cdot \frac{1}{8} - \frac{1}{3}$ **29.** $\left[\frac{1}{8} + \frac{1}{2}\left(\frac{3}{4} - \frac{1}{6}\right)\right]^2$ **30.** $-8 - 6 \cdot \frac{1}{3} \div \frac{4}{9}$

Using Technology: Fraction Arithmetic

Many scientific calculators include a mixed number key [Ab/c]. To enter a fraction, type the numerator, press [Ab/c], and type the denominator. When the [Ab/c] key is pressed, a ⌟ appears on the entry line.

EXAMPLE 20 Evaluate: $\left(\frac{3}{4}\right)\left(\frac{2}{9}\right)$

Press [3] [Ab/c] [4] [×] [2] [Ab/c] [9] [=] The answer is $\frac{1}{6}$.

The calculator simplifies the product to lowest terms.

Sometimes the result of fraction arithmetic is an improper fraction. Most scientific calculators automatically convert this into a mixed number. To rewrite the number as an improper fraction, press [2nd] [Ab/c]. The calculator uses the commands printed above the key: d/c or Ab/c ◀ ▶ d/e.

EXAMPLE 21 Evaluate: $\frac{4}{9} + \frac{7}{9}$

Press [4] [Ab/c] [9] [+] [7] [Ab/c] [9] [=]

The calculator displays 1⌞2⌟9 or 1⊔2/9, representing $1\frac{2}{9}$

Press [2nd] [Ab/c]. The calculator displays the mixed number as an improper fraction, using 11⌟9 or 11 / 9 to represent $\frac{11}{9}$.

To enter a fraction in a graphing calculator, use the [÷] key as the fraction bar. The calculator reports the answer as a decimal number.

EXAMPLE 22 Evaluate: $\frac{4}{9} + \frac{7}{9}$

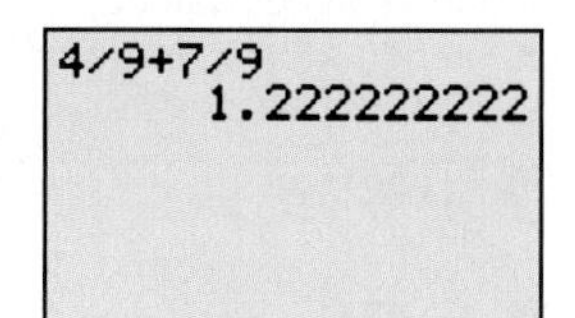

(a)

Press [4] [÷] [9] + [7] ÷ [9] [ENTER]

The calculator will show 1.22. . . .

To change this decimal number into a fraction, press [MATH]. The choice we need, Frac, is highlighted. Press [ENTER]. The screen shows Ans ▸ Frac. Press [ENTER]. The fraction that is equivalent to the decimal number appears on the screen.

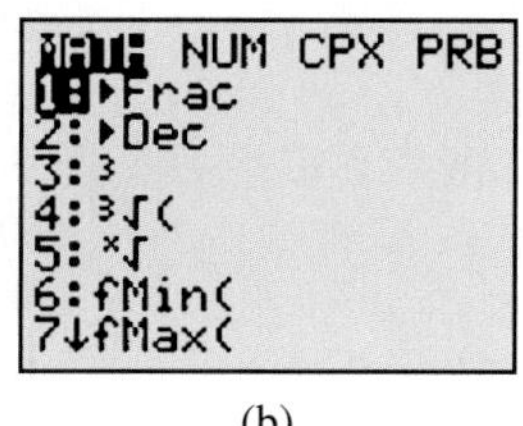

(b)

4/9+7/9
1.222222222
Ans▸Frac

(c)

4/9+7/9
1.222222222
Ans▸Frac
11/9

(d)

Practice Problems For problems 31–33, use a calculator to evaluate.

31. $\left(\frac{2}{15}\right)\left(\frac{20}{21}\right)$ **32.** $\frac{31}{21} - \frac{5}{67}$ **33.** $\frac{21}{40} \div \frac{12}{13}$

R.3 Exercises

Follow your instructor's guidelines for showing your work. Do not rewrite improper fractions as mixed numbers.

For exercises 1–12, simplify.

1. $\frac{14}{32}$ **2.** $\frac{21}{56}$

3. $\frac{48}{66}$ **4.** $\frac{84}{108}$

5. $\frac{54}{28}$ **6.** $\frac{72}{64}$

7. $\frac{21}{54}$ **8.** $\frac{21}{63}$

9. $\frac{5}{365}$ **10.** $\frac{5}{415}$

11. $\frac{34}{51}$ **12.** $\frac{38}{57}$

For exercises 13–22, rewrite the improper fraction as a mixed number.

13. $\frac{10}{7}$
14. $\frac{9}{7}$
15. $\frac{21}{2}$
16. $\frac{23}{2}$
17. $\frac{31}{9}$
18. $\frac{32}{9}$
19. $\frac{101}{51}$
20. $\frac{100}{51}$
21. $\frac{41}{6}$
22. $\frac{37}{6}$

For exercises 23–74, evaluate.

23. $\frac{1}{3}\cdot\frac{4}{9}$
24. $\frac{1}{9}\cdot\frac{2}{5}$
25. $\frac{3}{4}\cdot\frac{16}{21}$
26. $\frac{3}{5}\cdot\frac{20}{27}$
27. $\frac{4}{15}\cdot\frac{3}{8}$
28. $\frac{4}{9}\cdot\frac{3}{16}$
29. $\frac{1}{8}\cdot\frac{30}{7}$
30. $\frac{1}{6}\cdot\frac{50}{3}$
31. $\frac{3}{10}\cdot 4$
32. $\frac{5}{12}\cdot 6$
33. $8\cdot\frac{3}{4}$
34. $12\cdot\frac{5}{6}$
35. $\left(-\frac{1}{4}\right)\left(-\frac{5}{6}\right)$
36. $\left(-\frac{2}{3}\right)\left(-\frac{1}{2}\right)$
37. $\left(-\frac{8}{9}\right)\left(\frac{15}{16}\right)$
38. $\left(-\frac{20}{21}\right)\left(\frac{7}{8}\right)$
39. $\left(\frac{1}{3}\right)\left(-\frac{9}{10}\right)$
40. $\left(\frac{1}{4}\right)\left(-\frac{8}{9}\right)$
41. $\left(-\frac{15}{4}\right)\left(\frac{2}{3}\right)\left(\frac{1}{8}\right)$
42. $\left(-\frac{15}{4}\right)\left(\frac{3}{5}\right)\left(\frac{1}{6}\right)$
43. $\frac{1}{3}\div\frac{1}{8}$
44. $\frac{1}{4}\div\frac{1}{7}$
45. $\frac{1}{8}\div\frac{1}{3}$
46. $\frac{1}{7}\div\frac{1}{4}$
47. $\frac{2}{3}\div\frac{4}{27}$
48. $\frac{9}{10}\div\frac{3}{5}$
49. $\frac{11}{12}\div\frac{2}{3}$
50. $\frac{7}{20}\div\frac{2}{5}$
51. $9\div\frac{3}{4}$
52. $15\div\frac{3}{8}$
53. $\frac{3}{4}\div 12$
54. $\frac{3}{8}\div 15$
55. $-\frac{9}{10}\div\frac{1}{12}$
56. $-\frac{8}{9}\div\frac{1}{12}$
57. $-\frac{8}{15}\div\frac{3}{10}$
58. $-\frac{5}{6}\div\frac{4}{9}$
59. $\frac{1}{9}+\frac{4}{9}$
60. $\frac{1}{7}+\frac{5}{7}$
61. $\frac{3}{10}+\frac{5}{10}$
62. $\frac{4}{15}+\frac{2}{15}$
63. $\frac{9}{11}-\frac{3}{11}$
64. $\frac{12}{13}-\frac{7}{13}$
65. $\frac{9}{10}-\frac{3}{10}$
66. $\frac{7}{15}-\frac{4}{15}$
67. $-\frac{1}{6}-\frac{1}{6}$
68. $-\frac{1}{10}-\frac{1}{10}$
69. $\frac{4}{9}-\left(-\frac{2}{9}\right)$
70. $\frac{3}{14}-\left(-\frac{9}{14}\right)$
71. $-\frac{7}{8}+\frac{1}{8}$
72. $-\frac{9}{10}+\frac{1}{10}$
73. $-\frac{13}{12}+\frac{1}{12}$
74. $-\frac{7}{6}+\frac{1}{6}$

For exercises 75–80, rewrite the fraction as an equivalent fraction with the given denominator.

75. $\frac{4}{5}$; 100
76. $\frac{2}{7}$; 56
77. $\frac{3}{8}$; 24
78. $\frac{5}{6}$; 42
79. $\frac{4}{21}$; 63
80. $\frac{5}{18}$; 54

For exercises 81–96, evaluate.

81. $\frac{35}{12}-\frac{17}{18}$
82. $\frac{21}{16}-\frac{5}{24}$
83. $\frac{18}{5}-\frac{2}{15}$
84. $\frac{20}{3}-\frac{2}{21}$
85. $\frac{1}{30}+\frac{3}{20}$
86. $\frac{3}{50}+\frac{7}{20}$
87. $\frac{9}{20}-\frac{1}{25}$
88. $\frac{9}{10}-\frac{1}{16}$
89. $-\frac{1}{4}-\frac{5}{6}$
90. $-\frac{2}{3}-\frac{1}{2}$
91. $-\frac{8}{9}+\frac{2}{3}$
92. $-\frac{13}{16}+\frac{3}{4}$
93. $\frac{1}{8}-\left(-\frac{3}{16}\right)$
94. $\frac{1}{7}-\left(-\frac{3}{14}\right)$
95. $-\frac{6}{11}-\left(-\frac{3}{8}\right)$
96. $-\frac{7}{11}-\left(-\frac{2}{5}\right)$

For exercises 97–114, evaluate.

97. $\frac{1}{2}+\frac{1}{3}\left(\frac{1}{4}\right)$
98. $\frac{1}{4}+\frac{1}{3}\left(\frac{1}{2}\right)$
99. $\frac{3}{4}+\frac{1}{6}\div\frac{1}{2}-\frac{1}{3}$
100. $\frac{4}{5}+\frac{1}{10}\div\frac{1}{5}-\frac{1}{2}$
101. $\frac{3}{4}+\left(\frac{3}{5}-\frac{1}{10}\right)^2$
102. $\frac{8}{9}+\left(\frac{1}{2}-\frac{1}{6}\right)^2$
103. $\frac{8}{9}\div\frac{1}{3}\left(\frac{3}{5}+\frac{1}{4}\right)$
104. $\frac{9}{10}\div\frac{1}{2}\left(\frac{3}{8}+\frac{1}{6}\right)$
105. $\left[\frac{1}{10}+3\left(\frac{1}{2}\right)\right]^2$
106. $\left[\frac{1}{6}+2\left(\frac{1}{3}\right)\right]^2$

107. $-\frac{3}{8} \div \frac{1}{2} \cdot \frac{1}{5} + \frac{1}{4}$

108. $-\frac{2}{9} \div \frac{1}{6} \cdot \frac{1}{4} + \frac{1}{5}$

109. $\left(-\frac{1}{2}\right)^2 + \frac{1}{3} - \frac{5}{6}$

110. $\left(-\frac{3}{4}\right)^2 + \frac{1}{2} - \frac{5}{8}$

111. $-16 - 8\left(\frac{1}{2}\right)^2 - 12$

112. $-18 - 100\left(\frac{1}{5}\right)^2 - 30$

113. $-\frac{1}{2}(12 - 9) - \frac{1}{4}(15 - 6)$

114. $-\frac{1}{3}(11 - 7) - \frac{1}{9}(14 - 6)$

Technology

For exercises 115–118, evaluate. On a scientific calculator, use the fraction key. For graphing calculators, use the Frac command to rewrite the decimal answer as a fraction.

115. $\frac{5}{34} + \frac{8}{21}$

116. $\frac{5}{34} - \frac{8}{21}$

117. $\left(\frac{5}{34}\right)\left(\frac{8}{21}\right)$

118. $\frac{5}{34} \div \frac{8}{21}$

SECTION R.4 Decimal Numbers, Percent, and Order

Decimal Numbers

In our base 10 number system, each place value is 10 times bigger than the place value to its right. The value of a digit depends on its position in the number.

One billion	One hundred million	Ten million	One million	One hundred thousand	Ten-thousand	Thousand	Hundred	Ten	One	.	Tenth	Hundredth	Thousandth	Ten-thousandth
—	—	—	—	—	—	—	—	—	—	•	—	—	—	—

The decimal number 341.57 is in *place value notation* and represents 3 hundreds + 4 tens + 1 one + 5 tenths + 7 hundredths. In *expanded notation*, 341.57 is $300 + 40 + 1 + \frac{5}{10} + \frac{7}{100}$.

When writing a decimal number that is less than 1, we often write a **leading zero** in the "ones" place to ensure that the decimal point is noticed. For example, a medication order might be written as 0.2 milligram.

To rewrite a decimal number as a fraction, use the place value of the last digit in the number to identify the denominator. The digits of the decimal number are placed in the numerator. Do not include a leading zero in the numerator.

EXAMPLE 1 Rewrite 0.3 as a fraction.

SOLUTION ▶

0.3

$= \frac{3}{10}$ The place value of the last digit is the tenths place.

When rewriting a decimal number as a fraction, simplify the fraction to lowest terms.

EXAMPLE 2 Rewrite 1.54 as a fraction. Simplify the fraction to lowest terms.

SOLUTION ▶ 1.54

$= \frac{154}{100}$ The place value of the last digit is the hundredths place.

$= \frac{2 \cdot 77}{2 \cdot 50}$ Find common factors.

$= \frac{77}{50}$ Simplify.

To rewrite an improper fraction as a decimal number, divide the numerator by the denominator.

EXAMPLE 3 Rewrite $\frac{25}{2}$ as a decimal number.

SOLUTION ▶ $\frac{25}{2}$

$= 25 \div 2$ Divide the numerator by the denominator.

$= 12.5$ Simplify.

To rewrite a fraction that is less than 1 as a decimal number, divide the numerator by the denominator and write a leading 0 to the left of a decimal point.

EXAMPLE 4 Rewrite $\frac{1}{8}$ as a decimal number.

SOLUTION ▶ $\frac{1}{8}$

$= 1 \div 8$ Divide the numerator by the denominator.

$= 0.125$ Write a leading 0 to the left of the decimal point.

If the fraction is equal to a **repeating decimal number**, write a bar over the top of the repeating digit(s).

EXAMPLE 5 Rewrite $\frac{2}{3}$ as a decimal number.

SOLUTION ▶ $\frac{2}{3}$

$= 2 \div 3$ Divide the numerator by the denominator.

$= 0.6666\ldots$ The number 6 repeats.

$= 0.\overline{6}$ Draw a bar over the first 6.

If there is more than one repeating digit, write a bar over the top of the digits in the first repeat.

EXAMPLE 6 Rewrite $\frac{1}{7}$ as a decimal number.

SOLUTION ▶ $\frac{1}{7}$

$= 1 \div 7$ Divide the numerator by the denominator.

$= 0.14285714825714...$ The numbers 142857 repeat.

$= 0.\overline{142857}$ Draw a bar over the first 142857.

Practice Problems

For problems 1–4, rewrite the decimal number as a fraction.

1. 0.81 **2.** 3.999 **3.** 0.6 **4.** 0.0004

For problems 5–8, rewrite the fraction as a decimal number.

5. $\frac{3}{4}$ **6.** $\frac{22}{5}$ **7.** $\frac{5}{9}$ **8.** $\frac{3}{11}$

Rounding

When we **round a number to a place value**, we are writing an estimated value for the number. For example, we can round a ticket price of $52.68 to the nearest tenth. The original price is between $52.60 and $52.70. Since $52.68 is closer to $52.70 than it is to $52.60, we round to $52.70. To show that a number is rounded, use the approximately equal to sign, $\approx$. Rounded to the tenths place, $52.68 $\approx$ $52.70.

We can also use rules to round a number to a given place value. One set of rules depends on the digit just to the right of the given place value. This is the **rounding digit**.

Rounding a Decimal Number to a Given Place Value

1. Identify the rounding digit, which is just to the right of the given place value.
2. If the rounding digit is less than 5, don't change the digit in the given place value. Take off the rounding digit and any other digits to its right.
3. If the rounding digit is greater than or equal to 5, add 1 to the digit in the given place value. Remove the rounding digit and any other digits to its right.

EXAMPLE 7 Round 4.9376 to the nearest hundredth.

SOLUTION ▶

4.9376 The rounding digit, 7, is just to the right of the hundredths place.

$\approx$ **4.94** Since 7 is greater than 5, increase the digit in the hundredths place by 1; remove the rounding digit and all other digits to the right.

In the next example, the rounding digit is less than 5. The digit in the given place value does not change.

EXAMPLE 8 Round 51.392 to the nearest one.

SOLUTION ▶

51.392 The rounding digit, 3, is just to the right of the ones place.

$\approx$ **51** Since 3 is less than 5, the digit in the ones place does not change; remove the rounding digit and all other digits to the right. Since there are no digits to the right of the decimal point, the decimal point is not included.

In the next example, the digit in the given place value is 9. To round, this digit is increased by 1 to 10; write down a 0, and carry the 1 to the next place value to the left.

EXAMPLE 9 Round 21.6983 to the nearest hundredth.

SOLUTION ▶

21.6983 The rounding digit, 8, is just to the right of the hundredths place.

≈ **21.70** Since 8 is greater than 5, the digit in the hundredths place is increased by 1. Write down a 0, and carry the 1 to the tenths place. Remove the rounding digit and all other digits to the right. Usually, the remaining 0 in the given place value is included.

Practice Problems

9. Round 219 to the nearest ten.

10. Round 219 to the nearest hundred.

11. Round 31.749 to the nearest tenth.

12. Round 31.749 to the nearest hundredth.

13. Round 0.0595 to the nearest thousandth.

14. Round 0.0611 to the nearest tenth.

15. Round 34,916 to the nearest thousand.

16. Round 34,916 to the nearest ten thousand.

Percent

A percent represents the number of parts out of 100 total parts. The notation for percent is %. For example, 53% is equivalent to the fraction $\frac{53}{100}$ and to the decimal number 0.53.

EXAMPLE 10 Rewrite 41% as a decimal number.

SOLUTION ▶

41%

$= \frac{41}{100}$ Rewrite the percent as a fraction with a denominator of 100.

$= 0.41$ Divide the numerator by the denominator.

An interest rate may be represented by a percent. In using formulas that include interest rates, the interest rate is usually rewritten as a decimal number.

EXAMPLE 11 An interest rate is 3.75%. Rewrite this percent as a decimal number.

SOLUTION ▶

3.75%

$= \frac{3.75}{100}$ Rewrite the percent as a fraction with a denominator of 100.

$= 0.0375$ Divide the numerator by the denominator.

If an amount doubles, it has increased by 100%. If it more than doubles, it has increased by more than 100%.

EXAMPLE 12 Rewrite 150% as a decimal number.

SOLUTION ▶

150%

$= \frac{150}{100}$ Rewrite the percent as a fraction with a denominator of 100.

$= 1.5$ Divide the numerator by the denominator.

Practice Problems

For problems 17–20, rewrite the percent as a fraction. Simplify the fraction into lowest terms.

17. 3% **18.** 17% **19.** 60% **20.** 8%

For problems 21–24, rewrite the percent as a decimal number.

21. 18% **22.** 2% **23.** 6.5% **24.** 137%

Finding and Using a Percent

If 9 out of 10 students do their math homework in pencil, we can find the percent of students who do their math homework in pencil. There are 10 students in total; this is the "number of parts in the whole." There are 9 students who do their math homework in pencil; this is the "number of parts."

Percent

$$\text{percent} = \left(\frac{\text{number of parts}}{\text{total parts}}\right)(100)\%$$

EXAMPLE 13 If 9 out of 10 students do their math homework in pencil, find the percent of students who do their math homework in pencil.

SOLUTION ▶

$$\text{percent} = \left(\frac{\text{number of parts}}{\text{total parts}}\right)(100)\%$$

$$\text{percent} = \left(\frac{\mathbf{9}}{\mathbf{10}}\right)(100)\%$$ Replace the number of parts and total parts.

$$\text{percent} = \mathbf{90\%}$$ Follow the order of operations.

Given a percent and the total number of parts, we can find the number of parts that the percent represents. Write the percent as a decimal number and multiply by the total number of parts.

Number of Parts

$$\text{number of parts} = (\text{percent in decimal notation})(\text{total parts})$$

EXAMPLE 14 Find 6.5% of $40.

SOLUTION ▶ In decimal form, 6.5% is 0.065.

$$\text{number of parts} = (\text{percent})(\text{total parts})$$

$$\text{number of parts} = (0.065)(\$40)$$ Replace percent and total parts.

$$\text{number of parts} = \$2.60$$ Follow the order of operations.

Practice Problems

25. If 3 out of 5 trees in a park are oak trees, find the percent of trees in the park that are oak trees.

26. If 15 out of 24 voters voted yes, find the percent of voters who voted yes.

27. Find 12% of 50.

28. Find 4% of 150.

29. Find 2.5% of 60.

Order

The symbols $>$ (greater than) and $<$ (less than) are used to show whether a number is larger or smaller than another number. If a number is greater than another number, it is to the right of that number on the number line graph. For example, $6 > 2$.

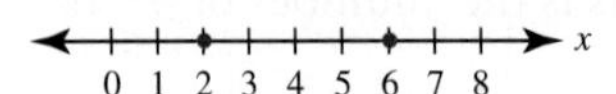

If a number is less than another number, it is to the left of that number on the number line graph. For example, $4 < 7$.

x
2 3 4 5 6 7 8 9

A negative number is less than a positive number. For example, $-3 < 1$.

x
−5 −4 −3 −2 −1 0 1 2 3

To order fractions, rewrite each fraction as a decimal number and then compare their values.

EXAMPLE 15 Make a true statement by writing a $>$ or $<$ sign between $\frac{3}{5}$ and $\frac{7}{10}$.

SOLUTION ▸ Since $\frac{3}{5} = 0.6$, $\frac{7}{10} = 0.7$, and $0.6 < 0.7$, $\frac{3}{5} < \frac{7}{10}$.

In the next example, both fractions are negative numbers.

EXAMPLE 16 Make a true statement by writing a $>$ or $<$ sign between $-\frac{3}{8}$ and $-\frac{9}{20}$.

SOLUTION ▸ Since $-\frac{3}{8} = -0.375$, $-\frac{9}{20} = -0.45$, and $-0.375 > -0.45$, $-\frac{3}{8} > -\frac{9}{20}$.

Practice Problems

For problems 30–40, write $>$ or $<$ between the numbers to make a true statement.

30. 8 13 **31.** 21 4 **32.** 8 −13 **33.** −21 4

34. −7 −2 **35.** −65 0 **36.** 0.2 0.6 **37.** 0.002 0.1

38. $\frac{5}{8}$ $\frac{3}{5}$ **39.** $-\frac{5}{8}$ $\frac{3}{5}$ **40.** $\frac{95}{2}$ $\frac{146}{3}$

R.4 Exercises

Follow your instructor's guidelines for showing your work.

For exercises 1–12, rewrite the decimal number as a fraction. Simplify the fraction to lowest terms.

1. 0.7 **2.** 0.9

3. 0.63 **4.** 0.11

5. 1.2 **6.** 1.8

7. 0.14 **8.** 0.16

9. 0.95 **10.** 0.85

11. 0.002 **12.** 0.004

For exercises 13–24, rewrite the fraction as a decimal number.

13. $\frac{4}{5}$ **14.** $\frac{3}{5}$

15. $\frac{8}{5}$ **16.** $\frac{9}{5}$

17. $\frac{3}{8}$ **18.** $\frac{5}{8}$

19. $\frac{1}{6}$ **20.** $\frac{1}{3}$

21. $\frac{1}{10}$ **22.** $\frac{3}{10}$

23. $\frac{3}{7}$ **24.** $\frac{2}{7}$

25. Round 351 to the nearest ten.

26. Round 461 to the nearest ten.

27. Round 0.87 to the nearest tenth.

28. Round 0.38 to the nearest tenth.

29. Round 0.2394 to the nearest hundredth.

30. Round 0.5492 to the nearest hundredth.

31. Round 65,432 to the nearest thousand.

32. Round 28,310 to the nearest thousand.

33. Round 5.496 to the nearest hundredth.

34. Round 7.298 to the nearest hundredth.

35. Round 0.0084 to the nearest thousandth.

36. Round 0.0053 to the nearest thousandth.

37. Round 8.01 to the nearest tenth.

38. Round 3.02 to the nearest tenth.

For exercises 39–46, rewrite the percent as a fraction. Simplify the fraction into lowest terms.

39. 7% **40.** 9%

41. 12% **42.** 16%

43. 85% **44.** 35%

45. 113% **46.** 117%

For exercises 47–58, rewrite the percent as a decimal number.

47. 2% **48.** 4%

49. 18% **50.** 15%

51. 8.4% **52.** 9.6%

53. 1.25% **54.** 3.25%

55. 200% **56.** 400%

57. 0.6% **58.** 0.3%

59. Copy and complete the table.

Fraction notation	Percent notation	Decimal number notation
$\frac{4}{5}$		
$\frac{61}{100}$		
	91%	
	7%	
	311%	
		0.01
		0.57

60. Copy and complete the table.

Fraction notation	Percent notation	Decimal number notation
$\frac{2}{5}$		
$\frac{71}{100}$		
	41%	
	1%	
	257%	
		0.03
		0.89

61. If 5 out of 20 shirts are T-shirts, find the percent of the shirts that are T-shirts.

62. If 4 out of 20 shirts are flannel shirts, find the percent of the shirts that are flannel shirts.

63. If 8 out of 200 bills are overdue, find the percent of the bills that are overdue.

64. If 6 out of 200 bills are overdue, find the percent of the bills that are overdue.

65. If 15 out of 80 pens are blue, find the percent of the pens that are blue.

66. If 60 out of 320 pens are black, find the percent of the pens that are black.

67. If 3 out of 300 people own a ferret, find the percent of the people that own a ferret.

68. If 2 out of 200 people own a gerbil, find the percent of the people that own a gerbil.

69. If 3 out of 3000 people own a ferret, find the percent of the people that own a ferret.

70. If 2 out of 2000 people own a gerbil, find the percent of the people that own a gerbil.

71. Find 5% of 60.

72. Find 5% of 40.

73. Find 20% of 60.

74. Find 20% of 40.

75. Find 32% of 50.

76. Find 24% of 75.

77. Find 2.5% of 30.

78. Find 2.5% of 18.

79. Find 125% of 45.

80. Find 125% of 65.

81. Find 1% of 300.

82. Find 1% of 200.

83. Find 6% of 21.

84. Find 8% of 14.

For exercises 85–108, write > or < between the numbers to make a true statement.

85. $12 \quad 5$

86. $18 \quad 7$

87. $-12 \quad 5$

88. $-18 \quad 7$

89. $-12 \quad -5$

90. $-18 \quad -7$

91. $0.4 \quad 0.9$

92. $0.6 \quad 0.8$

93. $-0.4 \quad -0.9$

94. $-0.6 \quad -0.8$

95. $0.4 \quad -0.9$

96. $0.6 \quad -0.8$

97. $-\frac{1}{4} \quad -\frac{3}{4}$

98. $-\frac{1}{3} \quad -\frac{2}{3}$

99. $\frac{5}{8} \quad \frac{2}{3}$

100. $\frac{3}{5} \quad \frac{2}{3}$

101. $\frac{3}{4} \quad \frac{5}{9}$

102. $\frac{3}{5} \quad \frac{5}{9}$

103. $\frac{2}{11} \quad \frac{3}{16}$

104. $\frac{3}{11} \quad \frac{2}{7}$

105. $0 \quad -\frac{1}{4}$

106. $0 \quad -\frac{1}{5}$

107. $\frac{35}{3} \quad \frac{49}{5}$

108. $\frac{28}{5} \quad \frac{47}{9}$

Study Plan for Review of Chapter R

SECTION R.1 Evaluating Expressions and the Order of Operations

Ask Yourself	Test Yourself	Help Yourself
Can I . . .	**Do Posttest Problems, page 34**	**See these Examples and Practice Problems**
write the order of operations? evaluate an expression, following the order of operations?	9	Ex. 1–9, PP 1–12
evaluate an expression that includes 0 or 1?	2, 20	Ex. 10, 11, PP 13–23

SECTION R.2 Signed Numbers

Ask Yourself	Test Yourself	Help Yourself
Can I . . .	**Do Posttest Problems, page 34**	**See these Examples and Practice Problems**
evaluate an expression that includes addition or subtraction of signed numbers?	1, 5, 7	Ex. 7–9, PP 10–19, 24–27
evaluate an expression that includes multiplication or division of signed numbers or an exponent?	3, 4, 13, 20	Ex. 10–13, PP 20–23, 28–30
evaluate an expression that includes signed numbers, following the order of operations?	10, 16	Ex. 14–16, PP 31–36

SECTION R.3 Fractions

Ask Yourself	Test Yourself	Help Yourself
Can I . . .	**Do Posttest Problems, page 34**	**See these Examples and Practice Problems**
identify the greatest common factor of two numbers? identify the numerator and denominator of a fraction? simplify a fraction into lowest terms?	12, 14, 15, 18	Ex. 1–5, PP 1–6
identify the difference between a proper and an improper fraction? simplify an improper fraction into lowest terms? rewrite an improper fraction as a mixed number?	17	Ex. 6, 7, PP 7–12
multiply fractions? identify the reciprocal of a number?	12, 18	Ex. 8–10, PP 13–16
divide fractions?	15	Ex. 11, 12, PP 17–20
add and subtract fractions?	6, 14, 17	Ex. 13–18, PP 21–27
evaluate an expression that includes fractions, following the order of operations?	19	Ex. 19, PP 28–30

SECTION R.4 Decimal Numbers, Percent, and Order

Ask Yourself	Test Yourself	Help Yourself
Can I . . .	**Do Posttest Problems, page 34**	**See these Examples and Practice Problems**
identify the place value of a digit in a base-10 number? rewrite a decimal number as a fraction and rewrite a fraction as a decimal number?	21, 22	Ex. 1–6, PP 1–8
round a number to a given place value?	21, 22	Ex. 7–9, PP 9–16
rewrite a percent as a fraction or decimal number? find or use a percent?	23–26	Ex 10–14, PP 17–29
make a true statement by writing a $>$ or $<$ sign between two numbers?	27–30	Ex. 15–16, PP 30–40

Chapter R Posttest

For problems 1–20, evaluate. All fractions should be simplified to lowest terms. Do not rewrite improper fractions as mixed numbers.

1. $-3 - (-11)$

2. $\frac{0}{4}$

3. -4^2

4. $(-4)^2$

5. $2 - 17$

6. $\frac{3}{8} + \frac{7}{12}$

7. $-21 - 18$

8. $18 \div 0$

9. $14 + 6 \div 2$

10. $-8 - (3 + 1)^2$

11. $196 \div 4$

12. $\frac{9}{10} \cdot \frac{5}{27}$

13. $(-8)(-7)$

14. $\frac{7}{12} - \frac{5}{12}$

15. $-\frac{8}{9} \div \frac{2}{15}$

16. $-16 \div 2 \cdot 4$

17. $3 - \frac{5}{8}$

18. $\left(\frac{3}{5}\right)\left(\frac{10}{21}\right)$

19. $\frac{1}{2} - \frac{2}{3} \div 8$

20. $(-6)(0)$

21. Round 3.4671 to the nearest hundredth.

22. Round 143.2 to the nearest ten.

23. Find 14% of 250.

24. Find 8% of 95.

25. What percent is 5 of 40?

26. What percent is 24 of 120?

For problems 27–30, write $>$ or $<$ between the numbers to make a true statement.

27. $-11 \quad -2$

28. $7 \quad -4$

29. $\frac{5}{6} \quad \frac{7}{9}$

30. $0.03 \quad -0.11$

The Real Numbers

1

SUCCESS IN COLLEGE MATHEMATICS

The Culture of College

Going to college is exciting, challenging, and sometimes confusing. You are in a world with its own rules, traditions, and culture. To be successful, you need to learn about this world and what it expects from you. You may need to develop new habits and new ways of learning. You may need to seek help from your advisor, your instructor, a college counselor, or other resources on campus. Orientation, workshops, and syllabi are important sources of information about your college.

The introduction to each chapter in this book focuses on strategies for being successful in college. Each exercise set includes follow-up questions about using these strategies. In this first chapter, these questions assess some of your knowledge about your college. Whether you are a commuter, live in a dorm, or take your classes on-line, you need to learn how to be successful in college.

1.1 The Real Numbers
1.2 Variables and Algebraic Expressions
1.3 Introduction to Problem Solving
1.4 Problem Solving with Percent
1.5 Units of Measurement and Formulas
1.6 Irrational Numbers and Roots

© Bikeriderlondon/Shutterstock.com

To do algebra, we must be able to talk about algebra, and this is possible only if we know vocabulary words that represent important terms and concepts. In the game of tennis, a collection of games is a set. In mathematics, a collection of numbers is also called a **set**. In this section, we will learn about different sets of numbers and other important vocabulary used in algebra.

SECTION 1.1

The Real Numbers

After reading the text, working the practice problems, and completing assigned exercises, you should be able to:

1. Use a number line graph to represent a number.
2. Identify whether a number is an element of the set of real numbers, rational numbers, irrational numbers, integers, and/or whole numbers.
3. Use the order of operations to evaluate an expression.

Whole Numbers and Integers

In mathematics, we often classify numbers or other objects into **sets**. The numbers or objects in the set are its **elements**. The set of **whole numbers** is $\{0, 1, 2, 3, 4, \ldots\}$. This set is in **roster notation**; the elements are separated with commas and are inside of **set brackets**. The three dots following the number 4 is an **ellipsis**. The ellipsis shows that this set of numbers is infinite, continuing on without end.

When numbers are multiplied together, the result is a **product** and the multiplied numbers are **factors**. A **prime number** is a whole number greater than 1. Its only factors are itself and 1. A **composite number** is a whole number greater than 1 that is not prime.

EXAMPLE 1 Use roster notation to represent the set of the first four prime numbers.

SOLUTION ▶ The whole numbers are $\{0, 1, 2, 3, 4, \ldots\}$. Since prime numbers are whole numbers greater than 1, the numbers 0 and 1 are not prime. Since the only factors of a prime number are itself and 1, the set of the first four prime numbers is $\{2, 3, 5, 7\}$.

A **notation** is a way of representing an operation or expression. For example, to show the multiplication of 3 and 5, we can write $3 \cdot 5$, $3(5)$, $(3)(5)$, or 3×5. We use the notation $6 \div 3$, $\frac{6}{3}$, or $3\overline{)6}$ to show the division of 6 by 3.

The number 0 is the **additive** identity. When 0 is added to or subtracted from a number, the number does not change. For example, $5 + 0 = 5$ and $2 - 0 = 2$. The number 1 is the **multiplicative identity**. When a number is multiplied or divided by 1, the number does not change. For example, $(5)(1) = 5$ and $\frac{5}{1} = 5$. The product of any number and 0 is 0. For example, $5(0) = 0$ and $5 \cdot 0 = 0$. If 0 is divided by any number (except 0), the quotient is 0. For example, $0 \div 9 = 0$ and $\frac{0}{9} = 0$. However, **division by zero** is **undefined**. So $9 \div 0$ is undefined, and $\frac{9}{0}$ is undefined.

To represent a whole number by graphing it on a number line, section a number line into equal distances with "tick marks." The number that appears below a tick mark shows its value. Together, these numbers are the **scale** of the number line (Figure 1). The x at the right end of the number line is included as preparation for more work with graphing in Chapter 3.

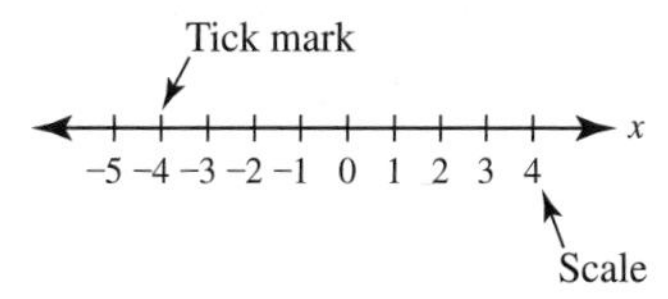

Figure 1

To use a number line graph to represent a number, draw a point directly on the number line at the position that corresponds to its value.

EXAMPLE 2 Use a number line graph to represent 4.

SOLUTION

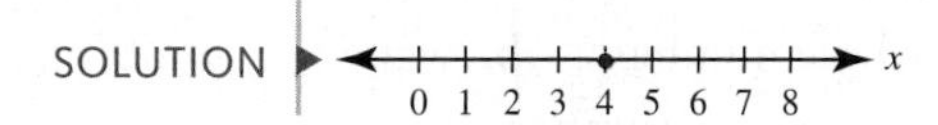

The numbers 3 and -3 are **opposites**. Opposites are the same distance from 0 on the number line graph but are on opposite sides of 0 (Figure 2). The set of integers includes the whole numbers and the **opposites** of the whole numbers. In roster notation, the set of **integers** is $\{\ldots, -5, -4, -3, -2, -1, 0, 1, 2, 3, 4, \ldots\}$.

−5 −4 −3 −2 −1 0 1 2 3 4 x

Figure 2

The notation for the opposite of negative 2 is $-(-2)$. Since the opposite of -2 is 2, $-(-2) = 2$ (Figure 3).

−5 −4 −3 −2 −1 0 1 2 3 4 x

Figure 3

A **base** raised to an **exponent** represents repeated multiplications. For example, 4^3 is an **exponential expression** that is equivalent to $4 \cdot 4 \cdot 4$. The base in this expression is 4. The exponent is 3. To **evaluate** an exponential expression, do the repeated multiplications.

EXAMPLE 3 Evaluate.

(a) 4^3

SOLUTION

4^3	The base is 4; the exponent is 3.
$= 4 \cdot 4 \cdot 4$	Rewrite as repeated multiplications.
$= 64$	Evaluate.

(b) $(-6)^2$

$(-6)^2$	The base is -6; the exponent is 2.
$= (-6)(-6)$	Rewrite as repeated multiplications.
$= 36$	Evaluate.

The **absolute value of a number** is its distance from 0 on a number line. The notation for absolute value is | |. Since distance is a number that is greater than or equal to 0, the absolute value of an integer is also greater than or equal to 0.

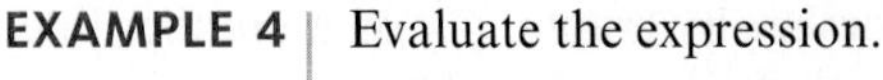

EXAMPLE 4 Evaluate the expression.

(a) $|3|$

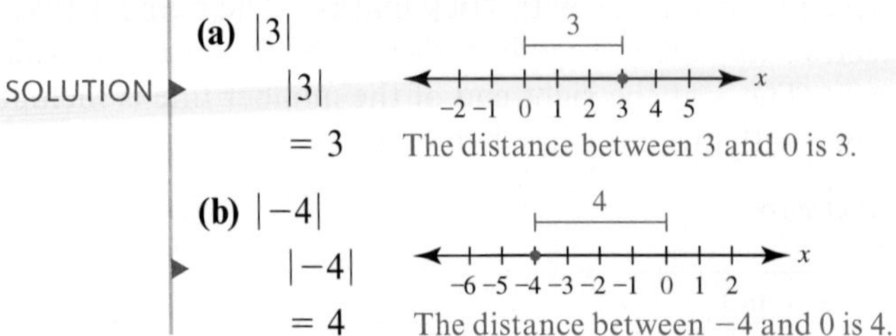

SOLUTION ▶ $|3|$

$= 3$ The distance between 3 and 0 is 3.

(b) $|-4|$

▶ $|-4|$

$= 4$ The distance between -4 and 0 is 4.

To review arithmetic with integers, read Section R.2 and do the Practice Problems and Exercises.

Practice Problems

1. Use roster notation to
   a. write the set of whole numbers. **b.** write the set of integers.
2. Explain why 6 is not a prime number.
3. Use a number line graph to represent 7.
4. What is the opposite of 7?
5. What is the opposite of -7?

For problems 6–11, evaluate.

6. $-(-7)$ **7.** 5^2 **8.** $(-5)^2$ **9.** $(-1)^3$ **10.** $|8|$ **11.** $|-8|$

Rational Numbers

In a fraction, the top number or expression is the **numerator**, and the bottom number or expression is the **denominator**. A **rational number** can be written as a fraction in which the numerator and the denominator are integers. Because division by 0 is undefined, the denominator of a fraction cannot be 0.

EXAMPLE 5 Rewrite the rational number as a fraction.

(a) 0.29

SOLUTION ▶ $= \frac{29}{100}$ Since 9 is in the hundredths place, the denominator is 100.

(b) 5

▶ $= \frac{5}{1}$ An integer can be written as a fraction with a denominator of 1.

(c) -7

▶ $= \frac{-7}{1}$ An integer can be written as a fraction with a denominator of 1.

$= -\frac{7}{1}$ The negative sign can be in the numerator, in the denominator, or in front of the fraction.

By tradition, arithmetic answers that are fractions are simplified into **lowest terms**. To simplify a fraction into lowest terms, identify common factors in the numerator and denominator, and simplify fractions that are equal to 1.

EXAMPLE 6 Simplify: $\frac{10}{15}$

SOLUTION ▶

$$\begin{aligned} \frac{10}{15} &= \frac{2 \cdot 5}{3 \cdot 5} && \text{Find common factors.} \\ &= \frac{2}{3} \cdot \mathbf{1} && \text{A fraction with the same numerator and denominator is equal to 1; } \frac{5}{5} = 1 \\ &= \frac{2}{3} && \text{The product of a number and 1 is the number.} \end{aligned}$$

When simplifying fractions, we may draw a line through the common factors and eliminate rewriting the fraction as 1 and multiplying by 1. This is possible because 1 is the multiplicative identity; the product of 1 and any number is that number.

EXAMPLE 7 Simplify: $\frac{150}{210}$

SOLUTION ▶

$$\begin{aligned} \frac{150}{210} &= \frac{5 \cdot \cancel{30}}{7 \cdot \cancel{30}} && \text{Find common factors.} \\ &= \frac{5}{7} && \text{Simplify; } \frac{30}{30} = 1 \text{ and } \frac{5}{7} \cdot 1 = \frac{5}{7} \end{aligned}$$

In the next example, we first rewrite the numerator as a product of itself and 1. When the fraction is simplified, the numerator is 1.

EXAMPLE 8 Simplify: $\frac{12}{36}$

SOLUTION ▶

$$\begin{aligned} \frac{12}{36} &= \frac{12 \cdot \mathbf{1}}{36} && 12 \cdot 1 = 12 \\ &= \frac{\cancel{12} \cdot 1}{\cancel{12} \cdot \mathbf{3}} && \text{Find common factors.} \\ &= \frac{1}{3} && \text{Simplify; } \frac{12}{12} = 1 \text{ and } \frac{1}{3} \cdot 1 = \frac{1}{3} \end{aligned}$$

To review arithmetic with fractions, read Section R.3 and do the Practice Problems and Exercises there.

Practice Problems

For problems 12–15, simplify.

12. $\frac{15}{35}$ **13.** $\frac{110}{170}$ **14.** $\frac{16}{48}$ **15.** $\frac{24}{80}$

Irrational Numbers and Real Numbers

A "nonrepeating, nonterminating" decimal number is an **irrational number**. An irrational number cannot be written as a fraction in which the numerator and the denominator are integers. The irrational number π (pronounced "pi") is 3.14159. . . ; the ellipsis shows that the digits do not stop (terminate) or repeat. Rounding to the nearest hundredth, $\pi \approx 3.14$. As we will see in Section 1.6, π is included in the formulas for finding the circumference and area of a circle.

The set of **real numbers** includes the rational numbers and the irrational numbers. Every point on the real number line (Figure 4) represents a real number.

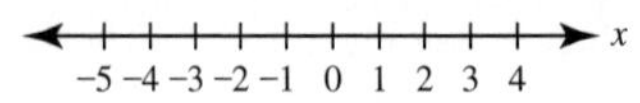

Figure 4

Sets of Real Numbers

Whole Numbers $\{0, 1, 2, 3, \ldots\}$

Integers $\{\ldots, -3, -2, -1, 0, 1, 2, \ldots\}$

Rational Numbers Numbers that can be written as a fraction in which the numerator and denominator are integers.

If a and b are integers and $b \neq 0$, $\frac{a}{b}$ is a rational number.

Irrational Numbers Numbers that cannot be written as a fraction in which the numerator and denominator are integers; nonrepeating, nonterminating decimal numbers.

Real Numbers The rational and irrational numbers.

EXAMPLE 9 Put an X in the box if the number is an element of the set at the top of the column.

SOLUTION ▶ All of the numbers are real numbers. The only irrational number is π. If a real number is not irrational, it is rational.

	Real numbers	Rational numbers	Irrational numbers	Integers	Whole numbers
7	X	X		X	X
-2	X	X		X	
0.2	X	X			
$\frac{2}{9}$	X	X			
$-\frac{2}{9}$	X	X			
0	X	X		X	X
π	X		X		

Practice Problems

16. Is every whole number a real number?

17. Is every rational number a whole number?

18. Copy the chart. Put an X in the box if the number is an element of the set at the top of the column.

	Real numbers	Rational numbers	Irrational numbers	Integers	Whole numbers
6					
$-\frac{5}{9}$					
π					
0.8					
-43					
$\frac{4}{5}$					

The Order of Operations

When evaluating expressions with more than one operation, follow the **order of operations**.

The Order of Operations

1. Working from left to right, evaluate expressions in parentheses or in other grouping symbols, whichever comes first.
2. Working from left to right, evaluate exponential expressions or roots, whichever comes first.
3. Working from left to right, do multiplication or division, whichever comes first.
4. Working from left to right, do addition or subtraction, whichever comes first.

In the next example, we evaluate $(7 - 10)^2$. Keep the parentheses around the base after evaluating inside the parentheses, $(-3)^2$.

EXAMPLE 10 Evaluate: $15 + 20 \div 5 \cdot 4 - (7 - 10)^2$

SOLUTION ▶

$15 + 20 \div 5 \cdot 4 - (7 - 10)^2$

$= 15 + 20 \div 5 \cdot 4 - \mathbf{(-3)^2}$ From left to right, evaluate inside parentheses.

$= 15 + 20 \div 5 \cdot 4 - \mathbf{9}$ From left to right, evaluate exponential expressions.

$= 15 + \mathbf{4} \cdot 4 - 9$ From left to right, multiply or divide.

$= 15 + \mathbf{16} - 9$ From left to right, multiply or divide.

$= \mathbf{31} - 9$ From left to right, add or subtract.

$= 22$ From left to right, add or subtract.

In the next example, we evaluate -4^2, which is equivalent to $-(4 \cdot 4)$. The base is *not* -4. The final expression is an improper fraction in which the numerator is greater than the denominator. Do not rewrite improper fractions as mixed numbers. For help with fraction arithmetic, read Section R.3.

EXAMPLE 11 Evaluate: $-4^2 \div \frac{1}{2} - \frac{3}{4}(2 + 1)$

SOLUTION

$-4^2 \div \frac{1}{2} - \frac{3}{4}(2 + 1)$

$= -4^2 \div \frac{1}{2} - \frac{3}{4}(\mathbf{3})$ From left to right, evaluate inside parentheses.

$= \mathbf{-16} \div \frac{1}{2} - \frac{3}{4}(3)$ From left to right, evaluate exponential expressions.

$= \frac{-16}{1} \cdot \frac{\mathbf{2}}{\mathbf{1}} - \frac{3}{4}(3)$ Rewrite division as multiplication by the reciprocal.

$= \mathbf{-32} - \frac{3}{4}(3)$ From left to right, multiply or divide.

$= -32 - \frac{\mathbf{9}}{\mathbf{4}}$ From left to right, multiply or divide.

$= -\frac{32}{\mathbf{1}} \cdot \frac{4}{4} - \frac{9}{4}$ The common denominator is 4; multiply by a fraction equal to 1.

$= -\frac{\mathbf{128}}{\mathbf{4}} - \frac{9}{4}$ From left to right, multiply or divide.

$= -\frac{137}{4}$ From left to right, add or subtract.

When an expression has more than one pair of grouping symbols, work from the inside out. The expression in the next example includes two types of grouping symbols, brackets and parentheses.

EXAMPLE 12 Evaluate: $8 - [9 - (-12 \div 6 \cdot 2)]^2$

SOLUTION

$8 - [9 - (-12 \div 6 \cdot 2)]^2$

$= 8 - [9 - (\mathbf{-2} \cdot 2)]^2$ From left to right, multiply or divide in parentheses.

$= 8 - [9 - (\mathbf{-4})]^2$ From left to right, multiply or divide in parentheses.

$= 8 - [9 \mathbf{+ 4}]^2$ The opposite of a negative number is a positive number.

$= 8 - [\mathbf{13}]^2$ From left to right, add or subtract in brackets.

$= 8 - \mathbf{169}$ From left to right, evaluate exponential expressions.

$= -161$ From left to right, add or subtract.

A fraction bar acts like a grouping symbol. In the next example, we apply the order of operations to the numerator and then to the denominator.

EXAMPLE 13 Evaluate: $\frac{-90 + 5 \cdot 2^3}{-50 \div 2}$

SOLUTION

$\frac{-90 + 5 \cdot 2^3}{-50 \div 2}$

$= \frac{-90 + 5 \cdot \mathbf{8}}{-50 \div 2}$ From left to right in numerator, evaluate exponential expressions.

$= \frac{-90 + \mathbf{40}}{-50 \div 2}$ From left to right in numerator, multiply or divide.

$= \dfrac{\mathbf{-50}}{-50 \div 2}$ From left to right in numerator, add or subtract.

$= \dfrac{-50}{\mathbf{-25}}$ From left to right in denominator, multiply or divide.

$= 2$ Divide the numerator by the denominator.

In the next example, the expression includes another grouping symbol, an absolute value.

EXAMPLE 14 Evaluate: $|5 - 13| + 8 \div 4$

SOLUTION ▶

$|5 - 13| + 8 \div 4$

$= |\mathbf{-8}| + 8 \div 4$ From left to right, evaluate inside the absolute value.

$= \mathbf{8} + 8 \div 4$ From left to right, evaluate absolute values.

$= 8 + \mathbf{2}$ From left to right, multiply or divide.

$= 10$ From left to right, add or subtract.

For more help with the order of operations, read the examples and do the practice problems and exercises in Sections R.1, R.2, and R.3 that use the order of operations.

Practice Problems

For problems 19–28, evaluate. Use the examples as models of how to show your work.

19. $18 - 15 \div 3 + 2$ **20.** $6 + 10 \div 2 + 8$ **21.** $6 + 10 \div (2 + 8)$

22. $20 \div 2^2 \cdot 5$ **23.** $(-6)^2 + |-48| \div 2 \cdot 3$

24. $-4^2 + 15 \div (3 + 2)(-6)$ **25.** $(-3 - 2)(5^2 - 16)^2$

26. $[24 - 18 \div 9 \cdot 3(8 - 3)]^2$ **27.** $\dfrac{-12 \div \frac{1}{2} \cdot \frac{1}{4}}{-3 + 1}$

28. $18 - [6 + (5 - 1)^2]$

1.1 Vocabulary Practice

Match the term with its description.

1. These numbers are elements of $\{\ldots, -3, -2, -1, 0, 1, 2, 3, \ldots\}$.
2. These numbers are elements of $\{0, 1, 2, 3, 4, \ldots\}$.
3. The numbers in the set that includes all of the elements of the set of rational numbers and all of the elements of the set of irrational numbers
4. The distance of a number from 0
5. Two numbers that are the same distance from zero but are on opposite sides of zero on the number line
6. All of the numbers in this set can be written as fractions in which the numerator and denominator are integers.
7. Three dots that follow a list of numbers to show that the list has no end
8. A real number that is not a rational number is an element of this set.
9. The number 1
10. The number 0

A. absolute value
B. additive identity
C. ellipsis
D. integers
E. irrational numbers
F. opposites
G. multiplicative identity
H. rational numbers
I. real numbers
J. whole numbers

1.1 Exercises

These exercises are opportunities for practice. A college basketball player spends hours shooting free throws and practicing layups. Successful math students practice in the same way. They practice until a skill is automatic and they can do it most of the time without mistakes.

Read the directions carefully for each group of problems. You can find the answers for many of the odd-numbered problems in the Answers to Selected Exercises. Check your answers. You should not practice doing something the wrong way. Get help if you do not understand how to do a problem.

Follow your instructor's guidelines for showing your work.

1. Use roster notation to represent the set of whole numbers.

2. Explain the purpose of ellipses in roster notation.

3. Use a number line graph to represent 6.

4. Use a number line graph to represent -6.

For exercises 5–22, evaluate.

5. $9 + 0$ **6.** $4 + 0$

7. $9 - 0$ **8.** $4 - 0$

9. $0 - 9$ **10.** $0 - 4$

11. $(1)(9)$ **12.** $(1)(4)$

13. $\frac{9}{1}$ **14.** $\frac{4}{1}$

15. $0 \cdot 9$ **16.** $0 \cdot 4$

17. $0 \div 9$ **18.** $0 \div 4$

19. $\frac{9}{0}$ **20.** $\frac{4}{0}$

21. $9 \div 0$ **22.** $4 \div 0$

23. Use roster notation to represent the set of integers.

24. The number 5 is the opposite of -5. Explain why.

For exercises 25–40, evaluate.

25. $-(-6)$ **26.** $-(-3)$

27. 10^2 **28.** 7^2

29. $(-10)^2$ **30.** $(-7)^2$

31. -10^2 **32.** -7^2

33. 2^3 **34.** 3^3

35. $(-2)^3$ **36.** $(-3)^3$

37. $|-25|$ **38.** $|-36|$

39. $|25|$ **40.** $|36|$

41. Explain why $|-4| = 4$. **42.** Explain why $|-6| = 6$.

43. a. Identify a number that is an element of the set of whole numbers and an element of the set of real numbers.

b. Are all whole numbers also real numbers?

44. a. Identify a number that is an element of the set of integers and an element of the set of real numbers.

b. Are all integers also real numbers?

For exercises 45–54, rewrite each number as a fraction in which the numerator and the denominator are integers.

45. 2 **46.** 8

47. -9 **48.** -6

49. 0.3 **50.** 0.7

51. 0.47 **52.** 0.31

53. 0.809 **54.** 0.709

55. Copy the chart below. Put an X in the box if the number is an element of the set at the top of the column.

56. Copy the chart on the next page. Put an X in the box if the number is an element of the set at the top of the column.

For exercises 57–114, evaluate. Do not rewrite improper fractions as mixed numbers.

57. $6 + 15 \div 3$ **58.** $8 + 20 \div 2$

59. $16 \div 8 \cdot 2$ **60.** $12 \div 6 \cdot 2$

61. $16 \div (8 \cdot 2)$ **62.** $12 \div (6 \cdot 2)$

63. $14 + 8 \div 2$ **64.** $10 + 6 \div 2$

65. $18 - 4(9)$ **66.** $24 - 3(10)$

67. $6 - 4^2 \div 2$ **68.** $5 - 4^2 \div 2$

69. $5 - |-9|$ **70.** $6 - |-10|$

71. $12 + |-8| \div (-2)$ **72.** $14 + |-12| \div (-3)$

73. $\frac{1}{4} + \frac{5}{8} \cdot \frac{2}{3}$ **74.** $\frac{1}{6} + \frac{2}{3} \cdot \frac{1}{8}$

Chart for exercise 55

	Real numbers	Rational numbers	Irrational numbers	Integers	Whole numbers
-6					
0					
π					
-0.75					
18					
$\frac{1}{3}$					

Chart for exercise 56

	Real numbers	Rational numbers	Irrational numbers	Integers	Whole numbers
-8					
π					
$\frac{9}{10}$					
-0.13					
0					
$-\frac{7}{8}$					

75. $\frac{1}{4} - \left(\frac{1}{3}\right)^2$

76. $\frac{1}{3} - \left(\frac{1}{2}\right)^2$

77. $\left(\frac{2}{3} + \frac{1}{2}\right)^2$

78. $\left(\frac{3}{4} + \frac{1}{2}\right)^2$

79. $\frac{1}{8} \div \frac{1}{2} - \frac{3}{4}$

80. $\frac{1}{10} \div \frac{1}{5} - \frac{3}{5}$

81. $3 - 4 \cdot 2 + 9$

82. $5 - 4 \cdot 3 + 8$

83. $-15 - (3 - 8)$

84. $-21 - (4 - 9)$

85. $\frac{14 - 2}{6 - 3}$

86. $\frac{15 - 3}{2 + 4}$

87. $\frac{14 - 2^2}{6 + 4}$

88. $\frac{15 - 3^2}{2 + 4}$

89. $12 + 4^2 \cdot 3 \div 2 + 4$

90. $15 + 6^2 \cdot 2 \div 3 + 9$

91. $9^2 \div 3 - 12(4 - 6)$

92. $8^2 \div 4 - 12(2 - 7)$

93. $-9^2 \div (3 - 12)(4 - 6)$

94. $-8^2 \div (4 - 12)(2 - 7)$

95. $5^3 + 3 \cdot 8 \div (-4 - 2)$

96. $4^3 + 4 \cdot 6 \div (-4 - 2)$

97. $\frac{3^2 - 5 \cdot 9}{-6^2}$

98. $\frac{6 \cdot 8 - 8^2}{-4^2}$

99. $\frac{4(5^2 - 6 \cdot 4)}{3 \cdot 4 - 6 \cdot 2}$

100. $\frac{3(4^2 - 5 \cdot 2)}{4 \cdot 8 - 2 \cdot 16}$

101. $\frac{1}{3} \cdot \frac{1}{8} \div \frac{5}{6}$

102. $\frac{1}{5} \cdot \frac{1}{12} \div \frac{5}{6}$

103. $\frac{1}{2}(12 - 7) - \frac{1}{4}(17 - 14)$

104. $\frac{1}{2}(11 - 4) - \frac{3}{4}(15 - 12)$

105. $\frac{2}{7} + \frac{2}{7} \div \left(-\frac{1}{3}\right)$

106. $\frac{2}{9} + \frac{2}{9} \div \left(-\frac{1}{5}\right)$

107. $-\frac{2}{7} + \frac{2}{7} \div \left(-\frac{1}{3}\right)^2$

108. $-\frac{4}{9} + \frac{2}{9} \div \left(-\frac{1}{5}\right)^2$

109. $(9 - 3^2) \div 18$

110. $(25 - 5^2) \div 10$

111. $[10 - 3(2 + 7) + 20]^2$

112. $[18 - 3(4 + 9) + 25]^2$

113. $2[21 \div 3(7 - 4)]$

114. $2[15 \div 3(5 - 1)]$

115. Write your own example of an expression that includes multiplication, division, and subtraction. When this expression is evaluated, it equals 5.

116. Write your own example of an expression that includes multiplication, division, and addition. When this expression is evaluated, it equals 6.

117. Write your own example of an expression that includes one set of parentheses and an exponent. When this expression is evaluated, it equals 5.

118. Write your own example of an expression that includes one set of parentheses and an exponent. When this expression is evaluated, it equals 4.

Technology

For exercises 119–122, use a calculator to evaluate the expression.

119. $3\left(\frac{1}{5} - \frac{2}{3}\right) - 4^2$

120. $4 + 20 - (7 + 12 \div (2 + 1))$

121. $(9 - 1)^2 - (4 + 3)^2$

122. $\frac{18 - 5 + 2}{1 - 4}$

Find the Mistake

For exercises 123–126, the completed problem has one mistake.

(a) Describe the mistake in words, or copy down the whole problem and highlight or circle the mistake.

(b) Do the problem correctly.

123. Problem: Evaluate: $100 - 25 \div 5$

Incorrect Answer: $100 - 25 \div 5$
$= 75 \div 5$
$= 15$

124. Problem: Evaluate: $4 + (7 - 1)^2$

Incorrect Answer: $4 + (7 - 1)^2$
$= 4 + 6^2$
$= 4 + 12$
$= 16$

125. Problem: Evaluate: $-6 - (-2)(5)$

Incorrect Answer: $-6 - (-2)(5)$
$= -6 - 10$
$= -16$

126. Problem: Evaluate: $-3^2 + 8 \div (-2)$

Incorrect Answer: $-3^2 + 8 \div (-2)$
$= 9 + 8 \div (-2)$
$= 9 + (-4)$
$= 5$

Review

For problems 127–130, evaluate. Do not use a calculator.

127. $921 + 587$

128. $921 - 587$

129. $(486)(29)$

130. $2484 \div 9$

SUCCESS IN COLLEGE MATHEMATICS

131. In college, you may need to communicate with your instructor in person, by e-mail, by text message, or by voice mail. Your instructors may vary in how they wish to be addressed. Some may prefer that you use their titles and last names. Others may prefer that you use only their first names. Describe how the instructor of your math class prefers to be addressed.

132. Your math class may be in a computer lab, in a regular classroom, in a large lecture hall, on-line, or some combination of these locations. You may use a printed text book or an e-book. Your instructor may require that you bring other materials to class. Describe the materials that you should have when you go to class or log in to your on-line class.

The word *algebra* comes from the title of a book by a Persian mathematician, Muhammad ibn Mūsā al-Khwārizmī, who was born around 780. His book, *Kitab al-Jabr wa-l-Muqabala*, describes how to solve linear and quadratic equations. A copy of this book traveled to Europe and was translated into Latin. Its Latin title was *Liber algebrae et almucabala*. *Algebrae* in Latin is "of algebra" in English.

Now, some 1200 years since the time of al-Khwārizmī, algebra is still the gateway to the study of mathematics beyond arithmetic. In this section, we begin by learning important vocabulary and concepts of algebra.

SECTION 1.2 Variables and Algebraic Expressions

After reading the text, working the practice problems, and completing assigned exercises, you should be able to:

1. Identify a coefficient, constant, variable, and term.
2. Simplify an algebraic expression.
3. Match an equation with the property it represents.

Variables, Terms, and Expressions

A **numerical expression** includes numbers and may include operations. We **evaluate** numerical expressions such as $12 + 15 \div 3$. A **variable** is a symbol that represents an unknown number. Variables are often letters such as x or y. An **algebraic expression** includes variables and may include numbers or operations. We **simplify** algebraic expressions.

A **term** is a variable, a number, or a product of a number and a variable. Terms are separated by addition or subtraction signs. In the expression $4x + 2y + 9$, there are three terms: $4x$, $2y$, and 9. A term such as 9 that is just a number is a **constant**. When a term is a product of a number and one or more variables, the number is a **coefficient**. In the term $4x$, the coefficient is 4 and the variable is x. In the term x, the coefficient is 1 because $x = 1x$.

EXAMPLE 1 An expression is $8a - 7b + c + 4$.

(a) How many terms are in this expression?

SOLUTION ▶ There are four terms in this expression: $8a$, $7b$, c, and 4.

(b) Identify the coefficients, constants, and variables in this expression.

▶ Since c can be rewritten as $1c$, the coefficients are 8, −7, and 1; the constant is 4; and the variables are a, b, and c.

Like terms that include variables have exactly the same variables and the same exponent on each variable. Constants are also like terms.

Like terms	**Not like terms**
If like terms include variables, the variables are the same. If a variable has an exponent, it is the same in each term. Constants are like terms.	If terms are not like terms, either a variable in the terms is not the same or the variables that are the same have different exponents.
$3x$ and $10x$	$3x$ and $10y$
$5x^2$ and x^2	$5x^2$ and x
$6ab$ and $9ab$	$6a^2b$ and $9ab^2$
4 and 7	

To simplify an algebraic expression, working from left to right, add or subtract the coefficients on the like terms; add or subtract constants. This process is **combining like terms.**

EXAMPLE 2 Simplify: $9a + a - 4a$

SOLUTION ▶

$9a + a - 4a$	
$= 9a + \mathbf{1}a - 4a$	Rewrite a as $1a$.
$= \mathbf{10a} - 4a$	Combine like terms.
$= 6a$	Combine like terms.

In the next example, the expression includes only one constant, 8.

EXAMPLE 3 Simplify: $5h + 2h + 9k - k + 8$

SOLUTION ▶

$5h + 2h + 9k - k + 8$	
$= 5h + 2h + 9k - \mathbf{1}k + 8$	Identify like terms; $-k = -1k$
$= 7h + 8k + 8$	Combine like terms.

Some students organize their work by drawing the same shape around like terms.

EXAMPLE 4 Simplify: $12x^2 + 3x - 9 + 4x^2 - 5x - 2$

SOLUTION ▶

$12x^2 + 3x - 9 + 4x^2 - 5x - 2$	
$= 12x^2 + 3x - 9 + 4x^2 - 5x - 2$	Identify like terms.
$= 16x^2 - 2x - 11$	Combine like terms.

In the next example, the coefficients of like terms are fractions. We can add or subtract only fractions that have the same (common) denominator. The least common denominator (LCD) for $\frac{2}{5}$ and $\frac{3}{8}$ is 40. To rewrite the fractions with this denominator, multiply each fraction by a fraction that is equal to 1. To rewrite $\frac{2}{5}$ with a denominator of 40, multiply by $\frac{8}{8}$. To rewrite $\frac{3}{8}$ with a denominator of 40, multiply by $\frac{5}{5}$. Multiplying by 1 does not change the value of the fraction.

EXAMPLE 5 Simplify: $\frac{2}{5}x + \frac{1}{9} + \frac{3}{8}x - \frac{4}{9}$

SOLUTION ▶

$\frac{2}{5}x + \frac{1}{9} + \frac{3}{8}x - \frac{4}{9}$

$= \frac{2}{5}x + \frac{1}{9} + \frac{3}{8}x - \frac{4}{9}$ Identify like terms.

$= \frac{2}{5}x \cdot \frac{8}{8} + \frac{1}{9} + \frac{3}{8}x \cdot \frac{5}{5} - \frac{4}{9}$ The LCD is 40; multiply by fractions equal to 1.

$= \frac{16}{40}x + \frac{1}{9} + \frac{15}{40}x - \frac{4}{9}$ Multiply numerators; multiply denominators.

$= \frac{31}{40}x - \frac{3}{9}$ Combine like terms.

$= \frac{31}{40}x - \frac{\cancel{3} \cdot 1}{\cancel{3} \cdot 3}$ To simplify $\frac{3}{9}$ into lowest terms, find common factors.

$= \frac{31}{40}x - \frac{1}{3}$ Simplify.

In a **proper fraction**, the numerator is less than the denominator. In an **improper fraction**, the numerator is greater than or equal to the denominator. When an expression includes an improper fraction, do not rewrite it as a mixed number.

EXAMPLE 6 Simplify: $\frac{4}{5}a - \frac{1}{3}b + \frac{3}{5}a - \frac{2}{9}b$

SOLUTION ▶

$\frac{4}{5}a - \frac{1}{3}b + \frac{3}{5}a - \frac{2}{9}b$

$= \frac{4}{5}a - \frac{1}{3}b + \frac{3}{5}a - \frac{2}{9}b$ Identify like terms.

$= \frac{4}{5}a - \frac{1}{3}b \cdot \frac{3}{3} + \frac{3}{5}a - \frac{2}{9}b$ The LCD of 3 and 9 is 9; multiply by a fraction equal to 1.

$= \frac{4}{5}a - \frac{3}{9}b + \frac{3}{5}a - \frac{2}{9}b$ Multiply numerators; multiply denominators.

$= \frac{7}{5}a - \frac{5}{9}b$ Combine like terms.

Practice Problems

For problems 1–2,

(a) identify the number of terms in the expression.

(b) copy the expression. Circle the coefficients, underline the variables, and draw a square around the constant.

1. $3x + 8$ **2.** $\frac{3}{4}x + 6y + 5$

For problems 3–8, simplify.

3. $8b + 15b - b$ **4.** $w - 10w - 3w$ **5.** $12m - 20m + 3p + 15p + 4$

6. $-\frac{1}{10}p + \frac{1}{8} - \frac{2}{5}p + \frac{3}{8}$ **7.** $\frac{5}{6}p + \frac{1}{3}p + \frac{1}{2}p$ **8.** $-\frac{1}{8}a + 9 - \frac{11}{12}a + 10$

Multiplying Terms

We can add and subtract only like terms. However, we can multiply unlike terms such as $4p$ and $3w$. Multiply the coefficients and multiply the variables. Rewrite many multiplications of the same base in exponential notation.

EXAMPLE 7 Simplify: $4p \cdot 3w$

SOLUTION ▶

$4p \cdot 3w$

$= \mathbf{12pw}$ Multiply coefficients; multiply variables.

In the next example, we can find and simplify common factors in the numerator and denominator before multiplying. The expression $h \cdot h$ is rewritten in exponential notation as h^2.

EXAMPLE 8 Simplify: $\left(-\frac{2}{3}h\right)\left(\frac{1}{6}hk\right)$

SOLUTION ▶

$\left(-\frac{2}{3}h\right)\left(\frac{1}{6}hk\right)$

$= \left(-\frac{\cancel{2}}{3}h\right)\left(\frac{1}{\cancel{2} \cdot 3}hk\right)$ Find common factors; $\frac{2}{2} = 1$

$= -\frac{\mathbf{1}}{\mathbf{9}}\mathbf{h^2k}$ Multiply coefficients; multiply variables; $(h)(h) = h^2$

Practice Problems

For problems 9–12, simplify.

9. $(5x)(3y)$ **10.** $(-6a)(-3b)$ **11.** $\left(\frac{1}{4}x\right)\left(\frac{5}{6}x\right)$ **12.** $3(3b)$

Properties

The **commutative properties**, the **associative properties**, and the **distributive property** are properties of the real numbers. The properties can be described by using words or **equations**. An equation includes an = sign. In the descriptions of the properties, the variables a, b, and c represent real numbers.

Properties of Real Numbers

The **commutative properties** state that we can change the *order* of addition or multiplication of real numbers without changing the result.

If a and b are real numbers,

$a + b = b + a$ **Commutative property of addition**

$ab = ba$ **Commutative property of multiplication**

The **associative properties** state that we can change the *grouping* of addition or multiplication without changing the result.

If a, b, and c are real numbers,

$a + (b + c) = (a + b) + c$ **Associative property of addition**

$a(bc) = (ab)c$ **Associative property of multiplication**

For the **distributive property**, if a, b, and c are real numbers,

$a(b + c) = ab + ac$ **Distributive property of multiplication over addition**

$ab + ac = a(b + c)$

Subtraction is not commutative or associative. If we change the order of the numbers we are subtracting, the result changes. For example, $8 - 5 = 3$ and $5 - 8 = -3$. In the same way, division is not commutative or associative. For example, $12 \div 6 = 2$ and $6 \div 12 = \frac{1}{2}$.

EXAMPLE 9 Identify the property represented by each equation.

(a) $(x + 4) + y = x + (4 + y)$

SOLUTION The *grouping* of terms on the left side is different than the *grouping* of terms on the right side; the *order* of terms is the same. This equation represents the associative property.

(b) $(x + 4) + y = (4 + x) + y$

The *grouping* of terms on the left side is the same as the *grouping* of terms on the right side; the *order* of terms is not the same. This equation represents the commutative property.

Practice Problems

For problems 13–18, match the equation with the property it represents.

A. Commutative property
B. Associative property
C. Distributive property

13. $3(2 + 9) = 3 \cdot 2 + 3 \cdot 9$ **14.** $4 + (2 + 9) = (4 + 2) + 9$

15. $8 \cdot 7 = 7 \cdot 8$ **16.** $4 + (3 + 2) = 4 + (2 + 3)$

17. $7(x + y) = 7x + 7y$ **18.** $3(xy) = (3x)y$

Simplifying Algebraic Expressions

To evaluate the numerical expression $3(6 + 4)$ following the order of operations, we first work from left to right inside parentheses, adding $6 + 4$. In the algebraic expression $3(6x + 4)$, we cannot combine $6x$ and 4 because they are not like terms.

We instead first use the distributive property, $a(b + c) = ab + ac$, to remove the parentheses.

EXAMPLE 10 Simplify: $3(6x + 4) + 9$

SOLUTION

$3(6x + 4) + 9$	
$= \mathbf{3(6x) + 3(4)} + 9$	Distributive property.
$= \mathbf{18x + 12} + 9$	Simplify; identify like terms.
$= 18x + \mathbf{21}$	Combine like terms.

To avoid incorrect signs, use parentheses when multiplying negative numbers rather than a multiplication dot.

EXAMPLE 11 Simplify: $4 - 8(w - 3)$

SOLUTION

$4 - 8(w - 3)$	
$= 4 \mathbf{- 8(w) - 8(-3)}$	Distributive property.
$= 4 - 8w \mathbf{+ 24}$	Simplify; identify like terms.
$= -8w \mathbf{+ 28}$	Combine like terms.

Like terms have the same variables and the same exponent on each variable. So $5x^2$ and $3x^2$ are like terms, but $5x^2$ and $7x$ are not like terms.

EXAMPLE 12 Simplify: $2(x^2 - 4x + 6) - 7x^2 + 5$

SOLUTION

$2(x^2 - 4x + 6) - 7x^2 + 5$	
$= \mathbf{2x^2 + 2(-4x) + 2(6)} - 7x^2 + 5$	Distributive property.
$= 2x^2 \mathbf{- 8x} + 12 - 7x^2 + 5$	Simplify; identify like terms.
$= -5x^2 - 8x + 17$	Combine like terms.

When multiplying an integer and a fraction, we can write the integer as a fraction with a denominator of 1. For example, $-8\left(\frac{1}{4}\right) = \frac{-8}{1}\left(\frac{1}{4}\right)$. Before multiplying fractions, identify and simplify common factors in the numerator and denominator.

EXAMPLE 13 Simplify: $\frac{1}{2}(2h - 6) - 8\left(\frac{1}{4}h - 1\right)$

SOLUTION

$\frac{1}{2}(2h - 6) - 8\left(\frac{1}{4}h - 1\right)$	
$= \mathbf{\frac{1}{2}}\left(\frac{2h}{1}\right) \mathbf{+ \frac{1}{2}}\left(\frac{-6}{1}\right) \mathbf{- \frac{8}{1}}\left(\frac{1}{4}h\right) \mathbf{- 8}(-1)$	Distributive property.
$= \frac{1}{\cancel{2}}\left(\frac{\cancel{2}h}{1}\right) + \frac{1}{\cancel{2}}\left(\frac{\mathbf{-\cancel{2}\cdot 3}}{1}\right) - \frac{\mathbf{2\cdot\cancel{4}}}{1}\left(\frac{1}{\cancel{4}}h\right) - 8(-1)$	Find common factors.
$= h - 3 - 2h + 8$	Simplify; identify like terms.
$= -h + 5$	Combine like terms.

In the next example, to make it easier to use the distributive property, we rewrite $-(2p - 9w)$ as $-1(2p - 9w)$.

EXAMPLE 14 Simplify: $3(5p - 6w) - (2p - 9w)$

SOLUTION

$3(5p - 6w) - (2p - 9w)$	
$= 3(5p - 6w) - \mathbf{1}(2p - 9w)$	Rewrite $-(2p - 9w)$ as $-1(2p - 9w)$.
$= \mathbf{3}(5p) + \mathbf{3}(-6w) - \mathbf{1}(2p) - \mathbf{1}(-9w)$	Distributive property.
$= 15p - 18w - 2p + 9w$	Simplify; identify like terms.
$= 13p - 9w$	Combine like terms.

Practice Problems

For problems 19–26, simplify.

19. $\frac{1}{2}(2x - 5)$ **20.** $3(4a + 9b) - 5(2b)$ **21.** $10 - 3(4z - 9)$

22. $4(2x - 3) - 8(6x + 1)$ **23.** $4(2x - 3) + 8(6y - 1)$

24. $5(3x - 1) - (2x - 7)$ **25.** $x^2 + 8x + 6 + x^2 + 3x + 1$

26. $2(a^2 - 3a - 9) + 4a^2 - a + 2$

1.2 Vocabulary Practice

Match the term with its description.

1. A symbol that represents an unknown number
2. A variable, a number, or a product of a number and a variable
3. $a(b + c) = ab + ac$
4. In the term $3x$, 3 is an example of this.
5. In the expression $5x + 7$, 7 is an example of this.
6. These terms have identical variables with the same exponents.
7. $a + b = b + a$
8. $a + (b + c) = (a + b) + c$
9. These numbers are elements of $\{\ldots, -3, -2, -1, 0, 1, 2, 3, \ldots\}$.
10. These numbers are elements of $\{0, 1, 2, 3, 4, \ldots\}$.

A. associative property of addition
B. coefficient
C. commutative property of addition
D. constant
E. distributive property
F. integers
G. like terms
H. term
I. variable
J. whole numbers

1.2 Exercises

Follow your instructor's guidelines for showing your work.

For exercises 1–4,
(a) identify the number of terms in the expression.
(b) identify the coefficient(s).
(c) identify the constant.
(d) identify the variable(s).

1. $8x + 2y + 5$
2. $7x + 9y + 3$
3. $8a - 6b$
4. $12h + 4k$

For exercises 5–32, simplify.

5. $14k + 9k - k$
6. $20u + 15u - u$
7. $b - 6b - 10b$
8. $n - 8n - 3n$
9. $\frac{1}{8}m - \frac{5}{8}m + \frac{3}{8}$
10. $\frac{1}{10}c - \frac{3}{10}c + \frac{7}{10}$
11. $\frac{3}{5}u + \frac{4}{9}u - \frac{8}{11}$
12. $\frac{5}{6}r + \frac{3}{5}r - \frac{2}{7}$
13. $4a - 5b + 6a - 9b + 1$
14. $9c - 8d + 2c - 5d + 3$
15. $-9w + c + 10 - 3w + c - 15$
16. $-5z + x + 8 - 7z + x - 20$
17. $-\frac{3}{5}p + \frac{1}{8} - \frac{1}{5}p + \frac{3}{8}$
18. $-\frac{3}{4}a + \frac{1}{9} - \frac{3}{4}a + \frac{2}{9}$
19. $\frac{1}{10}c + \frac{3}{5}d + \frac{3}{4}c - \frac{1}{2}d$
20. $\frac{1}{8}x + \frac{2}{3}y + \frac{2}{5}x - \frac{1}{2}y$

21. $\frac{7}{8}x + \frac{3}{10} + \frac{5}{8}x + \frac{9}{10}$

22. $\frac{5}{12}p + \frac{3}{10} + \frac{11}{12}p + \frac{1}{10}$

23. $-5u + 6y - \frac{3}{5}u - \frac{1}{3}y$

24. $-9h + 3k - \frac{3}{4}h - \frac{1}{2}k$

25. $6x^2 - 3x + 4x^2 - 1$

26. $7z^2 - 4z + 10z^2 - 5$

27. $c^2 + 5c + 9 + c^2 + 4c + 10$

28. $a^2 + 6a + 3 + a^2 + 5a + 4$

29. $k^2 - 5k - 6 + 2k^2 - 3k + 4$

30. $h^2 - 8h - 7 + 2h^2 - 4h + 3$

31. $2x^2 + x - 9 + 3x^2 - 2x + 15$

32. $3p^2 + p - 10 + 2p^2 - 2p + 13$

For exercises 33–40, match the equation with the property it represents.

A. Commutative property
B. Associative property
C. Distributive property

33. $6(3u) = (6 \cdot 3)u$

34. $8(2w) = (8 \cdot 2)w$

35. $n(7) = 7n$

36. $x(3) = 3x$

37. $6x + 4x = x(6 + 4)$

38. $3z + 5z = z(3 + 5)$

39. $3(x + 4) = 3x + 3(4)$

40. $4(c + 1) = 4c + 4(1)$

For exercises 41–48, simplify.

41. $(10y)(4z)$

42. $(5u)(3w)$

43. $(-3x)(-5x)$

44. $(-4h)(-7h)$

45. $(7a)(-3b)$

46. $(8c)(-2d)$

47. $-8(9m)$

48. $-6(5v)$

For exercises 49–82, simplify.

49. $9x + 2y - 14x + 3y$

50. $6x + 15y - 13x + 3y$

51. $100a + 8b + 4 + 250a + 16b - 51$

52. $5h + 200k + 11 + 8h + 130k - 17$

53. $3(2x + 1) - 8$

54. $4(3x + 5) - 32$

55. $10 - 6(a + 5)$

56. $12 - 8(c + 3)$

57. $10 - (a + 5)$

58. $12 - (c + 3)$

59. $-(x - 1)$

60. $-(z - 4)$

61. $3(2x + 1) - 7(4x - 9)$

62. $4(3x + 5) - 8(6x - 1)$

63. $3(2x + 1) - (4x - 9)$

64. $4(3x + 5) - (6x - 1)$

65. $12w - (7w + 1) - w$

66. $15d - (6d + 1) - d$

67. $-\frac{3}{4}(20w - 12)$

68. $-\frac{5}{6}(30k - 24)$

69. $-\frac{3}{4}\left(26a - \frac{1}{2}\right)$

70. $-\frac{5}{6}\left(15b - \frac{1}{4}\right)$

71. $\frac{9}{10}c - \frac{3}{4}c + \frac{2}{3}d - \frac{1}{8}d$

72. $\frac{11}{12}h - \frac{5}{8}h + \frac{4}{5}k - \frac{1}{9}k$

73. $6\left(\frac{2}{3}x + \frac{4}{9}\right)$

74. $12\left(\frac{3}{4}x + \frac{4}{15}\right)$

75. $\frac{2}{5}\left(\frac{1}{3}p + \frac{7}{10}w\right)$

76. $\frac{4}{9}\left(\frac{1}{7}f + \frac{5}{8}g\right)$

77. $12\left(\frac{2}{3}z - \frac{1}{4}\right) - 6$

78. $18\left(\frac{5}{6}h - \frac{1}{3}\right) - 7$

79. $\frac{1}{2}(6c + 9) - \frac{3}{4}(8c - 11)$

80. $\frac{1}{3}(12d + 6) - \frac{2}{5}(15d - 9)$

81. $\frac{2}{5}(8z - 6) + \frac{3}{7}(2z + 15)$

82. $\frac{3}{8}(15z - 7) + \frac{2}{9}(3z + 20)$

83. Explain why $6x$ and $5y$ are not like terms.

84. Explain why $9a$ and $3b$ are not like terms.

85. Explain why $2x^2y$ and $8xy^2$ are not like terms.

86. Explain why $7a^2b$ and $5ab^2$ are not like terms.

87. A student said that the expression $(5x)(3y)$ cannot be simplified because $5x$ and $3y$ are not like terms. Explain why the student is wrong.

88. A student said that the expression $(2a)(9b)$ cannot be simplified because $2a$ and $9b$ are not like terms. Explain why the student is wrong.

89. The expression $\left(\frac{3}{5}\right)\left(\frac{7}{9}x\right)$ can be simplified by multiplying the numerators, multiplying the denominators, and simplifying the product, $\frac{21}{45}x$. Or the expression can be simplified by finding and simplifying common factors in the numerator and denominator, $\left(\frac{\cancel{3}}{5}\right)\left(\frac{7}{\cancel{3} \cdot 3}\right)x$ and then multiplying the numerators and denominators. Explain the advantage of first finding and simplifying common factors in the numerator and denominator.

90. To add $\frac{5}{6}x$ and $\frac{1}{4}x$, the fractions can be rewritten with the common denominator 24, or they can be rewritten with the *least* common denominator, 12. Explain the advantage of using the least common denominator.

Find the Mistake

For exercises **91–94**, the completed problem has one mistake.

(a) Describe the mistake in words, or copy down the whole problem and highlight or circle the mistake.

(b) Do the problem correctly.

91. Problem: Simplify: $10 - 6(x - 3)$

Incorrect Answer: $10 - 6(x - 3)$
$= 4(x - 3)$
$= 4x - 12$

92. Problem: Simplify: $4 - (x + 9)$

Incorrect Answer: $4 - (x + 9)$
$= 4 - x + 9$
$= -x + 13$

93. Problem: Simplify: $8(4x - 1) - 3(2x - 5)$

Incorrect Answer: $8(4x - 1) - 3(2x - 5)$
$= 8(4x) + 8(-1) - 3(2x) - 3(5)$
$= 32x - 8 - 6x - 15$
$= 26x - 23$

94. Problem: Simplify: $12x + 15y + 3x + 2y$

Incorrect Answer: $12x + 15y + 3x + 2y$
$= 15x + 17y$
$= 32xy$

Review

95. Describe the order of operations.

96. Explain the difference between a rational number and an irrational number.

97. Evaluate: $5 \div 0$

98. Explain how to determine whether a fraction is in lowest terms.

SUCCESS IN COLLEGE MATHEMATICS

99. Some instructors in college do not admit students who arrive late. Others allow late students to enter if they do so quietly. Describe your instructor's policy about arriving late to class.

100. Some college students own a laptop, netbook, MP3 player, tablet, smartphone, cell phone, and/or other electronic devices. Describe your instructor's policy about the use of electronic devices such as these in class.

© Joe Ferrer/Shutterstock.com

In this section, you will solve many application problems. Think about the softball player who practices fielding ground balls for hours to improve her skills. In the same way, you need to do many problems to improve your skill in problem solving. You will continue to practice problem solving as you work through this textbook.

SECTION 1.3 Introduction to Problem Solving

After reading the text, working the practice problems, and completing assigned exercises, you should be able to:

Use the five steps to solve an application problem.

Problem Solving with Whole Numbers

Why solve application problems? People who can solve problems have an advantage in real life and in the workplace. Employers want employees who are confident problem solvers. Solving problems can also help us learn more mathematics.

There is no magic formula for becoming a good problem solver. But it does seem that successful problem solvers do a lot of problems; *they practice.* They don't give up when their first attempt doesn't work; *they persevere.* They don't always expect immediate success; *they are patient.*

THE FIVE STEPS

Step 1 Understand the problem.
Step 2 Make a plan.
Step 3 Carry out the plan.
Step 4 Look back.
Step 5 Report the solution.

In this book, we use **five steps** to solve many different kinds of problems in many different situations. The steps are based on the work of George Polya (1887–1985). Polya believed that students learn to solve problems the way they learn how to swim: with practice and coaching. To improve your problem-solving ability, you need to do many problems. As Dr. Polya wrote, "Mathematics is not a spectator sport."

Step 1 Understand the problem.

- Read the problem, perhaps more than once. Look up any unfamiliar words.
- Identify what you are trying to find, the *unknown*. **Assign a variable** that represents the unknown.
- Identify the information needed to solve the problem. Identify information that is not needed. This is called **extraneous information**.

Step 2 Make a plan.

- Identify the relationship between the unknown and the other information in the problem. For example, the relationship could be a difference, a sum, or a percent, or it could be described by a formula.
- If this problem is similar to a problem that you have done before, think about how you solved the earlier problem. The examples in the book and examples completed by your instructor can also help you make a plan to solve a new problem.
- Decide how to solve the problem. It may help to write a word equation that describes the relationship in the problem. Although there are other ways to solve a problem, we will most often use an equation or formula.

Step 3 Carry out the plan.

- Write and solve the equation or formula.

Step 4 Look back.

- Think about whether your answer solves the problem. Does it make sense? Is it reasonable? If it does not, make and carry out a new plan. Use *working backwards*, *estimating*, *in the ballpark*, or *doing the problem a different way* to check your work.
- Review your algebra or arithmetic, line by line, to make sure it is correct.
- Think about whether you have learned something new from this problem, something that you may use to solve similar problems in the future.

Step 5 Report the solution.

- How you report the solution depends on the audience. In a college mathematics class, writing the solution to the problem in a complete sentence is often what the audience (your instructor) wants.

THE FIVE STEPS

Step 1 Understand the problem.
Read the problem. Identify the unknown. Assign a variable. Identify extraneous information.

Step 2 Make a plan.
Identify the relationship between the unknown and the other information in the problem. Decide how to find the unknown: write and solve an equation or inequality, use a formula, draw and use a graph, evaluate a function, solve a system of equations.

Step 3 Carry out the plan.

Step 4 Look back.
Does the answer make sense? Is it reasonable?
Check for errors in arithmetic or algebra.
Think about what you have learned from solving this problem.

Step 5 Report the solution.
Write the answer to the problem in a complete sentence.

How much of this understanding, planning, and looking back should you write down? You need to ask your instructor for guidance. Since we will be using algebra to solve almost all of these problems, often the minimum amount of work that you must show is assigning a variable, writing and solving an equation, and writing the final answer in a complete sentence.

EXAMPLE 1 The rent in a rent-stabilized apartment in Brooklyn, New York, is $800 per month. When the lease is renewed, the landlord will raise the rent by 7.25%, which is $58 per month. Find the rent per year for this apartment after this increase. (*Source:* www.housingnyc.com)

SOLUTION ▶ **Step 1 Understand the problem.**

What is unknown? The amount of rent per year

Assign the variable. R = rent per year

Needed information. The previous rent is $800 per month; the increase is $58 per month.

Relationship. The amount per year is the number of months multiplied by the new monthly rent.

Extraneous information. The percent increase (7.25%) is not needed.

Step 2 Make a plan.

Add the increase to the previous rent and multiply by 12. A word equation that represents the relationship is

amount = (number of months)(previous rent + increase)

Step 3 Carry out the plan.

$R = (12)(\$800 + \$58)$ amount = (number of months)(previous rent + increase)

$R = \$10{,}296$ Follow the order of operations.

Step 4 Look back.

Is the answer reasonable? A reasonable answer should be "in the ballpark." In baseball, a ball in play can be anywhere in the ballpark—in the outfield, the infield, or in the catcher's mitt. An acceptable answer is also in some "ballpark." We are not checking to see whether the answer is exactly right. Rather, we are finding whether the answer is within an acceptable or expected range of answers. To find the ballpark in this problem, we find the rent per year if it had stayed at $800 per month and the rent per year if it had increased to $900 per month. A reasonable answer will be in the ballpark, somewhere between these amounts.

If the rent remained at $800 per month, the rent per year is 12($800), which is $9600. If the rent increased to $900 per month, the rent per year is 12($900), which equals $10,800. Since $858 is between $800 and $900 and $10,296 is between $9600 and $10,800, the answer is in the ballpark and seems reasonable.

What can we learn from this problem? If there is an *increase*, we often add.

Step 5 Report the solution.

To make sure that you are answering the problem, read both the problem and your solution to yourself or aloud. In this example, the problem is "Find the amount of rent per year after this increase." In a sentence, the final answer is "The rent per year after the increase is $10,296."

In media reports, we often see values described with a combination of numbers and word names. To change numbers that include word names such as *million*, *billion*, and *trillion* into numbers in place value notation, multiply by the number that the word name represents.

1 million	1,000,000
1 billion	1,000,000,000
1 trillion	1,000,000,000,000

EXAMPLE 2 The national debt is about \$15.9 trillion. Write this amount in place value notation.

SOLUTION ▶ \$15.9

$= (\$15.9)(1{,}000{,}000{,}000{,}000)$ 1 trillion = 1,000,000,000,000

$= \$15{,}900{,}000{,}000{,}000$ Simplify.

We will learn more about rewriting numbers using word names when we study changing units of measurement in Section 1.5.

EXAMPLE 3 Find the difference between 2008 and 1995 in the number of ER visits nationwide in which children were given CT scans.

The number of ER visits nationwide in which children were given CT scans surged from about 330,000 in 1995 to 1.65 million in 2008. (*Source:* www.usatoday.com, April 5, 2011)

SOLUTION ▶ **Step 1 Understand the problem.**

What is unknown? The difference in the number of ER visits.

Assign the variable. D = difference

Needed information. The number of ER visits in 2008 in which children were given CT scans and the number of ER visits in 1995 in which children were given CT scans.

Relationship. The difference is the number of visits in 2008 minus the number of visits in 1995.

Extraneous information. None.

Step 2 Make a plan.

To find a difference, subtract. A word equation is difference = number in 2008 − number in 1995. The amount in millions needs to be rewritten in place value notation: 1.65 million = 1.65(1,000,000) = 1,650,000.

Step 3 Carry out the plan.

$D = 1{,}650{,}000$ visits − 330,000 visits   difference = number in 2008 − number in 1995

$D =$ **1,320,000 visits**   Follow the order of operations.

Step 4 Look back.

Is the answer reasonable? Working backwards, since 1,320,000 visits + 330,000 visits equals 1,650,000 visits and that is the number of visits in 2008, the answer seems reasonable.

What can we learn from this problem? If there is a *difference*, we often subtract.

Step 5 Report the solution.

Between 1995 and 2008, the difference in the number of ER visits in which children were given CT scans was 1,320,000 visits.

When there is more than one way to solve a problem, we can check the reasonability of the solution by doing the problem in a different way.

EXAMPLE 4 The table shows the costs for attending Anne Arundel Community College in the academic year 2010–2011. For spring semester 2011, a student registered for 13 credits, including a biology course with a lab fee and a physical education course, but did not register for any telecourses. Find the total cost for the semester, including tuition and fees. (*Source:* www.aacc.edu)

Tuition and registration fees, 2010–2011	Cost
Tuition, per credit	\$88
Student activity fee, per credit	\$1
Educational services fee, per credit	\$9
Athletic fee, per credit	\$1
Registration fee, per semester	\$20
Physical education fee, per course	\$6
Telecourse fee, per course	\$35
Lab fee, per course	\$30

SOLUTION **Step 1 Understand the problem.**

What is unknown? The total cost for the semester for this student.

Assign the variable. $C =$ total cost

Needed information. Some costs depend on the number of credits such as tuition (\$88 per credit), student activity fee (\$1 per credit), educational services fee (\$9 per credit), and athletic fee (\$1 per credit). The student is taking 13 credits. Other fees do not depend on the number of credits, such as the registration fee (\$20), the physical education fee (\$6), and the lab fee (\$30).

Relationship. The total cost is the sum of the costs based on the number of credits, the registration fee, the cost for the physical education course, and the lab fee.

Extraneous information. The cost per telecourse.

Step 2 Make a plan.

Find the sum of the costs per credit and multiply this by the number of credits. Then add on the cost for the registration fee, the physical education fee, and the lab fee. A word equation is total cost = (cost per credit)(number of credits) + other costs. The total paid in fees per credit is \$88 + \$1 + \$9 + \$1 for a total of \$99 per credit.

Step 3 Carry out the plan.

$$C = \left(\frac{\$99}{1\ \cancel{\text{credit}}}\right)(13\ \cancel{\text{credits}}) + \$20 + \$6 + \$30$$ cost = (cost per credit)(number of credits) + other

$$C = \mathbf{\$1287} + \$20 + \$6 + \$30$$ Follow the order of operations.

$$C = \mathbf{\$1343}$$ Follow the order of operations.

Step 4 Look back.

Is the answer reasonable? Doing this problem a different way, we can separately multiply the different costs per credit and the number of credits, find their sum, and then add on the other costs. Since (\$88)(13) + (\$1)(13) + (\$9)(13) + (\$1)(13) + \$20 + \$6 + \$30 equals the original answer of \$1343, the answer seems reasonable. Or estimating, we can round the price per credit to \$100 and multiply by 13. The sum of \$1300 and the other costs of \$56 is \$1356, which is close to the original answer.

What can we learn from this problem? If we can do the problem a different way, we can use this to check the reasonability of the solution.

Step 5 Report the solution.

The total cost for the semester is \$1343.

Practice Problems

For problems 1–3, use the five steps.

1. Cedar Point Amusement Park in Ohio has 17 roller coasters. For customers age 3–61 years who are at least 48 inches tall, the price of admission is \$46.99. The price of admission for a senior over age 61 is \$21.00. The price for military personnel and their immediate family is \$36.99 per person. A family includes seven people. Each family member is over 3 years of age and at least 48 inches tall. One person is a senior citizen. None is associated with the military. Find the total cost of daily admission for this family. (*Source:* www.cedarpoint.com)
2. Find the difference between the predicted carbon dioxide emissions from coal in 2030 for China and the predicted emissions for India.

   For China alone, coal-related emissions are projected to grow by an average of 3.2 percent annually, from 4.3 billion metric tons in 2005 to 9.6 billion metric tons (51 percent of the world total) in 2030. India's carbon dioxide emissions from coal combustion are projected to total 1.4 billion metric tons in 2030, accounting for more than 7 percent of the world total. (*Source:* www.doe.gov, June 2008)
3. The city of Lawrence, Kansas, needs about \$2.1 million to replace two fire trucks and a rescue truck. Find the number of false alarms that the fire chief is using to make his estimate.

   Fire Chief Mark Bradford estimates that if the city charged \$25 per false alarm, the city could collect about \$18,000 per year. (*Source:* www.ljword.com, May 22, 2008)

Problem Solving with Negative Numbers and Fractions

In the next example, we use a negative number to represent a house payment.

EXAMPLE 5 The annual take-home pay of a computer programmer is \$45,000. Her monthly house payment is \$1400. Her monthly car payment is \$325. Find the amount of take-home pay that remains per year after making all of the house payments. (Use a negative number to represent a house payment.)

SOLUTION ▶ **Step 1 Understand the problem.**

The unknown is the remaining amount of take-home pay. The car payment amount is extraneous information.

L = remaining amount of take-home pay

Step 2 Make a plan.

Since there are 12 months in a year, the total amount of house payments equals 12 times the monthly house payment. The annual pay is a positive number; the total amount of house payments is a negative number. A word equation is remaining amount = annual pay + (12)(payment).

Step 3 Carry out the plan.

$L = \$45{,}000 + 12(-\$1400)$ amount = annual pay + 12(payment)

$L = \$45{,}000 - \mathbf{\$16{,}800}$ Follow the order of operations.

$L = \mathbf{\$28{,}200}$ Follow the order of operations.

Step 4 Look back.
If the monthly payment is \$1500, the remaining amount is \$45,000 + 12(−\$1500), which equals \$27,000. If the monthly payment is \$1300, the remaining amount is \$45,000 + 12(−\$1300), which equals \$29,400. Since the answer of \$28,200 is between (in the ballpark of) \$27,000 and \$29,400, it seems reasonable.

Step 5 Report the solution.
After making the house payments, \$28,200 will remain of the annual take-home pay.

In the next example, the equation includes a fraction. If this was not an application problem, we would write the answer as an improper fraction. In application problems, we often instead rewrite fractions as decimal numbers and round to a given place value. When rounding, use an approximately equal sign, ≈, instead of an equals sign, =.

EXAMPLE 6 Find the amount of water that the Metropolitan Water District will return to Westlands and San Luis in summer 2011. Write as a decimal number; round to the nearest hundred.

> The ability to reschedule (carry over) water in San Luis Reservoir from one contract year to the next has been available to the water service contractors south of the Delta since the early 1990s. Westlands and San Luis sent 110,692 acre · feet of their 2010 allocation to MWD (The Metropolitan Water District of South California) in late 2010, and MWD will return two-thirds of that amount to Westlands and San Luis in summer 2011. (*Source:* www.usbr.gov, Feb. 15, 2011)

SOLUTION ▶ **Step 1 Understand the problem.**
The unknown is the amount of water that will be returned to Westlands and San Luis in summer 2011.

A = amount of water to be returned

© Harris Shiffman/Shutterstock.com

Step 2 Make a plan.
When finding a fraction of an amount, multiply the fraction and the amount. A word equation is amount returned = (fraction returned)(2010 allocation).

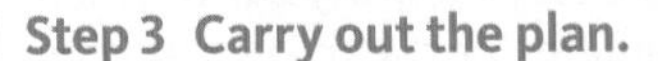

Step 3 Carry out the plan.

$A = \left(\frac{2}{3}\right)\left(\frac{110{,}692 \text{ acre} \cdot \text{feet}}{1}\right)$ amount returned = (fraction returned)(2010 allocation)

$A = \mathbf{73{,}794.66...}$ **acre · feet** Follow the order of operations; write as a decimal number.

$A \approx \mathbf{73{,}800}$ acre · feet Round to the nearest hundred.

Step 4 Look back.
Doing this problem another way, if we divide 110,692 acre · feet into three parts, each part is about 36,900 acre · feet. Since the amount returned is equal to two of these three parts of water, it is equal to 2(36,900 acre · feet), which is 73,800 acre · feet. Since this is equal to the rounded answer, the answer seems reasonable.

Step 5 Report the solution.
The amount of water to be returned to Westlands and San Luis is about 73,800 acre · feet.

In the next example, to find how many times an amount that is a fraction can be taken out of a whole number, we divide by the fraction and round down to the nearest whole number.

EXAMPLE 7 A recipe for chocolate chip cookies uses $\frac{3}{4}$ cup of sugar and 2 eggs. A container has 8 cups of sugar. Find the number of cookie recipes that can be made with this sugar. Round down to the nearest whole number.

SOLUTION ▶ **Step 1 Understand the problem.**
The unknown is the number of recipes that can be made with 8 cups of sugar. The number of eggs is extraneous information.

r = number of recipes

Step 2 Make a plan.
The number of recipes equals the total amount of sugar divided by the amount of sugar needed to make a recipe. A word equation is number of recipes = total sugar ÷ sugar in a recipe. Since cup ÷ cup is 1, we remove the units of measurement. We will learn more about simplifying units of measurement in Section 1.5.

Step 3 Carry out the plan.

$r = 8 \text{ cup} \div \frac{3}{4} \text{ cup}$	number of recipes = total sugar ÷ sugar in a recipe
$r = 8 \div \frac{3}{4}$	Remove the units.
$r = 8 \cdot \mathbf{\frac{4}{3}}$	Rewrite division as multiplication by the reciprocal.
$r = \mathbf{10.66...}$	Follow the order of operations; write as a decimal number.
$r \approx$ **10 recipes**	Include the units; round down to the nearest whole number.

Step 4 Look back.
If the recipe required 1 cup of sugar, we could make 8 recipes out of 8 cups of sugar. If the recipe required $\frac{1}{2}$ cup of sugar, we could make double this amount, or 16 recipes. Since $\frac{3}{4}$ cup is between $\frac{1}{2}$ cup and 1 cup and the answer of 10 recipes is between 8 recipes and 16 recipes, the answer is in the ballpark and seems reasonable.

Step 5 Report the solution.
Ten recipes of cookies can be made with this amount of sugar.

Practice Problems

For problems 4–7, use the five steps.

4. The balance in a student's bank account is $150. He uses his debit card to buy $70 of groceries, deposits $575, and uses his debit card to buy $228 of books. Find the balance in his bank account. Use negative numbers to represent withdrawals.
5. A board is 12 feet long. A builder is cutting it into pieces that are $\frac{3}{4}$ foot long. Find the number of pieces that can be cut from the board.
6. Three-fifths of the students in a class are using electronic textbooks. If the class has 60 students, find the number of students who are using an electronic textbook.
7. Find the number of the fuel assemblies (control rods) replaced. Round to the nearest whole number.

   Columbia Generating Station, the Northwest's only nuclear power plant, is scheduled to begin a refueling outage. . . . Approximately one-third of the 764 fuel assemblies in the reactor core will be replaced during the outage. (*Source:* www.energy-northwest.com, May 10, 2007)

1.3 Vocabulary Practice

Choose the term that best describes the measurement printed in *italics*.

A. reasonable answer
B. unreasonable answer

1. A dentist is injecting lidocaine into a mouth before filling a cavity. His calculations tell him to inject *2 liters* of lidocaine.
2. A hamburger costs $1.25. A student is buying five hamburgers. He estimates that he will need *at least $5 but less than $10* to pay for the hamburgers.
3. A student is 15 miles from school. He is traveling 60 miles an hour. He estimates that he will be there in *4 hours*.
4. A person needs to lose 40 pounds. The doctor recommends a weight loss of 2 pounds per week. At this rate, the person will lose the weight in about *5 months*.
5. The price of a 24-can case of soda is $8.99. The price per can is *$215.76*.
6. A football field is 120 yards long and 160 feet wide. The distance around the football field is *600 yards*.
7. The population of a country is 34,000,000 people. To give every person a gift of $0.50 will cost about $68,000,000.
8. An employee makes $8 per hour. His gross pay for two 40-hour weeks of work should be about *$640*.
9. An investment of $1000 is earning 5% interest a year. In 2 years, the value of the investment and interest will be about *$2000*.
10. Some friends are driving to a concert at a stadium. The stadium is 150 miles away, and they are averaging a speed of 60 miles per hour. It will take them *2 hours and 30 minutes* to get there.

1.3 Exercises

Follow your instructor's guidelines for showing your work.

For exercises 1–76, use the five steps.

1. The October electricity bill was $106.81. The November electricity bill was $126.56. Find the difference between the November bill and the October bill.
2. The rent on an apartment was increased from $475 a month to $540 a month. Find the amount that the rent was increased.
3. A contractor said that the cost to build a standard one-story home was about $122 per square foot. Find the cost to build a home with 1850 square feet.
4. The price of one drip coffee at a campus coffee shop is $1.25. A student buys about 180 drip coffees per school year. Find the cost to buy 180 drip coffees.
5. For students taking at least six credits, the college day-care center charges $350 a month for preschoolers, $400 a month for toddlers, and $450 a month for infants. Find the cost of daycare for 9 months for a family with one toddler and one infant.
6. For students taking at least six credits, the college day-care center charges $350 a month for preschoolers, $400 a month for toddlers, and $450 a month for infants. Find the cost of daycare for 9 months for one preschooler and one infant.
7. A student works 15 hours a week at a job that pays $7.25 per hour. Since he needs to make $225 per week to pay his bills, he is looking for another job. If he can find a job that pays $9 per hour, how many additional hours will he need to work? Round up to the nearest whole number.
8. A student works 12 hours a week at a job that pays $7.25 per hour. Since she needs to make $240 per week to pay her bills, she is looking for another job. If she can find a job that pays $10 per hour, how many additional hours will she need to work? Round up to the nearest whole number.
9. A basic wireless phone plan costs $39.99 a month. Each text message costs $0.20. A select plan costs $59.99 a month and has unlimited text messages. Find the number of text messages at which the basic plan and the select plan have the same monthly cost.
10. A select wireless phone plan costs $59.99 a month. Each megabyte of data costs $1.99. A connect wireless phone plan costs $69.99 a month. Data transfer is free. Find the number of megabytes of data at which the select plan and the connect plan have the same monthly cost. Round to the nearest whole number.
11. A gallon jug of brand name nonfat milk costs $2.39. A gallon jug of store brand nonfat milk costs $1.99. Find the savings in a year from buying the store brand if a family buys 3 gallons of nonfat milk per week (52 weeks = 1 year).
12. A 2010 Honda Civic with a 1.8-liter engine has an EPA highway rating of 34 miles per gallon. A 2010 Honda Civic with a 2.0-liter engine has an EPA highway rating of 29 miles per gallon. If a commuter drives 25,000 highway miles per year and gasoline is $3.10 per gallon, find the amount saved in a year by driving the car with the 1.8-liter engine. Round to the nearest whole number. (*Source:* www.fueleconomy.gov)

13. In October 2011, Citigroup announced a 74% change in earnings in the third quarter of $3.8 billion, an increase of $1.6 billion from the third quarter of 2010. Find the earnings in the third quarter of 2010. (*Source:* www.denverpost.com, Oct. 17, 2011)

14. Find the number of employees before the job cuts at General Tobacco's plant.

General Tobacco Co. will cut 31 jobs . . . the job cuts would trim the workforce to 95 full-time employees.
(*Source:* www.godanriver.com, Nov. 14, 2008)

15. The average weekly salary for a construction worker in May 2011 was $725. If a construction worker is employed for 50 weeks a year, find the employee's annual salary. (*Source:* www.bls.gov, July 2011)

16. The U.S. average weekly wage in the first quarter of May 2011 was $935. If an employee works at this wage for 50 weeks a year, find the employee's annual salary. (*Source:* www.bls.gov, Sept. 29, 2011)

For exercises 17–18, use the bar graph.

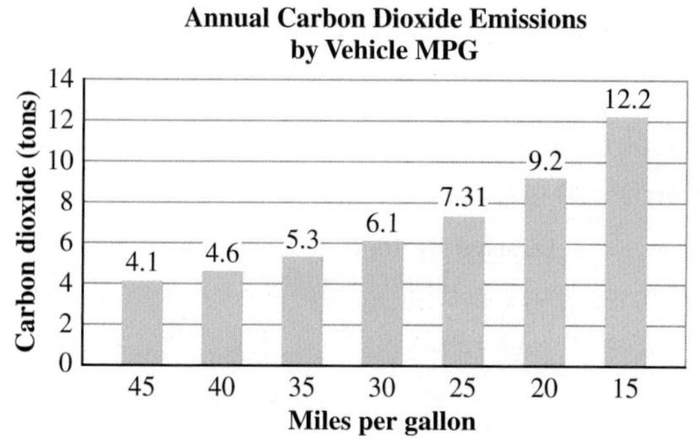

Source: www.fueleconomy.gov

17. For 6 years of driving, find the difference in carbon dioxide (CO_2) emissions of a car that gets 35 miles per gallon and of a car that gets 15 miles per gallon.

18. For 5 years of driving, find the difference in carbon dioxide (CO_2) emissions of a car that gets 40 miles per gallon and of a car that gets 20 miles per gallon.

For exercises 19–20, use the table below, which compares the estimated annual costs to attend college while living off campus during the 2010–2011 school year in California. Assume that the costs stay the same for 4 years and that the total annual cost includes fees, tuition, books, supplies, off-campus room and board, and miscellaneous expenses.

19. Find the difference in the total cost to attend a college in the California State University system for 4 years and the total cost to attend community college for 2 years and then transfer to a college in the California State University system for 2 years.

20. Find the difference in the total cost to attend a college in the University of California system for 4 years and the total cost to attend community college for 2 years and then transfer to a college in the University of California system for 2 years.

21. Find the population of China a decade ago. Write the answer using the word *million.*

For now, China remains the most populous nation, with 1.34 billion people. In the past decade it added 73.9 million, more than the population of France or Thailand.
(*Source:* www.chicagotribune.com, Oct. 15, 2011)

22. Find the amount of state funding UC received last year. Write the answer using the word *million.*

UC will receive about $2.37 billion in state funding this year, $650 million less than last year. (*Source:* www.latimes.com, Sept. 15, 2011)

23. Find the amount of manure produced by the herd of cows in 1 year (365 days). Write the answer using the word *million.*

"Can't you guys do something with this stuff—make a flowerpot or something?" Those were fateful words for brothers Ben and Matthew Freund, second-generation dairy farmers who at the time maintained a herd of 225 Holsteins in East Canaan. Each cow produces 120 pounds of manure daily.
(*Source:* www.nytimes.com, March 1, 2009)

24. New diets for cows that result in less belching can reduce greenhouse gas emissions. Find the maximum amount of methane gas that a herd of 60 average dairy cows produces by belching per year (1 year = 365 days). Write the answer using the word *million.*

The average dairy cow belches out about 100 to 200 liters of methane each day. (*Source:* www.reuters.com, July 9, 2007)

For exercises 25–26, use the information in the table.

Music downloads	2009	2010	Percent change
Number of downloaded singles	1.138 trillion	1.162 trillion	2.1%
Value of downloaded singles	$1.220 trillion	$1.367 trillion	12.0%

Source: www.riaa.com

25. Find the difference in the number of downloaded singles between 2009 and 2010. Write the answer in billions of downloaded singles.

Table for exercises 19–20

	Community colleges	California State University (CSU)	University of California (UC)
Fees and tuition	$864	$6489	$13,200
Books and supplies	$1656	$1652	$1500
Room and board (off campus)	$10,863	$11,379	$9500
Misc.	$4059	$4041	$4200

Source: www.californiacolleges.edu

26. Use the table on page 63 to find the difference in the value of downloaded singles between 2009 and 2010. Write the answer in billions of dollars.

27. Find the total amount the company is paying for pollution controls and the penalty. Write the answer in place value notation.

Last July, East Kentucky agreed to install pollution controls estimated to cost $650 million and to pay a $750,000 penalty to resolve violations of the new source review provisions of the Clean Air Act at the Dale facility and two other plants. (*Source:* www.yosemite.epa.gov, Sept. 20, 2007)

28. Find the total salary that the chief operating officer of Microsoft made in 2010. Write the answer in place value notation.

Chief Operating Officer Kevin Turner was the highest paid . . . $732,500 in salary, $1.9 million in bonus and $6.6 million in stock awards. (*Source:* news.cnet.com, Oct. 3, 2011)

29. A condominium in the East Village of New York City has two bedrooms, two bathrooms, and 1468 square feet of living area. The price of the condo is $1,575,000. Find the cost per square foot of this condo. Round to the nearest whole number.

30. A condominium in south Boston has three bedrooms, two baths, and 1807 square feet of living area. The price of the condo is $1,040,000. Find the cost per square foot of this condo. Round to the nearest whole number.

31. The regular price of a gallon of milk is $3.09. On sale, the milk is $2.39 per gallon. A daycare center uses 45 gallons of milk and 32 gallons of orange juice per week. Find the amount the daycare center will save per week when the milk is on sale.

32. The regular price of a gallon of orange juice is $3.89. On sale, the juice is $3.59 per gallon. A daycare center uses 45 gallons of milk and 32 gallons of orange juice per week. Find the amount the daycare center will save per week when the orange juice is on sale.

33. Find the number of preserved acres before the properties and easements were added in 2007.

In 2007, the Lowcountry Open Land Trust added 28 new protected properties and 2 additions to existing easements, helping landowners save a record 10,561 acres in perpetuity. With these easements, the Land Trust now has 57,579 preserved acres in South Carolina. (*Source:* www.lolt.org, Annual Report 2007)

34. Find the previous score of the Civic LX in *Consumer Reports*.

The redesigned Civic LX's score dropped a whopping 17 points to a mediocre 61. (*Source:* latimes.blog, Aug. 4, 2011)

35. A birthday party at a skating rink includes 15 children and 2 adults. Twelve of the children will rent skates, and no one will play hockey. Find the cost for the birthday party.

Cost to skate is $4 with $1 to rent skates. Parks and Rec also offers open hockey with separate times for both adults and youth at Packer. Hockey is $10 for adults, $5 for kids. (*Source:* www.austindailyherald.com, Dec. 6, 2008)

36. The cost to buy a phone with a 2-year service contract is $199. A single-line voice plan costs $69.99 per month. A 2 GB data plan is an additional $30 per month. An unlimited messaging plan is an additional $20 per month. A case is $35. Find the total cost to buy this phone and a case and to use the given plans for 2 years.

37. Find the number of nonfatal injuries in alcohol-related motor vehicle crashes that occur each day.

Alcohol-related motor vehicle crashes kill someone every 31 minutes and nonfatally injure someone every two minutes. (*Source:* www.cdc.gov)

38. Find the number of motor vehicle crashes that occur each day.

Every 12 minutes someone dies in a motor vehicle crash, every 10 seconds an injury occurs and every 5 seconds a crash occurs. (*Source:* www.osha.gov, 2005)

39. Find the number of additional hours each year a student will spend in school if the average school day is increased to 6.75 hours.

On average, U.S. students spend 6.5 hours per day in school, 180 days a year. That's far less than for students in many other developed nations. (*Source:* www.bnet.com, Jan. 28, 2008)

40. In Chicago in 2011–2012, most elementary students met from 9 a.m. to 2:45 p.m. for 170 days. If 10 days are added to the school year and 90 minutes are added to each school day, find the additional time a student will spend in school. (*Source:* www.chicagotribune.com, April 29, 2011)

For exercises 41–42, use the chart.

Personal Health Care Expenditures (billions of dollars)

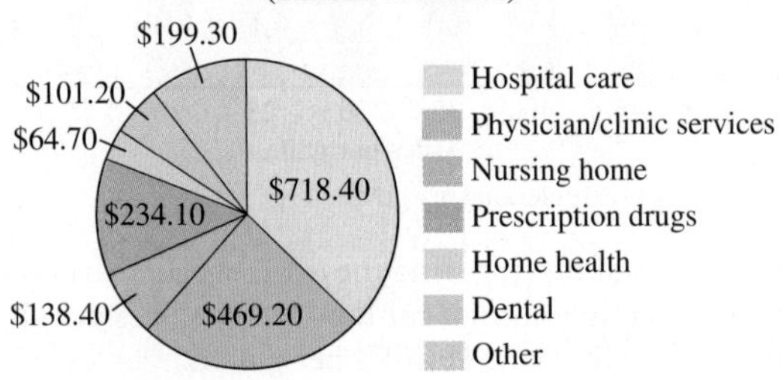

Source: www.cdc.gov, 2010

41. Find the total amount spent on hospital care, physician/clinic services, and prescription drugs. Write the answer in place value notation.

42. Find the total amount spent on hospital care, physician/clinic services, and nursing homes. Write the answer in place value notation.

For exercises 43–44, use the chart at the top of the next page.

43. Find the difference in the percent of employees that have access to medical benefits if there are 500 or more employees in the company and if there are 50 to 99 employees in the company.

44. Find the difference in the percent of employees that have access to medical benefits if there are 500 or more employees in the company and if there are 1 to 49 employees in the company.

45. The International Association for Wireless Communication predicted annual wireless data revenues in 2011 of $55.4 billion, compared to $280.8 million in 1996. Find the increase in revenues, writing the answer in place value notation. (*Source:* www.ctia.org, Oct. 11, 2011)

Chart for exercises 43–44

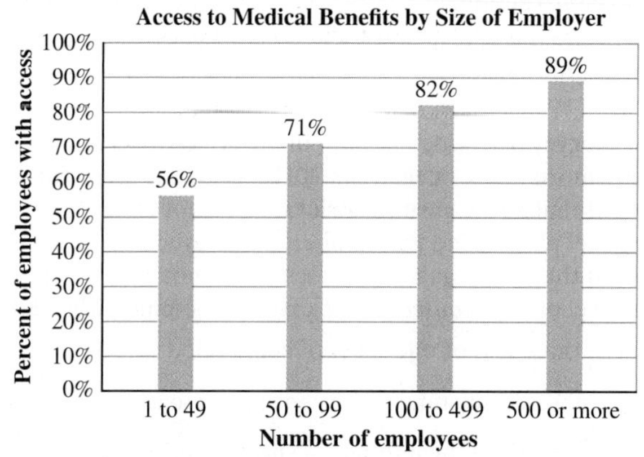

Source: www.bls.gov, March 2010

46. The International Association for Wireless Communication predicted 196.9 billion text messages in 2011, compared to 33.5 million text messages sent in 1996. Find the increase in the number of text messages, writing the answer in place value notation. (*Source:* www.ctia.org, Oct. 11, 2011)

47. The beginning balance on a bank account is \$285. A bank statement shows two deposits: \$650 and \$1324. It shows many debit card transactions: −\$39, −\$58, −\$61, −\$475, −\$173, and −\$81. Find the new balance in this account. Use negative numbers to represent debits.

48. An employee has her monthly paycheck of \$2080 deposited directly into her bank account. Her fixed monthly expenses include rent of \$425, car payment of \$320, cell phone payment of \$50, utilities payment of \$75, and cable bill payment of \$60. After paying fixed expenses, find the amount of money that she has left per month. Use negative numbers to represent payments.

49. A patient with heart disease needs to lose 96 pounds. The physician's assistant recommends an average loss of 3 pounds per week. Find the number of weeks needed to lose the weight. Use negative numbers to represent weight loss.

50. A runner is training for a marathon. She wants to reduce the time it takes her to run 6 miles by 3 minutes. She plans to reduce her time by 12 seconds every week. Find the number of weeks needed for her to meet her goal. Use negative numbers to represent loss of time.

51. On January 4, the high temperature in Minneapolis was 9°F. The low temperature was −4°F. Find the difference of these temperatures.

52. On January 23, the high temperature in Binghamton, New York, was 14°F. The low temperature was −5°F. Find the difference of these temperatures.

53. The price of a popular toy is \$49.99. A customer with \$765 in a checking account bought three of these toys with her debit card. Find the ending balance in the account. Use a negative number to represent a debit.

54. The price of a place setting of dishes is \$20.99. A customer with \$845 in a checking account bought eight place settings with her debit card. Find the ending balance in the account. Use a negative number to represent a debit.

55. The annual take-home pay of a teacher is \$34,000. His monthly house payment is \$976. After making all of his house payments for the year, find the remaining amount of his annual take-home pay. Use a negative number to represent a house payment.

56. The annual take-home pay of a chemist is \$69,000. Her monthly house payment is \$1255. After making all of her house payments for the year, find the remaining amount of her annual take-home pay. Use a negative number to represent a house payment.

57. Find the number of 2350 individual policyholders who will have their premiums raised.

Anthem Blue Cross, which plans to raise premiums for about four-fifths of the . . . individual policyholders . . . has outraged members, many of whom have . . . price increases of more than 30 percent. (*Source:* www.sfgate.com, Feb. 11, 2009)

58. A headline in the *New York Times* on July 16, 1916, was "Autos Increase, Horses Decrease." Find the number of 550 vehicles that were self-propelled and were not pulled by horses or other animals. Round to the nearest whole number.

Nearly seven-tenths of all vehicular traffic in . . . New York . . . is self-propelled. (*Source: New York Times*, July 16, 1916)

59. The width of a standard seam in home sewing is $\frac{5}{8}$ inch. A sewer decreases the width by $\frac{1}{4}$ inch. Find the new width of the seam.

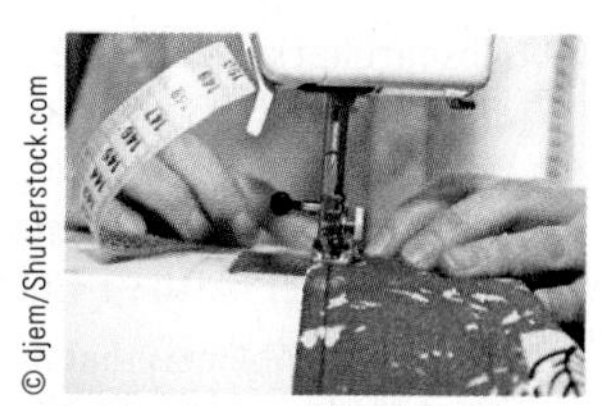

60. The ingredients of a recipe includes $\frac{3}{8}$ cup grated cheese. If an additional $\frac{1}{4}$ cup of cheese is used, find the total amount of cheese.

61. If two grandchildren each received $\frac{1}{32}$ of an estate and their mother received $\frac{1}{3}$ of the estate, find the total fraction of the estate received by the mother and her two children.

62. A building lot is $\frac{1}{6}$ acre. If it is combined with an adjoining lot that is $\frac{5}{32}$ acre, find the total area of the combined lots.

63. Ship decking for model ships is made by gluing together narrow strips of wood. Each strip is $\frac{3}{16}$ inch wide. Find the number of strips needed to create a sheet of decking that is 24 inches long.

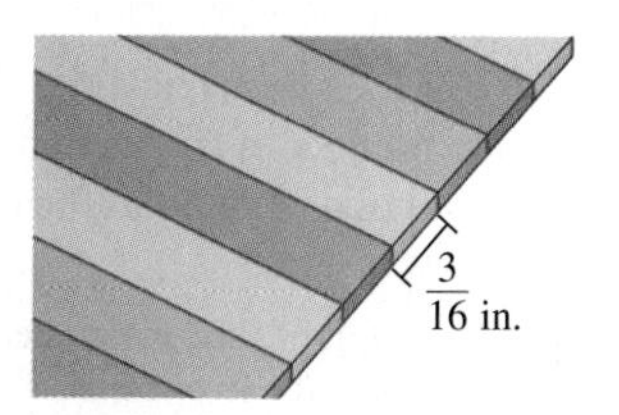

64. A cork board is made by gluing together wine corks. Each cork is $\frac{3}{4}$ inch wide. Find the number of corks needed to create a board that is 24 inches wide.

65. Find the number of school districts in which the program malfunctioned. Round to the nearest whole number.

A new online statewide math test was shut down this week, after the program malfunctioned for one-fourth of the 99 school districts using it. (*Source:* www.ap.org, April 26, 2007)

66. The U.S. Constitution states that a constitutional amendment requires approval by $\frac{3}{4}$ of the state legislatures. There are 50 states. Find the number of legislatures that must approve a constitutional amendment. Round to the nearest whole number.

67. A recipe for frosted cookies includes $\frac{1}{2}$ pound of butter for the cookies and $\frac{1}{8}$ pound of butter for the frosting. Find the total amount of butter in this recipe.

68. A sausage recipe includes $\frac{1}{2}$ pound of ground pork and $\frac{1}{3}$ pound of ground lamb. Find the total amount of ground meat in this recipe.

69. Researchers used placement test scores of 1788 freshmen at the University of Alaska to identify those who were not prepared to do college-level math and English. Find the number of freshmen who were not prepared to do college-level math or English.

A second problem is that many high-school graduates aren't prepared for college. A 2006 report estimated that as many as two-thirds of incoming UA freshmen weren't prepared to do college-level math and English. (*Source:* www.iser.uaa.alaska.edu, May 2008)

70. About 8.2 million Americans own recreational vehicles (RVs). Find the number of RV owners who plan to purchase another RV. Round to the nearest tenth of a million.

Two-thirds of current owners plan to purchase another RV. Among all U.S. households, nearly one quarter (23 percent) intend to purchase an RV in the future. (*Source:* www.rvia.org, Feb. 19, 2009)

71. The payroll for the New York Yankees in 2008 was about $215 million. Find the approximate combined payrolls in millions of dollars of the Kansas City Royals, the Oakland Athletics, and the Tampa Bay Rays.

The Yankees are in last place. They currently own a worse record than Kansas City. Worse than Oakland. Worse than Tampa Bay. We mention those three teams because their combined payrolls weigh barely three-fourths of the Yankees. (*Source:* www.usatoday.com, May 20, 2008)

72. In November 2007, U.S. airlines had 415,024 scheduled departures. Find the number of flights that landed close to their scheduled arrival time. Round to the nearest whole number. (*Source:* www.bts.gov)

U.S. airlines brought four-fifths of their November flights in close to schedule, making the month the second best of last year for on-time performance. (*Source:* www.usatoday.com, Jan. 8, 2008)

73. Find the number of worker fatalities in private industry in construction.

Out of 4,070 worker fatalities in private industry in calendar year 2010, one-fifth . . . were in construction. (*Source:* www.osha.gov, Aug. 2011)

74. A survey of 325 people who intended to travel on the Memorial Day weekend in 2011 found that three-fifths said that rising gasoline prices would not affect their travel plans. Find the number of surveyed people who said that rising gasoline prices would not affect their travel plans. (*Source:* www.aaanewsroom.net, May 2011)

75. The thickness of one sheet of corrugated cardboard is $\frac{3}{16}$ inch. The distance between shelves in a storage rack is 6 inches. Find the number of sheets of cardboard that can be stacked in this space.

76. The thickness of one sheet of card stock used in scrapbooking is $\frac{3}{64}$ inch. The distance between shelves in a storage rack is 6 inches. Find the number of sheets of card stock that can be stacked in this space.

Find the Mistake

For exercises 77–80, the completed problem has one mistake.

(a) Describe the mistake in words, or copy down the whole problem and highlight or circle the mistake.

(b) Do the problem correctly.

77. Problem: Complete the "Make a plan" step for this problem: The Government Accountability Office found cost overruns totaling $295 billion in 95 major government defense projects. Find the average amount of overrun per project. Round to the nearest tenth of a billion.

Incorrect Answer: The average amount is the product of the overruns and the number of projects. A word equation is average amount = (cost overruns)(number of projects).

78. Problem: Complete the "Understand the problem," "Make a plan," and "Carry out the plan" steps for this problem: Find the change in the maximum annual Pell Grant scholarship from 2010 to 2019.

The student-loan legislation would provide $40 billion to increase the maximum annual Pell Grant scholarship to $5,550 in 2010 and to $6,900 by 2019, from $5,350 now. (*Source:* www.nytimes.com, Sept. 17, 2009)

Incorrect Answer: *Understand the problem:* D = change in the maximum scholarship

Make a plan: The change is the difference between the amount in 2019 and the amount in 2010. A word equation is difference = amount in 2019 – amount in 2010.

Carry out the plan: $D = \$6900 - \$5350$

$D = \$1550$

79. Problem: Complete the "Understand the problem," "Make a plan," and "Carry out the plan" steps for this problem: Find the percent of community college students nationwide who can attend college full-time.

If students can attend full-time, they are four times as likely to complete as part-timers. But only 29 percent of California students can attend full-time. That's 12 percentage points below the national community college figure. (*Source:* www.communitycollegetimes.com, Aug. 3, 2007)

Incorrect Answer: *Understand the problem:* P = percent of community college students nationwide

Make a plan: The percent is the difference in the percent of California students and the national community college figure. A word equation is difference = California students − national community college students

Carry out the plan: $D = 29\% - 12\%$
$D = 17\%$

80. Problem: Complete the "Make a plan" step for this problem: Find the amount of the revenue that the state needs to raise in addition to this tax. Write the answer with the word *million*.

The tax is projected to raise about $688 million—or just over 5 percent—of the state's $13.4 billion in revenue for the coming budget year. (*Source:* www.chicagotribune.com, April 3, 2011)

Incorrect Answer: *Make a plan:* The percent is the sum of the tax and the revenue. A word equation is sum = tax + revenue.

Review

81. Find 15% of 6.

82. Find 8% of $48,000.

83. What percent is $5 of $80?

84. Rewrite 0.97 as a percent.

SUCCESS IN COLLEGE MATHEMATICS

85. Describe your instructor's policy about late assignments.

86. Some instructors include attendance, class participation, or participation in an on-line discussion board as part of your grade. Explain how, if at all, attendance or participation affects your grade.

© Cynthia Farmer/Shutterstock.com

In 1970, the average tuition and fees for one year of full-time study at a two-year public college was $178. By 2004, the average tuition and fees was $1702, an 856% increase. In this section, we will do problem solving with percents.

SECTION 1.4 Problem Solving with Percent

After reading the text, working the practice problems, and completing assigned exercises, you should be able to:

1. Rewrite a percent as a fraction and/or as a decimal number.
2. Solve application problems that involve percents.

Percent

A **percent** represents the number of parts out of 100 total parts. Since we can write a percent as a fraction in which the numerator and denominator are integers, a percent is a rational number.

EXAMPLE 1 Rewrite 3% as a fraction and as a decimal number.

SOLUTION ▶

3%	Percent notation; 3 parts out of 100 parts.
$= \frac{3}{100}$	Fraction notation; 3 parts out of 100 parts.
$= 0.03$	Decimal notation; 3 parts out of 100 parts.

To find $\frac{1}{2}$ of \$60, we multiply: $\frac{1}{2} \cdot \$60 = \$30$. To find 50% of 60, we also multiply, rewriting the percent in fraction or decimal notation.

EXAMPLE 2 Find 50% of \$60.

SOLUTION ▶

$(50\%)(\$60)$

$= (\mathbf{0.50})(\$60)$ Rewrite 50% in decimal notation as 0.50.

$= \$30$ Multiply.

To find the percent an amount is of a total, divide the amount by the total and multiply by 100%.

EXAMPLE 3 What percent is \$7.20 of \$160?

SOLUTION ▶

$\left(\frac{\$7.20}{\$160}\right)(100)\%$

$= 4.5\%$ Follow the order of operations.

Percent

$$\text{percent} = \left(\frac{\text{number of parts}}{\text{total parts}}\right)(100)\%$$

$$\text{number of parts} = (\text{percent in decimal notation})(\text{total parts})$$

In the next example, we use the five steps to solve the problem.

EXAMPLE 4 When crude oil is refined, gasoline is one of the products. Use the information in the graphic to find the cost in March 2011 of the crude oil used to make 1 gallon of gasoline. Round to the nearest hundredth.

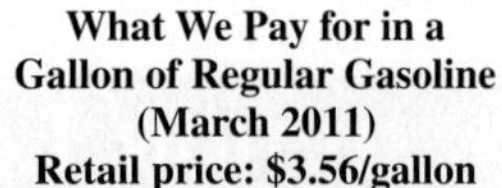

What We Pay for in a
Gallon of Regular Gasoline
(March 2011)
Retail price: \$3.56/gallon

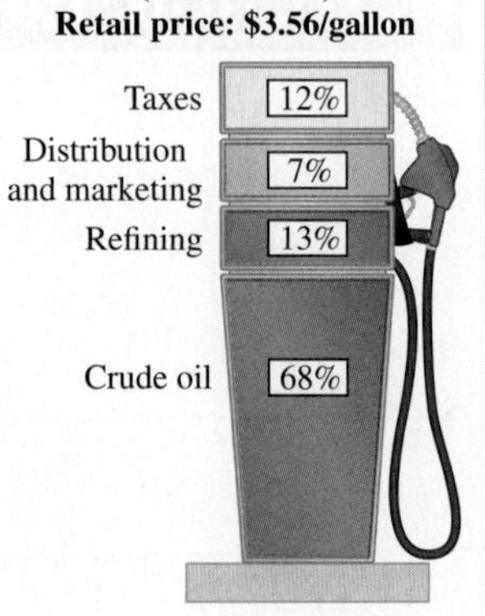

Source: www.eia.gov

SOLUTION ▶ **Step 1 Understand the problem.**
The unknown is the cost of the crude oil used to make 1 gallon of gasoline. The percent for taxes, distribution and marketing, and refining is extraneous information.

C = cost of crude oil

Step 2 Make a plan.
Since number of parts = (percent)(total parts), a word equation is

cost of crude oil = (percent crude oil)(total cost)

Step 3 Carry out the plan.

$C = (0.68)(\$3.56)$ cost of crude oil = (percent crude oil)(total cost)

$C = \mathbf{\$2.4208}$ Follow the order of operations.

$C \approx \mathbf{\$2.42}$ Round.

Step 4 Look back.
If crude oil is 60% of the cost of 1 gallon of gasoline, then the cost of crude oil would be (0.60)(\$3.56), which equals about \$2.14. If crude oil is 70% of the cost of

1 gallon of gasoline, then the cost of crude oil would be $(0.70)(\$3.56)$, which equals about \$2.49. Since 68% is between 60% and 70% and \$2.42 is between \$2.14 and \$2.49, the answer is in the ballpark and seems reasonable.

Step 5 Report the solution.
The cost of the crude oil to make 1 gallon of gas in March 2011 was about \$2.42.

Information in the media is often reported as a percent.

EXAMPLE 5 The Idaho State Senate approved a \$1.56 billion budget for the public schools. Find the percent of Senators who voted for the budget. Round to the nearest percent.

Senate lawmakers agreed to cut the public school budget for the second time in state history Tuesday . . . Senators approved the bill on a 30–5 vote. (*Source:* www.lmtribune.com, April 6, 2011)

SOLUTION ▶ **Step 1 Understand the problem.**
The unknown is the percent of senators who voted "yes." The amount of the budget is extraneous information.

$P =$ percent voting yes

Step 2 Make a plan.
Since percent $= \left(\dfrac{\text{number of parts}}{\text{total parts}}\right)(100)\%$, a word equation is

$$\text{percent voting yes} = \left(\frac{\text{number of yes votes}}{\text{total votes}}\right)(100)\%$$

The total votes is 30 votes + 5 votes, which equals 35 votes.

Step 3 Carry out the plan.

$P = \left(\dfrac{30 \cancel{\text{votes}}}{35 \cancel{\text{votes}}}\right)(100)\%$ — percent voting yes $= \left(\dfrac{\text{number of yes votes}}{\text{total votes}}\right)(100)\%$

$P = \mathbf{85.7\ldots\%}$ — Follow the order of operations; $\dfrac{\text{votes}}{\text{votes}} = 1$

$P \approx \mathbf{86\%}$ — Round.

Step 4 Look back.
Working backwards, since 86% of 35 votes is about 30 yes votes and that is the number of yes votes reported, the answer seems reasonable.

Step 5 Report the solution.
About 86% of the senators voted for the budget.

Information is often presented as a ratio in lowest terms. For example, a survey might report that four of five customers wanted a white car. To write a rate or ratio as a percent, divide the numerator (the number of parts) by the denominator (the total parts) and multiply by 100%.

EXAMPLE 6 Find the percent of health dollars in the United States spent on heart disease, stroke, and other cardiovascular diseases. Round to the nearest tenth of a percent.

Heart disease, stroke, and other cardiovascular (blood vessel) diseases are among the leading cause of death and now kill more than 800,000 adults in the U.S. each year. Of these, 150,000 are younger than age 65. These diseases are also two of the leading causes of health disparities in the U.S. Treatment of these diseases accounts for 1 in every 6 U.S. health dollars spent. (*Source:* www.cdc.gov, Feb. 1, 2011)

SOLUTION ▶ **Step 1 Understand the problem.**
The unknown is the percent of health dollars spent on cardiovascular diseases. The information about the number of adults killed by the disease and their age is extraneous.

P = percent of health dollars spent on cardiovascular diseases

Step 2 Make a plan.
Since percent $= \left(\dfrac{\text{number of parts}}{\text{total parts}}\right)(100)\%$, a word equation is

$$\text{percent spent} = \left(\frac{\text{amount spent on cardiovascular}}{\text{total amount spent}}\right)(100)\%$$

Step 3 Carry out the plan.

$P = \left(\dfrac{1 \text{ dollar}}{6 \text{ dollars}}\right)(100)\%$ — percent spent $= \left(\dfrac{\text{amount spent on cardiovascular}}{\text{total amount spent}}\right)(100)\%$

$P =$ **16.66. . . %** — Follow the order of operations; $\dfrac{\text{dollar}}{\text{dollars}} = 1$

$P \approx$ **16.7%** — Round.

Step 4 Look back.
Working backwards, since 16.7% of 6 is about 1, which is the original amount spent on cardiovascular diseases in the ratio, the answer seems reasonable.

Step 5 Report the solution.
About 16.7% of health dollars in the United States are spent on cardiovascular diseases.

Practice Problems

For problems 1–4, use the five steps.

1. The Pew Research Center's Internet & American Life Project surveyed 2011 adults, ages 18 and older. Find the number of these adults who own a cell phone. Round to the nearest whole number.

   Some 85% of adults own cell phones overall. Taking pictures (done by 76% of cell owners) and text messaging (done by 72% of cell owners) are the two non-voice functions that are widely popular among all cell phone users. (*Source:* www.pewinternet.org, Feb. 3, 2011)

2. When Chicago mayor Rahm Emanuel spoke about a $15 million investment in new security cameras, 1113 cameras out of 1500 cameras had been installed. Find the percent of cameras that had been installed. Round to the nearest percent. (*Source:* www.chicagotribune.com, Oct. 15, 2011)

3. Find the percent of women in Niger who will die from complications during pregnancy or delivery. Round to the nearest percent.

   Having a child remains one of the biggest health risks for women worldwide. Fifteen hundred women die every day while giving birth. . . . A woman in Niger has a one in seven chance of dying during the course of her lifetime from complications during pregnancy or delivery. That's in stark contrast to the risk for mothers in America, where it's one in 4,800 or in Ireland, where it's just one in 48,000. (*Source:* www.unicef.org, 2009)

4. The Smithfield Packing Plant in Tar Heel, North Carolina, processes about 30,000 hogs a day. About 4600 of its 5000 employees are eligible for union representation. Find the percent of voters who voted for union representation. Round to the nearest percent. (*Source:* www.reuters.com, Dec. 12, 2008)

   On Dec. 11, 2008, when the votes were counted in the same packing plant, 2,041 workers had voted to join the United Food and Commercial Workers (UFCW), while just 1,879 had voted against it. (*Source:* www.prospect.org, Dec. 17, 2008)

Percent Decrease and Increase

When an amount increases or decreases, we can use the relationship percent $= \left(\frac{\text{number of parts}}{\text{total parts}}\right)(100)\%$ and report the increase or decrease as a percent of the original amount. The number of total parts is the original amount, before the change.

EXAMPLE 7 In April 2011, the Recording Academy eliminated 31 categories from the Grammy Awards. Now performers can receive awards in only 78 categories. Find the percent change in the number of categories. Round to the nearest percent. (*Source:* twincities.com, April 6, 2011)

SOLUTION ▶ **Step 1 Understand the problem.**
The unknown is the percent decrease in the number of categories.

P = percent decrease in the number of categories.

Step 2 Make a plan.
Since percent $= \left(\frac{\text{number of parts}}{\text{total parts}}\right)(100)\%$, a word equation is

$$\text{percent decrease} = \left(\frac{\text{decrease in number of categories}}{\text{original number of categories}}\right)(100)\%$$

The decrease can be represented by a negative number: -31 categories. The original number of categories is 78 categories + 31 categories, which equals 109 categories.

Step 3 Carry out the plan.

$P = \left(\frac{-31 \text{ categories}}{109 \text{ categories}}\right)(100)\%$	percent decrease $= \left(\frac{\text{decrease in number of categories}}{\text{original number of categories}}\right)(100)\%$
$P = \mathbf{-28.4...\%}$	Follow the order of operations; $\frac{\text{categories}}{\text{categories}} = 1$
$P \approx \mathbf{-28\%}$	Round.

Step 4 Look back.
Estimating, if we round 31 to 30 and 109 to 100, the percent decrease is about 30%. Since this is close to the answer of a decrease of 28%, the answer seems reasonable.

Step 5 Report the solution.
The number of categories in the Grammy Awards decreased by about 28%.

In the next example, we need to find a difference in order to know the original amount.

EXAMPLE 8 The Port of Lewiston proposed increasing the length of a dock used for loading barges to 275 feet, an increase of 150 feet. Find the percent increase in the length of the dock. (*Source:* www.lmtribune.com, Oct. 15, 2011)

SOLUTION ▶ **Step 1 Understand the problem.**
The unknown is the percent increase in the length of the dock.

P = percent increase in the length of the dock

Step 2 Make a plan.
Since percent $= \left(\frac{\text{number of parts}}{\text{total parts}}\right)(100)\%$, a word equation is

$$\text{percent increase} = \left(\frac{\text{increase in length of dock}}{\text{original length of dock}}\right)(100)\%$$

The original amount is 275 feet – 150 feet, which equals 125 feet.

Step 3 Carry out the plan.

$P = \left(\frac{150 \text{ feet}}{125 \text{ feet}}\right)(100)\%$ — percent increase $= \left(\frac{\text{increase in length of dock}}{\text{original length of dock}}\right)(100)\%$

$P = \mathbf{120\%}$ — Follow the order of operations; $\frac{\text{feet}}{\text{feet}} = 1$

Step 4 Look back.
Working backwards, since 120% of 125 feet is 150 feet, which is equal to the amount of the increase, the answer seems reasonable.

Step 5 Report the solution.
The length of the dock will increase 120%.

Practice Problems

For problems 5–7, use the five steps.

5. Find the percent decrease in the median household income in the Portland area. Round to the nearest percent.

The median household income in the Portland area tumbled from $67,137 in 2007 to $53,078 in 2010, according to the American Community Survey. (*Source:* www.oregonlive.com, Oct. 15, 2011)

6. Find the percent decrease in the number of certified organic producers. Round to the nearest tenth of a percent.

The number of certified [organic] producers in the state declined by 18 in 2010 to 735, with five farms transitioning to organic. (*Source:* wsutoday.wsu.edu, March 23, 2011)

7. Find the percent increase in the amount of medical expenses associated with asthma. Round to the nearest tenth of a percent.

Medical expenses associated with asthma increased from $48.6 billion in 2002 to $50.1 billion in 2007. (*Source:* www.cdc.gov, May 20, 2011)

Retail Applications

When an item is on sale, its original (regular) price is reduced by an amount called the **discount**. The **sale price** of an item equals the original price minus the discount. The discount is often described as a percent of the original price.

EXAMPLE 9 The original price of a home theater speaker system is $499. The speaker system is on sale at a 35% discount. Find the sale price.

SOLUTION ▶ **Step 1 Understand the problem.**
The unknown is the sale price.

p = sale price

Step 2 Make a plan.
Since number of parts = (percent)(total parts), discount = (percent discount)(original price). The sale price equals the original price minus the discount. So sale price = original price − (percent discount)(original price).

Step 3 Carry out the plan.

$p = \$499 - (0.35)(\$499)$ — sale price = original price − (percent discount)(original price)

$p = \mathbf{\$324.35}$ — Follow the order of operations.

Step 4 Look back.
If the discount is 40%, the sale price is \$499 – (0.40)(\$499), which is \$299.40. If the discount is 30%, the sale price is \$499 – (0.30)(\$499), which is \$349.30. Since 35% is between 30% and 40% and \$324.35 is between \$299.40 and \$349.30, the answer is in the ballpark and seems reasonable.

Step 5 Report the solution.
The sale price of the stereo system is \$324.35.

Most states charge a sales tax on purchases. A percent of the purchase price is collected for the state government. To find the total cost of a purchase, the sales tax is calculated and added onto the price of the item.

EXAMPLE 10 The sales tax in a city is 7.65%. Find the total cost, including tax, of a computer with a price of \$499. Round to the nearest hundredth.

SOLUTION ▶ **Step 1 Understand the problem.**
The unknown is total cost of the computer.

C = total cost of the computer

Step 2 Make a plan.
Since number of parts = (percent)(total parts), a word equation for sales tax is sales tax = (percent tax)(price). A word equation for the total cost is total cost = price + sales tax or total cost = price + (percent tax)(price).

Step 3 Carry out the plan.

$C = \$499 + (0.0765)(\$499)$ total cost = price + (percent tax)(price)

$C = \mathbf{\$537.173...}$ Follow the order of operations.

$C \approx \mathbf{\$537.17}$ Round.

Step 4 Look back.
If the sales tax is 7%, the total cost is \$499 + (0.07)(\$499), which equals \$533.93. If the sales tax is 8%, the total cost is \$499 + (0.08)(\$499), which equals \$538.92. Since 7.65% is between 7% and 8% and \$537.17 is between \$533.93 and \$538.92, the answer is in the ballpark and seems reasonable.

Step 5 Report the solution.
The total cost of the computer is \$537.17.

Practice Problems

For problems 8–11, use the five steps.

8. The sales tax rate in a state is 6.5%. Find the sales tax to be collected on textbooks with a total cost of \$375. Round to the nearest hundredth.

9. A car with an original price of \$17,999 is marked down 15%. Find the sale price. Round to the nearest whole number.

10. The original price of a sweatshirt is \$39.99. It is marked down 35%. The sales tax is 8.25%. Find the total cost of the sweatshirt on sale, including tax.

11. A customer bought a set of knives for \$150. The sales tax on \$150 is \$12.30. Find the percent that the sales tax is of the purchase price. (This is the sales tax rate.)

1.4 Vocabulary Practice

Match the term with its description.

1. These numbers are elements of $\{\ldots, -3, -2, -1, 0, 1, 2, 3, \ldots\}$.
2. These numbers are elements of $\{0, 1, 2, 3, 4, \ldots\}$.
3. The elements of this set are either rational numbers or irrational numbers.
4. $\left(\dfrac{\text{number of parts}}{\text{total parts}}\right)(100)\%$
5. $(a + b) + c = a + (b + c)$
6. $a(b + c) = ab + ac$
7. Three dots that follow a list of numbers to show that the list continues on without repeating or stopping
8. $\dfrac{6}{0}$
9. The difference between the original price of an item and the sale price
10. $a + b = b + a$

A. associative property
B. commutative property
C. discount
D. distributive property
E. ellipsis
F. integers
G. percent
H. real numbers
I. undefined
J. whole numbers

1.4 Exercises

Follow your instructor's guidelines for showing your work.

For exercises 1–10, rewrite the percent as a decimal number.

1. 5%
2. 7%
3. 83%
4. 61%
5. 9.1%
6. 7.3%
7. 142%
8. 151%
9. 2.15%
10. 3.33%
11. Find 8% of 32.
12. Find 6% of 42.
13. Find 54% of 225.
14. Find 65% of 130.
15. Find 1.25% of 4500.
16. Find 1.25% of 5500.
17. Find 300% of 65.
18. Find 300% of 55.
19. What percent is 42 of 120?
20. What percent is 21 of 140?
21. What percent is 1.75 of 50?
22. What percent is 1.35 of 90?
23. What percent is 300 of 150?
24. What percent is 250 of 125?
25. What percent is 37.5 of 1500?
26. What percent is 42.5 of 1700?

For exercises 27–82, use the five steps.

27. Find the number of emergency room visits for nail gun injuries that involve workers.

© Christina Richards/Shutterstock.com

Nail guns are powerful, easy to operate, and boost productivity for nailing tasks. They are also responsible for an estimated 37,000 emergency room visits each year—68% of these involve workers and 32% involve consumers. (*Source:* www.cdc.gov, Oct 12, 2011)

28. In a sample of 920 children in Texas, ages 24–35 months, 32.3% had received the hepatitis A vaccine. Find the number of children who received the vaccine. Round to the nearest whole number. (*Source:* www.cdc.gov/nchs/nis.htm, 2003)

29. If 15% of the area of a park is playground and the area of the park is 6 acres, find the area of the playground.

30. If 8% of the receipts of a sporting event are donated to charity and the receipts total $48,000, find the amount donated to charity.

31. Find the percent of adults who wear a seatbelt on every trip. Round to the nearest percent.

One in 7 adults did wear a seat belt on every trip. (*Source:* www.cdc.gov, Jan. 4, 2011)

32. Find the percent of episodes of drinking and driving in 2010 for which men were responsible.

Men were responsible for 4 in 5 episodes . . . of drinking and driving in 2010. (*Source:* www.cdc.gov, Oct. 2011)

33. Find the percent of vehicles belonging to residents that can be parked in legal curb spaces. Round to the nearest percent.

The need for extra parking is clear: There are about 850 legal curb spaces in UCLA's North Village, but about 5,700 vehicles belonging to residents. (*Source:* www.latimes.com, June 28, 2011)

34. Faced with budget cuts from the state, the Minnesota Historical Society eliminated 19 jobs out of 325 total positions. Find the percent of jobs that were eliminated. Round to the nearest percent. (*Source:* www.startribune.com, Aug. 24, 2011)

For exercises 35–36, use the graph.

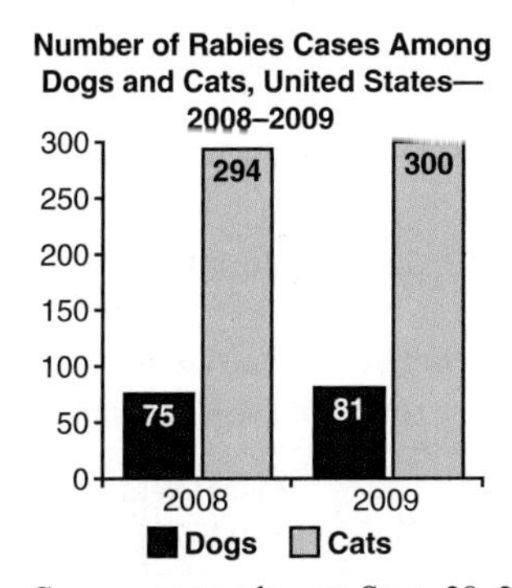

Source: www.cdc.gov, Sept. 28, 2011

35. Find the percent of rabies cases in dogs and cats in 2009 that were in cats. Round to the nearest percent.

36. Find the percent of rabies cases in dogs and cats in 2008 that were in dogs. Round to the nearest percent.

37. Find the minimum percent of streetlights that the Energy Committee recommends for removal. Round to the nearest percent.

A committee studying energy conservation in Jaffrey, N.H., is recommending that the town turn off more than a quarter of its streetlights. The Jaffrey Energy Committee says the town could save a substantial amount of money by removing at least 61 of its 225 streetlights and fitting the rest with lower wattage bulbs. (*Source:* www.amerlux.com, June 2008)

38. Find the percent that the fee revenues for each full-time equivalent student in California are of the national average. Round to the nearest percent.

California collects substantially lower amounts of tuition revenue per student than the national average. . . . Fee revenues per FTES (full-time equivalent student) in California were $1,441 in 2007, less than half the national average of $3,845. (*Source:* www.csus.edu/ihelp, Feb. 2009)

39. Find the percent of the budget cuts to the Arizona university system that are being cut from the Northern Arizona University budget. Round to the nearest percent.

Arizona Gov. Jan Brewer recently signed a bill that cuts $141.5 million from the budgets of the Arizona university system; $20.5 million will be cut from NAU (Northern Arizona University). (*Source:* www.jackcentral.com, Feb. 12, 2008)

40. Find the percent that the calories in a triple Whopper with cheese are of the recommended daily calorie intake for an adult woman. Round to the nearest percent.

At one Burger King restaurant . . . a triple Whopper with cheese has 1,230 calories . . . and a king-size chocolate shake has 1,260. The recommended daily calorie intake for an adult woman is about 1,800. (*Source:* www.nydailynews.com, July 7, 2007)

For exercises 41–42, use the bar graph at the top of the next column.

41. Find the percent of the total amount spent on pets in 2011 that was spent on food. Round to the nearest percent.

42. Find the percent of the total amount spent on pets in 2011 that was spent on vet care. Round to the nearest percent.

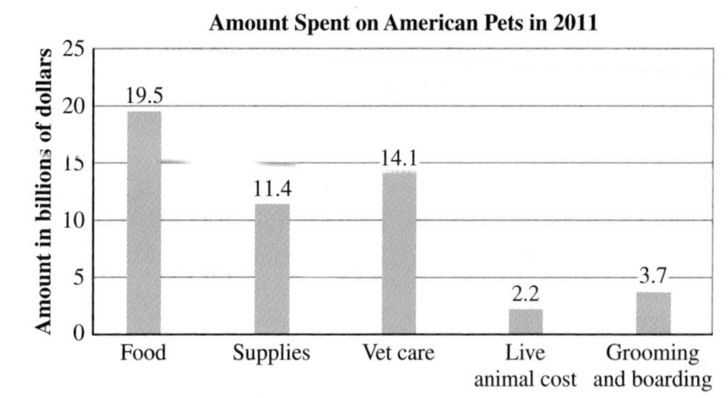

Source: www.americanpetproducts.org, 2011

For exercises 43–44, use the information about the 19th Amendment. Round to the nearest percent.

Advocates of woman suffrage won a victory in the Senate today when that body, by a vote of 56 to 25, adopted the Susan Anthony amendment to the Constitution. . . . The amendment, having already been passed by the House, where the vote was 304 to 89, now goes to the States for ratification. (*Source:* www.nytimes.com, June 5, 1919)

43. Find the percent of voting Senators who voted for the amendment.

44. Find the percent of voting members of the House who voted for the amendment.

For exercises 45–46, use the bar graph. Round to the nearest percent.

Armed Forces Personnel, Sept. 2011

600,000
500,000
400,000
300,000
200,000
100,000
0

565,463
325,123
201,157
333,370
43,251

Army Navy Marine Corps Air Force Coast Guard

Source: siadapp.dmdc.osd.mil/personnel, Oct. 31, 2011

45. Find the percent of total armed service personnel who were in the Marine Corps.

46. Find the percent of total armed service personnel who were in the Army.

For exercises 47–48, use the chart. Round to the nearest percent.

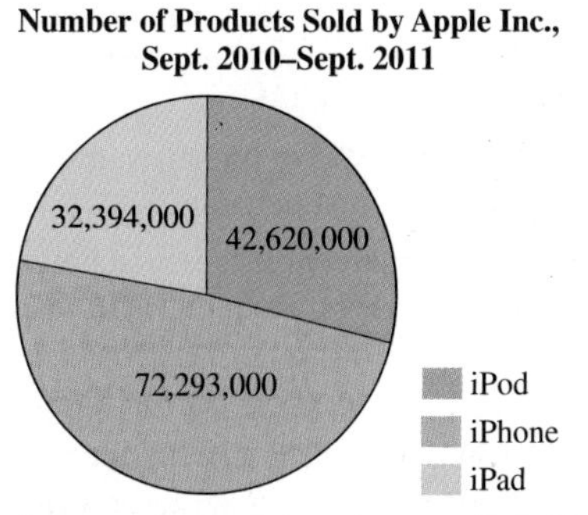

Source: www.appleinvestor.com, Oct. 26, 2011

47. Find the percent of the products that were iPods.

48. Find the percent of the products that were iPhones.

49. In 2005, Americans spent \$7.2 billion on vitamins. In 2010, Americans spent \$9.6 billion on vitamins. Find the percent increase in the spending on vitamins. Round to the nearest percent. (*Source:* www.lmtribune.com, Oct. 14, 2011)

50. In 2011, about 593 grizzly bears lived in Wyoming, Montana, and Idaho, a reduction from the 602 bears who lived there in 2010. Find the percent decrease. Round to the nearest tenth of a percent. (*Source:* www.lmtribune.com, Oct. 14, 2011)

For exercises 51–52, use the information in the bar graph.

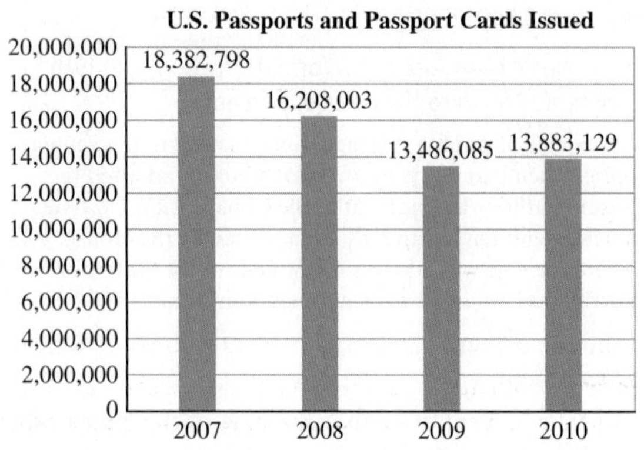

Source: travel.state.gov

51. Find the percent change between 2007 and 2010 in the number of passports and passport cards issued. Round to the nearest percent.

52. Find the percent change between 2008 and 2010 in the number of passports and passport cards issued. Round to the nearest percent.

53. The enrollment of a college in 2011 was 3000 students. In 2012, the enrollment increased by 600 students. Find the percent increase.

54. In 2011, there were 600 children playing in a soccer league. In 2012, the number playing increased by 150 children. Find the percent increase.

For exercises 55–56, use the information about hourly wages in Biloxi.

> "There's been this huge infusion of cash into the local economy (Biloxi). . . . And wages have gone up considerably given the demand for workers." That demand, Mr. Mahoney said, means he now pays a starting dishwasher \$8.50 an hour rather than \$6; a \$12-an-hour construction worker typically earns \$20 an hour or more now. (*Source:* www.gulfcoast.org, July 16, 2007)

55. Find the percent increase in hourly wages for dishwashers. Round to the nearest percent.

56. Find the minimum percent increase in hourly wages for construction workers. Round to the nearest percent.

57. In the second quarter of 2007, a biotechnology company that sells cancer drugs earned profits of \$747 million. In the second quarter of 2006, it earned profits of \$531 million. Find the percent increase in reported profits. Round to the nearest percent. (*Source:* www.nytimes.com, July 12, 2007)

58. In the second quarter of 2006, the revenue of a biotechnology company that sells cancer drugs was \$2.2 billion. In the second quarter of 2007, its revenue was \$3 billion. Find the percent increase in revenue. Round to the nearest percent. (*Source:* www.nytimes.com, July 12, 2007)

For exercises 59–60, use the information in the table.

Number of Daily Passengers on SEPTA Rail Line		
	Fiscal year 2009–2010	**Fiscal year 2010–2011**
Lansdale/ Doylestown Line	16,158	18,717
Airport Line	7454	6750

Source: www.philly.com, July 28, 2011

59. Find the percent change in the number of daily passengers on the Lansdale/Doylestown Line. Round to the nearest percent.

60. Find the percent change in the number of daily passengers on the Airport Line. Round to the nearest percent.

61. In 2001, there were 37 robberies in Central Park in New York City. In 2011, there were 17 robberies. Find the percent decrease in robberies. Round to the nearest percent. (*Source:* www.nytimes.com, Dec. 28, 2011)

62. In 2009, Jeff Gordon scored 6473 points and won \$6,476,460 in the NASCAR Sprint Cup Series. In 2010, he scored 6176 points and won \$5,703,710. Find the percent decrease in the number of points. Round to the nearest tenth of a percent. (*Source:* www.nascar.com)

63. The sales tax rate in a state is 6%. Find the sales tax on a purchase of \$750.

64. The sales tax rate in a state is 8%. Find the sales tax on a purchase of \$550.

65. A restaurant bill is \$48. If the tip is 15% of the bill, find the total cost of the meal including tip.

66. A restaurant bill is \$65. If the tip is 20% of the bill, find the total cost of the meal including tip.

67. The original price of a computer is \$899. If it is marked down 23%, find the sale price.

68. The original price of a phone is \$199. If it is marked down 32%, find the sale price.

69. The original cost of a meal at a fast-food restaurant is \$12.99. On sale, it is \$10.99. Find the percent discount. Round to the nearest percent.

70. The original price of a backpack carrier for dogs that weigh up to 22 pounds is \$69.97. On sale, it is \$39.95. Find the percent discount. Round to the nearest percent.

71. The original price of red bell peppers is \$2.99 per pound. If the peppers are on sale at 25% off, find the cost of 3 pounds of peppers. Round to the nearest hundredth.

72. The original price of a box of cereal is \$4.59. If the cereal is on sale at 18% off, find the cost of 4 boxes of cereal. Round to the nearest hundredth.

For exercises 73–74, use the information in the bar graph. Round to the nearest hundredth.

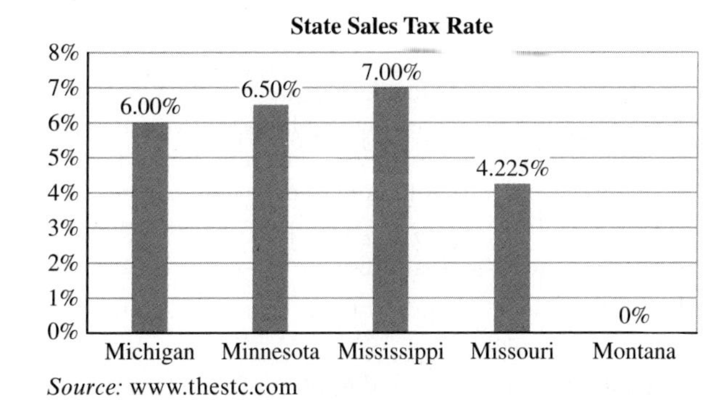

Source: www.thestc.com

73. The price of a used car in Missouri is \$13,500. Find the total cost of the car including sales tax.

74. The price of a 5.5-horsepower gas snow blower in Minnesota is \$699.99. Find the total cost of the snow blower including sales tax.

75. The regular price of a 14-piece cookware set is \$728. It is on sale at a discount of 35%. The sales tax is 6.5%. Find the final cost of the cookware including tax. Round to the nearest hundredth.

76. The regular price of an elliptical trainer is \$899.99. It is marked down 18%. The sales tax is 7.4%. Find the final cost of the elliptical trainer including tax. Round to the nearest hundredth.

77. The regular price of a portable GPS vehicle navigator is \$499.99. The sale price is \$389.99. Find the percent discount. Round to the nearest percent.

78. The regular price of a camp axe with a sheath is \$62.00. The sale price is \$30.52. Find the percent discount. Round to the nearest percent.

For problems 79–80, the percent markup compares the retail price of a product to the amount (cost) that the retailer paid for the product:

$$\text{percent markup} = \left(\frac{\text{price} - \text{cost}}{\text{cost}}\right)(100)\%$$

79. A furniture store bought a recliner from a wholesaler for \$150. It then sold the recliner for \$499. Find the percent markup. Round to the nearest percent.

80. A bookstore bought a used book from a student for \$20. It then resold the book as a used book for \$85. Find the percent markup.

81. An employee works for 8 hours at \$9.25 per hour. The employer withholds 7.65% of his gross wages for Social Security and Medicare taxes. Find the amount of wages after withholding.

82. An employee works for 8 hours at \$9.75 per hour. The employer withholds 7.65% of her gross wages for Social Security and Medicare taxes. Find the amount of wages after withholding.

Find the Mistake

For exercises 83–86, the completed problem has one mistake.
(a) Describe the mistake in words, or copy down the whole problem and highlight or circle the mistake.
(b) Do the problem correctly.

83. Problem: Complete the "Understand the problem," Make a plan," and "Complete the plan" steps. Use the information in the chart to find the percent decrease in violent crimes from 2009 to 2010. Round to the nearest percent.

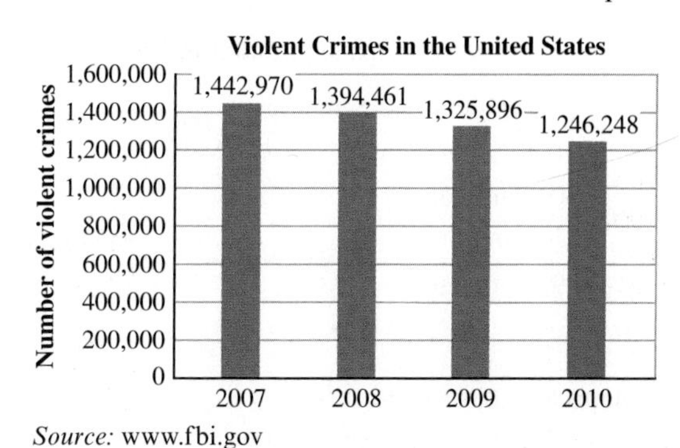

Source: www.fbi.gov

Incorrect Answer: The unknown is the percent decrease in violent crimes; P = percent decrease. A word equation is

$$\text{percent decrease} = \left(\frac{\text{decrease in crimes}}{\text{original number of crimes}}\right)(100)\%.$$

$$P = \left(\frac{1{,}246{,}248 \text{ crimes}}{1{,}325{,}896 \text{ crimes}}\right)(100)\%$$

$P \approx 94\%$ decrease

84. Problem: Complete the "Understand the problem," "Make a plan," and "Complete the plan" steps. A 20-acre field is going to be divided into building lots. The area of each lot is $\frac{2}{5}$ acre. Find the number of lots in the development.

Incorrect Answer: The unknown is the number of lots; L = number of lots. To find the number of lots, multiply the area by the number of lots.

$$L = \left(\frac{2}{5}\right)(20)$$

$L = 8$ lots

85. Problem: Complete the "Understand the problem," "Make a plan," and "Complete the plan" steps. The regular price of a 12-cup programmable coffee maker is \$79.99. The sale price is \$59.99. Find the percent discount. Round to the nearest percent.

Incorrect Answer: The unknown is the percent discount; P = percent discount. A word equation is

$$\text{percent decrease} = \left(\frac{\text{decrease in price}}{\text{sale price}}\right)(100)\%.$$

$$P = \left(\frac{\$79.99 - \$59.99}{\$59.99}\right)(100)\%$$

$P \approx 33\%$ decrease

86. **Problem:** Complete the "Understand the problem," "Make a plan," and "Complete the plan" steps. The population of the United States on July 4, 1776, was about 2.5 million people. In June 2008, the population had increased to about 304 million people. Find the percent increase in population. Round to the nearest percent. (*Source:* www.census.gov)

Incorrect Answer: The unknown is the percent increase; P = percent increase. A word equation is

$$\text{percent increase} = \left(\frac{\text{increase in population}}{\text{original population}}\right)(100)\%.$$

$$P = \frac{(304 \text{ million} - 2.5 \text{ million})}{304 \text{ million}}(100)\%$$

$$P = \frac{301.5 \text{ million}}{304 \text{ million}}(100)\%$$

$$P \approx 99\% \text{ increase}$$

Review

For problems 87–90, evaluate.

87. 5^2

88. 2^3

89. $(-2)^3$

90. $(-5)^2$

SUCCESS IN COLLEGE MATHEMATICS

91. If you use e-mail to communicate with your instructor, your instructor may prefer that you present the information more formally than you would if you were text messaging or using social media such as Twitter. Write an e-mail to your instructor that explains that you are too ill to come to class for a test. Ask what you should do about missing the exam. If your college does not have guidelines for writing e-mail, use the format shown below.

To: (Instructor e-mail address)

Subject: (include the name of your class and section number or class time)

Salutation: (use the title and name preferred by your instructor)

Body: (write in complete standard English sentences; check your spelling. Do not use smiley faces, other emoticons, or inappropriate language.)

Signature: (use your full name)

© Elena Elisseeva/Shutterstock.com

We can use a formula to find the distance traveled in a given time at a constant speed. The values we use in the formula may include units of measurement such as miles per hour. In this section, we will study both units of measurement and formulas.

SECTION 1.5 Units of Measurement and Formulas

After reading the text, working the practice problems, and completing assigned exercises, you should be able to:

1. Do arithmetic with measurements.
2. Use a conversion factor to change the units of a measurement.
3. Use a formula to solve an application problem.

Units of Measurement

Seconds, feet, meters, gallons, and degrees Fahrenheit are all **units of measurement**. We often just call them **units**.

People in the United States use two systems of measurement. The U.S. Customary System came to America from Europe. Its units have interesting histories. For example, an inch was originally the length of three grains of barley. The International System of Units (S.I.) was developed in France in the late 1700s. Instead of tradition, the units are based on multiples of 10. A unit of mass in S.I. is a **gram**. An M & M® milk chocolate candy piece has a mass of about 1 gram.

To create a bigger or smaller SI unit, write a **prefix** in front of the name of the standard unit. Table 1 lists some common prefixes.

A **deka**gram is 10 grams. A **kilo**gram is 1000 grams. In most countries, people measure their weight in kilograms rather than in pounds. A **deci**gram is 0.1 gram. A **milli**gram is 0.001 gram. Some medications are measured in milligrams.

There are abbreviations for many units of measurement. Table 2 lists many such abbreviations. The letters in the abbreviations are *not* variables.

Table 1 S.I. Prefixes

Multiple of ten	Prefix
1,000,000,000	giga-
1,000,000	mega-
1,000	kilo-
100	hecto-
10	deka-
0.1	deci-
0.01	centi-
0.001	milli-
0.000001	micro-
0.000000001	nano-
0.000000000001	pico-

Table 2 Abbreviations of Units

Units of length	meter	kilometer	centimeter	millimeter	mile	yard	foot	inch
Abbreviation	m	km	cm	mm	mi	yd	ft	in.

Units of volume	liter	milliliter	cubic centimeter	gallon	quart	pint	cup
Abbreviation	L	mL	cc or cm^3	gal	qt	pt	c

Units of mass/weight	gram	kilogram	milligram	microgram	pound	ounce
Abbreviation	g	kg	mg	μg	lb	oz

Units of time	hour	minute	second
Abbreviation	hr	min	s

When we add or subtract measurements with the same units, the units do not change.

EXAMPLE 1 Evaluate: 8 ft + 2 ft

SOLUTION ▶ 8 ft + 2 ft

= 10 ft Follow the order of operations.

When we multiply measurements, the units change. For example, the product of (ft)(ft) is ft^2, pronounced "feet squared" or "square feet."

EXAMPLE 2 Evaluate: (8 ft)(2 ft)

SOLUTION ▶ (8 ft)(2 ft)

= 16 ft^2 Follow the order of operations.

When we divide measurements, the units change. In the next example, since a fraction with the same numerator and denominator is equal to 1, $\frac{\text{ft}}{\text{ft}} = 1$.

EXAMPLE 3 Evaluate: $\frac{8 \text{ ft}}{2 \text{ ft}}$

SOLUTION ▶ $\frac{8 \cancel{\text{ft}}}{2 \cancel{\text{ft}}}$ Identify common factors.

= 4 Follow the order of operations; simplify units.

Practice Problems

For problems 1–5, evaluate.

1. $4 \text{ in.} + 7 \text{ in.}$ **2.** $7 \text{ in.} - 4 \text{ in.}$ **3.** $(4 \text{ in.})(7 \text{ in.})$ **4.** $\dfrac{6 \text{ in.}}{3 \text{ in.}}$

5. $(2 \text{ in.})(5 \text{ in.})(3 \text{ in.})$

Unit Conversion

The relationship of feet and inches is $1 \text{ ft} = 12 \text{ in.}$ Since a fraction with an equivalent numerator and denominator equals 1, both $\dfrac{1 \text{ ft}}{12 \text{ in.}} = 1$ and $\dfrac{12 \text{ in.}}{1 \text{ ft}} = 1$. Each of these fractions is a **conversion factor**. When we multiply a measurement by a conversion factor, we are multiplying by a fraction that is equal to 1. This does not change the value of the measurement.

To convert 2 ft into an equal measurement with a unit of inches, multiply by a conversion factor that is equal to 1. Choose the conversion factor $\dfrac{12 \text{ in.}}{1 \text{ ft}}$ so that the units of the product are inches. Notice that the unit "ft" is in the numerator of the original measurement and in the denominator of the conversion factor.

EXAMPLE 4 Convert 2 ft into inches.

SOLUTION ▸ 2 ft

$= \left(\dfrac{2 \cancel{\text{ft}}}{1}\right)\left(\dfrac{12 \text{ in.}}{1 \cancel{\text{ft}}}\right)$ Multiply by a conversion factor equal to 1; $\dfrac{12 \text{ in.}}{1 \text{ ft}} = 1$

$= 24 \text{ in.}$ Follow the order of operations; simplify units.

We can represent each relationship of two units of measurement in Table 3 as a conversion factor that is equal to 1.

Table 3 Relationships of Units of Measurement

S.I. units	U.S. customary units and S.I. units	U.S. customary units
1 kilometer = 1000 meters	1 inch = 2.54 centimeters	1 mile = 5280 feet
1 kilogram = 1000 grams	1 mile ≈ 1.6 kilometers	1 yard = 3 feet
1 liter = 1000 milliliters	2.2 pounds ≈ 1 kilogram*	1 foot = 12 inches
1 meter = 1,000,000,000 nanometers	1.06 quarts ≈ 1 liter	1 pound = 16 ounces
1 gram = 1,000,000 micrograms	1 teaspoon ≈ 5 milliliters	1 ton = 2000 pounds
1 tonne = 1000 kilograms		1 pint = 2 cups
1 meter = 100 centimeters		1 gallon = 4 quarts
1 milliliter = 1 cubic centimeter		1 quart = 2 pints
1 hour = 60 minutes		
1 minute = 60 seconds		

*The U.S. customary unit of pound measures force. The S.I. unit of kilogram measures mass. Your weight in pounds changes in different gravities. Your mass in kilograms does not change. The conversion factor here assumes that measurements are made at the surface of the earth.

To convert 45 ft into an equal measurement with a unit of yards, multiply by a conversion factor that is equal to 1. Choose the conversion factor $\dfrac{1 \text{ yd}}{3 \text{ ft}}$ so that the units of the product are yards. Notice that the unit "ft" is in the numerator of the original measurement and in the denominator of the conversion factor.

EXAMPLE 5 Convert 45 ft into yards.

SOLUTION ▶

45 ft

$\left(\frac{45\ \text{ft}}{1}\right)\left(\frac{1\ \text{yd}}{3\ \text{ft}}\right)$ Multiply by a conversion factor equal to 1; $\frac{1\ \text{yd}}{3\ \text{ft}} = 1$

$= 15$ yd Follow the order of operations; simplify units.

In the next example, to convert a measurement in meters into a measurement in kilometers $(1000\ \text{m} = 1\ \text{km})$, we can multiply by either $\frac{1000\ \text{m}}{1\ \text{km}}$ or $\frac{1\ \text{km}}{1000\ \text{m}}$. We choose $\frac{1\ \text{km}}{1000\ \text{m}}$ so that the unit "m" is in the numerator of the original measurement and in the denominator of the conversion factor. In working with S.I. units, the final answer is usually written as a decimal number rather than a fraction.

EXAMPLE 6 Convert 3500 m into kilometers.

SOLUTION ▶

3500 m

$= \left(\frac{3500\ \text{m}}{1}\right)\left(\frac{1\ \text{km}}{1000\ \text{m}}\right)$ Multiply by a conversion factor equal to 1; $\frac{1\ \text{km}}{1000\ \text{m}} = 1$

$= 3.5$ km Follow the order of operations; simplify units.

To convert a speed with units of $\frac{\text{km}}{\text{hr}}$ into units of $\frac{\text{km}}{\text{min}}$, multiply by the conversion factor $\frac{1\ \text{hr}}{60\ \text{min}}$. The unit "hr" is in the denominator of the original measurement and in the numerator of the conversion factor.

EXAMPLE 7 Convert $\frac{400\ \text{km}}{1\ \text{hr}}$ into $\frac{\text{km}}{\text{min}}$. Round to the nearest tenth.

SOLUTION ▶

$\frac{400\ \text{mi}}{1\ \text{hr}}$

$= \left(\frac{400\ \text{km}}{1\ \text{hr}}\right)\left(\frac{1\ \text{hr}}{60\ \text{min}}\right)$ Multiply by a conversion factor equal to 1; $\frac{1\ \text{hr}}{60\ \text{min}} = 1$

$= \frac{6.66\ldots\ \text{km}}{1\ \text{min}}$ Follow the order of operations; simplify units.

$\approx \frac{\mathbf{6.7}\ \text{km}}{1\ \text{min}}$ Round.

Practice Problems

6. Convert 8 in. into centimeters.

7. Convert 7 mi into kilometers.

8. Convert 6500 cm into kilometers.

9. Convert 12 cm into inches. Round to the nearest tenth.

10. Convert $\frac{3\ \text{L}}{1\ \text{hr}}$ into $\frac{\text{qt}}{\text{hr}}$.

Adding and Subtracting Measurements

We can add or subtract only measurements with the same "like" units. If the units are not the same, convert one of the measurements before adding or subtracting.

EXAMPLE 8 Evaluate 3 kg + 4 g. Write the answer in kilograms.

SOLUTION ▶ The relationship of kilograms and grams in Table 3 is 1 kg = 1000 g.

$3 \text{ kg} + 4 \text{ g}$

$= 3 \text{ kg} + \frac{4 \text{ g}}{1} \cdot \frac{1 \text{ kg}}{1000 \text{ g}}$ Multiply by a conversion factor equal to 1; $\frac{1 \text{ kg}}{1000 \text{ g}} = 1$

$= 3 \text{ kg} + \mathbf{0.004 \text{ kg}}$ Follow the order of operations; simplify units.

$= 3.004 \text{ kg}$ Follow the order of operations.

In the next example, we again evaluate 3 kg + 4 g. However, the answer is in grams instead of kilograms.

EXAMPLE 9 Evaluate 3 kg + 4 g. Write the answer in grams.

SOLUTION ▶ The relationship of kilograms and grams in Table 3 is 1 kg = 1000 g.

$3 \text{ kg} + 4 \text{ g}$

$= 3 \text{ kg} \cdot \frac{1000 \text{ g}}{1 \text{ kg}} + 4 \text{ g}$ Multiply by a conversion factor equal to 1; $\frac{1000 \text{ g}}{1 \text{ kg}} = 1$

$= \mathbf{3000 \text{ g}} + 4 \text{ g}$ Follow the order of operations; simplify units.

$= 3004 \text{ g}$ Follow the order of operations.

Practice Problems

11. Evaluate 2 mi + 1320 ft. Write the answer in feet.
12. Evaluate 5 kg + 13.2 lb. Write the answer in pounds.
13. Evaluate 6 km + 8 m. Write the answer in kilometers.
14. Evaluate 6 km + 8 m. Write the answer in meters.

Formulas for Distance and Simple Interest

$$(\text{speed})(\text{time}) = \frac{\text{mi}}{\text{hr}} \cdot \frac{\text{hr}}{1} = \text{mi}$$

A formula shows a relationship between variables. When we multiply an average speed in miles per hour by a time in hours, the product is a distance, miles. The formula for relationship of distance d, speed r, and time t is $d = rt$.

EXAMPLE 10 A student is driving to campus on a congested freeway at an average speed of $\frac{45 \text{ mi}}{1 \text{ hr}}$. She has 35 min to reach campus, which is 30 mi away. Will she make it to campus on time?

SOLUTION ▶ **Step 1 Understand the problem.**
The unknown is the distance she drives in 35 min.

d = distance she drives in 35 min.

Step 2 Make a plan.
Since we know rate (speed) and time, we can use the formula $d = rt$ to find the distance. Convert the measurement of 35 min into hours so that the units in the formula simplify to miles.

Step 3 Carry out the plan.

$d = rt$ distance = (rate)(time)

$d = \left(\frac{\mathbf{45 \text{ mi}}}{\mathbf{1 \text{ hr}}}\right)\left(\frac{\mathbf{35 \text{ min}}}{\mathbf{1}}\right)$ Replace the variables; include units.

$d = \left(\frac{45\text{ mi}}{1\text{ hr}}\right)\left(\frac{35\text{ min}}{1}\right)\left(\frac{1\text{ hr}}{60\text{ min}}\right)$ Multiply by a conversion factor equal to 1; $\frac{1\text{ hr}}{60\text{ min}} = 1$

$d = \mathbf{26.25\text{ mi}}$ Follow the order of operations; simplify units.

The student will travel 26.25 mi in 35 min. Since that is less than the distance to campus of 30 mi, the student will not make it to campus on time.

Step 4 Look back.
If the student's speed is $\frac{40\text{ mi}}{1\text{ hr}}$, she travels $\left(\frac{40\text{ mi}}{1\text{ hr}}\right)\left(\frac{35\text{ min}}{1}\right)\left(\frac{1\text{ hr}}{60\text{ min}}\right)$, which equals about 23 mi. If the student's speed is $\frac{50\text{ mi}}{1\text{ hr}}$, she travels $\left(\frac{50\text{ mi}}{1\text{ hr}}\right)\left(\frac{35\text{ min}}{1}\right)\left(\frac{1\text{ hr}}{60\text{ min}}\right)$, which equals about 29 mi. Since $\frac{45\text{ mi}}{1\text{ hr}}$ is between $\frac{40\text{ mi}}{1\text{ hr}}$ and $\frac{50\text{ mi}}{1\text{ hr}}$ and 26.25 mi is between 23 mi and 29 mi, the answer is in the ballpark and seems reasonable.

Step 5 Report the solution.
The student will not make it on time. She will travel only about 26 mi of the 30 mi to campus.

When someone borrows or invests money, interest is paid or earned. Interest is the fee paid for the use of money. **Simple interest** is calculated once a year, is not added to the original investment, and does not itself earn any interest. The formula $I = PRT$ represents the relationship of simple interest I, the amount invested P (the principal), the annual interest rate in decimal notation R, and the time in years T.

EXAMPLE 11 Find the simple interest earned on \$6000 invested for 4 years at an annual interest rate of 5.5%.

SOLUTION ▶ **Step 1 Understand the problem.**
The unknown is the amount of simple interest earned.

I = amount of interest

Step 2 Make a plan.
Since we know the principal, the rate, and the time, use the formula $I = PRT$.
The interest rate is 5.5% per year or $\frac{0.055}{1\text{ year}}$.

Step 3 Carry out the plan.

$I = PRT$ simple interest = (principal)(rate)(time)

$I = \left(\frac{\$6000}{1}\right)\left(\frac{0.055}{1\text{ year}}\right)\left(\frac{4\text{ years}}{1}\right)$ Replace the variables; include units.

$I = \mathbf{\$1320}$ Follow the order of operations; simplify units.

Step 4 Look back.
If the interest rate is 5%, the simple interest is $\left(\frac{\$6000}{1}\right)\left(\frac{0.05}{1\text{ year}}\right)\left(\frac{4\text{ years}}{1}\right)$, which equals \$1200. If the interest rate is 6%, the simple interest is $\left(\frac{\$6000}{1}\right)\left(\frac{0.06}{1\text{ year}}\right)\left(\frac{4\text{ years}}{1}\right)$, which equals \$1440. Since 5.5% is between 5% and 6% and \$1320 is between \$1200 and \$1440, the answer is in the ballpark and seems reasonable.

Step 5 Report the solution.
The investment earned \$1320 simple interest.

These formulas for distance and simple interest are two examples of the many formulas used in problem solving. To use a formula, we need to know the value of all of the variables in the formula except for the unknown. After replacing the variables, do any unit conversions required so that the units in the answer are correct.

Practice Problems

For problems 15–18, use the five steps.

15. The speed of a car is $\frac{65 \text{ mi}}{1 \text{ hr}}$. Find the distance it travels in 3.5 hr.

16. The ground speed of a plane is $\frac{512 \text{ mi}}{1 \text{ hr}}$. Find the distance it travels in 45 min.

17. Find the simple interest earned by an investment of $500 invested for 3 years at an annual simple interest rate of 3%.

18. Find the simple interest earned by an investment of $500 invested for 3 years at an annual simple interest rate of 8.4%.

Formulas for Perimeter, Area, and Volume

The distance around a rectangle, square, or triangle is its **perimeter**. For a rectangle with length L and width W, the formula for the perimeter P is $P = 2L + 2W$. For a triangle with sides a, b, and c, the formula for the perimeter is $P = a + b + c$.

When using a formula to find a perimeter, replace the variables with the measurements, including the units, and simplify.

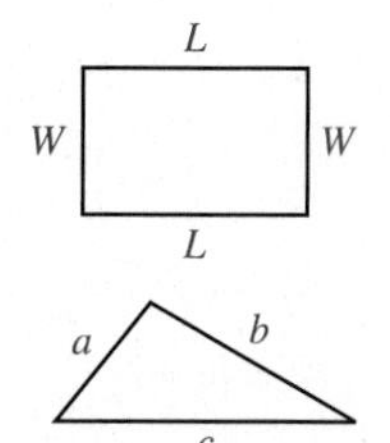

EXAMPLE 12 The width of a rectangle is 55 ft, and its length is 60 ft. Find the perimeter of the rectangle.

SOLUTION ▶

$P = 2W + 2L$	perimeter = 2(width) + 2(length)
$P = 2(\mathbf{55\text{ ft}}) + 2(\mathbf{60\text{ ft}})$	Replace the variables; include units.
$P = \mathbf{110\text{ ft}} + \mathbf{120\text{ ft}}$	Follow the order of operations.
$P = \mathbf{230\text{ ft}}$	Follow the order of operations.

To find the perimeter of the triangle in the next example, we need to change the units of the length of one side.

EXAMPLE 13 The sides of a triangle measure 95 cm, 74 cm, and 1.2 m. Find the perimeter of the triangle in centimeters. (1 m = 100 cm.)

SOLUTION ▶

$P = a + b + c$	perimeter = sum of the length of each side
$P = \mathbf{95\text{ cm}} + \mathbf{74\text{ cm}} + \mathbf{1.2\text{ m}}$	Replace the variables; include units.
$P = 95\text{ cm} + 74\text{ cm} + \left(\frac{1.2\cancel{\text{m}}}{1}\right)\left(\frac{100\text{ cm}}{1\cancel{\text{m}}}\right)$	Multiply by a fraction equal to 1.
$P = 95\text{ cm} + 74\text{ cm} + \mathbf{120\text{ cm}}$	Follow the order of operations; simplify units.
$P = \mathbf{289\text{ cm}}$	Follow the order of operations.

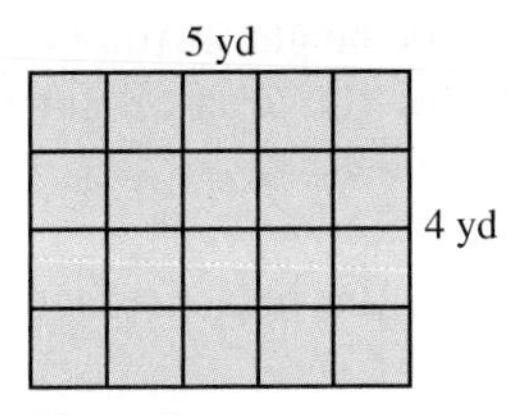

Figure 1

The perimeter of a rectangle or triangle is a measure of the distance around it. The **area** of a rectangle or triangle is a measure of the surface inside it. For example, the area of a square that is 1 yd long and 1 yd wide is 1 yd^2. Since a rectangle that is 5 yd long and 4 yd wide includes 20 of these squares (Figure 1), the area of the rectangle is 20 yd^2. Instead of counting squares to find the area of a rectangle A, we can use the formula $A = LW$, where L is the length and W is the width.

EXAMPLE 14 The width of a rectangle is 4 ft, and its length is 5 ft. Find the area of the rectangle.

SOLUTION

$A = LW$ area = (length)(width)

$A = \mathbf{(4\ ft)(5\ ft)}$ Replace the variables; include units.

$A = \mathbf{20\ ft^2}$ Follow the order of operations.

In the next example, the problem asks us to find the area of a building lot. We use the five steps to organize our work.

EXAMPLE 15 A building lot in a housing development in Fort Worth, Texas, is a rectangle with a width of 55 ft and a length of 110.3 ft. Find the area of this lot in acres. (1 acre = 43,560 ft^2.) Round to the nearest hundredth.

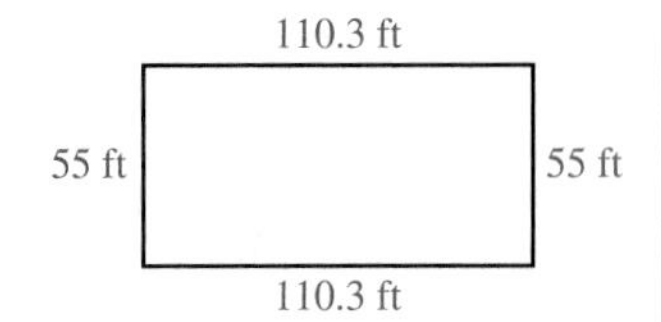

SOLUTION

Step 1 Understand the problem.
The unknown is the area of the lot in acres.

A = area of the lot

Step 2 Make a plan.
Since we know the length and width of a rectangle, we can use the formula $A = LW$. The units of area need to be converted into acres.

Step 3 Carry out the plan.

$A = LW$ area = (length)(width)

$A = \mathbf{(110.3\ ft)(55\ ft)}$ Replace variables; include units.

$A = \left(\frac{110.3\ \cancel{ft}}{1}\right)\left(\frac{55\ \cancel{ft}}{1}\right)\left(\frac{1\text{ acre}}{43{,}560\ \cancel{ft^2}}\right)$ Multiply by a fraction equal to 1; $\frac{1\text{ acre}}{43{,}560\text{ ft}^2} = 1$

$A = \mathbf{0.139\ldots\ acre}$ Follow the order of operations; simplify units; $\frac{\text{ft}\cdot\text{ft}}{\text{ft}^2} = 1$

$A \approx \mathbf{0.14}$ acre Round.

Step 4 Look back.
Estimating the length, width, and conversion factor, the area of the lot is about $(100\text{ ft})(60\text{ ft})\left(\frac{1\text{ acre}}{44{,}000\text{ ft}^2}\right)$, which equals about 0.14 acre. Since this is equal to the answer of 0.14 acre, the answer seems reasonable.

Step 5 Report the solution.
The area of the lot is about 0.14 acre.

If we draw a rectangle around a triangle with two equal sides (Figure 2), it appears that the area of the triangle is one-half of the area of the rectangle. Since the area of the rectangle is LW, the area of the triangle is $\frac{1}{2}LW$. The length of the rectangle

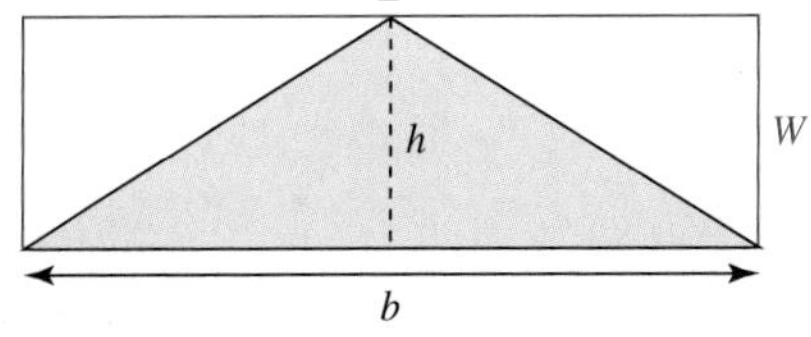

Figure 2

is the **base** of the triangle, b. The width of the rectangle is the **height** of the triangle, h. This relationship is true for all triangles. The formula for the area of a triangle is $A = \frac{1}{2}bh$.

In the next example, the formula includes a fraction but the measurements are decimal numbers. We write the area as a decimal number.

EXAMPLE 16 Find the area of the triangle. Write the answer as a decimal number.

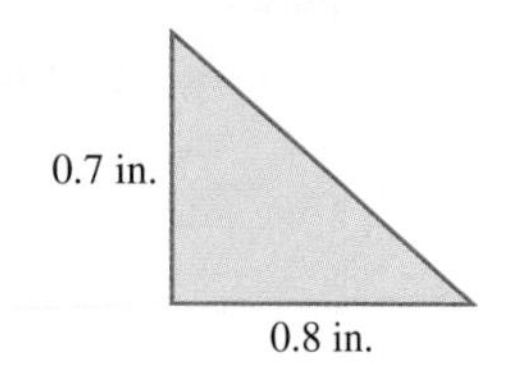

SOLUTION

$A = \frac{1}{2}bh$	area $= \frac{1}{2}$(base)(height)
$A = \frac{1}{2}\mathbf{(0.8\ in.)(0.7\ in.)}$	Replace the variables; include units.
$A = \mathbf{(0.4\ in.)}(0.7\ \text{in.})$	Follow the order of operations.
$A = \mathbf{0.28\ in.^2}$	Follow the order of operations.

We can use the formula for the area of a triangle to find the area of the gable end of a roof.

EXAMPLE 17 The gable end of a roof is a triangle with a base of 24 ft. The height is 6 ft. Find the area of plywood needed to cover the gable end.

SOLUTION **Step 1 Understand the problem.**

The unknown is the area of plywood.

A = area of plywood

Step 2 Make a plan.

Since the gable end is a triangle and we know its base and height, we can use the formula $A = \frac{1}{2}bh$ to find its area.

Step 3 Carry out the plan.

$A = \frac{1}{2}bh$	area $= \frac{1}{2}$(base)(height)
$A = \frac{1}{2}\mathbf{(24\ ft)(6\ ft)}$	Replace the variables; include units.
$A = \mathbf{(12\ ft)}(6\ \text{ft})$	Follow the order of operations.
$A = \mathbf{72\ ft^2}$	Follow the order of operations.

Step 4 Look back.

Doing this problem another way, a rectangle with a width that is equal to the height of the end of the gable (6 ft) and a length that is equal to the base of a triangle (24 ft) should have an area that is twice as much as the triangle. The area of this rectangle is (24 ft)(6 ft), which equals 144 ft^2. Since 144 ft^2 is twice as much as the area of the triangle, 72 ft^2, the answer seems reasonable.

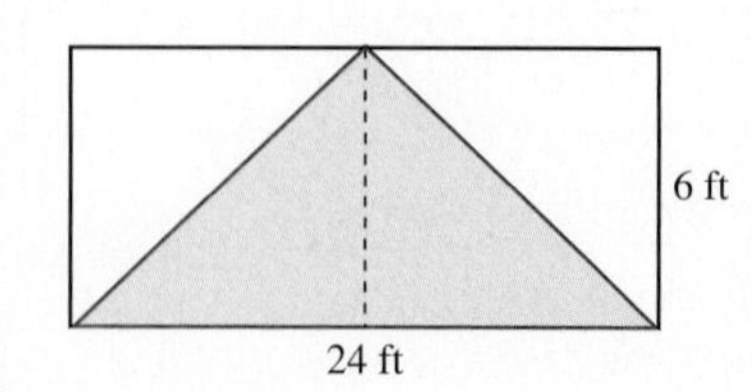

Step 5 Report the solution.

The area of the gable end of the roof is about 72 ft^2.

A box is an example of a **rectangular solid**. The volume of a solid is a measure of the space inside of it. The formula for the volume of a rectangular solid like a box is $V = LWH$, where V is volume, L is length, W is width, and H is height. In using this formula, the measurements are usually in the same units; an exception is the unit of volume often used for measuring water, acre · feet.

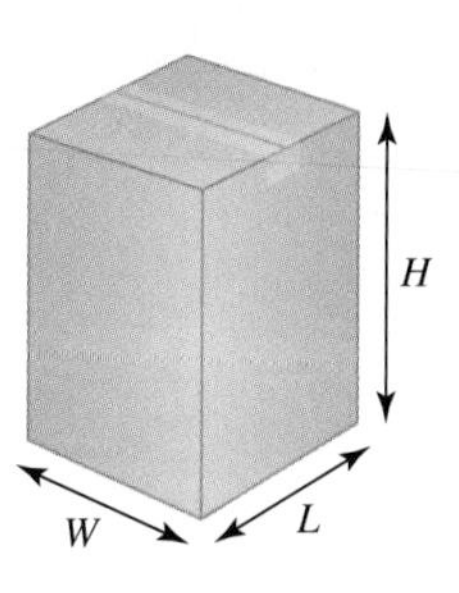

EXAMPLE 18 Find the volume in cubic inches of a rectangular solid that is 2 ft long, 15 in. wide, and 18 in. high. (1 ft = 12 in.)

SOLUTION ▶ Since the final answer is in cubic inches, we need to change the units of length into inches.

$V = LWH$	volume = (length)(width)(height)
$V = \mathbf{(2\ ft)(18\ in.)(15\ in.)}$	Replace the variables; include units.
$V = \left(\frac{2\ \cancel{ft}}{1}\right)\left(\frac{12\ in.}{1\ \cancel{ft}}\right)(18\ in.)(15\ in.)$	Multiply by a fraction equal to 1; $\frac{12\ in.}{1\ ft} = 1$
$V = \mathbf{6480\ in.^3}$	Follow the order of operations; simplify units.

To use other formulas for perimeter, area, and volume, replace the variables with the given measurements, and do any necessary unit conversions so that the units in the answer are correct.

Formulas for Perimeter, Area, and Volume

Rectangle with length L and width W

Perimeter: $P = 2L + 2W$

Area: $A = LW$

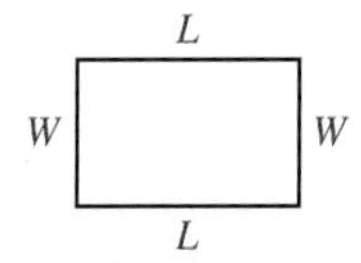

Triangle with sides of length a, b, and c

Perimeter: $P = a + b + c$

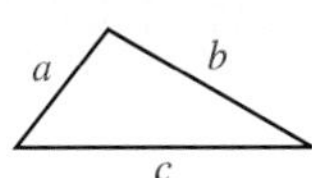

Triangle with height h and base b

Area of a triangle: $A = \frac{1}{2}bh$

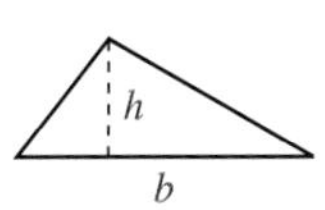

Rectangular solid with length L, width W, and height H

Volume: $V = LWH$

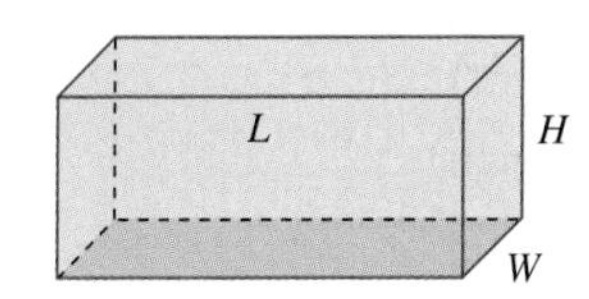

Practice Problems

19. Find the perimeter of a rectangle that is 3 ft long and 6 ft wide.
20. Find the area of a rectangle that is 3 ft long and 6 ft wide.
21. The sides of a triangle are 8 cm, 12.4 cm, and 10 cm. Find its perimeter.
22. The height of a triangle is 10 in., and its base is 24 in. Find its area.
23. A rectangle is 15 in. long and 2 ft wide. Find its area in square inches.

24. A rectangle is 18 in. long and 2.4 ft wide. Find its perimeter in feet.
25. Find the volume of a box that is 12 cm long, 4 cm wide, and 8 cm high.
26. Find the volume of a box that is 42 in. long, 30 in. wide, and 2 ft high. Write the answer in cubic inches.

For problems 27–28, use the five steps.

27. A camping tarp is 6 ft wide and 9 ft long. Find the perimeter of the tarp.
28. To make a croissant, a baker cuts a triangle of dough. The base of the triangle is 2.5 in., and the height of the triangle is 6 in. Find the area of this triangle.

1.5 Vocabulary Practice

Match the term with its description.

1. The product of the length, width, and height of a rectangular solid
2. The product of the length and width of a rectangle
3. The product of rate and time
4. The product of principal, interest rate, and time of investment
5. An S.I. prefix that means one-hundredth, 0.01
6. An S.I. prefix that means one-thousandth, 0.001
7. An S.I. prefix that means one thousand, 1000
8. $\approx$
9. A shape with three sides
10. A unit of area

A. approximately equal to
B. area
C. centi-
D. distance
E. kilo-
F. milli-
G. simple interest
H. square inches
I. triangle
J. volume

1.5 Exercises

Follow your instructor's guidelines for showing your work.

For exercises 1–8, evaluate.

1. $(16\text{ mm})(2\text{ mm})$
2. $(6\text{ ft})(8\text{ ft})$
3. $(16\text{ mm})(2\text{ mm})(4\text{ mm})$
4. $(6\text{ ft})(8\text{ ft})(2\text{ ft})$
5. $(32\text{ mm}^2)(4\text{ mm})$
6. $(48\text{ ft}^2)(2\text{ ft})$
7. $\dfrac{12\text{ km}}{2\text{ km}}$
8. $\dfrac{15\text{ lb}}{3\text{ lb}}$

For exercises 9–24, convert the measurement into the given unit. Use Examples 4–7 as a guide of how to show your work.

9. 5 mi; feet
10. 3 mi; feet
11. 6 gal; quarts
12. 8 gal; quarts
13. 3 km; meters
14. 9 km; meters
15. 12 kg; grams
16. 7 kg; grams
17. 2 qt; gallons
18. 1 qt; gallons
19. 9 in.; feet
20. 6 in.; feet
21. 5 mi; kilometers
22. 3 mi; kilometers
23. $\dfrac{60\text{ mi}}{1\text{ hr}}; \dfrac{\text{km}}{\text{hr}}$
24. $\dfrac{30\text{ mi}}{1\text{ hr}}; \dfrac{\text{km}}{\text{hr}}$

For exercises 25–30, convert the measurement into the given unit. Use Examples 4–7 as a guide of how to show your work. Round to the nearest tenth.

25. 18 km; miles
26. 14 km; miles
27. 18 lb; kilograms
28. 15 lb; kilograms
29. $\dfrac{60\text{ km}}{1\text{ hr}}; \dfrac{\text{miles}}{\text{hour}}$
30. $\dfrac{100\text{ km}}{1\text{ hr}}; \dfrac{\text{miles}}{\text{hour}}$
31. Evaluate 8 km + 4 m. Write the answer in meters.
32. Evaluate 12 kg + 54 g. Write the answer in grams.
33. Evaluate 8 km + 4 m. Write the answer in kilometers.
34. Evaluate 12 kg + 54 g. Write the answer in kilograms.

For exercises 35–40, find the amount of simple interest earned by the investment.

35. \$5000 earning 6% annual simple interest for 6 years.

36. \$9000 earning 5% annual simple interest for 8 years.

37. \$2500 earning 1.25% annual simple interest for 5 years.

38. \$2300 earning 1.75% annual simple interest for 7 years.

39. \$5000 earning 6% annual simple interest for 6 months. (1 year = 12 months.)

40. \$9000 earning 4% annual simple interest for 9 months. (1 year = 12 months.)

For exercises 41–48, find the distance that a car travels at the given speed in the given time. Round to the nearest whole number.

41. $\frac{65 \text{ mi}}{1 \text{ hr}}$; 4 hr

42. $\frac{55 \text{ mi}}{1 \text{ hr}}$; 7 hr

43. $\frac{65 \text{ mi}}{1 \text{ hr}}$; 20 min

44. $\frac{55 \text{ mi}}{1 \text{ hr}}$; 40 min

45. $\frac{45 \text{ mi}}{1 \text{ hr}}$; 35 min

46. $\frac{40 \text{ mi}}{1 \text{ hr}}$; 35 min

47. $\frac{55 \text{ mi}}{1 \text{ hr}}$; 75 min

48. $\frac{65 \text{ mi}}{1 \text{ hr}}$; 105 min

For exercises 49–50, find the perimeter of the rectangle.

49. Length: 4 in.; width: 8 in.

50. Length: 6 cm; width: 9 cm

For exercises 51–52, find the perimeter of the rectangle in inches.

51. Length: 2 ft, width: 5 in.

52. Length: 3 ft, width: 11 in.

For exercises 53–54, find the area of the rectangle.

53. Length: 6.5 in., width: 8 in.

54. Length: 3.5 in., width: 6 in.

For exercises 55–56, find the area of the rectangle in square inches.

55. Length: 2.5 ft, width: 9 in.

56. Length: 1.5: ft, width: 9 in.

For exercises 57–60, find the perimeter of the triangle.

57.

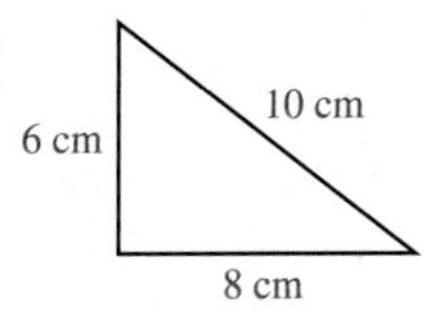

58.

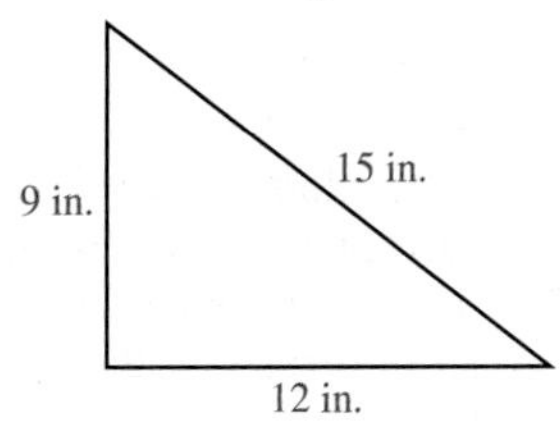

59. The lengths of the sides of a triangle are $\frac{1}{2}$ ft, $\frac{1}{4}$ ft, and $\frac{1}{3}$ ft.

60. The lengths of the sides of a triangle are $\frac{1}{4}$ ft, $\frac{1}{8}$ ft, and $\frac{1}{3}$ ft.

For exercises 61–64, find the area of the triangle with the given height and base. Do not rewrite fractions as decimal numbers.

61. Height: 12 ft, base: 15 ft

62. Height: 8 cm, base: 9 cm

63. Height: $\frac{1}{8}$ in., base: $\frac{1}{4}$ in.

64. Height: $\frac{1}{8}$ in., base: $\frac{1}{3}$ in.

For exercises 65–66, a square is a rectangle in which the length and width are equal. Find the area of the square.

65. The length of a square is 3 in.

66. The length of a square is 5 cm.

For exercises 67–68, a square is a rectangle in which the length and width are equal. Find the area of the square in *square yards*. Round to the nearest tenth.

67. The side of a square is 14 ft.

68. The side of a square is 8 ft.

69. The flag of the Philippines includes an equilateral triangle in which the lengths of the sides are equal. If the flag in the diagram is 5 ft long and 3 ft wide, find the perimeter of the triangle in the flag.

© Carsten Reisinger/Shutterstock.com

70. The flag of Puerto Rico includes an equilateral triangle in which the lengths of the sides are equal. If the flag in the diagram is 4 ft wide and 6 ft long, find the perimeter of the triangle in the flag.

© R. Gino Santa Maria/Shutterstock.com

For exercises 71–72, the formula for the density of a substance is $d = \frac{m}{v}$, where d is density, m is mass, and v is volume.

71. Find the density of a piece of wood with a mass of 12 g and a volume of 15 cm^3.

72. Find the density of a piece of gold with a mass of 3.50 g and a volume of 0.18 cm^3. Round to the nearest tenth.

For exercises 73–76, the Harris-Benedict formula represents the relationship of the basal metabolic rate of a human body at rest B (calories), height H (cm), weight W (kg), and age A (years). For men, the formula is $B = 664.730 + 13.7516W + 5.0033H - 6.7550A$. For women, the formula is $B = 655.0955 + 9.5634W + 1.8496H - 4.6756A$. Find the basal metabolic rate. Round to the nearest whole number. (*Source:* www.webmd.com)

73. The weight of a 22-year-old man is 160 lb, and his height is 72 in.

74. The weight of a 25-year-old man is 195 lb, and his height is 72 in.

75. The weight of a 22-year-old woman is 115 lb, and her height is 62 in.

76. The weight of a 22-year-old woman is 135 lb, and her height is 67 in.

77. The formula $P = \left(\frac{w}{w + i}\right)100$ represents the relationship of the percent of X-rays absorbed by an antiscatter grid P, the width of each grid strip w, and the width between the grid strips i. If the width of each grid strip is 50 micrometers and the width between the grid strips is 350 micrometers, find the percent of X-rays absorbed. Round to the nearest percent.

78. The intensity of X-ray radiation depends on the distance from the source of radiation. If the distance from the source of radiation changes from D_1 to D_2, the intensity changes from intensity I_1 to I_2. The formula for finding the new intensity is $I_2 = \frac{I_1 \cdot (D_1)^2}{(D_2)^2}$. When D_1 is 25 m, the intensity I_1 is 620 roentgen per hour. Find the intensity of radiation I_2 if D_2 is 5 m.

For exercises 79–80, use the formula $F = \frac{9}{5}C + 32$ to change the given temperature in degrees Celsius, C, into degrees Fahrenheit, F. Round to the nearest whole number.

79. The low temperature in Montreal, Canada, was $-10°C$.

80. The high temperature in Tijuana, Mexico, was 17°C.

For exercises 81–82, use the formula $C = \frac{5}{9}(F - 32)$ to change the given temperature in degrees Fahrenheit, F, into degrees Celsius, C. Round to the nearest whole number.

81. The low temperature in Albany, New York, was 15°F.

82. The high temperature in San Diego, California, was 65°F.

For exercises 83–84, a P/E ratio compares the price per share of stock of a company to the ability of the company to make money. In general, a stock with a lower P/E ratio is more valuable than a stock with a higher P/E ratio. The formula is

$$\text{P/E ratio} = \frac{\text{market value per share}}{\text{earnings per share}}$$

83. If the market value per share of Microsoft Corp. stock is \$30.16 and the earnings per share are \$1.39, find the P/E ratio. Round to the nearest hundredth.

84. If the market value per share of Apple, Inc. stock is \$138.91 and the earnings per share is \$3.16, find the P/E ratio. Round to the nearest hundredth.

For exercises 85–86, a 1917 pharmacy textbook describes rules for converting an adult dose of medication into the right dose for a child. (*Source:* Sollman, T.H., *A Manual of Pharmacology and Its Applications to Therapeutics and Toxicology*, 1917)

85. Clark's rule is "Multiply the adult dose with the weight of the child (in pounds) and divide by 150 (the average weight of an adult)." Assign the variables and write a formula that describes Clark's rule.

86. Fried's rule for infants is "Divide the age in months by 150 and multiply it by the adult dose." Assign the variables and write a formula that describes Fried's rule.

For exercises 87–88, the formula $V = \frac{LWH}{144 \text{ in.}^3}$ is used to find the volume V of lumber in *board feet*. The units of length L, width W, and height H are in inches. Find the total volume of the given boards in board feet. Round to the nearest tenth.

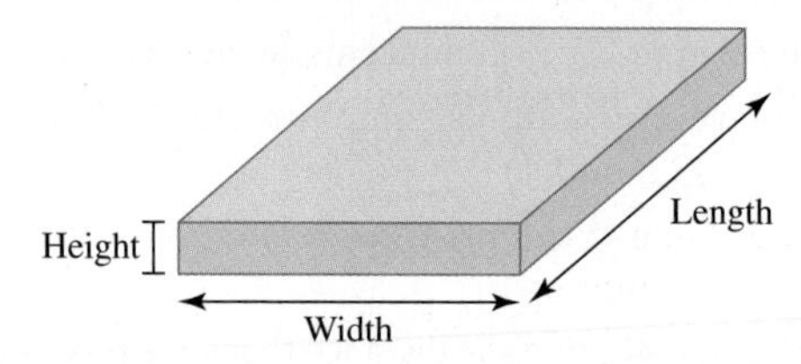

87. 50 boards that are each 10 ft long, 4 in. wide, and 2 in. thick.

88. 50 boards that are each 10 ft long, 6 in. wide, and 2 in. thick.

For exercises 89–92, a measure of central tendency is the **arithmetic mean** or average, $\overline{X}$. For a set of five test scores, $\overline{X} = \frac{x_1 + x_2 + x_3 + x_4 + x_5}{5}$. Find the arithmetic mean of the test scores.

89. 75, 77, 79, 71, 78

90. 71, 69, 73, 77, 72

91. 75, 77, 79, 71, 0

92. 71, 69, 73, 77, 97

For exercises 93–94, use the five steps. The formula for the area of a trapezoid is $A = \frac{1}{2}(b_1 + b_2)h$, where the bases are parallel line segments with lengths b_1 and b_2 and the height is h.

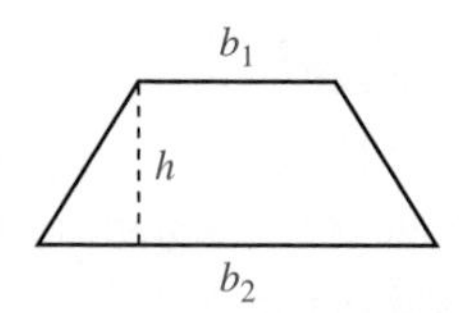

93. The shape of a building lot is a trapezoid with bases that measure 150 ft and 400 ft. The height is 220 ft. Find the area of the lot.

94. The shape of a table top is a trapezoid with bases that measure 25 in. and 60 in. The height is 30 in. Find the area of this table.

For exercises 95–96, use the five steps.

95. A quilter is buying fabric that is 44 in. wide to make a back for a rectangular quilt that is 102 in. wide and 120 in. long. The quilter needs to buy three strips of fabric that are as wide as the quilt plus two additional inches of fabric on each end of the strip. Find the amount of fabric in yards that the quilter should buy. Round *up* to the nearest tenth.

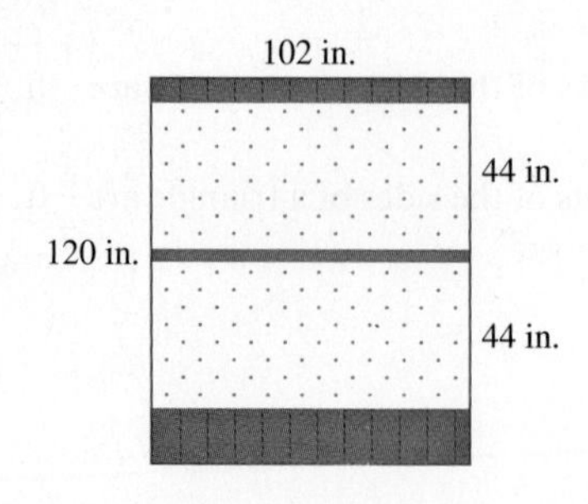

96. Roofing shingles are sold by the "square." A square can cover an area of roof that is 10 ft long and 10 ft wide. The area of a roof is 3450 ft^2. Find the number of squares of roofing shingles needed to cover this roof. Round *up* to the nearest whole number.

Find the Mistake

For exercises 97–100, the completed problem has one mistake.
(a) Describe the mistake in words, or copy down the whole problem and highlight or circle the mistake.
(b) Do the problem correctly.

97. Problem: Find the simple interest earned on \$300 invested for 5 years at an annual simple interest rate of 4.5%.

Incorrect Answer: $I = PRT$

$$I = (\$300)\left(\frac{4.5}{1\text{ year}}\right)(5\text{ years})$$

$$I = \$6750$$

98. Problem: The length of a rectangle is 2 ft, and its width is 9 in. Find its perimeter.

Incorrect Answer: $P = 2W + 2L$

$$P = 2(2\text{ ft}) + 2(9\text{ ft})$$

$$P = 22\text{ ft}$$

99. Problem: The length of a rectangle is 3 cm. Its width is 8 cm. Find its area.

Incorrect Answer: $A = LW$

$$A = 3 \cdot 8$$

$$A = 24$$

100. Problem: A Shinkansen high-speed train in Japan is traveling at a speed of 210 km per hour. Find the distance in kilometers that it travels in 15 min. Round to the nearest tenth.

Incorrect Answer: $d = rt$

$$d = \left(210\,\frac{\text{km}}{\text{hr}}\right)(15\text{ min})$$

$$d = 3150\text{ km}$$

Review

For problems 101–104, simplify each expression.

101. $\frac{2}{3}(6x - 8) - \frac{7}{10}x + 4$

102. $\frac{2}{15}h - \frac{5}{8}h + \frac{4}{5}k - \frac{1}{12}k$

103. $-6(3z - 7) - (2z + 1)$

104. $-8(9x - 5) - (4x + 1)$

SUCCESS IN COLLEGE MATHEMATICS

105. The syllabus for a course usually lists many of the expectations of the instructor. Explain how your grade is determined in this math class. What is the minimum grade needed to move on to your next math class?

106. Many colleges have a student code of conduct that describes the consequences of academic dishonesty (cheating). Some instructors also include this information in the class syllabus. What is the consequence in your math class for academic dishonesty?

About 4000 years ago, the Babylonians knew that the number equal to the quotient of the circumference of any circle and its diameter is about 3. Ancient Greek, Chinese, and Hindu mathematicians used an approximate value of this number to find the area of a circle. In 1706, English mathematician William Jones named this number π (pronounced "pi"). In this section, we will study π and other irrational numbers.

SECTION 1.6 Irrational Numbers and Roots

After reading the text, working the practice problems, and completing assigned exercises, you should be able to:

1. Use a formula that includes π to solve an application problem.
2. Evaluate a square root or cube root.
3. Use a formula that includes a square root to solve an application problem.

Rational Numbers, Irrational Numbers, and π

Terminating decimal numbers such as 0.7139 and repeating decimal numbers such as $0.\overline{3}$ are rational numbers that can be rewritten as fractions in which the numerator and denominator are integers. In rewriting a terminating decimal number as a fraction, the denominator is determined by the place value of the last digit in the fraction. For example, since 9 is in the ten-thousandths place, $0.7139 = \frac{7139}{10,000}$. Rewriting a repeating decimal as a fraction is a topic in Section 4.3.

To rewrite a fraction as a decimal number, divide the numerator by the denominator. If the decimal number repeats, write a bar over the digits that repeat.

EXAMPLE 1 Rewrite $\frac{4}{11}$ as a repeating decimal.

SOLUTION ▶ $\frac{4}{11}$

$= 0.363636...$ Divide the numerator by the denominator.

$= 0.\overline{36}$ The bar over the numbers shows the repeating digits.

An **irrational number** is a nonrepeating nonterminating decimal number that cannot be written as a fraction in which the numerator and denominator are integers. An important irrational number, 3.1415. . . is represented by a Greek letter, π, pronounced "pi." The quotient of the circumference of any circle (the distance around it) and its diameter (the distance across the circle, passing through the center), equals π. Because of this relationship, many formulas for shapes based on circles include π.

Formulas for Circles and Formulas That Include π

The radius of a circle, r, is the distance from the center to a point on the circle.

Diameter of a circle, D: $D = 2r$

Radius of a circle, r: $r = \frac{D}{2}$

Circumference of a circle, C: $C = \pi D$ or $C = 2\pi r$

Area of a circle, A: $A = \pi r^2$

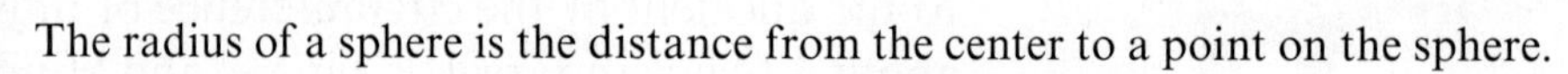

The radius of a sphere is the distance from the center to a point on the sphere.

Volume of a sphere: $V = \frac{4}{3}\pi r^3$

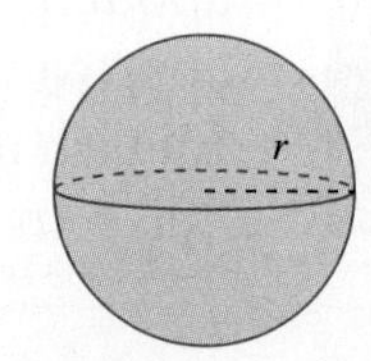

The radius of a cone or cylinder is the distance from the center of the circle that is its base to any point on the circle.

Volume of a cylinder: $V = \pi r^2 h$

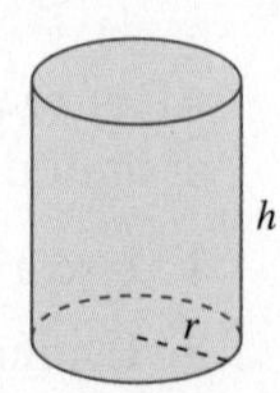

Volume of a cone: $V = \frac{1}{3}\pi r^2 h$

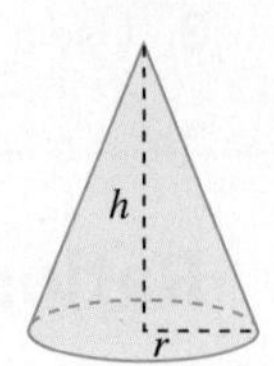

When using a formula that includes π, we use the rounded value of $\pi \approx 3.14$.

EXAMPLE 2 The diameter of a circle is 8 cm.

(a) Find the radius of the circle.

SOLUTION

$r = \frac{D}{2}$ radius = $\frac{\text{diameter}}{2}$

$r = \frac{\mathbf{8\ cm}}{2}$ Replace the variable.

$r = \mathbf{4\ cm}$ Follow the order of operations.

(b) Find the area of the circle. ($\pi \approx 3.14$.)

When replacing π in the formula with the rounded approximate value of 3.14, also replace the = sign in the formula with $\approx$.

$A = \pi r^2$ area = π (radius)^2

$A \approx \mathbf{(3.14)(4\ cm)^2}$ Replace the variable and π.

$A \approx (3.14)\mathbf{(16\ cm^2)}$ Follow the order of operations.

$A \approx \mathbf{50.24\ cm^2}$ Follow the order of operations.

(c) Find the circumference of the circle. ($\pi \approx 3.14$.)

$C = \pi D$ circumference = π (diameter)

$C \approx \mathbf{(3.14)(8\ cm)}$ Replace the variable and π.

$C \approx \mathbf{25.12\ cm}$ Follow the order of operations.

The volume of a **cylinder** is the area of its circular base multiplied by the height, $V = \pi r^2 h$.

EXAMPLE 3 The 5.7-L engine in a 2011 Dodge truck has eight cylinders. The diameter of each cylinder is 9.96 cm. The length of the piston stroke (the height of the cylinder) is 9.09 cm. Find the volume of one cylinder. ($\pi \approx 3.14$.) Round to the nearest whole number. (*Source:* www.ramtrucks.com)

Piston

SOLUTION

Step 1 Understand the problem.
The unknown is the volume of one cylinder.

V = volume of one cylinder

Step 2 Make a plan.
Use the formula for the volume of a cylinder, $V = \pi r^2 h$. The radius, r, equals the diameter of 9.96 cm divided by 2, which equals 4.98 cm.

Step 3 Carry out the plan.

$V = \pi r^2 h$ volume = π (radius)^2(height)

$V \approx \mathbf{(3.14)(4.98\ cm)^2(9.09\ cm)}$ Replace variables.

$V \approx (3.14)(\mathbf{24.8004\ cm^2})(9.09\ cm)$ Follow the order of operations.

$V \approx (\mathbf{77.873...\ cm^2})(9.09\ cm)$ Follow the order of operations.

$V \approx \mathbf{707.86...\ cm^3}$ Follow the order of operations; $cm^2 \cdot cm = cm^3$

$V \approx \mathbf{708\ cm^3}$ Round to the nearest whole number.

Step 4 Look back.
Estimating, if the volume of one cylinder is about 700 cm^3, then the volume of eight cylinders is about 5600 cm^3. Since the truck has a 5.7-L engine and $\frac{5600 \text{ cm}^3}{1} \cdot \frac{1 \text{ L}}{1000 \text{ cm}^3}$ equals 5.6 L, the answer seems reasonable.

Step 5 Report the solution.
The volume of one cylinder is about 708 cm^3.

A ball is an example of a **sphere**. Every point on the surface of the sphere is the same distance from its center. This distance is the radius of the sphere, r. The formula for the volume of a sphere is $V = \frac{4}{3}\pi r^3$.

EXAMPLE 4 Find the volume of a sphere with a radius of 0.5 in. ($\pi \approx 3.14$.) Round to the nearest tenth.

SOLUTION ▶

$V = \frac{4}{3}\pi r^3$ — volume $= \frac{4}{3}\pi(\text{radius})^3$

$V \approx \frac{\mathbf{4(3.14)(0.5\ in.)^3}}{3}$ — Replace the variable and π.

$V \approx \frac{4(3.14)(\mathbf{0.125\ in.^3})}{3}$ — Follow the order of operations.

$V \approx \frac{\mathbf{1.57\ in.^3}}{3}$ — Follow the order of operations.

$V \approx \mathbf{0.523\ldots}$ in.^{3} — Follow the order of operations.

$V \approx \mathbf{0.5}$ in.^{3} — Round.

Practice Problems

For problems 1–3, rewrite the fraction as a repeating decimal.

1. $\frac{2}{3}$ **2.** $\frac{5}{11}$ **3.** $\frac{6}{7}$

For problems 4–10, use $\pi \approx 3.14$. Round to the nearest tenth.

4. Find the circumference of a circle with a radius of 9 ft.
5. Find the area of a circle with a radius of 16 in.
6. Find the area of a circle with a diameter of 16 in.
7. Find the volume of a cylinder with a radius of 6 cm and a height of 15 cm.
8. Find the volume of a sphere with a diameter of 6 ft.
9. Find the volume of a cone with a radius of 8 in. and a height of 14 in.
10. Find the volume of a cylinder with a radius of 1.4 ft and a height of 11 in. Write the answer in cubic inches.

For problems 11–13, use the five steps. ($\pi \approx 3.14$.)

11. The diameter of a shot-put ring at a track meet is 7 ft. Find the area of the ring. Round to the nearest whole number.
12. The height of a cylindrical soup can is 6 in., and its diameter is 3 in. Find its volume. Round to the nearest tenth.
13. Find the volume of a spherical soap bubble with a diameter of 40 cm. Round to the nearest whole number.

Square Roots

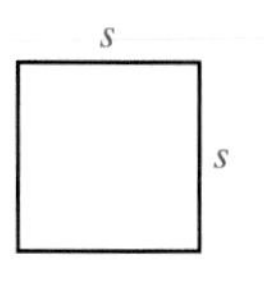

A square is a rectangle in which the length of each side s is equal. The area of a square equals s^2. Similarly, the **square of a number** n equals n^2. For example, the square of 5 is 5^2, which equals 25, and the square of 5 is $(\ 5)^2$, which also equals 25. To find the **square root** of a number, undo the process of squaring. For example, since $5 \cdot 5 = 25$ and $(-5)(-5) = 25$, the square roots of 25 are 5 and -5.

EXAMPLE 5 What are the square roots of 36?

SOLUTION ▸ Since $(6)(6) = 36$ and $(-6)(-6) = 36$, the square roots of 36 are 6 and -6.

The positive square root of a number is its **principal square root**. The notation for principal square root is $\sqrt{\ }$.

EXAMPLE 6 Evaluate: $\sqrt{36}$

SOLUTION ▸

$\sqrt{36}$ The principal square root of 36.

$= 6$ Evaluate; $6 \cdot 6 = 36$

If a number is a **perfect square**, its principal square root is a rational number that can be written as a fraction in which the numerator and denominator are integers. Since the principal square root of 36 equals 6 and 6 is a rational number, 36 is a perfect square. If the principal square root of a number is not a rational number, it is not a perfect square. For example, since the principal square root of 3 is an irrational number, 1.73205. . . , the number 3 is not a perfect square.

Before calculators were inexpensive, most algebra students learned a paper and pencil procedure for estimating irrational square roots. Now we most often use a calculator. A calculator displays only some of the digits of an irrational number. For example, on a calculator that shows ten digits, $\sqrt{6}$ is displayed as 2.449489743. The last digit is rounded.

EXAMPLE 7 Evaluate $\sqrt{8}$. Round to the nearest thousandth.

SOLUTION ▸

$\sqrt{8}$

$= 2.8284\ldots$ Evaluate; a nonterminating, nonrepeating decimal number.

≈ 2.828 Round.

We can also evaluate the square root of some units of measurement. For example, since ft · ft equals ft^2, $\sqrt{\text{ft}^2}$ equals ft.

EXAMPLE 8 Evaluate: $\sqrt{100 \text{ ft}^2}$

SOLUTION ▸

$\sqrt{100 \text{ ft}^2}$

$= 10 \text{ ft}$ Evaluate; $(10 \text{ ft})(10 \text{ ft}) = 100 \text{ ft}^2$

The *square root of a negative number is not a real number.* Since the product of two positive numbers is a positive number and the product of two negative numbers is also a positive number, the square of a real number cannot be a negative number. When asked to evaluate the square root of a negative number, do not say "no answer" or "can't do it." The correct answer is "not a real number."

EXAMPLE 9 Evaluate: $\sqrt{-16}$

SOLUTION ▸ $\sqrt{-16}$ is not a real number.

Practice Problems

14. What are the square roots of 9?

For problems 15–22, evaluate. If the number is irrational, round to the nearest tenth.

15. $\sqrt{81}$	**16.** $\sqrt{m^2}$	**17.** $\sqrt{81\text{ m}^2}$	**18.** $\sqrt{0}$
19. $\sqrt{1}$	**20.** $\sqrt{20}$	**21.** $\sqrt{5}$	**22.** $\sqrt{-25}$

Cube Roots

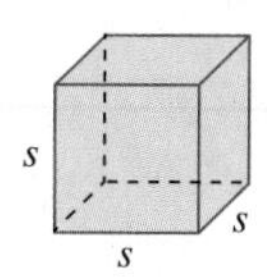

The volume of any rectangular solid such as a cube or box is the product of its length, width, and height: $V = LWH$. The **cube of a number** n equals n^3. For example, the cube of 4 is 4^3, which equals 64, and the cube of -4 is $(-4)^3$, which equals -64. To find the **principal cube root** of a number, undo the process of cubing. For example, since $(-3)(-3)(-3) = -27$, the cube root of -27 is -3. The notation for the principal cube root is $\sqrt[3]{\ }$. For the cube root $\sqrt[3]{64}$, 3 is the **index**.

EXAMPLE 10 Evaluate: $\sqrt[3]{-125}$

SOLUTION ▶

$\sqrt[3]{-125}$

$= -5$ Evaluate; $(-5)(-5)(-5) = -125$

Practice Problems

For problems 23–28, evaluate.

23. $\sqrt[3]{8}$	**24.** $\sqrt[3]{-8}$	**25.** $\sqrt[3]{1000\text{ ft}^3}$
26. $\sqrt[3]{0}$	**27.** $\sqrt[3]{1}$	**28.** $\sqrt[3]{-1}$

The Pythagorean Theorem and Other Applications

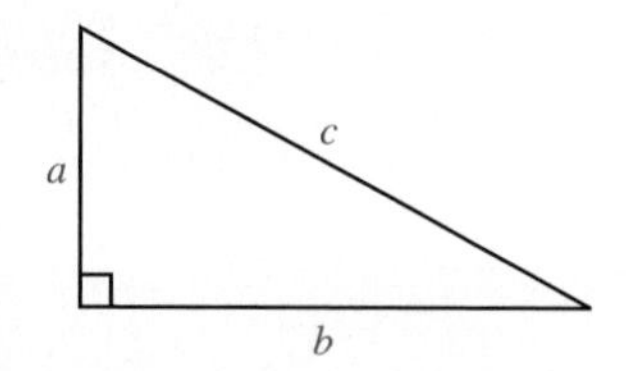

A right triangle has a 90-degree or **right angle**. The side that is opposite the right angle is the **hypotenuse**. The length of the hypotenuse is c. The lengths of the other two sides of the triangle (the **legs**) are a and b. The Pythagorean theorem states that $a^2 + b^2 = c^2$. In Chapter 8, we will learn that for values of c that are greater than 0, this can be rewritten as $c = \sqrt{a^2 + b^2}$. In the order of operations, a root is a grouping symbol. To evaluate $\sqrt{a^2 + b^2}$, replace the variables, evaluate the exponential expressions, find their sum, and then evaluate the square root.

EXAMPLE 11 The lengths of the legs of a right triangle are 3 cm and 7 cm. Find the length of the hypotenuse. Round to the nearest tenth.

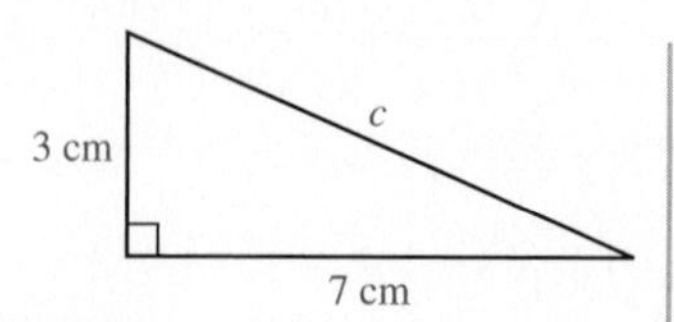

SOLUTION ▶

$c = \sqrt{a^2 + b^2}$

$c = \sqrt{(\mathbf{3\ cm})^2 + (\mathbf{7\ cm})^2}$ Replace the variables.

$c = \sqrt{\mathbf{9\ cm^2} + \mathbf{49\ cm^2}}$ Follow the order of operations.

$c = \sqrt{\mathbf{58\ cm^2}}$ Follow the order of operations.

$c = \mathbf{7.61...\ cm}$ Evaluate the square root.

$c \approx \mathbf{7.6}\text{ cm}$ Round.

Some formulas include roots. For example, to find the speed of a car before the brakes were applied, accident investigators may use the formula $S = \sqrt{30Dfn}$, where the average distance of the skid marks in feet is D, the drag factor of the road surface is f, the percent braking efficiency in decimal form is n, and the minimum speed in miles per hour of the car when the car began to skid is S. Since this formula includes a constant without units, 30, do not include the units when replacing the variables.

EXAMPLE 12 The average distance of skid marks at an accident is 109 ft, the asphalt road surface has a drag factor of 0.75, and the braking efficiency of the car is 0.50. Find the speed of the car in miles per hour before braking. Round to the nearest whole number.

SOLUTION ▸ **Step 1 Understand the problem.**

The unknown is the speed of the car before braking. Since the formula $S = \sqrt{30Dfn}$ includes a constant without units, the units for D, f, and n will not simplify to miles per hour. When replacing the variables, do not include the units.

S = speed before braking

Step 2 Make a plan.

Use the formula $S = \sqrt{30Dfn}$. The final units are miles per hour.

Step 3 Carry out the plan.

$S = \sqrt{30Dfn}$ — D = skid mark distance; f = drag factor; n = braking efficiency

$S = \sqrt{30\mathbf{(109)(0.75)(0.50)}}$ — Replace variables.

$S = \sqrt{\mathbf{1226.25}}$ — Follow the order of operations.

$S = \mathbf{35.017...}$ — Follow the order of operations.

$S \approx \frac{\mathbf{35\ mi}}{\mathbf{1\ hr}}$ — Round; the units are miles per hour.

Step 4 Look back.

If $D = 100$ ft, then $S = \sqrt{30(100)(0.75)(0.50)}$, which equals about $\frac{33.5\text{ mi}}{1\text{ hr}}$.

If $D = 125$ ft, then $S = \sqrt{30(125)(0.75)(0.50)}$, which equals $\frac{37.5\text{ mi}}{1\text{ hr}}$.

Since 109 ft is between 100 ft and 125 ft and $\frac{35\text{ mi}}{1\text{ hr}}$ is between $\frac{33.5\text{ mi}}{1\text{ hr}}$ and $\frac{37.5\text{ mi}}{1\text{ hr}}$, the answer is in the ballpark and seems reasonable.

Step 5 Report the solution.

The car was traveling about $\frac{35\text{ mi}}{1\text{ hr}}$ before braking.

Practice Problems

29. Find the length of the hypotenuse of the triangle.

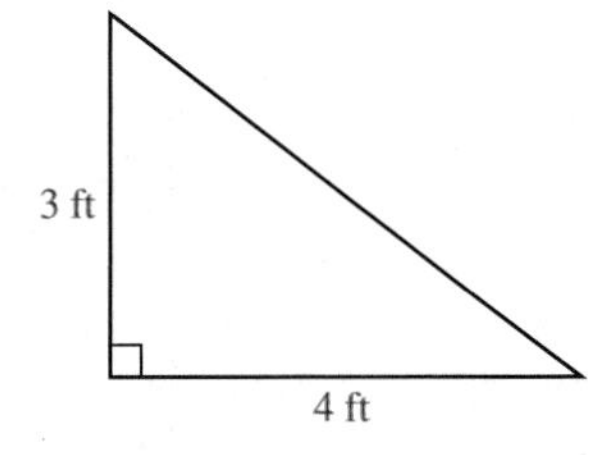

30. A doorway is 36 in. wide and 8 ft tall. Find the distance in inches of the diagonal from a lower corner to an upper corner. Round to the nearest tenth.

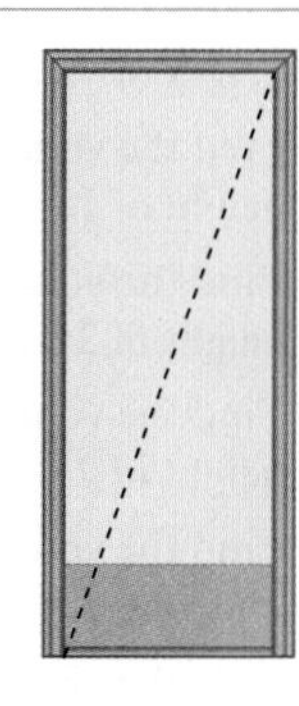

For problem 31, use the five steps.

31. The formula $S = 1.34\sqrt{L}$ represents the relationship of the maximum hull speed in knots, S, of a single-hull sailboat and the length of the hull at the waterline in feet, L. A single-hull sailboat is 22 ft long. Find its maximum speed. Round to the nearest whole number.

1.6 Vocabulary Practice

Match the term with its description.

1. $a^2 + b^2 = c^2$
2. The distance from any point on a circle to its center
3. The distance around a circle
4. A number that cannot be written as a fraction in which the numerator and denominator are integers
5. A repeating decimal number is this kind of number.
6. The quotient of the circumference of a circle and its diameter
7. Three dots that follow a list of numbers to show that the list does not end
8. The number 3 in $\sqrt[3]{8}$ is an example of this.
9. The side opposite the right angle in a right triangle
10. The square root of this is a rational number.

A. circumference
B. ellipsis
C. hypotenuse
D. index
E. irrational number
F. π
G. perfect square
H. Pythagorean theorem
I. radius
J. rational number

1.6 Exercises

Follow your instructor's guidelines for showing your work.

For all exercises in this section, use $\pi \approx 3.14$.

1. Explain the difference between a rational number and an irrational number.
2. Explain why 0.3 is a rational number.

For exercises 3–6, rewrite the fraction as a repeating decimal.

3. $\frac{5}{6}$ 4. $\frac{1}{6}$ 5. $\frac{3}{7}$ 6. $\frac{4}{7}$

For exercises 7–24, round to the nearest tenth.

7. Find the circumference of a circle with a diameter of 6 cm.
8. Find the circumference of a circle with a diameter of 13 cm.
9. Find the circumference of a circle with a radius of 0.5 in.
10. Find the circumference of a circle with a radius of 0.7 in.
11. Find the area of a circle with a radius of 6 cm.
12. Find the area of a circle with a radius of 10 cm.
13. Find the area of a circle with a diameter of 12 m.
14. Find the area of a circle with a diameter of 8 m.
15. Find the volume of a cylinder with a radius of 6 in. and height of 15 in.
16. Find the volume of a cylinder with a radius of 8 in. and height of 15 in.
17. Find the volume of a cylinder with a radius of 18 in. and height of 2 ft. Write the answer in cubic inches.
18. Find the volume of a cylinder with a radius of 16 in. and height of 2 ft. Write the answer in cubic inches.
19. Find the volume of a cylinder with a diameter of 15 in. and a height of 24 in.
20. Find the volume of a cylinder with a diameter of 11 in. and a height of 27 in.
21. Find the volume of a sphere with a diameter of 25 in.
22. Find the volume of a sphere with a diameter of 15 in.
23. Find the volume of a cone with a radius of 6 in. and a height of 11 in.
24. Find the volume of a cone with a radius of 6 in. and a height of 13 in.

For exercises 25–32, use the five steps.

25. The diameter of a circular bathroom sink is 15.75 in. Find its circumference. Round to the nearest hundredth.

26. The diameter of a circular saw blade is 7.25 in. Find its circumference. Round to the nearest hundredth.

27. The diameter of a circular bathroom sink is 15.75 in. Find its area. Round to the nearest tenth.
28. The diameter of a circular saw blade is 3.375 in. Find its area. Round to the nearest tenth.
29. The inner diameter of a cylindrical waste basket is 9.75 in. and its height is 14.75 in. Find its volume. Round to the nearest tenth.

30. The inner diameter of cylindrical mailing tube is 1.75 in. and its length is 9.5 in. Find its volume. Round to the nearest tenth.

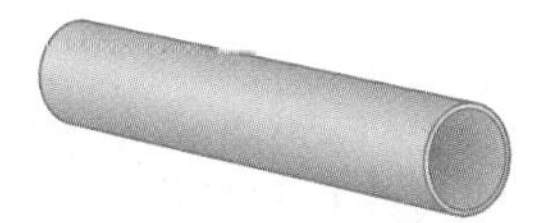

31. The International Basketball Federation rules (2006) state, "For all men's competitions in all categories, the circumference of the ball shall be no less than 749 mm and no more than 780 mm (size 7)." Find the maximum volume of the ball in cubic millimeters. Round to the nearest whole number. (*Source:* www.fiba.com)

32. The International Basketball Federation rules (2006) state, "For all women's competitions in all categories, the circumference of the ball shall be no less than 724 mm and no more than 734 mm (size 6)." Find the minimum radius of the ball in millimeters. Round to the nearest whole number. (*Source:* www.fiba.com)

For exercises 33–40, identify whether the number is a perfect square or is not a perfect square.

33. 81 **34.** 100 **35.** 144 **36.** 121

37. 6 **38.** 8 **39.** 17 **40.** 19

For exercises 41–42, the Herfindahl-Hirschman Index (HHI) is used by the U.S. Department of Justice in considering whether a merger of two companies violates antitrust law. Each company has a percent of the market called *market share*. The HHI equals the sum of the *squares* of the market share of each company, where the market share is written as a whole number. For example, if a market share is 20%, the square of the market share equals 20^2, or 400. (*Source:* U.S. District Court, District of Columbia, Case No.: 1:08-cv-00899)

41. A market consists of four companies. Company A has 30% of the market, Company B has 30% of the market, Company C has 20% of the market, and Company D has 20% of the market. Find the HHI for this market.

42. A market consists of four companies. Company A has 30% of the market, Company B has 10% of the market, Company C has 10% of the market, and Company D has 50% of the market. Find the HHI for this market.

For exercises 43–50, evaluate. If the square root is irrational, round to the nearest thousandth.

43. $\sqrt{49}$ **44.** $\sqrt{64}$

45. $\sqrt{15}$ **46.** $\sqrt{18}$

47. $\sqrt{49 \text{ km}^2}$ **48.** $\sqrt{64 \text{ km}^2}$

49. $\sqrt{-49}$ **50.** $\sqrt{-9}$

51. Explain why $\sqrt{-16}$ cannot be a real number.

52. Explain why $\sqrt{25}$ does not equal -5 even though $(-5)(-5) = 25$.

53. Given that $\sqrt{9} = 3$ and $\sqrt{16} = 4$, explain why $\sqrt{10}$ must be between 3 and 4.

54. Given that $\sqrt{16} = 4$ and $\sqrt{25} = 5$, explain why $\sqrt{18}$ must be between 4 and 5.

For exercises 55–56, the Bakhshali Manuscript was an early mathematical manuscript written on birchbark, discovered by a farmer in 1881 near Peshawar in what is now Pakistan. The manuscript includes a formula for approximating the square root of a number Q that is not a perfect square: $\sqrt{Q} = A + \frac{b}{2A} - \frac{\left(\frac{b}{2A}\right)^2}{2\left(A + \frac{b}{2A}\right)}$. In this formula, A^2 is the nearest perfect square smaller than Q, and $b = Q - A^2$.

55. a. For $\sqrt{12}$, $Q = 12$. Identify A^2, A, and b.

b. Use this method to approximate $\sqrt{12}$. Round to the nearest ten-thousandth.

c. Use the square root key on a calculator to evaluate $\sqrt{12}$. Round to the nearest ten-thousandth.

56. a. For $\sqrt{18}$, $Q = 18$. Identify A^2, A, and b.

b. Use this method to approximate $\sqrt{18}$. Round to the nearest ten-thousandth.

c. Use the square root key on a calculator to evaluate $\sqrt{18}$. Round to the nearest ten-thousandth.

For exercises 57–64, evaluate.

57. 6^3 **58.** 3^3 **59.** $\sqrt[3]{216}$

60. $\sqrt[3]{27}$ **61.** $\sqrt[3]{-216}$ **62.** $\sqrt[3]{-27}$

63. $\sqrt[3]{1000}$ **64.** $\sqrt[3]{-1000}$

65. Copy the chart below. Put an X in the box if the number is an element of the set at the top of the column.

Chart for exercise 65

	Real numbers	Rational numbers	Irrational numbers	Integers	Whole numbers
$\frac{3}{5}$					
π					
-8					
$\sqrt{7}$					
$\sqrt{100}$					
-0.4					
0					
$0.\overline{3}$					

66. Copy the chart below. Put an X in the box if the number is an element of the set at the top of the column.

For exercises 67–72, find the length of the hypotenuse of the right triangle. If the length is irrational, round to the nearest hundredth.

67.

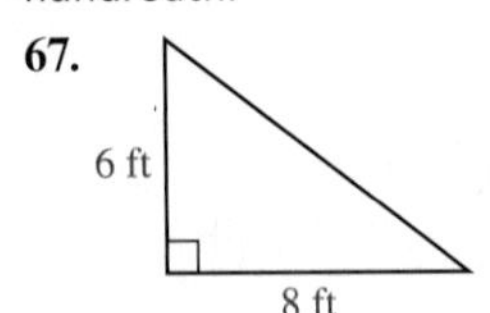

68.

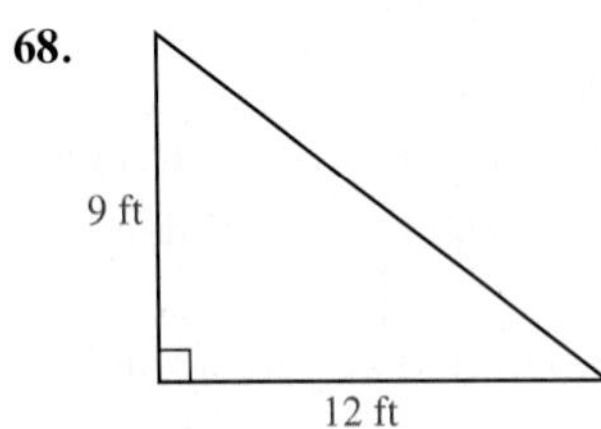

69.

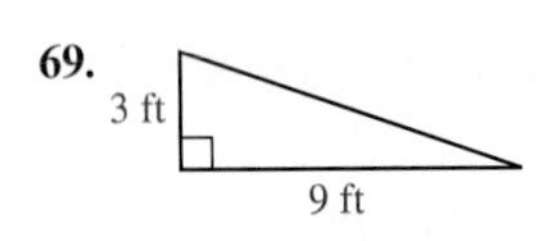

70.

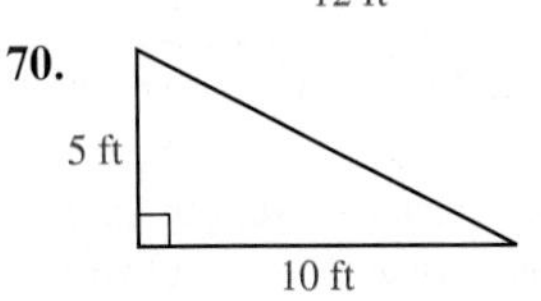

71.

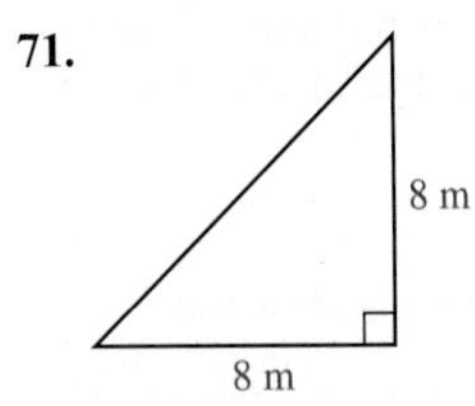

72.

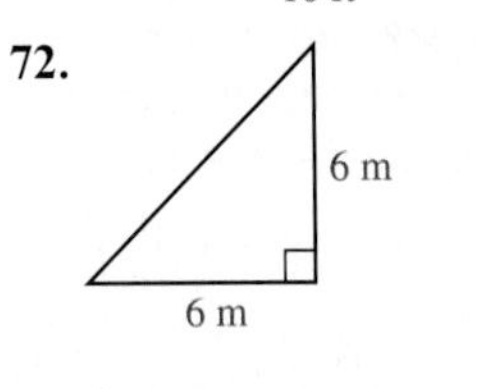

For exercises 73–74, the "square root formula," $V = \sqrt{P}$, was proposed by Poland in 2007 as a way to allocate the number of votes per country on the Council of Ministers of the European Union, V, where P is the population of the country.

73. The population of Germany in 2006 was about 82,422,300 people. Use the square root formula to allocate the number of Germany's votes on the Council. Round to the nearest whole number. (*Source:* www.infoplease.com)

74. The population of Denmark in 2006 was about 5,450,700. Use the square root formula to allocate the number of Denmark's votes on the Council. Round to the nearest whole number. (*Source:* www.infoplease.com)

For exercises 75–76, the formula $D = 1.17\sqrt{E}$ represents the relationship of the distance in nautical miles from a person in a boat to the horizon, D, and the height of the person's eye above the surface of the water in feet, E.

75. If the eye of a person standing on the deck of a boat is 12 ft above the surface of the water, find the distance of the person to the horizon. Round to the nearest tenth.

76. If the eye of a person sitting in a tower on a sport fishing boat is 20 ft above the surface of the water, find the distance of the person to the horizon. Round to the nearest tenth.

For exercises 77–80, the formula $V = 1.34\sqrt{L}$ represents the relationship of the speed in knots of a water wave in deep water, V, and its wavelength in feet, L. Round to the nearest tenth.

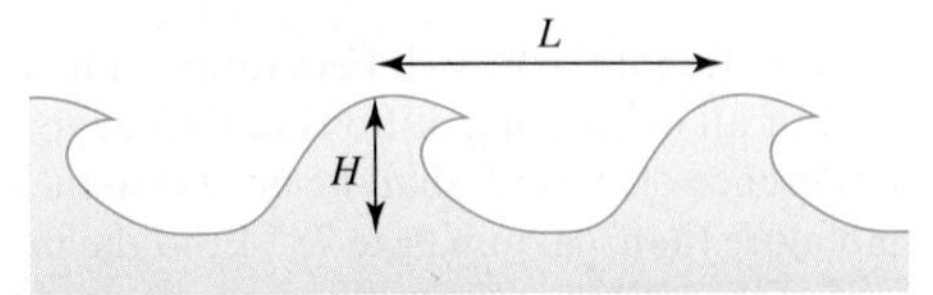

77. The wavelength of a wave in deep water is 10 ft. Find the speed of the wave.

78. The wavelength of a wave in deep water is 40 ft. Find the speed of the wave.

79. The wavelength of a wave in deep water is 30 ft. Find the speed of the wave in miles per hour. $\left(1 \text{ knot} \approx \dfrac{1.15 \text{ mi}}{1 \text{ hr}}.\right)$

80. The wavelength of a wave in deep water is 12 ft. Find the speed of the wave in miles per hour. $\left(1 \text{ knot} \approx \dfrac{1.15 \text{ mi}}{1 \text{ hr}}.\right)$

For exercises 81–82, when a layer of water is between the tires of an airplane and a runway, hydroplaning may occur. The formula $S = 9\sqrt{P}$ represents the relationship of the speed of the airplane in knots at which hydroplaning may occur, S, and the tire pressure in pounds per square inch, P. Find the speed at which hydroplaning may occur for an airplane with the given tire pressure. Round to the nearest whole number.

81. 192 pounds per square inch

82. 180 pounds per square inch

For exercises 83–84, the formula $A = \dfrac{1}{2}bh$ represents the relationship of the area of a triangle, A, the length of the base, b, and the height, h. Heron's formula,

$$A = \frac{\sqrt{(a^2 + b^2 + c^2)^2 - 2(a^4 + b^4 + c^4)}}{4},$$ represents

the relationship of the area of a triangle, A, and the length of each side, a, b, and c.

Chart for exercise 66

	Real numbers	Rational numbers	Irrational numbers	Integers	Whole numbers
$\sqrt{5}$					
$\sqrt{36}$					
-0.5					
$\frac{1}{2}$					
π					
-2					
0					
$0.\overline{7}$					

83. a. Use Heron's formula to find the area of the triangle.

b. Use $A = \frac{1}{2}bh$ to find the area of the triangle.

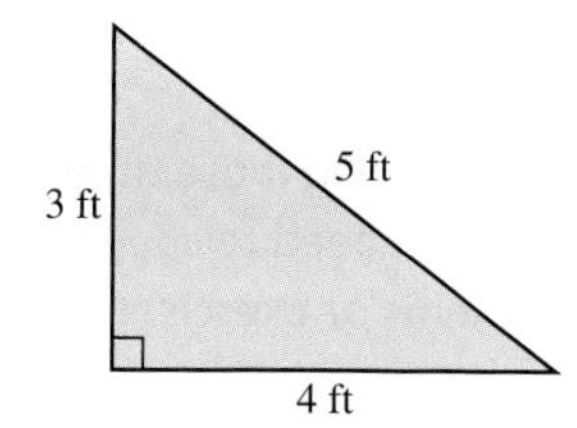

84. a. Use Heron's formula to find the area of the triangle.

b. Use $A = \frac{1}{2}bh$ to find the area of the triangle.

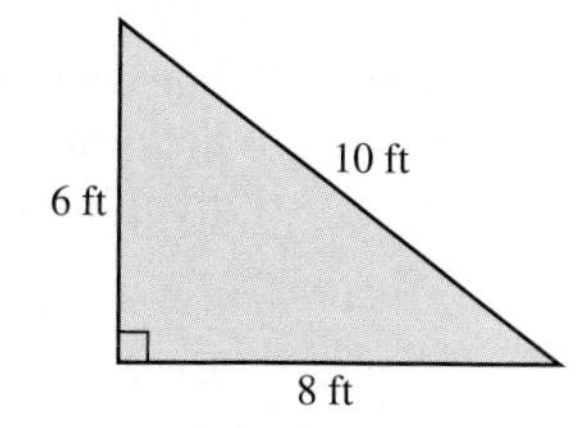

For exercises 85–86, use the five steps. The formula $G = 29.7D^2\sqrt{P}$ represents the relationship of the volume of water in gallons per minute delivered by a fire hose, G; the diameter of the nozzle on the hose in inches, D; and the water pressure at the nozzle in pounds per square inch, P.

85. Find the gallons per minute delivered by a fire hose with 64 pounds per square inch of pressure at a nozzle with a diameter of 2 in. Round to the nearest whole number.

86. Find the gallons per minute delivered by a fire hose with 90 pounds per square inch of pressure at a nozzle with a diameter of 1.5 in. Round to the nearest whole number.

For exercises 87–88, use the five steps. The formula $R = 1.42\sqrt{H}$ represents the relationship of the range in nautical miles, R, of a VHF marine antenna with a height of H ft. (*Source:* www.shakespeare-antennas.com)

87. The height of a VHF antenna on a sailboat is 64 ft. Find the range of this antenna. Round to the nearest tenth.

88. The height of a VHF antenna on a powerboat is 16 ft. Find the range of this antenna. Round to the nearest tenth.

Technology

For exercises 89–92, pressing the π key on a calculator enters the value of π with as many digits as the calculator can display. The last digit is rounded.

(a) Evaluate each expression using the π key on a calculator. Round to the nearest thousandth.

(b) Evaluate each expression using $\pi \approx 3.14$.

89. $\pi(15 \text{ m})^2$

90. $2\pi(8 \text{ ft})$

91. $\pi(5 \text{ in.})^2(8 \text{ in.})$

92. $\dfrac{4\pi(3 \text{ in.})^3}{3}$

Find the Mistake

For exercises 93–96, the completed problem has one mistake.

(a) Describe the mistake in words, or copy down the whole problem and highlight or circle the mistake.

(b) Do the problem correctly.

93. Problem: The diameter of a circle is 10 in. Using $\pi \approx 3.14$, find the area of the circle.

Incorrect Answer: $A = \pi r^2$

$A \approx (3.14)(10 \text{ in.})^2$

$A \approx (3.14)(100 \text{ in.}^2)$

$A \approx 314 \text{ in.}^2$

94. Problem: Evaluate: $\sqrt{-144}$

Incorrect Answer: $\sqrt{-144}$

$= -12$

95. Problem: Evaluate: $\sqrt[3]{-216}$

Incorrect Answer: $\sqrt[3]{-216}$ is not a real number.

96. Problem: Find the volume of a cylinder with a diameter of 14 in. and a height of 2 ft. Write the answer in cubic inches. Use $\pi \approx 3.14$.

Incorrect Answer: $V = \pi r^2 h$

$V \approx (3.14)(7 \text{ in.})^2(2 \text{ ft})$

$V \approx (3.14)(49 \text{ in.}^2)(2 \text{ ft})$

$V \approx 307.72 \text{ in.}^3$

Review

97. Use roster notation to represent the set of whole numbers.

98. Use roster notation to represent the set of integers.

99. Describe the purpose of an ellipsis.

100. Describe the order of operations.

SUCCESS IN COLLEGE MATHEMATICS

101. Explain how to add or drop a class at your college.

102. Explain how to get help if you cannot remember an important password for use on the college network.

Study Plan for Review of Chapter 1

To prepare for a test, you need to identify what you know and where you need to study more and practice. Complete the Review Exercises for each question. If you can do the Review Exercises correctly without referring to an example or your notes, you probably do not need to spend more time on this topic. If you cannot do the Review Exercises correctly without looking back, you need to spend more time learning by studying examples or your notes or by redoing Practice Problems or Exercises from the section.

SECTION 1.1 The Real Numbers

Ask Yourself	Test Yourself	Help Yourself
Can I...	**Do 1.1 Review Exercises**	**See these Examples and Practice Problems**
use roster notation to represent a set and describe the purpose of an ellipsis? identify a prime number?	1–3	Ex. 1, PP 1–2
use a number line graph to represent a number?	5	Ex. 2, PP 3
evaluate an exponential expression?	6–8	Ex. 3, PP 7–9
identify the opposite of a number? evaluate the absolute value of a number? evaluate expressions that include operations with 0 or 1?	9–19	Ex. 4, PP 4, 5, 10, 11
simplify a fraction into lowest terms?	20–25	Ex. 6–8, PP 12–15
classify a number as an element of the set of real numbers, irrational numbers, rational numbers, integers, and/or whole numbers?	26–27	Ex. 5, 9, PP 16–18
describe the order of operations and use the order of operations to evaluate an expression?	28–43	Ex. 10–14, PP 19–28

1.1 Review Exercises

1. Use roster notation to represent the set of
a. whole numbers.
b. integers.

2. A set is $\{1, 4, 7, 10, \ldots\}$. Explain what the ellipsis represents in this set.

3. Choose the numbers that are prime numbers.
a. 1 **b.** 3 **c.** 13 **d.** 10
e. $\frac{1}{2}$ **f.** 2 **g.** 51 **h.** 11

4. In $7 \cdot 5 = 35$,
a. What is the product?
b. What are the factors?

5. Graph the number 7 and its opposite on a real number line.

For exercises 6–19, evaluate.

6. 8^2
7. $(-8)^2$
8. -8^2
9. $|41|$
10. $|-73|$
11. $-(-4)$
12. $0(4)$
13. $\frac{0}{4}$
14. $4 + 0$
15. $0 - 4$
16. $4 \div 0$
17. $0 \div 4$
18. $\frac{4}{0}$
19. $(12)(1)$

For exercises 20–25, simplify the fraction into lowest terms.

20. $\frac{12}{30}$
21. $\frac{10}{90}$
22. $\frac{24}{72}$
23. $\frac{16}{10}$
24. $\frac{120}{305}$
25. $-\frac{54}{8}$

26. Explain why 0.13 is a rational number.

27. Copy the chart on the next page. Put an X in the box if a number is an element of the set at the top of the column.

28. Describe the order of operations.

29. To evaluate $12 \div 2 \cdot 3 + 1$ following the order of operations, what would you do first?

30. To evaluate $(12 \div 2) + 3(4)^2$ following the order of operations, what would you do first?

For exercises 31–43, evaluate.

31. $9 - 6(3 - 5)$
32. $24 \div 2 \cdot 3 \div 6$
33. $\frac{9(3)}{8 - 2 \cdot 4}$
34. $\frac{3}{5} + \frac{7}{8} \div \frac{1}{2}$
35. $(8 - 9)^2 - 16 \div 2 \cdot 3$
36. $5 - 4^2 \cdot 2 - (-7 - 3)$
37. $5(10 - 2)^2 - (6 + 1)$
38. $28 - 24 \div 2 \cdot 2^2$

Chart for exercise 27

	Real numbers	Rational numbers	Irrational numbers	Integers	Whole numbers
0.3					
$\frac{1}{8}$					
-9					
$\sqrt{11}$					
0					
π					

39. $\dfrac{7^2 - 49}{9 - 8 \cdot 3}$

40. $\dfrac{(15 \div 3 + 2)^2 - 6 \cdot 8}{(-7 + 6)^3}$

41. $2[5 - (1 + 9)^2 \div 10]$

42. $-5^2 - 45 \div 3$

43. $(3 \cdot 5 - 15) \div (21 + 3 \div 3)$

SECTION 1.2 Variables and Algebraic Expressions

Ask Yourself	Test Yourself	Help Yourself
Can I . . .	**Do 1.2 Review Exercises**	**See these Examples and Practice Problems**
identify the number of terms and the variables, coefficients, and constants in an expression?	44	Ex. 1, PP 1–2
simplify an algebraic expression by combining like terms?	45–47	Ex. 2–6, PP 3–8
multiply terms?	48–49	Ex. 7, 8, PP 9–12
write and identify examples of the commutative property, associative property, and distributive property?	50–56	Ex. 9, PP 13–18
simplify an algebraic expression, using the properties and the order of operations?	57–63	Ex. 10–14, PP 19–26

1.2 Review Exercises

44. An expression is $8x + 2y + 7z + 9$.
  a. Identify the coefficients.
  b. Identify the constant.
  c. Identify the variables.
  d. How many terms are in this expression?

For exercises 45–47, simplify.

45. $8p - 12p - p$

46. $2u + 3y - 9u + 8$

47. $x^2 + 8x - 9 + 3x^2 - 14x - 1$

For exercises 48–49, simplify.

48. $(-5w)(6z)$

49. $\left(\frac{3}{4}a\right)\left(\frac{2}{9}ab\right)$

50. a. Does $3 + (5 + 7) = 3 + (7 + 5)$ represent the commutative property or the associative property?
  b. Explain.

For exercises 51–56, match the equation with the property it represents.

51. $4(2 + 3) = 4 \cdot 2 + 4 \cdot 3$

52. $6 + 8 = 8 + 6$

53. $9 + (2 + 5) = 9 + (5 + 2)$

54. $(8 \cdot 7) \cdot 3 = 8 \cdot (7 \cdot 3)$

55. $x + (y + z) = (x + y) + z$

56. $3(m + 11) = 3m + 3(11)$

A. Associative property
B. Commutative property
C. Distributive property

For exercises 57–62, simplify.

57. $2(3a + 1) + 5$

58. $14 - 6(4x + 3) + x$

59. $8x + 5(3x - 2y) - 19(x + 2)$

60. $8p - (9p - 2) - 15$

61. $\frac{1}{6}x + \frac{4}{5}z - \frac{3}{10}x + \frac{7}{15}z$

62. $\frac{3}{4}(18z - 16) + \frac{5}{6}(15z + 18)$

63. Explain why it is not possible to first simplify within the parentheses of the expression $2(3x - 4) + 9$.

SECTION 1.3 Introduction to Problem Solving

Ask Yourself	Test Yourself	Help Yourself
Can I...	**Do 1.3 Review Exercises**	**See these Examples and Practice Problems**
use the five steps to solve an application problem?	64–73	Ex. 1, 3–7, PP 1–7

1.3 Review Exercises

64. Given the information below and asked to find the total calories of a meal that included a Five Guys Bacon Cheeseburger without mayo, a regular fries, and a large Coke, identify the information needed to solve the problem.

A Five Guys Bacon Cheeseburger has 920 calories and 30 grams of saturated fat ($1\frac{1}{2}$ days' worth) without toppings. . . . Add 620 calories for the regular fries or 1,460 calories for a large. (The large is as big as three large orders of fries at McDonald's.) Now your lunch of an unadorned Bacon Cheeseburger and large fries is up to 2,380 calories. Add 100 calories for every plop of mayo on your burger, another 300 for a large (32 oz.) Coke, and 300 more for every free refill. (*Source:* www.cspinet.org, June 2010)

65. A problem states, "Find the total losses for Hellenic Railways per year in U.S. dollars. Report the answer in billions of dollars; round to the nearest tenth."

Losses at Hellenic Railways, however, continue to mount— at the rate of 3 million euros ($3.8 million) a day. Its total debt has increased to $13 billion, or about 5 percent of Greece's gross domestic product. (*Source:* www.nytimes .com, July 20, 2010)

A student solved this problem by writing and solving the equation $L = \left(\frac{\$3.8 \text{ million}}{1 \text{ day}}\right)(365 \text{ days})$ and found the correct answer: "The total losses per year for Hellenic Railways will be $1.4 billion." Explain why this answer is reasonable.

66. A problem states, "The cost for Oregon to provide home and community based services to a low-income adult is $1500 per month. If these services are not available and the person ends up in a nursing home, the average cost to the state is $5900 a month. Find the difference in cost per year between providing care to 50 low-income adults with home and community-based services and providing care in a nursing home." (*Source:* www.nytimes.com, July 16, 2010)

A student solved this problem by writing and solving the equation

$$D = \left(\frac{\$5900 - \$1500}{1 \text{ month}}\right)(12 \text{ months})(50)$$

$$D = \$2,640,000$$

Write a complete and meaningful sentence to report this solution.

For exercises 67–73, use the five steps.

67. A community garden at Scottsdale Community College has an area of about 6 acres. Find the total cost for a new member to rent a garden plot for 2 years.

The annual cost to rent a 20- by 30-foot plot is $10 for insurance, $40 to rent the plot, plus $40 for water. New members pay a one-time equipment assessment of $62. (*Source:* www .azcentral.com, Nov. 20, 2009)

68. Find the cost per mile to create the light-rail line. Write the answer in place-value notation.

Just as difficult: juggling the money to pay for it all. The $1.4 billion spent to create Metro's 20-mile "starter line," which opens later this month, is a down payment. There will be more bills to pay once the trains start running. (*Source:* www.azcentral.com, Dec. 9, 2008)

69. The height of a steel windmill tower is 47 ft. It pumps water from a well with a depth of 35 ft. Using a negative number to represent depth, find the distance between the top of the tower and the bottom of the well.

70. If about two-fifths of students submit their financial aid FAFSA on or after July 1, find the number of 250,000 students applying for financial aid who submit their FAFSAs on or after July 1. (*Source:* www.nytimes.com, Jan. 22, 2010)

71. A piece of fabric is 45 in. wide. A rag rug maker is cutting strips that are $\frac{5}{8}$ in. wide from this fabric. How many strips can be cut from this fabric?

72. Find the amount of $250,000 spent on health care that is used in treating chronic diseases related to diet.

Three-fourths of every health care dollar is spent on treating chronic diseases related to diet. (*Source:* www.latimes.com, Feb. 1, 2011)

73. If $\frac{1}{8}$ cup olive oil is used for frying and $\frac{2}{3}$ cup olive oil is used for a salad dressing, find the total amount of olive oil used.

SECTION 1.4 Problem Solving with Percent

Ask Yourself	Test Yourself	Help Yourself
Can I...	**Do 1.4 Review Exercises**	**See these Examples and Practice Problems**
use the five steps to solve an application problem that includes percent?	77–82	Ex. 4–9, PP 1–11

1.4 Review Exercises

For exercises 74–75, rewrite the percent as a decimal number.

74. 7% **75.** 8.3%

76. Choose the expressions that are equivalent to 9%.

a. $\frac{9}{100}$ **b.** 0.09 **c.** 0.9 **d.** $\frac{18}{200}$ **e.** $\frac{90}{1000}$

For exercises 77–82, use the five steps.

77. Find the amount of popcorn consumed in homes. Round to the nearest tenth of a billion.

Americans consume 4 billion gallons of popcorn annually, totaling 13.5 gallons per person. . . . An estimated 70 percent . . . is consumed in homes, with the remaining 30 percent eaten at theaters, stadiums and schools. (*Source:* www.reuters.com, July 10, 2007)

78. Find the percent increase in number of students admitted to the Iowa Lakes Community College wind program. Round to the nearest percent.

Mesalands Community College . . . launched its wind program last fall with 32 students. GE offered to hire every qualified graduate for three years . . . At Iowa Lakes Community College . . . wind students have "two to three job offers each" by the time they complete the two-year program . . . The school will admit 102 students this fall, up from the 72 it had planned to take, because of surging demand. (*Source:* www.latimes.com, March 1, 2009)

79. Find the number of the surveyed institutions that said they would cut financial aid. Round to the nearest whole number.

In a survey of 372 institutions in December, the National Association of Independent Colleges and Universities found that 93 percent said they were moderately or greatly concerned about preventing enrollment declines. Only 8 percent said they would cut financial aid, compared with 50 percent that said they had stopped hiring. (*Source:* www.nytimes.com, Feb. 28, 2009)

80. Grocery stores in rural areas are closing because of competition from distant supercenters and relatively high operating costs. Use the information in the bar graph to find the number of low income rural residents who must travel 10 to 20 mi to buy groceries. Round to the nearest tenth of a million.

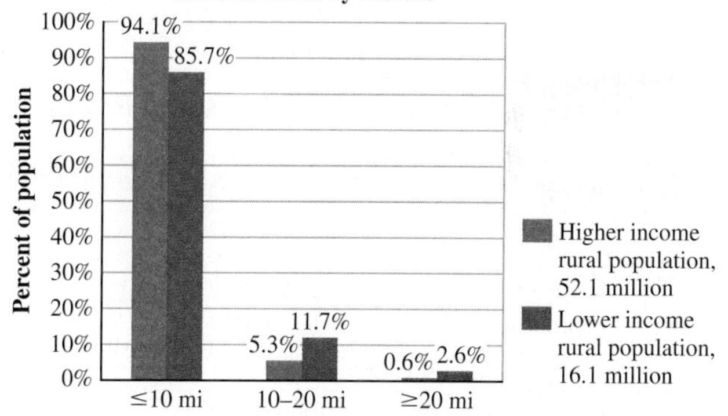

Source: www.ers.usda.gov, June 2009

81. Use the information in the table to find the percent of students who received a grade of C on a psychology test. Round to the nearest percent.

A	B	C	D	F
17	25	48	12	5

82. The regular price of one loaf of bread is \$4.25. Because it is one day old, it is marked down 35%. The sales tax on food is 6.5%. Find the total cost of one loaf of day-old bread including tax.

SECTION 1.5 Units of Measurement and Formulas

Ask Yourself	Test Yourself	Help Yourself
Can I...	**Do 1.5 Review Exercises**	**See these Examples and Practice Problems**
add, subtract, multiply, and divide measurements with the same unit?	83–86	Ex. 1–3, PP 1–5
use a conversion factor to change the units of a measurement into another unit?	88, 89, 91, 93	Ex 4–7, PP 6–10
add and subtract measurements in different units?	89	Ex. 8–9, PP 11–14
use a formula to solve an application problem including the formulas $d = rt$, $I = PRT$, $P = 2W + 2L$, $P = a + b + c$, $A = LW$, $A = \frac{1}{2}bh$, and $V = LWH$?	90–94, 96	Ex. 10–18, PP 15–28

1.5 Review Exercises

For exercises 83–86, evaluate.

83. 8 m + 12 m

84. $\frac{32 \text{ ft}}{4 \text{ ft}}$

85. (7 in.)(2 in.)(3 in.)

86. 129 kg − 30 kg

87. Using the relationship 1 kg = 1000 g, write two conversion factors that are equal to 1.

88. Change 3 qt into liters. (1 L ≈ 1.06 qt.) Round to the nearest tenth.

89. Find the sum in pounds (lb): 7 kg + 3 lb. Round to the nearest tenth. (1 kg ≈ 2.2 lb.)

90. Find the interest earned on $2850 invested for 8 years at a simple annual interest rate of 7.5%.

91. The speed of a car in rush hour traffic is $\frac{45 \text{ mi}}{1 \text{ hr}}$. Find the distance the car travels in 35 min.

92. Find the volume of a box in cubic inches that is 11 in. high, 14 in. long, and 8 in. wide.

93. Find the perimeter in centimeters of a rectangle that is 1.4 m long and 65 cm wide.

94. The base of a triangle is 3 in. and its height is 6 in. Find its area.

95. Choose the units of volume.
a. liter **b.** ft^2 **c.** cm^3
d. cubic inch **e.** meter

96. The Flesch-Kincaid Grade Level Readability Formula is $G = 0.39L + 11.8W - 15.59$, where G is the readability of a document by grade level in school, L is the average sentence length (the total words divided by the number of sentences), and W is the average number of syllables per word (total syllables divided by the number of words). For example, if G is 6.2, the readability of the document is the sixth grade in the United States. Find the readability of a document in which the average sentence length is 20.2 and the average number of syllables per word is 1.45. Round to the nearest tenth. (*Source:* www.utexas.edu)

SECTION 1.6 Irrational Numbers and Roots

Ask Yourself	Test Yourself	Help Yourself
Can I . . .	**Do 1.6 Review Exercises**	**See these Examples and Practice Problems**
rewrite a fraction as a repeating decimal?	98	Ex. 1, PP 1–3
use a formula that includes π?	100–103	Ex. 2–4, PP 4–13
identify the square roots of a real number?	104	Ex. 5, PP 14
evaluate a principal square root?	105–109, 112–113	Ex. 6–9, PP 15–22
evaluate a principal cube root?	110–111	Ex. 10, PP 23–28
given the length of the legs of a right triangle, use the Pythagorean theorem to find the length of the hypotenuse?	114	Ex. 11, PP 29, 30
use a formula that includes a square root?	115	Ex. 12, PP 31

1.6 Review Exercises

97. Rewrite 0.693 as a fraction.

98. Rewrite $\frac{7}{9}$ as a repeating decimal.

99. Explain the difference between a rational and an irrational number.

100. The diameter of a circle is 14 in. Find its circumference. ($\pi \approx 3.14$.) Round to the nearest whole number.

101. The diameter of a sphere is 8 cm. Find its volume. ($\pi \approx 3.14$.) Round to the nearest tenth.

102. Find the difference in the area of a paper plate with a diameter of 10 in. and the area of a paper plate with a diameter of 9 in. Round to the nearest tenth. ($\pi \approx 3.14$.)

103. A liquid chemical cylindrical storage tank has a diameter of 42 in. and a height of 48 in. The liquid must be at least 2 in. below the top of the tank. Use the five steps to find the volume in gallons of liquid chemicals that can be stored in this tank. ($\pi \approx 3.14$; 1 gal = 231 in.^3.) Round to the nearest whole number.

104. What are the square roots of 100?

For exercises 105–109, evaluate. If the number is irrational, round to the nearest hundredth.

105. $\sqrt{81}$

106. $\sqrt{0}$

107. $\sqrt{-4}$

108. $\sqrt{12}$

109. $\sqrt{51}$

For exercises 110–111, evaluate.

110. $\sqrt[3]{-27}$

111. $\sqrt[3]{8}$

112. A number is a perfect square. Will its square roots be rational numbers or irrational numbers?

113. Explain why $\sqrt{169}$ does not equal -13.

114. The legs of a right triangle are 6 ft and 8 ft long. Find the length of the hypotenuse.

115. Mosteller's formula $B = \sqrt{\frac{HW}{3131}}$ represents the relationship of the body surface area in square meters, B, the height in inches, H, and the weight in pounds, W. Find the body surface area of a person who is 4 ft tall and weighs 75 lb. Round to the nearest tenth.

Chapter 1 Test

1. Copy the chart below. Put an X in the box if the number is an element of the set at the top of the column.

For problems 2–10, evaluate.

2. $12 \div 4 \cdot 3 + 5^2$

3. $\dfrac{2 - 9(6)}{1 - 1^3}$

4. $\dfrac{14 - 2(7)}{3(1 - 6)}$

5. $9 - [-8 - (-15) - 3^2]^3$

6. $\dfrac{18 - 8 \div 2}{-3^2 + 2}$

7. $\dfrac{4}{15} \div \dfrac{2}{3} - \left(\dfrac{1}{2}\right)^2$

8. $5 - 2(-7) + 3$

9. $-4(9 - 15) - 10^2$

10. $\sqrt{144} - \sqrt[3]{-8}$

For problems 11–13, simplify.

11. $\dfrac{3}{4}(9h - 80) + \dfrac{5}{6}h$

12. $-19w - (-6w) + 5w$

13. $2(5p - 8) - (16p - 1)$

14. Explain why $5k$ and $4h$ are not like terms.

15. Describe the order of operations.

For problems 16–18, match the equation with the property it represents.

16. $9(4x + 3) = 9 \cdot 4x + 9 \cdot 3$

17. $4 \cdot 9 = 9 \cdot 4$

18. $6 + (2 + 1) = (6 + 2) + 1$

A. Commutative property
B. Associative property
C. Distributive property

19. The standard size of a croquet court is a rectangle that is 84 ft wide and 105 ft long. Find the area of this court. (*Source:* www.croquetamerica.com)

20. A weight-loss program assigns points to various foods, using the formula $P = \dfrac{C}{50} + \dfrac{F}{12} - \dfrac{R}{5}$, where the calories per serving is C, the grams of fat per serving is F, the grams of dietary fiber per serving is R, and the number of points assigned to the food is P. A Boston Market Pastry Top Chicken Pot Pie has 810 calories, 48 g of fat, and 4 g of fiber. Use the formula to find the points assigned to this chicken pot pie. Round to the nearest whole number. (*Source:* www.bostonmarket.com)

21. Find the volume in cubic meters of a box that is 90 cm high, 1.2 m wide, and 1.3 m long. (1 m = 100 cm.) Round to the nearest tenth.

22. The diameter of a circle is 15 in. Find its area. ($\pi \approx 3.14$.)

For problems 23–26, use the five steps.

23. The Board of Regents at the University of Hawaii Maui College announced an increase in tuition from \$97 per credit to \$101 per credit. Find the percent increase in tuition for a student taking 12 credits. Round to the nearest tenth of a percent. (*Source:* www.mauinews.com, Oct. 28, 2011)

24. Find the number of surveyed adults ages 18–29 who use their phones to send or receive text messages. Round to the nearest whole number. (*Source:* www.pewinternet.org, Aug. 15, 2011)

Number surveyed	321	535	572	430
Ages (years)	18–29	30–49	50–64	65+
Percent who send or receive text messages	95%	85%	58%	24%

25. In 2008, there were 460,000 children in foster care in the United States. In 2009, there were 424,000 children in foster care. Find the percent decrease in the number of children in foster care. Round to the nearest percent. (*Source:* www.childtrends.org, May 2011)

26. The regular price of a video game is \$39.99. If it is on sale for 20% off the regular price and the sales tax is 7.75%, find the price of the game, including tax.

Chart for problem 1

	Real numbers	Rational numbers	Irrational numbers	Integers	Whole numbers
-4					
$\sqrt{7}$					
$\dfrac{2}{9}$					
18					
π					
-0.13					

Solving Equations and Inequalities

2

SUCCESS IN COLLEGE MATHEMATICS

Time Management

As an adult, you are responsible for managing your time. This means that you need to balance and schedule the time needed for college, work, family, and relaxation. If this is the first time that you have been responsible for your own schedule, you need to develop a habit of organizing your time. Otherwise, you may find that the day or week has passed and you have not spent enough time on your college work.

Activities such as talking on the phone, surfing the web, shopping, gaming, watching videos, cooking meals, and doing housework are normal parts of American life. However, college students often have to change the way they use time. If you have non-college friends, you might not be able to do all of the things they can do. If you live with your family, you might not be able to do everything that you have done for them or with them in the past.

The first step in time management is to use a planner to build a weekly schedule. Then, as you use your schedule, you will need to adjust it to make it work for you. Initial estimates of how much time you spend on current activities and how much time you will need to spend on your college work are not always accurate.

2.1 The Properties of Equality
2.2 More Equations in One Variable
2.3 Variables on Both Sides
2.4 Solving an Equation for a Variable
2.5 Inequalities in One Variable

The pans of the balance scale in the picture are empty. The beam of the scale is horizontal. If we put a weight in the left pan, it will go down. To make the beam horizontal again, we must put an equal weight in the right pan. We can use the image of a balance scale to visualize equations and the process of solving an equation. In this section, we will use the properties of equality to solve equations in one variable.

SECTION 2.1

The Properties of Equality

After reading the text, working the practice problems, and completing assigned exercises, you should be able to:

1. Use the properties of equality to solve an equation in one variable.
2. Use an equation in one variable to solve an application problem.
3. Use a proportion in one variable to solve an application problem.

The Properties of Equality

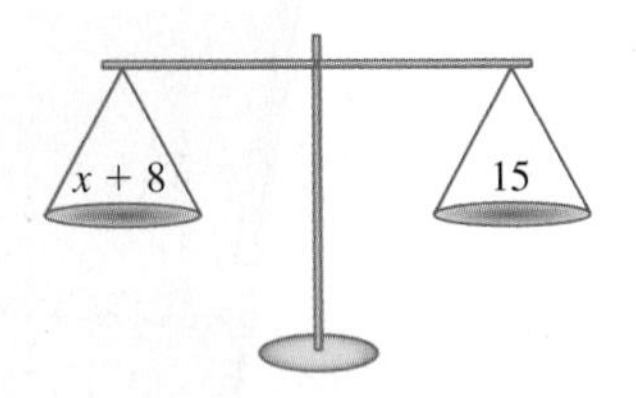

We can use the image of a balance scale to visualize the equation $x + 8 = 15$. The expression on the left side is equal to the expression on the right side. The equation is "balanced." If we add the same number to both sides or subtract the same number from both sides, the equation stays balanced. The new expression on the left side of the equation equals the new expression on the right side of the equation.

The properties of equality summarize these ideas.

> **The Properties of Equality**
>
> **The Addition and Subtraction Property of Equality** We can add a term to or subtract a term from both sides of an equation. The solution is not changed.
>
> **The Multiplication and Division Property of Equality** We can multiply both sides or divide both sides of an equation by a real number except for 0. The solution is not changed.

We can use the properties of equality to **isolate the variable** in an equation. If we isolate the variable, we know what number the variable represents. This number is the **solution** of the equation.

EXAMPLE 1 Solve: $x + 8 = 15$

SOLUTION ▸

$x + 8 = 15$

$x + 8 - 8 = 15 - 8$ Subtraction property of equality; subtract 8 from each side.

$\mathbf{x + 0 = 7}$ Simplify.

$\mathbf{x = 7}$ Simplify; the variable is isolated on the left side of the equation.

The solution of this equation is $x = 7$. If we replace x in the original equation $x + 8 = 15$ with the solution, 7, the equation is true: $7 + 8 = 15$ is true.

In the next example, we use the addition property of equality to isolate the variable.

EXAMPLE 2 Solve: $-12 = p - 20$

SOLUTION ▶

$$-12 = p - 20$$

$$-12 + 20 = p - 20 + 20$$ Addition property of equality; add 20 to each side.

$$\mathbf{8 = p + 0}$$ Simplify.

$$8 = \mathbf{p}$$ Simplify.

The isolated variable can be on either side of the equation. Since we are used to reading English from left to right, we often rewrite a solution such as $8 = p$ as the equivalent equation $p = 8$.

When using the addition or subtraction property of equality, we can also arrange the addition or subtraction of terms vertically.

EXAMPLE 3 Solve: $n - 6 = -13$

SOLUTION ▶

$$n - 6 = -13$$

$$\underline{+6 \qquad +6}$$ Addition property of equality.

$$\mathbf{n + 0 = -7}$$ Simplify.

$$\mathbf{n} = -7$$ Simplify.

To **check the solution** of an equation, replace the variable in the original equation with the solution. Simplify each side of the equation. If the solution is correct, the left side of the simplified equation will equal the right side.

EXAMPLE 4 **(a)** Solve: $-41 + y = 175$

SOLUTION ▶

$$-41 + y = 175$$

$$\underline{+41 \qquad\qquad +41}$$ Addition property of equality.

$$\mathbf{0} + y = \mathbf{216}$$ Simplify.

$$y = 216$$ Simplify.

(b) Check.

▶

$$-41 + y = 175$$ Use the original equation.

$$-41 + \mathbf{216} = 175$$ Replace the variable, y, with the solution, 216.

$$\mathbf{175} = 175$$ Evaluate; since the equation is true, the solution is correct.

If the variable term has a coefficient other than 1, the variable is not isolated. Use the division property of equality to change the coefficient to 1 and isolate the variable.

EXAMPLE 5 **(a)** Solve: $3k = -15$

SOLUTION ▶

$$3k = -15$$

$$\frac{3k}{3} = \frac{-15}{3}$$ Division property of equality.

$$\mathbf{1}k = \mathbf{-5}$$ Simplify.

$$k = -5$$ Simplify.

(b) Check.

▶

$$3k = -15$$ Use the original equation.

$$3(\mathbf{-5}) = -15$$ Replace the variable, k, with the solution, -5.

$$\mathbf{-15} = -15$$ Evaluate; since the equation is true, the solution is correct.

For some simple equations, it is possible to just look at the equation and figure out the solution. This is **solving by inspection**. However, showing work and using the properties of equality on simple equations is important preparation for solving more complicated equations in the future. Do not solve equations by inspection. Follow your instructor's guidelines for showing your work.

In the next example, we use the multiplication property of equality to change the coefficient of the variable term to 1 and isolate the variable.

EXAMPLE 6 **(a)** Solve: $\frac{h}{4} = 20$

SOLUTION

$$\frac{h}{4} = 20$$

$$\left(\frac{4}{1}\right)\frac{h}{4} = (4)20 \qquad \text{Multiplication property of equality.}$$

$$\mathbf{1}h = \mathbf{80} \qquad \text{Simplify; } \frac{4}{4} = 1$$

$$h = 80 \qquad \text{Simplify.}$$

(b) Check.

$$\frac{h}{4} = 20 \qquad \text{Use the original equation.}$$

$$\frac{\mathbf{80}}{4} = 20 \qquad \text{Replace the variable, } h\text{, with the solution, 80.}$$

$$\mathbf{20} = 20 \qquad \text{Evaluate; since the equation is true, the solution is correct.}$$

In the next example, the coefficient of the variable term is a fraction. To change the coefficient to 1, we multiply both sides of the equation by the reciprocal of the coefficient. In multiplying two fractions, the arithmetic is often easier if we simplify before multiplying the numerators and denominators. To review multiplication of fractions, go to Section R.3.

EXAMPLE 7 **(a)** Solve: $18 = -\frac{3}{4}x$

SOLUTION

$$18 = -\frac{3}{4}x$$

$$\left(-\frac{4}{3}\right)\left(\frac{18}{1}\right) = \left(-\frac{4}{3}\right)\left(-\frac{3}{4}x\right) \qquad \text{Multiplication property of equality.}$$

$$\left(-\frac{4}{\cancel{3}}\right)\left(\frac{\mathbf{6 \cdot \cancel{3}}}{1}\right) = \mathbf{1}x \qquad \text{Find common factors; simplify.}$$

$$-24 = x \qquad \text{Simplify.}$$

(b) Check.

$$18 = -\frac{3}{4}x \qquad \text{Use the original equation.}$$

$$18 = -\frac{3}{4}(\mathbf{-24}) \qquad \text{Replace the variable, } x\text{, with the solution, } -24.$$

$$18 = -\frac{3}{\cancel{4}}\left(\frac{\mathbf{-6 \cdot \cancel{4}}}{1}\right) \qquad \text{Find common factors.}$$

$$18 = \mathbf{18} \qquad \text{Evaluate; since the equation is true, the solution is correct.}$$

In the next example, we can use the division property of equality to solve $3k = -15$, dividing both sides by 3. Or we can use the multiplication property of equality, multiplying both sides by the reciprocal of 3, which is $\frac{1}{3}$. Multiplying both sides by $\frac{1}{3}$ is equivalent to dividing both sides by 3.

EXAMPLE 8 **(a)** Use the division property of equality to solve $3k = -15$.

SOLUTION

$3k = -15$

$\frac{3k}{3} = \frac{-15}{3}$ Division property of equality.

$1k = -5$ Simplify.

$k = -5$ Simplify.

(b) Use the multiplication property of equality to solve $3k = -15$.

$3k = -15$

$\left(\frac{1}{3}\right)3k = \left(\frac{1}{3}\right)(-15)$ Multiplication property of equality.

$1k = -5$ Simplify.

$k = -5$ Simplify.

(c) Check.

$3k = -15$ Use the original equation.

$3(-5) = -15$ Replace the variable, k, with the solution, -5.

$-15 = -15$ Evaluate; since the equation is true, the solution is correct.

To solve an equation, we use the properties of equality to isolate the variable. We can write directions for isolating the variable. As equations get more complicated, we will add more steps to these directions.

Solving an Equation in One Variable

1. Use the addition or subtraction property of equality to isolate the term that includes the variable. Any constant on the same side of the equation as the variable is changed to 0.
2. Use the multiplication or division property of equality to change the coefficient of the variable term to 1.
3. Check the solution in the original equation.

As we have seen before, a fraction with the same numerator and denominator is equal to 1 and multiplying a number by 1 does not change the number. Remember, a line drawn through common factors shows that the fraction is equal to 1.

EXAMPLE 9 (a) Solve: $\frac{x}{45} = \frac{2}{5}$

SOLUTION ▸

$\frac{x}{45} = \frac{2}{5}$

$\left(\frac{45}{1}\right)\frac{x}{45} = \left(\frac{45}{1}\right)\frac{2}{5}$ Multiplication property of equality.

$\frac{45}{1} \cdot \frac{x}{45} = \frac{\cancel{5} \cdot 9}{1} \cdot \frac{2}{\cancel{5}}$ Find common factors.

$1x = \mathbf{18}$ Simplify.

$\boldsymbol{x} = 18$ Simplify.

(b) Check.

▸ $\frac{x}{45} = \frac{2}{5}$ Use the original equation.

$\frac{\mathbf{18}}{45} = \frac{2}{5}$ Replace the variable, x, with the solution, 18.

$\frac{\mathbf{2 \cdot \cancel{9}}}{\mathbf{5 \cdot \cancel{9}}} = \frac{2}{5}$ Find common factors.

$\frac{\mathbf{2}}{\mathbf{5}} = \frac{2}{5}$ Evaluate; since the equation is true, the solution is correct.

If an equation includes a fraction and the solution is a fraction, we usually do not rewrite the solution as a decimal number or as a mixed number. Although the fraction $\frac{3}{4}$ exactly equals 0.75, $\frac{2}{7} = 0.\overline{285714}$, and writing the solution $x = \frac{2}{7}$ as $x = 0.29$ is not correct. Rounding changes the answer. (This does not apply to application problems; we often rewrite a fraction that is a solution of an application problem as a decimal number.)

EXAMPLE 10 (a) Solve: $50 = 8b$

SOLUTION ▸

$50 = 8b$

$\frac{50}{8} = \frac{8b}{8}$ Division property of equality.

$\frac{\mathbf{2 \cdot 25}}{\mathbf{\cancel{2} \cdot 4}} = \frac{8b}{8}$ Find common factors.

$\frac{\mathbf{25}}{\mathbf{4}} = \mathbf{1}b$ Simplify.

$\frac{25}{4} = \boldsymbol{b}$ Simplify; do not rewrite the solution as a mixed number.

(b) Check.

▸ $50 = 8b$ Use the original equation.

$50 = 8 \cdot \frac{\mathbf{25}}{\mathbf{4}}$ Replace the variable, x, with the solution, $\frac{25}{4}$.

$50 = \frac{\mathbf{\cancel{4} \cdot 2}}{1} \cdot \frac{25}{\cancel{4}}$ Find common factors.

$50 = \mathbf{50}$ Evaluate; since the equation is true, the solution is correct.

Practice Problems

For problems 1–12,
(a) solve.
(b) check.

1. $x + 6 = -21$ **2.** $18 = k - 21$ **3.** $14c = -168$

4. $\frac{d}{6} = 18$ **5.** $-20 = \frac{5}{6}y$ **6.** $-x = 9$

7. $\frac{1}{3} = \frac{p}{18}$ **8.** $\frac{h}{30} = \frac{4}{150}$ **9.** $\frac{x}{5} = \frac{8}{50}$

10. $\frac{2}{3}x = \frac{5}{6}$ **11.** $\frac{1}{2} + d = \frac{11}{15}$ **12.** $-4p = 34$

Equations with Decimal Numbers

If an equation includes decimal numbers and the solution is a decimal number, we usually do not rewrite the solution as a fraction or mixed number.

EXAMPLE 11 **(a)** Solve: $x - 7.28 = 15.69$

SOLUTION

$$x - 7.28 = 15.69$$

$$+7.28 \quad +7.28$$ Addition property of equality.

$$\mathbf{x + 0 = 22.97}$$ Simplify.

$$\boldsymbol{x} = 22.97$$ Simplify.

(b) Check.

$$x - 7.28 = 15.69$$ Use the original equation.

$$\mathbf{22.97} - 7.28 = 15.69$$ Replace the variable, x, with the solution, 22.97.

$$\mathbf{15.69} = 15.69$$ Evaluate; since the equation is true, the solution is correct.

If there are like terms on the same side of the equation, combine them before using a property of equality. In the next example, we rewrite p as $1p$ and then combine like terms.

EXAMPLE 12 **(a)** Solve: $p - 0.4p = 72$

SOLUTION

$$p - 0.4p = 72$$

$$\mathbf{1}p - 0.4p = 72$$ Rewrite p as $1p$.

$$\mathbf{0.6p} = 72$$ Combine like terms.

$$\frac{0.6p}{0.6} = \frac{72}{0.6}$$ Division property of equality.

$$\mathbf{1}p = \mathbf{120}$$ Simplify.

$$\boldsymbol{p} = 120$$ Simplify.

(b) Check.

$$p - 0.4p = 72$$ Use the original equation.

$$\mathbf{120} - 0.4(\mathbf{120}) = 72$$ Replace the variable, p, with the solution, 120.

$$120 - \mathbf{48} = 72$$ Follow the order of operations.

$$\mathbf{72} = 72$$ Evaluate; since the equation is true, the solution is correct.

Solving an Equation in One Variable

1. **Combine like terms.**
2. Use the addition or subtraction property of equality to isolate the term that includes the variable. Any constant on the same side of the equation as the variable is changed to 0.
3. Use the multiplication or division property of equality to change the coefficient of the variable term to 1.
4. Check the solution in the original equation.

Practice Problems

For problems 13–17,
(a) solve.
(b) check.

13. $c - 3.4 = 8.1$ **14.** $-9.5m = -2.85$ **15.** $\frac{h}{0.25} = 300$

16. $p - 0.25p = 300$ **17.** $p + 0.25p = 300$

When working with percent, we often use the relationship number of parts = (percent)(total parts). However, given the percent discount and the sale price, the discount is *not* equal to (percent discount)(sale price). Since the sale price is less than the original price, if we multiply the smaller sale price by the percent, the discount is too small. Instead, the discount equals (percent discount)(*original price*).

EXAMPLE 13 The sale price of a lawnmower is \$399. The percent discount is 20%. Find the original price.

SOLUTION ▶ **Step 1 Understand the problem.**
The unknown is the original price of the lawnmower.

p = original price

Step 2 Make a plan.
The discount is the amount that the original price is reduced. Since number of parts = (percent)(total parts), the discount equals (percent discount)(original price) or $0.20p$. A word equation is original price − discount = sale price.

Step 3 Carry out the plan.

$$p - 0.20p = \$399 \quad \text{Original price − discount = sale price}$$
$$\mathbf{1}p - 0.20p = \$399 \quad \text{Rewrite } p \text{ as } 1p.$$
$$\mathbf{0.80p} = \$399 \quad \text{Combine like terms; } 1p - 0.20p = 0.80p$$
$$\frac{0.80p}{0.80} = \frac{\$399}{0.80} \quad \text{Division property of equality.}$$
$$\mathbf{1}p = \mathbf{\$498.75} \quad \text{Simplify.}$$
$$p = \$498.75 \quad \text{Simplify.}$$

Step 4 Look back.
Working backwards, if the answer for the original price is correct, we should be able to use it and the percent discount to find the sale price. Since \$498.75 − (0.20)(\$498.75) equals the sale price of \$399, the answer seems reasonable.

In this problem, we used a relationship that may apply to future problems: *original + change = new.* In this problem, *original* is the original price, change is the discount, and new is the sale price.

Step 5 Report the solution.
The original price of the lawnmower was \$498.75.

In the next example, we again use the relationship *original* + *change* = *new*. In this problem, *original* is the number of minivans sold in 2009, *change* is the percent change in sales, and *new* is the number of minivans sold in 2010.

EXAMPLE 14 Find the number of minivans sold in 2009. Round to the nearest thousand.

Automakers were on pace to sell about 450,000 minivans last year [2010], a 9.3 percent increase from 2009 but far below the peak of 1.37 million in 2000. (*Source:* www.nytimes.com, Jan. 3, 2011)

SOLUTION **Step 1 Understand the problem.**
The unknown is the number of minivans sold in 2009. The number sold in 2000 is extraneous information.

N = number of minivans sold in 2009

Step 2 Make a plan.
Using the relationship *original* + *change* = *new*, the number sold in 2009 plus the increase equals the number sold in 2010. A word equation is number in 2009 + increase = number in 2010. Since number of parts = (percent)(total parts), the increase is (percent increase)(number in 2009), where the percent increase, 9.3%, is written as the decimal number 0.093.

Step 3 Carry out the plan.

$N + 0.093N = 450{,}000$ minivans	number in 2009 + increase = number in 2010
$\mathbf{1.093N} = 450{,}000$ minivans	Combine like terms; $1N + 0.093N = 1.093N$
$\dfrac{1.093N}{1.093} = \dfrac{450{,}000 \text{ minivans}}{1.093}$	Division property of equality.
$\mathbf{1N = 411{,}710.8\ldots}$ **minivans**	Simplify.
$N \approx \mathbf{412{,}000}$ minivans	Simplify; round.

Step 4 Look back.
Working backwards, if the answer for the number of minivans sold in 2009 is correct, we should be able to use it and the percent increase to find the number of minivans sold in 2010. Since 412,000 minivans + (0.093)(412,000 minivans) equals about 450,000 minivans, which is the number sold in 2010, the answer seems reasonable.

Step 5 Report the solution.
There were about 412,000 minivans sold in 2009.

In the next example, we again use the relationship number of parts = (percent)(total parts). Since we are given the percent and the number of parts, the unknown is the total parts.

EXAMPLE 15 Find the total number of households. Round to the nearest tenth of a million.

According to the 2011–2012 APPA National Pet Owners Survey, 62% of U.S. households own a pet, which equates to 72.9 million homes. (*Source:* www.americanpetproducts.org)

SOLUTION **Step 1 Understand the problem.**
The unknown is the total number of households.

N = total number of households

Step 2 Make a plan.
Use the relationship (percent)(total parts) = number of parts. The number of parts is 72.9 million homes; these are the households that own a pet. The unknown is the total number of households. A word equation is (percent of households with pets)(number of households) = households with pets.

Step 3 Carry out the plan.

$$0.62N = 72.9 \text{ million households} \quad \text{(percent)(total parts) = (number of parts)}$$

$$\frac{0.62N}{0.62} = \frac{72.9 \text{ million households}}{0.62} \quad \text{Division property of equality.}$$

$$\mathbf{1N = 117.58\ldots \text{ million households}} \quad \text{Simplify.}$$

$$N \approx \mathbf{117.6} \text{ million households} \quad \text{Simplify; round.}$$

Step 4 Look back.
Working backwards, and using the relationship percent $= \left(\dfrac{\text{number of parts}}{\text{total parts}}\right)$ $(100)\%$, $\left(\dfrac{72.9 \text{ million households}}{117.6 \text{ million households}}\right)(100)\%$ equals about 62%. Since this is the percent of households that own a pet, the answer seems reasonable.

Step 5 Report the solution.
The number of households used to find the percent was about 117.6 million households.

Practice Problems

For problems 18–20, use the five steps.

18. The sale price of a desktop computer is \$625.60. It is marked down 32%. Find the original price.

19. In 2010, Lincoln, Nebraska, police wrote 14,183 speeding tickets. This was a 6.1% increase over the number of tickets in 2009. Find the number of tickets in 2009. Round to the nearest whole number. (*Source:* www.lincoln.ne.gov/city/police)

20. Find the number of people killed on U.S. roads in 2009. Round to the nearest whole number.

The Transportation Department estimated Friday that 32,788 people were killed on U.S. roads in 2010, a decrease of about 3 percent from 2009. (*Source:* www.dallasnews.com, April 1, 2011)

A survey asked 80 students about their favorite morning beverage. Of these students, 61 chose coffee. We can write this as a rate: $\dfrac{61 \text{ students who prefer coffee}}{80 \text{ total students}}$.
A **proportion** is an equation with two equivalent rates. In the next example, the problem includes a *known rate* from a survey. To solve the problem, we write and solve a proportion, *unknown rate = known rate.* To explain why the solution to a proportion is reasonable, rewrite each rate as a decimal number. If the rates are equal or nearly equal, then the answer seems reasonable.

EXAMPLE 16 A short sale in real estate occurs when a buyer pays less for the property than the balance of the loan on the property. Assuming that each REALTOR® had one transaction, use a proportion to find the number of short sales in which the transaction closed.

According to a recent survey of 2,150 California REALTORS® who have assisted clients with a short sale, only three out of five transactions closed—even when there was an interested and qualified buyer. (*Source:* www.car.org, March 10, 2011)

SOLUTION ▶ **Step 1 Understand the problem.**
The unknown is the number of successful short sales.

N = number of successful short sales

Step 2 Make a plan.
In a proportion, unknown rate = known rate. The rate is $\dfrac{\text{number of successful short sales}}{\text{total number of short sales}}$.

Step 3 Carry out the plan.

$$\frac{N}{2150 \text{ total}} = \frac{3 \text{ successful}}{5 \text{ total}} \qquad \text{unknown rate = known rate}$$

$$\frac{N}{2150} = \frac{3}{5} \qquad \text{To solve, remove the units.}$$

$$\left(\frac{2150}{1}\right)\frac{N}{2150} = \left(\frac{2150}{1}\right)\frac{3}{5} \qquad \text{Multiplication property of equality.}$$

$$1N = \frac{\cancel{5} \cdot 430}{1} \cdot \frac{3}{\cancel{5}} \qquad \text{Simplify; find common factors.}$$

$$N = 1290 \text{ successful short sales} \qquad \text{Simplify; include the units.}$$

Step 4 Look back.
The unknown rate should equal the known rate. Since $\frac{1290}{2150} = 0.6$ and $\frac{3}{5} = 0.6$, the answer seems reasonable.

Step 5 Report the solution.
Out of 2150 short sales, 1290 were successful.

When solving an equation in which the coefficients and constants are integers or fractions and the solution is a fraction, we generally do not rewrite the solution as a decimal number. However, when we use an equation that includes fractions to solve an application problem, the most appropriate answer to the problem may be a decimal number. Whether we write the answer to an application as a fraction or a decimal number depends on the situation. In the next example, since the amount of an injection is usually written as a decimal number, the answer to the problem is a decimal number.

EXAMPLE 17 Lidocaine injections are used for local and regional anesthesia. A bottle of 250 mL of solution contains 2 g of lidocaine. If a patient needs an injection that contains 75 mg of lidocaine, find the amount of solution that the patient needs. Write the answer as a decimal number; round to the nearest hundredth.

SOLUTION ▶ **Step 1 Understand the problem.**
The unknown is the amount of solution that the patient needs.

A = amount of solution

Step 2 Make a plan.
Since the units of lidocaine in the bottle of solution are grams and the units of lidocaine that the patient needs are in milligrams, use the relationship 1 g = 1000 mg to change 2 g into milligrams: $\frac{2 \cancel{g}}{1} \cdot \frac{1000 \text{ mg}}{1 \cancel{g}}$ equals 2000 mg. The unknown rate is $\frac{A}{75 \text{ mg}}$. The known rate is $\frac{250 \text{ mL}}{2000 \text{ mg}}$. Use these rates to write a proportion.

Step 3 Carry out the plan.

$$\frac{A}{75\text{ mg}} = \frac{250\text{ mL}}{2000\text{ mg}}$$ unknown rate = known rate

$$\frac{A}{75} = \frac{250}{2000}$$ To solve, remove the units.

$$\left(\frac{75}{1}\right)\frac{A}{75} = \left(\frac{75}{1}\right)\frac{250}{2000}$$ Multiplication property of equality.

$$1A = 9.375$$ Simplify; write the answer as a decimal number.

$$A \approx 9.38\text{ mL}$$ Simplify; round; include the units.

Step 4 Look back.
If the answer is correct, the unknown rate equals the known rate. Since $\frac{9.38}{75} \approx 0.125$ and $\frac{250}{2000} = 0.125$, the answer seems reasonable.

Step 5 Report the solution.
The volume of solution is about 9.38 mL.

Practice Problems

For problems 21–24, use a proportion and the five steps.

21. Four out of five businesses with 100 or more employees will use social media marketing in 2011. Use a proportion to predict how many of 1250 businesses with 100 or more employees will use social media marketing. (*Source:* www.emarketer.com, Dec. 2010)

22. Predict the number of births in 2009 for 25,500 female teenagers. Round to the nearest whole number.

In 2009, the national teen birth rate was 39.1 births per 1,000 females, a 37% decrease from 61.8 births per 1,000 females in 1991 and the lowest rate ever recorded. (*Source:* www.cdc.gov, Apr. 8, 2011)

23. A solution contains 100 mg of antibiotic in 5 mL of solution. Find the amount of solution to give a child who needs 165 mg of antibiotic.

24. The owner of a Toyota Matrix drove 330 mi on 13.2 gal of gas. Find the number of miles that she can expect to drive on 5 gal of gas.

2.1 Vocabulary Practice

Match the term with its description.

1. A symbol that represents an unknown number
2. The product of a nonzero number and this is 1.
3. We can add a term to or subtract a term from both sides of an equality. The solution is not changed.
4. Solving an equation by just looking at it.
5. A number that is multiplied by a variable in a term
6. A number by itself in an expression or equation
7. We can multiply both sides or divide both sides of an equation by a real number except 0. The solution is not changed.
8. An example of this is $\frac{3\text{ mi}}{0.6\text{ gal}} = \frac{x}{40\text{ gal}}$.
9. Original price − discount
10. This includes an equals sign.

A. addition or subtraction property of equality
B. coefficient
C. constant
D. equation
E. multiplication or division property of equality
F. proportion
G. reciprocal
H. sale price
I. solving by inspection
J. variable

2.1 Exercises

Follow your instructor's guidelines for showing your work.

For exercises 1–10, solve.

1. $x + 8 = 24$
2. $x + 6 = 30$
3. $24 = x - 8$
4. $30 = x - 6$
5. $\frac{x}{8} = 24$
6. $\frac{x}{6} = 30$
7. $\frac{1}{8}x = 24$
8. $\frac{1}{6}x = 30$
9. $24 = 8x$
10. $30 = 6x$

For exercises 11–46,
(a) solve.
(b) check.

11. $-5p = 75$
12. $-8z = 120$
13. $c - \frac{3}{4} = \frac{17}{4}$
14. $d - \frac{1}{8} = \frac{23}{8}$
15. $\frac{3}{4}h = 15$
16. $\frac{4}{5}k = 60$
17. $0.8 = w + 1.9$
18. $0.29 = y + 0.81$
19. $\frac{a}{0.3} = 15$
20. $\frac{m}{0.7} = 12$
21. $\frac{2}{3}x = \frac{8}{9}$
22. $\frac{3}{4}x = \frac{9}{16}$
23. $\frac{1}{4} = m + \frac{5}{9}$
24. $\frac{1}{5} = d + \frac{4}{9}$
25. $0 = 6z$
26. $0 = 15b$
27. $-x = 40$
28. $-y = 50$
29. $p - 0.8p = 210$
30. $p - 0.4p = 259.2$
31. $x + 1.6 = 0$
32. $y + 9.7 = 0$
33. $\frac{5}{6} = k - \frac{3}{8}$
34. $\frac{14}{15} = w - \frac{3}{10}$
35. $\frac{a}{280} = \frac{3}{70}$
36. $\frac{b}{340} = \frac{3}{85}$
37. $\frac{5}{9} = \frac{v}{72}$
38. $\frac{4}{7} = \frac{u}{84}$
39. $\frac{75}{83} = \frac{n}{747}$
40. $\frac{92}{165} = \frac{m}{495}$
41. $0.31p = -0.155$
42. $0.27w = -0.108$
43. $80 = -p - 0.6p$
44. $120 = -p - 0.5p$
45. $\frac{x}{5} = \frac{3.5}{10}$
46. $\frac{x}{8} = \frac{2.5}{16}$
47. Explain why we cannot add a term to just one side of an equation.
48. The multiplication and division properties of equality do not allow both sides of an equation to be multiplied or divided by 0. Explain why you think zero cannot be used.

For exercises 49–82, use the five steps.

49. The sale price of a vacuum cleaner with a 7.1-amp motor is \$120. The percent discount is 25%. Find the original price.
50. The sale price of a 15-amp table saw and stand is \$438.98. The percent discount is 27%. Find the original price. Round to the nearest hundredth.
51. Find the median sales price in August. Round to the nearest whole number.

    The Median Sales Price was \$183,762 for the month of September. This is just 2.5% lower than the median price for August. (*Source:* rismedia.com, Oct. 17, 2011)
52. Find the amount of government payments in 2010. Round to the nearest tenth of a billion.

    Government payments [to farms] are forecast to be \$10.2 billion in 2011, a 17.7-percent decrease from 2010. (*Source:* www.ers.usda.gov, Aug. 30, 2011)
53. Find the number of people employed by Apple in 2010. Round to the nearest hundred.

    Apple now employs 63,300 people worldwide, up 28 percent from last year [2010]. (*Source:* gigaom.com, Oct 27, 2011)
54. Find the number of visitors to Las Vegas in August 2010. Round to the nearest tenth of a million.

    Las Vegas hosted nearly 3.3 million visitors in August, an increase of +2.8% over last August [2010]. (*Source:* www.lvcva.com, Aug. 2011)
55. *New Super Marios Bros* for the Nintendo Wii is marked down 12% to \$43.99. Find the original price. Round to the nearest hundredth.
56. *Just Dance* for the Nintendo Wii is marked down 35% to \$25.99. Find the original price. Round to the nearest hundredth.
57. A research firm interviewed 290 adults with a spinal cord injury or multiple sclerosis or who were recovering from polio. Four out of ten had experienced severe or excruciating pain in the previous 7 days. Find the number of the people interviewed who had experienced this kind of pain. (*Source:* www.unitedspinal.org)
58. In 2010, about seven of ten households in the United States experienced food insecurity, meaning that there was not always enough food for an active, healthy life for all household members. According to the 2010 Census, there were 116,716,292 households in the United States in 2010. Find the number of these households that experienced food insecurity. Round to the nearest thousand. (*Source:* www.ers.usda.gov, Sept. 2011; www.census.gov)
59. For 2011, the annual residential property tax rate in East Providence, Rhode Island, was \$19.41 per \$1000 of assessed valuation. Find the property tax on a home with a valuation of \$169,900. Round to the nearest whole number. (*Source:* www.muni-info.ri.gov)
60. For 2011, the annual residential property tax rate in Providence, Rhode Island, was \$30.38 per \$1000 of assessed valuation. Find the property tax on a home with a valuation of \$169,900. Round to the nearest whole number. (*Source:* www.muni-info.ri.gov)

61. A cancer incidence rate reports the number of new cases of cancer in a population per year. If the incidence rate of cancer in Kentucky is 516.2 cases of cancer per 100,000 people and the population is about 4,346,000 people, find the number of new cases of cancer in this population. Round to the nearest ten. (*Source:* 2010 Census; Siegel et al., *CA: A Cancer Journal for Clinicians*, July/Aug. 2011)

62. A cancer incidence rate reports the number of new cases of cancer in a population per year. If the incidence rate of cancer in Colorado is 439.4 cases of cancer per 100,000 people and the population is about 5,029,000 people, find the number of new cases of cancer in this population. Round to the nearest ten. (*Source:* 2010 Census; Siegel et al., *CA: A Cancer Journal for Clinicians*, July/Aug. 2011)

63. The Pew Research Center interviewed 688 American adults who own a smartphone. Find the number who use their smartphone for text messaging or taking pictures. Round to the nearest whole number.

One third of American adults (35%) own a smartphone of some kind, and these users take advantage of a wide range of their phones' capabilities. Fully nine in ten smartphone owners use text messaging or take pictures with their phones, while eight in ten use their phone to go online or send photos or videos to others. (*Source:* www.pewinternet.org, Aug. 15, 2011)

64. Find the number of 3850 adults with a checking, savings, or retirement account who had to withdraw money from a retirement account to pay their bills. Round to the nearest whole number.

Four-in-ten adults . . . who have a checking, savings or retirement account say that during the recession they have had to withdraw money from their savings account, 401(k) account or some other retirement account to pay their bills. Younger and middle-aged adults report having done this at higher rates than those ages 65 and older. (*Source:* www.pewsocialtrends.org, June 30, 2010)

65. Find the number of the Texan children who fell into poverty. Write the answer in place value notation.

Over the past decade, 2.4 million more U.S. kids fell into poverty, and one of every six of those kids are Texans. (*Source:* datacenter.kidscount.org, Aug. 17, 2011)

66. Find the number of jobs lost during the downturn that the state had gained back. Write the answer in place value notation. Round to the nearest thousand.

By July 2011, the most recent month for which data are available, the state [California] had gained back . . . just one out of six . . . of the nearly 1.4 million jobs the state lost during the downturn. (*Source:* www.cbp.org, Sept. 3, 2011)

67. A survey asked 1728 drivers who had driven in the last 30 days whether they had ever fallen asleep or nodded off while driving, even for just a second or two. Find the number who said that they had fallen asleep or nodded off. Round to the nearest whole number.

Two out of every five drivers . . . reported having ever fallen asleep or nodded off while driving, including 3.9% within the past month, 7.1% within the past 6 months, and 11.0% within the past 12 months. (*Source:* www.aaafoundation.org, Nov. 2010)

68. According to the 2010 Census, the population of Pennsylvania in 2010 was about 12,702,000 people. Find the number of people who lived in municipalities in fiscal distress.

Today, two out of every five Pennsylvanians live in a municipality in fiscal distress. (*Source:* issuespa.org, Feb. 22, 2011)

69. In a telephone poll of 1005 adults, one out of three said that they were not planning to get a flu shot. Find the number of these adults who said that they were not planning to get a flu shot. (*Source:* www.selectmedical.com, Oct. 26, 2011)

70. One out of five people living in Wisconsin is served by a Medicaid program. If the population in Wisconsin is about 5,687,000 people, find the number served by a Medicaid program. (*Source:* www.dhs.wisconsin.gov, Nov. 1, 2011)

71. A 20-10-10 fertilizer is 20% nitrogen. The recommended application rate is 5 lb of fertilizer for 1000 ft^2 of grass. Find the amount of fertilizer recommended for 725 ft^2 of grass. Round to the nearest tenth.

72. A 15-10-10 fertilizer is 15% nitrogen. The recommended application rate is 7.5 lb of fertilizer for 1000 ft^2 of grass. Find the amount of fertilizer recommended for 810 ft^2 of grass. Round to the nearest tenth.

73. A vial contains 50 mg Benadryl in 1 mL of solution. A child who weighs 14 kg needs 17.5 mg of Benadryl. Find the amount of solution that the child needs.

74. A vial contains 100 mg Synagis dissolved in 1 mL of solution. A child who weighs 3.4 kg should receive 15 mg of Synagis per kilogram. Find the amount of solution that the child needs.

75. The volume of the gas tank in a Ford Ranger V6 Super Cab is 19.5 gal. If the Ranger uses 19 gal to drive 304 mi, find the distance the owner can expect to drive on 8 gal of gas.

76. The volume of the gas tank in a Jeep Wrangler Rubicon V6 is 18.6 gal. If the Jeep uses 18 gal to drive 306 mi, find the distance the owner can expect to drive on 6 gal of gas.

For exercises 77–78, use the graph.

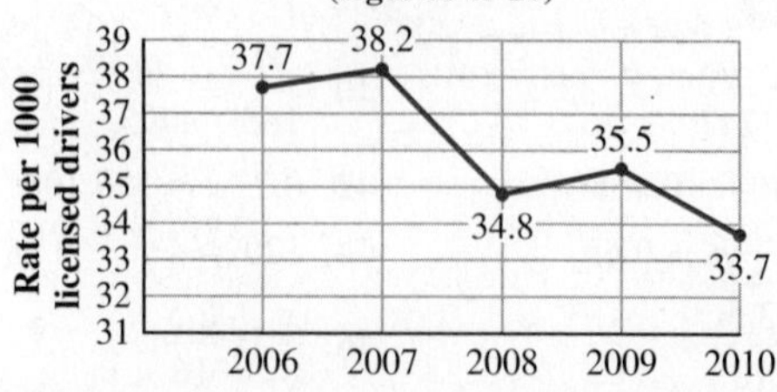

Source: www.dor.state.ne.us, 2010

77. In 2010, there were 112,436 licensed drivers in Nebraska, ages 16 to 20. Find the number of crashes with these drivers that involved fatalities or injuries. Round to the nearest whole number.

78. In 2009, there were 114,515 licensed drivers in Nebraska, ages 16 to 20. Find the number of crashes with these drivers that involved fatalities or injuries. Round to the nearest whole number.

79. In knitting, the *gauge* is the number of stitches a knitter completes in 1 in. A knitter's gauge is 4.5 stitches per inch. The total distance around a sweater is 40 in. Find the number of stitches needed to knit this width.

© Robyn Mackenzie/ Shutterstock.com

80. A baby blanket crochet project calls for superfine yarn and crochet hook size E. If a person averages 30 crochet stitches over 4 in. and the baby blanket is 26 in. wide, find the number of stitches needed to crochet this width.

81. There are 300 cal in a 60 g Hershey's milk chocolate bar. Find the number of calories in a piece of this chocolate that weighs 12 g.

82. There are 180 cal in a 41 g Hershey's Special Dark Chocolate bar. Find the number of calories in a piece of this chocolate that weighs 12 g. Round to the nearest whole number.

83. In 2008, one of every 31 adult Americans was in prison or jail or was on probation or parole. Predict how many of 65,000 adult Americans were in prison or jail or were on probation or parole. Round to the nearest hundred. (*Source:* www.nytimes.com, March 2, 2009)

84. If one in eleven telecommunications jobs is located in Texas, find the number of 844,200 telecommunications jobs that are located in Texas. Round to the nearest hundred. (*Source:* www.tstc.edu)

Problem Solving: Practice and Review

Follow your instructor's guidelines for using the five steps as outlined in Section 1.3, p. 55.

85. For the 2009–2010 school year, about \$29.99 billion in Pell Grants were awarded to undergraduate college students. This increased to \$34.7 billion in the 2010–2011 school year. Find the percent increase. Round to the nearest percent. (*Source:* trends.collegeboard.org, 2011)

86. Find the number of student borrowers who started repaying their loans in fiscal year 2009 and defaulted within the first 2 years. Write the answer in place value notation.

According to the U.S. Department of Education, 8.8 percent of the 3.6 million student borrowers who started repaying their loans in fiscal year 2009 defaulted within the first two years, up from 7 percent the year before. (*Source:* www.sfexaminer.com, Nov. 6, 2011)

87. Fidelity Investments partnered with Mr. Holland's Opus Foundation to donate instruments to an elementary school in Maryland. Find the percent of the instruments that were violins. Round to the nearest percent.

The students received 15 violins, five flutes, four clarinets, four trumpets, three bell kits and a concert bass drum. (*Source:* www.baltimoresun.com, Nov. 3, 2011)

88. A report by the National Foundation for Credit Counseling said that one-third of American adults, or 75.6 million people, have no savings other than retirement savings. Find the population of American adults used by this report. (*Source:* www.nfcc.org, March 2011)

Technology

For exercises 89–92, solve. Use a calculator to do the arithmetic.

89. $-24{,}598 + p = 89{,}457$

90. $\dfrac{c}{342} = -1840$

91. $291.35d = 14{,}864.677$

92. $-\dfrac{3}{8} = \dfrac{b}{160}$

Find the Mistake

For exercises 93–96, the completed problem has one mistake.

(a) Describe the mistake in words, or copy down the whole problem and highlight or circle the mistake.

(b) Do the problem correctly.

93. Problem: Solve: $\dfrac{x}{9} = 72$

Incorrect Answer:
$$\frac{x}{9} = 72$$
$$9 \cdot \frac{x}{9} = \frac{72}{9}$$
$$1x = \frac{72}{9}$$
$$x = 8$$

94. Problem: Solve: $x - 8 = 15$

Incorrect Answer:
$$x - 8 = 15$$
$$\underline{\quad -8 \quad -8}$$
$$x + 0 = 7$$
$$x = 7$$

95. Problem: Solve: $3x = 22$

Incorrect Answer:
$$3x = 22$$
$$\frac{3x}{3} = \frac{22}{3}$$
$$1x = 7.3$$
$$x = 7.3$$

96. Problem: Solve: $-4x - x = 21$

Incorrect Answer:
$$-4x - x = 21$$
$$-3x = 21$$
$$\frac{-3x}{-3} = \frac{21}{-3}$$
$$1x = -7$$
$$x = -7$$

Review

For exercises 97–100, simplify.

97. $7(x - 3) - (x + 4)$

98. $\frac{3}{4}(2x + 12)$

99. $\frac{2}{3}(x + 1) + \frac{4}{5}(x - 6)$

100. $9x - (3x - 8)$

SUCCESS IN COLLEGE MATHEMATICS

101. Copy the schedule below and mark when you will be *in a class, studying, or other* (including work, commuting, and personal activities).

102. How many hours do you plan to spend attending all of your classes and studying per week?

Chart for exercise 101

	Su	M	T	W	Th	F	Sa		Su	M	T	W	Th	F	Sa
12 mid								**12 noon**							
1 a.m.								**1 p.m.**							
2 a.m.								**2 p.m.**							
3 a.m.								**3 p.m.**							
4 a.m.								**4 p.m.**							
5 a.m.								**5 p.m.**							
6 a.m.								**6 p.m.**							
7 a.m.								**7 p.m.**							
8 a.m.								**8 p.m.**							
9 a.m.								**9 p.m.**							
10 a.m.								**10 p.m.**							
11 a.m.								**11 p.m.**							

Independent truckers own their own trucks. Analyzing their costs helps them make decisions that can improve their profit. In this section, we write and solve an equation that shows the relationship of cost per mile, the number of miles, fixed costs, and total costs. To solve this kind of equation, we use the distributive property and more than one property of equality.

SECTION 2.2

More Equations in One Variable

After reading the text, working the practice problems, and completing assigned exercises, you should be able to:

1. Use more than one property of equality to solve an equation in one variable.
2. Simplify an equation before solving.
3. Use an equation in one variable to solve an application problem.

Using More Than One Property of Equality

To solve an equation in one variable, we may need to use more than one property of equality to isolate the variable. For example, to solve $2x + 10 = 20$, we first subtract 10 from both sides. This isolates the term with the variable, $2x$. To finish isolating the variable, we then divide both sides by 2.

EXAMPLE 1 **(a)** Solve: $2x + 10 = 20$

SOLUTION

$$\begin{array}{rl} 2x + 10 = 20 & \\ -10 \quad -10 & \text{Subtraction property of equality.} \\ \hline \mathbf{2x + 0 = 10} & \text{Simplify.} \\ \dfrac{\mathbf{2x}}{2} = \dfrac{10}{2} & \text{Division property of equality.} \\ \mathbf{1}x = \mathbf{5} & \text{Simplify.} \\ \boldsymbol{x} = 5 & \text{Simplify.} \end{array}$$

(b) Check.

$$\begin{array}{rl} 2x + 10 = 20 & \text{Use the original equation.} \\ 2(\mathbf{5}) + 10 = 20 & \text{Replace the variable, } x\text{, with the solution, 5.} \\ \mathbf{10} + 10 = 20 & \text{Follow the order of operations.} \\ \mathbf{20} = 20 & \text{Evaluate; since the equation is true, the solution is correct.} \end{array}$$

To solve $2x + 10 = 20$, we isolated the term with the variable by *subtracting* 10 from both sides of the equation. In the next example, to solve $4 = \frac{p}{6} - 9$, we isolate the term with the variable by *adding* 9 to both sides.

EXAMPLE 2 Solve: $4 = \dfrac{p}{6} - 9$

SOLUTION

$$\begin{array}{rl} 4 = \dfrac{p}{6} - 9 & \\ +9 \qquad +9 & \text{Addition property of equality.} \\ \hline \mathbf{13} = \dfrac{p}{6} \mathbf{+ 0} & \text{Simplify.} \\ 13 = \dfrac{\boldsymbol{p}}{\mathbf{6}} & \text{Simplify.} \\ (6)(13) = \left(\dfrac{6}{1}\right)\left(\dfrac{p}{6}\right) & \text{Multiplication property of equality.} \\ \mathbf{78} = \mathbf{1}p & \text{Simplify.} \\ 78 = \boldsymbol{p} & \text{Simplify.} \end{array}$$

In the next example, we rewrite $-h$ as $-1h$ and then divide both sides of the equation by -1.

EXAMPLE 3 **(a)** Solve: $8 - h = 17$

SOLUTION

$$\begin{array}{rl} 8 - h = 17 & \\ -8 \qquad -8 & \text{Subtraction property of equality.} \\ \hline \mathbf{0} - h = \mathbf{9} & \text{Simplify.} \\ \boldsymbol{-h} = 9 & \text{Simplify.} \\ \mathbf{-1}h = 9 & \text{Rewrite } -h \text{ as } -1h. \\ \dfrac{-1h}{-1} = \dfrac{9}{-1} & \text{Division property of equality.} \\ \mathbf{1}h = \mathbf{-9} & \text{Simplify.} \\ \boldsymbol{h} = -9 & \text{Simplify.} \end{array}$$

(b) Check.

$8 - h = 17$	Use the original equation.
$8 - (\mathbf{-9}) = 17$	Replace the variable, h, with the solution, -9.
$8 + 9 = 17$	Simplify.
$\mathbf{17} = 17$	Evaluate; since the equation is true, the solution is correct.

Practice Problems

For problems 1–6,
(a) solve.
(b) check.

1. $6c + 3 = 21$ **2.** $28 = -4k - 12$ **3.** $\frac{x}{5} + 12 = -21$

4. $\frac{3}{4}x + 4 = 13$ **5.** $6 - w = 11$ **6.** $1.5x + 3.8 = -11.2$

Simplifying Equations

Before using the properties of equality to isolate the variable in an equation, combine like terms. If a solution is a fraction, do not rewrite it as a decimal number.

EXAMPLE 4 **(a)** Solve: $\frac{7}{8} = p + \frac{5}{9}p$

SOLUTION ▶

$\frac{7}{8} = p + \frac{5}{9}p$	
$\frac{7}{8} = \mathbf{1}p + \frac{5}{9}p$	Rewrite p as $1p$.
$\frac{7}{8} = (1p)\left(\frac{9}{9}\right) + \frac{5}{9}p$	Multiply by a fraction equal to 1.
$\frac{7}{8} = \frac{\mathbf{9}}{\mathbf{9}}p + \frac{5}{9}p$	Simplify.
$\frac{7}{8} = \frac{\mathbf{14}}{\mathbf{9}}p$	Combine like terms.
$\left(\frac{9}{14}\right)\left(\frac{7}{8}\right) = \left(\frac{9}{14}\right)\left(\frac{14}{9}p\right)$	Multiplication property of equality.
$\left(\frac{9}{\mathbf{2 \cdot 7}}\right)\left(\frac{7}{8}\right) = \left(\frac{9}{14}\right)\left(\frac{14}{9}p\right)$	Find common factors.
$\frac{\mathbf{9}}{\mathbf{16}} = \mathbf{1}p$	Simplify.
$\frac{9}{16} = \mathbf{p}$	Simplify.

(b) Check.

$\frac{7}{8} = p + \frac{5}{9}p$	Use the original equation.
$\frac{7}{8} = \frac{\mathbf{9}}{\mathbf{16}} + \frac{5}{9}\left(\frac{\mathbf{9}}{\mathbf{16}}\right)$	Replace the variable, p, with the solution, $\frac{9}{16}$.
$\frac{7}{8} = \frac{9}{16} + \frac{5}{9}\left(\frac{9}{16}\right)$	Find common factors.
$\frac{7}{8} = \frac{9}{16} + \frac{\mathbf{5}}{\mathbf{16}}$	Simplify.

$$\frac{7}{8} = \frac{\mathbf{14}}{\mathbf{16}}$$ Simplify.

$$\frac{7}{8} = \frac{\mathbf{2 \cdot 7}}{\mathbf{2 \cdot 8}}$$ Find common factors.

$$\frac{7}{8} = \frac{\mathbf{7}}{\mathbf{8}}$$ Evaluate; since the equation is true, the solution is correct.

In solving an equation, our goal is to find the number that the variable represents. We do this by isolating the variable. Before using the properties of equality, we may need to use the distributive property or combine like terms. In the next example, to solve the equation $-3(4a + 5) = 57$, we use the distributive property to rewrite $-3(4a + 5)$ as $-3(4a) - 3(5)$.

EXAMPLE 5 Solve: $-3(4a + 5) = 57$

SOLUTION

$-3(4a + 5) = 57$	
$\mathbf{-3}(4a) \mathbf{- 3}(5) = 57$	Distributive property.
$\mathbf{-12a - 15} = 57$	Simplify.
$+15 \quad +15$	Addition property of equality.
$-12a \mathbf{+ 0} = \mathbf{72}$	Simplify.
$\frac{\mathbf{-12a}}{-12} = \frac{72}{-12}$	Division property of equality.
$\mathbf{1}a = \mathbf{-6}$	Simplify.
$\mathbf{a} = -6$	Simplify.

Solving an Equation in One Variable

1. **Use the distributive property to rewrite the equation without parentheses.** Simplify; combine like terms.
2. Use the addition or subtraction property of equality to isolate the term that includes the variable. Any constant on the same side of the equation as the variable is changed to 0.
3. Use the multiplication or division property of equality to change the coefficient of the variable term to 1.
4. Check the solution in the original equation.

Multiplying an expression by 1 does not change the value of the expression. So in Example 6, we can rewrite $-(x - 8)$ as $-1(x - 8)$.

EXAMPLE 6 **(a)** Solve: $4x + 9 - (x - 8) = -10$

SOLUTION

$4x + 9 - (x - 8) = -10$	
$4x + 9 - \mathbf{1}(x - 8) = -10$	Multiply by 1.
$4x + 9 \mathbf{- 1}x \mathbf{- 1}(-8) = -10$	Distributive property.
$4x + 9 - 1x \mathbf{+ 8} = -10$	Simplify.
$\mathbf{3x + 17} = -10$	Combine like terms.
$-17 \quad -17$	Subtraction property of equality.
$3x \mathbf{+ 0} = \mathbf{-27}$	Simplify.
$\frac{\mathbf{3x}}{3} = \frac{-27}{3}$	Division property of equality.
$\mathbf{1}x = \mathbf{-9}$	Simplify.
$\mathbf{x} = -9$	Simplify.

(b) Check.

$$4x + 9 - (x - 8) = -10$$ Use the original equation.

$$4(-9) + 9 - (-9 - 8) = -10$$ Replace the variable, x, with the solution, -9.

$$4(-9) + 9 - (-17) = -10$$ Follow the order of operations.

$$-36 + 9 - (-17) = -10$$ Follow the order of operations.

$$-27 + 17 = -10$$ Follow the order of operations.

$$-10 = -10$$ Evaluate; since the equation is true, the solution is correct.

We can add or subtract only those fractions with the same (common) denominator. To change a fraction into an equivalent fraction with a different denominator, multiply by a fraction that is equal to 1.

EXAMPLE 7 Solve: $\frac{1}{4}(8z + 9) = \frac{7}{2}$

SOLUTION

$$\frac{1}{4}(8z + 9) = \frac{7}{2}$$

$$\frac{1}{4}\left(\frac{8z}{1}\right) + \frac{1}{4}\left(\frac{9}{1}\right) = \frac{7}{2}$$ Distributive property.

$$\frac{1}{4}\left(\frac{4 \cdot 2 \cdot z}{1}\right) + \frac{1}{4}\left(\frac{9}{1}\right) = \frac{7}{2}$$ Find common factors.

$$2z + \frac{9}{4} = \frac{7}{2}$$ Simplify.

$$-\frac{9}{4} \quad -\frac{9}{4}$$ Subtraction property of equality.

$$2z + 0 = \frac{7}{2} - \frac{9}{4}$$ Simplify.

$$2z = \frac{7}{2} \cdot \frac{2}{2} - \frac{9}{4}$$ Simplify; multiply by a fraction equal to 1.

$$2z = \frac{14}{4} - \frac{9}{4}$$ Multiply numerators; multiply denominators.

$$2z = \frac{5}{4}$$ Simplify.

$$\left(\frac{1}{2}\right)\left(\frac{2z}{1}\right) = \left(\frac{1}{2}\right)\left(\frac{5}{4}\right)$$ Multiplication property of equality.

$$1z = \frac{5}{8}$$ Simplify.

$$z = \frac{5}{8}$$ Simplify.

When an equation includes decimal numbers and the solution is a decimal number, do not rewrite the solution as a fraction or mixed number.

EXAMPLE 8 Solve: $5.6(3x - 4) - 13.5 = -12.38$

SOLUTION

$$5.6(3x - 4) - 13.5 = -12.38$$

$$5.6(3x) + 5.6(-4) - 13.5 = -12.38$$ Distributive property.

$$16.8x - 22.4 - 13.5 = -12.38$$ Simplify.

$16.8x \mathbf{- 35.9} = -12.38$	Combine like terms.
$+35.9 \quad +35.9$	Addition property of equality.
$16.8x + \mathbf{0} = \mathbf{23.52}$	Simplify.
$\dfrac{\mathbf{16.8x}}{16.8} = \dfrac{23.52}{16.8}$	Division property of equality.
$\mathbf{1}x = \mathbf{1.4}$	Simplify.
$\mathbf{x} = 1.4$	Simplify.

Practice Problems

For problems 7–11,
(a) solve.
(b) check.

7. $6(x + 9) = 18$ **8.** $-36 = 3(w + 1) - (w - 7)$ **9.** $\frac{3}{5}(10x + 4) = \frac{21}{5}$

10. $3.5(6.2x - 1) - 10.8 = 120.24$ **11.** $\frac{1}{5}(6z - 3) + \frac{1}{2}(z + 1) = \frac{3}{20}$

Applications

An independent trucker owns her semi-trailer truck. Her *fixed costs* do not depend on the number of miles she drives. These costs include the loan payment, insurance, licenses, permits, and accounting services. Her *nonfixed* costs do depend on the number of miles she drives. These costs include fuel, maintenance, repairs, lodging, meals, loading, unloading, scale fees, and taxes. The sum of the fixed costs and the nonfixed costs is the total cost.

EXAMPLE 9 The fixed annual cost for a trucker is \$27,600. The average nonfixed costs are \$0.45 per mile. If the total annual cost of operating the truck is limited to \$80,000, find the number of miles the trucker can drive. Round to the nearest hundred.

SOLUTION ▶ **Step 1 Understand the problem.**

The unknown is the number of miles that the trucker can drive.

N = number of miles that the trucker can drive

Step 2 Make a plan.

The fixed annual cost is \$27,600. The nonfixed cost equals $\left(\frac{\text{cost}}{\text{mile}}\right)$(number of miles). A word equation is

$$\text{fixed cost} + \left(\frac{\text{cost}}{\text{mile}}\right)(\text{number of miles}) = \text{total cost}$$

Step 3 Carry out the plan.

$\$27{,}600 + \left(\dfrac{\$0.45}{1 \text{ mi}}\right)N = \$80{,}000$	fixed costs + nonfixed costs = total cost
$27{,}600 + 0.45N = 80{,}000$	To solve, remove the units.
$-27{,}600 \quad -27{,}600$	Subtraction property of equality.
$\mathbf{0} + 0.45N = \mathbf{52{,}400}$	Simplify.
$\dfrac{\mathbf{0.45N}}{0.45} = \dfrac{52{,}400}{0.45}$	Division property of equality.
$\mathbf{1}N = \mathbf{116{,}444.44\ldots}$	Simplify.
$N \approx 116{,}400 \text{ mi}$	Simplify; round; include the units.

Step 4 Look back.

The sum of the fixed costs and the nonfixed costs should equal \$80,000. Since $\$27{,}600 + \left(\frac{\$0.45}{1\text{ mi}}\right)(116{,}400\text{ mi})$ equals \$79,980, which rounds to \$80,000, the answer seems reasonable. In this problem we used a relationship that may apply to future problems: *fixed amount* + *nonfixed amount* = *total amount.*

Step 5 Report the solution.

The trucker can drive about 116,400 mi.

We can use the general relationship in the last example, *fixed amount* + *nonfixed amount* = *total amount*, to solve a problem that includes a base salary and commission. For a salesperson who receives a base salary plus a commission which is a percent of sales, the *fixed amount* is the base salary, the *nonfixed amount* is the commission, and the *total amount* is the total salary. A word equation is base salary + (percent commission)(amount of sales) = total salary.

Practice Problems

For problems 12–13, use the five steps.

12. The fixed costs of owning a car include car payments, insurance, oil changes, car washes, repairs, and other routine maintenance. A car owner estimates that her fixed costs per year are \$5720. Her car averages 32 mi per gallon of gas. She wants to limit her total annual spending on her car to \$8000. If the cost of gas is \$4 per gallon, find the number of miles she can drive per year.

13. A door-to-door sales representative has an annual base salary of \$24,000. He makes \$20 commission on each product he sells. Find the number of products he must sell each year to have a total salary of \$60,000.

2.2 Vocabulary Practice

Match the term with its description.

1. In the term $7p$, 7 is an example of this.
2. We can multiply both sides or divide both sides of an equation by a real number that is not equal to 0. The solution is not changed.
3. We can add a term to or subtract a term from both sides of an equation. The solution is not changed.
4. A symbol that represents an unknown number
5. The product of a real number (except 0) and this is 1.
6. A number by itself in an expression or equation
7. $a + (b + c) = (a + b) + c$
8. $8x$ and $12x$ are examples of these.
9. $a(b + c) = ab + ac$
10. When we replace the variable in an equation with this, the equation is true.

A. addition or subtraction property of equality
B. associative property
C. coefficient
D. constant
E. distributive property
F. like terms
G. multiplication or division property of equality
H. reciprocal
I. solution of an equation
J. variable

2.2 Exercises

Follow your instructor's guidelines for showing your work.

For exercises 1–72,
(a) solve.
(b) check.

1. $2x + 10 = 24$
2. $3x + 15 = 36$
3. $2x - 10 = 24$
4. $3x - 15 = 36$
5. $18 = -6x + 48$
6. $24 = -6x + 36$
7. $-6 = 8x - 10$
8. $-4 = 6x - 7$
9. $-10x - 25 = 35$
10. $-10x - 15 = 25$
11. $-x + 8 = 11$
12. $-x + 9 = 14$
13. $8 - x = 11$
14. $9 - x = 14$
15. $4 = -5 - x$
16. $2 = -3 - x$
17. $\frac{x}{2} - 10 = 24$
18. $\frac{x}{3} - 15 = 36$
19. $\frac{1}{2}x - 10 = -24$
20. $\frac{1}{3}x - 15 = -36$
21. $10x + 16 = 106$
22. $18x + 30 = 174$
23. $-88 = -5p + 12$
24. $-43 = -4c + 21$
25. $8x - 9 = 63$
26. $7x - 12 = 44$
27. $-2x - 15 = 31$
28. $-3x - 18 = 45$
29. $9 + \frac{3}{4}w = 27$
30. $7 + \frac{2}{3}d = 29$
31. $\frac{2h}{9} - 3 = 15$
32. $\frac{3a}{5} - 8 = 16$
33. $6 - k = 32$
34. $9 - m = 27$
35. $5 = 6x + 3$
36. $11 = 7x + 2$
37. $8u - 9 = -2$
38. $-5v + 3 = -1$
39. $-z - 8 = 21$
40. $-p - 15 = 12$
41. $2x + 6 = 6$
42. $3x + 8 = 8$
43. $0.4b + 8 = 12.84$
44. $0.4w + 12.8 = 16.2$
45. $0.2x + 1.86 = 5.44$
46. $0.2x + 1.74 = 3.66$
47. $0.3x - 1.5 = 8.1$
48. $0.3x - 1.8 = 7.2$
49. $24 = -2(x - 1)$
50. $36 = -3(x - 1)$
51. $4(3z - 1) = 32$
52. $5(4p - 3) = 145$
53. $162 = -6(8k - 3)$
54. $49 = -7(2d - 3)$
55. $\frac{3}{4}(8x + 12) = 33$
56. $\frac{2}{5}(10x + 15) = 30$
57. $0.25(16a - 20) = 17$
58. $0.5(10c - 8) = 15$
59. $\frac{3}{2} = \frac{3}{8}(16h + 12)$
60. $\frac{2}{5} = \frac{4}{5}(15w + 2)$
61. $\frac{1}{4}(12p - 7) = \frac{3}{8}$
62. $\frac{1}{2}(6m - 9) = \frac{3}{4}$
63. $0.2(x - 3) + 0.4(x + 1) = 1.6$
64. $0.3(y - 2) + 0.2(y + 4) = 1.7$
65. $0 = 3(w + 4) - 5(w + 6)$
66. $0 = 9(b + 8) - 12(b + 2)$
67. $5(2y - 4) - (y + 3) = 58$
68. $4(3z - 9) - (z + 15) = -29$
69. $17 - (m + 8) = 21$
70. $13 - (p + 4) = 27$
71. $-47 = 20 - (6k - 5)$
72. $-63 = 15 - (8c - 10)$

For exercises 73–94, use the five steps.

73. A car owner estimates that the fixed costs per year to own his car are \$7400. His car averages 25 miles per gallon of gas. He wants to limit his annual car expenses to \$11,000. If gas costs \$4 per gallon, find the number of miles he can drive per year.

74. A granddad gave his granddaughter a used car that averages 20 miles per gallon of gas. The fixed costs per year are \$2300. The granddaughter wants to limit her annual car expenses to \$5000. If gas costs \$4 per gallon, find the number of miles she can drive per year.

75. The annual base salary for a sales representative is \$28,000. He makes \$4 commission on each product he sells. Find the number of products that he must sell per year to have a total annual salary of \$35,000.

76. The annual base salary for a sales representative is \$28,000. She makes \$50 commission on each product she sells. Find the number of products that she must sell per year to have a total annual salary of \$35,000.

77. The annual base salary for a sales position is \$45,000. An additional 12.5% commission is earned on any sales over \$750,000. Find the amount of additional sales needed for the total annual salary to be \$90,000.

78. The annual base salary for a sales position is \$54,000. An additional 4.5% commission is earned on any sales over \$500,000. Find the amount of additional sales needed for the total annual salary to be \$90,000. Round to the nearest whole number.

79. The cost for standard shipping of an order of textbooks from an on-line retailer includes a fixed cost of \$3 per shipment plus \$0.99 per book. Find the cost of shipping 12 textbooks.

80. The cost for standard shipping of an order of textbooks from an on-line retailer includes a fixed cost of \$3 per shipment plus \$0.99 per book. Find the cost of shipping 11 textbooks.

81. The cost for two-day shipping of an order of DVDs from an on-line retailer includes a fixed cost of \$5.99 per shipment plus \$1.99 per DVD. Find the cost of shipping eight DVDs.

82. The cost for two-day shipping of an order of DVDs from an on-line retailer includes a fixed cost of \$5.99 per shipment plus \$1.99 per DVD. Find the cost of shipping 15 DVDs.

83. The cost for next-day shipping of an order of video games from an on-line retailer includes a fixed cost of \$7.99 per shipment plus \$1.99 per game. Find the cost of shipping five games.

84. The cost for next-day shipping of an order of video games from an on-line retailer includes a fixed cost of \$7.99 per shipment plus \$1.99 per game. Find the cost of shipping six games.

85. A house cleaning company charges \$25 per visit plus \$18 per hour of cleaning. Find the number of hours the company will clean if the customer wants to limit the total charge to \$70.

86. A house cleaning company charges \$18 per visit plus \$16 per hour of cleaning. Find the number of hours the company will clean if the customer wants to limit the total charge to \$90.

87. Parking in a lot at a college in Texas costs \$3 for the first 2 hours and \$1.50 for each additional hour. A student has \$12. Find the number of additional hours that the student can park in the lot.

88. Parking in a garage in San Francisco costs \$4 for the first hour and \$2.50 for each additional hour. A shopper has \$14 to spend on parking. Find the number of additional hours that the shopper can park in the garage.

89. A taxi company charges a flat fee of \$2.40 for the first mile plus \$2.60 for each additional mile. Find how many additional miles a customer can travel for \$18.

90. A taxi company charges a flat fee of \$4 for the first mile plus \$1.87 for each additional mile. Find how many additional miles a customer can travel for \$22.70.

91. A professor has \$1250 to spend on fieldwork. The equipment for the fieldwork costs \$750. The reimbursement for driving during fieldwork is \$0.485 per mile. Find the number of miles that the professor can drive during fieldwork. Round to the nearest mile.

92. A costumer for a theater production has a \$1000 budget for tuxedos and dance costumes. She needs 12 tuxedos. She can buy used tuxedos on-line for \$30 each. A dance costume costs \$40. Find the number of dance costumes that she can buy.

For exercises 93–94, the cost to drive a car, including gas, is about $\frac{\$0.80}{1 \text{ mi}}$. A commuter must cross the Tacoma Narrows Toll Bridge to get to work. For drivers using cash, the toll is \$4. For drivers using an electronic pass, the toll is \$2.75.

© kathmanduphotdog/Shutterstock.com

93. If the commuter budgets \$10 for the one-way trip to work and uses the electronic pass, find the number of miles the commuter can live from work. Round to the nearest tenth.

94. If the commuter budgets \$12 for the one-way trip to work and uses the electronic pass, find the number of miles the commuter can live from work. Round to the nearest tenth.

Problem Solving: Practice And Review

Follow your instructor's guidelines for using the five steps as outlined in Section 1.3, p. 55.

95. In 2009, about 609,500 African American babies were born. Find the number of these babies who have sickle cell anemia.

In the United States, it's estimated that sickle cell anemia affects 70,000–100,000 people, mainly African Americans. The disease occurs in about 1 out of every 500 African American births. Sickle cell anemia also affects Hispanic Americans. The disease occurs in more than 1 out of every 36,000 Hispanic American births. (*Source:* www.nhlbi.nih.gov, Feb. 1, 2011)

96. A home in Lynchburg, Tennessee, has an appraised value of \$260,000. The assessed value of this property is 25% of the appraised value. If the annual property tax rate is \$2.41 per \$100 of assessed value, find the annual property tax on this home. (*Source:* www.mtida.org)

97. The Neutral Buoyancy Laboratory at the Johnson Space Center is a large water-filled tank used for training astronauts. It is 202 ft long, 102 ft wide, and 40 ft deep. Find its volume in gallons. ($1 \text{ ft}^3 \approx 7.48$ gal.) Round to the nearest whole number.

98. In 2010, there were 19,040 homeless students in Oregon schools. In 2011, this increased to 20,545 homeless students. Find the percent increase in homeless students. Round to the nearest percent. (*Source:* Children First for Oregon, 2011 Progress Report, Nov. 2011)

Technology

For exercises 99–102, solve the equation. Use a calculator to do arithmetic.

99. $350p + 10{,}817 = 16{,}750$ **100.** $\frac{c}{1063} + 945 = 2081$

101. $21{,}427 = 815(4d - 9)$ **102.** $\frac{3}{4}x + \frac{5}{9} = \frac{59}{50}$

Find the Mistake

For exercises 103–106, the completed problem has one mistake.
(a) Describe the mistake in words, or copy down the whole problem and highlight or circle the mistake.
(b) Do the problem correctly.

103. Problem: Solve: $3(x - 2) - (x + 5) = 1$

Incorrect Answer:
$$3(x - 2) - (x + 5) = 1$$
$$3x - 6 - x + 5 = 1$$
$$2x - 1 = 1$$
$$+1 \quad +1$$
$$2x + 0 = 2$$
$$\frac{2x}{2} = \frac{2}{2}$$
$$x = 1$$

104. Problem: Solve: $3x + 6 = 27$

Incorrect Answer:
$$3x + 6 = 27$$
$$+6 \quad +6$$
$$3x + 0 = 33$$
$$\frac{3x}{3} = \frac{33}{3}$$
$$x = 11$$

105. **Problem:** Solve: $\frac{1}{3}x - 5 = 2$

Incorrect Answer: $\frac{1}{3}x - 5 = 2$

$$3\left(\frac{1}{3}x - 5\right) = 3(2)$$
$$3 \cdot \frac{1}{3}x - 5 = 6$$
$$x - 5 = 6$$
$$\underline{\quad +5 \quad +5}$$
$$x + 0 = 11$$
$$x = 11$$

106. **Problem:** Solve: $3(x + 5) - 14 = 10$

Incorrect Answer: $3(x + 5) - 14 = 10$

$$3x + 5 - 14 = 10$$
$$3x - 9 = 10$$
$$\underline{\quad +9 \quad +9}$$
$$3x + 0 = 19$$
$$\frac{3x}{3} = \frac{19}{3}$$
$$x = \frac{19}{3}$$

Review

For exercises 107–110, the product of two whole numbers is a multiple. For example, 24 is a multiple of 8 because $8 \cdot 3 = 24$ and 24 is a multiple of 6 because $6 \cdot 4 = 24$; 12 is the least common multiple of 6 and 4. Find the least common multiple of the given numbers.

107. 8, 5, 2

108. 12, 15

109. 21, 24

110. 18, 30

SUCCESS IN COLLEGE MATHEMATICS

111. List the names of the classes you are taking. Using the schedule from Section 2.1, now identify how much time you will spend on each different class.

112. How many hours per week do you plan to use for completing your math homework?

113. When a test is coming up, you need to plan for time for extra studying, more than you would use just to finish your math homework. Where will you fit this in your schedule?

A business plan includes the costs of making a product and the revenue expected from selling the product. When the costs equal the revenue, the company is *breaking even.* In this section, we will use an equation with variables on both sides to find the number of products a company should make and sell to break even.

SECTION 2.3 Variables on Both Sides

After reading the text, working the practice problems, and completing assigned exercises, you should be able to:

1. Solve an equation in one variable in which the variable is on both sides of the equation.
2. Clear fractions or decimals from an equation before solving.
3. Identify an equation in one variable that is a contradiction or identity.
4. Use an equation with variables on both sides to solve an application problem.

Variables on Both Sides

In Sections 2.1 and 2.2, we used the properties of equality to solve equations in one variable. Often, the last step in isolating the variable is using the division property of equality to change the coefficient of the variable to 1.

$3x = 6$

$\frac{3x}{3} = \frac{6}{3}$ Division property of equality.

$\mathbf{1}x = \mathbf{2}$ Simplify.

$\mathbf{x} = 2$ Simplify.

From now on, we will usually combine two of the steps.

$\frac{3x}{3} = \frac{6}{3}$ Division property of equality.

$x = 2$ Simplify; the step of writing $1x = 2$ is not included.

When there are variables on both sides of an equation, we can use a property of equality to isolate the term with a variable.

EXAMPLE 1 **(a)** Solve: $2x + 8 = 6x - 20$

SOLUTION

$2x + 8 = 6x - 20$	
$-6x \quad\quad -6x$	Subtraction property of equality.
$\mathbf{-4x} + 8 = \mathbf{0} - 20$	Simplify.
$-4x + 8 = \mathbf{-20}$	Simplify; there are no variables on the right side.
$-8 \quad\quad -8$	Subtraction property of equality.
$-4x + \mathbf{0} = \mathbf{-28}$	Simplify.
$\frac{\mathbf{-4x}}{-4} = \frac{-28}{-4}$	Simplify; division property of equality.
$x = 7$	Simplify.

(b) Check.

$2x + 8 = 6x - 20$	Use the original equation.
$2(7) + 8 = 6(7) - 20$	Replace the variable, x, with the solution, 7.
$\mathbf{14} + 8 = \mathbf{42} - 20$	Follow the order of operations.
$22 = 22$	Evaluate; since the equation is true, the solution is correct.

In Example 1, we began by subtracting $6x$ from both sides. We can instead begin solving this equation by adding 20 to both sides. Either strategy results in the correct solution, $x = 7$.

EXAMPLE 2 Solve: $2x + 8 = 6x - 20$

SOLUTION

$2x + 8 = 6x - 20$	
$+20 \quad\quad +20$	Addition property of equality.
$2x + \mathbf{28} = 6x + \mathbf{0}$	Simplify.
$2x + 28 = \mathbf{6x}$	Simplify.
$-2x \quad\quad -2x$	Subtraction property of equality.
$\mathbf{0} + 28 = \mathbf{4x}$	Simplify.
$\frac{\mathbf{28}}{4} = \frac{4x}{4}$	Simplify; division property of equality.
$7 = x$	Simplify.

Solving an Equation in One Variable

1. Use the distributive property to rewrite the equation without parentheses. Simplify; combine like terms.
2. **Use the addition or subtraction property of equality so that only one side of the equation includes a term with a variable.**
3. Use the addition or subtraction property of equality to isolate the term that includes the variable. Any constant on the same side of the equation as the variable is changed to 0.
4. Use the multiplication or division property of equality to change the coefficient of the variable term to 1.
5. Check the solution in the original equation.

To simplify an expression such as $66p - (p - 78)$, we can rewrite the expression as $66p - 1(p - 78)$ and then use the distributive property.

EXAMPLE 3 Solve: $9(8p - 3) + 7 = 66p - (p - 78)$

SOLUTION ▶

$9(8p - 3) + 7 = 66p - (p - 78)$	
$9(8p - 3) + 7 = 66p - \mathbf{1}(p - 78)$	Multiply by 1.
$\mathbf{9}(8p) + \mathbf{9}(-3) + 7 = 66p - \mathbf{1}p - \mathbf{1}(-78)$	Distributive property.
$\mathbf{72p} - \mathbf{27} + 7 = 66p - \mathbf{p} + \mathbf{78}$	Simplify.
$72p - \mathbf{20} = \mathbf{65p} + 78$	Combine like terms.
$-65p \quad\quad -65p$	Subtraction property of equality.
$\mathbf{7p} - 20 = \mathbf{0} + 78$	Simplify.
$7p - 20 = \mathbf{78}$	Simplify.
$+20 \quad +20$	Addition property of equality.
$7p + \mathbf{0} = \mathbf{98}$	Simplify.
$\frac{\mathbf{7p}}{7} = \frac{98}{7}$	Simplify; division property of equality.
$p = 14$	Simplify.

There are often several paths to the solution of an equation. In Example 3, we subtracted $65p$ from both sides. We could have instead subtracted $72p$ from both sides. Some students prefer to always isolate the variable term on the left side of the equation. Others plan their strategy so that they work with fewer negative numbers. Each path, done correctly, has the same solution.

EXAMPLE 4 **(a)** Solve: $5(a + 1) - 2 = 4a + 3$

SOLUTION ▶

$5(a + 1) - 2 = 4a + 3$	
$\mathbf{5a} + \mathbf{5}(1) - 2 = 4a + 3$	Distributive property.
$\mathbf{5a} + \mathbf{5} - 2 = 4a + 3$	Simplify.
$5a + \mathbf{3} = 4a + 3$	Combine like terms.
$-3 \quad\quad -3$	Subtraction property of equality.
$5a + \mathbf{0} = 4a + \mathbf{0}$	Simplify.
$5a = 4a$	Simplify.
$-4a \quad -4a$	Subtraction property of equality.
$a = 0$	Simplify.

(b) Check.

$5(a + 1) - 2 = 4a + 3$	Use the original equation.
$5(\mathbf{0} + 1) - 2 = 4(\mathbf{0}) + 3$	Replace the variable, a, with the solution, 0.
$5(\mathbf{1}) - 2 = \mathbf{0} + 3$	Follow the order of operations.
$\mathbf{5} - 2 = \mathbf{3}$	Follow the order of operations.
$\mathbf{3} = 3$	Evaluate; since the equation is true, the solution is correct.

Practice Problems

For problems 1–3,
(a) solve.
(b) check.

1. $9w - 15 = 13w + 85$ **2.** $8(w + 1) - 15 = 20w - 31$
3. $2k - 8 = 4k$

Clearing Fractions or Decimals

Since addition and subtraction with fractions can take longer than arithmetic with integers, we may choose to "clear" the fractions from an equation. To **clear the fractions**, multiply both sides of the equation by the least common denominator of all of the fractions.

EXAMPLE 5 **(a)** Clear the fractions and solve $\frac{1}{3}z + \frac{2}{9} = \frac{4}{5}z - \frac{1}{15}$.

SOLUTION ▶ The least common denominator is 45. To clear the fractions, multiply both sides of the equation by 45. Use the distributive property to rewrite the equation without parentheses. Find common factors and simplify. The coefficients and constants are now integers; since there are no fractions in the equation, the fractions are "cleared."

$$\frac{1}{3}z + \frac{2}{9} = \frac{4}{5}z - \frac{1}{15}$$

$$\left(\frac{45}{1}\right)\left(\frac{1}{3}z + \frac{2}{9}\right) = \left(\frac{45}{1}\right)\left(\frac{4}{5}z - \frac{1}{15}\right) \quad \text{Multiplication property of equality.}$$

$$\left(\frac{\mathbf{45}}{\mathbf{1}}\right)\left(\frac{1}{3}z\right) + \left(\frac{\mathbf{45}}{\mathbf{1}}\right)\left(\frac{2}{9}\right) = \left(\frac{\mathbf{45}}{\mathbf{1}}\right)\left(\frac{4}{5}z\right) + \left(\frac{\mathbf{45}}{\mathbf{1}}\right)\left(-\frac{1}{15}\right) \quad \text{Distributive property.}$$

$$\left(\frac{\mathbf{3 \cdot 15}}{1}\right)\left(\frac{1}{3}z\right) + \left(\frac{\mathbf{5 \cdot 9}}{1}\right)\left(\frac{2}{9}\right) = \left(\frac{\mathbf{5 \cdot 9}}{1}\right)\left(\frac{4}{5}z\right) + \left(\frac{\mathbf{15 \cdot 3}}{1}\right)\left(-\frac{1}{15}\right) \quad \text{Find common factors.}$$

$15z + 10 = 36z - 3$	Simplify.
$-36z \qquad -36z$	Subtraction property of equality.
$\mathbf{-21z} + 10 = \mathbf{0} - 3$	Simplify.
$-21z + 10 = \mathbf{-3}$	Simplify.
$-10 \qquad -10$	Subtraction property of equality.
$-21z + \mathbf{0} = \mathbf{-13}$	Simplify.
$\frac{\mathbf{-21z}}{-21} = \frac{-13}{-21}$	Simplify; division property of equality.
$z = \frac{13}{21}$	Simplify.

The purpose of clearing fractions is to minimize the amount of fraction arithmetic in solving the equation. However, the solution itself may be a fraction. Do not rewrite the fraction as a decimal number.

(b) Check.

To check a solution that is a fraction, replace the variable(s) with the solution. Evaluate each side of the equation. It may be necessary to rewrite fractions with common denominators.

$$\frac{1}{3}z + \frac{2}{9} = \frac{4}{5}z - \frac{1}{15}$$ Use the original equation.

$$\frac{1}{3}\left(\mathbf{\frac{13}{21}}\right) + \frac{2}{9} = \frac{4}{5}\left(\mathbf{\frac{13}{21}}\right) - \frac{1}{15}$$ Replace the variable, z, with the solution, $\frac{13}{21}$.

$$\mathbf{\frac{13}{63}} + \frac{2}{9} = \mathbf{\frac{52}{105}} - \frac{1}{15}$$ Simplify.

$$\frac{13}{63} + \frac{2}{9}\left(\frac{7}{7}\right) = \frac{52}{105} - \frac{1}{15}\left(\frac{7}{7}\right)$$ Multiply by a fraction equal to 1.

$$\frac{13}{63} + \mathbf{\frac{14}{63}} = \frac{52}{105} - \mathbf{\frac{7}{105}}$$ Follow the order of operations.

$$\frac{27}{63} = \frac{45}{105}$$ Follow the order of operations.

$$\frac{3 \cdot \cancel{9}}{7 \cdot \cancel{9}} = \frac{3 \cdot \cancel{15}}{7 \cdot \cancel{15}}$$ Find common factors.

$$\frac{3}{7} = \frac{3}{7}$$ Evaluate; since the equation is true, the solution is correct.

In the next example, the equation includes parentheses.

EXAMPLE 6 Clear the fractions from $\frac{2}{5}(4c + 6) = -\frac{3}{8}(7c - 1)$. Do not solve.

SOLUTION

$$\frac{2}{5}(4c + 6) = -\frac{3}{8}(7c - 1)$$ Least common denominator is 40.

$$\left(\frac{40}{1}\right)\left(\frac{2}{5}\right)(4c + 6) = \left(\frac{40}{1}\right)\left(-\frac{3}{8}\right)(7c - 1)$$ Multiplication property of equality.

$$\frac{(\mathbf{8 \cdot \cancel{5}})}{1}\left(\frac{2}{\cancel{5}}\right)(4c + 6) = \left(\frac{\mathbf{\cancel{8} \cdot 5}}{1}\right)\left(-\frac{3}{\cancel{8}}\right)(7c - 1)$$ Find common factors.

$$\mathbf{16}(4c + 6) = \mathbf{-15}(7c - 1)$$ Simplify; the fractions are cleared.

We can also clear decimal numbers from an equation by multiplying both sides by a multiple of 10. When the decimals are cleared, the coefficients and constants are integers.

EXAMPLE 7 **(a)** Clear the decimals from $0.6x + 12 = 0.8x$. Do not solve.

SOLUTION The digit farthest to the right in 0.6 and in 0.8 is in the tenths place. Multiplying by 10 will change these coefficients to integers.

$$0.6x + 12 = 0.8x$$

$$(10)(0.6x + 12) = (10)(0.8x)$$ Multiplication property of equality.

$$\mathbf{10}(0.6x) + \mathbf{10}(12) = \mathbf{8x}$$ Distributive property; simplify.

$$\mathbf{6x + 120} = 8x$$ Simplify; the decimals are cleared.

(b) Clear the decimals from $0.05h = 2.65h + 12.7$. Do not solve.

The digit farthest to the right in 0.05, 2.65, and 12.7 is in the hundredths place. Multiplying by 100 will change these coefficients to integers.

$$0.05h = 2.65h + 12.7$$
$$(100)(0.05h) = (100)(2.65h + 12.7) \quad \text{Multiplication property of equality.}$$
$$\mathbf{5}h = \mathbf{100}(2.65h) + \mathbf{100}(12.7) \quad \text{Simplify; distributive property.}$$
$$5h = \mathbf{265}h + \mathbf{1270} \quad \text{Simplify; the decimals are cleared.}$$

Clearing an Equation of Fractions or Decimals

Fractions

1. Find the least common denominator of the fractions in the equation.
2. Multiply both sides of the equation by the least common denominator.
3. Simplify. The coefficients and constants in the equation are integers.

Decimals

1. Identify the multiple of ten needed to change each constant or coefficient to an integer.
2. Multiply both sides of the equation by this number.
3. Simplify. The coefficients and constants in the equation are integers.

Practice Problems

For problems 4–6,
(a) clear the fractions or decimals and solve.
(b) check.

4. $\frac{1}{4}x - \frac{1}{8} = \frac{5}{12}x + \frac{1}{6}$ **5.** $\frac{4}{9}(x + 12) = -20$ **6.** $0.2x - 15 = 0.04x + 3$

Identities and Contradictions

An equation in one variable with one solution is a **conditional equation**. An equation in one variable that has no solution is a **contradiction**.

EXAMPLE 8 Solve: $9x + 8 = 3(3x - 6)$

SOLUTION

$$9x + 8 = 3(3x - 6)$$
$$9x + 8 = \mathbf{3}(3x) + \mathbf{3}(-6) \quad \text{Distributive property.}$$
$$9x + 8 = \mathbf{9x - 18} \quad \text{Simplify.}$$
$$\underline{-9x \qquad -9x} \quad \text{Subtraction property of equality.}$$
$$\mathbf{0} + 8 = \mathbf{0} - 18 \quad \text{Simplify; there are no variables in the equation.}$$
$$8 = -18 \quad \text{Simplify; this equation is false.}$$

This equation has no solution. As we solve this equation, the variables are eliminated and the final equation is false. No number can replace the variable and create a true equation. Since there is no solution, we cannot do a check.

A linear equation in one variable that has **infinitely many solutions** is an **identity**. The solution of the equation is the set of real numbers. If we replace the variable in the equation with a real number, the equation is true.

EXAMPLE 9 **(a)** Solve: $9x + 8 = 3(3x - 6) + 26$

SOLUTION ▶

$9x + 8 = 3(3x - 6) + 26$	
$9x + 8 = \mathbf{3(3x) + 3(-6)} + 26$	Distributive property.
$9x + 8 = \mathbf{9x - 18} + 26$	Simplify.
$9x + 8 = 9x \mathbf{+ 8}$	Combine like terms.
$-9x \quad -9x$	Subtraction property of equality.
$\mathbf{0} + 8 = \mathbf{0} + 8$	Simplify; there are no variables in the equation.
$8 = 8$	Simplify; this equation is true.

The solution of this equation is the set of real numbers. As we solve this equation, the variables are eliminated and the final equation is true. Any real number can replace the variable and make this equation true.

(b) Check.

▶ If an equation has infinitely many solutions, at least two real numbers are solutions of the equation. We choose to check $x = 0$ and $x = 1$, since arithmetic with these numbers is often quicker than with other numbers.

Check: $x = 0$

$9x + 8 = 3(3x - 6) + 26$	Use the original equation to check.
$9(\mathbf{0}) + 8 = 3(3(\mathbf{0}) - 6) + 26$	Replace the variable, x, with a solution, 0.
$\mathbf{0} + 8 = 3(\mathbf{0} - 6) + 26$	Follow the order of operations.
$8 = 3(\mathbf{-6}) + 26$	Follow the order of operations.
$8 = \mathbf{-18} + 26$	Follow the order of operations.
$8 = \mathbf{8}$	Since the equation is true, the solution is correct.

This equation might only have one solution, $x = 0$. To confirm that it has infinitely many solutions, we check another real number, $x = 1$.

Check: $x = 1$

$9x + 8 = 3(3x - 6) + 26$	Use the original equation to check.
$9(\mathbf{1}) + 8 = 3(3(\mathbf{1}) - 6) + 26$	Replace the variable, x, with a solution, 1.
$\mathbf{9} + 8 = 3(\mathbf{3} - 6) + 26$	Follow the order of operations.
$\mathbf{17} = 3(\mathbf{-3}) + 26$	Follow the order of operations.
$17 = \mathbf{-9} + 26$	Follow the order of operations.
$17 = \mathbf{17}$	Since the equation is true, the solution is correct.

Since 0 and 1 are solutions of this equation, every number in the set of real numbers is a solution of this equation.

Practice Problems

For problems 7–9,
(a) solve.
(b) check.

7. $8x - (x + 2) = 7x - 19$ **8.** $3(4x - 3) = 12(x - 1) + 3$
9. $5(4x + 3) = 18x + 15$

Applications

A business that makes and sells a product "breaks even" when the costs of making a product are equal to the revenue from selling the product. If the revenue is *greater than* the costs, the business will make a **profit**. If the revenue is *less than* the costs, the business has a **loss**.

The total cost for making products equals the sum of the nonfixed costs and the fixed costs. To find the nonfixed costs, multiply the cost per product and the number of products. The fixed costs, also known as the **overhead**, include wages, utilities, rent, and interest payments.

A business earns revenue from selling products. To find the total revenue, multiply the revenue per product and the number of products.

Break Even

$$\text{costs} = \text{revenue}$$

$$\begin{pmatrix}\text{cost per}\\ \text{product}\end{pmatrix}\begin{pmatrix}\text{number of}\\ \text{products}\end{pmatrix} + \text{overhead} = \begin{pmatrix}\text{price per}\\ \text{product}\end{pmatrix}\begin{pmatrix}\text{number of}\\ \text{products}\end{pmatrix}$$

EXAMPLE 10 The cost to make a single product is \$5. The fixed overhead costs per month are \$4500. The price of the product is \$15. Find the number of products that the company should make and sell each month to break even.

SOLUTION ▶ **Step 1 Understand the problem.**
The unknown is the number of products that the company should make and sell each month to break even.

N = number of products

Step 2 Make a plan.
When the company breaks even, the costs equal the revenue. A word equation is (cost per product)(number of products) + overhead = (price per product)(number of products).

Step 3 Carry out the plan.

$\left(\dfrac{\$5}{1 \text{ product}}\right)N + \$4500 = \left(\dfrac{\$15}{1 \text{ product}}\right)N$	nonfixed costs + fixed costs = revenue
$5N + 4500 = 15N$	To solve, remove the units.
$-5N \qquad -5N$	Subtraction property of equality.
$\mathbf{0} + 4500 = \mathbf{10N}$	Simplify.
$\mathbf{4500} = 10N$	Simplify.
$\dfrac{4500}{10} = \dfrac{10N}{10}$	Division property of equality.
$450 \text{ products} = N$	Simplify; the units are products.

Step 4 Look back.
If the answer is correct, the costs to make 450 products should equal the revenue from selling 450 products. Since $\left(\dfrac{\$5}{1 \text{ product}}\right)(450 \text{ products}) + \$4500$ equals \$6750 and $\left(\dfrac{\$15}{1 \text{ product}}\right)(450 \text{ products})$ also equals \$6750, the answer seems reasonable.

Step 5 Report the solution.
To break even, the company should make and sell 450 products.

We can also use an equation with variables on both sides to compare the cost of a service such as a taxi ride.

EXAMPLE 11 If the standard meter rate for a taxi in Boston is a fee of \$1.45 plus \$0.30 for each $\frac{1}{8}$ mi and the standard meter rate for a taxi in Denver is a fee of \$2.25 plus \$0.25 for each $\frac{1}{8}$ mi, find the distance at which the cost for a taxi is the same in both cities.

SOLUTION **Step 1 Understand the problem.**

The unknown is the distance at which the cost for a taxi is the same in both cities. Since the fee in Boston is less than the fee in Denver but the cost per mile in Denver is less than the cost per mile in Boston, at some distance the cost for a taxi in both places is the same.

d = distance at which the cost is the same

Step 2 Make a plan.

The cost for a taxi equals the fee plus the cost for the distance traveled. A word equation is fee in Boston + $\left(\frac{\text{cost}}{1\text{ mi}}\right)$(distance) = fee in Denver + $\left(\frac{\text{cost}}{1\text{ mi}}\right)$(distance). Since the cost per mile is eight times the cost for $\frac{1}{8}$ mi, the cost per mile in Boston is $8\left(\frac{\$0.30}{1\text{ mi}}\right)$, which equals $\frac{\$2.40}{1\text{ mi}}$. The cost per mile in Denver is $8\left(\frac{\$0.25}{1\text{ mi}}\right)$, which equals $\frac{\$2.00}{1\text{ mi}}$.

Step 3 Carry out the plan.

$\$1.45 + \left(\frac{\$2.40}{1\text{ mi}}\right)d = \$2.25 + \left(\frac{\$2}{1\text{ mi}}\right)d$	fee + cost for distance = fee + cost for distance
$1.45 + 2.40d = 2.25 + 2d$	Remove the units.
$-2d \qquad -2d$	Subtraction property of equality.
$1.45 + \mathbf{0.40}\boldsymbol{d} = 2.25 + \mathbf{0}$	Simplify.
$-1.45 \qquad -1.45$	Subtraction property of equality.
$\mathbf{0} + 0.40d = \mathbf{0.80}$	Simplify.
$\frac{\mathbf{0.40}\boldsymbol{d}}{0.40} = \frac{0.80}{0.40}$	Division property of equality.
$d = 2\text{ mi}$	Simplify; the units are miles.

Step 4 Look back.

If the answer is correct, the cost for a 2-mi taxi trip in Boston is the same as the cost for a 2-mi trip in Denver. Since $\$1.45 + \left(\frac{\$2.40}{1\text{ mi}}\right)(2\text{ mi})$ equals \$6.25 and $\$2.25 + \left(\frac{\$2}{1\text{ mi}}\right)(2\text{ mi})$ also equals \$6.25, the answer seems reasonable.

Step 5 Report the solution.

When the distance of a taxi trip is 2 mi, the cost for the taxi in Boston is the same as the cost in Denver.

Practice Problems

For problems 10–12, use the five steps.

10. The cost to make a product is \$3200. The fixed overhead costs to make the product per month are \$27,000. The price of the product is \$5000. Find the number of products that should be made and sold each month to break even.

11. The owner of a long-arm quilting machine finishes quilts for other people. Her overhead is \$1525 per month. She uses about \$20 of materials for each quilt. Her average revenue per quilt is \$175. Find the number of quilts that she should complete per month to break even. *Round up* to the nearest quilt so that she will not lose money.

12. Company A charges \$75 for a service call plus \$40 per hour. Company B charges \$25 for a service call plus \$60 per hour. Find the time of the service call at which the charges by both companies are equal.

2.3 Vocabulary Practice

Match the term with its description.

1. $a(b + c) = ab + ac$
2. When we replace the variable in an equation with this, the left side of the equation equals the right side.
3. Amount of money collected from selling a product
4. Terms that can be added or subtracted
5. An equation in one variable with no solution
6. A number by itself in an expression or equation
7. In $7x$, 7 is an example of this.
8. A linear equation in one variable with infinitely many solutions
9. We can multiply both sides or divide both sides of an equation by a real number that is not equal to 0. The solution is not changed.
10. We can add a term to or subtract a term from both sides of an equation. The solution is not changed.

A. addition or subtraction property of equality
B. coefficient
C. constant
D. contradiction
E. distributive property
F. identity
G. like terms
H. multiplication or division property of equality
I. revenue
J. solution of an equation

2.3 Exercises

Follow your instructor's guidelines for showing your work.

For exercises 1–22,
(a) solve.
(b) check.

1. $8x + 12 = 5x + 24$
2. $9x + 15 = 6x + 27$
3. $2w - 9 = 6w + 31$
4. $3k - 4 = 8k + 26$
5. $9y - 20 = 5y$
6. $7z - 48 = 3z$
7. $15h = 21h + 42$
8. $2c = 5c + 36$
9. $-8k - 11 = -2k + 61$
10. $-12h - 19 = -6h + 65$
11. $x - 4 = 3x + 9$
12. $x - 6 = 4x + 11$
13. $7p + 5040 = 19p$
14. $6p + 8190 = 32p$
15. $3(2x + 9) = 7(x - 12)$
16. $5(3x + 12) = 2(8x - 16)$
17. $6(5a + 1) - 3(2a - 8) = 7(a + 9) + 3(5a - 6)$
18. $4(9d + 3) - 2(6d - 5) = 15(2d - 3) - 2(2d + 10)$
19. $18z - (z + 12) = 3(5z + 44)$
20. $15b - (b + 6) = 4(3b + 38)$
21. $9(v + 2) - 3 = 3(v + 5)$
22. $4(w + 7) - 6 = 11(w + 2)$

For exercises 23–54,
(a) clear the fractions and solve.
(b) check.

23. $\frac{5}{6}x + 2 = 17$
24. $\frac{3}{5}x + 2 = 17$
25. $\frac{1}{6}x + \frac{2}{3} = 8$
26. $\frac{1}{6}x + \frac{2}{3} = 5$
27. $\frac{1}{8}u + \frac{1}{3}u = 22$
28. $\frac{1}{6}v + \frac{3}{4}v = 22$

29. $\frac{1}{6}c + \frac{3}{4} = \frac{1}{2}$

30. $\frac{1}{6}c + \frac{3}{4} = \frac{1}{3}$

31. $\frac{2}{7}a + 1 = \frac{1}{3}$

32. $\frac{3}{7}a + 1 = \frac{1}{4}$

33. $2n - \frac{3}{4} = \frac{1}{2}$

34. $2n - \frac{1}{4} = \frac{3}{2}$

35. $\frac{3}{8} - \frac{2}{5}n = \frac{1}{4}$

36. $\frac{3}{8} - \frac{4}{5}n = \frac{1}{4}$

37. $\frac{7}{8} - p = \frac{4}{5}$

38. $\frac{5}{8} - p = \frac{4}{7}$

39. $\frac{3}{4}y + 10 = y - 9$

40. $\frac{3}{4}y + 12 = y - 9$

41. $\frac{2}{3}b + 12 = \frac{5}{8}b + 10$

42. $\frac{3}{4}u + 18 = \frac{5}{6}u + 9$

43. $\frac{1}{6}y + \frac{3}{4} = \frac{2}{9}y + \frac{2}{3}$

44. $\frac{1}{6}z + \frac{2}{9} = \frac{1}{2}z + \frac{1}{3}$

45. $\frac{3}{5}(4p + 1) = 15$

46. $\frac{2}{3}(8h + 2) = 12$

47. $\frac{5}{8}(x + 2) = \frac{1}{4}$

48. $\frac{5}{6}(m + 2) = \frac{1}{3}$

49. $-\frac{5}{8}(x + 2) = \frac{1}{4}$

50. $-\frac{5}{6}(m + 2) = \frac{1}{3}$

51. $\frac{1}{2}(x + 9) + \frac{2}{3}(2x - 4) = 11$

52. $\frac{7}{9}(2x + 6) - \frac{1}{2}(3x + 4) = 3$

53. $\frac{3}{5}a = \frac{5}{8}a - 6$

54. $\frac{4}{9}c = \frac{2}{3}c + 14$

For exercises 55–64,
(a) clear the decimals and solve.
(b) check.

55. $0.2x + 3 = 12$

56. $0.2x + 5 = 8$

57. $0.08x + 1 = 0.4$

58. $0.04y + 1 = 0.7$

59. $1.5w - 90 = 1.1w + 21$

60. $2.4g - 10.2 = 1.6g + 58$

61. $3.2n - 0.21 = 2.1n + 0.89$

62. $6.7r - 0.16 = 5.4r + 1.14$

63. $5(2.6k - 0.8) = 2(7.3k - 0.2)$

64. $8(2.4w - 0.4) = 2(5.1w + 0.2)$

For exercises 65–86,
(a) solve.
(b) check.

65. $3(2x - 9) = 6(x - 5)$

66. $4(3x - 8) = 2(6x - 1)$

67. $3(2x + 9) = 6(x + 5) - 3$

68. $4(3x - 8) = 2(6x - 1) - 30$

69. $3(2x - 9) = 6(x - 5) + x + 3$

70. $4(3x - 8) = 2(6x - 1) + x - 30$

71. $\frac{3}{4}(4a + 8) = 3(a + 2)$

72. $\frac{3}{4}(12c - 8) = 3(3c - 2)$

73. $\frac{2}{3}(p - 4) + \frac{1}{6} = p - \frac{5}{2}$

74. $\frac{2}{3}(n - 1) + \frac{1}{6} = n - \frac{1}{2}$

75. $8k - (k + 3) = 7k - 6$

76. $10u - (u + 4) = 9u - 2$

77. $15p - (8p + 2) = 3(p + 2) + 4(p - 1)$

78. $20n - (12n + 8) = 2(n + 8) + 6(n + 4)$

79. $12c - (10c + 4) = 6(c - 3) - 2(c - 7)$

80. $13w - (10w + 6) = 8(w - 2) + 5(w + 2)$

81. $\frac{3}{4}(2d + 1) = 5d - \frac{7}{2}\left(d - \frac{3}{14}\right)$

82. $\frac{3}{8}(4m + 6) = 4m - \frac{5}{2}\left(m - \frac{9}{10}\right)$

83. $9r + 8 = 3r - 11 + 6r$

84. $7a + 4 = a - 15 + 6a$

85. $9r + 8 = 3r + 10 + 5r$

86. $7a + 4 = a + 6 + 5a$

87. Write an equation in one variable that has no solution.

88. Write an equation in one variable that has infinitely many solutions, the set of real numbers.

89. Explain why you think an equation in one variable with no solution is called a contradiction.

90. Explain why you think an equation in one variable with infinitely many solutions is called an identity.

For exercises 91–98, use the five steps.

91. The cost to make a product is \$8. The fixed overhead costs to make the product per month are \$6600. The price of the product is \$30. Find the number of products that the company should make and sell per month to break even.

92. The cost to make a product is \$15. The fixed overhead costs to make the product per month are \$9500. The price of the product is \$40. Find the number of products that the company should make and sell per month to break even.

93. The cost to make a product is \$8. The fixed overhead costs to make the product per month are \$6600. The price of the product is \$60. Find the number of products that the company should make and sell per month to break even. *Round up* to the nearest product.

94. The cost to make a product is \$15. The fixed overhead costs to make the product per month are \$9500. The price of the product is \$80. Find the number of products that the company should make and sell per month to break even. *Round up* to the nearest product.

95. The charge for a taxi in St. Louis is a flat fee of \$2.30 plus \$0.20 for each $\frac{1}{10}$ mi. The charge for a taxi in Providence is a flat fee of \$2.00 plus \$0.25 for each $\frac{1}{10}$ mi. Find the distance at which the cost for a taxi is the same in both cities.

96. The charge for a taxi in Chicago is a flat fee of \$2.20 plus \$0.20 for each $\frac{1}{9}$ mi. The charge for a taxi in Salt Lake City is a flat fee of \$2.00 plus \$0.22 for each $\frac{1}{9}$ mi. Find the distance at which the cost for a taxi is the same in both cities. Round to the nearest tenth of a mile.

97. A lawn service charges a flat fee per visit of \$20 plus \$3.50 per hour. A different lawn service charges a flat fee per visit of \$10 plus \$5.00 per hour. Find the time in hours and minutes at which the cost for both lawn services is the same.

98. A cleaning company charges a monthly fee of \$40 plus \$12.00 per hour. A different company charges no monthly fee. Its hourly charge is \$16 per hour. Find the time in hours at which the cost to hire both companies is the same.

Problem Solving: Practice and Review

Follow your instructor's guidelines for using the five steps as outlined in Section 1.3, p. 55.

99. Find the percent increase in the price of a round bale of hay. Round to the nearest percent.

© holbox/Shutterstock.com

As hay prices continue to climb—an 800-pound round bale in Terrell was \$115 in mid-October, up from roughly \$60 last year—some ranchers have given up, selling off their herds and getting out of the cattle business. (*Source:* www.nytimes.com, Nov. 1, 2011)

100. Find the number of people in the surgery group who had strokes. Round to the nearest whole number.

An operation that doctors hoped would prevent strokes in people with poor circulation to the brain does not work. . . 97 received surgery and drugs, and 98 drugs alone. . . In the surgery group, 21 percent had strokes, compared with 22.7 percent in the medicine-only group. (*Source:* www.nytimes.com, Nov. 9, 2011)

101. Find the combined payroll of the Yankees and the Red Sox.

The Rays — whose \$43.8 million payroll is about one-fifth that of the rival Yankees and one-third that of the Red Sox — don't have the resources to recover from a poor trade or signing misstep, unlike their richer competitors. (*Source:* www.usatoday.com, Oct. 10, 2008)

102. A server at a restaurant made a total of \$95.00 in wages and tips in an 8-hr shift. The server's pay is \$6.25 per hour. Find the amount that the server collected in tips.

Technology

For exercises 103–106, solve the equation. Use a calculator to do the arithmetic.

103. $17{,}205c - 8510 = 12{,}485c + 19{,}810$

104. $901z + 302 = 2(492z + 3) - 83z$

105. $216p + 18(45p - 33) = 2000p - 594 - 974p$

106. $\frac{8}{15}x + \frac{10}{21} = \frac{2}{35}x + 1$

Find the Mistake

For exercises 107–110, the completed problem has one mistake.
(a) Describe the mistake in words, or copy down the whole problem and highlight or circle the mistake.
(b) Do the problem correctly.

107. Problem: Solve: $6x - 9 = 2x + 55$

Incorrect Answer:

$$\begin{aligned} 6x - 9 &= 2x + 55 \\ +2x \quad & \quad +2x \\ \hline 8x - 9 &= 0 + 55 \\ 8x - 9 &= 55 \\ +9 \quad & \quad +9 \\ \hline 8x + 0 &= 64 \\ \frac{8x}{8} &= \frac{64}{8} \\ x &= 8 \end{aligned}$$

108. Problem: Solve: $6x - 10 = 2(3x + 1) - 11$

Incorrect Answer:

$$\begin{aligned} 6x - 10 &= 2(3x + 1) - 11 \\ 6x - 10 &= 6x + 2 - 11 \\ 6x - 10 &= 6x - 9 \\ -6x \quad & \quad -6x \\ \hline 0 - 10 &= 0 - 9 \\ -10 &= -9 \end{aligned}$$

The solution is the set of real numbers.

109. Problem: Solve: $\frac{4}{5}(4x + 5) = x + 1$

Incorrect Answer:

$$\begin{aligned} \frac{4}{5}(4x + 5) &= x + 1 \\ 5 \cdot \frac{4}{5}(4x + 5) &= 5(x + 1) \\ 5 \cdot \frac{4}{5} \cdot 5(4x + 5) &= 5x + 5 \\ 4(20x + 25) &= 5x + 5 \\ 80x + 100 &= 5x + 5 \\ -5x \quad & \quad -5x \\ \hline 75x + 100 &= 0 + 5 \\ 75x + 100 &= 5 \\ -100 \quad & \quad -100 \\ \hline 75x + 0 &= -95 \\ \frac{75x}{75} &= \frac{-95}{75} \\ x &= -\frac{\cancel{5} \cdot 19}{\cancel{5} \cdot 15} \\ x &= -\frac{19}{15} \end{aligned}$$

110. Problem: Solve: $9x + 15 = 6x$

Incorrect Answer:

$$\begin{array}{rcl} 9x + 15 &=& 6x \\ -6x & & -6x \\ \hline 3x + 15 &=& 0 \\ +15 & & +15 \\ \hline \frac{3x}{3} &=& \frac{15}{3} \\ x &=& 5 \end{array}$$

Review

111. Change 45 min into hours.

112. Change 9.6 hr into minutes.

113. Simplify: $12\left(\frac{1}{4}x + \frac{2}{3}y\right)$

114. Simplify: $\dfrac{80 \text{ mi}}{40 \dfrac{\text{mi}}{\text{hr}}}$

SUCCESS IN COLLEGE MATHEMATICS

115. Math classes often have daily assignments, and it is essential to schedule regular times for doing homework. Other classes, especially those that involve projects, speeches, presentations, and papers, have assignments that are not due for several weeks or months. If a student procrastinates, the work for these other classes is often done at the last minute, and the regular planned time for math homework disappears. When in your weekly schedule do you work on longer-term assignments?

SUM ▾ X ✓ fx =B2/(C2*D2)

	A	B	C	D	E
1	Length	Volume	Width	Height	
2	=B2/(C2*D2)				
3					
4					

If a formula is solved for a variable, we can use a spreadsheet program to evaluate it. In this section, we will learn how to solve equations and formulas for a variable.

SECTION 2.4 Solving an Equation for a Variable

After reading the text, working the practice problems, and completing assigned exercises, you should be able to:

1. Solve an equation or formula for a variable.
2. Use a formula to solve an application problem.

Solving an Equation for a Variable

In this section, to prepare for work in Chapter 3 with equations in two variables we use the properties of equality to solve an equation for a variable. When an equation is solved for a variable, the variable is isolated on one side of the equation. In Examples 1–2, we solve equations in two variables, x and y, for y and simplify. We use the commutative property to write the terms in an order that matches the pattern $y = mx + b$, where m and b are real numbers. As we will learn in Chapter 3, $y = mx + b$ is an important pattern for linear equations in two variables.

EXAMPLE 1 Solve $3x + y = 8$ for y. Write the equation to match the pattern $y = mx + b$.

SOLUTION ▶

$$3x + y = 8$$

$$-3x \qquad -3x \qquad \text{Subtraction property of equality.}$$

$$\mathbf{0} + y = 8 - \mathbf{3x} \qquad \text{Simplify; 8 and } -3x \text{ are not like terms and cannot be combined.}$$

$$y = -3x + 8 \qquad \text{Simplify; change order of terms to match the pattern } y = mx + b.$$

In the next example, we divide $7 - 6x$ by -4. To match the pattern $y = mx + b$, we use the distributive property and divide *each term* by -4.

EXAMPLE 2 Solve $6x - 4y = 7$ for y. Write the equation to match the pattern $y = mx + b$.

SOLUTION ▶

$$6x - 4y = 7$$

$$\underline{-6x \qquad\quad -6x}$$ Subtraction property of equality.

$$\mathbf{0} - 4y = 7 - 6x$$ 7 and $-6x$ are not like terms and cannot be combined.

$$\mathbf{-4y} = 7 - 6x$$ Simplify.

$$\frac{-4y}{-4} = \frac{7}{-4} - \frac{6x}{-4}$$ Division property of equality; divide each term by -4.

$$y = -\frac{7}{4} + \frac{3 \cdot \cancel{2}x}{2 \cdot \cancel{2}}$$ Find common factors.

$$y = \frac{\mathbf{3x}}{\mathbf{2}} - \frac{7}{4}$$ Simplify; change order.

$$y = \frac{\mathbf{3}}{\mathbf{2}}\mathbf{x} - \frac{7}{4}$$ To match $y = mx + b$, rewrite $\frac{3x}{2}$ as $\frac{3}{2}x$.

In the next example, we clear the fractions before isolating the variable, x. Since we are not solving for y, we do not match the pattern $y = mx + b$. However, we continue to write the term with a variable to the left of the constant.

EXAMPLE 3 Solve $\frac{1}{4}x + \frac{2}{3}y = 5$ for x.

SOLUTION ▶ To clear the fractions, multiply both sides by the least common denominator, 12.

$$\frac{1}{4}x + \frac{2}{3}y = 5$$

$$\left(\frac{12}{1}\right)\left(\frac{1}{4}x + \frac{2}{3}y\right) = (12)(5)$$ Multiplication property of equality.

$$\frac{\mathbf{12}}{\mathbf{1}}\left(\frac{1}{4}x\right) + \frac{\mathbf{12}}{\mathbf{1}}\left(\frac{2}{3}y\right) = \mathbf{60}$$ Distributive property; simplify.

$$\frac{\mathbf{3 \cdot \cancel{4}}}{1}\left(\frac{1}{\cancel{4}}x\right) + \frac{\mathbf{\cancel{3} \cdot 4}}{1}\left(\frac{2}{\cancel{3}}y\right) = 60$$ Find common factors.

$$\mathbf{3x + 8y} = 60$$ Simplify; the fractions are cleared.

$$\underline{-8y \qquad -8y}$$ Subtraction property of equality.

$$3x + \mathbf{0} = 60 \mathbf{- 8y}$$ Simplify.

$$\frac{\mathbf{3x}}{3} = \frac{60}{3} - \frac{8y}{3}$$ Division property of equality.

$$x = -\frac{8}{3}y + 20$$ Simplify; change order.

Practice Problems

For problems 1–3, solve for the given variable.

1. $3x + 4y = 12$ for x **2.** $-5x + 2y = 15$ for y

3. $\frac{2}{9}x + \frac{1}{12}y = 20$ for y

Solving a Formula for a Variable

When using a formula to solve an application problem, we often first solve the formula for a variable.

EXAMPLE 4 The formula $D = \frac{M}{V}$ represents the relationship of density D, mass M, and volume V. Solve for M.

SOLUTION ▶

$D = \frac{M}{V}$ We need to isolate M.

$(V)(D) = \left(\frac{\cancel{V}}{1}\right)\left(\frac{M}{\cancel{V}}\right)$ Multiplication property of equality.

$\mathbf{VD = M}$ Simplify; M is isolated.

Since we read from left to right, we often rewrite the formula so that the isolated variable is on the left side and change the order on the right side so that the variables are arranged in alphabetical order.

$M = VD$ Rewrite with the isolated variable on the left.

$M = \mathbf{DV}$ Change the order; D comes before V in the alphabet.

Many formulas include subscripts. A **subscript** is a small number or letter written at the lower right of a variable that is part of the name of the variable. For example, in the formula $V_f = V_i + at$, the variable V_f represents a final velocity, and V_i represents an initial velocity. The subscripts show that they are two different variables. We say, "V sub f" or "$V f$."

In the next example, we divide both sides of the formula by a. When we divide $V_f - V_i$ by a, we can write the quotient as a single fraction $\frac{V_f - V_i}{a}$ or as the difference of two fractions, $\frac{V_f}{a} - \frac{V_i}{a}$. To make it easier to enter a formula into a spreadsheet or other software program, we often write expressions as single fractions.

EXAMPLE 5 The formula $V_f = V_i + at$ represents the relationship of constant acceleration a, time t, initial velocity V_i, and final velocity V_f. Solve this formula for t.

SOLUTION ▶

$V_f = V_i + at$

$-V_i \quad -V_i$ Subtraction property of equality.

$\mathbf{V_f - V_i = 0 + at}$ Simplify.

$\frac{V_f - V_i}{a} = \frac{at}{a}$ Division property of equality.

$\frac{V_f - V_i}{a} = \mathbf{t}$ Simplify.

$t = \frac{V_f - V_i}{a}$ Rewrite with the isolated variable on the left.

The formula $d = rt$ represents the relationship of distance d, rate r, and time t. In the next example, we solve this formula for t.

EXAMPLE 6 **(a)** Solve $d = rt$ for t.

SOLUTION ▸

$d = rt$

$\frac{d}{r} = \frac{rt}{r}$ Division property of equality.

$\frac{d}{r} = \boldsymbol{t}$ Simplify.

$t = \frac{d}{r}$ Rewrite with the isolated variable on the left.

(b) Use $t = \frac{d}{r}$ to find the time in hours and minutes to travel 80 mi at a speed of $\frac{45\text{ mi}}{1\text{ hr}}$. Round to the nearest minute.

▸

$t = \frac{d}{r}$

$t = \dfrac{\mathbf{80\ mi}}{\dfrac{\mathbf{45\ mi}}{\mathbf{1\ hr}}}$ Replace the variables with the given values.

$t = 80\text{ mi} \div \dfrac{\mathbf{45\ mi}}{\mathbf{1\ hr}}$ Rewrite the fraction as division.

$t = 80\,\cancel{\text{mi}} \cdot \dfrac{\mathbf{1\ hr}}{\mathbf{45}\,\cancel{\mathbf{mi}}}$ Rewrite division as multiplication by the reciprocal.

$t = \mathbf{1.77\ldots\ hr}$ Simplify; $\frac{\text{mi}}{\text{mi}} = 1$

$t = \mathbf{1\ hr + 0.77\ldots\ hr}$ Write as a sum of 1 hour + part of an hour.

$t = 1\text{ hr} + \left(\frac{0.77\ldots\,\cancel{\text{hr}}}{1}\right)\left(\frac{60\text{ min}}{1\,\cancel{\text{hr}}}\right)$ Multiply by a conversion factor equal to 1.

$t = 1\text{ hr} + \mathbf{46.66\ldots\ min}$ Simplify; $\frac{\text{hr}}{\text{hr}} = 1$

$t \approx 1\text{ hr}\ \mathbf{47\ min}$ Round to the nearest minute.

When we invest money at a given interest rate, the value of the money grows. The original amount of money invested is the **principal**, and the amount of money in the investment after a given time is its **future value**. A formula for finding the future value A of an amount P invested at an annual interest rate R for T years is $A = P + PRT$. To solve this formula for T, divide both sides by a product of two variables, PR. The fraction $\frac{PR}{PR}$ equals 1.

EXAMPLE 7 **(a)** Solve $A = P + PRT$ for T.

SOLUTION ▸

$$\begin{array}{r} A = P + PRT \\ -P \quad -P \\ \hline \end{array}$$ Subtraction property of equality.

$\mathbf{A - P = 0 + PRT}$ Simplify.

$\frac{A - P}{PR} = \frac{\mathbf{PRT}}{PR}$ Division property of equality.

$\frac{A - P}{PR} = \boldsymbol{T}$ Simplify; $\frac{PR}{PR} = 1$

$T = \frac{A - P}{PR}$ Rewrite with the isolated variable on the left.

(b) An investment of \$5000 earns 6.5% simple annual interest. Find the time when the future value of the investment is \$7600.

Write the interest rate in decimal form, $\frac{0.065}{1\text{ year}}$.

$$T = \frac{A - P}{PR}$$

$$T = \frac{\$7600 - \$5000}{\left(\frac{\$5000}{1}\right)\left(\frac{0.065}{1\text{ year}}\right)}$$ Replace the variables A, P, and R.

$$T = \frac{\$2600}{\left(\frac{\$325}{1\text{ year}}\right)}$$ Simplify.

$$T = \$2600 \div \frac{\$325}{1\text{ year}}$$ Rewrite the fraction as division.

$$T = \frac{\$2600}{1} \cdot \frac{1\text{ year}}{\$325}$$ Rewrite division as multiplication by the reciprocal.

$$T = 8\text{ years}$$ Simplify.

Practice Problems

4. Solve $x = v_1 t$ for t.

5. Solve $PV = nRT$ for V.

6. a. Solve $A = LW$ for W.

b. The length of a rectangle is 123 ft, and its area is 9963 ft^2. Find the width of the rectangle.

7. a. Solve $P = 2L + 2W$ for L.

b. The width of a rectangle is 152 m, and its perimeter is 1290 m. Find the length of the rectangle.

2.4 Vocabulary Practice

Match the term with its description.

1. This operation is equivalent to multiplication by the reciprocal of the fraction.
2. The original value of an investment
3. The combined value of the principal of an investment and the interest earned
4. The terms y and $5y$ are examples of this.
5. The distance around a rectangle
6. The product of a real number (except 0) and this is always 1.
7. $\approx$
8. The value of a fraction in which the numerator and the denominator are equivalent
9. We can multiply both sides or divide both sides of an equation by a real number that is not equal to 0. The solution is not changed.
10. We can add a term to or subtract a term from both sides of an equation. The solution is not changed.

A. addition or subtraction property of equality
B. approximately equal to
C. division by a fraction
D. future value
E. like terms
F. multiplication or division property of equality
G. one
H. perimeter
I. principal
J. reciprocal

2.4 Exercises

Follow your instructor's guidelines for showing your work.

For exercises 1–28, solve the equation for y. Write the equation to match the pattern $y = mx + b$.

1. $3x + y = 20$

2. $9x + y = 40$

3. $3x - y = 20$

4. $9x - y = 40$

5. $3x - 6y = 20$

6. $9x - 10y = 40$

7. $7x + 2y = 15$

8. $8x + 3y = 25$

9. $\frac{1}{3}x + 4y = -8$

10. $\frac{1}{4}x + 5y = -10$

11. $\frac{2}{3}x + \frac{4}{5}y = 12$

12. $\frac{3}{8}x + \frac{2}{9}y = 16$

13. $y - 6 = 3(x - 9)$

14. $y - 8 = 5(x - 7)$

15. $3x = 4y$

16. $5x = 6y$

17. $-3x = \frac{1}{4}y$

18. $-5x = \frac{1}{6}y$

19. $y - 4 = \frac{3}{4}(x - 9)$

20. $y - 5 = \frac{7}{8}(x - 3)$

21. $-40x + 51y = 12$

22. $-28x + 45y = 63$

23. $2y = 9x$

24. $3y = 10x$

25. $\frac{5}{6}x + \frac{9}{20}y = 30$

26. $\frac{8}{15}x + \frac{3}{10}y = 20$

27. $-\frac{1}{2}x + \frac{3}{5}y = \frac{1}{4}$

28. $-\frac{1}{2}x + \frac{3}{7}y = \frac{1}{4}$

For exercises 29–36, solve the equation for x.

29. $3x - y = 20$

30. $9x - y = 40$

31. $3x - 6y = 20$

32. $9x - 10y = 40$

33. $\frac{2}{3}x + \frac{4}{5}y = 12$

34. $\frac{3}{8}x + \frac{2}{9}y = 16$

35. $3x = 4y$

36. $5x = 6y$

37. Solve $8A + 20B = 100$ for A.

38. Solve $6C + 14D = 112$ for C.

39. Solve $1000P - 3500R = 7000$ for P.

40. Solve $2000A + 4500B = 18{,}000$ for A.

41. Solve $-F - G = 300$ for G.

42. Solve $-H - K = 800$ for K.

43. Solve $\frac{R}{2} + \frac{W}{5} = 21$ for R.

44. Solve $\frac{M}{3} + \frac{Y}{7} = 4$ for M.

45. Solve $A = LW$ for L.

46. Solve $d = rt$ for r.

47. Solve $PV = nRT$ for n.

48. Solve $PV = nRT$ for T.

49. Solve $A = P + PRT$ for R.

50. Solve $P = 2L + 2W$ for W.

51. Solve $V = LWH$ for H.

52. Solve $V = LWH$ for W.

53. Solve $A = \frac{1}{2}bh$ for b.

54. Solve $A = \frac{1}{2}bh$ for h.

55. Solve $F = \frac{9}{5}C + 32$ for C.

56. Solve $C = \frac{5}{9}(F - 32)$ for F.

57. Solve $V = \frac{LWT}{144}$ for T.

58. Solve $V = \frac{LWT}{144}$ for W.

59. Solve $A = \frac{x_1 + x_2 + x_3}{3}$ for x_1.

60. Solve $A = \frac{x_1 + x_2 + x_3}{3}$ for x_3.

61. Solve $C = 2\pi r$ for r.

62. Solve $C = \pi D$ for D.

63. Solve $P = \frac{1}{3}bh$ for b.

64. Solve $P = \frac{1}{3}bh$ for h.

65. Solve $L = C - H - \frac{1}{5}T$ for H.

66. Solve $L = C - H - \frac{1}{5}T$ for C.

67. Solve $R = \frac{E + C}{L}$ for C.

68. Solve $R = \frac{E + C}{L}$ for E.

69. Each of the three sides of an equilateral triangle is the same length. If P is the perimeter of the triangle and L is the length of a side, write a formula in P and L for the perimeter of an equilateral triangle and solve this formula for L.

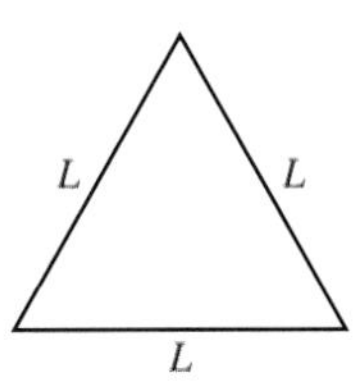

70. Each of the four sides of a square is the same length. If P is the perimeter of the square and S is the length of a side, write a formula in P and S for the perimeter of a square and solve this formula for S.

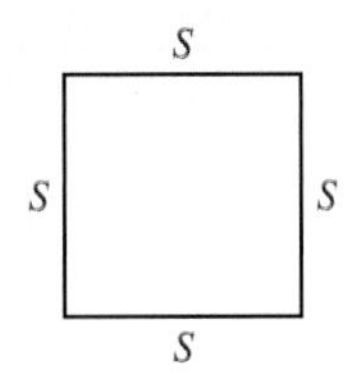

71. a. Solve $d = rt$ for r.
b. Find the speed in miles per hour needed to travel 240 mi in 3 hr 20 min.

72. a. Solve $d = rt$ for r.
b. Find the speed in miles per hour needed to travel 350 mi in 5 hr 20 min. Round to the nearest whole number.

73. a. Solve $A = P + PRT$ for R.
b. An investment of $500 grows to a future value of $650 in 5 years. Find the simple interest rate.

74. a. Solve $A = P + PRT$ for R.
b. An investment of $600 grows to a future value of $900 in 5 years. Find the simple interest rate.

75. a. Solve $D = \frac{M}{V}$ for V.
b. The density D of gasoline is $\frac{0.737 \text{ g}}{1 \text{ mL}}$, and the mass M is 50 g. Find the volume, V. Round to the nearest tenth.

76. a. Solve $D = \frac{M}{V}$ for V.
b. The density D of antifreeze is $\frac{1.1132 \text{ g}}{1 \text{ mL}}$, and the mass M is 12 g. Find the volume V. Round to the nearest tenth.

For exercises 77–78, $V = \frac{LWT}{144 \text{ in.}^3}$ where V is the volume of lumber in board feet, L is the length, W is the width, and T is the thickness. The units of length, width, and thickness are inches.

77. a. Solve $V = \frac{LWT}{144 \text{ in.}^3}$ for L. Include the units of measurement.
b. The thickness of each board in a pile is 1.5 in., and the width of each board is 7.25 in. The volume of the boards is 348 board feet. Find the total length of the boards.
c. Change this length into feet.

78. a. Solve $V = \frac{LWT}{144 \text{ in.}^3}$ for L. Include the units of measurement.
b. The thickness of each board in a pile is 1.5 in., and the width of each board is 3.5 in. The volume of the boards is 182 board feet. Find the total length of the boards.
c. Change this length into feet.

79. a. Solve $V_f = V_i + at$ for a.
b. The initial speed V_i of a 2007 Nissan 350Z is $\frac{0 \text{ ft}}{1 \text{ s}}$. The final speed V_f is $\frac{88 \text{ ft}}{1 \text{ s}}$. If the time t is 5.3 s, find the acceleration, a. Round to the nearest tenth.

80. a. Solve $V_f = V_i + at$ for a.
b. The initial speed V_i of a 2008 Ford Mustang is $\frac{0 \text{ ft}}{1 \text{ s}}$. The final speed V_f is $\frac{88 \text{ ft}}{1 \text{ s}}$. If the time t is 7.6 s, find the acceleration, a. Round to the nearest tenth.

81. a. Solve $A = \frac{h(b_1 + b_2)}{2}$ for b_1.
b. One of the bases b_1 of a trapezoid is 10 in., the height is 4 in., and the area is 70 in.^{2}. Find the other base b_2.

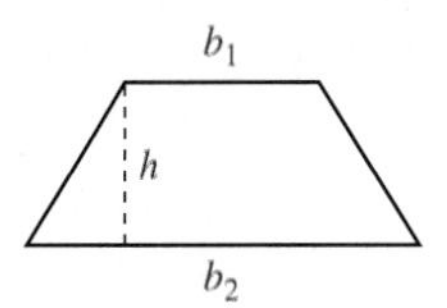

82. a. Solve $V = \frac{hw(b_1 + b_2)}{2}$ for b_2.
b. One of the bases b_2 of a trapezoidal prism is 12 in., the height is 4 in., the width is 10 in., and the volume is 400 in.^{2}. Find the other base, b_1.

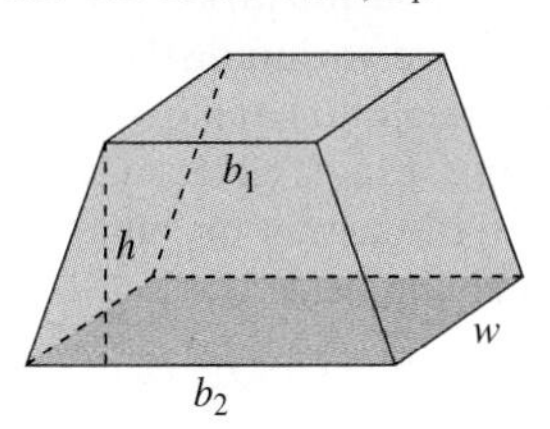

For exercises 83–84, a rectangular solid has C corners, E edges, and F faces.

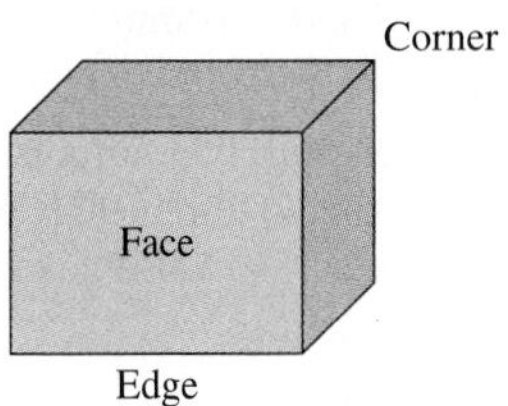

83. a. Solve $C - E + F = 2$ for E.
b. A rectangular shoebox has eight corners (C) and six faces (F). Find the number of edges E.

84. a. Solve $C - E + F = 2$ for F.
b. A rectangular book has eight corners (C) and 12 edges (E). Find the number of faces (F).

Problem Solving: Practice and Review

Follow your instructor's guidelines for using the five steps as outlined in Section 1.3, p. 55.

85. The cost of one double orchid lei from The Hawaiian Lei Company is \$25.08. Find the total cost for an order of three double orchid leis sent to the mainland, including the FedEx shipping charge and the shipping surcharge.

Number of leis	FedEx shipping charge	Surcharge
1	\$17.92	\$8
2	\$21.28	\$8
3	\$25.52	\$8
4	\$26.88	\$8

Source: www.hawaiianleicompany.com

86. The 2010 First College Year Study surveyed 201,818 U.S. full-time college freshmen. Find the number of students who received grants and scholarships to attend college. Round to the nearest hundred.

Higher levels in the use of loans that we reported in 2009 continue in 2010, with 53.1% of incoming students using loans as part of the financial package needed to attend college . . . more students also reported receiving grants and scholarships to attend college, as this figure moved up . . . to 73.4% in 2010. (*Source:* www.heri.ucla.edu, Jan. 2011)

87. The production cost for a tool is \$15.75 per tool. The weekly overhead is \$25,000. The price of the tool is \$19.99. Find the number of tools that should be made and sold to break even. Round to the nearest whole number.

88. The rectangular playing field of a pool table is 50 in. wide and 100 in. long. The top of the entire table, including the playing field and the area of the table that surrounds the playing field, is 62 in. wide and 112 in. long. Find the area of the table that surrounds the playing field.

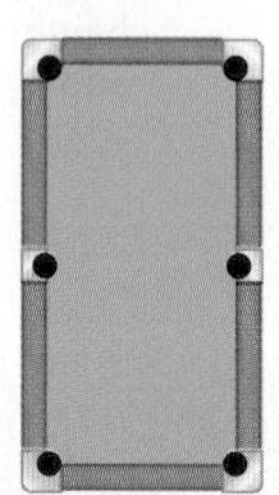

Technology

For problems 89–92, do the arithmetic with a calculator.

89. The area A of a rectangular playground is 28,800 ft^2. The length L is 180 ft. Solve the formula $A = LW$ for L, and use it to find the width W of the playground.

90. The volume V of an atrium in a hotel is 648,000 ft^3. The rectangular floor of the atrium is 90 ft wide and 120 ft long. Solve the formula $V = LWH$ for H, and use it to find the height of the atrium.

91. The principal P of an investment is \$1,650,900. The annual simple interest rate R is 6.25%. Use the formula $A = P + PRT$ to find the future value of the investment in 25 years.

92. The volume V of a beaker of turpentine is 325 mL, and the mass M of the turpentine is 283.16 g. Use the formula $D = \frac{M}{V}$ to find the density of the turpentine. Round to the nearest thousandth.

Find the Mistake

For exercises 93–96, the completed problem has one mistake.
(a) Describe the mistake in words, or copy down the whole problem and highlight or circle the mistake.
(b) Do the problem correctly.

93. Problem: Solve $12x - 19y = 38$ for y.

Incorrect Answer:

$$\begin{aligned} 12x - 19y &= 38 \\ \underline{-12x \qquad\quad} &\;\; \underline{-12x} \\ 19y &= -12x + 38 \\ \frac{19y}{19} &= \frac{-12x}{19} + \frac{38}{19} \\ y &= -\frac{12}{19}x + 2 \end{aligned}$$

94. Problem: Solve $15x - 6y = 45$ for x.

Incorrect Answer:

$$\begin{aligned} 15x - 6y &= 45 \\ \underline{-6y} &\;\; \underline{-6y} \\ 15x &= -6y + 45 \\ \frac{15x}{15} &= -\frac{6y}{15} + \frac{45}{15} \\ x &= -\frac{2}{5}y + 3 \end{aligned}$$

95. Problem: Solve $8x - 2y = 54$ for x.

Incorrect Answer:

$$\begin{aligned} 8x - 2y &= 54 \\ \underline{+2y} &\;\; \underline{+2y} \\ 8x + 0 &= 56 \\ \frac{8x}{8} &= \frac{56}{8} \\ x &= 7 \end{aligned}$$

96. Problem: Solve $V_f - V_i = at$ for V_i.

Incorrect Answer:

$$\begin{aligned} V_f - V_i &= at \\ \underline{-V_f \qquad} &\;\; \underline{-V_f} \\ V_i &= at - V_f \end{aligned}$$

Review

For problems 97–100, the symbol > means "greater than," and the symbol < means "less than." Replace the blank ____ with < or > to create a true statement.

97. 15 ____ 2

98. −15 ____ 2

99. −3 ____ −8

100. −12 ____ 0

SUCCESS IN COLLEGE MATHEMATICS

101. Monitoring or participating in social networks, texting, surfing the Internet, and using other technology can affect your concentration. You might need to limit multitasking to do effective studying. List the other tasks that you can manage at the same time that you are doing math homework or studying for a math test.

102. Family members, roommates, or friends can affect your concentration. Describe how you manage to study effectively and still maintain your relationships with family members, roommates, or friends.

© Daboost/Shutterstock.com

If an amusement park lets only people who are *at least* 54 in. tall ride a roller coaster, height is a constraint on riding the roller coaster. We can represent this constraint with an inequality, $x \geq 54$ in. In this section, we will write and solve inequalities and constraints.

SECTION 2.5 Inequalities in One Variable

After reading the text, working the practice problems, and completing assigned exercises, you should be able to:

1. Use a number line graph to represent an inequality.
2. Solve an inequality; check the solution.
3. Use an inequality to solve an application problem.

Inequalities in One Variable

The inequality $x \leq 3$ represents the words "x is less than or equal to 3." The number 3 is a **boundary value**. To represent this inequality with a number line graph, use a solid dot or a square bracket to show that the boundary value is a solution of the inequality. Use an arrowed line to represent the solutions of the inequality that are less than 3.

EXAMPLE 1 Use a number line graph to represent $x \leq 3$.

SOLUTION

−2 −1 0 1 2 3 4 5 x

A square bracket is graphed at the boundary value, 3.

−2 −1 0 1 2 3 4 5 x

A solid dot is graphed at the boundary value, 3.

The inequality $x > 6$ represents the words "x is greater than 6." The number 6 is a boundary value. To represent this inequality with a number line graph, use an open circle or a parenthesis to show that the boundary value is *not* a solution of the inequality. Use an arrowed line to represent the solutions of the inequality that are greater than 6.

EXAMPLE 2 Use a number line graph to represent $x > 6$.

SOLUTION

4 5 6 7 8 9 10 11 x

A parenthesis is graphed at the boundary value, 6.

4 5 6 7 8 9 10 11 x

An open circle is graphed at the boundary value, 6.

Since representing a boundary value with either a parenthesis or a square bracket is preparation for later work in algebra with interval notation, this is the notation that is used in the rest of the examples. Ask your instructor how you should represent boundary values on number line graphs.

Practice Problems

For problems 1–4, use a number line graph to represent the inequality.

1. $x > 3$ **2.** $x < -4$ **3.** $x \geq -1$ **4.** $x \leq 0$

Solving an Inequality in One Variable

To solve an equation, we use the properties of equality to isolate the variable by adding, subtracting, multiplying, or dividing both sides of the equation. To solve an inequality, we use the **properties of inequality** to isolate the variable.

The Properties of Inequality

The Addition and Subtraction Property of Inequality We can add a term to or subtract a term from both sides of an inequality. The solution is not changed.

The Multiplication and Division Property of Inequality We can multiply both sides or divide both sides of an inequality by a real number (except for 0). If the value of the real number is *less than* 0, **reverse the direction of the inequality sign**. If the value of the real number is *greater than* 0, the direction of the inequality sign does not change.

To solve an equation in one variable, we use the addition or subtraction property of equality to isolate the variable term and use the multiplication or division property of equality to change the coefficient of the variable term to 1. We apply the properties of inequality in the same way.

EXAMPLE 3 **(a)** Solve: $x + 4 < 7$

SOLUTION

$$x + 4 < 7$$

$$\underline{\;-4 \quad -4\;}$$ Subtraction property of inequality.

$$\mathbf{x + 0 < 3}$$ Simplify.

$$\mathbf{x} < 3$$ Simplify.

(b) Use a number line graph to represent the solution.

−2 −1 0 1 2 3 4 5 x

Every real number less than 3 is a solution of this inequality. Each point on the arrowed line represents a solution. The parenthesis shows that the boundary value, 3, is not a solution.

EXAMPLE 4 **(a)** Solve: $4a - 20 \geq -80$

SOLUTION

$$4a - 20 \geq -80$$

$$\underline{\;+20 \quad +20\;}$$ Addition property of inequality.

$$\mathbf{4a + 0 \geq -60}$$ Simplify.

$$\frac{\mathbf{4a}}{4} \geq \frac{-60}{4}$$ Division property of inequality.

$$a \geq -15$$ Simplify.

(b) Use a number line graph to represent the solution.

−17 −15 −13 −11 a

Every real number that is greater than or equal to −15 is a solution of this inequality. Since −15 is a solution, use a square bracket. Notice that the number line is labeled with the variable in the inequality, *a*.

To check whether the inequality sign of the solution is correct, check a solution that is not the boundary value.

EXAMPLE 5 **(a)** Solve: $3x + 15 < 2x + 8$

SOLUTION

$3x + 15 < 2x + 8$	
$-15 \quad -15$	Subtraction property of inequality.
$3x + \mathbf{0} < 2x - \mathbf{7}$	Simplify.
$-2x \quad -2x$	Subtraction property of inequality.
$\mathbf{x} + 0 < \mathbf{0} - 7$	Simplify.
$x < -7$	Simplify.

(b) Use a number line graph to represent the solution.

−11 −9 −7 −5 x

(c) Check the inequality sign in the solution.

Choose a solution of the inequality that is not a boundary value and that is less than −7. In this example, we choose to check −8, the first integer less than −7.

$3x + 15 < 2x + 8$	Use the original inequality.
$3(\mathbf{-8}) + 15 < 2(\mathbf{-8}) + 8$	Replace the variable, x, with a solution of the inequality, −8.
$\mathbf{-24} + 15 < \mathbf{-16} + 8$	Follow the order of operations.
$-9 < -8$	Since the inequality is true, the inequality sign is correct.

If we multiply both sides of the inequality $4 < 12$ by a positive number, the new inequality is true. If we multiply both sides of the inequality $4 < 12$ by a negative number, the new inequality is false. This is why the multiplication and division property of inequality says to reverse the direction of the inequality sign when multiplying or dividing both sides of the inequality by a negative number.

If we multiply or divide both sides of the inequality $4 < 12$ by a positive number, the inequality is true.	$4 < 12$ $2 \cdot 4 < 2 \cdot 12$ $8 < 24$ True.	$4 < 12$ $\frac{4}{2} < \frac{12}{2}$ $2 < 6$ True.
If we multiply or divide both sides of the inequality $4 < 12$ by a negative number, the inequality is **false**.	$4 < 12$ $-2 \cdot 4 < -2 \cdot 12$ $-8 < -24$ False.	$4 < 12$ $\frac{4}{-2} < \frac{12}{-2}$ $-2 < -6$ False.
If we multiply or divide both sides of the inequality $4 < 12$ by a negative number *and reverse the direction of the inequality sign*, the inequality is true.	$4 < 12$ $-2 \cdot 4 > -2 \cdot 12$ $-8 > -24$ True.	$4 < 12$ $\frac{4}{-2} > \frac{12}{-2}$ $-2 > -6$ True.

EXAMPLE 6 **(a)** Solve: $-3x - 3 > 15$

SOLUTION

$-3x - 3 > 15$	
$+3 \quad +3$	Addition property of inequality.
$-3x + \mathbf{0} > \mathbf{18}$	Simplify.
$\frac{\mathbf{-3x}}{-3} < \frac{18}{-3}$	Division property of inequality; reverse the inequality sign.
$x < -6$	Simplify.

(b) Use a number line graph to represent the solution.

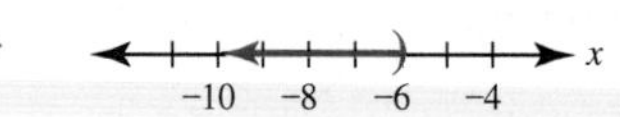

(c) Check the inequality sign in the solution.

Choose a solution of the inequality that is not the boundary value and that is less than −6. In this example, we choose to check −7.

$-3x - 3 > 15$	Use the original inequality.
$-3(\mathbf{-7}) - 3 > 15$	Replace x with a solution of the inequality, −7.
$\mathbf{21} - 3 > 15$	Follow the order of operations.
$\mathbf{18} > 15$	Since the inequality is true, the inequality sign is correct.

Solving an Inequality in One Variable

1. Use the distributive property to simplify the inequality. Combine like terms.
2. Use the addition or subtraction property of inequality so that there is only one term with a variable in the inequality.
3. Use the addition or subtraction property of inequality to isolate the term that includes the variable. Any constant on the same side as the variable is changed to 0.
4. Use the multiplication or division property of inequality to change the coefficient of the variable term to 1. If the inequality is multiplied or divided on both sides by a negative number, reverse the direction of the inequality sign.
5. Check the inequality sign in the solution.

EXAMPLE 7 **(a)** Solve: $18 \geq -2(x - 12)$

SOLUTION

$18 \geq -2(x - 12)$	
$18 \geq \mathbf{-2}(x) - \mathbf{2}(-12)$	Distributive property.
$18 \geq \mathbf{-2x + 24}$	Simplify.
$\underline{-24 \qquad -24}$	Subtraction property of inequality.
$\mathbf{-6} \geq -2x + \mathbf{0}$	Simplify.
$\dfrac{-6}{-2} \leq \dfrac{\mathbf{-2x}}{-2}$	Division property of inequality; reverse the inequality sign.
$3 \leq x$	Simplify.

We can rewrite $3 \leq x$ with the variable isolated on the left side as $x \geq 3$. The statement "3 is less than or equal to x" is equivalent to "x is greater than or equal to 3." Notice that the inequality sign "points" at 3 in both inequalities.

(b) Use a number line graph to represent the solution.

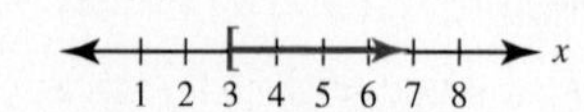

(c) Check the inequality sign in the solution.

Choose a solution of the inequality that is not the boundary value and that is greater than 3. In this example, we choose to check 4.

$18 \geq -2(x - 12)$ Use the original inequality.

$18 \geq -2(\mathbf{4} - 12)$	Replace the variable with a solution, 4.
$18 \geq -2(\mathbf{-8})$	Follow the order of operations.
$18 \geq \mathbf{16}$	Since the inequality is true, the inequality sign is correct.

Practice Problems

For problems 5–10,
(a) solve.
(b) use a number line graph to represent the solution.
(c) check the direction of the inequality sign.

5. $4x + 20 > 12$ **6.** $4x + 20 \geq -12$ **7.** $-4x + 20 < 12$
8. $-4x + 20 > -12$ **9.** $3x - 25 > -2x + 35$
10. $5(x + 2) + 4 \leq 7x + 14$

Inequalities in One Variable with Fractions or Decimals

To clear the fractions in an inequality, multiply both sides of the inequality by the least common denominator of the fractions and simplify.

EXAMPLE 8 **(a)** Solve: $\frac{3}{4}x + \frac{1}{2} \geq 6$

SOLUTION ▶ The least common denominator is 4.

$$\frac{3}{4}x + \frac{1}{2} \geq 6$$

$$\left(\frac{4}{1}\right)\left(\frac{3}{4}x + \frac{1}{2}\right) \geq (4)(6) \quad \text{Multiplication property of equality.}$$

$$\frac{\mathbf{4}}{\mathbf{1}}\left(\frac{3}{4}x\right) + \frac{\mathbf{4}}{\mathbf{1}}\left(\frac{1}{2}\right) \geq \mathbf{24} \quad \text{Distributive property; simplify.}$$

$$\frac{\cancel{4}}{1}\left(\frac{3}{\cancel{4}}x\right) + \frac{\mathbf{2 \cdot 2}}{1}\left(\frac{1}{2}\right) \geq 24 \quad \text{Find common factors.}$$

$$\mathbf{3x + 2} \geq 24 \quad \text{Simplify; the fractions are cleared.}$$

$$\underline{\quad -2 \quad -2} \quad \text{Subtraction property of inequality.}$$

$$3x + \mathbf{0} \geq \mathbf{22} \quad \text{Simplify.}$$

$$\frac{\mathbf{3x}}{3} \geq \frac{22}{3} \quad \text{Simplify; division property of inequality.}$$

$$x \geq \frac{22}{3} \quad \text{Simplify; do not rewrite as a mixed or decimal number.}$$

(b) Use a number line graph to represent the solution.

▶ We can choose a scale that is divided into one-thirds.

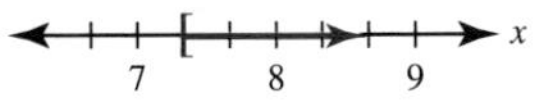

Or, we can we choose a scale with integers and estimate the location of $\frac{22}{3}$.

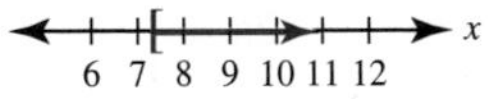

(c) Check the inequality sign in the solution.

▶ Since $\frac{22}{3} \approx 7.3$, choose to check 8, which is the first integer greater than 7.3.

$\frac{3}{4}x + \frac{1}{2} \geq 6$	Use the original inequality.
$\frac{3}{4}\left(\frac{\mathbf{8}}{\mathbf{1}}\right) + \frac{1}{2} \geq 6$	Replace the variable, x, with a solution, 8.
$\frac{3}{\not{4}}\left(\frac{\mathbf{2 \cdot \not{4}}}{1}\right) + \frac{1}{2} \geq 6$	Find common factors.
$\mathbf{6} + \frac{1}{2} \geq 6$	Follow the order of operations.
$\frac{6}{1}\cdot\frac{2}{2} + \frac{1}{2} \geq \frac{6}{1}\cdot\frac{2}{2}$	Multiply by a fraction equal to 1.
$\frac{\mathbf{12}}{\mathbf{2}} + \frac{1}{2} \geq \frac{\mathbf{12}}{\mathbf{2}}$	Follow the order of operations.
$\frac{\mathbf{13}}{\mathbf{2}} \geq \frac{12}{2}$	Since the inequality is true, the inequality sign is correct.

When solving inequalities that include decimals, either clear the decimals or do decimal arithmetic. If the original inequality includes decimals, the solution is usually also written as a decimal number rather than a fraction.

EXAMPLE 9 **(a)** Solve: $0.3x + 0.12 \geq 0.06x$

SOLUTION ▶ The digit farthest to the right in 0.3, 0.12, and 0.06 is in the hundredths place. Multiplying both sides of the equation by 100 will change these coefficients to integers.

$0.3x + 0.12 \geq 0.06x$	
$(100)(0.3x + 0.12) \geq (100)(0.06x)$	Multiplication property of equality.
$\mathbf{100}(0.3x) + \mathbf{100}(0.12) \geq \mathbf{6x}$	Distributive property; simplify.
$\mathbf{30x + 12} \geq 6x$	Simplify; the decimals are cleared.
$-30x \quad -30x$	Subtraction property of inequality.
$\mathbf{0} + 12 \geq \mathbf{-24x}$	Simplify.
$\frac{\mathbf{12}}{-24} \leq \frac{-24x}{-24}$	Division property of inequality; reverse the sign.
$-0.5 \leq x$	Simplify.

(b) Use a number line graph to represent the solution.

▶ The inequality $-0.5 \leq x$ is equivalent to $x \geq -0.5$.

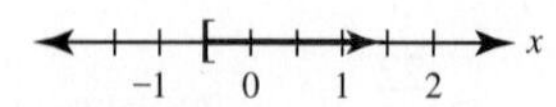

Practice Problems

For problems 11–15,
(a) clear the fractions or decimals and solve.
(b) use a number line graph to represent the solution.
(c) check the inequality sign in the solution.

11. $\frac{2}{3}x + \frac{1}{5} < -2x + 9$ **12.** $-\frac{3}{4}x + \frac{1}{2} \geq \frac{5}{8}$ **13.** $0.2x + 0.9 \leq -1.7$

14. $-0.4x + 8 > 0.6x + 20$ **15.** $\frac{1}{2}(3x + 6) - 8 \geq 0$

Applications

An inequality that represents a situation in which there is a maximum or minimum amount is a **constraint**.

EXAMPLE 10 To pass a math class, a student must score at least 72% on the final exam. Assign the variable, and write an inequality that represents this constraint.

SOLUTION ▶

m = score on final exam — Assign the variable.

$m \geq 72\%$ — score $\geq$ 72%

We can use inequalities to represent situations described by the phrases "no more than," "less than," "at least," "nearly," "limited to," and "greater than."

EXAMPLE 11 A family is applying for a mortgage of $120,000. Assign the variable, and write an inequality that represents the constraint on the total commission and additional fees that can be charged.

The total commission and additional fees charged by lenders and others involved in the mortgage process will effectively be limited to a maximum 3 percent of the loan amount. (*Source:* www.nytimes.com, July 7, 2010)

SOLUTION ▶

T = total commission and additional fees — Assign the variable.

$T \leq (0.03)(\$120{,}000)$ — total $\leq$ 3% of the loan amount

We can use an inequality to solve a problem situation that includes a maximum or a minimum value.

EXAMPLE 12 The adult population of the United States in 2010 was about 234,564,000 people. Find the number of these adults who report binge drinking. (*Source:* www.census .gov, 2011)

More than 15% of US adults report binge drinking. It is most common in men, adults in the 18–34 age range, and people with household incomes of $75,000 or more. (*Source:* www.cdc.gov, Oct. 2010)

SOLUTION ▶ **Step 1 Understand the problem.**

The unknown is the number of adults who report binge drinking.

N = number of adults who report binge drinking

Step 2 Make a plan.

Since number of parts = (percent) (total parts) and the number of parts is *more than* 15% of the total parts, a word inequality is number who report binge drinking > (percent binge drinking) (number of adults).

Step 3 Carry out the plan.

$N > (0.15)(234{,}564{,}000 \text{ people})$ — number of parts > (part)(total parts)

$N > \mathbf{35{,}184{,}600 \text{ people}}$ — Simplify.

Step 4 Look back.

Working backwards, $\dfrac{35{,}184{,}600 \text{ people}}{234{,}564{,}000 \text{ people}}$ equals 0.15, or 15%. Since more than this percent of people report binge drinking, the answer seems reasonable.

Step 5 Report the solution.

Of 234,564,000 adults, more than 35,184,600 adults report binge drinking.

In the next example, the inequality is a proportion that includes a known rate and an unknown rate.

EXAMPLE 13 In 2009, the population of a city included 18,000 youths age 13–17 years. Find the number of these youths who are vaccinated against meningitis.

Public Health recommends that all adolescents receive an annual influenza vaccine and that 11–12 year olds receive a meningococcal conjugate vaccine, along with a booster dose between 16 and 18 years of age. In 2009, fewer than two out of three youth 13–17 years of age had been vaccinated against meningitis. (*Source:* publichealth.lacounty.gov, April 14, 2011)

SOLUTION **Step 1 Understand the problem.**

The unknown is the number of youths who are vaccinated against meningitis.

N = number of youths vaccinated against meningitis

Step 2 Make a plan.

The rate is $\frac{\text{youths vaccinated against meningitis}}{\text{total number of youths}}$. The unknown rate is $\frac{N}{18{,}000 \text{ youths}}$.

The known rate is $\frac{2 \text{ youths}}{3 \text{ youths}}$. The unknown rate is less than the known rate.

Step 3 Carry out the plan.

$$\frac{N}{18{,}000 \text{ youths}} < \frac{2 \text{ youths}}{3 \text{ youths}} \quad \text{unknown rate} < \text{known rate}$$

$$\frac{N}{18{,}000} < \frac{2}{3} \quad \text{Remove units.}$$

$$\left(\frac{18{,}000}{1}\right)\left(\frac{N}{18{,}000}\right) < \left(\frac{18{,}000}{1}\right)\left(\frac{2}{3}\right) \quad \text{Multiplication property of inequality.}$$

$$N < \left(\frac{\cancel{3} \cdot 6000}{1}\right)\left(\frac{2}{\cancel{3}}\right) \quad \text{Simplify; find common factors.}$$

$$N < \mathbf{12{,}000 \text{ youths}} \quad \text{Simplify; the units are youths.}$$

Step 4 Look back.

Since 12,000 youths is the boundary value, then the unknown rate of $\frac{11{,}999 \text{ youths}}{18{,}000 \text{ youths}}$ should be less than the known rate, $\frac{2 \text{ youths}}{3 \text{ youths}}$. Since $\frac{11{,}999}{18{,}000} = 0.66661\ldots$, $\frac{2}{3} = 0.66666\ldots$, and $0.66661\ldots < 0.66666\ldots$, the answer seems reasonable.

Step 5 Report the solution.

Of 18,000 youths, fewer than 12,000 are vaccinated against meningitis.

Practice Problems

For problems 16–18, use the five steps.

16. A building lot is 4200 ft^2. Find the area that can be covered by built structures.

The city's backyard-cottage legislation includes a number of restrictions: Cottages can be no larger than 800 square feet. . . . The lot must be in a single-family residence zone and be at least 4,000 square feet in size. . . . the total area of built structures can cover no more than 35 percent of the lot. (*Source:* www.seattletimes.com, Oct. 28, 2011)

17. Find the number of 2500 sales of existing homes that are foreclosed or about to be foreclosed.

The large number of foreclosed (or about to be foreclosed) houses on the market . . . account for no less than four out of 10 sales of existing homes. (*Source:* www.slate.com, May 11, 2011)

18. In Huntington Beach, California, it costs \$10 to park all day in a city beach parking lot. The parking meters on the Pacific Coast Highway cost \$0.25 per 10 min. Find the number of minutes for which the cost of using a parking meter is less than or equal to the cost of using a parking lot.

2.5 Vocabulary Practice

Match the term with its description.

1. When we multiply or divide both sides of an inequality by a negative number, we must do this to the inequality sign.
2. When we multiply or divide both sides of an inequality by a positive number, we must do this to the inequality sign.
3. In $x > 6$, 6 is an example of this.
4. $\geq$
5. $<$
6. $>$
7. $\leq$
8. On a number line graph, this indicates greater than and less than.
9. On a number line graph, this indicates greater than or equal to and less than or equal to.
10. $\approx$

A. approximately equal to
B. boundary value
C. bracket
D. greater than
E. greater than or equal to
F. keep the direction the same
G. less than
H. less than or equal to
I. parenthesis
J. reverse the direction

2.5 Exercises

Follow your instructor's guidelines for showing your work.

For exercises 1–8, use a number line graph to represent the inequality.

1. $x < 8$
2. $x < 6$
3. $x \geq -4$
4. $x \geq -5$
5. $x \leq \frac{1}{2}$
6. $x \leq \frac{3}{4}$
7. $5 < x$
8. $-7 < x$

For exercises 9–36,
(a) solve.
(b) check the direction of the inequality sign.

9. $p + 12 > 3$
10. $z + 14 > 5$
11. $p + 12 \geq -3$
12. $z + 14 \geq -5$
13. $-p + 12 > 3$
14. $-z + 14 > 5$
15. $p + 12 > -3$
16. $z + 14 > -5$
17. $2c - 6 < 40$
18. $4a - 12 < 60$
19. $-2c - 6 \leq 40$
20. $-4a - 12 \leq 60$
21. $2c - 6 < -40$
22. $4a - 12 < -60$
23. $8z - 10 \geq 30$
24. $3p - 15 \geq 18$
25. $24 \leq 4x - 8$
26. $18 \leq 3x - 12$
27. $35 \leq -3w + 34$
28. $41 \leq -5h + 40$
29. $50 > -2a + 60$
30. $20 > -2u + 40$
31. $5x + 10 < 3x - 20$
32. $6x + 14 < 4x - 52$
33. $2z + 9 > 7z - 21$
34. $3p + 19 > 10p - 2$
35. $3y - 8 \leq 5y + 15$
36. $5z - 11 \leq 9z - 8$

For exercises 37–52,
(a) solve.
(b) use a number line graph to represent the solution.
(c) check the direction of the inequality sign.

37. $8x + 9 < 73$
38. $7x + 3 < 52$
39. $-12x - 15 \geq 21$
40. $-11x - 17 \geq 49$
41. $-x + 8 \leq -17$
42. $-x + 7 \leq -15$
43. $8x - 9 \geq 6x - 6$
44. $12x - 3 \geq 10x - 4$
45. $7x - 12 > 6x - 12$
46. $5x - 34 > 4x - 34$
47. $4(2x - 6) \leq 5(x - 9)$
48. $3(4x - 1) \leq 9(x - 3)$
49. $\frac{x}{12} \geq \frac{3}{4}$
50. $\frac{x}{30} \geq \frac{2}{15}$
51. $\frac{1}{16} > \frac{x}{80}$
52. $\frac{1}{12} > \frac{x}{96}$

For exercises 53–62,
(a) clear the fractions or decimals and solve.
(b) check the direction of the inequality sign.

53. $\frac{1}{4}m + \frac{1}{6} > 1$
54. $\frac{5}{6}n + \frac{1}{4} > 1$
55. $\frac{1}{3}k - 4 \leq \frac{3}{5}k - 6$
56. $\frac{1}{4}h + 3 \leq \frac{2}{3}h + 2$
57. $0.2w + 5 < 8.6$
58. $0.3z + 9 < 15.6$
59. $-0.98p + 4 \geq -0.18p + 8$
60. $-1.04h + 0.6 \geq -1.02h + 0.8$
61. $\frac{3}{4}(2u + 8) < \frac{5}{6}(3u - 12)$
62. $\frac{1}{2}(3v - 4) < \frac{4}{5}(2v - 10)$

63. Describe the difference between the properties of equality and the properties of inequality.

64. Describe the difference in the solution of $x < 9$ and the solution of $x \leq 9$.

For exercises 65–72, assign a variable, and write an inequality that represents the constraint.

65. The number of passengers in a Nissan NV3500 HD van

The NV3500 HD can hold up to 12 passengers. (*Source:* www.chicagotribune.com, Nov. 13, 2011)

66. The number of ballots affected by a technical problem in mailing

A technical problem was affecting the mailing of up to 21,000 ballots. (*Source:* www.seattletimes.com, Nov. 7, 2011)

67. The number of customers who had no electricity

More than 2.3 million customers from Pennsylvania through New England had no electricity, according to reports, as the region was lashed by surprisingly high winds and the snowdrifts piled up. (*Source:* www.nytimes.com, Oct. 31, 2011)

68. The number of Americans who will not be allowed to vote because of a criminal conviction

Next November more than 5 million Americans will not be allowed to vote because of a criminal conviction in their past. (*Source:* www.nytimes.com, Nov. 7, 2011)

69. The maximum amount of protein allowed per day for a patient on dialysis is 84 g. (*Source:* www.akp.org)

70. The maximum amount of fat allowed per day in a special diet is 200 cal.

71. At Six Flags Fiesta Texas, riders on the Superman Krypton Coaster must be at least 54 in. tall.

72. To qualify for a group discount, a group must have at least 14 people.

For exercises 73–84, use the five steps and an inequality.

73. The maximum fuel economy of a car is $\frac{28 \text{ mi}}{1 \text{ gal}}$. Find the maximum distance it can travel on a full tank of 14 gal of gasoline.

74. The maximum fuel economy of a car is $\frac{32 \text{ mi}}{1 \text{ gal}}$. Find the maximum distance it can travel on a full tank of 12 gal of gasoline.

75. An employee is working a temporary job for $8 per hour. His employer withholds 7.65% of his wages for Social Security and Medicare. His car payment is $350 per month. Find the number of hours the employee must work to earn enough to at least pay his next three car payments.

76. An employee is working a temporary job for $9 per hour. Her employer withholds 7.65% of her wages to pay for Social Security and Medicare. Her rent payment is $420 per month. Find the number of hours the employee must work to earn enough to at least pay her next two rent payments.

77. In the first five meets of the season, a collegiate gymnast had vault scores of 9.625, 9.750, 9.825, 9.725 and 9.550. Find the score she needs in the next meet to have an average score of at least 9.700.

78. An athlete needs a final grade of at least 82 (B) in his chemistry class to maintain his athletic eligibility. His final grade is the sum of 55% of his test score average, 25% of his lab score average, and 20% of his final exam score. Before the final, his test score average is 75, and his lab score average is 88. Find the score on the final that he needs to have a final grade of at least 82. Round to the nearest whole number.

79. The average cost to keep a person in prison for one day in the United States is $78.95. Find the maximum cost of supervising a person who is on probation for one day. Round to the nearest hundredth.

1 day in prison costs more than 10 days on parole or 22 days on probation. (*Source:* The Pew Charitable Trusts, "One is 31," March 2009)

80. In August 2010, the price of a barrel of crude oil was about $73. Find the maximum value of the oil that gushed from BP's well.

Nearly five million barrels of oil have gushed from the BP's well since the Deepwater Horizon spill began on April 20. (*Source:* www.nytimes.com, Aug. 2, 2010)

81. In 2009, Americans generated 243 million tons of municipal solid waste. Food waste was more than 14% of this total. Find the minimum amount of food waste generated. Round to the nearest million. (*Source:* U.S. Statistical Abstract 2012; www.epa.gov)

82. If the population of a city includes 75,000 people who are not immune to measles, find the minimum number of people who will get measles if they are exposed to the virus.

Before measles immunization was available, nearly everyone in the U.S. got measles. . . . Measles is one of the most infectious diseases in the world and is frequently imported into the U.S. . . . More than 90 percent of people who are not immune will get measles if they are exposed to the virus. (*Source:* www.cdc.gov, Aug. 2010)

83. In Sept. 2010, 1.4 million people received cash assistance from the California Work Opportunity and Responsibility to Kids (CalWORKS) program. More than three out of four of these recipients were children. Find the minimum number of recipients who were children. (*Source:* www.lafla.org, Jan 2011)

84. An estimated 3.5 million people in southeast Asia are infected with HIV. More than two out of three in need of antiretroviral treatment in this region do not receive it. Find the minimum number of people who need and do not receive the antiretroviral treatment. Round to the nearest tenth of a million. (*Source:* www.searo.who.int, Sept. 8, 2011)

Problem Solving: Practice and Review

Follow your instructor's guidelines for using the five steps outlined in Section 1.3, p. 55.

85. Find the percent decrease in the number of active Hopi weavers. Round to the nearest percent.

The 75th Hopi Festival of Arts and Culture is the oldest Hopi venue in the world. . . . To date, there are only about 20 active Hopi weavers, but back in the 1930s the MNA (Museum of North Arizona) recorded approximately 213 Hopi weavers, all men. (*Source:* nativetimes.com, July 2008)

86. The cost to rent a hardwood floor polisher is $18.25 for 3 hr and $4 for each additional hour. Find the cost to rent the polisher for 10 hr.

87. The charge for a taxi is a flat fee of $2.50 plus $0.40 for each $\frac{1}{5}$ mi. Find the cost to travel 3 mi.

88. Find the enrollment at Oregon State University before the increase. Round to the nearest whole number.

Oregon State University saw the biggest growth, a 5.3 percent jump of 1,302 students. Portland State University grew by only 1.5 percent, but remains the state's largest university with 28,958 students. (*Source:* www.oregonlive.com, Nov. 10. 2011)

Technology

For exercises 89–92, solve. Use a calculator to do arithmetic.

89. $21{,}000h + 38{,}450 > 47{,}802$

90. $0.35x - 0.226 \geq 0.33x + 0.784$

91. $\frac{8}{15}x + \frac{10}{21} < \frac{2}{35}x + 1$

92. $\frac{1}{9}x + \frac{5}{21} \leq -\frac{2}{9}x - \frac{12}{21}$

Find the Mistake

For exercises 93–96, the completed problem has one mistake.

(a) Describe the mistake in words, or copy down the whole problem and highlight or circle the mistake.

(b) Do the problem correctly.

93. Problem: Solve: $-2x + 12 > 36$

Incorrect Answer:

$$\begin{aligned} -2x + 12 &> 36 \\ -12 \quad & \quad -12 \\ -2x + 0 &> 24 \\ \frac{-2x}{-2} &> \frac{24}{-2} \\ x &> -12 \end{aligned}$$

94. Problem: Solve: $3p + 21 > -45$

Incorrect Answer:

$$\begin{aligned} 3p + 21 &> -45 \\ -21 \quad & \quad -21 \\ 3p + 0 &> -66 \\ \frac{3p}{3} &> \frac{-66}{3} \\ p &< -22 \end{aligned}$$

95. Problem: Use a number line graph to represent $x \geq 2$.

Incorrect Answer: (number line from 0 to 7 labeled x)

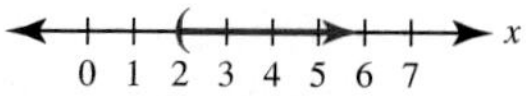

96. Problem: Use a number line graph to represent $x < 3$.

Incorrect Answer: (number line from −2 to 5 labeled x)

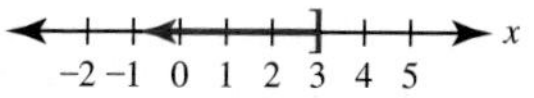

Review

97. Copy the chart below. Put an X in the box if the number is an element of the set at the top of the column.

98. Solve $3x + 2y = -8$ for y.

99. Solve: $7(2c - 1) = 15(c - 3) - c$

100. Solve: $2(n + 3) = 4n + 6$

SUCCESS IN COLLEGE MATHEMATICS

101. The first time that you create a schedule to help you manage your time, you may find that your plans are unrealistic. It may take longer to finish assignments than you planned; you may find that you need more time with your family; you may realize that you need more breaks in order to be able to study effectively; you may spend more time commuting than you hoped. In this situation, students sometimes cope by reducing the time they spend sleeping. What is the minimum amount of sleep you need each night to be able to pay attention in class and learn?

Chart for exercise 97

	Real numbers	**Rational numbers**	**Irrational numbers**	**Integers**	**Whole numbers**
$\sqrt{12}$					
$\frac{3}{11}$					
−25					
0.68					

Study Plan for Review of Chapter 2

SECTION 2.1 The Properties of Equality

Ask Yourself	Test Yourself	Help Yourself
Can I . . .	**Do 2.1 Review Exercises**	**See these Examples and Practice Problems**
use a property of equality to solve an equation in one variable and check the solution? identify the reciprocal of a number?	1–15	Ex. 1–10, PP 1–12
use the properties of equality to solve an equation that includes decimal numbers?	14–15	Ex. 11–12, PP 13–17
find the original amount, given the new amount and the percent change? use a percent to solve an application problem?	16–18, 21	Ex. 13–15, PP 18–20
use a proportion to solve an application problem?	19–20	Ex. 16–17, PP 21–24

2.1 Review Exercises

For exercises 1–15,
(a) solve.
(b) check.

1. $x + 40 = 22$

2. $\frac{x}{5} = 15$

3. $5a = 15$

4. $b - 9 = -41$

5. $x - \frac{2}{3} = \frac{5}{12}$

6. $4x - 9x = 6$

7. $-15 + h - 7 = 38$

8. $\frac{k}{9} = 0$

9. $\frac{5}{6} = \frac{p}{216}$

10. $\frac{z}{150} = \frac{1}{600}$

11. $2x + 6 = 6$

12. $\frac{2}{3}z + 5 = 11$

13. $10 - \frac{3}{4}a = -20$

14. $0.72 = 0.2x + 0.4$

15. $1.3 - 0.2a = 4$

For exercises 16–21, use the five steps.

16. A promotion for an apartment complex includes a 30% discount on the first month's rent. The reduced rent is $625 per month. Find the rent before the discount. Round to the nearest dollar.

17. The sale price of a dishwasher is $544. It is marked down 20%. Find the original price.

18. The amount spent for magazine advertising in the top 12 categories was $22,051,798,030. Find the total spent on magazine advertising. Round to the nearest million.

The top 12 advertising categories account for 87% of total magazine (advertising) spending, according to the Publishers Information Bureau (PIB). (*Source:* www.magazine.org, 2008)

19. A study of announcements on the New York subway system was done by 51 volunteers who made 6600 observations of in-car announcement opportunities. Find the number of delays in which either no announcement or a garbled inaudible announcement was played. (*Source:* www.straphangers.org)

In three out of five delays, no announcement played, or only a garbled, inaudible one. (*Source:* www.nytimes.com, March 30, 2011)

20. Predict the number of fatal work injuries for 430,000 workers in 2007. Round to the nearest whole number.

The rate of fatal work injury for U.S. workers in 2010 was 3.5 per 100,000 full-time equivalent (FTE) workers, (*Source:* www.bls.gov, Aug. 25, 2011)

21. The Bureau of Labor Statistics predicts a total of 190,000 jobs in psychology in 2018. Find the number of these psychologists who will be self-employed.

About 34 percent of psychologists are self-employed, mainly as private practitioners and independent consultants. (*Source:* Occupational Outlook Handbook, 2010–2011 edition, www.bls.gov)

SECTION 2.2 More Equations in One Variable

Ask Yourself	Test Yourself	Help Yourself
Can I...	**Do 2.2 Review Exercises**	**See these Examples and Practice Problems**
use more than one property of equality to solve an equation in one variable and check the solution?	22–27	Ex. 1–3, PP 1–6
combine like terms and use the distributive property to simplify an equation in one variable?	25–27	Ex. 4–8, PP 7–11
use an equation in one variable to solve an application problem?	28–30	Ex. 9, PP 12–13

2.2 Review Exercises

For exercises 22–27,
(a) solve.
(b) check.

22. $6x - 15 = 39$

23. $\frac{m}{2} + 18 = 30$

24. $9 = -5b + 9$

25. $9(2z + 14) - 16z = -58$

26. $4(7x - 9) - (24x - 8) = 30$

27. $13 = 6x - (x + 1) - 1$

For exercises 28–30, use the five steps.

28. A student buys a 2-year cell phone contract for $60 per month. For every person he refers who then signs a contract, the company reduces his monthly rate by $1.25. Find the number of people he needs to refer who then sign a contract so that his monthly rate is $42.50 per month.

29. The headline of a news release by a market research company reads, "Four out of Five Cell Phones to Integrate GPS by End of 2011." Predict the number of 1.2 million cell phones at the end of 2011 that had GPS. Write the answer in place value notation. (*Source:* www.isuppli.com, July 16, 2010)

30. An individual cell phone plan that includes 450 anytime minutes and unlimited texting is $59.99 per month. The cost for additional anytime calls is $0.45 per minute. Find the cost per year to use this plan if each month an average of 60 extra minutes is used.

SECTION 2.3 Variables on Both Sides

Ask Yourself	Test Yourself	Help Yourself
Can I...	**Do 2.3 Review Exercises**	**See these Examples and Practice Problems**
use the properties of equality to solve an equation in one variable with variables on both sides and check the solution?	31–37	Ex. 1–4, PP 1–3
clear the fractions from an equation in one variable?	38, 40–41	Ex. 5–6, PP 4–5
clear the decimals from an equation in one variable?	39	Ex. 7, PP 6
identify an equation with no solution or infinitely many solutions and check whether the solution is the set of real numbers?	33, 35, 41–43	Ex. 8–9, PP 7–8
use an equation with variables on both sides to solve an application problem?	44–45	Ex. 10–11, PP 10–12

2.3 Review Exercises

For exercises 31–37,
(a) solve.
(b) check.

31. $8a - 9 = 13a + 61$

32. $15x - (x + 3) = 16x - 2$

33. $12(b + 8) + 4b = 4(4b + 20) + 16$

34. $2(2x - 3) + 6x = 7(x + 4) + 2(x - 1)$

35. $8w - 11 = 4(2w - 5) + 8$

36. $2x = 6x - 4$

37. $3x = 7x$

For exercises 38–41,
(a) clear the fractions or decimals and solve.
(b) check.

38. $\frac{5}{8}(2z - 16) = 20$

39. $0.08d + 1.6 = 0.06d - 2.4$

40. $\frac{1}{9}k + \frac{1}{4} = \frac{3}{8}$

41. $\frac{3}{4}(14x + 8) = 10x + \frac{1}{2}(x - 9)$

42. Choose the equation(s) with no solution.
- **a.** $x - 4 = x + 4$
- **b.** $2x + 6 = 2(x + 3)$
- **c.** $2x + 6 = 2(x - 3)$
- **d.** $8n = 2n$

43. Choose the equation(s) with infinitely many solutions.
- **a.** $a + 4 = 2a + 8$
- **b.** $4p + 12 = 4(p + 3)$
- **c.** $7c = 7c$
- **d.** $4d = 0$
- **e.** $8 - d = -d + 8$

For exercises 44–45, use the five steps.

44. The current total monthly payment on a home mortgage is $890 a month. Closing costs to refinance the loan at a lower interest rate are $2950. The new payment after refinancing will be $654 a month. Find the number of months for the closing costs to equal the savings from the lower monthly payments. Round *up* to the nearest month.

45. The cost to create a single product is $100. The fixed overhead costs per month to make the product are $7500. The price of the product is $125. Find the number of products that the company must make and sell to break even.

SECTION 2.4 Solving an Equation for a Variable

Ask Yourself	Test Yourself	Help Yourself
Can I...	**Do 2.4 Review Exercises**	**See these Examples and Practice Problems**
use the properties of equality to solve an *equation* for a variable?	46–47	Ex. 1–3, PP 1–3
use the properties of equality to solve a *formula* for a variable?	48–51	Ex. 4–7, PP 4–7
use a formula to solve an application problem?	50–51	Ex. 6–7, PP 6–7

2.4 Review Exercises

46. Solve $3x - 2y = 9$ for y.

47. Solve $3x - 2y = 9$ for x.

48. Solve $D = \frac{NS}{A}$ for S.

49. Solve $d = \frac{D(a + 1)}{24}$ for a.

50. When a car is equipped with larger than standard tires, the speed shown on the speedometer is lower than the actual speed of the car. The formula $S = \frac{AD}{N}$ represents the relationship of the actual speed A, the observed speed on the speedometer S, the new tire diameter N, and the standard tire diameter for the car, D.
- **a.** Solve this formula for A.
- **b.** A car has a standard tire diameter of 28 in. If the tires on the car are changed to a diameter of 32 in. and the speedometer reads $\frac{55 \text{ mi}}{1 \text{ hr}}$, find the actual speed of the car. Round to the nearest whole number.

51. The formula $C = R(P - W)$ represents the relationship of the commission paid to a salesperson C, the percent commission in decimal form R, the selling price of an item P, and the wholesale cost of the item, W.
- **a.** Solve this formula for R.
- **b.** The wholesale price of a lawnmower is $250. The selling price of the lawnmower is $325. Find the percent commission if the commission is $1.50.

SECTION 2.5 Inequalities in One Variable

Ask Yourself	Test Yourself	Help Yourself
Can I...	**Do 2.5 Review Exercises**	**See these Examples and Practice Problems**
identify the meaning of the symbols $>$, $<$, $\geq$, and $\leq$? use a number line graph to represent an inequality in one variable?	52–60, 64–65	Ex. 1–9, PP 1–4
use the properties of inequality to solve an inequality in one variable?	54–63	Ex. 4–15, PP 5–15
check the inequality sign in the solution of an inequality in one variable?	54–58	Ex. 5–8, PP 5–15
clear the fractions or decimals from an inequality in one variable?	59–60	Ex. 8–9, PP 11–15
assign a variable and write an inequality that describes a constraint?	66–69	Ex. 10–13, PP 16–18
use an inequality in one variable to solve an application problem?	68–69	Ex. 12–13, PP 16–18

2.5 Review Exercises

For exercises 52–53, use a number line graph to represent the inequality.

52. $x > 5$ **53.** $x \leq \frac{3}{2}$

For exercises 54–58,
(a) solve.
(b) use a number line graph to represent the solution.
(c) check the direction of the inequality sign.

54. $5x + 8 < 2x + 29$

55. $-8x - 9 \geq 15$

56. $2(6x + 4) < 15x - 10$

57. $x - 3 > 2x - 3$

58. $1 - x > 6$

For exercises 59–60,
(a) clear the fractions or decimals and solve.
(b) use a number line graph to represent the solution.

59. $-\frac{3}{8}x + \frac{1}{2} \leq \frac{7}{8}$

60. $0.2x + 1.4 > 0.4x - 3.8$

61. Choose the numbers that are solutions of $x + 8 < 15$.
a. 2 **b.** -11
c. 7 **d.** 8
e. 0 **f.** $\frac{1}{4}$

62. Choose the numbers that are solutions of $x - 1 \geq 4$.
a. 2 **b.** 7
c. 5 **d.** 21
e. 0 **f.** 4

63. Choose the inequalities with a solution of $x < -2$.
a. $2x < -4$ **b.** $-4x > 8$
c. $-6x < 12$ **d.** $3x > -6$

64. Choose the number line graph that represents $x > 9$.

a. 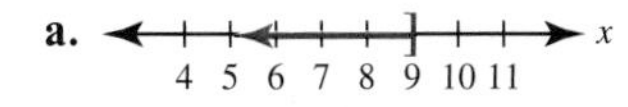

b. 4 5 6 7 8 9 10 11 x

c. 7 8 9 10 11 12 13 14 x

d. 7 8 9 10 11 12 13 14 x

65. The number 5 is a solution of an inequality. Choose the number line graphs that could represent this inequality.

a. 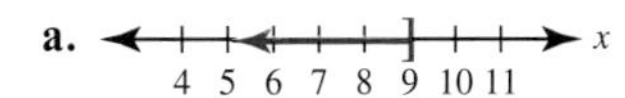

b. 4 5 6 7 8 9 10 11 x

c. 7 8 9 10 11 12 13 14 x

d. 7 8 9 10 11 12 13 14 x

For exercises 66–67, assign a variable and write an inequality that represents the constraint.

66. The number of pieces of art to which there is comprehensive access on the website

The Metropolitan Museum of Art has relaunched its website. . . . Key features of the expanded and redesigned site include comprehensive access to more than 340,000 works of art in the Museum's encyclopedic collections. (*Source:* www.metmuseum.org, Sept. 26, 2011)

67. The area median income level for a household that is a family of four

The one-, two- and three-bedroom apartments are limited to households earning no more than 60 percent of area median income. . . . For a single person, the maximum income level is \$53,220. For a family of four, it is \$75,960. (*Source:* www.nytimes.com, Nov. 11, 2011)

For exercises 68–69, use an inequality and the five steps.

68. A customer purchased 9.5 gal of E15 fuel for a car manufactured in 2011. Find the maximum amount of ethanol that is in this fuel. (*Source:* www.epa.gov)

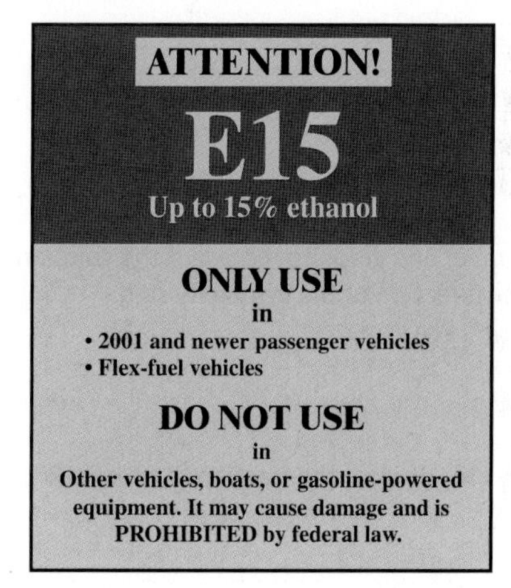

69. To make the cut in a golf tournament, the average score for three rounds must be no more than 80 strokes. A golfer's first-round score was 84 strokes, and her second-round score was 79 strokes. Find the score in the third round so that the golfer at least makes the cut.

Chapter 2 Test

Follow your instructor's guidelines for showing your work.

For problems 1–11,
(a) solve.
(b) check.

1. $-8x + 9 = 41$
2. $\frac{5}{12}p - 7 = 3$
3. $-3(2z + 5) = 6(-z + 4)$
4. $10c - (c - 18) = -9$
5. $2(2w - 10) = 9w + 30$
6. $9 - 7(2x + 8) + 2 = -17x$
7. $\frac{d}{3} - 4 = -1$
8. $\frac{3}{8} = \frac{a}{36}$
9. $24 - x = 3(x + 9) - 4x - 3$
10. $p - 0.2p = 5$
11. $8n = 5n$

For problems 12–13,
(a) solve.
(b) use a number line graph to represent the solution.
(c) check the direction of the inequality sign.

12. $8x - 5 \leq 12x + 15$
13. $-\frac{3}{4}x + 2 > -1$

For problems 14–15, clear the fractions or decimals and solve.

14. $\frac{9}{10}x + \frac{1}{2} = \frac{1}{5}$
15. $1.3h + 2.9 = 0.5(3h + 9)$
16. The sale price of a boat is \$17,500. The percent discount is 20% off the original price. Explain why the original price does not equal $\$17,500 + (0.20)(\$17,500)$.
17. Solve $12x - 5y = 40$ for y. Write the equation to match the pattern $y = mx + b$.
18. Solve the formula $D = A - 2B + 2C$ for C.
19. In the formula $A = 1.06\left(\frac{LW}{9}\right)$, A is the area in square yards of carpet to buy for a rectangular room, L is the length of the room in feet, and W is the width of the room in feet. Solve this formula for W.
20. The maximum Pell Grant award for the 2011–2012 award year was \$5550. Assign a variable, and write an inequality that represents this constraint.

For exercises 21–25, use the five steps.

21. In 2010, there were 151.8 million male Americans. Find the number of these Americans who will develop lung cancer. Round to the nearest tenth of a million. (*Source:* www.census.gov)

    The average lifetime chance that a man will develop lung cancer is about 1 in 13. For a woman it is 1 in 16. These numbers include both smokers and non-smokers. For smokers the risk is much higher, while for non-smokers the risk is lower. (*Source*: www.cancer.gov)

22. Find the number of people who traveled 50 mi or more for Thanksgiving in 2010. Round to the nearest tenth of a million.

    AAA forecasts 42.5 million Americans will travel 50 miles or more from home during the Thanksgiving holiday weekend, a four percent increase from . . . one year ago [2010]. (*Source:* www.newsroom.aaa.com, Nov. 17, 2011)

23. The cost to create a product is \$20. The fixed overhead costs to make the product per month are \$9500. The price of the product is \$35. Find the number of products that the company must make and sell to break even. Round up to the nearest whole number.
24. The sale price of a chair is \$556. The discount is 20%. Find the original price of the chair.
25. A rice and grain mix that does not contain salt or water and is labeled "organic" weighs 16 oz. Find the minimum weight of the organic ingredients in the box.

    Products labeled "organic" must consist of at least 95 percent organically produced ingredients (excluding water and salt). (*Source:* www.ams.usda.gov)

Linear Equations and Inequalities in Two Variables

3

SUCCESS IN COLLEGE MATHEMATICS

Learning Preferences

Learning preferences describe the surroundings in which you best learn and study, in which you most easily transfer information from your short-term memory into your long-term memory. Learning preferences might not be the same for all subjects.

Having a learning preference does not mean that you cannot learn in other ways. In college, students must adapt to a wide range of teaching methods and learning environments. However, students can often control the environment in which they study and practice what they have been taught.

If your college has a counseling center, the counselors might be able to help you identify your learning preferences and plan your study environment.

3.1 Ordered Pairs and Solutions of Linear Equations
3.2 Standard Form and Graphing with Intercepts
3.3 Average Rate of Change and Slope
3.4 Slope-Intercept Form
3.5 Point-Slope Form and Writing the Equation of a Line
3.6 Linear Inequalities in Two Variables
3.7 Relations and Functions

© Georgios Kollidas/Shutterstock.com

A legend says that a French boy named René Descartes was sick for a long time and needed to stay in bed. Watching a spider on the ceiling, he thought about how to describe its location. He imagined a grid on the ceiling and described the spider's location on the grid. This grid became what we now call the Cartesian or rectangular coordinate system. Whether or not the legend is true, Mr. Descartes did invent analytic geometry, the bridge between geometry and algebra. He lived from 1596 to 1650 and was an important philosopher and scientist as well as a mathematican. In this section, we will learn how to draw a Cartesian coordinate system, describe the location of points graphed on it, and graph a linear equation in two variables.

SECTION 3.1

Ordered Pairs and Solutions of Linear Equations

After reading the text, working the practice problems, and completing assigned exercises, you should be able to:

1. Graph ordered pairs on an xy-coordinate system.
2. Complete a table of solutions for a linear equation.
3. Use a table of solutions to graph a linear equation.

Ordered Pairs

We can represent a **whole number** by graphing it on a number line (Figure 1). The number line is sectioned into equal distances with tick marks. The number that appears below a tick mark shows its value. These numbers are the **scale** of the number line.

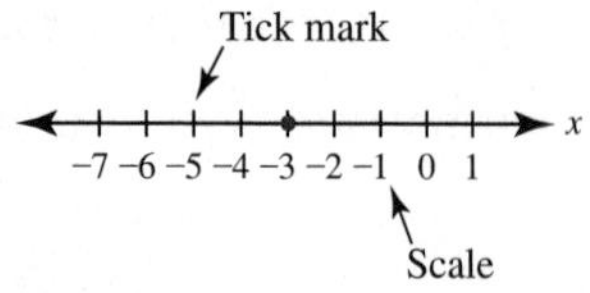

Figure 1

To graph an **ordered pair**, we use a **Cartesian coordinate system** (Figure 2), also called a rectangular coordinate system or just a coordinate system. A coordinate system includes a horizontal number line and a vertical number line. Each number line is an **axis**. When we use the variables x and y, the horizontal number line is usually called the x-axis and the vertical number line is called the y-axis. Each axis is labeled with its variable and a scale.

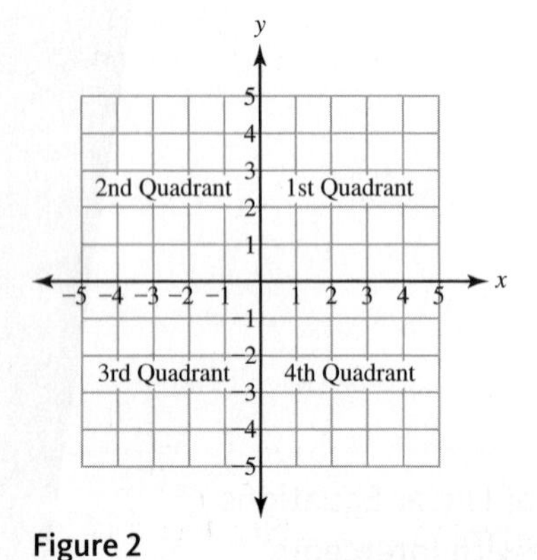

Figure 2

The scales do not have to be the same on both axes. (The plural of axis is "axes.") Ask your instructor about how scales should be shown on the graph. Some instructors prefer that only a few numbers be shown. Others want each tick mark labeled.

The intersection of the axes is the **origin**. The axes divide the coordinate system into four regions. Each region is a **quadrant** ("quad" means "four").

To represent an ordered pair, we graph a **point** on a coordinate system. In Figure 3, the point $(0, 0)$ is at the origin. The point that represents the ordered pair $(3, 1)$ is in the 1st quadrant, located 3 units to the right of the origin and 1 unit up from the origin.

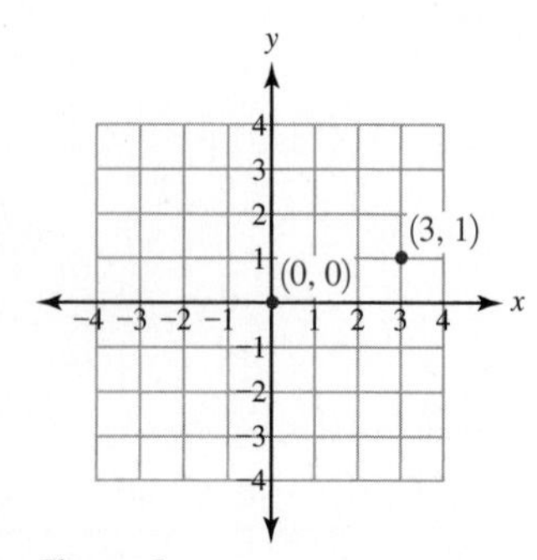

Figure 3

In Figure 4, a point in the 2nd quadrant that is 2 units to the left of the origin and 3 units up from the origin represents the ordered pair $(-2, 3)$.

In Figure 5, a point in the 3rd quadrant that is 3 points to the left of the origin and 1 unit down from the origin represents the ordered pair $(-3, -1)$. The ***x*-coordinate** of this ordered pair is -3. The ***y*-coordinate** is -1.

The distance between tick marks on an axis does not always equal 1. In Figure 6, the labeled point in the 4th quadrant represents the ordered pair $(20, -3)$. The distance between tick marks on the x-axis is 5. The distance between tick marks on the y-axis is 1.

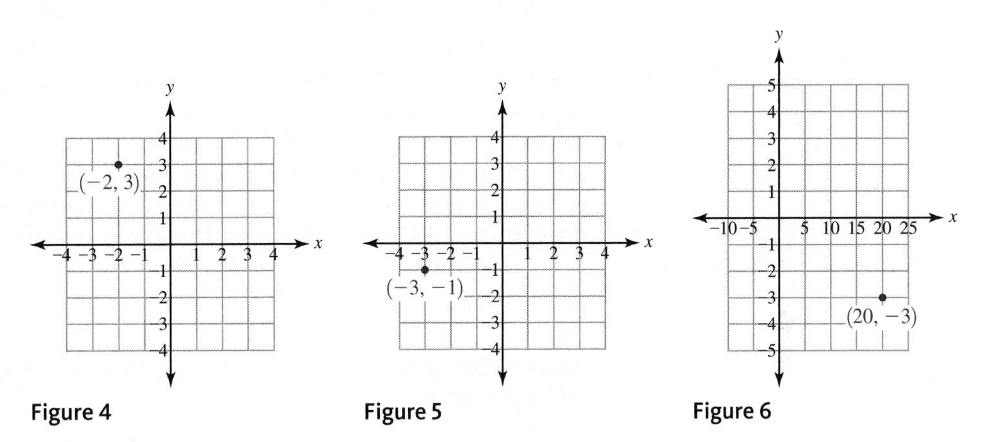

Figure 4 **Figure 5** **Figure 6**

Since the scales on each axis tell us the coordinates of a point, labeling the point with the ordered pair it represents can be optional. Ask your instructor whether you should label points.

EXAMPLE 1 Graph the ordered pairs $(0, 3)$, $(4, 0)$, and $(-1, -2)$. Label the axes; write a scale for each axis.

SOLUTION ▶ Draw a coordinate system, labeling the x-axis with x and the y-axis with y. Write a scale on each axis. The graph is a reasonable size if the distance between tick marks is 1. Graph the point that represents each ordered pair.

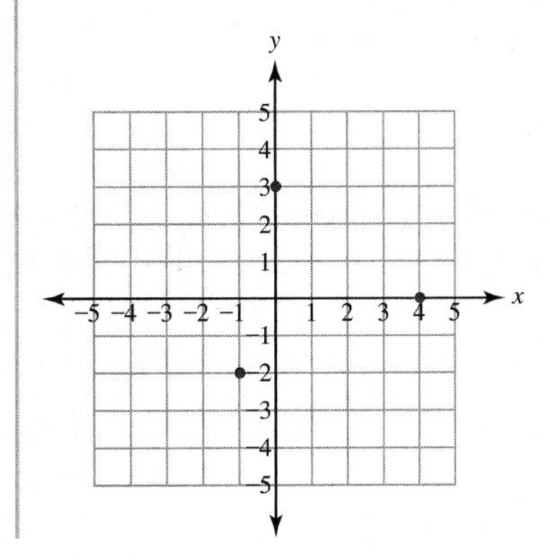

In the next example, the ordered pairs are organized in a table. The x-coordinate of the ordered pair is in the column to the left; the y-coordinate is in the column to the right. Since the ordered pairs include units, each axis is labeled with its unit. To make the graph a reasonable size, the distance between tick marks on each axis is 50 mi.

EXAMPLE 2 Graph the ordered pairs in the table. Label the axes, including the units of measurement; write a scale for each axis.

x (miles)	y (miles)
200	550
300	450
600	150

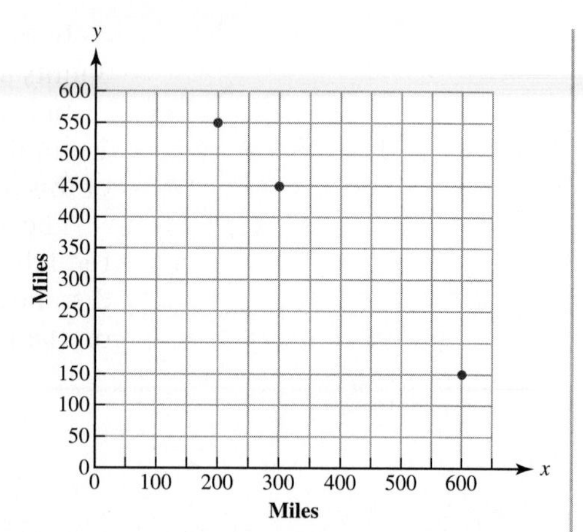

SOLUTION ▶ Since distances are positive and the ordered pairs are only in the first quadrant, draw only that part of the graph. To keep the size of the graph reasonable, the distance between tick marks on both axes is 50 mi. Write the units of measurement under the horizontal axis and to the left of the vertical axis. Graph the point that represents each ordered pair.

To create a graph of ordered pairs with x-coordinates that are much larger than the y-coordinates, we can use different scales on each axis. In the next example, the distance between tick marks on the x-axis is 5 and the distance between tick marks on the y-axis is 1.

EXAMPLE 3 Graph $(-20, 3)$, $(40, 0)$, and $(0, -1)$. Label the axes; write a scale for each axis.

SOLUTION ▶

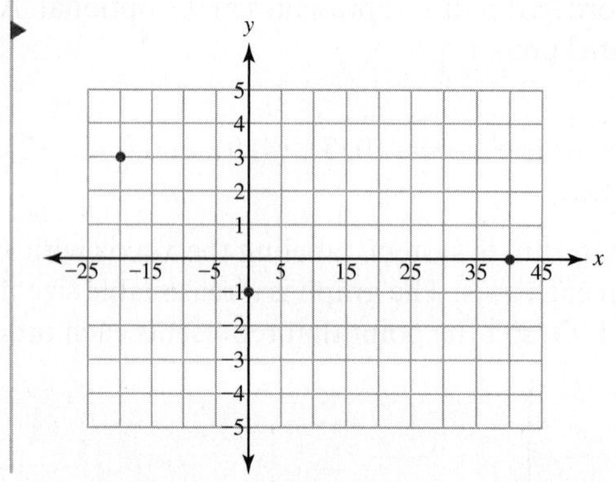

To graph an ordered pair that includes a fraction, we can use a scale in which the distance between tick marks is equal to a fraction, or we can use a whole number scale and estimate the locations of the points.

EXAMPLE 4 Graph $\left(\frac{1}{2}, 2\right)$ and $\left(4, -\frac{3}{2}\right)$. Label the axes; write a scale for each axis.

SOLUTION ▶ The point $\left(\frac{1}{2}, 2\right)$ is $\frac{1}{2}$ unit to the right of the origin and 2 units up from the origin. Estimate the distance to the right of the origin. The point $\left(4, -\frac{3}{2}\right)$ is 4 units to the right of the origin and $\frac{3}{2}$ unit down from the origin. Estimate the distance down from the origin.

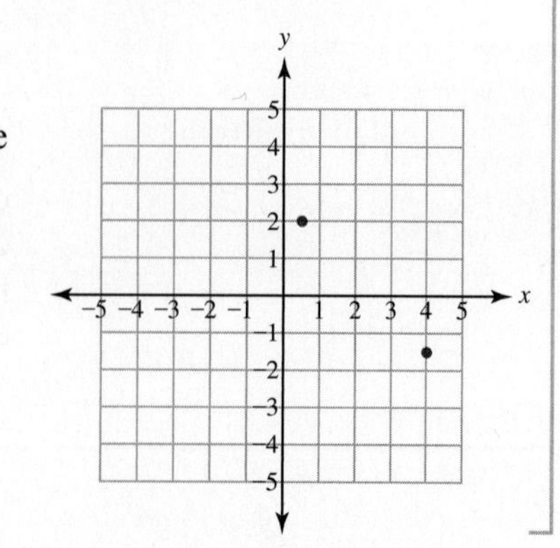

Practice Problems

For problems 1–4,
(a) identify the ordered pair represented by the point.
(b) identify the quadrant in which the point is located.

1.

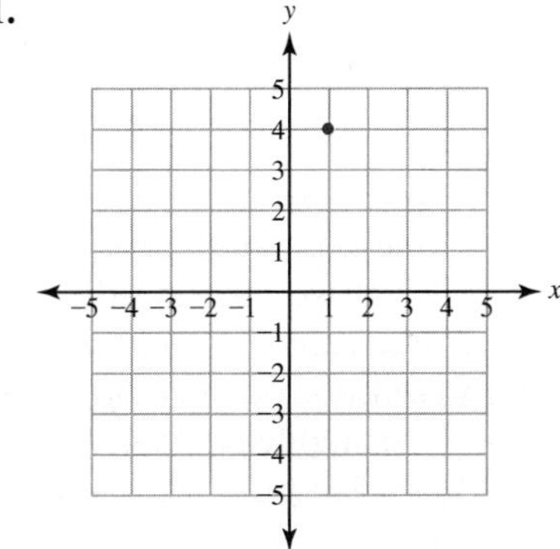

2.

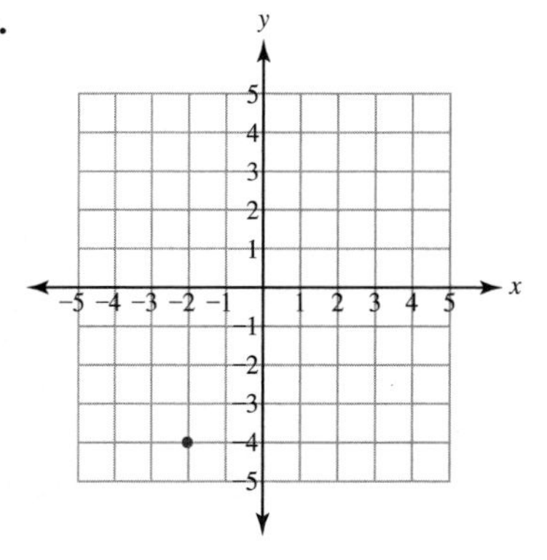

3.

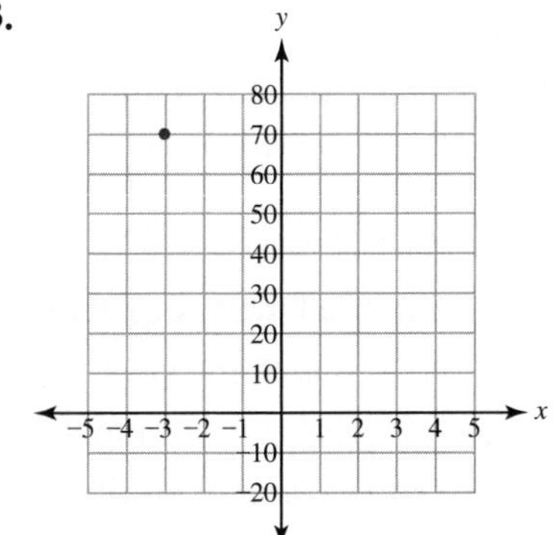

4.

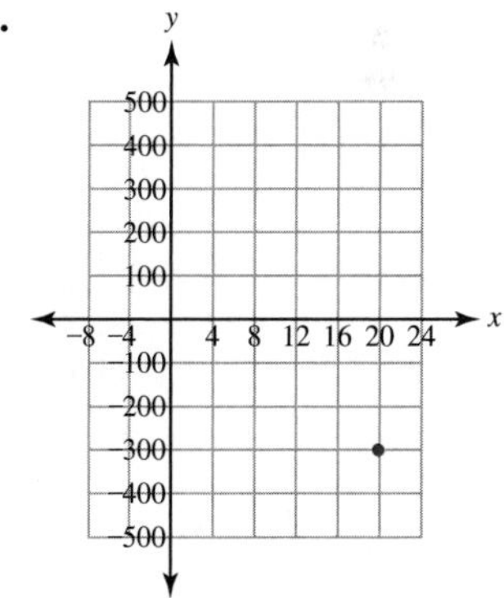

For problems 5–11, graph the ordered pairs. Label the axes, including any units of measurement; write a scale for each axis.

5. $(0, 0)$; $(4, 5)$; $(-6, -3)$

6. $(9, 2)$; $(-7, 5)$; $(0, 4)$; $(2, 0)$

7. $(2, 100)$; $(-3, 150)$; $(-5, -200)$

8. $(8000, 3)$; $(-7500, -6)$; $(4500, -5)$

9.

x (days)	y (lb)
0	8
2	12
3	16

10.

x	y
0	-6
-4	0
3	$\frac{3}{2}$

11.

x	y
0	5
2	$\frac{13}{2}$
3	$-\frac{3}{2}$

Linear Equations in Two Variables

If a student spends a total of \$100 on food and/or gasoline, the amount spent on food plus the amount spent on gasoline equals \$100. Describing this situation in words is a **verbal representation**. We can also assign two variables and create a **symbolic representation** of the relationship, an equation. If x represents the amount spent on food and y represents the amount spent on gasoline, then $x + y = \$100$. If we can replace the variables in the equation with the coordinates of an ordered pair and the equation is true, the ordered pair is a **solution** of the equation.

EXAMPLE 5 Use the table of ordered pairs to graph the equation $x + y = \$100$.

x (dollars)	y (dollars)
0	100
100	0
75	25

SOLUTION ▶ The ordered pairs in the table are three of the infinitely many solutions of this equation. To graph these ordered pairs, draw a rectangular coordinate system, labeling the horizontal axis with x and the vertical axis with y. Write the units of measurement on each axis.

To keep the size of the graph reasonable, the distance between tick marks on both axes is \$10. Because the values for x and for y are greater than or equal to 0, we can limit the graph to just the 1st quadrant of the coordinate system. To represent the other solutions of this equation, use a straightedge to draw a line from (\$0, \$100) to (\$100, 0). Each point on this line is a solution of $x + y = \$100$.

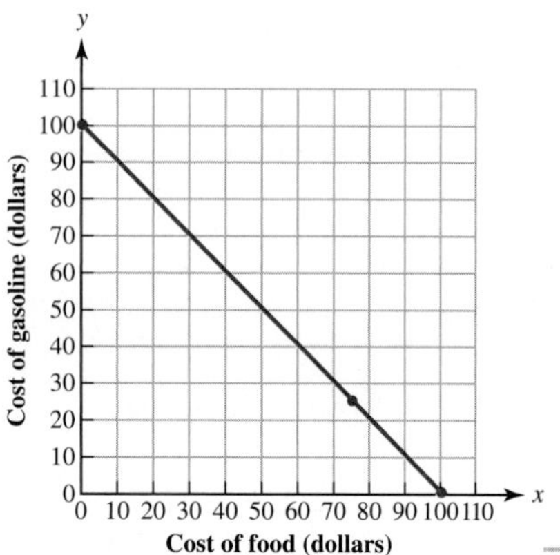

The graph of a **linear equation** like $x + y = \$100$ is a straight line. To find three or more solutions of a linear equation, we can use the properties of equality.

Finding a Solution of a Linear Equation in Two Variables

1. Choose a value for one of the variables.
2. Replace the variable with this number.
3. Solve the equation for the remaining variable.
4. Write the solution as an ordered pair.

EXAMPLE 6 A linear equation is $x + 2y = 6$.

(a) Complete the table of solutions.

x	y
0	
	0
−2	

SOLUTION ▶ To find y when x is 0, replace x in the equation with 0 and use the properties of equality to solve for y.

$$x + 2y = 6$$
$$\mathbf{0} + 2y = 6 \quad \text{Replace } x \text{ with 0.}$$
$$2y = 6 \quad \text{Simplify.}$$
$$\frac{2y}{2} = \frac{6}{2} \quad \text{Division property of equality.}$$
$$y = 3 \quad \text{Simplify.}$$

A solution of this equation is (0, 3).

To find x when y is 0, replace y in the equation with 0 and use the properties of equality to solve for x.

$$x + 2y = 6$$
$$x + 2(\mathbf{0}) = 6 \quad \text{Replace } y \text{ with 0.}$$
$$x + \mathbf{0} = 6 \quad \text{Simplify.}$$
$$x = 6 \quad \text{Simplify.}$$

A solution of this equation is (6, 0).

To find y when x is -2, replace x in the equation with -2 and use the properties of equality to solve for y.

$x + 2y = 6$

$\mathbf{-2} + 2y = 6$ Replace x with -2.

$+2 \quad +2$ Addition property of equality.

$\mathbf{0} + 2y = \mathbf{8}$ Simplify.

$\frac{\mathbf{2y}}{2} = \frac{8}{2}$ Division property of equality.

$y = 4$ Simplify.

A solution of this equation is $(-2, 4)$.

The table of solutions:

x	y
0	3
6	0
−2	4

(b) Graph the equation.

Draw a coordinate system, labeling the axes and writing a scale for each axis. Since the x-coordinates are -2, 0, and 6 and the y-coordinates are 0, 3, and 4, the distance between tick marks on each axis can be 1. Graph the point that represents each solution. Use a straightedge to draw a line through the points.

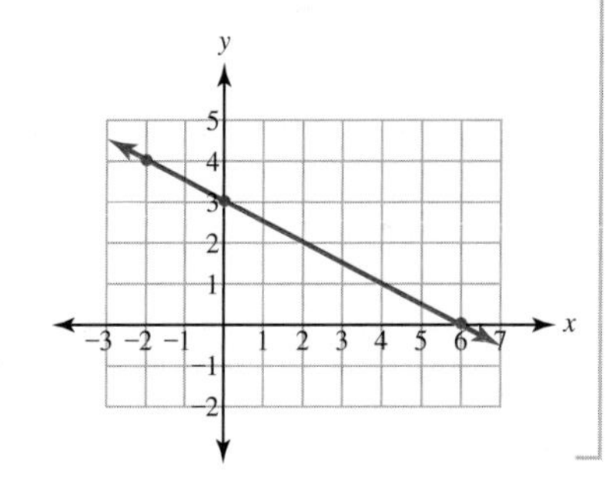

Because rounding a decimal changes its value, do not write solutions that are fractions as decimal numbers. To help estimate the location of the point on the graph, write an approximate value of the number in parentheses in the table.

EXAMPLE 7 A linear equation is $3x + 5y = 20$.

(a) Complete the table of solutions.

x	y
0	
	0
3	

SOLUTION

$3x + 5y = 20$

$3(\mathbf{0}) + 5y = 20$ Replace x with 0.

$\mathbf{0} + 5y = 20$ Simplify.

$\frac{\mathbf{5y}}{5} = \frac{20}{5}$ Division property of equality.

$y = 4$ Simplify.

A solution of this equation is $(0, 4)$.

$3x + 5y = 20$

$3x + 5(\mathbf{0}) = 20$ Replace y with 0.

$3x + \mathbf{0} = 20$ Simplify.

$\frac{\mathbf{3x}}{3} = \frac{20}{3}$ Division property of equality.

$x = \frac{20}{3}$ Simplify.

A solution of this equation is $\left(\frac{20}{3}, 0\right)$.

$$3x + 5y = 20$$

$$3(3) + 5y = 20 \quad \text{Replace } x \text{ with 3.}$$

$$\mathbf{9} + 5y = 20 \quad \text{Simplify.}$$

$$\underline{-9 \qquad -9} \quad \text{Subtraction property of equality.}$$

$$\mathbf{0} + 5y = \mathbf{11} \quad \text{Simplify.}$$

$$\frac{\mathbf{5y}}{5} = \frac{11}{5} \quad \text{Division property of equality.}$$

$$y = \frac{11}{5} \quad \text{Simplify.}$$

A solution of this equation is $\left(3, \frac{11}{5}\right)$.

The table of solutions:

x	y
0	4
$\frac{20}{3}$ ($\approx$6.7)	0
3	$\frac{11}{5}$ (=2.2)

(b) Graph the equation.

Draw a coordinate system, labeling the axes and writing a scale for each axis. Graph the point that represents each solution, estimating the location when necessary. Use a straight-edge to draw a line through the points.

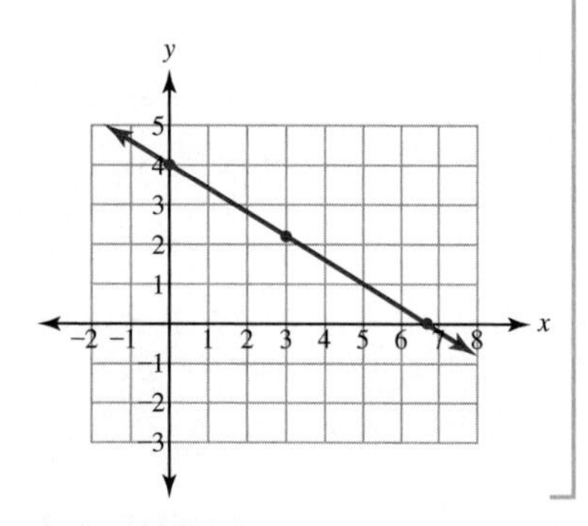

In the next example, the distance between tick marks on the x-axis is 10. The distance between tick marks on the y-axis is 1.

EXAMPLE 8 A linear equation is $-2x + 25y = 100$.

(a) Complete the table of solutions.

x	y
0	
	0
25	

SOLUTION

$$-2x + 25y = 100$$

$$-2(\mathbf{0}) + 25y = 100 \quad \text{Replace } x \text{ with 0.}$$

$$\mathbf{0} + 25y = 100 \quad \text{Simplify.}$$

$$\frac{\mathbf{25y}}{25} = \frac{100}{25} \quad \text{Division property of equality.}$$

$$y = 4 \quad \text{Simplify.}$$

A solution of this equation is $(0, 4)$.

$$-2x + 25y = 100$$

$$-2x + 25(\mathbf{0}) = 100 \quad \text{Replace } y \text{ with 0.}$$

$$-2x + \mathbf{0} = 100 \quad \text{Simplify.}$$

$$\frac{-2x}{-2} = \frac{100}{-2}$$ Division property of equality.

$$x = -50$$ Simplify.

A solution of this equation is $(-50, 0)$.

$$-2x + 25y = 100$$

$$-2(\mathbf{25}) + 25y = 100$$ Replace x with 25.

$$\mathbf{-50} + 25y = 100$$ Simplify.

$$+50 \qquad +50$$ Addition property of equality.

$$\mathbf{0} + 25y = \mathbf{150}$$ Simplify.

$$\frac{\mathbf{25y}}{25} = \frac{150}{25}$$ Division property of equality.

$$y = 6$$ Simplify.

A solution of this equation is $(25, 6)$.

The table of solutions:

x	y
0	4
−50	0
25	6

(b) Graph the equation.

Draw a coordinate system, labeling the axes and writing a scale for each axis. Graph the point that represents each solution, estimating the location when necessary. Use a straightedge to draw a line through the points.

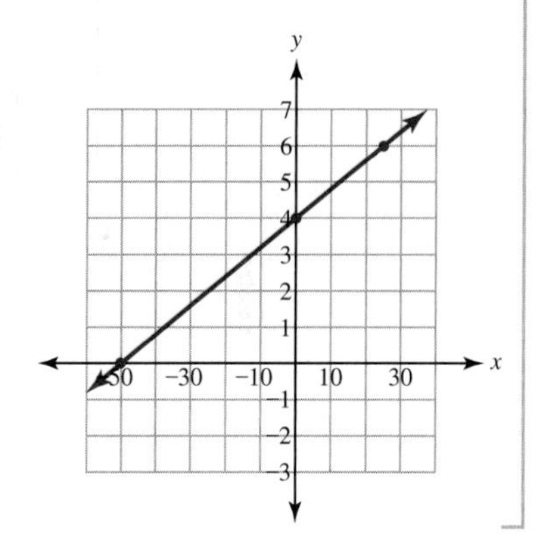

When an equation is solved for y, it is often easier to find solutions by choosing values only for x and solving for y. In the next example, the coefficient of the term $\frac{3}{4}x$ is $\frac{3}{4}$. Choosing values for x that are multiples of 4 makes the arithmetic easier.

EXAMPLE 9 A linear equation is $y = \frac{3}{4}x - 6$.

(a) Complete the table of solutions.

x	y
0	
4	
−4	

SOLUTION

$$y = \frac{3}{4}x - 6$$

$$y = \frac{3}{4}(\mathbf{0}) - 6$$ Replace x with 0.

$$y = \mathbf{0} - 6$$ Simplify.

$$y = \mathbf{-6}$$ Simplify.

A solution of this equation is $(0, -6)$.

$y = \frac{3}{4}x - 6$

$y = \frac{3}{4}(\mathbf{4}) - 6$ Replace x with 4.

$y = \mathbf{3} - 6$ Simplify.

$y = \mathbf{-3}$ Simplify.

A solution of this equation is $(4, -3)$.

$y = \frac{3}{4}x - 6$

$y = \frac{3}{4}(\mathbf{-4}) - 6$ Replace x with -4.

$y = \mathbf{-3} - 6$ Simplify.

$y = \mathbf{-9}$ Simplify.

A solution of this equation is $(-4, -9)$.

The table of solutions:

x	y
0	−6
4	−3
−4	−9

(b) Graph the equation.

Draw a coordinate system, labeling the axes and including a scale on each axis. Graph the point that represents each ordered pair. Use a straightedge to draw a line through the points.

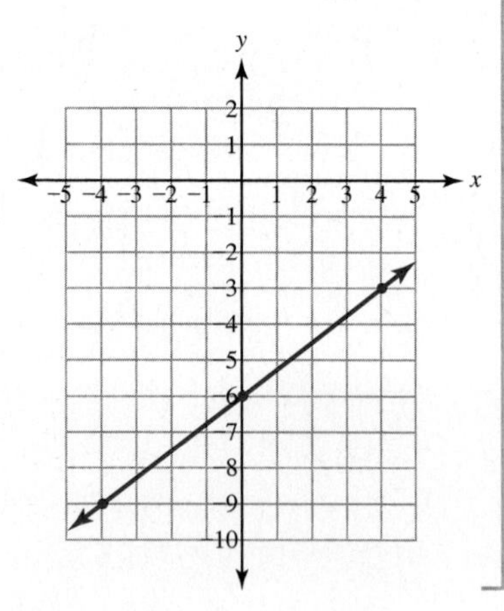

Practice Problems

For problems 12–20,

(a) complete the table of solutions.

(b) graph the equation.

12. $x + y = 6$

x	y
0	
	0
−3	

13. $x - y = 6$

x	y
0	
	0
−3	

14. $x + 2y = -6$

x	y
0	
	0
2	

15. $2x + 3y = 18$

x	y
0	
	0
	2

16. $3x - y = 15$

x	y
0	
	0
	−6

17. $2x + 3y = -9$

x	y
0	
	0
−3	

18. $y = \frac{2}{3}x - 4$

x	y
3	
0	
−3	

19. $y = \left(\frac{\$2}{1 \text{ pound}}\right)x + \$3$

x (lb)	y (\$)
0	
2	
4	

20. $y = -\frac{5}{6}x + 2$

x	y
6	
0	
−6	

Using Technology: Building a Table of Solutions

To build a table of solutions of a linear equation, use the TABLE SETUP screen and the TABLE screen.

EXAMPLE 10 Use a calculator to build a table of solutions of the equation $y = -\frac{3}{4}x + 6$ with a beginning x-value of 10 and an interval of 2 between x-values.

Press Y=. Type the equation. Go to the TBLSET screen; press 2nd WINDOW. Replace 0 with 10. This is the beginning x-value. Press ENTER.

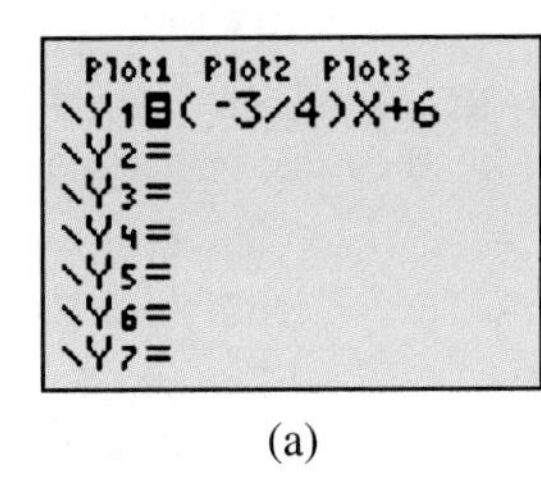

(a)

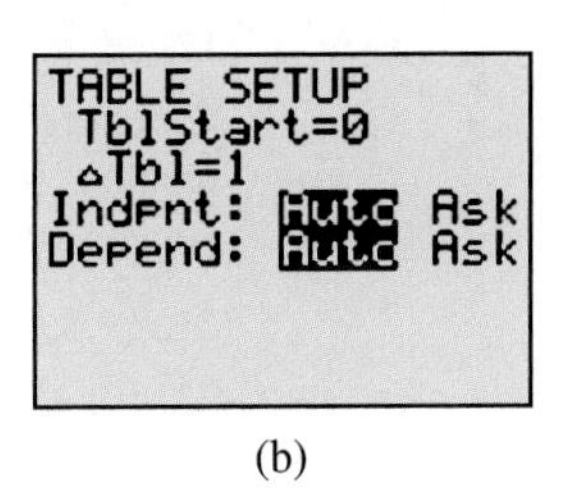

(b)

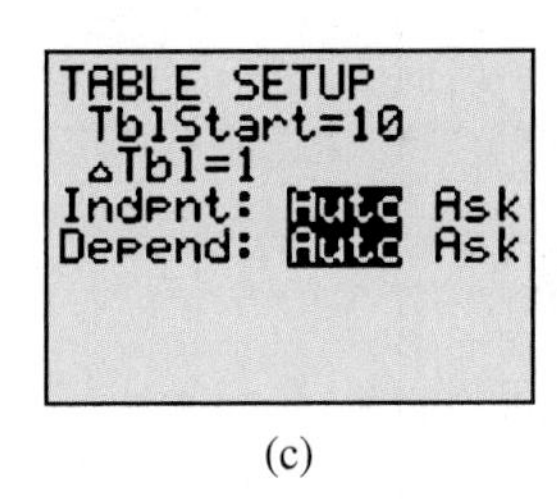

(c)

The symbol Δ means "change in." So ΔTbl means the change or interval between each x-value in the table. Replace 1 by typing 2. To see the table, press 2nd GRAPH to go to the TBL screen.

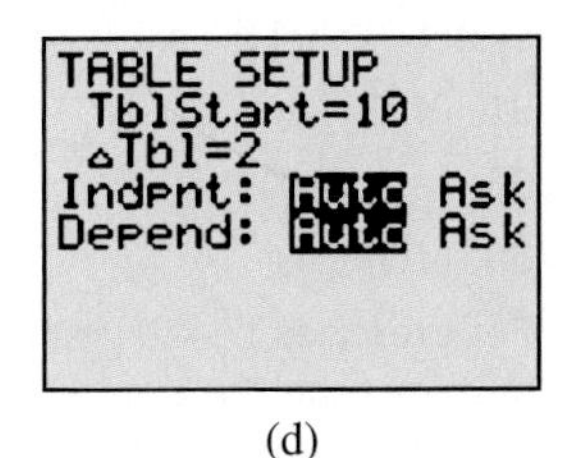

(d)

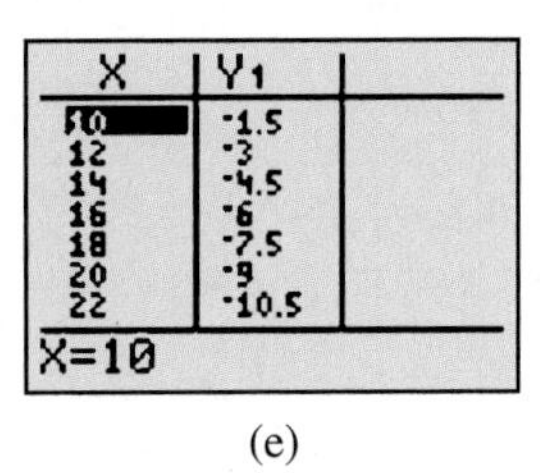

(e)

Practice Problems For problems 21–23, use a calculator to build a table of solutions of $y = 10x - 40$ with the given beginning x-value and interval between x-values. Write a table that includes the first five solutions.

21. $x = 5$, interval = 2 **22.** $x = 0$, interval = 20 **23.** $x = -20$, interval = 5

3.1 Vocabulary Practice

Match the term with its description.

1. A horizontal number line on an xy-coordinate system
2. A vertical number line on an xy-coordinate system
3. The graph of an ordered pair
4. A solution of a linear equation in two variables
5. This shows the value of the distance between tick marks on a number line.
6. $\{\ldots, -2, -1, 0, 1, 2, 3, \ldots\}$
7. In (3, 4), the number 3 is this.
8. In (3, 4), the number 4 is this.
9. A coordinate system is divided by the axes into these four regions.
10. (0, 0)

A. ordered pair
B. origin
C. point
D. quadrants
E. scale
F. set of integers
G. x-axis
H. x-coordinate
I. y-axis
J. y-coordinate

3.1 Exercises

Follow your instructor's guidelines for showing your work and for labeling points.

1. What ordered pair represents the origin?
2. What is the name of the numbers that are written next to the axes on a coordinate system?

For exercises 3–18, use the graph.

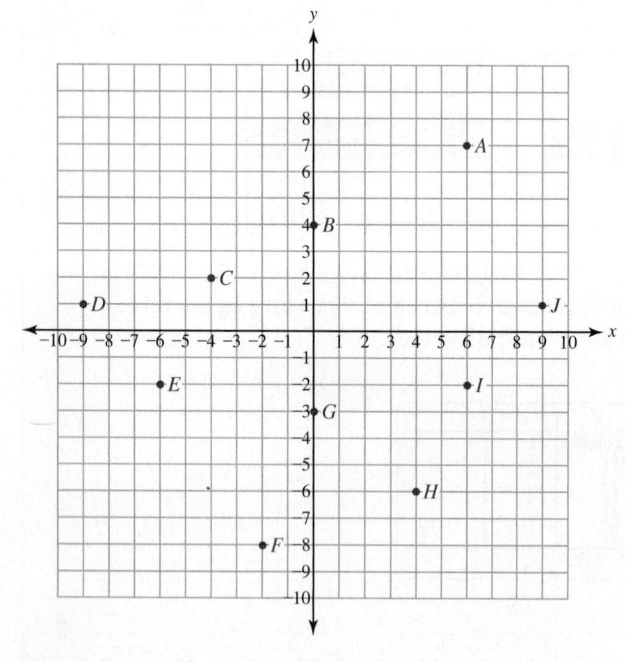

For exercises 3–10, identify the ordered pair represented by the point.

3. A
4. B
5. C
6. D
7. E
8. F
9. G
10. H

For exercises 11–18, identify the quadrant in which the point is located.

11. A
12. J
13. C
14. D
15. E
16. F
17. H
18. I

For exercises 19–30, graph each ordered pair on the same coordinate system. Label the axes; write a scale for each axis.

19. (1, 2)
20. (1, 5)
21. (−3, 6)
22. (−3, 2)
23. (0, 4)
24. (0, 2)
25. (−3, 0)
26. (−6, 0)
27. (−3, −7)
28. (−1, −4)
29. (6, −5)
30. (4, −3)

For exercises 31–36, graph each ordered pair on a coordinate system. Label the axes; write a scale for each axis.

31. (1, 30)
32. (2, 25)
33. (−3, 100)
34. (−5, 90)
35. (1000, 4)
36. (900, 3)

For exercises 37–42, graph each ordered pair on a coordinate system. Label the axes, including the units of measurement; write a scale for each axis.

37. (3 years, \$1000)
38. (4 years, \$500)
39. (50 min, 2 gal)
40. (40 min, 3 L)
41. (20 cm, 10 cm)
42. (15 ft, 10 ft)

For exercises 43–48, graph the ordered pair on a coordinate system. Label the axes; write a scale for each axis.

43. $\left(3, \frac{1}{2}\right)$
44. $\left(4, \frac{1}{2}\right)$

45. $\left(\frac{5}{2}, 6\right)$

46. $\left(\frac{7}{2}, 5\right)$

47. $\left(\frac{10}{3}, \frac{11}{2}\right)$

48. $\left(\frac{13}{3}, \frac{15}{2}\right)$

For exercises 49–84,
(a) complete the table of solutions.
(b) graph the equation.

49. $x + y = \$10$

x ($)	y ($)
0	
	0
3	

50. $x + y = \$8$

x ($)	y ($)
0	
	0
4	

51. $x + y = 50$ kg

x (kg)	y (kg)
0	
	0
20	

52. $x + y = 60$ cm

x (cm)	y (cm)
0	
	0
20	

53. $2x + y = -10$

x	y
0	
	0
3	

54. $2x + y = -8$

x	y
0	
	0
3	

55. $y - 3x = 12$

x	y
0	
	0
3	

56. $y - 4x = 12$

x	y
0	
	0
3	

57. $3x + 5y = 15$

x	y
0	
	0
−3	

58. $5x + 3y = 15$

x	y
0	
	0
−2	

59. $3x - 5y = 15$

x	y
0	
	0
−5	

60. $5x - 3y = 15$

x	y
0	
	0
−3	

61. $x - 6y = 9$

x	y
0	
	0
3	

62. $x - 4y = 10$

x	y
0	
	0
6	

63. $6x - y = 9$

x	y
0	
	0
−3	

64. $4x - y = 10$

x	y
0	
	0
−3	

65. $7x + 4y = 21$

x	y
0	
	0
−3	

66. $9x + 5y = 18$

x	y
0	
	0
−2	

67. $x + y = 30$

x	y
0	
	0
3	

68. $x + y = 40$

x	y
0	
	0
4	

69. $x + 60y = 120$

x	y
0	
	0
−60	

70. $x + 70y = 140$

x	y
0	
	0
−70	

71. $-3x + 50y = 150$

x	y
0	
	0
50	

72. $-3x + 40y = 120$

x	y
0	
	0
40	

73. $y = x + 2$

x	y
0	
2	
4	

74. $y = x + 3$

x	y
0	
2	
4	

75. $y = \left(\frac{\$2}{1 \text{ ft}}\right)x + \$10$

x (ft)	y ($)
0	
5	
10	

76. $y = \left(\frac{\$2}{1 \text{ ft}}\right)x + \$5$

x (ft)	y ($)
0	
5	
10	

77. $y = -2x + 1$

x	y
−2	
0	
2	

78. $y = -3x + 1$

x	y
−2	
0	
2	

79. $y = 3x - 4$

x	y
−4	
0	
4	

80. $y = 2x - 5$

x	y
−4	
0	
4	

81. $y = \frac{2}{3}x - 6$

x	y
−3	
0	
3	

82. $y = \frac{2}{3}x - 7$

x	y
−3	
0	
3	

83. $y = -\frac{5}{6}x + 8$

x	y
−6	
0	
6	

84. $y = -\frac{5}{6}x + 4$

x	y
−6	
0	
6	

Problem Solving: Practice and Review

Follow your instructor's guidelines for using the five steps as outlined in Section 1.3, p. 55.

85. Yummy Meats is an on-line retailer that sells meat in packs. Find the total price for an order of three pork packs, one chicken pack, and one beef pack. (*Source:* www.yummymeats.com)

Meat	Price per pack
Beef	\$389
Pork	\$319
Chicken	\$289
Seafood	\$369

86. Find the total amount of assets in the fund at the end of January 2011. Round to the nearest million.

Illinois' prepaid tuition program, a 12-year-old financial plan enabling children to attend state colleges at today's prices when they have grown up, has the deepest shortfall of any such fund in the United States and is plowing money into unconventional—and some financial experts say high-risk—investments to close the gap. . . . At the start of 2009, its assets were virtually all stocks and bonds; as of the end of January, the fund held \$419 million, or 38%, in such alternatives as hedge funds, real estate and private equity. (*Source:* www.chicagobusiness.com, March 7, 2011)

87. The base rate to rent a large RV in July for 2 weeks is \$1806. There is an additional charge of \$0.32 per mile. A family is going to take a two-week trip in the RV from Anaheim, California to Arches National Park in Utah, a round-trip distance of 1484 mi. Find the total cost to rent the RV for this trip.

88. The Monster pizza at Pizza Perfection in Pullman, Washington is circular with a diameter of 26 in. A large pizza has a diameter of 16 in. Find the difference in area between the Monster pizza and a large pizza. ($\pi \approx 3.14$.) (*Source:* www.pizzaperfection.net)

Technology

For exercises 89–92, use a calculator to build a table of solutions of $y = 4x - 6$ with the given beginning x-value and interval between x-values. Write a table that includes the first five solutions.

89. $x = 1$, interval $= 5$

90. $x = 3$, interval $= 2$

91. $x = -8$, interval $= 3$

92. $x = -12$, interval $= 0.5$

Find the Mistake

For exercises 93–96, the completed problem has one mistake.

(a) Describe the mistake in words, or copy down the whole problem and highlight or circle the mistake.

(b) Do the problem correctly.

93. Problem: Identify the ordered pair represented by the point.

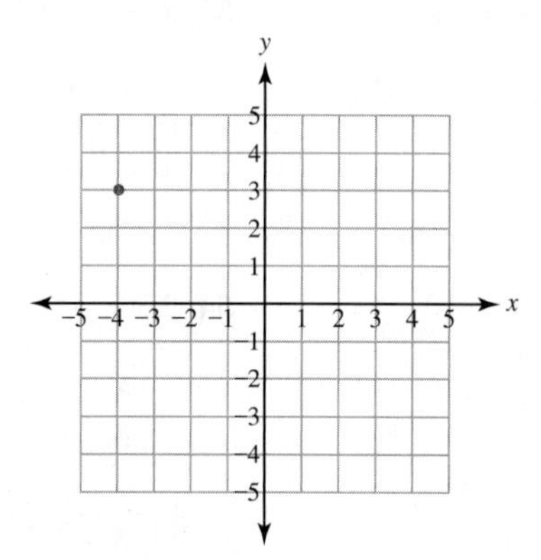

Incorrect Answer: $(3, -4)$

94. Problem: Graph: $(-1, -2)$

Incorrect Answer:

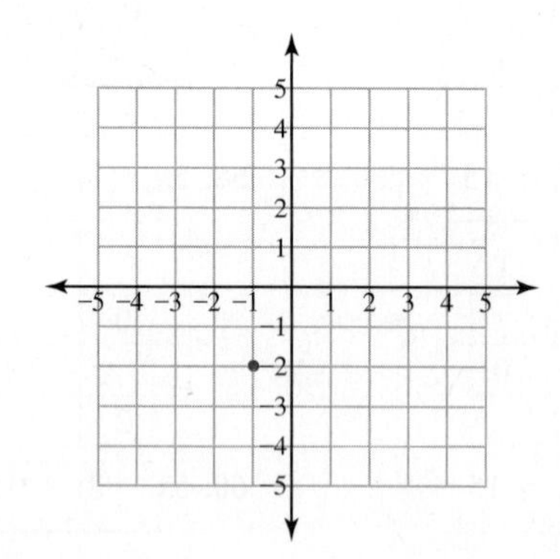

95. Problem: Use the solutions in the table to graph $2x + y = 6$.

x	y
0	6
2	2
3	0

Incorrect Answer:

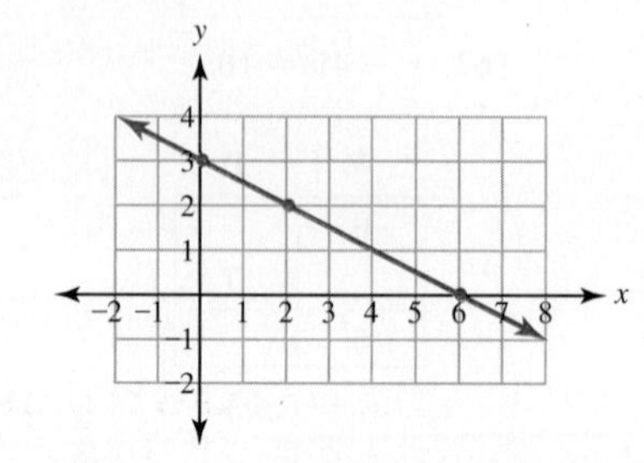

96. Problem: Identify the ordered pair represented by the point.

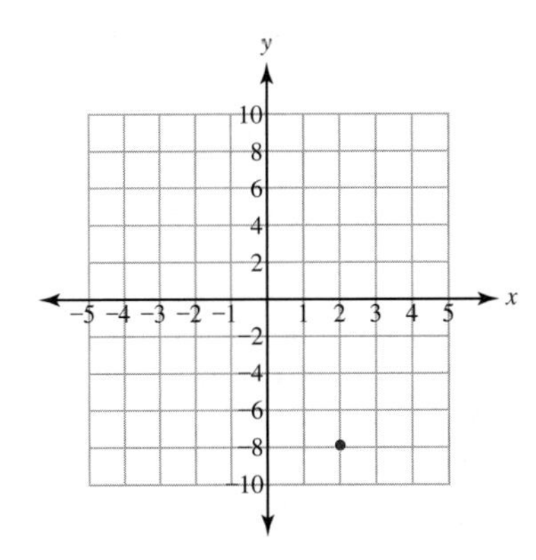

Incorrect Answer: (2, −4)

Review

For exercises 97–100, evaluate.

97. $\frac{2}{5} \cdot 15$

98. $\frac{4}{3} \cdot 15$

99. $-\frac{1}{8} \cdot 56$

100. $-\frac{1}{9} \cdot 54$

SUCCESS IN COLLEGE MATHEMATICS

101. Learning preferences describe the surroundings in which you best learn and study. Choose all the conditions that most closely match your preferences when learning or practicing math.

A. Quiet
B. Background noise (music, TV, family, friends)
C. Bright light shining on your work
D. Dim lighting
E. Sitting at a desk or table
F. Being relaxed on a couch or bed
G. Eating and/or drinking while you learn
H. Only eating or drinking during breaks
I. Working hard without breaks
J. Taking frequent breaks
K. Being in the library, math study center, or somewhere else at college
L. Being where you live

102. Do you now study in conditions that match your learning preferences?

If a family drives 750 mi in 2 days, the number of miles the family drives on the first day plus the number of miles the family drives on the second day equals 750 mi. If x equals the number of miles driven on the first day and y equals the number of miles driven on the second day, the linear equation $x + y = 750$ mi is a symbolic representation of the relationship of x and y. This equation is written in standard form. In this section, we will rewrite linear equations in standard form and use intercepts to graph these equations.

SECTION 3.2 Standard Form and Graphing with Intercepts

After reading the text, working the practice problems, and completing assigned exercises, you should be able to:

1. Rewrite a linear equation in standard form.
2. Given a linear equation or its graph, identify the intercepts.
3. Find three solutions of a linear equation and graph the equation.
4. Write the equation of a horizontal or vertical line that passes through a given point and graph the line.

Standard Form of a Linear Equation

Each solution of a **linear equation** in two variables can be represented by a point on a graph. All of these points lie on the same straight line.

Standard Form of a Linear Equation

The standard form of a linear equation in two variables is $ax + by = c$. The values of a, b, and c are real numbers; a and b cannot both be 0.

Note: There are variations in what is required for standard form. Some mathematicians write standard form for linear equations with rational coefficients and constants as $Ax + By = C$, where A, B, and C are integers.

We can use the properties of equality to rewrite a linear equation in standard form and clear the fractions.

EXAMPLE 1 Rewrite $y = -6x + 3$ in standard form.

SOLUTION

$$y = -6x + 3$$

$$+6x \qquad +6x \qquad \text{Addition property of equality.}$$

$$\mathbf{6x + y = 0 + 3} \qquad \text{Simplify.}$$

$$6x + y = \mathbf{3} \qquad \text{Simplify; the equation matches the form } ax + by = c.$$

When a linear equation includes fractions, it may be easier to clear the fractions first and then use the properties of equality to rewrite the equation in standard form.

EXAMPLE 2 Clear the fractions, and rewrite $y - 8 = \frac{3}{5}(x - 2)$ in standard form.

SOLUTION

$$y - 8 = \frac{3}{5}(x - 2)$$

$$5(y - 8) = (5)\left(\frac{3}{5}\right)(x - 2) \qquad \text{Clear fractions; multiplication property of equality.}$$

$$5(y - 8) = \mathbf{3}(x - 2) \qquad \text{Simplify.}$$

$$\mathbf{5y + 5(-8) = 3x + 3(-2)} \qquad \text{Distributive property.}$$

$$5y \mathbf{- 40} = 3x \mathbf{- 6} \qquad \text{Simplify.}$$

$$+40 \qquad +40 \qquad \text{Addition property of equality.}$$

$$5y \mathbf{+ 0} = 3x \mathbf{+ 34} \qquad \text{Simplify.}$$

$$-3x \qquad -3x \qquad \text{Subtraction property of equality.}$$

$$\mathbf{-3x} + 5y = \mathbf{0} + 34 \qquad \text{Simplify.}$$

$$-3x + 5y = \mathbf{34} \qquad \text{Simplify; this equation matches the form } ax + by = c.$$

Practice Problems

1. Rewrite $y = 7x - 5$ in standard form.
2. Rewrite $y - 6 = 2(x - 4)$ in standard form.
3. Clear the fractions and rewrite $\frac{1}{2}x = 9 - \frac{1}{6}y$ in standard form.
4. Clear the fractions and rewrite $y - \frac{2}{7} = \frac{5}{6}(x - 1)$ in standard form.

Intercepts

The graph (Figure 1) of $2x + 3y = 6$ includes the points $(0, 2)$ and $(3, 0)$. These points are the **intercepts** of the line. The ***y*-intercept** is the point on the line that is on the y-axis. The x-coordinate of the y-intercept is always 0. The ***x*-intercept** is the point on the line that is on the x-axis. The y-coordinate of the x-intercept is always 0.

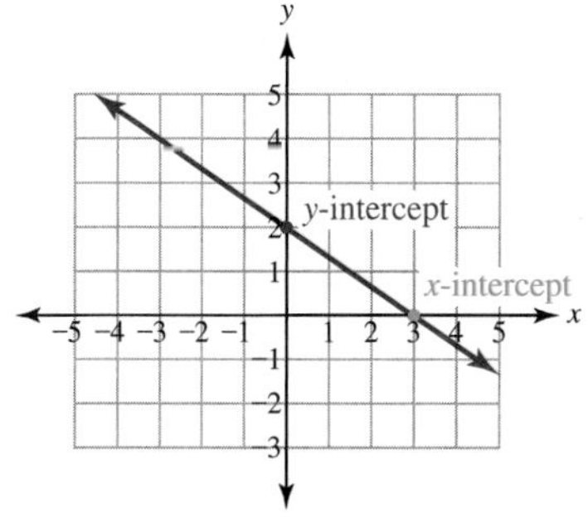

Figure 1

EXAMPLE 3 The graph represents a linear equation.

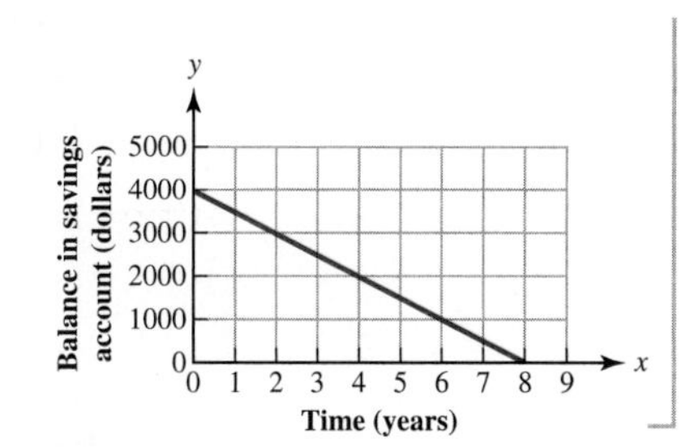

(a) Identify the y-intercept.

SOLUTION ▶ The y-intercept is (0 years, \$4000).

(b) Identify the x-intercept.

▶ The x-intercept is (8 years, \$0).

To graph a linear equation in standard form, we often graph both of the intercepts and a third point. The third point is a check; since the equation is a *linear* equation, all three points should be on a straight line. If the graph of the solutions of a linear equation is not a straight line, one or more of the solutions is not correct.

Intercept Graphing of a Linear Equation

1. To find the y-intercept, replace x in the equation of the line with 0 and solve for y.
2. To find the x-intercept, replace y in the equation of the line with 0 and solve for x.
3. Find a third solution of the equation.
4. Graph the ordered pairs. Draw a straight line through the points.

EXAMPLE 4 A linear equation is $-5x + 3y = 20$.

(a) Find the y-intercept.

SOLUTION ▶ Since the x-coordinate of the y-intercept is 0, replace x with 0 and solve for y.

$$-5x + 3y = 20$$

$$-5(\mathbf{0}) + 3y = 20 \quad \text{Replace } x \text{ with 0.}$$

$$\mathbf{0} + 3y = 20 \quad \text{Simplify.}$$

$$\frac{\mathbf{3y}}{3} = \frac{20}{3} \quad \text{Division property of equality.}$$

$$y = \frac{20}{3} \quad \text{Simplify.}$$

x	y
0	?

The y-intercept is $\left(0, \frac{20}{3}\right)$.

(b) Find the x-intercept.

Since the y-coordinate of the x-intercept is 0, replace y with 0 and solve for x.

$-5x + 3y = 20$

$-5x + 3(\mathbf{0}) = 20$ Replace y with 0.

$-5x + \mathbf{0} = 20$ Simplify.

$\dfrac{-5x}{-5} = \dfrac{20}{-5}$ Division property of equality.

$x = -4$ Simplify.

x	y
0	$\frac{20}{3}$
?	0

The x-intercept is $(-4, 0)$.

(c) Find a third solution of the equation.

Choose any value for x or y. Since it is easier to graph an accurate line if the three points are spread out, choose a value for x that is greater than 0, such as $x = 2$.

$-5x + 3y = 20$

$-5(\mathbf{2}) + 3y = 20$ Replace x with 2.

$\mathbf{-10} + 3y = 20$ Simplify.

$+10 \qquad +10$ Addition property of equality.

$\mathbf{0} + 3y = \mathbf{30}$ Simplify.

$\dfrac{3y}{3} = \dfrac{30}{3}$ Division property of equality.

$y = 10$ Simplify.

x	y
0	$\frac{20}{3}$
-4	0
2	?

A third solution of the equation is $(2, 10)$.

(d) Graph the equation.

Graph the three solutions. Use a straightedge to draw a line through the points. Each point on this line represents an ordered pair that is a solution of the equation $-5x + 3y = 20$.

x	y
0	$\frac{20}{3}$ (≈ 6.7)
-4	0
2	10

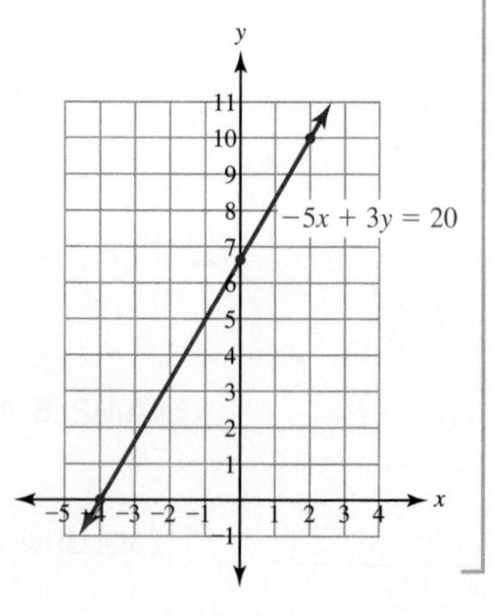

In the next example, we graph the equation $x + y = 750$ mi. Since distances are greater than or equal to 0, the graph is limited to the first quadrant.

EXAMPLE 5 A linear equation is $x + y = 750$ mi.

(a) Find the y-intercept.

SOLUTION

$x + y = 750$ mi

$\mathbf{0} + y = 750$ mi Replace x with 0.

$y = 750$ mi Simplify.

x (mi)	y (mi)
0	?

The y-intercept is (0 mi, 750 mi).

(b) Find the x-intercept.

$x + y = 750$ mi

$x + \mathbf{0} = 750$ mi   Replace y with 0.

$\mathbf{x} = 750$ mi   Simplify.

x (mi)	y (mi)
0	750
?	0

The x-intercept is $(750 \text{ mi}, 0 \text{ mi})$.

(c) Find a third solution of the equation.

Choose a value for x or y. Since the values for x must be between 0 mi and 750 mi, choose $x = 300$ mi.

$x + y = 750$ mi

$\mathbf{300\ mi} + y = 750$ mi   Replace x with 300 mi.

-300 mi -300 mi   Subtraction property of equality.

$\mathbf{0} + y = \mathbf{450\ mi}$ Simplify.

$y = 450$ mi   Simplify.

x (mi)	y (mi)
0	750
750	0
300	?

A third solution of the equation is $(300 \text{ mi}, 450 \text{ mi})$.

(d) Graph the equation.

Graph the three solutions. Use a straightedge to draw a line through the points. Each point on this line represents an ordered pair that is a solution of the equation $x + y = 750$ mi.

x (mi)	y (mi)
0	750
750	0
300	450

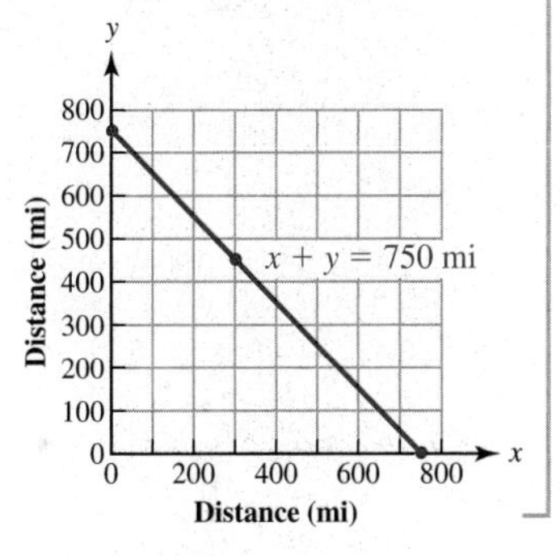

Practice Problems

For problems 5–11,
(a) find the y-intercept.
(b) find the x-intercept.
(c) find a third solution of the equation.
(d) graph the equation.

5. $x + y = 5$
6. $2x + y = 6$
7. $x - 3y = 9$
8. $4x - 5y = 20$
9. $x + 5y = 40$
10. $-3x + 2y = 9$
11. $2x + y = \$800$

Graphing Linear Equations with an Isolated Variable

To graph a linear equation in standard form, we often graph the x-intercept, the y-intercept, and one other point. The linear equation $y = -2x + 5$ is not in standard form; the variable y is isolated. To find three solutions of this equation, it is easier to choose three different values for x and solve for y.

EXAMPLE 6 A linear equation is $y = -2x + 5$.

(a) Find three solutions of the equation.

SOLUTION ▶ Since y is isolated on the left side of the equation, choose values for x and solve for y. Choose values for x that spread the points out across the graph. If the points are too close together, it is difficult to draw an accurate line.

$y = -2x + 5$

$y = -2(\mathbf{0}) + 5$ Replace x with 0.

$y = \mathbf{0} + 5$ Simplify.

$y = \mathbf{5}$ Simplify.

x	y
0	?

A solution of the equation is $(0, 5)$; this is the y-intercept.

$y = -2x + 5$

$y = -2(\mathbf{2}) + 5$ Replace x with 2.

$y = \mathbf{-4} + 5$ Simplify.

$y = \mathbf{1}$ Simplify.

x	y
0	5
2	?

A solution of the equation is $(2, 1)$.

$y = -2x + 5$

$y = -2(\mathbf{-2}) + 5$ Replace x with -2.

$y = \mathbf{4} + 5$ Simplify.

$y = \mathbf{9}$ Simplify.

x	y
0	5
2	1
−2	?

A solution of the equation is $(-2, 9)$.

(b) Graph the equation.

▶ Draw a coordinate system, labeling the axes and writing a scale for each axis. Graph the point that represents each ordered pair. Use a straightedge to draw a line through the points.

x	y
0	5
2	1
−2	9

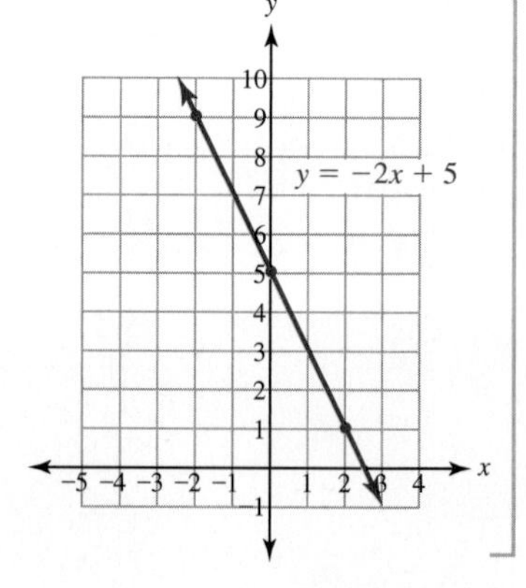

The next equation, $y = \frac{2}{5}x - 4$, includes a fraction with a denominator of 5. When finding solutions of this equation, the arithmetic is easier if we choose values for x that are multiples of 5.

EXAMPLE 7 A linear equation is $y = \frac{2}{5}x - 4$.

(a) Find three solutions of the equation.

SOLUTION ▶ $y = \frac{2}{5}x - 4$

$y = \frac{2}{5}(\mathbf{0}) - 4$ Replace x with 0.

$y = \mathbf{0} - 4$ Simplify.

$y = \mathbf{-4}$ Simplify.

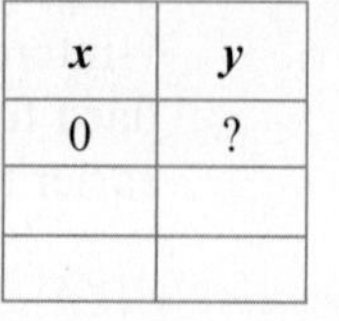

x	y
0	?

A solution of the equation is $(0, -4)$; this is the y-intercept.

$y = \frac{2}{5}x - 4$

$y = \frac{2}{5}(\mathbf{5}) - 4$ Replace x with 5.

$y = \mathbf{2} - 4$ Simplify.

$y = \mathbf{-2}$ Simplify.

x	y
0	−4
5	?

A solution of the equation is $(5, -2)$.

$y = \frac{2}{5}x - 4$

$y = \frac{2}{5}(\mathbf{-5}) - 4$ Replace x with −5.

$y = \mathbf{-2} - 4$ Simplify.

$y = \mathbf{-6}$ Simplify.

x	y
0	−4
5	−2
−5	?

A solution of the equation is $(-5, -6)$.

(b) Graph the equation.

Draw a coordinate system, labeling the axes and writing a scale for each axis. Graph the point that represents each ordered pair. Use a straightedge to draw a line through the points.

x	y
0	−4
5	−2
−5	−6

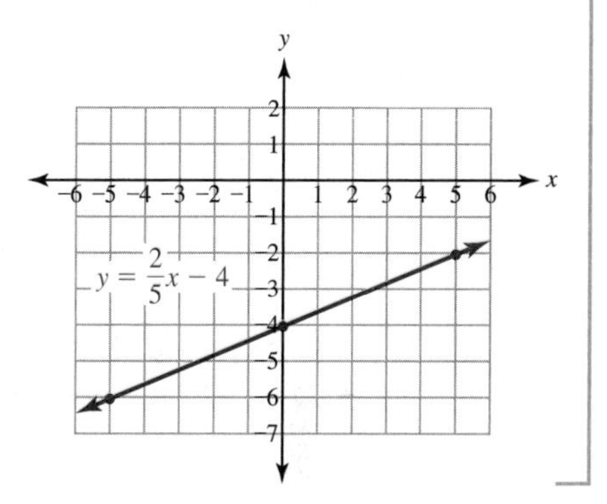

Graphing a Linear Equation with an Isolated Variable

1. Replace the variable that is not isolated with a number. Solve for the remaining variable.
2. Repeat this process to find three ordered pairs that are solutions of the equation.
3. Graph these points on a coordinate system. Draw a straight line through the points.

Practice Problems

For problems 12–14,
(a) find three solutions of the equation.
(b) graph the equation.

12. $y = 3x + 1$ **13.** $y = -4x + 9$ **14.** $y = \frac{2}{3}x - 5$

Horizontal and Vertical Lines

All of the points on a horizontal line have the same y-coordinate. The graph in Figure 2 represents the equation $y = 3$. The y-coordinate of every point on the line is 3. In standard form, the equation is $0x + 1y = 3$, which simplifies to $y = 3$.

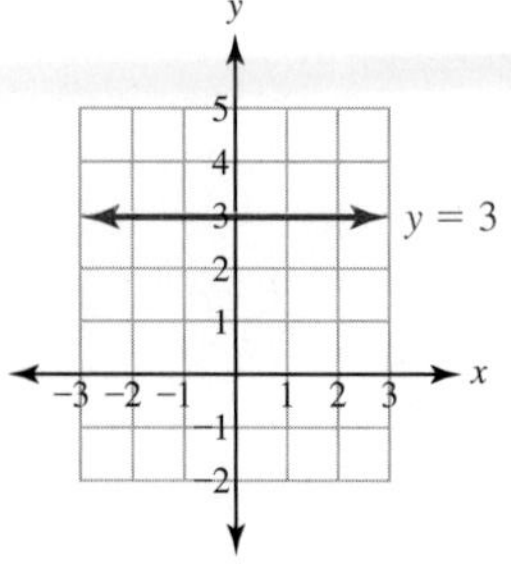

Figure 2

EXAMPLE 8 A horizontal line passes through the point $(2, -1)$.

(a) Write the equation of the line.

SOLUTION ▶ The y-coordinate of every point on this line is the same, -1. The equation is $y = -1$.

(b) Graph the equation.

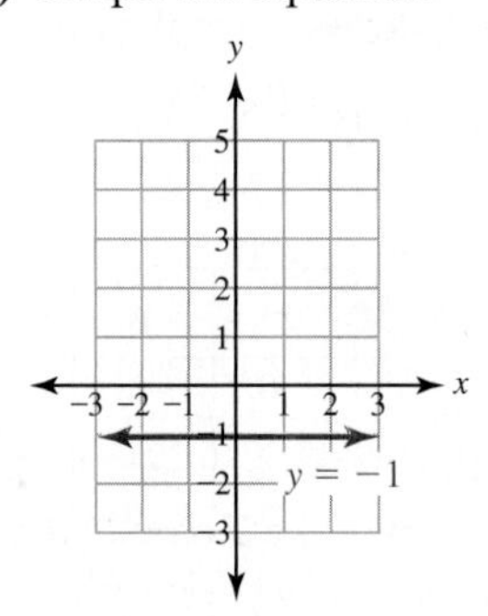

All of the points on a vertical line have the same x-coordinate. The graph in Figure 3 represents the equation $x = -2$. The x-coordinate of every point on the line is -2. In standard form, the equation is $1x + 0y = -2$, which simplifies to $x = -2$.

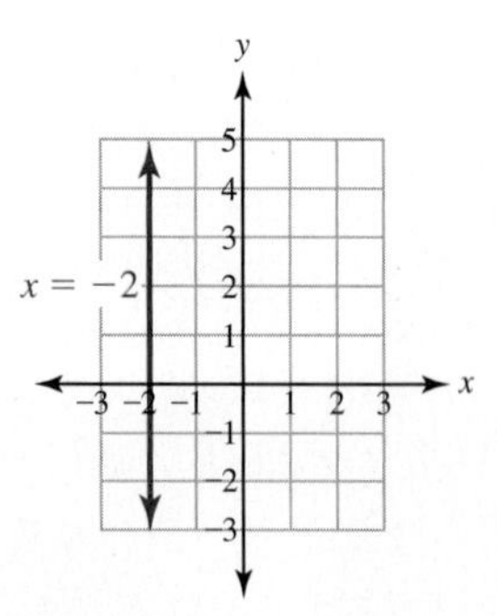

Figure 3

EXAMPLE 9 A vertical line passes through the point $(2, -3)$.

(a) Write the equation of the line.

SOLUTION ▶ The x-coordinate of every point on the line is the same, 2. The equation is $x = 2$.

(b) Graph the equation.

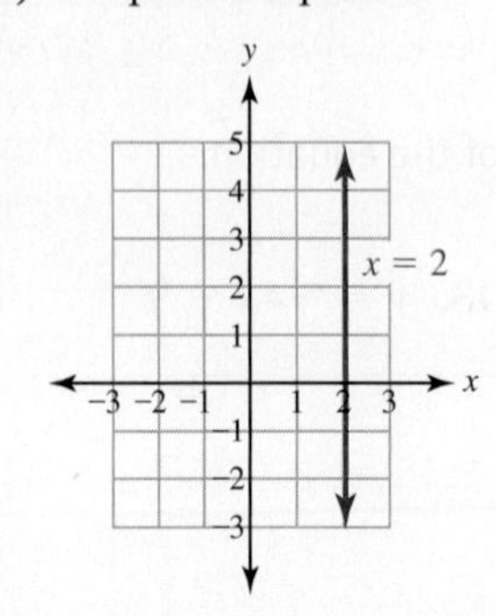

Practice Problems

15. A horizontal line passes through the point $(4,\ 5)$.
   a. Write the equation of the line.
   b. Graph the equation.

16. A vertical line passes through the point $(4, -5)$.
   a. Write the equation of the line.
   b. Graph the equation.

Using Technology: Graphing a Linear Equation on a Graphing Calculator

On a graphing calculator, use the WINDOW screen or the ZOOM screen to adjust the scale of the graph. The **window** describes the part of the graph that is visible on the calculator screen.

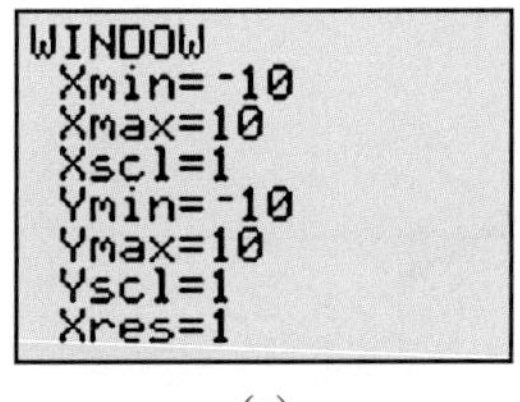

(a)

Press WINDOW. A **standard window** shows a graph from −10 to 10 on the x-axis and from −10 to 10 on the y-axis. To change a window, type in new values for X_{min}, X_{max}, Y_{min}, and Y_{max}.

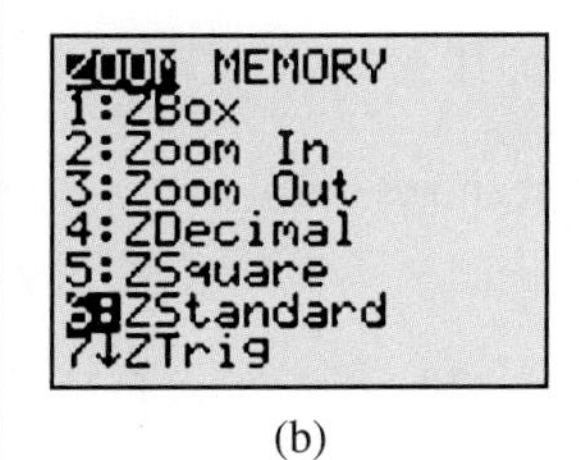

(b)

Or, press ZOOM. This opens the ZOOM screen. The **Zoom In** command decreases the amount of the graph in the window. The **Zoom Out** command increases the amount of the graph. Choosing **ZStandard** returns the graph to the standard window.

Choose a window that shows important features of the graph. When graphing a linear equation, we often choose a window that shows both the x-intercept and the y-intercept.

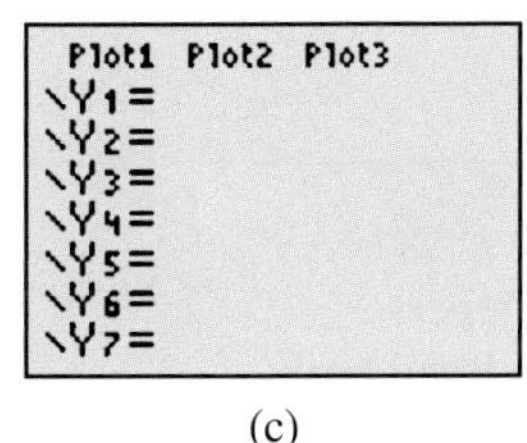

(c)

On the Y= screen, we can only type equations in which y is isolated on the left side. Press Y=. The subscripts are numbers to the lower right of the number. These subscripts are part of the name of the variable.

EXAMPLE 10 Graph $y = 3x + 6$. Sketch the graph; describe the window.

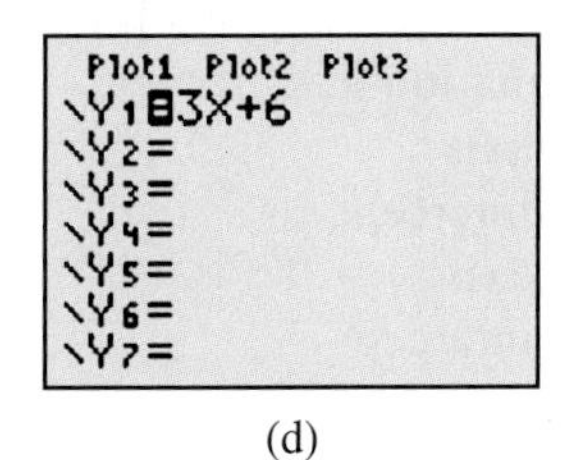

(d)

The Y= is already typed on the screen. Beginning with 3, type the rest of the equation, pressing the X,T,θ,n key for the variable. Press ENTER.

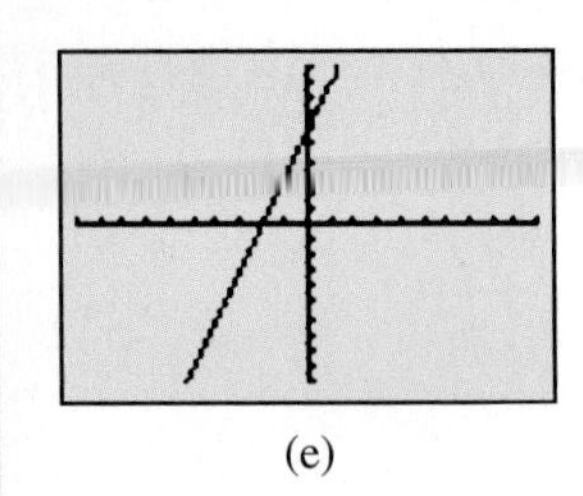
(e)

Press GRAPH. The graph of the equation appears. Since this is a standard window, the x-axis extends from $x = -10$ to $x = 10$. The y-axis extends from $y = -10$ to $y = 10$. The distance between tick marks on each axis is 1. In this window, we can see both the y-intercept and the x-intercept.

To *sketch the graph*, draw a rectangle. Draw a coordinate system inside the rectangle. Draw the line, being careful to place the intercepts in the correct positions. Some instructors prefer that you label the scale and the axes. This is one way to "describe the window." Other instructors prefer that you describe the window by writing the window setting inside of brackets. For a standard window, this is [−10, 10, 1; −10, 10, 1].

EXAMPLE 11 Graph $y = -\frac{3}{4}x + 6$. Sketch the graph; describe the window.

When typing the equation, enclose the fraction inside of parentheses.

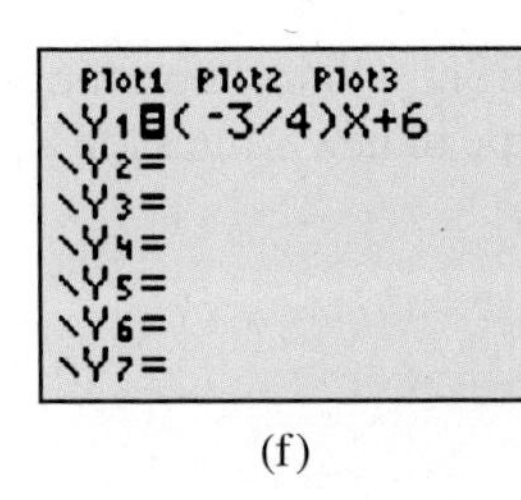

(f)

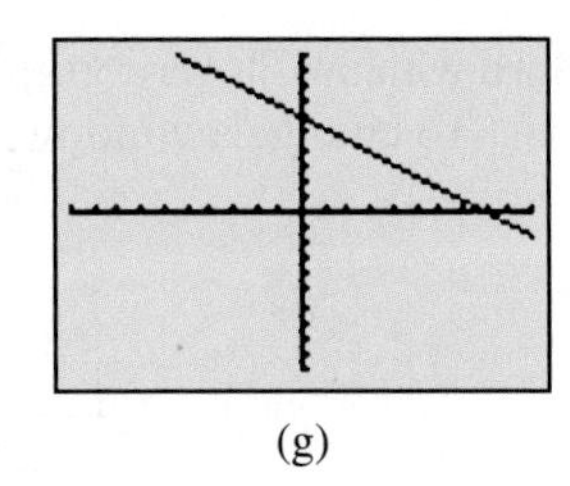
(g)

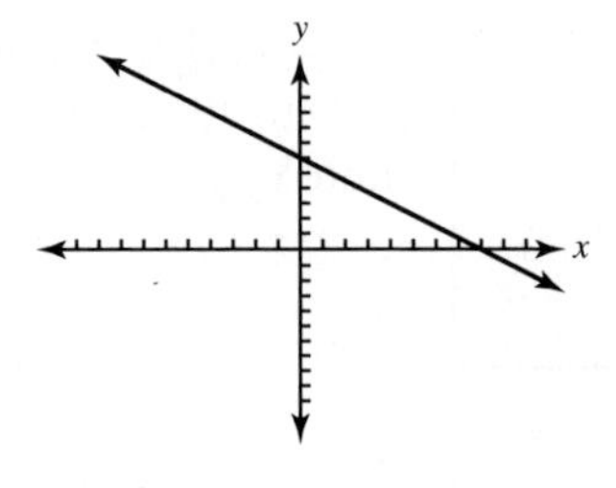

[−10, 10, 1; −10, 10, 1]

Practice Problems For problems 17–20, graph the equation. Choose a window that shows both intercepts. Sketch the graph; describe the window.

17. $y = 5x + 2$ **18.** $y = \frac{3}{5}x - 8$ **19.** $y = 3$ **20.** $y = -3x + 15$

3.2 Vocabulary Practice

Match the term with its description.

1. A horizontal number line on an xy-coordinate system
2. A vertical number line on an xy-coordinate system
3. The graph of an ordered pair
4. A solution of a linear equation in two variables
5. This shows the value of the distance between tick marks on a number line.
6. $\{\ldots, -2, -1, 0, 1, 2, 3, \ldots\}$
7. $ax + by = c$
8. The x-coordinate of this point is always 0.
9. The y-coordinate of this point is always 0.
10. $(0, 0)$

A. integers
B. ordered pair
C. origin
D. point
E. scale
F. standard form of a linear equation
G. x-axis
H. x-intercept
I. y-axis
J. y-intercept

3.2 Exercises

Follow your instructor's guidelines for showing your work.

For exercises 1–10, clear the fractions, and rewrite each equation in standard form.

1. $y = 6x - 2$

2. $y = 5x - 9$

3. $\frac{3}{8}y = -\frac{2}{5}x + 20$

4. $\frac{5}{6}y = -\frac{2}{7}x + 2$

5. $y - 3 = 2(x + 4)$

6. $y - 8 = 9(x + 6)$

7. $y = \frac{3}{4}x - 11$

8. $y = \frac{2}{3}x - 13$

9. $y - 3 = \frac{5}{8}(x + 1)$

10. $y - 4 = \frac{2}{3}(x + 5)$

For exercises 11–20,
(a) identify the y-intercept of the graphed line.
(b) identify the x-intercept of the graphed line.

11.

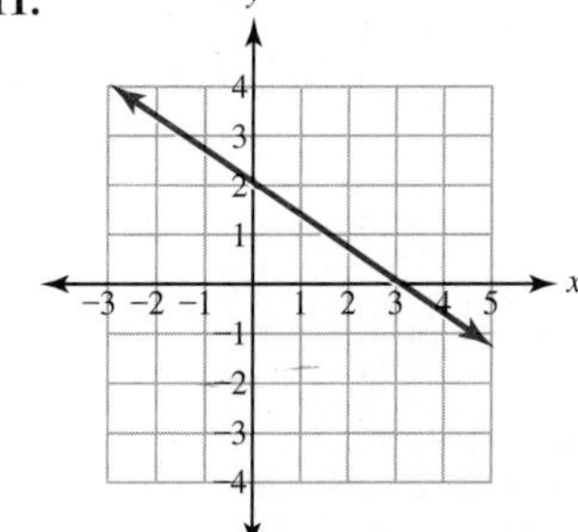

12.

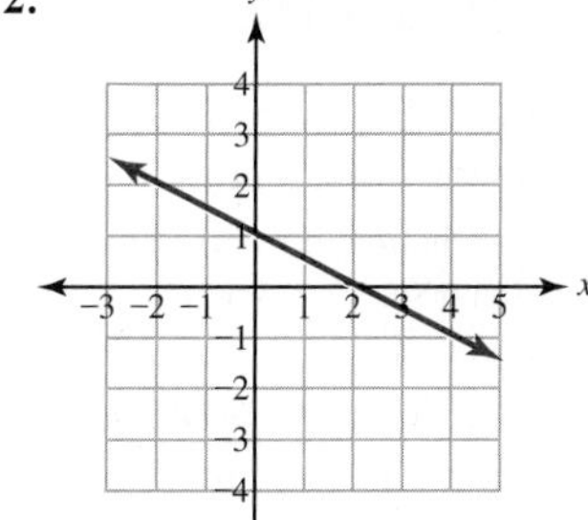

13.

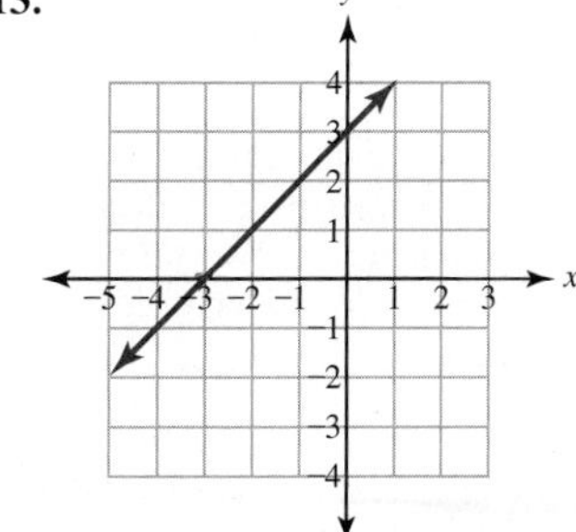

14.

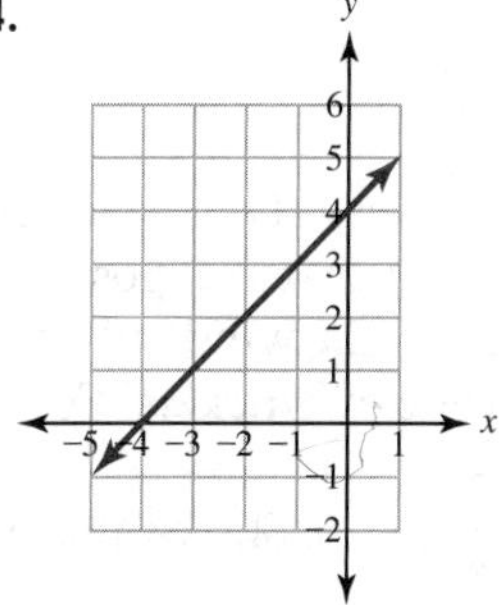

15.

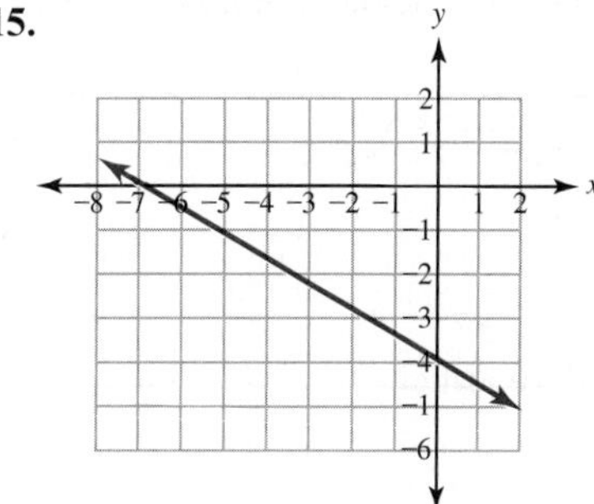

16.

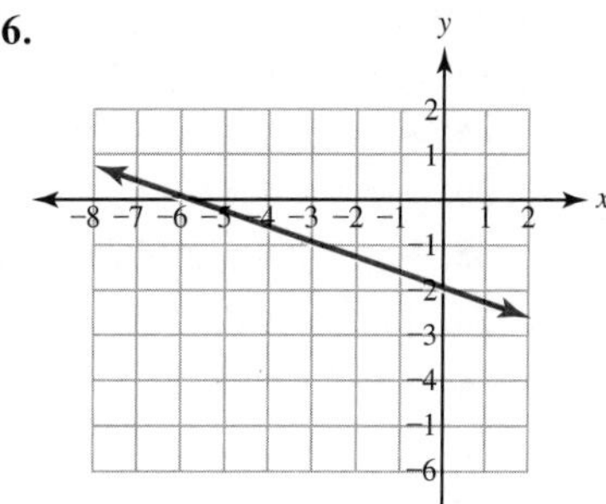

17.

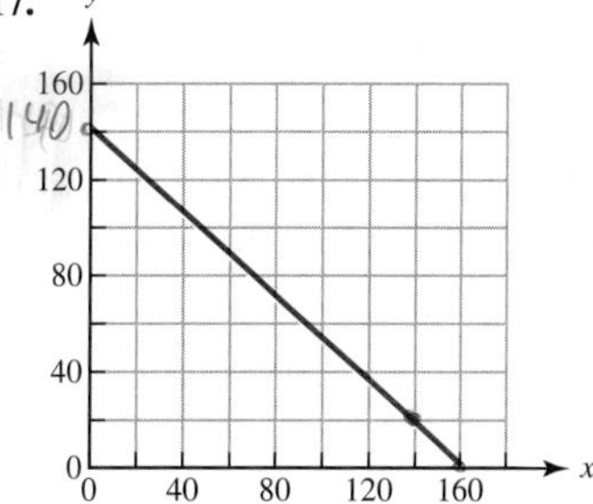

18.

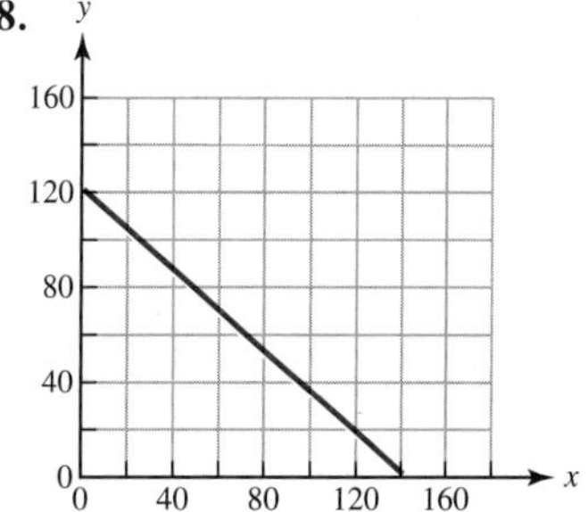

19.

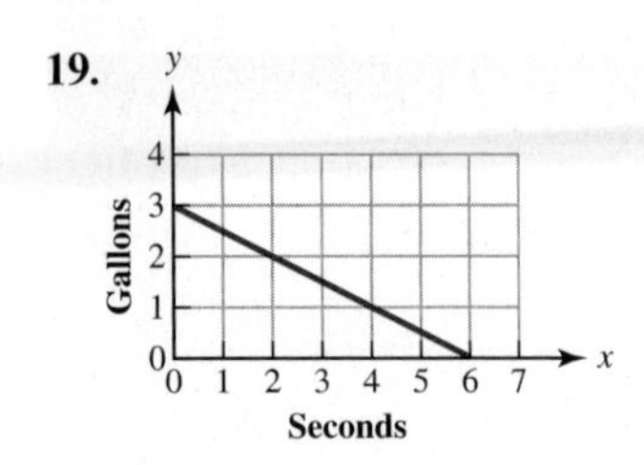

20.

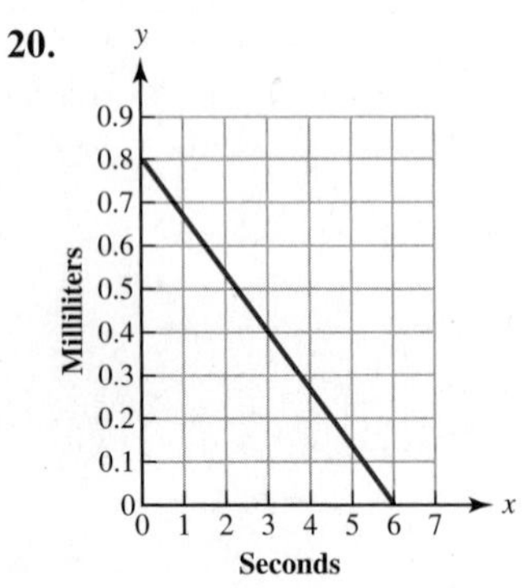

For exercises 21–64,
(a) find the y-intercept.
(b) find the x-intercept.
(c) find a third solution of the equation.
(d) graph the equation.

21. $x + y = 6$ **22.** $x + y = 7$
23. $x + y = -5$ **24.** $x + y = -3$
25. $x - y = 6$ **26.** $x - y = 7$
27. $x - y = -5$ **28.** $x - y = -3$
29. $-x - y = 9$ **30.** $-x - y = 8$
31. $-x + y = -1$ **32.** $-x + y = -2$
33. $3x + y = 12$ **34.** $4x + y = 12$
35. $x + 3y = 12$ **36.** $x + 4y = 12$
37. $7x + 2y = 28$ **38.** $2x + 7y = 28$
39. $2x - 5y = 20$ **40.** $5x - 2y = 20$
41. $-6x + 5y = 60$ **42.** $-5x + 6y = 60$
43. $2x - 9y = -27$ **44.** $9x - 2y = -27$
45. $8x + 3y = 24$ **46.** $3x + 8y = 24$
47. $-8x + 3y = 24$ **48.** $-3x + 8y = 24$
49. $4x + 3y = 20$ **50.** $3x + 4y = 20$
51. $3x + 2y = -15$ **52.** $2x + 3y = -15$
53. $x + y = 400$ **54.** $x + y = 200$
55. $x + y = -1000$ **56.** $x + y = -2000$
57. $20x + y = 500$ **58.** $30x + y = 600$
59. $100x - y = -500$ **60.** $100x - y = -600$
61. $-10x + 3y = 300$ **62.** $-10x + 5y = 400$
63. $2x + 3y = 4$ **64.** $3x + 2y = 4$

For exercises 65–76,
(a) find three solutions of the equation.
(b) graph the equation.

65. $y = 3x - 5$ **66.** $y = 4x - 6$
67. $y = -x + 4$ **68.** $y = -x + 7$
69. $y = 20x + 60$ **70.** $y = 10x + 50$
71. $y = x$ **72.** $y = -x$
73. $y = \frac{2}{5}x - 3$ **74.** $y = \frac{2}{3}x - 4$
75. $y = -\frac{1}{3}x + 5$ **76.** $y = -\frac{1}{4}x + 3$

77. Explain why the x-intercept, the y-intercept, and one other point are not three different solutions of $y = 2x$.

78. Explain why we use three points instead of two points to draw the graph of a linear equation.

For exercises 79–82,
(a) write the equation of the vertical line that passes through the point.
(b) graph the equation.

79. $(4, 5)$ **80.** $(6, 3)$
81. $(-2, -1)$ **82.** $(-3, -2)$

For exercises 83–86,
(a) write the equation of the horizontal line that passes through the point.
(b) graph the equation.

83. $(4, 5)$ **84.** $(6, 3)$
85. $(-2, -1)$ **86.** $(-3, -2)$

Problem Solving: Practice and Review

Follow your instructor's guidelines for using the five steps as outlined in Section 1.3, p. 55.

87. A large garden is a collection of 24 raised beds. Each bed is a rectangle that is 8 ft long and 4 ft wide. Find the total area of the raised beds.

© Laura Bracken

88. Find the operating budget of the Texas Education Agency before the reductions. Round to the nearest million.

It [the Texas Education Agency] is reducing its operating budget by \$48 million, or 36 percent. (*Source:* www.thestatesman.com, July 12, 2011)

89. According to the Pew Research Center, the Millennial Generation in the United States includes about 50 million people born after 1980. Find the number of this generation who say that "you can't be too careful" when dealing with people. Round to the nearest tenth of a million.

Whether as a by-product of protective parents, the age of terrorism or a media culture that focuses on dangers, they [Millennials] cast a wary eye on human nature. Two-thirds say "you can't be too careful" when dealing with people. (*Source:* www.pewresearch.org/millennials, 2010)

90. A high-speed Shinkansen train in Japan travels at a speed of $\frac{270 \text{ km}}{1 \text{ hr}}$ for 18 min. Find the distance it travels.

© Hiroshi Ichikawa/Shutterstock.com

Technology

For exercises 91–94, use a graphing calculator to graph each equation. Choose a window that shows the x-intercept and y-intercept. Sketch the graph; describe the window.

91. $y = -2x + 5$

92. $y = 2x + 5$

93. $y = -4$

94. $y = \frac{1}{2}x - 8$

Find the Mistake

For exercises 95–98, the completed problem has one mistake.

(a) Describe the mistake in words, or copy down the whole problem and highlight or circle the mistake.

(b) Do the problem correctly.

95. Problem: Find the y-intercept of $6x + 5y = 30$.

Incorrect Answer:
$$\begin{aligned} 6x + 5y &= 30 \\ 6x + 5(0) &= 30 \\ 6x &= 30 \\ \frac{6x}{6} &= \frac{30}{6} \\ x &= 5 \end{aligned}$$

The y-intercept is $(5, 0)$.

96. Problem: Find the x-intercept of $9x + 2y = 36$.

Incorrect Answer:
$$\begin{aligned} 9x + 2y &= 36 \\ 9(0) + 2y &= 36 \\ 2y &= 36 \\ \frac{2y}{2} &= \frac{36}{2} \\ y &= 18 \end{aligned}$$

The x-intercept is $(0, 18)$.

97. Problem: Write the equation of the vertical line that passes through $(6, 3)$.

Incorrect Answer: $y = 3$

98. Problem: Graph: $x = 2$

Incorrect Answer:

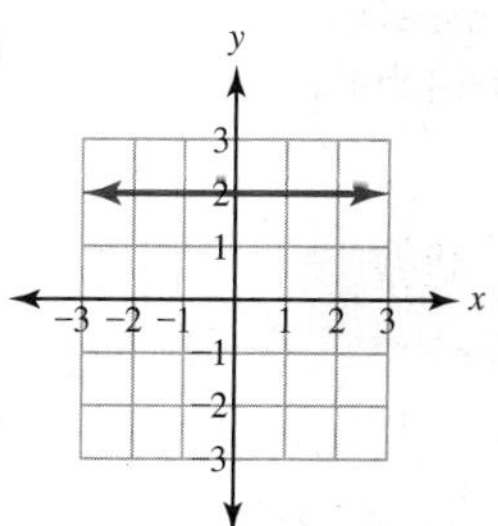

Review

For problems 99–102, solve.

99. $m = \frac{10 - 4}{5 - 3}$

100. $m = \frac{-10 - 4}{-5 - (-3)}$

101. $m = \frac{6 - 6}{4 - 9}$

102. $m = \frac{4 - 9}{6 - 6}$

SUCCESS IN COLLEGE MATHEMATICS

For exercises 103–104, some learning preferences describe how you prefer to receive, think about, and learn new information. These preferences include *visual* learning, *auditory* learning, and *kinesthetic* learning. Many students use more than one of these categories as they learn mathematics.

- ***Visual learners*** prefer to *see* information. Although you definitely listen to your instructor, you also like to see the example on a white board or screen. You may be able to recall a process by visualizing it in your mind; you may learn better by organizing information in charts, tables, diagrams, or pictures. You may prefer the use of colored markers instead of just black.
- ***Auditory learners*** prefer to *hear* information. Although you definitely watch what your instructor is doing, you also like your instructor to explain things aloud as he or she works. You may find it difficult to take notes because you cannot concentrate enough on what is being said while you write. You may learn better if you have the chance to explain things to others.
- ***Kinesthetic learners*** prefer to *do*. You may find it difficult to sit still and just watch and listen; you want to be trying it out. You may find that you must take notes in order to learn. If you only watch and listen, you may understand the concept but not remember it after you leave the classroom. You often learn better if you can show others how to do things.

103. Do you have a strong preference for visual, auditory, or kinesthetic learning?

104. Have you noticed anything that your instructor does while teaching that you find helps you remember what has been taught?

Between 2007 and 2010, the percent of U.S. households with only wireless telephone service increased by about 4.3% per year. This increase per year is an example of an **average rate of change**. In this section, we will study average rates of change and slope.

SECTION 3.3

Average Rate of Change and Slope

After reading the text, working the practice problems, and completing assigned exercises, you should be able to:

1. Find an average rate of change.
2. Given the equation or graph of a line, find the slope of the line.
3. Identify the slope of a horizontal or vertical line.

Rates of Change

The average of a group of test scores is the sum of the scores divided by the number of scores. The average speed of a car equals the distance the car travels divided by the time spent traveling. Although at times the car may travel faster and slower than the average speed, the product of the average speed and the time spent traveling is the total distance. Since a speed compares one measurement (the distance) to another measurement (the time), it is an **average rate of change**. We often use the variable m to represent an average rate of change.

EXAMPLE 1 After driving 2 hr on a highway, a car is at mile marker 60. After 5 hr, the car is at mile marker 300.

(a) Represent the information as two ordered pairs.

SOLUTION ▶ If the ordered pairs are (time, distance), then this information is (2 hr, 60 mi) and (5 hr, 300 mi).

(b) Graph the ordered pairs.

▶

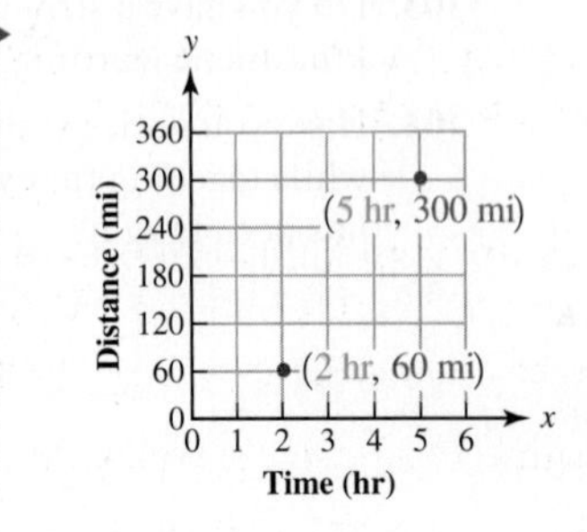

(c) Find the average rate of change, m.

$$\text{average rate of change} = \frac{\text{change in distance}}{\text{change in time}}$$

$$m = \frac{300 \text{ mi} - 60 \text{ mi}}{5 \text{ hr} - 2 \text{ hr}} \qquad m = \text{average rate of change}$$

$$m = \frac{\mathbf{240 \text{ mi}}}{\mathbf{3 \text{ hr}}} \qquad \text{Simplify.}$$

$$m = \frac{\mathbf{80 \text{ mi}}}{\mathbf{1 \text{ hr}}} \qquad \text{Simplify.}$$

The average rate of change is $\frac{80 \text{ mi}}{1 \text{ hr}}$.

In the next example, we find the average rate of change in the number of violent crimes per year. Since the number of violent crimes per year is decreasing, the average rate of change is negative.

EXAMPLE 2 In Mesa, Arizona, 2289 violent crimes were reported in 2008 and 1790 violent crimes were reported in 2010. (*Source:* www.fbi.gov, 2010)

(a) Represent the information as two ordered pairs.

SOLUTION If the ordered pairs are (time, number of crimes), then this information is (2008, 2289 crimes) and (2010, 1790 crimes).

(b) Find the average rate of change, m. Round to the nearest integer.

$$\text{average rate of change} = \frac{\text{change in number of violent crimes}}{\text{change in time}}$$

$$m = \frac{1790 \text{ crimes} - 2289 \text{ crimes}}{2010 - 2008} \qquad m = \text{average rate of change}$$

$$m = \frac{\mathbf{-499 \text{ crimes}}}{\mathbf{2 \text{ years}}} \qquad \text{Simplify.}$$

$$m = -\frac{\mathbf{249.5 \text{ crimes}}}{\mathbf{1 \text{ year}}} \qquad \text{Simplify.}$$

$$m \approx -\frac{\mathbf{250 \text{ crimes}}}{\mathbf{1 \text{ year}}} \qquad \text{Round.}$$

The average rate of change is about $-\frac{250 \text{ violent crimes}}{1 \text{ year}}$.

In the next example, since the percent of households with only wireless telephone service is increasing, the average rate of change is positive.

EXAMPLE 3 In 2007, 13.6% of American households had only wireless telephone service. By 2010, 26.6% of American households had only wireless telephone service. (*Source:* National Center for Health Statistics, Dec. 2010)

(a) Represent the information as two ordered pairs.

SOLUTION If the ordered pairs are (time, percent with only wireless), then this information is (2007, 13.6%) and (2010, 26.6%).

(b) Find the average rate of change, m. Round to the nearest tenth of a percent.

$$\text{average rate of change} = \frac{\text{change in percent}}{\text{change in time}}$$

$$m = \frac{26.6\% - 13.6\%}{2010 - 2007} \qquad m = \text{average rate of change}$$

$$m = \frac{\mathbf{13\%}}{\mathbf{3\ years}} \qquad \text{Simplify.}$$

$$m = \frac{\mathbf{4.33\ldots\%}}{\mathbf{1\ year}} \qquad \text{Simplify.}$$

$$m \approx \frac{\mathbf{4.3\%}}{\mathbf{1\ year}} \qquad \text{Round.}$$

The average rate of change is about $\frac{4.3\%}{1 \text{ year}}$.

Practice Problems

For problems 1–3,

(a) represent the information as two ordered pairs.

(b) find the average rate of change, m.

1. In 2004, producers in the United States made 8.87 billion lb of cheese. In 2010, production increased to 10.1 billion lb of cheese. Round to the nearest hundredth of a billion. (*Source:* www.eatwisconsincheese.com)
2. In 1998, Americans consumed 9.5 lb of carrots per person. In 2011, this decreased to 7.6 lb of carrots per person. Round to the nearest tenth. (*Source:* www.usda.gov)
3. In the United States in 2007, there were 69.5 births per 1000 women (age 15 years to 44 years). In 2010, there were 64.7 births per 1000 women in this age group. (*Source:* www.cdc.gov)

Slope

In Example 1, we graphed the ordered pairs (2 hr, 60 mi) and (5 hr, 300 mi). We then found the average speed by dividing the change in distance by the change in time: $m = \frac{300 \text{ mi} - 60 \text{ mi}}{5 \text{ hr} - 2 \text{ hr}}$. On the graph (Figure 1), the change in distance is the change in the y-coordinates of the two points. This is the rise. The change in time is the change in the x-coordinates of the two points. This is the run. The ratio of $\frac{\text{rise}}{\text{run}}$ is the **slope** of the line between the two points. So the average rate of change is equal to the slope of the line that passes through the two points.

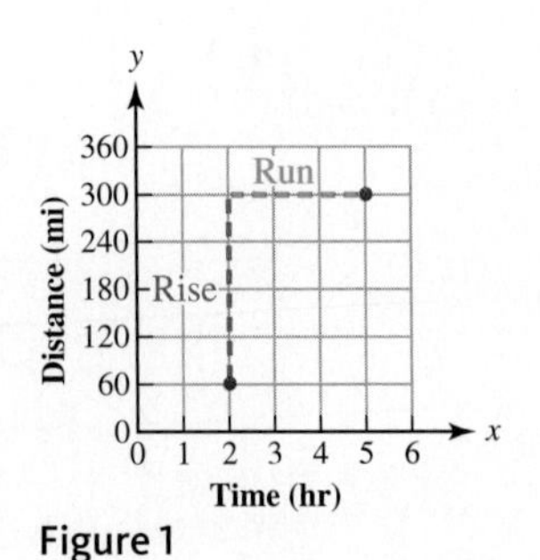

Figure 1

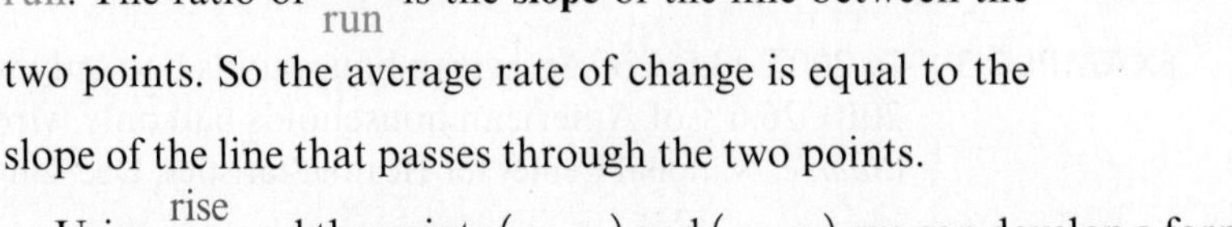

Using $\frac{\text{rise}}{\text{run}}$ and the points (x_1, y_1) and (x_2, y_2), we can develop a formula for finding the slope of a line that passes through any two points.

Slope of a Line

If m is the variable that represents slope, for a line that passes through the points (x_1, y_1) and (x_2, y_2),

$$m = \frac{\text{rise}}{\text{run}}$$

$$m = \frac{\text{change in } y}{\text{change in } x}$$

$$m = \frac{y_2 - y_1}{x_2 - x_1}$$

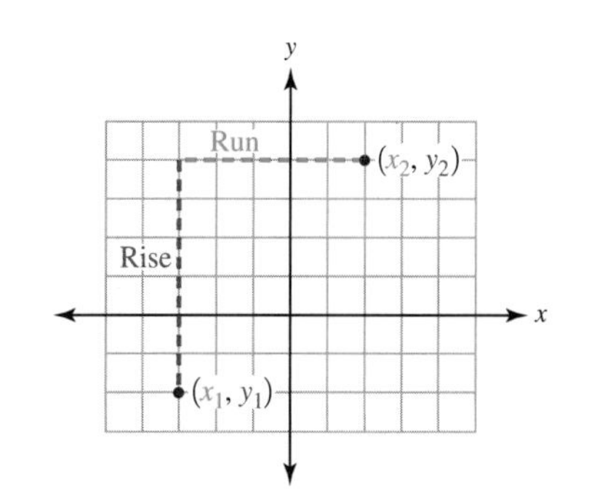

The formula for finding the **slope of a line** that passes through the points (x_1, y_1) and (x_2, y_2) is $m = \frac{y_2 - y_1}{x_2 - x_1}$.

EXAMPLE 4 Use the slope formula to find the slope of the line that passes through (2 hr, 60 mi) and (5 hr, 300 mi).

SOLUTION ▶

$\text{slope} = \frac{\text{rise}}{\text{run}}$ $\quad m = \frac{y_2 - y_1}{x_2 - x_1}$

$m = \frac{300 \text{ mi} - 60 \text{ mi}}{5 \text{ hr} - 2 \text{ hr}}$ $\quad (x_1, y_1) = (2 \text{ hr}, 60 \text{ mi}); (x_2, y_2) = (5 \text{ hr}, 300 \text{ mi})$

$m = \frac{\mathbf{240 \text{ mi}}}{\mathbf{3 \text{ hr}}}$ $\quad$ Simplify.

$m = \frac{\mathbf{80 \text{ mi}}}{\mathbf{1 \text{ hr}}}$ $\quad$ Simplify.

In construction, the pitch or slope of a roof is the change in height divided by the change in length. Between any two points on the roof, the change in height of the roof is the *rise*. The change in length of the roof is the *run*.

EXAMPLE 5 Use the slope formula to find the slope of the roof.

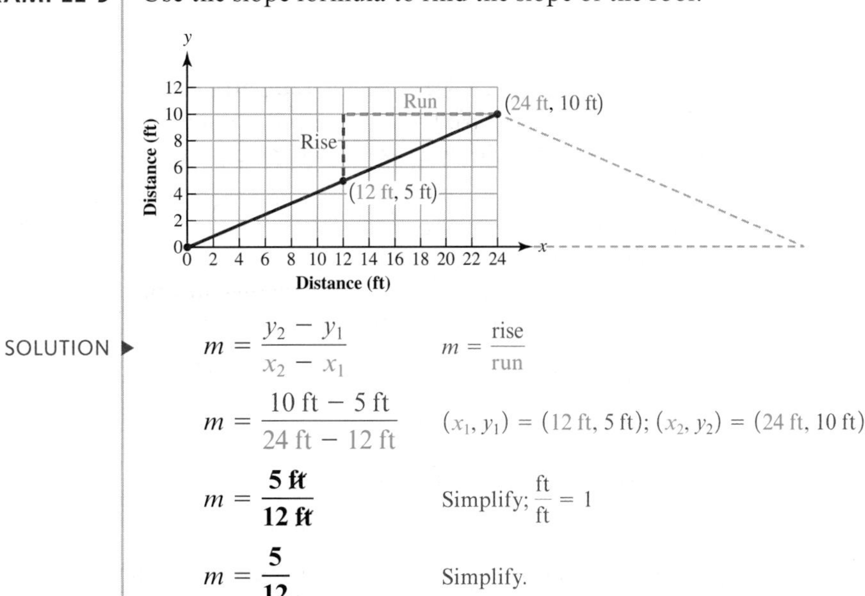

SOLUTION ▶

$m = \frac{y_2 - y_1}{x_2 - x_1}$ $\quad m = \frac{\text{rise}}{\text{run}}$

$m = \frac{10 \text{ ft} - 5 \text{ ft}}{24 \text{ ft} - 12 \text{ ft}}$ $\quad (x_1, y_1) = (12 \text{ ft}, 5 \text{ ft}); (x_2, y_2) = (24 \text{ ft}, 10 \text{ ft})$

$m = \frac{\mathbf{5 \text{ ft}}}{\mathbf{12 \text{ ft}}}$ $\quad$ Simplify; $\frac{\text{ft}}{\text{ft}} = 1$

$m = \frac{\mathbf{5}}{\mathbf{12}}$ $\quad$ Simplify.

The slope or pitch of the roof is $\frac{5}{12}$.

Moving from left to right on a graph, a line with a negative slope "goes down."

EXAMPLE 6 A line passes through (8, 2) and (6, 9).

(a) Graph (8, 2) and (6, 9), and draw a line through the points.

SOLUTION

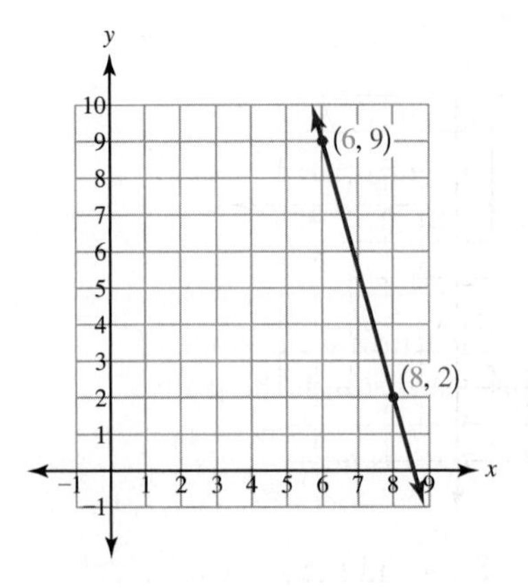

(b) Use the graph to find the slope of the line.

Begin at (6, 9). Move vertically down 7 units to the level of the next point, (8, 2). Since we moved *down*, the rise is *negative* 7. Move horizontally to the right 2 units and arrive at the point (8, 2). Since we moved to the *right*, the run is *positive* 2. The slope, $\frac{\text{rise}}{\text{run}}$, is $\frac{-7}{2}$ or $-\frac{7}{2}$.

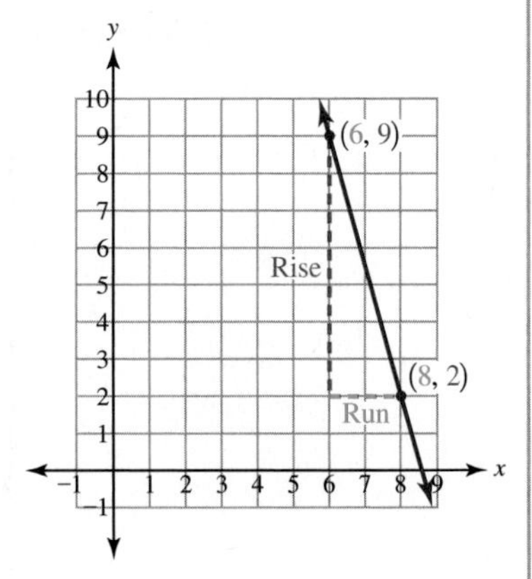

(c) If $(x_1, y_1) = (8, 2)$ and $(x_2, y_2) = (6, 9)$, use the slope formula to find the slope of the line.

$$m = \frac{y_2 - y_1}{x_2 - x_1}$$

$$m = \frac{9 - 2}{6 - 8} \quad (x_1, y_1) = (8, 2); (x_2, y_2) = (6, 9)$$

$$m = \frac{\mathbf{7}}{\mathbf{-2}} \quad \text{Simplify.}$$

$$m = -\frac{\mathbf{7}}{\mathbf{2}} \quad \text{Simplify.}$$

(d) Rename the points $(x_1, y_1) = (6, 9)$ and $(x_2, y_2) = (8, 2)$. Use the slope formula to find the slope of the line.

$$m = \frac{y_2 - y_1}{x_2 - x_1}$$

$$m = \frac{2 - 9}{8 - 6} \quad (x_1, y_1) = (6, 9); (x_2, y_2) = (8, 2)$$

$$m = \frac{\mathbf{-7}}{\mathbf{2}} \quad \text{Simplify.}$$

$$m = -\frac{\mathbf{7}}{\mathbf{2}} \quad \text{Simplify.}$$

No matter which point we choose to be (x_1, y_1), the slope is the same, $m = -\frac{7}{2}$.

Moving from left to right on a graph, a line with a positive slope "goes up."

EXAMPLE 7 A line passes through $(-2, -6)$ and $(0, -3)$.

(a) Graph $(-2, -6)$ and $(0, -3)$, and draw a line through the points.

SOLUTION ▶

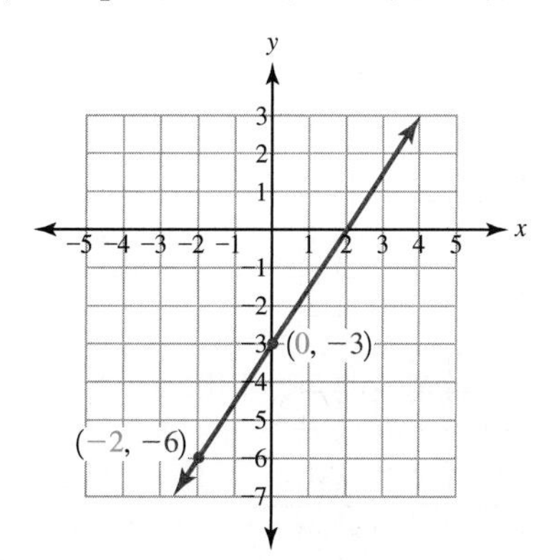

(b) Use the graph to find the slope of the line.

▶ Begin at $(-2, -6)$. Move vertically *up* 3 units to the level of the next point, $(0, -3)$. The rise is *positive* 3. Move horizontally to the *right* 2 units and arrive at the point $(0, 3)$. The run is *positive* 2. The slope, $\frac{\text{rise}}{\text{run}}$, is $\frac{3}{2}$.

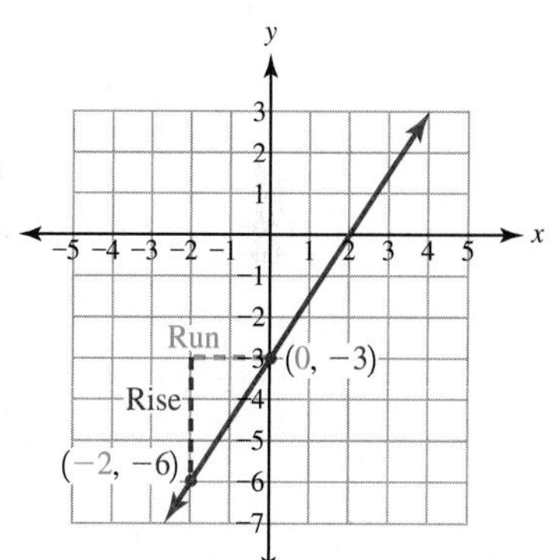

(c) Use the slope formula to find the slope of the line.

▶

$$m = \frac{y_2 - y_1}{x_2 - x_1}$$

$$m = \frac{-3 - (-6)}{0 - (-2)} \qquad (x_1, y_1) = (-2, -6); (x_2, y_2) = (0, -3)$$

$$m = \frac{-3 + 6}{0 + 2} \qquad \text{Simplify.}$$

$$m = \frac{3}{2} \qquad \text{Simplify. Do not rewrite the fraction as a decimal.}$$

When the scale of a graph includes units of measurement, the slope should also include units of measurement.

EXAMPLE 8 The graph represents the relationship of the number of months, x, and the cost per pound, y.

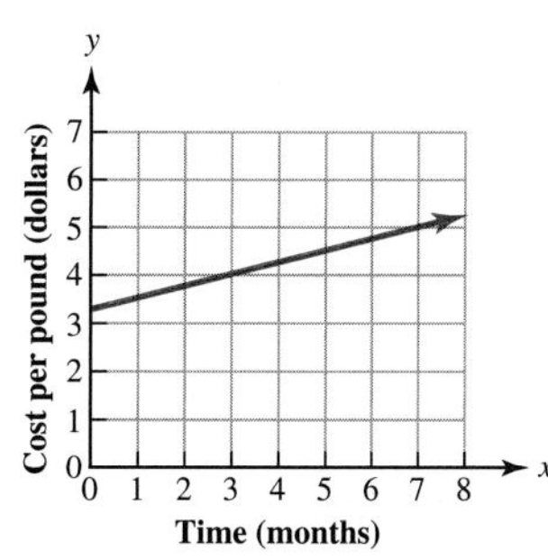

(a) Identify two points on the line.

SOLUTION ▶ Look for two points that have integer coordinates such as (3 months, \$4) and (7 months, \$5).

(b) Use the slope formula to find the slope of the line.

$$m = \frac{y_2 - y_1}{x_2 - x_1}$$

$$m = \frac{\$5 - \$4}{7 \text{ months} - 3 \text{ months}}$$ $(x_1, y_1) = (3 \text{ months}, \$4); (x_2, y_2) = (7 \text{ months}, \$5)$

$$m = \frac{\$1}{4 \text{ months}}$$ Simplify.

$$m = \frac{\$0.25}{1 \text{ month}}$$ Divide; write amounts of money as a decimal number.

If we change the scale of one of the axes on a graph, the appearance of the graph changes. However, *the slope does not change.* For example, the graphs in Figures 2 and 3 represent the relationship of the number of on-line classes, y, and the number of years since 2004, x. The slope is the average rate of change in the number of on-line classes per year. Although it may appear that the slope of the line in Figure 2 is greater than the slope of the line in Figure 3, it is not. The slope in both graphs is $\frac{20 \text{ on-line classes}}{1 \text{ year}}$. In preparing graphs for presentations or when reading media reports, it is important to realize that the scales on a graph can change the perception of the slope of a line.

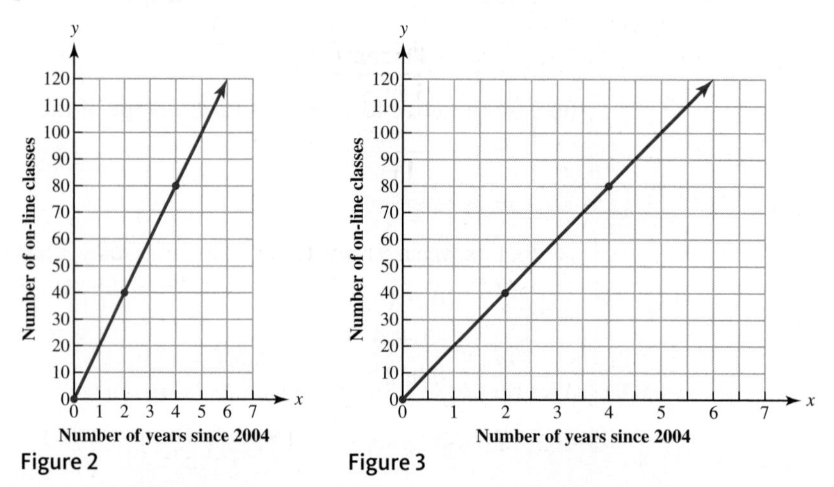

Figure 2 Figure 3

EXAMPLE 9 The graph represents the relationship of the number of reams of paper in a stock room, y, and the number of days, x.

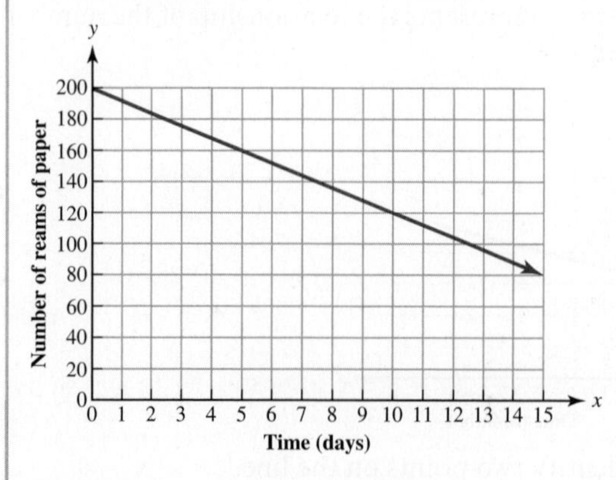

(a) Identify two points on the line.

SOLUTION ▸ Look for two points with integer coordinates such as (5 days, 160 reams) and (10 days, 120 reams).

(b) Use the graph to find the slope of the line.

The scale on each axis is different, and the distance between tick marks on the y-axis is not equal to 1. Begin at (5 days, 160 reams). Move vertically *down* 40 reams to the level of the next point, (10 days, 120 reams). The rise is *negative* 40 reams. Move horizontally to the *right* 5 days and, arrive at the point (10 days, 120 reams). The run is *positive* 5 days. The slope, $\frac{\text{rise}}{\text{run}}$, is $\frac{-40 \text{ reams}}{5 \text{ days}}$, which simplifies to $\frac{-8 \text{ reams}}{1 \text{ day}}$ or $-\frac{8 \text{ reams}}{1 \text{ day}}$.

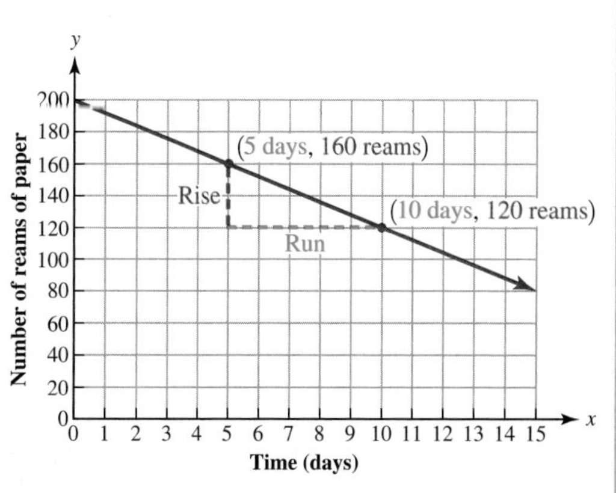

(c) Use the slope formula to find the slope of the line.

$$m = \frac{y_2 - y_1}{x_2 - x_1}$$

$$m = \frac{120 \text{ reams} - 160 \text{ reams}}{10 \text{ days} - 5 \text{ days}} \qquad (x_1, y_1) = (5 \text{ days}, 160 \text{ reams}); (x_2, y_2) = (10 \text{ days}, 120 \text{ reams})$$

$$m = \frac{\mathbf{-40 \text{ reams}}}{\mathbf{5 \text{ days}}} \qquad \text{Simplify.}$$

$$m = -\frac{\mathbf{8 \text{ reams}}}{\mathbf{1 \text{ day}}} \qquad \text{Simplify.}$$

Just as we can rewrite the fraction $\frac{a}{b}$ as $a \div b$, we can rewrite the slope $\frac{3}{\frac{2}{5}}$ as $3 \div \frac{2}{5}$.

EXAMPLE 10 Use the slope formula to find the slope of the line that passes through $\left(\frac{1}{5}, 1\right)$ and $\left(\frac{3}{5}, 4\right)$.

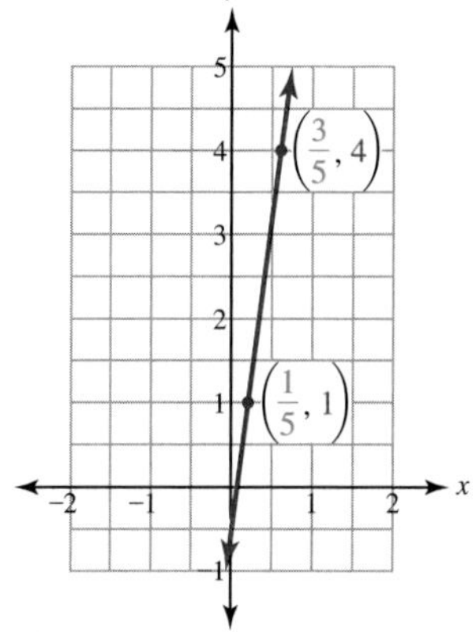

SOLUTION

$$m = \frac{y_2 - y_1}{x_2 - x_1}$$

$$m = \frac{4 - 1}{\frac{3}{5} - \frac{1}{5}} \qquad (x_1, y_1) = \left(\frac{1}{5}, 1\right); (x_2, y_2) = \left(\frac{3}{5}, 4\right)$$

$$m = \frac{\mathbf{3}}{\frac{\mathbf{2}}{\mathbf{5}}} \qquad \text{Simplify the numerator and denominator.}$$

$$m = 3 \div \frac{\mathbf{2}}{\mathbf{5}} \qquad \text{Rewrite the fraction as division.}$$

$$m = \frac{3}{1} \cdot \frac{\mathbf{5}}{\mathbf{2}} \qquad \text{Rewrite division as multiplication by the reciprocal.}$$

$$m = \frac{\mathbf{15}}{\mathbf{2}} \qquad \text{Simplify.}$$

The vertical distance between these specific points on the line is 3, and the horizontal distance is $\frac{2}{5}$. But a fraction with a numerator of 3 and a denominator of $\frac{2}{5}$ is awkward to write. Instead, we simplify and write the slope as $\frac{15}{2}$. If we move vertically up 15 units from $\left(\frac{3}{5}, 4\right)$ and then move horizontally 2 units to the right, we are at the location of another point on the same line.

In the next example, we use the equation of the line to find its intercepts and then use the slope formula to find the slope.

EXAMPLE 11 The equation of a line is $5x + 2y = 20$.

(a) Find the y-intercept.

SOLUTION ▶

$$5x + 2y = 20$$

$$5(\mathbf{0}) + 2y = 20 \quad \text{Replace } x \text{ with 0.}$$

$$\mathbf{0} + 2y = 20 \quad \text{Simplify.}$$

$$\frac{\mathbf{2y}}{2} = \frac{20}{2} \quad \text{Division property of equality.}$$

$$y = 10 \quad \text{Simplify.}$$

The y-intercept is $(0, 10)$.

(b) Find the x-intercept.

▶

$$5x + 2y = 20$$

$$5x + 2(\mathbf{0}) = 20 \quad \text{Replace } y \text{ with 0.}$$

$$5x + \mathbf{0} = 20 \quad \text{Simplify.}$$

$$\frac{\mathbf{5x}}{5} = \frac{20}{5} \quad \text{Division property of equality.}$$

$$x = 4 \quad \text{Simplify.}$$

The x-intercept is $(4, 0)$.

(c) Use the slope formula to find the slope of the line.

▶

$$m = \frac{y_2 - y_1}{x_2 - x_1}$$

$$m = \frac{0 - 10}{4 - 0} \quad (x_1, y_1) = (0, 10); (x_2, y_2) = (4, 0)$$

$$m = \frac{\mathbf{-10}}{\mathbf{4}} \quad \text{Simplify.}$$

$$m = -\frac{\mathbf{5 \cdot 2}}{\mathbf{2 \cdot 2}} \quad \text{Find common factors.}$$

$$m = -\frac{\mathbf{5}}{\mathbf{2}} \quad \text{Simplify.}$$

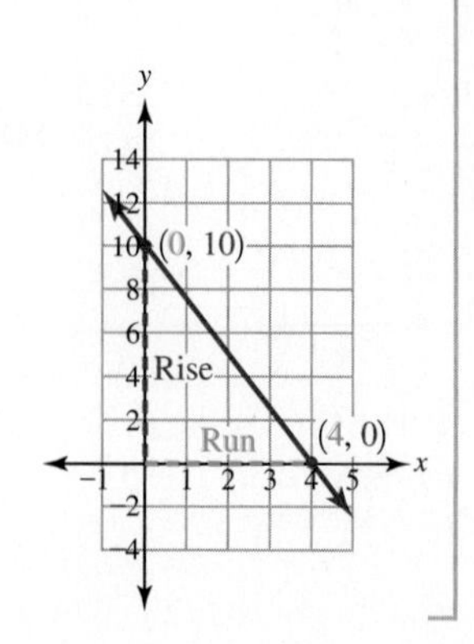

Methods for Finding the Slope of a Line, m

- Given two ordered pairs (x_1, y_1) and (x_2, y_2), use the slope formula, $m = \frac{y_2 - y_1}{x_2 - x_1}$.
- Given the equation of the line, find two ordered pairs such as the y-intercept and the x-intercept that are solutions of the equation. Label these (x_1, y_1) and (x_2, y_2). Use the slope formula.
- Given the graph of a line, find two points on the line. Label these (x_1, y_1) and (x_2, y_2). Use the slope formula.
- Given the graph of a line, find two points on the line. Find the vertical distance (the rise) and the horizontal distance (the run) between the two points. The slope is $\frac{\text{rise}}{\text{run}}$.

Practice Problems

For problems 4–7,
(a) graph the given points and draw a line through the points.
(b) use the graph to find the slope of the line.
(c) use the slope formula to find the slope of the line.

4. $(3, 8)\ (1, 14)$
5. $(5, 9)\ (1, 6)$
6. $(5, -9)\ (-1, -6)$
7. $(-8, 0)\ (0, 5)$

For problems 8–9,
(a) identify two points on the line.
(b) use the graph to find the slope of the graphed line.

8.

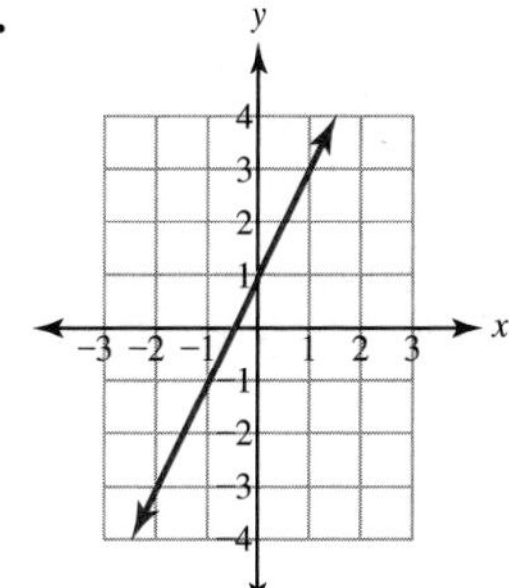

9.

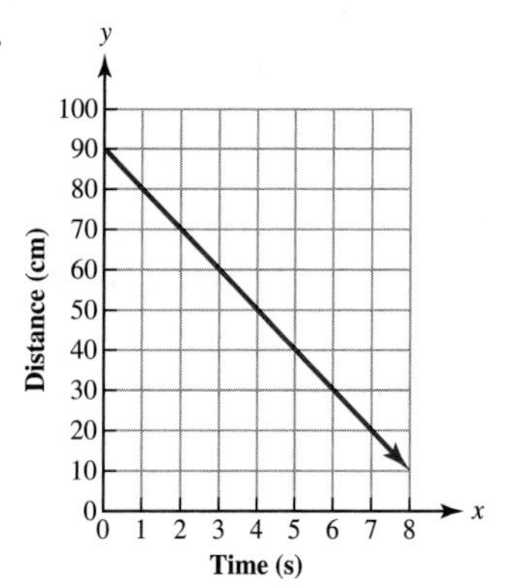

10. Use the slope formula to find the slope of the line that passes through $\left(\frac{1}{2}, \frac{5}{6}\right)$ and $\left(\frac{9}{2}, \frac{1}{6}\right)$.

11. The equation of a line is $3x + 4y = 24$.
a. Find the y-intercept.
b. Find the x-intercept.
c. Use the slope formula to find the slope of the line.

Horizontal and Vertical Lines

Since the rise (change in height) from one end of a table to the other is 0 inches, the slope of the tabletop is 0. The slope of a horizontal line that runs along the edge of a tabletop is also 0.

EXAMPLE 12 Use the slope formula to find the slope of the line that passes through $(-1, 3)$ and $(2, 3)$.

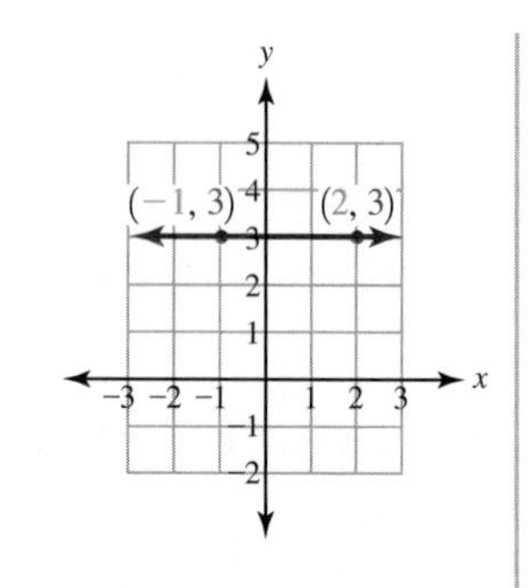

SOLUTION ▶

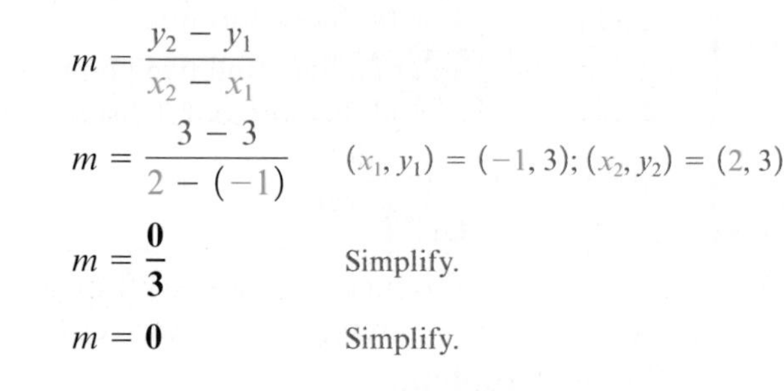

$$m = \frac{y_2 - y_1}{x_2 - x_1}$$

$$m = \frac{3 - 3}{2 - (-1)} \qquad (x_1, y_1) = (-1, 3); (x_2, y_2) = (2, 3)$$

$$m = \frac{\mathbf{0}}{\mathbf{3}} \qquad \text{Simplify.}$$

$$m = \mathbf{0} \qquad \text{Simplify.}$$

The slope of this horizontal line is 0.

When the slope of a line is 0, do not say that it has "no slope." The meaning of "no slope" is not precise enough to use in mathematics.

Division by zero is undefined. In the slope formula, the run is in the denominator, slope $= \frac{\text{rise}}{\text{run}}$. Since the run from ground level to the top of the building (Figure 4) is 0, the slope of the wall of the building is undefined. The slope of a vertical line that runs along the edge of the building is also undefined.

Figure 4

EXAMPLE 13 Use the slope formula to find the slope of the line that passes through $(-3, -2)$ and $(-3, 4)$.

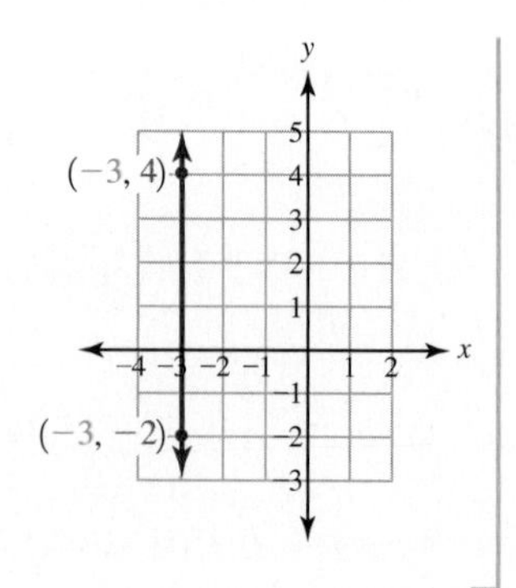

SOLUTION ▶

$$m = \frac{y_2 - y_1}{x_2 - x_1}$$

$$m = \frac{4 - (-2)}{-3 - (-3)} \qquad (x_1, y_1) = (-3, -2); (x_2, y_2) = (-3, 4)$$

$$m = \frac{\mathbf{6}}{\mathbf{0}} \qquad \text{Simplify.}$$

The slope of this vertical line is undefined.

Practice Problems

For problems 12–13,

(a) use the slope formula to find the slope of the line that passes through the given points.

(b) identify the line as horizontal or vertical.

12. $(3, 14)$ and $(1, 14)$ **13.** $(-1, 9)$ and $(-1, -6)$

Using Technology: Using a Graph to Find Ordered Pairs

EXAMPLE 14 The equation of a line is $y = -2x + 9$. What is y when $x = 3$?

To graph the equation, press Y=. Type the equation. Press WINDOW to check the window. If it is standard, press GRAPH. If not, change the window to standard.

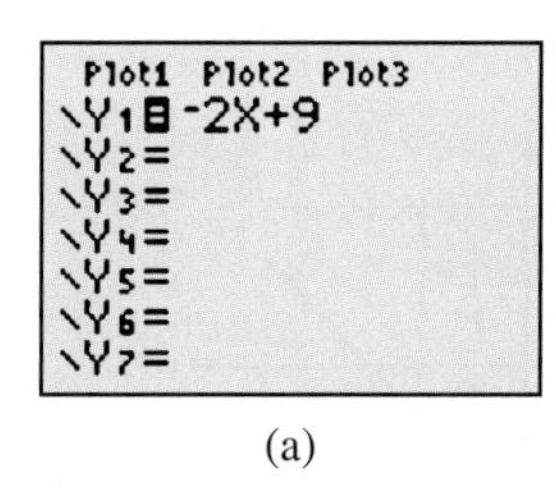

(a)

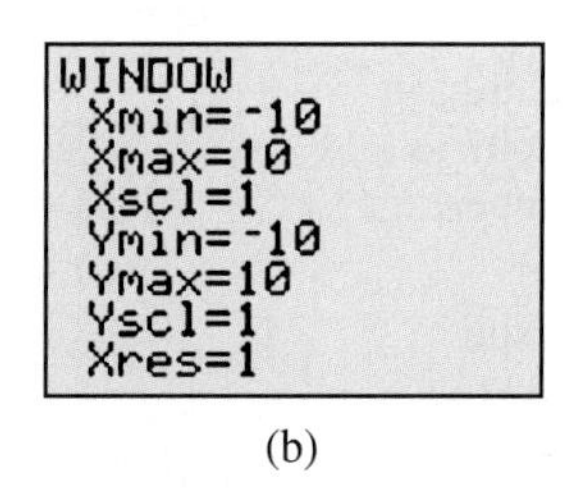

(b)

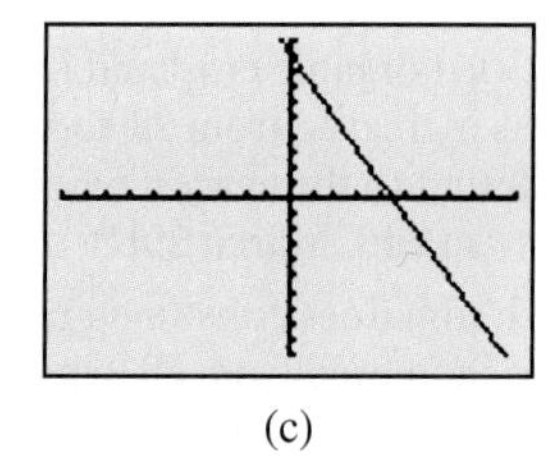

(c)

To open the CALCULATE screen, press 2nd TRACE.

In the CALCULATE screen, the **Value** command is highlighted. Press 1 or, since 1 is already highlighted, press ENTER. The cursor is now next to X=. Type −3. Press ENTER. The screen tells us that when $x = -3$, $y = 15$.

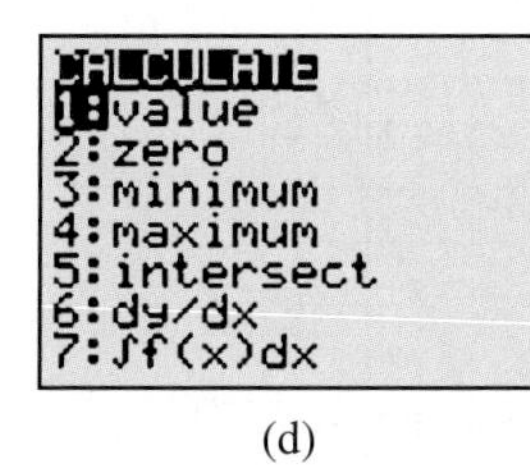

(d)

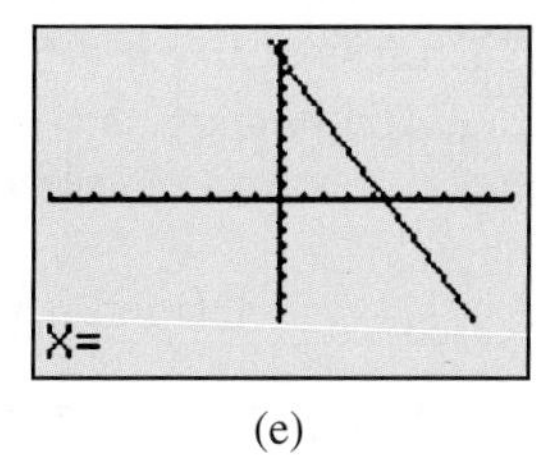

(e)

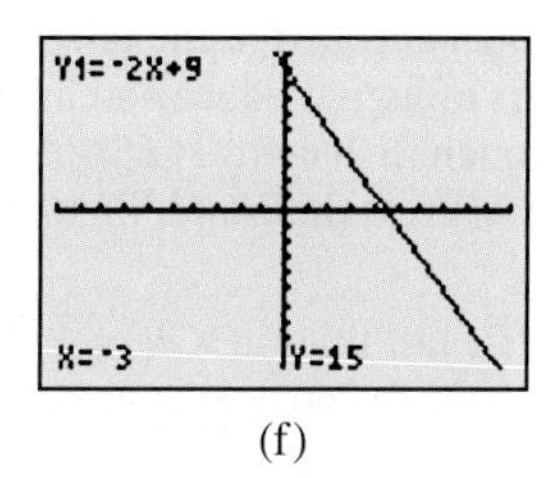

(f)

Practice Problems

14. Graph $y = 3x + 1$. Sketch the graph; describe the window.
15. Use the graph of $y = 3x + 1$ and the **Value** command to find y when $x = 2$.
16. Without changing the window, try to use the **Value** command to find y when $x = 17$. Describe what happens.
17. Change the window so that $X_{max} = 20$.
    a. Sketch the graph; describe the window.
    b. Find y when $x = 17$.

3.3 Vocabulary Practice

Match the term with its description.

1. On an xy-coordinate system, the number 9 in (4, 9) is an example of this.
2. On an xy-coordinate system, the number 4 in (4, 9) is an example of this.
3. A letter often used to represent slope.
4. $ax + by = c$
5. The point where a line intersects the y-axis.
6. $\{\ldots, -2, -1, 0, 1, 2, 3, \ldots\}$
7. A solution of a linear equation in two variables.
8. $m = \dfrac{y_2 - y_1}{x_2 - x_1}$
9. 0
10. Undefined

A. set of integers
B. m
C. ordered pair
D. slope formula
E. slope of a horizontal line
F. slope of a vertical line
G. standard form of a linear equation
H. x-coordinate
I. y-coordinate
J. y-intercept

3.3 Exercises

Follow your instructor's guidelines for showing your work.

For exercises 1–8,
(a) represent the information as two ordered pairs.
(b) find the average rate of change, m.

1. The amount of certified organic cropland in Washington State planted in peas increased from 28 acres in 2007 to 252 acres in 2010. Round to the nearest whole number. (*Source:* www.tfrec.wsu.edu, March 2011)
2. The amount of fresh tomatoes consumed per person in the United States increased from 89.9 lb in 2009 to 93.5 lb in 2011. (*Source:* www.ers.usda.gov, Dec. 15, 2011)
3. The number of men enrolled in the fall in degree-granting institutions of higher education increased from 7,575,000 men in 2006 to 8,770,000 men in 2009. Round to the nearest thousand. (*Source:* nces.ed.gov, 2011)
4. The number of women enrolled in the fall in degree-granting institutions of higher education increased from 10,184,000 women in 2006 to 11,658,000 women in 2009. Round to the nearest thousand. (*Source:* nces .ed.gov, 2011)
5. The number of traffic fatalities in Kentucky decreased from 985 deaths in 2005 to 760 deaths in 2010. (*Source:* www-nrd.nhtsa.dot.gov)
6. The number of traffic fatalities in Missouri decreased from 1257 deaths in 2005 to 819 deaths in 2010. Round to the nearest whole number. (*Source:* www-nrd.nhtsa .dot.gov)
7. The estimated number of wireless connections in the United States increased from 207,896,198 connections in 2005 to 302,859,674 connections in 2010. Round to the nearest thousand. (*Source:* www.ctia.org)
8. The estimated 12-month total service revenues for wireless service in the United States increased from \$113,538,221 in 2005 to \$159,929,648 in 2010. Round to the nearest thousand. (*Source:* www.ctia.org)

For exercises 9–20,
(a) graph the given points, and draw a line through the points.
(b) use the graph to find the slope of the line.
(c) use the slope formula to find the slope of the line.

9. $(1, 4); (3, 10)$
10. $(1, 5); (3, 13)$
11. $(1, 4); (3, -6)$
12. $(1, 5); (3, -3)$
13. $(-2, -5); (1, 3)$
14. $(-3, -4); (1, 3)$
15. $(-1, -3); (-4, -1)$
16. $(-2, -4); (-5, -2)$
17. $(-20, -50); (10, 30)$
18. $(-30, -40); (10, 30)$
19. $(0, -7); (3, 0)$
20. $(0, -5); (2, 0)$

For exercises 21–30,
(a) identify two points on the line.
(b) use the graph to find the slope of the graphed line.

21.

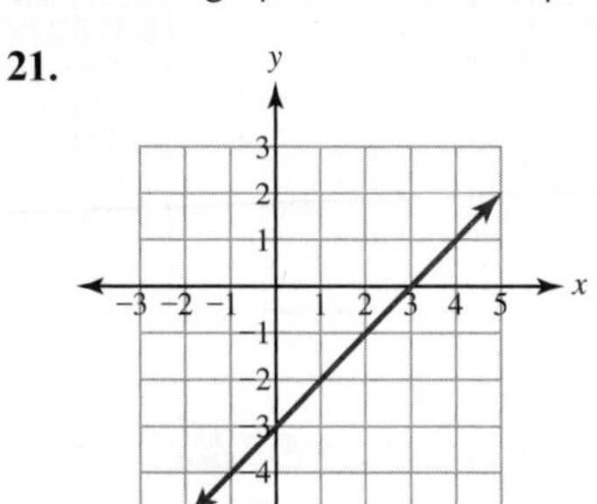

22.

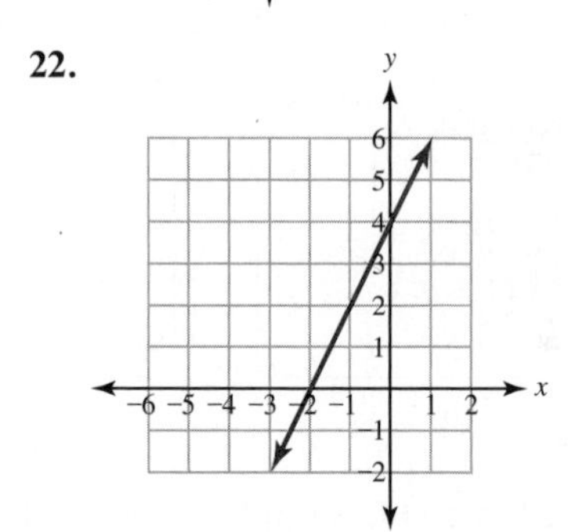

23.

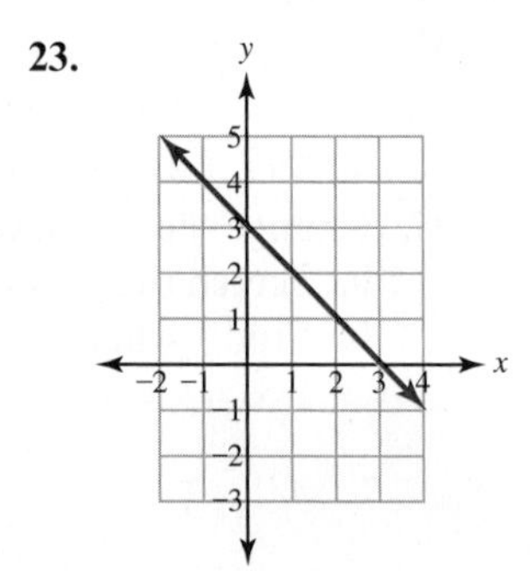

24.

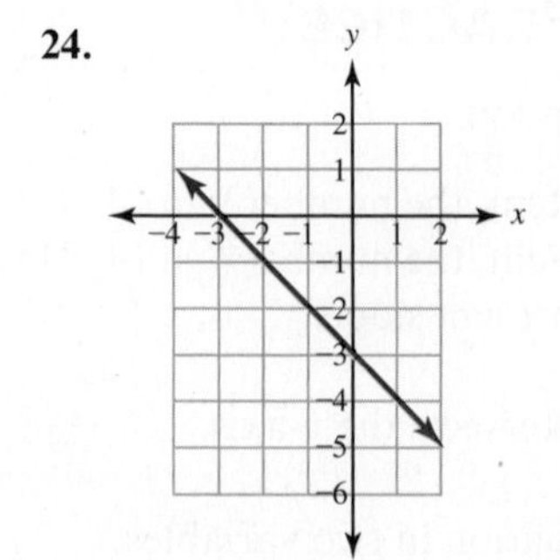

25.

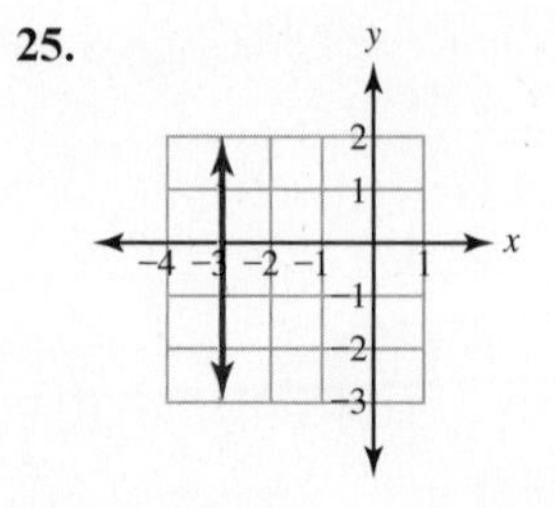

26.

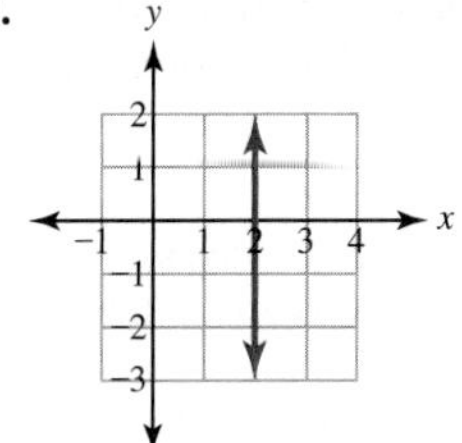

27.

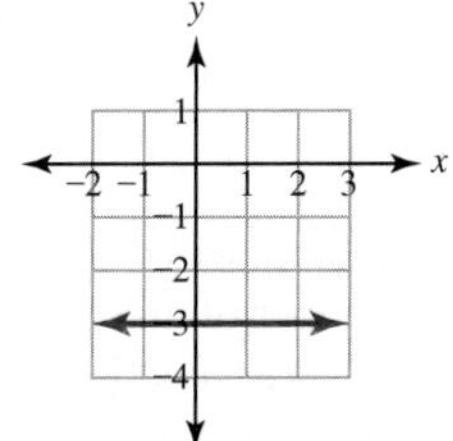

28.

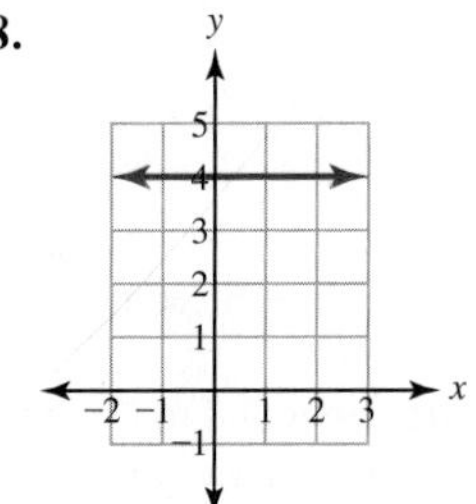

29.

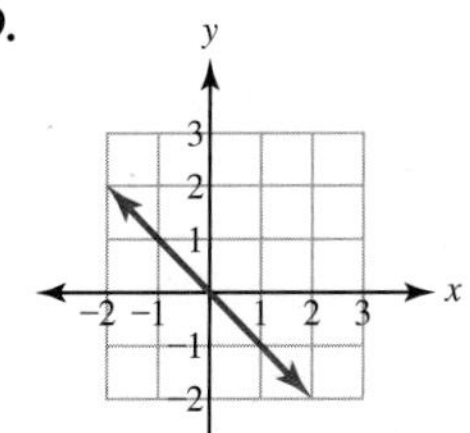

30.

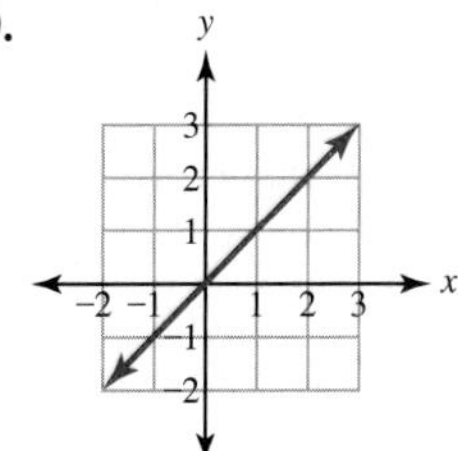

For exercises 31–36,
(a) identify two points on the line.
(b) use the graph to find the slope of the graphed line.

31.

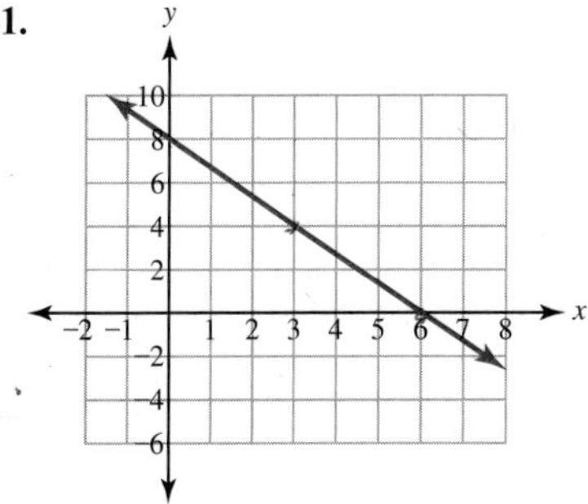

32.

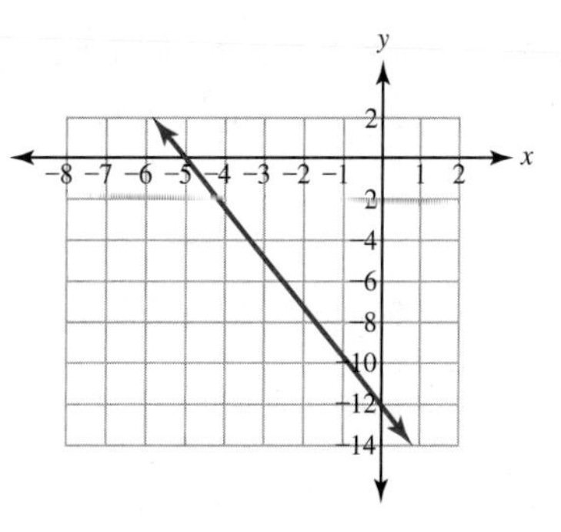

33.

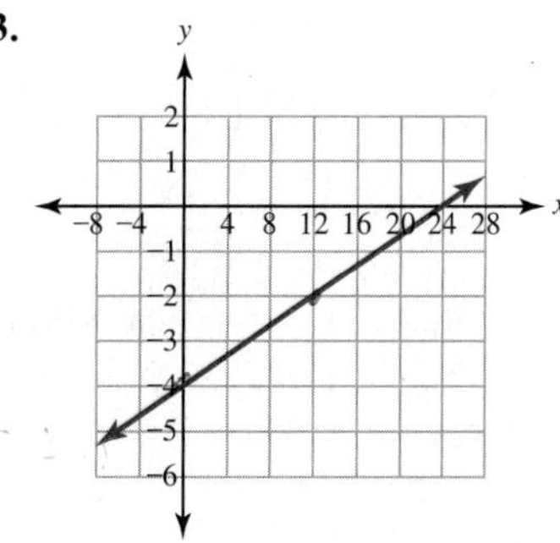

34.

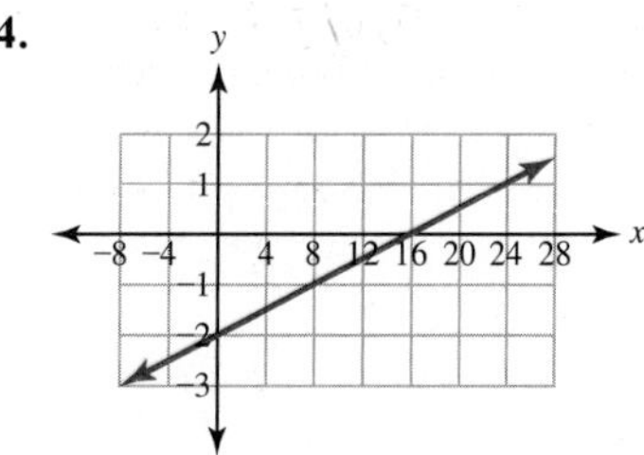

35.

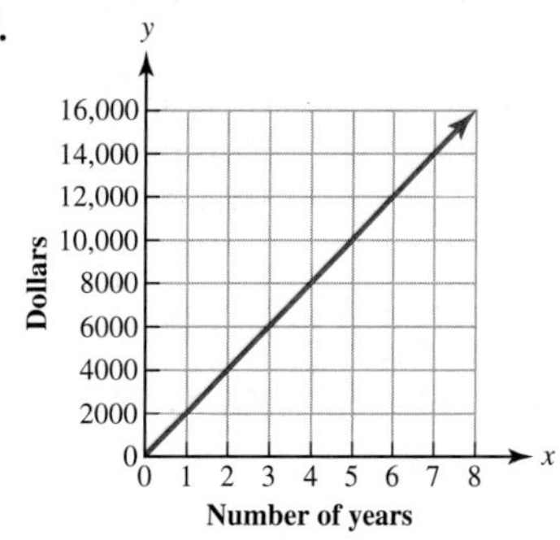

36.

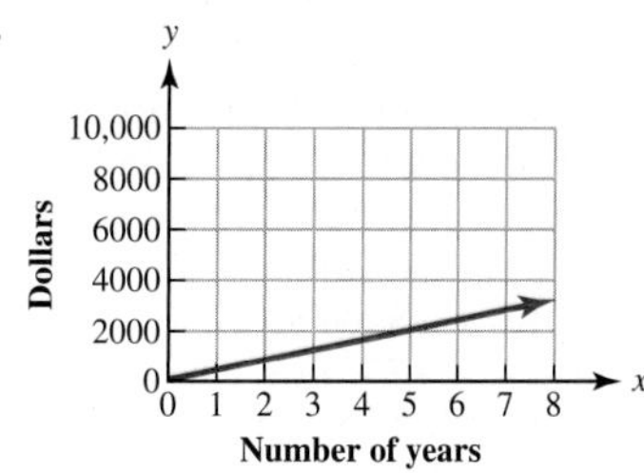

For exercises 37–66, use the slope formula to find the slope of the line that passes through the points.

37. $(1, 9); (3, 14)$ **38.** $(1, 8); (3, 15)$
39. $(5, 10); (7, 16)$ **40.** $(3, 11); (7, 39)$
41. $(-3, 8); (11, 23)$ **42.** $(-1, 5); (6, 13)$
43. $(-3, 8); (-11, -23)$ **44.** $(-1, 5); (-6, -13)$
45. $(0, -4); (-3, 0)$ **46.** $(0, -9); (-2, 0)$

47. $(-20, 45); (-15, 90)$

48. $(-30, 55); (-5, 80)$

49. $(-6, 3); (-9, 2)$

50. $(-5, 4); (-9, 3)$

51. $(-6, 3); (-9, -2)$

52. $(-5, 4); (-9, -3)$

53. $(-6, -3); (-9, -2)$

54. $(-5, -4); (-9, -3)$

55. $\left(0, \frac{3}{4}\right); \left(\frac{1}{4}, 0\right)$

56. $\left(0, \frac{4}{5}\right); \left(\frac{1}{5}, 0\right)$

57. $\left(\frac{1}{3}, \frac{4}{5}\right); \left(\frac{5}{3}, \frac{2}{5}\right)$

58. $\left(\frac{1}{3}, \frac{9}{7}\right); \left(\frac{5}{3}, \frac{3}{7}\right)$

59. $\left(\frac{1}{8}, 7\right); \left(\frac{7}{8}, 9\right)$

60. $\left(\frac{1}{6}, 8\right); \left(\frac{5}{6}, 11\right)$

61. $\left(-6, \frac{1}{2}\right); \left(10, \frac{3}{4}\right)$

62. $\left(-8, \frac{1}{4}\right); \left(16, \frac{1}{2}\right)$

63. $\left(\frac{5}{6}, \frac{1}{4}\right); \left(\frac{1}{6}, \frac{7}{4}\right)$

64. $\left(\frac{5}{8}, \frac{1}{6}\right); \left(\frac{3}{8}, \frac{5}{6}\right)$

65. (4 hr, \$36); (6 hr, \$54)

66. (3 hr, \$45); (7 hr, \$105)

For exercises 67–78,
(a) find the y-intercept.
(b) find the x-intercept.
(c) use the slope formula to find the slope of the line.

67. $4x + 9y = 72$

68. $3x + 10y = 60$

69. $x - 3y = 27$

70. $x - 4y = 48$

71. $-5x + 2y = 40$

72. $-8x + 3y = 48$

73. $2x + 5y = 15$

74. $2x + 3y = 9$

75. $x - y = 8$

76. $x - y = 7$

77. $y = -x + 3$

78. $y = -x + 2$

79. Explain why the slope of a vertical line is undefined.

80. Explain why the slope of a horizontal line is always zero.

Problem Solving: Practice and Review

Follow your instructor's guidelines for using the five steps as outlined in Section 1.3, p. 55.

81. An employee in Maine has two jobs that pay minimum wage. He works 28 hr per week at one job and 18 hr per week at the other job. Find the difference in his pay per week between October 2007 and October 2009.
October 1, 2007—Minimum Wage is \$7.00 per hour
October 1, 2008—Minimum Wage is \$7.25 per hour
October 1, 2009—Minimum Wage is \$7.50 per hour
(*Source:* www.maine.gov/labor/posters/minimumwage.pdf)

82. Balanced Rock in Arches National Park is 55 ft tall and weighs 3500 tons. Find its height in meters. Round to the nearest tenth. (1 m ≈ 3.2808 ft.) (*Source:* www.desertusa.com)

© Laura Bracken

83. Find the volume of the contaminated groundwater plume in gallons. (1 U.S. gal ≈ 0.133 ft³.) Write the answer in billions of gallons. Round to the nearest hundredth of a billion.

Over a century of ASARCO's pollution has created a 233 million cubic foot contaminated groundwater plume around the smelter. The remediation plan calls for monitoring and extraction wells to be placed around the ASARCO facility in an attempt to prevent contaminants from migrating into my community's drinking water. (*Source:* Senator Eliot Shapleigh, Feb. 27, 2009, www.shapleigh.org)

84. The New York City Feedback Citywide Customer Survey Report was published in 2008. The response rate describes the percent of mailed questionnaires that were completed and returned. Find the number of mailed questionnaires. Round to the nearest whole number.

- 24,339 questionnaires were completed and returned
- 18% response rate

(*Source:* www.nyc.gov)

Technology

For exercises 85–88,
(a) use a graphing calculator to graph $y = \frac{3}{4}x - 5$. Sketch the graph; describe the window.
(b) for the given value of x, use the **Value** command to find the value of y. If it is necessary to change the window, describe the changed window.

85. $x = 2$

86. $x = -5$

87. $x = -15$

88. $x = 20$

Find the Mistake

For exercises 89–92, the completed problem has one mistake.
(a) Describe the mistake in words, or copy down the whole problem and highlight or circle the mistake.
(b) Do the problem correctly.

89. Problem: Find the slope of the line that passes through (7, 1) and (9, 4).

Incorrect Answer: $m = \frac{y_2 - y_1}{x_2 - x_1}$

$m = \frac{9 - 7}{4 - 1}$

$m = \frac{2}{3}$

90. Problem: Find the slope of the graphed line.

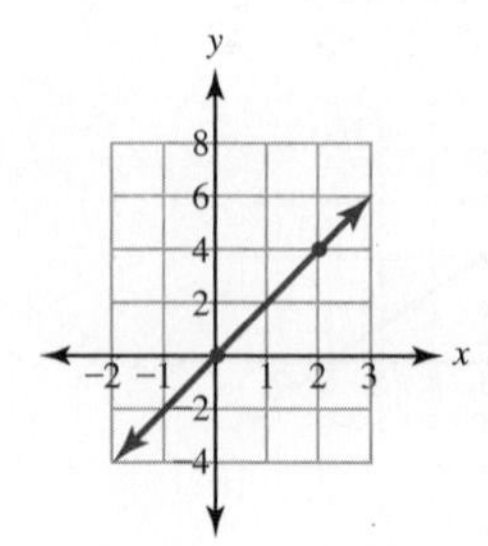

Incorrect Answer: $(x_1, y_1) = (0, 0); (x_2, y_2) = (2, 2)$

$$m = \frac{y_2 - y_1}{x_2 - x_1}$$

$$m = \frac{2 - 0}{2 - 0}$$

$$m = \frac{2}{2}$$

$$m = 1$$

91. Problem: Use the slope formula to find the slope of the line that passes through $(5, -9)$ and $(-2, -11)$.

Incorrect Answer: $m = \frac{y_2 - y_1}{x_2 - x_1}$

$$m = \frac{-11 - 9}{-2 - 5}$$

$$m = \frac{-20}{-7}$$

$$m = \frac{20}{7}$$

92. Problem: Use the slope formula to find the slope of the line that passes through $(6, 2)$ and $(6, 7)$.

Incorrect Answer: $m = \frac{y_2 - y_1}{x_2 - x_1}$

$$m = \frac{6 - 6}{7 - 2}$$

$$m = \frac{0}{5}$$

$$m = 0$$

Review

For exercises 93–96, identify the y-intercept of each line. Write the y-intercept as an ordered pair.

93.

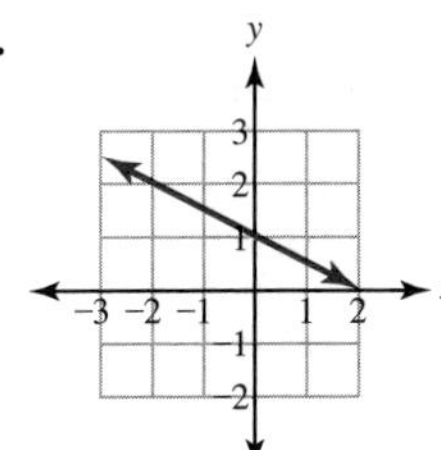

94.

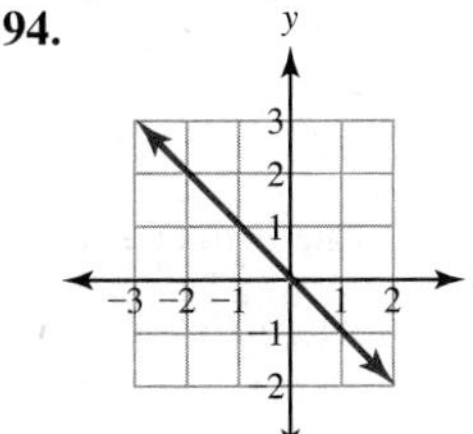

95.

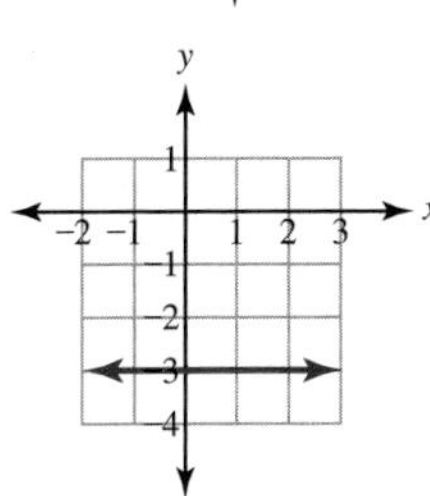

96.

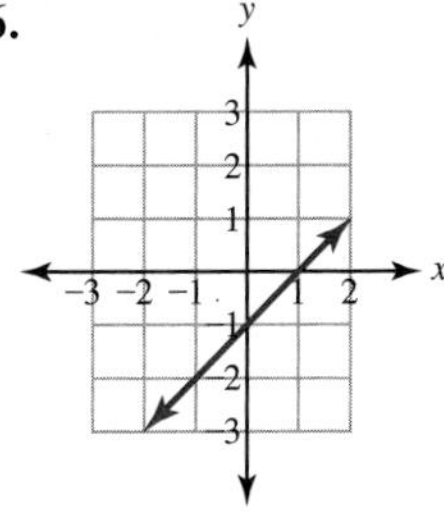

SUCCESS IN COLLEGE MATHEMATICS

For exercises 97–98, some students find it helpful to use their learning preferences as a guide in how to study.

Visual Learner

- Take detailed notes during class. Use colored pens and highlighters.
- Reorganize and rewrite notes after class; draw diagrams that summarize what you have learned.
- Read your book; watch the videos or DVDs for this text.
- Use flash cards for memory work.
- Sit where you can see everything in the classroom. Turn your phone or tablet off so that you are not distracted.

Auditory Learner

- With permission, record your class. Take only brief notes of the big ideas and examples. After class, listen to the recording. Complete your notes. Restate the main ideas aloud to yourself. Use videos and DVDs to fill in anything you missed in class.
- Talk to yourself as you do your homework. Explain each step to yourself.
- Do memory work by repeating definitions aloud. Listen to a recording of the words and definitions. Create songs that help you remember a definition.
- Sit where you can hear everything. Turn your phone or tablet off so that you are not distracted.

Kinesthetic Learner

- With permission, record your class. Take brief notes of the big ideas and examples. After class, listen to the recording. Complete your notes. Draw pictures. Use videos and DVDs to fill in anything you missed during class.
- With your finger, trace diagrams and graphs. Do not just look at them.
- Imagine symbols such as variables as three-dimensional objects or even cartoon characters. Imagine yourself counting them, combining them, or subtracting them.
- Do memory work as you exercise or walk to your car. Walk around your room as you repeat definitions. You may find it helpful to come up with physical motions and/or a song that correspond to a procedure.
- If your class is mostly lecture, prepare yourself mentally before you walk into class to concentrate and not daydream. Turn your phone or tablet off so that you are not distracted.

97. Identify any of the strategies listed above that you currently use to study math.

98. Identify any strategies listed that you don't currently use but you think might be helpful.

If a patient is receiving a solution from an IV bag at a constant rate and we know the beginning volume of solution, we can write an equation that describes the relationship of time and the amount of IV solution in the bag. This equation is an example of a linear model. In this section, we will learn about linear models and the slope-intercept form of a linear equation.

SECTION 3.4 Slope-Intercept Form

After reading the text, working the practice problems, and completing assigned exercises, you should be able to:

1. Write a linear model.
2. Describe what the slope or y-intercept of a linear model represent.
3. Given the slope and the y-intercept, write the equation of a line.
4. Given the equation of a line, rewrite the equation in slope-intercept form and identify the slope and y-intercept.
5. Given the graph of a line, write its equation in slope-intercept form.
6. Use slope-intercept graphing to graph a linear equation in two variables.

Linear Models

A linear equation can describe or "model" some problem situations. The y-intercept of this equation is an ordered pair that describes the situation at the beginning, when time equals 0.

EXAMPLE 1 A bag of IV solution contains 500 mL. After 10 min of the IV delivering the solution to a patient at a constant rate, the bag contains 450 mL.

(a) Write ordered pairs (time, volume) that represent this information.

SOLUTION ▶ At the beginning, the volume is 500 mL; (0 min, 500 mL).
After 10 min, the volume is 450 mL; (10 min, 450 mL).

(b) Graph the ordered pairs, and draw a line beginning at the y-intercept.

▶

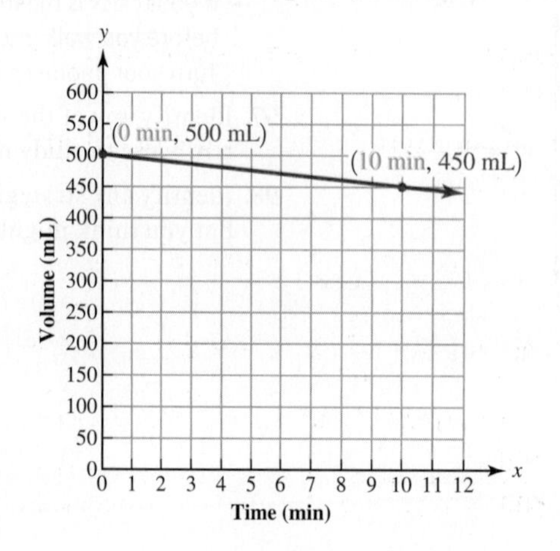

(c) Identify the y-intercept of the line, and describe what the y-coordinate of the y-intercept represents.

The y-intercept is (0 min, 500 mL). The y coordinate represents the amount of solution in the IV bag at the beginning.

(d) Use the slope formula to find the slope of the line, and describe what the slope represents.

$(x_1, y_1) = (0 \text{ min}, 500 \text{ mL}) \qquad (x_2, y_2) = (10 \text{ min}, 450 \text{ mL})$

$$m = \frac{450 \text{ mL} - 500 \text{ mL}}{10 \text{ min} - 0 \text{ min}} \qquad m = \frac{y_2 - y_1}{x_2 - x_1}$$

$$m = \frac{\mathbf{-50 \text{ mL}}}{\mathbf{10 \text{ min}}} \qquad \text{Simplify.}$$

$$m = -\frac{\mathbf{5 \text{ mL}}}{\mathbf{1 \text{ min}}} \qquad \text{Simplify.}$$

The slope, $-\frac{5 \text{ mL}}{1 \text{ min}}$, represents the average rate of change of the volume of solution in the bag.

(e) Write an equation that represents the relationship of the volume of solution in the bag, y, and the time, x.

To find the volume in the bag, we need to know how much solution has been delivered to the patient and the beginning volume. The amount of solution delivered is equal to the product of the average rate of change and the time. The average rate of change is the slope of the line.

$$\text{volume} = (\text{average rate of change})(\text{time}) + \text{beginning volume}$$
$$y = (\text{slope})x + y\text{-coordinate of the } y\text{-intercept}$$
$$y = \left(-\frac{\mathbf{5 \text{ mL}}}{\mathbf{1 \text{ min}}}\right)x + \mathbf{500 \text{ mL}}$$

(f) Use the equation to find the volume of solution after 60 min.

$$y = \left(-\frac{5 \text{ mL}}{1 \text{ min}}\right)x + 500 \text{ mL}$$

$$y = \left(-\frac{5 \text{ mL}}{1 \cancel{\text{min}}}\right)(\mathbf{60} \cancel{\textbf{min}}) + 500 \text{ mL} \qquad \text{Replace } x \text{ with 60 min.}$$

$$y = \mathbf{-300 \text{ mL}} + 500 \text{ mL} \qquad \text{Simplify.}$$

$$y = \mathbf{200 \text{ mL}} \qquad \text{Simplify.}$$

After 60 min, the volume of the solution is 200 mL.

(g) Find the x-intercept, and describe what the x-coordinate of the x-intercept represents.

The y-coordinate of the x-intercept is 0. To find the x-intercept, replace y with 0 and solve for x.

$$y = \left(-\frac{5 \text{ mL}}{1 \text{ min}}\right)x + 500 \text{ mL}$$

$$\mathbf{0 \text{ mL}} = \left(-\frac{5 \text{ mL}}{1 \text{ min}}\right)x + 500 \text{ mL} \qquad \text{Replace } y \text{ with 0 mL.}$$

$$0 = -5x + 500 \qquad \text{Remove the units.}$$

$$\underline{-500 \qquad\quad -500} \qquad \text{Subtraction property of equality.}$$

$$\mathbf{-500} = -5x + \mathbf{0} \qquad \text{Simplify.}$$

$\frac{-500}{-5} = \frac{\mathbf{-5x}}{-5}$ Division property of equality.

$100 = x$ Simplify.

$100 \textbf{ min} = x$ Replace units.

The x-intercept is (100 min, 0 mL). The x-coordinate of the x-intercept represents the time when the bag is empty, 100 min.

The equation $y = \left(-\frac{5 \text{ mL}}{1 \text{ min}}\right)x + 500 \text{ mL}$ is an example of a **linear model**. It represents the relationship of volume, y, and time, x. In the next example, the linear model represents the relationship of an amount of money, y, and time, x.

EXAMPLE 2 A student earned $8000 during the summer fighting wildfires. When he returned to school in the fall, he used $500 of this money per month to pay for his share of rent and utilities.

(a) Write an equation that represents the relationship of the amount of firefighting money, y, and the time, x.

SOLUTION ▶ The average rate of change is the amount spent per month, $-\frac{\$500}{1 \text{ month}}$. The amount of firefighting money, y, equals the amount spent plus the beginning amount of money, $8000. The amount spent is equal to the product of the average rate of change and the time, $\left(-\frac{\$500}{1 \text{ month}}\right)x$.

amount of money = (average rate of change)(time) + beginning amount

$$y = \left(-\frac{\$500}{1 \text{ month}}\right)x + \$8000$$

(b) Use this equation to find the amount of firefighting money after 5 months.

▶ $y = \left(-\frac{\$500}{1 \text{ month}}\right)x + \$8000$

$y = \left(-\frac{\$500}{1 \cancel{\text{month}}}\right)(\mathbf{5 \cancel{months}}) + \$8000$ Replace x with the time, 5 months.

$y = \mathbf{-\$2500} + \$8000$ Simplify.

$y = \mathbf{\$5500}$ Simplify.

There is $5500 of firefighting money left after 5 months.

Practice Problems

1. A student is downloading a program that is 2100 megabits. After 200 seconds (s), 1500 megabits of software remain to be downloaded.
   a. Write ordered pairs that represent this information.
   b. Graph the ordered pairs, and draw a line beginning at the y-intercept.
   c. Identify the y-intercept of the line, and describe what the y-coordinate of the y-intercept represents.
   d. Use the slope formula to find the slope of the line, and describe what the slope represents.
   e. Write an equation that represents the relationship of the amount of software to be downloaded, y, and the time, x.
   f. Use the equation to find the amount of software to be downloaded after 600 s.
   g. Find the x-intercept, and describe what the x-coordinate of the x-intercept represents.

2. A construction project had 2400 bricks placed before funding was stopped. When funding was resumed, a new crew of bricklayers was hired. This crew can place 60 bricks per hour. (*Source:* S. Geddes and J. Williams, *Estimating for Building and Civil Engineering Works*, 1996)
   a. Write an equation that represents the relationship of the total number of bricks placed in the project, y, and the time that the new crew has worked, x.
   b. Use the equation to find the number of bricks placed after 120 hr of work by the new crew.

Slope-Intercept Form

In Example 1, we used the slope and the y-coordinate of the y-intercept to write a linear model that describes the relationship between x and y. If $\boldsymbol{m}$ represents the slope and $\boldsymbol{b}$ represents the y-coordinate of the y-intercept, then $y = \boldsymbol{m}x + \boldsymbol{b}$. This is the **slope-intercept form** of a linear equation.

Forms of a Linear Equation in Two Variables, x and y

Standard Form $ax + by = c$, where a, b, and c are real numbers

Slope-Intercept Form $y = mx + b$, where m and b are real numbers

The slope of the line is m; the y-intercept is $(0, b)$.

EXAMPLE 3 If the slope of a line is $\frac{3}{4}$ and its y-intercept is $(0, -2)$, write its equation in slope-intercept form.

SOLUTION

$y = mx + b$ Slope-intercept form.

$y = \mathbf{\frac{3}{4}}x + (\mathbf{-2})$ Replace m and b; $m = \frac{3}{4}$; $b = -2$

$y = \frac{3}{4}x - \mathbf{2}$ Rewrite addition of a negative number as subtraction.

To rewrite a linear equation in slope-intercept form, use the properties of equality. If it is necessary to use the division property of equality, divide each term so that the simplified equation matches slope-intercept form.

EXAMPLE 4 A linear equation in standard form is $9x - 4y = 12$.

(a) Rewrite $9x - 4y = 12$ in slope-intercept form.

SOLUTION

$9x - 4y = 12$

$\underline{-9x \qquad\qquad -9x}$ Subtraction property of equality.

$\mathbf{0} - 4y = 12 \mathbf{- 9x}$ Simplify.

$\frac{\mathbf{-4y}}{-4} = \frac{12}{-4} - \frac{9x}{-4}$ Division property of equality; divide each term.

$y = -3 + \frac{9}{4}x$ Simplify.

$y = \mathbf{\frac{9}{4}x - 3}$ Change order.

(b) Identify the slope.

In $y = mx + b$, the slope is m. For $y = \frac{9}{4}x - 3$, the slope is $\frac{9}{4}$.

(c) Identify the y-intercept.

In $y = mx + b$, the y-intercept is $(0, b)$. For $y = \frac{9}{4}x - 3$, the y-intercept is $(0, -3)$.

(d) Find the x-intercept.

To find the x-intercept, replace y with 0 and solve for x.

$$9x - 4y = 12$$
$$9x - 4(\mathbf{0}) = 12 \quad \text{Replace } y \text{ with 0.}$$
$$9x - \mathbf{0} = 12 \quad \text{Simplify.}$$
$$\frac{\mathbf{9x}}{9} = \frac{12}{9} \quad \text{Division property of equality.}$$
$$x = \frac{\cancel{3} \cdot 4}{\cancel{3} \cdot 3} \quad \text{Find common factors.}$$
$$x = \frac{4}{3} \quad \text{Simplify.}$$

The x-intercept is $\left(\frac{4}{3}, 0\right)$.

In the next example, the coefficients and constant in the equation are fractions. To rewrite the equation in slope-intercept form, it is easier to first clear the fractions.

EXAMPLE 5 A linear equation is $\frac{3}{4}x + \frac{5}{8}y = -\frac{2}{3}$.

(a) Clear the fractions, and rewrite the equation in slope-intercept form.

SOLUTION

$$\frac{3}{4}x + \frac{5}{8}y = -\frac{2}{3} \quad \text{Least common denominator is 24.}$$
$$24\left(\frac{3}{4}x + \frac{5}{8}y\right) = 24\left(-\frac{2}{3}\right) \quad \text{Multiplication property of equality.}$$
$$\frac{\mathbf{24}}{\mathbf{1}}\left(\frac{3}{4}x\right) + \frac{\mathbf{24}}{\mathbf{1}}\left(\frac{5}{8}y\right) = \frac{24}{\mathbf{1}}\left(-\frac{2}{3}\right) \quad \text{Distributive property.}$$
$$\frac{6 \cdot \cancel{4}}{1} \cdot \frac{3}{\cancel{4}}x + \frac{3 \cdot \cancel{8}}{1} \cdot \frac{5}{\cancel{8}}y = \frac{8 \cdot \cancel{3}}{1}\left(-\frac{2}{\cancel{3}}\right) \quad \text{Find common factors.}$$
$$18x + 15y = -16 \quad \text{Simplify; the fractions are cleared.}$$
$$\underline{-18x \qquad\qquad -18x} \quad \text{Subtraction property of equality.}$$
$$\mathbf{0} + 15y = -16 - \mathbf{18x} \quad \text{Simplify.}$$
$$\frac{\mathbf{15y}}{15} = -\frac{16}{15} - \frac{18x}{15} \quad \text{Division property of equality.}$$
$$\frac{15y}{15} = -\frac{16}{15} - \frac{\cancel{3} \cdot 6 \cdot x}{\cancel{3} \cdot 5} \quad \text{Find common factors.}$$
$$y = -\frac{16}{15} - \frac{\mathbf{6}}{\mathbf{5}}x \quad \text{Simplify.}$$
$$y = -\frac{\mathbf{6}}{\mathbf{5}}x - \frac{\mathbf{16}}{\mathbf{15}} \quad \text{Change order.}$$

(b) Identify the slope.

In $y = mx + b$, the slope is m. For $y = -\frac{6}{5}x - \frac{16}{15}$, the slope is $-\frac{6}{5}$.

(c) Identify the y-intercept.

In $y = mx + b$, the y-intercept is $(0, b)$. For $y = -\frac{6}{5}x - \frac{16}{15}$, the y-intercept is $\left(0, -\frac{16}{15}\right)$.

(d) Find the x-intercept.

To find the x-intercept, use the equation in which the fractions have been cleared. Replace y with 0 and solve for x.

$$18x + 15y = -16$$
$$18x + 15(\mathbf{0}) = -16 \quad \text{Replace } y \text{ with 0.}$$
$$18x + \mathbf{0} = -16 \quad \text{Simplify.}$$
$$\frac{\mathbf{18x}}{18} = \frac{-16}{18} \quad \text{Division property of equality.}$$
$$\frac{18x}{18} = \frac{\mathbf{-8 \cdot 2}}{\mathbf{9 \cdot 2}} \quad \text{Find common factors.}$$
$$x = -\frac{8}{9} \quad \text{Simplify.}$$

The x-intercept is $\left(-\frac{8}{9}, 0\right)$.

To write the equation of a graphed line, identify the y-intercept, the rise, the run, and the slope. Replace m and b in slope-intercept form of a line, $y = mx + b$. Since the precision of graphs and our ability to identify points are limited, the equation is an estimate. In the next example, the y-intercept certainly appears to be $(0, -2)$. However, it could instead be $(0, -1.99)$ or $(0, -2.01)$.

EXAMPLE 6 The graph represents a linear equation.

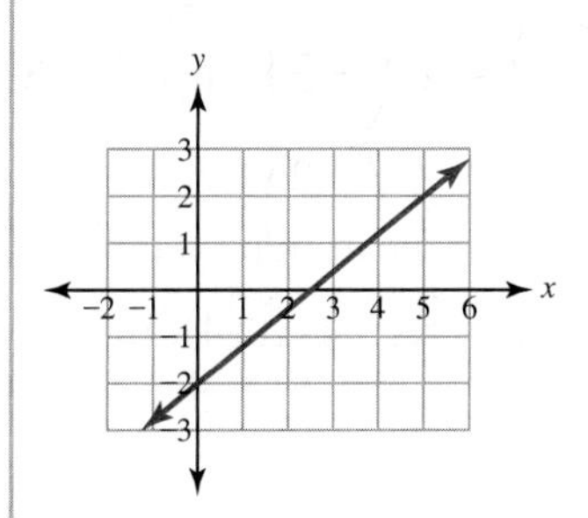

(a) Identify the y-intercept.

SOLUTION The y-intercept is $(0, -2)$.

(b) Identify the slope.

Another point on the line with integer coordinates appears to be $(5, 2)$. Beginning from $(0, -2)$, move vertically up 4 units to the level of $(5, 2)$; the rise is positive 4. Move horizontally to the right 5 units; the run is positive 5. The slope, $\frac{\text{rise}}{\text{run}}$, is $\frac{4}{5}$.

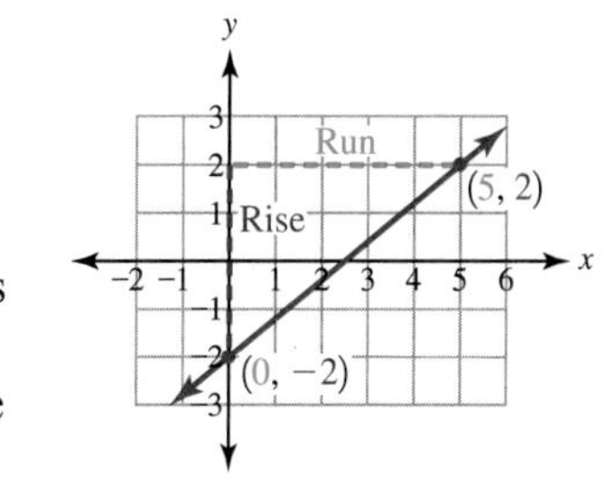

(c) Write the equation of the line in slope-intercept form.

The slope is $\frac{4}{5}$; the y-intercept is $(0, -2)$. In slope-intercept form, $y = mx + b$, the equation of the line is $y = \frac{4}{5}x - 2$.

In the next example, as we move from left to right from one point to the next, the vertical distance or rise between the points is negative. The horizontal distance between the points is positive. Since $\frac{\text{negative number}}{\text{positive number}}$ = negative number, the slope is negative.

EXAMPLE 7 The graph represents a linear equation.

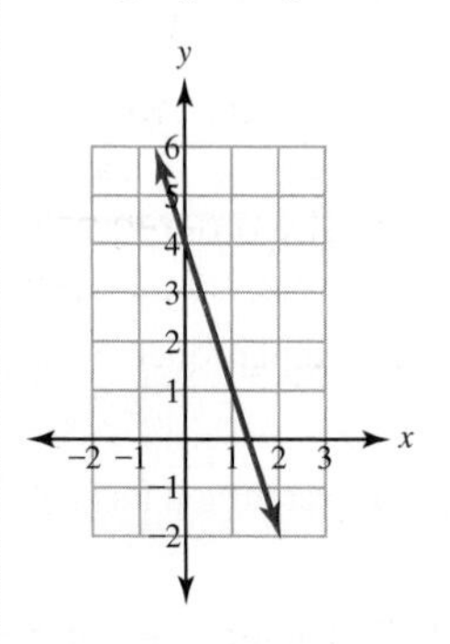

(a) Identify the y-intercept.

SOLUTION ▶ The y-intercept is $(0, 4)$.

(b) Identify the slope.

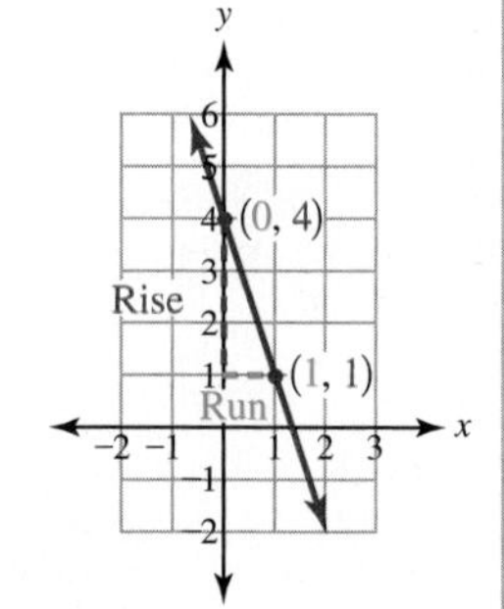

▶ Another point with integer coordinates appears to be $(1, 1)$. Beginning from $(0, 4)$, move vertically down 3 units to the level of $(1, 1)$; the rise is negative 3. Move horizontally to the right 1 unit; the run is positive 1. The slope, $\frac{\text{rise}}{\text{run}}$, is $\frac{-3}{1}$ which simplifies to -3. If we choose a different second point on the grid such as $(2, -2)$, the rise is -6 and the run is 2. The slope, $\frac{\text{rise}}{\text{run}}$, is $\frac{-6}{2}$ which again simplifies to -3. We can choose any two points on the line; the slope of the line is the same.

(c) Write the equation of the line in slope-intercept form.

▶ The slope is -3; the y-intercept is $(0, 4)$. In slope-intercept form, $y = mx + b$, the equation of the line is $y = -3x + 4$.

Methods for Writing the Equation of a Line in Slope-Intercept Form

- Given the slope $\boldsymbol{m}$ and y-intercept $(0, \boldsymbol{b})$, replace m and b in the form $y = \boldsymbol{m}x + \boldsymbol{b}$.
- Given a linear equation, use the properties of equality to rewrite the equation in the form $y = mx + b$.
- Given the graph of a linear equation, identify the y-intercept. The y-coordinate of the y-intercept is $\boldsymbol{b}$. To find the slope, identify another point on the line. Find the rise and the run between these two points; the slope $\boldsymbol{m}$ is $\frac{\text{rise}}{\text{run}}$. Replace m and b in the form $y = mx + b$.

Practice Problems

For problems 3–6, write the equation of the line in slope-intercept form.

3. slope $= 5$; y-intercept: $(0, 9)$ **4.** slope $= -8$; y-intercept: $(0, -5)$

5. $m = \frac{4}{3}$; $b = -9$ **6.** $m = 7$; $b = \frac{1}{2}$

For problems 7–8,
(a) rewrite the equation in slope-intercept form.
(b) identify the slope.
(c) identify the y-intercept. Write the ordered pair, not just the y-coordinate.
(d) find the x-intercept. Write the ordered pair, not just the x-coordinate.

7. $9x - 2y = 36$ **8.** $3x + 5y = -12$

For problems 9–10,
(a) clear the fractions, and rewrite the equation in slope-intercept form.
(b) identify the slope.
(c) identify the y-intercept. Write the ordered pair, not just the y-coordinate.
(d) find the x-intercept. Write the ordered pair, not just the x-coordinate.

9. $\frac{2}{9}x - \frac{4}{3}y = \frac{1}{2}$ **10.** $y - 8 = \frac{3}{7}(x - 2)$

For problems 11–13, the graph represents a linear equation.
(a) Identify the y-intercept. Write the ordered pair, not just the y-coordinate.
(b) Identify the slope.
(c) Write the equation of the line in slope-intercept form.

11.

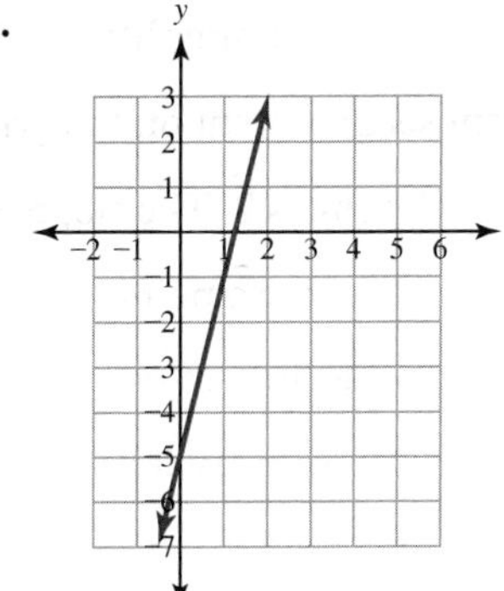

12.

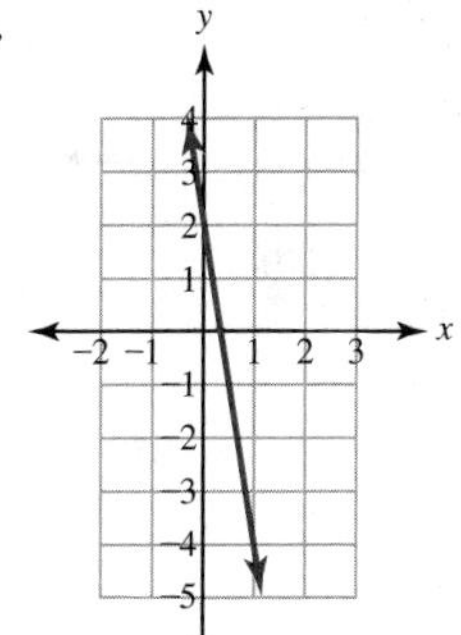

13.

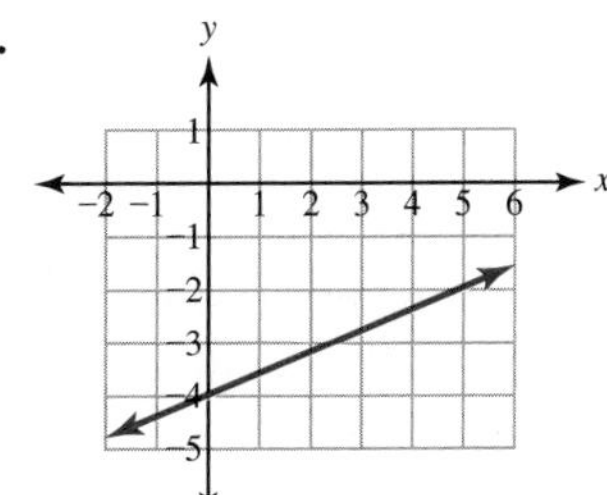

Horizontal and Vertical Lines

The slope of a horizontal line is 0. Every point on the line has the same y-coordinate. The equation in slope-intercept form of a horizontal line with a y-intercept of b is $y = 0x + b$, which simplifies to $y = b$.

EXAMPLE 8 A horizontal line passes through $(2, 3)$.

(a) Graph the line.

SOLUTION ▶

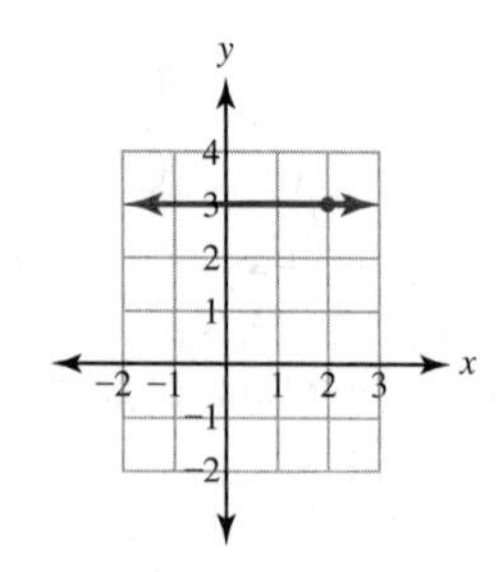

(b) Write the equation of the line.

▶ Every y-coordinate on this line is 3; its slope is 0. Its equation is $y = 0x + 3$, which simplifies to $y = 3$.

The slope of a vertical line is undefined. Every point on the line has the same x-coordinate. Since the slope of a vertical line is undefined, *we cannot write the equation of a vertical line in slope-intercept form.* In standard form, the equation of a vertical line is $1x + 0y = c$, which simplifies to $x = c$.

EXAMPLE 9 A vertical line passes through $(2, 3)$.

(a) Graph the line.

SOLUTION ▶

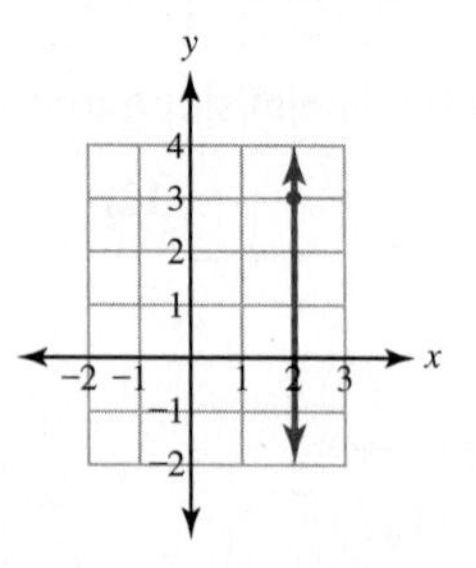

(b) Write the equation of the line.

▶ The x-coordinate on this line is 2; its slope is undefined. We cannot write its equation in slope-intercept form. In standard form, its equation is $1x + 0y = 2$, which simplifies to $x = 2$.

Practice Problems

14. A horizontal line passes through $(7, 1)$.
 a. Graph the line.
 b. Write the equation of the line.

15. A vertical line passes through $(7, 1)$.
 a. Graph the line.
 b. Write the equation of the line.

Slope-Intercept Graphing

We can graph a linear equation by finding three solutions of the equation, graphing the points, and drawing a line through the points. Or if the equation is in slope-intercept form, we can use the **slope-intercept method** to graph it.

Slope-Intercept Graphing of a Linear Equation

1. Write the equation in slope intercept form, $y = mx + b$.
2. Graph the y-intercept, $(0, b)$.
3. The slope, m, is $\frac{\text{rise}}{\text{run}}$. Beginning at the y-intercept, move in the vertical direction a distance equal to the rise. (If the slope of a line is negative and the negative sign is assigned to the numerator, move vertically downward a distance equal to the rise.) Then move in the horizontal direction a distance equal to the run. At this location, graph a second point.
4. Draw a straight line through the two points.

EXAMPLE 10 Use slope-intercept graphing to graph $y = \frac{7}{3}x - 4$.

SOLUTION ▶ $y = \frac{7}{3}x - 4$ $\quad m = \frac{7}{3}$; $b = -4$; rise $= 7$; run $= 3$

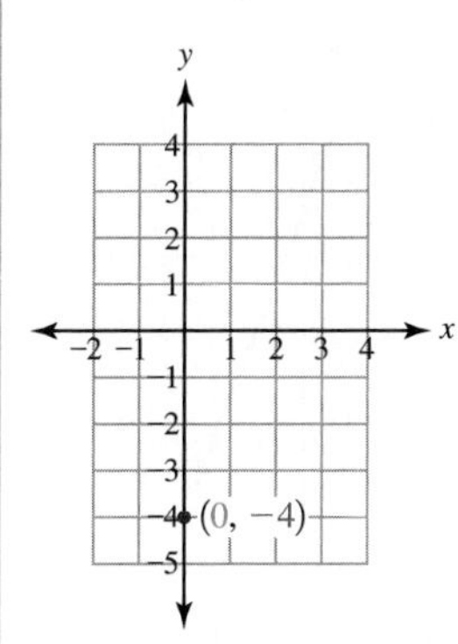

Graph the y-intercept.

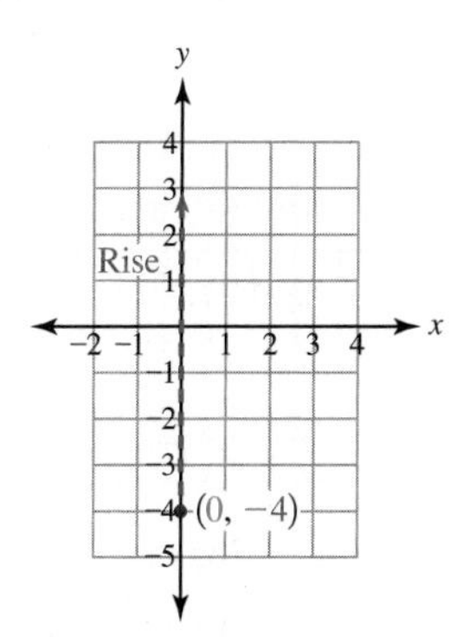

Rise $= 7$; move up 7.

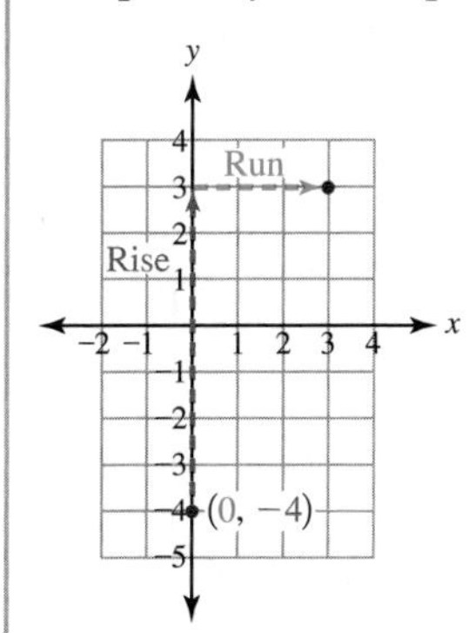

Run $= 3$; move right 3. Graph the second point.

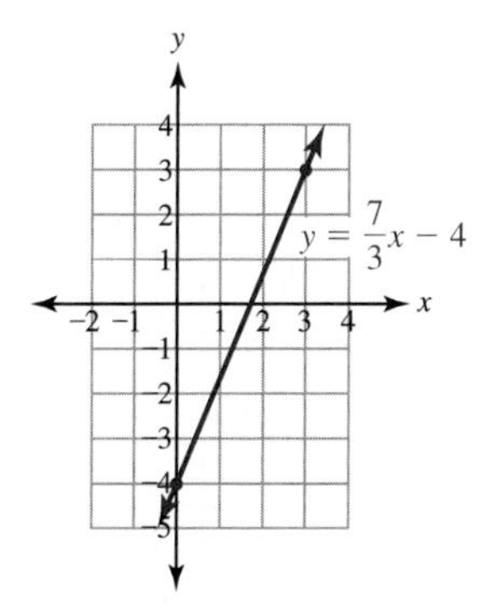

Draw the line.

When the slope of a line is negative, we can assign the negative sign to the numerator. Since the "rise" is negative, move down from the y-intercept to find the location of another point on the line.

EXAMPLE 11 Use slope-intercept graphing to graph $y = -\frac{5}{2}x + 1$.

SOLUTION $y = -\frac{5}{2}x + 1 \qquad m = -\frac{5}{2} = \frac{-5}{2}$; $b = 1$; rise $= -5$; run $= 2$

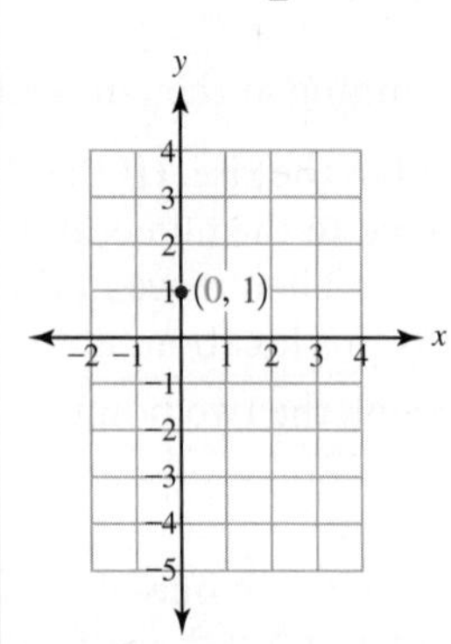

Graph the y-intercept.

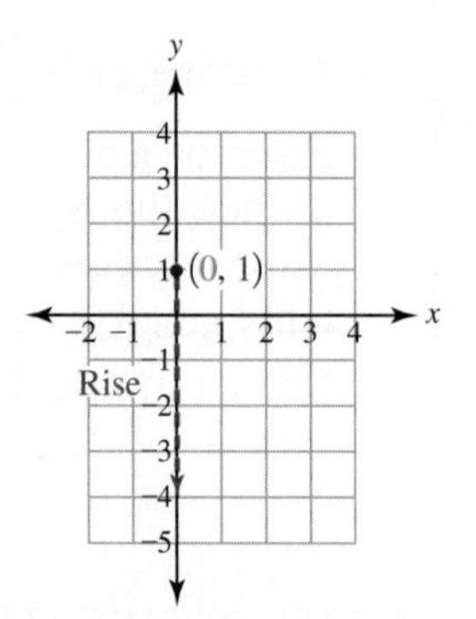

Rise $= -5$; move down 5.

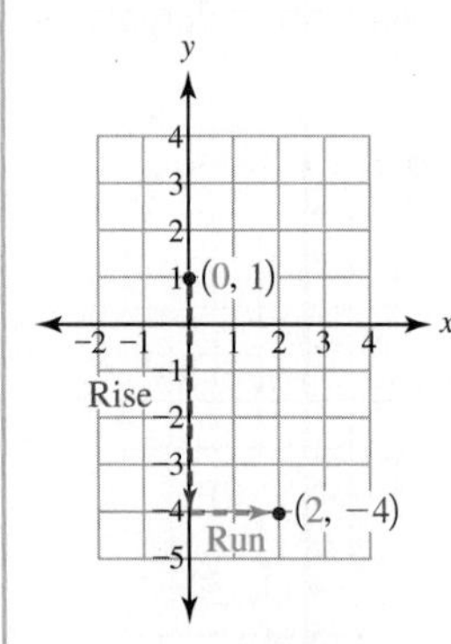

Run $= 2$; move right 2.
Graph the second point.

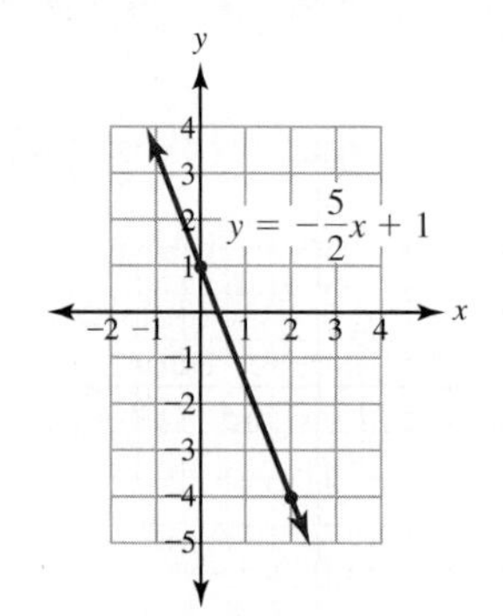

Draw the line.

If we assign the negative sign to the run, the graph is the same.

$y = -\frac{5}{2}x + 1 \qquad m = -\frac{5}{2} = \frac{5}{-2}$; $b = 1$; rise $= 5$; run $= -2$

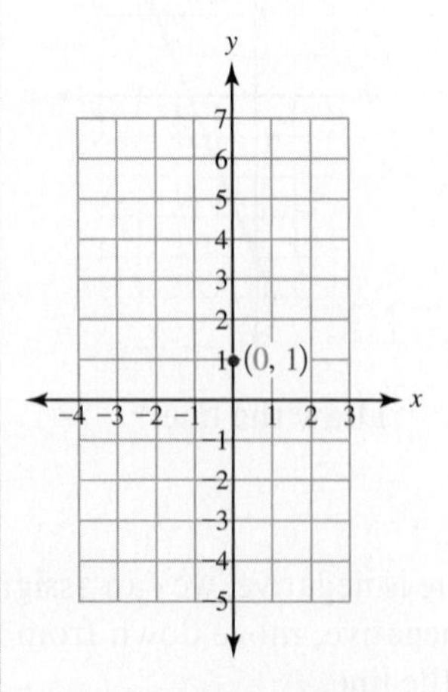

Graph the y-intercept.

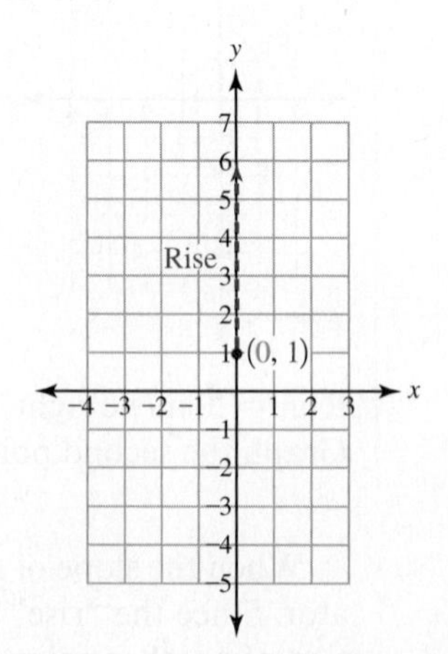

Rise $= 5$; move up 5.

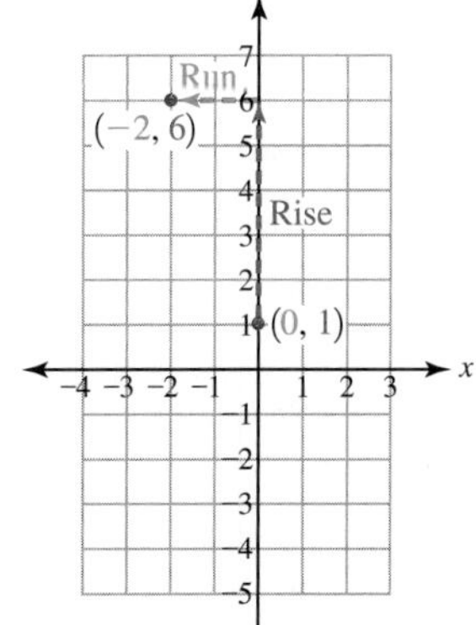

Run = −2; move left 2.
Graph the second point.

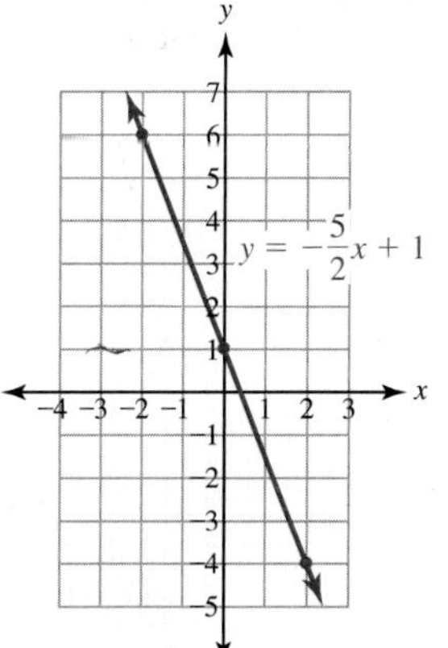

Draw the line.
The graph is the same.

In the next example, notice that the scales on the axes are not the same.

EXAMPLE 12 Use slope-intercept graphing to graph $y = \frac{3}{20}x - 1$.

SOLUTION $y = \frac{3}{20}x - 1$ $\quad m = \frac{3}{20}$; $b = -1$; rise = 3; run = 20

The distance between tick marks on the x-axis is 5. The distance between tick marks on the y-axis is 1.

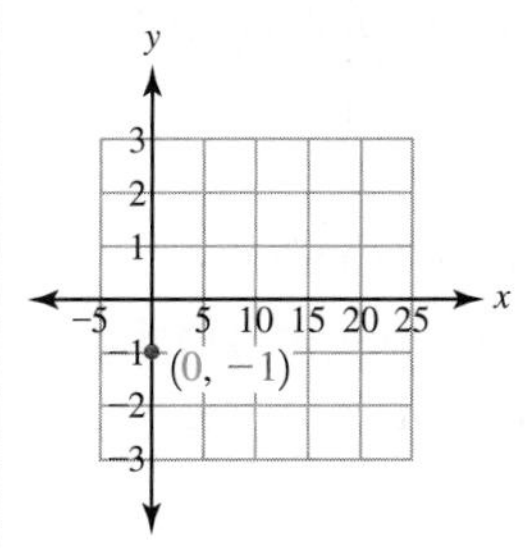

Graph the y-intercept.

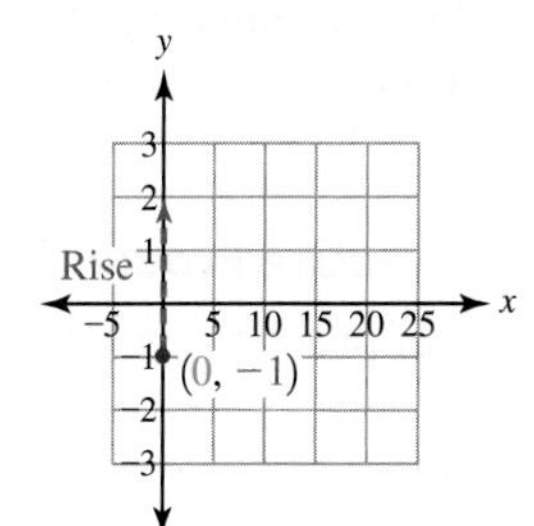

Rise = 3; move up 3 tick marks.

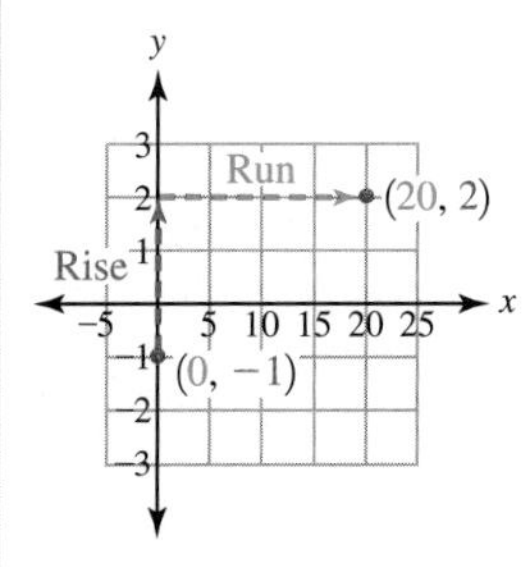

Run = 20; move right 4 tick marks. Graph the second point.

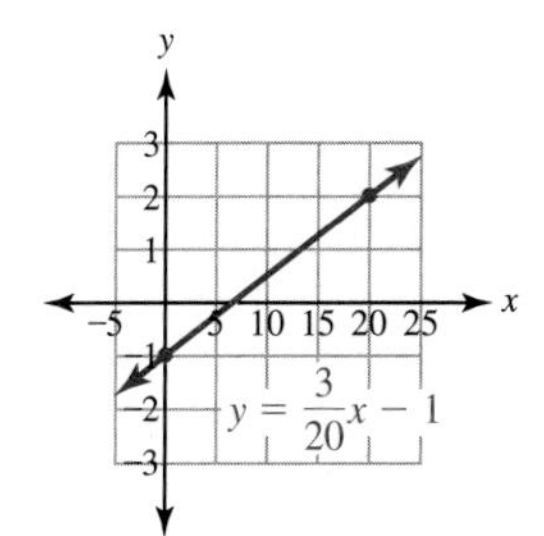

Draw the line.

In the next example, the slope is 2, which can be rewritten as $\frac{2}{1}$. The rise is 2, and the run is 1.

EXAMPLE 13 Use slope-intercept graphing to graph $y = 2x - 3$.

SOLUTION ▶ $y = 2x - 3 \quad m = \frac{2}{1}; b = -3; \text{rise} = 2; \text{run} = 1$

The distance between tick marks on the x-axis is 1. The distance between tick marks on the y-axis is 1.

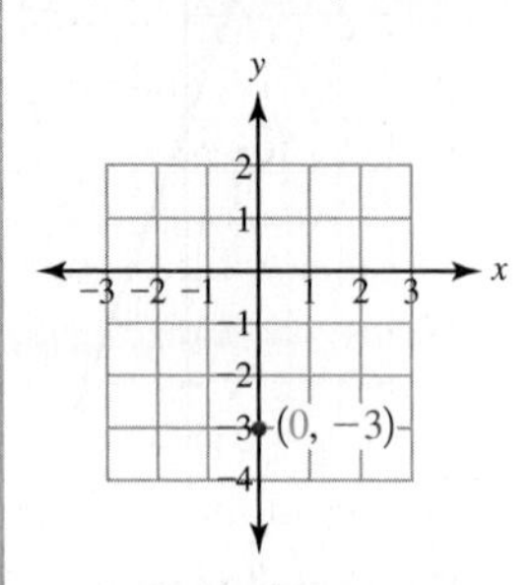

Graph the y-intercept.

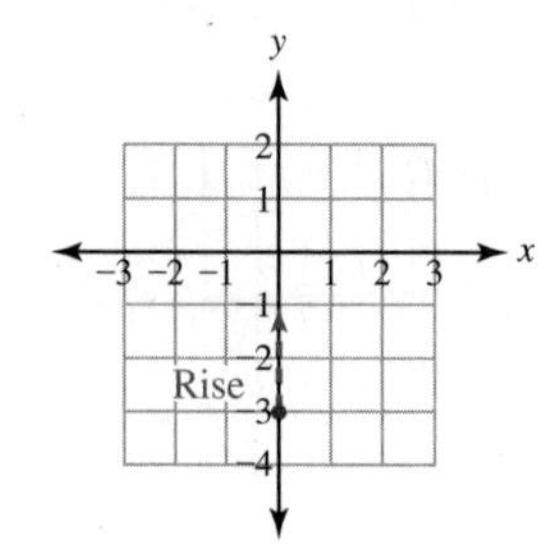

Rise = 2; move up 2 tick marks.

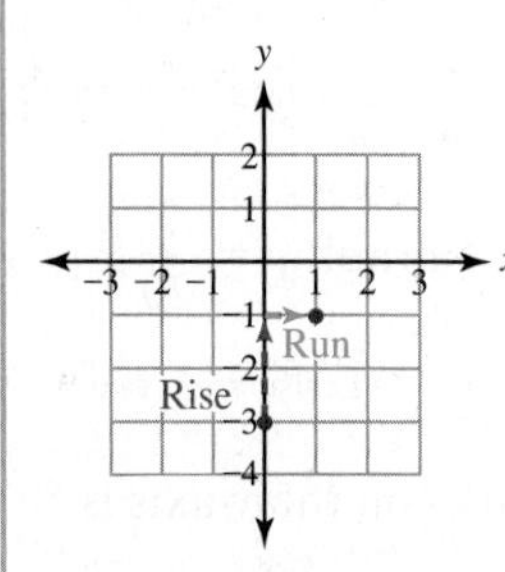

Run = 1; move right 1 tick mark.
Graph the second point.

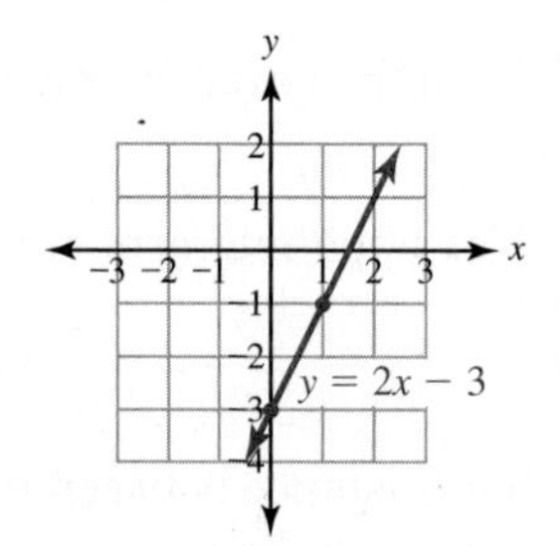

Draw the line.

Practice Problems

For problems 16–19, use slope-intercept graphing to graph the equation.

16. $y = \frac{3}{4}x + 2$ **17.** $y = -\frac{3}{4}x + 2$ **18.** $y = 5x - 7$ **19.** $y = -5x + 4$

For problems 20–21,
(a) rewrite the equation in slope-intercept form.
(b) use slope-intercept graphing to graph the equation.

20. $-2x + 3y = -12$
21. $6x + 5y = 40$

3.4 VOCABULARY PRACTICE

Match the term with its description.

1. $y = mx + b$
2. $ax + by = c$
3. In the ordered pair that represents the y-intercept, this is always 0.
4. In the ordered pair that represents the x-intercept, this is always 0.
5. The point where a line intersects the y-axis
6. The intersection of the x-axis and the y-axis
7. A solution of a linear equation in two variables
8. $m = \frac{y_2 - y_1}{x_2 - x_1}$
9. 0
10. Undefined

A. ordered pair
B. origin
C. slope formula
D. slope of a horizontal line
E. slope of a vertical line
F. slope-intercept form of a linear equation
G. standard form of a linear equation
H. x-coordinate
I. y-coordinate
J. y-intercept

3.4 Exercises

Follow your instructor's guidelines for showing your work.

1. The volume of diesel in a tank is 24 gal. After 2 hr of driving at a constant speed, the tank contains 18 gal of diesel.
- **a.** Write ordered pairs that represent this information.
- **b.** Graph the ordered pairs, and draw a line beginning at the y-intercept.
- **c.** Identify the y-intercept of the line, and describe what the y-coordinate of the y-intercept represents.
- **d.** Use the slope formula to find the slope of the line, and describe what the slope represents.
- **e.** Write an equation that represents the relationship of the volume of diesel in the tank, y, and the time, x.
- **f.** Use the equation to find the amount of diesel in the tank after 3 hr of driving at a constant speed.
- **g.** Find the x-intercept, and describe what the x-coordinate of the x-intercept represents.

2. The volume of gasoline in a tank is 14 gal. After 3 hr of driving at a constant speed, the tank contains 8 gal of gasoline.
- **a.** Write ordered pairs that represent this information.
- **b.** Graph the ordered pairs, and draw a line beginning at the y-intercept.
- **c.** Identify the y-intercept of the line, and describe what the y-coordinate of the y-intercept represents.
- **d.** Use the slope formula to find the slope of the line, and describe what the slope represents.
- **e.** Write an equation that represents the relationship of the volume of gasoline in the tank, y, and the time, x.
- **f.** Use the equation to find the amount of gasoline in the tank after 5 hr of driving at a constant speed.
- **g.** Find the x-intercept, and describe what the x-coordinate of the x-intercept represents.

3. After 3 hr, a potato chip factory has produced 6000 lb of potato chips. After 5 hr, the factory has produced 10,000 lb of potato chips.
- **a.** Write ordered pairs that represent this information.
- **b.** Graph the ordered pairs, and draw a line beginning at the y-intercept.
- **c.** Identify the y-intercept of the line, and describe what the y-coordinate of the y-intercept represents.
- **d.** Use the slope formula to find the slope of the line, and describe what the slope represents.
- **e.** Write an equation that represents the relationship of the amount of potato chips, y, and the time, x.
- **f.** Use the equation to find the amount of potato chips produced after 7 hr.
- **g.** Find the x-intercept, and describe what the x-coordinate of the x-intercept represents.

4. After 3 hr, a cheesecake factory has produced 9000 cheesecakes. After 5 hr, the factory has produced 15,000 cheesecakes.
- **a.** Write ordered pairs that represent this information.
- **b.** Graph the ordered pairs, and draw a line beginning at the y-intercept.
- **c.** Identify the y-intercept of the line, and describe what the y-coordinate of the y-intercept represents.
- **d.** Use the slope formula to find the slope of the line, and describe what the slope represents.
- **e.** Write an equation that represents the relationship of the number of cheesecakes, y, and the time, x.
- **f.** Use the equation to find the number of cheesecakes produced after 6 hr.
- **g.** Find the x-intercept, and describe what the x-coordinate of the x-intercept represents.

5. At an on-line retailer, the price to ship an order of books is \$3 per shipment plus \$0.99 per book.
- **a.** Write an equation that represents the relationship of the price of shipping, y, and the number of books shipped, x.
- **b.** Use the equation to find the price of shipping 25 books.

6. At an on-line retailer, the price of shipping for toys is \$3.99 per shipment plus \$0.85 per pound.
- **a.** Write an equation that represents the relationship of the price of shipping, y, and the number of pounds of toys shipped, x.
- **b.** Use the equation to find the price of shipping an electronic dump truck that weighs 6 pounds.

7. A premature baby needs 6 grams (g) of protein per day. The baby receives 0.25 g of protein per hour.
- **a.** Write an equation that represents the relationship of the amount of protein the baby still needs on a particular day, y, and the number of hours that have passed on that day, x.
- **b.** Use the equation to find the amount of protein the baby still needs after 8 hr have passed.
- **c.** Find the x-intercept.
- **d.** Describe what the x-coordinate of the x-intercept represents.

8. To prepare for a colonoscopy, a patient needs to drink 4000 mL of CoLyte solution. The patient drinks 250 mL every 0.25 hr.
- **a.** Write an equation that represents the relationship of the amount of solution remaining to drink, y, and the time since first drinking the solution, x.
- **b.** Use the equation to find the amount of solution the patient still needs to drink after 3 hr.
- **c.** Find the x-intercept.
- **d.** Describe what the x-coordinate of the x-intercept represents.

9. The cost of wild turkey shipped to Portland, Oregon, from an on-line retailer is \$8.90 per pound plus a shipping cost of \$9.95.
- **a.** Write an equation that represents the relationship of the cost, y, and the weight of the turkey, x.
- **b.** Use the equation to find the cost for a 5-lb turkey.

10. The cost of French roast coffee beans is \$12.95 per pound plus a shipping cost of \$9.95.

a. Write an equation that represents the relationship of the cost, y, and the weight of the coffee, x.

b. Use the equation to find the cost for 6 lb of coffee.

11. An editorial assistant paid a \$1200 broker's fee to sign a lease on an apartment in Brooklyn. His share of the rent is \$800 a month.

a. Write an equation that represents the relationship of the total amount (including the broker's fee) that he will pay for housing, y, and the time, x.

b. Use the equation to find the total amount that he will pay for housing over 5 months.

12. A rain barrel attached to a downspout in Seattle contains 20 gal of water. When it starts to rain, the rain from the roof flows into the barrel at a speed of $\frac{4\text{ gal}}{1\text{ hr}}$.

a. Write an equation that represents the relationship of the amount of water in the barrel, y, and the time, x. Assume that the rain continues to fall at a steady rate.

b. Use the equation to find the amount of water in the barrel after 6 hr.

For exercises 13–14, an asphalt machine paves a road at a speed of $\frac{40\text{ ft}}{1\text{ min}}$. The machine has already paved 8000 ft. The equation $y = \left(\frac{40\text{ ft}}{1\text{ min}}\right)x + 8000\text{ ft}$ represents the relationship of the total length of road paved, y, and the additional time that the machine paves. (*Source:* www.apellc.com)

13. Find the length paved after 3 *hours.*

14. Find the length paved after 2 *hours.*

For exercises 15–16, an asphalt grinder digs a trench at a speed of $\frac{1000\text{ ft}}{1\text{ hr}}$. A grinder is digging a communications cable trench with a length of 158,400 ft. The equation $y = \left(\frac{-1000\text{ ft}}{1\text{ hr}}\right)x + 158{,}400\text{ ft}$ represents the relationship between the length of trench remaining to dig, y, and the time the grinder digs, x. (*Source:* www.asphaltzipper.com)

15. Find the length left to be dug *in miles* after 8 hr. (1 mi = 5280 ft.) Round to the nearest tenth.

16. Find the length left to be dug *in miles* after 40 hr. (1 mi = 5280 ft.) Round to the nearest tenth.

For exercises 17–24, write the equation of the line in slope-intercept form.

17. slope: -4; y-intercept: $(0, 3)$

18. slope: -6; y-intercept: $(0, 2)$

19. slope: 5; y-intercept: $(0, 0)$

20. slope: 6; y-intercept: $(0, 0)$

21. slope: $\frac{2}{3}$; y-intercept: $(0, 4)$

22. slope: $\frac{5}{9}$; y-intercept: $(0, -2)$

23. $m = \frac{30\text{ mi}}{1\text{ hr}}$; $b = 28\text{ mi}$

24. $m = \frac{40\text{ mi}}{1\text{ hr}}$; $b = 35\text{ mi}$

For exercises 25–28,
(a) identify the slope.
(b) identify the y-intercept. Write the ordered pair, not just the y-coordinate.
(c) find the x-intercept. Write the ordered pair, not just the x-coordinate.

25. $y = 3x + 2$

26. $y = 2x + 3$

27. $y = -5x + 10$

28. $y = -4x + 12$

For exercises 29–44,
(a) rewrite the equation in slope-intercept form.
(b) identify the slope.
(c) identify the y-intercept. Write the ordered pair, not just the y-coordinate.
(d) find the x-intercept. Write the ordered pair, not just the x-coordinate.

29. $7x + 9y = -63$

30. $8x + 9y = -72$

31. $7x - 9y = -63$

32. $8x - 9y = -72$

33. $x + y = 4$

34. $x + y = 7$

35. $x - y = 6$

36. $x - y = 2$

37. $x + y = 0$

38. $x - y = 0$

39. $-4x + 3y = 24$

40. $-2x + 15y = 30$

41. $2x - 9y = 27$

42. $5x - 8y = 64$

43. $9x - 2y = 27$

44. $8x - 5y = 64$

For exercises 45–52,
(a) clear the fractions, and rewrite the equation in slope-intercept form.
(b) identify the slope.
(c) identify the y-intercept. Write the ordered pair, not just the y-coordinate.
(d) find the x-intercept. Write the ordered pair, not just the x-coordinate.

45. $-\frac{3}{4}x + \frac{2}{3}y = 8$

46. $-\frac{3}{8}x + \frac{2}{5}y = 10$

47. $y - 3 = \frac{1}{2}(x - 4)$

48. $y - 1 = \frac{1}{4}(x - 8)$

49. $y + 3 = \frac{4}{9}(x - 10)$

50. $y + 4 = \frac{5}{8}(x - 12)$

51. $y - \frac{5}{6} = 2\left(x - \frac{3}{8}\right)$

52. $y - \frac{3}{8} = 3\left(x - \frac{5}{6}\right)$

For exercises 53–70,

(a) identify the y-intercept. Write the ordered pair, not just the y-coordinate.

(b) identify the slope.

(c) write the equation of the line in slope-intercept form.

53.

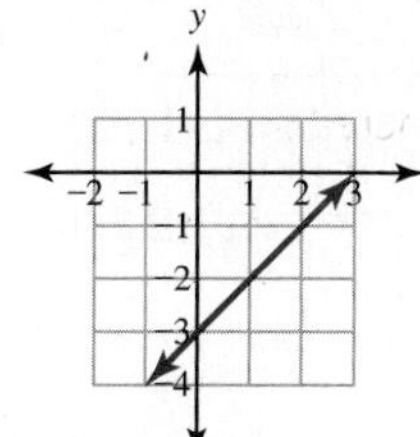

54.

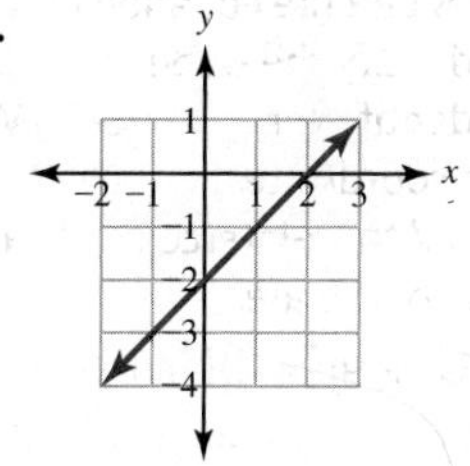

55.

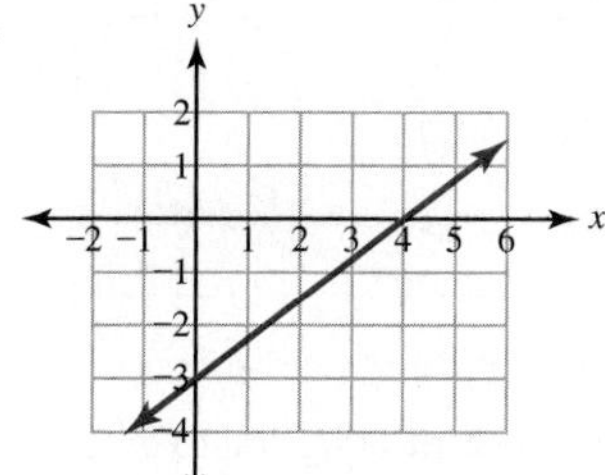

56.

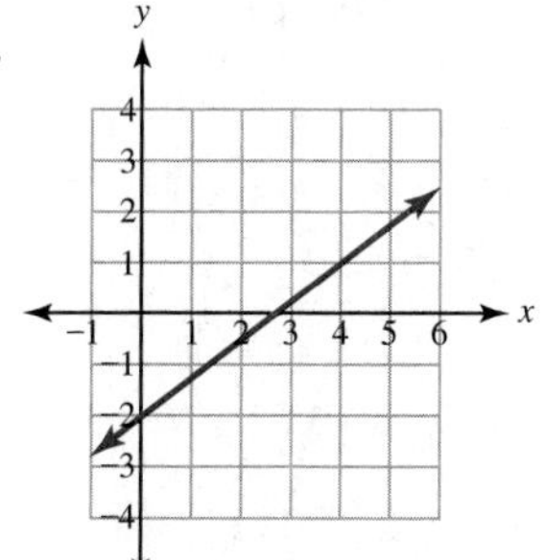

57.

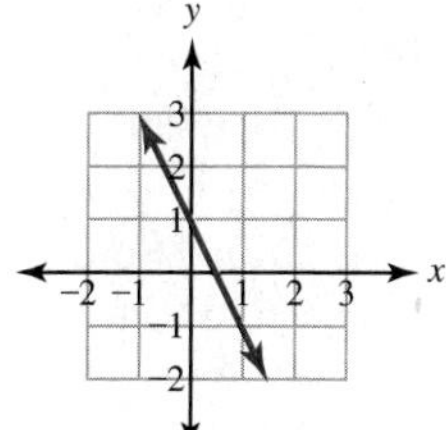

58.

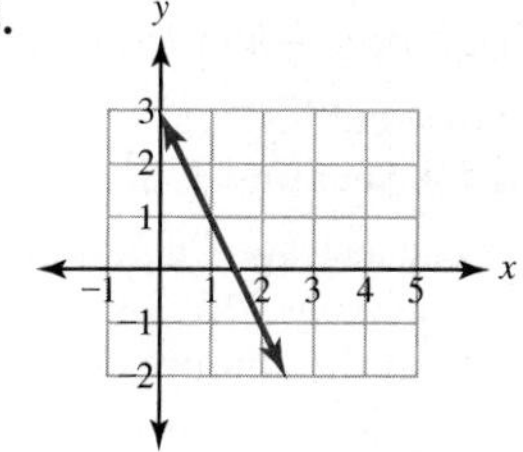

59.

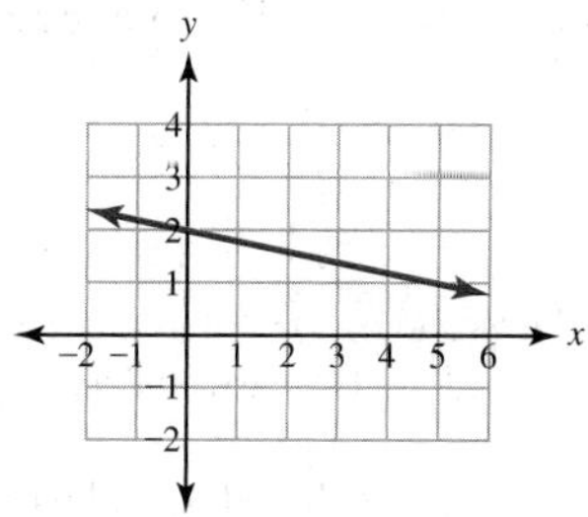

60.

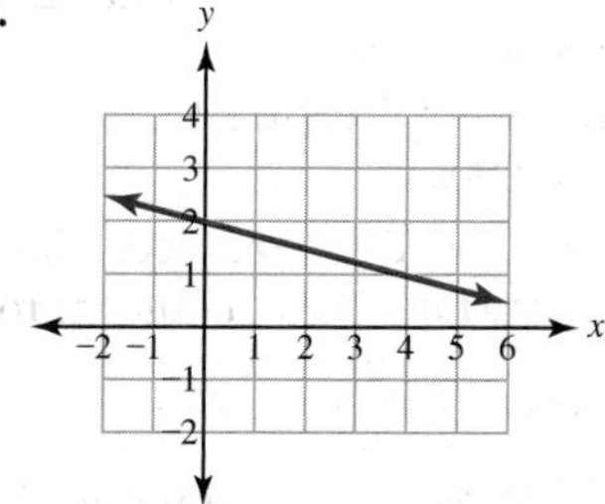

61.

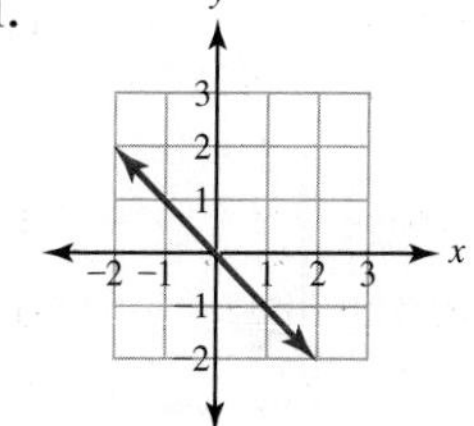

62.

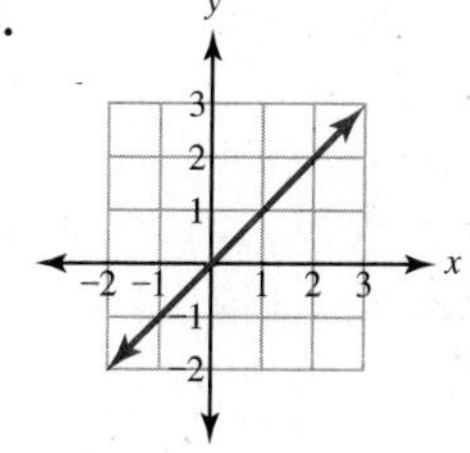

63.

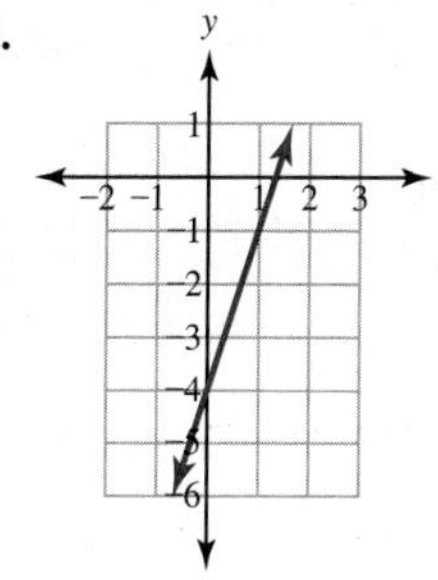

64.

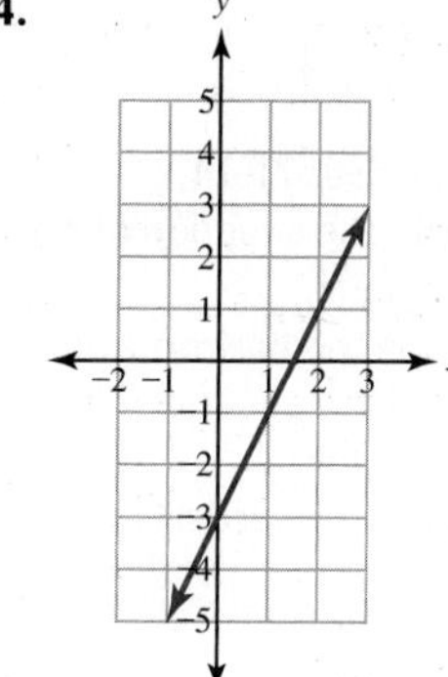

65.

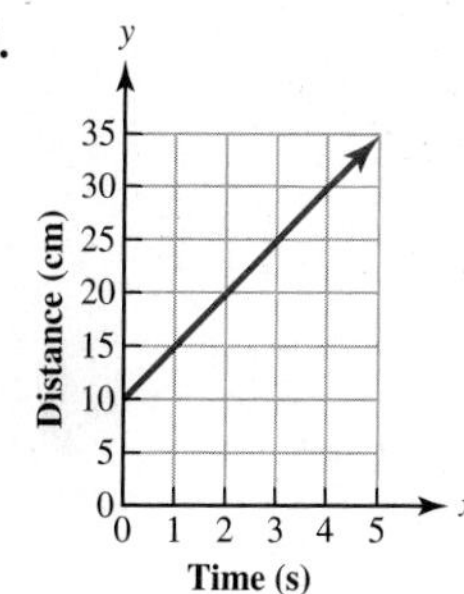

66.

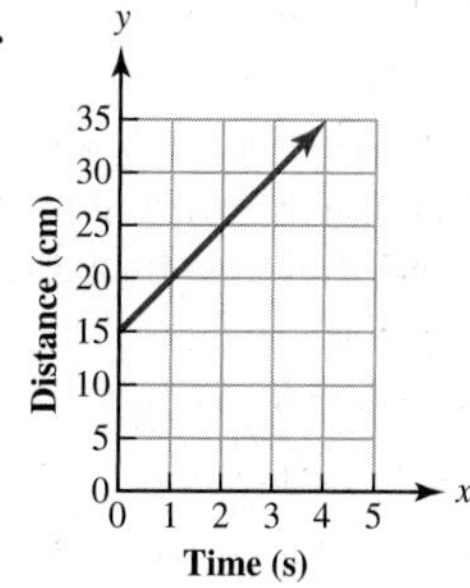

67.

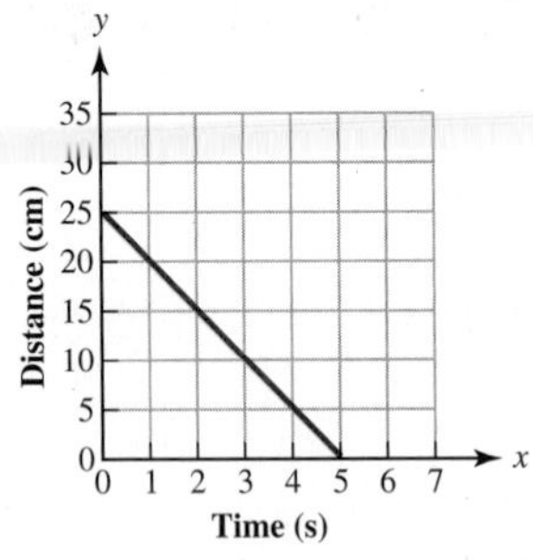

68.

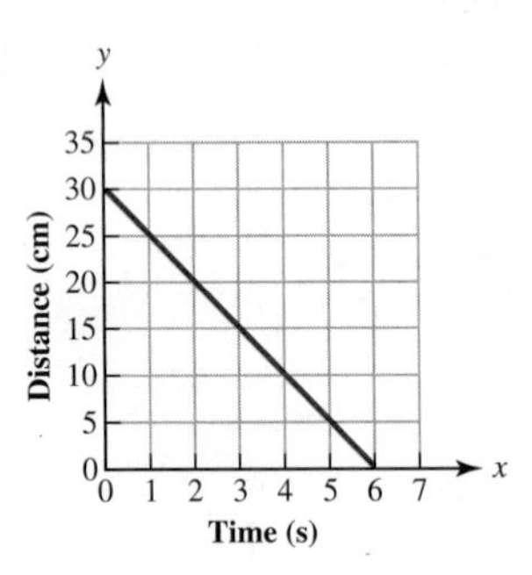

69.

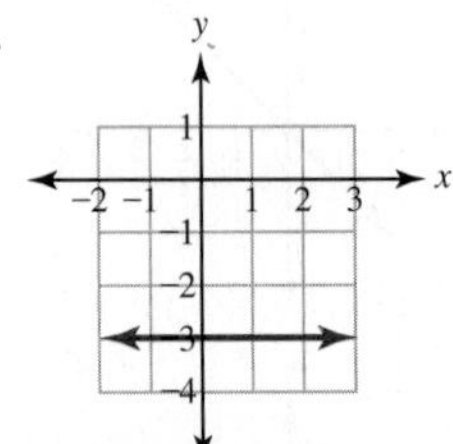

70.

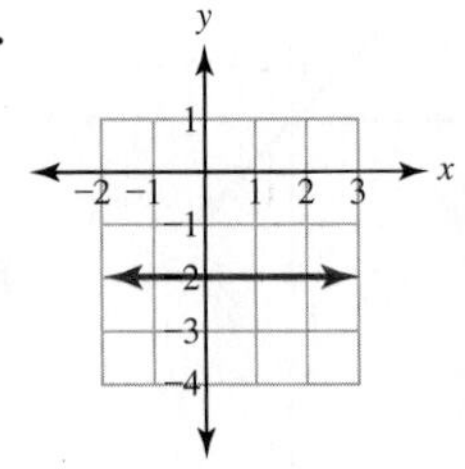

For exercises 71–74,

(a) write the equation of the horizontal line that passes through the point.

(b) graph the horizontal line.

(c) write the equation of the vertical line that passes through the point.

(d) graph the vertical line.

71. (4, 5)

72. (9, 2)

73. (−4, −1)

74. (−3, −8)

For exercises 75–78,

(a) identify the slope of the line.

(b) write the equation of the line.

75.

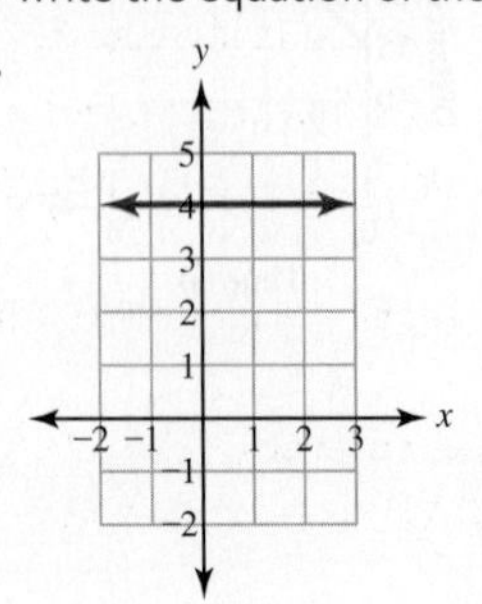

76.

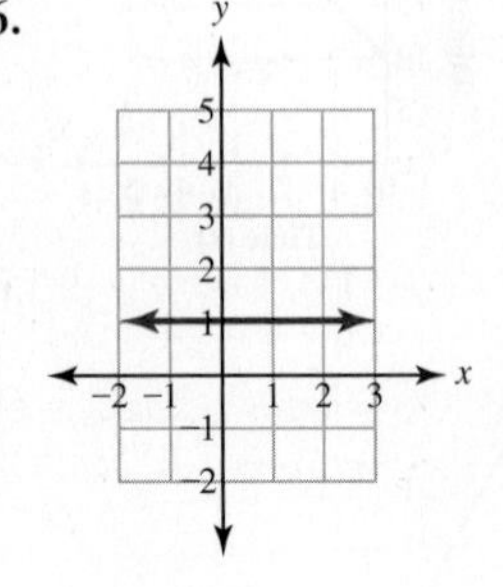

77.

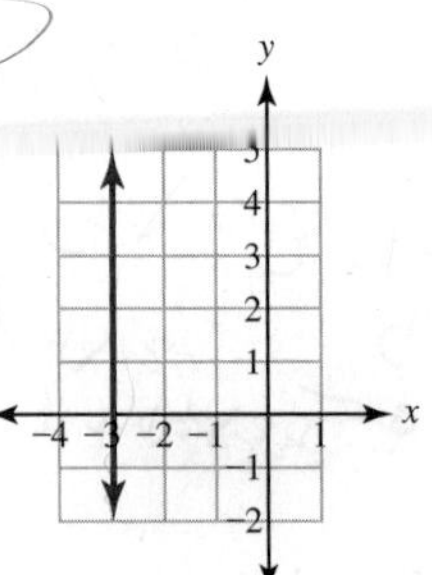

78.

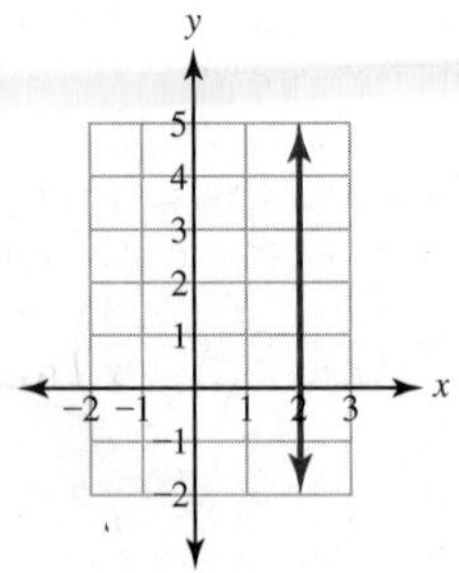

79. Copy and fill in the table.

Equation	Slope	y-intercept
$y = -\frac{5}{6}x + 2$		
$y = 8$		
$y = 8x$		
$x = -12$		

80. Copy and fill in the table.

Equation	Slope	y-intercept
$y = -\frac{7}{8}x + 5$		
$y = 10$		
$y = 10x$		
$x = -7$		

For exercises 81–92, use slope-intercept graphing to graph the equation.

81. $y = \frac{2}{5}x - 1$

82. $y = \frac{2}{5}x - 4$

83. $y = -\frac{2}{5}x + 6$

84. $y = -\frac{2}{5}x + 4$

85. $y = 3x - 2$

86. $y = 2x - 3$

87. $y = -4x + 3$

88. $y = -4x + 2$

89. $y = 6x$

90. $y = 5x$

91. $y = -\frac{1}{50}x + 20$

92. $y = -\frac{1}{40}x + 10$

Problem Solving: Practice and Review

Follow your instructor's guidelines for using five steps as outlined in Section 1.3, p. 55.

93. The average single-family home in the United States consumes 12,773 kilowatt · hour of electricity per year. Find the amount of carbon dioxide released in generating the electricity needed to power the average single family home. Assume that 7 percent of electricity generated is lost during transmission and distribution to homes. Round to the nearest hundred. (1 megawatt · hour = 1000 kilowatt · hour.)

The national average carbon dioxide output rate for electricity generated . . . was 1,329 lbs CO_2 per megawatt · hour. (*Source:* www.epa.gov)

94. Water from the Columbia River system is used for irrigation in Eastern Washington. A center pivot irrigation sprinkler waters a circular area. The radius of the circle is equal to the length of the sprinkler. A sprinkler is 1320 ft long. Find the area that it irrigates. ($\pi \approx 3.14$.)

© Laura Bracken

95. The following media report describes the results of a study of 28,962 rescue and recovery workers. Find the number of workers who were still suffering serious mental health effects three years after the disaster. Round to the nearest ten.

Thousands of World Trade Center rescue and recovery workers were still suffering serious mental health effects three years after the disaster, the Health Department reported today. New findings released from the World Trade Center Health Registry show that one in eight rescue and recovery workers . . . likely had post-traumatic stress disorder when they were interviewed in 2003 and 2004. (*Source:* www.nyc.gov, Aug. 29, 2007)

96. Find the amount of simple interest earned by an investment of \$6500 invested for 8 years at an annual interest rate of 4%.

Find the Mistake

For exercises 97–100, the completed problem has one mistake.

(a) Describe the mistake in words, or copy down the whole problem and highlight or circle the mistake.

(b) Do the problem correctly.

97. Problem: Write the equation in slope-intercept form.

Incorrect Answer:

$$6x - 5y = 30$$
$$-5y = -6x + 30$$
$$\frac{-5y}{-5} = \frac{-6x}{-5} + \frac{30}{-5}$$
$$y = -\frac{6}{5}x - 6$$

98. Problem: Find the slope and y-intercept of the line that represents $x = 3$.

Incorrect Answer: The slope is undefined. The y-intercept is $(0, 3)$.

99. Problem: Use slope-intercept graphing to graph $y = \frac{3}{4}x - 2$.

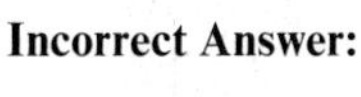

Incorrect Answer:

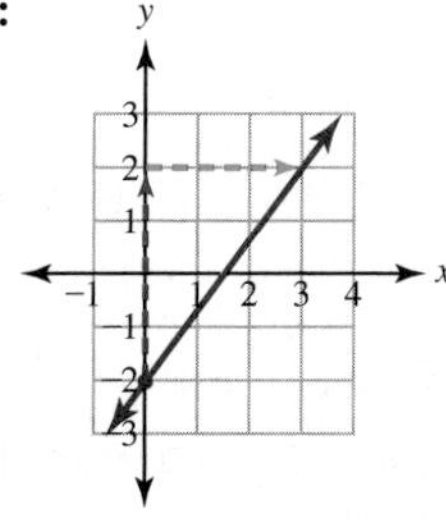

100. Problem: Write $y - 6 = 2(x - 9)$ in slope-intercept form.

Incorrect Answer:

$$y - 6 = 2(x - 9)$$
$$y - 6 = 2x - 18$$
$$\underline{-6 \qquad -6}$$
$$y = 2x - 24$$

Review

For exercises 101–104, two perpendicular lines intersect each other at a 90-degree (right) angle. Two parallel lines never intersect. Are the lines in the graph parallel, perpendicular, or neither parallel nor perpendicular?

101. **102.**

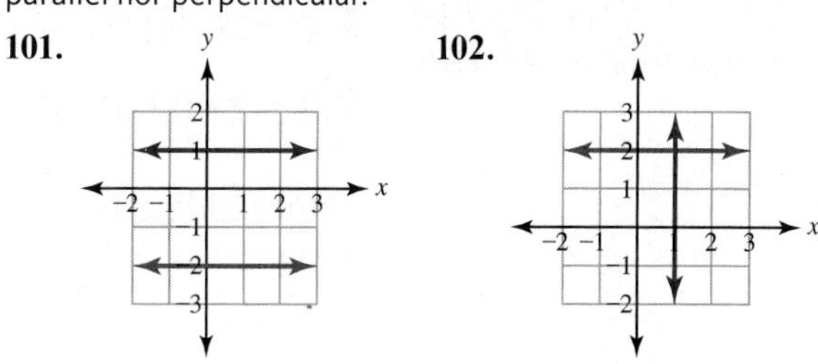

103. **104.**

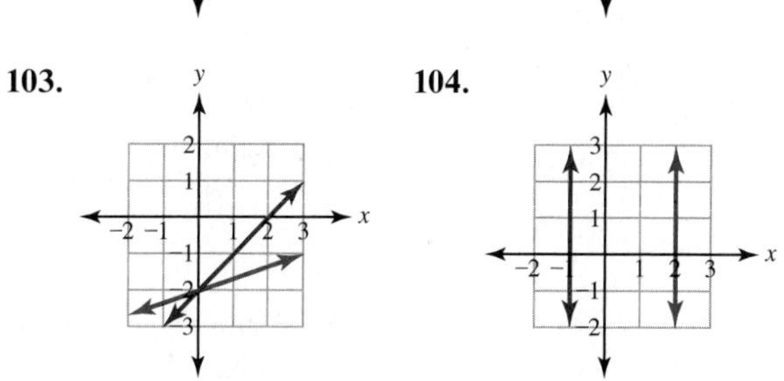

SUCCESS IN COLLEGE MATHEMATICS

105. Learning preferences can include the *social setting* in which you study and learn mathematics. Choose all of the conditions that most closely match your preferences when learning or practicing difficult math.

A. Work by myself.

B. Work with close friends.

C. Work with a study group who are not friends outside of class.

D. If I need help, I prefer to get it from my friends or my study group.

E. If I need help, I prefer to get it from my spouse or significant other or one of my children.

F. If I need help, I prefer to go to a tutoring center.

G. If I need help, I prefer to get it from my instructor.

H. I do not like to stop my work to help others who do not understand.

I. I do not mind stopping my work to help others when they do not understand.

106. Do you now study in conditions that match these learning preferences? If not, explain why.

A linear model describes the relationship of two variables such as the percent of paper waste that is recycled, y, and the time, x. The **line of best fit** is the straight line that best represents the relationship. In this section, we will write equations of lines of best fit. We will also use point-slope form to write the equation of a line.

SECTION 3.5

Point-Slope Form and Writing the Equation of a Line

After reading the text, working the practice problems, and completing assigned exercises, you should be able to:

1. Given the slope and one point on a line, write its equation in slope-intercept form.
2. Given two points on a line, write its equation in slope-intercept form.
3. Given the graph of a line, write its equation in slope-intercept form.
4. Write a linear model that represents a line of best fit.
5. Write the equation in slope-intercept form of a line that is parallel to a given line.
6. Write the equation in slope-intercept form of a line that is perpendicular to a given line.

Point-Slope Form

The slope formula is $m = \dfrac{y_2 - y_1}{x_2 - x_1}$. If we change the subscripts on the slope formula to $m = \dfrac{y - y_1}{x - x_1}$ and use the properties of equality, the result is the **point-slope form** of the equation of a line.

$$m = \frac{y_2 - y_1}{x_2 - x_1} \quad \text{Slope formula.}$$

$$m = \frac{\boldsymbol{y} - y_1}{\boldsymbol{x} - x_1} \quad \text{Change the subscripts on the formula.}$$

$$(x - x_1)m = \frac{\cancel{(x - x_1)}}{1} \cdot \frac{y - y_1}{\cancel{x - x_1}} \quad \text{Clear fractions; multiplication property of equality.}$$

$$(x - x_1)m = \boldsymbol{y - y_1} \quad \text{Simplify.}$$

$$y - y_1 = (x - x_1)m \quad \text{Switch sides.}$$

$$y - y_1 = \boldsymbol{m(x - x_1)} \quad \text{Change order; this is point-slope form.}$$

Forms of a Linear Equation in Two Variables, x and y

Standard Form $ax + by = c$, where a, b, and c are real numbers and a and b are not both 0

Slope-Intercept Form $y = mx + b$, where m and b are real numbers
The slope of the line that represents this equation is m. The y-intercept of this line is $(0, b)$.

Point-Slope Form $y - y_1 = m(x - x_1)$, where m, x_1, and y_1 are real numbers
The slope of the line that represents this equation is m. A point on this line is (x_1, y_1).

If we know the slope and a point on the line but do not know the y-intercept, we can write the equation of the line by replacing m, x_1, and y_1 in point-slope form. We then usually rewrite the equation in slope-intercept form.

EXAMPLE 1 The slope of the graphed line is 2, and it passes through $(5, -1)$.

(a) Write the equation of the line in slope-intercept form.

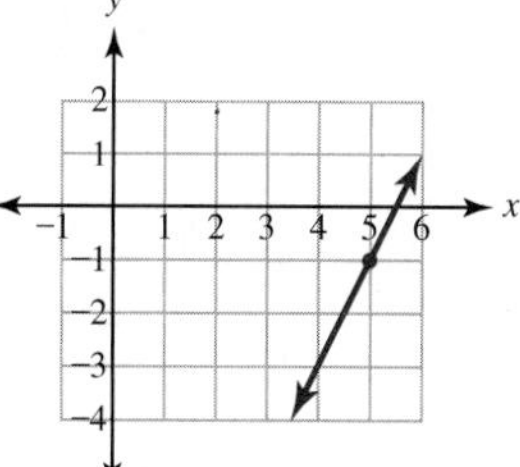

SOLUTION ▶ Since the y-intercept is not known, begin with point-slope form. Replace m with the slope, 2, and replace x_1 and y_1 with the coordinates of the point, $(5, -1)$. Then use the properties of equality to rewrite the equation in slope-intercept form.

$y - y_1 = m(x - x_1)$	Point-slope form.
$y-(-1) = 2(x - 5)$	Replace m, x_1, and y_1.
$y + \mathbf{1} = \mathbf{2x + 2(-5)}$	Simplify; distributive property.
$y + 1 = 2x - \mathbf{10}$	Simplify.
$-1 \qquad -1$	Subtraction property of equality.
$y + \mathbf{0} = 2x - \mathbf{11}$	Simplify.
$\mathbf{y} = 2x - 11$	Simplify.

The equation $y = 2x - 11$ is in slope-intercept form, $y = mx + b$.

(b) Identify the y-intercept.

▶ In $y = mx + b$, the y-intercept is $(0, b)$. For $y = 2x - 11$, the y-intercept is $(0, -11)$.

Writing the Equation of a Line in Slope-Intercept Form, Given the Slope and a Point on the Line

1. Replace m, x_1, and y_1 with the given values in point-slope form, $y - y_1 = m(x - x_1)$.
2. Use the distributive property and the properties of equality to rewrite the equation in slope-intercept form, $y = mx + b$.

In the next example, we are given two points but do not know the slope. We use the slope definition to find the slope and then use point-slope form to write the equation of the line. Even though the problem asks for the equation in slope-intercept form, we first write the equation in point-slope form.

EXAMPLE 2 A line passes through $(14, -3)$ and $(-6, -4)$. Write the equation of the line in slope-intercept form.

SOLUTION ▶ Use the slope formula to find the slope of the line.

$m = \dfrac{y_2 - y_1}{x_2 - x_1}$ Slope formula.

$m = \dfrac{-4 - (-3)}{-6 - 14}$ $(x_1, y_1) = (14, -3)$; $(x_2, y_2) = (-6, -4)$

$m = \dfrac{\mathbf{-1}}{\mathbf{-20}}$ Simplify.

$m = \dfrac{\mathbf{1}}{\mathbf{20}}$ Simplify; $\dfrac{-1}{-20} = \dfrac{1}{20}$

Write the equation of the line in point-slope form.

$y - y_1 = m(x - x_1)$ Point-slope form.

$y - (-3) = \dfrac{\mathbf{1}}{\mathbf{20}}(x - 14)$ Replace m, x_1, and y_1; $(x_1, y_1) = (14, -3)$

$\mathbf{y + 3} = \dfrac{1}{20}(x - 14)$ Simplify.

Rewrite the equation in slope-intercept form.

$y + 3 = \dfrac{\mathbf{1}}{\mathbf{20}}x + \dfrac{\mathbf{1}}{\mathbf{20}}\left(-\dfrac{14}{1}\right)$ Distributive property.

$y + 3 = \dfrac{1}{20}x + \dfrac{1}{\mathbf{2} \cdot 10}\left(-\dfrac{\mathbf{2} \cdot 7}{1}\right)$ Find common factors.

$y + 3 = \dfrac{1}{20}x - \dfrac{\mathbf{7}}{\mathbf{10}}$ Simplify.

$\underline{-3 \qquad\qquad -3}$ Subtraction property of equality.

$y + \mathbf{0} = \dfrac{1}{20}x - \dfrac{7}{10} \mathbf{- 3}$ Simplify.

$\mathbf{y} = \dfrac{1}{20}x - \dfrac{7}{10} - \dfrac{3}{\mathbf{1}} \cdot \dfrac{10}{10}$ Multiply by a fraction equal to 1.

$y = \dfrac{1}{20}x - \dfrac{7}{10} - \dfrac{\mathbf{30}}{\mathbf{10}}$ Simplify.

$y = \dfrac{1}{20}x - \dfrac{\mathbf{37}}{\mathbf{10}}$ Simplify.

In Example 2, we used both of the ordered pairs to find the slope but used only one of the points, $(14, -3)$, to write the equation of the line. No matter which point we choose, the equation of the line in slope-intercept form is the same.

EXAMPLE 3 Write the equation of the graphed line in slope-intercept form.

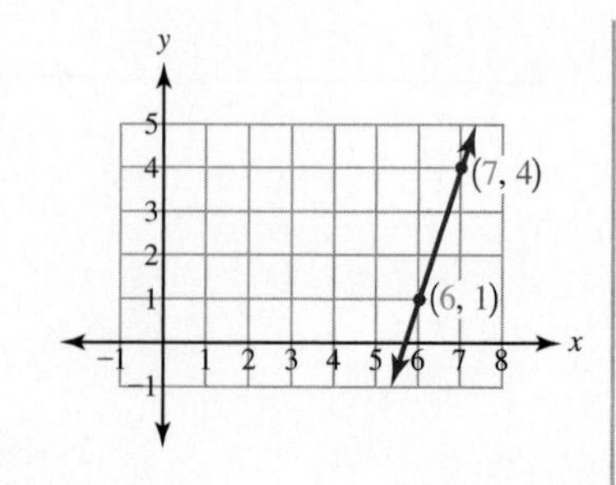

SOLUTION ▶ Identify two points on the line: $(6, 1)$ and $(7, 4)$.

Use the slope formula to find the slope of the line.

$m = \dfrac{y_2 - y_1}{x_2 - x_1}$ Slope formula.

$m = \dfrac{4 - 1}{7 - 6}$ $(x_1, y_1) = (6, 1)$; $(x_2, y_2) = (7, 4)$

$m = \dfrac{\mathbf{3}}{\mathbf{1}}$ Simplify.

Write the equation of the line in point-slope form.

$y - y_1 = m(x - x_1)$ Point-slope form.

$y - 1 = \mathbf{3}(x - \mathbf{6})$ Replace m, x_1, y_1; $(x_1, y_1) = (6, 1)$

Rewrite the equation in slope-intercept form.

$y - 1 = \mathbf{3x + 3(-6)}$ Distributive property.

$y - 1 = 3x - \mathbf{18}$ Simplify.

$\underline{+1 \qquad +1}$ Addition property of equality.

$y - \mathbf{0} = 3x - \mathbf{17}$ Simplify.

$y = 3x - 17$ Simplify.

Writing the Equation of a Line in Slope-Intercept Form, Given Two Points or the Graph of a Line

1. Use the slope formula to find the slope of the line.
2. Use point-slope form and either one of the points to write the equation of the line.
3. Rewrite the equation in slope-intercept form.

EXAMPLE 4 Write the equation in slope-intercept form of a line that passes through $(-1, 5)$ and $(-4, 9)$.

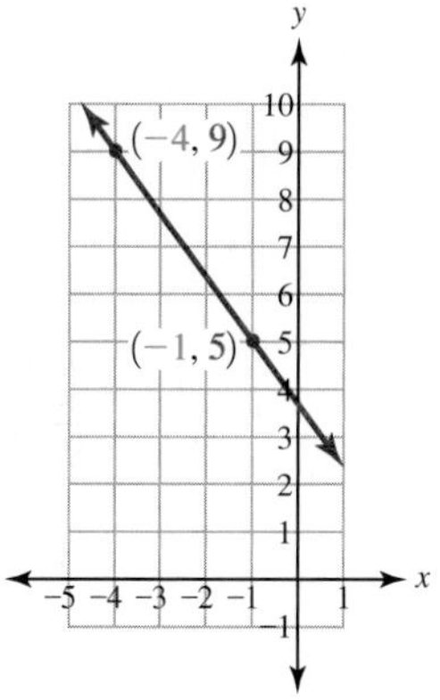

SOLUTION ▶ Use the slope formula to find the slope of the line.

$m = \frac{y_2 - y_1}{x_2 - x_1}$ Slope formula.

$m = \frac{9 - 5}{-4 - (-1)}$ $(x_1, y_1) = (-1, 5)$; $(x_2, y_2) = (-4, 9)$

$m = -\frac{4}{3}$ Simplify.

Write the equation of the line in point-slope form.

$y - y_1 = m(x - x_1)$ Point-slope form.

$y - 5 = -\frac{4}{3}(x - (-1))$ Replace m, x_1, y_1; $(x_1, y_1) = (-1, 5)$

$y - 5 = -\frac{4}{3}(x + \mathbf{1})$ Simplify.

Rewrite the equation in slope-intercept form.

$y - 5 = -\frac{4}{3}x - \frac{4}{3} \cdot 1$ Distributive property.

$y - 5 = -\frac{4}{3}x - \frac{4}{3}$ Simplify.

$\underline{+5 \qquad +5}$ Addition property of equality.

$y + \mathbf{0} = -\frac{4}{3}x - \frac{4}{3} + \mathbf{5}$ Simplify.

$y = -\frac{4}{3}x - \frac{4}{3} + \frac{5}{\mathbf{1}} \cdot \frac{3}{3}$ Simplify; multiply by a fraction equal to 1.

$y = -\frac{4}{3}x - \frac{4}{3} + \frac{\mathbf{15}}{\mathbf{3}}$ Simplify.

$y = -\frac{4}{3}x + \frac{\mathbf{11}}{\mathbf{3}}$ Simplify.

The equation of the line in slope-intercept form is $y = -\frac{4}{3}x + \frac{11}{3}$.

Practice Problems

1. The slope of a line is 3, and it passes through $(5, 12)$.
   a. Write the equation of the line in slope-intercept form.
   b. Identify the y-intercept.
2. The slope of a line is $-\frac{5}{8}$, and it passes through the point $(24, -4)$. Write the equation of the line in slope-intercept form.
3. Write the equation in slope-intercept form of the line that passes through $(8, 15)$ and $(10, 29)$.
4. Write the equation in slope-intercept form of the line that passes through $(8, -1)$ and $(13, -4)$.
5. Write the equation in slope-intercept form of the graphed line.

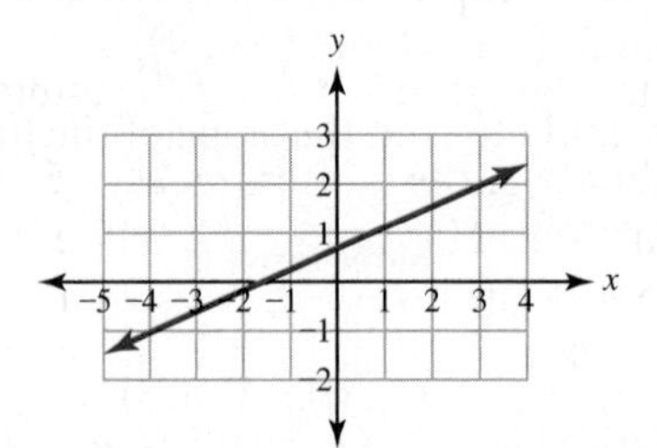

Writing the Equation of a Line of Best Fit

A linear model represents the relationship of two variables. To create a linear model, we collect data and organize it as a list of ordered pairs. If the relationship is perfectly linear, all of the points that represent the ordered pairs will be on the same straight line. In real-life situations, the relationship between two variables might not be perfectly linear. The **line of best fit** is the straight line that best represents the relationship of the two variables. This line may pass through some, none, or all of the points. Mathematicians use *regression methods* to find the line of best fit. We will not learn regression here. However, we will use information from lines of best fit to write linear models.

Writing the Equation of a Line of Best Fit in Slope-Intercept Form

1. Use slope-intercept form, $y = mx + b$.
2. Replace m with the slope of the line of best fit, including the units of measurement.
3. Replace b with the y-coordinate of the y-intercept of the line of best fit, including the units of measurement.

EXAMPLE 5 The points on the graph represent the ordered pairs in the table. The equation of the line of best fit for these points is a linear model of the relationship of Texas public school enrollment, y, and the number of years since 2001, x.

Texas Public School Enrollment

Number of students: 0; 1,000,000; 2,000,000; 3,000,000; 4,000,000; 5,000,000; 6,000,000

Number of years since 2001: 0 1 2 3 4 5 6 7 8

$m = \frac{85{,}166 \text{ students}}{1 \text{ year}}$

$b = 4{,}163{,}191$ students

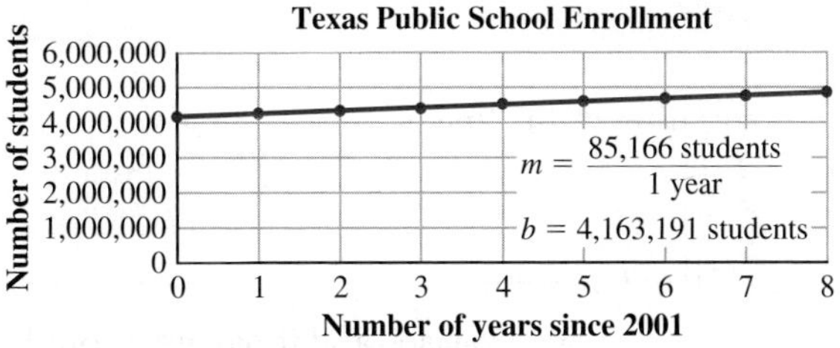

Source: datacenter.kidscount.org

x, number of years since 2001	y, public school enrollment
0	4,160,968
1	4,255,821
2	4,328,028
3	4,400,644
4	4,525,394
5	4,594,942
6	4,671,493
7	4,749,571
8	4,847,844

(a) Write the equation of the line of best fit.

SOLUTION ▶ In slope-intercept form, $y = mx + b$, m is the slope of the line of best fit, $\frac{85{,}166 \text{ students}}{1 \text{ year}}$, and b is the y-coordinate of the y-intercept of the line of best fit, 4,163,191 students. The equation is $y = \left(\frac{85{,}166 \text{ students}}{1 \text{ year}}\right)x + 4{,}163{,}191$ students.

(b) Describe what the slope represents.

▶ The slope, $\frac{85{,}166 \text{ students}}{1 \text{ year}}$, is the average increase in enrollment per year.

(c) Describe what the y-coordinate of the y-intercept represents.

▶ The y-coordinate of the y-intercept is 4,163,191 students, which is the estimated enrollment in 2001. Notice that the y-intercept in the table of ordered pairs and the y-intercept of the line of best fit are not the same. Points on the line of best fit might not exactly match the points that represent actual data.

(d) Use the equation to find the enrollment in 2016. Round to the nearest thousand.

▶ $x = 2016 - 2001$ — x is the number of years since 2001.

$x = 15$ years — Simplify.

$y = \left(\frac{85{,}166 \text{ students}}{1 \text{ year}}\right)x + 4{,}163{,}191 \text{ students}$

$y = \left(\frac{85{,}166 \text{ students}}{1 \cancel{\text{year}}}\right)(\mathbf{15}\ \cancel{\textbf{years}}) + 4{,}163{,}191 \text{ students}$ — Replace x.

$y = \mathbf{1{,}277{,}490\ students} + 4{,}163{,}191 \text{ students}$ — Simplify.

$y = \mathbf{5{,}440{,}681\ students}$ — Simplify.

$y \approx \mathbf{5{,}441{,}000\ students}$ — Round.

In 2015, the enrollment will be about 5,441,000 students.

Extrapolation is the process of using the equation or graph of a linear model to make predictions. A model is only as accurate as the data used to create it, and a model can predict the future only if circumstances stay essentially the same. For example, the linear model in Example 5 is based on the enrollment from 2001 to 2009. When we evaluate this function to find the enrollment in 2016, we are extrapolating.

Practice Problems

6. The graph represents the relationship of the U.S. gross domestic product in trillions of dollars, y, and the number of years since 1998, x.

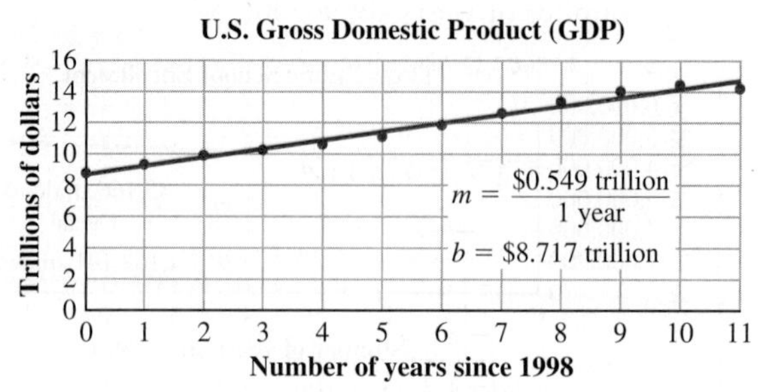

Source: www.bea.gov

a. Write the equation of the line of best fit.
b. Describe what the slope represents.
c. Describe what the y-coordinate of the y-intercept represents.
d. Use the equation to find the gross domestic product in 2012.
e. The data point for 2008 is (10 years, \$14.441 trillion), and the data point for 2009 is (11 years, \$14.256 trillion). Describe what was happening in the U.S. economy between 2008 and 2009.

Parallel Lines

If two lines do not intersect, they are parallel lines. Because they do not intersect, their slopes are equal. For example, if the slope of a line is -2, then the slope of a line that is parallel to it is also -2.

EXAMPLE 6 The slope of line A is -2 and passes through $(1, 4)$. A parallel line (line B) passes through $(-3, 2)$.

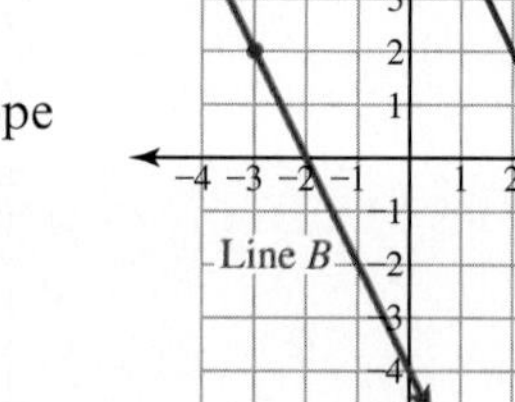

(a) Identify the slope of line B.

SOLUTION ▶ Since line A and line B are parallel, their slopes are equal. Since the slope of line A is -2, the slope of the parallel line (line B) is also -2.

(b) Write the equation of line B in slope-intercept form.

▶ Replace m, x_1, and y_1 in point-slope form, and rewrite in slope-intercept form.

$y - y_1 = m(x - x_1)$	Point-slope form.
$y - 2 = \mathbf{-2}(x - (\mathbf{-3}))$	$m = -2, (x_1, y_1) = (-3, 2)$
$y - 2 = -2(x + \mathbf{3})$	Simplify.
$y - 2 = \mathbf{-2x - 2 \cdot 3}$	Distributive property.
$y - 2 = -2x - \mathbf{6}$	Simplify.
$\quad +2 \qquad +2$	Addition property of equality.
$y + \mathbf{0} = -2x - \mathbf{4}$	Simplify.
$y = -2x - 4$	Simplify.

In slope-intercept form, the equation of line B is $y = -2x - 4$. Notice that the given point on line A, $(1, 4)$, is extraneous information that is not needed to write the equation of line B.

In the next example, we compare the slopes of two lines, line A and line B. So that we can use m to represent both slopes but also refer specifically to the slope of each line, we use subscripts, small letters to the lower right of m. The slope of line A is m_A, and the slope of line B is m_B.

EXAMPLE 7 Line A passes through $(-3, -4)$ and $(1, 2)$. A parallel line (line B) passes through $(6, 2)$.

(a) Find the slope of line A.

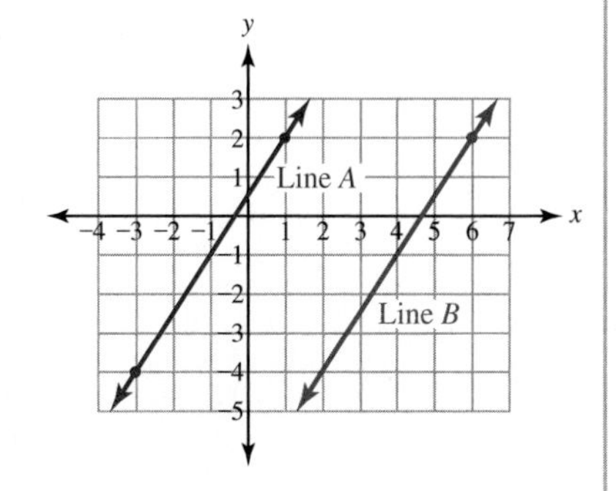

SOLUTION

$m = \dfrac{y_2 - y_1}{x_2 - x_1}$ The slope formula.

$m_A = \dfrac{2 - (-4)}{1 - (-3)}$ $(x_1, y_1) = (-3, -4)$; $(x_2, y_2) = (1, 2)$

$m_A = \dfrac{6}{4}$ Simplify.

$m_A = \dfrac{3 \cdot 2}{2 \cdot 2}$ Find common factors.

$m_A = \dfrac{3}{2}$ Simplify.

(b) Identify the slope of line B.

Since line A and line B are parallel, their slopes are equal. Since $m_A = \dfrac{3}{2}$, $m_B = \dfrac{3}{2}$.

(c) Write the equation of line B in slope-intercept form.

Replace m, x_1, and y_1 in point-slope form, and rewrite in slope-intercept form.

$y - y_1 = m(x - x_1)$ Point-slope form.

$y - 2 = \dfrac{3}{2}(x - 6)$ $m = \dfrac{3}{2}$, $(x_1, y_1) = (6, 2)$

$y - 2 = \dfrac{3}{2}x + \dfrac{3}{2}\left(-\dfrac{6}{1}\right)$ Distributive property.

$y - 2 = \dfrac{3}{2}x + \dfrac{3}{2}\left(-\dfrac{2 \cdot 3}{1}\right)$ Find common factors.

$y - 2 = \dfrac{3}{2}x - 9$ Simplify.

$ + 2 \qquad\quad + 2$ Addition property of equality.

$y + 0 = \dfrac{3}{2}x - 7$ Simplify.

$y = \dfrac{3}{2}x - 7$ Simplify.

In slope-intercept form, the equation of line B is $y = \dfrac{3}{2}x - 7$.

Practice Problems

7. Line A is $y = 7x - 9$. A parallel line (line B) passes through $(-6, 1)$.
   a. Identify the slope of line A.
   b. Identify the slope of line B.
   c. Write the equation of line B in slope-intercept form.

8. Line A passes through $(-2, -7)$ and $(1, 2)$. A parallel line (line B) passes through $(5, -4)$.
a. Find the slope of line A.
b. Identify the slope of line B.
c. Write the equation of line B in slope-intercept form.

9. The slope of a line is -4. Write the equation in slope-intercept form of a parallel line that passes through $(1, 3)$.

Perpendicular Lines

When perpendicular lines intersect, they form a right angle with a measure of 90 degrees. A right triangle includes a 90-degree angle. To show a right angle in a drawing, draw a small square in the angle. Since **perpendicular lines** form a right angle, the legs of a right triangle are perpendicular. For the right triangle in Figure 1, the slope of the vertical leg is undefined, and the slope of the horizontal leg is 0. However, if the legs of a right triangle are not horizontal and vertical, the product of their slopes is -1.

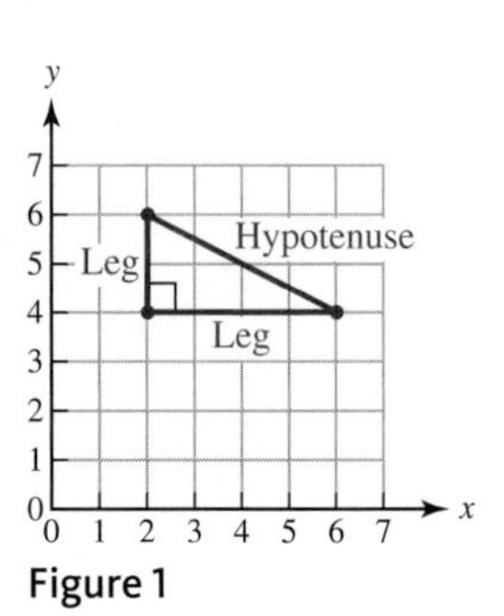

Figure 1

EXAMPLE 8 The vertices of the right triangle are $(2, 4)$, $(4, 6)$, and $(5, 1)$.

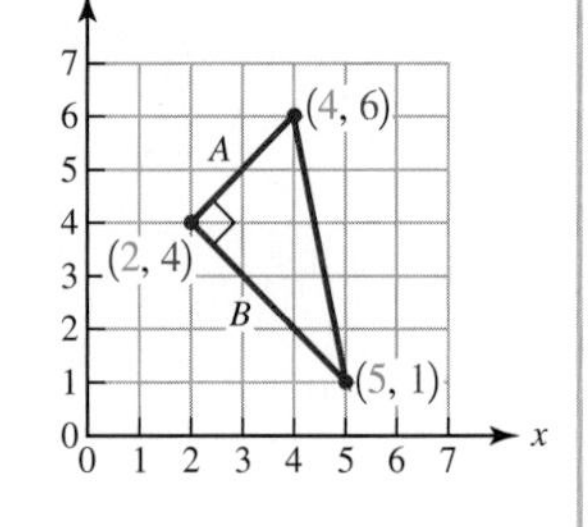

(a) Find the slope of leg A, m_A.

SOLUTION

$m_A = \dfrac{y_2 - y_1}{x_2 - x_1}$ The slope formula.

$m_A = \dfrac{6 - 4}{4 - 2}$ $(x_1, y_1) = (2, 4); (x_2, y_2) = (4, 6)$

$m_A = \mathbf{\dfrac{2}{2}}$ Simplify.

$m_A = \mathbf{1}$ The slope of leg A.

(b) Find the slope of leg B, m_B.

$m_B = \dfrac{y_2 - y_1}{x_2 - x_1}$ The slope formula.

$m_B = \dfrac{1 - 4}{5 - 2}$ $(x_1, y_1) = (2, 4); (x_2, y_2) = (5, 1)$

$m_B = \mathbf{\dfrac{-3}{3}}$ Simplify.

$m_B = \mathbf{-1}$ The slope of leg B.

(c) Find the product of the slopes of leg A and leg B.

$m_A \cdot m_B$

$= (1)(-1)$ Replace m_A and m_B.

$= -1$ Simplify.

Just as the product of the slopes of the perpendicular legs of the right triangle is -1, for all perpendicular lines with defined slopes, the product of the slopes equals -1.

EXAMPLE 9 The graph represents two linear equations.

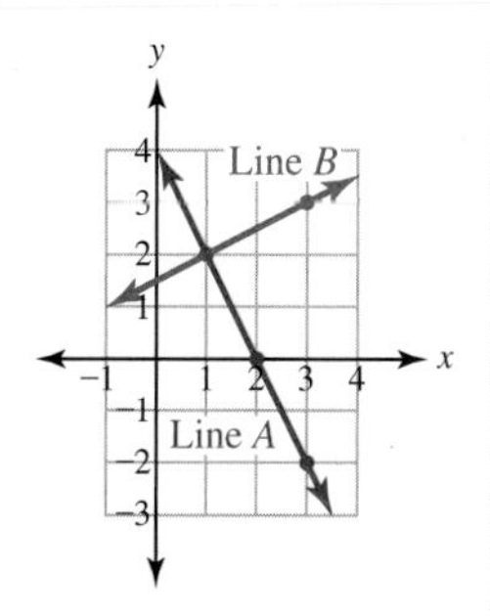

(a) Find the slope of line A.

SOLUTION ▶ Identify two points on line A: $(2, 0)$ and $(3, -2)$.

$m_A = \dfrac{y_2 - y_1}{x_2 - x_1}$ The slope formula.

$m_A = \dfrac{-2 - 0}{3 - 2}$ $(x_1, y_1) = (2, 0); (x_2, y_2) = (3, -2)$

$m_A = -\dfrac{\mathbf{2}}{\mathbf{1}}$ Simplify.

(b) Find the slope of line B.

▶ Identify two points on line B: $(1, 2)$ and $(3, 3)$.

$m_B = \dfrac{y_2 - y_1}{x_2 - x_1}$ The slope formula.

$m_B = \dfrac{3 - 2}{3 - 1}$ $(x_1, y_1) = (1, 2); (x_2, y_2) = (3, 3)$

$m_B = \dfrac{\mathbf{1}}{\mathbf{2}}$ Simplify.

(c) Find the product of the slopes of line A and line B.

▶ $m_A \cdot m_B$ The product of the slopes of line A and line B.

$= -\dfrac{2}{1} \cdot \dfrac{1}{2}$ $m_A = -\dfrac{2}{1}; m_B = \dfrac{1}{2}$

$= -1$ Simplify.

(d) Are line A and line B perpendicular?

▶ Since the product of their slopes is -1, line A and line B are perpendicular.

In Example 9, the slopes of the perpendicular lines are **opposite reciprocals,** $-\dfrac{2}{1}$ and $\dfrac{1}{2}$. The product of opposite reciprocals is -1. In the next example, the slope of line A is $-\dfrac{5}{6}$. The slope of a perpendicular line is the opposite reciprocal, $\dfrac{6}{5}$.

EXAMPLE 10 The equation of line A is $y = -\dfrac{5}{6}x + 3$. A perpendicular line (line B) passes through $(8, 4)$.

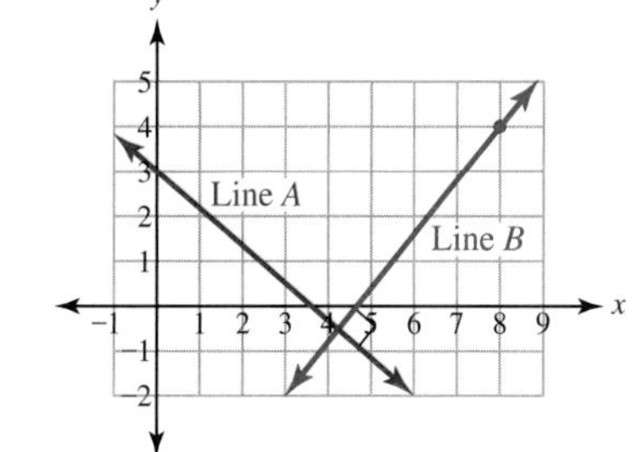

(a) Identify the slope of line A.

SOLUTION ▶ The equation of line A is in slope-intercept form, $y = mx + b$; $m_A = -\dfrac{5}{6}$.

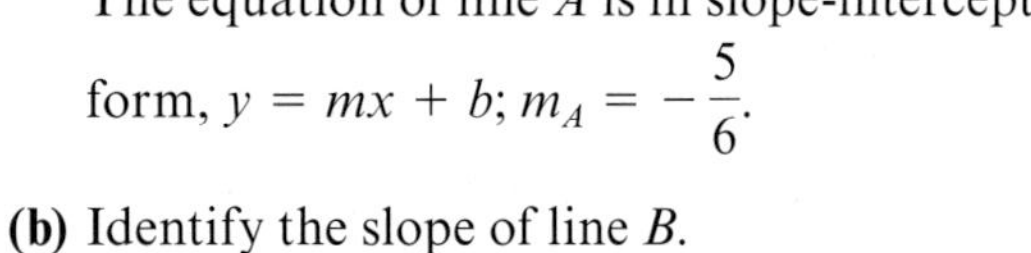

(b) Identify the slope of line B.

▶ Since line B is perpendicular to line A, the slope of line B is the opposite reciprocal of $-\dfrac{5}{6}$; $m_B = \dfrac{6}{5}$.

(c) Write the equation of line B in slope-intercept form.

▶ Replace m, x_1, and y_1 in point-slope form, and rewrite in slope-intercept form.

$y - y_1 = m(x - x_1)$	Point-slope form.
$y - 4 = \frac{6}{5}(x - 8)$	$m = \frac{6}{5}$, $(x_1, y_1) = (8, 4)$
$y - 4 = \mathbf{\frac{6}{5}}x + \mathbf{\frac{6}{5}}\left(-\frac{8}{1}\right)$	Distributive property.
$y - 4 = \frac{6}{5}x - \mathbf{\frac{48}{5}}$	Simplify.
$+4 \qquad\qquad +4$	Addition property of equality.
$y + \mathbf{0} = \frac{6}{5}x - \frac{48}{5} \mathbf{+ 4}$	Simplify; common denominator of the constants is 5.
$\mathbf{y} = \frac{6}{5}x - \frac{48}{5} + \frac{4}{1}\cdot\frac{5}{5}$	Simplify; multiply by a fraction equal to 1.
$y = \frac{6}{5}x - \frac{48}{5} + \mathbf{\frac{20}{5}}$	Simplify.
$y = \frac{6}{5}x - \mathbf{\frac{28}{5}}$	Simplify.

In slope-intercept form, the equation of line B is $y = \frac{6}{5}x - \frac{28}{5}$.

Parallel Lines and Perpendicular Lines

Parallel lines have the same slope and never intersect. When two perpendicular lines intersect, they form a right (90-degree) angle. The product of the defined slopes of two perpendicular lines is -1. If the slopes of two perpendicular lines are defined, then the slopes are opposite reciprocals.

EXAMPLE 11 Line A passes through (2, 3) and (6, 1). Write the equation in slope-intercept form of the perpendicular line (line B) that passes through $(-4, 5)$.

(a) Find the slope of line A.

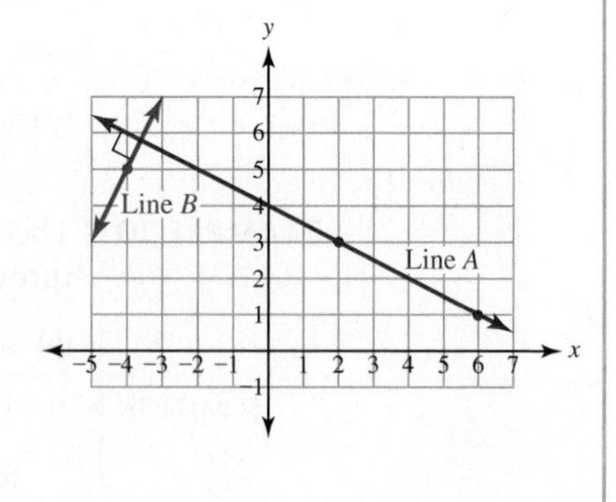

SOLUTION ▶

$m_A = \frac{y_2 - y_1}{x_2 - x_1}$	The slope formula.
$m_A = \frac{1 - 3}{6 - 2}$	$(x_1, y_1) = (2, 3)$; $(x_2, y_2) = (6, 1)$
$m_A = \mathbf{\frac{-2}{4}}$	Simplify.
$m_A = \mathbf{\frac{-1\cdot 2}{2\cdot 2}}$	Find common factors.
$m_A = -\mathbf{\frac{1}{2}}$	Simplify.

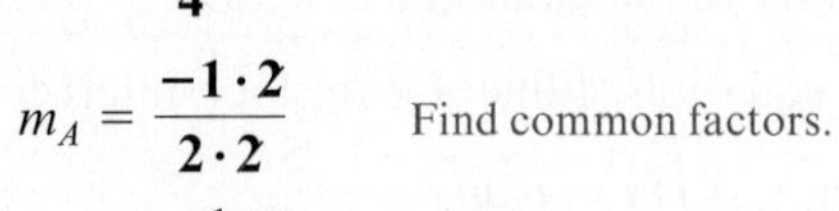

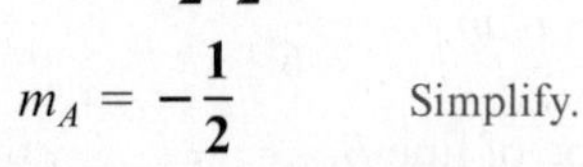

(b) Identify the slope of line B.

▶ Since line A and line B are perpendicular, the slope of line B is the opposite reciprocal of $-\frac{1}{2}$, and $m_B = 2$.

(c) Write the equation of line B in slope-intercept form.

▶ Replace m, x_1, and y_1 in point-slope form, and rewrite in slope-intercept form.

$y - y_1 = m(x - x_1)$	Point-slope form.
$y - 5 = \mathbf{2}(x - (\mathbf{-4}))$	$m = 2$, $(x_1, y_1) = (-4, 5)$
$y - 5 = 2(x + \mathbf{4})$	Simplify.
$y - 5 = \mathbf{2}x + \mathbf{2 \cdot 4}$	Distributive property.
$y - 5 = 2x + \mathbf{8}$	Simplify.
$\quad +5 \qquad +5$	Addition property of equality.
$y + \mathbf{0} = 2x + \mathbf{13}$	Simplify.
$y = 2x + 13$	Simplify.

In slope-intercept form, the equation of line B is $y = 2x + 13$.

Practice Problems

10. Line A is $y = 2x - 9$. Line B is perpendicular to line A and passes through $(-6, 1)$.
  a. Identify the slope of line A.
  b. Identify the slope of line B.
  c. Write the equation of line B in slope-intercept form.

11. Line A passes through $(-2, -7)$ and $(1, 2)$. Line B is perpendicular to line A and passes through $(5, 4)$.
  a. Find the slope of line A.
  b. Identify the slope of line B.
  c. Write the equation of line B in slope-intercept form.

12. The slope of line A is $\frac{5}{6}$. The slope of line B is $\frac{6}{5}$. Are these lines perpendicular? Explain.

Using Technology: Trace, Value, and ZoomFit

On a graphing calculator, use the **Trace** or **Value** command to extrapolate.

EXAMPLE 12 The equation $y = \left(\frac{\$536{,}509}{1 \text{ year}}\right)x + \$8{,}610{,}545$ represents the relationship of the U.S. gross domestic production in millions of dollars, y, and the number of years since 1998, x. Graph this equation. Find the gross domestic product in 2010. Round to the nearest tenth of a trillion.

The number of years since 1998 x, is 2010 − 1998, which equals 12 years. Since the y-intercept is (0 years, \$8,610,545), a standard window is not a good choice. To find a good window, use **ZoomFit**. Type the equation on the Y= screen. Press ZOOM. Scroll down to **ZoomFit**. Press ENTER. The graph appears.

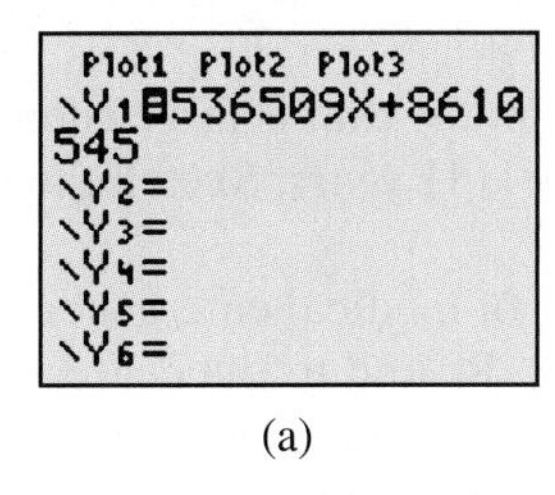

(a)

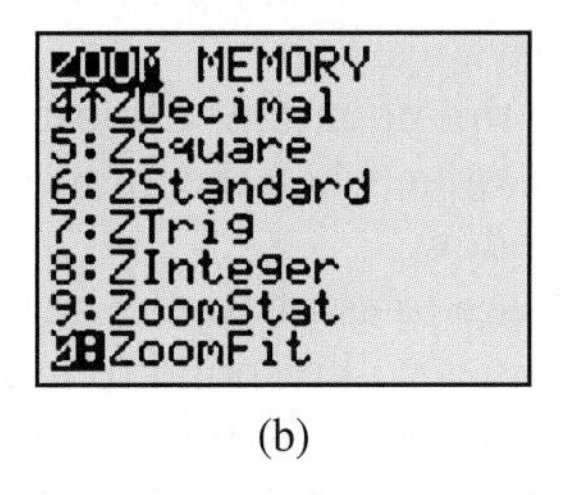

(b)

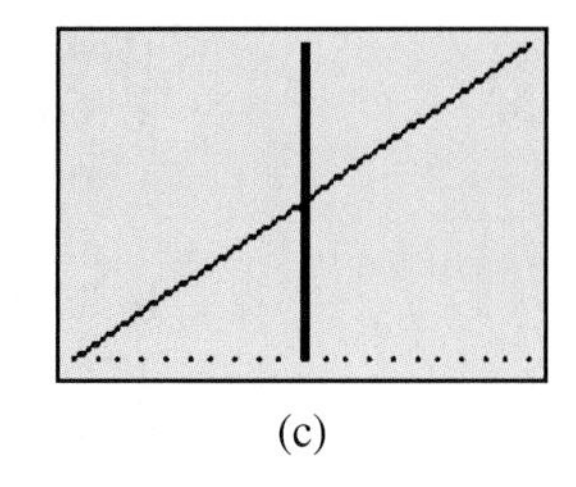
(c)

Using **ZoomFit** often results in a window that shows the intercept(s). We can then change the window settings to fit the problem situation. In this example, since only x-values or y-values greater than 0 make sense, change X_{min} and Y_{min} to 0. Since the goal is to find y when $x = 12$, X_{max} should be greater than 12. Choose 15. Because Y_{scl} is 1, there are many tick marks on the graph, and the y-axis in the graph is thick. Change Y_{scl} to 1,000,000 to reduce the number of tick marks. Press GRAPH. Although this window is better, the point (12, ?) is still not visible.

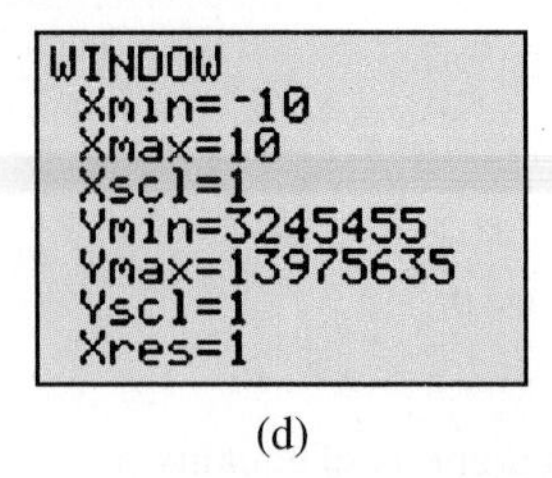

(d)

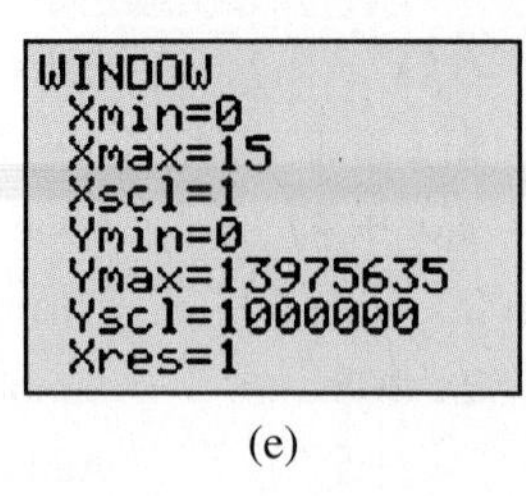

(e)

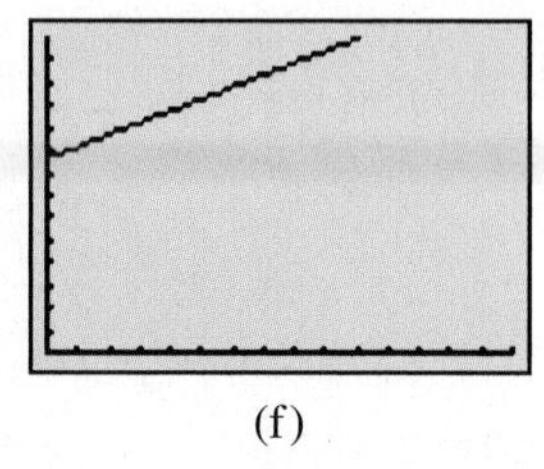
(f)

Increase Y_{max} to 20,000,000. Press GRAPH. The point (12, ?) is now visible. This is a good window for finding y when $x = 12$.

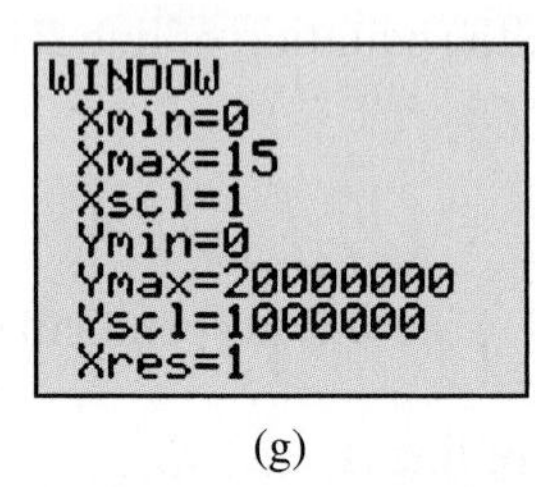

(g)

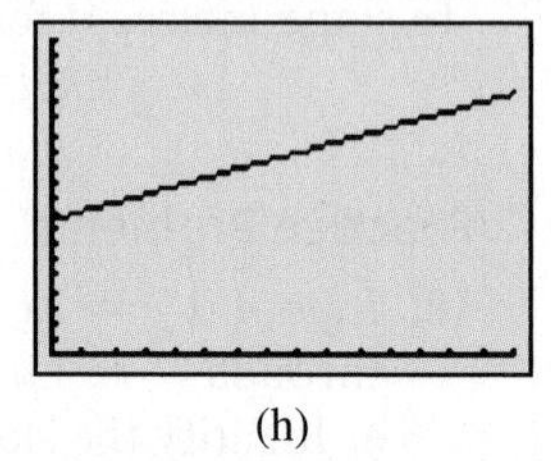
(h)

Press TRACE. Use the right arrow key to move the cursor. Since the calculator moves one pixel on the screen at a time, the cursor cannot fall exactly on $x = 12$ and the value of y is an estimate. However, this value may be precise enough to use for a purpose such as budget planning. So an estimate of the U.S. gross domestic product for 2010 is $15,117,144 million, or about $15.1 trillion.

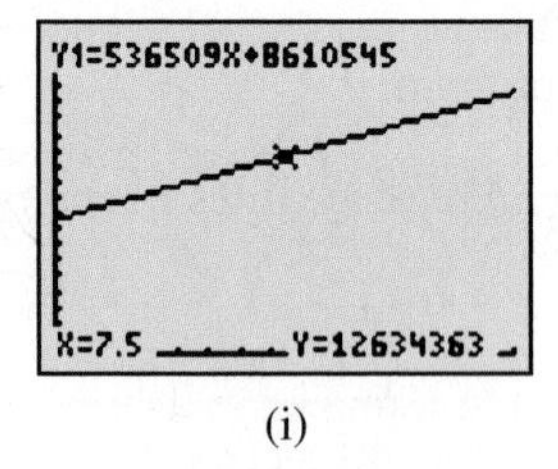

(i)

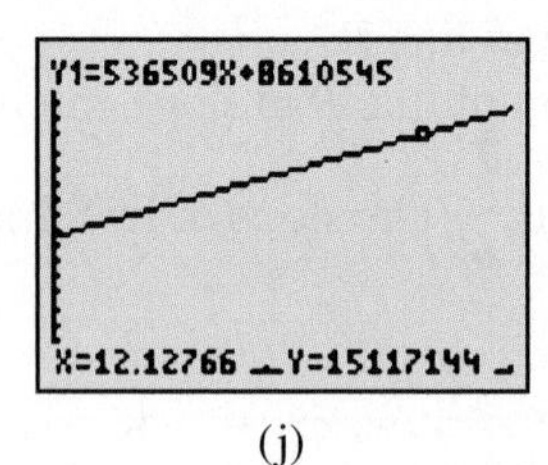

(j)

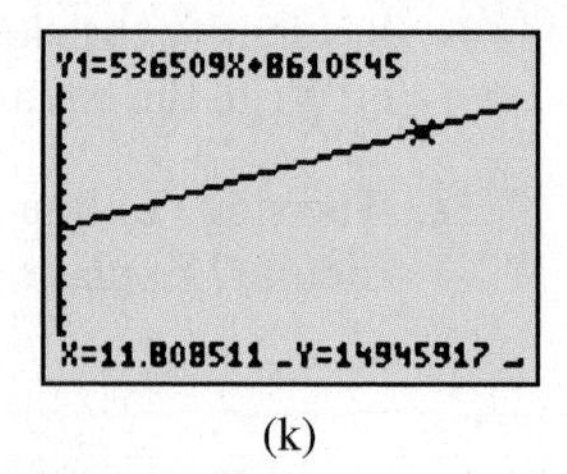

(k)

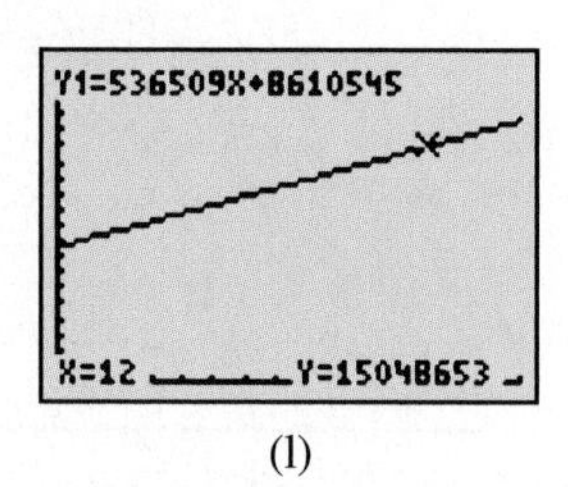

(l)

To use the **Value** command in the same window, press CALC. Choose 1: value. Type 12. Press ENTER. The extrapolated value is $15,048,653, or about $15.0 trillion.

This equation is based on the GDP before the recession of 2009 and 2010. Because the rate of growth was affected by the recession, the equation might not be a good predictor.

Practice Problems 13. In the Bronx, fire engine companies respond to selected life-threatening emergencies as first responders. The equation $y = \left(\frac{2426 \text{ medical emergencies}}{1 \text{ year}}\right)x + 27{,}556$ medical emergencies represents the relationship of the number of medical emergencies, y, and the number of years since 2001, x.
(*Source:* www.nyc.gov)

a. Graph this equation. Choose a window that is large enough to find the output value when the input value is 11 years. Sketch the graph; describe the window.

b. Use **Trace** to estimate the number of medical emergencies that fire engine companies will respond to in 2012. Round to the nearest ten.

c. Use **Value** to estimate the number of medical emergencies that fire engine companies will respond to in 2012. Round to the nearest ten.

3.5 VOCABULARY PRACTICE

Match the term with its description.

1. Two numbers that are the same distance from 0 but on opposite sides of zero on the number line
2. Two lines that have the same slope
3. The result of multiplication
4. $ax + by = c$
5. $y - y_1 = m(x - x_1)$
6. $y = mx + b$
7. Two lines that intersect at a 90-degree angle
8. The straight line that best represents the relationship of two variables
9. The product of a nonzero real number and this equals 1.
10. The measure of this angle is 90 degrees.

A. line of best fit
B. opposites
C. parallel lines
D. perpendicular lines
E. point-slope form of a linear equation
F. product
G. reciprocal
H. right angle
I. slope-intercept form of a linear equation
J. standard form of a linear equation

3.5 Exercises

Follow your instructor's guidelines for showing your work.

For exercises 1–20, a line with the given slope passes through the given point. Write the equation of the line in slope-intercept form.

1. slope $= 8$; $(2, -9)$
2. slope $= 6$; $(3, -5)$
3. slope $= -4$; $(3, 1)$
4. slope $= -5$; $(4, 1)$
5. slope $= -2$; $(-9, -5)$
6. slope $= -3$; $(-8, -6)$
7. slope $= \frac{1}{4}$; $(8, 12)$
8. slope $= \frac{1}{4}$; $(12, 8)$
9. slope $= -\frac{2}{3}$; $(3, -6)$
10. slope $= -\frac{2}{3}$; $(6, -9)$
11. slope $= \frac{2}{3}$; $(6, 8)$
12. slope $= \frac{1}{2}$; $(5, 10)$
13. slope $= -\frac{3}{4}$; $(5, 1)$
14. slope $= -\frac{2}{3}$; $(2, 1)$
15. slope $= 8$; $\left(\frac{1}{2}, 15\right)$
16. slope $= 6$; $\left(\frac{1}{2}, 12\right)$
17. slope $= 0$; $(6, -3)$
18. slope $= 0$; $(8, -1)$
19. slope $= 0.2$; $(1, 6.4)$
20. slope $= 0.4$; $(2, 1.6)$

For exercises 21–40, a line passes through the given points.
(a) Find the slope of the line.
(b) Write the equation of the line in slope-intercept form.

21. $(6, 1)$; $(8, 11)$
22. $(5, 9)$; $(7, 17)$
23. $(9, 12)$; $(3, 36)$
24. $(10, 15)$; $(6, 27)$
25. $(1, -3)$; $(-2, -18)$
26. $(-2, -5)$; $(3, 20)$
27. $(-3, 15)$; $(5, 21)$
28. $(-6, 30)$; $(-14, 24)$
29. $(5, -9)$; $(16, -9)$
30. $(3, -11)$; $(20, -11)$
31. $(3, 3)$; $(7, 7)$
32. $(4, 4)$; $(9, 9)$
33. $(3, -3)$; $(7, -7)$
34. $(4, -4)$; $(9, -9)$
35. $\left(\frac{1}{2}, \frac{3}{5}\right)$; $\left(\frac{3}{2}, \frac{8}{5}\right)$
36. $\left(\frac{1}{2}, \frac{3}{8}\right)$; $\left(\frac{3}{2}, \frac{11}{8}\right)$
37. $\left(\frac{1}{6}, \frac{2}{9}\right)$; $\left(\frac{5}{6}, \frac{5}{9}\right)$

38. $\left(\frac{1}{4}, \frac{1}{8}\right); \left(\frac{3}{4}, \frac{7}{8}\right)$

39. $\left(4, \frac{3}{8}\right); \left(6, \frac{9}{10}\right)$

40. $\left(3, \frac{5}{12}\right); \left(5, \frac{9}{10}\right)$

For exercises 41–44,
(a) Identify the two points on the line.
(b) Identify the slope of the line.
(c) Write the equation of the line in slope-intercept form.

41.

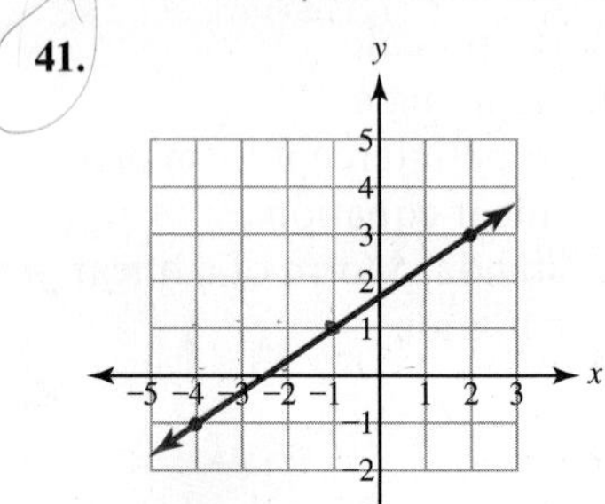

42.

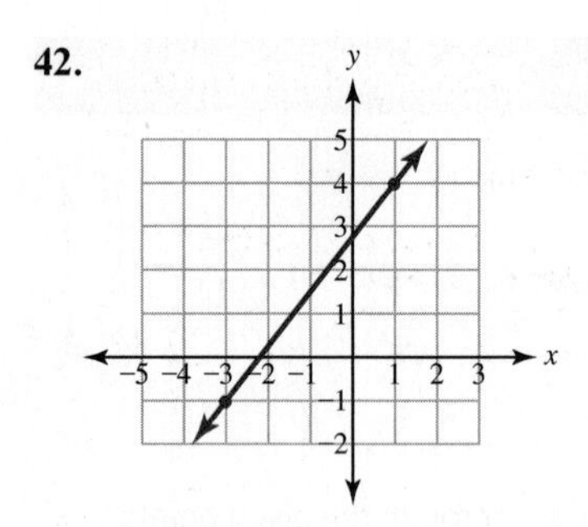

43.

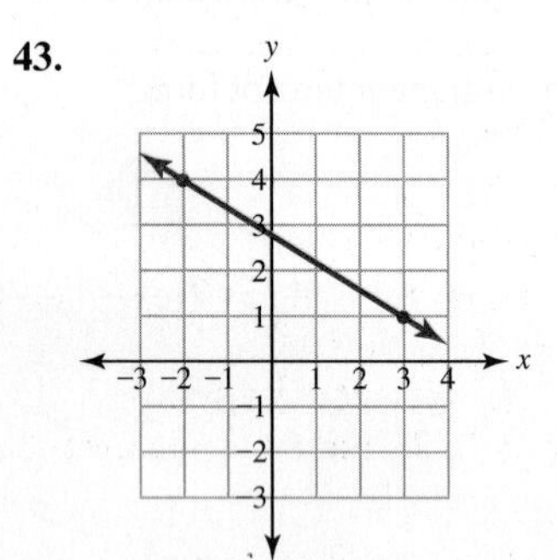

44.

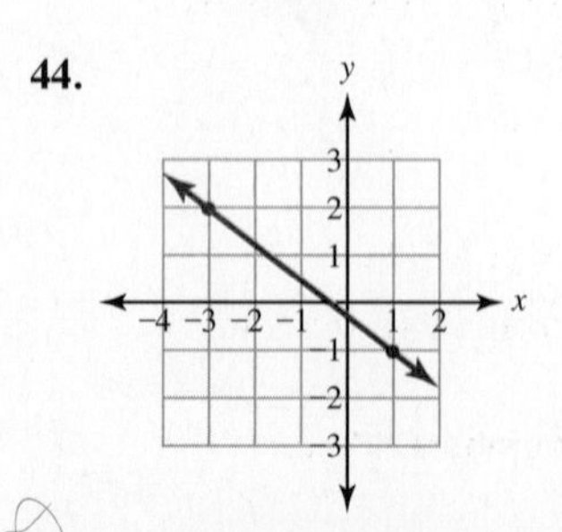

45. The graph represents the relationship of the population of Hispanics in the United States, y, and the number of years since 1970, x.
a. Write the equation of the line of best fit.
b. Describe what the slope represents.
c. Describe what the y-coordinate of the y-intercept represents.
d. Use the equation to find the population of Hispanics in the United States in 2020. Round to the nearest tenth of a million.

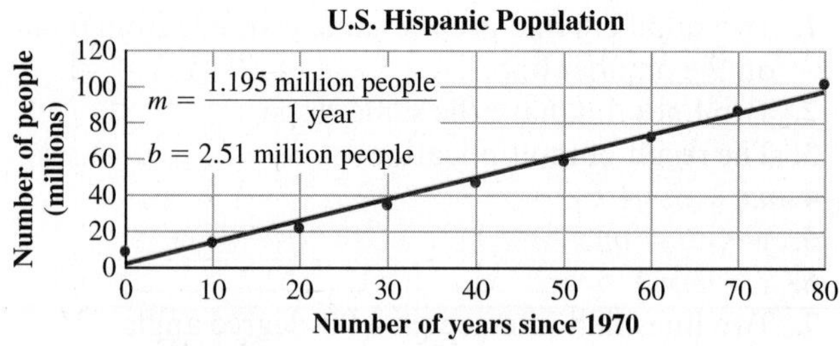

Source: www.census.gov

46. The graph represents the relationship of the number of health care employees, y, and the number of years since 2000, x.

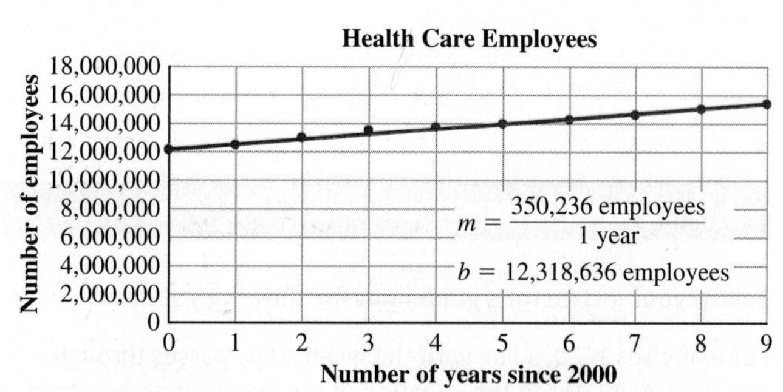

Source: www.cdc.gov

a. Write the equation of the line of best fit.
b. Describe what the slope represents.
c. Describe what the y-coordinate of the y-intercept represents.
d. Use the equation to find the number of health care employees in 2016. Round to the nearest thousand.

47. The graph represents the relationship of the number of recreational boat registrations in Delaware, y, and the number of years since 1996, x.

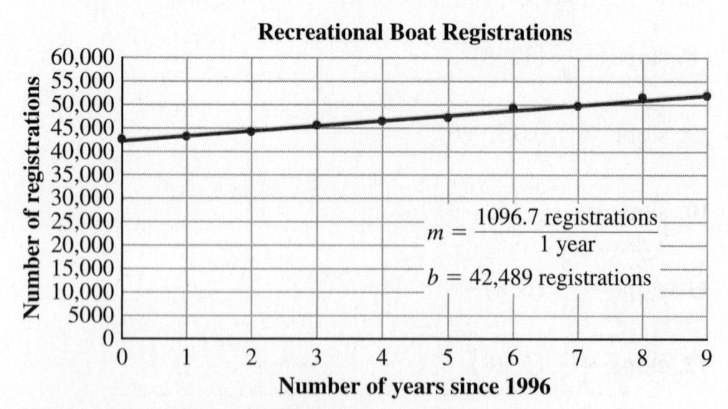

Source: www.nmma.org, 2006

a. Write the equation of the line of best fit.
b. Describe what the slope represents.
c. Describe what the y-coordinate of the y-intercept represents.
d. Use the equation to find the number of recreational boat registrations in 2013. Round to the nearest hundred.

48. The graph represents the relationship of the population of Puerto Rico, y, and the number of years since 1950, x.

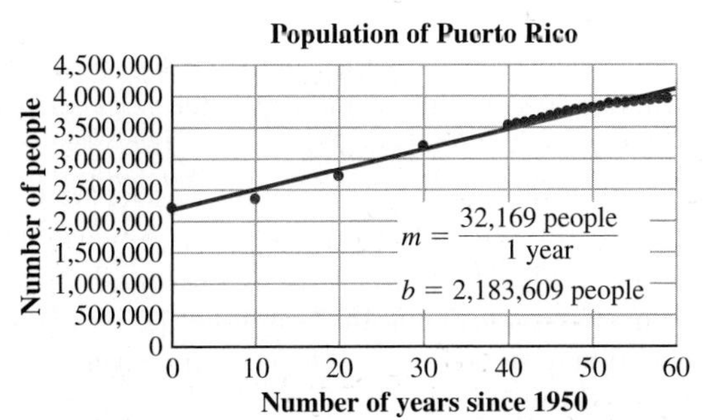

Source: www.census.gov

a. Write the equation of the line of best fit.
b. Describe what the slope represents.
c. Describe what the y-coordinate of the y-intercept represents.
d. Use the equation to find the population of Puerto Rico in 2015. Round to the nearest hundred.

49. The graph represents the relationship of the average cinema ticket price, y, and the number of years since 2001, x.

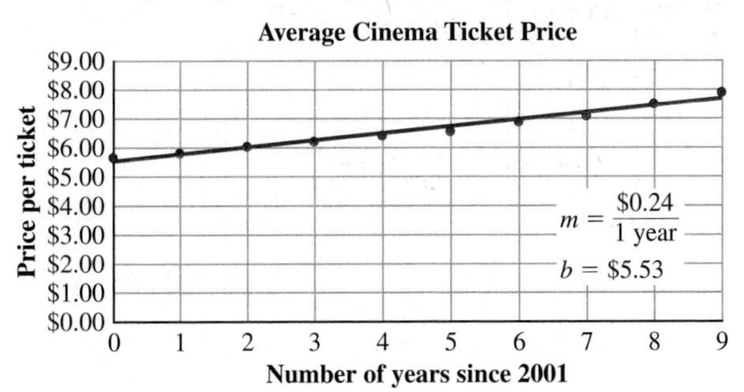

Source: www.mpaa.org, 2010

a. Write the equation of the line of best fit.
b. Describe what the slope represents.
c. Describe what the y-coordinate of the y-intercept represents.
d. Use the equation to find the average cinema ticket price in 2014. Round to the nearest hundredth.

50. The graph represents the relationship of the number of pounds of broiler chicken sold, y, and the number of years since 1992, x.

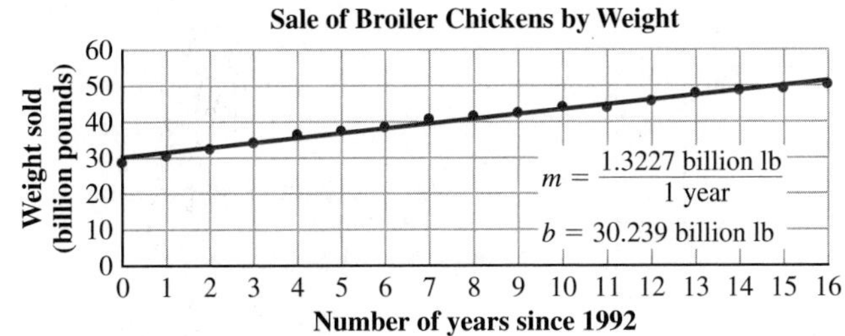

Source: www.census.gov

a. Write the equation of the line of best fit.
b. Describe what the slope represents.
c. Describe what the y-coordinate of the y-intercept represents.
d. Use the equation to find the average number of pounds of broiler chicken sold in 2014. Round to the nearest tenth of a billion.

51. The graph represents the relationship of the number of full-time elementary and secondary teachers in the United States, y, and the number of years since 1983, x.

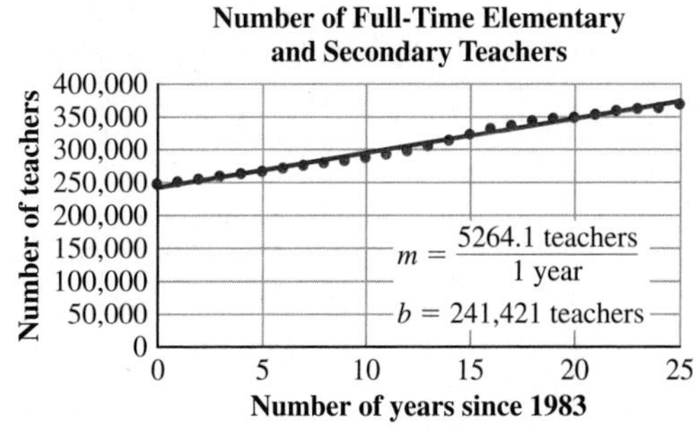

Source: www.census.gov

a. Write the equation of the line of best fit.
b. Describe what the slope represents.
c. Describe what the y-coordinate of the y-intercept represents.
d. Use the equation to find the number of full-time elementary and secondary teachers in 2013. Round to the nearest hundred.

52. The graph represents the relationship of the percent of paper and paperboard recovered from municipal waste, y, and the number of years since 2004, x.

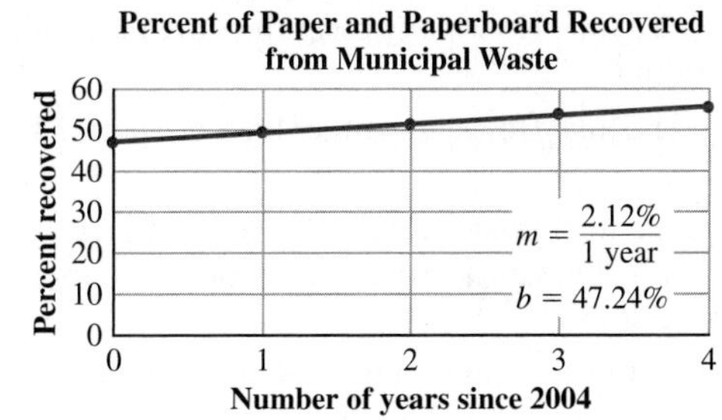

Source: www.census.gov

a. Write the equation of the line of best fit.
b. Describe what the slope represents.
c. Describe what the y-intercept represents.
d. Use the equation to find the percent of paper and paperboard recovered in 2013. Round to the nearest percent.

For exercises 53–56, identify the slope of a line that is:

53. parallel to the line $y = 8x - 11$.

54. parallel to the line $y = 3x - 14$.

55. perpendicular to the line $y = 8x - 11$.

56. perpendicular to the line $y = 3x - 14$.

57. Identify the product of two reciprocals that are opposites.

58. Identify the product of two numbers that are reciprocals.

For exercises 59–62, the equation of line *A* is given. Write the equation in slope-intercept form of the line (line *B*) that is parallel to line *A* and that passes through the given point.

59. $y = 9x + 2$; $(5, -13)$

60. $y = 5x - 16$; $(7, -15)$

61. $y = \frac{3}{4}x + 8$; $(-2, -1)$

62. $y = \frac{3}{8}x + 2$; $(-6, -1)$

For exercises 63–66, the equation of line *A* is given. Write the equation in slope-intercept form of the line (line *B*) that is perpendicular to line *A* and that passes through the given point.

63. $y = \frac{5}{8}x + 10$; $(10, -11)$

64. $y = \frac{3}{4}x + 2$; $(6, -15)$

65. $y = -\frac{2}{3}x + \frac{1}{2}$; $(-4, 6)$

66. $y = -\frac{6}{5}x + \frac{3}{8}$; $(-8, 12)$

67. Point-slope form is $y - y_1 = m(x - x_1)$. We can change the subscripts and rewrite this form as $y_2 - y_1 = m(x_2 - x_1)$. Use the properties of equality to rewrite $y_2 - y_1 = m(x_2 - x_1)$ as the slope definition, $m = \frac{y_2 - y_1}{x_2 - x_1}$.

68. A line has a slope of m and passes through the point $(0, b)$.
   a. Write the equation of this line in point-slope form.
   b. Rewrite this equation in slope-intercept form.

69. Imagine a line. How many other lines are perpendicular to this line?

70. Imagine a line. How many other lines are parallel to this line?

71. Imagine a line. Imagine a point that is not on this line. How many lines pass through this point that are parallel to the first line?

72. Imagine a line. Imagine a point that is not on this line. How many lines pass through this point that are perpendicular to the first line?

73. Line *A* passes through $(1, 8)$ and $(3, 18)$.
   a. Find the slope of line *A*.
   b. Line *B* is perpendicular to line *A*. Identify the slope of line *B*.
   c. Line *B* passes through the point $(4, -6)$. Write the equation of line *B* in slope-intercept form.

74. Line *A* passes through $(4, 10)$ and $(6, 14)$.
   a. Find the slope of line *A*.
   b. Line *B* is perpendicular to line *A*. Identify the slope of line *B*.
   c. Line *B* passes through the point $(-1, 5)$. Write the equation of line *B* in slope-intercept form.

75. Line *A* passes through $(1, 8)$ and $(3, 18)$.
   a. Find the slope of line *A*.
   b. Line *B* is parallel to line *A*. Identify the slope of line *B*.
   c. Line *B* passes through the point $(4, -6)$. Write the equation of line *B* in slope-intercept form.

76. Line *A* passes through $(3, 10)$ and $(6, 19)$.
   a. Find the slope of line *A*.
   b. Line *B* is parallel to line *A*. Identify the slope of line *B*.
   c. Line *B* passes through the point $(-1, 5)$. Write the equation of line *B* in slope-intercept form.

77. A line passes through $(3, 10)$ and $(6, 14)$. Write the equation in slope-intercept form of the *perpendicular* line that passes through $(-1, 5)$.

78. A line passes through $(14, 9)$ and $(17, 11)$. Write the equation in slope-intercept form of the *perpendicular* line that passes through $(-3, 2)$.

79. A line passes through $(24, 15)$ and $(21, 17)$. Write the equation in slope-intercept form of the *parallel* line that passes through $(-4, 3)$.

80. A line passes through $(38, 7)$ and $(44, 2)$. Write the equation in slope-intercept form of the *parallel* line that passes through $(-10, 9)$.

For exercises 81–88,

(a) write the equation of the graphed line in slope-intercept form.

(b) write the equation of the line that is parallel to this line and passes through the graphed point.

81.

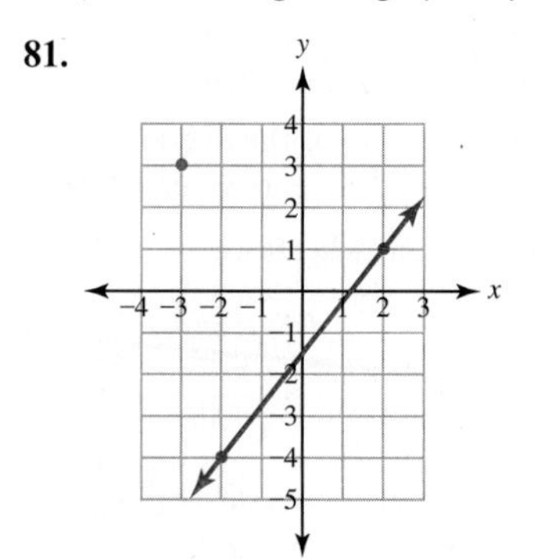

82.

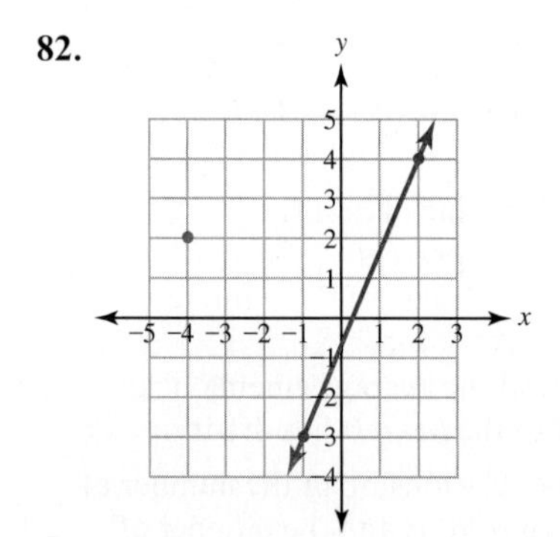

83.

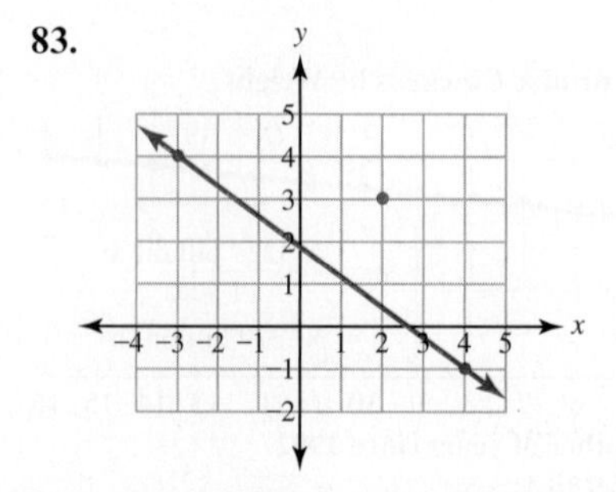

84. 

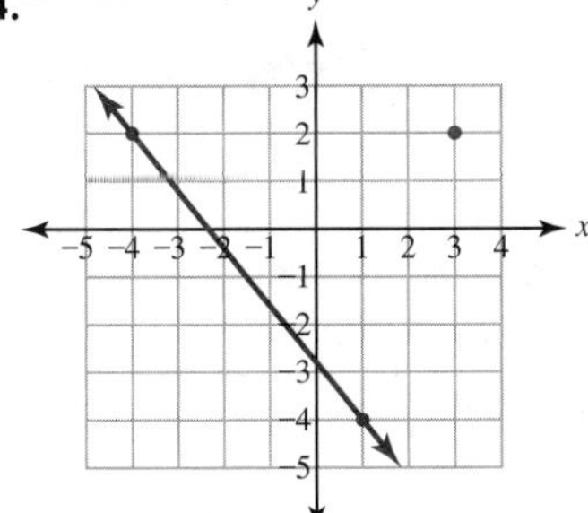

85.

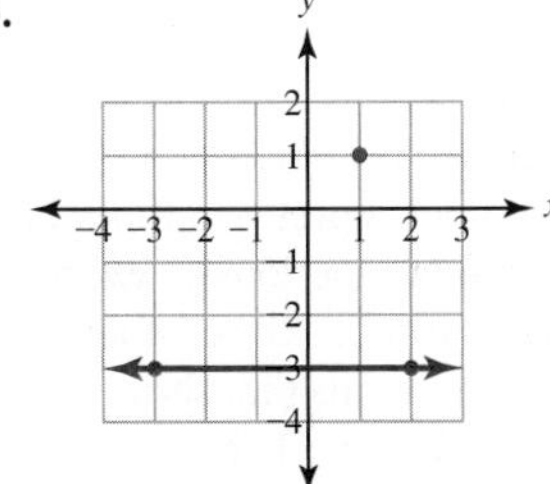

86.

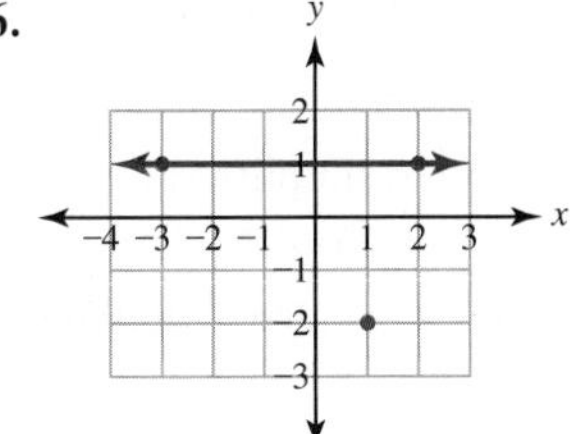

87.

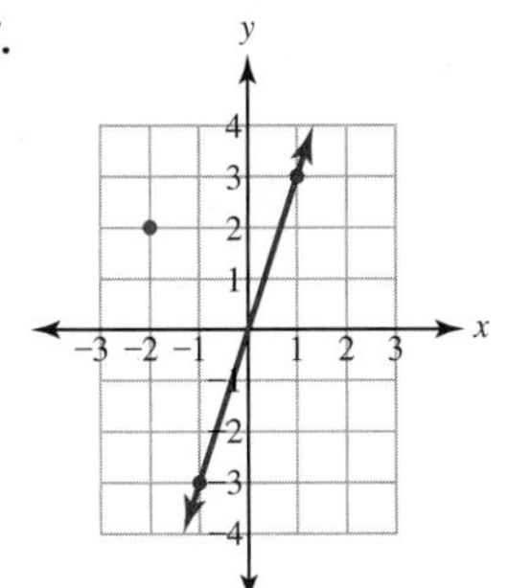

88.

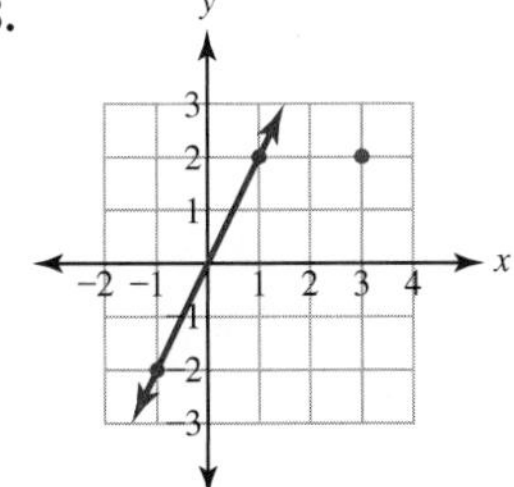

Problem Solving: Practice and Review

Follow your instructor's guidelines for using the five steps as outlined in Section 1.3, p. 55.

89. The average weight of a tomato on a garden plant is 5 oz. Over the season, a gardener harvests 45 tomatoes. Predict the number of pounds of tomatoes harvested. (1 lb = 16 oz.) Round to the nearest tenth of a pound.

90. The rectangular white board in a classroom is 16 ft long and 4 ft wide.
a. Find its perimeter.
b. Find its area.

91. The table lists the rates for parking in a downtown parking garage in Chicago. Find the difference in the price per minute to park for 25 min or to park for 3 hr. Round to the nearest hundredth.

Time	Rate (dollars)
First 20 min	6
20 min–40 min	11
40 min–60 min	14
1 hr–1 hr 20 min	17
1 hr 20 min–1 hr 40 min	18
1 hr 40 min–2 hr	20
2 hr–2 hr 20 min	22
2 hr 20 min–2 hr 40 min	24
2 hr 40 min–3 hr	26

92. Supplies are brought to the town of Stehekin on Lake Chelan in Washington State by boat. Each passenger can bring 75 lb of freight without charge. For freight exceeding 75 lb, the charge is \$0.08 per pound with a minimum charge of \$6. Find the price for a passenger to bring four 25-lb boxes of tomatoes and three 25-lb boxes of lettuce to Stehekin. (*Source:* www.ladyofthelake.com)

© Laura Bracken

Technology

For exercises 93–94, use the **Trace** or **Value** command to find the value of *y* in the given year for the model in the given window. Sketch the graph; describe the window.

93. The equation $y = \left(\frac{1900 \text{ people}}{1 \text{ year}}\right)x + 63{,}180 \text{ people}$ represents the relationship of the population of a county, *y*, and number of years since 2009, *x*. Find the population in 2013. Round to the nearest hundred. Window: [0, 20, 2; 0, 200,000, 10,000]

94. The equation $y = \left(\frac{8 \text{ birds}}{1 \text{ year}}\right)x + 81 \text{ birds}$ represents the relationship of the number of species on a birder's lifetime list, *y*, and the number of years since 2010, *x*. Find the number of birds on the list in 2014. Round to the nearest whole number. Window: [0, 10, 1; 0, 250, 10]

Find the Mistake

For exercises 95–98, the completed problem has one mistake.

(a) Describe the mistake in words, or copy down the whole problem and highlight or circle the mistake.
(b) Do the problem correctly.

95. Problem: Write the equation in slope-intercept form of a line with a slope of 6 that passes through (1, 3).

Incorrect Answer: $y = mx + b$

$y = 6x + 3$

96. Problem: Write the equation of the line in slope-intercept form that passes through (2, 9) and (1, 6).

Incorrect Answer:

$$m = \frac{y_2 - y_1}{x_2 - x_1} \qquad y - y_1 = m(x - x_1)$$

$$m = \frac{6 - 9}{1 - 2} \qquad y - 9 = 3(x - 1)$$

$$y - 9 = 3x - 3$$

$$+9 \quad +9$$

$$m = \frac{-3}{-1} \qquad y = 3x + 6$$

$$m = 3$$

97. Problem: The equation of a line is $y = \frac{2}{3}x + 8$. Write the equation in slope-intercept form of the perpendicular line that passes through (10, 6).

Incorrrect Answer: $y - y_1 = m(x - x_1)$

$$y - 6 = \frac{3}{2}(x - 10)$$

$$y - 6 = \frac{3}{2}x + \frac{3}{2}(-10)$$

$$y - 6 = \frac{3}{2}x - \frac{3 \cdot 2 \cdot 5}{2}$$

$$y - 6 = \frac{3}{2}x - 15$$

$$+6 \qquad +6$$

$$y = \frac{3}{2}x - 9$$

98. Problem: The equation of a line is $y = 5x - 8$. Write the equation in slope-intercept form of the parallel line that passes through (−6, 3).

Incorrect Answer: $y - y_1 = m(x - x_1)$

$$y - 3 = 5(x - 6)$$

$$y - 3 = 5x + 5(-6)$$

$$y - 3 = 5x - 30$$

$$+3 \qquad +3$$

$$y = 5x - 27$$

Review

For exercises 99–100, use the set {(5, 2), (3, 4), (9, 8), (15, 1)}.

99. Write a set in roster notation whose elements are the first coordinate of each ordered pair.

100. Write a set in roster notation whose elements are the second coordinate of each ordered pair.

101. An equation is $y = \frac{3}{4}x - 9$. Find y when $x = 8$.

102. An equation is $y = -\frac{3}{4}x - 9$. Find y when $x = 8$.

SUCCESS IN COLLEGE MATHEMATICS

103. Learning preferences can include how you prefer to organize information. Some students prefer to think in steps, similar to the step-by-step instructions given in procedure boxes found in this textbook. Other students prefer to draw pictures or concept maps that show the relationship between different concepts. The steps below use mathematical vocabulary. Rewrite these steps in your own words, as you would explain them to another student.

Solving an Inequality in One Variable

1. Use the distributive property to simplify the inequality. Combine any like terms.
2. Use the addition or subtraction property of inequality so that there is only one term with a variable in the inequality.
3. Use the addition or subtraction property of inequality to isolate the term that includes the variable.
4. Use the multiplication or division property of inequality to change the coefficient of the variable term to 1.
5. Check whether the inequality sign is correct.

104. A concept web or concept map such as the one shown below is another way to organize information. Do you prefer to write a list of steps such as the steps in exercise 103 or draw a concept map? Explain.

Figure for exercise 104

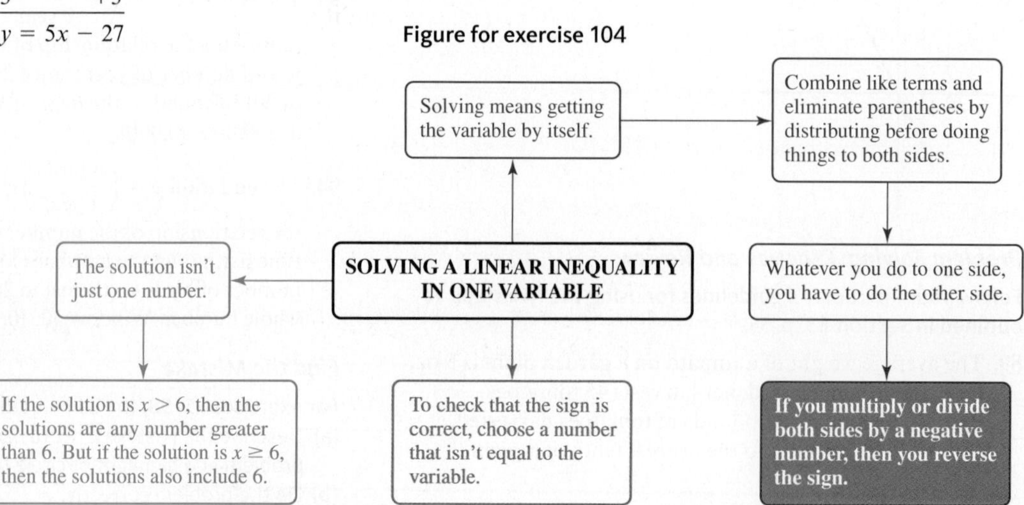

The FAA requires that spectators at an air show must be kept a horizontal distance of at least 500 ft away from planes in normal flight. We can represent this distance with a linear constraint. In this section, we will write and graph linear inequalities and constraints. (*Source:* www.faa.gov)

SECTION 3.6 Linear Inequalities in Two Variables

After reading the text, working the practice problems, and completing assigned exercises, you should be able to:

1. Graph a linear inequality.
2. Write the linear inequality represented by a graph.
3. Write and graph a linear constraint.

Graphing a Linear Inequality

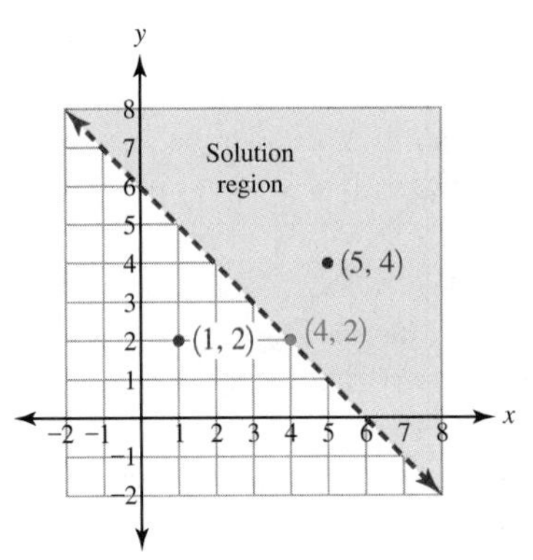

Figure 1

The symbols used to write inequalities include $\geq$ (greater than or equal to), $\leq$ (less than or equal to), $>$ (greater than), and $<$ (less than). Inequalities that include $>$ or $<$ are **strict inequalities**. The **linear inequality** $x + y > 6$ is a symbolic representation of "the sum of two numbers is *greater than* 6." This is a strict inequality; the sum cannot be equal to 6.

On the graph of $x + y > 6$ (Figure 1), the points in the shaded **solution region** represent the ordered pairs that are solutions of $x + y > 6$. For example, the point $(5, 4)$ is in the solution region; the inequality $5 + 4 > 6$ is true. Points outside the solution region do not represent solutions of the inequality. For example, $(1, 2)$ is not in the solution region; the inequality $1 + 2 > 6$ is false. The **dashed boundary line** on the graph represents the equation $x + y = 6$. The line is dashed because the points on the line do not represent solutions of $x + y > 6$. For example, $(4, 2)$ is a point on the dashed line; the inequality $4 + 2 > 6$ is false.

An inequality that includes $\geq$ or $\leq$ is **not strict**. The boundary line on the graph of an inequality that is not strict is solid, and the points on the boundary line represent solutions of the inequality.

EXAMPLE 1 Graph: $x + y \leq 7$

SOLUTION ▶ Use a table of ordered pairs to graph the solid boundary line, $x + y = 7$.

x	y
0	7
7	0
3	4

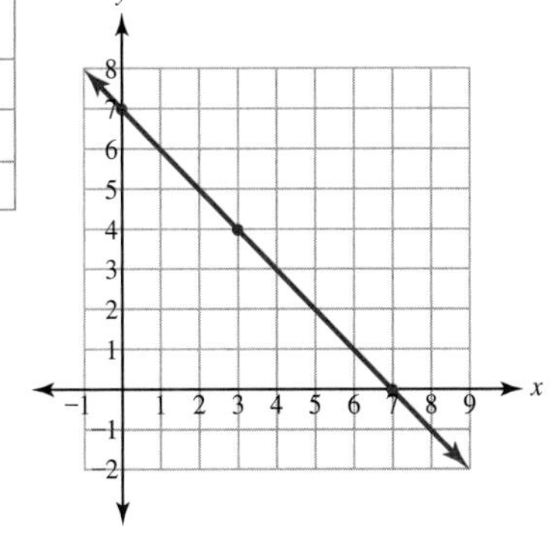

Every point on this line is a solution of the inequality $x + y \leq 7$. Since the inequality is not strict ($\leq$), the boundary line is solid.

▶ Use a test point to identify the solution region.

Choose a **test point** that is not on the boundary line such as $(8, 4)$. Replace the variables in the inequality with the coordinates of the test point. If the inequality is true, the test point and all of the points in this area are solutions of the inequality. If the inequality is false, neither the test point nor any of the points in this area are solutions of the inequality.

$$x + y \leq 7$$

$$\mathbf{8} + \mathbf{4} \leq 7 \quad \text{Replace variables with (8, 4).}$$

$$\mathbf{12} \leq 7 \quad \text{False.} \quad \text{Evaluate.}$$

Since $(8, 4)$ is *not* a solution of the inequality, the solutions are the points in the area on the other side of the boundary line. To show that this is true, test a point on the other side of the boundary line, such as $(0, 0)$.

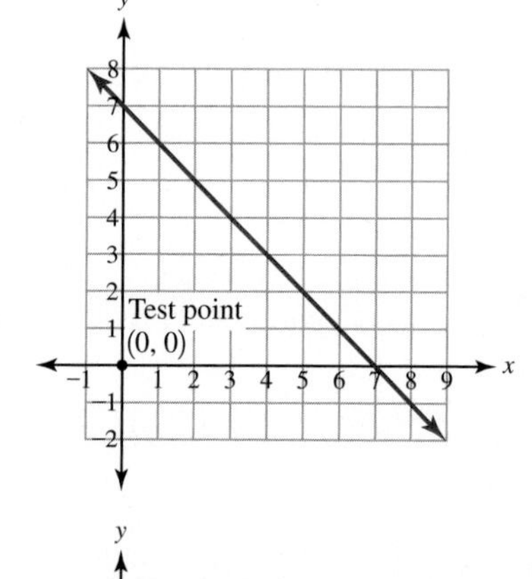

$$x + y \leq 7$$

$$\mathbf{0} + \mathbf{0} \leq 7 \quad \text{Replace the variables with (0, 0).}$$

$$\mathbf{0} \leq 7 \quad \text{True.} \quad \text{Evaluate.}$$

▶ Shade the solution region.

Since $(0, 0)$ is a solution of $x + y < 7$, every point on the side of the boundary line that includes $(0, 0)$ is also a solution. Shade the area of the graph that includes $(0, 0)$.

Ask your instructor whether you should label the test points and/or solution region and how to show shading.

The graph of $x + y \leq 7$ includes all of the points on the solid boundary line and all of the points in the shaded solution region.

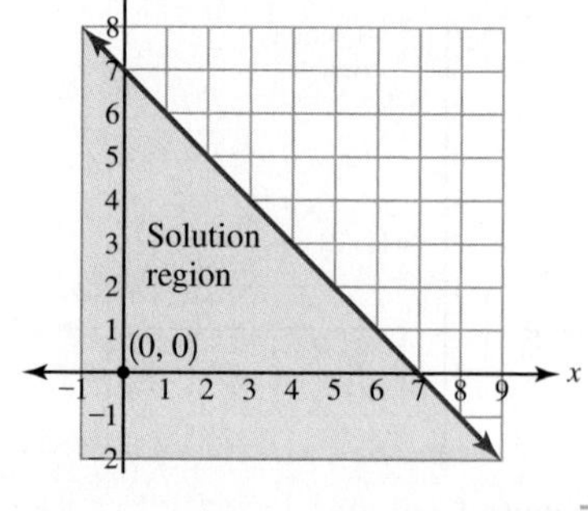

In the next example, since the graph represents a strict inequality, the boundary line is dashed.

EXAMPLE 2 Graph: $y > 2x - 6$

SOLUTION ▶ Use slope-intercept graphing to graph the dashed boundary line, $y = 2x - 6$.

The y-intercept is $(0, -6)$. The slope is $\frac{2}{1}$. Graph the y-intercept; from the y-intercept, move vertically up 2 units (the rise) and then move horizontally to the right 1 unit (the run). This is the location of another point on the line. Draw a line through the points. Since this is a strict inequality (>), the boundary line is dashed.

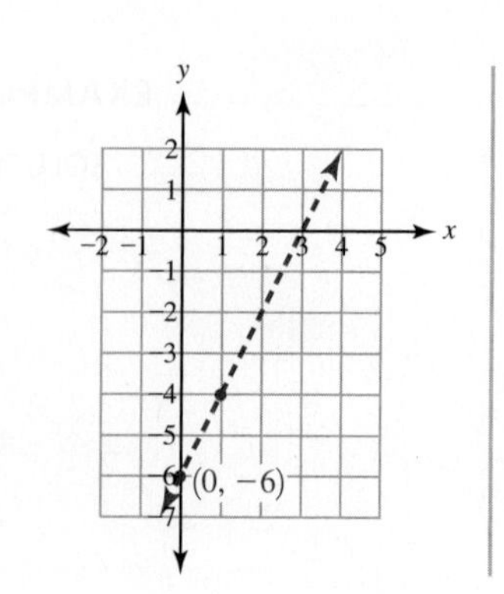

▶ Use a test point to identify the solution region.

Choose a test point that is not on the boundary line, (2, 1).

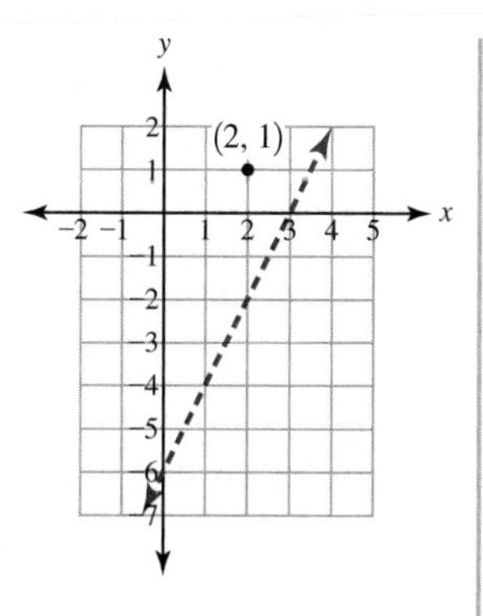

$y > 2x - 6$

$\mathbf{1} > 2(\mathbf{2}) - 6$ Replace variables with (2, 1).

$1 > \mathbf{4} - 6$ Evaluate.

$1 > \mathbf{-2}$ True. Evaluate.

▶ Shade the solution region.

Since (2, 1) is a solution of $y > 2x - 6$, every point on the side of the boundary line that includes (2, 1) is also a solution. Shade the area of the graph that includes (2, 1).

The graph of $y > 2x - 6$ is limited to the points in the shaded solution region. The points on the dashed boundary line are not part of the graph.

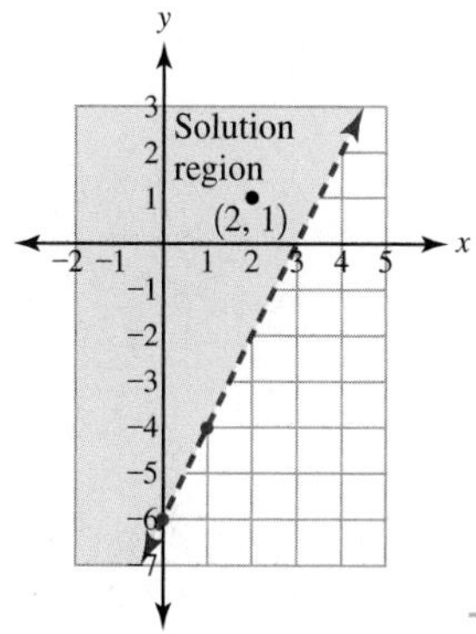

Graphing a Linear Inequality

1. Graph the boundary line. If the inequality includes a $\geq$ or $\leq$, the boundary line will be a solid line. If the inequality includes a $>$ or $<$ (a strict inequality), the boundary line will be dashed.
2. Select a test point that is not on the boundary line. Replace the variables in the inequality with the coordinates of the test point. The resulting inequality will be true or false.
3. If the inequality is true, the test point is in the solution region. Shade the area on this side of the boundary line. If the inequality is false, the test point is *not* in the solution region. Shade the area on the *other* side of the boundary line.

We can also graph inequalities in one variable on an *xy*-coordinate system. Only one of the coordinates of the test point is needed to find the solution region.

EXAMPLE 3 Graph: $x < 3$

SOLUTION ▶ Graph the dashed boundary line, $x = 3$.

The equation $x = 3$ is a vertical line in which every x-coordinate is 3. Since the inequality is strict (<), the boundary line is dashed.

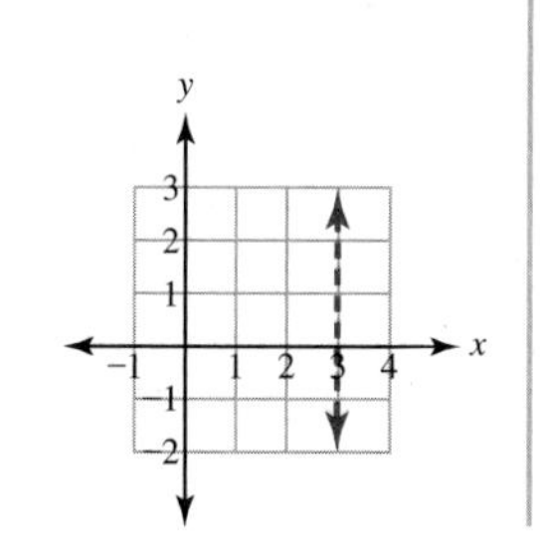

▶ Use a test point to identify the solution region.

Choose a test point that is not on the boundary line, (1, 2). Since the inequality has only one variable, x, use only the x-coordinate of the test point.

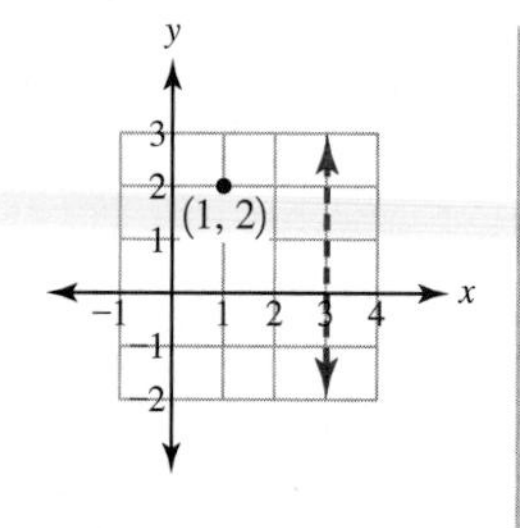

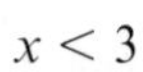

$x < 3$

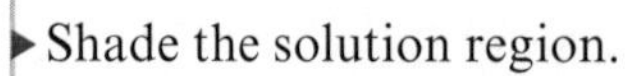

$1 < 3$ True. Replace variable with x-coordinate of (1, 2).

▶ Shade the solution region.

Since (1, 2) is a solution of $x < 3$, every point on the side of the boundary line that includes (1, 2) is also a solution. Shade the area of the graph that includes (1, 2).

The graph of $x < 3$ is limited to the points in the shaded solution region. The points on the dashed boundary line are not part of the graph.

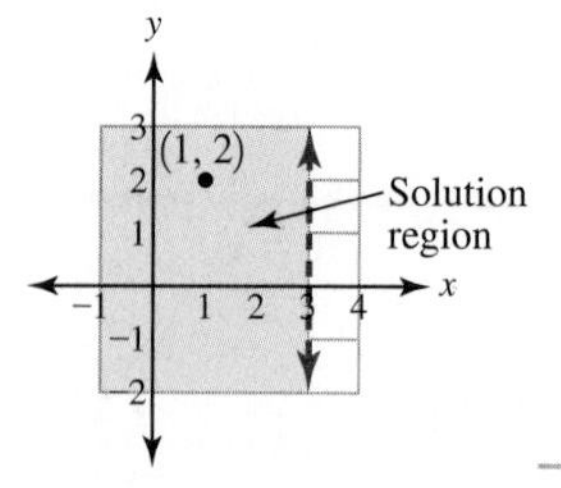

In the next example, the boundary line is a horizontal line.

EXAMPLE 4 Graph: $y \geq -2$

SOLUTION ▶ Graph the solid boundary line, $y = -2$.

The equation $y = -2$ is a horizontal line in which every y-coordinate is -2. Since the inequality is not strict ($\geq$), the boundary line is solid.

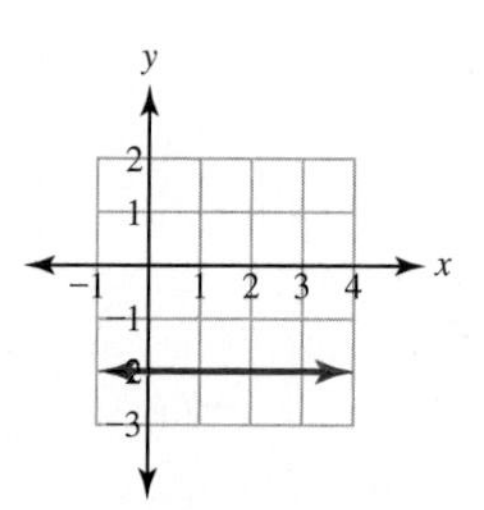

▶ Use a test point to identify the solution region.

Choose a test point that is not on the boundary line, (3, 1). Since the inequality has only one variable, y, use only the y-coordinate of the test point.

$y \geq -2$

$1 \geq -2$ True. Replace variable with y-coordinate of (3, 1).

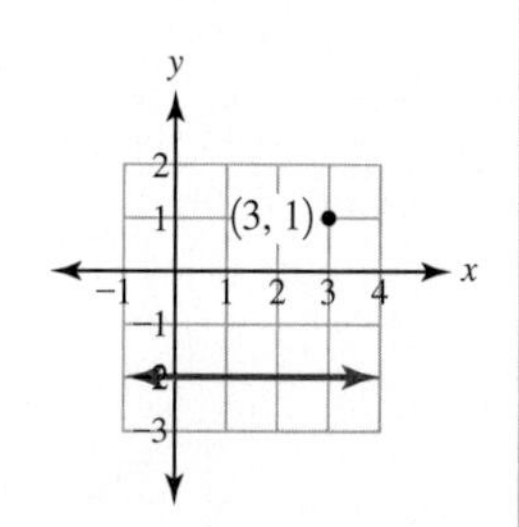

▶ Shade the solution region.

Since (3, 1) is a solution of $y \geq -2$, every point on the side of the boundary line that includes (3, 1) is also a solution. Shade the area of the graph that includes (3, 1).

The graph of $y \geq -2$ includes the points in the shaded solution region and the points on the boundary line.

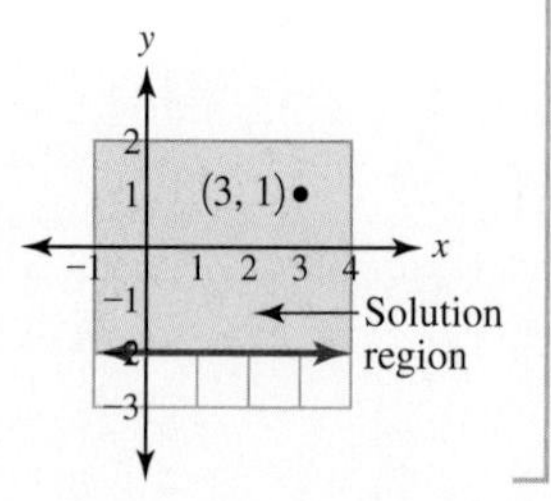

Practice Problems

For problems 1–4, graph the inequality.

1. $6x + 7y < 42$ **2.** $y \geq -3x + 8$ **3.** $5x - 4y < 20$ **4.** $y < 7$

Writing a Linear Inequality

To write the inequality represented by a graph, we need to find the equation of the boundary line and choose the correct inequality sign.

EXAMPLE 5 Write the inequality represented by the graph.

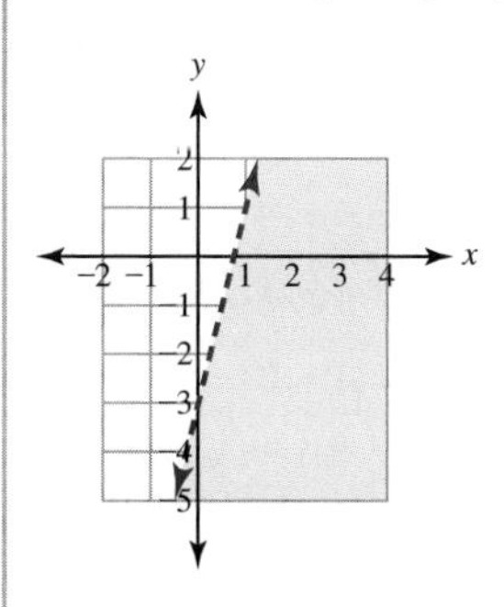

SOLUTION ▶ Write the equation of the boundary line.

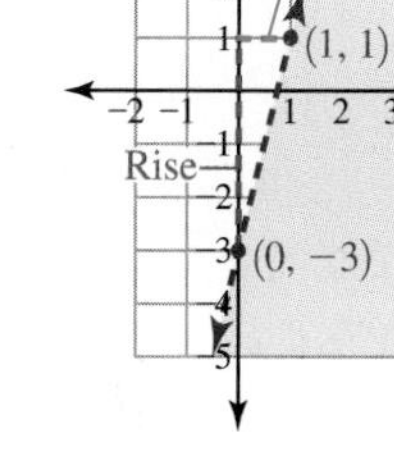

The y-intercept of this boundary line is $(0, -3)$. To find the slope, choose two points on the boundary line, $(0, -3)$ and $(1, 1)$. Moving from left to right, the vertical rise between the points is positive 4, and the horizontal run between the points is positive 1. The slope, $\frac{\text{rise}}{\text{run}}$, is $\frac{4}{1}$.

In slope-intercept form, $y = mx + b$, the boundary line is $y = 4x - 3$.

▶ Choose the inequality sign and its direction.

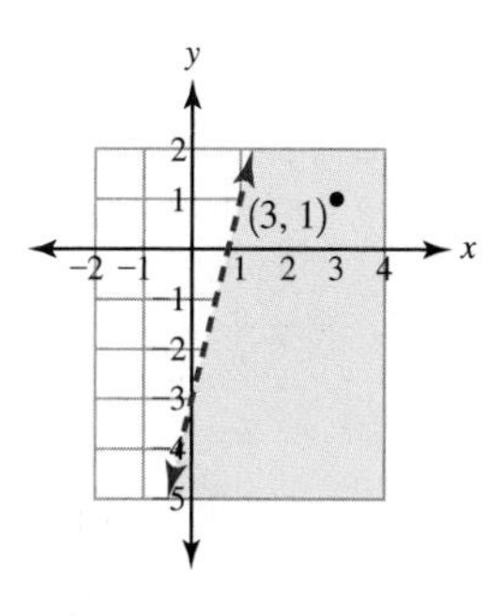

Since the boundary line is dashed, this is a strict inequality, and the sign will be $<$ or $>$. Choose a sign, and use a point from the solution region to find out whether this sign is correct.

Choose a sign: $>$. Choose a point in the solution region: $(3, 1)$.

$y > 4x - 3$	Choose $>$ as the inequality sign.
$\mathbf{1} > 4(\mathbf{3}) - 3$	Replace variables with (3, 1).
$1 > \mathbf{9}$ False.	Evaluate.

Since the inequality is false, the sign we chose, $>$, is incorrect. The correct sign is $<$. This graph represents the inequality $y < 4x - 3$.

If the y-intercept is difficult to identify on the graph, do not use slope-intercept form to write the equation of the boundary line. Instead, identify two other points on the boundary line, find the slope of the line, and use point-slope form to write its equation.

EXAMPLE 6 Write the inequality represented by the graph.

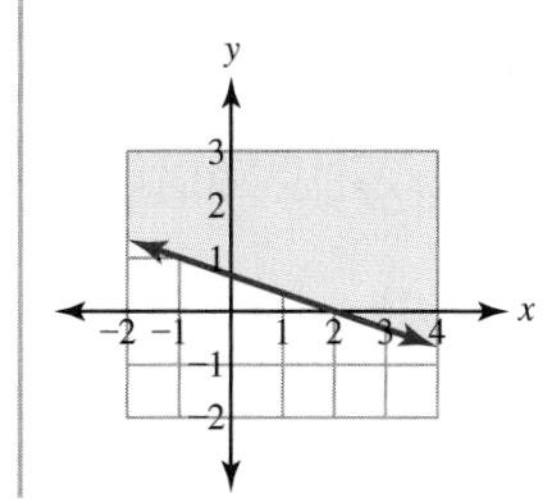

SOLUTION ▶ Find the slope of the boundary line.

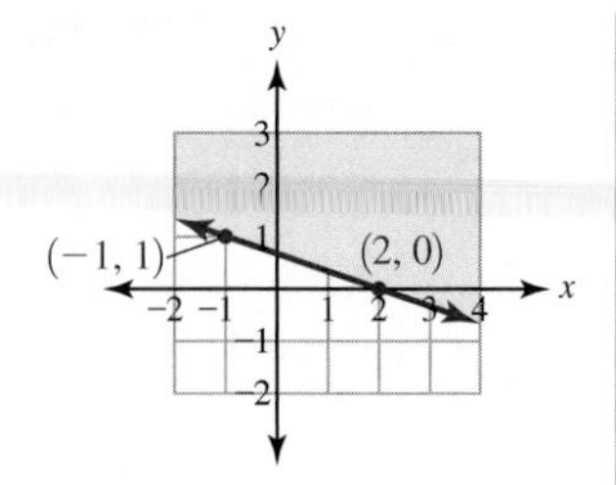

Identify two points on the line with integer coordinates, $(-1, 1)$ and $(2, 0)$. We can use the graph to identify the slope or use the slope formula.

$m = \frac{y_2 - y_1}{x_2 - x_1}$ The slope formula.

$m = \frac{0 - 1}{2 - (-1)}$ $(x_1, y_1) = (-1, 1)$; $(x_2, y_2) = (2, 0)$

$m = -\frac{1}{3}$ Simplify.

▶ Write the equation of the boundary line.

$y - y_1 = m(x - x_1)$ Use point-slope form.

$y - 1 = -\frac{1}{3}(x - (-1))$ Replace m, x_1, and y_1.

$y - 1 = -\frac{1}{3}(x + 1)$ Simplify.

$y - 1 = -\frac{1}{3}x - \frac{1}{3} \cdot 1$ Distributive property.

$y - 1 = -\frac{1}{3}x - \frac{1}{3}$ Simplify.

$+1 \quad\quad +1$ Addition property of equality.

$y + 0 = -\frac{1}{3}x - \frac{1}{3} + 1$ Simplify; common denominator is 3.

$y = -\frac{1}{3}x - \frac{1}{3} + 1 \cdot \frac{3}{3}$ Simplify; multiply by a fraction equal to 1.

$y = -\frac{1}{3}x - \frac{1}{3} + \frac{3}{3}$ Simplify.

$y = -\frac{1}{3}x + \frac{2}{3}$ Simplify.

▶ Choose the inequality sign and its direction.

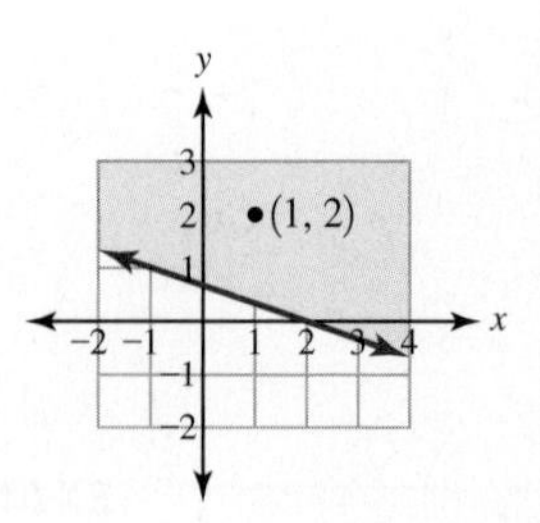

Since the boundary line is solid, this is not a strict inequality, and the sign will be ≤ or ≥. Choose a sign, and use a point from the solution region to find out whether this sign is correct.

Choose a sign: ≥. Choose a point in the solution region: $(1, 2)$.

$y \geq -\frac{1}{3}x + \frac{2}{3}$

$2 \geq -\frac{1}{3} \cdot 1 + \frac{2}{3}$ Replace variables with (1, 2).

$2 \geq -\frac{1}{3} + \frac{2}{3}$ Evaluate.

$2 \geq \frac{1}{3}$ True. Evaluate.

Since the inequality is true, the sign we chose, ≥, is correct. This graph represents the inequality $y \geq -\frac{1}{3}x + \frac{2}{3}$.

Writing the Inequality Represented by a Graph

1. Find the equation of the boundary line.
2. If the boundary line is dashed, the inequality is strict, and the sign is $>$ or $<$. If the boundary line is solid, the inequality is not strict, and the sign is $\geq$ or $\leq$.
3. Choose an inequality sign, and use this sign to write the inequality. Choose a test point in the solution region. Determine whether this test point is a solution of the inequality. If it is a solution, the sign is correct. If it is not a solution, reverse the direction of the sign.

Practice Problems

For problems 5–7, write the inequality represented by the graph.

5.

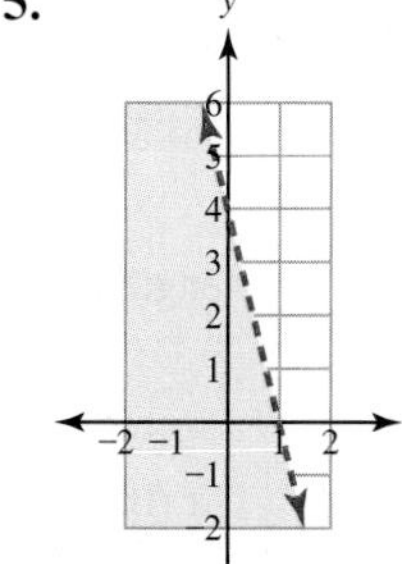

6.

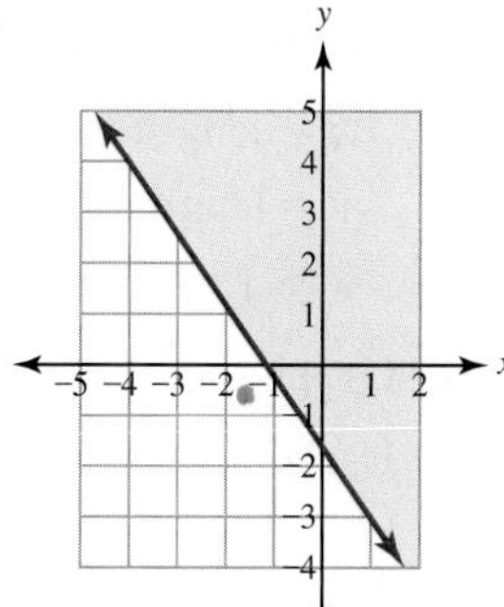

7.

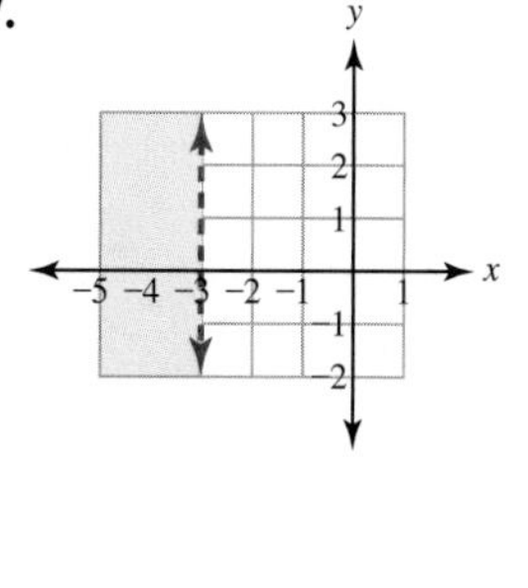

Constraints

When a parent puts a child in a car seat, the movement of the child is *constrained*. An inequality in one or two variables can represent some constraints.

EXAMPLE 7 The amount of carbohydrates is a constraint on snacks in vending machines in Alabama. A snack may contain no more than 30 grams (g) of carbohydrates. Assign one variable, and write an inequality that represents the constraint. (*Source:* www.latimes.com, Sept. 26, 2011)

SOLUTION ▶

x = amount of carboyhydrates   Assign the variable.

$x \leq 30$ g   amount of carbohydrates ≤ 30 g

In the next example, the constraint is represented by an inequality in two variables.

EXAMPLE 8 The sum of the Critical Reading section and the Math section on the SAT is a constraint on admission of freshman students to the University of North Carolina. The combined score must be at least 700. Assign two variables, and write an inequality that represents the constraint.

SOLUTION ▶

x = score on Critical Reading section   Assign the variables.

y = score on Math section

$x + y \geq 700$ critical reading score + math score ≥ 700

A constraint may include a percent.

EXAMPLE 9 The amount of sugar is a constraint on snacks in vending machines in California. Assign two variables, and write an inequality that represents the constraint.

In California, snacks, including those in vending machines, can't derive more than 35% of their calories from fat (including a limit of 10% of calories from saturated fats) or contain more than 35% sugar. (*Source:* www.latimes.com, Sept. 26, 2011)

SOLUTION ▶ $x =$ amount of snack Assign the variables.

$y =$ amount of sugar

$y \leq 0.35x$ amount of sugar $\leq$ (percent)(amount of snack)

When graphing a constraint, the solution region may instead be called the **feasible region**. The word *feasible* means "possible" or "can be done." A point in the feasible region represents an ordered pair that is within the limits of the constraint.

EXAMPLE 10 The sum of the length of a truck-tractor and the length of its semi-trailer is a constraint on the combined length of a vehicle in Michigan. The combined length must be no more than 50 ft. (*Source:* www.michigan.gov)

(a) Assign two variables, and write an inequality that represents the constraint.

SOLUTION ▶ $x =$ length of a truck-tractor Assign the variables.

$y =$ length of a semi-trailer

$x + y \leq 50$ ft length of tractor + length of trailer $\leq$ 50 ft

(b) Graph the inequality.

▶ Use a table of ordered pairs to graph the solid boundary line, $x + y = 50$ ft.

Since the inequality is not strict, $\leq$, the line is solid.

x (ft)	*y* (ft)
0	50
50	0
25	25

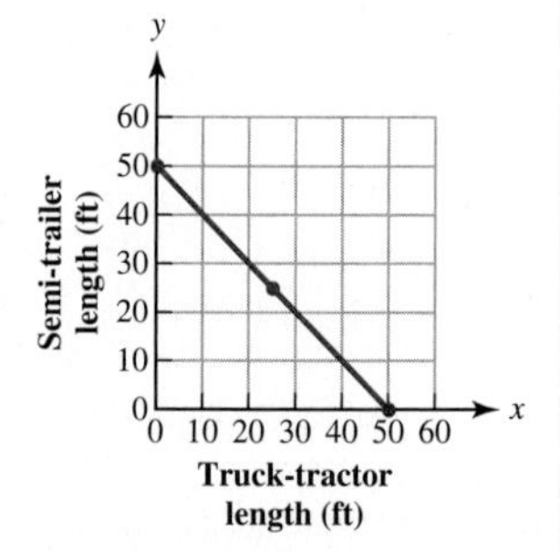

▶ Use a test point to identify the solution region.

Choose a test point that is not on the boundary line, (10 ft, 20 ft).

$x + y \leq 50$ ft

$\mathbf{10\ ft} + \mathbf{20\ ft} \leq 50$ ft Replace the variables.

$\mathbf{30\ ft} \leq 50$ ft True. Evaluate.

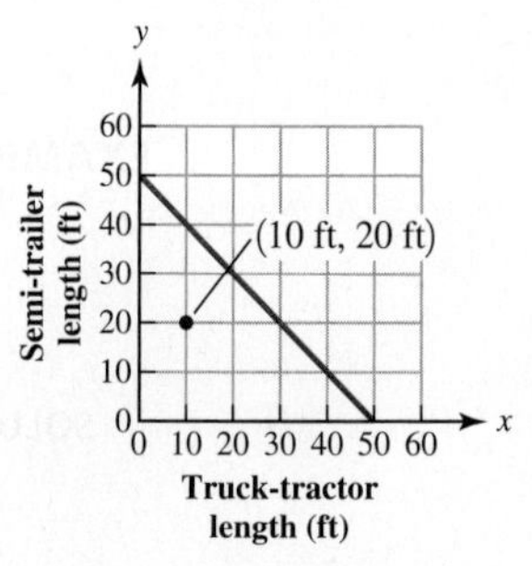

▶ Shade the feasible region.

The coordinates of the points in the feasible region must be greater than or equal to 0 ft. Shade the area of the graph that includes (10 ft, 20 ft) and does not extend past the x-axis or y-axis.

The graph of $x + y \leq 50$ ft includes the points in the shaded feasible region and the points on the boundary line.

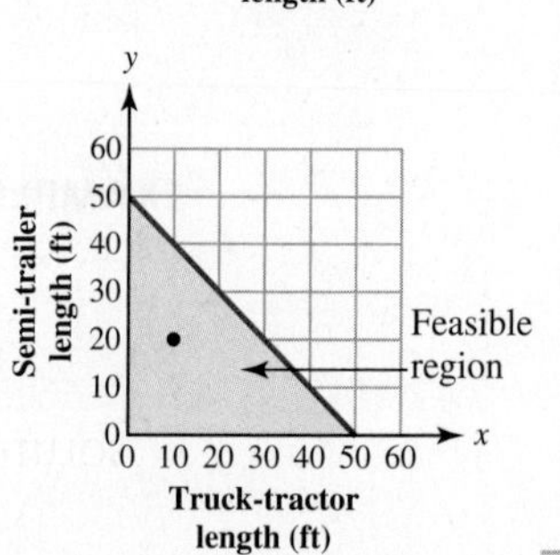

In the next example, since it is not practical to shade each individual point that represents a solution of the constraint, the entire feasible region is shaded. However, only ordered pairs with whole number coordinates are feasible solutions.

EXAMPLE 11 The total amount spent on hats and posters is a constraint on a political campaign event. The cost of a hat is \$4; the cost of a poster is \$1.25. The amount spent on hats and posters must be less than or equal to \$3000.

(a) Assign two variables, and write an inequality that represents this constraint.

SOLUTION ▶

x = number of hats — Assign the variables.

y = number of posters

$\left(\frac{\$4}{1 \text{ hat}}\right)x + \left(\frac{\$1.25}{1 \text{ poster}}\right)y \leq \$3000$ — amount hats + amount posters ≤ \$3000

(b) Graph the inequality.

▶ Use a table of ordered pairs to graph the solid boundary line, $\$4x + \$1.25y = \$3000$.

The inequality is not strict, ≤; the line is solid. However, since the campaign can give away only whole hats or posters, only the ordered pairs with whole number coordinates are actually solutions of the constraint.

x (hats)	y (posters)
0	2400
750	0
200	1760

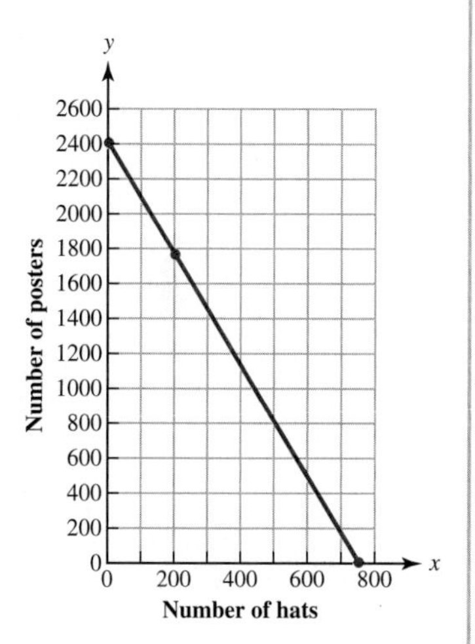

▶ Use a test point to identify the solution region.

Choose a test point that is not on the boundary line, (300 hats, 600 posters).

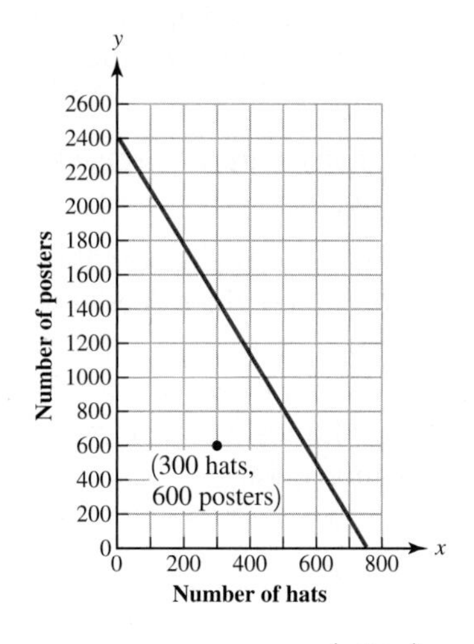

$$\left(\frac{\$4}{1 \text{ hat}}\right)x + \left(\frac{\$1.25}{1 \text{ poster}}\right)y \leq \$3000$$

$$\left(\frac{\$4}{1 \text{ hat}}\right)(\mathbf{300 \text{ hats}}) + \left(\frac{\$1.25}{1 \text{ poster}}\right)(\mathbf{600 \text{ posters}}) \leq \$3000$$ Replace variables.

$$\mathbf{\$1200 + \$750} \leq \$3000$$ Evaluate.

$$\mathbf{\$1950} \leq \$3000 \quad \text{True.}$$ Evaluate.

▶ Shade the feasible region.

The coordinates of the points in the solution region must be greater than or equal to 0 hats or 0 posters. Shade the area of the graph that includes (300 hats, 600 posters) and does not extend past the x-axis or y-axis.

The graph of $\left(\frac{\$4}{1\text{ hat}}\right)x + \left(\frac{\$1.25}{1\text{ poster}}\right)y \le \$3000$ includes the points in the shaded feasible region and the points on the boundary line that have whole number coordinates.

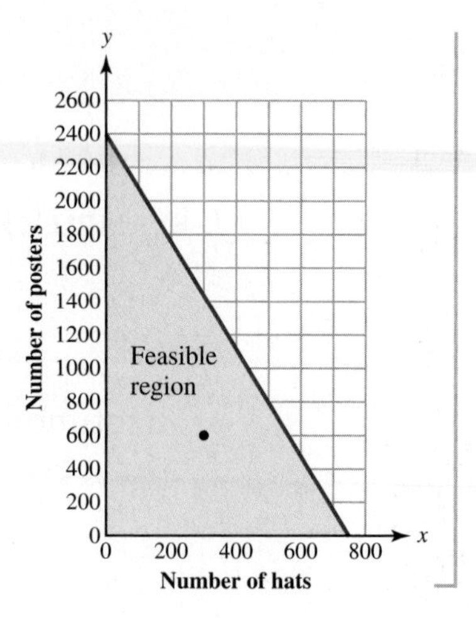

Practice Problems

For problems 8–9, assign one variable, and write an inequality that represents the constraint.

8. An egress window for a bedroom in a basement must have a minimum net clear opening of 5.7 ft^2.
9. The maximum amount of rent a student can afford is \$550 per month.

For problems 10–11, assign two variables, and write an inequality that represents the constraint.

10. A buyer will order a total of no more than 7000 lb of Guatemala Antigua coffee and Colombia Nariño coffee.
11. To reduce the expense of benefits, a state required that employee pension contributions rise by at least 1 percent of the employee's total salary.
12. The maximum volume of garbage per household collected by the city is 90 gal. Assign one variable, and write an inequality that represents the constraint.

For problems 13–14,
(a) assign two variables and write an inequality that represents the constraint.
(b) graph the inequality.

13. A multimedia company creates wedding presentations that contain digital photo files and/or digital video files. The presentation can include no more than 100 total files.
14. The total amount spent for gifts at a company's booth at a convention is limited to \$4000. The cost of a computer USB drive is \$5; the cost of an insulated coffee mug is \$8.

Using Technology: Graphing Inequalities

To use a graphing calculator to graph an inequality in two variables, first enter the boundary line equation. After using a test point, use the "graph style icon" to shade the solution region.

EXAMPLE 12 Graph $y \leq -4x + 9$. Sketch the graph; describe the window.

The window is standard. Type the equation on the Y= screen. Press GRAPH.

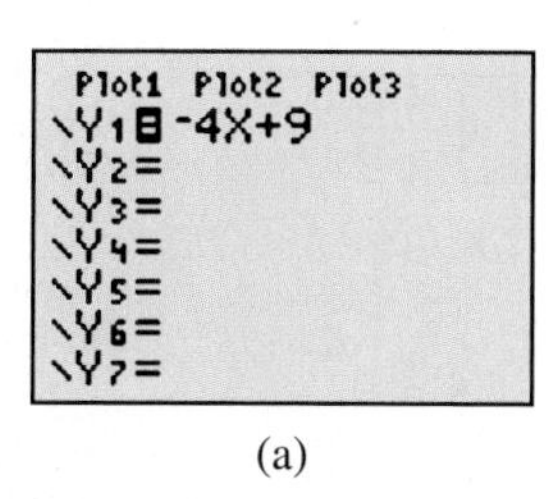

(a)

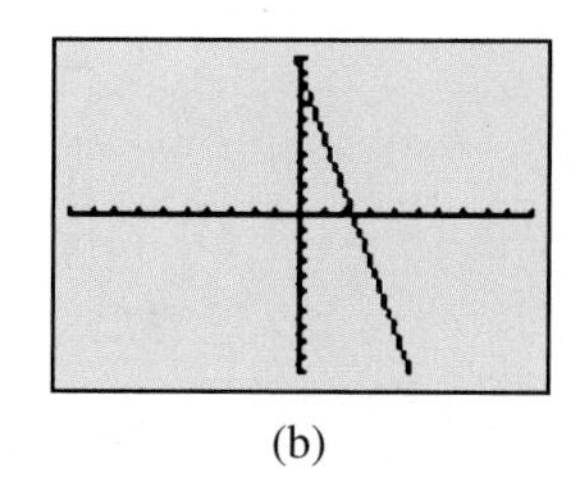

(b)

Test point: (0, 0)

$y \leq -4x + 9$

$\mathbf{0} \leq -4(\mathbf{0}) + 9$

$0 \leq \mathbf{9}$ True.

The "graph style icon" draws the line as dashed or solid or it shades one side of the line. It cannot do both at the same time. To shade the solution region, move the cursor to the left of the equals sign. Press ENTER repeatedly until the "shaded below" icon appears. Press GRAPH.

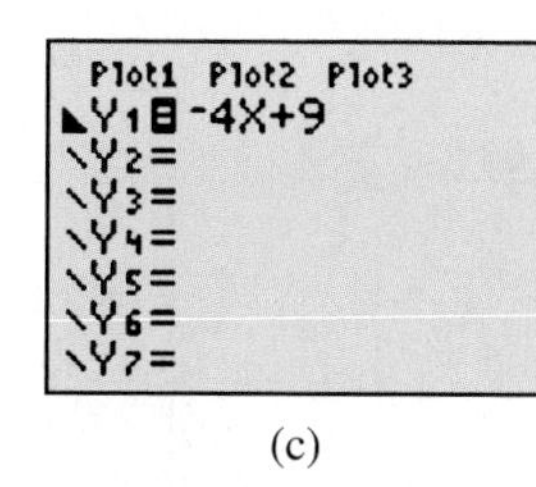

(c)

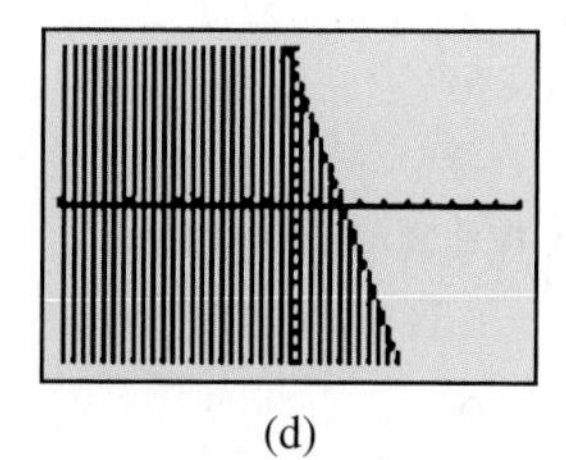

(d)

Practice Problems For problems 15–17, graph the inequality. Use the graph style icon to shade the solution region. Sketch the graph; describe the window. Use a dashed line to represent the boundary line of a strict inequality.

15. $y \geq 2x - 7$ **16.** $y < -\frac{3}{4}x + 8$ **17.** $y > 6$

3.6 Vocabulary Practice

Match the term with its description.

1. $<$
2. $>$
3. $\leq$
4. $\geq$
5. We use this to find the shaded area on the graph of a linear inequality.
6. This line marks the edge of the solution region on the graph of a linear inequality.
7. On the graph of a linear inequality, this area includes all the points that represent solutions of the inequality.
8. On the graph of a constraint, this area includes all the points that represent the ordered pairs that are within the limits of the constraint.
9. A boundary line used to represent a strict inequality is this.
10. A boundary line used to represent an inequality that is not strict is this.

A. boundary line
B. dashed line
C. feasible region
D. greater than
E. greater than or equal to
F. less than
G. less than or equal to
H. solid line
I. solution region
J. test point

3.6 Exercises

Follow your instructor's guidelines for showing your work.

For exercises 1–28, graph the inequality.

1. $y > 4x + 3$

2. $y > 2x + 5$

3. $y \le 2x - 6$

4. $y \le 3x - 7$

5. $y < -5$

6. $y < 4$

7. $x \ge 6$

8. $x \ge -5$

9. $2x + 9y \le 18$

10. $3x + 5y \le 15$

11. $2x - 9y \ge 18$

12. $3x - 5y \ge 15$

13. $y > x$

14. $y < x$

15. $y \le -x$

16. $y \ge -x$

17. $y < -\frac{9}{4}x + 7$

18. $y < -\frac{8}{5}x + 6$

19. $x - y \ge 6$

20. $x - y \ge 3$

21. $5x + 2y > -15$

22. $3x + 2y > -9$

23. $y \ge -4x$

24. $y \ge -5x$

25. $x > 0$

26. $x < 0$

27. $y \le 0$

28. $y \ge 0$

For exercises 29–40, write the inequality that represents the graph.

29.

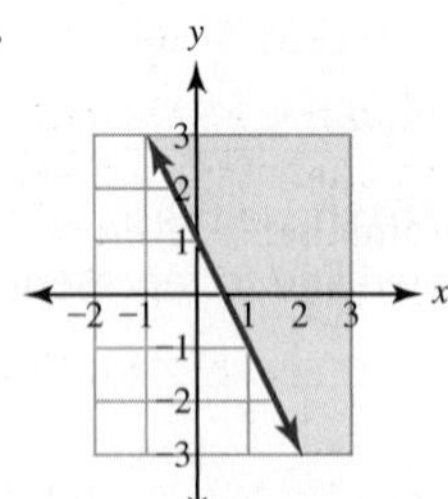

30.

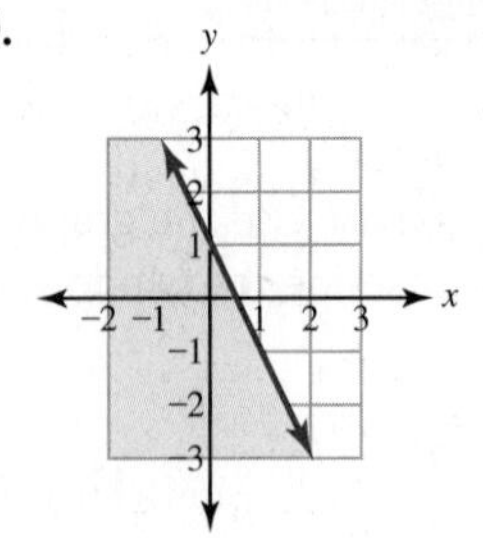

31.

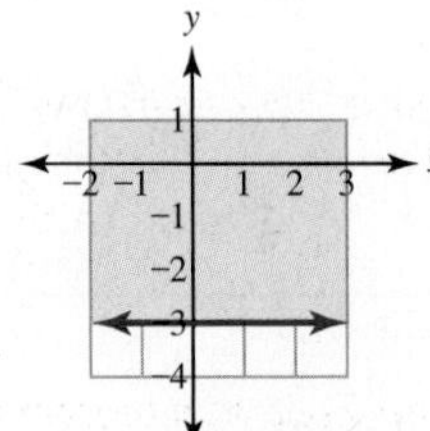

32.

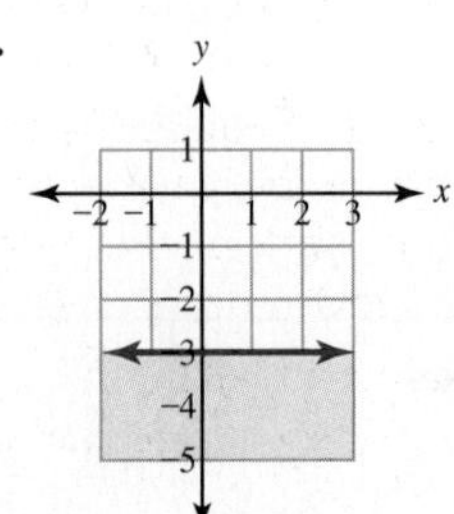

33.

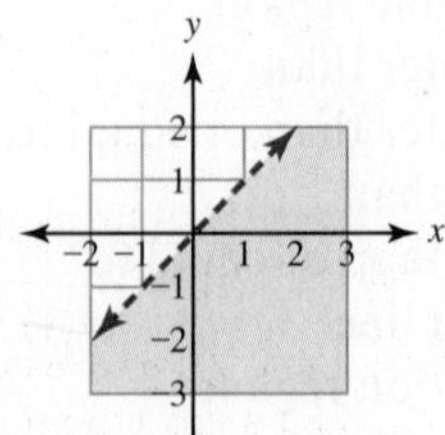

34.

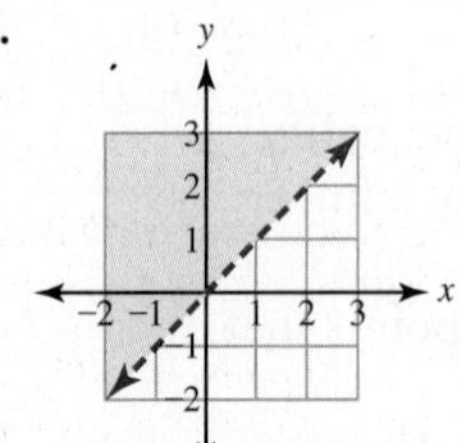

35.

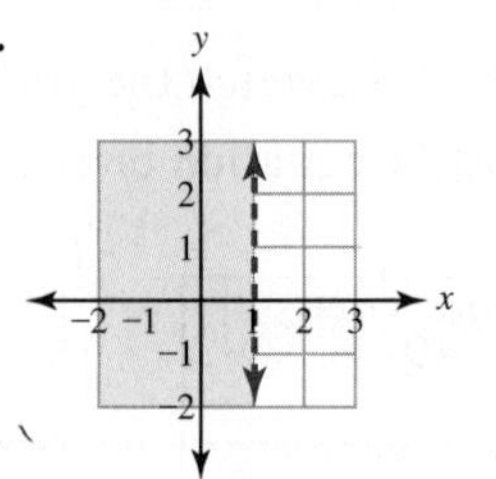

36.

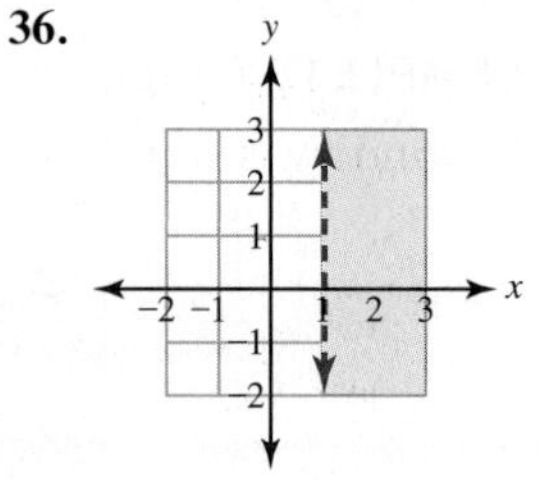

37.

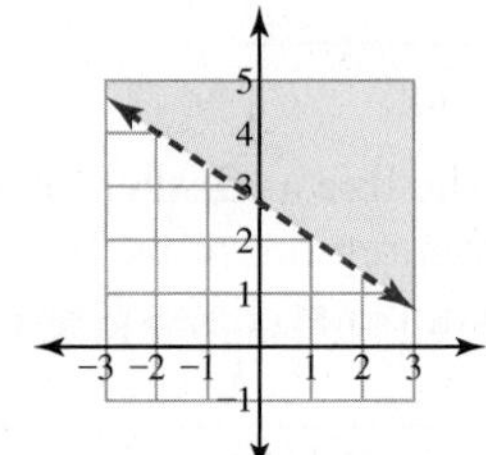

38.

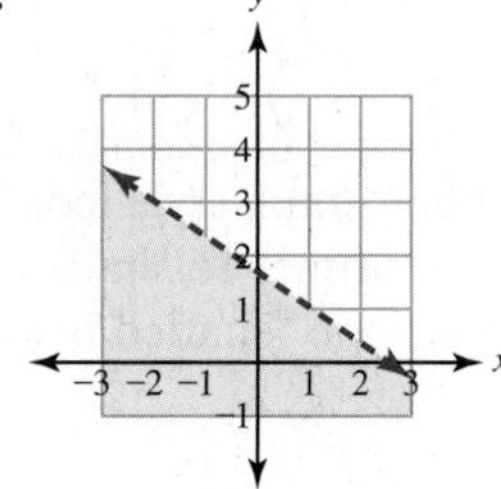

39.

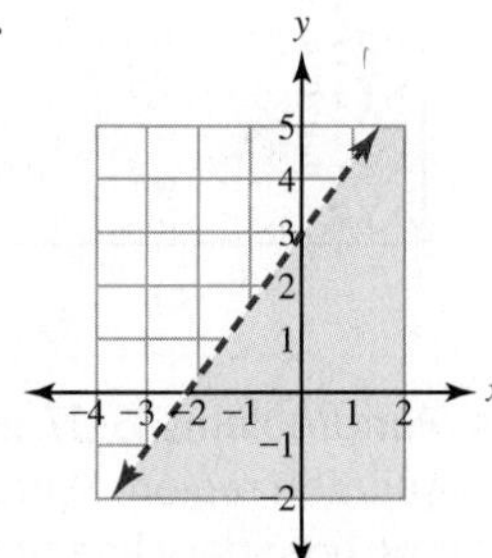

40.

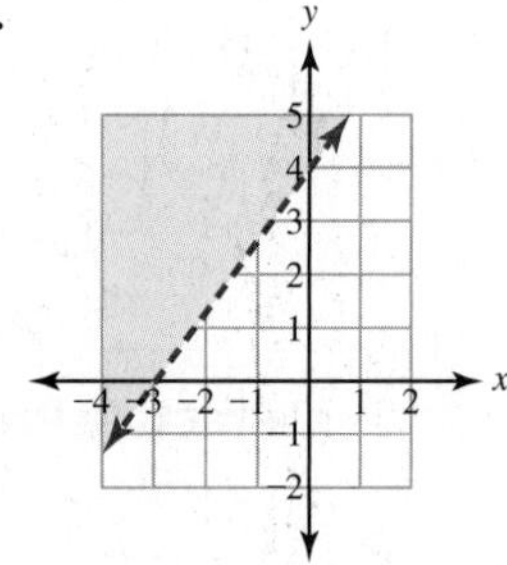

41. Explain how to determine whether the boundary line on the graph of an inequality should be solid or dashed.

42. Explain how to determine whether an inequality is a strict inequality.

43. Explain why (0, 0) is often a good choice for a test point when graphing an inequality.

44. Describe a situation in which (0, 0) cannot be used as a test point when graphing an inequality.

45. Explain how to write the equation of a line in slope-intercept form given two points on the line.

46. Explain how to write the equation of a line in slope-intercept form given the slope and one point on the line.

For exercises 47–54, assign one variable, and write an inequality that represents the constraint.

47. In the East Orchards of Lewiston, Idaho, at least $\frac{1}{2}$ acre of land is required to keep a horse. (*Source:* www.cityoflewiston.org)

48. The area of land designated as forest in Albemarle County in Virginia must be at least 20 acres. (*Source:* www.albemarle.org)

49. Directors on the board of directors of a corporation must acquire and maintain a minimum ownership of 2500 shares of the corporation's stock.

50. The minimum initial investment in the Vanguard Windsor Admiral Fund is \$100,000. (*Source:* www.vanguard.com)

51. The maximum limit of employee exposure to flour dust over an 8-hr period is $\frac{10 \text{ mg}}{1 \text{ m}^3}$.

52. The maximum short-term limit of employee exposure to flour dust in a 15-min period is $\frac{30 \text{ mg}}{1 \text{ m}^3}$.

53. A woman of childbearing age should ingest at least 400 micrograms of folate per day. (*Source:* ods.nih.gov)

54. A pregnant woman, 19 years of age or older, should ingest no more than 10,000 international units of preformed vitamin A per day. (*Source:* ods.nih.gov)

For exercises 55–76, assign two variables, and write an inequality that represents the constraint.

55. To qualify for free checking, the minimum combined balance of deposits and loans is \$1000.

56. The maximum combined contributions of an employee and an employer to a 403(b) retirement plan is \$44,000.

57. At the iTunes Store, a single music download costs \$0.99, and an episode of a television show costs \$1.99. A customer wants to spend no more than \$35 on singles and television shows. (*Source:* www.apple.com/itunes)

58. A picnic planner can spend no more than \$100 on baked beans and macaroni salad. The cost of baked beans is $\frac{\$2.99}{1 \text{ lb}}$, and the cost of macaroni salad is $\frac{\$3.40}{1 \text{ lb}}$.

59. In 2011–2012, federal law limited individual political contributions to each candidate or candidate committee per election. The sum of two contributions must be less than or equal to \$2500. (*Source:* www.fec.gov)

60. In 2011–2012, federal law limited individual political contributions to a national party committee per calendar year. The sum of two contributions must be less than or equal to \$30,800. (*Source:* www.fec.gov)

61. If a state employee qualifies for the Rule of 90, a state allows early retirement without reduction of benefits. The sum of the number of years of employment and the age of the employee in years must be at least 90 years.

62. The American Public Transportation Association sponsors an International Bus "Roadeo." In 2011, the maximum possible combined score of the maintenance team and the bus operator was 2925 points. (*Source:* www.apta.com)

63. The maximum benefit level for a retirement plan for a person is eight times the final average salary.

64. The maximum benefit level for a retirement plan for a person is three times the final average salary.

65. The maximum benefit level for a retirement plan for a person is eight times the final average salary plus \$500,000.

66. The maximum benefit level for a retirement plan for a person is three times the final average salary plus \$455,999.

67. An investor will transfer a total of no more than \$15,500 into a stock fund and into a bond fund.

68. An investor will transfer a total of no more than \$20,500 into a stock fund and into a bond fund.

69. Financial aid can pay for a maximum of 125% of the credits required for a degree or certificate program.

70. A loan guarantee for agricultural production is a maximum of 50% of the total loan amount.

71. No more than 80% of a loan can be guaranteed by the federal government.

72. No more than 60% of a loan can be guaranteed by the federal government.

73. Admission to a master's degree program in Business Administration at the University of Texas Permian Basin requires a minimum Entrance Score of at least 1120. The Entrance Score is the sum of the product of 200 and the undergraduate GPA and the score on the Graduate Management Admission Test (GMAT). (*Source:* bus.utpb.edu)

74. Admission to a master's degree program in Exercise Science at Florida State University requires a minimum combined score of 1000 on the verbal and quantitative sections of the Graduate Record Examination (GRE). (*Source:* www.chs.fsu.edu)

75. Children's and adult jewelry sold by certain retail companies in California must contain no more than 0.03% cadmium. (*Source:* www.latimes.com, Sept. 7, 2011)

76. Homework in the Los Angeles Unified School District will count for no more than 10% of a student's academic achievement grade. (*Source:* panoramahs.org, Policy Bulletin BUL-5502.0, May 20, 2011)

For exercises 77–78, assign one variable and write an inequality that represents the constraint.

77. The FAA requires that at air shows and races, spectators must be kept at least 500 ft away from planes in normal flight. (*Source:* www.faa.gov)

78. The FAA requires that at air shows and races, spectators must be kept a horizontal distance of at least 1500 ft from planes performing aerobatic maneuvers. (*Source:* www.faa.gov)

For exercises 79–82,

(a) assign two variables and write an inequality that represents the constraint.

(b) graph the inequality.

79. For two pieces of luggage checked at EasyJet Airline, the sum of the weights of the luggage can be no more than 20 kg. (*Source:* www.easyjet.com)

80. The sum of the annual contributions to a Roth IRA and a Traditional IRA can be no more than \$5000.

81. The total amount spent on ground beef and ground lamb to make meatballs for a reception can be no more than \$90. The cost of ground beef is $\frac{\$3}{1 \text{ lb}}$ and the cost of ground lamb is $\frac{\$4}{1 \text{ lb}}$.

82. The total amount spent on raw carrots and broccoli for a reception can be no more than \$45. The cost of carrots is $\frac{\$0.90}{1 \text{ lb}}$, and the cost of broccoli is $\frac{\$1.50}{1 \text{ lb}}$.

Problem Solving: Practice and Review

Follow your instructor's guidelines for using the five steps as outlined in Section 1. 3, p. 55.

83. An asparagus packing box is 19 in. long, $10\frac{3}{4}$ in. wide, and $10\frac{1}{4}$ in. high. Find the volume of this box in cubic inches. Round to the nearest whole number. (*Source:* www.caspack.com)

84. Find the total complaint load. Round to the nearest whole number.

Private fair housing groups continue to investigate the highest number of complaints—18,665, or 65 percent of the total complaint load, although there are fewer organizations operating than in 2009. (*Source:* www.nationalfairhousing.org, April 29, 2011)

85. In 2010, box-office receipts for movies shown in China were \$1.5 billion. This was a 64% increase over receipts in 2009. Find the receipts in 2009. Round to the nearest tenth of a billion. (*Source:* www.latimes.com, Aug. 24, 2011)

86. The Elkhorn Slough is an area of tidal salt marsh on the California coastline. Find the amount of greenhouse gas an acre soaks up per hour. (365 days = 1 year.) Round to the nearest tenth.

Every acre of the slough's . . . vegetation soaks up 870 kilograms of greenhouse gas annually, roughly equal to the emissions of driving 2,280 miles. (*Source:* www.Santacruzsentinel.com. Dec. 30, 2008)

Technology

For exercises 87–90, use a graphing calculator to graph the inequality. Choose a window that shows the *y*-intercept. Use the graph style icon to shade the solution region. Sketch the graph; describe the window. Use a dashed line to represent the boundary line of a strict inequality.

87. $y > 3x - 15$

88. $y < -2x + 18$

89. $y \leq 2500x + 8000$

90. $y \geq 4000x + 12{,}000$

Find the Mistake

For exercises 91–94, the completed problem has one mistake.

(a) Describe the mistake in words, or copy down the whole problem and highlight or circle the mistake.

(b) Do the problem correctly.

91. Problem: Graph: $y < -\frac{1}{3}x + 2$

Incorrect Answer:

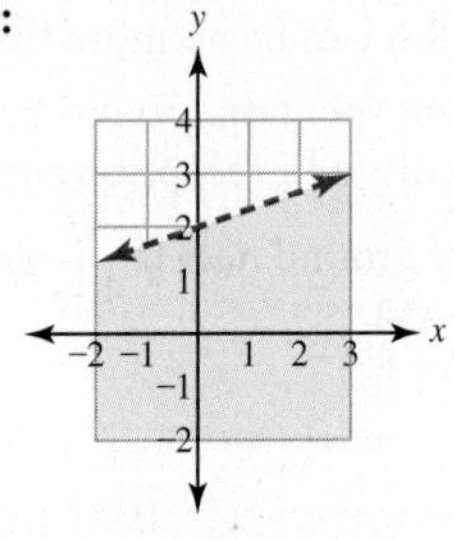

92. Problem: Graph: $y > -3x + 2$

Incorrect Answer:

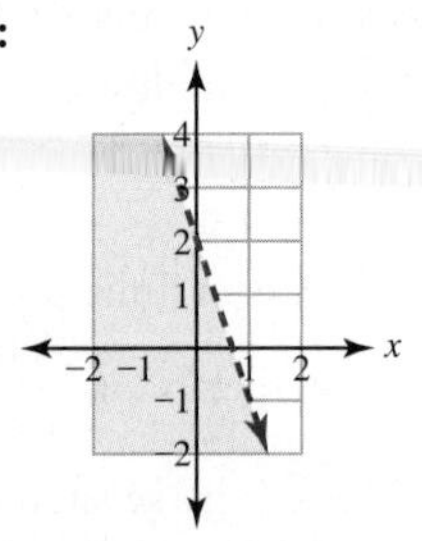

93. Problem: Graph: $5x + 2y < 10$

Incorrect Answer:

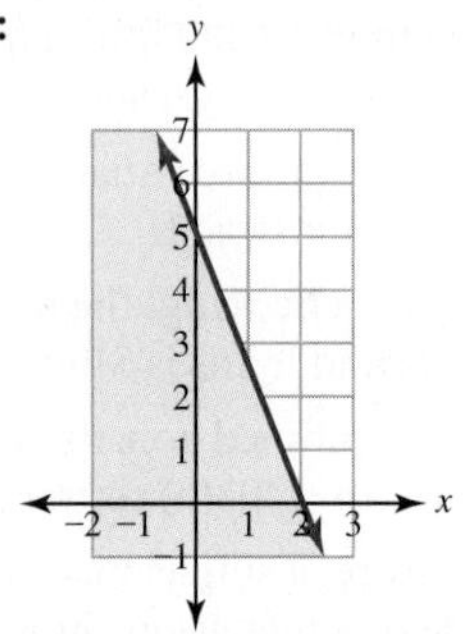

94. Problem: Graph: $y > 3$

Incorrect Answer:

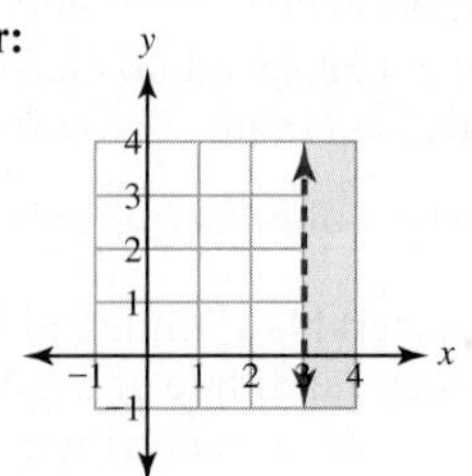

Review

95. The cost to create a product is \$6. The fixed overhead costs per month to make the product are \$7500. The price of the product is \$10. Assume that the company will sell all the products. Find the number of products that must be sold to break even.

96. Two lines are parallel. The slope of one line is 5. Identify the slope of the other line.

97. Two lines are perpendicular. The slope of one line is 8. Identify the slope of the other line.

98. Two lines are perpendicular. The slope of one line is 0. Identify the slope of the other line.

SUCCESS IN COLLEGE MATHEMATICS

99. Learning mathematics includes understanding new concepts, mastering procedures, and memorizing key terms and properties. Your learning preferences for memorizing may be different from your learning preferences for practicing procedures and answering questions about concepts. Describe the method that you use to learn a list of definitions or to memorize a property.

100. Is this method effective? Explain.

© Courtesy Logan Fowler/Lewis-Clark State College

The roster of the Lewis-Clark State College baseball team includes the name of each player and his position. The set of ordered pairs (name, position) is a relation. In this section, we will study relations and functions.

SECTION 3.7

Relations and Functions

After reading the text, working the practice problems, and completing assigned exercises, you should be able to:

1. Determine whether a relation is a function.
2. Use the vertical line test to determine whether a graph represents a function.
3. Evaluate a function.
4. Identify the domain or range of a function.
5. Graph a function.

Relations

On the roster of the Lewis-Clark State College men's baseball team, the name of each player is matched with his position. These matches are a **correspondence**. We can also represent the correspondence as a set of ordered pairs: {(Fassold, pitcher), (Miller, designated hitter), (Casini, second base), (Sanchez, pitcher)}. A set of ordered pairs is a **relation**.

Name	Position
Fassold	Pitcher
Miller	Designated hitter
Casini	Second base
Sanchez	Pitcher

The first coordinate in an ordered pair in a relation is an **input value**. The second coordinate is an **output value**. We often say just "input" and "output." For the relation $y = x + 3$, the input value is the number that replaces x. The output value, y, is the sum of the input value and 3. The relation is the set of ordered pairs that are solutions of the equation $y = x + 3$.

EXAMPLE 1 Determine whether the set is a relation.

(a) $\{(94\%, \text{A}), (75\%, \text{C}), (86\%, \text{B})\}$

SOLUTION ▶ Since this is a set of ordered pairs, it is a relation.

(b) $\{2, 4, 6, 8\}$

▶ This is a set but it is not a set of ordered pairs; it is not a relation.

(c) $y = 5x - 1$

▶ Since this equation represents a set of ordered pairs (x, y) in which the input value is x and the output value is y, there are infinitely many ordered pairs in this set. Since this is a set of ordered pairs, it is a relation.

A **function** is a special relation in which each input value corresponds to exactly one output value. No two ordered pairs have the same input value.

EXAMPLE 2 Determine whether the relation is a function.

(a) $A = \{(0, 4), (1, 6), (2, 8)\}$

SOLUTION ▶ Since each input value corresponds to exactly one output value (there are no input values that repeat), it is a function.

(b) $Q = \{(0, 0), (2, 4), (-2, 4)\}$

▶ Since each input value corresponds to exactly one output value, this is a function. It does not matter whether output values repeat.

(c) $P = \{(3, 2), (5, 8), (3, 1)\}$

▶ Since the input value 3 corresponds to the output value 2 and the input value 3 also corresponds to the output value 1, this is not a function.

(d) The solutions of $y = 5x + 3$. The input value is x, and the output value is y.

▶ Each solution of this equation is an ordered pair. If the output value changes, the input value also changes. Since an input value corresponds to exactly one output value, this is a function.

(e) The correspondence between the given fruits and their color.

▶ 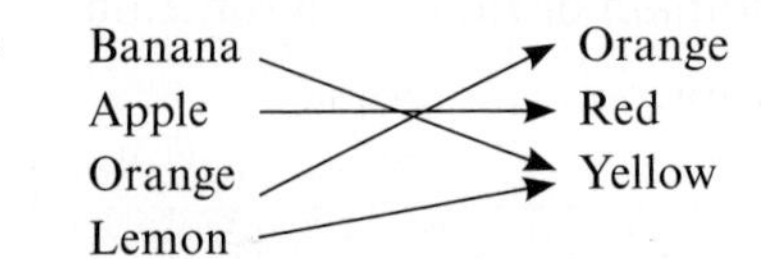

In roster notation, this relation is {(banana, yellow), (apple, red), (orange, orange), (lemon, yellow)}. Since each input value corresponds to exactly one output value, this is a function.

(f) The correspondence of the set of the name of each American (inputs) and the set of each American's Social Security number (outputs).

▶ Since some Americans have the same names, each input value does not correspond to exactly one output value. This is not a function.

(g) The correspondence of the set of Social Security numbers in the United States (inputs) and the set of the name of the person assigned to each number (outputs).

▶ Since each U.S. Social Security number is unique, each input value corresponds to exactly one output value. This is a function.

(h) The result of pushing a turn signal lever down.

▶ There is one ordered pair. The input value is pushing the turn signal lever down, and the output value is the blinking of the indicator lights on the left side of the car. Since each input value corresponds to exactly one output value, it is a function.

The function that is the correspondence of pushing a turn signal lever down and the blinking of the indicator lights on the left side of the car illustrates that functions are *predictable.* If pushing the turn signal lever down sometimes resulted in the left indicator lights blinking and sometimes resulted in the right indicator lights blinking, this correspondence would be a relation but not a function. The action of pressing the turn signal lever down would not be predictable or useful.

If we can represent a relation with a graph, we can use the **vertical line test** to determine whether it is a function. If a vertical line crosses a graph in more than one point, at least two ordered pairs on the graph have the same input value. The relation has repeating input values, and it is not a function. To use the vertical line test, draw a *dashed* vertical line. Just as a dashed boundary line on the graph of an inequality does not include solutions of the inequality, a dashed vertical line is not part of the graph of a function.

Vertical Line Test

If every vertical line crosses the graph of a relation in at most one point, the graph represents a function.

EXAMPLE 3 Use the vertical line test to determine whether each graph represents a function.

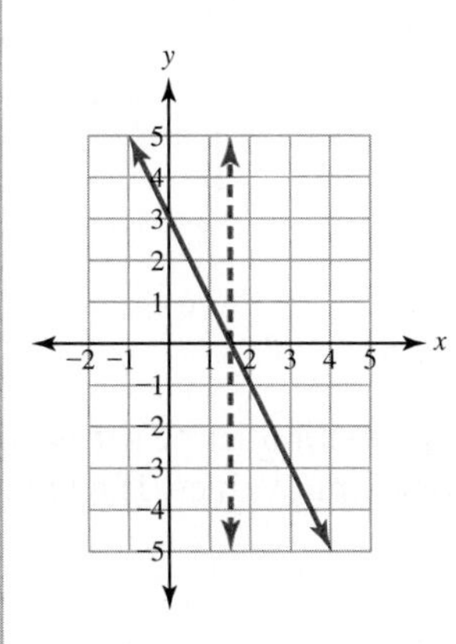

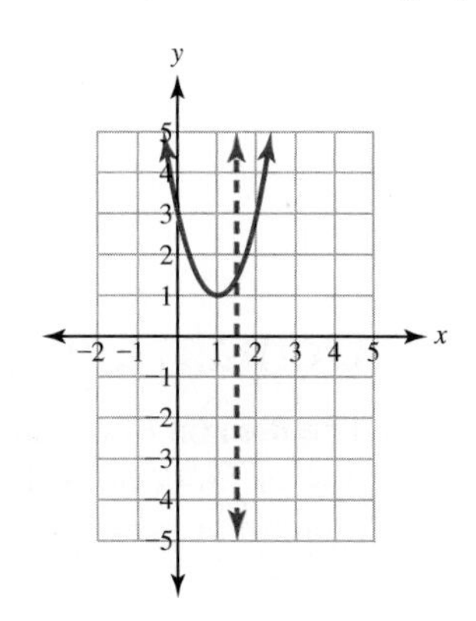

SOLUTION ▶ Every vertical line crosses the graph in at most one point. This is a function.

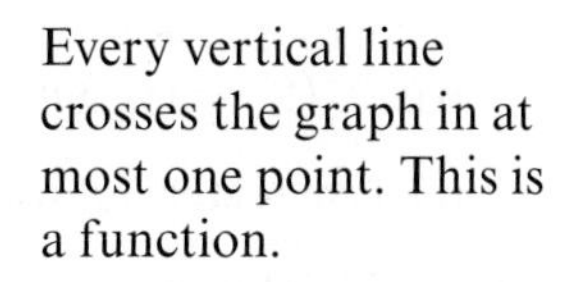
Every vertical line crosses the graph in at most one point. This is a function.

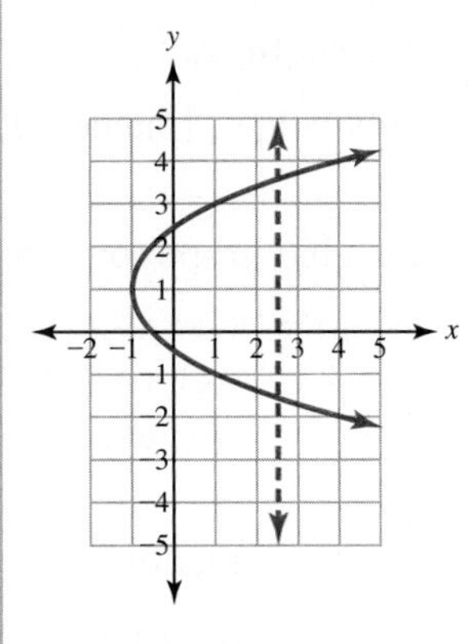

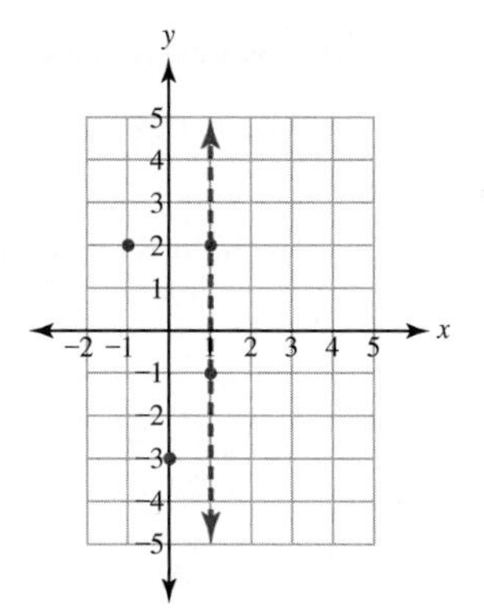

▶ At least one vertical line crosses the graph in more than one point. This is not a function.

At least one vertical line crosses the graph in more than one point. This is not a function.

Practice Problems

For problems 1–4, identify the set as a relation, a function, both a relation and a function, or neither a relation nor a function.

1. $\{(8, 0), (9, -1), (10, -2)\}$

2. $\{\ldots, -3, -2, -1, 0, 1, 2, 3, 4, \ldots\}$

3. The correspondence between the ID number of each student at a college (the input) and the age of the student rounded to the nearest year (the output).

4. The correspondence between the age of each student at a college rounded to the nearest year (the input) and the ID number of the student (the output).

For problems 5–7, use the vertical line test to determine whether the graph represents a function.

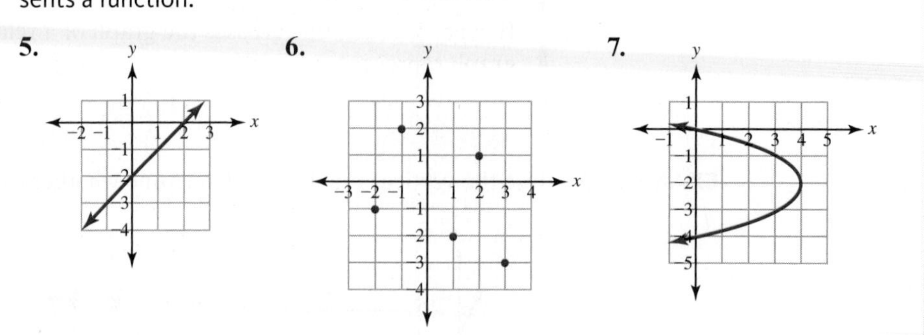

Domain and Range

The **domain** of a relation or function is the set of input values. The **range** of a relation or function is the set of output values. If a value repeats, list it only once in the domain or range.

EXAMPLE 4 A relation is $R = \{$(Fassold, pitcher), (Miller, designated hitter), (Casini, second base), (Sanchez, pitcher)$\}$.

(a) Use roster notation to represent the domain.

SOLUTION ▶ The domain is the set of input values, {Fassold, Miller, Casini, Sanchez}.

(b) Use roster notation to represent the range.

▶ The range is the set of output values, {pitcher, designated hitter, second base}.

In the next example, we use a bar graph to represent a function.

EXAMPLE 5 The function $N = \{$(AA, 242), (AAT, 44), (AS, 205), (AAS, 931)$\}$ represents the relationship of the type of degree and the number of degrees awarded in 2010 at Austin Community College.

(a) Use a bar graph to represent this function.

SOLUTION ▶ The label on the horizontal axis under each bar is the input value; the height of the bar on the vertical axis is the corresponding output value.

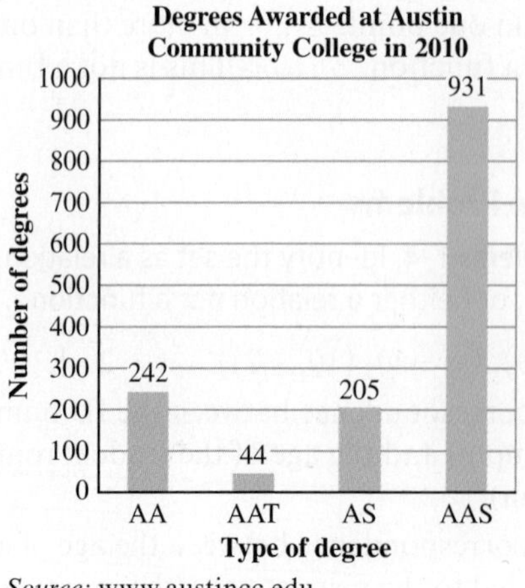

Source: www.austincc.edu

(b) Use roster notation to represent the domain.

The domain is the set of input values, {AA, AAT, AS, AAS}.

(c) Use roster notation to represent the range.

The range is the set of output values, {242 degrees, 44 degrees, 205 degrees, 931 degrees}.

For the function $y = 2x + 1$, the input value is x, and the output value is y. Each point on the graph of this equation represents an ordered pair (x, y) in this relation. The input value, x, is the **independent variable**. Since the value of the output value, y, depends on x, it is the **dependent variable**.

EXAMPLE 6 The graph represents the relation $y = 2x + 1$. Each point on the graph represents an ordered pair in the relation.

$y = 2x + 1$

(a) Identify the independent variable.

SOLUTION The independent variable is x.

(b) Identify the dependent variable.

The dependent variable is y.

(c) Identify the domain.

Since any real number can be an input value, the domain is the set of real numbers.

(d) Identify the range.

Each solution of this equation is an ordered pair. Each real number input value corresponds to a unique real number output value. The range is the set of real numbers.

Practice Problems

8. A relation is {(Yankees, NY), (Twins, MN), (Mets, NY), (Mariners, WA)}.
a. Use roster notation to represent the domain.
b. Use roster notation to represent the range.

9. The number of electoral college votes per state is equal to the sum of the number of its U.S. representatives and the number of its U.S. Senators. For the 2012 election, four states had more than 20 electoral college votes. (*Source:* www.eac.gov)
a. Use a bar graph to represent the relation V = {(CA, 55 votes), (FL, 29 votes), (NY, 29 votes), (TX, 38 votes)}.
b. Use roster notation to represent the domain.
c. Use roster notation to represent the range.

10. A relation is $y = -2x + 4$.
a. Identify the independent variable.
b. Identify the dependent variable.
c. Identify the domain.
d. Identify the range.

Function Notation

To **evaluate** a function, find the output value that corresponds to a given input value.

EXAMPLE 7 Evaluate $y = 5x + 3$ when $x = 8$.

SOLUTION ▶

$y = 5x + 3$

$y = 5 \cdot \mathbf{8} + 3$ Replace x with the input value, 8.

$y = \mathbf{40} + 3$ Simplify.

$y = \mathbf{43}$ Simplify.

When the input value is 8, the output value is 43.

Instead of writing $y = 5x + 3$, we can use **function notation** and write $f(x) = 5x + 3$, where $y = f(x)$. The notation $f(x)$ means "the output value of the function f when the input value is x." The expression $5x + 3$ is the **function rule**. Since "The output value of the function f when the input value is x" is a long phrase, we often just say "f of x." The f in $f(x)$ is *not* a variable. We are not multiplying f and x; we cannot divide both sides by f or x. To **evaluate** $f(x) = 5x + 3$ for a given input value, replace x with the input value, and follow the order of operations.

EXAMPLE 8 If $f(x) = 5x + 3$, evaluate $f(1)$.

SOLUTION ▶

$f(x) = 5x + 3$

$f(\mathbf{1}) = 5 \cdot \mathbf{1} + 3$ Replace x with the input value, 1.

$f(1) = \mathbf{5} + 3$ Simplify.

$f(1) = \mathbf{8}$ Simplify.

Since $f(1) = 8$, when the input value is 1, the output value is 8. This function is a set of ordered pairs that includes the ordered pair $(1, 8)$.

Functions are also named with other letters or symbols. For example, the notation $R(x)$ means "the output value of the function R when the input value is x." We use letters to name functions because they are easy to write and talk about. We could name a function "smiley face," ☺$(x) = 5x + 3$, but this is not as convenient as a letter name.

In the next example, the input value is a negative number. When replacing the variable in the function rule, use parentheses.

EXAMPLE 9 If $R(x) = 6x^2 - 9x + 1$, evaluate $R(-2)$.

SOLUTION ▶

$R(x) = 6x^2 - 9x + 1$

$R(\mathbf{-2}) = 6(\mathbf{-2})^2 - 9(\mathbf{-2}) + 1$ Replace x with the input value, -2; use parentheses.

$R(-2) = 6(\mathbf{4}) - 9(-2) + 1$ Simplify.

$R(-2) = \mathbf{24} + \mathbf{18} + 1$ Simplify.

$R(-2) = \mathbf{43}$ Simplify.

Every ordered pair in the function $f(x) = 11$ has an output value of 11. When we evaluate this function for any input value that is a real number, the output value is 11.

EXAMPLE 10 If $f(x) = 11$, evaluate $f(500)$.

SOLUTION ▶ For every input value, the output value is 11. So $f(500) = 11$.

In the next example, the function is written in roster notation and does not have a numerical rule. To evaluate the function for an input value, find the ordered pair with this input value, and identify the corresponding output value.

EXAMPLE 11 If $R = \{$(Fassold, pitcher), (Miller, designated hitter), (Casini, second base), (Sanchez, pitcher)$\}$, evaluate R(Sanchez).

SOLUTION ▶ One of the ordered pairs is (Sanchez, pitcher). When the input is *Sanchez*, the output is *pitcher*. So R(Sanchez) = pitcher.

To use a graph to evaluate a function, estimate the coordinates of the ordered pair with the given input value.

EXAMPLE 12 If the graph represents the function f, evaluate $f(3)$.

SOLUTION ▶ It appears that the point on this graph with an input value of 3 is $(3, 1)$. The y-coordinate of this point is 1. Since $y = f(x)$, $f(3) = 1$.

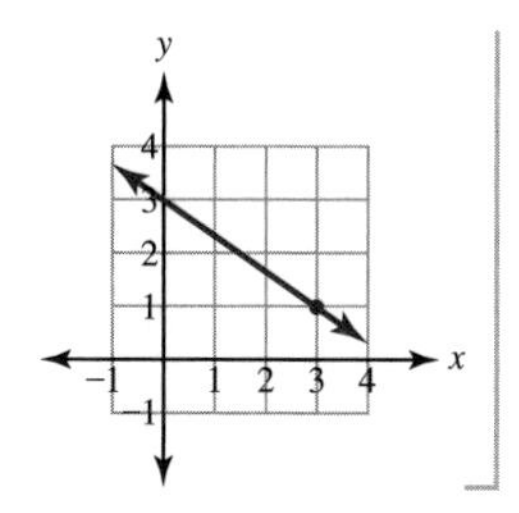

Practice Problems

11. Evaluate $y = -\frac{3}{4}x - 8$ when $x = 12$.

12. If $f(x) = 4x^2 - 5x + 2$, evaluate $f(-3)$.

13. If $N = \{$(George, Clooney), (Brad, Pitt), (Matt, Damon), (Julia, Roberts)$\}$, evaluate N(Brad).

14. Use the graph of the function f to evaluate $f(3)$.

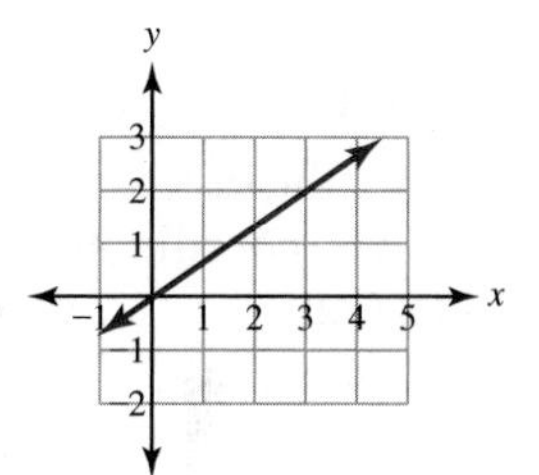

Graphing Functions

Since $y = f(x)$, the graph of $f(x) = 5x + 3$ is the same as the graph of $y = 5x + 3$. To graph a function, build and graph a table of ordered pairs, or use slope-intercept graphing.

EXAMPLE 13 Graph: $f(x) = 5x + 3$

(a) Find three ordered pairs that are solutions of the equation. Graph the points, and draw a straight line through them.

SOLUTION ▶

x	y
0	3
1	8
−1	−2

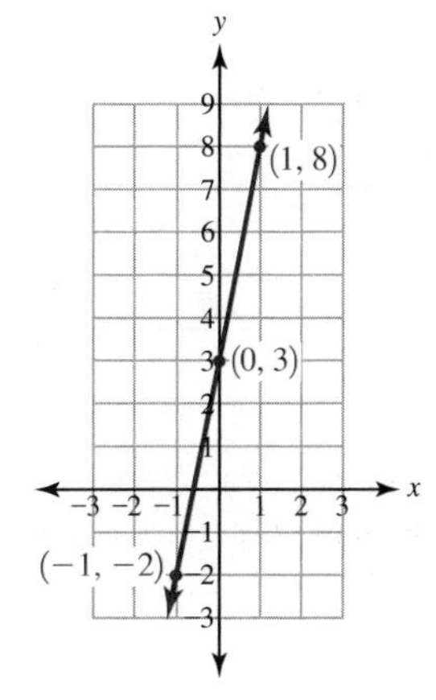

(b) Use slope-intercept graphing.

For $f(x) = 5x + 3$, $m = \frac{5}{1}$, and the y-intercept is $(0, 3)$.

Graph the y-intercept. The rise is 5; move vertically up 5 units. The run is 1; move horizontally to the right 1 unit. Graph the second point. Draw a line through the points.

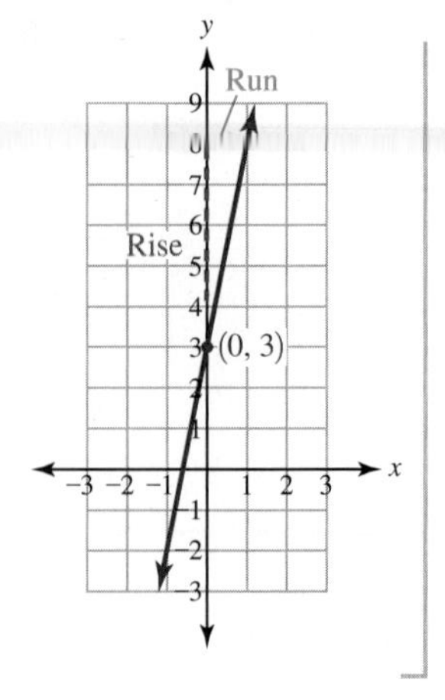

Practice Problems

For problems 15–16, graph the function.

15. $f(x) = -2x + 5$ **16.** $h(x) = 3$

Using Technology: Finding Outputs and Zoom Out

EXAMPLE 14 If $f(x) = x + 9$, evaluate $f(8)$.

Rewrite the function as $y = x + 9$. Type the function in the Y= screen. Press GRAPH. The point on the graph with an x-coordinate of 8 is not visible in a standard window. To make the window larger, press ZOOM. Choose **Zoom Out**.

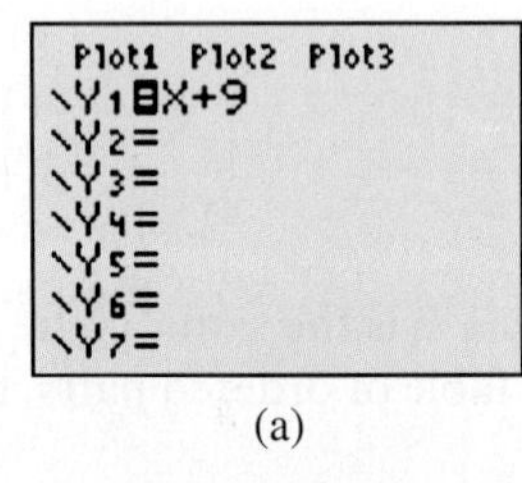

(a)

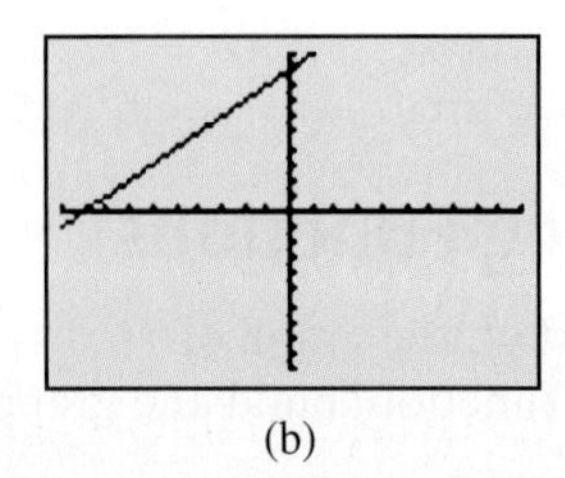
(b)

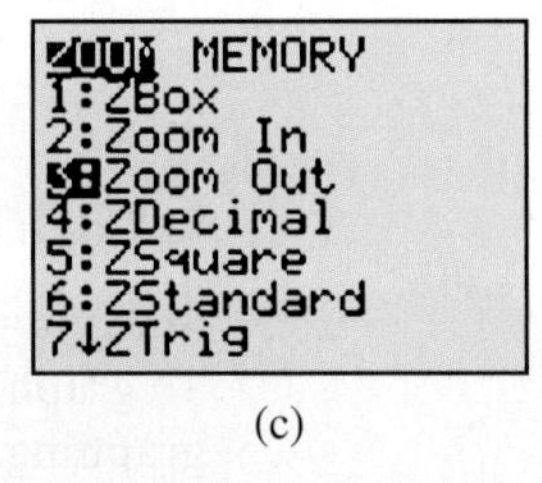

(c)

Press ENTER. The equations X=0 and Y=0 appear at the bottom of the screen, showing that the current center of the graph is (0, 0). To keep (0, 0) as the center, press ENTER. The graph reappears in a new window. Press WINDOW. Since the x-axis now extends from $x = -40$ to $x = 40$, the point on the line with an x-coordinate of 8 is now included in the window.

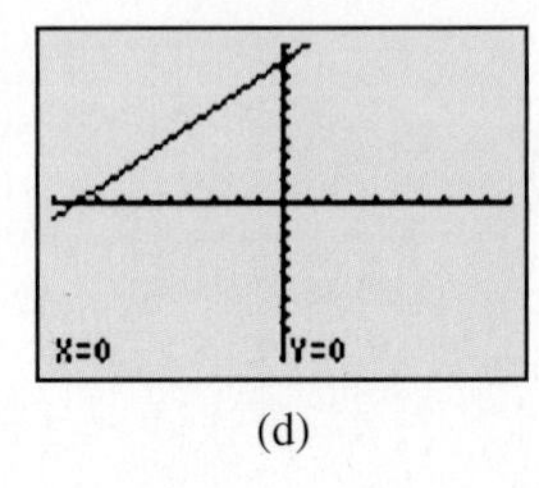

(d)

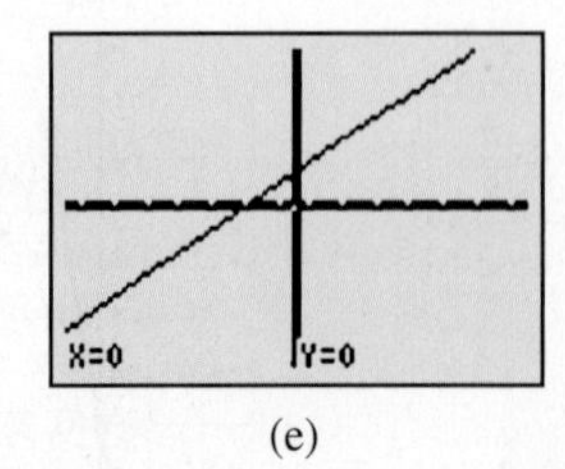

(e)

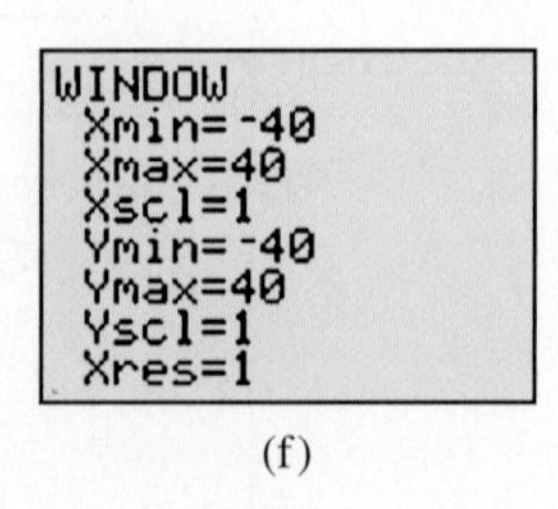

(f)

Go to the CALC screen. Choose 1: value. Press ENTER. Type "8." Press ENTER.

CALCULATE
1:value
2:zero
3:minimum
4:maximum
5:intersect
6:dy/dx
7:∫f(x)dx

(g)

X=■

(h)

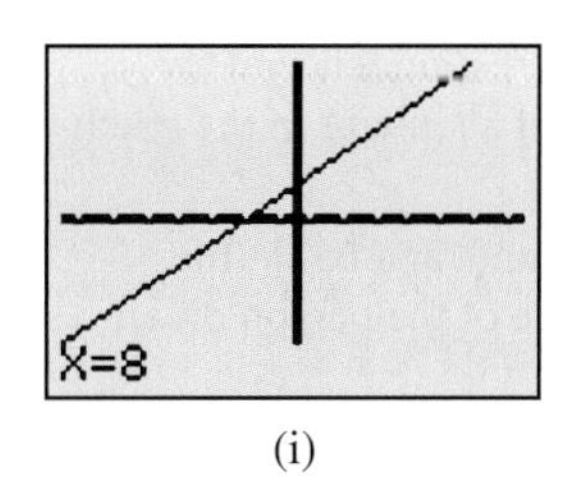

(i)

Y1=X+9
X=8 Y=17

(j)

The cursor moves to $(8, 17)$. The coordinates of the point are shown at the bottom of the screen, X=8 and Y=17. Since an input value of 8 corresponds to an output value of 17, $f(8) = 17$.

Practice Problems For problems 17–19,

(a) graph the function. To evaluate the function for the given input value, it might be necessary to change the window. Sketch the graph; describe the window.

(b) evaluate the function for the given input value.

17. $f(x) = 2x - 8; f(5)$ **18.** $f(x) = 2x - 8; f(14)$

19. $f(x) = 2x - 10; f(-3)$

3.7 Vocabulary Practice

Match the term with its description.

1. The output value of the function f when the input value is x
2. The set of the input values of a function
3. The set of the output values of a function
4. A set that includes all of the rational numbers and all of the irrational numbers
5. Finding the output value in a function that corresponds to a given input value
6. In an ordered pair in a function, this is the first value.
7. In an ordered pair in a function, this is the second value.
8. A set of ordered pairs
9. $y = mx + b$
10. Each vertical line crosses the graph of a function in at most one point.

A. domain
B. evaluating a function
C. $f(x)$
D. input
E. output
F. range
G. real numbers
H. relation
I. slope-intercept form of a linear equation
J. vertical line test

3.7 Exercises

Follow your instructor's guidelines for showing your work.

For exercises 1–6, determine whether the set is a relation.

1. $\{(7, 2), (8, 2), (5, 3)\}$ **2.** $\{(9, 4), (13, 4), (6, 5)\}$

3. $\{0, 2, 4, 6, 8, \ldots\}$ **4.** $\{1, 3, 5, 7, \ldots\}$

5. $y = \frac{3}{4}x - 2$ **6.** $y = \frac{5}{8}x - 1$

For exercises 7–20, identify the set as a relation, a function, or both a relation and a function.

7. $\{(5, 4), (9, 2), (13, 0)\}$

8. $\{(8, 20), (13, 29), (18, 37)\}$

9. $\{(0, 0), (1, 1), (2, 2), (3, 3), \ldots\}$

10. $\{(0, 0), (-1, -1), (-2, -2), (-3, -3), \ldots\}$

11. $\{(0, -1), (1, -1), (2, -1), (3, -1)\}$

12. $\{(0, 1), (1, 1), (2, 1), (3, 1)\}$

13. $\{(2, 1), (2, 2), (2, 3), (2, 4)\}$

14. $\{(3, 1), (3, 2), (3, 3), (3, 4)\}$

15. $y = -6x + 7$

16. $y = -11x + 2$

17. $\{(\text{MN, state}), (\text{VA, state}), (\text{LWS, airport}), (\text{SEA, airport})\}$

18. {(Adam, 130 lb), (Brown, 300 lb), (Ayanbadejo, 230 lb), (Hill, 230 lb)}

19. For an elementary school, the correspondence of the name of a teacher and the name of a student in the teacher's class.

20. For a person with two bank accounts at a bank, the correspondence between the name of the person on an account and the account number.

21. Describe the difference between a relation and a function.

22. Explain why not every set is a relation. Include an example in the explanation.

23. Create a set with six ordered pairs that is a function.

24. Create a set with six ordered pairs that is a relation but is not a function.

For exercises 25–26, the graph represents a relation.
(a) Write a set of ordered pairs in roster notation that represents this relation.
(b) Is this relation a function? Explain.

25.

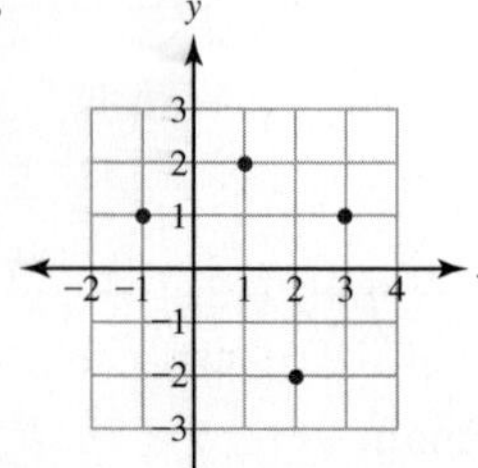

26.

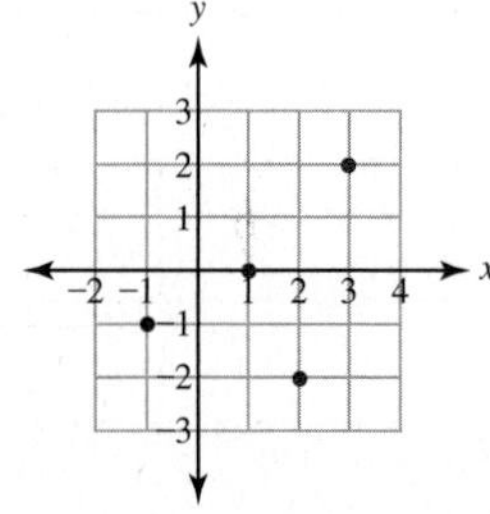

For exercises 27–30, use the vertical line test to determine if the graph represents a function.

27.

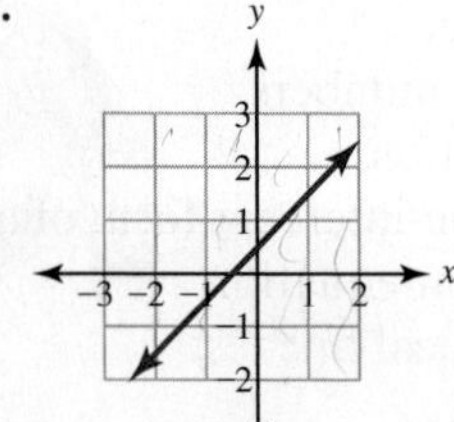

28.

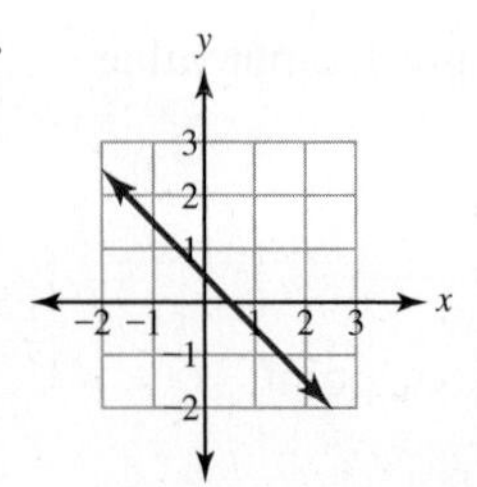

29.

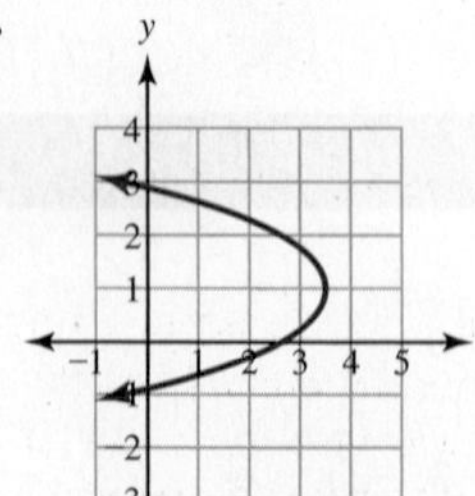

30.

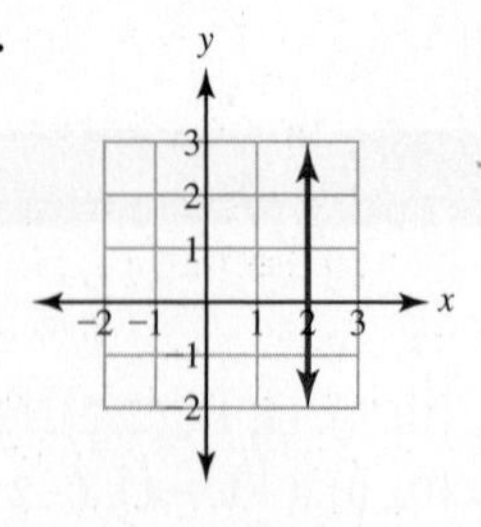

For exercises 31–32,
(a) Graph the relation.
(b) Is the relation a function? Explain.

31. $y = -\frac{5}{6}x + 8$

32. $y = -\frac{3}{4}x + 7$

For exercises 33–40,
(a) use roster notation to represent the domain.
(b) use roster notation to represent the range.

33. {(Buddy, dog), (Joey, dog), (Skunkface, goat), (Oscar, cat)}

34. {(orange, 72 calories), (peach, 37 calories), (apple, 72 calories), (banana, 105 calories)}

35. {(June 6, cloudy), (June 7, partly cloudy), (June 8, clear), (June 9, cloudy)}

36. {(St. Paul, MN), (Austin, TX), (Eugene, OR), (White Bear Lake, MN)}

37. For an order of 20 to 120 buckets, $C(x) = \left(\frac{\$4.80}{1 \text{ bucket}}\right)x$ represents the relationship of the number of buckets ordered, x, and the cost of the order, $C(x)$.

38. For an order of 4 to 11 cigarette waste cans, $C(x) = \left(\frac{\$61.50}{1 \text{ can}}\right)x$ represents the relationship of the number of waste cans ordered, x, and the cost of the order, $C(x)$.

39. The correspondence of the name (in English) of each day of the week and the number of letters in the name.

40. The correspondence of the name (in English) of each day of the week and the first letter of the name.

41. The correspondence of the name of a broadband Internet service plan and the cost per month is {(Road Runner, \$45), (Performance, \$60), (Express, \$30), (Charter, \$55)}.
- **a.** Use a bar graph to represent this relation.
- **b.** Use roster notation to represent the domain.
- **c.** Use roster notation to represent the range.

42. The correspondence of the name of a parking lot and the cost for a parking permit for the lot is {(Blue, \$75), (Gold, \$325), (Red, \$175), (Purple, \$125)}.
- **a.** Use a bar graph to represent this relation.
- **b.** Use roster notation to represent the domain.
- **c.** Use roster notation to represent the range.

43. A relation is $y = -5x + 1$.
- **a.** Identify the independent variable.
- **b.** Identify the dependent variable.
- **c.** Identify the domain.
- **d.** Identify the range.

44. A relation is $y = -4x + 3$.
- **a.** Identify the independent variable.
- **b.** Identify the dependent variable.
- **c.** Identify the domain.
- **d.** Identify the range.

For exercises 45–66,
(a) identify the domain.
(b) identify the range.

45. $f(x) = 9x - 2$

46. $f(x) = 8x - 5$

47. Evaluate $y = 9x + 5$ when $x = -2$.

48. Evaluate $y = 8x + 7$ when $x = -3$.

49. Evaluate $y = 6x - 8$ when $x = 0$.

50. Evaluate $y = 9x - 2$ when $x = 0$.

51. Evaluate $y = \frac{2}{3}x + 8$ when $x = 18$.

52. Evaluate $y = \frac{3}{4}x + 9$ when $x = 24$.

53. If $f(x) = 8x - 9$, evaluate $f(3)$.

54. If $f(x) = 8x - 9$, evaluate $f(4)$.

55. If $g(x) = 10x$, evaluate $g(40)$.

56. If $g(x) = 10x$, evaluate $g(50)$.

57. If $g(x) = 3x^2 - 6x + 5$, evaluate $g(2)$.

58. If $g(x) = 3x^2 - 4x + 6$, evaluate $g(2)$.

59. If $g(x) = 3x^2 - 6x + 5$, evaluate $g(-3)$.

60. If $g(x) = 3x^2 - 4x + 6$, evaluate $g(-3)$.

61. If $f(x) = 8x - 9$, evaluate $f\left(\frac{2}{3}\right)$.

62. If $f(x) = 8x - 9$, evaluate $f\left(\frac{3}{5}\right)$.

63. If $k(x) = 3$, evaluate $k(100)$.

64. If $k(x) = 9$, evaluate $k(200)$.

65. If $f(x) = 0$, evaluate $f(-8)$.

66. If $f(x) = 0$, evaluate $f(-9)$.

For exercises 67–68, C = {(Alachua, Gainesville), (Citrus, Inverness), (Clay, Green Cove Springs), (Escambia, Pensacola), (Marion, Ocala)}.

67. Evaluate C(Marion).

68. Evaluate C(Clay).

For exercises 69–70, E = {(Denver, 5280 ft), (Lewiston, 802 ft), (McCall, 5280 ft), (Moscow, 2583 ft)}.

69. Evaluate E(Denver).

70. Evaluate E(Lewiston).

For exercises 71–72, the graph represents the function f.

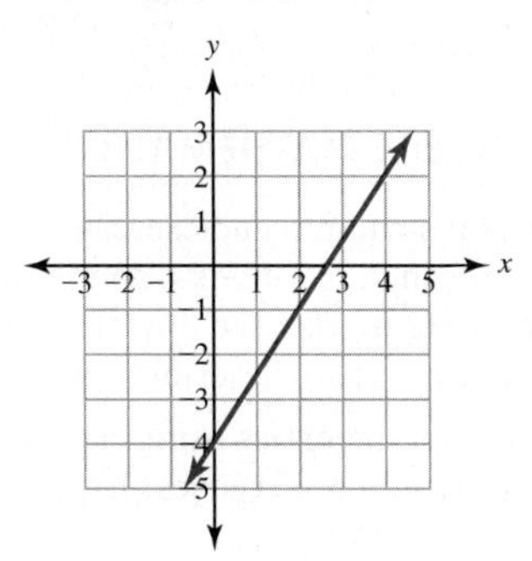

71. Evaluate $f(2)$.

72. Evaluate $f(4)$.

For exercises 73–74, the function

$$N(x) = \left(\frac{17.6 \text{ million devices}}{1 \text{ year}}\right)x + 164.3 \text{ million devices}$$

represents the relationship between the number of mobile computing devices sold in the United States, $N(x)$, and the number of years since 2006, x. (*Source:* www.epa.gov, May 2011)

73. Evaluate N(5 years). **74.** Evaluate N(4 years).

For exercises 75–76, the bar graph represents the function P.

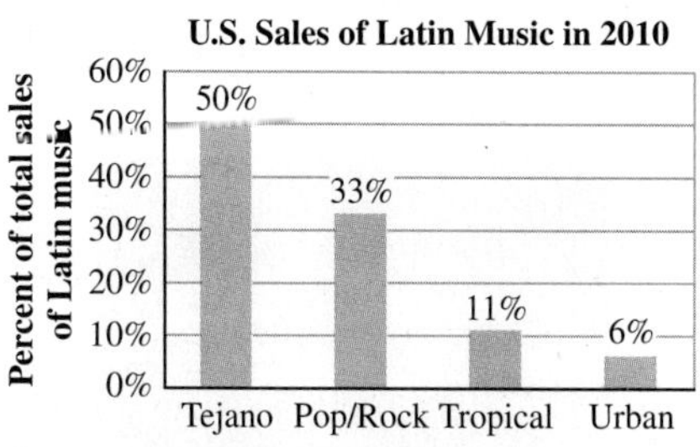

Source: www.riaa.com, 2010

75. Evaluate P(Tejano).

76. Evaluate P(Tropical).

For exercises 77–78, the bar graph represents the function N.

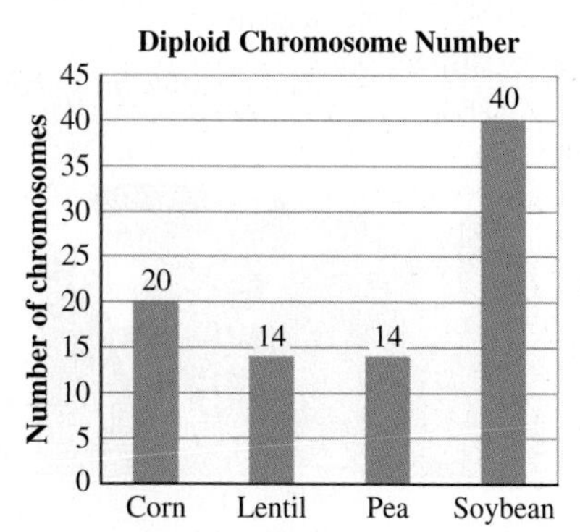

Source: morgan.rutgers.edu

77. Evaluate N(lentil).

78. Evaluate N(soybean).

For exercises 79–84, graph the function.

79. $f(x) = \frac{2}{5}x - 3$

80. $f(x) = \frac{5}{6}x - 8$

81. $h(x) = 6$

82. $w(x) = -4$

83. $\{(-3, 5), (-1, 6), (0, 6), (4, -2)\}$

84. $\{(-4, 1), (-3, 4), (0, 4), (4, -2)\}$

Problem Solving: Practice and Review

Follow your instructor's guidelines for using the five steps as outlined in Section 1.3, p. 55.

85. Groupon, Inc. offers daily deals on a variety of products and services. In Chicago, National Louis University sponsored a Groupon deal for a graduate-level three-credit Introduction to Teaching course. The regular tuition is \$2232. Using the Groupon deal, the tuition is \$950. Find the percent discount in tuition. Round to the nearest percent. (*Source:* www.chicagotribune.com, Sept. 5, 2011)

86. The horizontal distance from the base of the stairs to a point under the edge of the landing is 14 ft. The vertical distance from the base of the stairs to the landing is 12 ft. Find the slope of these stairs.

© Laura Bracken

For exercises 87–88, fenbendazole is used for stomach worm control in goats. The concentration of fenbendazole in Safe-Guard® Goat is $\frac{100 \text{ mg fenbendazole}}{1 \text{ mL suspension}}$. The recommended dosage is $\frac{2.9 \text{ mL suspension}}{125 \text{ lb}}$. (*Source:* www.drugs.com/vet)

© Laura Bracken

87. Spot the goat weighs 110 lb. Find the amount of suspension to give to Spot. Round to the nearest tenth.

88. Petunia the goat weighs 80 lb. Find the amount of fenbendazole (in milligrams) to give to Petunia.

Technology

For exercises 89–92,

(a) graph the function on a graphing calculator. Sketch the graph; describe the window.

(b) evaluate the function for the given input value.

89. $f(x) = 3x - 4$; $f(2)$

90. $f(x) = 3x - 4$; $f(-4)$

91. $f(x) = 3x + 2$; $f(8)$

92. $f(x) = -3x + 28$; $f(1)$

Find the Mistake

For exercises 93–96, the completed problem has one mistake.

(a) Describe the mistake in words, or copy down the whole problem and highlight or circle the mistake.

(b) Do the problem correctly.

93. Problem: Use roster notation to represent the domain of $\{(8, 2), (2, 8), (0, 0), (5, -3)\}$.

Incorrect Answer: The domain is $\{2, 8, 0, -3\}$.

94. Problem: If $f(x) = x^2 - x + 9$, evaluate $f(-6)$.

Incorrect Answer:
$$f(x) = x^2 - x + 9$$
$$f(x) = (-6)^2 - 6 + 9$$
$$f(x) = 36 - 6 + 9$$
$$f(x) = 39$$

95. Problem: Graph: $p(x) = 2$

Incorrect Answer:

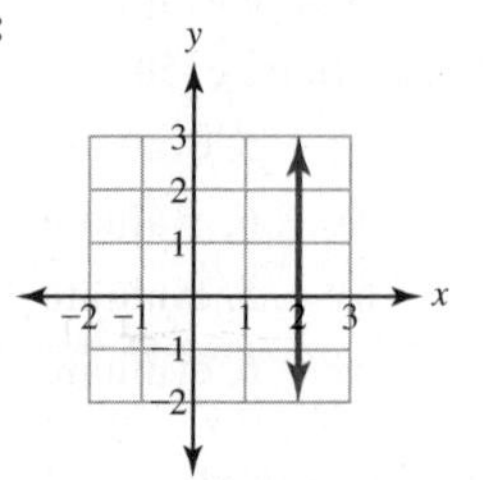

96. Problem: Use the vertical line test to determine whether the graph represents a function.

Incorrect Answer: This graph represents a function. A vertical line can be drawn that does not cross it more than once.

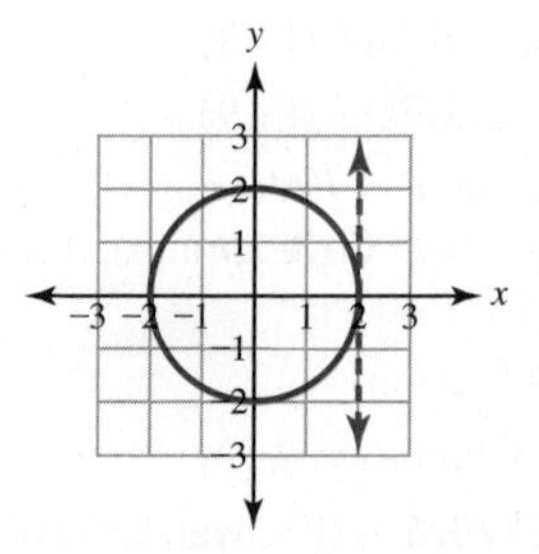

Review

97. Solve: $2(x - 9) = 5x - 3(x + 6)$

98. Solve: $2(x - 9) = 5x + 15$

99. Solve: $2(x - 9) = 4x - 3(x + 6)$

100. Solve: $2(x - 9) = 5x - 3(x + 7)$

SUCCESS IN COLLEGE MATHEMATICS

101. Mathematics instruction and exercises are now available on the Internet, in e-books, and in on-line homework systems. Have you used any of these types of electronic instruction? If so, describe your experience.

102. Does learning mathematics using electronic instruction match your learning preferences? Explain.

Study Plan for Review of Chapter 3

SECTION 3.1 Ordered Pairs and Solutions of Linear Equations

Ask Yourself	Test Yourself	Help Yourself
Can I...	**Do 3.1 Review Exercises**	**See these Examples and Practice Problems**
graph an ordered pair on a correctly labeled xy-coordinate system?	1	Ex. 1–4, PP 1–11
complete a table with three solutions of a linear equation and graph the equation?	2–5	Ex. 5–9, PP 12–20

3.1 Review Exercises

1. Graph each ordered pair on the same coordinate system. Label the axes; write a scale for each axis.

a. $(4, -3)$ **b.** $(-2, 5)$
c. $(-6, -7)$ **d.** $(0, 0)$
e. $(1, 3)$

For exercises 2–5,
(a) complete the table of solutions.
(b) graph the equation.

2. $2x - 5y = 10$

x	y
0	
	0
−5	

3. $3x + 2y = -9$

x	y
0	
	0
3	

4. $6x - y = -6$

x	y
0	
	0
2	

5. $x + 2y = 16$

x (gigabytes)	y (gigabytes)
0	
	0
8	

SECTION 3.2 Standard Form and Graphing with Intercepts

Ask Yourself	Test Yourself	Help Yourself
Can I...	**Do 3.2 Review Exercises**	**See these Examples and Practice Problems**
write the standard form of a linear equation? clear the fractions and rewrite a linear equation in standard form?	6	Ex. 1–2, PP 1–4
given a linear equation or its graph, identify the intercepts?	7–14	Ex. 3–5, PP 5–11
find three solutions of a linear equation in standard form and graph the equation?	12–14, 18	Ex. 3–5, PP 5–11
find three solutions of a linear equation with an isolated variable and graph the equation?	15–17	Ex. 6–7, PP 12–14
write the equation of a horizontal or vertical line that passes through a given point and graph the equation?	20–21	Ex. 8–9, PP 15–16

3.2 Review Exercises

6. Clear any fractions and rewrite the equation in standard form.

a. $y - 6 = -9(x - 1)$ **b.** $2x = \frac{1}{8}y + \frac{4}{5}$

For exercises 7–11,
(a) identify the y-intercept.
(b) identify the x-intercept.

7.

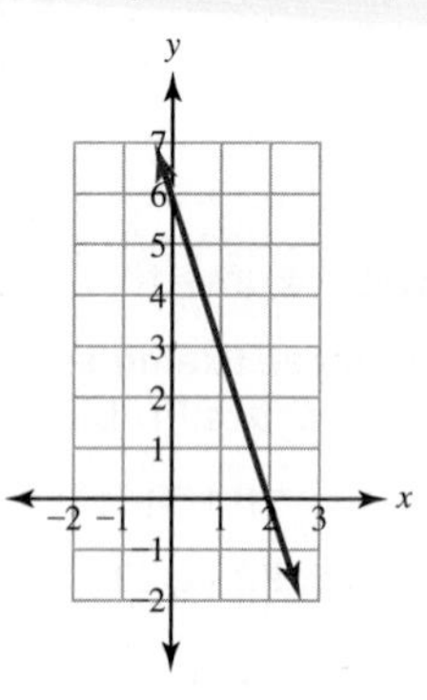

8.

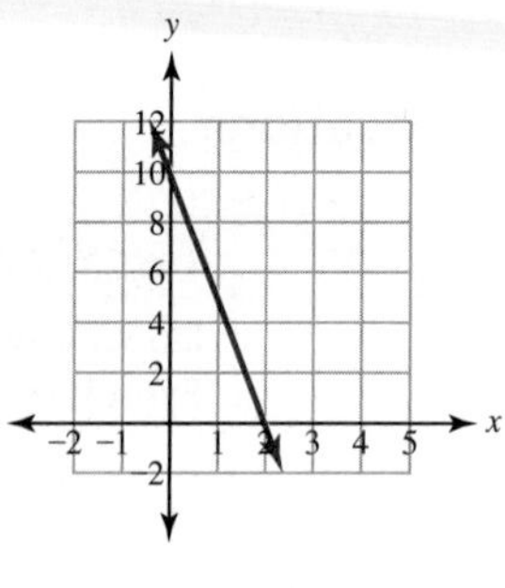

9.

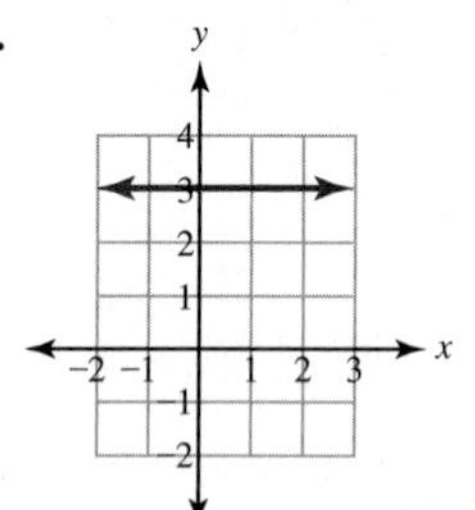

10.

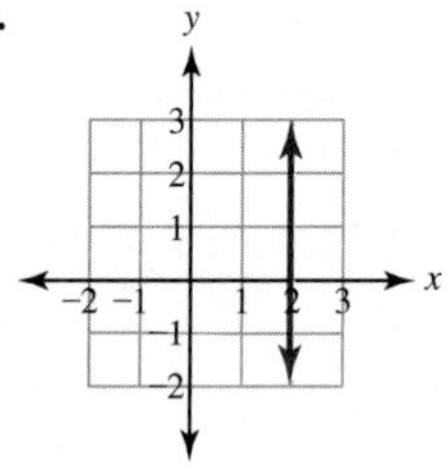

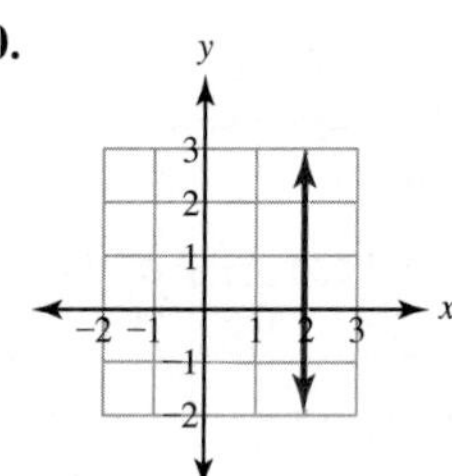

11.

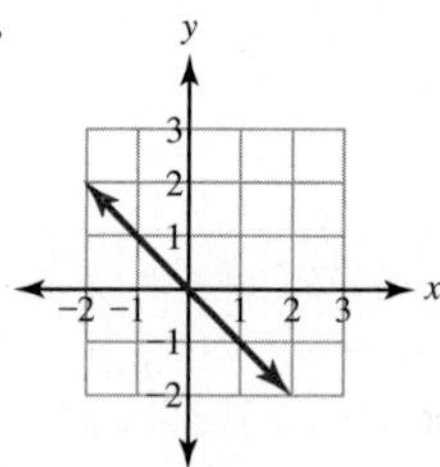

For exercises 12–14,
(a) Find the y-intercept.
(b) Find the x-intercept.
(c) Find a third solution of the equation.
(d) Graph the equation.

12. $6x + y = -12$

13. $6x - 5y = 180$

14. $-5x + 2y = 15$

For exercises 15–16,
(a) find three solutions of the equation.
(b) graph the equation.

15. $y = -3x + 7$

16. $y = \frac{2}{3}x - 8$

17. To graph $y = -\frac{5}{6}x + 1$, a student chose values for x of 0, 6, and 12. Explain why these are good choices for graphing this equation.

18. Identify the solutions of $7x - 3y = -21$. More than one answer may be correct.

a. $(0, -7)$

b. $(0, 7)$

c. $(-3, 0)$

d. $\left(-\frac{3}{7}, 6\right)$

e. $\left(4, \frac{7}{3}\right)$

19. The equation of a line is $2x + 3y = 6$. How many solutions are there for this equation?

20. A vertical line passes through the point $(8, 2)$.

a. Write the equation of the line.

b. Graph the equation.

21. A horizontal line passes through the point $(8, 2)$.

a. Write the equation of the line.

b. Graph the equation.

SECTION 3.3 Average Rate of Change and Slope

Ask Yourself	Test Yourself	Help Yourself
Can I...	**Do 3.3 Review Exercises**	**See these Examples and Practice Problems**
find an average rate of change?	22–24	Ex. 1–3, PP 1–3
write the slope formula? given two points on a line, use the slope formula to find its slope?	25–26, 31–33	Ex. 4–13, PP 4–7, 10–13
use the graph of a linear equation to find its slope?	27–30	Ex. 6–7, 9, PP 4–9
given a linear equation, find the slope of the line?	31	Ex. 11, PP 11
identify the slope of a horizontal or vertical line?	32–33	Ex. 12–13, PP 12–13

3.3 Review Exercises

22. A group of employees share a large 55-cup coffee urn. Two hours after the coffee was made, there were 30 cups of coffee left in the urn.

a. Represent the information as two ordered pairs.

b. Find the average rate of change in the amount of coffee in the urn.

23. Fifteen minutes after an IV is started, an IV bag contains 800 mL of solution. Twenty minutes after the IV is started, the bag contains 750 mL of solution.

a. Represent the information as two ordered pairs.

b. Find the average rate of change in the amount of solution in the IV bag.

24. At the beginning of a trip, there are 12 gal of gas in the gas tank. After 3 hr, there are 5.5 gal of gas in the tank. Will the average rate of change in the volume of gas in the gas tank be a positive or a negative number? Explain.

For exercises 25–26, find the slope of the line that passes through the given points.

25. $(5, -8); (2, 11)$

26. $(9, -2); (7, -16)$

For exercises 27–30,

(a) identify two points on the line.

(b) use the graph to find the slope of the line.

27.

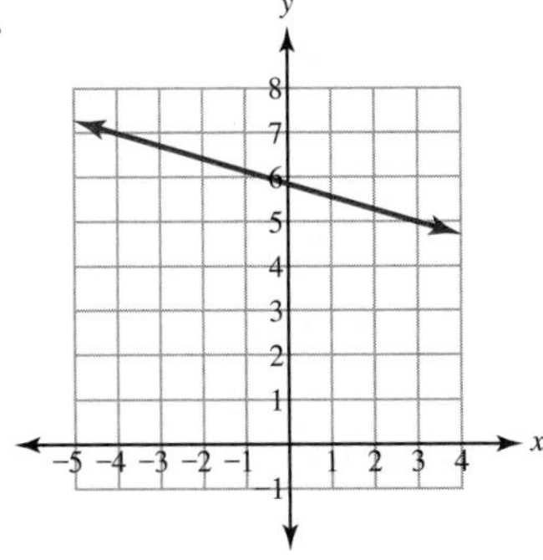

28.

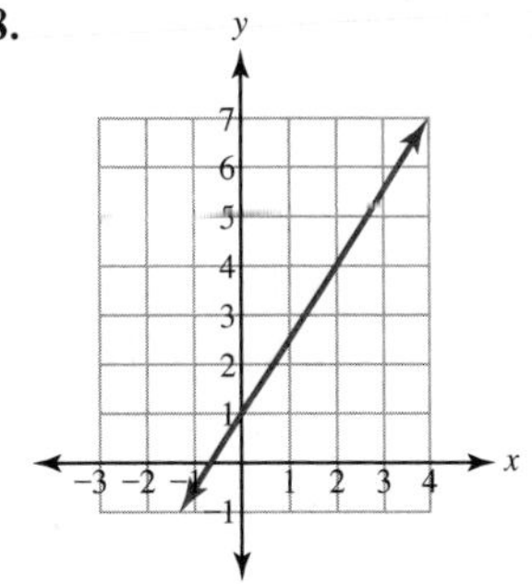

29.

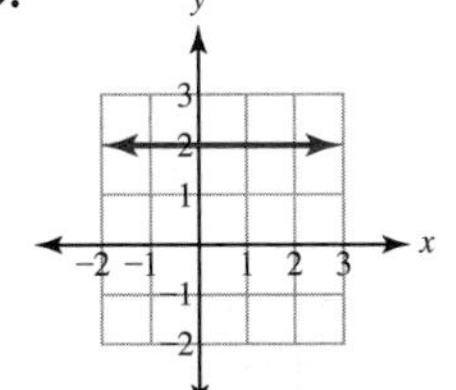

30.

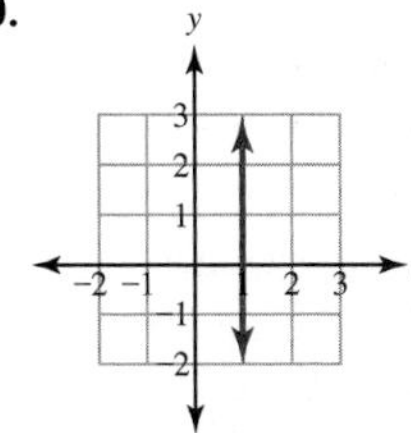

31. Find the slope of the line represented by $2x - 15y = 30$.

32. Find the slope of the line that passes through $(3, -15)$, and $(3, 2)$.

33. Find the slope of the line that passes through $(-1, 4)$ and $(-8, 4)$.

SECTION 3.4 Slope-Intercept Form

Ask Yourself	Test Yourself	Help Yourself
Can I...	**Do 3.4 Review Exercises**	**See these Examples and Practice Problems**
write and evaluate a linear equation that is a model? identify the slope and intercepts of a linear model and describe what they represent?	34–35	Ex. 1–2, PP 1–2
write the slope-intercept form of a linear equation? given the slope and y-intercept, write the equation of a line in slope-intercept form?	36	Ex 3, PP 3–6
rewrite a linear equation in slope-intercept form?	37–40	Ex 4–5, PP 7–10, 20–21
given the equation of a line in slope-intercept form, identify the slope, y-intercept and x-intercept?	37–40	Ex 4–5, PP 7–10
given the graph of a linear equation, write the equation in slope-intercept form?	41–43, 47	Ex 6–7, PP 11–13
given a point on a horizontal or vertical line, write its equation and graph the line?	44–45	Ex. 8–9, PP 14–15
use slope-intercept graphing to graph a linear equation?	46	Ex. 10–13, PP 16–21

3.4 Review Exercises

34. The relationship of the number of full-time students enrolled at Coastal Carolina University, y, and the number of years since 2000, x, is
$y = \left(\frac{407.13 \text{ students}}{1 \text{ year}}\right)x + 4059.2$ students.
(*Source:* www.che.sc.gov, 2010)

   a. Use the equation to predict the number of full-time students in 2014. Round to the nearest whole number.
   b. Identify the y-intercept of the line, and describe what the y-coordinate of the y-intercept represents.
   c. Identify the slope of the line, and describe what the slope represents.

35. The capacity of a landfill is 2,000,000 tons of compressed garbage. The average rate of garbage deposited into the landfill is 4177 tons per day.

a. Write an equation that represents the relationship of the remaining capacity of the landfill, y, and the time, x.

b. Use the equation to find the remaining capacity of the landfill after 90 days. Round to the nearest thousand.

c. Find the x-intercept of this model, rounding the x-coordinate to the nearest whole number, and describe what the x-coordinate of the x-intercept represents.

d. Identify the y-intercept of this model, and describe what the y-coordinate of the y-intercept represents.

36. The slope of a line is –6. Its y-intercept is $(0, 8)$. Write the equation of this line in slope-intercept form.

For exercises 37–40,
(a) rewrite the equation in slope-intercept form. Clear any fractions.
(b) identify the slope.
(c) identify the y-intercept. Write the ordered pair, not just the y-coordinate.
(d) find the x-intercept. Write the ordered pair, not just the x-coordinate.

37. $3x - 8y = 21$

38. $\frac{3}{4}x + \frac{2}{3}y = 6$

39. $-y = 7$

40. $x = y$

For exercises 41–43,
(a) identify the y-intercept.
(b) identify the slope.
(c) write the equation of the graphed line in slope-intercept form.

41.

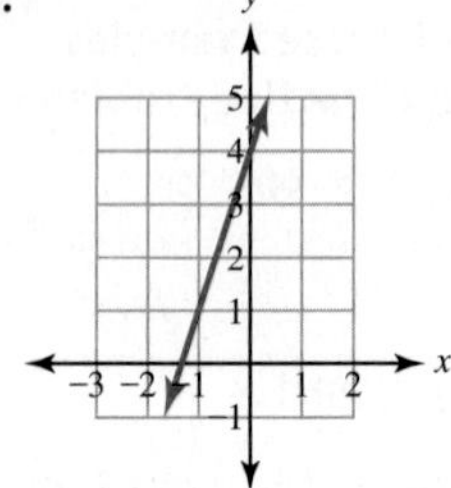

42.

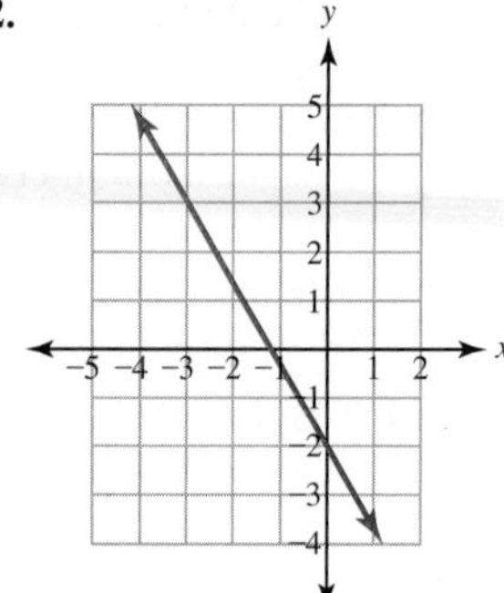

43.

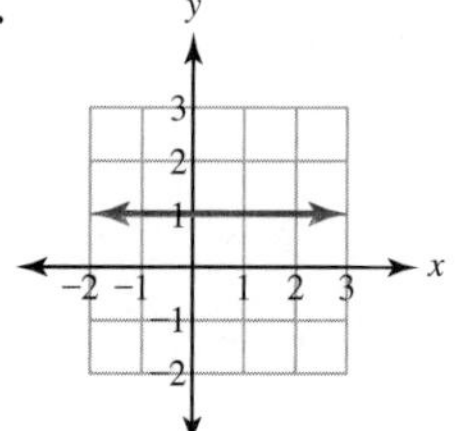

44. A vertical line passes through $(-1, 4)$.

a. Graph the line.

b. Write the equation of the line.

45. A horizontal line passes through $(-1, 4)$.

a. Graph the line.

b. Write the equation of the line.

46. Use slope-intercept graphing to graph the equation.

a. $y = \frac{4}{7}x - 5$

b. $y = -3x + 8$

47. Choose the equation(s) that represent the graph.

a. $5x + 3y = 9$

b. $y = \frac{5}{3}x + 3$

c. $y = -\frac{5}{3}x + 3$

d. $y = -\frac{3}{5}x + 3$

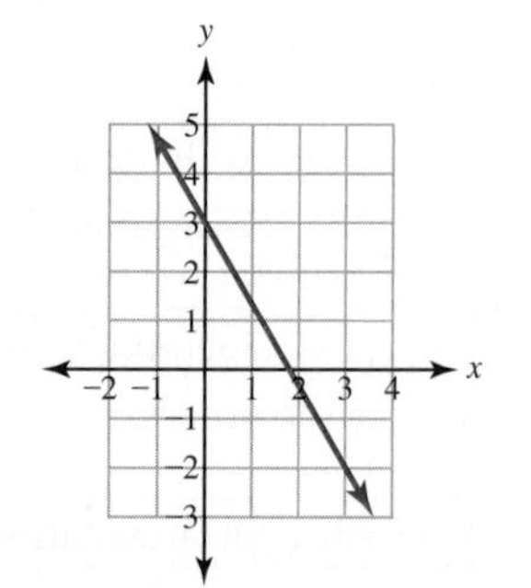

SECTION 3.5 Point-Slope Form and Writing the Equation of a Line

Ask Yourself	Test Yourself	Help Yourself
Can I...	**Do 3.5 Review Exercises**	**See these Examples and Practice Problems**
write the point-slope form of a linear equation? given the slope and a point on a line, write the equation of the line in slope-intercept form?	48–51	Ex. 1, PP 1–2
given two points on a line or the graph of a line, write the equation of the line in slope-intercept form?	52–58	Ex. 2–4, PP 3–5
write the equation of a line of best fit and use the equation to solve an application problem?	59	Ex. 5, PP 6

Ask Yourself	Test Yourself	Help Yourself
Can I...	Do 3.5 Review Exercises	See these Examples and Practice Problems
identify the slope of a parallel or perpendicular line?	60–65	Ex. 6–11, PP 7–11
write the equation in slope-intercept form of a parallel or perpendicular line that passes through a given point?	63–65	Ex. 6–7, 10–11, PP 7–11
given the graph of two lines, determine whether they are parallel, perpendicular, or neither?	66	Ex. 8–9

3.5 Review Exercises

For exercises 48–51, write the equation in slope-intercept form of the line with the given slope that passes through the given point.

48. slope $= 8$; $(15, -29)$

49. slope $= \frac{11}{2}$; $(-6, 3)$

50. slope $= -9$; $(0, 2)$

51. slope $= 0$; $(0, -5)$

For exercises 52–57, write the equation in slope-intercept form of the line that passes through the given points.

52. $(15, 21)$; $(18, 15)$

53. $(-2, 8)$; $(6, 10)$

54. $(6, -2)$; $(5, -2)$

55. $(3, 3)$; $(8, 8)$

56. $(7, 2)$; $(0, 4)$

57. $(0, 1)$; $(8, 1)$

58. Write the equation of the graphed line in slope-intercept form.

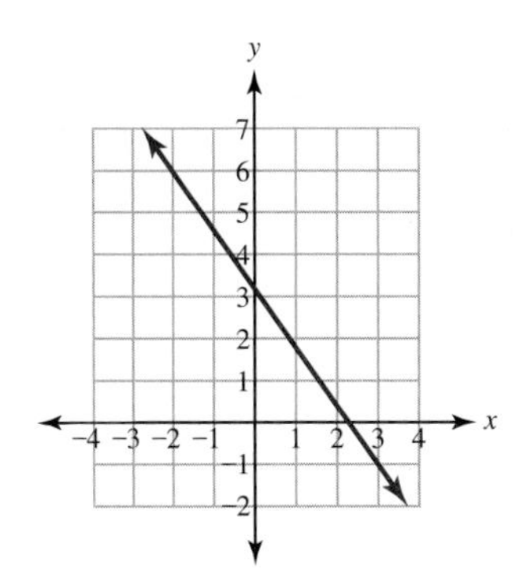

59. The graph shows the relationship between the number of violent crimes committed in California, y, and the number of years since 2001, x.

a. Use the equation of the line of best fit to predict the number of violent crimes in 2013.

b. Identify the slope of the line, and describe what the slope represents.

c. Identify the y-intercept of the line, and describe what the y-coordinate of the y-intercept represents.

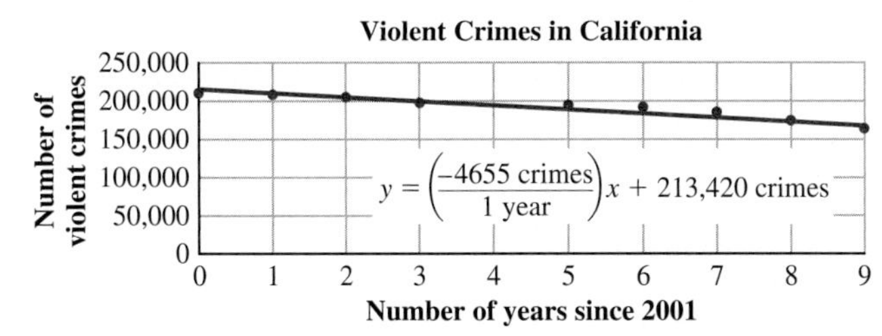

Source: oag.ca.gov, 2010

60. The equation of a line is $y = -\frac{5}{6}x + 1$.

a. Identify the slope of a line that is parallel to this line.

b. Identify the slope of a line that is perpendicular to this line.

61. The equation of a line is $9x + 5y = 20$. Identify the slope of a line that is parallel to this line.

a. 5 **b.** 9

c. $\frac{9}{5}$ **d.** $\frac{5}{9}$

e. $-\frac{9}{5}$ **f.** 0

62. The slope of a line is $\frac{3}{4}$. Identify the equation(s) of the line that is perpendicular to this line.

a. $y = -\frac{4}{3}x + 8$ **b.** $y = -\frac{3}{4}x - 1$

c. $y = \frac{4}{3}x + 2$ **d.** $4x + 3y = 21$

63. The slope of a line is 2.

a. Write the equation in slope-intercept form of a line that is parallel to this line and passes through $(5, 1)$.

b. Write the equation in slope-intercept form of a line that is perpendicular to this line and passes through $(5, 1)$.

64. The equation of a line is $2x + 3y = 6$.

a. Write the equation in slope-intercept form of a line that is parallel to this line and passes through $(9, 4)$.

b. Write the equation in slope-intercept form of a line that is perpendicular to this line and passes through $(9, 4)$.

65. A line passes through $(2, 8)$ and $(4, 20)$.

a. Write the equation in slope-intercept form of a line that is parallel to this line and passes through $(7, 1)$.

b. Write the equation in slope-intercept form of a line that is perpendicular to this line and passes through $(12, 18)$.

66. **a.** Identify the slope of line A in the graph.

b. Identify the slope of line B.

c. Explain why the slopes show that these lines are not perpendicular.

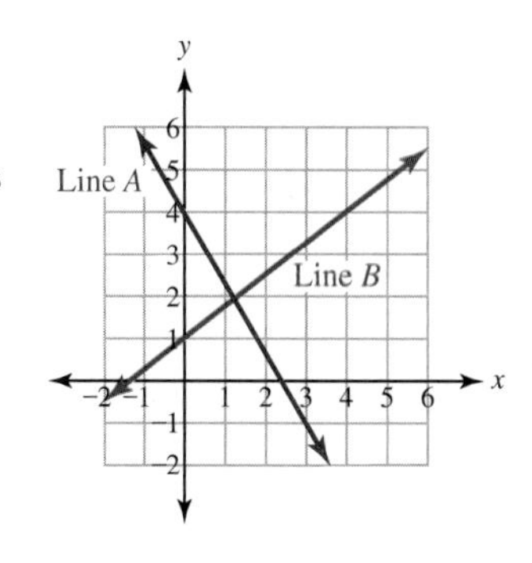

SECTION 3.6 Linear Inequalities in Two Variables

Ask Yourself	Test Yourself	Help Yourself
Can I...	**Do 3.6 Review Exercises**	**See These Examples and Practice Problems**
describe the meaning of ≥, ≤, >, and <? determine whether the boundary line of the graph of an inequality is solid or dashed? Identify whether an inequality is strict or not strict? use a test point to determine the shaded region of the graph of an inequality? graph a linear inequality?	67–73	Ex. 1–3, PP 1–4
write the inequality represented by a graph?	74–76	Ex. 4–6, PP 5–7
write a linear inequality that represents a constraint?	77–79	Ex. 7–9, PP 8–12
graph a linear inequality that represents a constraint?	78	Ex. 10–11, PP 13–14

3.6 Review Exercises

67. Is the boundary line of the graph of the inequality solid or dashed?
- **a.** $3x + 5y > 15$
- **b.** $y < -2x + 3$
- **c.** $x \leq 4$
- **d.** $y > 6$
- **e.** $x + y \geq 8$

68. Choose the sentence that is *not* true for a strict linear inequality.
- **a.** A point in the solution region is a solution of the inequality.
- **b.** A point on the boundary line is a solution of the inequality.
- **c.** If the inequality represents a constraint, the solution region is also called the feasible region.

69. The graph represents the inequality $y \leq -\frac{1}{2}x + 3$.
- **a.** Choose a test point in the solution region of the graphed inequality.
- **b.** Use the test point to show that the shading of the solution region is correct.

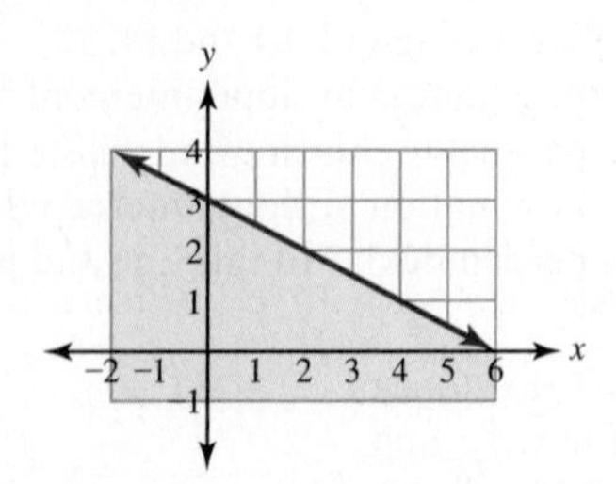

For exercises 70–73, graph the inequality.

70. $5x - 9y > -45$

71. $y \leq \frac{1}{2}x - 5$

72. $x < -6$

73. $y \geq 0$

For exercises 74–76, write the inequality represented by the graph.

74.

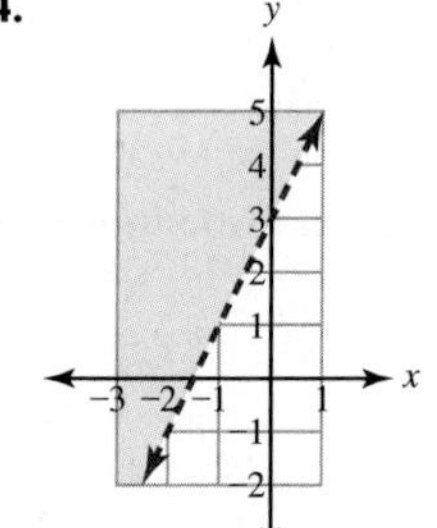

75.

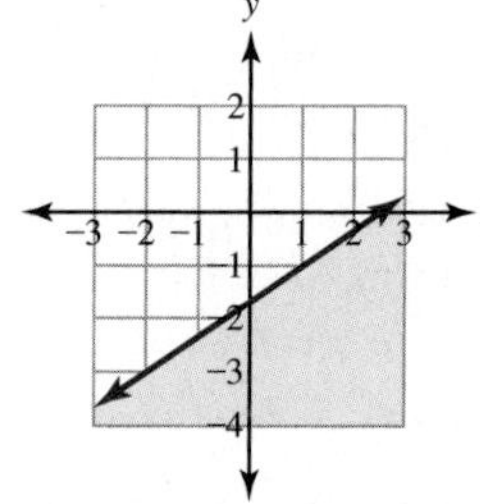

76.

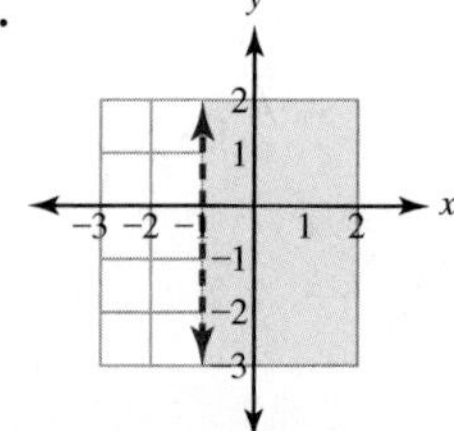

77. When peach trees are pruned with a central leader, an orchard should have no more than 201 trees per acre. Assign one variable, and write an inequality that describes this constraint. (*Source:* www.omafra.gov.on.ca, 2011)

78. The sum of the arithmetic reasoning score and the word knowledge and paragraph comprehension score on the Armed Services Vocational Aptitude Battery must be greater than or equal to 100. (*Source:* www.us-army-info.com)
- **a.** Assign two variables, and write an inequality that represents the constraint.
- **b.** Graph this equality.

79. A state required employers to pay workers who earn tips at least 60% of the minimum wage. Assign two variables, and write an inequality that represents the constraint.

SECTION 3.7 Relations and Functions

Ask Yourself	Test Yourself	Help Yourself
Can I . . .	**Do 3.7 Review Exercises**	**See these Examples and Practice Problems**
write a definition of relation? determine whether a set is a relation?	80	Ex. 1, PP 1–4
write a definition of function? determine whether a set is a function?	81–86	Ex. 2, PP 1–4
use the vertical line test to determine whether a graph represents a function?	85–86	Ex. 3, PP 5–7
evaluate a function?	87–91	Ex. 7–12, PP 11–14
graph a relation or function?	93–94	Ex. 5, 13, PP 9, 15–16
identify the domain and range of a relation or function?	92, 95	Ex. 4–6, PP 8–10
identify the independent and dependent variable in a relation or function?	92	Ex. 6, PP 10

3.7 Review Exercises

80. Explain why the set of whole numbers is not a relation.

For exercises 81–84, is the relation a function? Explain why.

81. $\{(1, 5), (2, 6), (3, 7)\}$

82. $\{(8, 5), (9, 5), (10, 3), (11, 3)\}$

83. $\{(5, 8), (5, 9), (3, 10), (3, 11)\}$

84. The correspondence between the shoe sizes of the students in a class and their heights. Some students with different heights have the same shoe sizes. Some students have the same heights.

For exercises 85–86, does the graph represent a function?

85.

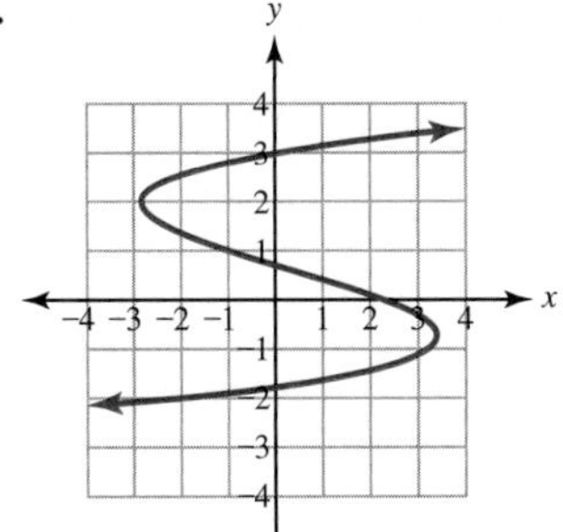

86.

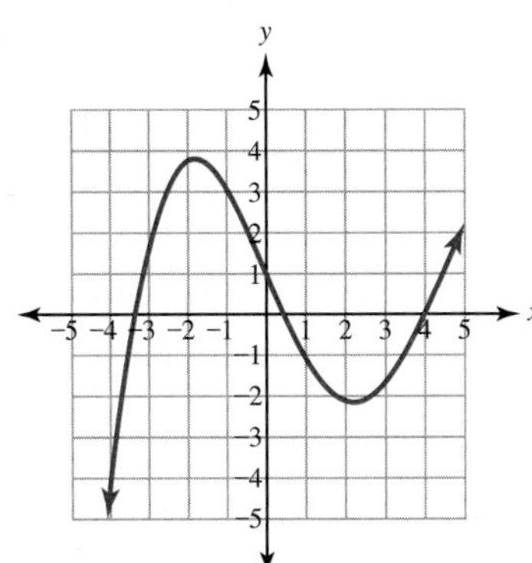

87. If $f(x) = x^2 + 8x - 3$, evaluate $f(-2)$.

88. Evaluate $y = \frac{3}{4}x - 18$ when $x = 20$.

89. If $f(x) = 4$, evaluate $f(7)$.

90. If $H = \left\{\left(\text{John}, 7\frac{3}{8}\right), \left(\text{Bob}, 7\frac{1}{4}\right), \left(\text{Mike}, 7\frac{5}{8}\right)\right\}$, evaluate $H(\text{Bob})$.

91. The graph represents the function f.

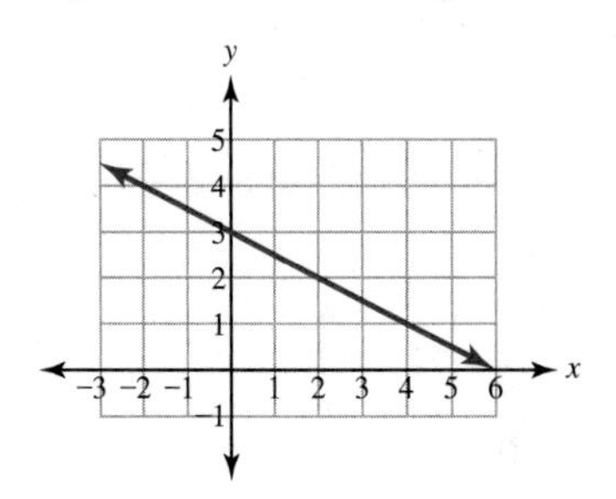

a. Evaluate $f(2)$.
b. If $f(x) = 4$, what is x?

92. A function is $y = \frac{5}{6}x - 4$.

a. Identify the independent variable.
b. Identify the dependent variable.
c. Identify the domain.
d. Identify the range.

93. Graph: $f(x) = -5x - 2$

94. The U.S. Department of Agriculture divides food spending into four categories, ranging from "thrifty" to "liberal." The chart shows the relationship of each category and the monthly cost for food at home for a family of two adults, 19–50 years old. Use a bar graph to represent this function.

Category	Monthly cost for food at home
Thrifty	\$361
Low-cost	\$459
Moderate	\$569
Liberal	\$711

Source: www.lmtribune.com, March 15, 2009

95. A function is {(Laurie, 21 years), (Judy, 23 years), (Jennifer, 24 years), (Jane, 24 years)}.
  a. Use roster notation to represent the domain.
  b. Use roster notation to represent the range.

Chapter 3 Test

1. The graph represents a linear equation.
  a. Identify the y-intercept.
  b. Identify the x-intercept.

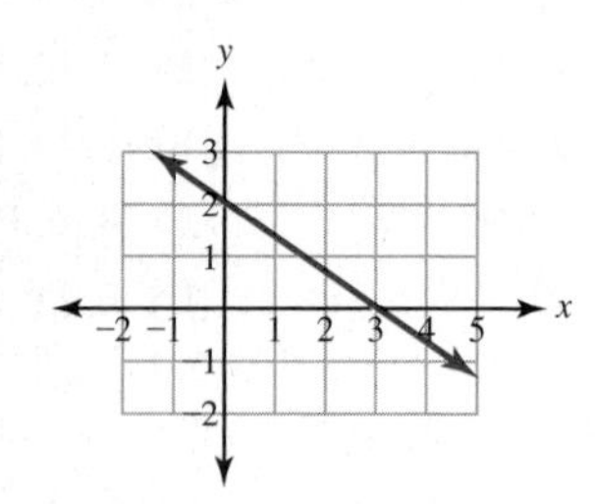

2. An equation is $9x - 4y = 36$.
  a. Find the y-intercept.
  b. Find the x-intercept.
  c. Find a third solution of the equation.
  d. Graph the equation.

3. Use slope-intercept graphing to graph $y = -\frac{4}{3}x + 5$.

For problems 4–9, graph the equation or inequality.

4. $x = 5$

5. $2x + 3y = -9$

6. $y \leq -4x + 10$

7. $y = -3$

8. $y = \left(\frac{3 \text{ mi}}{1 \text{ hr}}\right)x + 7 \text{ mi}$

9. $\$2.50x + \$3.00y < \$150$

10. Find the slope of the line that passes through $(-15, 2)$ and $(-7, 12)$.

11. Find the x-intercept of the graph of $-7x + 4y = 28$.

12. Write the equation in slope-intercept form of the line that passes through $(4, 9)$ and $(8, 11)$.

13. Write the equation of the horizontal line that passes through the point $(150, 800)$.

14. Write the equation in slope-intercept form of a line that passes through the point $(1, 2)$ and is perpendicular to the line $y = \frac{1}{3}x + 9$.

15. Write the inequality represented by the graph.

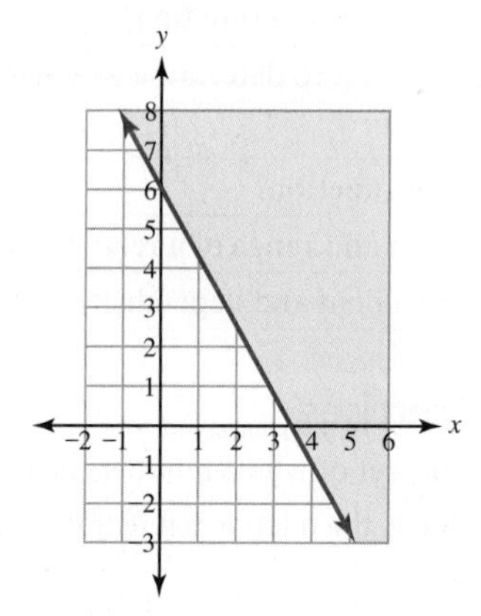

16. Explain how to determine whether the graph of a relation represents a function.

17. A function is $f(x) = -x + 9$. Evaluate $f(-15)$.

18. A function is $B = \{(2, 3), (4, 0), (6, -3), (8, -6)\}$. Evaluate $B(6)$.

19. Rewrite $y - 6 = 5(x + 9)$ in standard form.

20. The graph shows the relationship of the population of the United States, y, and the number of years since 1990, x.

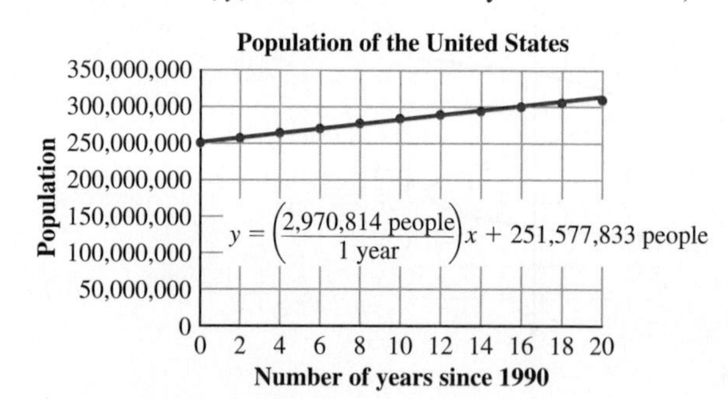

Source: www.census.gov, 2011

  a. Use the equation of the line of best fit to predict the population of the United States in 2015. Round to the nearest million.
  b. Identify the average rate of increase in population per year.

Cumulative Review Chapters 1–3

Follow your instructor's directions for showing your work.

1. Use roster notation to describe the set of integers.
2. Evaluate: $-24 \div 3 \cdot 2 - (3 - 7)^2$
3. Simplify: $\frac{5}{6}(12x - 9) - \frac{1}{3}(15x + 8)$
4. Simplify: $-8(x^2 - 3x + 2) - (x - 4)$
5. Name the property illustrated by the equation $3 + 7 = 7 + 3$.
6. Write the distributive property.
7. Find the perimeter of a rectangle that is 6 in. long and 2.5 in. wide.
8. Find the simple interest earned on \$250 invested for 3 years at 3.2%.
9. Find the area of a circle with a diameter of 6.2 in. ($\pi \approx 3.14$.) Round to the nearest tenth.
10. A box has a height of 3 ft, a width of 16 in., and a length of 4 ft. Find its volume in cubic inches.
11. Evaluate: $\sqrt{121}$
12. The legs of a right triangle are 3 in. and 4 in. Find the length of the hypotenuse.

For exercises 13–18,
(a) solve.
(b) check.

13. $\frac{1}{2}p - \frac{1}{3} = \frac{1}{4}$
14. $6(2x - 9) - 5x = 8x - 3(x + 4)$
15. $17 - 3x = 5(4x + 1) - 23x$
16. $7a = -3a$
17. $85w + 1575 = 90w$
18. $\frac{8}{15} = \frac{n}{195}$

For exercises 19–21,
(a) solve.
(b) check.

19. $\frac{3}{4}x - 9 < -30$
20. $-8v + 14 \geq -6v - 30$
21. $a < 3a - 10$
22. Use a number line graph to represent $x > 4$.
23. Use a number line graph to represent $x \leq -2$.
24. Solve $8x - 9y = 15$ for y.
25. The Khine formula for finding the diameter of a tracheal tube used in intubation is $D = \frac{A}{4} + 3$, where D is the diameter in millimeters and A is the patient's age in years. Solve for A.
26. Find the x-intercept and the y-intercept of the graph of $7x - 4y = 56$.

For exercises 27–30, graph.

27. $y = 3$
28. $x = -5$
29. $y \geq 4x - 7$
30. $5x - 3y > 20$
31. A linear equation is $-2x + 5y = 20$.
    a. Find three solutions of the equation.
    b. Graph the equation.
32. Use slope-intercept graphing to graph $y = -\frac{5}{6}x + 9$.
33. Write the equation of the graphed line.

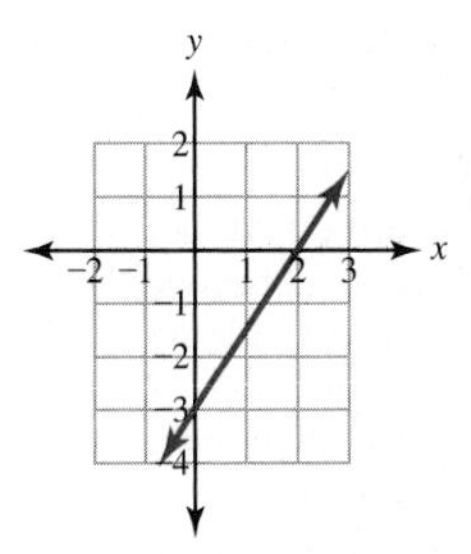

For exercises 34–37, find the slope of the line that passes through the given points.

34. $(4, 9), (4, -12)$
35. $(-2, -6), (-5, 10)$
36. $(8, 21), (-2, 21)$
37. $(4, -8), (2, -30)$
38. Write the equation of the horizontal line that passes through $(2, 7)$.
39. Write the equation in slope-intercept form of the line that passes through $(1, 4)$ and $(12, 5)$.
40. Write the equation in slope-intercept form of the line that is perpendicular to $y = -\frac{1}{3}x - 9$ and passes through $(-5, 2)$.
41. A model of the height in inches of rain in a rain gauge, y, after x hours of a steady rain is $y = \left(\frac{0.25\text{ in.}}{1\text{ hr}}\right)x + 1.75\text{ in.}$ Use this model to find the height in 7 hr.
42. A set of ordered pairs is {(Buddy, mutt), (Sam, schnauzer), (Baxter, bichon frise), (Harry, mutt)}.
    a. Is this set a function?
    b. Explain.
43. A function is $f(x) = x^2 - 6x + 2$. Evaluate $f(-4)$.

44. a. Does the graph represent a function?
b. Explain.

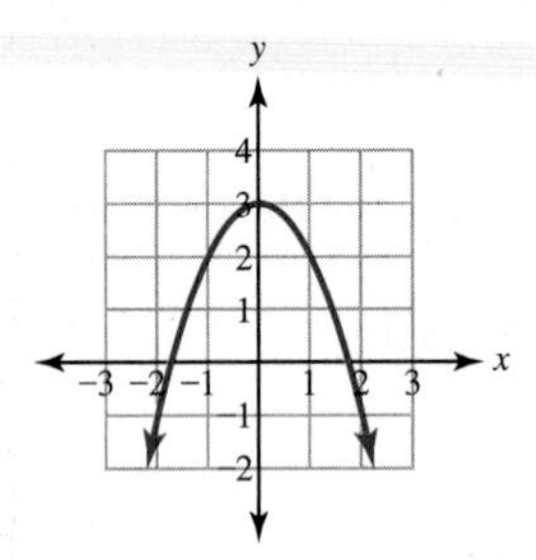

For exercises 45–53, use the five steps.

45. Of the \$339 million awarded in 2012 by the CDC for HIV prevention activities, Maryland received \$12.5 million, a decrease of \$604,000 from the amount received in 2011. Find the percent of the total awarded to Maryland in 2012. Round to the nearest percent. (*Source:* www.baltimoresun.com, Jan. 4, 2012)

46. Find the difference between the weekday circulation of *The Chronicle* and the combined circulation of the other newspapers.

The Chronicle (weekday circulation 339,000 last year) dominates in San Francisco, but beyond the city it trucks papers long distances to compete with the *San Jose Mercury News* (224,000) to the south, the *Contra Costa Times* (181,000) and the *Oakland Tribune* (92,000) to the east. (*Source:* www.nytimes.com, March 29, 2009)

47. In 2011, the Florida Department of Agriculture and Consumer Affairs received 41,061 complaints, a 9% increase over the previous year [2010]. Find the number of complaints in 2010. Round to the nearest whole number. (*Source:* www.miamiherald.com, Jan. 5, 2012)

48. Find the maximum number of Americans who can get all of their toilet paper for 1 year from the pulp of one eucalyptus tree. Round to the nearest whole number.

The pulp from one eucalyptus tree . . . produces as many as 1,000 rolls of toilet tissue. Americans use an average of 23.6 rolls per capita a year. (*Source:* www.nytimes.com, Feb. 25, 2009)

49. Find the number of 6500 air travelers whose top five frustrations with flying were directly related to the TSA passenger screening process.

Four out of air travelers' top five frustrations with flying are directly related to the TSA passenger screening process. (*Source:* www.ustravel.org, Nov. 16, 2011)

50. Find the number of surveyed adults who use social media platforms.

Two-thirds of online adults . . . use social media platforms such as Facebook, Twitter, MySpace or LinkedIn. . . . the results reported here are based on a national telephone survey of 2,277 adults. (*Source:* pewinternet.org, Nov. 15, 2011)

51. Find the distance traveled by a commuter driving at a speed of $\frac{62 \text{ mi}}{1 \text{ hr}}$ for 13 min. Round to the nearest tenth.

52. Find the percent increase in the population of the United States over the 15-month period. Round to the nearest tenth of a percent.

The United States as a whole saw its population increase by 2.8 million over the 15-month period, to 311.6 million. (*Source:* www.census.gov, Dec. 21, 2011)

53. The formula $W = 4.22 + 0.19D + 0.108M + 0.374S + 0.06T$ represents the relationship for one dairy cow of the recommended daily water intake W in gallons, the amount of dry matter feed per day D in pounds, the amount of milk produced per day M in pounds, the amount of sodium consumed per day S in ounces, and the average minimum weekly temperature T in degrees Fahrenheit. Find the recommended daily water intake for a herd of 140 dairy cows with an average daily production of 65 lb of milk per cow and average daily consumption of 40 lb of dry matter feed and 3 oz of sodium per cow at an average minimum weekly temperature of 50°F. Round to the nearest whole number. (*Source:* www.extension.umn.edu)

Systems of Equations and Inequalities

4

SUCCESS IN COLLEGE MATHEMATICS

Using Your Textbook

Like many stories, a chapter in a math textbook includes new vocabulary words and new ideas. The main idea is described in the section title. The supporting ideas are described by the subsection headings.

Reading your textbook before class builds a foundation for what your instructor will teach you in class. If you read your textbook before class, you are much less likely to feel lost during class, and you will be better prepared to ask questions.

Using the textbook during class can help you connect your instructor's examples to similar examples in the book. Mark what your instructor thinks is particularly important, and ask whether you are responsible for topics not discussed in class.

Reading the textbook after class can help you transfer what you learned in class into your long-term memory. As soon as possible after class, rewrite your notes and review any work done in class. Refer to your textbook for help on anything you do not understand. Before starting your homework exercises, do at least one practice problem from each subsection, and check your answer with the answers in the back of the book. Read the steps in the procedure boxes aloud, and make sure that you understand what is meant by each step. Do the vocabulary exercises. As you do the homework exercises assigned by your instructor, check your answers to make sure that you are doing them correctly. This is not cheating.

Using your textbook before, during, and after class can help you be successful in this math class. Although it may take more time each day to study in this way, you will need less time to prepare right before a test.

4.1 Systems of Linear Equations
4.2 The Substitution Method
4.3 The Elimination Method
4.4 More Applications
4.5 Systems of Linear Inequalities

© .shock/Shutterstock.com

If one line on a graph represents the costs to make a product and another line on the graph represents the price of the product, the break-even point is the point of intersection of the two lines. The equations for the lines are a system of linear equations. In this section, we will graph and solve systems of linear equations.

SECTION 4.1

Systems of Linear Equations

After reading the text, working the practice problems, and completing assigned exercises, you should be able to:

1. Given the graph of a system of linear equations, identify the number of solutions of the system.
2. Given the graph of a system of linear equations, find the solution(s) of the system.
3. Solve a system of linear equations by graphing.
4. Use a system of linear equations and graphing to solve an application problem.

A System of Linear Equations

The standard form of a linear equation in two variables is $ax + by = c$. A solution of this linear equation is an ordered pair. Each ordered pair can be represented as a point. The graph of the equation is the line formed by these points.

A **system of linear equations** is two or more linear equations. An ordered pair that is a **solution of the system** is a solution of all of the equations. On a graph, the solution of the system is the intersection point(s) of the lines that represent each equation.

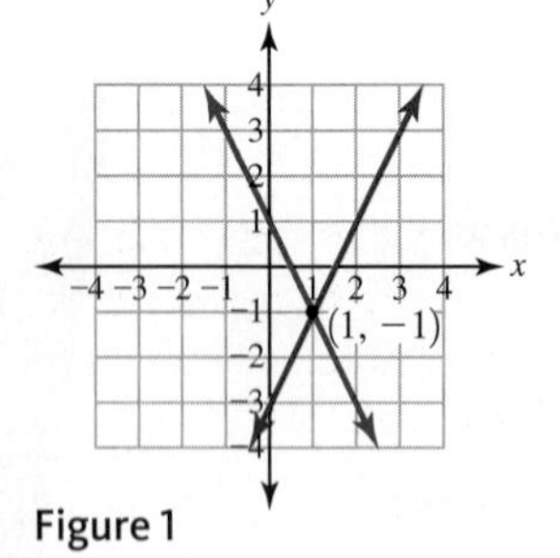

Figure 1

The graph in Figure 1 represents the system $\begin{matrix} y = 2x - 3 \\ y = -2x + 1 \end{matrix}$. The intersection point and solution of this system is $(1, -1)$. It is the only point shared by both lines. This system has *one solution.*

The graph in Figure 2 represents the system $\begin{matrix} y = \dfrac{1}{2}x - 1 \\ y = \dfrac{1}{2}x + 2 \end{matrix}$. These are parallel lines that do not intersect. This system has *no solution.*

The graph in Figure 3 represents the system $\begin{matrix} y = 3x + 2 \\ -3x + y = 2 \end{matrix}$. Since the solutions of these different equations are the same, the graphs are **coinciding lines**. The solution of this system is the infinite set of all of the ordered pairs that represent points on these lines. This system has *infinitely many solutions.*

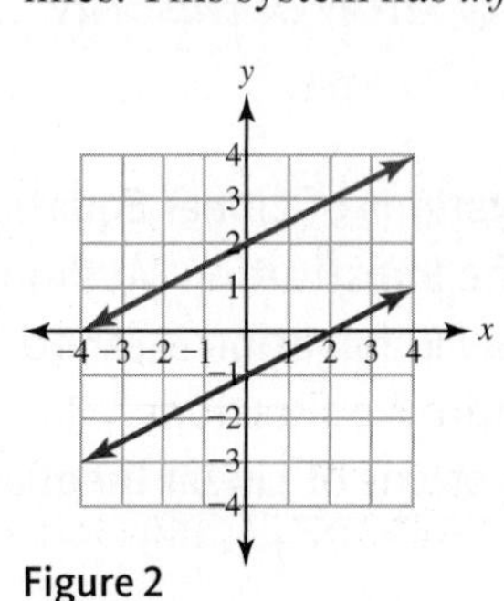

Figure 2

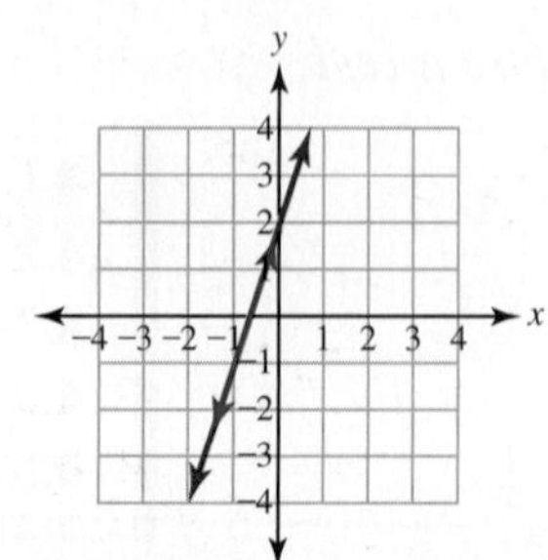

Figure 3

The Graph of a System of Two Linear Equations

Lines That Intersect in One Point The intersection point is the solution of the system.

Parallel Lines No intersection point; no solution.

Coinciding Lines Infinitely many intersection points; infinitely many solutions.

Practice Problems

For problems 1–4, identify the number of solutions of the graphed system of equations.

1.

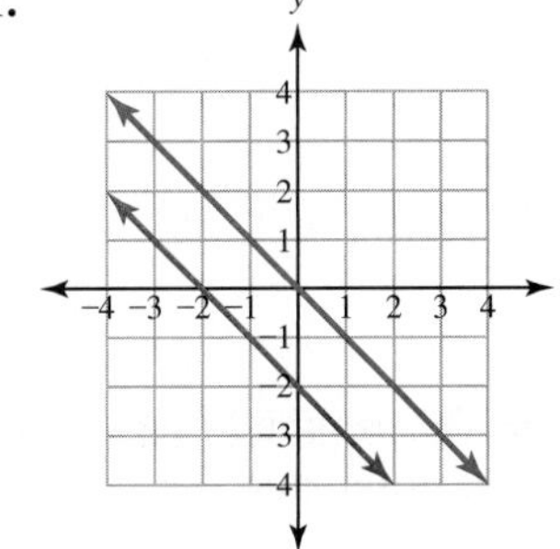

2.

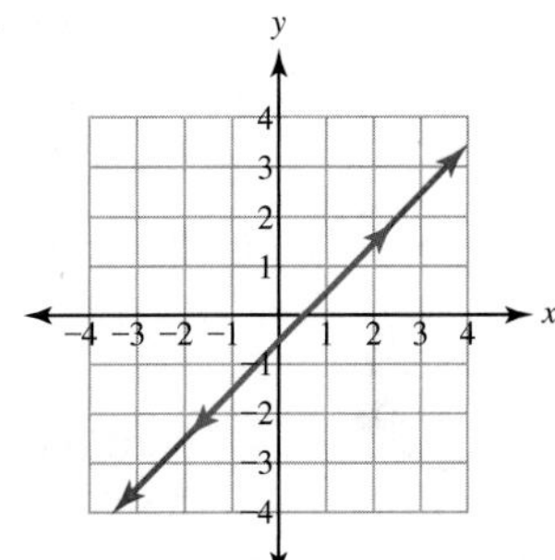

3.

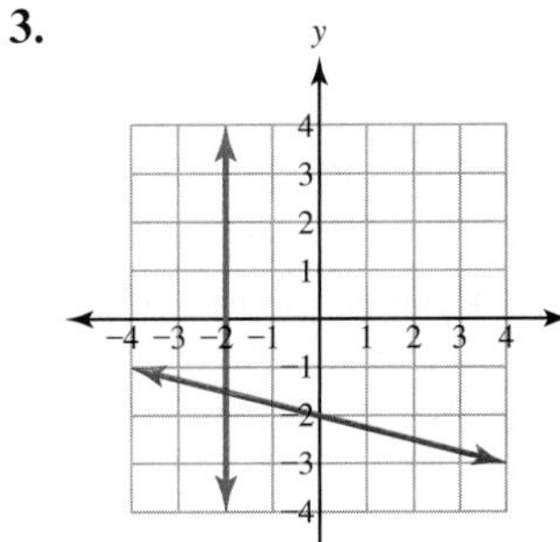

4.

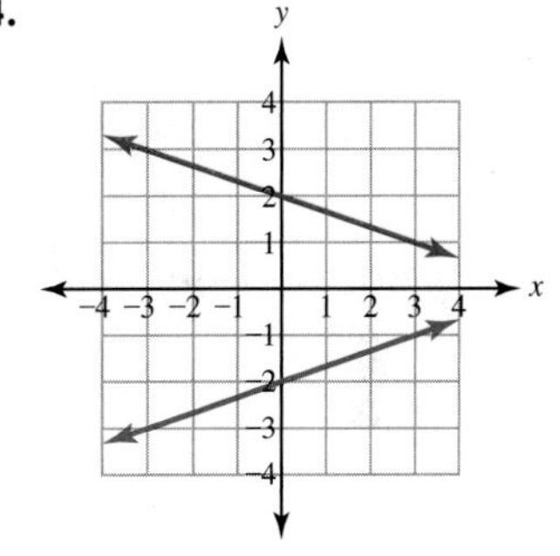

Solving a System of Linear Equations by Graphing

To find the solution of a system of linear equations by graphing, graph each line. If there is a single intersection point, this is the solution of the system. To check the solution, replace the variables in each equation with the coordinates of the intersection point. If the solution is correct, then both equations are true.

EXAMPLE 1 **(a)** Solve $\begin{aligned} y &= x - 6 \\ 3x + 2y &= 3 \end{aligned}$ by graphing.

SOLUTION ▶ Since $y = x - 6$ is in slope-intercept form, use slope-intercept graphing. The slope is $\frac{1}{1}$. Graph the y-intercept, $(0, -6)$. From this point, move up 1 unit, and then move to the right 1 unit. This is the location of another point on the line, $(1, -5)$.

x	y
0	$\frac{3}{2}$
1	0
5	-6

Since $3x + 2y = 3$ is in standard form, find the intercepts and one other point. Graph the points, and draw a straight line through them. The solution of the system is the point of intersection, $(3, -3)$.

(b) Check.

▸ Show that $(3, -3)$ is a solution of both equations in the system.

$$y = x - 6 \qquad\qquad 3x + 2y = 3$$
$$\mathbf{-3} = \mathbf{3} - 6 \qquad\qquad 3(\mathbf{3}) + 2(\mathbf{-3}) = 3$$
$$-3 = \mathbf{-3} \quad \text{True.} \qquad\qquad \mathbf{9 - 6} = 3$$
$$\mathbf{3} = 3 \quad \text{True.}$$

EXAMPLE 2 Solve $\begin{matrix} x = 4 \\ y = 2 \end{matrix}$ by graphing.

SOLUTION ▸ Graph the vertical line $x = 4$. Graph the horizontal line $y = 2$. The solution of the system is the point of intersection, $(4, 2)$.

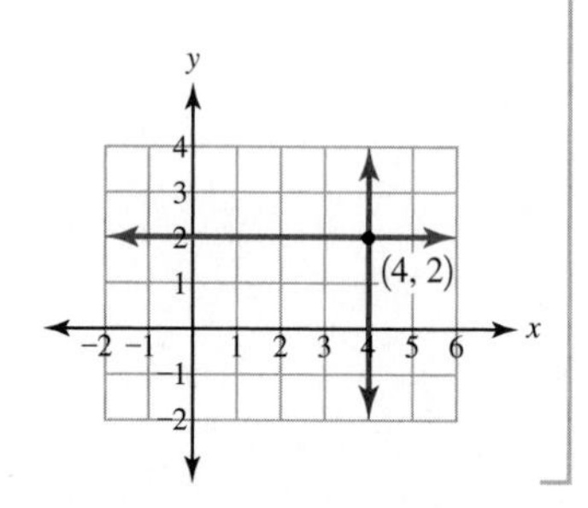

In the next example, the graphed lines appear to be parallel. If they are parallel, the slopes of the lines are equal and the system has no solution.

EXAMPLE 3 Solve $\begin{matrix} y = \frac{3}{2}x - 4 \\ 3x - 2y = -2 \end{matrix}$ by graphing.

SOLUTION ▸ To graph $y = \frac{3}{2}x - 4$, use slope-intercept graphing. The y-intercept is $(0, -4)$; the slope is $\frac{3}{2}$. We can choose to rewrite $3x - 2y = -2$ in slope-intercept form and use slope-intercept graphing.

$$3x - 2y = -2$$
$$\underline{-3x \qquad\quad -3x} \qquad \text{Subtraction property of equality.}$$
$$\mathbf{0} - 2y = \mathbf{-3x} - 2 \qquad \text{Simplify.}$$
$$\frac{\mathbf{-2y}}{-2} = \frac{-3x}{-2} - \frac{2}{-2} \qquad \text{Division property of equality.}$$
$$y = \frac{3}{2}x + 1 \qquad \text{Simplify; the } y\text{-intercept is } (0, 1)\text{; the slope is } \frac{3}{2}.$$

Since the slope of both lines is the same, $m = \frac{3}{2}$, and parallel lines have the same slope, the lines are parallel. Since the lines never intersect, this system has no solution.

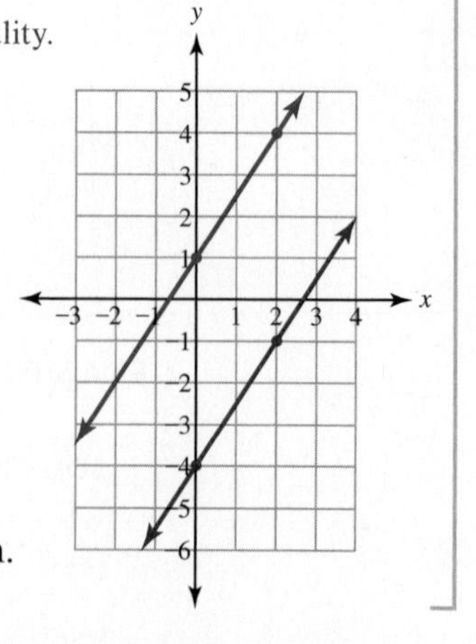

In the next example, the graphed lines that represent the system of equations coincide and share infinitely many points. This system has infinitely many solutions.

EXAMPLE 4 Solve $\begin{array}{l} y = -2x + 5 \\ 6x + 3y = 15 \end{array}$ by graphing.

SOLUTION ▶ Use slope-intercept graphing to graph $y = -2x + 5$. The y-intercept is $(0, 5)$. The slope is -2. Rewrite $6x + 3y = 15$ in slope-intercept form.

$$6x + 3y = 15$$

$$\underline{-6x \qquad -6x} \qquad \text{Subtraction property of equality.}$$

$$\mathbf{0} + 3y = \mathbf{-6x} + 15 \qquad \text{Simplify.}$$

$$\frac{\mathbf{3y}}{3} = \frac{-6x}{3} + \frac{15}{3} \qquad \text{Division property of equality.}$$

$$y = -2x + 5 \qquad \text{Simplify.}$$

In slope-intercept form, the equations are the same. The lines are coinciding with infinitely many intersection points. The solution of this system is an infinite set of ordered pairs that includes all the points on the line. The solution is *not* the set of real numbers. To describe the solution precisely, we use **set-builder notation**: $\{(x, y) \mid y = -2x + 5\}$. This notation tells us that the solution of the system of linear equations is "the set of ordered pairs, (x, y), such that y equals $-2x + 5$." The vertical line in this notation represents the words "such that."

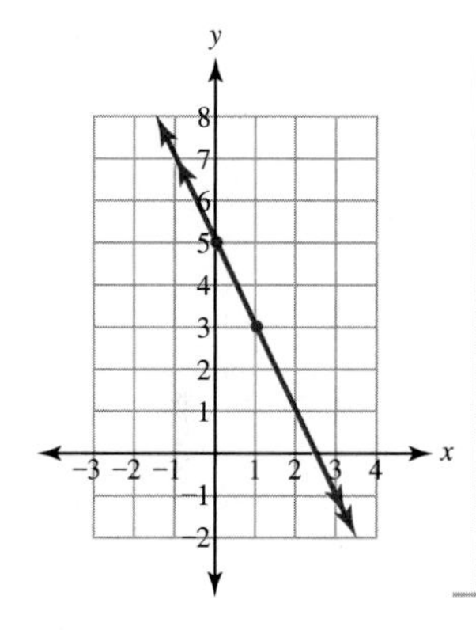

When writing a solution in set-builder notation, we can use either equation to describe the ordered pairs that are the solutions.

EXAMPLE 5 The system $\begin{array}{l} y = \frac{3}{4}x - 1 \\ 3x - 4y = 4 \end{array}$ has infinitely many solutions. Use set-builder notation to represent this solution.

SOLUTION ▶ The solution of this system is the set of ordered pairs, (x, y), such that $y = \frac{3}{4}x - 1$. In set-builder notation, the solution is $\left\{(x, y) \mid y = \frac{3}{4}x - 1\right\}$.

Ask your instructor how to describe the solution of a system with infinitely many solutions. The answers in the back of the book use set-builder notation. Some instructors prefer that students write "infinitely many solutions."

Practice Problems

For problems 5–10, solve by graphing.

5. $\begin{array}{l} y = 3x - 4 \\ 2x + y = 1 \end{array}$

6. $\begin{array}{l} y = \frac{1}{3}x + 4 \\ x + y = 8 \end{array}$

7. $\begin{array}{l} x = 5 \\ y = 2x - 7 \end{array}$

8. $\begin{array}{l} 2x + 5y = 20 \\ x + 3y = 12 \end{array}$

9. $\begin{array}{l} x = 3 \\ y = -2 \end{array}$

10. $\begin{array}{l} 2x + 3y = 6 \\ y = -\frac{2}{3}x + 2 \end{array}$

Applications

A business "breaks even" when the total costs of making a product is equal to the revenue from selling a product. If the revenue is greater than the costs, the business makes a **profit**. If the revenue is less than the costs, the business has a **loss**.

We can graph a system of equations in which one equation represents the relationship of the number of products sold, x, and the total cost to make the products, y, and the other equation represents the relationship of the number of products sold, x, and the revenue from selling the products, y. The solution of this system of equations is the **break-even point** represented by the point (x, y). The x-coordinate of the break-even point is the number of products when the business breaks even. The y-coordinate of the break-even point is equal to the costs and the revenue when the business breaks even.

Writing a System of Linear Equations for the Break-Even Point

1. Write an equation that represents the costs, y, of making x products.

   $y = (\text{cost per product})x + \text{fixed overhead costs}$

2. Write an equation that represents the revenue, y, from selling x products.

   $y = (\text{price per product})x$

Notice that both equations are in slope-intercept form, $y = mx + b$. The slope of the costs equation, m, represents the cost per product. The y-coordinate of the y-intercept, b, represents the fixed overhead costs. The slope of the revenue equation, m, represents the price per product. The y-coordinate of the y-intercept is \$0.

A graph of the cost equation, the revenue equation, and the break-even point can have a powerful visual impact. For example, if the owner of a small business is presenting an application to a bank officer for a loan to expand the business, a graph might be an excellent way to show the expected break-even point.

EXAMPLE 6 The cost to make a product is \$5. The fixed overhead costs per month are \$4500. A model of the total cost, y, to make x products is $y = \left(\frac{\$5}{1\text{ product}}\right)x + \$4500$.

(a) If the price of the product is \$10, a model of the total revenue, y, from the sale of x products is $y = \left(\frac{\$10}{1\text{ product}}\right)x$. Find the break-even point by graphing.

SOLUTION ▶

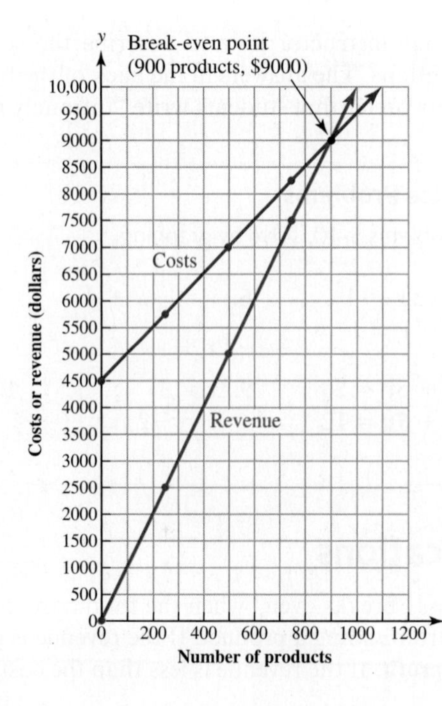

Costs $y = \left(\frac{\$5}{1\text{ product}}\right)x + \$4500$

x (products)	y (dollars)
0	4500
250	5750
500	7000
750	8250
900	9000

Revenue $y = \left(\frac{\$10}{1\text{ product}}\right)x$

x (products)	y (dollars)
0	0
250	2500
500	5000
750	7500

The intersection point of the two lines is (900 products, \$9000). If 900 products are made and sold, the costs to make these products equals the revenue collected from selling them.

(b) If the price of the product is \$15, a model of the total revenue, y, from the sale of x products is $y = \left(\frac{\$15}{1 \text{ product}}\right)x$. Find the break-even point by graphing.

▶ If the price of the product increases and the cost to make the product stays the same, the slope of the line that represents the revenue increases. The break-even point changes. Fewer products must be sold to break even.

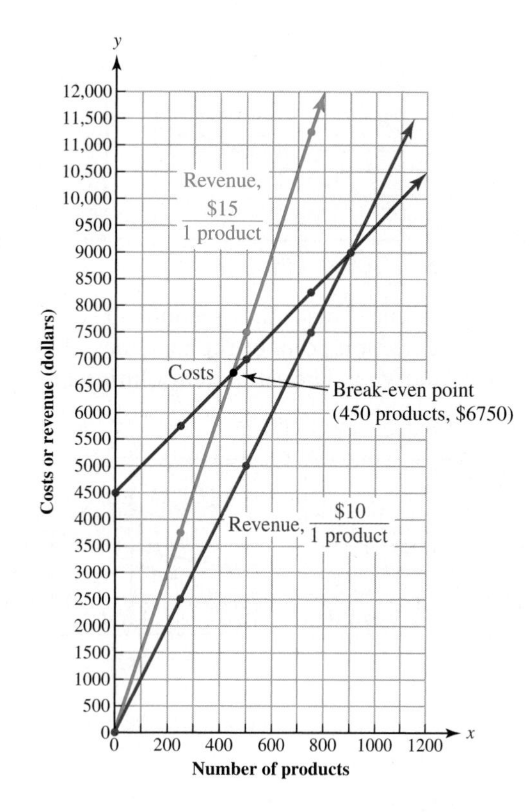

Revenue $y = \left(\frac{\$15}{1 \text{ product}}\right)x$

x (products)	y (dollars)
0	0
250	3750
500	7500
750	11,250

The intersection point of the two lines is (450 products, \$6750). If 450 products are made and sold, the costs to make these products equals the revenue collected from selling them.

Practice Problems

For problems 11–12, the cost to make a product is \$0.50. The fixed overhead costs per month are \$3000. A model of the total cost, y, to make x products is $y = \left(\frac{\$0.50}{1 \text{ product}}\right)x + \$3000$.

11. If the price of each product is \$2, a model of the total revenue, y, from the sale of x products is $y = \left(\frac{\$2}{1 \text{ product}}\right)x$. Find the break-even point by graphing.

12. If the price of each product is \$3.50,

a. write an equation that is a model of the revenue y when x products are sold.

b. find the break-even point by graphing.

Using Technology: Systems of Linear Equations

A graphing calculator can estimate the intersection point of two lines.

EXAMPLE 7 Solve $\begin{aligned} y &= \frac{3}{5}x - 2 \\ 2x - 7y &= 3 \end{aligned}$ by graphing.

Rewrite $2x - 7y = 3$ in slope-intercept form: $y = \frac{2}{7}x - \frac{3}{7}$. Press Y=. Type in the equations. Press GRAPH.

The intersection point is visible in a standard window. On the CALC screen, choose **Intersect**. The cursor appears on one of the graphed lines. This is the "first curve." Press ENTER.

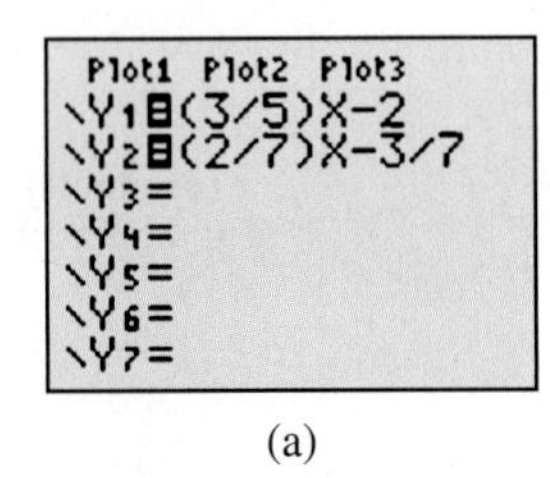

(a)

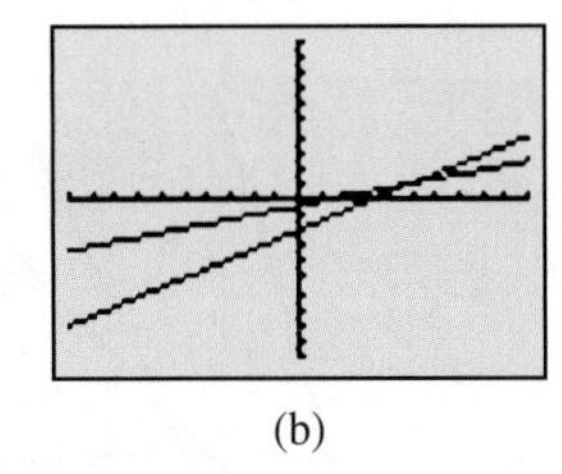

(b)

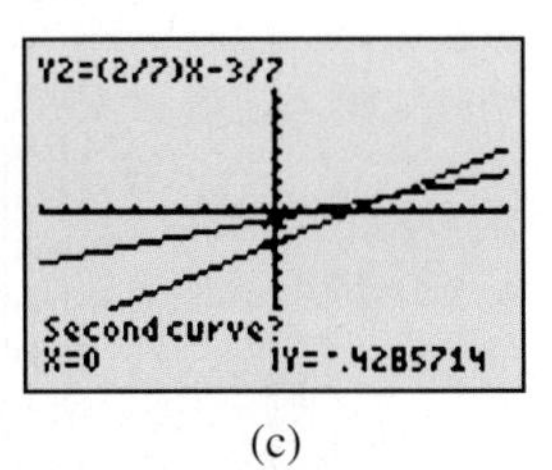

(c)

The cursor has moved to the other graphed line. This is the "second curve." Press ENTER. The calculator asks, "Guess?" Move the cursor close to the intersection point of the two lines. Press ENTER. The calculator displays the estimated point of intersection: X=5 and Y=1. The estimated solution of the system of equations is the ordered pair (5, 1).

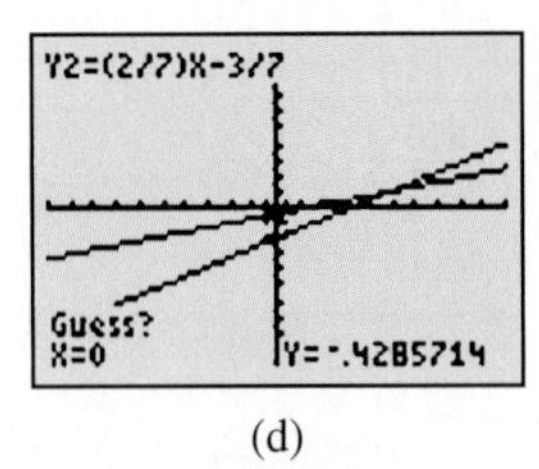

(d)

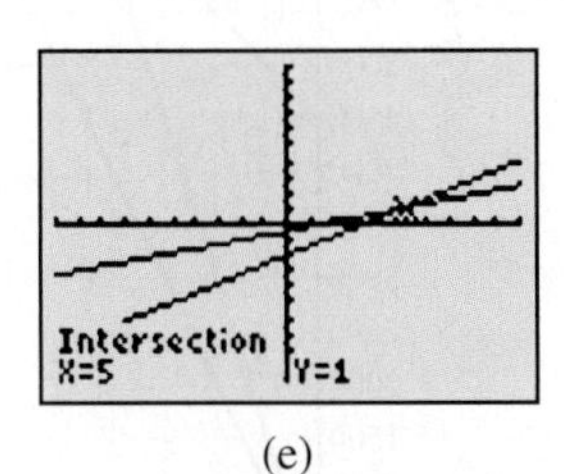

(e)

Practice Problems For problems 13–15, solve by graphing. Sketch the graph; describe the window; identify the solution.

13. $\begin{aligned} y &= -4x + 7 \\ y &= 6x - 3 \end{aligned}$

14. $\begin{aligned} y &= 3x + 15 \\ y &= -4x + 8 \end{aligned}$

15. $\begin{aligned} y &= \frac{2}{3}x - 10 \\ y &= -x \end{aligned}$

4.1 Vocabulary Practice

Match the term with its description.

1. The solution of a system of equations represented by parallel lines
2. Lines with exactly the same points
3. The number of solutions of a system of coinciding lines
4. $ax + by = c$
5. $y = mx + b$
6. $\{\ldots, -2, -1, 0, 1, 2, 3, \ldots\}$
7. $m = \frac{y_2 - y_1}{x_2 - x_1}$
8. The x-coordinate of this point is always 0.
9. The y-coordinate of this point is always 0.
10. The x-coordinate of this point represents the number of products that must be made and sold so that costs and revenue are equal.

A. break-even point
B. coinciding lines
C. infinitely many solutions
D. integers
E. no solution
F. slope formula
G. slope-intercept form of a linear equation
H. standard form of a linear equation
I. x-intercept
J. y-intercept

4.1 Exercises

Follow your instructor's guidelines for showing your work.

For exercises 1–6, identify the number of solutions of the graphed system of equations.

1.

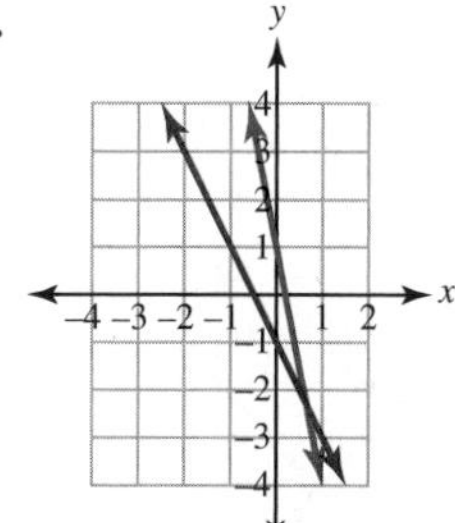

2.

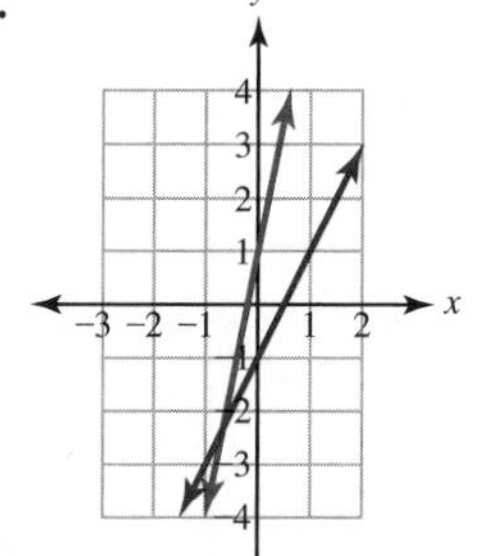

3.

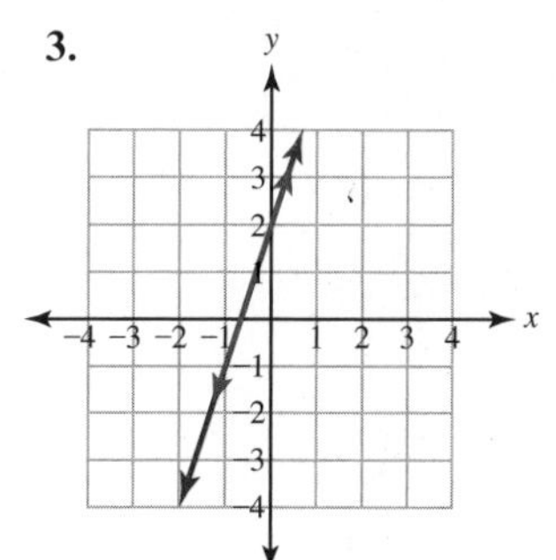

4.

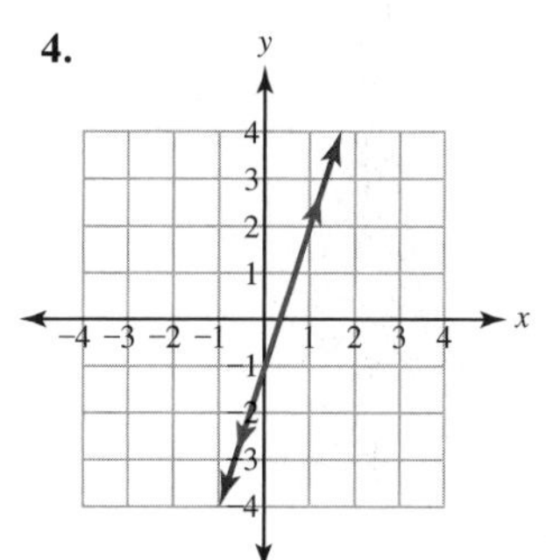

5.

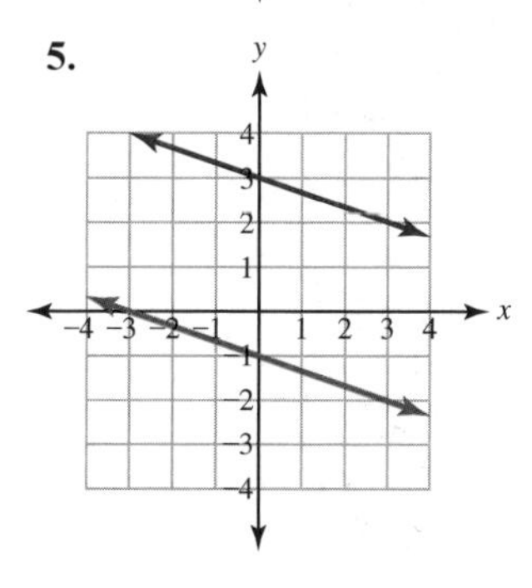

6.

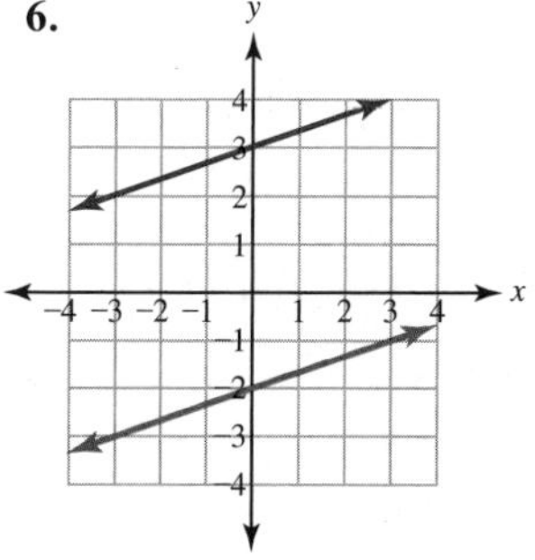

For exercises 7–26, the graph represents a system of linear equations. Use the graph to solve.

7.

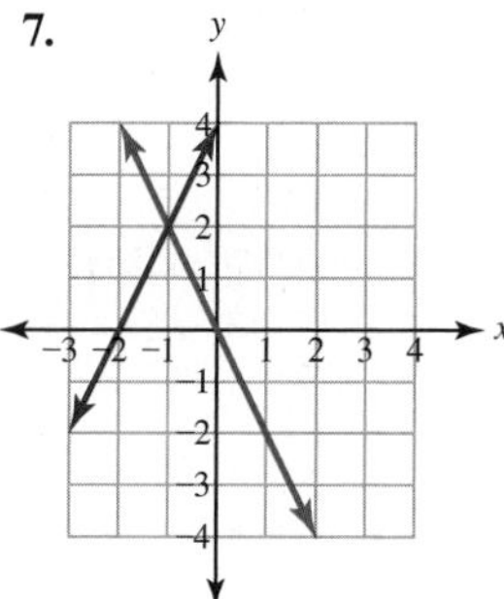

8.

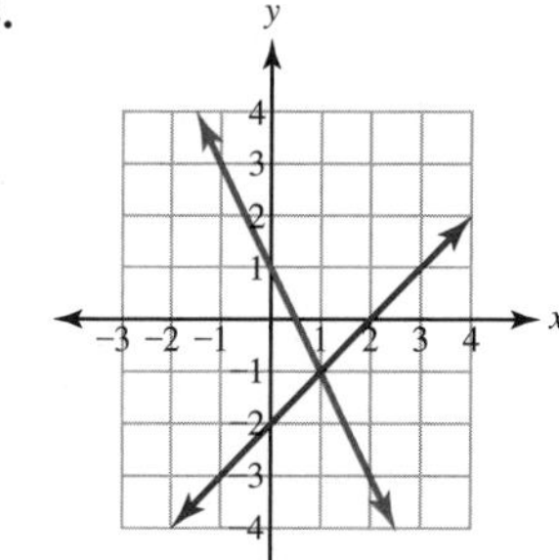

9.

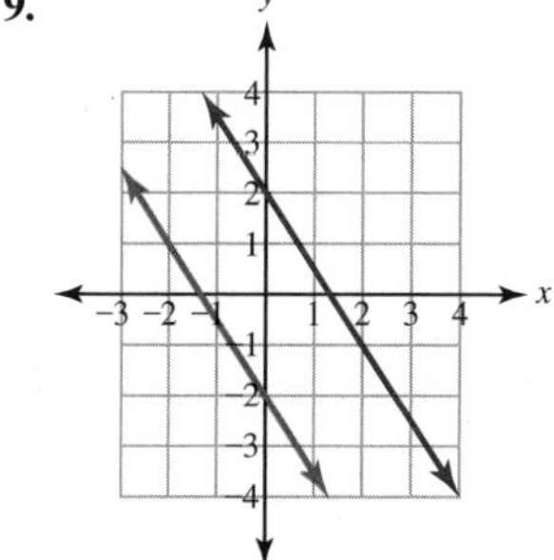

10.

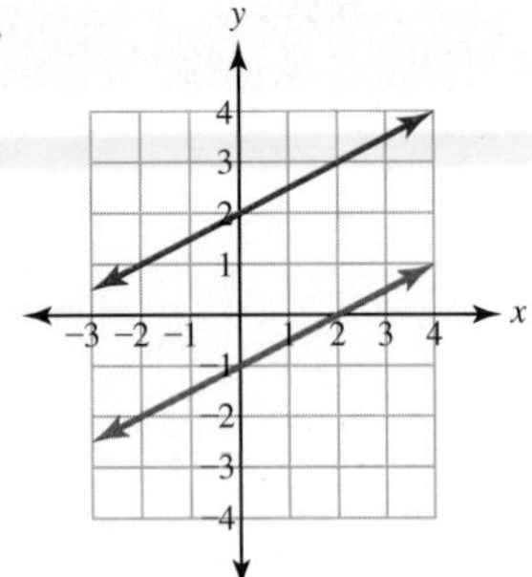

11.

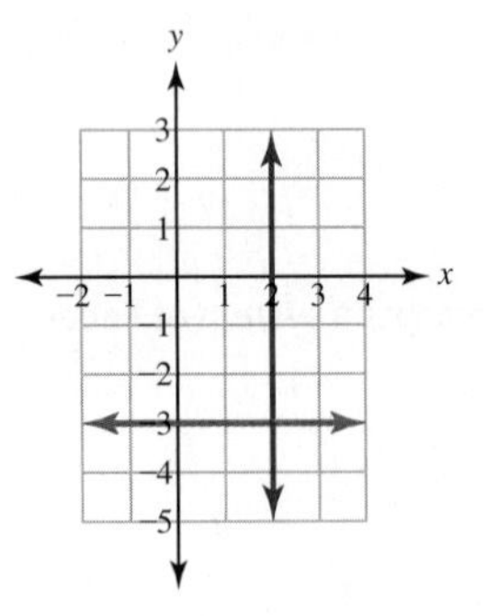

12.

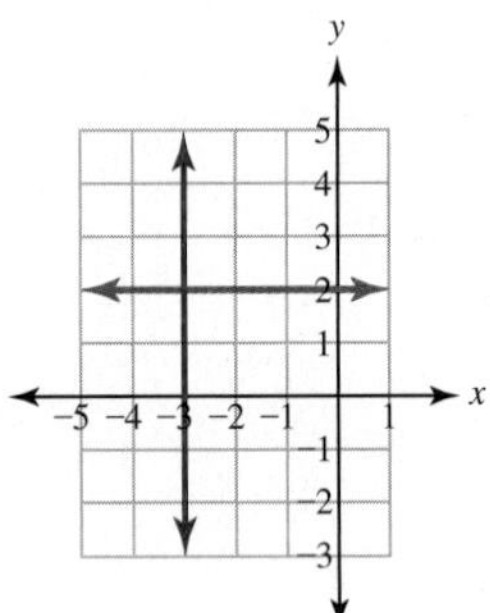

13.

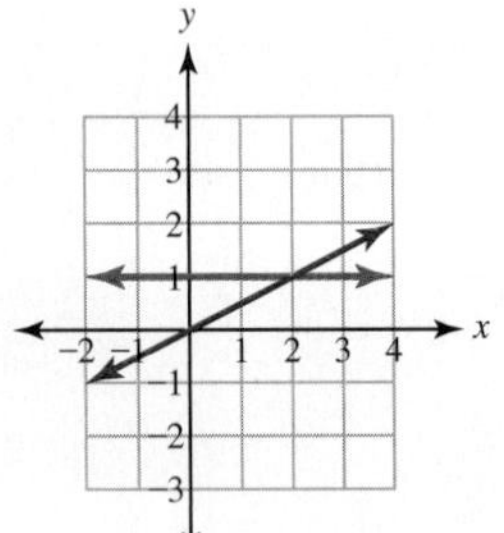

14.

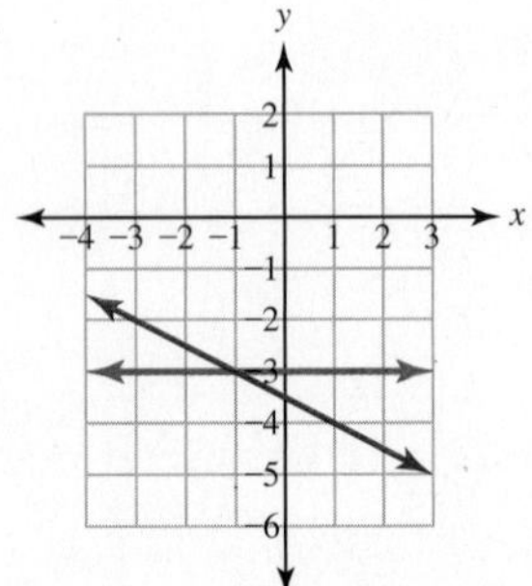

15.

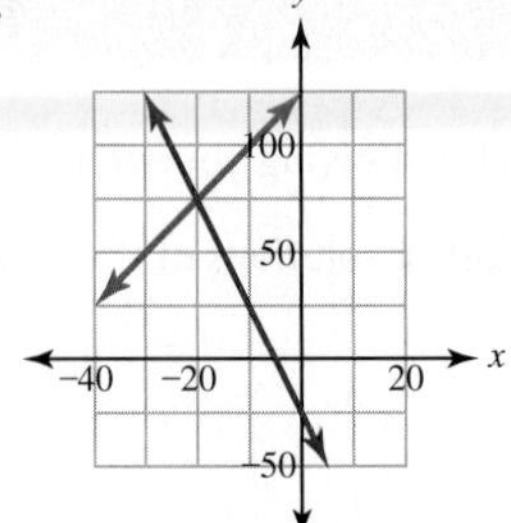

16.

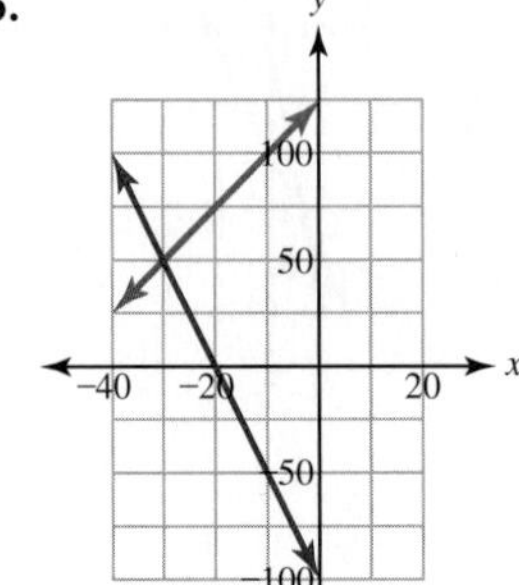

17.

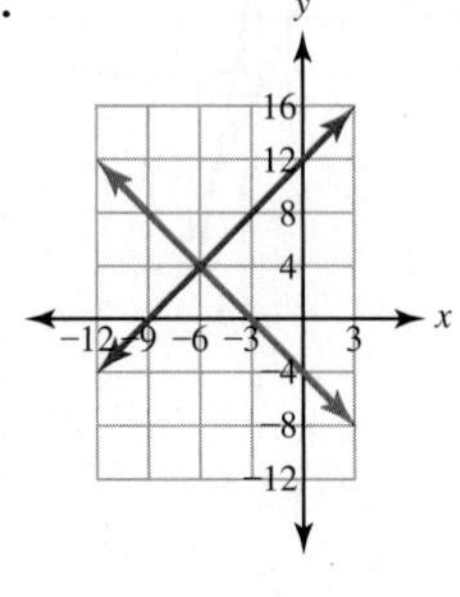

18.

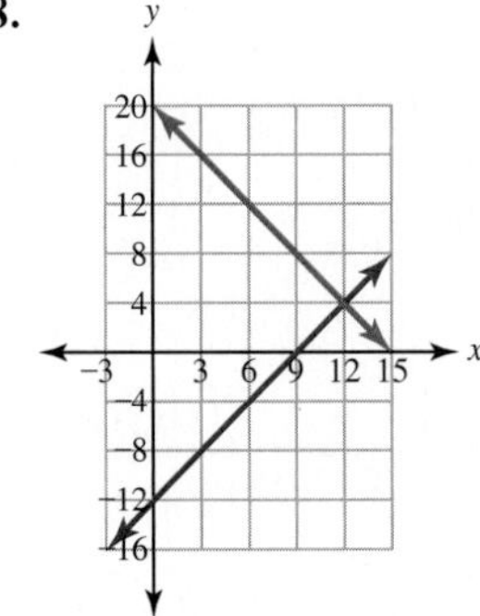

19.

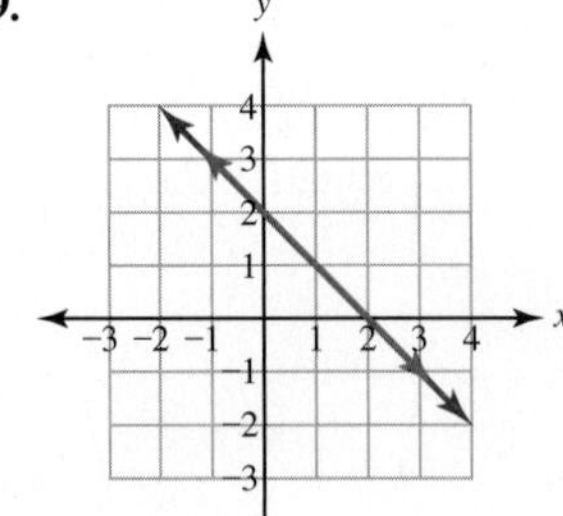

20.

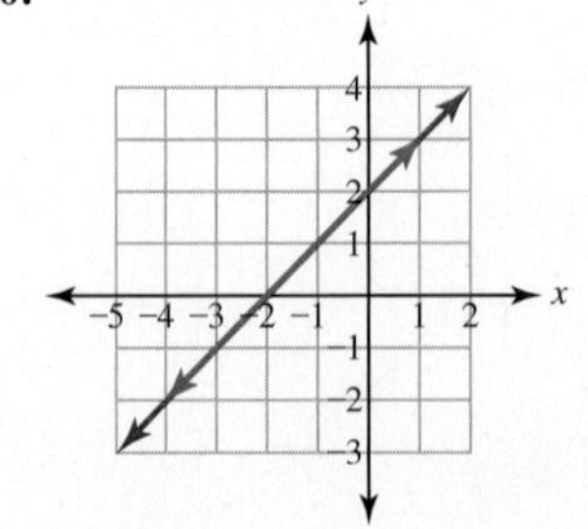

21.

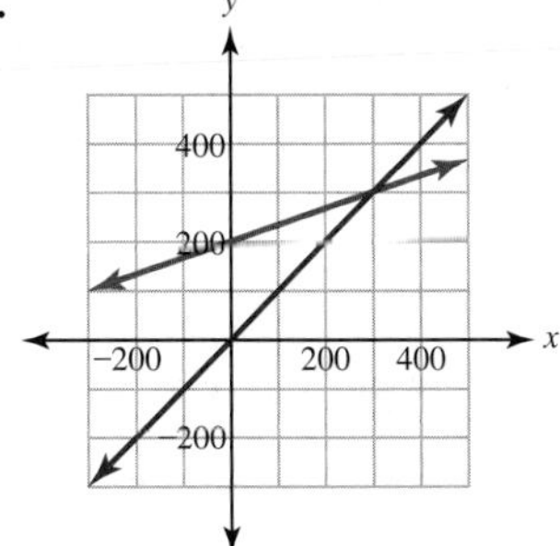

22.

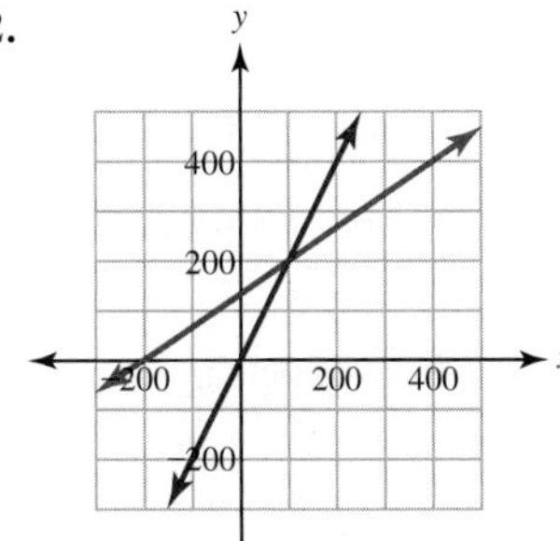

23.

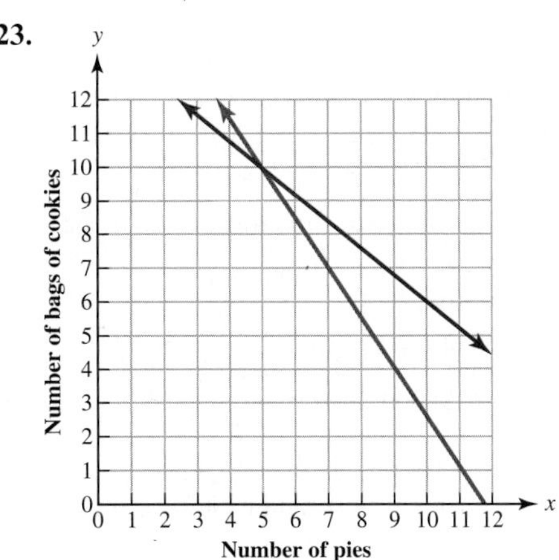

24.

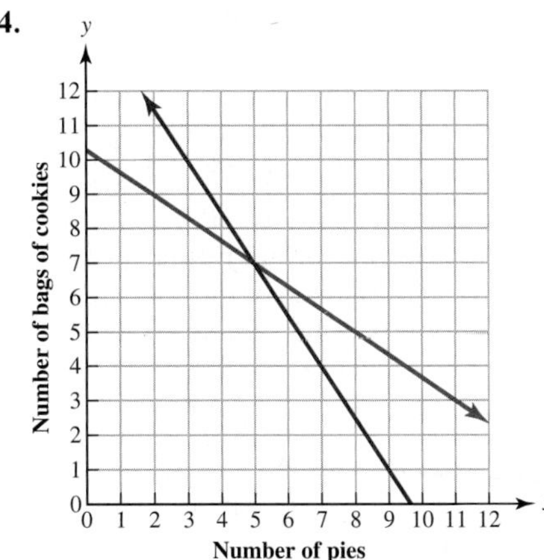

25.

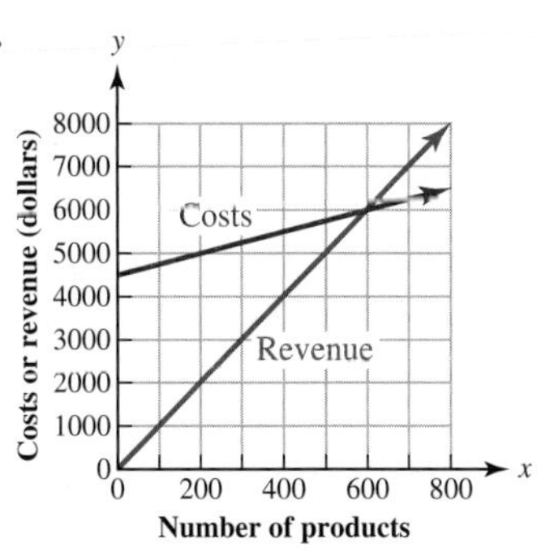

26.

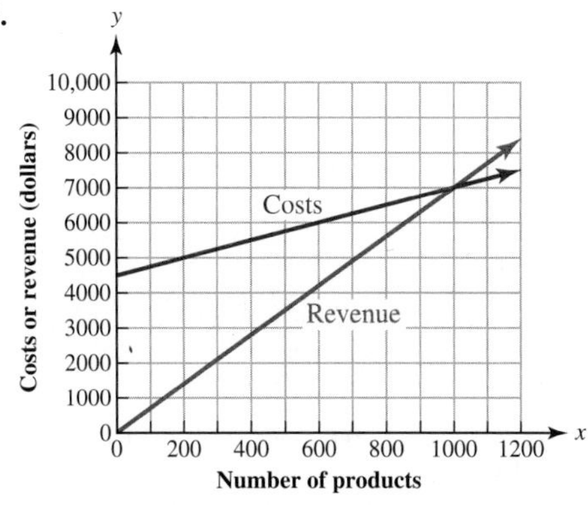

For exercises 27–68, solve by graphing.

27. $y = 3x + 1$
$y = -2x + 6$

28. $y = 2x + 1$
$y = -3x + 6$

29. $y = 2x - 7$
$y = 3x - 9$

30. $y = 2x - 7$
$y = 3x - 10$

31. $y = 3x$
$y = 2x$

32. $y = 4x$
$y = 2x$

33. $y = 2x + 5$
$y = -3x + 5$

34. $y = 3x + 4$
$y = -2x + 4$

35. $y = \frac{1}{2}x - 3$
$y = -2x + 7$

36. $y = \frac{1}{2}x + 6$
$y = 2x + 9$

37. $y = \frac{3}{4}x - 7$
$y = -\frac{1}{2}x - 2$

38. $y = \frac{2}{3}x - 5$
$y = -\frac{1}{3}x - 2$

39. $y = \frac{1}{4}x + 3$
$y = \frac{1}{4}x - 2$

40. $y = \frac{1}{2}x + 4$
$y = \frac{1}{2}x - 1$

41. $x + y = 8$
$-2x + y = -10$

42. $x + y = 6$
$-2x + y = -9$

43. $3x + y = 7$
$2x + y = 4$

44. $3x + y = 9$
$2x + y = 5$

45. $x + y = 0$
$-x + y = 4$

46. $x + y = 0$
$-x + y = 2$

47. $x + 2y = 6$
$y = 4x - 6$

48. $x + 2y = 9$
$y = 5x - 12$

49. $4x + 3y = 24$
$y = \frac{1}{3}x - 7$

50. $5x + 2y = 20$
$y = \frac{1}{2}x - 8$

51. $x = 9$
$y = -2$

52. $x = 7$
$y = -4$

53. $7x + 2y = 8$
$y = -\frac{7}{2}x - 6$

54. $3x + 8y = 16$
$y = -\frac{3}{8}x - 5$

55. $y = x$
$5x - 2y = -6$

56. $y = x$
$7x - 2y = -20$

57. $x - 9y = 18$
$y = \frac{1}{9}x - 2$

58. $x - 5y = 30$
$y = \frac{1}{5}x - 6$

59. $y = -5x$
$y = \frac{3}{8}x$

60. $y = -6x$
$y = \frac{5}{9}x$

61. $y = \frac{9}{10}x - 8$
$x = 10$

62. $y = \frac{8}{9}x - 6$
$x = 9$

63. $5x + 2y = 20$
$-3x - 4y = -12$

64. $4x + 3y = 12$
$-5x + 2y = -15$

65. $y = \frac{3}{4}x - 2$
$y = -x + 5$

66. $y = \frac{3}{5}x - 1$
$y = -x + 7$

67. $800x + 300y = -2400$
$-800x - 300y = 4800$

68. $900x + 200y = -3600$
$-900x - 200y = 1800$

For exercises 69–70,
(a) complete the table of solutions for each equation.
(b) solve the system of equations by finding a solution in the table that is shared by both equations.

69. $x + 2y = 7$ $\qquad$ $y = 2x - 9$

x	y
−1	
0	
1	
2	
3	
4	
5	
6	

x	y
−1	
0	
1	
2	
3	
4	
5	
6	

70. $x + 3y = -4$ $\qquad$ $y = 4x - 23$

x	y
−1	
0	
1	
2	
3	
4	
5	
6	

x	y
−1	
0	
1	
2	
3	
4	
5	
6	

For exercises 71–74, the cost to make each product is \$2 and the fixed overhead costs per month to make products are \$4800. A model of the cost per month, y, to make x products is $y = \left(\frac{\$2}{1\text{ product}}\right)x + \$4800$.

71. What does the slope of the graph of $y = \left(\frac{\$2}{1\text{ product}}\right)x + \$4800$ represent?

72. What does the y-coordinate of the y-intercept of the graph of $y = \left(\frac{\$2}{1\text{ product}}\right)x + \$4800$ represent?

73. The price of each product is \$10. A model of the total revenue, y, from the sale of x products is $y = \left(\frac{\$10}{1\text{ product}}\right)x$. Find the break-even point by graphing.

74. The price of each product is \$6. A model of the total revenue, y, from the sale of x products is $y = \left(\frac{\$6}{1\text{ product}}\right)x$. Find the break-even point by graphing.

For exercises 75–78, the cost to make a product is \$200. The fixed overhead costs per month to make the product are \$6000. A model of the cost per month, y, to make x products is $y = \left(\frac{\$200}{1\text{ product}}\right)x + \$6000$.

75. What does the y-coordinate of the y-intercept of the graph of $y = \left(\frac{\$200}{1\text{ product}}\right)x + \$6000$ represent?

76. What does the slope of the graph of $y = \left(\frac{\$200}{1\text{ product}}\right)x + \$6000$ represent?

77. If the price of each product is \$300,
a. write an equation that is a model of the revenue, y, when x products are sold.
b. find the break-even point by graphing.

78. If the price of each product is \$250,
a. write an equation that is a model of the revenue, y, when x products are sold.
b. find the break-even point by graphing.

79. The balance in a retirement account is \$18,000, with x dollars in short-term investments and y dollars in long-term investments. If twice as much of the balance is in short-term investments as in long-term investments, then a model of this situation is $x + y = \$18{,}000$, $y = \frac{1}{2}x$. Find the amount in short-term investments and the amount in long-term investments by graphing.

80. An investor with \$50,000 has x dollars invested in real estate and y dollars invested in stocks. If three times as much is invested in real estate as in stock, then a model of this situation is $x + y = \$50{,}000$, $y = \frac{1}{3}x$. Find the amount invested in real estate and the amount invested in stocks by graphing.

81. A concession stand sells x large bags of popcorn for \$3 each and y small bags of popcorn for \$1 each. If a total of 3000 bags are sold and the total sales are \$4200, then a model of this situation is $\begin{matrix} \left(\frac{\$3}{1 \text{ large bag}}\right)x + \left(\frac{\$1}{1 \text{ small bag}}\right)y = \$4200 \\ x + y = 3000 \end{matrix}$. Find the number of large bags and the number of small bags of popcorn sold by graphing.

82. A store sells x small packages of toilet paper for \$3 each and y large packages of toilet paper for \$6 each. If a total of 750 packages are sold and the total sales are \$3750, then a model of this situation is $\begin{matrix} \left(\frac{\$3}{1 \text{ small pkg}}\right)x + \left(\frac{\$6}{1 \text{ large pkg}}\right)y = \$3750 \\ x + y = 750 \text{ packages} \end{matrix}$. Find the number of small packages and the number of large packages of toilet paper sold by graphing.

Problem Solving: Practice and Review

Follow your instructor's guidelines for using the five steps as outlined in Section 1.3, p. 55.

83. A pedestrian is crossing a four-lane highway at a stoplight. Because of a disability, this person walks slowly, at a speed of $\frac{0.5 \text{ mi}}{1 \text{ hr}}$. The distance across the highway is 80 ft. Find the time in *seconds* for the person to cross the road. (1 mi = 5280 ft.) Round to the nearest whole number.

84. Find the number of FM high-definition radio stations that transmitted iTunes-tagging UFID codes. Round to the nearest whole number.

Of about 1,600 FM HD Radio stations in early 2009, most delivered song and artist metadata, and about two-thirds of those transmitted iTunes-tagging UFID codes.
(*Source:* www.twice.com, March 9, 2009)

85. The sale price of a mountain bike is \$439. The regular price is \$489. Find the percent discount. Round to the nearest percent.

86. Find the percent increase in daily pay for substitute teachers in the Wayland, Massachusetts, for the sixth day of teaching. (*Source:* www.wayland.k12.ma.us)

Massachusetts-certified substitutes brought in from outside the school system covering for teachers

A. *Per diem substitute* for one day or several days
  1. \$75.00 per day
  2. No benefits

B. *Long term substitute* in one position for six or more days but less than or equal to eight weeks (≤40 consecutive teaching days).
  1. \$75.00 for first five days
  2. \$120.00 for sixth day and subsequent consecutive days
  3. No benefits

Technology

For exercises 87–90, solve by graphing. Sketch the graph; describe the window; identify the solution.

87. $y = x + 5$
$y = 3x + 9$

88. $y = -4x + 20$
$y = 10x + 6$

89. $y = \frac{3}{4}x - 10$
$y = \frac{1}{2}x - 8$

90. $y = x + 6$
$y = -5x$

Find the Mistake

For exercises 91-94, the completed problem has one mistake.
(a) Describe the mistake in words, or copy down the whole problem and highlight or circle the mistake.
(b) Do the problem correctly.

91. Problem: Solve $\begin{matrix} y = -\frac{1}{3}x + 2 \\ x + 3y = 6 \end{matrix}$ by graphing.

Incorrect Answer:

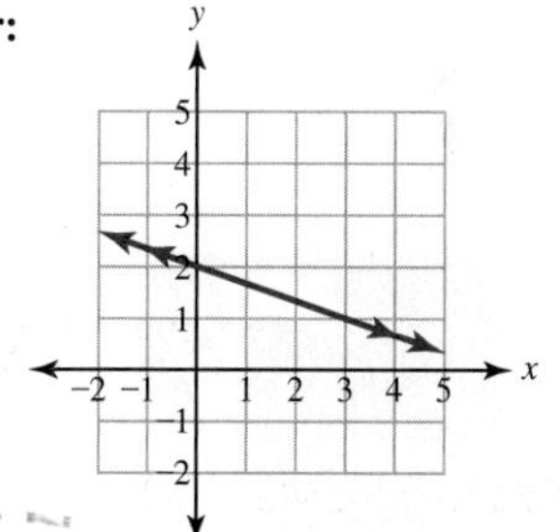

The solution is the set of real numbers.

92. Problem: Solve $\begin{matrix} y = -2x + 3 \\ x = 1 \end{matrix}$ by graphing.

Incorrect Answer:

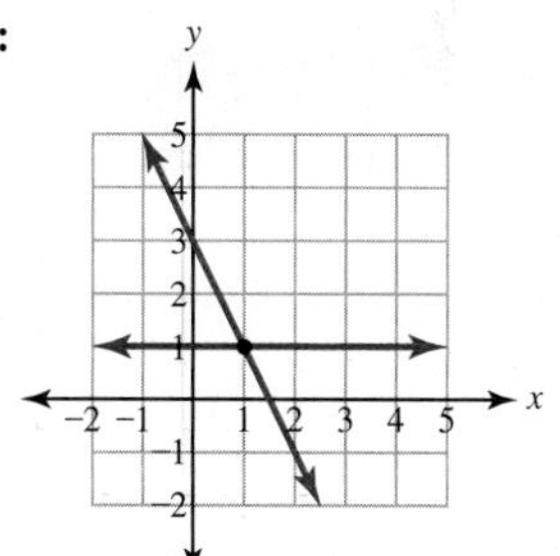

The solution is (1, 1).

93. Problem: Solve $\begin{matrix} y = -x \\ x + y = 4 \end{matrix}$ by graphing.

Incorrect Answer:

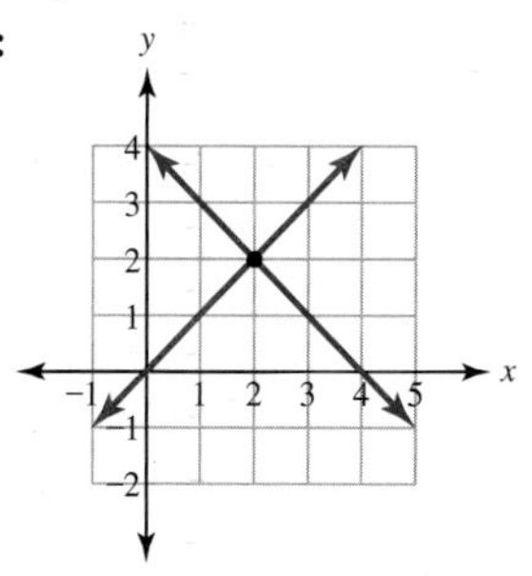

The solution is (2, 2).

94. Problem: Check whether $(2, -8)$ is a solution of $\begin{aligned} 3x + 7y &= -50 \\ x - y &= -6 \end{aligned}$.

Incorrect Answer:

$$\begin{aligned} 3x + 7y &= -50 \\ 3(2) + 7(-8) &= -50 \\ 6 - 56 &= -50 \\ -50 &= -50 \quad \text{True.} \end{aligned} \qquad \begin{aligned} x - y &= -6 \\ 2 - 8 &= -6 \\ -6 &= -6 \quad \text{True.} \end{aligned}$$

Since it is a solution of each equation in the system, $(2, -8)$ is a solution.

Review

For exercises 95–98,
(a) solve the equation.
(b) check.

95. $4x - (7x + 2) = -17$

96. $4x - 3(7x + 2) = 28$

97. $4x - 6\left(\frac{2}{3}x + 5\right) = -30$

98. $x - \left(\frac{3}{4}x - 9\right) = \frac{73}{8}$

SUCCESS IN COLLEGE MATHEMATICS

99. Turn to the first page of this section. At the beginning of the section is a list of objectives. The section is then broken into subsections, each with a heading at the beginning and practice problems at the end. For each of the following objectives, write the heading of the subsection in which this objective is taught.

1. Given the graph of a system of linear equations, identify the number of solutions of the system.
2. Given the graph of a system of linear equations, find the solution(s) of the system.
3. Solve a system of linear equations by graphing.
4. Use a system of linear equations and graphing to solve an application problem.

100. Look at practice problems 1–4 in this section. If you can complete these questions correctly, what objective have you learned?

USDA

If a food bank manager receives a donation of 600 lb of food and decides to give four times as much food to families as she gives to a shelter, a system of linear equations can represent the relationship of the amounts of food going to the shelter and to families. In this section, we will use the substitution method to solve this and other systems of linear equations.

SECTION 4.2 The Substitution Method

After reading the text, working the practice problems, and completing assigned exercises, you should be able to:

1. Solve a system of two linear equations by substitution.
2. Identify a system of linear equations with no solution or infinitely many solutions.
3. Use the substitution method and a system of equations to solve an application problem.

Solving a System of Equations by Substitution

The **substitution method** is an algebraic method for finding the solution of a system of equations. When a system of equations has one solution, the graphs of the equations intersect at one point. The lines on the graph (Figure 1) represent the equations $y = 3x - 7$ and $y = -2x + 8$. The equations share a solution, $(3, 2)$, which is the intersection point of their graphs.

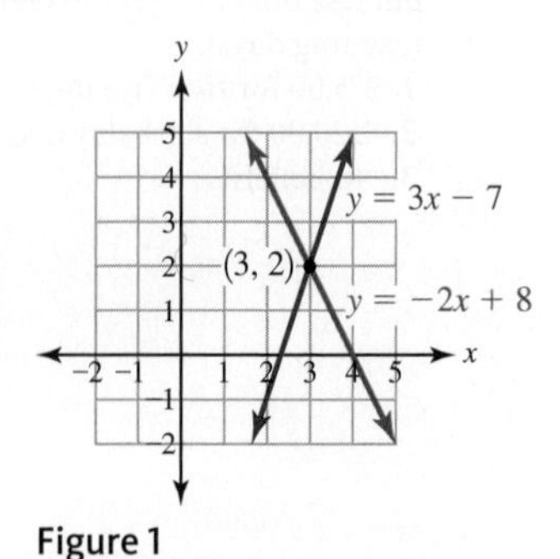

Figure 1

At this point, the value of y in $y = 3x - 7$ equals the value of y in $y = -2x + 8$. Since $y = y$, then $3x - 7 = -2x + 8$. If we solve this equation to find x, we can then use this value for x to find y.

EXAMPLE 1 The graph represents $\begin{matrix} y = 3x - 7 \\ y = -2x + 8 \end{matrix}$.

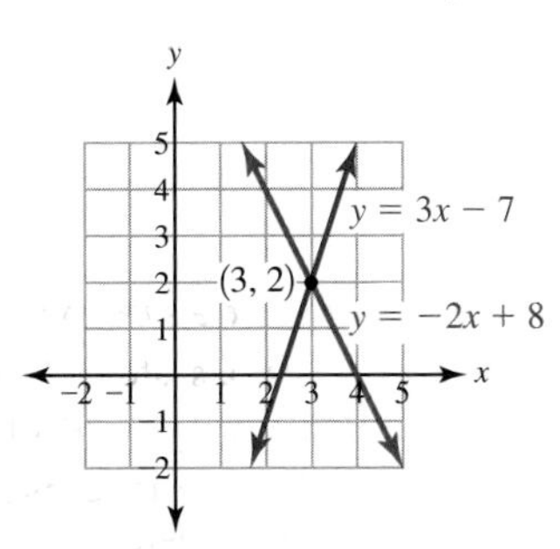

(a) Solve by substitution.

SOLUTION ▶ Find the x-coordinate of the intersection point.

$y = y$	At the intersection point, the y-values are equal.
$3x - 7 = -2x + 8$	Replace y on both sides of the equation.
$+7 \quad\quad +7$	Addition property of equality.
$3x + \mathbf{0} = -2x + \mathbf{15}$	Simplify.
$+2x \quad\quad +2x$	Addition property of equality.
$\mathbf{5x} + 0 = \mathbf{0} + 15$	Simplify.
$\frac{5x}{5} = \frac{15}{5}$	Division property of equality.
$x = 3$	Simplify; this is the x-coordinate of the intersection point.

Use this value for x to find the y-coordinate of the intersection point.

$y = 3x - 7$	Choose either of the original equations.
$y = 3(\mathbf{3}) - 7$	Replace x with the x-coordinate of the intersection point, 3.
$y = \mathbf{9} - 7$	Simplify.
$y = \mathbf{2}$	Simplify; this is the y-coordinate of the intersection point.

The solution is $(3, 2)$.

(b) Check.

▶ The solution of a system of equations is a solution of each equation in the system.

$y = 3x - 7$	Use the original equations.	$y = -2x + 8$
$\mathbf{2} = 3(\mathbf{3}) - 7$	Replace the variables.	$\mathbf{2} = -2(\mathbf{3}) + 8$
$2 = \mathbf{9} - 7$	Simplify.	$2 = \mathbf{-6} + 8$
$2 = \mathbf{2}$ True.	The solution is correct.	$2 = \mathbf{2}$ True.

In the next example, one of the equations is solved for a variable, $y = 5x - 3$. At the intersection point, each line has the same y-coordinate, so $y = y$. Since $y = 5x - 3$, we can replace y in the other equation with the **substitution expression**, $5x - 3$, and solve for x.

EXAMPLE 2 The graph represents $\begin{matrix} y = 5x - 3 \\ 6x - 7y = -8 \end{matrix}$.

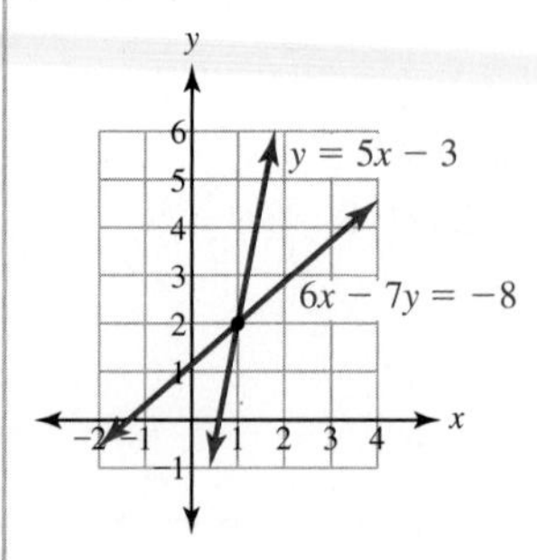

(a) Solve by substitution.

SOLUTION ▶ Find the x-coordinate of the intersection point.

$6x - 7y = -8$	
$6x - 7(5x - 3) = -8$	Replace y with the substitution expression, $5x - 3$.
$6x \mathbf{- 35x + 21} = -8$	Distributive property.
$\mathbf{-29x} + 21 = -8$	Combine like terms.
$-21 \quad -21$	Subtraction property of equality.
$-29x + \mathbf{0} = \mathbf{-29}$	Simplify.
$\dfrac{\mathbf{-29x}}{-29} = \dfrac{-29}{-29}$	Division property of equality.
$x = 1$	Simplify; this is the x-coordinate of the intersection point.

Use this value for x to find the y-coordinate of the intersection point.

$y = 5x - 3$	Choose either of the original equations.
$y = 5(\mathbf{1}) - 3$	Replace x with the x-coordinate of the intersection point, 1.
$y = \mathbf{5} - 3$	Simplify.
$y = \mathbf{2}$	Simplify; this is the y-coordinate of the intersection point.

The solution is $(1, 2)$.

(b) Check.

▶ $y = 5x - 3$	Use the original equations.	$6x - 7y = -8$
$\mathbf{2} = 5(\mathbf{1}) - 3$	Replace the variables.	$6(\mathbf{1}) - 7(\mathbf{2}) = -8$
$2 = \mathbf{5} - 3$	Simplify.	$\mathbf{6 - 14} = -8$
$2 = \mathbf{2}$ True.	The solution is correct.	$\mathbf{-8} = -8$ True.

In the next example, the equation $x = \frac{1}{2}y + 3$ is solved for x. The substitution expression is $\frac{1}{2}y + 3$.

EXAMPLE 3 The graph represents $\begin{array}{l} x = \frac{1}{2}y + 3 \\ 8x + 3y = -11 \end{array}$. Solve by substitution.

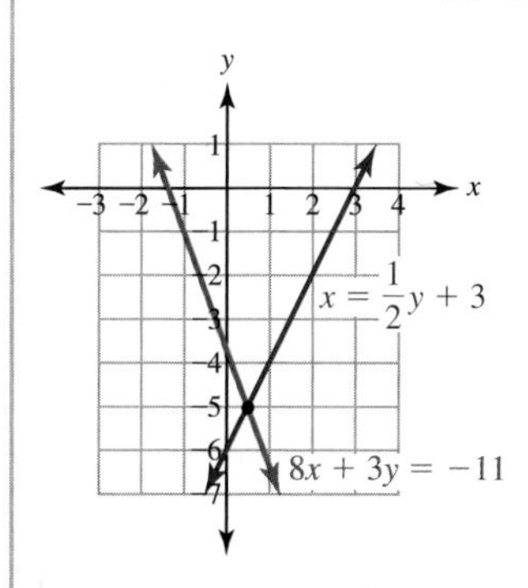

SOLUTION ▶ Find the y-coordinate of the intersection point.

$$8x + 3y = -11$$

$$8\left(\frac{1}{2}y + 3\right) + 3y = -11 \quad \text{Replace } x \text{ with the substitution expression, } \frac{1}{2}y + 3.$$

$$\mathbf{8}\left(\frac{1}{2}y\right) + \mathbf{8}(3) + 3y = -11 \quad \text{Distributive property.}$$

$$\mathbf{4y + 24} + 3y = -11 \quad \text{Simplify.}$$

$$\mathbf{7y} + 24 = -11 \quad \text{Combine like terms.}$$

$$-24 \quad -24 \quad \text{Subtraction property of equality.}$$

$$7y + \mathbf{0} = \mathbf{-35} \quad \text{Simplify.}$$

$$\frac{\mathbf{7y}}{7} = \frac{-35}{7} \quad \text{Division property of equality.}$$

$$y = -5 \quad \text{Simplify; this is the } y\text{-coordinate of the intersection point.}$$

Use this value for y to find the x-coordinate of the intersection point.

$$x = \frac{1}{2}y + 3 \quad \text{Choose either of the original equations.}$$

$$x = \frac{1}{2}(\mathbf{-5}) + 3 \quad \text{Replace } y \text{ with the } y\text{-coordinate of the intersection point, } -5.$$

$$x = -\frac{\mathbf{5}}{\mathbf{2}} + 3 \quad \text{Simplify.}$$

$$x = -\frac{5}{2} + \frac{3}{\mathbf{1}} \cdot \frac{2}{2} \quad \text{The LCD is 2; multiply by a fraction equal to 1.}$$

$$x = -\frac{5}{2} + \frac{\mathbf{6}}{\mathbf{2}} \quad \text{Simplify.}$$

$$x = \frac{\mathbf{1}}{\mathbf{2}} \quad \text{Simplify; this is the } x\text{-coordinate of the intersection point.}$$

The solution is $\left(\frac{1}{2}, -5\right)$.

The variables in the next example are a and b. When writing the solution as an ordered pair, write the variables in alphabetical order, (a, b).

EXAMPLE 4 The graph represents $\begin{matrix} 5a + 2b = 46 \\ a = 3b - 1 \end{matrix}$. Solve by substitution.

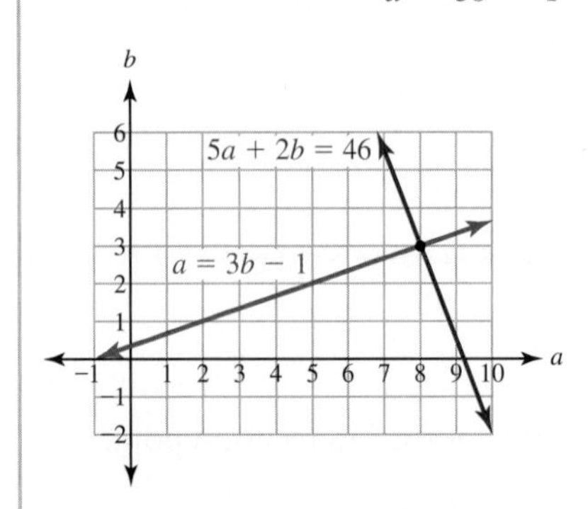

SOLUTION ▶ Find the b-coordinate of the intersection point.

$5a + 2b = 46$	
$5(3b - 1) + 2b = 46$	Replace a with the substitution expression, $3b - 1$.
$\mathbf{5(3b) + 5(-1)} + 2b = 46$	Distributive property.
$\mathbf{15b - 5} + 2b = 46$	Simplify.
$\mathbf{17b} - 5 = 46$	Combine like terms.
$+5 \quad +5$	Addition property of equality.
$17b + \mathbf{0} = \mathbf{51}$	Simplify.
$\frac{\mathbf{17b}}{17} = \frac{51}{17}$	Division property of equality.
$b = 3$	Simplify; this is the b-coordinate of the intersection point.

Use this value for b to find the a-coordinate of the intersection point.

$a = 3b - 1$	Choose either of the original equations.
$a = 3(\mathbf{3}) - 1$	Replace b with the b-coordinate of the intersection point, 3.
$a = \mathbf{9} - 1$	Simplify.
$a = \mathbf{8}$	Simplify; this is the a-coordinate of the intersection point.

The solution is $(8, 3)$.

To use substitution when both equations are in standard form, $ax + by = c$, first solve one of the equations for a variable.

EXAMPLE 5 The graph represents $\begin{matrix} 3x + 5y = -25 \\ 4x - 7y = -6 \end{matrix}$.

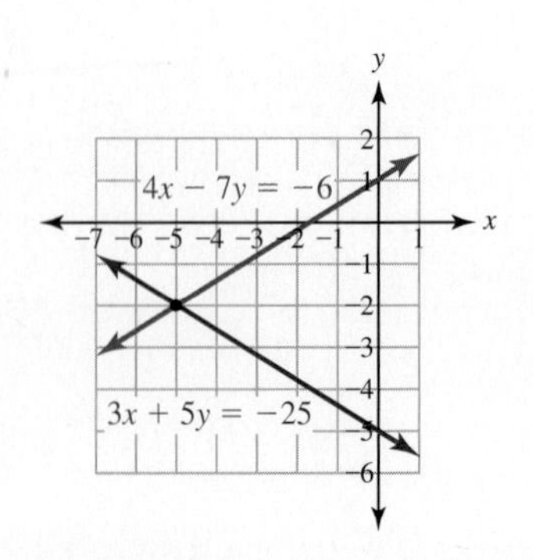

(a) Solve by substitution.

SOLUTION ▶ Choose one of the equations to solve for either x or y. We will solve $3x + 5y = -25$ for x.

$3x + 5y = -25$

$-5y \qquad -5y$ — Subtraction property of equality.

$3x + \mathbf{0} = -25 - \mathbf{5y}$ — Simplify.

$\frac{3x}{3} = \frac{-25}{3} - \frac{5y}{3}$ — Division property of equality.

$x = -\frac{25}{3} - \frac{5y}{3}$ — Simplify. The substitution expression is $-\frac{25}{3} - \frac{5y}{3}$.

Find the y-coordinate of the intersection point.

$4x - 7y = -6$

$4\left(-\frac{25}{3} - \frac{5y}{3}\right) - 7y = -6$ — Replace x with the substitution expression, $-\frac{25}{3} - \frac{5y}{3}$.

$\mathbf{4}\left(-\frac{25}{3}\right) + \mathbf{4}\left(-\frac{5y}{3}\right) - 7y = -6$ — Distributive property; simplify.

$-\frac{\mathbf{100}}{\mathbf{3}} - \frac{\mathbf{20y}}{\mathbf{3}} - 7y = -6$ — Simplify.

$3\left(-\frac{100}{3} - \frac{20y}{3} - 7y\right) = 3(-6)$ — Clear fractions; multiplication property of equality.

$\cancel{3}\left(-\frac{100}{\cancel{3}}\right) + \cancel{3}\left(-\frac{20y}{\cancel{3}}\right) + 3(-7y) = \mathbf{-18}$ — Distributive property; simplify.

$\mathbf{-100 - 20y - 21y} = -18$ — Simplify.

$-100 - \mathbf{41y} = -18$ — Combine like terms.

$+100 \qquad +100$ — Addition property of equality.

$\mathbf{0} - 41y = \mathbf{82}$ — Simplify.

$\frac{\mathbf{-41y}}{-41} = \frac{82}{-41}$ — Division property of equality.

$y = -2$ — Simplify; this is the y-coordinate of the intersection point.

Use this value for y to find the x-coordinate of the intersection point.

$3x + 5y = -25$ — Choose either of the original equations.

$3x + 5(\mathbf{-2}) = -25$ — Replace y with the y-coordinate of the intersection point, -2.

$3x - \mathbf{10} = -25$ — Simplify.

$+10 \qquad +10$ — Addition property of equality.

$3x - \mathbf{0} = \mathbf{-15}$ — Simplify.

$\frac{\mathbf{3x}}{3} = \frac{-15}{3}$ — Division property of equality.

$x = -5$ — Simplify; this is the x-coordinate of the intersection point.

The solution is $(-5, -2)$.

(b) Check.

$3x + 5y = -25$	Use the original equations.	$4x - 7y = -6$
$3(\mathbf{-5}) + 5(\mathbf{-2}) = -25$	Replace the variables.	$4(\mathbf{-5}) - 7(\mathbf{-2}) = -6$
$\mathbf{-15 - 10} = -25$	Simplify.	$\mathbf{-20 + 14} = -6$
$\mathbf{-25} = -25$ True.	The solution is correct.	$\mathbf{-6} = -6$ True.

Solving a System of Two Linear Equations by Substitution

1. Solve at least one equation for a variable. The expression that is equal to the variable is the substitution expression.
2. Replace the variable in the other equation with the substitution expression.
3. Solve for the remaining variable.
4. Replace the variable in one of the original equations with the value found in step 3. Solve for the other variable.
5. To check, replace the variables in each equation with the solution.

The substitution method finds the exact solution of a system of equations. For some systems, a graph might not be precise enough to identify the solution.

EXAMPLE 6 The graph represents $\begin{aligned} y &= \frac{3}{4}x - 5 \\ 2x + 7y &= -14 \end{aligned}$. Solve by substitution.

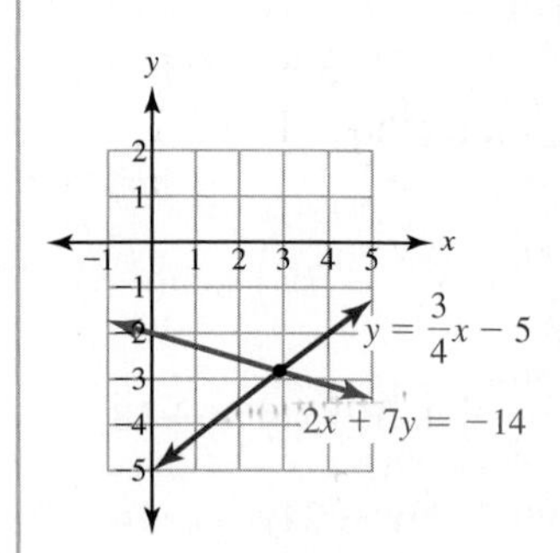

SOLUTION ▶ Find the x-coordinate of the intersection point.

$$2x + 7y = -14$$

$$2x + 7\left(\frac{3}{4}x - 5\right) = -14 \quad \text{Replace } y \text{ with the substitution expression, } \frac{3}{4}x - 5.$$

$$2x + \mathbf{7}\left(\frac{3}{4}x\right) + \mathbf{7(-5)} = -14 \quad \text{Distributive property.}$$

$$2x + \mathbf{\frac{21}{4}x - 35} = -14 \quad \text{Simplify.}$$

$$\frac{2x}{\mathbf{1}} \cdot \frac{4}{4} + \frac{21}{4}x - 35 = -14 \quad \text{The LCD is 4; multiply by a fraction equal to 1.}$$

$$\mathbf{\frac{8}{4}}x + \frac{21}{4}x - 35 = -14 \quad \text{Simplify.}$$

$$\mathbf{\frac{29}{4}x} - 35 = -14 \quad \text{Combine like terms.}$$

$$+35 \quad +35 \quad \text{Addition property of equality.}$$

$$\frac{29}{4}x + \mathbf{0} = \mathbf{21} \quad \text{Simplify.}$$

$$\frac{4}{29} \cdot \frac{\mathbf{29}}{\mathbf{4}}x = \frac{4}{29} \cdot \frac{21}{1} \quad \text{Simplify; multiplication property of equality.}$$

$$x = \frac{84}{29} \quad \text{Simplify; this is the } x\text{-coordinate of the intersection point.}$$

Use this value for x to find the y-coordinate of the intersection point.

$y = \frac{3}{4}x - 5$	Choose either of the original equations.
$y = \frac{3}{4} \cdot \mathbf{\frac{84}{29}} - 5$	Replace x with the x-coordinate of the intersection point, $\frac{84}{29}$.
$y = \frac{3}{\cancel{4}} \cdot \frac{\mathbf{\cancel{4} \cdot 21}}{29} - 5$	Find common factors.
$y = \mathbf{\frac{63}{29}} - \frac{5}{\mathbf{1}} \cdot \frac{29}{29}$	Simplify; the LCD is 29; multiply by a fraction equal to 1.
$y = \frac{63}{29} - \mathbf{\frac{145}{29}}$	Simplify.
$y = \mathbf{-\frac{82}{29}}$	Simplify; this is the y-coordinate of the intersection point.

The solution is $\left(\frac{84}{29}, -\frac{82}{29}\right)$. The graph is not precise enough to identify this solution.

In Section 4.3, we will learn another algebraic method for solving a system of equations, the elimination method. That method is often a more efficient way than substitution for solving a system of two equations that are in standard form.

Practice Problems

For problems 1–5, solve by substitution.

1. $y = -2x + 9$
$y = 6x - 7$

2. $y = 3x + 1$
$x - 2y = -7$

3. $x = 5y + 1$
$x - y = -7$

4. $2a - 3b = 23$
$4a + b = 11$

5. $6x + 5y = 33$
$y = 2x - 6$

Parallel Lines and Coinciding Lines

When we use the substitution method to solve a system of linear equations, we are assuming that there is an intersection point. When the lines that represent a system of equations are parallel, there is no intersection point and the system has no solution. Using the substitution method on this system of equations results in a false equation with no variables.

EXAMPLE 7 The graph represents $\begin{array}{l} y = 2x + 1 \\ y = 2x - 5 \end{array}$. Solve by substitution.

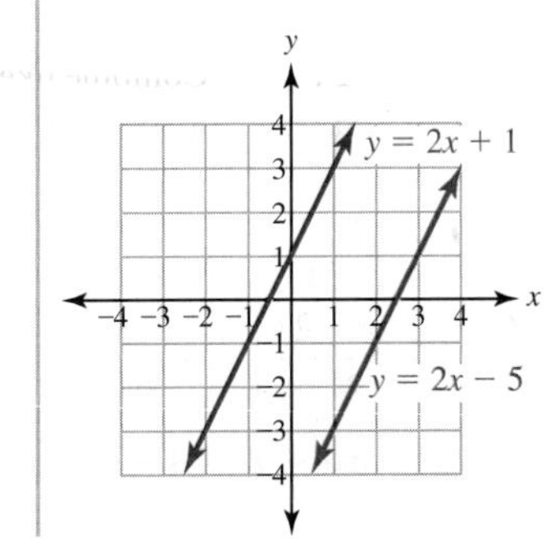

SOLUTION ▶ Find the x-coordinate of the intersection point.

$y = 2x - 5$	
$2x + 1 = 2x - 5$	Replace y with the substitution expression, $2x + 1$.
$-1 \quad -1$	Subtraction property of equality.
$2x + \mathbf{0} = 2x - \mathbf{6}$	Simplify.
$-2x \quad -2x$	Subtraction property of equality.
$\mathbf{0} + 0 = \mathbf{0} - 6$	Simplify.
$0 = -6$	Simplify; this equation is false.

The result of the substitution is a false equation with no variables. The graph of this system is two parallel lines, which do not intersect. The system has no solution.

When lines are coinciding, *every* point on either of the lines is an intersection point. The system has infinitely many solutions. Using the substitution method on a system with infinitely many solutions results in a true equation with no variables.

EXAMPLE 8 The graph represents $\begin{matrix} y = -2x + 5 \\ 6x + 3y = 15 \end{matrix}$. Solve by substitution.

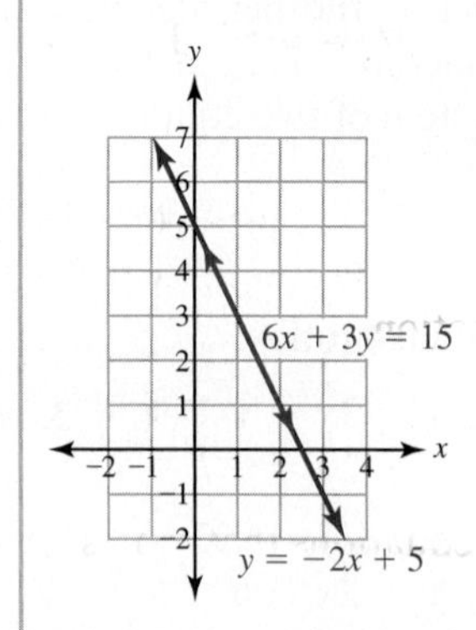

SOLUTION ▶ Find the x-coordinate of the intersection point.

$6x + 3y = 15$	
$6x + 3(-2x + 5) = 15$	Replace y with the substitution expression, $-2x + 5$.
$6x + \mathbf{3(-2x)} + \mathbf{3 \cdot 5} = 15$	Distributive property.
$6x + \mathbf{-6x} + \mathbf{15} = 15$	Simplify.
$\mathbf{0} + 15 = 15$	Simplify.
$\mathbf{15} = 15$	Simplify; this equation is true.

The result of substitution is a true equation with no variables. The graph of this system is two coinciding lines, which share infinitely many points. This system has infinitely many solutions. In set-builder notation, the solution is $\{(x, y) \mid y = -2x + 5\}$.

Practice Problems

For problems 6–9, solve by substitution.

6. $15x + 3y = 3$
$y = -5x + 10$

7. $y = \frac{3}{4}x + 1$
$3x + 4y = 4$

8. $x = 5y$
$x + y = 0$

9. $y = 4x + 2$
$4x - y = -2$

Applications

When the total costs to make products and the revenue from selling the products are equal, a company is breaking even. If a system of equations includes a model of the costs and a model of the revenue, the intersection of the graphs of the equations is the break-even point. In Section 4.1, we found the break-even point by graphing. We can also find the break-even point by substitution. The intersection point and the solution of the system are the same ordered pair.

Writing a System of Linear Equations for the Break-Even Point

1. Write an equation that represents the costs, y, of making x products.

   $y = (\text{cost per product})x + \text{fixed overhead costs}$

2. Write an equation that represents the revenue, y, from selling x products.

   $y = (\text{price per product})x$

EXAMPLE 9 The cost to make a product is \$5. The fixed overhead costs per month are \$4500. The price of the product is \$10. Most of the people who buy this product are over age 25. Find the break-even point.

SOLUTION **Step 1 Understand the problem.**
The two unknowns are the number of products that the company must make and sell to break even and the costs/revenue at the break-even point. The age of the customers is extraneous information.

x = number of products $\quad$ y = costs or revenue at the break-even point

Step 2 Make a plan.
Write and solve a system of equations that represent the costs and the revenue.

The word equations are costs $= \left(\dfrac{\text{cost}}{1 \text{ product}}\right)$(number of products) + overhead and revenue $= \left(\dfrac{\text{price}}{1 \text{ product}}\right)$(number of products).

Step 3 Carry out the plan.
For the costs, $y = \left(\dfrac{\$5}{1 \text{ product}}\right)x + \$4500$. For the revenue, $y = \left(\dfrac{\$10}{1 \text{ product}}\right)x$.

Remove the units and solve $\begin{aligned} y &= 5x + 4500 \\ y &= 10x \end{aligned}$ by substitution.

Find the x-coordinate of the solution.

$$y = 10x$$

$$5x + 4500 = 10x \quad \text{Replace } y \text{ with the substitution expression, } 5x + 4500.$$

$$-5x \qquad -5x \quad \text{Subtraction property of equality.}$$

$$\mathbf{0} + 4500 = \mathbf{5x} \quad \text{Simplify.}$$

$$\frac{\mathbf{4500}}{5} = \frac{5x}{5} \quad \text{Division property of equality.}$$

$$900 = x \quad \text{Simplify; this is the } x\text{-coordinate of the solution.}$$

Use this value for x to find the y-coordinate of the solution.

$y = 10x$	Choose either of the original equations.
$y = 10(\mathbf{900})$	Replace x with the x-coordinate of the solution, 900.
$y = \mathbf{9000}$	Simplify; this is the y-coordinate of the solution.

Including the units, the solution is (900 products, $9000).

Step 4 Look back.
The cost to make 900 products is $\left(\frac{\$5}{1 \text{ product}}\right)$(900 products) + $4500, which equals $9000. The revenue from selling 900 products, $\left(\frac{\$10}{1 \text{ product}}\right)$(900 products), is also $9000. Since the costs equal the revenue, the answer seems reasonable.

Step 5 Report the solution.
The break-even point is (900 products, $9000). To break even, the company must make and sell 900 products. The costs and revenue at the break-even point are $9000.

In the next example, the amount of food given to families, x, is four times the amount of food given to a senior food program, y. The amount given to the senior food program must be multiplied by 4 to equal the amount given to families.

EXAMPLE 10 A donor gave 600 lb of food and $500 for fuel bills relief to a community agency. The agency will distribute four times as much of the donated food to families as to the senior food program. Find the amount of food for families and the amount for the senior food program.

SOLUTION ▶ **Step 1 Understand the problem.**
The two unknowns are the amount of food for families and the amount of food for the senior food program. The amount given for fuel bill relief is extraneous information.

x = amount of food for families $\quad$ y = amount of food for senior food program

Step 2 Make a plan.
Write and solve a system of equations. One equation represents the sum of the amount of food for families and the amount of food for the senior food program. The other equation compares the amount of food for families to the amount of food for the senior food program. Word equations are amount families + amount program = total and amount families = 4(amount program).

Step 3 Carry out the plan.
The equations are $x + y = 600$ lb and $x = 4y$. Solve this system $\begin{aligned} x + y &= 600 \text{ lb} \\ x &= 4y \end{aligned}$ by substitution.

Find the y-coordinate of the solution.

$x + y = 600 \text{ lb}$	
$4y + y = 600 \text{ lb}$	Replace x with the substitution expression, $4y$.
$\mathbf{5y} = 600 \text{ lb}$	Combine like terms.
$\frac{5y}{5} = \frac{600 \text{ lb}}{5}$	Division property of equality.
$y = 120 \text{ lb}$	Simplify; this is the y-coordinate of the solution.

Use this value for y to find the x-coordinate of the solution.

$x = 4y$	Choose either of the original equations.
$x = 4(\mathbf{120\ lb})$	Replace y with the y-coordinate of the solution, 120 lb.
$x = \mathbf{480\ lb}$	Simplify; this is the x-coordinate of the solution.

The solution is (480 lb, 120 lb).

Step 4 Look back.
Since 480 lb plus 120 lb equals the total amount of 600 lb and 480 lb equals four times 120 lb, the answer seems reasonable.

Step 5 Report the solution.
The manager should give 480 lb to families and 120 lb to the senior food program.

Practice Problems

For problems 10–11, use the five steps, a system of equations, and the substitution method.

10. The cost to make a product is $3500. The fixed overhead costs per month are $200,000. If the price of the product is $4300, find the number of products the company must make and sell to break even. Find the costs and revenue at the break-even point.
11. A tree farm has 75 acres of mature trees. The grower decides to sell four times as many acres of trees this year as next year. About 45% of the trees are Scotch pine, and 55% of the trees are balsam fir. Find the number of acres of trees to sell this year and the number of acres of trees to sell next year.

4.2 Vocabulary Practice

Match the term with its description.

1. Two lines that never intersect
2. The number of solutions of a system of coinciding lines
3. A solution found by solving a system of equations using graphing
4. A solution found by solving a system of equations using an algebraic method
5. This method of solving a system assumes that the lines that represent the equations intersect in one point.
6. $ax + by = c$
7. Money received from selling products
8. Money required to make a product
9. The slopes of two parallel lines
10. The slopes of two intersecting lines

A. costs
B. different slopes
C. estimated solution
D. exact solution
E. infinitely many solutions
F. parallel lines
G. revenue
H. same slopes
I. standard form of a linear equation
J. substitution method

4.2 Exercises

Follow your instructor's guidelines for showing your work.

For exercises 1–2,
(a) find the x-coordinate of the intersection point.
(b) find the y-coordinate of the intersection point.
(c) identify the solution of the system.
(d) check.

1. $y = 3x + 2$
   $y = x + 6$

2. $y = 5x + 3$
   $y = x + 7$

For exercises 3–4,
(a) identify the substitution expression.
(b) find the x-coordinate of the intersection point.
(c) find the y-coordinate of the intersection point.
(d) identify the solution of the system.
(e) check.

3. $5x + y = 9$
   $y = 2x + 2$

4. $6x + y = 13$
   $y = 3x + 4$

For exercises 5–40,
(a) solve by substitution.
(b) if there is one solution, check.

5. $y = 3x - 5$
$3x + 2y = 17$

6. $y = 2x - 5$
$2x + 5y = 23$

7. $y = 6x - 21$
$y = -2x + 3$

8. $y = 3x - 19$
$y = -4x + 16$

9. $y = 4x - 1$
$y = 4x + 3$

10. $y = 6x - 2$
$y = 6x + 7$

11. $x + 2y = 6$
$y = 4x - 6$

12. $x + 2y = 9$
$y = 5x - 12$

13. $y = \frac{1}{3}x - 7$
$4x + 3y = 24$

14. $y = \frac{1}{2}x - 8$
$5x + 2y = 20$

15. $y = -\frac{7}{2}x - 6$
$7x + 2y = 8$

16. $y = -\frac{3}{8}x - 5$
$3x + 8y = 16$

17. $5x - 2y = -6$
$y = x$

18. $7x - 2y = -20$
$y = x$

19. $x - 9y = 18$
$y = \frac{1}{9}x - 2$

20. $x - 5y = 30$
$y = \frac{1}{5}x - 6$

21. $y = -5x$
$y = \frac{3}{8}x$

22. $y = -6x$
$y = \frac{5}{9}x$

23. $y = \frac{9}{10}x - 8$
$x = 20$

24. $y = \frac{8}{9}x - 6$
$x = 18$

25. $y = \frac{3}{4}x - 2$
$y = -x + 5$

26. $y = \frac{3}{5}x - 1$
$y = -x + 7$

27. $y = 6x - 3$
$y = 6x + 1$

28. $y = 5x + 8$
$y = 5x - 3$

29. $x = 8y + 2$
$3x - 4y = 16$

30. $x = 10y + 3$
$5x - 20y = 30$

31. $y = -8x$
$12x + 5y = 7$

32. $y = -6x$
$9x + 2y = 1$

33. $y = 10x - 5$
$y = 2x - 1$

34. $y = 8x - 4$
$y = 6x - 3$

35. $4x + 8y = 7$
$y = x + \frac{1}{2}$

36. $3x + 9y = 16$
$y = x + \frac{4}{3}$

37. $y = \frac{3}{4}x + 2$
$-3x + 4y = 8$

38. $y = \frac{3}{8}x + 2$
$-3x + 8y = 16$

39. $y = -\frac{8}{9}x + \frac{7}{9}$
$-x + 2y = \frac{1}{6}$

40. $y = -\frac{9}{8}x + \frac{7}{8}$
$-x + 2y = \frac{2}{3}$

For exercises 41–48, solve by substitution.

41. $a = 2b - 15$
$3a + 4b = 55$

42. $k = -3h + 100$
$4h + 9k = 210$

43. $g = 3f - 2$
$f = g + 1$

44. $c = 2b - 1$
$b = 2c + 1$

45. $-6p + 11w = 16$
$w = 14p - 12$

46. $-8c + 5d = 7$
$d = 17c - 14$

47. $k = \frac{4}{9}h + 8$
$-4h + 9k = 70$

48. $w = \frac{3}{8}v + 5$
$-3v + 8w = 41$

For exercises 49–60, solve by substitution. Include the units of measurement in the solution.

49. $y = \left(\frac{\$40}{1 \text{ product}}\right)x + \$8000$
$y = \left(\frac{\$60}{1 \text{ product}}\right)x$

50. $y = \left(\frac{\$25}{1 \text{ product}}\right)x + \$10{,}000$
$y = \left(\frac{\$50}{1 \text{ product}}\right)x$

51. $y = \left(\frac{\$40}{1 \text{ product}}\right)x + \$8000$
$y = \left(\frac{\$50}{1 \text{ product}}\right)x$

52. $y = \left(\frac{\$25}{1 \text{ product}}\right)x + \$10{,}000$
$y = \left(\frac{\$75}{1 \text{ product}}\right)x$

53. $\left(\frac{\$6}{1 \text{ lb}}\right)x + \left(\frac{\$10}{1 \text{ lb}}\right)y = \$3700$
$x + y = 450 \text{ lb}$

54. $\left(\frac{\$7}{1 \text{ lb}}\right)x + \left(\frac{\$8}{1 \text{ lb}}\right)y = \$920$
$x + y = 125 \text{ lb}$

55. $x + y = 5000 \text{ acres}$
$y = 4x$

56. $x + y = 7500 \text{ acres}$
$y = 9x$

57. $y = \left(\frac{60 \text{ mi}}{1 \text{ hr}}\right)x + 80 \text{ mi}$
$y = \left(\frac{65 \text{ mi}}{1 \text{ hr}}\right)x$

58. $y = \left(\frac{55 \text{ mi}}{1 \text{ hr}}\right)x + 300 \text{ mi}$
$y = \left(\frac{75 \text{ mi}}{1 \text{ hr}}\right)x$

59. $\left(\frac{\$10}{1 \text{ adult ticket}}\right)x + \left(\frac{\$5}{1 \text{ youth ticket}}\right)y = \$950$
$x + y = 150 \text{ tickets}$

60. $\left(\frac{\$10}{1 \text{ adult ticket}}\right)x + \left(\frac{\$4}{1 \text{ youth ticket}}\right)y = \$1700$
$x + y = 275 \text{ tickets}$

For exercises 61–72,
(a) solve one of the equations for a variable and solve by substitution.
(b) if there is one solution, check.

61. $3x + y = 2$
$2x + 3y = 20$

62. $4x + y = -5$
$2x + 3y = 15$

63. $7x + y = 8$
$7x + y = 13$

64. $9x + y = 15$
$9x + y = -2$

65. $x + 2y = -8$
$6x + y = 18$

66. $x + 3y = -7$
$10x + y = 17$

67. $3x - 4y = -8$
$x + y = 2$

68. $5x - 6y = -18$
$x + y = 3$

69. $x + 2y = 4$
$\frac{1}{2}x + y = 2$

70. $x + 3y = 6$
$\frac{1}{3}x + y = 2$

71. $x + y = 8$
$2x + y = 8$

72. $x + y = 6$
$2x + y = 6$

For exercises 73–76, use the five steps, a system of equations, and the substitution method to find the break-even point.

73. The cost to make a product is $8. The fixed costs per month to make the product are $4800. The price of each product is $10.

74. The cost to make a product is $6. The fixed costs per month to make the product are $3600. The price of each product is $12.

75. The cost to make a product is $2000. Forty percent of these products will be sold to women. The fixed costs per month to make the product are $60,000. The price of each product is $2500.

76. The cost to make a product is $4000. Seventy percent of these products will be sold to men. The fixed costs per month to make the product are $90,000. The price of each product is $4500.

For exercises 77–82, use the five steps, a system of equations, and the substitution method.

77. A chef is creating a stew that will include a total of 28 lb of potatoes and carrots. If the amount of potatoes is to be three times the amount of carrots, find the amount of potatoes and carrots to put in the stew.

78. A grass seed mixture will include a total of 32 lb of perennial grass seed and annual grass seed. If the amount of perennial grass seed is to be three times the amount of annual grass seed, find the amount of perennial grass seed and annual grass seed in the mixture.

79. A grandchild received an inheritance of $32,000, and his parents received an inheritance of $45,000. The grandchild will invest an amount that is three times the amount he uses to pay debts. Find the amount he will invest and the amount he will use to pay debts.

80. A college foundation receives a gift of $750,000 from an estate. About 70% of the income from the investments is for scholarships. A financial advisor suggests that the foundation invests three times as much in stocks as in real estate. Find the amount to invest in stocks and the amount to invest in real estate.

81. A librarian has $45,000 to spend on print journals and on-line access to journals. If one-fourth as much money will be spent on print journals as on on-line access to journals, find the amount that will be spent on print journals and the amount that will be spent on on-line access to journals.

82. A programmer is writing code to represent a total of 180 math problems in which the answers are either true/false or multiple choice. If there will be one-fifth as many answers that are true/false as there are multiple choice answers, find the number of answers that will be true/false and the number of answers that will be multiple choice.

Problem Solving: Practice and Review

Follow your instructor's guidelines for using the five steps as outlined in Section 1.3, p. 55.

83. The sale price of a flat screen television is $995. The percent discount is 18%. Find the regular price of the television. Round to the nearest whole number.

84. The arithmetic mean (also called the average) of a collection of test scores is the sum of the test scores divided by the total number of test scores. Find the arithmetic mean of these scores: 85, 70, 62, 79, 65, 81. Round to the nearest whole number.

85. A line between two points (x_1, y_1) and (x_2, y_2) is a line segment. The midpoint of this line segment is $\left(\frac{x_1 + x_2}{2}, \frac{y_1 + y_2}{2}\right)$. Find the midpoint of the line segment created by connecting the points $(1, -2)$ and $(-3, 8)$.

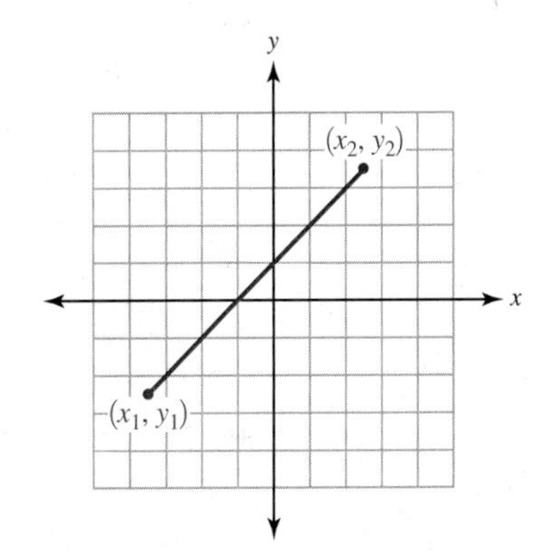

86. A model of the amount of e-commerce revenue for rental and leasing services companies, y, in x years after 2005 is $y = \left(\frac{\$8.71 \text{ billion}}{1 \text{ year}}\right)x + \$6.01 \text{ billion}$. Use the model to find the amount of e-commerce revenue in 2012. (*Source:* www.census.gov, 2011)

Technology

For exercises 87–90, solve by graphing. Sketch the graph; describe the window; identify the solution.

87. $y = 3x + 2$
$y = x + 6$

88. $y = 5x + 3$
$y = x + 7$

89. $y = x + 13$
$y = -2x + 19$

90. $y = x + 8$
$y = \frac{3}{4}x + 5$

Find the Mistake

For exercises 91–94, the completed problem has one mistake.
(a) Describe the mistake in words, or copy down the whole problem and highlight or circle the mistake.
(b) Do the problem correctly.

91. Problem: Solve $\begin{matrix} y = -2x + 4 \\ 3x - y = 21 \end{matrix}$ by substitution.

Incorrect Answer:
$$\begin{aligned} 3x - y &= 21 \\ 3x - 2x + 4 &= 21 \\ x + 4 &= 21 \\ x &= 17 \end{aligned}$$
$$\begin{aligned} y &= -2(17) + 4 \\ y &= -34 + 4 \\ y &= -30 \end{aligned}$$
The solution is $(17,-30)$.

92. Problem: Solve $\begin{matrix} y = x - 9 \\ 2x - 3y = 22 \end{matrix}$ by substitution.

Incorrect Answer:
$$\begin{aligned} 2x - 3y &= 22 \\ 2x - 3(x - 9) &= 22 \\ 2x - 3x - 9 &= 22 \\ -x - 9 &= 22 \\ +9 \quad &\ \ +9 \\ -x &= 31 \\ \frac{-x}{-1} &= \frac{31}{-1} \\ x &= -31 \end{aligned}$$
$$\begin{aligned} y &= x - 9 \\ y &= -31 - 9 \\ y &= -40 \end{aligned}$$
The solution is $(-31,-40)$.

93. Problem: Solve $\begin{matrix} y = 2x + 3 \\ -x + \frac{1}{2}y = \frac{3}{2} \end{matrix}$ by substitution.

Incorrect Answer:
$$\begin{aligned} -x + \frac{1}{2}y &= \frac{3}{2} \\ -x + \frac{1}{2}(2x + 3) &= \frac{3}{2} \\ -x + x + \frac{3}{2} &= \frac{3}{2} \\ 0 + \frac{3}{2} &= \frac{3}{2} \\ \frac{3}{2} &= \frac{3}{2} \end{aligned}$$
The solution is the set of real numbers.

94. Problem: Solve $\begin{matrix} y = 3x - 4 \\ -3x + y = 15 \end{matrix}$ by substitution.

Incorrect Answer:
$$\begin{aligned} -3x + y &= 15 \\ -3x + 3x - 4 &= 15 \\ -4 &= 15 \end{aligned}$$
The solution is $(-4, 15)$.

Review

For exercises 95–96, clear the fractions, and rewrite the equation in standard form.

95. $y = -\frac{3}{4}x + 9$

96. $y = -\frac{1}{2}x - \frac{3}{4}$

97. Write a definition and give an example of the *opposite of a real number.*

98. What is the sum of two numbers that are opposites?

SUCCESS IN COLLEGE MATHEMATICS

99. Following Example 5 in this section, there is a procedure box that describes the substitution method. Read these steps. Explain the meaning of the "substitution expression."

100. Do practice problem 4. To the right of each line, write the number of the step as given in the procedure box to show the step that corresponds to your work. For example, write, "Used step 1."

At an on-line video retailer, downloads for children cost \$9.99 or \$14.99, and the cost to download eight movies is \$94.92. We can use a system of linear equations to represent these relationships. In this section, we will write and solve systems of linear equations like this one.

SECTION 4.3

The Elimination Method

After reading the text, working the practice problems, and completing assigned exercises, you should be able to:

1. Rewrite a repeating decimal as a fraction.
2. Solve a system of linear equations by elimination.
3. Identify a system of linear equations with no solution or infinitely many solutions.
4. Use the elimination method and a system of equations to solve an application problem.

Addition and Subtraction of Equations

The scale in Figure 1 is balanced. The scale in Figure 2 is also balanced. The weight in the left pan equals the weight in the right pan. Now imagine that we combine the weights from the first two scales on another scale (Figure 3). The weight in the left pan equals the weight in the right pan. The scale is balanced.

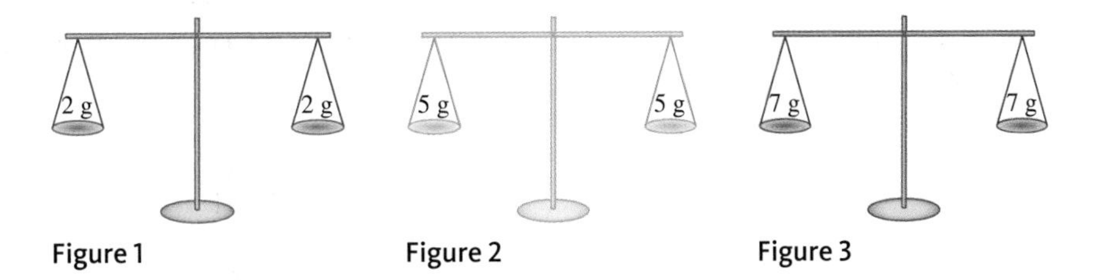

Figure 1 Figure 2 Figure 3

This idea also applies to linear equations. *If we add two linear equations together, the sum is a new equation. The left side of this new equation equals the right side* (Figure 4). We are using the addition property of equality to add equivalent expressions to both sides of an equation. This does not change the solution of the equation.

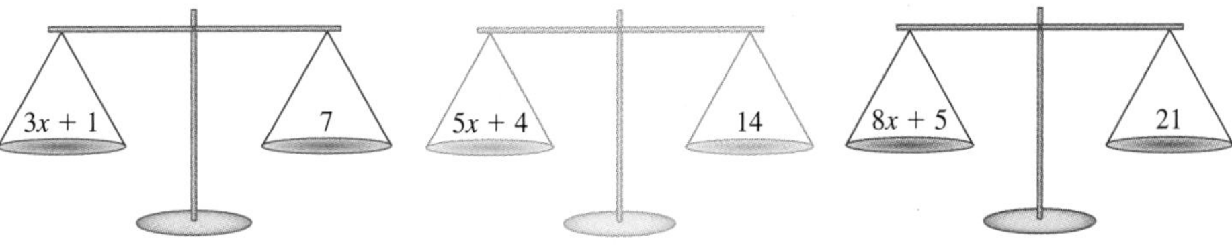

Figure 4

Similarly, we can use the subtraction property of equality and subtract two equations. In the next example, to rewrite a repeating decimal as a fraction, we subtract two equations.

EXAMPLE 1 Rewrite $0.\overline{6}$ as a fraction.

SOLUTION ▶

$x = 0.\overline{6}$

$x = \mathbf{0.666...}$ Rewrite $0.\overline{6}$ as $0.666....$

$(10)x = (10)0.666...$ Multiplication property of equality.

$10x = 6.666...$ Simplify.

Subtract $x = 0.666...$ from $10x = 6.666....$

$10x = 6.666...$

$-\quad x = 0.666...$ Subtraction property of equality.

$9x = 6$ Simplify.

$\frac{9x}{9} = \frac{6}{9}$ Division property of equality.

$x = \frac{2}{3}$ Simplify into lowest terms.

Since $x = 0.\overline{6}$ and $x = \frac{2}{3}$, $0.\overline{6} = \frac{2}{3}$. To check, rewrite $\frac{2}{3}$ as a decimal number.

$\frac{2}{3}$

$= 2 \div 3$ Rewrite the fraction as division.

$= 0.666...$ Divide; the quotient is a repeating decimal.

$= 0.\overline{6}$ Rewrite with a bar over the repeating digit.

To rewrite a repeating decimal with a two-digit repeat as a fraction, such as $0.\overline{47}$, multiply $x = 0.\overline{47}$ on both sides by 100 and subtract the equations.

EXAMPLE 2 Rewrite $0.\overline{47}$ as a fraction.

SOLUTION ▶

$x = 0.\overline{47}$

$x = \mathbf{0.47474747...}$ Rewrite $0.\overline{47}$ as $0.4747....$

$(100)x = (100)0.474747...$ Multiplication property of equality.

$100x = 47.4747...$ Simplify.

Subtract $x = 0.4747...$ from $100x = 47.4747....$

$100x = 47.4747...$

$-\quad x = \ \ 0.4747...$ Subtraction property of equality.

$99x = 47$ Simplify.

$\frac{99x}{99} = \frac{47}{99}$ Division property of equality.

$x = \frac{47}{99}$ Simplify into lowest terms.

Since $x = 0.\overline{47}$ and $x = \frac{47}{99}$, $0.\overline{47} = \frac{47}{99}$.

Practice Problems

For problems 1–4, rewrite the repeating decimal as a fraction.

1. $0.\overline{3}$ **2.** $0.\overline{5}$ **3.** $0.\overline{1}$ **4.** $0.\overline{13}$

Solving a System of Equations by Elimination

The substitution method is an algebraic method for solving a system of equations. Another algebraic method for solving a system of equations is the **elimination method**. When the equations in the system are added together, one of the variables is eliminated. Since the equations are added together, this method is also known as the addition method.

EXAMPLE 3 The graph represents $\begin{matrix} 5x + 8y = 31 \\ -5x + 6y = -3 \end{matrix}$.

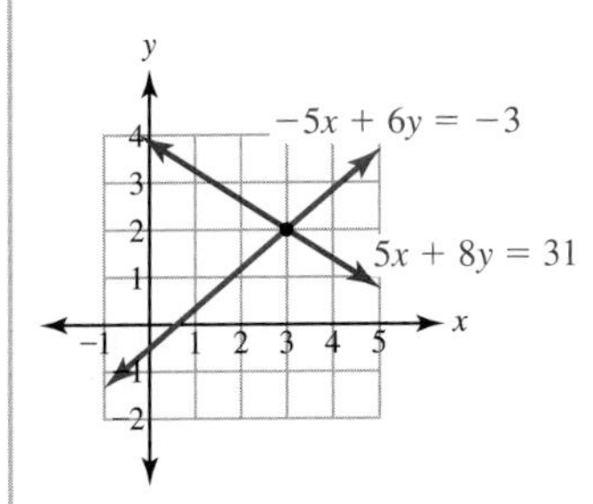

(a) Solve by elimination.

SOLUTION ▶ To find the y-coordinate of the intersection point, add the equations together.

$5x + 8y = 31$	
$+\ -5x + 6y = -3$	Add the equations; addition property of equality.
$0 + 14y = 28$	Simplify.
$\frac{\mathbf{14y}}{14} = \frac{28}{14}$	Division property of equality.
$y = 2$	Simplify; this is the y-coordinate of the intersection point.

Use this value for y to find the x-coordinate of the solution.

$5x + 8y = 31$	Choose either of the original equations.
$5x + 8(\mathbf{2}) = 31$	Replace y with the y-coordinate of the intersection point, 2.
$5x + \mathbf{16} = 31$	Simplify.
$-16 \quad -16$	Subtraction property of equality.
$5x + \mathbf{0} = \mathbf{15}$	Simplify.
$\frac{\mathbf{5x}}{5} = \frac{15}{5}$	Division property of equality.
$x = 3$	Simplify; this is the x-coordinate of the intersection point.

The solution is $(3, 2)$.

(b) Check.

$5x + 8y = 31$		Use the original equations.	$-5x + 6y = -3$
$5(\mathbf{3}) + 8(\mathbf{2}) = 31$		Replace variables.	$-5(\mathbf{3}) + 6(\mathbf{2}) = -3$
$\mathbf{15 + 16} = 31$		Simplify.	$\mathbf{-15 + 12} = -3$
$\mathbf{31} = 31$	True.	The solution is correct.	$\mathbf{-3} = -3$ True.

In Example 3, the coefficients of the x-terms are opposites, 5 and -5. When the equations are added, the sum of the x-terms is 0. If the coefficients of at least one of the variables in the system are not opposites, use the multiplication property of equality to change the coefficients of one of the variables to opposites. In the next example, the coefficients of the x-terms are 2 and 7. When we multiply both sides of one equation by 7 and both sides of the other equation by -2, the coefficients of the x-terms are opposites, $14x$ and $-14x$, and the sum of these terms is 0.

EXAMPLE 4 The graph represents $\begin{matrix} 2x + 5y = -7 \\ 7x + 6y = 10 \end{matrix}$. Solve by elimination.

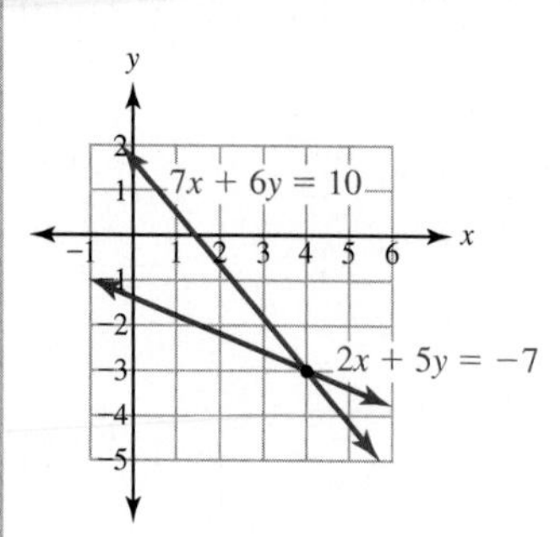

SOLUTION ▶ Choose either x or y to eliminate. In this example, we will eliminate x. To create x-terms that are opposites, multiply $2x + 5y = -7$ by 7, and multiply $7x + 6y = 10$ by -2.

$7(2x + 5y) = 7(-7)$	Multiplication property of equality.
$14x + 35y = -49$	Simplify; the x-term is now $14x$.
$-2(7x + 6y) = -2(10)$	Multiplication property of equality.
$-14x - 12y = -20$	Simplify; the x-term is now $-14x$.

Find the y-coordinate of the intersection point.

$14x + 35y = -49$	
$+\ -14x - 12y = -20$	Add the equations; addition property of equality.
$0 + 23y = -69$	Simplify.
$\frac{23y}{23} = \frac{-69}{23}$	Division property of equality.
$y = -3$	Simplify; this is the y-coordinate of the intersection point.

Use this value for y to find the x-coordinate of the intersection point.

$2x + 5y = -7$	Choose either of the original equations.
$2x + 5(\mathbf{-3}) = -7$	Replace y with the y-coordinate of the intersection point, -3.
$2x - \mathbf{15} = -7$	Simplify.
$+15 \quad +15$	Addition property of equality.
$2x + \mathbf{0} = \mathbf{8}$	Simplify.
$\frac{2x}{2} = \frac{8}{2}$	Division property of equality.
$x = 4$	Simplify; this is the x-coordinate of the intersection point.

The solution is $(4, -3)$.

Solving a System of Two Linear Equations by Elimination

1. Write each equation in standard form, $ax + by = c$. Choose the variable to eliminate.
2. Multiply both sides of either one or both of the equations by a constant so that addition of the equations results in elimination of the chosen variable.

3. Add the equations.
4. Solve the equation from step 3.
5. Replace the variable in one of the original equations with the value found in step 4. Solve for the remaining variable.
6. To check, replace the variables in each equation with the solution.

When using the elimination method, either the x-terms or y-terms can be eliminated. In the next example, we choose to eliminate the y-terms because only one of the equations needs to be multiplied on both sides by a constant to create y-terms with opposite coefficients.

EXAMPLE 5 The graph represents $\begin{matrix} 12x + y = 12 \\ 4x - 3y = -16 \end{matrix}$. Solve by elimination.

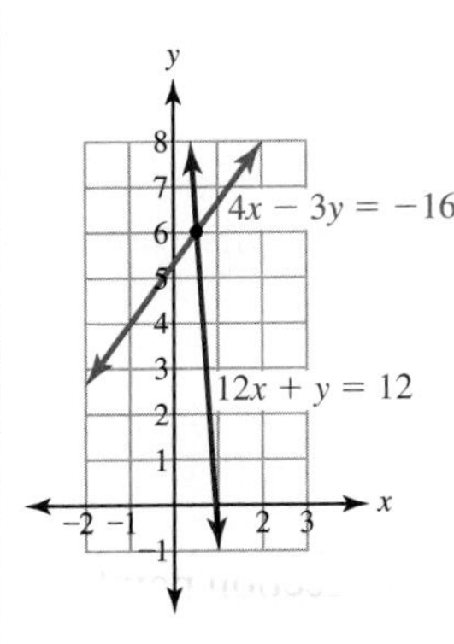

SOLUTION ▶ Choose either x or y to eliminate. In this example, we will eliminate y. To create y-terms that are opposites, multiply $12x + y = 12$ by 3.

$3(12x + y) = 3(12)$ Multiplication property of equality.

$36x + 3y = 36$ Distributive property; simplify.

Add the equations together to find the x-coordinate of the intersection point.

$36x + 3y = 36$

$+\ 4x - 3y = -16$ Add the equations; addition property of equality.

$40x + 0 = 20$ Simplify.

$\frac{\mathbf{40x}}{40} = \frac{20}{40}$ Division property of equality.

$x = \frac{1}{2}$ Simplify; this is the x-coordinate of the intersection point.

Use this value for x to find the y-coordinate of the intersection point.

$12x + y = 12$ Choose either of the original equations.

$12\left(\frac{\mathbf{1}}{\mathbf{2}}\right) + y = 12$ Replace x with the x-coordinate of the intersection point, $\frac{1}{2}$.

$\mathbf{6} + y = 12$ Simplify.

$-6 \quad\quad -6$ Subtraction property of equality.

$\mathbf{0} + y = \mathbf{6}$ Simplify.

$y = 6$ Simplify; this is the y-coordinate of the intersection point.

The solution is $\left(\frac{1}{2}, 6\right)$.

Before using the elimination method, write the equations in the system in standard form, $ax + by = c$, so that like terms line up in the same column. Since adding integers is easier than adding fractions, also clear any fractions from the equations.

EXAMPLE 6 The graph represents $\begin{aligned} \frac{1}{2}x + \frac{3}{4}y &= -35 \\ 2y &= 5x - 30 \end{aligned}$. Solve by elimination.

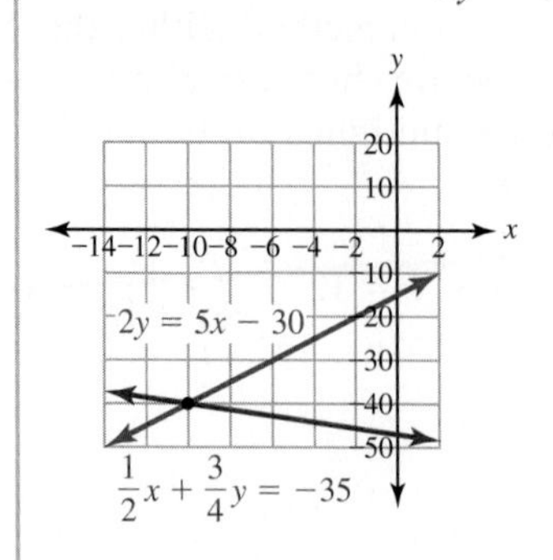

SOLUTION ▶ Choose either x or y to eliminate. In this example, we will eliminate x. Clear the fractions from $\frac{1}{2}x + \frac{3}{4}y = -35$.

$$\frac{1}{2}x + \frac{3}{4}y = -35$$

$$4\left(\frac{1}{2}x + \frac{3}{4}y\right) = 4(-35) \quad \text{Clear fractions; multiplication property of equality.}$$

$$\mathbf{4} \cdot \frac{1}{2}x + \mathbf{4} \cdot \frac{3}{4}y = \mathbf{-140} \quad \text{Distributive property; simplify.}$$

$$\mathbf{2x + 3y} = -140 \quad \text{Simplify.}$$

Rewrite $2y = 5x - 30$ in standard form.

$$2y = 5x - 30$$

$$\underline{-5x \qquad -5x} \quad \text{Subtraction property of equality.}$$

$$\mathbf{-5x} + 2y = \mathbf{0} - 30 \quad \text{Simplify.}$$

$$-5x + 2y = \mathbf{-30} \quad \text{Simplify.}$$

To create x-terms that are opposites, multiply $2x + 3y = -140$ by 5, and multiply $-5x + 2y = -30$ by 2.

$$5(2x + 3y) = 5(-140) \quad \text{Multiplication property of equality.}$$

$$10x + 15y = -700 \quad \text{Distributive property; simplify.}$$

$$2(-5x + 2y) = 2(-30) \quad \text{Multiplication property of equality.}$$

$$-10x + 4y = -60 \quad \text{Distributive property; simplify.}$$

Find the y-coordinate of the intersection point.

$$10x + 15y = -700$$

$$\underline{+ \; -10x + 4y = -60} \quad \text{Add the equations; addition property of equality.}$$

$$0 + 19y = -760 \quad \text{Simplify.}$$

$$\frac{\mathbf{19y}}{19} = \frac{-760}{19} \quad \text{Division property of equality.}$$

$$y = -40 \quad \text{Simplify; this is the } y\text{-coordinate of the intersection point.}$$

Use this value for y to find the x-coordinate of the intersection point.

$2y = 5x - 30$	Choose either of the original equations.
$2(\mathbf{-40}) = 5x - 30$	Replace y with the y-coordinate of the intersection point, -40.
$\mathbf{-80} = 5x - 30$	Simplify.
$+30 \qquad +30$	Addition property of equality.
$\mathbf{-50} = 5x - \mathbf{0}$	Simplify.
$\frac{-50}{5} = \frac{\mathbf{5x}}{5}$	Division property of equality.
$-10 = x$	Simplify; this is the x-coordinate of the intersection point.

The solution is $(-10, -40)$.

Practice Problems

For problems 5–11,
(a) solve by elimination.
(b) check.

5. $2x + 5y = 12$
$-2x + 3y = 4$

6. $x + y = 1$
$2x + 3y = 5$

7. $5x + 4y = -2$
$3x + 3y = -3$

8. $x + 6y = -5$
$9x + 4y = -20$

9. $4x + 7y = 19$
$3x - 5y = 45$

10. $\frac{1}{3}x - \frac{2}{5}y = 4$
$5y = -2x - 13$

11. $3p + 5w = -17$
$-2p + 6w = -26$

Parallel Lines and Coinciding Lines

If we use the elimination method to solve $\begin{matrix} -2x + y = 1 \\ -2x + y = -5 \end{matrix}$, we are assuming that the lines that represent these equations intersect in a single point. Since these lines are parallel and do not intersect, using the elimination method results in a false equation with no variables.

EXAMPLE 7 The graph represents $\begin{matrix} -2x + y = 1 \\ -2x + y = -5 \end{matrix}$. Solve by elimination.

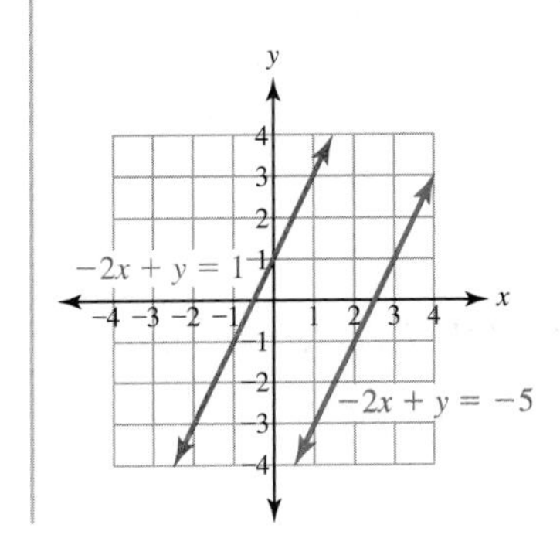

SOLUTION ▶ Choose either x or y to eliminate. In this example, we will eliminate y.

To create y-terms that are opposites, multiply $-2x + y = 1$ by -1.

$-1(-2x + y) = -1(1)$ Multiplication property of equality.

$2x - y = -1$ Distributive property; simplify.

Find the x-coordinate of the intersection point.

$$\begin{array}{r} 2x - y = -1 \\ + \; -2x + y = -5 \\ \hline 0 + 0 = -6 \\ \mathbf{0} = -6 \end{array}$$

Add the equations; addition property of equality.
Simplify.
Simplify; this equation is false.

The result of elimination is a false equation with no variables. The graphs of these equations are two parallel lines, which do not intersect. The system has no solution.

When lines are coinciding, every point on either of the lines is an intersection point. The system has infinitely many solutions. Using the elimination method on a system with infinitely many solutions results in a true equation with no variables. The solution is the set of all of the ordered pairs that are solutions of the equation; the solution is *not* the set of real numbers.

EXAMPLE 8 The graph represents $\begin{array}{l} 2x + y = 5 \\ 6x + 3y = 15 \end{array}$. Solve by elimination.

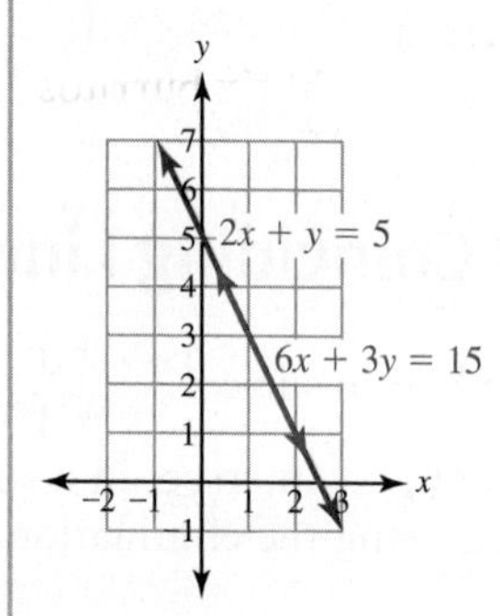

SOLUTION ▶ Choose either x or y to eliminate. In this example, we will eliminate y.

To create y-terms that are opposites, multiply $2x + y = 5$ by -3.

$-3(2x + y) = -3(5)$ Multiplication property of equality.

$-6x - 3y = -15$ Distributive property; simplify.

Find the x-coordinate of the intersection point.

$$\begin{array}{r} -6x - 3y = -15 \\ + \quad 6x + 3y = 15 \\ \hline 0 + 0 = 0 \\ \mathbf{0} = 0 \end{array}$$

Add the equations; addition property of equality.
Simplify.
Simplify; this equation is true.

The result of elimination is a true equation with no variables. The graphs of these equations are two coinciding lines, which share infinitely many points. This system has infinitely many solutions. In set-builder notation, the solution is $\{(x, y) | 2x + y = 5\}$.

The substitution method and the elimination method find the exact solution of a system of two linear equations. In general, the substitution method is the most efficient method to use when one or both of the equations is solved for a variable, and the elimination method is the most efficient method to use when both equations are in standard form.

Practice Problems

For problems 12–15, solve by elimination.

12. $\begin{aligned} -3x + y &= 5 \\ 3x - y &= 9 \end{aligned}$

13. $\begin{aligned} 6x - y &= 9 \\ -12x + 2y &= -18 \end{aligned}$

14. $\begin{aligned} -\frac{3}{4}x + y &= 2 \\ 3x - 4y &= 24 \end{aligned}$

15. $\begin{aligned} 10x + y &= 4 \\ 20x + 2y &= 1 \end{aligned}$

Applications

If an application problem has two unknowns, we may be able to use a system of two linear equations to represent the problem.

EXAMPLE 9 In 2010, there were about 460 mobile food trucks selling takeout food in Portland, Oregon. At a taco truck, Customer A bought two tacos and a quesadilla for \$6. Customer B bought five tacos and two quesadillas for \$13.25. Customer C bought six burritos for \$25.50. Find the cost of a taco and a quesadilla. (*Source:* www.oregonbusiness.com, Jan. 2010)

© Laura Bracken

SOLUTION ▶ **Step 1 Understand the problem.**

The two unknowns are the cost of a taco and the cost of a quesadilla. The cost of six burritos and the number of food trucks are extraneous information.

x = cost of one taco $\quad$ y = cost of one quesadilla

Step 2 Make a plan.

Write and solve a system of equations representing the relationship of the number of tacos, the number of quesadillas, and the total cost. Both word equations are

$$(\text{number of tacos})\left(\frac{\text{cost}}{1\text{ taco}}\right) + (\text{number of quesadillas})\left(\frac{\text{cost}}{1\text{ quesadilla}}\right) = \text{total cost.}$$

Step 3 Carry out the plan.

For Customer A, $(2\text{ tacos})x + (1\text{ quesadilla})y = \$6$. For Customer B, $(5\text{ tacos})x + (2\text{ quesadillas})y = \$13.25$. Since both equations are in standard form, remove the units and solve the system $\begin{aligned} 2x + 1y &= 6 \\ 5x + 2y &= 13.25 \end{aligned}$ by elimination.

To create y-terms that are opposites, multiply $2x + 1y = 6$ by -2.

$$\begin{aligned} 2x + 1y &= 6 \\ -2(2x + 1y) &= -2(6) && \text{Multiplication property of equality.} \\ \mathbf{-2}(2x) \mathbf{- 2}(1y) &= -2(6) && \text{Distributive property.} \\ -4x - 2y &= -12 && \text{Simplify.} \end{aligned}$$

Find the x-coordinate of the solution.

$-4x - 2y = -12$	
$+\ 5x + 2y = 13.25$	Add equations; addition property of equality.
$1x + 0 = 1.25$	Simplify.
$\mathbf{x} = 1.25$	Simplify; the x-coordinate of the solution is 1.25.

Use this value for x to find the y-coordinate of the solution.

$2x + 1y = 6$	Choose one of the original equations.
$2(\mathbf{1.25}) + y = 6$	Replace x with the x-coordinate of the solution, 1.25.
$\mathbf{2.50} + y = 6$	Simplify.
$-2.50 \quad -2.50$	Subtraction property of equality.
$\mathbf{0} + y = \mathbf{3.50}$	Simplify.
$\mathbf{y} = 3.50$	Simplify; this is the y-coordinate of the solution.

Including the units, the solution is $\left(\dfrac{\$1.25}{1\text{ taco}}, \dfrac{\$3.50}{1\text{ quesadilla}}\right)$.

Step 4 Look back.
Since 2($1.25) + 1($3.50) equals the bill for Customer A, $6, and 5($1.25) + 2($3.50) equals the bill for Customer B, $13.25, the answer seems reasonable.

Step 5 Report the solution.
A taco costs $1.25, and a quesadilla costs $3.50.

In Example 9, we knew the number of tacos and quesadillas and the total cost for two different purchases. We found the cost of a taco and the cost of a quesadilla. In the next example, we are given the cost of honey and the cost of jam, the total cost, and the total number of jars. We need to find the number of jars of jam and honey sold.

EXAMPLE 10 A stall at a farmer's market sells honey, jam, and bread. If 82 jars of honey and jam sold for $325, find the number of jars of honey and the number of jars of jam sold.

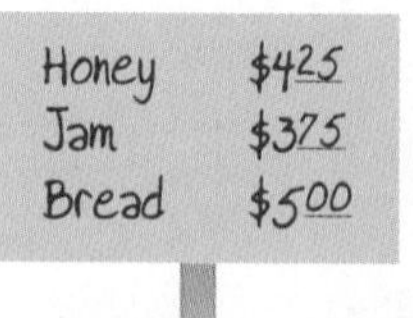

SOLUTION ▶ **Step 1 Understand the problem.**
The two unknowns are the number of jars of honey sold and the number of jars of jam sold. The cost of bread is extraneous information.

x = number of jars of honey $\qquad$ y = number of jars of jam

Step 2 Make a plan.
Write and solve a system of equations in which one equation represents the relationship of the number of jars and the other represents the relationship of the costs. The word equations are number of honey + number of jam = total number and $\left(\dfrac{\text{cost}}{1\text{ honey}}\right)(\text{number of honey}) + \left(\dfrac{\text{cost}}{1\text{ jam}}\right)(\text{number of jam}) = \text{total cost}$.

Step 3 Carry out the plan.
The relationship of the number of jars, is $x + y = 82$ jars, and the relationship of the costs is $\left(\dfrac{\$4.25}{1\text{ honey}}\right)x + \left(\dfrac{\$3.75}{1\text{ jam}}\right)y = \$325$. Since both equations are in standard form, remove the units and solve the system $\begin{matrix} x + y = 82 \\ 4.25x + 3.75y = 325 \end{matrix}$ by elimination.

To create x-terms that are opposites, multiply $x + y = 82$ by -4.25.

$x + y = 82$	
$(-4.25)(x + y) = (-4.25)(82)$	Multiplication property of equality.
$-4.25x - 4.25y = -348.50$	Distributive property; simplify.

Find the y-coordinate of the solution.

$-4.25x - 4.25y = -348.50$	
$+\ 4.25x + 3.75y = 325$	Add equations; addition property of equality.
$0 - 0.50y = -23.50$	Simplify.
$\dfrac{\mathbf{-0.50y}}{-0.50} = \dfrac{-23.50}{-0.50}$	Division property of equality.
$y = 47$	Simplify; this is the y-coordinate of the solution.

Use this value for y to find the x-coordinate of the solution.

$x + y = 82$	Choose one of the original equations.
$x + \mathbf{47} = 82$	Replace y with the y-coordinate of the solution, 47.
$-47 \quad -47$	Subtraction property of equality.
$x + \mathbf{0} = \mathbf{35}$	Simplify.
$\boldsymbol{x} = 35$	Simplify; this is the x-coordinate of the solution.

Including the units, the solution is (35 jars of honey, 47 jars of jam).

Step 4 Look back.
Since 35 jars + 47 jars equals 82 jars and 35($4.25) + 47($3.75) equals $325, the answer seems reasonable.

Step 5 Report the solution.
The stall sold 35 jars of honey and 47 jars of jam.

In Example 10, the coefficients of the equation $4.25x + 3.75y = 325$ are decimal numbers. If we had cleared the decimals by multiplying both sides of the equation by 100 before using the elimination method, the solution of the system would have been the same.

Practice Problems

For problems 16–17, use the five steps, a system of equations, and the elimination method.

16. The volume of 13 large cups of soda and 6 small cups of soda is 154 oz. The volume of 2 large cups of soda and 15 small cups of soda is 80 oz. Find the volume of a large cup of soda. Find the volume of a small cup of soda.

17. If downloaded movies for children cost $9.99 or $14.99 and 8 downloaded movies cost $94.92, find the number of each kind of movie that was downloaded.

4.3 VOCABULARY PRACTICE

Match the term with its description.

1. In this method, the equations in a system are added together.
2. In this method, a variable is replaced by an expression.
3. $ax + by = c$
4. $y = mx + b$
5. Two lines with the same slope but different y-intercepts
6. Two lines with the same slope and the same y-intercept
7. $\{(x, y) | y = 2x + 5\}$ is an example of this kind of notation.
8. We can multiply both sides of an equation by a real number except for 0 without changing its solution(s).
9. We can add a real number to both sides of an equation without changing its solution(s).
10. We can subtract a real number from both sides of an equation without changing its solution(s).

A. addition property of equality
B. coinciding lines
C. elimination method
D. multiplication property of equality
E. parallel lines
F. set-builder notation
G. slope-intercept form of a linear equation
H. standard form of a linear equation
I. substitution method
J. subtraction property of equality

4.3 Exercises

Follow your instructor's guidelines for showing your work.

For exercises 1–4, rewrite the repeating decimal as a fraction.

1. $0.\overline{4}$ **2.** $0.\overline{7}$

3. $0.\overline{38}$ **4.** $0.\overline{51}$

For exercises 5–36,
(a) solve by elimination.
(b) if there is one solution, check.

5. $3x + y = 1$; $-3x + y = 7$ **6.** $2x + y = -2$; $-2x + y = 10$

7. $x + 2y = -4$; $x - 2y = 8$ **8.** $x + 3y = -3$; $x - 3y = 9$

9. $2x + 2y = 8$; $3x + 2y = 9$ **10.** $3x + 3y = 9$; $3x + 2y = 7$

11. $x + y = 8$; $3x + 2y = 21$ **12.** $x + y = 9$; $3x + 2y = 22$

13. $x - y = 5$; $5x + 2y = -17$ **14.** $x - y = 7$; $4x + 3y = -28$

15. $x + y = 5$; $3x - 2y = -10$ **16.** $x + y = 4$; $5x - 3y = -12$

17. $2x + y = 7$; $4x - 3y = -16$ **18.** $4x + y = 5$; $2x - 3y = -8$

19. $x + y = 5$; $x + y = 4$ **20.** $x + y = 6$; $x + y = 1$

21. $5x - 6y = 28$; $7x + 6y = 68$ **22.** $4x - 7y = 10$; $3x + 7y = 32$

23. $-2x + 9y = -21$; $2x - 5y = 17$ **24.** $-5x + 4y = -33$; $5x - 3y = 31$

25. $4x + 3y = -6$; $5x + 2y = -11$ **26.** $3x + 2y = -6$; $4x + 5y = -1$

27. $x + y = 5$; $2x + 5y = 13$ **28.** $x + y = 4$; $3x + 2y = 11$

29. $-5x + 6y = 3$; $5x - 6y = -7$ **30.** $-7x + 8y = 4$; $7x - 8y = -9$

31. $15x + 4y = 8$; $20x - 7y = -14$ **32.** $3x + 5y = 20$; $5x - 8y = -32$

33. $-2x + y = -3$; $4x - 2y = 6$ **34.** $-5x + y = -6$; $10x - 2y = 12$

35. $x - 5y = 8$; $-2x + 10y = -16$ **36.** $x - 6y = 9$; $-2x + 12y = -18$

For exercises 37–44,
(a) solve by elimination.
(b) if there is one solution, check.

37. $3a - 5b = 2$; $7a + 8b = -15$ **38.** $2h - 5k = 9$; $7h + 3k = -30$

39. $4m - 6z = 3$; $-2m + 3z = 6$ **40.** $10c - 16n = 3$; $-5c + 8n = 9$

41. $2a + 9b = -5$; $3a + 5b = 1$ **42.** $3h + 4k = -5$; $6h + 5k = -4$

43. $-5p + w = -15$; $2p - 3w = 6$ **44.** $-4c + d = -8$; $5c - 3d = 10$

45. In a system of linear equations with two variables, x and y, explain how to decide whether to eliminate the x-terms or the y-terms.

46. In solving a system of linear equations with two variables, one student chooses to eliminate the x-terms. Another student chooses to eliminate the y-terms. Will they get the same solution? Explain.

47. To eliminate the x-terms in the system $\begin{aligned} 3x + 4y &= 11 \\ 5x + 7y &= 19 \end{aligned}$, each equation is multiplied by a constant. Explain how to decide which constants to use.

48. To eliminate the y-terms in the system $\begin{aligned} 3x + 4y &= 11 \\ 5x + 7y &= 19 \end{aligned}$, each equation is multiplied by a constant. Explain how to decide which constants to use.

For exercises 49–56, solve by either elimination or substitution.

49. $\begin{aligned} x + y &= 9 \\ 3x - y &= -5 \end{aligned}$

50. $\begin{aligned} x + y &= 6 \\ 2x - y &= -3 \end{aligned}$

51. $\begin{aligned} x + y &= 5000 \\ -4x + y &= 0 \end{aligned}$

52. $\begin{aligned} x + y &= 6000 \\ -3x + y &= 0 \end{aligned}$

53. $\begin{aligned} 2x + y &= 6 \\ 2x + y &= -8 \end{aligned}$

54. $\begin{aligned} 4x + y &= 12 \\ 4x + y &= -8 \end{aligned}$

55. $\begin{aligned} -3x + y &= 9 \\ 12x - 4y &= -36 \end{aligned}$

56. $\begin{aligned} -4x + y &= 1 \\ 8x - 2y &= -2 \end{aligned}$

57. Would you use the substitution method or the elimination method to solve $\begin{aligned} x + y &= 6 \\ 3x - 5y &= -14 \end{aligned}$? Explain.

58. Would you use the substitution method or the elimination method to solve $\begin{aligned} 2x + y &= 5 \\ 5x + y &= 14 \end{aligned}$? Explain.

For exercises 59–68, solve by substitution or elimination.

59. $\begin{aligned} y - 5 &= \frac{1}{2}(x - 6) \\ y - 5 &= -2(x - 6) \end{aligned}$

60. $\begin{aligned} y - 4 &= \frac{1}{4}(x - 3) \\ y - 4 &= 2(x - 3) \end{aligned}$

61. $\begin{aligned} y - \frac{3}{4} &= \frac{3}{2}x \\ 3x - \frac{1}{2}y &= \frac{3}{4} \end{aligned}$

62. $\begin{aligned} y - \frac{5}{6} &= \frac{1}{2}x \\ 2x - 2y &= \frac{4}{3} \end{aligned}$

63. $\begin{aligned} \frac{1}{2} &= 3x - y \\ 2x + 3y &= \frac{5}{4} \end{aligned}$

64. $\begin{aligned} \frac{1}{2} &= 4x - y \\ 3x - 2y &= -\frac{3}{2} \end{aligned}$

65. $\begin{aligned} \frac{1}{2}x + \frac{1}{3}y &= 4 \\ \frac{1}{4}x + \frac{1}{2}y &= 3 \end{aligned}$

66. $\begin{aligned} \frac{1}{2}x + \frac{1}{5}y &= 3 \\ \frac{1}{4}x + \frac{3}{4}y &= 8 \end{aligned}$

67. $\begin{aligned} \frac{1}{2}x + \frac{1}{7}y &= 9 \\ \frac{1}{3}x + \frac{1}{3}y &= 11 \end{aligned}$

68. $\begin{aligned} \frac{1}{4}x + \frac{1}{5}y &= 5 \\ \frac{1}{2}x + \frac{1}{3}y &= 9 \end{aligned}$

69. Use graphing to solve $\begin{aligned} x &= 6 \\ y &= -2 \end{aligned}$.

70. Use graphing to solve $\begin{aligned} x &= -4 \\ y &= 3 \end{aligned}$.

For exercises 71–74,
(a) solve by elimination.
(b) check by graphing.

71. $\begin{aligned} x + y &= 12 \\ 2x - y &= 9 \end{aligned}$

72. $\begin{aligned} x + y &= 10 \\ 2x - y &= 8 \end{aligned}$

73. $\begin{aligned} 3x + y &= -3 \\ x - 2y &= -8 \end{aligned}$

74. $\begin{aligned} 5x + y &= -9 \\ x - 2y &= -4 \end{aligned}$

For exercises 75–80, use the five steps, a system of equations, and the elimination method.

75. The cost of two hamburgers and three orders of fries is \$6.25. The cost of five hamburgers and one order of fries is \$9.84. The cost of two orders of chicken fingers is \$7.50. Find the cost of one hamburger. Find the cost of one order of fries.

76. The cost of three scones and four sweet rolls is \$23.85. The cost of three loaves of bread and one muffin is \$12.75. The cost of two loaves of bread and five muffins is \$15. Find the cost of one loaf of bread. Find the cost of one muffin.

77. The cost of 320 tickets for a comedy show is \$1165. Student tickets cost \$3.50. All other tickets cost \$5. Find the number of student tickets sold. Find the number of other tickets sold.

78. A roadside stand sold 190 lb of peaches and pears. Peaches cost \$0.79 per pound. Pears cost \$1.29 per pound. The total sales were \$210.60. Find the number of pounds of peaches sold. Find the number of pounds of pears sold.

79. A group of students at a table in front of a grocery store sold 100 chocolate bars and boxes of thin mints. The price of an 8-oz chocolate bar was \$2. The price of a 10-oz box of thin mints was \$3. The total amount of money collected was \$225. Find the number of chocolate bars sold. Find the number of boxes of thin mints sold.

80. A scientific calculator weighs 4 oz and costs \$13. A graphing calculator weighs 10 oz and costs \$75. The total weight of 50 calculators is 278 oz. Find the number of scientific calculators and the number of graphing calculators.

Problem Solving: Practice and Review

Follow your instructor's guidelines for using the five steps as outlined in Section 1.3, p. 55.

81. A sign on a gas pump offers a rebate of \$0.18 per gallon up to \$50 for gas purchased using a credit card issued by the company that supplies the gasoline. Find the number of gallons of gas to purchase to earn a rebate of \$50. Round *up* to the nearest whole number.

82. Fruit boxes are stacked at a warehouse. Each individual fruit box is 4 ft wide, 4 ft long, and 2 ft high. If a stack is 4 boxes high, 8 boxes wide, and 12 boxes deep, find the total volume of the stack.

83. According to an article on safety tips for blogging, two out of three teenagers provide their age on their blog, three out of five reveal their location, and one out of five reveals their full name. Find the number of 3675 teenagers who reveal their location on their blog. (*Source:* www.microsoft.com)

84. Use the information in the circle graph to find the percent of the Washington Metropolitan Transit Authority 2012 budget that is federal funds. Round to the nearest percent.

Sources of Funds for FY2012 Budget (in millions)

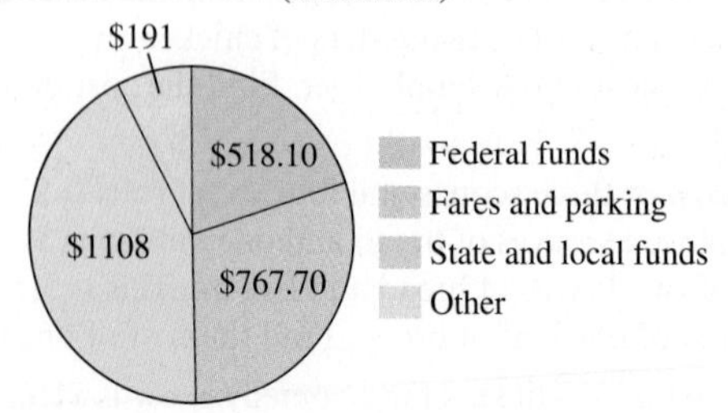

Source: www.wmata.com, June 23, 2011

Technology

For exercises 85–88, solve by graphing. Sketch the graph; describe the window; identify the solution.

85. $y = \frac{3}{4}x + 9$
$y = \frac{1}{2}x + 10$

86. $y = \frac{3}{4}x + 9$
$y = -\frac{1}{2}x + 14$

87. $y = 10x - 100$
$y = x + 8$

88. $y = 2x + 5$
$y = -x + 20$

Find the Mistake

For exercises 89–92, the completed problem has one mistake.
(a) Describe the mistake in words, or copy down the whole problem and highlight or circle the mistake.
(b) Do the problem correctly.

89. Problem: Solve $\begin{aligned} 3x + 2y &= 10 \\ 5x + 2y &= 14 \end{aligned}$ by elimination.

Incorrect Answer:
$$\begin{aligned} 3x + 2y &= 10 \\ + \; 5x + 2y &= 14 \\ \hline 8x + 0 &= 24 \\ \frac{8x}{8} &= \frac{24}{8} \\ x &= 3 \end{aligned}$$
$$\begin{aligned} 3x + 2y &= 10 \\ 3(3) + 2y &= 10 \\ 9 + 2y &= 10 \\ -9 \quad & \quad -9 \\ \hline 2y &= 1 \\ \frac{2y}{2} &= \frac{1}{2} \\ y &= \frac{1}{2} \end{aligned}$$
The solution is $\left(3, \frac{1}{2}\right)$.

90. Problem: Solve $\begin{aligned} x + y &= 8 \\ 2x - 3y &= -9 \end{aligned}$ by elimination.

Incorrect Answer:
$$\begin{aligned} x + y &= 8 \\ 3(x + y) &= 3 \cdot 8 \\ 3x + 3y &= 24 \end{aligned}$$
$$\begin{aligned} 3x + 3y &= 24 \\ + \; 2x - 3y &= -9 \\ \hline x + 0 &= 15 \\ x &= 15 \end{aligned}$$
$$\begin{aligned} x + y &= 8 \\ 15 + y &= 8 \\ -15 \quad & \quad -15 \\ \hline 0 + y &= -7 \\ y &= -7 \end{aligned}$$
The solution is $(15, -7)$.

91. Problem: Solve $\begin{aligned} -3x + 4y &= 8 \\ 6x - 8y &= -16 \end{aligned}$ by elimination.

Incorrect Answer:
$$\begin{aligned} -3x + 4y &= 8 \\ 2(-3x + 4y) &= 2(8) \\ -6x + 8y &= 16 \end{aligned}$$
$$\begin{aligned} -6x + 8y &= 16 \\ + \; 6x - 8y &= -16 \\ \hline 0 + 0 &= 0 \end{aligned}$$
The solution is the set of real numbers.

92. Problem: Solve $\begin{aligned} 2x + 3y &= 18 \\ x + 2y &= 11 \end{aligned}$ by elimination.

Incorrect Answer:
$$-2(2x + 3y) = -2(18) \qquad 3(x + 2y) = 3(11)$$
$$-4x - 6y = -36 \qquad 3x + 6y = 33$$
$$\begin{aligned} -4x - 6y &= -36 \\ + \; 3x + 6y &= 33 \\ \hline -x + 0 &= 3 \\ \frac{-x}{-1} &= \frac{3}{-1} \\ x &= -3 \end{aligned}$$
$$\begin{aligned} x + 2y &= 11 \\ -3 + 2y &= 11 \\ 2y &= 14 \\ \frac{2y}{2} &= \frac{14}{2} \\ y &= 7 \end{aligned}$$
The solution is $(-3, 7)$.

Review

93. Solve $d = rt$ for t.

94. A glass contains 6 oz of cranberry juice drink. This drink is 24% cranberry juice. Find the amount of pure cranberry juice in this glass.

95. A chemist is going to mix an acid solution with pure water. The acid solution is 35% acid. What is the percent of acid in pure water?

96. Solve $d = rt$ for r.

SUCCESS IN COLLEGE MATHEMATICS

97. To prepare for your next class meeting, look ahead to Section 4.4. Examples 1–4 are often called *mixture problems.* The orange boxes in this section include two important relationships that are used in solving mixture problems. What are these relationships?

98. Read Example 1 in Section 4.4. Copy the example, and use a highlighter to mark any steps in the example that you do not understand or find difficult. This should help you prepare to ask questions about the example or a similar problem in your next class.

© Charles Petersen

Imagine a group of slow hikers who start on a trail early in the morning. Another group of faster hikers starts the trail later. The fast group will eventually catch up to the slow group. In this section, we will learn to write and solve systems of equations that are models of application problems.

SECTION 4.4 More Applications

After reading the text, working the practice problems, and completing assigned exercises, you should be able to:

Use a system of equations to solve an application problem.

Percent, Mixtures, and Systems of Equations

In Section 4.3, we used the five steps and solved a system of equations to find the price of a taco and a quesadilla. Although it might be quicker to find the price by looking at the menu, this does not develop the critical thinking skills needed to solve other problems that occur at work or in personal life. In this section, you will continue to improve your critical thinking skills as you use the five steps. Look for similarities between problems. Use your previous experience to develop strategies for solving new problems. Do not expect to always write the correct equations on the first try. Problem solving often involves dead ends or false starts. Be persistent.

A percent represents the number of parts out of 100 total parts.

Percent

$$\text{percent} = \left(\frac{\text{number of parts}}{\text{total parts}}\right)(100)\%$$

$$\text{number of parts} = (\text{percent in decimal notation})(\text{total parts})$$

When pure juice is in a mixture with other liquids, the percent juice in the liquid represents the amount of the mixture that is juice. The volume of juice in a mixture is the product of the percent that is juice and the total volume of the mixture. For example, if a mixture is 50% juice, the volume of juice in the mixture equals (0.50)(total volume). If x represents the total volume, then the volume of juice in the mixture is $0.50x$.

Volume of an Ingredient in a Mixture

volume of an ingredient = (percent ingredient)(total volume of the mixture)

EXAMPLE 1 Most cranberry juice drinks are between 20% and 30% cranberry juice. Drink A is 27% cranberry juice. Drink B is 23% cranberry juice. Find the amount of each drink needed to make 5000 gal of a new drink that is 24% cranberry juice. Round to the nearest whole number.

SOLUTION **Step 1 Understand the problem.**

The two unknowns are the amount of Drink A and the amount of Drink B.

x = amount of Drink A $\quad$ y = amount of Drink B

Step 2 Make a plan.

Since number of parts = (percent)(total parts), the amount of cranberry juice in a drink equals (percent cranberry juice)(amount of the drink). Write and solve a system of equations in which one equation represents the relationship of the amounts and the other equation represents the amount of cranberry juice in each drink. The word equations are amount A + amount B = amount new and (percent)(amount A) + (percent)(amount B) = (percent)(amount new).

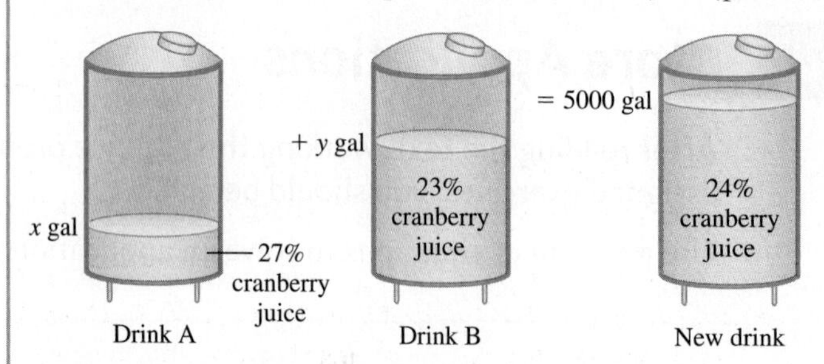

Step 3 Carry out the plan.

The equation for the relationship of the amounts is $x + y = 5000$ gal. The equation for the relationship of the percents is $0.27x + 0.23y = (0.24)(5000 \text{ gal})$, which simplifies to $0.27x + 0.23y = 1200$ gal. Since both equations are in standard form, remove the units and solve $\begin{array}{c} x + y = 5000 \\ 0.27x + 0.23y = 1200 \end{array}$ by elimination.

To create y-terms that are opposites, multiply $x + y = 5000$ by -0.23.

$$x + y = 5000$$

$$(-0.23)(x + y) = (-0.23)(5000) \quad \text{Multiplication property of equality.}$$

$$-0.23x - 0.23y = -1150 \quad \text{Distributive property; simplify.}$$

Find the x-coordinate of the solution.

$$\begin{array}{r} -0.23x - 0.23y = -1150 \\ +\ 0.27x + 0.23y = 1200 \\ \hline 0.04x + 0 = 50 \end{array} \quad \begin{array}{l} \\ \text{Add the equations; addition property of equality.} \\ \text{Simplify.} \end{array}$$

$$\frac{\mathbf{0.04x}}{0.04} = \frac{50}{0.04} \quad \text{Division property of equality.}$$

$$x = 1250 \quad \text{Simplify; this is the } x\text{-coordinate of the solution.}$$

Use this value for x to find the y-coordinate of the solution.

$x + y = 5000$	Choose either of the original equations.
$\mathbf{1250} + y = 5000$	Replace x with the x-coordinate of the solution.
$-1250 \quad -1250$	Subtraction property of equality.
$\mathbf{0} + y = \mathbf{3750}$	Simplify.
$y = 3750$	Simplify; this is the y-coordinate of the solution.

Including the units, the solution is (1250 gal, 3750 gal).

Step 4 Look back.
The amount of Drink A, 1250 gal, plus the amount of Drink B, 3750 gal, equals the amount of the new drink, 5000 gal. The amount of juice in Drink A is (0.27)(1250 gal), which equals 337.5 gal; the amount of juice in Drink B is (0.23)(3750 gal), which equals 862.5 gal; and the amount of juice in the new drink is (0.24)(5000 gal), which equals 1200 gal. Since 337.5 gal plus 862.5 gal equals 1200 gal, the answer seems reasonable.

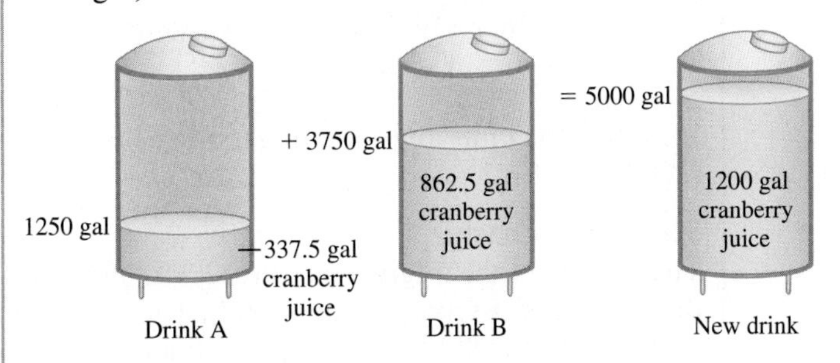

Step 5 Report the solution.
To make the new drink, 1250 gal of Drink A and 3750 gal of Drink B are needed.

To decrease the percent of an ingredient in a mixture, we can dilute the mixture by adding water. The percent of an ingredient in pure water is 0%.

EXAMPLE 2 A drum contains a liquid mixture that is 6.5% weed killer. Find the amount of this mixture and the amount of water needed to make 450 L of a new mixture that is 3% weed killer. Round to the nearest whole number.

SOLUTION ▸ **Step 1 Understand the problem.**
The two unknowns are the amount of the original mixture and the amount of water.

x = amount of original mixture $\qquad y$ = amount of water

Step 2 Make a plan.
Write and solve a system of equations in which one equation represents the relationship of the amounts and the other equation represents the amount of weed killer in each mixture. The word equations are amount original + amount water = amount new and (percent)(amount original) + (percent)(amount water) = (percent)(amount new).

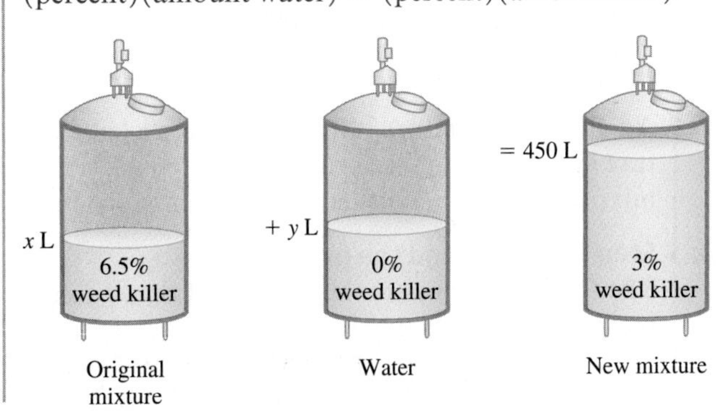

Step 3 Carry out the plan.
The equation for the relationship of the amounts is $x + y = 450$ L. The equation for the amount of weed killer in each mixture is $0.065x + 0y = (0.03)(450\text{ L})$, which simplifies to $0.065x = 13.5\text{ L}$. Since $0.065x = 13.5\text{ L}$ can be solved for x, remove the units, and solve $\begin{matrix} 0.65x = 13.5 \\ x + y = 450 \end{matrix}$ by substitution.

$$0.065x = 13.5$$

$$\frac{0.065x}{0.065} = \frac{13.5}{0.065} \qquad \text{Division property of equality.}$$

$$x = 207.6\ldots \qquad \text{Simplify; this is the } x\text{-coordinate of the solution.}$$

Use this value for x to find the y-coordinate of the solution.

$x + y = 450$	Choose either of the original equations; remove units.
$\mathbf{207.6\ldots} + y = 450$	Replace x with the x-coordinate of the solution, 207.6. . . .
$-207.6\ldots \qquad -207.6\ldots$	Subtraction property of equality.
$\mathbf{0} + y = \mathbf{242.3\ldots}$	Simplify.
$y = 242.3\ldots$	Simplify; this is the y-coordinate of the solution.

Including the units and rounding to the nearest whole number, the solution is about (208 L, 242 L).

Step 4 Look back.
The amount of the original mixture, 208 L, plus the amount of water, 242 L, equals the amount of the new mixture, 450 L. The amount of weed killer in the original mixture is $(0.065)(208\text{ L})$, which equals about 13.5 L; the amount of weed killer in water is 0 L; and the amount of weed killer in the new mixture is $(0.03)(450\text{ L})$, which equals 13.5 L. Since 13.5 L plus 0 L equals 13.5 L, the answer seems reasonable.

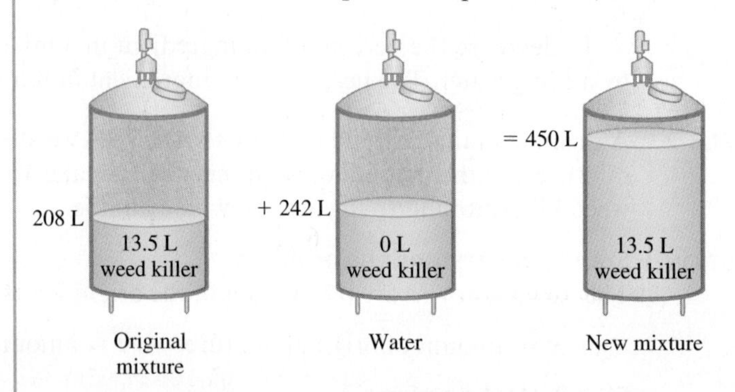

Step 5 Report the solution.
To make the new mixture, about 208 L of the original mixture and 242 L of water are needed.

In Example 2, the water added to the original mixture contains 0% weed killer. In the next example, pure (100%) recycled crushed glass is added to a mixture. In decimal form, 100% = 1.

EXAMPLE 3 An aggregate mixture is 5% crushed recycled glass. Find the amount of this mixture and the amount of crushed recycled glass needed to make 7500 lb of a new mixture that is 8% crushed recycled glass. Round to the nearest whole number.

SOLUTION ▶ **Step 1 Understand the problem.**
The two unknowns are the amount of the aggregate mixture and the amount of recycled glass.

x = amount of aggregate mixture $\qquad y$ = amount of recycled glass

Step 2 Make a plan.
Write and solve a system of equations in which one equation represents the relationship of the amounts and the other equation represents the amount of glass in each mixture. The word equations are amount aggregate + amount recycled glass = amount new and (percent)(amount aggregate) + (percent)(amount recycled glass) = (percent)(amount new).

x lb + y lb = 7500 lb

5% crushed glass	100% crushed glass	8% crushed glass
Original aggregate	Crushed glass	New mixture

Step 3 Carry out the plan.
The equation for the total amount is $x + y = 7500$ lb. The equation for the amount of crushed recycled glass in each mixture is $0.05x + 1y = (0.08)(7500 \text{ lb})$, which simplifies to $0.05x + y = 600$ lb. Since both equations are in standard form, remove the units and solve $\begin{array}{r} x + y = 7500 \\ 0.05x + y = 600 \end{array}$ by elimination.

To create y-terms that are opposites, multiply $x + y = 7500$ by -1.

$$x + y = 7500$$

$$-1(x + y) = -1(7500) \quad \text{Multiplication property of equality.}$$

$$\mathbf{-1x - 1y = -7500} \quad \text{Distributive property; simplify.}$$

$$-x - y = -7500 \quad \text{Simplify.}$$

Find the x-coordinate of the solution.

$$-x - y = -7500$$

$$+\ 0.05x + y = 600 \quad \text{Add equations; addition property of equality.}$$

$$-0.95x + 0 = -6900 \quad \text{Simplify.}$$

$$\frac{-0.95x}{-0.95} = \frac{-6900}{-0.95} \quad \text{Division property of equality.}$$

$$x = 7263.1\ldots \quad \text{Simplify; this is the } x\text{-coordinate of the solution.}$$

Use this value for x to find the y-coordinate of the solution.

$$x + y = 7500 \quad \text{Choose one of the original equations.}$$

$$\mathbf{7263.1\ldots} + y = 7500 \quad \text{Replace } x \text{ with the } x\text{-coordinate of the solution, } 7263.1\ldots.$$

$$-7263.1\ldots \qquad -7263.1\ldots \quad \text{Subtraction property of equality.}$$

$$\mathbf{0} + y = \mathbf{236.8\ldots} \quad \text{Simplify.}$$

$$\mathbf{y} = 236.8\ldots \quad \text{Simplify; this is the } y\text{-coordinate of the solution.}$$

Including the units and rounding, the solution is about (7263 lb aggregate, 237 lb crushed recycled glass).

Step 4 Look back.
The sum of 7263 lb of aggregate and 237 lb of pure crushed recycled glass equals the total amount of 7500 lb. The amount of recycled glass in the original aggregate is (0.05)(7263 lb), which equals about 363 lb; the amount of pure glass is 237 lb; and the amount of glass in the new mixture is (0.08)(7500 lb), which equals 600 lb. Since 363 lb plus 237 lb equals 600 lb, the answer seems reasonable.

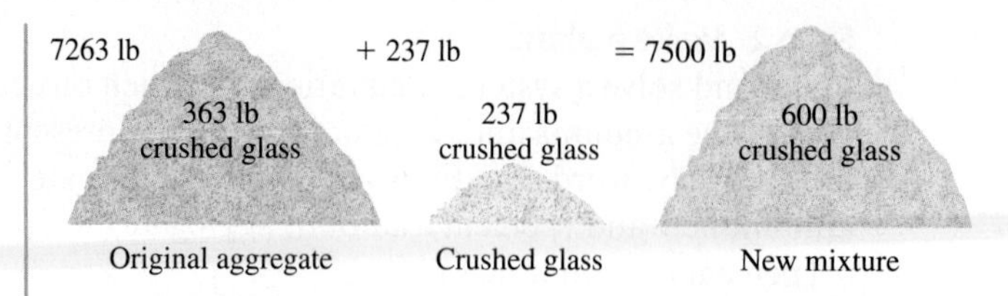

Step 5 Report the solution.
About 237 lb of crushed recycled glass needs to be mixed with 7263 lb of the original mixture.

In Example 3, the unknowns are the amounts of the original mixture and the amount of the pure substance being added. In the next example, given the amount of the original mixture, the unknowns are the amount of pure substance to add to this mixture and the amount of the new mixture.

EXAMPLE 4 Trucks need different mixtures of diesel fuels for summer and winter driving. A fuel distributor has 7500 gal of fuel that is 65% No. 2 Diesel. Find the amount of pure No. 2 Diesel to add to this mixture to make a new mixture that is 80% No. 2 Diesel, and find the total amount of the new mixture.

SOLUTION ▶ **Step 1 Understand the problem.**
The two unknowns are the amount of pure diesel and the amount of the new mixture.

x = amount of pure diesel $\quad y$ = amount of new mixture

Step 2 Make a plan.
Write and solve a system of equations in which one equation represents the relationship of the amounts and the other equation represents the amount of diesel in each mixture. The word equations are amount original + amount pure diesel = amount new and (percent)(amount original) + (percent)(amount pure diesel) = (percent)(amount new).

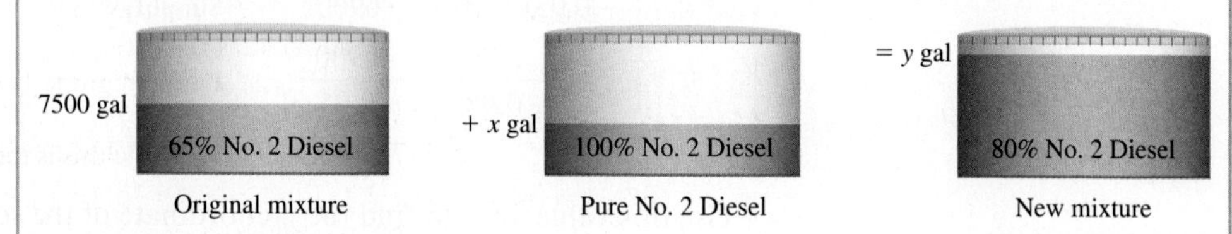

Step 3 Carry out the plan.
The equation for the relationship of the amounts is $7500 \text{ gal} + x = y$. The equation for the amount of diesel in each mixture is $0.65(7500 \text{ gal}) + 1.00x = 0.80y$, which simplifies to $4875 \text{ gal} + x = 0.8y$. Since one of the equations is solved for a variable, remove the units and solve $\begin{matrix} 7500 + x = y \\ 4875 + x = 0.8y \end{matrix}$ by substitution.

Find the x-coordinate of the solution.

$4875 + x = 0.8y$

$4875 + x = 0.8(7500 + x)$ Replace y with the substitution expression, $7500 + x$.

$4875 + x = \mathbf{0.8}(7500) + \mathbf{0.8x}$ Distributive property.

$4875 + x = \mathbf{6000} + 0.8x$ Simplify.

$-4875 \qquad -4875$ Subtraction property of equality.

$\mathbf{0} + x = \mathbf{1125} + 0.8x$ Simplify.

$-0.8x \qquad -0.8x$ Subtraction property of equality.

$\mathbf{0.2x} = 1125 + \mathbf{0}$ Simplify.

$$\frac{0.2x}{0.2} = \frac{\mathbf{1125}}{0.2}$$ Simplify; division property of equality.

$$x = 5625$$ Simplify; this is the x-coordinate of the solution.

Use this value for x to find the y-coordinate of the solution.

$$7500 + x = y$$ Use either of the original equations.

$$7500 + \mathbf{5625} = y$$ Replace x with the x-coordinate of the solution, 5625.

$$\mathbf{13{,}125} = y$$ Simplify; this is the y-coordinate of the solution.

Including the units, the solution is (5625 gal No. 2 Diesel, 13,125 gal new mixture).

Step 4 Look back.
The amount of original fuel, 7500 gal, plus the amount of pure No. 2 Diesel, 5625 gal, equals the amount of new mixture, 13,125 gal. The amount of No. 2 Diesel in the original fuel is (0.65)(7500 gal), which equals 4875 gal; the amount of pure No. 2 Diesel added is 5625 gal; and the amount of No. 2 Diesel in the new mixture is (0.80)(13,125 gal), which equals 10,500 gal. Since 4875 gal plus 5625 gal equals 10,500 gal, the answer seems reasonable.

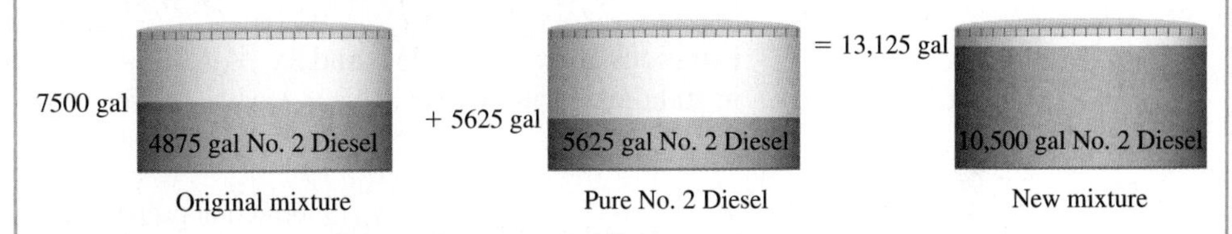

Step 5 Report the solution.
The distributor should add 5625 gal of No. 2 Diesel to the original fuel. The total amount of the new mixture is 13,125 gal.

Practice Problems

For problems 1–4, use the five steps and a system of equations.

1. Mixture A is 69.5% nitric acid. Mixture B is 4% nitric acid. Find the amount of Mixture A and the amount of Mixture B needed to make 2 L of a new mixture that is 30% nitric acid. Round to the nearest tenth.
2. The mixture in a bottle is 32% insecticide. Find the amount of this mixture and the amount of water needed to make 256 oz of a new mixture that is 5% insecticide.
3. A mixture is 45% antifreeze. Find the amount of this mixture and the amount of pure antifreeze needed to make 6 L of a new mixture that is 60% antifreeze. Round to the nearest tenth.
4. A tank contains 200 L of a mixture that is 26.5% salt. Find the amount of water to add to the tank so that the new mixture is 12.5% salt. Find the amount of the new mixture in the tank.

Cost and Systems of Equations

In the next example, we know the total number of items sold, the price of each item, and the total cost. We can use a system of equations to find the number of each item sold.

EXAMPLE 5 At a fundraiser, the cost of a box of cookies is \$3 and the cost of a candy bar is \$1.50. If a total of 750 boxes and bars are sold for \$1537.50, find the number of boxes of cookies and the number of candy bars sold.

SOLUTION ▶ **Step 1 Understand the problem.**
The unknowns are the number of boxes of cookies sold and the number of candy bars sold.

x = number of boxes of cookies sold $\quad$ y = number of candy bars sold

Step 2 Make a plan.
Write and solve a system of equations in which one equation represents the total number of items sold and the other equation represents the costs. The word equations are number of boxes of cookies + number of candy bars = total and

$$\left(\frac{\text{cost}}{\text{1 box cookies}}\right)(\text{number boxes}) + \left(\frac{\text{cost}}{\text{1 candy bar}}\right)(\text{number candy bars}) = \text{total cost.}$$

Step 3 Carry out the plan.
The equation for the number of items sold is $x + y = 750$ items, and the equation for the costs is $\left(\frac{\$3}{1\text{ box}}\right)x + \left(\frac{\$1.50}{1\text{ bar}}\right)y = \$1537.50$. Removing the units, the equations are $x + y = 750$, and $3x + 1.50y = 1537.50$. Although these equations are in standard form, we can solve $x + y = 750$ for y and solve this system by substitution.

$$x + y = 750$$

$$\underline{-x \qquad -x} \quad \text{Subtraction property of equality.}$$

$$\mathbf{0} + y = \mathbf{750 - x} \quad \text{Simplify.}$$

$$\mathbf{y} = 750 - x \quad \text{Simplify.}$$

The system is $\begin{matrix} y = 750 - x \\ 3x + 1.50y = 1537.50 \end{matrix}$. Find the x-coordinate of the solution.

$$3x + 1.50y = 1537.50$$

$$3x + 1.50(750 - x) = 1537.50 \quad \text{Replace } y \text{ with the substitution expression, } 750 - x.$$

$$3x + \mathbf{1125 - 1.50x} = 1537.50 \quad \text{Distributive property.}$$

$$\mathbf{1.5x} + 1125 = 1537.50 \quad \text{Combine like terms.}$$

$$\underline{-1125 \quad -1125} \quad \text{Subtraction property of equality.}$$

$$1.5x + \mathbf{0} = \mathbf{412.50} \quad \text{Simplify.}$$

$$\frac{\mathbf{1.5x}}{1.5} = \frac{412.50}{1.5} \quad \text{Division property of equality.}$$

$$x = 275 \quad \text{Simplify; this is the } x\text{-coordinate of the solution.}$$

Use this value for x to find the y-coordinate of the solution.

$$x + y = 750 \quad \text{Use either of the original equations.}$$

$$\mathbf{275} + y = 750 \quad \text{Replace } x \text{ with the } x\text{-coordinate of the solution, 275.}$$

$$\underline{-275 \qquad -275} \quad \text{Subtraction property of equality.}$$

$$\mathbf{0} + y = \mathbf{475} \quad \text{Simplify.}$$

$$\mathbf{y} = 475 \quad \text{Simplify; this is the } y\text{-coordinate of the solution.}$$

Including the units, the solution is (275 boxes of cookies, 475 candy bars).

Step 4 Look back.
Since 275 + 475 equals 750, which is the number of items sold, and (275)(\$3) + (475)(\$1.50) equals \$1537.50, which is the amount of total sales, the answer seems reasonable.

Step 5 Report the solution.
The group sold 275 boxes of cookies and 475 candy bars.

When writing a system of equations, use the units of measurement to check whether each equation makes sense. The simplified units on the left side of the equation should equal the simplified units on the right side of the equation.

EXAMPLE 6 A produce broker pays \$3.99 per pound for shiitake mushrooms and \$2.37 per pound for oyster mushrooms. The broker wants a mixture of 800 lb of mushrooms at a price of \$3.25 per pound. Find the amount of each kind of mushroom that the broker should buy. Round to the nearest whole number. (*Source:* www.americanmushroom.org)

© Rebecca Craven

SOLUTION ▶ **Step 1 Understand the problem.**

The two unknowns are the amount of shiitake mushrooms and the amount of oyster mushrooms.

x = amount of shiitake mushrooms $\quad$ y = amount of oyster mushrooms

Step 2 Make a plan.

Write and solve a system of equations in which one equation represents the total amount of mushrooms and the other equation represents the costs. The word equations are amount of shiitake + amount of oyster = total amount and

$$\left(\frac{\text{cost}}{\text{1 lb shiitake}}\right)(\text{amount shiitake}) + \left(\frac{\text{cost}}{\text{1 lb oyster}}\right)(\text{amount oyster}) = \left(\frac{\text{cost}}{\text{1 lb mixed}}\right)(\text{amount mixed}).$$

Step 3 Carry out the plan.

The equation for the total amount of mushrooms is $x + y = 800$ lb, and the equation for the costs is $\left(\frac{\$3.99}{1\text{ lb}}\right)x + \left(\frac{\$2.37}{1\text{ lb}}\right)y = \left(\frac{\$3.25}{1\text{ lb}}\right)(800\text{ lb})$, which simplifies to $\left(\frac{\$3.99}{1\text{ lb}}\right)x + \left(\frac{\$2.37}{1\text{ lb}}\right)y = \$2600$. Remove the units, and solve $\begin{aligned} x + y &= 800 \\ 3.99x + 2.37y &= 2600 \end{aligned}$ by elimination.

Create y-terms that are opposites by multiplying $x + y = 800$ by -2.37.

$x + y = 800$	
$-2.37(x + y) = -2.37(800)$	Multiplication property of equality.
$-2.37x - 2.37y = -1896$	Distributive property; simplify.

Find the x-coordinate of the solution.

$-2.37x - 2.37y = -1896$	
$+\ 3.99x + 2.37y = 2600$	Add equations; addition property of equality.
$1.62x + 0 = 704$	Simplify.
$\frac{\mathbf{1.62x}}{1.62} = \frac{704}{1.62}$	Division property of equality.
$x = 434.56\ldots$	Simplify; this is the x-coordinate of the solution.

Use this value for x to find the y-coordinate of the solution.

$x + y = 800$	Choose one of the original equations.
$\mathbf{434.56\ldots} + y = 800$	Replace x with the x-coordinate of the solution, 434.56. . . .
$-434.56\ldots \quad -434.56\ldots$	Subtraction property of equality.
$0 + y = \mathbf{365.43\ldots}$	Simplify.
$y = 365.43\ldots$	Simplify; this is the y-coordinate of the solution.

Including the units and rounding, the solution is about (435 lb shiitake, 365 lb oyster).

Step 4 Look back.
The sum of 435 lb and 365 lb is 800 lb, which equals the total amount of mushrooms. The cost of shiitake mushrooms is $\left(\frac{\$3.99}{1\text{ lb}}\right)(435\text{ lb})$, which equals about \$1736; the cost of oyster mushrooms is $\left(\frac{\$2.37}{1\text{ lb}}\right)(365\text{ lb})$, which equals about \$865; and the cost of 800 lb is $\left(\frac{\$3.25}{1\text{ lb}}\right)(800\text{ lb})$, which equals \$2600. Since \$1736 plus \$865 equals \$2601, the answer seems reasonable.

Step 5 Report the solution.
The broker should buy 435 lb of shiitake mushrooms and 365 lb of oyster mushrooms.

Practice Problems

For problems 5–6, use the five steps and a system of equations.

5. A report requires white copy paper that costs \$3.99 per ream and colored copy paper that costs \$8.89 per ream. The total cost of buying 9 reams of paper is \$60.41. Find the number of reams of colored paper and the number of reams of white paper.
6. The kitchen staff at a senior citizens center is buying 250 lb of ham and turkey for a holiday dinner. The cost of ham is \$1.25 per pound, and the cost of turkey is \$0.49 per pound. If the budget allows for a cost of \$0.97 per pound, find the amount of ham and the amount of turkey to buy. Round to the nearest whole number.

Formulas and Systems of Equations

The formula for the perimeter of a rectangle is $P = 2L + 2W$. If we know the perimeter of a rectangle and the relationship of the length and width, we can write and solve a system of linear equations to find the length and the width.

EXAMPLE 7 The perimeter of a rectangle is 48 in. The length plus eight times the width is 87 in. Find the length and width.

SOLUTION ▶ **Step 1 Understand the problem.**
The two unknowns are the length and width of the rectangle.

L = length $\quad W$ = width

L W W L

Step 2 Make a plan.
Write a system of equations in which one equation represents the perimeter and the other equation represents the relationship of the length and width. The word equations are 2(length) + 2(width) = perimeter and length + 8(width) = 87 in.

Step 3 Carry out the plan.
The equation that represents the perimeter is $2L + 2W = 48$ in. The equation that represents the relationship of the length and width is $L + 8W = 87$ in. Since the equations are in standard form, remove the units and solve $\begin{matrix} 2L + 2W = 48 \\ L + 8W = 87 \end{matrix}$ by elimination. Choose either L or W to eliminate. In this example, we will eliminate L.

To create L-terms that are opposites, multiply $L + 8W = 87$ by -2.

$L + 8W = 87$

$(-2)(L + 8W) = (-2)(87)$ Multiplication property of equality.

$-2L - 16W = -174$ Distributive property; simplify.

Find the W-coordinate of the solution.

$$\begin{aligned} 2L + 2W &= 48 \\ + \; -2L - 16W &= -174 \\ \hline 0L - 14W &= -126 \end{aligned}$$

Add equations; addition property of equality.

Simplify.

$$\frac{-14W}{-14} = \frac{-126}{-14}$$ Division property of equality.

$$W = 9$$ Simplify; the W-coordinate of the solution is 9.

Use this value for W to find the L-coordinate of the solution.

$L + 8W = 87$	Choose either of the original equations.
$L + 8(\mathbf{9}) = 87$	Replace W with the W-coordinate of the solution, 9.
$L + \mathbf{72} = 87$	Simplify.
$-72 \quad -72$	Subtraction property of equality.
$L + \mathbf{0} = \mathbf{15}$	Simplify.
$L = 15$	Simplify; the L-coordinate of the solution is 15.

Including the units, the solution is (15 in., 9 in.).

Step 4 Look back.
The given perimeter is 48 in. and 2(15 in.) + 2(9 in.) also equals 48 in. Since the length plus eight times the width equals 87 in. and 15 in. + 8(9 in.) also equals 87 in., the answer seems reasonable.

Step 5 Report the solution.
The length of the rectangle is 15 in. and the width of the rectangle is 9 in.

The next example includes the relationship of distance, rate, and time, $d = rt$. The solution of a system of equations that includes time and distance is not written in the usual alphabetical order but is instead written (time, distance) or (t, d).

EXAMPLE 8 A group of eight cyclists leaves a parking lot traveling $\frac{15 \text{ mi}}{1 \text{ hr}}$. After 35 min, another group of cyclists leaves the parking lot traveling the same route at $\frac{24 \text{ mi}}{1 \text{ hr}}$. Find the time in minutes when the second group will catch up with the first group. Round to the nearest minute. Find the distance traveled by each group. Round to the nearest tenth.

SOLUTION ▶ **Step 1 Understand the problem.**
When the fast group catches up to the slow group, both groups will have traveled the same distance. The two unknowns are the time traveled by the fast group and the total distance traveled by each group. The number of cyclists is extraneous information.

t = time traveled by the fast group $\qquad$ d = distance traveled by each group

Step 2 Make a plan.
Since the time is in minutes and the answer is to be in minutes, change the units of speed into $\frac{\text{mi}}{\text{min}}$.

$$\frac{15 \text{ mi}}{1 \text{ hr}} \cdot \frac{1 \text{ hr}}{60 \text{ min}}$$ Speed of the slow group; multiply by a fraction equal to 1.

$$= \frac{0.25 \text{ mi}}{1 \text{ min}}$$ Simplify.

$$\frac{24 \text{ mi}}{1 \text{ hr}} \cdot \frac{1 \text{ hr}}{60 \text{ min}}$$ Speed of the fast group; multiply by a fraction equal to 1.

$$= \frac{0.4 \text{ mi}}{1 \text{ min}}$$ Simplify.

For the slow group, a word equation is total distance = distance traveled in 35 min + distance traveled in unknown time. For the fast group, a word equation is total distance = distance traveled in unknown time. To write an expression for a distance, use $d = rt$.

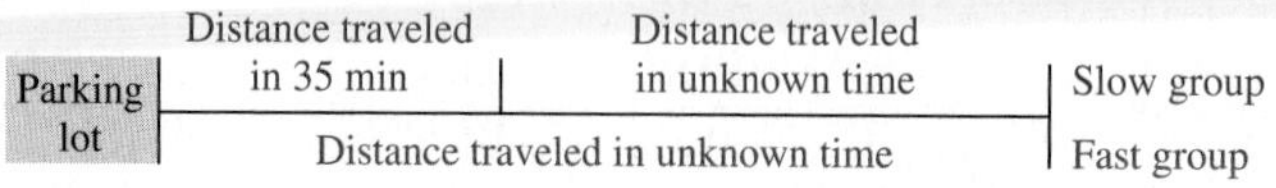

Parking lot | Distance traveled in 35 min | Distance traveled in unknown time | Slow group

Distance traveled in unknown time | Fast group

Step 3 Carry out the plan.
The equation for the slow group is $d = \left(\frac{0.25\text{ mi}}{1\text{ min}}\right)(35\text{ min}) + \left(\frac{0.25\text{ mi}}{1\text{ min}}\right)t$, which simplifies to $d = 8.75\text{ mi} + \left(\frac{0.25\text{ mi}}{1\text{ min}}\right)t$. The equation for the fast group is $d = \left(\frac{0.4\text{ mi}}{1\text{ min}}\right)t$.

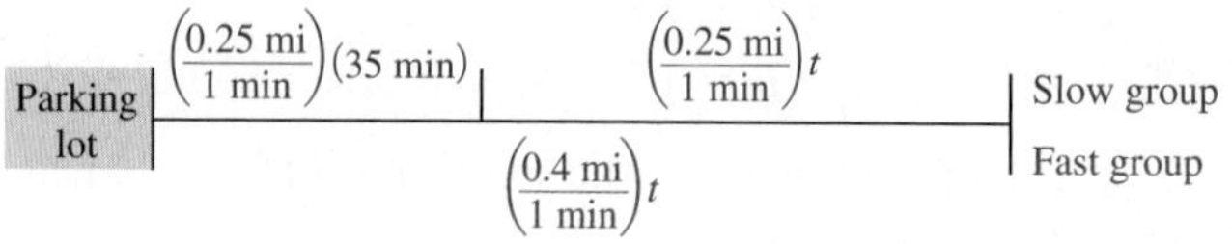

Parking lot | $\left(\frac{0.25\text{ mi}}{1\text{ min}}\right)(35\text{ min})$ | $\left(\frac{0.25\text{ mi}}{1\text{ min}}\right)t$ | Slow group

$\left(\frac{0.4\text{ mi}}{1\text{ min}}\right)t$ | Fast group

Since both equations are solved for a variable, remove the units and solve $d = 8.75 + 0.25t$, $d = 0.4t$ by substitution.

Find the t-coordinate of the solution.

$d = 0.4t$	
$8.75 + 0.25t = 0.4t$	Replace d with the substitution expression, $8.5 + 0.25t$.
$-0.25t \quad -0.25t$	Subtraction property of equality.
$8.75 + \mathbf{0} = \mathbf{0.15t}$	Simplify.
$\frac{\mathbf{8.75}}{0.15} = \frac{0.15t}{0.15}$	Division property of equality.
$58.33... = t$	Simplify; this is the t-coordinate of the solution.

Use this value for t to find the d-coordinate of the solution.

$d = 0.4t$	Choose either of the original equations.
$d = 0.4(\mathbf{58.3...})$	Replace t with the t-coordinate of the solution, 58.33. . . min.
$d = \mathbf{23.33...}$	Simplify.

Rounding the time to the nearest whole number and the distance to the nearest tenth and including the units, the solution is about (58 min, 23.3 mi).

Step 4 Look back.
In the first 35 min, the slow group travels 8.75 mi. In the next 58 min, the slow group travels about $\left(\frac{0.25\text{ mi}}{1\text{ min}}\right)(58\text{ min})$, or 14.5 mi. The total distance traveled by the slow group is 23.25 mi. The fast group travels a total distance of $\left(\frac{0.4\text{ mi}}{1\text{ min}}\right)(58\text{ min})$, which is 23.2 mi. Since both groups travel about the same total distance, the answer seems reasonable. The small difference in distance is due to rounding.

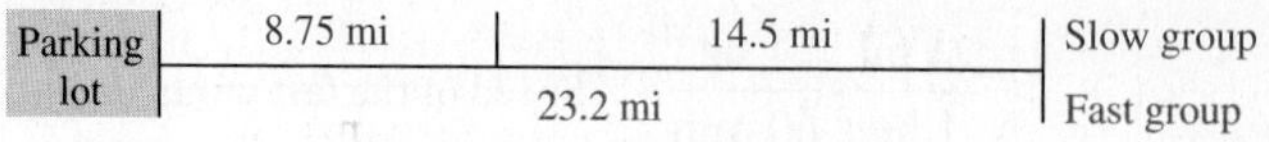

Parking lot | 8.75 mi | 14.5 mi | Slow group

23.2 mi | Fast group

Step 5 Report the solution.
The fast group will catch up to the slow group after about 58 min. Both groups will travel about 23.2 mi.

Practice Problems

For problems 7–9, use the five steps and a system of equations.

7. The perimeter of a rectangle is 32 in. The length plus three times the width is 28 in. Find the length and width of the rectangle.
8. The perimeter of a rectangle is 84 in. Seven times the length plus the width is 192 in. Find the length of the rectangle. Find the width of the rectangle.
9. A truck leaves a town traveling at $\frac{60 \text{ mi}}{1 \text{ hr}}$. After 30 min, a car leaves town, following the same route, traveling at $\frac{70 \text{ mi}}{1 \text{ hr}}$. Find the time when the car will catch up with the truck. Find the distance traveled by the truck and the distance traveled by the car.

4.4 Vocabulary Practice

Match the term with its description.

1. What L represents in the formula $P = 2L + 2W$
2. The distance around a rectangle
3. In this method for solving a system of equations, the equations in a system are added together.
4. In this method for solving a system of equations, a variable is replaced by an expression.
5. The graphs of the linear equations in a system with no solution are this.
6. The product of rate and time
7. $ax + by = c$
8. We can multiply both sides of an equation by a real number except for 0 without changing its solution(s).
9. We can add a real number to both sides of an equation without changing its solution(s).
10. Speed is an example of this.

A. addition property of equality
B. distance
C. elimination method
D. length of a rectangle
E. multiplication property of equality
F. parallel lines
G. perimeter
H. rate
I. standard form of a linear equation
J. substitution method

4.4 Exercises

Follow your instructor's guidelines for showing your work.

For exercises 1–40, use the five steps and a system of equations.

1. Drink A is 15% orange juice. Drink B is 4% orange juice. Find the amount of each drink needed to make 800 gal of a new drink that is 7% orange juice. Round to the nearest whole number.
2. Drink A is 13% grapefruit juice. Drink B is 2% grapefruit juice. Find the amount of each drink needed to make 900 gal of a new drink that is 5% grapefruit juice. Round to the nearest whole number.
3. Mixture A is 27% acid. Find the amount of this mixture and the amount of water needed to make 320 L of a new mixture that is 21% acid. Round to the nearest whole number.
4. Mixture A is 31% acid. Find the amount of this mixture and the amount of water needed to make 410 L of a new mixture that is 24% acid. Round to the nearest whole number.

5. Mixture A is 13% merlot grape juice. Find the amount of pure merlot grape juice to add to this mixture to make 59 gal of a new mixture that is 44% merlot grape juice. Round to the nearest whole number.

6. Mixture A is 9% merlot grape juice. Find the amount of pure merlot grape juice to add to this mixture to make 59 gal of a new mixture that is 21% merlot grape juice. Round to the nearest whole number.

7. The cost of regular gasoline is $3.85 per gallon and the cost of premium gasoline is $4.05 per gallon. If a total of 1200 gal of gasoline is sold for $4675, found the amount of regular gasoline and the amount of premium gasoline sold.

8. The cost of playground grass seed mix is $2.16 per pound, and the cost of shade grass seed mix is $3.60 per pound. If a total of 500 lb of grass seed is sold for $1728, find the amount of playground grass seed mix and the amount of shade grass seed mix sold.

9. A daycare center needs 50 lb of apples and oranges. Apples cost $1.49 per pound, and oranges cost $0.89 a pound. If the budget allows for a cost of $1.05 per pound, find the amount of apples and the amount of oranges to buy. Round to the nearest whole number.

10. A caterer needs 40 lb of deli roast beef and deli turkey. Deli roast beef costs $7.99 per pound, and deli turkey costs $5.59 per pound. If the budget allows for a cost of $6.29 per pound, find the amount of deli roast beef and deli turkey to buy. Round to the nearest whole number.

11. The perimeter of a rectangle is 46 in. The length plus three times the width is 37 in. Find the length and width.

12. The perimeter of a rectangle is 42 in. The length plus two times the width is 27 in. Find the length and width.

13. The perimeter of a rectangle is 26 cm. Three times the length minus the width is 23 cm. Find the length and width.

14. The perimeter of a rectangle is 32 cm. Three times the length minus the width is 24 cm. Find the length and width.

15. A truck leaves a town traveling at a constant speed of $\frac{55 \text{ mi}}{1 \text{ hr}}$. After 20 min, a car follows the same route traveling at a constant speed of $\frac{60 \text{ mi}}{1 \text{ hr}}$. Find the time in minutes when the car will catch up with the truck. Find the distance traveled by the truck and the distance traveled by the car.

16. A truck leaves a town traveling at a constant speed of $\frac{65 \text{ mi}}{1 \text{ hr}}$. After 25 min, a car follows the same route traveling at a constant speed of $\frac{70 \text{ mi}}{1 \text{ hr}}$. Find the time in minutes when the car will catch up with the truck. Find the distance traveled by each vehicle. Round to the nearest whole number.

17. A prep cook at a restaurant in a national park gets off work at 8 a.m. At 8:30 a.m., she starts hiking on a trail at a speed of 2.5 mi per hour. A server gets off work at 10 a.m. At 10:30 a.m., she starts hiking on the same trail at a speed of 4.5 mi per hour. Find the time in minutes when the server will catch up with the cook. Find the distance traveled by each person.

18. A tugboat leaves a port pushing two barges, traveling at an average speed of 6 mi per hour. Four hours later, a tugboat without barges leaves the port, traveling at an average speed of 8 mi per hour. Find the time after the fast tugboat leaves port needed for the fast tugboat to catch up with the slower tugboat. Find the distance that the boats travel.

19. Snack Mix A is 18% almonds. Snack Mix B is 6% almonds. Find the amount of Snack Mix A and Snack Mix B needed to make 40 lb of a new snack mix that is 9% almonds.

20. Snack Mix A is 17% raisins. Snack Mix B is 8% raisins. Find the amount of Snack Mix A and Snack Mix B needed to make 30 pounds of a new snack mix that is 12% raisins. Round to the nearest whole number.

21. Mixture A is 15% sugar. Mixture B is 18% sugar. Find the amounts of Mixture A and water needed to make 20 L of a new mixture that is 12% sugar.

22. Mixture A is 12% acid. Mixture B is 17% acid. Find the amounts of Mixture A and water needed to make 30 L of a new mixture that is 8% acid.

23. An adult ticket to an event cost $7. A child's ticket cost $3.50. A total of 300 tickets were sold for $1648.50. The concessions stand sold 500 bags of popcorn. Find the number of each kind of ticket that were sold.

24. An adult ticket to an event cost $7. A child's ticket cost $3.50. A total of 300 tickets were sold for $1501.50. The concessions stand sold 200 hotdogs. Find the number of each kind of ticket that were sold.

25. Drink A is 5% orange juice. Drink B is 2% orange juice. Eight ounces of Drink B contains 27 g of sugar. Find the amount of each drink needed to make 1400 gal of a new drink that is 3% orange juice. Round to the nearest whole number.

26. Blackberries and blueberries are among the fruits with the highest amount of antioxidants. Drink A is 22% blackberry juice. Drink B is 27% blackberry juice. Find the amount of each mixture needed to make 8000 gal of a new drink that is 25% blackberry juice.

27. A mixture is 26.5% salt. Find the amount of this mixture and the amount of pure water needed to make 300 L of a new mixture for curing feta cheese that is 12.5% salt. Round to the nearest whole number.

28. An antifreeze mixture is 45% antifreeze. Find the amount of this mixture and the amount of pure water needed to make 6.2 L of a new mixture that is 40% antifreeze. Round to the nearest tenth.

For exercises 29–34, a *karat* describes the percent gold in an alloy (a mixture of metals).

Name of alloy	Percent gold
10-karat gold	41.7%
14-karat gold	58.3%
18-karat gold	75%
20-karat gold	83.3%
24-karat gold	100%

29. Find the amount of 14-karat gold and the amount of 20-karat gold to combine to make 8 oz of 18-karat gold. Round to the nearest hundredth.

30. Find the amount of 10-karat gold and 20-karat gold to combine to make 3 g of 14-karat gold. Round to the nearest hundredth.

31. Find the amount of 24-karat gold and the amount of copper to mix to make 15 oz of 10-karat gold. Round to the nearest hundredth.

32. Find the amount of 24-karat gold and the amount of silver to mix to make 8 oz of 10-karat gold. Round to the nearest hundredth.

33. If 6 g of 20-karat gold jewelry is melted down, find the amount of 14-karat gold to add to the melted jewelry to create a new alloy that is 18-karat gold. Find the amount of the new alloy. Round to the nearest tenth.

34. If 9 oz of 14-karat gold jewelry is melted down, find the amount of 20-karat gold to add to the melted jewelry to create a new alloy that is 18-karat gold. Find the amount of the new alloy. Round to the nearest tenth.

35. Logging trucks use a winter fuel that is 60% No. 2 Diesel. Find the amount of winter fuel and the amount of pure No. 2 Diesel needed to make 7100 gal of a new spring fuel mixture that is 80% No. 2 Diesel.

© Laura Bracken

36. A fuel is 20% No. 1 Diesel. Find the amount of this fuel and the amount of pure No. 1 Diesel needed to make 8000 gallons that is 40% No. 1 Diesel.

37. A campsite is 6 mi from a trailhead. One group of hikers begins at the trailhead, hikes on the trail to this campsite, and spends the night. At 8 a.m. the next morning, they start hiking on the trail again at a speed of 3 mi per hour. At 10 a.m. the same day, another group of hikers begins the trail at the trailhead, hiking at a speed of 5 mi per hour. Find the time and distance hiked by the second group when they catch up with the first group. (Assume that the groups do not take breaks or stop.)

38. A tugboat leaves a port on the Columbia River at a speed of 10 mi per hour. One hour later, a powerboat leaves the port traveling at a speed of 24 mi per hour. Find the time and distance traveled by the powerboat when it catches up with the barge. Round to the nearest minute.

39. At a food co-op where co-op members are required to work 18 times per year, almonds cost $9.46 per pound and Brazil nuts cost $7.31 per pound. Find the amount of almonds and Brazil nuts needed to make 40 lb of a mixture that costs $8 per pound. Round to the nearest tenth.

40. If pecan halves cost $11.75 per pound, cashews cost $8.90 per pound, and English walnuts cost $10.20 per pound, find the amount of pecans and English walnuts needed to make 50 lb of a mixture that costs $11 per pound. Round to the nearest tenth.

For exercises 41–76, use the five steps and a system of equations.

41. The cost of two cans of chili and three cans of soup is $7.78. The cost of five cans of chili and four cans of soup is $12.45. Find the cost of one can of chili. Find the cost of one can of soup.

42. The cost of two jars of peanut butter and three jars of mayonnaise is $10.87. The cost of five jars of peanut butter and four jars of mayonnaise is $19.16. Find the cost of one jar of peanut butter. Find the cost of one jar of mayonnaise.

43. An assistant is buying 300 reams of white and colored paper for the office. He wants five times as many reams of white paper as colored paper. Find the number of reams of white paper and the number of reams of colored paper that he should buy.

44. A caterer is buying 40 lb of oranges and apples. She wants four times as many pounds of apples as oranges. Find the amount of oranges and the amount of apples that she should buy.

45. A winemaker has 11,328 gal of wine. Three times as much wine will be bottled as will be stored in barrels. Find the amount of wine that will be bottled. Find the amount of wine that will be stored in barrels.

46. A quilter has 108 yd of fabric. She wants to donate five times as much fabric to the Quilts for Children project as she keeps. Find the number of yards that she should donate. Find the number of yards that she should keep.

47. The cost to make a product is $9.50. The fixed overhead costs per month to make the product are $2400. The price of each product is $15.75. Find the break-even point for this product.

48. The cost to make a product is $21.50. The fixed overhead costs per month to make the product are $21,450. The price of each product is $29.75. Find the break-even point for this product.

49. The perimeter of a rectangle is 50 in. The length is four times the width. Find the length and width.

50. The perimeter of a rectangle is 66 in. The length is twice as long as the width. Find the length and width.

51. The perimeter of a rectangle is 78 in. The length is 9 in. longer than the width. Find the length and the width.

52. The perimeter of a rectangle is 60 in. The length is 14 in. longer than the width. Find the length and the width.

53. Concentrated stop bath for photography is 28% acetic acid. Find the amount of concentrated stop bath and the amount of water needed to make 3 L of a new mixture that is 2% acetic acid. Round to the nearest tenth.

54. A mixture in a stock room contains 37% formaldehyde. Find the amount of this mixture and pure water needed to make 40 mL of a new mixture that is 4% formaldehyde. Round to the nearest tenth.

55. The costs to deliver a course at a college include $1100 per section plus an additional $10 in materials per student. The revenue from tuition is about $150 per student. Find the number of students at which the cost of delivering a course is equal to the revenues from tuition, and find the revenues from tuition. Round *up* to the nearest student, and round the tuition to the nearest hundred.

56. The costs to deliver a course at a college include $2300 per section plus an additional $20 in materials per student. The revenue from tuition is about $175 per student. Find the number of students at which the cost of delivering a course is equal to the revenues from tuition, and find the revenues from tuition. Round *up* to the nearest student, and round the tuition to the nearest hundred.

57. Distillers grains (30% protein), corn gluten feed (20% protein), and cornstalks (5% protein) are used as cattle feed. Find the amount of distillers grains and cornstalks that should be combined to make 1500 lb of feed that is 18% protein. (*Source:* www.extension.iastate.edu)

58. Distillers grains (30% protein), corn gluten feed (20% protein), and cornstalks (5% protein) are used as cattle feed. Find the amount of corn gluten feed and cornstalks that should be combined to make 1500 lb of feed that is 18% protein. (*Source:* www.extension.iastate.edu)

59. A canoe traveling 4 miles per hour leaves a portage on one end of Saganaga Lake. Another faster canoe traveling 5 mi per hour begins the same route 15 min later. The distance to the next portage is 9 mi. Find the time in minutes when the faster canoe will catch up with the slower canoe. Find the distance traveled by each canoe.

© Tom Bracken

60. A canoe traveling 3 mi per hour leaves a portage on one end of Knife Lake. Another faster canoe traveling 5 mi per hour begins the same route 20 min later. The distance to the next portage is 9.5 mi. Find the time in minutes when the faster canoe will catch up with the slower canoe. Find the distance traveled by each canoe.

61. A timber bolt with a diameter of $\frac{5}{8}$ in. weighs 1.2 lb and costs $3.65. A hex bolt with a diameter of 1 in. weighs 3 lb and costs $7.82. The total weight of 20 bolts is 51 lb. Find the number of timber bolts and the number of hex bolts.

62. A timber bolt with a diameter of $\frac{1}{2}$ in. weighs 0.9 lb and costs $3.73. A hex bolt with a diameter of 1 in. weighs 3 lb and costs $7.82. The total weight of 30 bolts is 69 lb. Find the number of timber bolts and the number of hex bolts.

63. Pigeon Feed A contains popcorn, whole milo, Canadian peas, whole wheat, maple peas, Austrian peas, and oat groats and is 14% protein. Pigeon Feed B contains popcorn, milo, wheat, oat groats, and Red Proso Millet and is 17% protein. Find the amount of each feed to mix together to make 50 lb of a new feed that is 16.2% protein. Round to the nearest tenth.

64. Pigeon Feed A contains popcorn, whole milo, Canadian peas, whole wheat, maple peas, Austrian peas, and oat groats and is 14% protein. Pigeon Feed B contains popcorn, milo, wheat, oat groats, and Red Proso Millet and is 17% protein. Find the amount of each feed to mix together to make 40 lb of a new feed that is 16.2% protein. Round to the nearest tenth.

65. A total of $17,000 is invested in two mutual funds for 1 year. The return on Mutual Fund A is 3% per year, the return on Mutual Fund B is 2% per year, and the total return is $407.50. Find the amount invested in Mutual Fund A and the amount invested in Mutual Fund B.

66. A total of $16,000 is invested in two mutual funds for 1 year. The return on Mutual Fund A is 2% per year, the return on Mutual Fund B is 3% per year, and the total return is $441.50. Find the amount invested in Mutual Fund A and the amount invested in Mutual Fund B.

67. A boat travels 60 mi upstream in a river, against the current, in 3 hr. The boat travels 60 mi downstream, with the current, in 2 hr. Find the speed of the boat in still water. Find the speed of the current.

68. A boat travels 120 mi upstream in a river, against the current, in 4 hr. The boat travels 120 mi downstream, with the current, in 3 hr. Find the speed of the boat in still water. Find the speed of the current.

69. If 18 lb of bird seed mix has four times as much Golden German Finch Millet in it as Red Proso Millet, find the amount of each kind of millet in the mix.

70. If 16 lb of bird seed mix has four times as much Golden German Finch Millet in it as Red Proso Millet, find the amount of each kind of millet in the mix.

71. Jelly beans cost \$3 per pound, chocolate-covered espresso beans cost \$6 per pound, and chocolate-covered pomegranate seeds cost \$7 per pound. Find the amount of jelly beans and the amount of chocolate-covered pomegranate seeds needed to make 40 lb of a mixture with a total value of \$265.

72. Jelly beans cost \$3 per pound, chocolate-covered espresso beans cost \$6 per pound, and chocolate-covered pomegranate seeds cost \$7 per pound. Find the amount of jelly beans and the amount of chocolate-covered espresso beans needed to make 60 lb of a mixture with a total value of \$240.

73. A new hotel will have 520 rooms. There must be seven times as many rooms that are designated as nonsmoking as rooms that are designated as smoking. Find the number of nonsmoking rooms. Find the number of smoking rooms.

74. A fisheries department needs a total of 5760 kokanee trout and rainbow trout for stocking lakes. They want three times as many kokanee trout as rainbow trout. Find the number of kokanee trout and the number of rainbow trout that they need.

75. The cost of one bag of dog food and four bags of sweet feed for livestock is \$83. The cost of two bags of dog food and one bag of sweet feed is \$82. Find the cost of one bag of dog food. Find the cost of one bag of sweet feed.

76. At an office supply store, three reams of paper and two ink cartridges cost \$56. Seven reams of paper and five ink cartridges cost \$139. Find the cost of one ream of paper. Find the cost of one ink cartridge.

Problem Solving: Practice and Review

Follow your instructor's guidelines for using the five steps as outlined in Section 1.3, p. 55.

77. A 2010 report on homelessness said that there were 1.6 million homeless children in the United States and that 1 in 45 children in the United States are homeless in a year. According to this report, how many children live in the United States? (*Source:* www.homelesschildrenamerica.org, 2010)

78. The maximum combined length and girth of a package that is mailed at the priority rate with the U.S. Postal Service is 108 in. The length is the measure of the longest side of the package, and the girth is the distance measured around the thickest part of the parcel. A box is 3 in. high, 15 in. long, and 8 in. wide. Find its girth.

79. Researchers surveyed 3111 drivers in 14 large cities with red light camera programs. Find the number of surveyed drivers who favor the use of cameras for red light enforcement.

Among drivers in the 14 cities with red light camera programs, two-thirds favor the use of cameras for red light enforcement. (*Source:* www.iihs.org, June 2011)

80. At the end of the last quarter of 2010, about 11.1 million residential properties with a mortgage were underwater, with a balance on the mortgage that was more than the home was worth. At end of the third quarter of 2011, 10.7 million residential mortgages were underwater. Find the percent decrease in the number of underwater mortgages. Round to the nearest tenth of a percent. (*Source:* www.brookings.edu, Dec. 13, 2011)

Technology

For exercises 81–84, solve by graphing. If necessary, round to the nearest tenth. Sketch the graph; describe the window; identify the solution.

81. $y = \left(\dfrac{\$25}{1 \text{ product}}\right)x + \$1500$

$y = \left(\dfrac{\$40}{1 \text{ product}}\right)x$

82. $y = \left(\dfrac{29 \text{ degrees}}{1 \text{ year}}\right)x + 520 \text{ degrees}$

$y = \left(\dfrac{74 \text{ degrees}}{1 \text{ year}}\right)x + 195 \text{ degrees}$

83. $y = \left(-\dfrac{6.95 \text{ heart disease deaths}}{1 \text{ year}}\right)x +$ 325.4 heart disease deaths

$y = \left(-\dfrac{1.99 \text{ cancer deaths}}{1 \text{ year}}\right)x + 217.9 \text{ cancer deaths}$

84. $y = \left(\dfrac{0.16\%}{1 \text{ year}}\right)x + 12.11\%$

$y = \left(-\dfrac{0.46\%}{1 \text{ year}}\right)x + 33.88\%$

Find The Mistake

For exercises 85–88, the completed problem has one mistake.

(a) Describe the mistake in words, or copy down the whole problem and highlight or circle the mistake.

(b) Do the problem correctly.

85. **Problem:** Write a system of equations that represents the relationship of the variables.

Mixture A is 6% juice. Mixture B is 15% juice. Mixture B contains 50% of the RDA of vitamin C. Find the amount of each mixture needed to make 80 gal of a new drink that is 11% juice. Round to the nearest whole number.

x = amount of Mixture A
y = amount of Mixture B

Incorrect Answer: $x + y = 80 \text{ gal}$

$0.06x + 0.50y = (0.11)(80 \text{ gal})$

86. **Problem:** Write a system of equations that represents the relationship of the variables.

Each pound of Cereal A contains 0.5 oz of almonds and 2 oz of dried cranberries. Each pound of Cereal B contains 2 oz of almonds and 1 oz of dried cranberries. Find the amount of each cereal needed to make 400 lb of a new cereal that contains 0.75 oz of almonds per pound of cereal. (1 lb = 16 oz.)

x = amount of Cereal A
y = amount of Cereal B

Incorrect Answer: $x + y = 400 \text{ lb}$

$\left(\dfrac{0.5 \text{ oz}}{1 \text{ lb}}\right)x + \left(\dfrac{2 \text{ oz}}{1 \text{ lb}}\right)y = 0.75 \text{ oz}$

87. Problem: Write a system of equations that represents the relationship of the variables.

The perimeter of a rectangle is 54 in. The length is 15 in. more than the width. Find the length and the width.

x = width
y = length

Incorrect Answer: $x + y = 54$ in.
$y = x + 15$ in.

88. Problem: Write a system of equations that represents the relationship of the variables.

At 8 a.m., a motor home with three people leaves a restaurant traveling an average speed of 50 mi per hour. At 9 a.m., a motorcycle with two people leaves the restaurant traveling the same road at an average speed of 65 mi per hour. Find the number of hours and the distance traveled when the motorcycle catches up to the motor home. (Assume that neither vehicle stops.)

x = time when the vehicles meet
y = distance when the vehicles meet

Incorrect Answer: $y = \left(\frac{50 \text{ mi}}{1 \text{ hr}}\right)x$
$y = \left(\frac{65 \text{ mi}}{1 \text{ hr}}\right)x$

Review

For exercises 89–92, graph.

89. $y > 3x - 5$

90. $y \leq -\frac{2}{3}x + 7$

91. $x - y > 4$

92. $3x + 2y \geq -6$

SUCCESS IN COLLEGE MATHEMATICS

93. Step 1 of the five steps for problem solving (Section 1.3) is to understand the problem, step 2 is to make a plan, and step 3 is to carry out the plan. Looking closely at Example 4 in this section, describe in detail what has to be done to carry out the plan.

94. The practice problems in this section are very similar to the examples. Some of the exercises are also very similar to the examples. Identify the example in this section that is similar to an exercise assigned as homework by your instructor.

Home equity lines of credit can be used to pay for college expenses or home improvement. However, the total amount spent cannot be *greater than* the line of credit. This limit is an example of a constraint. In this section, we will study systems of inequalities and constraints.

SECTION 4.5 Systems of Linear Inequalities

After reading the text, working the practice problems, and completing assigned exercises, you should be able to:

1. Graph a system of linear inequalities.
2. Identify the vertices of a bounded solution region.
3. Determine whether an ordered pair is a solution of a system of inequalities.
4. Write linear inequalities that represent constraints.

Systems of Linear Inequalities

The solution region of the graph of a linear inequality (Section 3.6) includes all the points that represent ordered pairs that are solutions of the inequality. If the inequality is strict ($>$ or $<$), the boundary line is dashed, and the points on the boundary line are not included in the solution region. If the inequality is not strict ($\geq$ or $\leq$), the boundary line is solid, and the points on the boundary line are included in the solution region.

Graphing a Linear Inequality

1. Graph the boundary line. If the inequality is not strict and includes a $\geq$ or $\leq$, the boundary line is a solid line. If the inequality is strict and includes a $>$ or $<$, the boundary line is a dashed line.
2. Select a test point that is *not* on the boundary line. Replace the variables in the inequality with the coordinates of the test point. The resulting inequality will be true or false.
3. If the inequality is true, the test point is in the solution region. Shade the area on this side of the boundary line. If the inequality is false, the test point is *not* in the solution region. Shade the area on the *other* side of the boundary line.

A **system of linear inequalities** includes two or more inequalities. Each point in the solution region of the system must be in the solution region of every inequality in the system.

EXAMPLE 1 Graph $\begin{matrix} y \leq 2x - 4 \\ 8x + 5y \geq 40 \end{matrix}$. Label the solution region.

SOLUTION ▶ Graph: $y \leq 2x - 4$

The solid boundary line is $y = 2x - 4$. Since this equation is in slope-intercept form, use slope-intercept graphing (Section 3.4); the y-intercept is $(0, -4)$, and the slope is 2.

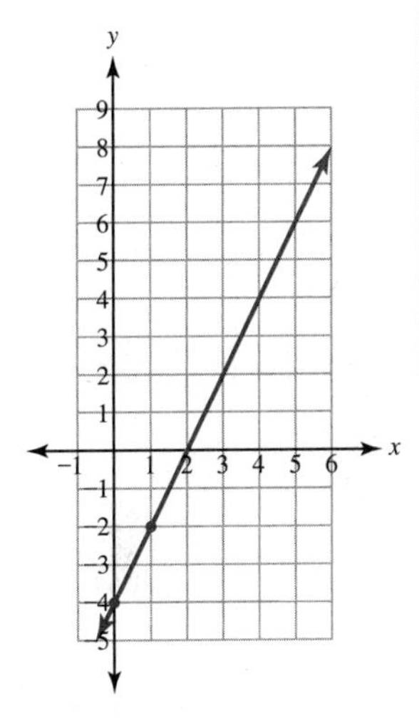

To identify the solution region, choose a test point, $(0, 0)$. If the test point is a solution of the inequality, it is in the solution region.

$y \leq 2x - 4$

$\mathbf{0} \leq 2(\mathbf{0}) - 4$ Replace the variables.

$0 \leq \mathbf{-4}$ False. Simplify.

Since (0, 0) is not a solution of the inequality, the test point is not in the solution region. Shade the solution region on the side of the boundary line that does not include $(0, 0)$.

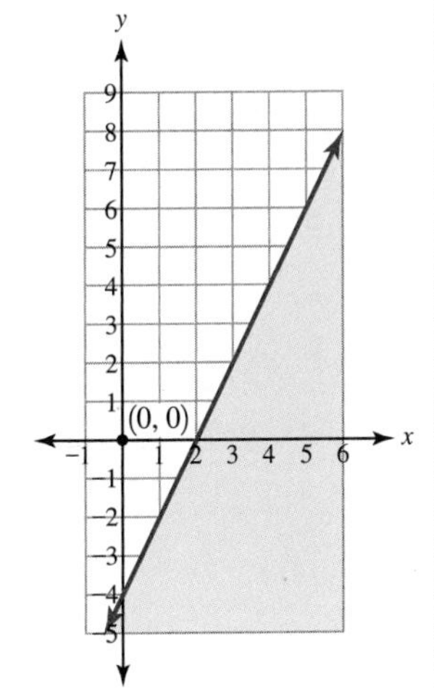

Graph: $8x + 5y \geq 40$

The solid boundary line is $8x + 5y = 40$. Since this equation is in standard form, graph the intercepts (Section 3.2) and one other point.

x	y
0	8
5	0
2	$\frac{24}{5}$

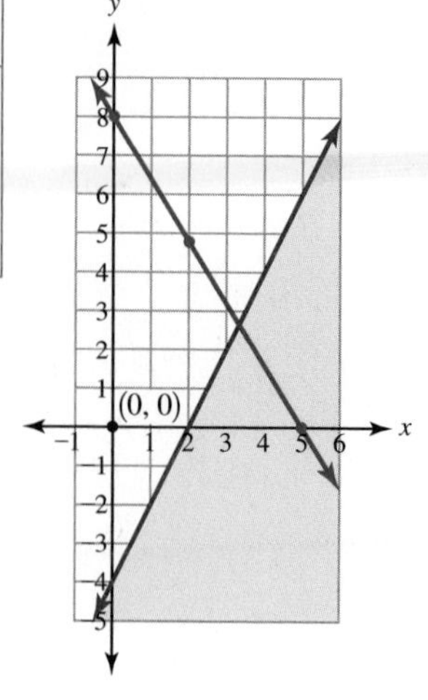

To identify the solution region, choose a test point, (0, 0).

$$8x + 5y \geq 40$$

$$8(\mathbf{0}) + 5(\mathbf{0}) \geq 40 \quad \text{Replace the variables.}$$

$$\mathbf{0} \geq 40 \quad \text{False.} \quad \text{Simplify.}$$

Since (0, 0) is not a solution of the inequality, the test point is not in the solution region. Shade the solution region on the side of the boundary line that does not include (0, 0). Label the solution region of the system, the region that is shaded by both inequalities.

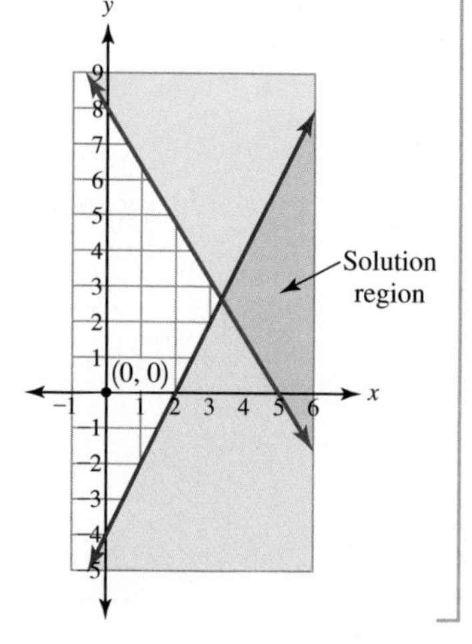

A test point can be any point that is not on a boundary line. We often choose (0, 0) as the test point because it results in easier arithmetic.

EXAMPLE 2 Graph $\begin{aligned} y &\leq -3x + 7 \\ x &\geq 1 \\ y &\geq 2 \end{aligned}$. Label the solution region.

SOLUTION ▶ Graph: $y \leq -3x + 7$

The solid boundary line is $y = 3x - 7$. Use slope-intercept graphing; the y-intercept is (0, 7), and the slope is -3.

Choose a test point: (0, 0)

$$y \leq -3x + 7$$

$$\mathbf{0} \leq -3(\mathbf{0}) + 7 \quad \text{Replace the variables.}$$

$$0 \leq \mathbf{7} \quad \text{True.} \quad \text{Simplify.}$$

Since (0, 0) is a solution of the inequality, the test point is in the solution region. Shade the solution region on the side of the boundary line that includes (0, 0).

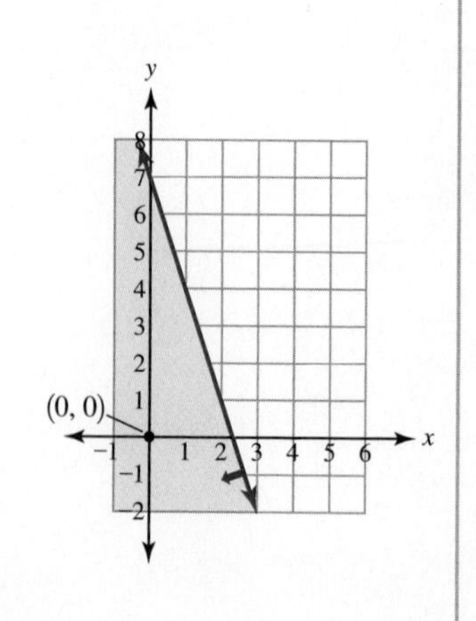

Graph: $x \geq 1$

The solid vertical boundary line is $x = 1$.

Choose a test point: $(0, 0)$

$x \geq 1$

$\mathbf{0} \geq 1$ False. Replace the variable; simplify.

Since $(0, 0)$ is not a solution of the inequality, the test point is not in the solution region. Shade the solution region on the side of the boundary line that does not include $(0, 0)$.

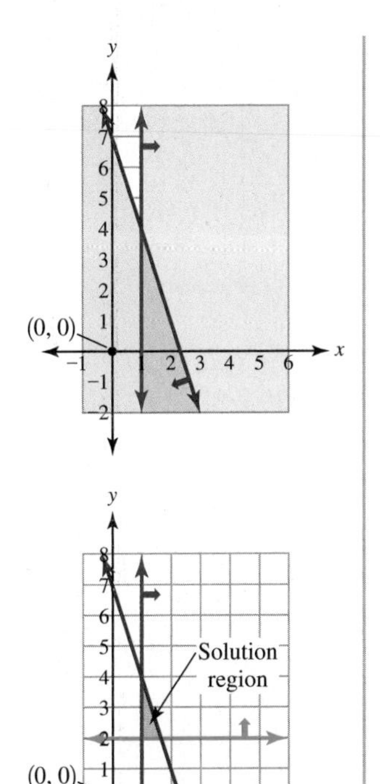

Graph: $y \geq 2$

The solid horizontal boundary line is $y = 2$.

Choose a test point: $(0, 0)$

$y \geq 2$

$\mathbf{0} \geq 2$ False. Replace the variable; simplify.

Since $(0, 0)$ is not a solution of the inequality, the test point is not in the solution region. Shade the solution region on the side of the boundary line that does not include $(0, 0)$. Label the solution region of the system.

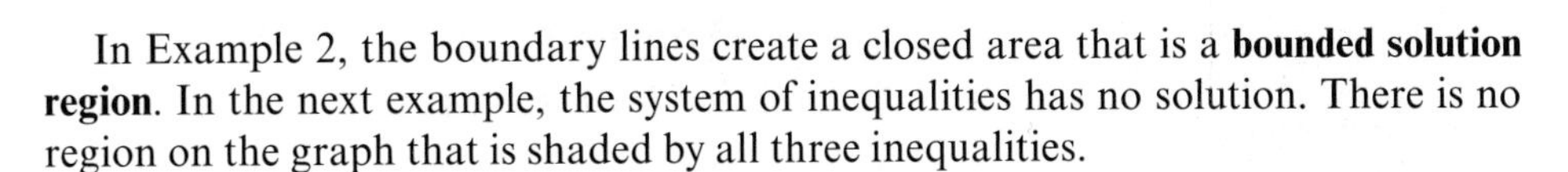

In Example 2, the boundary lines create a closed area that is a **bounded solution region**. In the next example, the system of inequalities has no solution. There is no region on the graph that is shaded by all three inequalities.

EXAMPLE 3 Graph $\begin{matrix} y > x \\ y < 3x - 4 \\ y < -2 \end{matrix}$. Label the solution region.

SOLUTION ▶ Graph: $y > x$

The dashed boundary line is $y = x$. Use slope-intercept graphing; the y-intercept is $(0, 0)$, and the slope is 1. Since $(0, 0)$ is on the boundary line, it cannot be a test point.

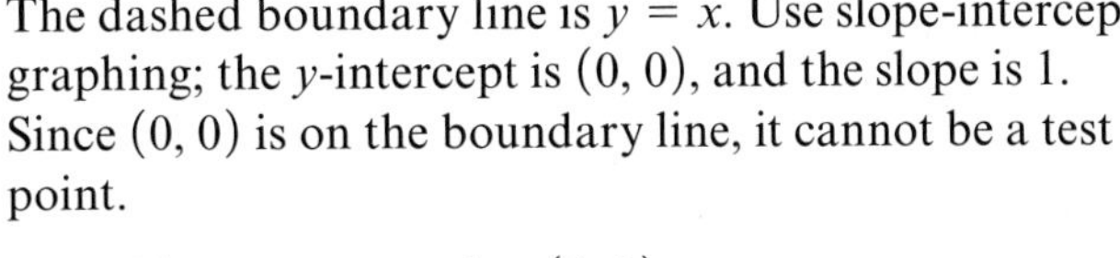

Choose a test point: $(1, 2)$

$y > x$

$\mathbf{2} > \mathbf{1}$ True. Replace the variables; simplify.

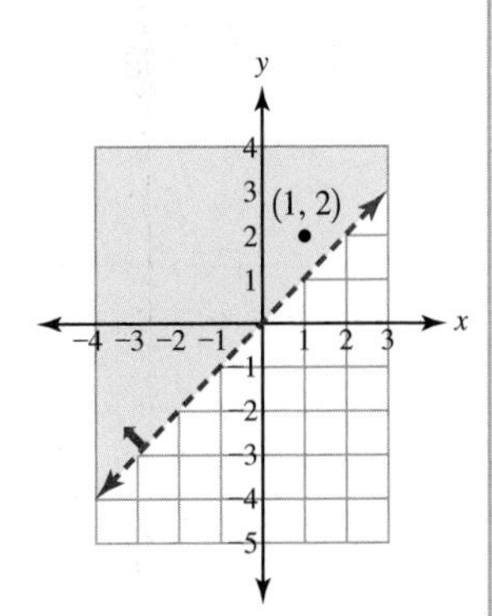

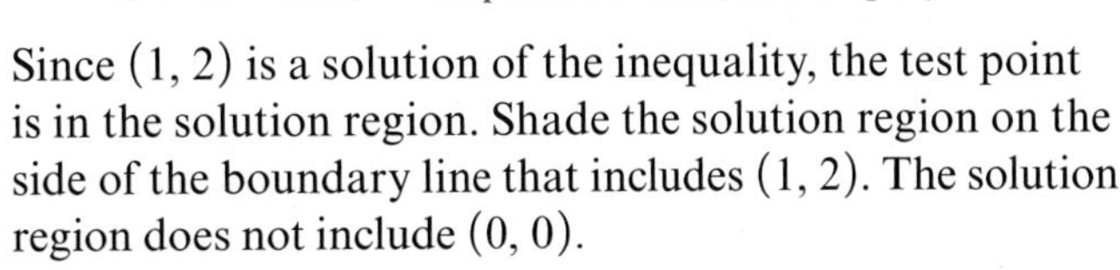

Since $(1, 2)$ is a solution of the inequality, the test point is in the solution region. Shade the solution region on the side of the boundary line that includes $(1, 2)$. The solution region does not include $(0, 0)$.

Graph: $y < 3x - 4$

The dashed boundary line is $y = 3x - 4$. Use slope-intercept graphing; the y-intercept is $(0, -4)$, and the slope is $\frac{3}{1}$.

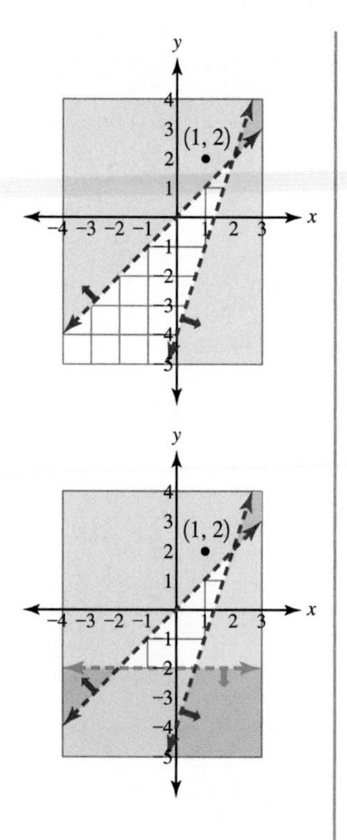

Choose a test point: $(1, 2)$

$y < 3x - 4$

$\mathbf{2} < 3(\mathbf{1}) - 4$ Replace the variables.

$2 < \mathbf{-1}$ False. Simplify.

Since $(1, 2)$ is not a solution of the inequality, the test point is not in the solution region. Shade the solution region on the side of the boundary line that does not include $(1, 2)$.

Graph: $y < -2$

The horizontal dashed boundary line is $y = -2$.

Choose a test point: $(1, 2)$

$y < -2$

$\mathbf{2} < \mathbf{-2}$ False. Replace the variable; simplify.

Since $(1, 2)$ is not a solution of the inequality, the test point is not in the solution region. Shade the solution region on the side of the boundary line that does not include $(1, 2)$.

There is no shaded region that is common to all three inequalities. This system of inequalities has no solution.

Practice Problems

For problems 1–5, graph. Label the solution region.

1. $y < -3x + 4$
$y > 2x - 5$

2. $x + y \geq 5$
$y \leq 2x - 3$

3. $2x + y < 6$
$x > 1$

4. $y \geq \frac{1}{2}x + 4$
$y \leq \frac{1}{2}x - 3$

5. $x + y > 0$
$y < -3x + 8$

For problems 6–7, graph. Label the solution region.

6. $x + y \geq 6$
$x \geq 0$
$y \geq 0$

7. $y \geq \frac{1}{2}x - 2$
$y \leq -x + 4$
$x \geq -2$

Identifying Solutions

Given the graph of a system of linear inequalities, it may be clear that a given point is not in the shaded solution region. In Figure 1, $(4, 3)$ is clearly not in the shaded solution region and is not a solution of the system. Similarly, $(1, 2)$ is clearly in the shaded solution region and is a solution of the system. To determine whether a point that is close to a boundary line is a solution of the system, replace the variables in each inequality with the coordinates of the point. If all of the inequalities are true, the point is a solution of the system.

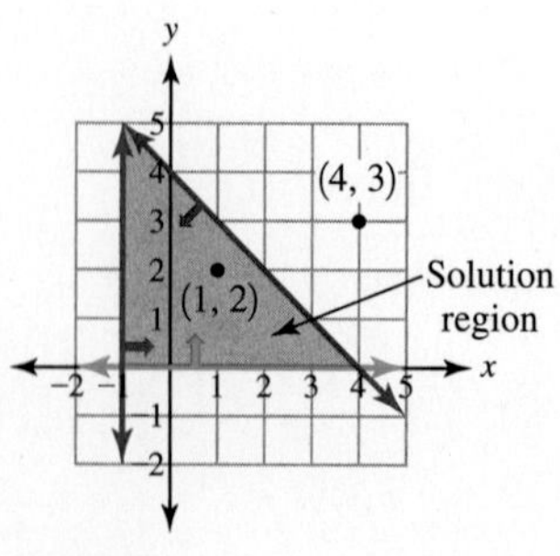

Figure 1

EXAMPLE 4 The graph represents $\begin{aligned} y &\le -\frac{4}{5}x + 2 \\ x &\ge 0 \\ y &\ge -3 \end{aligned}$. Determine whether $(6, -3)$ is a solution of this system.

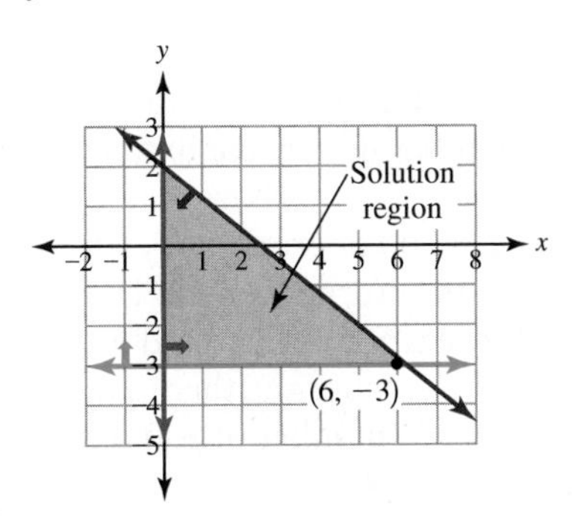

SOLUTION ▸ Check $(6, -3)$ in each inequality.

$y \le -\frac{4}{5}x + 2$ — Choose an inequality in the system.

$\mathbf{-3} \le -\frac{4}{5}(\mathbf{6}) + 2$ — Replace the variables.

$-3 \le -\mathbf{\frac{24}{5}} + 2$ — Simplify.

$-\frac{3}{1}\cdot\frac{5}{5} \le -\frac{24}{5} + \frac{2}{1}\cdot\frac{5}{5}$ — The LCD is 5; multiply by a fraction equal to 1.

$-\mathbf{\frac{15}{5}} \le -\mathbf{\frac{14}{5}}$ True. — Simplify.

$x \ge 0$ — Choose another inequality in the system.

$\mathbf{6} \ge 0$ True. — Replace the variable; simplify.

$y \ge -3$ — Choose another inequality in the system.

$\mathbf{-3} \ge -3$ True. — Replace the variable; simplify.

Since $(6, -3)$ is a solution of each inequality, it is a solution of the system.

Practice Problems

For problems 8–9, determine whether the ordered pair is a solution of the system.

8. $(1, -5)$; $\begin{aligned} y &\le -6x + 10 \\ -4x + y &\ge -9 \\ y &\le 3 \end{aligned}$

9. $(5, 7)$; $\begin{aligned} x + 2y &< 18 \\ x &\ge 2 \\ y &\ge 3 \end{aligned}$

Constraints

An inequality in two variables can represent a constraint. In Section 3.6, we assumed that the variables represented positive numbers. In the next example, we will instead use an additional constraint, $x \ge 0$, to ensure that the values for x are positive.

EXAMPLE 5 The maximum tax-free amount of gifts from an individual to another individual is \$13,000. Let x = amount of gifts. Write two inequalities that represent the constraints. (*Source:* www.irs.gov, Nov. 21, 2011)

SOLUTION ▶ Since the amount of a gift must be less than or equal to \$13,000, a constraint is $x \le \$13{,}000$. Since a gift cannot be less than \$0, another constraint is $x \ge \$0$.

In the next example, the constraints represent the relationship between two unknowns: the number of new computers and the number of software licenses.

EXAMPLE 6 A professor has \$25,000 in grant money to buy and install new computers and software in the math tutoring center. The cost of each computer is \$1040. The cost of a software license is \$850. The total installation cost is \$1500. There must be at least as many new computers as there are software licenses. Let x = number of computers, and let y = number of software licenses. Write four inequalities that represent the constraints.

SOLUTION ▶ Since the budget must be less than or equal to \$25,000, a constraint is $\left(\frac{\$1040}{1 \text{ computer}}\right)x + \left(\frac{\$850}{1 \text{ license}}\right)y + \$1500 \le \$25{,}000$. Since there must be at least as many new computers as there are software licenses, a constraint is $x \ge y$. Since there must be at least 0 computers and at least 0 software licenses, other constraints are $x \ge 0$ and $y \ge 0$.

We graph a system of constraints just as we graph a system of inequalities. Since *feasible* means *possible*, the solution region of a system of constraints is called the **feasible region**.

EXAMPLE 7 A student can spend no more than \$225 this month on food and gas. She must spend at least \$25 per month on gas. Let x = amount spent on food, and let y = amount spent on gas.

(a) Write three inequalities that represent the constraints.

SOLUTION ▶

$x + y \le \$225$ The total spent must be less than or equal to \$225.

$x \ge \$0$ The minimum amount spent on food is \$0.

$y \ge \$25$ The minimum amount spent on gas is \$25.

(b) Graph the inequalities that represent the constraints. Label the feasible region.

▶ Graph: $x + y \le \$225$

The solid boundary line is $x + y \le \$225$. Since this equation is in standard form, graph the intercepts and one other point.

x	y
0	\$225
\$225	0
\$100	\$125

To identify the feasible region, choose a test point, (\$100, \$50).

$x + y \le \$225$

\$100 + **\$50** $\le \$225$ Replace the variables.

\$150 $\le \$225$ True. Simplify.

Since (\$100, \$50) is a solution of the inequality, the test point is in the feasible region. Shade the feasible region on the side of the boundary line that includes (\$100, \$50).

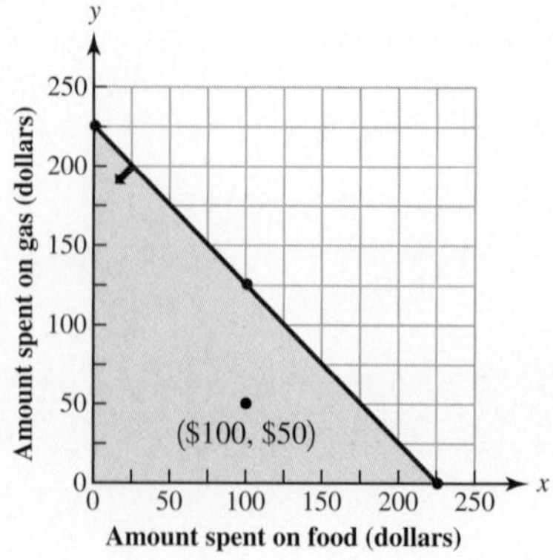

Graph: $x \geq \$0$

The solid vertical boundary line is $x = 0$. To identify the feasible region, choose a test point, (\$100, \$50).

$x \geq \$0$

$\mathbf{\$100} \geq \$0$ True. Replace the variable; simplify.

Since (\$100, \$50) is a solution of the inequality, the test point is in the feasible region. Shade the feasible region on the side of the boundary line that includes (\$100, \$50).

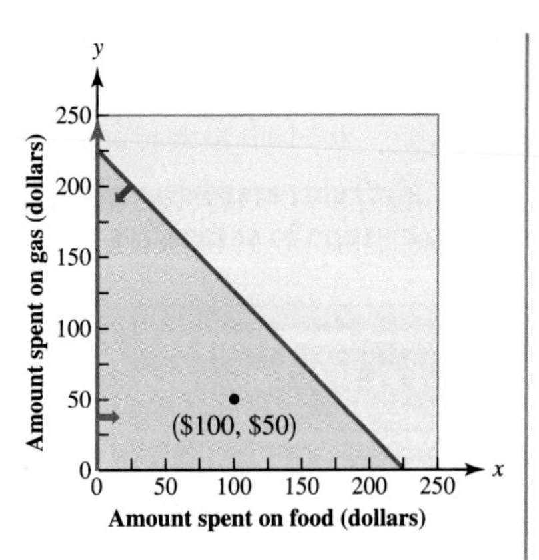

Graph: $y \geq \$25$

The solid horizontal boundary line is $y = 25$. To identify the feasible region, choose a test point, (\$100, \$50).

$y \geq \$25$

$\mathbf{\$50} \geq \$25$ True. Replace the variable; simplify.

Since (\$100, \$50) is a solution of the inequality, the test point is in the feasible region. Shade the feasible region on the side of the boundary line that includes (\$100, \$50). Label the feasible region.

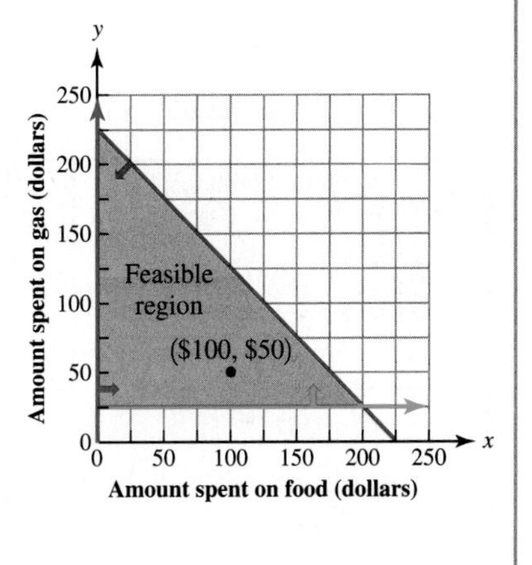

Writing constraints and identifying the feasible region are part of *optimization*, an important process used in making decisions in business.

Practice Problems

10. The minimum weight of a blue marlin in the 2011 North Carolina Saltwater Fishing Tournament is 400 lb. If x = weight of a blue marlin, write an inequality that represents the constraint. (*Source:* portal.ncdenr.org)

11. A caterer is preparing a maximum of 30 lunch boxes that either include gluten or do not include gluten. If x = lunch boxes with gluten and y = lunch boxes without gluten, write three inequalities that represent the constraints.

12. It takes 10 min to wash a window and 15 min to caulk a window. A homeowner has no more than 180 min for washing and caulking. At least 8 windows need to be caulked. At least 4 windows need to be washed. If x = number of windows washed and y = number of windows caulked,

a. write three inequalities that represent the constraints.

b. graph the system of constraints and label the feasible region.

Using Technology: Graphing Systems of Linear Inequalities

To graph an inequality, graph the boundary line, use a test point to find out whether the shading is "below" or "above" the line, move the cursor to the left of the equals sign, and choose the correct shading icon.

EXAMPLE 8 Graph $\begin{aligned} y &\le 3x \\ y &\le -2x + 10 \\ y &\ge x - 8 \end{aligned}$. Identify the vertices of the solution region.

To graph $y \le 3x$, type the equation of the boundary line, $y = 3x$, on the Y= screen. Press ENTER. Press GRAPH. Use a test point to determine the shaded solution region.

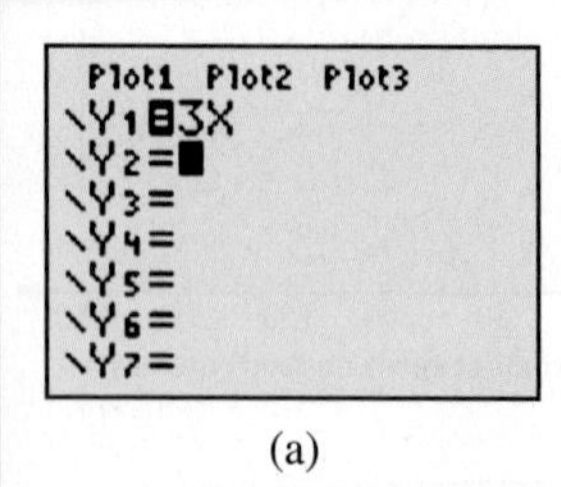

(a)

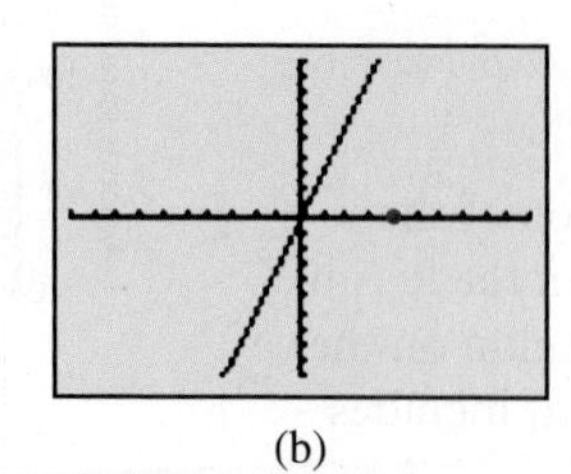

(b)

Choose a test point: (4, 0)

$y \le 3x$

$0 \le 3(4)$

$0 \le 12$ True.

Since (4, 0) is a solution of $y \le 3x$, the shading includes (4, 0). This point is "below" the boundary line. To shade, move the cursor to the left of the equals sign. Press ENTER repeatedly until the "shaded below" icon appears. Press GRAPH.

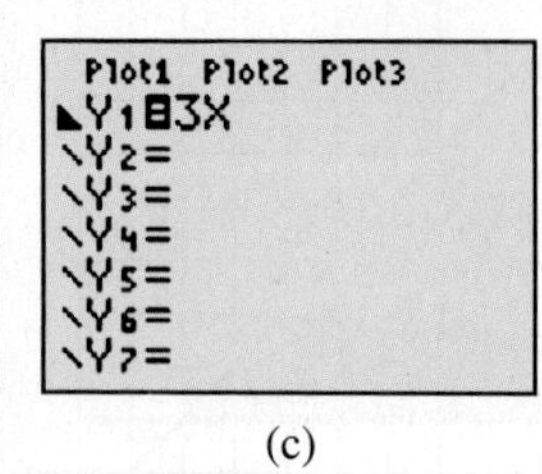

(c)

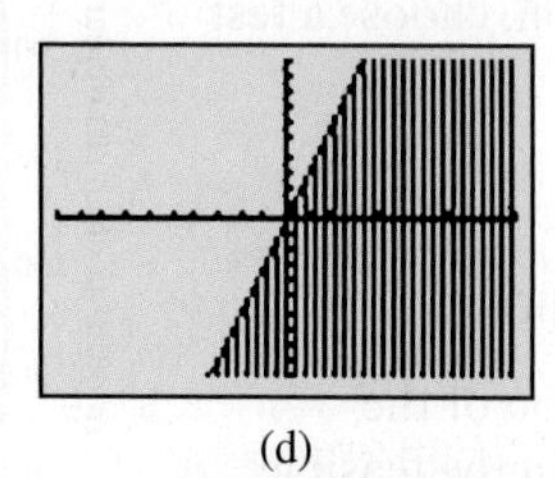

(d)

The shading includes the test point, (4, 0).

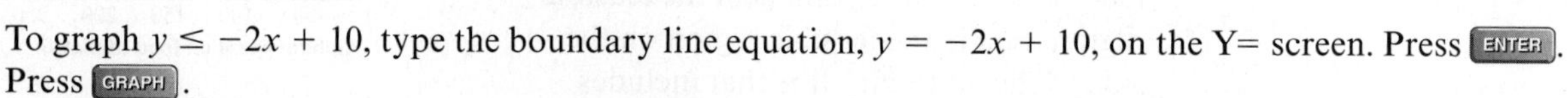

To graph $y \le -2x + 10$, type the boundary line equation, $y = -2x + 10$, on the Y= screen. Press ENTER. Press GRAPH.

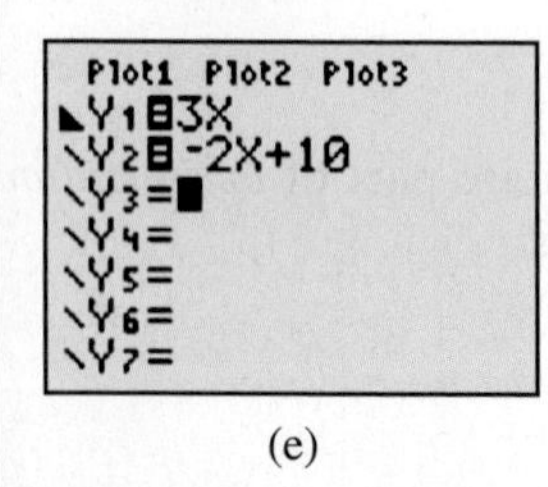

(e)

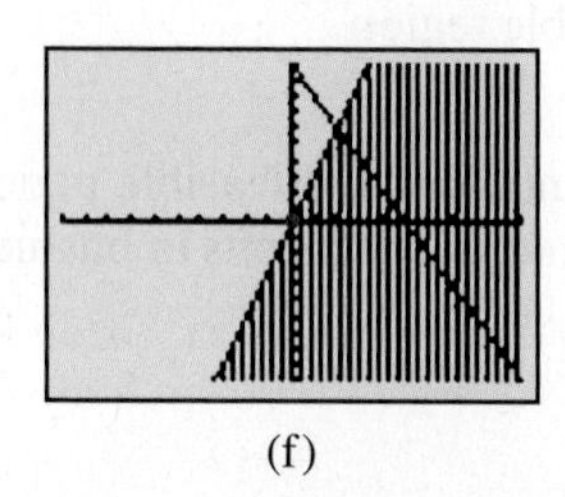

(f)

Choose a test point: (0, 0)

$y \le -2x + 10$

$0 \le -2(0) + 10$

$0 \le 10$ True.

Since (0, 0) is a solution of $y \le -2x + 10$, the shading includes (0, 0). This point is "below" the boundary line. To shade, move the cursor to the left of the equals sign. Press ENTER repeatedly until the "shaded below" icon appears. Press GRAPH.

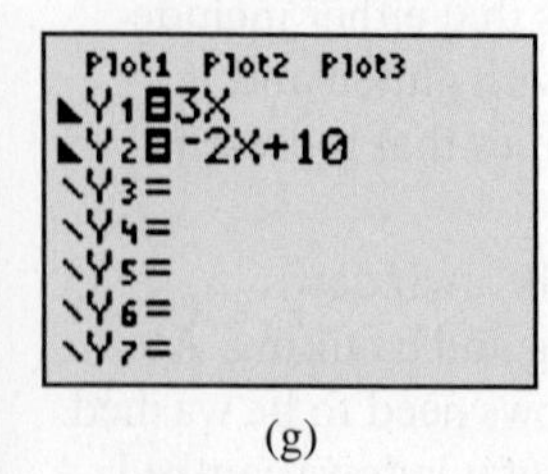

(g)

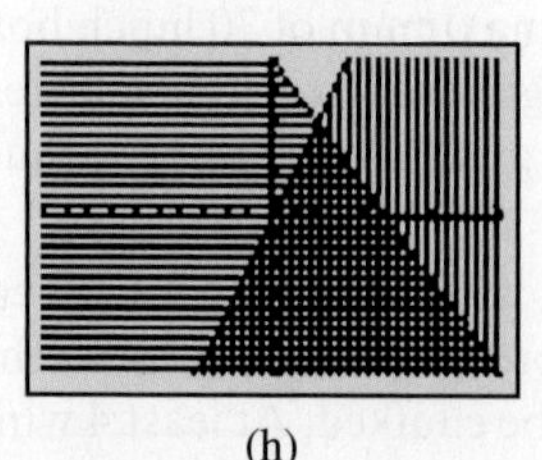

(h)

The shading includes the test point. The area with the shading from both inequalities is the solution region.

To graph $y \ge x - 8$, type the boundary line equation, $y = x - 8$, on the Y= screen. Press ENTER. Press GRAPH. Use a test point to determine the shaded solution region.

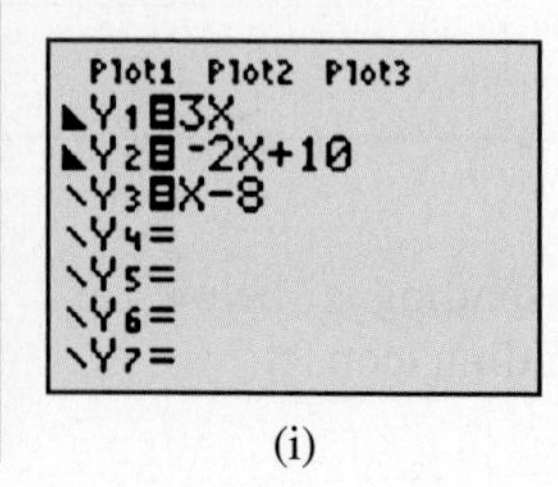

(i)

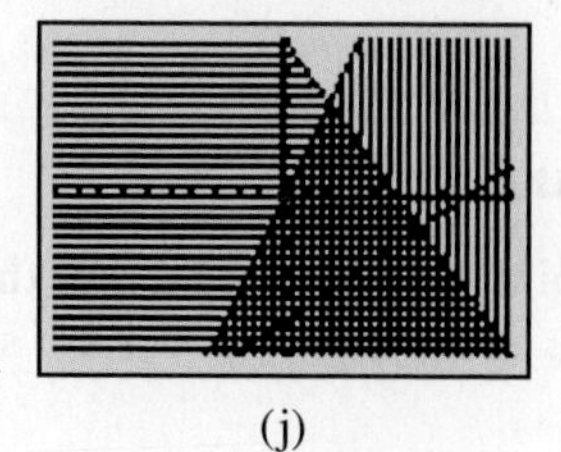

(j)

Choose a test point: (0, 0)

$y \ge x - 8$

$0 \ge 0 - 8$

$0 \ge -8$ True.

Since $(0, 0)$ is a solution of $y \geq x - 8$, the shading includes $(0, 0)$. This point is "above" the boundary line. To shade, move the cursor to the left of the equals sign. Press ENTER repeatedly until the "shaded above" icon appears. Press GRAPH.

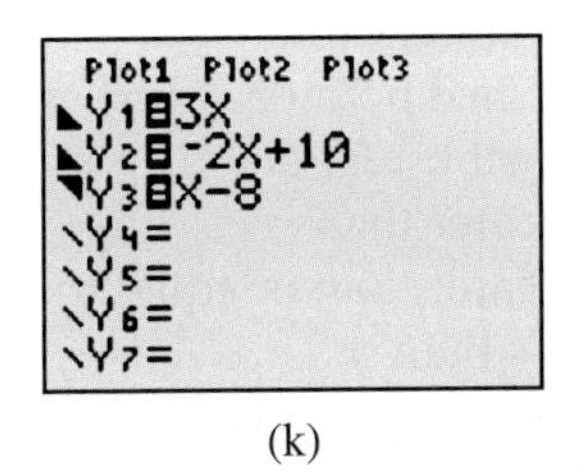

(k)

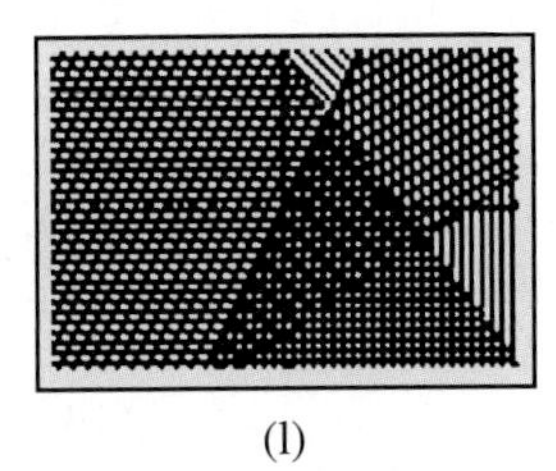

(l)

The area that is shaded by all three inequalities is the solution region.

To identify the vertices, change the window so that the lower left vertex is visible. Press WINDOW. Change the Y_{min} value to -15. Press GRAPH.

WINDOW
Xmin=-10
Xmax=10
Xscl=1
Ymin=-10
Ymax=10
Yscl=1
Xres=1

(m)

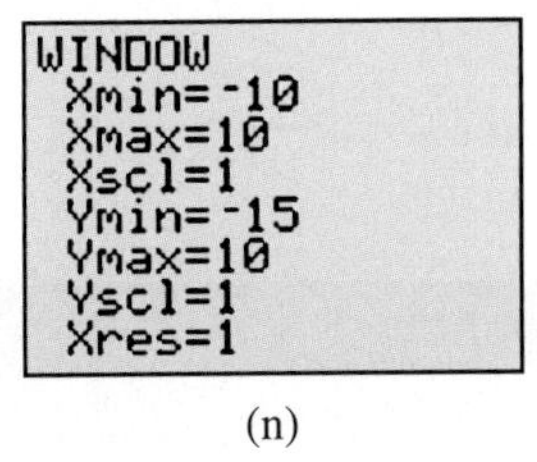

(n)

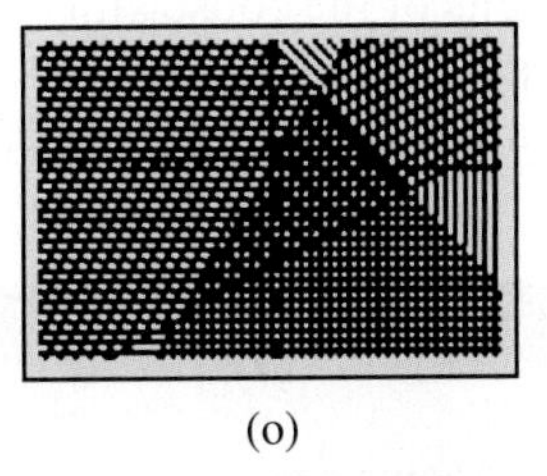

(o)

To find a vertex, go to the CALC screen. Choose 5: intersect. Press ENTER.

CALCULATE
1:value
2:zero
3:minimum
4:maximum
5:intersect
6:dy/dx
7:∫f(x)dx

(p)

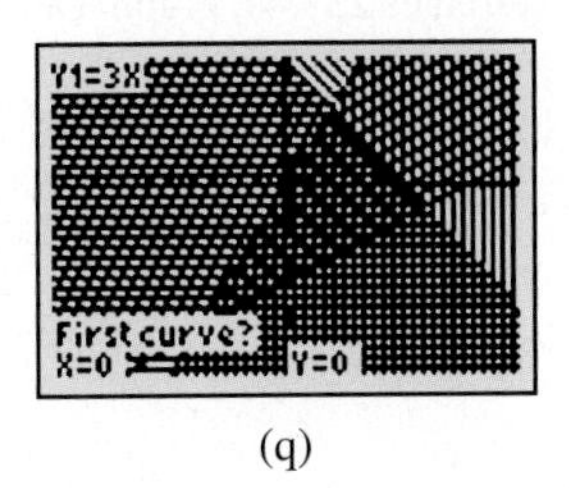

(q)

The calculator asks, "First curve?" The upper left-hand corner tells us that the cursor is on the line $y = 3x$. Press ENTER.

The calculator asks, "Second curve?" The cursor is on the line $y = -2x + 10$. Press ENTER. The calculator asks, "Guess?" Move the cursor to the intersection point of these two lines. (The vertex may be difficult to see; it will be blinking.) Press ENTER. This vertex is (2, 6).

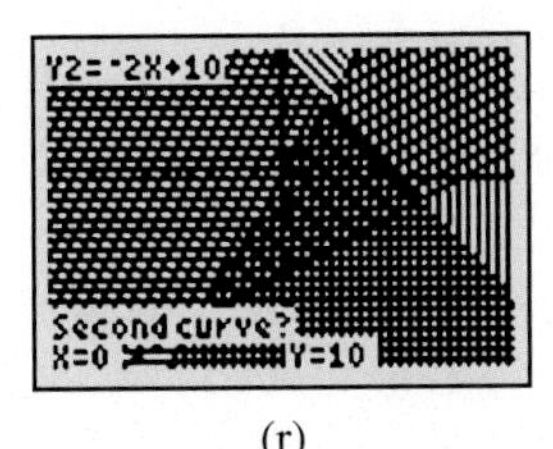

(r)

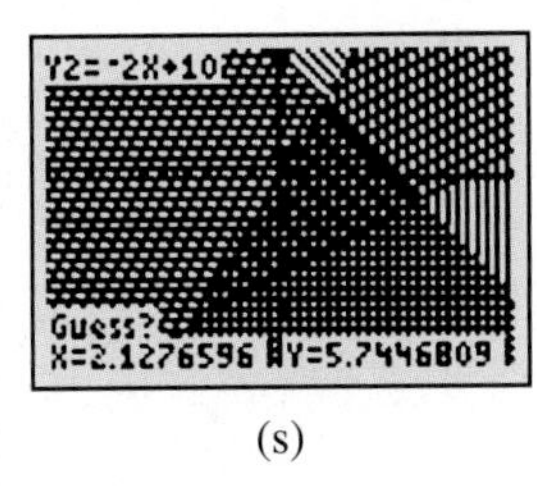

(s)

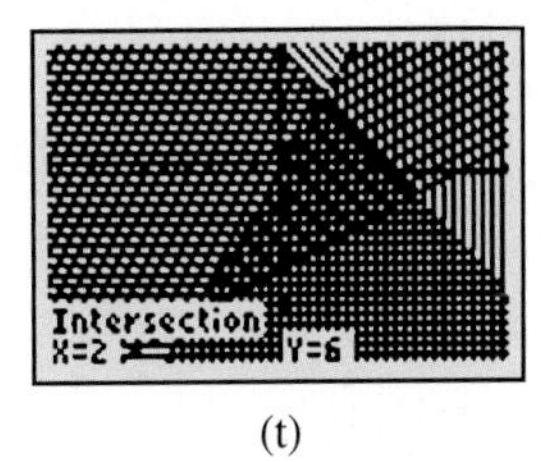

(t)

To find the other vertices, repeat this process for the other intersection points.

Practice Problems 13. A system of inequalities is $\begin{array}{l} y \leq 4x \\ y \geq x - 5 \\ y \leq -2x + 12 \end{array}$. Graph. Choose a window that shows all the vertices of the solution region. Sketch the graph; describe the window. Label the solution region.

4.5 VOCABULARY PRACTICE

Match the term with its description.

1. $<$
2. $>$
3. $\leq$
4. $\geq$
5. We use this as we determine the shaded area on the graph of a linear inequality.
6. This line marks the edge of the solution region on the graph of a linear inequality.
7. On the graph of a linear inequality, this area includes all the points that represent solutions of the inequality.
8. On the graph of a constraint, this area includes all the points that represent the ordered pairs that are within the limits of the constraint.
9. A boundary line used to represent a strict inequality
10. A boundary line used to represent an inequality that is not strict

A. boundary line
B. dashed boundary line
C. feasible region
D. greater than
E. greater than or equal to
F. less than
G. less than or equal to
H. solid boundary line
I. solution region
J. test point

4.5 Exercises

Follow your instructor's guidelines for showing your work.

1. Describe when the boundary line of the graph of a linear inequality is solid and when it is dashed.

2. Explain how to use a test point to identify the solution region of the graph of a linear inequality.

For exercises 3–22, graph. Label the solution region.

3. $x + y \leq 8$, $x \geq 2$

4. $x + y \leq 10$, $x \geq 3$

5. $2x + 3y > 12$, $y < 2x + 8$

6. $3x + 2y > 18$, $y < 4x + 5$

7. $2x + 3y < 12$, $y < 2x + 8$

8. $3x + 2y < 18$, $y < 4x + 5$

9. $2x + 3y < 12$, $y > 2x + 8$

10. $3x + 2y < 18$, $y > 4x + 5$

11. $y \geq 4x - 6$, $y \geq -4x - 6$

12. $y \geq 5x - 7$, $y \geq -5x - 7$

13. $y \leq x$, $y > 6$

14. $y \leq -x$, $y > 2$

15. $2x + 5y < 20$, $2x + 5y > 10$

16. $2x + 3y > 12$, $2x + 3y < 18$

17. $x \geq 1$, $x \leq 9$, $y \geq 3$, $y \leq 11$

18. $x \geq 2$, $x \leq 8$, $y \geq 1$, $y \leq 15$

19. $y \geq -\frac{1}{5}x$, $y \leq \frac{1}{5}x$, $x < 7$

20. $y \geq -\frac{1}{3}x$, $y \leq \frac{1}{3}x$, $x < 8$

21. $y \leq x$, $y \leq -x$, $y \geq -8$

22. $y \geq x$, $y \geq -x$, $y \leq 6$

For exercises 23–44, graph. Label the solution region.

23. $x + 3y \leq 9$, $x \geq 0$, $y \geq 0$

24. $x + 2y \leq 16$, $x \geq 0$, $y \geq 0$

25. $x + 3y \leq 9$, $x \geq 4$, $y \geq 1$

26. $x + 2y \leq 16$, $x \geq 3$, $y \geq 2$

27. $y \leq \frac{3}{4}x + 1$, $y \leq -\frac{5}{6}x + 8$, $x \geq 1$, $y \geq 0$

28. $y \leq \frac{4}{5}x + 2$, $y \leq -\frac{2}{3}x + 7$, $x \geq 0$, $y \geq 1$

29. $y \geq \frac{3}{4}x + 1$, $y \leq -\frac{5}{6}x + 8$, $x \geq 0$

30. $y \geq \frac{4}{5}x + 2$, $y \leq -\frac{2}{3}x + 7$, $x \geq 0$

31. $-3x + 5y \leq 15$, $-3x + 5y \geq -10$, $x \geq 0$, $x \leq 10$

32. $-3x + 4y \leq 20$, $-3x + 4y \geq 4$, $x \geq 4$, $x \leq 8$

33. $-3x + 5y \leq 15$, $-3x + 5y \geq -10$, $y \geq 0$, $y \leq 5$

34. $-3x + 4y \leq 20$, $-3x + 4y \geq 4$, $y \geq 0$, $y \leq 4$

35. $1200x + 500y \geq 20{,}000$, $x + y \leq 40$, $x \geq 0$, $y \geq 0$

36. $1500x + 600y \geq 18{,}000$, $x + y \leq 30$, $x \geq 0$, $y \geq 0$

37. $1200x + 500y \geq 20{,}000$, $x + y \leq 40$, $y \leq x$, $x \geq 0$, $y \geq 0$

38. $1500x + 600y \geq 18{,}000$, $x + y \leq 30$, $y \leq x$, $x \geq 0$, $y \geq 0$

39. $y \geq 2x$
$x + y \leq 50$
$y \leq 5x + 15$
$x \geq 0$

40. $y \geq 3x$
$x + y \leq 60$
$y \leq 6x + 10$
$x \geq 0$

41. $y \leq -5x + 40$
$y \geq -5x + 15$
$y \leq x$
$y \geq 0$

42. $y \leq -10x + 80$
$y \geq -10x + 30$
$y \leq x$
$y \geq 0$

43. $x \geq 10$
$x \leq 100$
$y \geq 20$
$y \leq 80$

44. $x \geq 60$
$x \leq 150$
$y \geq 10$
$y \leq 100$

For exercises 45–48, determine whether the ordered pair is a solution of the system.

45. $(2, 4)$; $2x + y \leq 30$
$y \geq 2x$
$x \geq 1$

46. $(5, 46)$; $6x + y \leq 80$
$y \geq 9x$
$x \geq 2$

47. $(2, 25)$;
$-2x + 9y \geq -90$
$15x + y \geq 50$
$y \leq -4x + 30$

48. $(24, 12)$;
$-7x + 8y \geq -200$
$x + 2y \geq 57$
$y \geq -6x + 45$

For exercises 49–50,
(a) Write five inequalities that represent the constraints.
(b) Graph the inequalities that represent the constraints. Label the feasible region.

49. Dependent students in their first year of college can receive up to \$5500 in Stafford loans. Of this amount, a maximum of \$3500 can be federally subsidized Stafford loans, and a maximum of \$2000 can be unsubsidized Stafford loans. Let x = amount in federally subsidized Stafford loans, and let y = amount in unsubsidized Stafford loans. (*Source:* www.staffordloan.com, 2011)

50. Independent students in their first year of college can receive up to \$9500 in Stafford loans. Of this amount, a maximum of \$3500 can be federally subsidized Stafford loans, and a maximum of \$6000 can be unsubsidized Stafford loans. Let x = amount in federally subsidized Stafford loans, and let y = amount in unsubsidized Stafford loans.

For exercises 51–52,
(a) Write five inequalities that represent the constraints.
(b) Graph the inequalities that represent the constraints. Label the feasible region.

51. Of the protein in beef cattle feed, 30 to 40 percent is available bypass protein, and 60 to 70 percent is rumen soluble protein. Let x = percent of protein that is rumen soluble protein, and let y = percent of protein that is available bypass protein. (*Source:* www.omafra.gov.on.ca, Jan. 2010)

52. The dry matter of dairy cow feed contains both crude protein and carbohydrates. Crude protein should be 14 to 19 percent of the feed. Carbohydrates should be 65 to 75 percent of the feed. Let x = percent of feed that is crude protein, and let y = percent of feed that is carbohydrate. (*Source:* Grant, Rick J., "G91-1027 Protein and Carbohydrate Nutrition of High Producing Dairy Cows," 1991)

For exercises 53–54,
(a) Write three inequalities that represent the constraints.
(b) Graph the inequalities that represent the constraints. Label the feasible region.

53. A bird feed mixture of sunflower seeds and safflower seeds can weigh no more than 50 lb and must contain at least 35 lb of sunflower seeds. Let x = amount of sunflower seeds, and let y = amount of safflower seeds.

54. The volume of wine in a barrel is no more than 60 gal. To be entered in a competition, the wine must contain at least 37.5 gal of merlot grape juice. Let x = amount of merlot grape juice, and let y = amount of other grape juice.

For exercises 55–58,
(a) Write four inequalities that represent the constraints.
(b) Graph the inequalities that represent the constraints. Label the feasible region.

55. An investor will put a maximum of \$25,500 in foreign investments and domestic investments with at least four times as much in domestic investments as in foreign investments and a minimum of \$2000 in foreign investments. Let x = amount in foreign investments, and let y = amount of domestic investments.

56. An investor will put a maximum of \$60,000 in growth funds and income funds with at least three times as much in income funds as in growth funds and a minimum of \$5000 in growth funds. Let x = amount in growth funds, and let y = amount of income funds.

57. The maximum amount that can be spent on almonds and peanuts for a reception is \$120. Each can of almonds weighs 6 oz and costs \$2.50. Each jar of peanuts weighs 16 oz and costs \$2. There should be at least three times as many cans of almonds as jars of peanuts, and there should be at least five jars of peanuts. Let x = cans of roasted almonds, and let y = jars of roasted peanuts.

58. The maximum amount that can be spent on sliced ham and sliced turkey for a wedding reception is \$500. Each pound of ham costs \$5. Each pound of turkey costs \$6. There should be no more than two times the amount of turkey as the amount of ham and there should be at least 20 lb of turkey. Let x = pounds of ham, and let y = pounds of turkey.

Problem Solving: Practice and Review

Follow your instructor's guidelines for using the five steps as outlined in Section 1.3, p. 55.

59. A gutter company charges a builder \$1.75 per foot to make gutters. If the builder buys a seamless gutter machine for \$7000, the cost to make gutters is \$0.89 per foot. Find the length of gutter at which the builder's cost for paying for and using the seamless gutter machine is equal to the amount charged by the gutter company. Round to the nearest whole number.

60. If the population in Colorado is about 5,100,000 people, find the number of people who had no health insurance. (*Source:* www.census.gov)

One of every six Coloradans has no health insurance. (*Source:* www.gazette.com, Nov. 15, 2011)

61. Find the percent increase in the composite price of a pound of coffee from 2009 to 2010. Round to the nearest percent. (*Source:* www.ico.org, Dec. 2011)

Annual Composite Coffee Prices on the N.Y. Market Colombian (Mild Arabicas) Price in U.S. Cents per Pound

2007	2008	2009	2010
126.74	145.85	180.87	223.76

62. The size of cylindrical cans is described by using two three-digit numbers. The first number describes the diameter, and the second number describes the height. The first digit in each number is the number of whole inches, and the second two digits are the number of sixteenths of an inch. For example, a 303 by 407 can has a diameter of $3\frac{3}{16}$ in. and is $4\frac{7}{16}$ in. high. The formula for the volume V of a cylinder is $V = \pi r^2 h$, where r is the radius and h is the height. Find the volume of a 200 by 503 beverage can. Round to the nearest whole number. (*Source:* www.ellenskitchen.com)

Technology

For exercises 63–66, graph. Sketch the graph; describe the window.

63. $y \geq -x - 5$
$y \leq 2x + 4$
$y \leq -4x + 9$

64. $y \leq \frac{3}{4}x + 5$
$y \geq 2x - 8$
$y \leq 4x + 5$

65. $y \leq x + 8$
$y \geq 4x + 8$
$y \geq 2$

66. $y \leq -2x + 10$
$y \leq 3x - 8$
$y \geq -6$

Find the Mistake

For exercises 67–70, the completed problem has one mistake.

(a) Describe the mistake in words, or copy down the whole problem and highlight or circle the mistake.

(b) Do the problem correctly.

67. Problem: Graph $\begin{matrix} y \leq -x + 9 \\ y \leq 2x - 8 \\ x \geq 0 \\ y \geq 0 \end{matrix}$. Label the solution region.

Incorrect Answer:

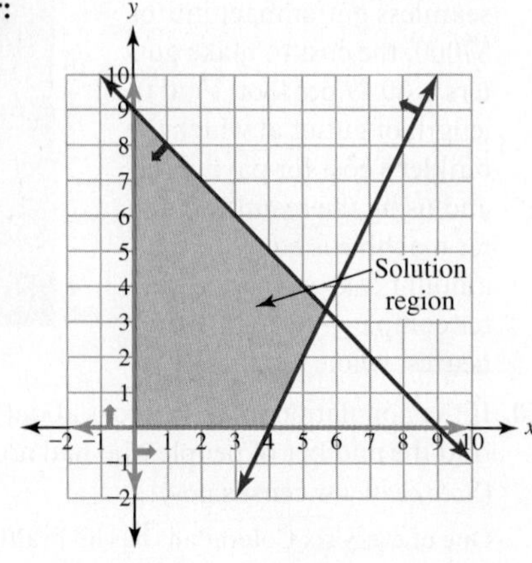

68. Problem: Graph $\begin{matrix} x + y \leq 6 \\ y \geq 2 \\ x \geq 1 \end{matrix}$. Label the solution region.

Incorrect Answer:

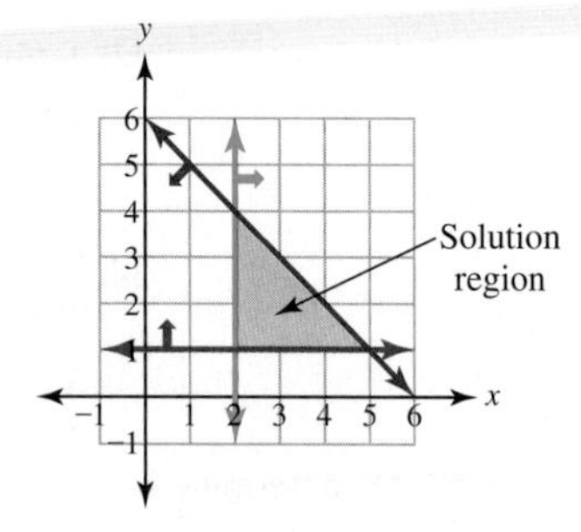

69. Problem: Graph $\begin{matrix} y \geq x \\ y \geq -x \end{matrix}$. Label the solution region.

Incorrect Answer:

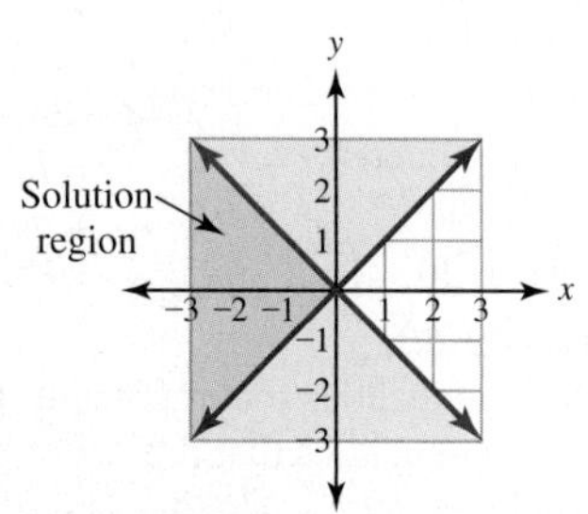

70. Problem: Graph $\begin{matrix} y \leq x \\ y \geq -x - 6 \\ y \leq 0 \\ x \leq 0 \end{matrix}$. Label the solution region.

Incorrect Answer:

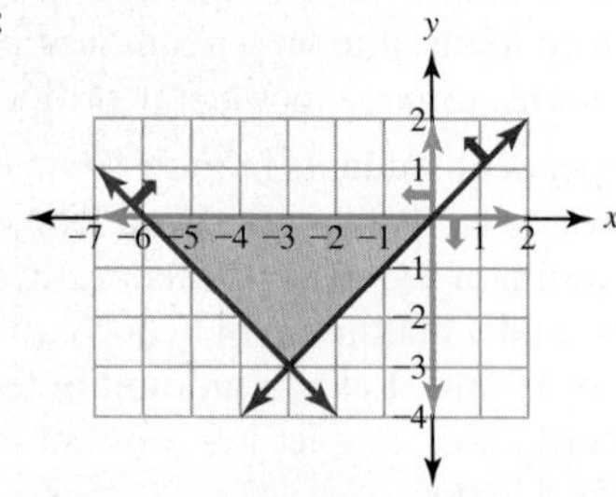

Review

71. Identify the coefficient in the expression $5x + 9$.

72. Identify the constant in the expression $8x - 2$.

73. Simplify: $5(3x - 2) - (7x + 1)$

74. Simplify: $5(3x - 2) - (7x - 1)$

SUCCESS IN COLLEGE MATHEMATICS

75. List the vocabulary words from the Vocabulary Practice in this section. Highlight the words that your instructor used when teaching about this section today in class, that you remember from reading the text, or that you had to use when completing an activity.

76. The first exercise in this section asks you to "describe" something. Identify the location in this section of the information needed to complete exercise 1.

Study Plan for Review of Chapter 4

SECTION 4.1 Systems of Linear Equations

Ask Yourself	Test Yourself	Help Yourself
Can I...	**Do 4.1 Review Exercises**	**See these Examples and Practice Problems**
identify the number of solutions of a system of equations, given its graph?	1–6, 15	Ex. 1–4, PP 1–12
solve a system of linear equations by graphing? if a system of equations has one solution, check the solution?	7–13	Ex. 1–4, PP 5–12
use set-builder notation to represent the solution of a system of equations that has infinitely many solutions?	14	Ex. 4–5, PP 10
use a system of equations and graphing to solve an application problem?	16–17	Ex. 6, PP 11–12

4.1 Review Exercises

For exercises 1–3, use the graph of the system of linear equations to identify the number of solutions.

1. 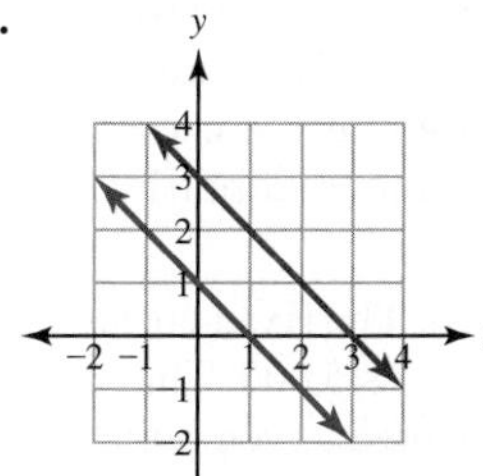

2. 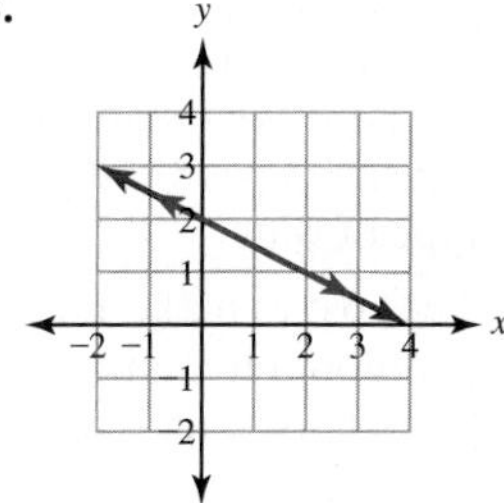

3. 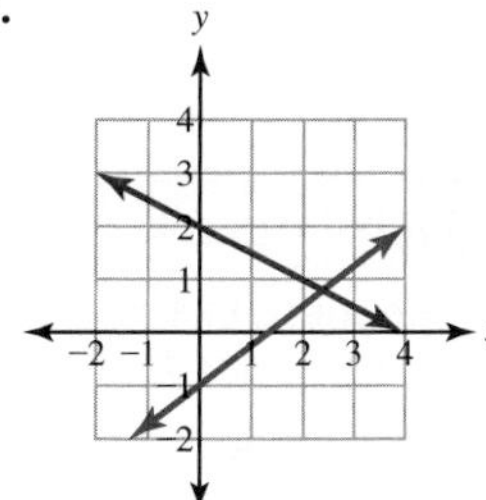

For exercises 4–6, the graph represents a system of linear equations. Use the graph to solve.

4. 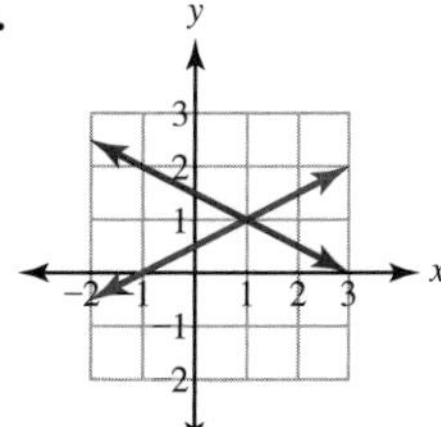

5. 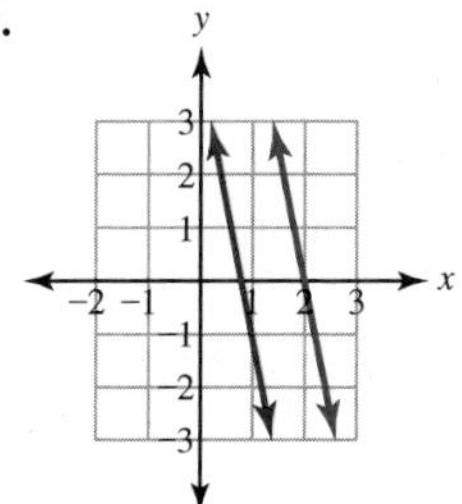

6. 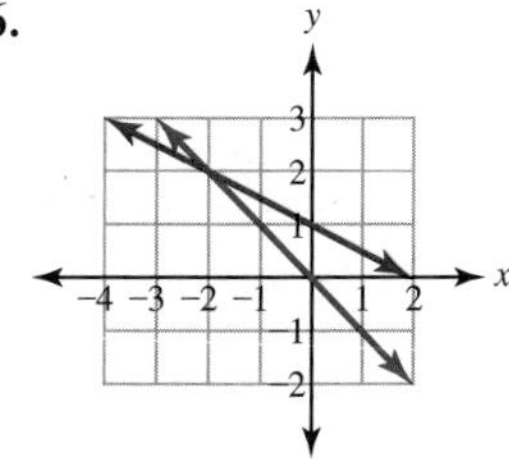

For exercises 7–12, solve by graphing.

7. $y = -2x + 1$
   $y = \frac{4}{3}x - 9$

8. $9x - y = 7$
   $2x + y = 4$

9. $y = -2x - 3$
   $6x + 3y = -9$

10. $2x + 5y = 10$
    $-2x + 5y = 30$

11. $3x + y = 12$
    $x = 2$

12. $x = 6$
    $y = -2$

13. Explain why $(1, -3)$ is not a solution of $\begin{matrix} 4x - 2y = 10 \\ x - y = -2 \end{matrix}$.

14. The system $\begin{matrix} y = 3x + 2 \\ -9x + 3y = 6 \end{matrix}$ has infinitely many solutions. Use set-builder notation to represent the solution of this system.

15. Describe or sketch the graph of a system of linear equations with
    a. one solution.
    b. no solution.
    c. infinitely many solutions.

16. The cost to make a product is \$50. The fixed costs per month to make the product are \$10,000. For x products, a model of the costs per month, y, is $y = \left(\frac{\$50}{1 \text{ product}}\right)x + \$10{,}000$. The price of each product is \$300. A model of the total revenue, y, from the sale of x products is $y = \left(\frac{\$300}{1 \text{ product}}\right)x$. Find the break-even point by graphing.

17. Describe what it means when a company breaks even.

SECTION 4.2 The Substitution Method

Ask Yourself	Test Yourself	Help Yourself
Can I . . .	**Do 4.2 Review Exercises**	**See these Examples and Practice Problems**
solve a system of linear equations by substitution and check the solution? determine whether an ordered pair is a solution of a system of equations?	18–25	Ex. 1–8, PP 1–9
use a system of equations and substitution to solve an application problem?	26–27	Ex. 9–10, PP 10–11

4.2 Review Exercises

For exercises 18–23,
(a) solve by substitution.
(b) if there is one solution, check.

18. $2x + 3y = 9$
$y = x - 7$

19. $-8x + 2y = 15$
$y = 4x + 3$

20. $y = 8x - 5$
$y = -3x + 6$

21. $y = \frac{1}{2}x - \frac{3}{4}$
$-2x + 4y = -3$

22. $y = -3$
$2x + 5y = -27$

23. $3x + 8y = 20$
$-5x + 16y = -26$

24. Choose the ordered pair that is a solution of
$-2x - 5y = 19$
$y = -\frac{1}{2}x - 4$

a. $(2, -3)$ **b.** $(-2, 3)$
c. $(-2, -3)$ **d.** $(2, -6)$
e. $(0, -4)$

25. Choose the ordered pairs that are solutions of
$-3x + 4y = 2$
$y = \frac{3}{4}x + \frac{1}{2}$

a. $(2, 2)$ **b.** $(6, 5)$
c. $\left(0, \frac{1}{2}\right)$ **d.** $\left(-\frac{2}{3}, 0\right)$
e. $(-6, -4)$

For exercises 26–27, use the five steps, a system of equations, and the substitution method.

26. A caterer is putting 480 bottles of carbonated water and noncarbonated water into tubs of ice. There are twice as many bottles of carbonated water as there are of noncarbonated water. Find the number of bottles of each type of water.

27. The cost to make a product is \$25. The fixed costs per month to make the product are \$15,000. The price of each product is \$325. Find the break-even point.

SECTION 4.3 The Elimination Method

Ask Yourself	Test Yourself	Help Yourself
Can I . . .	**Do 4.3 Review Exercises**	**See these Examples and Practice Problems**
rewrite a repeating decimal as a fraction?	28–29	Ex. 1–2, PP 1–4
use elimination to solve a system of linear equations and check the solution?	30–39	Ex. 3–8, PP 5–15
clear the fractions from a linear equation?	36	Ex. 6, PP 10, 14
use a system of equations and elimination to solve an application problem?	40–41	Ex. 9–10, PP 16–17

4.3 Review Exercises

For exercises 28–29, rewrite the repeating decimal as a fraction.

28. $0.\overline{5}$ **29.** $0.\overline{12}$

30. If a student adds two equations together, one of the equations is $7x - 3y = 49$, and the sum of the equations is $9x = 63$, what is the other equation?

For exercises 31–36,
(a) solve by elimination.
(b) if there is one solution, check.

31. $2x + 5y = 41$
$3x + 2y = 1$

32. $9x - 12y = -10$
$3x - 4y = -12$

33. $9x + 2y = 16$
$-x + 4y = 32$

34. $-x + y = -2$
$3x - 3y = 6$

35. $7a + 4b = 33$
$11a + 20b = 45$

36. $\frac{1}{3}x = -\frac{2}{5}y - 1$
$5x - 2y = 65$

37. A student solving $\begin{aligned} 5x + 2y &= 3 \\ 6x - 7y &= 13 \end{aligned}$ by elimination decides to multiply both sides of $5x + 2y = 3$ by 7 and both sides of $6x - 7y = 13$ by 2. Identify the variable that the student is eliminating.

38. A student solving $\begin{array}{l} 3x + 5y = 2 \\ -4x + y = -18 \end{array}$ by elimination decides to multiply both sides of $-4x + y = -18$ by 5 and not change $3x + 5y = 2$. Explain what is wrong with this thinking.

39. A linear equation in one variable such as $2(x + 4) = 2x + 8$ has infinitely many solutions. The solution is the set of real numbers. A system of linear equations such as $\begin{array}{l} x + y = 3 \\ 2x + 2y = 6 \end{array}$ has infinitely many solutions, but the solution is not the set of real numbers. Explain why.

For exercises 40–41, use the five steps, a system of equations, and the elimination method.

40. A book is published in paperback, in hardback, and as an audiobook. The price of one paperback and one hardback is \$24.66. The price of eight paperbacks and three hardbacks is \$115.83. The price of two paperbacks and one audiobook is \$39.81. Find the price of a paperback. Find the price of a hardback.

41. The cost of 100 lattes and mochas is \$372.20. The cost of a mocha is \$4.25. The cost of a latte is \$3.59. Find the number of mochas sold. Find the number of lattes sold.

SECTION 4.4 More Applications

Ask Yourself	Test Yourself	Help Yourself
Can I. . .	**Do 4.4 Review Exercises**	**See these Examples and Practice Problems**
write a system of linear equations to solve an application problem?	42–48	Ex. 1–8, PP 1–9

4.4 Review Exercises

For exercises 42–48, use the five steps and a system of equations.

42. The length of a rectangle is 3 in. more than twice the width. The perimeter of the rectangle is 48 in. Find the length and width of the rectangle.

43. Mixture A is 0.25% fertilizer, and Mixture B is 9% fertilizer. Find the amount of Mixture A and the amount of Mixture B needed to make 2 gal of a new mixture that is 1.5% fertilizer. Round to the nearest tenth.

44. At 60 degrees Fahrenheit, 500 gal of saturated brine solution is 26.4% salt. Find the amount of water to add to make a new mixture that is 20% salt. Find the amount of the new mixture.

45. Brass is an alloy of copper and zinc. Brass A is 65% copper. Brass B is 80% copper. Brass C is 68% copper. Find the amount of Brass A and Brass B needed to make 500 kg of Brass C.

46. A credit card user is shifting his balances. His total debt is \$1950. He wants four times as much debt on his VISA® card as on his MasterCard®. Find the amount of debt on each card.

47. Three pounds of lemon drops and 5 lb of chocolates cost \$14. Seven pounds of lemon drops and 9 lb of chocolates cost \$26. Find the cost of a pound of lemon drops and the cost of a pound of chocolates.

48. The perimeter of a rectangle is 44 cm. The width is 6 cm shorter than the length. Find the length and the width of the rectangle.

SECTION 4.5 Systems of Linear Inequalities

Ask Yourself	Test Yourself	Help Yourself
Can I. . .	**Do 4.5 Review Exercises**	**See these Examples and Practice Problems**
graph a system of inequalities and label the solution region?	49–55	Ex. 1–3, PP 1–7
determine whether an ordered pair is a solution of a system of linear inequalities?	55	Ex. 4, PP 8–9
write and graph a system of linear constraints and label the feasible region?	56	Ex. 5–7, PP 10–12

4.5 Review Exercises

For exercises 49–52, graph. Label the solution region.

49. $y > 2x + 5$
$3x + 5y < 15$
$x \geq -5$

50. $y < x$
$x > 2$
$x < 4$

51. $y \leq 3x$
$y \leq -2x + 10$
$y \geq -2$

52. $y \leq 3x + 4$
$y \geq -\frac{2}{3}x - 7$
$x \leq -2$

For exercises 53–54, graph. Label the solution region.

53. $x \geq 3$
$x \leq 8$
$y \geq -2$
$y \leq 6$

54. $x + y \leq 4$
$x \geq 1$
$y \geq 6$

55. A system of inequalities is $y \geq 2x - 6$, $y \leq -6x + 10$, $x \geq -2$.
   a. Graph. Label the solution region.
   b. Use an algebraic method to determine whether $(1, -2)$ is a solution of the system.

56. A woman over age 19 who is eating 2000 cal per day of fat and carbohydrate should eat a minimum of 400 cal per day of fat, a maximum of 700 cal per day of fat, a minimum of 900 cal per day of carbohydrate, and a maximum of 1300 cal per day of carbohydrate. Let x = calories from fat, and let y = calories from carbohydrates. (*Source:* www.cnpp.usda.gov, 2010)
   a. Write five constraints that describe this situation.
   b. Graph the constraints and label the feasible region.

Chapter 4 Test

1. Solve $\begin{array}{l} x + y = 5 \\ 3x + 2y = 12 \end{array}$ by graphing.

2. Solve $\begin{array}{l} y = -7x + 2 \\ 2x - 5y = -158 \end{array}$ by substitution.

3. Solve $\begin{array}{l} 7x - 2y = -9 \\ 8x + 3y = 32 \end{array}$ by elimination.

For problems 4–6, solve by substitution or elimination.

4. $y = 2x - 9$
$-2x + y = 15$

5. $y = 8x + 15$
$y = \frac{4}{3}x + \frac{50}{3}$

6. $y = \frac{5}{6}x - \frac{3}{4}$
$-10x + 12y = -9$

7. Describe or sketch the graph of a system of two linear equations with
   a. infinitely many solutions.
   b. no solution.
   c. one solution.

8. Show that $(-3, 1)$ is a solution of the system
$2x + 7y = 1$
$-4x + 3y = 15$

For problems 9–12, use a system of two linear equations and the five steps.

9. A stand sells slices of pizza and soft drinks. Each slice of pizza is \$2.50. Each soft drink is \$1.00. A total of 300 soft drinks and slices are sold. The total receipts are \$562.50. Find the number of pizza slices sold. Find the number of soft drinks sold.

10. A daycare provider has a cleaning solution for cleaning the diaper pail that is 4.7% household bleach. Find the amount of diaper pail solution and the amount of water needed to make 2 gal of a new solution for presoaking laundry that is 1.5% household bleach. Round to the nearest tenth.

11. The cost to make a product is \$75. The fixed costs per month to make the product are \$25,000. The price of each product is \$80. Assume that the company will sell all the products it makes. Find how many products the company needs to make to break even. Find the costs and revenues at the break-even point.

12. A homeowner is buying sheep and goats. If the pasture can feed 18 animals and the homeowner wants five times as many sheep as goats, find the number of sheep and the number of goats that the homeowner should buy.

13. Graph $\begin{array}{l} 2x + 5y \leq 20 \\ y \geq 2 \\ x \geq 0 \end{array}$. Label the solution region.

14. The weekly budget for buying boxes of cereal (\$4) and bottles of milk (\$3) for a preschool is \$600. Each week, the preschool needs at least 20 boxes of cereal and at least 40 bottles of milk. Let x = boxes of cereal and let y = bottles of milk.
   a. Write three constraints that represent this situation.
   b. Graph the constraints and label the feasible region.

Exponents and Polynomials

5

SUCCESS IN COLLEGE MATHEMATICS

Taking and Using Notes

In many math classes, there are times when the instructor lectures. Sometimes the instructor writes key words, examples, and explanations on a whiteboard or document projector. Sometimes the instructor uses presentation software like PowerPoint. The examples may be the same or different than the examples in the textbook.

As a student, you need to listen, understand, ask questions, and take notes. For many learners, the very act of taking notes helps them retain the material. Notes also are a record of what your instructor thinks is most important and a source of help for doing homework.

When you are taking notes, you are writing, listening, and thinking about new material at the same time. If you search the Internet for note taking in math, you will find many different systems for organizing and using your notes. What you choose to do depends on whether your instructor expects you to always take notes, how fast you can write, how much you need to watch (not just listen) to learn, and how your instructor presents the information. For most students, the system they use is not as important as having a time as soon after class as possible in which they rewrite and organize their notes, connecting the notes to the material in the textbook.

5.1 Exponential Expressions and the Exponent Rules
5.2 Scientific Notation
5.3 Introduction to Polynomials
5.4 Multiplication of Polynomials
5.5 Division of Polynomials

$\text{base}^{\text{exponent}}$

We can rewrite *repeated additions* of a number as multiplication: $5 + 5 + 5 + 5 = (4)(5)$. We can rewrite *repeated multiplications* of a number as an exponential expression: $6 \cdot 6 \cdot 6 = 6^3$. In this section, we will use the exponent rules to simplify exponential expressions.

SECTION 5.1

Exponential Expressions and the Exponent Rules

After reading the text, working the practice problems, and completing assigned exercises, you should be able to:

1. Use the exponent rules to simplify an expression.
2. Simplify an expression that includes an exponent equal to 0.
3. Simplify an expression that includes an exponent that is a negative integer.

Exponential Expressions

We use **exponential notation** to represent repeated multiplications of the same number, variable, or term. The expression 3^4 equals $3 \cdot 3 \cdot 3 \cdot 3$. The **base** of this expression is 3, and the **exponent** or **power** is 4. The base of an exponential expression can also be a variable: $x^5 = x \cdot x \cdot x \cdot x \cdot x$. When writing $4y \cdot 4y \cdot 4y$ as an exponential expression, put the base $4y$ in parentheses: $(4y)^3$. Since $4y \cdot 4y \cdot 4y$ equals $64y^3$, unless $y = 0$, $(4y)^3$ does not equal $4y^3$.

EXAMPLE 1 Use exponential notation to represent $y \cdot y \cdot y$, $4 \cdot y \cdot y \cdot y$, and $4y \cdot 4y \cdot 4y$.

SOLUTION ▶

$y \cdot y \cdot y$	$4 \cdot y \cdot y \cdot y$	$4y \cdot 4y \cdot 4y$
$= y^3$	$= 4y^3$	$= (4y)^3$

Instead of simplifying exponential expressions by first rewriting them as repeated multiplications, we can use the **exponent rules**. The **addition and subtraction rule of exponents** restates what we know about combining like terms.

> **Addition and Subtraction Rule of Exponents**
> To simplify exponential expressions, add or subtract terms that have exactly the same variable(s) with the same exponent(s).

EXAMPLE 2 Simplify: $3x^4 - 8x^2 + 9x + x^2 - 13x$

SOLUTION ▶

$3x^4 - 8x^2 + 9x + x^2 - 13x$ Identify like terms.

$= 3x^4 - 7x^2 - 4x$ Combine like terms.

Practice Problems

For problems 1–3, use exponential notation to represent the expression.

1. $n \cdot n \cdot n \cdot n$ **2.** $7 \cdot n \cdot n \cdot n \cdot n$ **3.** $7n \cdot 7n \cdot 7n \cdot 7n$

For problems 4–6, simplify.

4. $9z^3 - 17z^3$ **5.** $6x + 5x^2 - 10x - 9$ **6.** $x - 8x - 3x$

The Product Rule, Quotient Rule, and Power Rule of Exponents

To multiply exponential expressions, we can rewrite each expression as repeated multiplications, or we can use the **product rule of exponents**.

Product Rule of Exponents

$$x^m \cdot x^n = x^{m+n}$$

where x is a real number, m and n are integers, and x and $m + n$ cannot both be 0.

To multiply exponential expressions with the same base, add the exponents and keep the same base.

We can show that the product rule of exponents is true by simplifying $x^2 \cdot x^3$ using repeated multiplications and using the product rule of exponents. The result is the same.

EXAMPLE 3 Simplify: $x^2 \cdot x^3$

SOLUTION

Using repeated multiplications:

$x^2 \cdot x^3$

$= x \cdot x \cdot x \cdot x \cdot x$

$= x^5$

Using the product rule of exponents:

$x^2 \cdot x^3$

$= x^{2+3}$

$= x^5$

Using either repeated multiplications or the product rule results in the same simplified expression, x^5.

The commutative property of multiplication states that changing the order of factors in multiplication does not change the product. In the next example, we use the commutative property to rewrite $(6p^4)(-3p^5)$ as $(6)(-3)(p^4)(p^5)$.

EXAMPLE 4 Simplify: $(6p^4)(-3p^5)$

SOLUTION

$(6p^4)(-3p^5)$

$= (6)(-3)(p^4)(p^5)$ Commutative property of multiplication.

$= -18p^{(4+5)}$ Simplify; product rule of exponents; add the exponents.

$= -18p^9$ Simplify.

We can use the product rule only when multiplying exponential expressions with the same base. If the bases are different numbers, first evaluate each exponential expression by rewriting as repeated multiplications.

EXAMPLE 5 Evaluate: $2^3 \cdot 5^2$

SOLUTION ▶

$2^3 \cdot 5^2$ — The bases 2 and 5 are not the same.

$= 2 \cdot 2 \cdot 2 \cdot 5 \cdot 5$ — Rewrite as repeated multiplications.

$= 200$ — Multiply.

We can use the product rule of exponents to simplify the product of exponential expressions with the same base. To simplify a quotient of exponential expressions with the same base, we can use the **quotient rule of exponents.**

Quotient Rule of Exponents

$$\frac{x^m}{x^n} = x^{m-n}, \quad x \neq 0$$

where x is a real number and m and n are integers.

To divide exponential expressions with the same base, subtract the exponents and keep the same base.

We can show that the quotient rule of exponents is true by simplifying $\frac{x^5}{x^2}$ using repeated multiplications and using the quotient rule of exponents. The result is the same.

EXAMPLE 6 Simplify: $\frac{x^5}{x^2}$

SOLUTION ▶

Using repeated multiplications:

$$\frac{x^5}{x^2}$$

$$= \frac{x \cdot x \cdot x \cdot x \cdot x}{x \cdot x}$$

$$= \frac{x}{x} \cdot \frac{x}{x} \cdot x \cdot x \cdot x$$

$$= 1 \cdot 1 \cdot x^3$$

$$= x^3$$

Using the quotient rule of exponents:

$$\frac{x^5}{x^2}$$

$$= x^{5-2}$$

$$= x^3$$

Using either repeated multiplications or the quotient rule results in the same simplified expression, x^3.

In the next example, we simplify $\frac{10}{8}$ into lowest terms and use the quotient rule of exponents to simplify $\frac{h^{10}}{h^3}$.

EXAMPLE 7 Simplify: $\frac{10h^{10}}{8h^3}$

SOLUTION ▶

$$\frac{10h^{10}}{8h^3}$$

$= \frac{10h^{10-3}}{8}$ — Quotient rule of exponents; subtract the exponents.

$= \dfrac{5 \cdot 2 \cdot h^7}{4 \cdot 2}$ Find common factors.

$= \dfrac{5h^7}{4}$ Simplify.

The rule for *multiplying* exponential expressions is the *product* rule of exponents. The rule for *dividing* exponential expressions is the *quotient* rule of exponents. The rule for raising an exponential expression to a *power* is the *power* rule of exponents.

Power Rule of Exponents

$$(x^m)^n = x^{mn}$$

where x is a real number, m and n are integers, and x and mn cannot both be 0.

When raising an exponential expression to a power, multiply the exponents and keep the same base.

We can show that the power rule of exponents is true by simplifying $(x^3)^2$ using repeated multiplications and using the power rule of exponents. The result is the same.

EXAMPLE 8 Simplify: $(x^3)^2$

SOLUTION

Using repeated multiplications:

$(x^3)^2$

$= (x \cdot x \cdot x)(x \cdot x \cdot x)$

$= x^6$

Using the power rule of exponents:

$(x^3)^2$

$= x^{(3)(2)}$

$= x^6$

Using either repeated multiplications or the power rule results in the same simplified expression, x^6.

In the next example, we use both the power rule and the quotient rule.

EXAMPLE 9 Simplify: $\dfrac{(a^2)^4}{2a^3}$

SOLUTION

$\dfrac{(a^2)^4}{2a^3}$

$= \dfrac{a^{(2)(4)}}{2a^3}$ Power rule of exponents; multiply the exponents.

$= \dfrac{a^8}{2a^3}$ Simplify.

$= \dfrac{a^{8-3}}{2}$ Quotient rule of exponents; subtract the exponents.

$= \dfrac{a^5}{2}$ Simplify.

Practice Problems

For problems 7–11, simplify.

7. $(3x^5)(5x^4)$ **8.** $(p^8)(p)$ **9.** $-\dfrac{18x^{11}}{9x^7}$ **10.** $\dfrac{30a^9}{18a^4}$ **11.** $(m^5)^9$

Many Bases Rule for a Product and Quotient

If the base of an exponential expression is a product, we can simplify this expression using repeated multiplications, or we can use the **many bases rule for a product**.

Many Bases Rule for a Product

$$(xy)^m = x^m \cdot y^m$$

where x and y are real numbers, m and n are integers, and x or y and m cannot both be 0.

When raising the product of two or more bases to a power, raise each base to the power.

We can show that the many bases rule for a product is true by simplifying $(xy)^2$ using repeated multiplications and using the many bases rule for a product. The result is the same.

EXAMPLE 10 Simplify: $(xy)^2$

SOLUTION

Using repeated multiplications:

$(xy)^2$
$= (xy)(xy)$
$= x \cdot x \cdot y \cdot y$
$= x^2y^2$

Using the many bases rule for a product:

$(xy)^2$
$= x^2y^2$

Using either repeated multiplications or the many bases rule for a product results in the same simplified expression, x^2y^2.

If a constant such as 4 is one of the many bases that is raised to a power, use repeated multiplications to evaluate it.

EXAMPLE 11 Simplify: $(4a^2b)^3$

SOLUTION

$(4a^2b)^3$
$= 4^3(a^2)^3b^3$ Many bases rule for a product; raise each base to the power.
$= \mathbf{4 \cdot 4 \cdot 4} \cdot (a^2)^3b^3$ Rewrite 4^3 as repeated multiplications.
$= \mathbf{64}a^{\mathbf{(2)(3)}}b^3$ Simplify; power rule of exponents.
$= 64a^6b^3$ Simplify.

If the base of an exponential expression is a quotient, we can simplify this expression using repeated multiplications, or we can use the **many bases rule for a quotient**.

Many Bases Rule for a Quotient

$$\left(\frac{x}{y}\right)^m = \frac{x^m}{y^m}$$

where x and y are real numbers, $y \neq 0$, m is an integer, and x and m cannot both be 0.

When raising the quotient of two or more bases to a power, raise each base to the power.

We can show that the many bases rule for a quotient is true by simplifying $\left(\frac{x}{y}\right)^2$ using repeated multiplications and using the many bases rule for a quotient. The result is the same.

EXAMPLE 12 Simplify: $\left(\frac{x}{y}\right)^2$

SOLUTION ▶

Using repeated multiplications:

$$\left(\frac{x}{y}\right)^2$$
$$= \left(\frac{x}{y}\right)\left(\frac{x}{y}\right)$$
$$= \frac{x \cdot x}{y \cdot y}$$
$$= \frac{x^2}{y^2}$$

Using the many bases rule for a quotient:

$$\left(\frac{x}{y}\right)^2$$
$$= \frac{x^2}{y^2}$$

Using either repeated multiplications or the many bases rule for a quotient results in the same simplified expression, $\frac{x^2}{y^2}$.

To simplify the expression in the next example, we use the many bases rule for a quotient and the many bases rule for a product.

EXAMPLE 13 Simplify: $\left(\frac{3x}{5py}\right)^2$

SOLUTION ▶

$$\left(\frac{3x}{5py}\right)^2$$
$$= \frac{(3x)^2}{(5py)^2}$$ Many bases rule for a quotient; raise each base to a power.
$$= \frac{3^2x^2}{5^2p^2y^2}$$ Many bases rule for a product; raise each base to a power.
$$= \mathbf{\frac{9x^2}{25p^2y^2}}$$ Simplify.

In the next example, we use the many bases rule for a quotient, the many bases rule for a product, and the power rule of exponents.

EXAMPLE 14 Simplify: $\left(\frac{8h^3}{5k^5n^7}\right)^2$

SOLUTION ▶

$$\left(\frac{8h^3}{5k^5n^7}\right)^2$$
$$= \frac{(8h^3)^2}{(5k^5n^7)^2}$$ Many bases rule for a quotient; raise each base to a power.
$$= \frac{8^2 \cdot (h^3)^2}{5^2 \cdot (k^5)^2(n^7)^2}$$ Many bases rules for a product; raise each base to a power.
$$= \frac{64h^{(3)(2)}}{25k^{(5)(2)}n^{(7)(2)}}$$ Simplify; power rule of exponents; multiply the exponents.
$$= \frac{64h^6}{25k^{10}n^{14}}$$ Simplify.

In the next example, we simplify $(-4x^3)(-5x^2)$ and $(-4x^3)(-5x)^2$. Since the parentheses are in a different position in each expression, the simplified expressions are not the same.

EXAMPLE 15 **(a)** Simplify: $(-4x^3)(-5x^2)$

SOLUTION

$(-4x^3)(-5x^2)$

$= (-4)(-5) \cdot x^3 \cdot x^2$ Commutative property of multiplication.

$= \mathbf{20}x^{\mathbf{3+2}}$ Simplify; product rule of exponents.

$= 20x^{\mathbf{5}}$ Simplify.

(b) Simplify: $(-4x^3)(-5x)^2$

$(-4x^3)(-5x)^2$

$= (-4x^3)(-5)^{\mathbf{2}}x^{\mathbf{2}}$ Many bases rule for a product.

$= (-4x^3)(\mathbf{25})x^2$ Simplify.

$= (-4)(25) \cdot x^3 \cdot x^2$ Commutative property of multiplication.

$= \mathbf{-100}x^{\mathbf{3+2}}$ Simplify; product rule of exponents.

$= -100x^{\mathbf{5}}$ Simplify.

Practice Problems

For problems 12–16, simplify.

12. $(hk)^3$ **13.** $\left(\frac{w}{p}\right)^8$ **14.** $(5x^3)^2$

15. $\left(\frac{-3a}{b^4}\right)^2$ **16.** $\left(\frac{4a}{5b^4}\right)^2$

Zero as an Exponent

We can use repeated multiplications and the quotient rule of exponents to show that if x is not equal to 0, $x^0 = 1$.

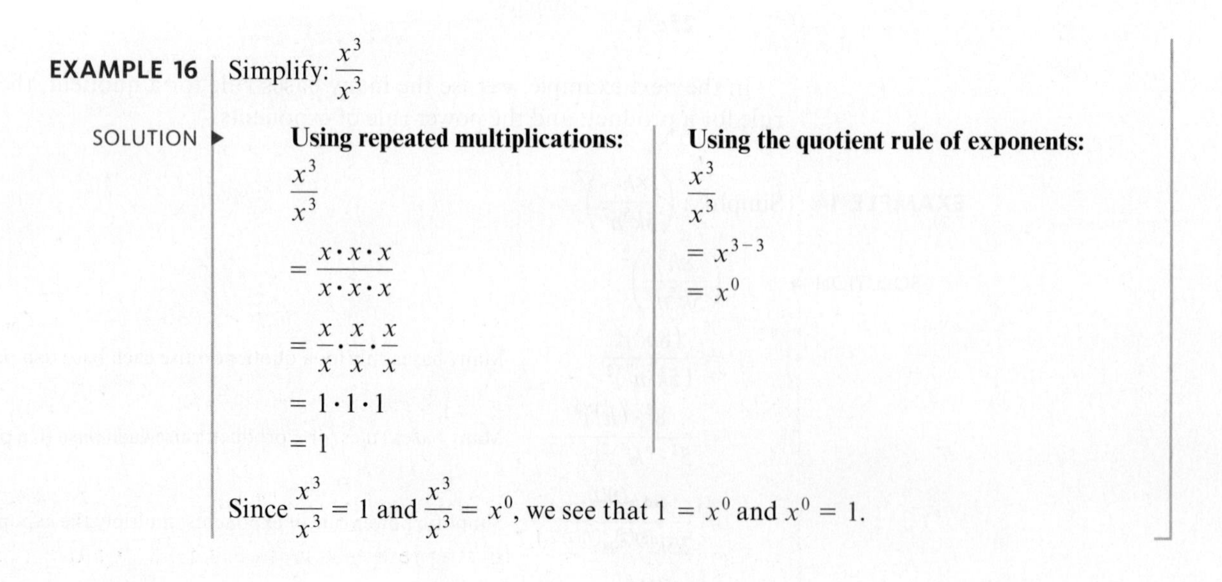

EXAMPLE 16 Simplify: $\frac{x^3}{x^3}$

SOLUTION

Using repeated multiplications:

$\frac{x^3}{x^3}$

$= \frac{x \cdot x \cdot x}{x \cdot x \cdot x}$

$= \frac{x}{x} \cdot \frac{x}{x} \cdot \frac{x}{x}$

$= 1 \cdot 1 \cdot 1$

$= 1$

Using the quotient rule of exponents:

$\frac{x^3}{x^3}$

$= x^{3-3}$

$= x^0$

Since $\frac{x^3}{x^3} = 1$ and $\frac{x^3}{x^3} = x^0$, we see that $1 = x^0$ and $x^0 = 1$.

Zero as an Exponent

If x is a real number and x is not equal to 0, $x^0 = 1$.

In the next example, an expression that is not equal to 0 is raised to the zero power. The simplified expression equals 1.

EXAMPLE 17 If $\frac{3a}{7b} \neq 0$, simplify $\left(\frac{3a}{7b}\right)^0$.

SOLUTION ▶

$\left(\frac{3a}{7b}\right)^0$

$= 1$ Simplify.

Practice Problems

For problems 17–19, simplify.

17. 5^0 **18.** c^0 **19.** $\left(\frac{4n}{3p}\right)^0$

Negative Exponents

By tradition, simplified expressions include only positive exponents. We can use repeated multiplications and the quotient rule of exponents to show that x^{-3} and $\frac{1}{x^3}$ are equivalent expressions. We can then use this relationship to rewrite expressions with negative exponents as equivalent expressions with only positive exponents.

EXAMPLE 18 Simplify: $\frac{x^2}{x^5}$

SOLUTION ▶

Using repeated multiplications:

$\frac{x^2}{x^5}$

$= \frac{x \cdot x}{x \cdot x \cdot x \cdot x \cdot x}$

$= \frac{x}{x} \cdot \frac{x}{x} \cdot \frac{1}{x \cdot x \cdot x}$

$= 1 \cdot 1 \cdot \frac{1}{x^3}$

$= \frac{1}{x^3}$

Using the quotient rule of exponents:

$\frac{x^2}{x^5}$

$= x^{2-5}$

$= x^{-3}$

Since $\frac{x^2}{x^5} = \frac{1}{x^3}$ and $\frac{x^2}{x^5} = x^{-3}$, we see that $\frac{1}{x^3} = x^{-3}$ and $x^{-3} = \frac{1}{x^3}$.

From Example 18, we see that $\frac{1}{x^m} = x^{-m}$ and $x^{-m} = \frac{1}{x^m}$. We can use the relationship $x^{-m} = \frac{1}{x^m}$ to rewrite an expression with a negative exponent as an equivalent expression with a positive exponent.

EXAMPLE 19 Simplify: $\frac{3a^5b^7}{a^9}$

SOLUTION ▶ $\frac{3a^5b^7}{a^9}$

$= 3a^{5-9}b^7$ Quotient rule of exponents; subtract the exponents.

$= 3a^{-4}b^7$ Simplify.

$= 3 \cdot \frac{1}{a^4} \cdot b^7$ Use the relationship $x^{-m} = \frac{1}{x^m}$.

$= \frac{3b^7}{a^4}$ Simplify.

We can use the relationship $\frac{1}{x^m} = x^{-m}$ to rewrite an expression that includes a base with a negative exponent in the denominator as an equivalent expression with only positive exponents.

EXAMPLE 20 Simplify: $\frac{5}{x^{-2}}$

SOLUTION ▶ $\frac{5}{x^{-2}}$

$= 5x^{-(-2)}$ Use the relationship $\frac{1}{x^m} = x^{-m}$.

$= 5x^2$ Simplify.

Since $\frac{5}{x^{-2}} = 5x^2$, we see that $\frac{1}{x^{-m}} = x^m$.

Negative Exponents

If x and m are real numbers and x is not equal to 0, $x^{-m} = \frac{1}{x^m}$ and $\frac{1}{x^{-m}} = x^m$.

When using the quotient rule of exponents to simplify an expression in which the exponent in the denominator is a negative number, we subtract a negative number. For example, $\frac{b^3}{b^{-2}}$ equals $b^{3-(-2)}$.

EXAMPLE 21 Simplify: $\frac{a^{-6}b^3c}{a^9b^{-2}c^8}$

SOLUTION ▶ $\frac{a^{-6}b^3c}{a^9b^{-2}c^8}$

$= a^{-6-9}b^{3-(-2)}c^{1-8}$ Quotient rule of exponents.

$= a^{-6-9}b^{3+2}c^{1-8}$ Rewrite $3 - (-2)$ as $3 + 2$.

$= a^{-15}b^5c^{-7}$ Simplify.

$= \frac{b^5}{a^{15}c^7}$ Rewrite with positive exponents.

In the next example, the base is a number instead of a variable. We rewrite the expression with a positive exponent, write the exponential expression as repeated multiplications, and multiply.

EXAMPLE 22 Evaluate: 10^{-3}

SOLUTION ▶

10^{-3}

$= \frac{1}{10^3}$ Rewrite with a positive exponent.

$= \frac{1}{\mathbf{10 \cdot 10 \cdot 10}}$ Rewrite as repeated multiplications.

$= \frac{1}{\mathbf{1000}}$ Multiply.

Practice Problems

For problems 20–24, simplify.

20. $\frac{x^7}{x^{12}}$ **21.** $(y^{-5})^3$ **22.** $a^{-2}a^{-7}a^9$ **23.** $p^{-4}z^3p^{-2}z$ **24.** $\frac{5f^{-3}}{f}$

Simplifying Exponential Expressions

To simplify more complicated exponential expressions, we use several of the exponent rules. In the final expression, the exponents must be positive, and any rational numbers must be in lowest terms.

EXAMPLE 23 Simplify: $\frac{(2x^5)^3(4x^3)^2}{100x^{16}}$

SOLUTION ▶

$\frac{(2x^5)^3(4x^3)^2}{100x^{16}}$

$= \frac{2^3(x^5)^3(4^2)(x^3)^2}{100x^{16}}$ Many bases rule for a product.

$= \frac{\mathbf{2 \cdot 2 \cdot 2} \cdot x^{\mathbf{(5)(3)}} \cdot \mathbf{16}x^{\mathbf{(3)(2)}}}{100x^{16}}$ Rewrite as repeated multiplications; power rule of exponents.

$= \frac{\mathbf{8}x^{\mathbf{15}} \cdot 16x^{\mathbf{6}}}{100x^{16}}$ Simplify.

$= \frac{\mathbf{128}x^{\mathbf{15+6}}}{100x^{16}}$ Simplify; product rule of exponents; add the exponents.

$= \frac{\mathbf{32 \cdot 4} \cdot x^{\mathbf{21}}}{\mathbf{25 \cdot 4} \cdot x^{16}}$ Simplify; find common factors.

$= \frac{32x^{21}}{25x^{16}}$ Simplify.

$= \frac{32x^{\mathbf{21-16}}}{25}$ Quotient rule of exponents.

$= \frac{32x^{\mathbf{5}}}{25}$ Simplify.

In the next example, we can first use the quotient rule of exponents to simplify inside the parentheses or first use the many bases rule for a quotient. The simplified expression is the same.

EXAMPLE 24 Simplify: $\left(\frac{3z^5}{2z^9}\right)^2$

SOLUTION ▶ $\left(\frac{3z^5}{2z^9}\right)^2$

$= \left(\frac{3z^{5-9}}{2}\right)^2$ Quotient rule of exponents.

$= \left(\frac{3z^{-4}}{2}\right)^2$ Simplify.

$= \frac{(3z^{-4})^2}{2^2}$ Many bases rule for a quotient.

$= \frac{3^2(z^{-4})^2}{2^2}$ Many bases rule for a product.

$= \frac{\mathbf{3 \cdot 3} \cdot (z^{-4})^2}{\mathbf{2 \cdot 2}}$ Rewrite as repeated multiplications.

$= \frac{\mathbf{9}z^{(-4)(2)}}{\mathbf{4}}$ Simplify; power rule of exponents.

$= \frac{9z^{-8}}{4}$ Simplify.

$= \frac{9}{4z^8}$ Rewrite with a positive exponent.

In the next example, we can first rewrite the expression with a positive exponent and then use the many bases rule for a product, or we can first use the many bases rule for a product and then rewrite the expression with a positive exponent. The simplified expression is the same.

EXAMPLE 25 Simplify: $(7p)^{-2}$

SOLUTION ▶ $(7p)^{-2}$

$= \frac{1}{(7p)^2}$ Rewrite with a positive exponent.

$= \frac{1}{7^2p^2}$ Many bases rule for a product.

$= \frac{1}{\mathbf{7 \cdot 7} \cdot p^2}$ Rewrite as repeated multiplications.

$= \frac{1}{\mathbf{49}p^2}$ Simplify.

Although we will not do any application problems with exponents in this section, we will use the exponent rules in Section 5.2 as we work with measurements written in scientific notation. Scientists use scientific notation to show the precision of measurements and to speed up their work with very small and very large numbers.

Practice Problems

For problems 25–27, simplify.

25. $\frac{(3m^3)^2}{12m^4}$ **26.** $(4p^3)^2(3p^4)^2$ **27.** $\left(\frac{4w}{3y}\right)^2(w^3)^2(2y^2)$

5.1 Vocabulary Practice

Match the term with its description.

1. In the expression x^3, 3 is an example of this.
2. In the expression x^3, x is an example of this.
3. $(x^m)^n = x^{mn}$
4. $\frac{x^m}{x^n} = x^{m-n}$
5. $\left(\frac{x}{y}\right)^n = \frac{x^n}{y^n}, y \neq 0$
6. $x^m \cdot x^n = x^{m+n}$
7. $(xy)^m = x^m y^m$
8. The value of 5^0
9. Division by zero
10. In the expression $\frac{x^n}{y^m}$, y cannot be equal to this.

A. base
B. exponent
C. many bases rule for a product
D. many bases rule for a quotient
E. one
F. power rule of exponents
G. product rule of exponents
H. quotient rule of exponents
I. undefined
J. zero

5.1 Exercises

Follow your instructor's guidelines for showing your work.

For exercises 1–82, simplify. Assume that denominators and bases that are raised to the zero power do not equal 0. Simplified expressions must be in lowest terms, and exponents must be positive numbers.

1. $5x^2 + 3 - 7x^2 - 9$
2. $7y^2 + 5 - 9y^2 - 8$
3. $3x^4 + 5y^2 + 8x^4 + 11y^2$
4. $8p^2 + 9w^3 + 2p^2 + 13w^3$
5. $\frac{1}{8}x + \frac{4}{9} - \frac{7}{8}x - \frac{7}{9}$
6. $\frac{1}{10}a + \frac{5}{12} - \frac{9}{10}a - \frac{1}{12}$
7. $\frac{1}{2}h^2 - \frac{3}{4}h + \frac{5}{6}h^2 + \frac{1}{3}h$
8. $\frac{1}{5}k^2 - \frac{3}{10}k + \frac{5}{8}k^2 + \frac{1}{4}k$
9. $k^3 - 4k^2 - k^2 - k + 6$
10. $x^3 - 5x^2 - x^2 - x + 1$
11. $b^{11}b^4$
12. $c^{10}c^3$
13. $(9x^3)(4x^5)$
14. $(5y^7)(6y^3)$
15. $\frac{x^9}{x^2}$
16. $\frac{z^8}{z^3}$
17. $\frac{32h^9}{8h^4}$
18. $\frac{40p^9}{8p^3}$
19. $\frac{60x^{10}}{34z^3}$
20. $\frac{50z^{12}}{14y^8}$
21. $-\frac{14z^6}{30z}$
22. $-\frac{18p^7}{40p}$
23. $(u^2)^7$
24. $(r^3)^5$
25. $(cd)^3$
26. $(xy)^5$
27. $(a^3b)^2$
28. $(c^4d)^2$
29. $(3a)^2$
30. $(4f)^2$
31. $(-3a)^3$
32. $(-4f)^3$
33. $(-6n^3)^2$
34. $(-7f^4)^2$
35. $\left(\frac{h}{k}\right)^7$
36. $\left(\frac{c}{d}\right)^8$
37. $\left(\frac{x^2}{y^5}\right)^2$
38. $\left(\frac{w^3}{q^5}\right)^2$
39. $\left(\frac{3x^2}{4y^5}\right)^2$
40. $\left(\frac{4w^3}{5q^5}\right)^2$
41. $(2z^3)^3(4z)$
42. $(3p^4)^2(5p)$
43. 4^0
44. 5^0
45. c^0
46. a^0
47. $(5x)^0$
48. $(12y)^0$
49. $\left(\frac{8p^4}{56p^6}\right)^0$
50. $\left(\frac{9n^3}{54n^8}\right)^0$
51. $(y^{-3})^2$
52. $(b^{-4})^2$
53. 8^{-2}
54. 5^{-2}
55. 10^{-4}
56. 10^{-5}
57. $\left(\frac{3x^2}{y^5}\right)^2\left(\frac{x}{y}\right)$
58. $\left(\frac{4w^3}{q^5}\right)^2\left(\frac{w}{q}\right)$
59. $\left(\frac{5p}{6}\right)\left(\frac{3}{p}\right)^2$
60. $\left(\frac{5a}{12}\right)\left(\frac{3}{a}\right)^2$
61. $(6x)^2(2x)^3$
62. $(9a)^2(2a)^3$
63. $(-x^2)^3$
64. $(-z^7)^3$
65. $(-x^4)^2(-x^2)^3$
66. $(-z^5)^2(-z^7)^3$
67. $(3x)^{-1}$
68. $(4a)^{-1}$
69. $(3x)^{-2}$
70. $(4a)^{-2}$
71. $\left(-\frac{3x^4}{4y^9}\right)\left(\frac{2x^7}{9y^6}\right)$
72. $\left(-\frac{5a^3}{6b^5}\right)\left(\frac{2a^8}{15b^4}\right)$
73. $\left(\frac{2a^3}{7b^8}\right)\left(\frac{21b^{12}}{10a}\right)$
74. $\left(\frac{8z^5}{15w^9}\right)\left(\frac{3w^{19}}{20z}\right)$
75. $\left(\frac{3x^4}{5z^8}\right)\left(\frac{5z^8}{3x^4}\right)$
76. $\left(\frac{2a^5}{5b^7}\right)\left(\frac{5b^7}{2a^5}\right)$

77. $\dfrac{(4w^3)^2}{32w}$

78. $\dfrac{(3x^4)^2}{180x}$

79. $\left(\dfrac{2p}{3q}\right)^2(4p)^2$

80. $\left(\dfrac{3a}{5b}\right)^2(2a)^2$

81. $6z(z^2) + (6z)(8z) + 6z(4)$

82. $9a(8a^2) + 9a(4a) + 9a(6)$

83. Simplify: $\left(\dfrac{x^9}{x^3}\right)^2$

a. Use the quotient rule first. Then use the power rule.
b. Use the many bases rule for a quotient first. Next, use the power rule. Finally, use the quotient rule.
c. Compare the simplified expressions from part a and part b.

84. Simplify: $\left(\dfrac{z^7}{z^4}\right)^2$

a. Use the quotient rule first. Then use the power rule.
b. Use the many bases rule for a quotient first. Next, use the power rule. Finally, use the quotient rule.
c. Compare the simplified expressions from part a and part b.

Problem Solving: Practice and Review

Follow your instructor's guidelines for using the five steps as outlined in Section 1.3, p. 55.

85. Write a linear equation that represents the relationship between the number of hours to do an inspection, x, and the total cost of the inspection, y.

In North Dakota, which has about 200 potato farms, GAP inspections are handled by the state Seed Department. Ken Bertsch, the state seed commissioner, said there is a flat fee of $50 besides charges of $75 per hour. (*Source:* Times-News, www.magicvalley.com, Oct. 1, 2007)

86. Find last year's (2010) enrollment.

The district's 2011 head count, released Friday, shows this year's enrollment is about 1,500 students higher than last—a 3 percent increase. (*Source:* www.nwsource.com, Oct. 16, 2011)

87. A periodontist removes a rectangular piece from the roof of a patient's mouth to use as a gum graft. The graft is 1.6 cm long, 9 mm wide, and 2 mm thick. Find the volume of the graft in cubic millimeters. (1 cm = 10 mm.) (*Source:* Bradley Morlock, DDS)

88. A community has 37,500 low-wage workers. Predict how many do not receive paid sick days.

Three out of four low-wage workers in the private sector do not have employer-provided health insurance, while eight out of nine do not participate in a pension plan. Three-fourths of low-wage workers do not receive paid sick days, so if they need to miss two days work because they are sick or their child is sick, they receive no pay for those days and often risk getting fired. (*Source:* Greenhouse, Steven, *The Big Squeeze: Tough Times for the American Worker*, Knopf, 2008)

Technology

For exercises 89–92, evaluate. Use the exponent key on a scientific calculator.

89. 6^5

90. $(-7)^5$

91. $(-5)^{11}$

92. $(10)^7$

Find the Mistake

For exercises 93–96, the completed problem has one mistake.
(a) Describe the mistake in words, or copy down the whole problem and highlight or circle the mistake.
(b) Do the problem correctly.

93. Problem: Simplify: $(6n^3)^2$

Incorrect Answer: $(6n^3)^2$
$= 6^2(n^3)^2$
$= 36n^5$

94. Problem: Simplify: $(7n^4)(9n^6)$

Incorrect Answer: $(7n^4)(9n^6)$
$= 63n^{(4)(6)}$
$= 63n^{24}$

95. Problem: Simplify: $\dfrac{15n^{12}}{3n^4}$

Incorrect Answer: $\dfrac{15n^{12}}{3n^4}$
$= 5n^{12 \div 4}$
$= 5n^3$

96. Problem: Simplify: $\dfrac{(4x^3)^3}{(8x)^2}$

Incorrect Answer: $\dfrac{(4x^3)^3}{(8x)^2}$
$= \dfrac{(2 \cdot 2 \cdot x^3)^3}{(2 \cdot 2 \cdot 2 \cdot x)^2}$
$= \dfrac{(x^3)^3}{(2x)^2}$
$= \dfrac{x^{(3)(3)}}{2^2x^2}$
$= \dfrac{x^9}{4x^2}$
$= \dfrac{x^7}{4}$

Review

97. Write the distributive property.

98. Write the commutative property of addition.

99. Simplify: $6(3x - 9) - (x - 15)$

100. Simplify: $-3(2x - 11) - 4(5x - 8)$

SUCCESS IN COLLEGE MATHEMATICS

101. Describe how you currently take notes and what your notes look like.

102. Describe how you use your notes.

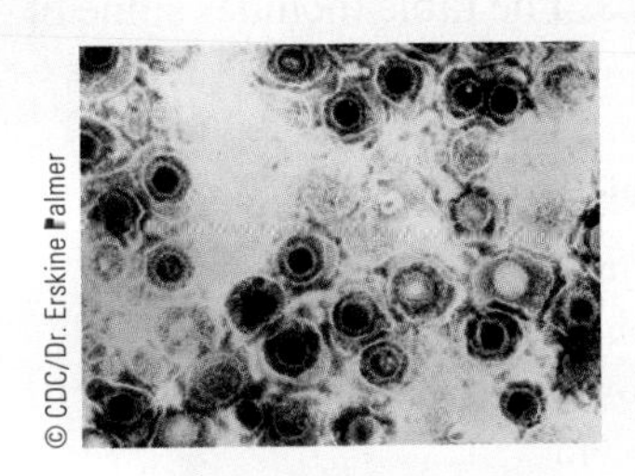

The herpes simplex virus causes cold sores. It is spherical in shape. The diameter of these spheres ranges from 1.2×10^{-7} meter to 2×10^{-7} meter. These measurements are written in **scientific notation**. Scientific notation is especially useful for writing very small and very large measurements. In this section, we will learn to use scientific notation.

SECTION 5.2 Scientific Notation

After reading the text, working the practice problems, and completing assigned exercises, you should be able to:

1. Write a measurement in scientific notation.
2. Multiply and divide measurements in scientific notation.
3. Add and subtract measurements in scientific notation.
4. Solve an equation that includes measurements in scientific notation.

Powers of Ten and Scientific Notation

A power of ten is an exponential expression with a base of ten. For example, since 100 is equal to $10 \cdot 10$, we can rewrite it in exponential notation as a power of ten, 10^2.

> **Power of Ten**
>
> A power of ten, 10^p, is an exponential expression with a base of 10 that is raised to a power (an exponent), p, where p is an integer.

When the exponent on a power of ten is a whole number, the value of the power of ten is greater than or equal to 1.

EXAMPLE 1 Write 10^0, 10^1, 10^2, and 10^3 in place value notation.

SOLUTION ▶

10^0	10^1	10^2	10^3
$= 1$	$= 10$	$= 10 \cdot 10$	$= 10 \cdot 10 \cdot 10$
		$= 100$	$= 1000$

When the exponent on a power of ten is an integer that is less than 0, the value of the power of ten is less than 1.

EXAMPLE 2 Write 10^{-1}, 10^{-2}, 10^{-3}, and 10^{-4} in place value notation.

SOLUTION ▶ Since $x^{-m} = \frac{1}{x^m}$, we can rewrite 10^{-p} as a fraction, $\frac{1}{10^p}$, and then rewrite the fraction in place value notation.

10^{-1}	10^{-2}	10^{-3}	10^{-4}
$= \frac{1}{10^1}$	$= \frac{1}{10^2}$	$= \frac{1}{10^3}$	$= \frac{1}{10^4}$
$= \frac{1}{10}$	$= \frac{1}{100}$	$= \frac{1}{1000}$	$= \frac{1}{10,000}$
$= 0.1$	$= 0.01$	$= 0.001$	$= 0.0001$

We can also describe powers of ten using word names. The table includes some of these names.

Power of ten	Place value notation	Word name
10^{12}	1,000,000,000,000	One trillion
10^{9}	1,000,000,000	One billion
10^{6}	1,000,000	One million
10^{3}	1000	One thousand
10^{-2}	0.01	One hundredth
10^{-6}	0.000001	One millionth

Instead of word names, scientists often use scientific notation to describe numbers and measurements.

Scientific Notation

$n \times 10^p$

A number in **scientific notation** is the product of a real number n that is at least 1 and that is less than 10 and a power of ten, 10^p, where p is an integer. The **mantissa** is n.

EXAMPLE 3 Use scientific notation to represent the speed of the space shuttle in miles per hour.

SOLUTION ▶ To achieve orbit, the shuttle must accelerate from zero to a speed of almost 28,968 km per hour (18,000 mi per hour), a speed nine times as fast as the average rifle bullet. (*Source:* www.NASA.gov)

© John A Davis/Shutterstock.com

$18{,}000\,\frac{\text{mi}}{\text{hr}}$ Original measurement in place value notation.

$= \mathbf{1.8 \times 10{,}000}\,\frac{\text{mi}}{\text{hr}}$ The mantissa is at least 1 and less than 10.

$= 1.8 \times \mathbf{10^4}\,\frac{\text{mi}}{\text{hr}}$ Rewrite 10,000 as a power of ten.

To quickly rewrite this measurement in scientific notation, move the decimal point so that the mantissa is at least 1 and less than 10. The exponent on the power of ten is positive and equals the number of places the decimal point moves.

$18{,}000.\,\frac{\text{mi}}{\text{hr}}$ Original measurement in place value notation.

$= \mathbf{1.8 \times 10^4}\,\frac{\text{mi}}{\text{hr}}$ Move the decimal point left 4 place values; the exponent is 4.

In the next example, the power of ten has a negative exponent.

EXAMPLE 4 The herpes simplex virus is a sphere with a diameter of 0.00000012 m. Write this measurement in scientific notation.

SOLUTION ▶

0.00000012 m Original measurement in place value notation.

$= \mathbf{1.2 \times 0.0000001}$ m The mantissa is at least 1 and less than 10.

$= 1.2 \times \mathbf{10^{-7}}$ m Rewrite 0.0000001 as a power of 10.

To quickly rewrite this measurement, move the decimal point to the right so that the mantissa is at least 1 and less than 10. The exponent on the power of ten is negative.

0.00000012 m	Original measurement in place value notation.
$= \mathbf{1.2 \times 10^{-7}}$ m	Move the decimal point right 7 place values; the exponent is −7.

When rewriting numbers in scientific notation, count the number of **place values** that the decimal point moves. *Do not count the number of zeros.*

Write 5,000,000 in scientific notation.

5,000,000.	Move 6 place values to the left.
$= 5 \times 10^6$	Positive exponent.

Write 0.0000005 in scientific notation.

0.0000005	Move 7 place values to the right.
$= 5 \times 10^{-7}$	Negative exponent.

EXAMPLE 5 In fiscal year 2011, the amount of federal education money loaned out and not yet paid back was about $745.5 billion. Rewrite $745.5 billion in scientific notation. (*Source:* www.finaid.com, 2011)

SOLUTION ▶

$745.5 billion	
$= \$745.5 \times \mathbf{10^9}$	1 billion = 10^9
$= \mathbf{\$7.455 \times 10^2} \times 10^9$	Rewrite mantissa; $\$745.5 = \$7.455 \times 10^2$
$= \$7.455 \times \mathbf{10^{2+9}}$	Product rule of exponents.
$= \$7.455 \times \mathbf{10^{11}}$	Simplify.

Writing a Number in Scientific Notation

1. Move the decimal point in the original number until the number is at least 1 and less than 10. This number is the mantissa.
2. Multiply the mantissa by a power of ten. If the original number is greater than 1, the exponent on the power of ten is equal to the number of place values that the decimal point moves. If the original number is less than 1, the exponent on the power of ten is equal to the opposite of the number of place values that the decimal point moves.

Practice Problems

For problems 1–6, write the measurement in scientific notation.

1. 0.000000085 m **2.** $30{,}000{,}000{,}000\ \frac{\text{cm}}{\text{s}}$ **3.** 9800 kg

4. 0.000063 L **5.** $2 billion **6.** $3.5 trillion

Multiplication and Division in Scientific Notation

To multiply measurements in scientific notation, multiply the mantissas, use the product rule of exponents to multiply the powers of ten, and multiply any units of measurement.

EXAMPLE 6 Evaluate: $(2 \times 10^{-2}\text{ cm})(1 \times 10^{-2}\text{ cm})(4 \times 10^{-2}\text{ cm})$

SOLUTION ▶

$(2 \times 10^{-2}\text{ cm})(1 \times 10^{-2}\text{ cm})(4 \times 10^{-2}\text{ cm})$	
$= (2 \times 1 \times 4) \times (10^{-2} \times 10^{-2} \times 10^{-2}) \times (\text{cm} \times \text{cm} \times \text{cm})$	Multiply mantissas, powers of ten, and units.
$= 8 \times 10^{-2+(-2)+(-2)} \times \text{cm}^3$	Simplify; product rule of exponents.
$= 8 \times \mathbf{10^{-6}}\ \text{cm}^3$	Simplify.

In the next example, the product is 40×10^9 cm^2. Since the mantissa, 40, is not less than 10, this measurement is not in scientific notation. The product is rewritten as 4.0×10^{10} cm^2. Decreasing the value of the mantissa from 40 to 4.0 results in an increase in the power of ten from 9 to 10.

EXAMPLE 7 Evaluate: $(5 \times 10^2 \text{ cm})(8 \times 10^7 \text{ cm})$

SOLUTION ▶

$(5 \times 10^2 \text{ cm})(8 \times 10^7 \text{ cm})$	
$= (5 \times 8) \times (10^2 \times 10^7) \times (\text{cm} \times \text{cm})$	Multiply mantissas, powers of ten, and units.
$= \mathbf{40} \times 10^{2+7} \times \text{cm}^2$	Simplify; product rule of exponents.
$= 40 \times 10^9 \times \text{cm}^2$	Simplify; mantissa is not between 1 and 10.
$= \mathbf{4.0 \times 10^1} \times 10^9 \times \text{cm}^2$	Rewrite the mantissa; $40 = 4.0 \times 10^1$
$= 4.0 \times 10^{1+9} \times \text{cm}^2$	Product rule of exponents.
$= 4.0 \times 10^{10} \text{ cm}^2$	Simplify.

To divide measurements written in scientific notation, use the quotient rule of exponents, $\frac{x^m}{x^n} = x^{m-n}$.

EXAMPLE 8 Evaluate: $\frac{8.4 \times 10^6 \text{ m}^2}{2 \times 10^3 \text{ m}}$

SOLUTION ▶

$\frac{8.4 \times 10^6 \text{ m}^2}{2 \times 10^3 \text{ m}}$	
$= \frac{8.4}{2} \times \frac{10^6}{10^3} \times \frac{\text{m}^2}{\text{m}}$	Divide mantissas, powers of ten, and units.
$= \mathbf{4.2} \times 10^{6-3} \times \text{m}^{2-1}$	Quotient rule of exponents.
$= 4.2 \times 10^3 \text{ m}$	Simplify.

In the next example, the quotient is $0.25 \times 10^{-3} \frac{\text{g}}{\text{cm}^3}$. Since the mantissa, 0.25, is less than 1, this measurement is not in scientific notation. The quotient is rewritten as $2.5 \times 10^{-4} \frac{\text{g}}{\text{cm}^3}$. Increasing the value of the mantissa results in a decrease in the power of ten from −3 to −4.

EXAMPLE 9 Evaluate: $\frac{2.075 \times 10^{-4} \text{ g}}{8.3 \times 10^{-1} \text{ cm}^3}$

SOLUTION ▶

$\frac{2.075 \times 10^{-4} \text{ g}}{8.3 \times 10^{-1} \text{ cm}^3}$	
$= \frac{2.075}{8.3} \times \frac{10^{-4}}{10^{-1}} \times \frac{\text{g}}{\text{cm}^3}$	Divide mantissas, powers of ten, and units.
$= \mathbf{0.25} \times 10^{-4-(-1)} \times \frac{\text{g}}{\text{cm}^3}$	Simplify; quotient rule of exponents.
$= 0.25 \times 10^{-3} \times \frac{\text{g}}{\text{cm}^3}$	Simplify; the mantissa is less than 1.
$= \mathbf{2.5 \times 10^{-1}} \times 10^{-3} \times \frac{\text{g}}{\text{cm}^3}$	Rewrite the mantissa; $0.25 = 2.5 \times 10^{-1}$
$= 2.5 \times 10^{-1+(-3)} \times \frac{\text{g}}{\text{cm}^3}$	Product rule of exponents.
$= 2.5 \times 10^{-4} \frac{\text{g}}{\text{cm}^3}$	Simplify.

Multiplying or Dividing Measurements in Scientific Notation

1. Multiply or divide the mantissas.
2. Apply the product rule or the quotient rule to the powers of ten.
3. Multiply or divide the units of measurement.
4. If the mantissa is not at least 1 and less than 10, rewrite the measurement in scientific notation. In this process, if the value of the mantissa increases, the exponent decreases, and if the value of the mantissa decreases, the exponent increases.

When raising a measurement in scientific notation to a power, use the many bases rule for a product, and apply the power to each part of the measurement. In the next example, the product is $729 \times 10^{-9}\ \text{m}^3$. Since the mantissa, 729, is not less than 10, this measurement is not in scientific notation. The product is rewritten as $7.29 \times 10^{-7}\ \text{m}^3$. Decreasing the value of the mantissa from 729 to 7.29 results in an increase in the power of ten from -9 to -7.

EXAMPLE 10 Evaluate: $(9 \times 10^{-3}\ \text{m})^3$

SOLUTION

$(9 \times 10^{-3}\ \text{m})^3$	
$= \mathbf{9^3} \times \mathbf{(10^{-3})^3} \times \mathbf{m^3}$	Many bases rule for a product.
$= \mathbf{9 \cdot 9 \cdot 9} \times (10^{-3})^3 \times \text{m}^3$	Rewrite as repeated multiplications.
$= \mathbf{729} \times 10^{(-3)(3)} \times \text{m}^3$	Simplify; power rule of exponents.
$= 729 \times 10^{-9} \times \text{m}^3$	Simplify; the mantissa is greater than 10.
$= \mathbf{7.29 \times 10^2} \times 10^{-9} \times \text{m}^3$	Rewrite the mantissa; $729 = 7.29 \times 10^2$
$= 7.29 \times 10^{2+(-9)} \times \text{m}^3$	Product rule of exponents.
$= 7.29 \times 10^{-7}\ \text{m}^3$	Simplify.

Raising a Measurement in Scientific Notation to a Power

1. Apply the many bases rule for a product to the expression.
2. Evaluate the mantissa.
3. Apply the power rule of exponents to the power of ten.
4. If needed, use the power rule of exponents to simplify the units.
5. If the mantissa is not at least 1 and less than 10, rewrite the measurement in scientific notation. In this process, if the value of the mantissa increases, the exponent decreases, and if the value of the mantissa decreases, the exponent increases.

Practice Problems

For problems 7–12, evaluate. The final answer must be in scientific notation.

7. $(3.4 \times 10^8\ \text{km})(2 \times 10^7\ \text{km})$

8. $(3.4 \times 10^8\ \text{km})(5 \times 10^7\ \text{km})$

9. $(3.4 \times 10^{-8}\ \text{km})(5 \times 10^{-7}\ \text{km})$

10. $\dfrac{9 \times 10^{-2}\ \text{g}}{3 \times 10^1\ \text{mL}}$

11. $\dfrac{4.2 \times 10^4\ \text{m}^3}{6 \times 10^1\ \text{m}^2}$

12. $(4 \times 10^2\ \text{m})^3$

Addition and Subtraction in Scientific Notation

We can add or subtract measurements in scientific notation only if the powers of ten and the units are the same. To add or subtract, add the mantissas, and leave the power of ten and the units the same. If the mantissa in the sum or difference is not at least 1 and less than 10, rewrite in scientific notation.

EXAMPLE 11 The mass of the earth is 5.97×10^{24} kg. The mass of Venus is 4.869×10^{24} kg. Find the total mass of the earth and Venus.

SOLUTION ▶

$5.97 \times 10^{24}\text{ kg} + 4.869 \times 10^{24}\text{ kg}$	The powers of ten are the same.
$= (5.97 + 4.869) \times 10^{24}\text{ kg}$	Add the mantissas.
$= \mathbf{10.839} \times 10^{24}\text{ kg}$	The mantissa is greater than 10.
$= \mathbf{1.0839 \times 10^{1}} \times 10^{24}\text{ kg}$	Rewrite the mantissa; $10.839 = 1.0839 \times 10^{1}$
$= 1.0839 \times 10^{\mathbf{1+24}}\text{ kg}$	Product rule of exponents.
$= 1.0839 \times 10^{\mathbf{25}}\text{ kg}$	Simplify.

In the next example, since the powers of ten are not the same, we rewrite one of the measurements, 6.42×10^{23} kg, as 0.642×10^{24} kg. Decreasing the value of the mantissa from 6.42 to 0.642 results in an increase in the power of ten from 23 to 24.

EXAMPLE 12 The mass of the earth is 5.97×10^{24} kg. The mass of Mars is 6.42×10^{23} kg. Find the total mass of the earth and Mars. Round the mantissa to the nearest hundredth.

SOLUTION ▶

$5.97 \times 10^{24}\text{ kg} + 6.42 \times 10^{23}\text{ kg}$	The powers of ten are different.
$= 5.97 \times 10^{24}\text{ kg} + \mathbf{0.642 \times 10^{1}} \times 10^{23}\text{ kg}$	Rewrite the mantissa; $6.42 = 0.642 \times 10^{1}$
$= 5.97 \times 10^{24}\text{ kg} + 0.642 \times 10^{\mathbf{1+23}}\text{ kg}$	Product rule of exponents.
$= 5.97 \times 10^{24}\text{ kg} + 0.642 \times 10^{\mathbf{24}}\text{ kg}$	Simplify.
$= (5.97 + 0.642) \times 10^{24}\text{ kg}$	The powers of ten are the same; add the mantissas.
$= \mathbf{6.612} \times 10^{24}\text{ kg}$	Simplify.
$\approx \mathbf{6.61} \times 10^{24}\text{ kg}$	Round mantissa to the nearest hundredth.

Adding and Subtracting Measurements in Scientific Notation

1. If the powers of ten in the measurements are not the same, rewrite one of the measurements so that the powers of ten are the same. After rewriting, the measurement will no longer be in scientific notation.
2. Add the mantissas; the powers of ten and the units of measurement do not change.
3. If the mantissa is not at least 1 and less than 10, rewrite the measurement in scientific notation. In this process, if the value of the mantissa increases, the exponent decreases, and if the value of the mantissa decreases, the exponent increases.

Practice Problems

For problems 13–18, evaluate. The final answer must be in scientific notation.

13. $6 \times 10^{5}\text{ km} + 2 \times 10^{5}\text{ km}$
14. $8 \times 10^{5}\text{ m} + 4 \times 10^{5}\text{ m}$
15. $8 \times 10^{5}\text{ km} + 3.1 \times 10^{6}\text{ km}$
16. $6.4 \times 10^{-8}\text{ m} - 1.9 \times 10^{-8}\text{ m}$
17. $3 \times 10^{-4}\text{ km} + 8 \times 10^{-5}\text{ km}$
18. $9 \times 10^{-2}\text{ s} + 5 \times 10^{-4}\text{ s}$

Applications

Formulas and measurements in scientific notation are common in college science courses. The units of measurement include **combined units** such as $\frac{\text{kg}\cdot\text{m}}{\text{s}^2}$.

EXAMPLE 13 Solve: $P = \left(2 \times 10^4 \frac{\text{kg}\cdot\text{m}}{\text{s}^2}\right)\left(3 \times 10^2 \frac{\text{m}}{\text{s}}\right)$

SOLUTION

$$P = \left(2 \times 10^4 \frac{\text{kg}\cdot\text{m}}{\text{s}^2}\right)\left(3 \times 10^2 \frac{\text{m}}{\text{s}}\right)$$

$$P = 2\cdot 3 \times 10^4\cdot 10^2 \times \frac{\text{kg}\cdot\text{m}}{\text{s}^2}\cdot\frac{\text{m}}{\text{s}}$$ Multiply mantissas, powers of ten, and units.

$$P = \mathbf{6} \times 10^{4+2} \times \frac{\text{kg}\cdot\text{m}^{1+1}}{\text{s}^{2+1}}$$ Product rule of exponents.

$$P = 6 \times 10^6 \frac{\text{kg}\cdot\text{m}^2}{\text{s}^3}$$ Simplify.

In the next example, we begin by using the many bases rule for a product to simplify $(4 \times 10^{-2}\,\text{m})^2$.

EXAMPLE 14 Solve: $W = (0.5)\left(80\frac{\text{kg}}{\text{s}^2}\right)(4 \times 10^{-2}\,\text{m})^2$

SOLUTION

$$W = (0.5)\left(80\frac{\text{kg}}{\text{s}^2}\right)(4 \times 10^{-2}\,\text{m})^2$$

$$W = (0.5)\left(80\frac{\text{kg}}{\text{s}^2}\right)(4^2 \times (10^{-2})^2\,\text{m}^2)$$ Many bases rule for a product.

$$W = (0.5)\left(80\frac{\text{kg}}{\text{s}^2}\right)(16 \times 10^{(-2)(2)}\,\text{m}^2)$$ Simplify; power rule of exponents.

$$W = (0.5)\left(80\frac{\text{kg}}{\text{s}^2}\right)(16 \times 10^{-4}\,\text{m}^2)$$ Simplify.

$$W = (0.5)(80)(16) \times 10^{-4} \times \frac{\text{kg}\cdot\text{m}^2}{\text{s}^2}$$ Multiply mantissas and units.

$$W = \mathbf{640} \times 10^{-4}\frac{\text{kg}\cdot\text{m}^2}{\text{s}^2}$$ Simplify.

$$W = \mathbf{6.4 \times 10^2} \times 10^{-4}\frac{\text{kg}\cdot\text{m}^2}{\text{s}^2}$$ Rewrite the mantissa; $640 = 6.4 \times 10^2$

$$W = 6.4 \times 10^{2+(-4)}\frac{\text{kg}\cdot\text{m}^2}{\text{s}^2}$$ Product rule of exponents

$$W = 6.4 \times 10^{-2}\frac{\text{kg}\cdot\text{m}^2}{\text{s}^2}$$ Simplify.

In the next example, we round the mantissa to a given place value.

EXAMPLE 15 Solve $C = \frac{1.0 \times 10^{-14}\,\text{mole}}{5.2 \times 10^{-7}\,\text{L}}$. Round the mantissa to the nearest tenth.

SOLUTION

$$C = \frac{1.0 \times 10^{-14}\,\text{mole}}{5.2 \times 10^{-7}\,\text{L}}$$

$C = \frac{1.0}{5.2} \times \frac{10^{-14}}{10^{-7}} \times \frac{\text{mole}}{\text{L}}$	Divide mantissas, powers of ten, and units.
$C = \mathbf{0.192\ldots} \times 10^{-14-(-7)} \frac{\text{mole}}{\text{L}}$	Simplify; quotient rule of exponents.
$C = 0.192\ldots \times 10^{-7} \frac{\text{mole}}{\text{L}}$	Simplify.
$C = \mathbf{1.92\ldots \times 10^{-1}} \times 10^{-7} \frac{\text{mole}}{\text{L}}$	Rewrite the mantissa; $0.192\ldots = 1.92\ldots \times 10^{-1}$
$C = 1.92\ldots \times 10^{-1+(-7)} \frac{\text{mole}}{\text{L}}$	Product rule of exponents.
$C = 1.92\ldots \times 10^{-8} \frac{\text{mole}}{\text{L}}$	Simplify.
$C \approx \mathbf{1.9} \times 10^{-8} \frac{\text{mole}}{\text{L}}$	Round the mantissa to the nearest tenth.

Practice Problems

For problems 19–20, solve.

19. $E = \frac{1}{2}(6.8 \times 10^1 \text{ kg})\left(4 \frac{\text{m}}{\text{s}}\right)^2$

20. $v = -\left(\frac{1.5 \times 10^{-1} \text{ kg}}{5.0 \times 10^1 \text{ kg}}\right)\left(4 \times 10^1 \frac{\text{m}}{\text{s}}\right)$

Using Technology: Scientific Notation

For arithmetic in place value notation, a scientific calculator is in Normal or Floating Point mode. To enter numbers and have answers in scientific notation, we can change the mode to **Scientific Notation**. On some calculators, look for the letters SCI printed above a key. Press [2nd]; press this key. The calculator will now present answers in scientific notation.

On other calculators, press the [MODE] key. The entry screen will show different modes. Move the cursor to select SCI. Or, the entry screen will show a list of different modes paired with numbers. Press the number paired with SCI.

Most scientific calculators allow the entry of numbers in scientific notation even when it is not in scientific notation mode. *Do not use the exponent keys,* [y^x] *or* [^] *to enter numbers in scientific notation.* Instead, use the [EE] key.

EXAMPLE 16 Enter 3×10^5 on a scientific calculator.

Press [3] [EE] [5]

Notice that the base, 10, is not typed. EE means $\times 10^?$ and will appear on the calculator screen as E. Press the key for the exponent after pressing EE.

EXAMPLE 17 Enter 6.02×10^{23} on a scientific calculator.

If necessary, change the mode of the calculator to scientific notation.

Press [6] [.] [0] [2] [EE] [2] [3]

Practice Problems For problems 21–23, use a scientific calculator in scientific notation mode to evaluate.

21. $(3.4 \times 10^8 \text{ km})(2.5 \times 10^7 \text{ km})$ **22.** $\frac{3.3 \times 10^{-2} \text{ g}}{1.5 \times 10^1 \text{ mL}}$ **23.** $(2.5 \times 10^2 \text{ m})^3$

5.2 Vocabulary Practice

Match the term with its description.

1. $(x^m)^n = x^{mn}$
2. $\dfrac{x^m}{x^n} = x^{m-n}, x \neq 0$
3. $\left(\dfrac{x}{y}\right)^n = \dfrac{x^n}{y^n}, y \neq 0$
4. $x^m \cdot x^n = x^{m+n}$
5. A notation in which we write a number as a product of a number that is at least 1 and less than 10 and a power of ten
6. In 6×10^7, 10^7 is an example of this.
7. $\times$
8. The value of an integer other than 0 raised to the zero power
9. $(xy)^m = x^m y^m$
10. In 5×10^3, 5 is an example of this.

A. many bases rule for a product
B. many bases rule for a quotient
C. mantissa
D. notation for multiplication
E. one
F. power of ten
G. power rule of exponents
H. product rule of exponents
I. quotient rule of exponents
J. scientific notation

5.2 Exercises

Follow your instructor's guidelines for showing your work.

For exercises 1–10, write the measurement or number in scientific notation.

1. The virus that causes dengue fever is spherical with a diameter of about 0.00000004 m.
2. The minimum length of the virus that causes measles is 0.000001 m.

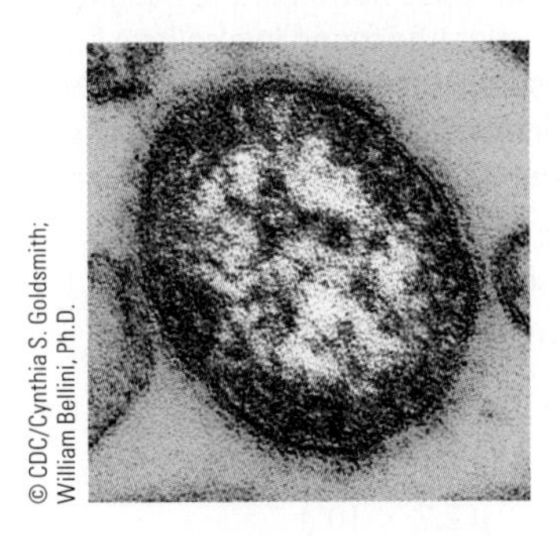

© CDC/Cynthia S. Goldsmith; William Bellini, Ph.D.

3. The genome of the virus that causes rabies contains a maximum of 15,000 nucleotides.
4. The genome of the virus that causes rubella contains a maximum of 11,800 nucleotides.
5. Avogadro's number is 602,000,000,000,000,000,000,000.
6. The speed of light in a vacuum is about 299,000,000 $\dfrac{\text{m}}{\text{s}}$.
7. Planck's constant is 0.0000000000000000000000000000000006626 $\dfrac{\text{m}^2 \cdot \text{kg}}{\text{s}}$.
8. Newton's constant of gravitation is 0.0000000000667 $\dfrac{\text{m}^3}{\text{kg} \cdot \text{s}^2}$.
9. The diameter of Jupiter is about 88,800 mi.
10. The diameter of Uranus is about 31,800 mi.
11. A measurement is 34×10^{12} km. When this measurement is rewritten in scientific notation, will the exponent be greater or less than 12?
12. A measurement is 560×10^{14} km. When this measurement is rewritten in scientific notation, will the exponent be greater or less than 14?
13. A measurement is 0.0005×10^7 g. When this measurement is rewritten in scientific notation, will the exponent be greater or less than 7?
14. A measurement is 0.000003×10^4 g. When this measurement is rewritten in scientific notation, will the exponent be greater or less than 4?

For exercises 15–16, use the information in the table.

Holders of United States Treasury securities, Oct. 2011	Amount
China, mainland	\$1134.1 billion
Japan	\$979.0 billion
United Kingdom	\$408.2 billion

Source: www.treasury.gov, Dec. 15, 2011

15. Write the amount of securities held by mainland China in scientific notation.
16. Write the amount of securities held by Japan in scientific notation.

For exercises 17–18, use the information in the table.

Public domain land acquisition (year acquired)	Area (acres)
Louisiana Purchase (1803)	236,825,600
Oregon Compromise (1846)	183,386,240
Gadsden Purchase (1853)	18,988,800
Alaska Purchase (1867)	378,242,560

Source: www.blm.gov, 2010

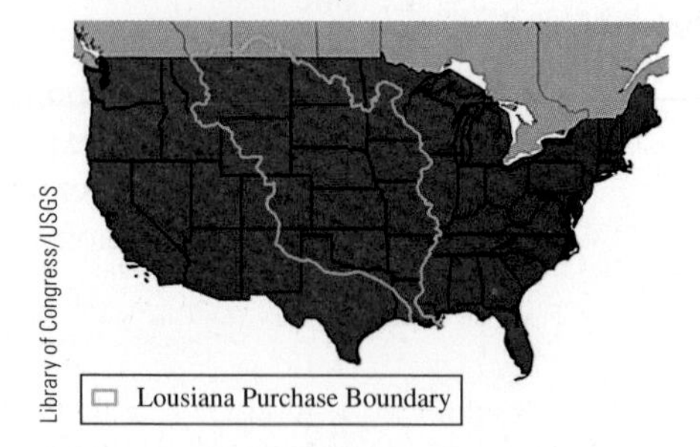

Library of Congress/USGS

17. Round the amount of land acquired in the Oregon Compromise to the nearest million, and write in scientific notation.

18. Round the amount of land acquired in the Louisiana Purchase to the nearest million, and write in scientific notation.

For exercises 19–50, evaluate. The final answer must be in scientific notation.

19. $(4 \times 10^3)(2 \times 10^4)$

20. $(3 \times 10^4)(2 \times 10^5)$

21. $(6 \times 10^3)(2 \times 10^4)$

22. $(8 \times 10^4)(2 \times 10^5)$

23. $(6 \times 10^{-3})(2 \times 10^{-4})$

24. $(8 \times 10^{-4})(2 \times 10^{-5})$

25. $\dfrac{6 \times 10^8}{2 \times 10^3}$

26. $\dfrac{8 \times 10^9}{2 \times 10^3}$

27. $\dfrac{6 \times 10^8}{2 \times 10^{-3}}$

28. $\dfrac{8 \times 10^9}{2 \times 10^{-3}}$

29. $\dfrac{6 \times 10^{-8}}{2 \times 10^3}$

30. $\dfrac{8 \times 10^{-9}}{2 \times 10^3}$

31. $\dfrac{164 \times 10^7}{2 \times 10^5}$

32. $\dfrac{146 \times 10^8}{2 \times 10^5}$

33. $\dfrac{164 \times 10^7}{2 \times 10^{-5}}$

34. $\dfrac{146 \times 10^8}{2 \times 10^{-5}}$

35. $\dfrac{164 \times 10^{-7}}{2 \times 10^5}$

36. $\dfrac{146 \times 10^{-8}}{2 \times 10^5}$

37. $(9 \times 10^3)^2$

38. $(6 \times 10^4)^2$

39. $(9 \times 10^{-3})^2$

40. $(6 \times 10^{-4})^2$

41. $7 \times 10^3 + 2 \times 10^3$

42. $5 \times 10^4 + 2 \times 10^4$

43. $7 \times 10^3 - 2 \times 10^3$

44. $5 \times 10^4 - 2 \times 10^4$

45. $9 \times 10^2 + 5 \times 10^2$

46. $9 \times 10^3 + 4 \times 10^3$

47. $8 \times 10^5 + 3 \times 10^4$

48. $7 \times 10^6 + 2 \times 10^5$

49. $8 \times 10^{-5} + 3 \times 10^{-4}$

50. $7 \times 10^{-6} + 2 \times 10^{-5}$

For exercises 51–80, evaluate. The final answer must be in scientific notation.

51. $(2.4 \times 10^5\ \text{cm})(3 \times 10^8\ \text{cm})$

52. $(1.5 \times 10^4\ \text{cm})(4 \times 10^7\ \text{cm})$

53. $(8 \times 10^4\ \text{in.})(2 \times 10^3\ \text{in.})(2 \times 10^4\ \text{in.})$

54. $(7 \times 10^3\ \text{in.})(4 \times 10^2\ \text{in.})(3 \times 10^2\ \text{in.})$

55. $(3.14)(2.4 \times 10^{-8}\ \text{km})^2$

56. $(3.14)(1.8 \times 10^{-5}\ \text{km})^2$

57. $(3.14)(3 \times 10^{-1}\ \text{cm})^2(4 \times 10^{-1}\ \text{cm})$

58. $(3.14)(4 \times 10^{-1}\ \text{cm})^2(2 \times 10^{-1}\ \text{cm})$

59. $\dfrac{9 \times 10^8\ \text{cm}^2}{2 \times 10^5\ \text{cm}}$

60. $\dfrac{7 \times 10^9\ \text{cm}^2}{2 \times 10^4\ \text{cm}}$

61. $\dfrac{9 \times 10^{-8}\ \text{cm}^2}{2 \times 10^{-5}\ \text{cm}}$

62. $\dfrac{7 \times 10^{-9}\ \text{cm}^2}{2 \times 10^{-4}\ \text{cm}}$

63. $\dfrac{4 \times 10^{-2}\ \text{g}}{5 \times 10^1\ \text{mL}}$

64. $\dfrac{3 \times 10^{-2}\ \text{g}}{5 \times 10^2\ \text{mL}}$

65. $\dfrac{(8 \times 10^3\ \text{in.})^2}{4 \times 10^2\ \text{in.}}$

66. $\dfrac{(6 \times 10^3\ \text{in.})^2}{6 \times 10^4\ \text{in.}}$

67. $(6 \times 10^{-4}\ \text{m})^3$

68. $(8 \times 10^{-5}\ \text{m})^3$

69. $(6 \times 10^4\ \text{m})^3$

70. $(8 \times 10^5\ \text{m})^3$

71. $\dfrac{1}{4 \times 10^8}$

72. $\dfrac{1}{5 \times 10^6}$

73. $5.2 \times 10^{-8}\ \text{g} + 1.3 \times 10^{-8}\ \text{g}$

74. $3.9 \times 10^{-6}\ \text{m} + 4.2 \times 10^{-6}\ \text{m}$

75. $5.2 \times 10^{-9}\ \text{g} + 1.3 \times 10^{-8}\ \text{g}$

76. $3.9 \times 10^{-7}\ \text{m} + 4.2 \times 10^{-6}\ \text{m}$

77. $9.3 \times 10^3\ \text{kg} + 4.8 \times 10^3\ \text{kg}$

78. $7.9 \times 10^4\ \text{km} + 6.8 \times 10^4\ \text{km}$

79. $9.3 \times 10^3\ \text{kg} + 4.8 \times 10^4\ \text{kg}$

80. $7.9 \times 10^4\ \text{km} + 6.8 \times 10^5\ \text{km}$

For exercises 81–96, the solution of the equation is the answer to a problem that might be found in a college science textbook. The units of measurement are not included. Solve. If the number is less than 1 or greater than or equal to 10, write the answer in scientific notation. If necessary, round the mantissa to the nearest hundredth.

81. $C = \dfrac{7.7 \times 10^{-13}}{5 \times 10^{-6}}$

82. $C = \dfrac{5 \times 10^{-8}}{2 \times 10^{-2}}$

83. $n = \dfrac{(1 \times 10^2)(2 \times 10^{-4})}{(8.31)(2.93 \times 10^2)}$

84. $F = \dfrac{(7.5 \times 10^1)(2.7 \times 10^1)}{1 \times 10^{-2}}$

85. $M = \dfrac{(4)(3.14)(1 \times 10^{-7})(6)}{2(3.14)(5 \times 10^{-3})}$

86. $L = \dfrac{9.45 \times 10^6}{(5 \times 10^1)(9.8)(2)}$

87. $m = \dfrac{(9.8)(6.38 \times 10^6)^2}{6.67 \times 10^{-11}}$

88. $f = \dfrac{(1.34 \times 10^1)^2}{(5 \times 10^1)(9.8)}$

89. $v = \dfrac{5 \times 10^4}{(4 \times 10^3)(7.99 \times 10^{-1})}$

90. $d = \dfrac{(3.5 \times 10^5)}{(1 \times 10^3)(9.8)(5.7)}$

91. $h = \dfrac{2(7.3 \times 10^{-2})}{(1 \times 10^3)(9.8)(6 \times 10^{-3})}$

92. $r = \dfrac{(3.14)(3 \times 10^{-4})^4(5.15 \times 10^3)}{8(2.7 \times 10^{-3})(2.0 \times 10^{-2})}$

93. $v = \dfrac{(2.6 \times 10^{-3})(3 \times 10^3)}{(1.05 \times 10^3)(2 \times 10^{-3})}$

94. $K_a = \dfrac{(1.34 \times 10^{-3})(1.34 \times 10^{-3})}{9.87 \times 10^{-2}}$

95. $C = \dfrac{(1.2 \times 10^{-5})(1.5 \times 10^{-1})}{1.8 \times 10^{-5}}$

96. $f = \dfrac{(6.67 \times 10^{-11})(3 \times 10^{-1})(9 \times 10^{-1})}{(4 \times 10^{-1})^2}$

For exercises 97–104, solve. If the number is less than 1 or greater than or equal to 10, write the answer in scientific notation. Include the units of measurement. If necessary, round the mantissa to the nearest hundredth.

97. $P = \dfrac{3.5 \times 10^2 \text{ joule}}{2.5 \times 10^{-1} \text{ s}}$

98. $R = \dfrac{1.5 \times 10^{-6} \text{ ohm} \cdot \text{m}}{3.6 \times 10^{-7} \text{ m}^2}$

99. $P = (5.0 \times 10^{-2} \text{ kg})\left(4.4 \times 10^1 \dfrac{\text{m}}{\text{s}}\right)$

100. $P = (3.0 \times 10^{-2} \text{ kg})\left(5.4 \times 10^1 \dfrac{\text{m}}{\text{s}}\right)$

101. $E = \dfrac{1}{2}(5.2 \times 10^2 \text{ kg})\left(3.8 \dfrac{\text{m}}{\text{s}}\right)^2$

102. $E = \dfrac{1}{2}(2.5 \times 10^3 \text{ kg})\left(1.2 \dfrac{\text{m}}{\text{s}}\right)^2$

103. $P = 3.9 \times 10^3 \dfrac{\text{kg} \cdot \text{m}}{\text{s}} - \left(-2.25 \times 10^4 \dfrac{\text{kg} \cdot \text{m}}{\text{s}}\right)$

104. $P = 4.3 \times 10^3 \dfrac{\text{kg} \cdot \text{m}}{\text{s}} - \left(-3.6 \times 10^4 \dfrac{\text{kg} \cdot \text{m}}{\text{s}}\right)$

For exercises 105–106, solve. An A.U. is a unit of distance, the astronomical unit. The distance from the earth to the sun is 1 A.U. Round the mantissa to the nearest hundredth.

105. $D = \left(\dfrac{6 \text{ A.U.}}{1}\right)\left(\dfrac{1.49 \times 10^8 \text{ km}}{1 \text{ A.U.}}\right)\left(\dfrac{1 \text{ mi}}{1.61 \text{ km}}\right)$

106. $D = \left(\dfrac{6 \text{ light-years}}{1}\right)\left(\dfrac{9.47 \times 10^{12} \text{ km}}{1 \text{ light-year}}\right)\left(\dfrac{1 \text{ mi}}{1.61 \text{ km}}\right)$

Problem Solving: Practice and Review

Follow your instructor's guidelines for using the five steps as outlined in Section 1.3, p. 55.

107. Find the number of gallons delivered per day from the wells to Brantley Reservoir.

> The Seven Rivers well field . . . and an additional three private wells can produce about 20,000 gallons of water per minute that is delivered through a 12-mile pipeline network directly to Brantley Reservoir located about 12 miles north of Carlsbad. (*Source:* www.currentargus.com, Sept. 10, 2009)

108. Find the number of boxes of apples that needed to be sold in 1998 to make a profit of $20,000. Round to the nearest whole number.

> Last year, wholesale prices went as low as $5 a box; it takes $9.50 a box to break even, the growers say. This year [1998], apples are selling at an average of $14 a box. (*Source:* www.nytimes.com, Oct. 14, 1998)

For exercises 109–110, according to the United States Bowling Congress, a regulation ten-pin bowling ball shall not have a circumference of more than 27.002 in. nor less than 26.704 in.

© Nikkytok/Shutterstock.com

109. Find the maximum diameter of a regulation bowling ball. ($\pi \approx 3.14$.) Round to the nearest tenth of an inch. (*Source:* www.bowl.com)

110. Find the maximum volume in cubic inches of a regulation bowling ball. ($\pi \approx 3.14$.) Round to the nearest tenth.

Technology

For exercises 111–114, evaluate the expression or solve the equation. Use a scientific calculator in scientific notation mode. If necessary, round the mantissa to the nearest hundredth.

111. $\dfrac{4 \times 10^{-2} \text{ g}}{5 \times 10^1 \text{ mL}}$

112. $\dfrac{3 \times 10^{-2} \text{ g}}{5 \times 10^2 \text{ mL}}$

113. $K_a = \dfrac{(1.34 \times 10^{-3})(1.34 \times 10^{-3})}{9.87 \times 10^{-2}}$

114. $C = \dfrac{(1.2 \times 10^{-5})(1.5 \times 10^{-1})}{1.8 \times 10^{-5}}$

Find the Mistake

For exercises 115–118, the completed problem has one mistake.
(a) Describe the mistake in words, or copy down the whole problem and highlight or circle the mistake.
(b) Do the problem correctly.

115. Problem: Rewrite 0.00000079 in scientific notation.

Incorrect Answer: 7.9×10^7

116. Problem: Evaluate $(6.4 \times 10^5)(3 \times 10^8)$. Write the answer in scientific notation.

Incorrect Answer: 19.2×10^{13}

117. Problem: Evaluate $(2 \times 10^3)(4 \times 10^5)$. Write the answer in scientific notation.

Incorrect Answer: 8×10^{15}

118. **Problem:** Evaluate $(9 \times 10^{-5})(3 \times 10^{-7})$. Write the answer in scientific notation.

Incorrect Answer: $(9 \times 10^{-5})(3 \times 10^{-7})$

$= 27 \times 10^{-12}$

$= 2.7 \times 10^{-13}$

Review

119. Explain how to identify like terms.

120. Are $5x^2$ and $5x^3$ like terms? Explain.

121. Simplify: $8x - 9x + 12x - 15x$

122. Simplify: $-3x^2 - 5x^2 + x^2 - 7x^2$

SUCCESS IN COLLEGE MATHEMATICS

123. Search the Internet for different note-taking systems. Describe two of these.

124. Do you think that you would find either of these note-taking systems to be better than what you are using now? If so, why? If not, why not?

© Sourabh/Shutterstock.com

A *mono*rail is a train that runs on *one* rail. A *bi*cycle has *two* wheels. A *tri*angle has *three* angles. A *poly*syllabic word has *many* syllables. The prefixes in these words help us understand their meanings. In this section, we will learn how these prefixes are used to classify some algebraic expressions.

SECTION 5.3 Introduction to Polynomials

After reading the text, working the practice problems, and completing assigned exercises, you should be able to:

1. Determine whether an expression is a polynomial.
2. Identify polynomial expressions that are monomials, binomials, or trinomials.
3. Identify the degree of a polynomial expression in one variable.
4. Write a polynomial expression in descending order.
5. Add or subtract polynomial expressions.
6. Multiply a polynomial expression by a monomial.

Polynomial Expressions

A **variable** is a letter or other symbol that represents an unknown number. A **constant** is a number. When we multiply a number and a variable, the number is a **coefficient**. A **term** is a variable, a constant, or the product of a number and a variable(s).

EXAMPLE 1 Identify the number of terms, the variable, the coefficient, and the constant in the expression $4x + 6$.

SOLUTION ▸ This expression has two terms, $4x$ and 6. The variable is x. The coefficient is 4. The constant is 6.

A **polynomial expression** is a single term or a sum or difference of terms. In a polynomial expression, a variable can never be an exponent, be in a denominator, or be under a radical sign such as a square root. If a variable is raised to a power, this power is always an integer greater than or equal to 0.

EXAMPLE 2 Is the expression a polynomial? If not, explain why.

SOLUTION **(a)** $3x + 9$

Yes.

(b) $2x - \frac{4}{x}$

No, there is a variable in a denominator.

(c) $x^2 + 5x - 8$

Yes.

(d) $2^x + 3^x$

No, a variable is an exponent.

(e) 9

Yes, a polynomial can be a constant.

(f) $x^2 + xy + y^2$

Yes.

When a polynomial in one variable is in **descending order**, the exponents on the variables decrease from left to right and the coefficient of the first term is the **lead coefficient**. For example, $2x^4 + 7x^2 - 9x + 8$ is in descending order, and the lead coefficient is 2.

EXAMPLE 3 **(a)** Write the polynomial $8a + 9a^2 + 5a^4 - 6$ in descending order.

SOLUTION $5a^4 + 9a^2 + 8a - 6$

(b) Identify the lead coefficient in $5a^4 + 9a^2 + 8a - 6$.

The lead coefficient is 5.

The **degree** of a polynomial expression in one variable is the value of the largest exponent on the variable. For example, the degree of $p^2 - 7p + 1$ is 2. The polynomial expression 7 has no variable. However, since $x^0 = 1$, we can rewrite this expression as $7x^0$, and its degree is 0.

EXAMPLE 4 Identify the degree of each polynomial.

SOLUTION **(a)** $8x^3 - 5x^2 + 4$

Since the greatest exponent is 3, the degree is 3.

(b) $-9x$

Since $-9x$ is equivalent to $-9x^1$, the degree is 1.

(c) 5

Since 5 is equivalent to $5x^0$, the degree is 0.

To create special names for polynomials with one, two, or three terms, we combine a prefix with the root word "nomial."

Prefix	Meaning of prefix	Name of expression	Number of terms	Example
mono-	one	monomial	1	$-9x$
bi-	two	binomial	2	$3x + 4$
tri-	three	trinomial	3	$x^2 + 5x + 6$

Polynomial Expression

A polynomial expression is a sum or difference of terms or a single term. A polynomial expression in one variable can be written in the form $a_n x^n + a_{n-1}x^{n-1} + \cdots + a_2 x^2 + a_1 x + a_0$, where x is a variable, n is an integer ≥ 0, and $a_0, a_1, a_2, \ldots, a_n$ are real numbers. The **degree** of this expression is n. The **lead coefficient** is a_n.

Practice Problems

1. For $7x^2 + 8x + 1$, identify the
   a. number of terms **b.** variable **c.** constant
2. Identify the expressions that are polynomials.
   a. $6z^3 + 8z - 9$ **b.** $\dfrac{2b + 1}{b - 3}$ **c.** x^2 **d.** -2 **e.** $7x^2 + 11$ **f.** $7^x + 8$
3. Explain the difference between a monomial, a binomial, and a trinomial, and write an example of each.
4. Write $8x + 9x^4 - 14x^2 + 12 - x^3$ in descending order, and identify the lead coefficient.
5. Identify the degree of $8x^3 - 5x^2 + 9$.
6. Identify the degree of 7.

Adding and Subtracting Polynomial Expressions

In the next example, we simplify a polynomial expression by combining like terms.

EXAMPLE 5 Simplify: $3p^2 + 8p - 9 + 2p^2 - p + 15$

SOLUTION ▶

$3p^2 + 8p - 9 + 2p^2 - p + 15$

$= 3p^2 + 2p^2 + 8p - p - 9 + 15$ Change order; commutative property.

$= 5p^2 + 7p + 6$ Combine like terms.

To add polynomial expressions, write the terms in descending order, arrange the polynomials horizontally or vertically, and combine like terms.

EXAMPLE 6 Simplify: $(10a^2 + 9a + 3) + (a^2 + 4a + 1)$

SOLUTION ▶

Horizontal Addition

$(10a^2 + 9a + 3) + (a^2 + 4a + 1)$

$= 10a^2 + a^2 + 9a + 4a + 3 + 1$

$= 11a^2 + 13a + 4$

Vertical Addition

$$\begin{array}{r} 10a^2 + 9a + 3 \\ + \quad a^2 + 4a + 1 \\ \hline 11a^2 + 13a + 4 \end{array}$$

When adding polynomials horizontally, some students find it helpful to use a shape to mark like terms.

$10a^2 + 9a + 3 + a^2 + 4a + 1$

$= 11a^2 + 13a + 4$

In the next example, one of the polynomials does not have a w-term. In vertical addition, we include $0w$ in this expression to act as a placeholder.

EXAMPLE 7 Simplify: $(5w^2 - 6w - 9) + (2w^2 - 10)$

SOLUTION

Horizontal Addition

$(5w^2 - 6w - 9) + (2w^2 - 10)$

$= 5w^2 + 2w^2 - 6w - 9 - 10$

$= 7w^2 - 6w - 19$

Vertical Addition

$$\begin{array}{r} 5w^2 - 6w - 9 \\ + \; 2w^2 + \mathbf{0w} - 10 \\ \hline 7w^2 - 6w - 19 \end{array}$$

In subtracting two polynomials, the second polynomial is inside parentheses. When subtracting polynominals horizontally, use the distributive property to simplify. In vertical subtraction, remember that each term is being subtracted.

EXAMPLE 8 Simplify: $(6c + 9) - (2c - 10)$

SOLUTION

Horizontal Subtraction

$(6c + 9) - (2c - 10)$

$= (6c + 9) \mathbf{- 1}(2c - 10)$

$= (6c + 9) \mathbf{- 1}(2c) \mathbf{- 1}(-10)$

$= 6c + 9 - 2c + 10$

$= 4c + 19$

Vertical Subtraction

$$\begin{array}{r} 6c + 9 \\ - \; (2c - 10) \\ \hline 4c + 19 \end{array}$$

In the next example, we again use a placeholder, $0y$, in the vertical subtraction.

EXAMPLE 9 Simplify: $(y^3 + 6y^2 - 4y + 2) - (5y^3 + 9y^2 - 8)$

SOLUTION

Horizontal Subtraction

$(y^3 + 6y^2 - 4y + 2) - (5y^3 + 9y^2 - 8)$

$= (y^3 + 6y^2 - 4y + 2) \mathbf{- 1}(5y^3 + 9y^2 - 8)$

$= (y^3 + 6y^2 - 4y + 2) \mathbf{- 1}(5y^3) \mathbf{- 1}(9y^2) \mathbf{- 1}(-8)$

$= y^3 + 6y^2 - 4y + 2 \mathbf{- 5y^3 - 9y^2 + 8}$

$= y^3 - 5y^3 + 6y^2 - 9y^2 - 4y + 2 + 8$

$= -4y^3 - 3y^2 - 4y + 10$

Vertical Subtraction

$$\begin{array}{r} y^3 + 6y^2 - 4y + 2 \\ - \; (5y^3 + 9y^2 + 0y - 8) \\ \hline -4y^3 - 3y^2 - 4y + 10 \end{array}$$

We can also add and subtract polynomials that include more than one variable. For two terms to be like terms, each variable and its exponent must be exactly the same.

EXAMPLE 10 Simplify: $(9x^2 + 12xy + 4y^2) - (x^2 - xy + y^2)$

SOLUTION

Horizontal Subtraction

$(9x^2 + 12xy + 4y^2) - (x^2 - xy + y^2)$

$= (9x^2 + 12xy + 4y^2) \mathbf{- 1}(x^2 - xy + y^2)$

$= (9x^2 + 12xy + 4y^2) \mathbf{- 1}(x^2) \mathbf{- 1}(-xy) \mathbf{- 1}(y^2)$

$= 9x^2 + 12xy + 4y^2 - x^2 + xy - y^2$

$= 9x^2 - x^2 + 12xy + xy + 4y^2 - y^2$

$= 8x^2 + 13xy + 3y^2$

Vertical Subtraction

$$\begin{array}{r} 9x^2 + 12xy + 4y^2 \\ - \; (x^2 - xy + y^2) \\ \hline 8x^2 + 13xy + 3y^2 \end{array}$$

When adding polynomial expressions in which the coefficients or constants are fractions, identify like terms and find a common denominator for the coefficients of like terms. Do not rewrite improper fractions as mixed numbers. We can clear fractions only from an equation in which we can multiply both sides by the same constant. Fractions cannot be cleared from an expression.

EXAMPLE 11 Simplify: $\left(\frac{1}{6}a^2 + \frac{1}{2}a + 9\right) + \left(\frac{2}{5}a^2 + \frac{3}{4}a\right)$

SOLUTION ▶ $\left(\frac{1}{6}a^2 + \frac{1}{2}a + 9\right) + \left(\frac{2}{5}a^2 + \frac{3}{4}a\right)$

$= \frac{1}{6}a^2 + \frac{2}{5}a^2 + \frac{1}{2}a + \frac{3}{4}a + 9$ — Commutative property; change order.

$= \frac{1}{6}\cdot\frac{\mathbf{5}}{\mathbf{5}}\cdot a^2 + \frac{2}{5}\cdot\frac{\mathbf{6}}{\mathbf{6}}\cdot a^2 + \frac{1}{2}\cdot\frac{\mathbf{2}}{\mathbf{2}}\cdot a + \frac{3}{4}a + 9$ — Multiply by fractions equal to 1.

$= \frac{5}{30}a^2 + \frac{12}{30}a^2 + \frac{2}{4}a + \frac{3}{4}a + 9$ — Simplify.

$= \frac{17}{30}a^2 + \frac{5}{4}a + 9$ — Combine like terms.

Practice Problems

For problems 7–12, simplify.

7. $8d^3 - 2d^2 - 12d^3 + d - d^2 + 9$ **8.** $(2x^2 + 15x - 9) + (7x^2 + x - 4)$

9. $(8x^2 + 15x - 9) - (7x^2 + x - 4)$ **10.** $(6x^2 + 5x + 1) - (4x^2 + 10)$

11. $(8h^3 + 3h^2k - 2hk^2 - 9) + (4h^3 - 5hk^2 - 17)$

12. $\left(\frac{1}{8}x^2 + \frac{2}{3}x - 9\right) + \left(\frac{2}{5}x^2 - \frac{1}{4}x + 2\right)$

Multiplication of a Polynomial by a Monomial

A monomial is a polynomial with one term. To multiply two monomials, multiply the coefficients and use the product rule of exponents, $x^m \cdot x^n = x^{m+n}$, to multiply the variables.

EXAMPLE 12 Simplify: $(9z^5)(3z^2)$

SOLUTION ▶ $(9z^5)(3z^2)$

$= 9\cdot 3\cdot z^{(5+2)}$ — Multiply coefficients; product rule of exponents.

$= 27z^7$ — Simplify.

A binomial is a polynomial with two terms. When multiplying a monomial and a binomial, use the distributive property, $a(b + c) = ab + ac$.

EXAMPLE 13 Simplify: $4x(6x - 9)$

SOLUTION ▶ $4x(6x - 9)$

$= \mathbf{4x}(6x) + \mathbf{4x}(-9)$ — Distributive property.

$= 4x^{\mathbf{1}}(6x^{\mathbf{1}}) + 4x(-9)$ — $4x = 4x^1$

$= 24x^{(1+1)} - \mathbf{36}x$ — Simplify; product rule of exponents.

$= 24x^{\mathbf{2}} - 36x$ — Simplify.

In the next example, the binomial has two variables.

EXAMPLE 14 Simplify: $(7a^5)(3a^2 - 8b)$

SOLUTION ▶ $(7a^5)(3a^2 - 8b)$

$= \mathbf{7a^5}(3a^2) + \mathbf{7a^5}(-8b)$ — Distributive property.

$= \mathbf{21}a^{(5+2)} - \mathbf{56}a^5b$ — Simplify; product rule of exponents.

$= 21a^7 - 56a^5b$ — These are not like terms and cannot be combined.

In the next example, we use the distributive property to multiply a monomial and a polynomial with more than two terms.

EXAMPLE 15 Simplify: $-2x(x^2 - 8x + 5)$

SOLUTION

$$-2x(x^2 - 8x + 5)$$
$$= \mathbf{-2x}(x^2) \mathbf{- 2x}(-8x) \mathbf{- 2x}(5) \quad \text{Distributive property.}$$
$$= -2x^{\mathbf{1+2}} + \mathbf{16x^{1+1}} - \mathbf{10x} \quad \text{Product rule of exponents.}$$
$$= -2x^3 + 16x^2 - 10x \quad \text{Simplify.}$$

When multiplying fractions, look for common factors and simplify.

EXAMPLE 16 Simplify: $\frac{3}{4}x\left(\frac{5}{6}x - \frac{2}{7}\right)$

SOLUTION

$$\frac{3}{4}x\left(\frac{5}{6}x - \frac{2}{7}\right)$$
$$= \left(\frac{\mathbf{3}}{\mathbf{4}}\mathbf{x}\right)\left(\frac{5}{6}x\right) + \left(\frac{\mathbf{3}}{\mathbf{4}}\mathbf{x}\right)\left(-\frac{2}{7}\right) \quad \text{Distributive property.}$$
$$= \left(\frac{\cancel{3}}{4}x\right)\left(\frac{5}{\mathbf{2 \cdot \cancel{3}}}x\right) + \left(\frac{3}{\mathbf{2 \cdot 2}}x\right)\left(-\frac{\cancel{2}}{7}\right) \quad \text{Find common factors.}$$
$$= \frac{5}{8}x^2 - \frac{3}{14}x \quad \text{Simplify.}$$

In the next example, we first use the distributive property and simplify. To combine like terms in which the coefficients are fractions, we find common denominators.

EXAMPLE 17 Simplify: $\frac{3}{4}(2x + 8) - \frac{1}{2}x(5x - 12)$

SOLUTION

$$\frac{3}{4}(2x + 8) - \frac{1}{2}x(5x - 12)$$
$$= \left(\frac{\mathbf{3}}{\mathbf{4}}\right)(2x) + \left(\frac{\mathbf{3}}{\mathbf{4}}\right)(8) - \frac{\mathbf{1}}{\mathbf{2}}x(5x) - \frac{\mathbf{1}}{\mathbf{2}}x(-12) \quad \text{Distributive property.}$$
$$= \left(\frac{3}{\mathbf{2 \cdot 2}}\right)(2x) + \left(\frac{3}{\cancel{4}}\right)(\mathbf{2 \cdot \cancel{4}}) - \frac{1}{2}x(5x) - \frac{1}{2}x(\mathbf{-6 \cdot \cancel{2}}) \quad \text{Find common factors.}$$
$$= \frac{3}{2}x + 6 - \frac{5}{2}x^2 + 6x \quad \text{Simplify.}$$
$$= -\frac{5}{2}x^2 + \frac{3}{2}x + 6x + 6 \quad \text{Commutative property.}$$
$$= -\frac{5}{2}x^2 + \frac{3}{2}x + \frac{6}{1}x \cdot \frac{\mathbf{2}}{\mathbf{2}} + 6 \quad \text{Multiply by a fraction equal to 1.}$$
$$= -\frac{5}{2}x^2 + \frac{3}{2}x + \frac{\mathbf{12}}{\mathbf{2}}x + 6 \quad \text{Simplify.}$$
$$= -\frac{5}{2}x^2 + \frac{\mathbf{15}}{\mathbf{2}}x + 6 \quad \text{Combine like terms.}$$

Practice Problems

For problems 13–19, simplify.

13. $6(3x - 10)$

14. $6x^2(3x - 10)$

15. $7a^3(4a^2 + 5a - 2)$

16. $6x^2(3x - 10y)$

17. $\frac{2}{3}c\left(\frac{1}{4}c - \frac{6}{5}\right)$

18. $\frac{5}{6}n^2\left(\frac{3}{10}n^2 + \frac{2}{3}n\right)$

19. $\frac{1}{2}(3x + 6) + \frac{3}{8}(2x + 16)$

5.3 Vocabulary Practice

Match the term with its description.

1. The result of multiplication
2. This is equal to the value of the largest exponent in a polynomial expression in one variable.
3. When a polynomial in one variable is in this arrangement, the value of the exponents decrease from left to right.
4. A polynomial with three terms
5. $3xy$ and $7xy$
6. A polynomial with one term
7. $a(b + c) = ab + ac$
8. The value of any real number (except 0) raised to the 0 power
9. $x^m \cdot x^n = x^{m+n}$
10. A polynomial with two terms

A. binomial
B. degree
C. descending order
D. distributive property
E. like terms
F. monomial
G. one
H. product
I. product rule of exponents
J. trinomial

5.3 Exercises

Follow your instructor's guidelines for showing your work.

For exercises 1–12,
(a) identify the degree of the polynomial.
(b) identify the type of polynomial (monomial, binomial, or trinomial).

1. $9x + 1$
2. $3x + 17$
3. $4x^3 + 8x^2 - 9$
4. $2x^3 - 8x + 1$
5. $-7x$
6. $-5x$
7. 15
8. 19
9. $x^2 - 1$
10. $x^2 - 9$
11. $x^2 + 5x + 6$
12. $x^2 + 5x + 4$
13. The prefix *tri* means "three." State an English word besides *trinomial* that includes the prefix *tri*. Explain the meaning of the word.
14. The prefix *mono* means "one." State an English word besides *monomial* that includes the prefix *mono*. Explain the meaning of the word.
15. A student simplified $3x + 8x^2$ as $11x^3$. Explain why this is not correct.
16. A student simplified $3x + 7y$ as $10xy$. Explain why this is not correct.

For exercises 17–30,
(a) is the expression a polynomial?
(b) if it is not a polynomial, explain why it is not.

17. $m^2 + 8m + 12$
18. $a^2 + 9a + 20$
19. 150
20. 160
21. $\frac{1}{2}d$
22. $\frac{1}{3}p$
23. $\frac{1}{2d}$
24. $\frac{1}{3p}$
25. $\frac{3}{4}x^2 + \frac{5}{8}x - \frac{1}{2}$
26. $\frac{5}{6}x^2 + \frac{1}{3}x - \frac{3}{4}$
27. $5^x + 10$
28. $3^x + 15$
29. $x^2 + x^{\frac{3}{4}} - 8$
30. $x + x^{\frac{1}{2}} + 30$

For exercises 31–34,
(a) rewrite each polynomial in descending order.
(b) identify the lead coefficient.

31. $3x^2 + 7 - 2x - 4x^3$
32. $5x^2 + 21 - 8x - 2x^3$
33. $-216 + 7a + a^3$
34. $-8 + 2r + r^3$
35. Explain why we want polynomials to be in descending order when we do vertical addition or subtraction.
36. The term 15 does not include a variable. Explain why 15 is an example of a polynomial.

For exercises 37–86, simplify.

37. $(3x + 15) + (4x - 9)$
38. $(7x + 15) + (2x - 3)$
39. $(3x + 15) - (4x - 9)$
40. $(7x + 15) - (2x - 3)$
41. $(6x - 1) - (8x - 4)$
42. $(3x - 4) - (9x - 6)$
43. $(3x^2 - 15) + (4x - 9)$
44. $(7x^2 - 15) + (2x - 3)$
45. $(3x^2 - 15x) - (4x - 9)$
46. $(7x^2 - 15x) - (2x - 3)$
47. $(4y^2 + 6y - 13) + (9y^2 - 5y - 2)$
48. $(2h^2 + 7h - 17) + (8h^2 - 6h - 3)$
49. $(4y^2 + 6y - 13) - (9y^2 - 5y - 2)$
50. $(2h^2 + 7h - 17) - (8h^2 - 6h - 3)$
51. $(x^3 - x^2 - 4) - (x^3 - x - 4)$
52. $(w^3 - w - 5) - (w^3 - w^2 - 5)$
53. $\left(\frac{5}{8}p + \frac{2}{9}\right) + \left(\frac{1}{6}p + \frac{1}{3}\right)$
54. $\left(\frac{5}{9}a + \frac{3}{8}\right) + \left(\frac{1}{6}a + \frac{1}{4}\right)$

55. $\left(8b^2 - \frac{7}{10}b + 5\right) - \left(6b^2 - \frac{1}{10}b + 3\right)$

56. $\left(10c^2 - \frac{9}{10}c + 8\right) - \left(7c^2 - \frac{1}{10}c + 2\right)$

57. $(5c - 6) - (9c - 8) - (4c + 11)$

58. $(2x - 9) - (3x - 4) - (7x + 8)$

59. $(9 - 4r) - (5r + 7r^2) - (3r^2 + 1)$

60. $(5 - 8k) - (k + 3k^2) - (4k^2 + 1)$

61. $(5m^2)(3m^9)$

62. $(6c^2)(9c^7)$

63. $(-4a^3)(-a^5)$

64. $(-3z^4)(-z^2)$

65. $\frac{3}{4}(8x - 9)$

66. $\frac{5}{6}(12x - 7)$

67. $3x(4x - 9)$

68. $7x(2x - 3)$

69. $-p(8p - 1)$

70. $-w(4w - 1)$

71. $-9c^2(3c - 4)$

72. $-5k^2(2k - 6)$

73. $\frac{2}{3}x(x^2 - 6x + 9)$

74. $\frac{3}{5}z(z^2 - 10z + 15)$

75. $-\frac{3}{5}a\left(10a + \frac{1}{2}\right)$

76. $-\frac{5}{6}b\left(12b + \frac{1}{4}\right)$

77. $8k(2k^2 - k + 5)$

78. $9h(3h^2 - h + 7)$

79. $x(6x + 5y)$

80. $a(4a + 3b)$

81. $x\left(\frac{5}{6}x - 8\right) + 2\left(\frac{3}{4}x - 9\right)$

82. $m\left(\frac{3}{4}m - 2\right) + 7\left(\frac{1}{3}m - 5\right)$

83. $x(4x + 8) - 2(3x - 9)$

84. $m(9m - 2) - 7(4m - 5)$

85. $10c^2(c - 9) + c(c^2 + 12)$

86. $20d^2(d - 6) + d(d^2 + 15)$

Problem Solving: Practice and Review

Follow your instructor's guidelines for using the five steps as outlined in Section 1.3, p. 55.

For exercises 87–88, www.ESPN.com uses Major League Baseball statistics formulas. The table includes some of these formulas. Notice that an abbreviation with two letters does not mean multiplication. When multiplication is needed, the formula includes the word *times*.

G = games played
H = number of hits
2B = number of doubles
3B = number of triples
HR = number of home runs
AB = number of at-bats
BA = batting average = H divided by AB
TB = total bases = H + 2B + (3B times 2) + (HR times 3)
Slg = slugging percentage in decimal form
= TB divided by AB
IsoP = isolated power
= slugging percentage − batting average

87. Find the slugging percentage for Albert Pujols of the St. Louis Cardinals. Round to the nearest thousandth.

Pujols	2011 regular season
G	147
AB	579
R	105
H	173
2B	29
3B	0
HR	37

Source: www.ESPN.com

88. Find the isolated power for Ian Kinsler of the Texas Rangers.

Kinsler	2011 regular season
G	155
AB	620
R	121
H	158
2B	34
3B	4
HR	32

Source: www.ESPN.com

89. Find the number of Haitians who live in poverty.

Haiti occupies an area roughly the size of Maryland on the Caribbean island of Hispaniola, which it shares with the Dominican Republic. Nearly all of the 8.7 million residents are of African descent and speak Creole and French. The capital is Port-au-Prince. The country is the poorest in the Western Hemisphere, with four out of five people living in poverty and more than half in abject poverty. (*Source:* www.nytimes.com, Dec. 27, 2011)

90. The board designated unrestricted endowment at the Walker Art Center in Minneapolis increased from \$22,140,129 in 2009 to \$27,590,778 in 2010. Find the percent increase. Round to the nearest percent. (*Source:* www.walkerart.org)

Find the Mistake

For exercises 91–94, the completed problem has one mistake.

(a) Describe the mistake in words, or copy down the whole problem and highlight or circle the mistake.

(b) Do the problem correctly.

91. Problem: Simplify: $(6x^2 + 8x - 4) - (3x^2 + 5x - 1)$

Incorrect Answer: $(6x^2 + 8x - 4) - (3x^2 + 5x - 1)$
$= 6x^2 - 3x^2 + 8x - 5x - 4 - 1$
$= 3x^2 + 3x - 5$

92. Problem: Simplify: $3x(x - 4) + 8x(9x - 3)$

Incorrect Answer: $3x(x - 4) + 8x(9x - 3)$
$= 3x^2 - 12x + 72x^2 - 3$
$= 75x^2 - 12x - 3$

93. Problem: Simplify: $(5x^2 - 8) - (3x - 9)$

Incorrect Answer: $(5x^2 - 8) - (3x - 9)$
$= 5x^2 - 8 - 3x + 9$
$= 2x^2 + 1$

94. Problem: Simplify: $(3p^3)(4p^3 + p^2)$

Incorrect Answer: $(3p^3)(4p^3 + p^2)$
$= 12p^9 + 3p^6$

Review

For exercises 95–98, simplify.

95. $\dfrac{x^9}{x^5}$ **96.** $\dfrac{110x^9}{300x^5}$

97. $\dfrac{x^5}{x^9}$ **98.** $\dfrac{5p^8}{100p^{13}}$

SUCCESS IN COLLEGE MATHEMATICS

99. Attach a copy of your notes for today's class to your completed homework assignment.

100. Rewriting notes as soon as possible after class is often recommended to improve your memory of what you learned and your understanding. To rewrite notes, copy anything that you cannot read well, and fill in words or steps that you did not have time to write. At the bottom of the notes, write summary sentences such as "The main topic of today's class was . . . ," "It is important to know that . . . ," and "The most common mistake made in these kinds of problems is . . ." Rewrite your notes from today's class. If you did not take notes, explain why.

We can describe the relationship of the length and width of a rectangle. For example, a parking lot might be five times longer than it is wide. In this section, we will use polynomial expressions to describe the area of rectangles and other geometric figures.

SECTION 5.4

Multiplication of Polynomials

After reading the text, working the practice problems, and completing assigned exercises, you should be able to:

1. Use the distributive property to multiply polynomial expressions.
2. Use a pattern to multiply polynomial expressions.
3. Use a polynomial expression to describe the perimeter and area of a geometric figure.

Multiplication

We can use a rectangle to visualize the multiplication of two binomials. The area of a rectangle equals the product of its length and its width. One of the binomials represents the length. The other binomial represents the width. The area of the rectangle is equal to the product of the binomials.

EXAMPLE 1 Use a rectangle to simplify $(x + 3)(x + 2)$.

SOLUTION ▶ The area of the rectangle equals the product of the length and the width. The length of the rectangle is $x + 3$. The width is $x + 2$. The area of the entire rectangle is A.

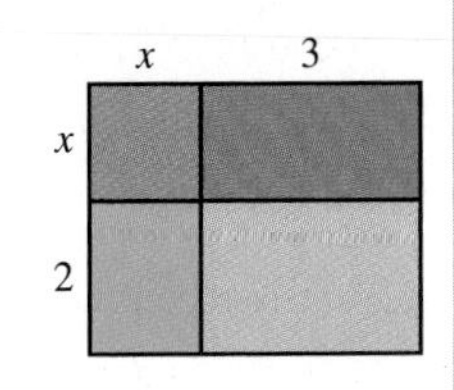

$$\text{Area of the rectangle} = (\text{length})(\text{width})$$
$$A = (x + 3)(x + 2)$$

The area of the rectangle also equals the sum of the four different areas.

area $= x^2$ area $= 3x$ area $= 2x$ area $= 2 \cdot 3 = 6$

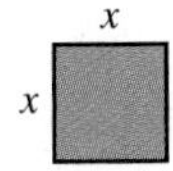

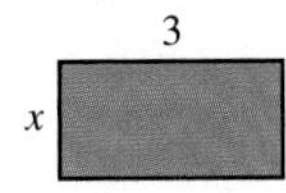

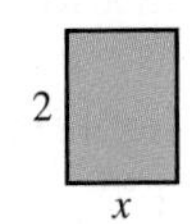

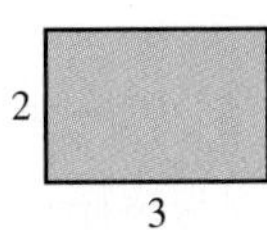

$$\text{Area of the rectangle} = \text{sum of the four areas shown in the diagram.}$$
$$A = x^2 + 3x + 2x + 6$$
$$A = x^2 + \mathbf{5x} + 6$$

Since $A = (x + 3)(x + 2)$ and $A = x^2 + 5x + 6$, $(x + 3)(x + 2)$ equals $x^2 + 5x + 6$.

Instead of using a rectangle to find the product of two binomials, we can use the distributive property and multiply the second binomial by each term in the first binomial. We then use the distributive property again.

EXAMPLE 2 Simplify: $(x + 3)(x + 4)$

SOLUTION ▶

$(x + 3)(x + 4)$	
$= \mathbf{x}(x + 4) + \mathbf{3}(x + 4)$	Distributive property.
$= \mathbf{x}(x) + \mathbf{x}(4) + \mathbf{3}(x) + \mathbf{3}(4)$	Use the distributive property again.
$= x^2 + 4x + 3x + 12$	Simplify.
$= x^2 + \mathbf{7x} + 12$	Combine like terms.

The directions "simplify" in Example 2 and in the next examples tell us to multiply the binomials and then simplify the product by combining like terms.

EXAMPLE 3 Simplify: $(2m - 7)(3m - 4)$

SOLUTION ▶

$(2m - 7)(3m - 4)$	
$= \mathbf{2m}(3m - 4) \mathbf{- 7}(3m - 4)$	Distributive property.
$= \mathbf{2m}(3m) + \mathbf{2m}(-4) \mathbf{- 7}(3m) \mathbf{- 7}(-4)$	Distributive property.
$= 6m^2 - 8m - 21m + 28$	Simplify.
$= 6m^2 \mathbf{- 29m} + 28$	Combine like terms.

The product in the next example is a "difference of squares," $h^2 - 16$.

EXAMPLE 4 Simplify: $(h - 4)(h + 4)$

SOLUTION ▶

$(h - 4)(h + 4)$	
$= \mathbf{h}(h + 4) \mathbf{- 4}(h + 4)$	Distributive property.
$= \mathbf{h}(h) + \mathbf{h}(4) \mathbf{- 4}(h) \mathbf{- 4}(4)$	Distributive property.
$= h^2 + 4h - 4h - 16$	Simplify.

$= h^2 + \mathbf{0h} - 16$ Simplify.

$= h^2 - 16$ Combine like terms.

Since the terms are subtracted, this is a *difference*. Since h^2 is a *square* and 16 is another *square*, 4^2, this expression is a **difference of squares**.

Instead of thinking about the distributive property, some students use the phrase "**F**irsts, **O**uters, **I**nners, **L**asts (**FOIL**)" to remember how to multiply binomials. However, this phrase is limited to binomials. It cannot be used to multiply a binomial and a trinomial. Unlike FOIL, the distributive property is not limited to multiplication of binomials.

EXAMPLE 5 Simplify: $(x + 3)(x^2 + 5x + 6)$

SOLUTION ▶

$(x + 3)(x^2 + 5x + 6)$

$= \mathbf{x}(x^2 + 5x + 6) + \mathbf{3}(x^2 + 5x + 6)$ Distributive property.

$= \mathbf{x}(x^2) + \mathbf{x}(5x) + \mathbf{x}(6) + \mathbf{3}(x^2) + \mathbf{3}(5x) + \mathbf{3}(6)$ Distributive property.

$= x^3 + 5x^2 + 6x + 3x^2 + 15x + 18$ Simplify.

$= x^3 + \mathbf{8x^2} + \mathbf{21x} + 18$ Combine like terms.

To multiply two exponential expressions with the same base, use the product rule of exponents, $x^m \cdot x^n = x^{m+n}$. For example, $5a(a^2)$ equals $5a^{1+2}$, which simplifies to $5a^3$.

EXAMPLE 6 Simplify: $(a^2 + 5a - 2)(a^2 - 4a + 3)$

SOLUTION ▶

$(a^2 + 5a - 2)(a^2 - 4a + 3)$

$= \mathbf{a^2}(a^2 - 4a + 3) + \mathbf{5a}(a^2 - 4a + 3) - \mathbf{2}(a^2 - 4a + 3)$

$= \mathbf{a^2}(a^2) + \mathbf{a^2}(-4a) + \mathbf{a^2}(3) + \mathbf{5a}(a^2) + \mathbf{5a}(-4a) + \mathbf{5a}(3) - \mathbf{2}(a^2)$
$- \mathbf{2}(-4a) - \mathbf{2}(3)$

$= a^{\mathbf{2+2}} - 4a^{\mathbf{2+1}} + 3a^2 + 5a^{\mathbf{1+2}} - 20a^{\mathbf{1+1}} + 15a - 2a^2 + 8a - 6$

$= a^4 - 4a^3 + 3a^2 + 5a^3 - 20a^2 + 15a - 2a^2 + 8a - 6$

$= a^4 + \mathbf{a^3} - \mathbf{19a^2} + \mathbf{23a} - 6$

When we are finding the product of more than two polynomials, the commutative property allows us to change the order of the factors. However, you may find it helpful to develop the habit of working from left to right.

EXAMPLE 7 Simplify: $2a(3a - 5)(a + 1)$

SOLUTION ▶

$2a(3a - 5)(a + 1)$

$= (\mathbf{2a} \cdot 3a + \mathbf{2a} \cdot -5)(a + 1)$ Distributive property.

$= (\mathbf{6a^2} - \mathbf{10a})(a + 1)$ Simplify.

$= \mathbf{6a^2}(a + 1) - \mathbf{10a}(a + 1)$ Distributive property.

$= \mathbf{6a^2}(a) + \mathbf{6a^2}(1) - \mathbf{10a}(a) - \mathbf{10a}(1)$ Distributive property.

$= 6a^3 + 6a^2 - 10a^2 - 10a$ Simplify.

$= 6a^3 - \mathbf{4a^2} - 10a$ Combine like terms.

Practice Problems

For problems 1–6, simplify.

1. $2k(5k + 4)$ **2.** $(2k + 8)(5k + 4)$ **3.** $(3a + 4)(a^2 + 6a + 2)$

4. $5c(8c - 7)$ **5.** $(2d - 5)(d + 6)$ **6.** $(2w)(w - 5)(3w + 4)$

Patterns

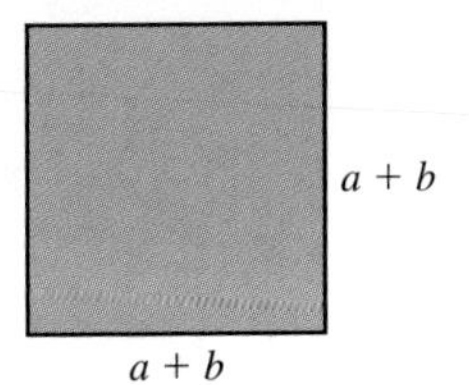

When we multiply a binomial by itself, we are **squaring** it. The product is a **perfect square trinomial**. In the next example, we find a pattern for the product of a binomial and itself.

EXAMPLE 8 Simplify: $(a + b)(a + b)$

SOLUTION ▶

$(a + b)(a + b)$	
$= \boldsymbol{a}(a + b) + \boldsymbol{b}(a + b)$	Distributive property.
$= \boldsymbol{a}(a) + \boldsymbol{a}(b) + \boldsymbol{b}(a) + \boldsymbol{b}(b)$	Distributive property.
$= a^2 + ab + ab + b^2$	Simplify; commutative property.
$= a^2 + \mathbf{2ab} + b^2$	Combine like terms.

The result of Example 8 is a **perfect square trinomial pattern**: $(a + b)^2 = a^2 + 2ab + b^2$. It is a perfect square pattern because $(a + b)^2$ is a perfect square, and it is a trinomial pattern because the product is a trinomial. To use this perfect square trinomial pattern to simplify an expression such as $(2x + 3)^2$, first identify what a and b represent. Matching $(a + b)^2$ and $(2x + 3)^2$ gives $a = 2x$ and $b = 3$. Then replace a with $2x$ and b with 3 in the pattern $a^2 + 2ab + b^2$ and simplify.

EXAMPLE 9 Use a pattern to simplify $(2x + 3)^2$.

SOLUTION ▶ Use the pattern $(a + b)^2 = a^2 + 2ab + b^2$. For $(2x + 3)^2$, $a = 2x$ and $b = 3$.

$(2x + 3)^2$	$(a + b)^2$
$= (2x)^2 + 2(2x)(3) + 3^2$	Replace a and b in the pattern $a^2 + 2ab + b^2$.
$= 2^2 \cdot x^2 + 12x + 9$	Many bases rule for a product; simplify.
$= \mathbf{4}x^2 + 12x + 9$	Simplify.

In the next example, we develop a pattern for $(a - b)^2$.

EXAMPLE 10 Simplify: $(a - b)^2$

SOLUTION ▶

$(a - b)^2$	
$(a - b)(a - b)$	Rewrite as multiplication.
$= \boldsymbol{a}(a - b) - \boldsymbol{b}(a - b)$	Distributive property.
$= \boldsymbol{a}(a) + \boldsymbol{a}(-b) - \boldsymbol{b}(a) - \boldsymbol{b}(-b)$	Distributive property.
$= a^2 - ab - ab + b^2$	Simplify; commutative property.
$= a^2 - \mathbf{2ab} + b^2$	Simplify.

This is another perfect square trinomial pattern, $(a - b)^2 = a^2 - 2ab + b^2$.

Whether we use the distributive property or a pattern to multiply polynomials, the result is the same. In some situations, it is quicker to use a pattern than to use the distributive property. In Chapter 6, we will reverse the patterns and use them to factor polynomials.

EXAMPLE 11 Use a pattern to simplify $(5p - 4)^2$.

SOLUTION ▶ Use the pattern $(a - b)^2 = a^2 - 2ab + b^2$. For $(5p - 4)^2$, $a = 5p$ and $b = 4$.

$(5p - 4)^2$	$(a - b)^2$
$= (5p)^2 - 2(5p)(4) + 4^2$	Replace a and b in the pattern $a^2 - 2ab + b^2$.
$= 5^2 \cdot p^2 - 40p + 16$	Many bases rule for a product; simplify.
$= \mathbf{25}p^2 - 40p + 16$	Simplify.

In the next example, we develop a pattern for the product $(a - b)(a + b)$.

EXAMPLE 12 Simplify: $(a - b)(a + b)$

SOLUTION ▶

$(a - b)(a + b)$	
$= \mathbf{a}(a + b) - \mathbf{b}(a + b)$	Distributive property.
$= \mathbf{a}(a) + \mathbf{a}(b) - \mathbf{b}(a) - \mathbf{b}(b)$	Distributive property.
$= a^2 + ab - ab - b^2$	Simplify; commutative property.
$= a^2 + \mathbf{0} - b^2$	Combine like terms.
$= a^2 - b^2$	Simplify.

Because a^2 is a perfect square, b^2 is a perfect square, and a subtraction is a difference, $(a - b)(a + b) = a^2 - b^2$ is the **difference of squares** pattern.

Patterns for Multiplying Polynomials

$(a + b)(a + b) = a^2 + 2ab + b^2$	*Perfect square trinomial pattern*
$(a - b)(a - b) = a^2 - 2ab + b^2$	*Perfect square trinomial pattern*
$(a + b)(a - b) = a^2 - b^2$	*Difference of squares pattern*

EXAMPLE 13 Use a pattern to simplify $(y + 8)(y - 8)$.

SOLUTION ▶ Use the pattern $(a - b)(a + b) = a^2 - b^2$. For $(y + 8)(y - 8)$, $a = y$ and $b = 8$.

$(y + 8)(y - 8)$	$(a - b)(a + b)$
$= y^2 - 8^2$	Replace a and b in the pattern $a^2 - b^2$.
$= y^2 - \mathbf{64}$	Simplify.

Practice Problems

For problems 7–12, use a pattern to simplify.

7. $(x + 5)(x + 5)$ **8.** $(3x + 5)(3x + 5)$ **9.** $(x - 5)(x - 5)$
10. $(3x - 5)(3x - 5)$ **11.** $(x - 5)(x + 5)$ **12.** $(3x - 5)(3x + 5)$

Writing Polynomial Expressions

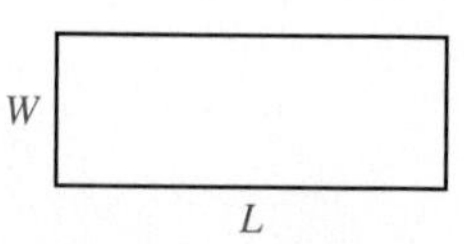

In Section 1.5, we used the formula for the perimeter of a rectangle, $P = 2L + 2W$, and the formula for the area of a rectangle, $A = LW$, where L is the length and W is the width. If we know the relationship of the length and the width, we can write a polynomial expression with only one variable that represents the perimeter or the area. To describe a polynomial expression such as $6W + 10$ that includes only one variable, W, we often say that the expression is "in" that variable. So $6W + 10$ is a polynomial expression "in W."

EXAMPLE 14 The length of a rectangle is 5 ft more than twice its width.

(a) If W = width, write a polynomial expression in W that represents the length, and draw a diagram of the rectangle. Do not include the units.

SOLUTION ▶ Since the length is twice the width plus 5 ft, the length, L, equals $2W + 5$.

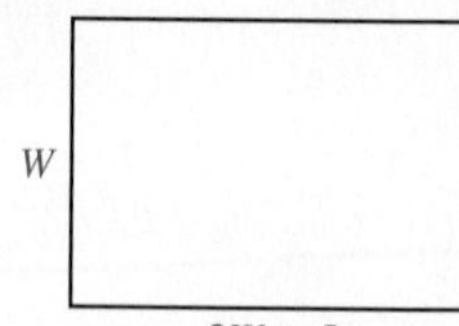

(b) Write a polynomial expression in W that represents the perimeter.

$2L + 2W$	Perimeter of a rectangle.
$= 2(\mathbf{2W + 5}) + 2W$	Replace L with $2W + 5$.
$= \mathbf{2}(2W) + \mathbf{2}(5) + 2W$	Distributive property.
$= \mathbf{4W + 10} + 2W$	Simplify.
$= \mathbf{6W} + 10$	Simplify.

(c) Write a polynomial expression in W that represents the area.

LW	Area of a rectangle.
$= (\mathbf{2W + 5})W$	Replace L with $2W + 5$.
$= (2W)\mathbf{W} + (5)\mathbf{W}$	Distributive property.
$= 2W^2 + 5W$	Simplify.

In Section 1.5, we used the formula for the perimeter of a triangle, $P = a + b + c$, where the length of the sides of the triangle are a, b, and c. If we know the relationship of the lengths of the sides, we can write a polynomial expression with only one variable that represents the perimeter.

EXAMPLE 15 The length of one side of a triangle is a ft. The other sides of the triangle are 3 ft longer and 7 ft shorter than this side.

(a) If a = length of one side, write polynomial expressions in a that represent the lengths of the other sides, and draw a diagram of the triangle. Do not include the units.

SOLUTION Since one side, b, is 3 ft longer, its length equals $a + 3$. Since another side, c, is 7 ft shorter, its length is $a - 7$.

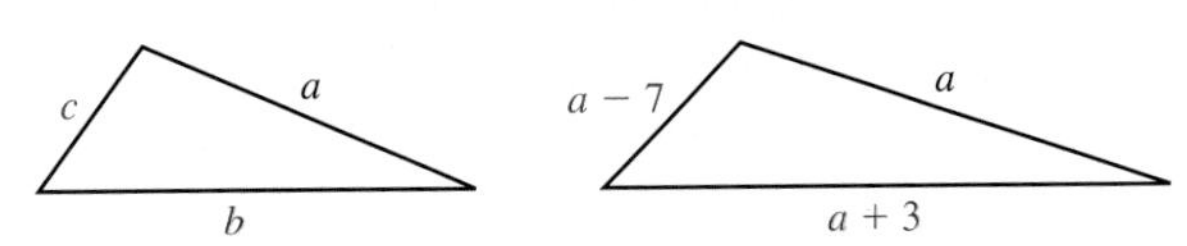

(b) Write a polynomial expression in a that represents the perimeter.

$a + b + c$	Perimeter of a triangle.
$= a + a + 3 + a - 7$	Replace b with $a + 3$, and replace c with $a - 7$.
$= 3a - 4$	Simplify.

In Section 1.5, we used the formula for the area of a triangle, $A = \frac{1}{2}bh$, where b is the base and h is the height. If we know the relationship of the base and the height, we can write a polynomial expression with only one variable that represents the area.

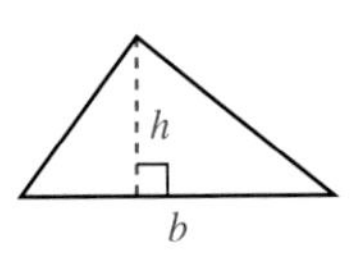

EXAMPLE 16 The base of a triangle is 6 in. less than twice the height.

(a) If h = height, write a polynomial expression in h that represents the base, and draw a diagram of the triangle. Do not include the units.

SOLUTION Since the base is 6 in. less than twice the height, the base, b, equals $2h - 6$.

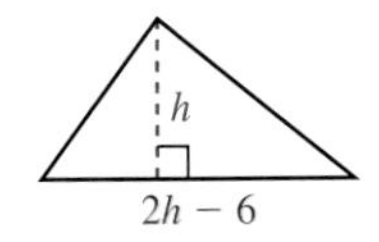

(b) Write a polynomial expression in h that represents the area.

$\frac{1}{2}bh$	Area of a triangle.
$= \frac{1}{2}(2h - 6)(h)$	Replace b with $2h - 6$.
$= \left(\frac{1}{2} \cdot 2h + \frac{1}{2}(-6)\right)(h)$	Distributive property.
$= (h - 3)(h)$	Simplify.
$= h^2 - 3h$	Distributive property.

Practice Problems

13. The length of a rectangle is 4 cm more than the width.

a. If W = width, write a polynomial expression in W that represents the length, and draw a diagram of the rectangle. Do not include the units.

b. Write a polynomial expression in W that represents the perimeter.

c. Write a polynomial expression in W that represents the area.

14. The length of one side of a triangle is a ft. The other sides of the triangle are 6 ft longer and 2 ft shorter than this side.

a. If a = length of one side, write polynomial expressions in a that represent the lengths of the other sides, and draw a diagram of the triangle. Do not include the units.

b. Write a polynomial expression in a that represents the perimeter.

15. The height of a triangle is 10 cm less than twice the base.

a. If b = base, write a polynomial expression in b that represents the height, and draw a diagram of the triangle. Do not include the units.

b. Write a polynomial expression in b that represents the area.

5.4 Vocabulary Practice

Match the term with its description.

1. $(a + b)(a + b) = a^2 + 2ab + b^2$
2. $(a + b)(a - b) = a^2 - b^2$
3. $a(b + c) = ab + ac$
4. The product of the length and the width of a rectangle
5. The distance around a triangle
6. A polynomial with three terms
7. A polynomial with two terms
8. The result of multiplication
9. A polynomial with one term
10. The exponents in a polynomial arranged this way decrease from left to right.

A. area
B. binomial
C. descending order
D. difference of squares pattern
E. distributive property
F. monomial
G. perfect square trinomial pattern
H. perimeter
I. product
J. trinomial

5.4 Exercises

Follow your instructor's guidelines for showing your work.

For exercises 1–50, simplify. Do not use a pattern.

1. $h(h + 2)$
2. $z(z + 7)$
3. $(h + 4)(h + 2)$
4. $(z + 9)(z + 3)$
5. $(h - 4)(h + 2)$
6. $(z - 9)(z + 3)$
7. $(h - 4)(h - 2)$
8. $(z - 9)(z - 3)$
9. $(n + 3)(n - 4)$
10. $(m + 8)(m - 5)$
11. $(q - 1)(q - 1)$
12. $(n - 1)(n - 1)$
13. $(q - 1)(q + 1)$
14. $(n - 1)(n + 1)$
15. $(h + k)(h + k)$
16. $(c + d)(c + d)$
17. $(3x + 5)(x + 4)$
18. $(2z + 3)(z + 6)$
19. $(a - 1)(4a + 2)$
20. $(c - 1)(5c + 3)$
21. $(2x - 7)(2x - 3)$
22. $(3x - 4)(3x - 5)$
23. $(7z - 8)(z + 6)$
24. $(5p - 9)(p + 8)$
25. $(5h - 4)(3h - 2)$
26. $(8z - 9)(3z - 7)$
27. $(m - n)(m + 3n)$
28. $(x - y)(x + 3y)$
29. $(3k - 1)(2k + 5)$
30. $(8h - 1)(3h + 2)$
31. $(7b + 3)(b - 9)$
32. $(9w + 5)(w - 7)$
33. $8u(10u^2 - u + 3)$
34. $5v(10v^2 - v + 4)$
35. $-c(c^2 - c + 6)$
36. $-z(z^2 - z + 8)$
37. $(2a + 5)(3a^2 + 7a - 9)$
38. $(2w + 7)(3w^2 + 9w - 5)$
39. $(x + 4)(x^3 + 5x^2 + 8x - 1)$
40. $(x + 2)(x^3 + 3x^2 + 8x - 1)$
41. $(k + 1)(k^2 + 5k + 3)$
42. $(w + 1)(w^2 + 3w + 5)$
43. $(p - 4)(p^2 + 5p - 1)$
44. $(n - 7)(n^2 + 2n - 1)$
45. $(b - 4)(b^2 - 3b + 5)$
46. $(a - 3)(a^2 - 4a + 6)$
47. $(2r - 3)(r^2 + r - 6)$
48. $(2v - 3)(v^2 + 5v - 4)$
49. $(3x + 5)(x^2 + 7x + 2)$
50. $(3x + 2)(x^2 + 5x + 3)$

51. a. Simplify: $(x + 5)(x + 8)$
b. Simplify: $(x - 5)(x - 8)$
c. Describe the difference in the products.

52. a. Simplify: $(x + 3)(x + 5)$
b. Simplify: $(x - 3)(x - 5)$
c. Describe the difference in the products.

53. a. Simplify: $(x - 5)(x + 8)$
b. Simplify: $(x + 5)(x - 8)$
c. Describe the difference in the products.

54. a. Simplify: $(x - 3)(x + 5)$
b. Simplify: $(x + 3)(x - 5)$
c. Describe the difference in the products.

For exercises 55–60, simplify.

55. $2(x + 3)(x + 1)$
56. $3(x + 5)(x + 1)$
57. $3a(a - 4)(a + 2)$
58. $2z(z - 6)(z + 3)$
59. $4w(w - 5)(3w - 8)$
60. $4c(c - 3)(2c - 5)$

For exercises 61–68, use a pattern to simplify.

61. $(x + 5)(x + 5)$
62. $(x + 6)(x + 6)$
63. $(x - 4)(x - 4)$
64. $(x - 3)(x - 3)$
65. $(x - 9)(x + 9)$
66. $(x - 5)(x + 5)$
67. $(5x - 6)(5x + 6)$
68. $(3x - 7)(3x + 7)$

For exercises 69–72,
(a) each of these equations is a pattern. Simplify the left side of each equation; use the distributive property; combine like terms.
(b) Is the simplified expression on the left side the same as the right side?

69. $(a + b)(a + b) = a^2 + 2ab + b^2$
70. $(a - b)(a - b) = a^2 - 2ab + b^2$
71. $(a + b)(a - b) = a^2 - b^2$
72. $(a - b)(a + b) = a^2 - b^2$

73. Explain why "difference of squares" is a good name for the pattern $(a - b)(a + b) = a^2 - b^2$.

74. The pattern for the difference of squares is given as $(a - b)(a + b) = a^2 - b^2$. Is this equivalent to the pattern $(a + b)(a - b) = a^2 - b^2$? Explain.

75. The length of a rectangle is 8 in. longer than the width.
  a. If W = width, write a polynomial expression in W that represents the length, and draw a diagram of the rectangle. Do not include the units.
  b. Write a polynomial expression in W that represents the perimeter.
  c. Write a polynomial expression in W that represents the area.

76. The width of a rectangle is 10 cm shorter than the length.
  a. If L = length, write a polynomial expression in L that represents the width, and draw a diagram of the rectangle. Do not include the units.
  b. Write a polynomial expression in L that represents the perimeter.
  c. Write a polynomial expression in L that represents the area.

77. The width of a rectangle is 8 in. shorter than three times the length.
  a. If L = length, write a polynomial expression in L that represents the width, and draw a diagram of the rectangle. Do not include the units.
  b. Write a polynomial expression in L that represents the perimeter.
  c. Write a polynomial expression in L that represents the area.

78. The length of a rectangle is 3 cm longer than twice the width.
  a. If W = width, write a polynomial expression in W that represents the length, and draw a diagram of the rectangle. Do not include the units.
  b. Write a polynomial expression in W that represents the perimeter.
  c. Write a polynomial expression in W that represents the area.

79. The length of one side of a triangle is a ft. The other sides of the triangle are 4 ft longer and 3 ft shorter than this side.
  a. If a = length of one side, write polynomial expressions in a that represent the lengths of the other sides, and draw a diagram of the triangle. Do not include the units.
  b. Write a polynomial expression in a that represents the perimeter.

80. The length of one side of a triangle is a ft. The other sides of the triangle are 6 ft longer and 9 ft shorter than this side.
  a. If a = length of one side, write polynomial expressions in a that represent the lengths of the other sides, and draw a diagram of the triangle. Do not include the units.
  b. Write a polynomial expression in a that represents the perimeter.

81. The height of a triangle is 6 cm less than the base.
  a. If b = base, write a polynomial expression in b that represents the height, and draw a diagram of the triangle. Do not include the units.
  b. Write a polynomial expression in b that represents the area.

82. The height of a triangle is 12 cm less than the base.
  a. If b = base, write a polynomial expression in b that represents the height, and draw a diagram of the triangle. Do not include the units.
  b. Write a polynomial expression in b that represents the area.

83. The base of a triangle is 4 in. longer than twice the height.
  a. If h = height, write a polynomial expression in h that represents the base, and draw a diagram of the triangle. Do not include the units.
  b. Write a polynomial expression in h that represents the area.

84. The base of a triangle is 6 in. longer than twice the height.
  a. If h = height, write a polynomial expression in h that represents the base, and draw a diagram of the triangle. Do not include the units.
  b. Write a polynomial expression in h that represents the area.

85. A square is a rectangle in which the lengths of all four sides are equal. If s = length of a side, write a polynomial expression in s that represents the perimeter.

86. A square is a rectangle in which the lengths of all four sides are equal. If s = length of a side, write a polynomial expression in s that represents the area.

87. A drafter is making enlargements of a rectangular drawing that preserve the relative width and length of the drawing. The length of the drawing is five-fourths of the width.
  a. If W = width, write a polynomial expression in W that represents the length, and draw a diagram of the rectangle. Do not include the units.
  b. Write a polynomial expression in W that represents the perimeter.
  c. Write a polynomial expression in W that represents the area.

88. A quilter is making enlargements of a rectangular machine-quilting pattern that preserve the relative width and length of the pattern. The width of the pattern is two-thirds of the length.

  a. If L = length, write a polynomial expression in L that represents the width, and draw a diagram of the rectangle. Do not include the units.

b. Write a polynomial expression in L that represents the perimeter.

c. Write a polynomial expression in L that represents the area.

89. A rectangular parking lot is three times as wide as it is long.

a. If L = length, write a polynomial expression in L that represents the width, and draw a diagram of the rectangle. Do not include the units.

b. Write a polynomial expression in L that represents the perimeter.

c. Write a polynomial expression in L that represents the area.

90. A rectangular parking lot is three times as long as it is wide.

a. If W = width, write a polynomial expression in W that represents the length, and draw a diagram of the rectangle. Do not include the units.

b. Write a polynomial expression in W that represents the perimeter.

c. Write a polynomial expression in W that represents the area.

Problem Solving: Practice and Review

Follow your instructor's guidelines for using the five steps as outlined in Section 1.3, p. 55.

91. Concentrated sulfuric acid is 98% sulfuric acid. The liquid mixture in lead batteries is 33.5% sulfuric acid. Use a system of linear equations to find the amount of concentrated sulfuric acid and the amount of water needed to make 5 L of the liquid mixture for a lead battery. Round to the nearest tenth.

92. Find the difference between the amount earned in 2011 and the amount earned in 2012 by an employee working a 40-hr week in Washington State for minimum wage.

Washington's minimum wage will increase to $9.04 per hour beginning Jan. 1, 2012, the Department of Labor & Industries announced today. L&I calculates the state's minimum wage each year as required by Initiative 688, approved by Washington voters in 1998. The 37-cent increase reflects a 4.258 percent increase in the Consumer Price Index for Urban Wage Earners and Clerical Workers (CPIW) since August 2010. (*Source:* www.lni.wa.gov, Sept. 30, 2011)

93. A veterinary clinic charges $51 for the first X-ray of a dog. Each additional X-ray is $41.50.

a. Write a linear equation that represents the relationship of the total cost, C, of n additional X-rays.

b. Use this equation to find the cost of eight *total* X-rays.

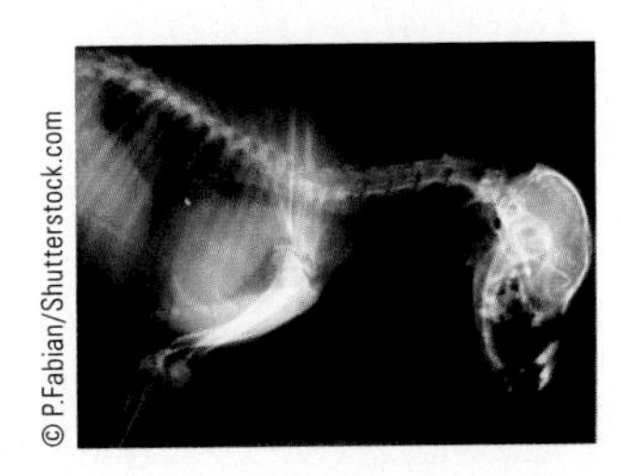
© P.Fabian/Shutterstock.com

94. A wireless gaming headset is marked down 30% to a sale price of $139.99. Find the original price. Round to the nearest hundredth.

Find the Mistake

For exercises 95–98, the completed problem has one mistake.

(a) Describe the mistake in words, or copy down the whole problem and highlight or circle the mistake.

(b) Do the problem correctly.

95. Problem: Simplify: $(5x - 2)(x - 3)$

Incorrect Answer: $(5x - 2)(x - 3)$
$= 5x(x) + 5x(-3) - 2x - 2(3)$
$= 5x^2 - 15x - 2x - 6$
$= 5x^2 - 17x - 6$

96. Problem: Simplify: $(2p - 9)(3p^2 + 6p - 5)$

Incorrect Answer:
$(2p - 9)(3p^2 + 6p - 5)$
$= 2p(3p^2) + 2p(6p) + 2p(-5) - 9(3p^2) - 9(6p) - 9(-5)$
$= 6p^3 + 12p^2 + 10p - 27p^2 - 54p + 45$
$= 6p^3 - 15p^2 - 44p + 45$

97. Problem: Simplify: $(4x - 1)(6x + 1)$

Incorrect Answer: $(4x - 1)(6x + 1)$
$= 4x(6x) + 4x(1) - 1(6x) - 1(1)$
$= 24x^2 + 4x - 6x - 1$
$= 24x^2 + 2x - 1$

98. Problem: Simplify: $(3p)(p + 4)(p + 2)$

Incorrect Answer: $(3p)(p + 4)(p + 2)$
$= 3p(p + 4) + 3p(p + 2)$
$= 3p(p) + 3p(4) + 3p(p) + 3p(2)$
$= 3p^2 + 12p + 3p^2 + 6p$
$= 6p^2 + 18p$

Review

For exercises 99–101, $20 \div 4 = 5$. Identify the

99. quotient

100. divisor

101. dividend

102. Rewrite $\frac{18}{3}$ using a $\div$ sign instead of a fraction bar.

SUCCESS IN COLLEGE MATHEMATICS

103. Instructors often make comments such as "This is really important, and it will be on the test" or "Make sure that you can work a problem like number 45 for the test." Describe where you put these kinds of comments in your notes.

104. There are many vocabulary words in mathematics that you must know so that you can understand your instructor, the book, and the directions on a test. You may also have to write definitions of these words on tests. Describe where you write definitions of vocabulary words in your notes.

$$\begin{array}{r} 13 \\ 10\overline{)138} \\ -10 \\ \hline 38 \\ -30 \\ \hline 8 \end{array}$$

When we divide a whole number dividend by a whole number divisor, the result is a quotient. For example, $138 \div 10 = 13\text{ R}(8)$. In this section, we will extend the process for long division of whole numbers to the long division of polynomials.

SECTION 5.5 Division of Polynomials

After reading the text, working the practice problems, and completing assigned exercises, you should be able to:

1. Divide a polynomial by a monomial.
2. Use long division to divide a polynomial by a binomial.
3. Check the result of division.

Division of a Polynomial by a Monomial

To simplify $\frac{26m^9p^7}{2m^3p}$, we use the quotient rule of exponents, $\frac{x^m}{x^n} = x^{m-n}$.

EXAMPLE 1 Simplify: $\frac{26m^9p^7}{2m^3p}$

SOLUTION ▶

$\frac{26m^9p^7}{2m^3p}$

$= \frac{13 \cdot \cancel{2} \cdot m^{9-3}p^{7-1}}{1 \cdot \cancel{2}}$ Find common factors; quotient rule of exponents.

$= \frac{13m^6p^6}{1}$ Simplify.

$= 13m^6p^6$ Simplify.

We can rewrite a fraction as the sum of two fractions with the same denominator. For example, we can rewrite $\frac{5}{11}$ as $\frac{2}{11} + \frac{3}{11}$. To divide a polynomial with more than one term by a monomial, we rewrite it as a sum or difference of fractions and then simplify.

EXAMPLE 2 Simplify: $\frac{10x^3 - 65x - 5}{5x}$

SOLUTION ▶

$\frac{10x^3 - 65x - 5}{5x}$

$= \frac{10x^3}{5x} - \frac{65x}{5x} - \frac{5}{5x}$ Rewrite as a difference of fractions.

$= \frac{2 \cdot \cancel{5} \cdot x^{3-1}}{\cancel{5}} - \frac{\cancel{5} \cdot 13 \cdot \cancel{x}}{\cancel{5} \cdot \cancel{x}} - \frac{\cancel{5} \cdot 1}{\cancel{5}x}$ Find common factors; quotient rule of exponents.

$= 2x^2 - 13 - \frac{1}{x}$ Simplify.

In division, the result of dividing a **dividend** by a **divisor** is the **quotient**. Since a fraction bar represents division, we can rewrite $\frac{10x^3 - 65x - 5}{5x}$ as $(10x^3 - 65x - 5) \div (5x)$. To check that a quotient is correct, multiply the quotient by the divisor. The product should be the dividend.

EXAMPLE 3 **(a)** Simplify: $(12x^4 - 15x^3) \div (3x)$

SOLUTION

$(12x^4 - 15x^3) \div (3x)$

$= \frac{12x^4 - 15x^3}{3x}$ Rewrite division as a fraction.

$= \frac{12x^4}{3x} - \frac{15x^3}{3x}$ Rewrite as a difference of fractions.

$= \frac{\cancel{3} \cdot 4 \cdot x^{4-1}}{\cancel{3} \cdot 1} - \frac{5 \cdot \cancel{3} \cdot x^{3-1}}{\cancel{3} \cdot 1}$ Find common factors; quotient rule of exponents.

$= \frac{4x^3}{1} - \frac{5x^2}{1}$ Simplify.

$= 4x^3 - 5x^2$ Simplify.

(b) Check.

If the quotient is correct, then (quotient)(divisor) equals the dividend, $12x^4 - 15x^3$.

$(4x^3 - 5x^2)(3x)$ (quotient)(divisor)

$= 4x^3\mathbf{(3x)} - 5x^2\mathbf{(3x)}$ Distributive property.

$= 12x^{3+1} - 15x^{2+1}$ Multiply coefficients; product rule of exponents.

$= 12x^{\mathbf{4}} - 15x^{\mathbf{3}}$ This equals the dividend; the quotient is correct.

Practice Problems

For problems 1–3,
(a) identify the dividend and divisor in each problem.
(b) simplify.
(c) check.

1. $\frac{28a^6b^{11}c}{7a^2bc}$ **2.** $\frac{21x^3 - 27x^2 + 6x - 9}{3x}$ **3.** $\frac{28a^6b^{11}c - 28a^5bc + 35}{7a^2bc}$

Division by a Polynomial

We can use the process of **long division** to find the quotient of whole numbers. In long division, the dividend is inside the division box, and the divisor is to the left of the division box.

EXAMPLE 4 **(a)** Use long division to evaluate $358 \div 2$.

SOLUTION

The dividend is 358, and the divisor is 2.

$2\overline{)358}$ Dividend is inside the division box; divisor is to the left.

$\begin{array}{r} \mathbf{1} \\ 2\overline{)358} \end{array}$ Look at the first digit in the dividend, 3. What should we multiply the divisor by so that the product is the largest number that is less than or equal to 3? The answer is 1.

$$\begin{array}{r} 1 \\ 2\overline{)358} \\ -\mathbf{2} \\ \hline \mathbf{1} \end{array}$$

Multiply the divisor by 1: 1(2) = 2. Write this product under the first digit in the dividend. Subtract. The difference is 1.

$$\begin{array}{r} 1 \\ 2\overline{)358} \\ -2\downarrow \\ \hline \mathbf{15} \end{array}$$

Bring down 5 from the dividend. What should we multiply the divisor by so that the product is the largest number that is less than or equal to 15? The answer is 7.

$$\begin{array}{r} 17 \\ 2\overline{)358} \\ -2 \\ \hline 15 \\ -\mathbf{14} \\ \hline \mathbf{1} \end{array}$$

Multiply the divisor by 7: 7(2) = 14. Write this product below 15. Subtract. The difference is 1.

$$\begin{array}{r} 17 \\ 2\overline{)358} \\ -2 \\ \hline -14\downarrow \\ \hline \mathbf{18} \end{array}$$

Bring down 8 from the dividend. What should we multiply the divisor by so that the product is the largest number that is less than or equal to 18? The answer is 9.

$$\begin{array}{r} 179 \\ 2\overline{)358} \\ -2 \\ \hline 15 \\ -14 \\ \hline 18 \\ -\mathbf{18} \\ \hline 0 \end{array}$$

Multiply the divisor by 9: 9(2) = 18. Write this product below 18. Subtract. The final difference is the remainder, 0: 358 ÷ 2 = 179.

(b) Check.

If the answer is correct, then (quotient)(divisor) + remainder equals the dividend.

$(179)(2) + 0$ (quotient)(divisor) + remainder

$= 358$ This equals the dividend; the quotient is correct.

If the terms in the dividend and the divisor are in descending order, we can also use long division to find the quotient of polynomials.

EXAMPLE 5 **(a)** Use long division to simplify $(x^2 + 7x + 10) \div (x + 5)$.

SOLUTION The dividend is $x^2 + 7x + 10$, and the divisor is $x + 5$.

$x + 5\overline{)x^2 + 7x + 10}$ Dividend is inside the division box; divisor is to the left.

$$\begin{array}{r} \boldsymbol{x} \\ x + 5\overline{)x^2 + 7x + 10} \end{array}$$

Look at the first term in the dividend, x^2. What should we multiply the divisor by so that the first term in the product is x^2? The answer is x.

Practice Problems

For problems 4–7,
(a) use long division to simplify.
(b) check.

4. $(x^2 + 12x + 20) \div (x + 2)$
5. $(x^2 - 8x + 15) \div (x - 5)$
6. $(15x^2 + 14x - 8) \div (5x - 2)$
7. $(x^2 - 64) \div (x - 8)$

Remainders

In the next example, the quotient is 13 plus a remainder of 8, often written as 13 R(8). Do not write additional zeros after the decimal point to continue the division.

EXAMPLE 8 **(a)** Use long division to evaluate $138 \div 10$.

SOLUTION

$$\begin{array}{r} 13 \\ 10\overline{)138} \\ -10\downarrow \\ \hline 38 \\ -30 \\ \hline 8 \end{array} \leftarrow \text{Remainder}$$

The final difference is the remainder. The remainder of this quotient is 8.

The quotient is 13 R(8).

(b) Check.

If the answer is correct, then (quotient)(divisor) + remainder equals the dividend.

$(13)(10) + 8$	(quotient)(divisor) + remainder
$= \mathbf{130} + 8$	Evaluate.
$= 138$	This equals the dividend, the quotient is correct.

The degree of a polynomial in one variable is the value of the largest exponent. When the quotient of two polynomials includes a remainder, continue the process of long division until the degree of the remainder is less than the degree of the divisor or the remainder is 0.

EXAMPLE 9 **(a)** Use long division to simplify $(x^2 + 9x + 15) \div (x + 2)$.

SOLUTION

$x + 2\overline{)x^2 + 9x + 15}$ — The degree of the divisor, $x + 2$, is 1, since $x = x^1$.

$$\begin{array}{r} \mathbf{x} \\ x + 2\overline{)x^2 + 9x + 15} \end{array}$$

Look at the first term in the dividend, x^2. What should we multiply the divisor by so that the first term in the product is x^2? The answer is x.

$$\begin{array}{r} x \\ x + 2\overline{)x^2 + 9x + 15} \\ \mathbf{-(x^2 + 2x)} \\ \hline \mathbf{7x} \end{array}$$

Multiply the divisor by x: $x(x + 2) = x^2 + 2x$. Write this product under the dividend, aligning like terms. Subtract. The difference is $7x$.

$$\begin{array}{r} x \mathbf{+ 7} \\ x + 2\overline{)x^2 + 9x + 15} \\ -(x^2 + 2x) \quad\downarrow \\ \hline 7x \mathbf{+ 15} \end{array}$$

Bring down 15 from the dividend. What should we multiply the divisor by so that the first term in the product is $7x$? The answer is 7.

$$\begin{array}{r} x + 7 \\ x + 2\overline{)x^2 + 9x + 15} \\ -(x^2 + 2x) \\ \hline 7x + 15 \\ -(7x + 14) \\ \hline 1 \end{array}$$

Multiply the divisor by 7: $7(x + 2) = 7x + 14$. Write this product under the dividend, aligning like terms. Subtract. The final difference, the remainder, is 1. Since we can rewrite 1 as $1x^0$, the degree of the remainder is 0. Since 0 is less than the degree of the divisor, 1, the division is done: $(x^2 + 9x + 15) \div (x + 2) = (x + 7)\ \text{R}(1)$.

(b) Check.

If the answer is correct, then (quotient)(divisor) + remainder equals the dividend.

$(x + 7)(x + 2) + 1$	(quotient)(divisor) + remainder
$= x(x + 2) + 7(x + 2) + 1$	Distributive property.
$= x(x) + x(2) + 7x + 7(2) + 1$	Distributive property.
$= x^2 + 2x + 7x + 14 + 1$	Commutative property; simplify.
$= x^2 + 9x + 15$	This equals the dividend; the quotient is correct.

In the next example, the remainder is a negative number.

EXAMPLE 10 **(a)** Use long division to simplify $(x^2 + 10x + 12) \div (x + 6)$.

SOLUTION

$$x + 6\overline{)x^2 + 10x + 12}$$

The degree of the divisor, $x + 6$, is 1, since $x = x^1$.

$$\begin{array}{r} x \\ x + 6\overline{)x^2 + 10x + 12} \end{array}$$

Look at the first term in the dividend, x^2. What should we multiply the divisor by so that the first term in the product is x^2? The answer is x.

$$\begin{array}{r} x \\ x + 6\overline{)x^2 + 10x + 12} \\ -(x^2 + 6x) \\ \hline 4x \end{array}$$

Multiply the divisor by x: $x(x + 6) = x^2 + 6x$. Write this product under the dividend, aligning like terms. Subtract. The difference is $4x$.

$$\begin{array}{r} x + 4 \\ x + 6\overline{)x^2 + 10x + 12} \\ -(x^2 + 6x) \downarrow \\ \hline 4x + 12 \end{array}$$

Bring down 12 from the dividend. What should we multiply the divisor by so that the first term in the product is $4x$? The answer is 4.

$$\begin{array}{r} x + 4 \\ x + 6\overline{)x^2 + 10x + 12} \\ -(x^2 + 6x) \\ \hline 4x + 12 \\ -(4x + 24) \\ \hline -12 \end{array}$$

Multiply the divisor by 4: $4(x + 6) = 4x + 24$. Write this product under the dividend, aligning like terms. Subtract. The final difference, the remainder, is -12. Since we can rewrite -12 as $-12x^0$, the degree of the remainder is 0. Since 0 is less than the degree of the divisor, 1, the division is done: $(x^2 + 10x + 12) \div (x + 6) = (x + 4)\ \text{R}(-12)$.

(b) Check.

If the answer is correct, then (quotient)(divisor) + remainder equals the dividend.

$(x + 4)(x + 6) + (-12)$	(quotient)(divisor) + remainder
$= x(x + 6) + 4(x + 6) + (-12)$	Distributive property.
$= x(x) + x(6) + 4x + 4(6) + (-12)$	Distributive property.
$= x^2 + 6x + 4x + 24 + (-12)$	Commutative property; simplify.
$= x^2 + 10x + 12$	This equals the dividend; the quotient is correct.

In the next example, we rewrite the dividend with a placeholder, $x^2 + 0x + 8$.

EXAMPLE 11 Use long division to simplify $(x^2 + 8) \div (x + 5)$.

SOLUTION ▶ Rewrite the dividend as $x^2 + 0x + 8$.

$$x + 5\overline{)x^2 + 0x + 8}$$

The degree of the divisor, $x + 5$, is 1, since $x = x^1$.

$$\begin{array}{r} \mathbf{x} \\ x + 5\overline{)x^2 + 0x + 8} \end{array}$$

Look at the first term in the dividend, x^2. What should we multiply the divisor by so that the first term in the product is x^2? The answer is x.

$$\begin{array}{r} x \\ x + 5\overline{)x^2 + 0x + 8} \\ \mathbf{-(x^2 + 5x)} \\ \hline \mathbf{-5x} \end{array}$$

Multiply the divisor by x: $x(x + 5) = x^2 + 5x$. Write this product under the dividend, aligning like terms. Subtract. The difference is $-5x$.

$$\begin{array}{r} x \mathbf{- 5} \\ x + 5\overline{)x^2 + 0x + 8} \\ -(x^2 + 5x) \downarrow \\ \hline -5x \mathbf{+ 8} \end{array}$$

Bring down 8 from the dividend. What should we multiply the divisor by so that the first term in the product is $-5x$? The answer is -5.

$$\begin{array}{r} x - 5 \\ x + 5\overline{)x^2 + 0x + 8} \\ -(x^2 + 5x) \\ \hline -5x + 8 \\ \mathbf{-(-5x - 25)} \\ \hline 33 \end{array}$$

Multiply the divisor by -5: $-5(x + 5) = -5x - 25$. Write this product under the dividend, aligning like terms. Subtract. The final difference, the remainder, is 33. Since we can rewrite 33 as $33x^0$, the degree of the remainder is 0. Since 0 is less than the degree of the divisor, 1, the division is done: $(x^2 + 8) \div (x + 5) = (x - 5)$ R(33).

Practice Problems

For problems 8–11,

(a) use long division to simplify.

(b) check.

8. $(x^2 + 4x + 20) \div (x + 3)$

9. $(x^2 + 3x + 8) \div (x + 2)$

10. $(x^2 + 48) \div (x - 3)$

11. $(x^2 + 16) \div (x + 4)$

Using Technology: Checking Division

EXAMPLE 12 The result of simplifying $(x^2 - 5x + 6) \div (x - 3)$ is $x - 2$. Check.

If the answer is correct, then the graphs of $y = \dfrac{x^2 - 5x + 6}{x - 3}$ and $y = x - 2$ should appear to be the same. On the Y= screen, type $\dfrac{x^2 - 5x + 6}{x - 3}$, placing parentheses around the numerator and around the denominator. Press ENTER. Sketch the graph; describe the window.

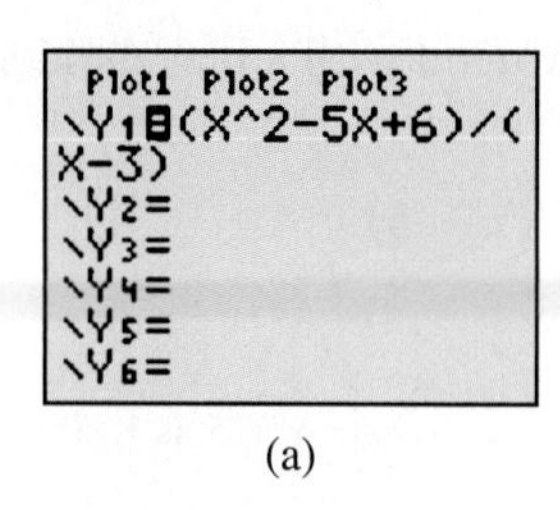

(a)

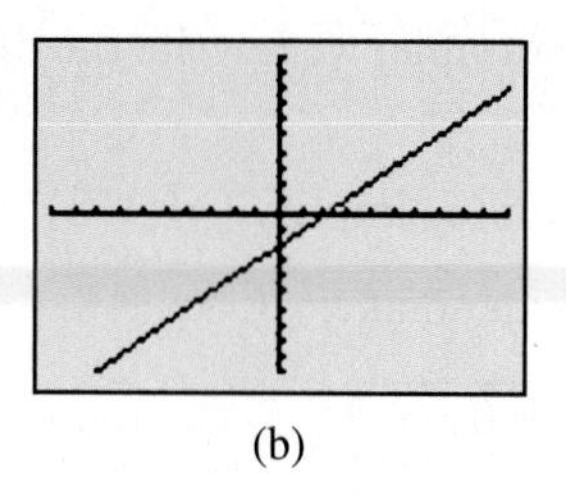
(b)

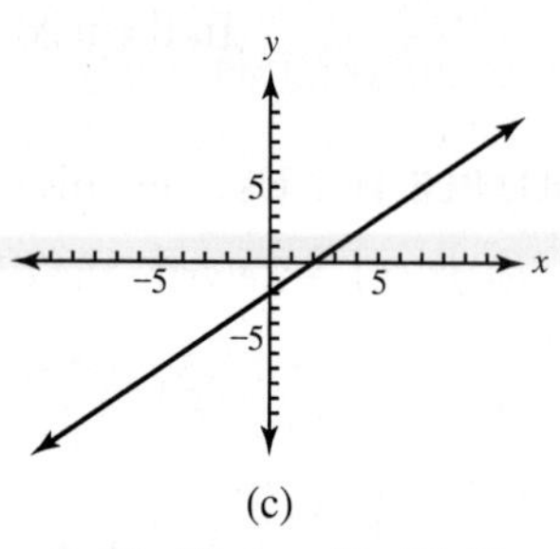

(c)

[−10, 10, 1; −10, 10, 1]

Now repeat this process with $y = x - 2$.

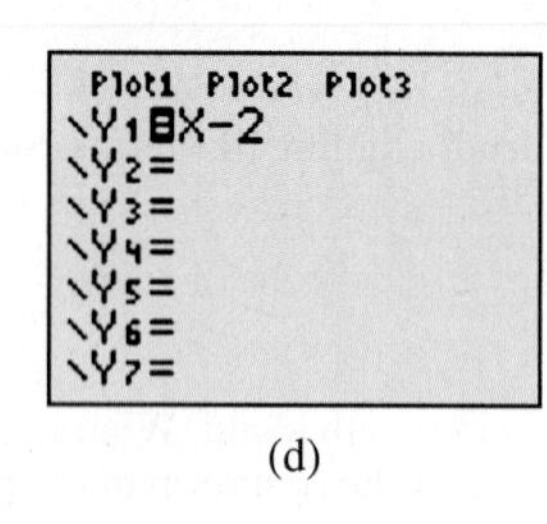

(d)

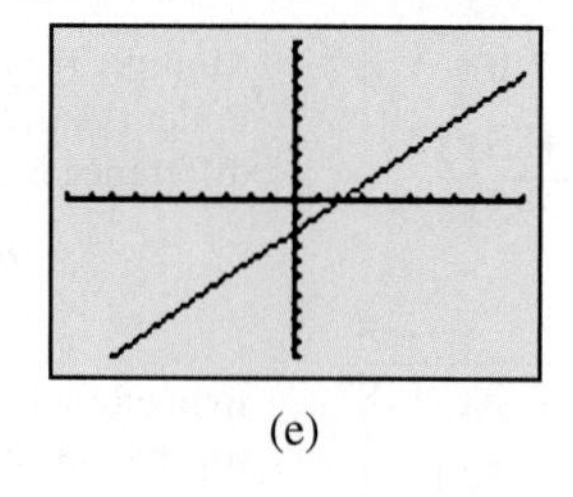
(e)

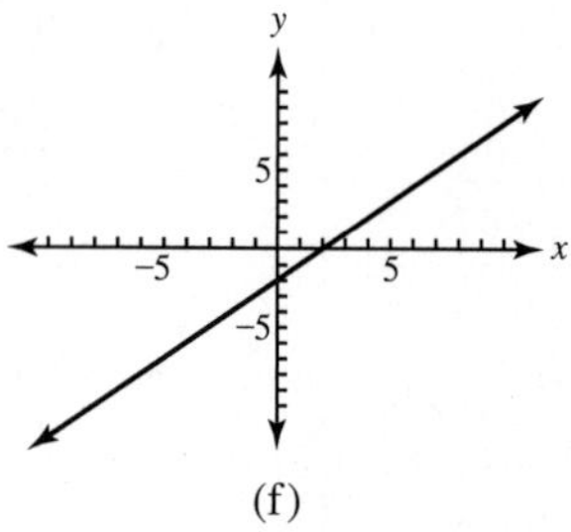

(f)

[−10, 10, 1; −10, 10, 1]

The graphs appear to be the same; the answer is probably correct. Since division by zero is undefined, there actually is a small difference in these graphs that cannot be seen in this window. The graph of $y = \dfrac{x^2 - 5x + 6}{x - 3}$ has a hole at $x = 3$ because this value causes division by zero.

Practice Problems For problems 12–14,
(a) use long division to simplify.
(b) use a graphing calculator to check.

12. $\dfrac{x^2 - 11x + 18}{x - 2}$ **13.** $\dfrac{x^2 + 7x + 12}{x + 4}$ **14.** $\dfrac{x^2 - 25}{x - 5}$

5.5 Vocabulary Practice

Match the term with its description.

1. The result of division
2. The product of the quotient and the divisor
3. In the expression $3x + 5$, 3 is an example of this.
4. In the expression $3x + 5$, 5 is an example of this.
5. A polynomial with one term
6. A polynomial with two terms
7. A polynomial with three terms
8. When a polynomial in one variable is in this arrangement, the value of the exponents decreases from left to right.
9. In $14 \div 2 = 7$, 2 is an example of this.
10. $x^m \cdot x^n = x^{m+n}$

A. binomial
B. coefficient
C. constant
D. descending order
E. dividend
F. divisor
G. monomial
H. product rule of exponents
I. quotient
J. trinomial

5.5 Exercises

Follow your instructor's guidelines for showing your work.

For exercises 1–4,
(a) identify the dividend.
(b) identify the divisor.

1. $\frac{x^2 + 5x}{x}$

2. $\frac{x^2 + 9x + 20}{x + 4}$

3. $(x^2 - 25) \div (x + 5)$

4. $(x^2 - 100) \div (x + 10)$

For exercises 5–30, simplify.

5. $(8x^3 + 14x^2 + 6x) \div (2x)$

6. $(15y^3 + 21y^2 + 54y) \div (3y)$

7. $(12a^4 + 6a^3 - 3a^2) \div 3$

8. $(18b^5 + 12b^2 - 6b) \div 6$

9. $(21a^3 - 7a) \div (7a)$

10. $(24c^3 - 8c) \div (8c)$

11. $(35d^4 - 20d^3 - 5d^2 + 10d) \div (5d)$

12. $(54z^4 - 36z^3 - 6z^2 + 12z) \div (6z)$

13. $(56u^5 - 64u^3 + 72u^2) \div (8u^2)$

14. $(72v^6 - 81v^4 + 54v^2) \div (9v^2)$

15. $(14h^3 - 6h^2 + 12h) \div (-2h)$

16. $(20k^3 - 24k^2 + 8k) \div (-2k)$

17. $(5w^4 - 15w^2 + 60w + 20) \div (5w)$

18. $(4p^4 - 16p^2 + 60p + 20) \div (4p)$

19. $(5w^4 - 15w^2 + 60w + 20) \div (-5w)$

20. $(4p^4 - 16p^2 + 60p + 20) \div (-4p)$

21. $(h^8 - 4h^2 + 100h + 20) \div 4$

22. $(k^7 - 5k^3 + 100k + 20) \div 5$

23. $(h^8 - 4h^2 + 100h + 20) \div (4h^2)$

24. $(k^7 - 5k^3 + 100k + 20) \div (5k^2)$

25. $(m^2 - 4m + 8) \div (2m)$

26. $(y^2 - 6y + 9) \div (3y)$

27. $(6x^2 + 3x - 9) \div (3x)$

28. $(14x^2 + 7x - 21) \div (7x)$

29. $(12y^3 - 10y^2 + 6y) \div (5y)$

30. $(18p^3 - 14p^2 + 5p) \div (7p)$

For exercises 31–36,
(a) use long division to evaluate.
(b) check.

31. $408 \div 2$

32. $609 \div 3$

33. $1416 \div 4$

34. $2524 \div 4$

35. $26{,}520 \div 85$

36. $10{,}120 \div 23$

For exercises 37–54,
(a) use long division to simplify.
(b) check.

37. $(x^2 + 10x + 16) \div (x + 8)$

38. $(x^2 + 12x + 32) \div (x + 4)$

39. $(x^2 - 10x + 16) \div (x - 8)$

40. $(x^2 - 12x + 32) \div (x - 4)$

41. $(2x^2 + 17x + 21) \div (x + 7)$

42. $(2x^2 + 11x + 15) \div (x + 3)$

43. $(2x^2 + 17x + 21) \div (2x + 3)$

44. $(2x^2 + 11x + 15) \div (2x + 5)$

45. $(2x^2 - 11x - 21) \div (x - 7)$

46. $(2x^2 - x - 15) \div (2x + 5)$

47. $(y^2 - 81) \div (y - 9)$

48. $(c^2 - 36) \div (c - 6)$

49. $(a^2 - 144) \div (a + 12)$

50. $(d^2 - 169) \div (d + 13)$

51. $(9z^2 + 12z + 4) \div (3z + 2)$

52. $(25w^2 + 20w + 4) \div (5w + 2)$

53. $(9x^2 - 6x + 1) \div (3x - 1)$

54. $(16x^2 - 8x + 1) \div (4x - 1)$

For exercises 55–70,
(a) use long division to simplify.
(b) check.

55. $(x^2 + 8x + 10) \div (x + 2)$

56. $(x^2 + 10x + 18) \div (x + 2)$

57. $(x^2 - 8x + 10) \div (x + 2)$

58. $(x^2 - 10x + 18) \div (x + 2)$

59. $(x^2 + 11x + 15) \div (x + 3)$

60. $(x^2 + 13x + 18) \div (x + 3)$

61. $(x^2 + 11x + 15) \div (x - 3)$

62. $(x^2 + 13x + 18) \div (x - 3)$

63. $(2x^2 + 9x + 20) \div (x + 4)$

64. $(2x^2 + 7x + 30) \div (x + 2)$

65. $(3x^2 + 20x + 4) \div (x + 1)$

66. $(3x^2 + 19x + 5) \div (x + 1)$

67. $(a^2 - 21) \div (a + 5)$

68. $(p^2 - 19) \div (p + 6)$

69. $(c^2 + 8) \div (c + 6)$

70. $(w^2 + 9) \div (w + 2)$

Problem Solving: Practice and Review

Follow your instructor's guidelines for using the five steps as outlined in Section 1.3, p. 55.

For exercises 71–72, a car salesperson is paid a commission based on a percentage of the profit the car dealership makes on a sale.

Profit made by dealership	Percent of profit paid in commission to salesperson
\$0–\$749	20%
\$750–\$1249	25%
> \$1250	30%

71. Find the commission made by a salesperson on the sale of a car on which the dealership made a profit of \$950.

72. a. Write a linear equation that represents the total commission, C, on a profit made by dealership, p, that is less than \$749.

b. Find the commission paid to the salesperson when the dealership makes a profit of \$500.

73. A customer borrows \$8500 for 6 years at a simple interest rate of 11.3%. Find the simple interest.

74. Find the percent increase in price when paying the "1 hr Express" price for 2500 copies instead of the "Standard" price for 2500 copies.

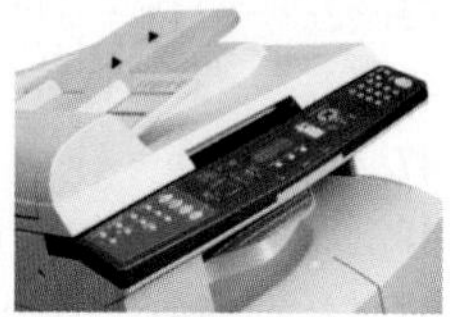

© R-O-M-A/Shutterstock.com

Black & White Copying, Letter (8.5 × 11)	1 hr Express	Standard
1–99	13¢ ea.	10¢ ea.
100–499	10¢ ea.	8¢ ea.
500–999	8¢ ea.	6¢ ea.
1,000–9,999	6¢ ea.	5¢ ea.
10,000–19,999	5¢ ea.	4¢ ea.

Source: www.staplescopyandprint.com, Jan. 2012

Technology

For exercises 75–78,

(a) use long division to simplify.

(b) use a graphing calculator to check. Sketch the graph; describe the window.

75. $(x^2 + 5x - 24) \div (x + 8)$

76. $(x^2 + 7x - 18) \div (x + 9)$

77. $(x^2 + 27x + 180) \div (x + 12)$

78. $(40x^2 + 122x + 55) \div (2x + 5)$

Find the Mistake

For exercises 79–82, the completed problem has one mistake.

(a) Describe the mistake in words, or copy down the whole problem and highlight or circle the mistake.

(b) Do the problem correctly.

79. Problem: Simplify: $(24x^3 + 8x^2 - 2x) \div (4x)$

Incorrect Answer: $\dfrac{24x^3 + 8x^2 - 2x}{4x}$

$= \dfrac{6\cdot \cancel{4}\cdot \cancel{x}\cdot x^2 + 2\cdot 4x^2 - 2x}{\cancel{4}\cdot \cancel{x}}$

$= 6x^2 + 8x^2 - 2x$

$= 14x^2 - 2x$

80. Problem: Use long division to simplify $(x^2 - 8x + 12) \div (x - 4)$.

Incorrect Answer:

$$\begin{array}{r} x - 12 \\ x - 4\overline{)x^2 - 8x + 12} \\ \underline{-(x^2 - 4x)} \\ -12x + 12 \\ \underline{-(-12x + 48)} \\ -36 \end{array}$$

The answer is $(x - 12)\ \mathrm{R}(-36)$.

81. Problem: Use long division to simplify $(2x^2 + 5x + 3) \div (x - 8)$.

Incorrect Answer:

$$\begin{array}{r} x + 13 \\ x - 8\overline{)2x^2 + 5x + 3} \\ \underline{-(x^2 - 8x)} \\ 13x + 3 \\ \underline{-(13x - 104)} \\ 107 \end{array}$$

The answer is $(x + 13)\ \mathrm{R}(107)$.

82. Problem: Simplify: $(6x^2 + 8x - 3) \div (3x)$

Incorrect Answer: $(6x^2 + 8x - 3) \div (3x)$

$= \dfrac{6x^2}{3x} + \dfrac{8x}{3x} - \dfrac{3}{3x}$

$= 2x + \dfrac{8}{3} - x$

$= x + \dfrac{8}{3}$

Review

83. Define *function.*

84. Define *domain.*

85. Describe the purpose of the vertical line test.

86. Is the set {(John, A), (Mary, A), (Miguel, A), (Lee, A)} a function? Explain.

SUCCESS IN COLLEGE MATHEMATICS

87. This section includes long division of polynomials. Looking at your notes from this section, identify the main ideas that your instructor presented.

88. Looking at your notes from this section, identify the topic that you found most difficult or that you do not yet understand.

Study Plan for Review of Chapter 5

SECTION 5.1 Exponential Expressions and the Exponent Rules

Ask Yourself	Test Yourself	Help Yourself
Can I...	**Do 5.1 Review Exercises**	**See these Examples and Practice Problems**
rewrite an expression with repeated multiplications using exponential notation? simplify a polynomial expression by combining like terms?	1, 2	Ex. 1, 2, PP 1–6
use the product rule of exponents?	3, 6, 15	Ex. 3, 4, 15, PP 7, 8, 22–24, 26, 27
use the quotient rule of exponents?	4, 5, 11, 13, 14	Ex. 6, 7, 9, 16, 18, 19, 21, PP 9, 10, 20, 24, 25
use the power rule of exponents?	8–15, 17	Ex. 8, 9, 11, 14, PP 11, 14–16, 21, 25–27
use the many bases rule for a product?	9–15, 17, 18	Ex. 10, 11, 13–15, 23–25, PP 12, 14–16, 25–27
use the many bases rule for a quotient?	11, 13	Ex. 12–14, 24, PP 13, 15, 16, 27
simplify an expression that includes an exponent of 0?	7, 16	Ex. 16, 17, PP 17–19
rewrite an expression with a negative exponent as an equivalent expression with positive exponents?	18	Ex. 18–22, PP 20–24

5.1 Review Exercises

For exercises 1–15, simplify. Simplified expressions must be in lowest terms, and all exponents must be positive.

1. $3w + 8p - 9w + 15p - 4$
2. $5x^2 + 8x - 13x^2 - 3(x - 2)$
3. $(8x^5)(6x^5)$
4. $\dfrac{35a^5b^3}{7z^4b}$
5. $\dfrac{28p^9q^3}{52p^4q^{10}}$
6. $(2x^3y)(-9x^4y)$
7. x^0
8. $(a^7)^9$
9. $(3x)^2$
10. $(-4x^3y^4)^2$
11. $\left(\dfrac{5a^3}{4b^5}\right)^2$
12. $(9w^{-3})^2$
13. $\left(\dfrac{2d^9}{5f^2}\right)^2\left(\dfrac{5f^3}{8d^{20}}\right)$
14. $\dfrac{(4x^3y)^2}{8xy}$
15. $(-6x^8y^9)^2(5x^2y)$
16. Choose the expressions that are equal to 1.
    a. h^0 b. $\dfrac{d^2}{d^2}$
    c. $\left(\dfrac{8ab}{9a^3b^2}\right)^0$ d. $\dfrac{7}{7}$
    e. $9k^0 - 8k^0$
17. A student rewrote $(7n^3)^2$ as 7^2n^6 and then simplified this to $14n^6$. Explain what is wrong with this work.
18. Choose the expressions that are equivalent to $(5a)^{-2}$.
    a. $5^{-2}a^{-2}$ b. $\dfrac{1}{(5a)^2}$
    c. $\dfrac{1}{-25a^2}$ d. $\dfrac{-25}{a^2}$
    e. $\dfrac{1}{25a^2}$

SECTION 5.2 Scientific Notation

Ask Yourself	Test Yourself	Help Yourself
Can I...	**Do 5.2 Review Exercises**	**See these Examples and Practice Problems**
write a measurement in scientific notation?	19–21, 30	Ex. 3–5, PP 1–6
multiply measurements in scientific notation?	22, 27, 28	Ex. 6, 7, PP 7–9
divide measurements in scientific notation?	24, 28	Ex. 8, 9, PP 10, 11
raise a measurement in scientific notation to a power?	23, 27	Ex. 10, PP 12
add or subtract measurements in scientific notation?	25, 26	Ex. 11, 12, PP 13–18
solve an equation that includes numbers or measurements in scientific notation?	28	Ex. 13–15, PP 19, 20

5.2 Review Exercises

For exercises 19–21, write the measurement in scientific notation. Include the units of measurement.

19. The volume of an ink drop is 0.000000000003 L.

20. The estimated number of hydrogen atoms in a human body is 4,220,000,000,000,000,000,000,000,000 atoms. (*Source:* www.foresight.org)

21. The total national debt on Dec. 29, 2011, was about $15,125,000,000,000. (*Source:* www.treasurydirect.gov)

For exercises 22–27, evaluate. Write in scientific notation, and include the units of measurement. If necessary, round the mantissa to the nearest hundredth.

22. $\left(6.02 \times 10^{23}\ \frac{\text{atoms}}{\text{mole}}\right)(5 \times 10^{3}\ \text{moles})$

23. $(5.3 \times 10^{-4}\ \text{m})^2$

24. $\frac{2 \times 10^4\ \text{kg}}{2.2 \times 10^4\ \text{L}}$

25. $8 \times 10^3\ \text{L} + 5 \times 10^2\ \text{L}$

26. $5 \times 10^{-3}\ \text{g} + 7 \times 10^{-4}\ \text{g}$

27. $\frac{1}{2}(6.0\ \text{kg})\left(5 \times 10^3\ \frac{\text{m}}{\text{s}}\right)^2$

28. Solve $E = \frac{(3.2 \times 10^4)(7.2 \times 10^1)}{(4)(2.7 \times 10^7)}$. Round the mantissa to the nearest hundredth.

29. Some exponential expressions are called "powers of ten." Write an example of a power of ten.

30. A student said that 0.0004 is equal to 4×10^{-3} because there are three zeros in 0.0004. Explain what is wrong with this thinking.

SECTION 5.3 Introduction to Polynomials

Ask Yourself	Test Yourself	Help Yourself
Can I...	**Do 5.3 Review Exercises**	**See these Examples and Practice Problems**
identify the number of terms, the variable(s), the coefficient(s), and the constant in a polynomial expression? identify a monomial, binomial, or trinomial?	38–42	Ex. 1, PP 1, 3
determine whether an expression is a polynomial?	31–37	Ex. 2, PP 2
write a polynomial in descending order and identify the lead coefficient and its degree?	41, 42	Ex. 3, 4, PP 4–6
simplify a polynomial expression by combining like terms?	43	Ex. 5, PP 7
add or subtract polynomial expressions?	44–46	Ex. 6–11, PP 8–12
multiply polynomial expressions?	47–50, 52	Ex. 12–17, PP 13–19

5.3 Review Exercises

For exercises 31–37,
(a) is the expression a polynomial?
(b) if it is not a polynomial, explain why.

31. $5x^2 - 8x + 10$

32. $\frac{3}{4}x^5 - 9x$

33. $\frac{4}{x - 9}$

34. 2^x

35. x^2

36. 7

37. $\sqrt{x + 8}$

For exercises 38–40, write a polynomial expression that is an example of

38. a monomial.

39. a binomial.

40. a trinomial.

For exercises 41–42,
(a) rewrite each expression in descending order.
(b) identify the degree of the expression.
(c) identify the lead coefficient.
(d) identify the constant.
(e) identify the number of terms.

41. $21x^2 + 15 - x^5 + 9x + 6x^3$

42. $\frac{5}{6} + \frac{3}{4}x^3 - x + \frac{2}{3}x^2$

For exercises 43–50, simplify.

43. $7x^2 - 9x + 2 + 15x^2 - 41x - 8$

44. $(7x^2 - 9x + 2) - (15x^2 - 41x - 8)$

45. $(8x^2 - 14) - (5x^2 + 7x + 14)$

46. $\left(\frac{5}{8}y^2 + 3y - \frac{1}{12}\right) + \left(\frac{2}{3}y^2 + 5y - \frac{5}{12}\right)$

47. $(9p)(5p^2)$

48. $\frac{1}{2}x(7x^2 + 6x - 4)$

49. $-x(6x^3 - x^2 + 9x - 8)$

50. $-2x(6x^3 - x^2 + 9x - 8)$

51. Given the expression $8x^2(2x^5 + 3)$, a student used the distributive property and a rule of exponents to rewrite it as $16x^7 + 24x^2$. What rule of exponents did the student use?

52. Choose the expressions that are equivalent to $-5a(3a - 4)$.
a. $-5a(3a) - 5a(4)$
b. $-5a(3a) - 5a(-4)$
c. $(3a - 4) - 5a$
d. $(3a - 4)(-5a)$

SECTION 5.4 Multiplication of Polynomials

Ask Yourself	Test Yourself	Help Yourself
Can I...	**Do 5.4 Review Exercises**	**See these Examples and Practice Problems**
use the distributive property to multiply polynomial expressions?	53–61, 67	Ex. 2–8, 10, 12, PP 1–6
use a pattern to multiply polynomial expressions?	62–67	Ex. 9, 11, 13, PP 7–12
write a polynomial expression that represents the perimeter or area of a rectangle or triangle?	68–70	Ex. 14–16, PP 13–15

5.4 Review Exercises

For exercises 53–61, simplify. Do not use a pattern.

53. $(x - 4)(x + 9)$

54. $(x - 10)(x - 6)$

55. $(7x + 2)(x - 9)$

56. $(3a + 11)(2a + 7)$

57. $(9x - 5)(9x + 5)$

58. $(x - 4)(x^2 - 5x + 9)$

59. $(x + 3y)(x - 7y)$

60. $(2w - 11)(7w^2 + 8w - 3)$

61. $(2p)(3p - 1)(p + 6)$

For exercises 62–65, use a pattern to simplify.

62. $(x + 8)(x + 8)$

63. $(a - 10)(a + 10)$

64. $(4y - 5)(4y - 5)$

65. $(10z - y)(10z + y)$

66. a. The linear equation $5x + 2y = 20$ is in standard form, $ax + by = c$. Identify a and b for this equation.
b. The expression $(3h + 4)(3h + 4)$ matches the pattern $(a + b)(a + b) = a^2 + 2ab + b^2$. Identify a and b in this expression.

67. Choose the expressions that are equivalent to $(x + 7)^2$.
a. $x^2 + 14x + 49$
b. $x^2 + 7^2$
c. $x(x + 7) + 7(x + 7)$
d. $(x + 7)(x + 7)$

68. The length of a rectangle is 6 ft longer than three times its width.
a. If W = width, write a polynomial expression in W that represents the length, and draw a diagram of the rectangle. Do not include the units.
b. Write a polynomial expression in W that represents the perimeter.
c. Write a polynomial expression in W that represents the area.

69. The width of a rectangle is one-half of the length.

a. If L = length, write a polynomial expression in L that represents the width, and draw a diagram of the rectangle. Do not include the units.

b. Write a polynomial expression in L that represents the perimeter.

c. Write a polynomial expression in L that represents the area.

70. The height of a triangle is equal to three times the base plus 5.

a. If b = base, write a polynomial expression in b that represents the height. Draw a diagram of this triangle. Do not include the units.

b. Write a polynomial expression in b that represents the area.

SECTION 5.5 Division of Polynomials

Ask Yourself	Test Yourself	Help Yourself
Can I...	**Do 5.5 Review Exercises**	**See these Examples and Practice Problems**
divide a polynomial by a monomial?	71, 72	Ex. 1–3, PP 1–3
use long division to divide a polynomial by a binomial?	73–79	Ex. 5–7, 9–11, PP 4–11
check the result of division?	71–76	Ex. 3–10, PP 1–11

5.5 Review Exercises

For exercises 71–72,
(a) simplify.
(b) check.

71. $(20x^3 + 12x^2 + 3x + 8) \div 2$

72. $(9x^{10} - 6x^7 + 2x^3 + 8) \div (2x)$

For exercises 73–76,
(a) use long division to simplify.
(b) check.

73. $(x^2 + 15x + 56) \div (x + 7)$

74. $(x^2 + x - 56) \div (x - 7)$

75. $(x^2 + 12x + 50) \div (x + 8)$

76. $(x^2 + 9x + 15) \div (x + 6)$

77. The divisor in a division problem is $a + 5$, and the quotient is $3a - 4$. Identify the dividend.

78. The divisor in a division problem is $n - 3$, and the quotient is $(n + 8)\ \mathrm{R}(31)$. What is the dividend?

79. The quotient in a division problem is $w + 4$, and the dividend is $w^2 + 11w + 28$. What is the divisor?

Chapter 5 Test

For problems 1–18, simplify.

1. $(5x^3)^2(4xy)$

2. $\dfrac{38x^5y^9}{4x^8y^3}$

3. $(-9a^5b^2c)^2$

4. 5^0

5. $(5p^{-4})^2$

6. $\left(\dfrac{7w}{6c}\right)^2$

7. $(5xy^3)^{-2}$

8. $\dfrac{1.8 \times 10^{-3}\ \mathrm{g}}{2 \times 10^2\ \mathrm{mL}}$

9. $3.4 \times 10^{20}\ \mathrm{kg} + 9.7 \times 10^{19}\ \mathrm{kg}$

10. $(6 \times 10^{-9}\ \mathrm{m})^3$

11. $4h^3(h^2 - 9h + 5)$

12. $(8x - 5)(x + 1)$

13. $(x - 7)(x^2 + 4x - 3)$

14. $3x(x - 1)(x + 4)$

15. $(5x^2 + 7x - 18) + (3x^2 - 4x - 2)$

16. $(5p^2 - 4p - 3) - (7p^2 - 8p + 10)$

17. $(5.2 \times 10^8\ \mathrm{m})(2 \times 10^9\ \mathrm{m})$

18. $(3x - 7)(3x + 7)$

19. Choose the expressions that are polynomials.

a. $\dfrac{3}{4}x + 1$

b. $\dfrac{3}{4x} + 1$

c. $\left(\dfrac{3}{4}\right)^x + 1$

d. $\dfrac{3}{4}x^2 - 9x + 7$

e. $\sqrt{\dfrac{3}{4}x} + 1$

20. a. Explain the difference between a trinomial and a monomial.

b. Write an example of a trinomial.

c. Write an example of a binomial.

21. A polynomial is $4x + 12 - 9x^2 + x^3$.
    a. Write this polynomial in descending order.
    b. Identify the degree of this polynomial.
    c. Identify the number of terms in this polynomial.

22. A pattern for multiplying polynomials is $(a + b)(a - b) = a^2 - b^2$.
    a. What is the name of this pattern?
    b. Use this pattern to multiply $(7x - 2)(7x + 2)$.

23. The length of a rectangle is 6 ft more than five times its width.
    a. If W = width, write a polynomial expression in W that represents the length, and draw a diagram of the rectangle. Do not include the units.
    b. Write a polynomial expression in W that represents the perimeter.
    c. Write a polynomial expression in W that represents the area.

24. The recommended dose of a drug is 0.000002 g per kilogram per minute. Write this measurement in scientific notation.

25. The mass of Jupiter is 1,898,600,000,000,000,000,000,000,000 kg. Write this measurement in scientific notation.

26. Find E when $E = (7.5 \times 10^2 \text{ kg})\left(3.0 \frac{\text{m}}{\text{s}}\right)^2$. Write the answer in scientific notation.

27. Simplify: $(12x^4 - 24x^3 + 4x - 36) \div (4x)$

28. a. Use long division to simplify $(x^2 - 11x + 30) \div (x - 6)$.
    b. Check.

29. a. Use long division to simplify $(x^2 + 8x + 14) \div (x + 2)$.
    b. Check.

Factoring Polynomials

6

SUCCESS IN COLLEGE MATHEMATICS

Learning from Tests and Quizzes

Quizzes and tests are "summative" or "formative." A summative test is used to assess student progress. When you take a summative test such as a final exam, you often do not get the test back, and you do not use the test for more learning. A formative test has two purposes. It is a measure of your progress and usually is assigned a grade of some kind. It also is an opportunity for you to learn from your mistakes and to identify where you need to do more work.

When you get a graded formative test back, it is important to correct your mistakes. Sometimes it is difficult to find the time to do this, since you are starting the next chapter. However, if you do not learn how to do the problems you missed, you still don't know important information needed for the next chapters. Just looking over the problem is not enough. You need to start over, do the problem from the beginning, and then ask your instructor, a tutor, or a person in your class whether it is now correct.

Tests also are important resources for studying for a final exam. The questions that you missed are important to review before taking the final. You can also use the questions from your chapter tests to make a practice final.

6.1 Introduction to Factoring
6.2 Factoring Trinomials
6.3 Patterns
6.4 Factoring Completely
6.5 The Zero Product Property

With many computer software programs, we can "undo" our last action by choosing a command in a menu or pressing a combination of keys on the keyboard. With some polynomials, we can "undo" multiplication by rewriting the polynomial as a product of factors. In this section, we begin our study of factoring.

SECTION 6.1 Introduction to Factoring

After reading the text, working the practice problems, and completing assigned exercises, you should be able to:

1. Find the greatest common factor of two or more terms.
2. Use the greatest common factor strategy to factor a polynomial.
3. Use the grouping strategy to factor a polynomial.

Greatest Common Factor and Prime Factorization

The result of multiplication is a product. The multiplied numbers are factors. For example, since $1 \cdot 6 = 6$ and $2 \cdot 3 = 6$, the factors of 6 are 1, 2, 3, and 6. When two or more numbers share a factor, it is a **common factor**. The greatest of these shared factors is the **greatest common factor**, sometimes abbreviated as the **GCF**. To find the greatest common factor of two numbers, list all of the factors of each number and identify the greatest factor that appears in both lists.

EXAMPLE 1 Find the greatest common factor of 24 and 60.

SOLUTION ▶

Factors of 24: 1, 2, 3, 4, 6, 8, (12), 24

Factors of 60: 1, 2, 3, 4, 5, 6, 10, (12), 15, 20, 30, 60

The greatest common factor is 12.

In the next example, the greatest common factor includes variables. The exponent on each variable in the greatest common factor is the smallest exponent on this variable in both terms.

EXAMPLE 2 Find the greatest common factor of x^2y^5z and x^3y^4.

SOLUTION ▶

Factors of x^2y^5z: $(x \cdot x) \cdot (y \cdot y \cdot y \cdot y) \cdot y \cdot z$

Factors of x^3y^4: $(x \cdot x) \cdot x \cdot (y \cdot y \cdot y \cdot y)$

The greatest common factor is $x \cdot x \cdot y \cdot y \cdot y \cdot y$. In exponential notation, it is x^2y^4. The exponent on x is 2, and the exponent on y is 4. These are the smallest exponents on these variables in both terms.

In the next example, we find the greatest common factor of terms that include numbers and variables.

EXAMPLE 3 Find the greatest common factor of $42y^3$ and $63x^2y^2$.

SOLUTION ▶

Factors of 42: 1, 2, 3, 6, 7, 14, (21), 42   Factors of y^3: $y \cdot (y \cdot y)$

Factors of 63: 1, 3, 7, 9, (21), 63   Factors of x^2y^2: $x \cdot x \cdot (y \cdot y)$

The greatest common factor is $21y^2$. Since x is not a factor in both terms, it cannot be a factor in the greatest common factor. The exponent on y is the smallest exponent on the y in both terms.

The set of whole numbers is {0, 1, 2, 3, 4, . . .}. A **prime number** is a whole number greater than 1 with only two factors, itself and 1. The set of the first ten prime numbers is {2, 3, 5, 7, 11, 13, 17, 19, 23, 29}. For any whole number greater than 1, there is only one combination of prime number factors with a product equal to this number. This expression is the **prime factorization** of the number.

The Fundamental Theorem of Arithmetic

Every whole number greater than 1 can be represented in exactly one way other than rearrangement as a product of one or more prime numbers.

EXAMPLE 4 Write the prime factorization of 60.

SOLUTION ▶

60

$= 6 \cdot 10$ Choose any two factors of 60.

$= 2 \cdot 3 \cdot 2 \cdot 5$ If a factor is not prime, rewrite it as a product.

$= \mathbf{2^2} \cdot 3 \cdot 5$ Rewrite repeated factors in exponential notation; $2 \cdot 2 = 2^2$

The prime factorization of 60 is $2^2 \cdot 3 \cdot 5$.

Writing the Prime Factorization of a Whole Number

1. Write the whole number as a product of two factors.
2. If a factor is not prime, rewrite it as the product of two factors.
3. Continue this process until all of the factors are prime numbers.
4. Write the factors in increasing order.
5. Use exponential notation to represent any repeated factors.

We can also use prime factorization to find the greatest common factor.

Using Prime Factorizations to Find the Greatest Common Factor

1. Write the prime factorization of the number in each term.
2. Choose each numerical factor that appears in each of the prime factorizations and the smallest exponent for this factor that appears in the factorizations.
3. Choose each variable that appears in each of the prime factorizations and the smallest exponent for this variable that appears in the factorizations.
4. The product of these numbers and variables is the greatest common factor.

EXAMPLE 5 Find the greatest common factor of $20a^3b^2c^3$, $72a^3bc^2$, and $180a^3c^2$.

SOLUTION ▶ Write the prime factorization of $20a^3b^2c^3$.

$20a^3b^2c^3$

$= \mathbf{2 \cdot 2 \cdot 5} \cdot a^3 \cdot b^2 \cdot c^3$ Rewrite as a product of primes.

$= \mathbf{2^2} \cdot 5 \cdot a^3 \cdot b^2 \cdot c^3$ Rewrite in exponential notation.

Write the prime factorization of $72a^3bc^2$.

$72a^3bc^2$

$= \mathbf{2 \cdot 2 \cdot 2 \cdot 3 \cdot 3} \cdot a^3 \cdot b \cdot c^2$ Rewrite as a product of primes.

$= \mathbf{2^3 \cdot 3^2} \cdot a^3 \cdot b \cdot c^2$ Rewrite in exponential notation.

Write the prime factorization of $180a^3c^2$.

$180a^3c^2$

$= \mathbf{2 \cdot 2 \cdot 3 \cdot 3 \cdot 5} \cdot a^3 \cdot c^2$ Rewrite as a product of primes.

$= \mathbf{2^2 \cdot 3^2} \cdot 5 \cdot a^3 \cdot c^2$ Rewrite using exponents.

Write the greatest common factor.

$2^2a^3c^2$ The product of common factors with the smallest exponent.

$= \mathbf{4}a^3c^2$ Simplify; $2^2 = 4$

Practice Problems

For problems 1–6, find the greatest common factor.

1. $28xy^2$; $36x^4y$ **2.** $32x^2y^2$; $60x^3y^2$ **3.** $30xz^2$; $45x^4y^2$
4. $90a^2c^3$; $108d^2f$ **5.** $30x^2y$; $12x^2yz$; $75xy^2z$ **6.** $15a^2d^3$; $28a^3d$; $32a^2d$

Factoring Polynomials

The polynomial $3x^2 + 6y$ is *simplified*. We can rewrite $3x^2 + 6y$ as the product of two factors, 3 and $x^2 + 2y$. Rewriting a polynomial as a product of factors is called **factoring**. We are using the distributive property to "undo" a multiplication. If the only factors of a polynomial are itself and 1, it is a **prime polynomial**. For example, the polynomial $3p - 5w$ is prime. The only factors of this polynomial are $3p - 5w$ and 1.

To factor some polynomials, we **factor out the greatest common factor**. In the next example, we look for a greatest common factor of all of the terms in the polynomial. To factor the polynomial, we rewrite it as the product of the greatest common factor and a new polynomial.

EXAMPLE 6 **(a)** Factor: $3ab + 2ac$

SOLUTION ▶

$3ab + 2ac$

$= a(\_\_\_ + \_\_\_)$ Factor out the greatest common factor, a.

$= a(\mathbf{3b} + \_\_\_)$ The product of a and the first term equals $3ab$.

$= a(3b + \mathbf{2c})$ The product of a and the second term equals $2ac$.

(b) Check.

▶ To check factoring, use the distributive property to multiply the factors. If the factoring is correct, the product is equivalent to the original polynomial.

$a(3b + 2c)$

$= \mathbf{a}(3b) + \mathbf{a}(2c)$ Distributive property.

$= 3ab + 2ac$ Simplify; the factoring is correct.

In the next example, the polynomial has three terms. The greatest common factor is a factor of all three terms.

EXAMPLE 7 **(a)** Factor: $6x^2y + 21xy^2 + 3xy$

SOLUTION ▶

$6x^2y + 21xy^2 + 3xy$

$= 3xy(\_\_\_ + \_\_\_ + \_\_\_)$ Factor out the greatest common factor, $3xy$.

$= 3xy(\mathbf{2x} + \_\_\_ + \_\_\_)$ The product of $3xy$ and the first term equals $6x^2y$.

$= 3xy(2x + \mathbf{7y} + \_\_\_)$ The product of $3xy$ and the second term equals $21xy^2$.

$= 3xy(2x + 7y + \mathbf{1})$ The product of $3xy$ and the third term equals $3xy$.

(b) Check.

$3xy(2x + 7y + 1)$
$= \mathbf{3xy}(2x) + \mathbf{3xy}(7y) + \mathbf{3xy}(1)$ Distributive property.
$= 6x^2y + 21xy^2 + 3xy$ Simplify; the factoring is correct.

In the next example, we combine some of the steps.

EXAMPLE 8 Factor: $48x + 30y - 12z$

SOLUTION

$48x + 30y - 12z$
$= 6(\_\_ + \_\_ - \_\_)$ Factor out the greatest common factor, 6.
$= 6(\mathbf{8x + 5y - 2z})$ Complete the factoring.

A greatest common factor can also be a binomial such as $x + 2$.

EXAMPLE 9 Factor: $3(x + 2) + y(x + 2)$

SOLUTION

$3(x + 2) + y(x + 2)$
$= (x + 2)(\_\_ + \_\_)$ Factor out the greatest common factor, $x + 2$.
$= (x + 2)(\mathbf{3 + y})$ Complete the factoring; both of the factors are binomials.

EXAMPLE 10 Factor: $7x + 6y + 8z$

SOLUTION This polynomial is prime; the only factors of this polynomial are $7x + 6y + 8z$ and 1.

Practice Problems

For problems 7–12,
(a) factor out the greatest common factor. Identify any prime polynomials.
(b) check.

7. $24cd^4 + 18c^3d$ **8.** $8a^2c^5 + 30ac$ **9.** $36a^3b^4 - 12ab$
10. $7m^3p^2 + 3m^2p^4 + 5m^2p^3$ **11.** $20xy + 27z$ **12.** $x^3 + 3x^2 - 5x$

Factor by Grouping

Although the four terms in $4a + 4c + ab + bc$ do not have a common factor, this polynomial is not prime. In the next example, we factor this polynomial **by grouping**. We use parentheses to group pairs of terms that have a common factor, factor out this common factor, and then factor again.

EXAMPLE 11 **(a)** Factor: $4a + 4c + ab + bc$

SOLUTION

$4a + 4c + ab + bc$ These terms have no common factor.
$= (4a + 4c) + (ab + bc)$ Group terms.
$= 4(\_\_ + \_\_) + b(\_\_ + \_\_)$ Identify a common factor of each group.
$= 4(a + c) + b(a + c)$ Complete the factoring of each group.
$= (a + c)(\_\_ + \_\_)$ Factor out a binomial, $a + c$.
$= (a + c)(4 + b)$ Complete the factoring.

(b) Check.

After using the distributive property, we use the commutative property of multiplication to change the order of the variables in three of the terms and then use the commutative property of addition to change the order of the terms in the polynomial to match the original polynomial.

$(a + c)(4 + b)$	
$= \mathbf{a}(4 + b) + \mathbf{c}(4 + b)$	Distributive property.
$= \mathbf{a}(4) + \mathbf{a}(b) + \mathbf{c}(4) + \mathbf{c}(b)$	Distributive property.
$= \mathbf{4a} + ab + \mathbf{4c} + \mathbf{bc}$	Commutative property of multiplication.
$= 4a + \mathbf{4c} + \mathbf{ab} + bc$	Commutative property of addition. Change order; the factoring is correct.

The greatest common factor of a group of terms may be the number 1.

EXAMPLE 12 Factor: $2wx^2 - 4wy + x^2 - 2y$

SOLUTION ▶

$2wx^2 - 4wy + x^2 - 2y$	These terms have no common factor.
$= (2wx^2 - 4wy) + (x^2 - 2y)$	Group terms.
$= 2w(\_\_ - \_\_) + 1(\_\_ - \_\_)$	Identify a common factor of each group.
$= 2w(x^2 - 2y) + 1(x^2 - 2y)$	Complete the factoring of each group.
$= (x^2 - 2y)(\_\_ + \_\_)$	Factor out a binomial, $x^2 - 2y$.
$= (x^2 - 2y)(2w + 1)$	Complete the factoring.

We can rewrite subtraction as addition of a negative number. For example, $6 - 4 = 6 + (-4)$. When a polynomial with four terms includes subtraction between the middle two terms, rewrite this subtraction as addition of a negative number before factoring by grouping.

EXAMPLE 13 Factor: $6x^2 + 3xz - 2xy - yz$

SOLUTION ▶

$6x^2 + 3xz - 2xy - yz$	These terms have no common factor.
$= 6x^2 + 3xz \mathbf{+ -}2xy - yz$	Rewrite subtraction.
$= (6x^2 + 3xz) + (-2xy - yz)$	Group terms.
$= 3x(\_\_ + \_\_) + -1y(\_\_ + \_\_)$	Identify a common factor of each group.
$= 3x(2x + z) + -1y(2x + z)$	Complete the factoring of each group.
$= (2x + z)(\_\_ + \_\_)$	Factor out a binomial, $2x + z$.
$= (2x + z)(3x + -1y)$	Complete the factoring.
$= (2x + z)(3x \mathbf{- y})$	Rewrite $3x + -1y$ as $3x - y$.

We factored $-1y$ from the second group. If we had instead factored out $1y$ from the second group, the factoring would be $3x(2x + z) + y(-2x - z)$. Since $3x(2x + z)$ and $y(-2x - z)$ do not have a common factor, we could not have completed the factoring.

The greatest common factor of a group of terms may be the number -1.

EXAMPLE 14 Factor: $7ac + 7bc - a - b$

SOLUTION ▶

$7ac + 7bc - a - b$	These terms have no common factor.
$= (7ac + 7bc) \mathbf{+} (\mathbf{-}a - b)$	Rewrite subtraction; group terms.
$= 7c(\_\_ + \_\_) + -1(\_\_ + \_\_)$	Identify a common factor of each group.
$= 7c(a + b) + -1(a + b)$	Complete the factoring of each group.
$= (a + b)(\_\_ + \_\_)$	Factor out a binomial, $a + b$.
$= (a + b)(7c + -1)$	Complete the factoring.
$= (a + b)(7c \mathbf{- 1})$	Rewrite $7c + -1$ as $7c - 1$.

To factor some polynomials by grouping, we need to change the order of the middle terms.

EXAMPLE 15 Factor: $-18p + 15w + 10pw - 27$

SOLUTION

$-18p + 15w + 10pw - 27$	These terms have no common factor.
$= (-18p + 15w) + (10pw - 27)$	Group terms.
$= -3(\_\_ - \_\_) + 1(\_\_ - \_\_)$	Identify a common factor of each group.
$= -3(\mathbf{6p - 5w}) + 1(\mathbf{10pw - 27})$	Complete the factoring of each group.

The groups do not have a common factor. Change the order of the middle terms, group terms, and factor.

$-18p + 15w + 10pw - 27$	These terms have no common factor.
$= -18p + \mathbf{10pw + 15w} - 27$	Change the order of the middle terms.
$= (-18p + 10pw) + (15w - 27)$	Group terms.
$= 2p(\_\_ + \_\_) + 3(\_\_ - \_\_)$	Identify a common factor of each group.
$= 2p(-9 + 5w) + 3(5w - 9)$	Complete the factoring of each group.
$= 2p(5w - 9) + 3(5w - 9)$	Change the order of terms in the first group.
$= (5w - 9)(\_\_ + \_\_)$	Factor out a binomial, $5w - 9$.
$= (5w - 9)(2p + 3)$	Complete the factoring.

Factoring a Four-Term Polynomial by Grouping

1. If the polynomial includes subtraction between the middle terms, rewrite subtraction as addition of a negative number.
2. Group the first two terms and the second two terms.
3. Factor the greatest common factor from each group. This factor may be 1 or -1.
4. If the groups now have a common binomial factor, factor again.
5. If the groups do not have a common binomial factor, change the order of the middle terms and repeat the process.
6. To check, use the distributive property to multiply the factors. If the factoring is correct, the product is equivalent to the original polynomial.

The common factor of two groups can be negative or positive. The sign we choose affects the final factors.

EXAMPLE 16 **(a)** Factor: $-3y + 2y^2 + 15x - 10xy$

SOLUTION

Factor y from the first group of terms:	**Factor $-y$ from the first group of terms:**
$-3y + 2y^2 + 15x - 10xy$	$-3y + 2y^2 + 15x - 10xy$
$= (-3y + 2y^2) + (15x - 10xy)$	$= (-3y + 2y^2) + (15x - 10xy)$
$= y(\_\_ + \_\_) + -5x(\_\_ + \_\_)$	$= -y(\_\_ - \_\_) + 5x(\_\_ - \_\_)$
$= y(-3 + 2y) + -5x(-3 + 2y)$	$= -y(3 - 2y) + 5x(3 - 2y)$
$= (-3 + 2y)(y - 5x)$	$= (3 - 2y)(-y + 5x)$

The signs in the final factors are different, but the factorings are equivalent, and both are correct. When we multiply the factors, the products are the same.

(b) Check.

$$\begin{aligned} &(-3 + 2y)(y - 5x) \\ &= \mathbf{-3}(y - 5x) + \mathbf{2y}(y - 5x) \\ &= -3y + 15x + 2y^2 - 10xy \\ &= -3y + 2y^2 + 15x - 10xy \end{aligned}$$

$$\begin{aligned} &(3 - 2y)(-y + 5x) \\ &= \mathbf{3}(-y + 5x) - \mathbf{2y}(-y + 5x) \\ &= -3y + 15x + 2y^2 - 10xy \\ &= -3y + 2y^2 + 15x - 10xy \end{aligned}$$

When you check your factoring with the answers given in this textbook, the signs on the terms in your factors may be opposites of those given in the answers. Though the signs on each set of factors may be opposites, both sets of factors are correct.

Many polynomials are prime. The only factors of a prime polynomial are itself and 1.

EXAMPLE 17 Factor: $15xy - 2x^2 + 40y - 8x$

SOLUTION

$15xy - 2x^2 + 40y - 8x$	These terms have no common factor.
$= (15xy - 2x^2) + (40y - 8x)$	Group terms.
$= x(\_\_\_ - \_\_\_) + 8(\_\_\_ - \_\_\_)$	Identify a common factor of each group.
$= x(\mathbf{15y - 2x}) + 8(\mathbf{5y - x})$	Complete the factoring of each group.

The groups do not have a common factor. Change the order of the middle terms, regroup, and factor.

$15xy + 40y - 2x^2 - 8x$	Change the order of the terms.
$= (15xy + 40y) + (-2x^2 - 8x)$	Rewrite subtraction; group terms.
$= 5y(\_\_\_ + \_\_\_) + -2x(\_\_\_ + \_\_\_)$	Identify a common factor of each group.
$= 5y(\mathbf{3x + 8}) + -2x(\mathbf{x + 4})$	Complete the factoring of each group.

The groups again do not have a common factor. The polynomial $15xy - 2x^2 + 40y - 8x$ is prime.

Practice Problems

For problems 13–18,
(a) factor by grouping. Identify any prime polynomials.
(b) check.

13. $12x^2 + 8xy + 3xz + 2yz$
14. $10az - 15bz + 6ac - 9bc$
15. $35hk - 24 + 40k - 21h$
16. $7x^2 + 35xy^2 - 3xy - 15y^3$
17. $12xy - 4xz - 15y + 5z$
18. $p^2 + 6p + 6p + 36$

6.1 Vocabulary Practice

Match the term with its description.

1. The result of multiplication
2. The only factors of this polynomial are itself and 1.
3. This operation can be rewritten as addition of a negative number.
4. The largest factor shared by two numbers
5. In the term $3xy$, 3 is an example of this.
6. The process of rewriting a polynomial as the product of factors
7. $a + b = b + a$
8. $a(b + c) = ab + ac$
9. Numbers that are the same distance from zero on the number line but have different signs
10. $a + (b + c) = (a + b) + c$

A. associative property of addition
B. coefficient
C. commutative property of addition
D. distributive property
E. factoring
F. greatest common factor
G. opposites
H. prime polynomial
I. product
J. subtraction

6.1 Exercises

Follow your instructor's guidelines for showing your work.

For exercises 1–10, find the greatest common factor of the terms.

1. 48; 54
2. 48; 56
3. $48x^2y^3$; $60xy^2$
4. $48x^2y^3$; $80x^2y$
5. $48x^5y^3$; $60x^2y$
6. $48x^7y^2$; $80xy^4$
7. $18ab$; $21cd^2$
8. $15pw$; $36c^2d$
9. $62x^2y^5$; $27xz^4$
10. $34a^4b^7$; $15ad^3$

For exercises 11–40,
(a) factor out the greatest common factor. Identify any prime polynomials.
(b) check.

11. $14p + 21q$
12. $18c + 24d$
13. $14p + 21q + 29r$
14. $18c + 24d + 31f$
15. $4x^2 + 20x$
16. $6y^2 + 30y$
17. $56x - 35z$
18. $21p - 56w$
19. $60x^2 - 100x$
20. $70y^2 - 30y$
21. $60x^2 - 60x$
22. $70y^2 - 70y$
23. $15a^3b^6 - 90ab^7 + 35a^2b^5$
24. $45hk^{15} - 27h^3k^9 + 18hk^{10}$
25. $-20x^2 - 6xy + 14xz + 2x$
26. $-30a^2 - 15ab + 33ac + 3a$
27. $50m^2n^2 + 20m^2n + 30mn^2 + 100$
28. $80h^2k^2 + 24h^2k + 64hk^2 + 240$
29. $x^4y^7 + x^5y^5 + x^3y^6 + x^2y^5$
30. $a^5b^{11} + a^4b^{10} + a^3b^9 + a^3b^5$
31. $26mn + 78m - 39n$
32. $22xy - 55x + 132y$
33. $9x^2 + 16y^2$
34. $4a^2 + 25b^2$
35. $30u^3 + 18u^2 + 54u$
36. $48v^3 + 56v^2 + 32v$
37. $10w^3 - 32w^2 + 8w$
38. $12z^3 - 20z^2 + 18z$
39. $p^4 - p^3 - p^2 - p$
40. $w^4 - w^3 - w^2 - w$

For exercises 41–66,
(a) factor by grouping. Identify any prime polynomials.
(b) check.

41. $2p^2 - 10p + 3pw - 15w$
42. $5x^2 - 35x + 2xy - 14y$
43. $30f^2 + 40f - 3fg - 4g$
44. $45a^2 + 63a - 5ab - 7b$
45. $30cd - 28 + 35d - 24c$
46. $3hk - 10 + 15k - 2h$
47. $2x^2 - 5x + 9xy - 27y$
48. $7w^2 - 34w + 3pw - 15p$
49. $3m^2 + 3mv + m + v$
50. $8u^2 + 8uz + u + z$
51. $3m^2 + 3mv - m - v$
52. $8u^2 + 8uz - u - z$
53. $3m^2 - 3mv - m - v$
54. $8u^2 - 8uz - u - z$
55. $2fj + gh + 2gj + fh$
56. $3ad + bc + ac + 3bd$
57. $x^3 - x^2z + x - z$
58. $2f^3 - f^2g + 2f - g$
59. $21uw - 6ku - 35hw + 10hk$
60. $33xz - 22ax - 21yz + 14ay$
61. $15cm + 12dm + 10ac + 8ad$
62. $8cz + 10dz + 12ac + 15ad$
63. $3ac + bd - 3ad - bc$
64. $2px - 2hx - pv + hv$
65. $3h^2 - 3hk - hk^2 - k^3$
66. $5x^2 - 5xz - xz^2 - z^3$

For exercises 67–82, factor by grouping. Do not combine like terms before factoring.

67. $x^2 + 7x + 4x + 28$
68. $x^2 + 9x + 4x + 36$
69. $p^2 - 9p + 6p - 54$
70. $p^2 - 8p + 7p - 56$
71. $c^2 - 8c - 3c + 24$
72. $d^2 - 7d - 6d + 42$
73. $10a^2 + 15a - 4a - 6$
74. $12b^2 - 21b + 20b - 35$
75. $36x^2 - 9x - 20x + 5$
76. $40x^2 - 35x - 8x + 7$
77. $a^2 + ab - ab - b^2$
78. $x^2 + xy - xy - y^2$
79. $a^2 + ab + ab + b^2$
80. $x^2 + xy + xy + y^2$
81. $25x^2 + 15x + 15x + 9$
82. $49z^2 + 14z + 14z + 4$

For exercises 83–88, either factor out the greatest common factor or factor by grouping.

83. $2xy + 4ab + 6cd$
84. $8ab + 2cd + 6cf$
85. $ax + 2y + ay + 2x$
86. $by + 3z + bz + 3y$
87. $a^3 + a^4 + a^5 + a^7$
88. $c^2 + c^4 + c^5 + c^6$

Problem Solving: Practice and Review

Follow your instructor's guidelines for using the five steps as outlined in Section 1.3, p. 55.

89. Predict the number of 240 death penalty sentences that were overturned on appeal.

The most far-reaching study of the death penalty in the United States has found that two out of three sentences were overturned on appeal, mostly because of serious errors by incompetent defense lawyers or overzealous police officers and prosecutors who withheld evidence. (*Source:* www.nytimes.com, June 12, 2009)

90. An e-mail from an airline promoted a new way to earn mileage plan miles. The chart shows the relationship of mileage plan miles earned, y, and the multiples of $10,000 of refinanced mortgage, x.

Buy, sell, or finance your home and earn Mileage Plan Miles at no additional cost after closing.

Refinance mortgage	1,250 miles per $10,000 financed + 15,000 Bonus Miles
New mortgage	1,250 miles per $10,000 financed + 5,000 Bonus Miles
Home equity loan	1,250 miles per $10,000 financed + 5,000 Bonus Miles

a. Write a linear model that represents the relationship of x and y.

b. Use this linear model to find the amount of miles earned for the refinance of a $275,000 loan.

91. Find the number of albums sold at the same point last year (2011). Round to the nearest tenth of a million.

Year to date album sales stand at 5.80 million, up 7% compared to the same total at this point last year. (*Source:* www.billboard.com, Jan. 11, 2012)

92. In 2000, 30,530,000 tons of yard waste went into the municipal waste stream. In 2010, 33,400,000 tons of yard waste went into the municipal waste stream. Find the average rate of change of the amount of yard waste going into the municipal waste stream per year.
(*Source:* www.epa.gov, Nov. 2011)

© imageegami/Shutterstock.com

Find the Mistake

For exercises 93–96, the completed problem has one mistake.
(a) Describe the mistake in words, or copy down the whole problem and highlight or circle the mistake.
(b) Do the problem correctly.

93. Problem: Factor: $60x^5y^4 + 36x^3y^2$
Incorrect Answer: $6x^3y^2(10x^2y^2 + 6)$

94. Problem: Factor: $16x^2 + 12xy + 4$
Incorrect Answer: $16x^2 + 12xy + 4$
$= 4x(4x + 3y + 1)$

95. Problem: Factor: $(-6x^2 + 4xz)$
Incorrect Answer: $(-6x^2 + 4xz)$
$= -2x(3x + 2z)$

96. Problem: Factor $6xy - 2xp + 3y - p$ by grouping.
Incorrect Answer: $6xy - 2xp + 3y - p$
$= 2x(3y - p) + (3y - p)$
$= 2x(3y - p)$

Review

For exercises 97–100, the sum and the product of two unknown numbers is given. Find the numbers.

97. sum = 17; product = 72

98. sum = −1; product = −56

99. sum = −4; product = −96

100. sum = 18; product = 77

SUCCESS IN COLLEGE MATHEMATICS

101. When you get a test back, do you redo the problems that you missed? Why or why not?

102. What percent of your final grade comes from quizzes, chapter tests, and the final exam?

$ax^2 + bx + c$

A trinomial is a polynomial with three terms. In this section, we will use the guess and check method and the *ac* method to factor trinomials.

SECTION 6.2 Factoring Trinomials

After reading the text, working the practice problems, and completing assigned exercises, you should be able to:

1. Use the guess and check method to factor a trinomial.
2. Use the *ac* method to factor a trinomial.
3. Use the discriminant to determine whether a quadratic trinomial is prime.

The Guess and Check Method

A trinomial that is degree 2 and matches the pattern $ax^2 + bx + c$ is a **quadratic trinomial**. The lead coefficient is a, the coefficient of the middle term is b, and the constant is c. The factors of some quadratic trinomials $ax^2 + bx + c$, where the lead coefficient is 1, are two binomials. The first term in each binomial factor is x. The second term in each binomial factor is a number. If both b and c are greater than 0, write the factors with a blank for the numbers: $(x + \_\_)(x + \_\_)$. To fill the blanks in the factors (the guess), find two numbers whose sum is equal to b and whose product is equal to c. To check that the factors are correct, use the distributive property to multiply.

EXAMPLE 1 Use the guess and check method to factor $x^2 + 9x + 8$.

SOLUTION ▸ Guess.

Since $b = 9$ and $c = 8$, use the factors $(x + \_\_)(x + \_\_)$. The sum of the numbers in each factor is 9, and the product is 8. Guess that the numbers are 1 and 8.

$x^2 + 9x + 8$

$= (x + \_\_)(x + \_\_)$ Since $x \cdot x = x^2$, the first term in both factors is x.

$= (x + \mathbf{1})(x + \mathbf{8})$ Guess: The numbers are 1 and 8.

Check.

$(x + 1)(x + 8)$

$= \mathbf{x}(x + 8) + \mathbf{1}(x + 8)$ Distributive property.

$= \mathbf{x}(x) + \mathbf{x}(8) + \mathbf{1}(x) + \mathbf{1}(8)$ Distributive property.

$= x^2 + 8x + 1x + 8$ Simplify.

$= x^2 + \mathbf{9x} + 8$ Simplify; the factoring is correct.

For a quadratic trinomial in which $a = 1$, the sum of the numbers in the correct factors equals b and the product of the numbers in the factors equals c. In Example 1, we factored $x^2 + 9x + 8$. In the next example, we factor $x^2 + 6x + 8$. Although the constant c in both trinomials is 8, the value of b is different. Changing the value of b changes the factors.

EXAMPLE 2 Use the guess and check method to factor $x^2 + 6x + 8$.

SOLUTION ▶ Guess.

Since $b = 6$ and $c = 8$, use the factors $(x + \_\_)(x + \_\_)$. The sum of the numbers is 6, and the product is 8. Guess that the numbers are 2 and 4.

$x^2 + 6x + 8$	
$= (x + \_\_)(x + \_\_)$	Since $x \cdot x = x^2$, the first term in both factors is x.
$= (x + \mathbf{2})(x + \mathbf{4})$	Guess: the numbers are 2 and 4.

Check.

$(x + 2)(x + 4)$	
$= \mathbf{x}(x + 4) + \mathbf{2}(x + 4)$	Distributive property.
$= \mathbf{x}(x) + \mathbf{x}(4) + \mathbf{2}(x) + \mathbf{2}(4)$	Distributive property.
$= x^2 + 4x + 2x + 8$	Simplify.
$= x^2 + \mathbf{6x} + 8$	Simplify; the factoring is correct.

Factoring a Quadratic Trinomial Using the Guess and Check Method

1. For a trinomial $ax^2 + bx + c$, where $a = 1$,

   If $b > 0$ and $c > 0$, the factors are $(x + \_\_)(x + \_\_)$.
   If $c < 0$, the factors are $(x + \_\_)(x - \_\_)$.
   If $b < 0$ and $c > 0$, the factors are $(x - \_\_)(x - \_\_)$.

   The sum of the numbers in the factors equals b, and the product of the numbers in the factors equals c.
2. Guess the numbers to fill in the blanks, and check by multiplying the factors. If the guess is incorrect, try a different pair or arrangement of numbers.

In the next example, both b and c are less than 0. Since the product of a negative number and a positive number is a negative number, the factors are $(p + \_\_)(p - \_\_)$.

EXAMPLE 3 Use the guess and check method to factor $p^2 - p - 12$.

SOLUTION ▶ Guess.

Since $b = -1$ and $c = -12$, use the factors $(p + \_\_)(p - \_\_)$. The sum of the numbers is -1 and the product is -12. Guess that the numbers are 4 and -3.

$p^2 - p - 12$	
$= (p + \_\_)(p - \_\_)$	Since $p \cdot p = p^2$, the first term in both factors is p.
$= (p + \mathbf{4})(p - \mathbf{3})$	Guess: The numbers are 4 and -3.

Check.

$(p + 4)(p - 3)$	
$= \mathbf{p}(p - 3) + \mathbf{4}(p - 3)$	Distributive property.
$= \mathbf{p}(p) + \mathbf{p}(-3) + \mathbf{4}p + \mathbf{4}(-3)$	Distributive property.
$= p^2 - 3p + 4p - 12$	Simplify.
$= p^2 + \mathbf{p} - 12$	Simplify; the factoring is *not* correct.

Guess.

Since the factoring is not correct, make another guess. Guess that the numbers are 3 and -4.

$p^2 - p - 12$	
$= (p + \_\_\_)(p - \_\_\_)$	Since $p \cdot p = p^2$, the first term in both factors is p.
$= (p + 3)(p - 4)$	New guess: The numbers are 3 and -4.

Check.

$(p + 3)(p - 4)$	
$= \mathbf{p}(p - 4) + \mathbf{3}(p - 4)$	Distributive property.
$= \mathbf{p}(p) + \mathbf{p}(-4) + \mathbf{3}(p) + \mathbf{3}(-4)$	Distributive property.
$= p^2 - 4p + 3p - 12$	Simplify.
$= p^2 \mathbf{- p} - 12$	Simplify; the factoring is correct.

In the next example, b is greater than 0, and c is less than 0. Since the product of a negative number and a positive number is a negative number, the factors are $(w + \_\_\_)(w - \_\_\_)$.

EXAMPLE 4 Use the guess and check method to factor $w^2 + 9w - 22$.

SOLUTION ▶ Guess.

Since $b = 9$ and $c = -22$, use the factors $(w + \_\_\_)(w - \_\_\_)$. The sum of the numbers is 9, and the product is -22. Guess that the numbers are 11 and -2.

$w^2 + 9w - 22$	
$= (w + \_\_\_)(w - \_\_\_)$	Since $w \cdot w = w^2$, the first term in both factors is w.
$= (w + \mathbf{11})(w - \mathbf{2})$	Guess: The numbers are 11 and -2.

Check.

$(w + 11)(w - 2)$	
$= \mathbf{w}(w - 2) + \mathbf{11}(w - 2)$	Distributive property.
$= \mathbf{w}(w) + \mathbf{w}(-2) + \mathbf{11}(w) + \mathbf{11}(-2)$	Distributive property.
$= w^2 - 2w + 11w - 22$	Simplify.
$= w^2 + \mathbf{9w} - 22$	Simplify; the factoring is correct.

In the next example, b is less than 0 and c is greater than 0. Since the product of a negative number and a negative number is a positive number, the factors are $(k - \_\_\_)(k - \_\_\_)$.

EXAMPLE 5 Use the guess and check method to factor $k^2 - 9k + 14$.

SOLUTION ▶ Guess.

Since $b = -9$ and $c = 14$, use the factors $(k - \_\_\_)(k - \_\_\_)$. The sum of the numbers is -9, and the product is 14. Guess that the numbers are -2 and -7.

$k^2 - 9k + 14$	
$= (k - \_\_\_)(k - \_\_\_)$	Since $k \cdot k = k^2$, the first term in both factors is k.
$= (k - \mathbf{2})(k - \mathbf{7})$	Guess: The unknown numbers are -2 and -7.

Check.

$(k - 2)(k - 7)$
$= \mathbf{k}(k - 7) - \mathbf{2}(k - 7)$ Distributive property.
$= \mathbf{k}(k) + \mathbf{k}(-7) - \mathbf{2}(k) - \mathbf{2}(-7)$ Distributive property.
$= k^2 - 7k - 2k + 14$ Simplify.
$= k^2 - \mathbf{9k} + 14$ Simplify; the factoring is correct.

In the next example, we use the guess and check method to factor a trinomial in which the lead coefficient a is not 1. When we fill the blanks of the factors with numbers, the product of the numbers is c, but the sum of the numbers is not b.

EXAMPLE 6 Use the guess and check method to factor $2x^2 + 7x + 3$.

SOLUTION ▶ Guess.

Since $a = 2$, use the factors $(2x + \_\_)(x + \_\_)$. Since $b = 7$ and $c = 3$, the product of the numbers in the factors equals 3, and the sum of *two times* one of the numbers and the other number equals 7.

$2x^2 + 7x + 3$
$= (2x + \_\_)(x + \_\_)$ Since $2x \cdot x = 2x^2$, the first term in one factor is $2x$.
$= (2x + \mathbf{3})(x + \mathbf{1})$ Guess: The unknown numbers are 3 and 1.

Check.

$(2x + 3)(x + 1)$
$= \mathbf{2x}(x + 1) + \mathbf{3}(x + 1)$ Distributive property.
$= \mathbf{2x}(x) + \mathbf{2x}(1) + \mathbf{3}(x) + \mathbf{3}(1)$ Distributive property.
$= 2x^2 + 2x + 3x + 3$ Simplify.
$= 2x^2 + \mathbf{5x} + 3$ Simplify; the factoring is *not* correct.

Guess.

Since the factoring is not correct, make another guess. Guess that the numbers are 1 and 3.

$2x^2 + 7x + 3$
$= (2x + \mathbf{1})(x + \mathbf{3})$ Guess: The unknown numbers are 1 and 3.

Check.

$(2x + 1)(x + 3)$
$= \mathbf{2x}(x + 3) + \mathbf{1}(x + 3)$ Distributive property.
$= \mathbf{2x}(x) + \mathbf{2x}(3) + \mathbf{1}(x) + \mathbf{1}(3)$ Distributive property.
$= 2x^2 + 6x + 1x + 3$ Simplify.
$= 2x^2 + \mathbf{7x} + 3$ Simplify; the factoring is correct.

Practice Problems

For problems 1–4, use guess and check to factor. Identify any prime polynomials.

1. $x^2 + 13x + 36$ **2.** $z^2 - 9z + 18$ **3.** $w^2 + 3w - 54$
4. $3x^2 + 29x + 40$

The *ac* Method

Another method for factoring trinomials is the *ac* method. To use the *ac* method, first identify the values of a, b, and c for any quadratic trinomial of the form $ax^2 + bx + c$, and find the product ac. Next, find a pair of factors whose product is equal to ac and whose sum is equal to b. Then rewrite the middle term, bx, of the trinomial as two terms with coefficients using the pair of factors you found whose product is ac and whose sum is b. Finally, factor by grouping. This method is usually more efficient than guess and check for factoring quadratic trinomials in which $a > 1$. Although we still need to find the numbers with a product of ac and a sum of b, we do not need to guess the factors of the trinomial.

EXAMPLE 7 Use the *ac* method to factor $2x^2 + 7x + 3$. Check.

SOLUTION ▶

$2x^2 + 7x + 3$ $a = 2,\ b = 7,\ c = 3,\ ac = 6$	
Factor pairs of 6 Both numbers are positive.	**Sum of factor pair**
$1 \cdot 6$	7
$2 \cdot 3$	5

Rewrite $2x^2 + 7x + 3$ as $2x^2 + \mathbf{1x} + \mathbf{6x} + 3$, and factor by grouping.

$$
\begin{aligned}
&2x^2 + 1x + 6x + 3 && \text{Rewrite } 7x \text{ as } 1x + 6x.\\
&= (2x^2 + 1x) + (6x + 3) && \text{Group terms.}\\
&= x(\_\_ + \_\_) + 3(\_\_ + \_\_) && \text{Identify a common factor of each group.}\\
&= x(2x + 1) + 3(2x + 1) && \text{Complete the factoring of each group.}\\
&= (2x + 1)(\_\_ + \_\_) && \text{Factor out a binomial, } 2x + 1.\\
&= (2x + 1)(x + 3) && \text{Complete the factoring.}
\end{aligned}
$$

Check.

$$
\begin{aligned}
&(2x + 1)(x + 3)\\
&= \mathbf{2x}(x + 3) + \mathbf{1}(x + 3) && \text{Distributive property.}\\
&= \mathbf{2x}(x) + \mathbf{2x}(3) + \mathbf{1}(x) + \mathbf{1}(3) && \text{Distributive property.}\\
&= 2x^2 + 6x + x + 3 && \text{Simplify.}\\
&= 2x^2 + \mathbf{7x} + 3 && \text{Simplify; the factoring is correct.}
\end{aligned}
$$

In the next example, *ac* equals 72, a number with six different factor pairs. However, only one of these factor pairs has a sum that is equal to b, 17.

EXAMPLE 8 Use the *ac* method to factor $6x^2 + 17x + 12$.

SOLUTION ▶

$6x^2 + 17x + 12$ $a = 6,\ b = 17,\ c = 12,\ ac = 72$	
Factor pairs of 72 Both numbers are positive.	**Sum of factor pair**
$1 \cdot 72$	73
$2 \cdot 36$	38
$3 \cdot 24$	27
$4 \cdot 18$	22
$6 \cdot 12$	18
$8 \cdot 9$	17

Rewrite $6x^2 + 17x + 12$ as $6x^2 \mathbf{+ 8x + 9x} + 12$, and factor by grouping.

$6x^2 + 8x + 9x + 12$	Rewrite $17x$ as $8x + 9x$.
$= (6x^2 + 8x) + (9x + 12)$	Group terms.
$= 2x(\_\_\_ + \_\_\_) + 3(\_\_\_ + \_\_\_)$	Identify a common factor of each group.
$= 2x(3x + 4) + 3(3x + 4)$	Complete the factoring of each group.
$= (3x + 4)(\_\_\_ + \_\_\_)$	Factor out a binomial, $3x + 4$.
$= (3x + 4)(2x + 3)$	Complete the factoring.

The product of a negative number and a positive number is a negative number. When *ac* is a negative number, one of the numbers in the factor pair is negative, and the other is positive.

EXAMPLE 9 Use the *ac* method to factor $2x^2 - 11x - 6$.

SOLUTION ▶

$2x^2 - 11x - 6$ $a = 2, b = -11, c = -6, ac = -12$			
Factor pairs of −12 **One number is negative; one number is positive.**		**Sum of factor pair**	
$(-1)(12)$	$(1)(-12)$	11	−11
$(-2)(6)$	$(2)(-6)$	4	−4
$(-3)(4)$	$(3)(-4)$	1	−1

Rewrite $2x^2 - 11x - 6$ as $2x^2 \mathbf{+ 1x + -12x} - 6$, and factor by grouping.

$2x^2 + 1x + -12x - 6$	Rewrite $-11x$ as $1x + -12x$.
$(2x^2 + 1x) + (-12x - 6)$	Group terms.
$= x(\_\_\_ + \_\_\_) + -6(\_\_\_ + \_\_\_)$	Identify a common factor of each group.
$= x(2x + 1) + -6(2x + 1)$	Complete the factoring of each group.
$= (2x + 1)(\_\_\_ + \_\_\_)$	Factor out a binomial, $2x + 1$.
$= (2x + 1)(x + -6)$	Complete the factoring.
$= (2x + 1)(x - 6)$	Rewrite $x + -6$ as $x - 6$.

The product of a negative number and a negative number is a positive number. When *ac* is a positive number and *b* is a negative number, both of the numbers in the factor pair are negative.

EXAMPLE 10 Use the *ac* method to factor $3x^2 - 32x + 45$.

SOLUTION ▶

$3x^2 - 32x + 45$ $a = 3, b = -32, c = 45, ac = 135$	
Factor pairs of 135 **Both numbers are negative.**	**Sum of factor pair**
$(-1)(-135)$	−136
$(-3)(-45)$	−48
$(-5)(-27)$	−32

Rewrite $3x^2 - 32x + 45$ as $3x^2 \mathbf{- 5x + -27x} + 45$, and factor by grouping.

$3x^2 - 5x + -27x + 45$	Rewrite $-32x$ as $-5x + -27x$.
$= (3x^2 - 5x) + (-27x + 45)$	Group terms.

$= x(\_\_\_ - \_\_\_) + -9(\_\_\_ - \_\_\_)$	Identify a common factor of each group.
$= x(3x - 5) + -9(3x - 5)$	Complete the factoring of each group.
$= (3x - 5)(\_\_\_ + \_\_\_)$	Factor out a binomial, $3x - 5$.
$= (3x - 5)(x + -9)$	Complete the factoring.
$= (3x - 5)(x - 9)$	Rewrite $x + -9$ as $x - 9$.

The *ac* Method for Factoring Quadratic Trinomials

1. For the quadratic trinomial $ax^2 + bx + c$, identify a, b, c, and ac.
2. Identify the factor pairs of ac. Choose the factor pair with a sum that equals b.
3. Rewrite the middle term of the trinomial as two terms. The coefficients of the new terms are the factor pair from step 2.
4. Factor by grouping.

In the next example, $ac = -54$ and $b = -25$. Although all of the factor pairs are identified, we can stop looking once we find the correct pair, $(2)(-27)$.

EXAMPLE 11 Use the *ac* method to factor $6x^2 - 25x - 9$.

SOLUTION ▶

$6x^2 - 25x - 9$

$a = 6$, $b = -25$, $c = -9$, $ac = -54$

Factor pairs of −54 One number is negative; one number is positive.		Sum of factor pair	
$(-1)(54)$	$(1)(-54)$	53	−53
$(-2)(27)$	$(2)(-27)$	25	−25
$(-3)(18)$	$(3)(-18)$	15	−15
$(-6)(9)$	$(6)(-9)$	3	−3

Rewrite $6x^2 - 25x - 9$ as $6x^2 + \mathbf{2x} + \mathbf{-27x} - 9$, and factor by grouping.

$6x^2 + 2x + -27x - 9$	Rewrite $-25x$ as $2x + -27x$.
$= (6x^2 + 2x) + (-27x - 9)$	Group terms.
$= 2x(\_\_\_ + \_\_\_) + -9(\_\_\_ + \_\_\_)$	Identify a common factor of each group.
$= 2x(3x + 1) + -9(3x + 1)$	Complete the factoring of each group.
$= (3x + 1)(\_\_\_ + \_\_\_)$	Factor out a binomial, $3x + 1$.
$= (3x + 1)(2x + -9)$	Complete the factoring.
$= (3x + 1)(2x - 9)$	Rewrite $2x + -9$ as $2x - 9$.

The *ac* method can also be used to factor quadratic trinomials in which $a = 1$, although the guess and check method might be more efficient.

EXAMPLE 12 Use the *ac* method to factor $k^2 - 9k + 14$.

SOLUTION ▶

$k^2 - 9k + 14$

$a = 1$, $b = -9$, $c = 14$, $ac = 14$

Factor pairs of 14 Both numbers are negative.	Sum of factor pair
$(-1)(-14)$	−15
$(-2)(-7)$	−9

Rewrite $k^2 - 9k + 14$ as $k^2 \mathbf{- 2k + -7k} + 14$, and factor by grouping.

$k^2 - 2k + -7k + 14$	Rewrite $-9k$ as $-2k + -7k$.
$= (k^2 - 2k) + (-7k + 14)$	Group terms.
$= k(\_\_\_ - \_\_\_) + -7(\_\_\_ - \_\_\_)$	Identify a common factor of each group.
$= k(k - 2) + -7(k - 2)$	Complete the factoring of each group.
$= (k - 2)(\_\_\_ + \_\_\_)$	Factor out a binomial, $k - 2$.
$= (k - 2)(k + -7)$	Complete the factoring.
$= (k - 2)(k \mathbf{- 7})$	Rewrite $k + -7$ as $k - 7$.

We have now used four methods to factor polynomials. We will use factoring later in this chapter to solve polynomial equations.

Methods for Factoring Polynomials

- Greatest common factor
- Grouping
- Guess and check
- *ac* method

Practice Problems

For problems 5–8, use the *ac* method to factor. Identify any prime polynomials.

5. $2x^2 + 17x + 30$ **6.** $2x^2 - 17x + 30$ **7.** $6x^2 + 13x + 5$
8. $12x^2 - 5x - 2$

The Discriminant and Quadratic Trinomials

A rational number can be written as a fraction in which the numerator and denominator are integers. If a number is a perfect square, its square root is a rational number. For example, since $\sqrt{484} = 22$ and 22 is a rational number, 484 is a perfect square. Since $\sqrt{13} \approx 3.6055\ldots$, an irrational number, 13 is not a perfect square.

The standard form of a quadratic trinomial in one variable is $ax^2 + bx + c$. The **discriminant** of a quadratic trinomial is $b^2 - 4ac$. If the discriminant is a perfect square, then we can factor the trinomial. If the discriminant is not a perfect square or if the discriminant is less than 0, the trinomial is prime. The only factors of a prime trinomial are itself and 1.

Discriminant Test for Quadratic Trinomials

For a quadratic trinomial $ax^2 + bx + c$, where a, b, and c have no common factor, if $b^2 - 4ac$ is a perfect square, the trinomial can be factored. If $b^2 - 4ac$ is not a perfect square or if $b^2 - 4ac < 0$, the trinomial is prime.

EXAMPLE 13 Determine whether $3x^2 - 32x + 45$ is a prime polynomial.

(a) Find the discriminant.

SOLUTION ▶

$b^2 - 4ac$	
$= (\mathbf{-32})^2 - 4 \cdot \mathbf{3} \cdot \mathbf{45}$	$a = 3, b = -32, c = 45$
$= 1024 - 540$	Simplify.
$= 484$	Simplify.

(b) Use the discriminant to determine whether the polynomial is prime.

The discriminant is 484 and $\sqrt{484}$ is 22. Since 22 is a rational number, 484 is a perfect square, and this polynomial is not prime. We factored it in Example 10: $3x^2 - 32x + 45 = (3x - 5)(x - 9)$.

The discriminant is part of the quadratic formula. We will study the quadratic formula in Chapter 9 and see why we can use the discriminant to identify prime quadratic trinomials.

EXAMPLE 14 Determine whether $2x^2 - x + 6$ is a prime polynomial.

(a) Find the discriminant.

SOLUTION

$b^2 - 4ac$

$= (\mathbf{-1})^2 - 4 \cdot \mathbf{2} \cdot \mathbf{6}$ — $a = 2, b = -1, c = 6$

$= 1 - 48$ — Simplify.

$= -47$ — Simplify.

(b) Use the discriminant to determine whether the polynomial is prime.

Since the discriminant, -47, is less than zero, this polynomial is prime.

Practice Problems

For problems 9–11,
(a) find the discriminant.
(b) use the discriminant to determine if the polynomial is prime.

9. $2x^2 + 3x + 1$ **10.** $6x^2 - x + 12$ **11.** $21x^2 - 19x - 12$

Factoring Trinomials That Are Not Quadratic

The degree of a quadratic trinomial, $ax^2 + bx + c$, is 2. We can use the *ac* method to factor some trinomials that are not quadratic.

EXAMPLE 15 Use the *ac* method to factor $x^6 - x^3 - 20$. Check.

SOLUTION

$x^6 - 1x^3 - 20$
$a = 1,\ b = -1,\ c = -20,\ ac = -20$

Factor pairs of −20 One number is negative; one number is positive.		Sum of factor pair	
$(-1)(20)$	$(1)(-20)$	19	-19
$(-2)(10)$	$(2)(-10)$	8	-8
$(-4)(5)$	$(4)(-5)$	1	-1

The middle term, $-x^3$, can be rewritten as $4x^3 - 5x^3$ or as $-5x^3 + 4x^3$. Rewrite $x^6 - x^3 - 20$ as $x^6 \mathbf{- 5x^3 + 4x^3} - 20$, and factor by grouping.

$x^6 - 5x^3 + 4x^3 - 20$ — Rewrite $-x^3$ as $-5x^3 + 4x^3$.

$= (x^6 - 5x^3) + (4x^3 - 20)$ — Group terms.

$= x^3(\_\_\_ - \_\_\_) + 4(\_\_\_ - \_\_\_)$ — Identify a common factor of each group.

$= x^3(x^3 - 5) + 4(x^3 - 5)$ — Complete the factoring of each group.

$= (x^3 - 5)(\_\_\_ + \_\_\_)$ — Factor out a binomial, $x^3 - 5$.

$= (x^3 - 5)(x^3 + 4)$ — Complete the factoring.

Check.

$(x^3 - 5)(x^3 + 4)$

$= \mathbf{x^3}(x^3 + 4) - \mathbf{5}(x^3 + 4)$ — Distributive property.

$= \mathbf{x^3}(x^3) + \mathbf{x^3}(4) - \mathbf{5}(x^3) - \mathbf{5}(4)$ — Distributive property.

$= x^{3+3} + 4x^3 - 5x^3 - 20$ — Simplify; product rule of exponents.

$= \mathbf{x^6} - \mathbf{x^3} - 20$ — Simplify; the factoring is correct.

Practice Problems

For problems 12–15, use the *ac* method to factor. Identify any prime polynomials.

12. $x^6 + 9x^3 + 20$ **13.** $a^8 - 6a^4 - 16$ **14.** $2x^4 - 9x^2 - 5$
15. $6x^6 - x^3 - 12$

6.2 Vocabulary Practice

Match the term with its description.

1. The result of addition
2. The result of multiplication
3. $x^m \cdot x^n = x^{m+n}$
4. A polynomial with three terms
5. The coefficient of the first term of a quadratic trinomial
6. The only factors of this polynomial are itself and 1.
7. The largest factor of two numbers
8. $b^2 - 4ac$
9. In the polynomial $5x^2 + 3x + 7$, 7 is an example of this.
10. $a(b + c) = ab + ac$

A. constant
B. discriminant
C. distributive property
D. greatest common factor
E. lead coefficient
F. prime polynomial
G. product
H. product rule of exponents
I. sum
J. trinomial

6.2 Exercises

Follow your instructor's guidelines for showing your work.

For exercises 1–22, use the guess and check method to factor. Identify any prime polynomials.

1. $x^2 + 16x + 63$
2. $y^2 + 15y + 54$
3. $p^2 - 9p + 20$
4. $w^2 - 11w + 24$
5. $h^2 + 11h + 42$
6. $k^2 + 15k + 40$
7. $z^2 - 7z - 18$
8. $a^2 - 4a - 21$
9. $z^2 + 7z - 18$
10. $a^2 + 4a - 21$
11. $x^2 - 14x + 49$
12. $y^2 - 16y + 64$
13. $x^2 + 14x + 49$
14. $y^2 + 16y + 64$
15. $x^2 + 14x - 49$
16. $y^2 + 16y - 64$
17. $2c^2 - 17c - 9$
18. $2f^2 - 13f - 7$
19. $3v^2 + 8v + 4$
20. $3u^2 + 14u + 8$
21. $5m^2 - 2m - 3$
22. $7a^2 - 5a - 2$

For exercises 23–50, use the *ac* method to factor. Check the factoring. Identify any prime polynomials.

23. $2c^2 - c - 21$
24. $2x^2 + 9x - 35$
25. $2j^2 + 7j - 39$
26. $2m^2 + 3m - 44$
27. $2j^2 - 7j - 39$
28. $2m^2 - 3m - 44$
29. $2b^2 - 11b - 20$
30. $2c^2 - 9c - 36$
31. $3v^2 + 8v + 4$
32. $3u^2 + 14u + 8$
33. $5y^2 - 19y - 4$
34. $7b^2 + b - 6$
35. $3b^2 + 28b + 32$
36. $3n^2 + 26n + 35$

37. $9w^2 - 18w + 5$

38. $9d^2 - 27d + 8$

39. $4x^2 + 20x + 25$

40. $4x^2 + 12x + 9$

41. $2f^2 - 5f - 88$

42. $2h^2 - 7h - 72$

43. $x^2 + 20x + 99$

44. $y^2 + 15y + 56$

45. $p^2 - 10p + 24$

46. $w^2 - 12w + 27$

47. $z^2 - 6z - 27$

48. $a^2 - 4a - 32$

49. $z^2 + 2z - 48$

50. $a^2 + 7a - 18$

51. A student is using the *ac* method to factor $6x^2 - 11x - 35$. Since $a = 6$, $b = -11$, $c = -35$, and $ac = -210$, the student needs to find two numbers whose sum is -11 and whose product is -210. The student does not identify the factors paired with 1, 2, 3, and 5 because she thinks that the sum of these factors cannot be -11. Is the student correct? Explain.

52. A student is factoring a quadratic trinomial in which the lead coefficient and the constant are perfect squares. He wonders whether he can always use the square root of the lead coefficient and the square root of the constant to write the factors. Write a quadratic trinomial in which this method does not work.

For exercises 53–58, use the *ac* method to factor. Check the factoring. Identify any prime polynomials.

53. $y^4 - 4y^2 - 32$

54. $j^4 + 4j^2 - 21$

55. $x^6 + 11x^3 + 18$

56. $b^6 + 11b^3 + 30$

57. $z^8 - 11z^4 + 30$

58. $p^8 - 7p^4 + 10$

For exercises 59–74, factor. Either factor out the greatest common factor, factor by grouping, use the guess and check method, or use the *ac* method.

59. $3b^2 + 28b + 32$

60. $3n^2 + 26n + 35$

61. $6y^2 + 11y - 72$

62. $6x^2 + 5x - 56$

63. $8d^2 + 26d + 21$

64. $8c^2 + 26c + 15$

65. $4k^2 - 36k + 81$

66. $4y^2 - 28y + 49$

67. $5q^2 + 9q + 3$

68. $4u^2 + 10u + 5$

69. $2x^2 + 6x + 2$

70. $3x^2 + 9x + 3$

71. $6mp + 3pw - 8m - 4w$

72. $15ab + 5ac - 6b - 2c$

73. $h^2 + 4h - 3$

74. $w^2 + 6h - 5$

75. Use the *ac* method to factor $x^2 + 0x - 36$.

76. Use the *ac* method to factor $x^2 + 0x - 64$.

For exercises 77–84,
(a) find the discriminant.
(b) use the discriminant to determine whether the trinomial is prime.

77. $1x^2 + 7x + 12$

78. $1x^2 + 8x + 12$

79. $88x^2 + 28x - 5$

80. $91x^2 + 16x - 4$

81. $200x^2 - 90x + 9$

82. $600x^2 - 110x + 3$

83. $x^2 - x - 182$

84. $x^2 - x - 156$

Problem Solving: Practice and Review

Follow your instructor's guidelines for using the five steps as outlined in Section 1.3, p. 55.

85. A standard rectangular city block in Manhattan is about 264 ft by 900 ft. One acre of land is 43,560 ft². Find the size of a standard city block in acres. Round to the nearest tenth.

86. Police in Lakeland, Florida, believe that installation of red light cameras are the reason that red light violations have decreased. Find the percent decrease in red light violators. Round to the nearest percent.

There has been a decrease in red-light violators from 7,522 over a six-month period in 2010 to 4,666 over a same six-month period in 2011. (*Source:* www.theledger.com, Jan. 13, 2012)

87. The new floating bridge across Lake Washington in Seattle will include 21 large bridge pontoons. Find the total volume of these pontoons.

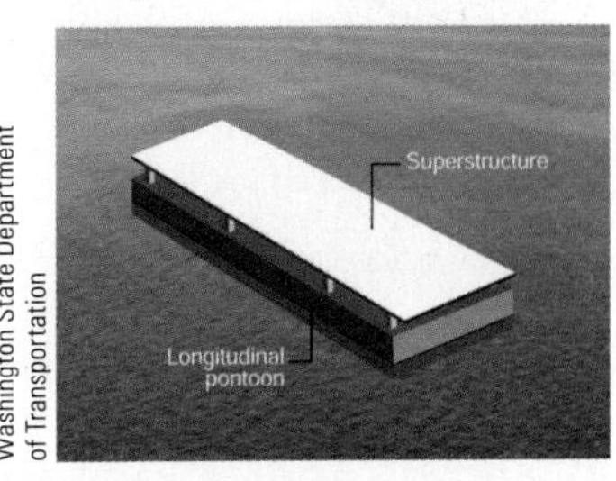

New SR 520 bridge pontoons will be approximately 28 feet tall, 75 feet wide and 360 feet long—as long as a football field. (*Source:* www.wsdot.wa.gov)

88. The average cost for a community to fluoridate its water is about \$0.62 per person per year in large communities and \$3.90 per person per year in small communities. Fluoridated water significantly reduces cavities. A dentist charges \$125 to fill a cavity. Find the number of years in a small community at which the cost for one person to drink fluoridated water is equal to the cost of filling one cavity. Round to the nearest whole number.

Find the Mistake

For exercises 89–92, the completed problem has one mistake.
(a) Describe the mistake in words, or copy down the whole problem and highlight or circle the mistake.
(b) Do the problem correctly.

89. **Problem:** Use the guess and check method to factor $x^2 - 5x - 6$.

Incorrect Answer: $x^2 - 5x - 6$
$= (x + 3)(x - 2)$

90. Problem: Use the guess and check method to factor $2a^2 - 11a + 12$.

Incorrect Answer: $2a^2 - 11a + 12$
$= (2a - 4)(a - 3)$

91. Problem: Use the *ac* method to factor $6c^2 - 5c - 4$.

Incorrect Answer: $6c^2 - 5c - 4$
$= 6c^2 + 8c - 3c - 4$
$= (6c^2 + 8c) + (-3c - 4)$
$= 2c(\_\_\_ + \_\_\_) + -1(\_\_\_ + \_\_\_)$
$= 2c(3c + 4) - 1(3c + 4)$
$= (3c + 4)(2c - 1)$

92. Problem: Use factor by grouping to factor $3x^2 - 21x - x - 7$.

Incorrect Answer: $3x^2 - 21x - x - 7$
$= (3x^2 - 21x) - (x - 7)$
$= 3x(x - 7) - 1(x - 7)$
$= (x - 7)(3x - 1)$

Review

93. Use the distributive property to multiply $(4x + 5)(4x + 5)$.

94. Use a pattern to multiply $(4x + 5)(4x + 5)$.

95. Use a pattern to multiply $(6x + 7)(6x - 7)$.

96. Use the distributive property to multiply $(6x + 7)(6x - 7)$.

SUCCESS IN COLLEGE MATHEMATICS

97. Look at your last test from this class.
a. How many total points are on this test?
b. How many points did you miss because of arithmetic mistakes including sign errors?
c. What percent of the points that you missed were because of arithmetic mistakes including sign errors?
d. Are arithmetic mistakes a major problem for you on tests?

98. What can you do while taking a test to reduce the number of arithmetic mistakes that you make?

A log cabin quilt block is made by sewing together strips of fabric. The arrangement of the strips in the block follows the log cabin pattern. In different log cabin quilts, the colors of the strips of fabric change, but the pattern is the same. In this section, we will use patterns to factor polynomials.

SECTION 6.3 Patterns

After reading the text, working the practice problems, and completing assigned exercises, you should be able to:

Use a pattern to factor a polynomial.

Perfect Square Trinomials

A rational number can be written as a fraction in which the numerator and denominator are integers. The square root of a perfect square is a rational number. For example, 9 is a perfect square because $\sqrt{9} = 3$ and 3 is a rational number.

Perfect Squares

$1^2 = \mathbf{1}$	$2^2 = \mathbf{4}$	$3^2 = \mathbf{9}$	$4^2 = \mathbf{16}$	$5^2 = \mathbf{25}$
$6^2 = \mathbf{36}$	$7^2 = \mathbf{49}$	$8^2 = \mathbf{64}$	$9^2 = \mathbf{81}$	$10^2 = \mathbf{100}$
$11^2 = \mathbf{121}$	$12^2 = \mathbf{144}$	$13^2 = \mathbf{169}$	$14^2 = \mathbf{196}$	$15^2 = \mathbf{225}$

In Section 5.4, we used patterns to *multiply* polynomials. If we reverse the direction of the patterns, we can use them to *factor* polynomials. For example, $(a + b)(a + b) = a^2 + 2ab + b^2$ is a pattern for *multiplying* $(a + b)(a + b)$, and $a^2 + 2ab + b^2 = (a + b)(a + b)$ is a pattern for *factoring* $a^2 + 2ab + b^2$.

Perfect Square Trinomial Patterns

$$a^2 + 2ab + b^2 = (a + b)(a + b)$$
$$a^2 - 2ab + b^2 = (a - b)(a - b)$$

To use the perfect square trinomial patterns, look for perfect squares in the first term and the last term of the trinomial. For example, in the trinomial $9x^2 + 24x + 16$, $9x^2$ is a perfect square, since $(3x)(3x) = 9x^2$, and 16 is a perfect square, since $(4)(4) = 16$. Identify a and b, and write the factors, matching the pattern. Rewrite the factors in exponential notation.

Notice that in the standard form of a quadratic trinomial, $ax^2 + bx + c$, the lead coefficient is a and the middle term is b, but in the perfect square trinomial pattern, $a^2 + 2ab + b^2 = (a + b)(a + b)$, the lead coefficient is not a, and the coefficient of the middle term is not b. In mathematics, we often use the same letter to represent different things.

EXAMPLE 1 Use a pattern to factor $9x^2 + 24x + 16$.

SOLUTION ▶ Since $9x^2$ is a perfect square and 16 is a perfect square, the pattern may be $a^2 + 2ab + b^2 = (a + b)(a + b)$. Since $9x^2 + 24x + 16$ is equivalent to $(3x)^2 + 2(3x)(4) + 4^2$, $a = 3x$ and $b = 4$.

$9x^2 + 24x + 16$

$= (3x)^2 + 2 \cdot 3x \cdot 4 + 4^2$ — Match the pattern: $a = 3x$; $b = 4$

$= (3x + 4)(3x + 4)$ — Write the factors: $(a + b)(a + b)$.

$= (3x + 4)^2$ — Rewrite the factors in exponential notation.

In a perfect square trinomial, the first and last terms are perfect squares. The middle term is $2ab$ or $-2ab$. When the middle term is $-2ab$, use the pattern $a^2 - 2ab + b^2 = (a - b)(a - b)$.

EXAMPLE 2 Use a pattern to factor $36w^2 - 12w + 1$. Check.

SOLUTION ▶ Since $36w^2$ is a perfect square and 1 is a perfect square, the pattern may be $a^2 - 2ab + b^2 = (a - b)(a - b)$. Since $36w^2 - 12w + 1$ is equivalent to $(6w)^2 - 2(6w)(1) + 1^2$, $a = 6w$ and $b = 1$.

$36w^2 - 12w + 1$

$= (6w)^2 - 2 \cdot 6w \cdot 1 + 1^2$ — Match the pattern: $a = 6w$; $b = 1$

$= (6w - 1)(6w - 1)$ — Write the factors: $(a - b)(a - b)$.

$= (6w - 1)^2$ — Rewrite the factors in exponential notation.

To check, multiply the factors.

$(6w - 1)(6w - 1)$

$= \mathbf{6w}(6w - 1) - \mathbf{1}(6w - 1)$ — Distributive property.

$= \mathbf{6w}(6w) + \mathbf{6w}(-1) - \mathbf{1}(6w) - \mathbf{1}(-1)$ — Distributive property.

$= 36w^2 - 6w - 6w + 1$ — Simplify.

$= 36w^2 - \mathbf{12w} + 1$ — Simplify. The factoring is correct.

In the next example, the lead coefficient and the end constant are perfect squares. However, the middle term does not match the perfect square trinomial pattern. The polynomial is prime.

EXAMPLE 3 Use a pattern to factor $4p^2 + 58p + 49$.

SOLUTION ▶ Since $4p^2$ is a perfect square and 49 is a perfect square, the pattern may be $a^2 + 2ab + b^2 = (a + b)(a + b)$. If $a = 2p$ and $b = 7$, then $2ab = 2 \cdot 2p \cdot 7$. But $2 \cdot 2p \cdot 7 = 28p$, and the middle term of this polynomial is $58p$. The polynomial does not match the pattern $a^2 + 2ab + b^2 = (a + b)(a + b)$.

Perhaps this quadratic trinomial is prime. If the discriminant is a perfect square and it is greater than 0, then the trinomial can be factored. If it is not, it is prime.

$b^2 - 4ac$	The discriminant.
$= \mathbf{58}^2 - 4 \cdot \mathbf{4} \cdot \mathbf{49}$	$a = 4, b = 58, c = 49$
$= 2580$	Simplify.
$\sqrt{2580}$	Is the discriminant a perfect square?
$= 50.793...$	Irrational number; 2580 is not a perfect square.

This polynomial is prime. Its only factors are itself and 1.

Practice Problems

For problems 1–3, use a pattern to factor. Check. Identify any prime polynomials.

1. $x^2 + 12x + 36$ **2.** $4h^2 + 12h + 9$ **3.** $9w^2 + 20w + 1$

Perfect Square Trinomials That Are Not Quadratic

We can use a pattern and the power rule of exponents, $(x^m)^n = x^{mn}$, to factor some trinomials that are not quadratic. In the next example, since $x^8 = (x^4)^2$, the first term, x^8, is a perfect square. If the coefficient of an exponential expression is 1 and its exponent is divisible by 2, the expression is a perfect square.

EXAMPLE 4 Use a pattern to factor $x^8 + 18x^4 + 81$.

SOLUTION ▶ Since x^8 is a perfect square and 81 is a perfect square, the pattern may be $a^2 + 2ab + b^2 = (a + b)(a + b)$. Since $x^8 + 18x^4 + 81$ is equivalent to $(x^4)^2 + 2 \cdot x^4 \cdot 9 + 9^2$, $a = x^4$ and $b = 9$.

$x^8 + 18x^4 + 81$	
$= (x^4)^2 + 2 \cdot x^4 \cdot 9 + 9^2$	Match the pattern: $a = x^4$; $b = 9$
$= (x^4 + 9)(x^4 + 9)$	Write the factors: $(a + b)(a + b)$.
$= (x^4 + 9)^2$	Rewrite the factors in exponential notation.

Practice Problems

For problems 4–5, use a pattern to factor. Check. Identify any prime polynomials.

4. $49h^6 - 28h^3 + 4$ **5.** $36x^{10} + 60x^5 + 25$

The Difference of Squares

In Section 5.4, we used the difference of squares pattern to multiply binomials. If we reverse this pattern, we can use it for factoring.

Difference of Squares Pattern

$$a^2 - b^2 = (a + b)(a - b)$$

EXAMPLE 5 Use a pattern to factor $169v^2 - 25$.

SOLUTION ▸ Since $169v^2$ is a perfect square and 25 is a perfect square, the pattern may be $a^2 - b^2 = (a + b)(a - b)$. Since $169v^2 - 25$ is equivalent to $(13v)^2 - 5^2$, $a = 13v$ and $b = 5$.

$169v^2 - 25$

$= (13v)^2 - 5^2$ Match the pattern: $a = 13v$; $b = 5$

$= (13v + 5)(13v - 5)$ Write the factors: $(a + b)(a - b)$.

In the next example, we use the power rule of exponents, $(x^m)^n = x^{mn}$, to rewrite the perfect squares.

EXAMPLE 6 Use a pattern to factor $16k^{10} - 81y^4$. Check.

SOLUTION ▸ Since $16k^{10}$ is a perfect square and $81y^4$ is a perfect square, the pattern is $a^2 - b^2 = (a + b)(a - b)$. Since $16k^{10} - 81y^4$ is equivalent to $(4k^5)^2 - (9y^2)^2$, $a = 4k^5$ and $b = 9y^2$.

$16k^{10} - 81y^4$

$= (4k^5)^2 - (9y^2)^2$ Match the pattern: $a = 4k^5$; $b = 9y^2$

$= (4k^5 + 9y^2)(4k^5 - 9y^2)$ Write the factors: $(a + b)(a - b)$.

Check.

$(4k^5 + 9y^2)(4k^5 - 9y^2)$

$= \mathbf{4k^5}(4k^5 - 9y^2) + \mathbf{9y^2}(4k^5 - 9y^2)$ Distributive property.

$= \mathbf{4k^5}(4k^5) + \mathbf{4k^5}(-9y^2) + \mathbf{9y^2}(4k^5) + \mathbf{9y^2}(-9y^2)$ Distributive property.

$= 16k^{10} - 36k^5y^2 + 36k^5y^2 - 81y^4$ Simplify.

$= 16k^{10} - 81y^4$ Simplify; the factoring is correct.

A polynomial that is the sum of two squares in which the terms have no common factors is prime. For example, $x^2 + 4$ and $9a^2 + 25$ are prime polynomials.

Practice Problems

For problems 6–9, use a pattern to factor. Check. Identify any prime polynomials.

6. $x^2 - 36$ **7.** $25d^2 - f^2$ **8.** $64p^4 + 49$ **9.** $121m^8 - 16n^4$

The Sum and Difference of Cubes

The volume of a cube is equal to the product of its length, width, and height. Since the length, width, and height of a cube are equal, the formula for the volume of a cube is $V = s^3$, where s is the length of any side. A cube in a polynomial is a number, variable, or term raised to the third power.

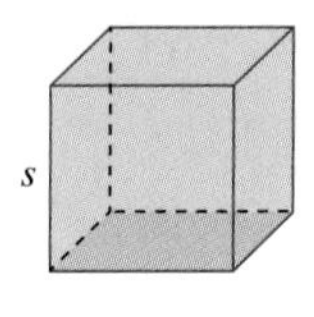

Perfect Cubes

$1^3 = \mathbf{1}$	$2^3 = \mathbf{8}$	$3^3 = \mathbf{27}$	$4^3 = \mathbf{64}$	$5^3 = \mathbf{125}$
$6^3 = \mathbf{216}$	$7^3 = \mathbf{343}$	$8^3 = \mathbf{512}$	$9^3 = \mathbf{729}$	$10^3 = \mathbf{1000}$

To factor a sum or difference of cubes, use a pattern.

Sum or Difference of Cubes Pattern

$a^3 + b^3 = (a + b)(a^2 - ab + b^2)$ *Sum of cubes pattern*

$a^3 - b^3 = (a - b)(a^2 + ab + b^2)$ *Difference of cubes pattern*

EXAMPLE 7 Use a pattern to factor $h^3 + 8$. Check.

SOLUTION ▶ Since h^3 is a perfect cube and 8 is a perfect cube, the pattern is $a^3 + b^3 = (a + b)(a^2 - ab + b^2)$. Since $h^3 + 8$ is equivalent to $(h)^3 + (2)^3$, $a = h$ and $b = 2$.

$h^3 + 8$	
$= h^3 + 2^3$	Match the pattern: $a = h$; $b = 2$
$= (h + 2)(h^2 - h \cdot 2 + 2^2)$	Write the factors: $(a + b)(a^2 - ab + b^2)$.
$= (h + 2)(h^2 - 2h + 4)$	Simplify.

Check.

$(h + 2)(h^2 - 2h + 4)$	
$= \mathbf{h}(h^2 - 2h + 4) + \mathbf{2}(h^2 - 2h + 4)$	Distributive property.
$= \mathbf{h}(h^2) + \mathbf{h}(-2h) + \mathbf{h}(4) + \mathbf{2}(h^2) + \mathbf{2}(-2h) + \mathbf{2}(4)$	Distributive property.
$= h^3 - 2h^2 + 4h + 2h^2 - 4h + 8$	Simplify.
$= h^3 + 8$	Simplify; the factoring is correct.

In the next example, we use the difference of cubes pattern.

EXAMPLE 8 Use a pattern to factor $512p^3 - 27w^3$.

SOLUTION ▶ Since $512p^3$ is a perfect cube and $27w^3$ is a perfect cube, the pattern is $a^3 - b^3 = (a - b)(a^2 + ab + b^2)$. Since $512p^3 - 27w^3$ is equivalent to $(8p)^3 - (3w)^3$, $a = 8p$ and $b = 3w$.

$512p^3 - 27w^3$	
$= (8p)^3 - (3w)^3$	Match the pattern: $a = 8p$; $b = 3w$
$= (8p - 3w)((8p)^2 + 8p \cdot 3w + (3w)^2)$	Write the factors: $(a - b)(a^2 + ab + b^2)$.
$= (8p - 3w)\mathbf{(64p^2 + 24pw + 9w^2)}$	Simplify.

We have now used five methods for factoring polynomials.

Methods for Factoring Polynomials

- Greatest common factor
- Grouping
- Guess and check
- *ac* method
- Pattern

Practice Problems

For problems 10–13, use a pattern to factor. Check. Identify any prime polynomials.

10. $x^3 - 729$ **11.** $x^3 + 729$ **12.** $8h^3 + 125k^3$ **13.** $8h^3 - 125k^3$

6.3 Vocabulary Practice

Match the term with its description.

1. The coefficient of the first term of a quadratic trinomial
2. The only factors of this polynomial are itself and 1.
3. $a^3 - b^3 = (a - b)(a^2 + ab + b^2)$
4. $a^3 + b^3 = (a + b)(a^2 - ab + b^2)$
5. $a^2 - b^2 = (a + b)(a - b)$
6. $a^2 + 2ab + b^2 = (a + b)(a + b)$
7. $a(b + c) = ab + ac$
8. In a polynomial with one variable, this is equal to the largest exponent.
9. $x^m \cdot x^n = x^{m+n}$
10. $(x^m)^n = x^{mn}$

A. degree
B. difference of cubes pattern
C. difference of squares pattern
D. distributive property
E. lead coefficient
F. perfect square trinomial pattern
G. power rule of exponents
H. prime polynomial
I. product rule of exponents
J. sum of cubes pattern

6.3 Exercises

Follow your instructor's guidelines for showing your work.

For exercises 1–20, use a pattern to factor. Check. Identify any prime polynomials.

1. $x^2 + 8x + 16$
2. $c^2 + 16x + 64$
3. $y^2 - 12y + 36$
4. $a^2 - 10a + 25$
5. $u^2 + 20u + 100$
6. $v^2 + 18v + 81$
7. $y^2 - 6y + 9$
8. $c^2 - 14c + 49$
9. $k^2 + 16k + 64$
10. $y^2 + 18y + 81$
11. $9k^2 + 48k + 64$
12. $4y^2 + 36y + 81$
13. $9k^2 + 48km + 64m^2$
14. $4y^2 + 36yz + 81z^2$
15. $25c^2 - 60c + 36$
16. $16w^2 - 56w + 49$
17. $y^2 - 6yz + 9z^2$
18. $c^2 - 14cd + 49d^2$
19. $y^{20} - 6y^{10}z^{10} + 9z^{20}$
20. $c^{18} - 14c^9d^9 + 49d^{18}$
21. Show that x^{20} is a perfect square.
22. Show that x^{30} is a perfect square.

For exercises 23–34, use a pattern to factor. Check. Identify any prime polynomials.

23. $f^2 - 25$
24. $m^2 - 4$
25. $w^2 - 100$
26. $v^2 - 81$
27. $4f^2 - 25$
28. $49m^2 - 36$
29. $100f^2 - 9h^2$
30. $64m^2 - 49p^2$
31. $36 - y^2$
32. $9 - d^2$
33. $x^{18} - y^{50}$
34. $x^{32} - y^{90}$
35. Is 9 a perfect cube? Explain.
36. Is x^9 a perfect cube? Explain.

For exercises 37–46, use a pattern to factor. Check. Identify any prime polynomials.

37. $x^3 + 125$
38. $x^3 + 216$
39. $h^3 - 27$
40. $k^3 - 64$
41. $p^3 - 1000$
42. $k^3 - 729$
43. $8h^3 - 27k^3$
44. $27k^3 - 64n^3$
45. $64m^3 + 1$
46. $1000n^3 + 1$

For exercises 47–56, use a pattern to factor. Check. Identify any prime polynomials.

47. $x^2 - 1$
48. $y^2 - 400$
49. $c^2 + 2c + 1$
50. $h^2 + 4h + 4$
51. $n^3 + 1000$
52. $m^3 + 216$
53. $81w^2 + 36w + 4$
54. $64c^2 + 48c + 9$
55. $9p^2 - w^8$
56. $100j^2 - k^{12}$

For exercises 57–84, use any of the factoring methods to factor. Identify any prime polynomials.

57. $n^2 - 7n - 60$
58. $u^2 - 6u - 91$

59. $m^2 + 64$
60. $p^2 + 121$
61. $6a^2 + 48a + 60$
62. $4b^2 + 36b + 64$
63. $18x^2 + 5x - 7$
64. $16r^2 + 10r - 9$
65. $pv - 9p + 6v - 54$
66. $wx - 8w + 7x - 56$
67. $15px - 9rx + 10py - 6ry$
68. $14vx - 10px + 21vz - 15pz$
69. $25u^4 - 81z^6$
70. $36h^6 - 49j^{10}$
71. $v^2 + 18v + 81$
72. $q^2 + 20q + 100$
73. $42p^5 - 28p^4 + 56p^3 - 70p^2 + 21p$
74. $36u^6 - 21u^5 + 45u^4 + 30u^3 - 9u^2$
75. $9w^2 - 18w + 5$
76. $9d^2 - 27d + 8$
77. $9m^2 + 81n^2$
78. $4p^2 + 16r^2$
79. $3b^2 + 28b + 32$
80. $3n^2 + 26n + 35$
81. $14x^2 + 7x - 49$
82. $6x^2 + 3x - 81$
83. $4c^2 + 8c - 5$
84. $4p^2 + 4p - 3$

85. In this section, the difference of squares pattern is $a^2 - b^2 = (a + b)(a - b)$. We can rewrite this pattern as $a^2 - b^2 = (a - b)(a + b)$. What property of the real numbers allows us to do this?

86. We can write the difference of squares pattern as $a^2 - b^2 = (a + b)(a - b)$ or as $a^2 - b^2 = (a - b)(a + b)$. Show that $(a + b)(a - b) = (a - b)(a + b)$.

Problem Solving: Practice and Review

Follow your instructor's guidelines for using the five steps as outlined in Section 1.3, p. 55.

87. A pickup truck with 45 ft^3 of compost in its bed can be emptied with 3 loads of a garden cart and 10 loads of a wheelbarrow, or it can be emptied with 6 loads of a garden cart and 5 loads of a wheelbarrow. Use a system of two linear equations to find the volume of compost carried by the garden cart and by the wheelbarrow.

88. Harris Interactive conducted an on-line poll of 1434 U.S. community college students. Find the number of students who were taking courses that required completing homework, problems, or quizzes on-line. Round to the nearest whole number.

About three-quarters of students surveyed were taking courses that required completing homework, problems, or quizzes online. (*Source:* www.pearsonfoundation.org, Feb. 9, 2011)

89. A model of the gross receipts for Broadway shows, y, for x years since 2002 is $y = \left(\frac{\$44{,}388{,}300}{1 \text{ year}}\right)x + \$682{,}573{,}500$. Use this model to find the gross receipts for Broadway in 2013. Round to the nearest thousand. (*Source:* www.broadwayworld.com, Jan. 2012)

© Stuart Monk/Shutterstock.com

90. A semester grade equals 20% of the average homework score, 55% of the average test score, and 25% of the final exam score. Just before the final exam, a student has a average test score of 75 and an average homework score of 74. Find the minimum score that the student needs on the final exam to earn at least a B (a semester grade greater than or equal to 80) in the class.

Find the Mistake

For exercises 91–94, the completed problem has one mistake.
(a) Describe the mistake in words, or copy down the whole problem and highlight or circle the mistake.
(b) Do the problem correctly.

91. **Problem:** Use a pattern to factor $16x^2 + 40x + 25$.
Incorrect Answer: Since the pattern is $a^2 + 2ab + b^2 = (a + b)(a + b)$, $a = 8x$ and $b = 5$, and the factored polynomial is $(8x + 5)(8x + 5)$.

92. **Problem:** Use a pattern to factor $9x^2 - 6x + 1$.
Incorrect Answer: Since the pattern is $a^2 + 2ab + b^2 = (a + b)(a + b)$, $a = 3x$ and $b = 1$, and the factored polynomial is $(3x + 1)(3x + 1)$.

93. **Problem:** Use a pattern to factor $25z^2 + 9$.
Incorrect Answer: Since the pattern is $a^2 + b^2 = (a + b)(a + b)$, $a = 5z$ and $b = 3$, and the factored polynomial is $(5z + 3)(5z + 3)$.

94. **Problem:** Use a pattern to factor $p^3 + 8w^3$.
Incorrect Answer: Since the pattern is $a^3 + b^3 = (a + b)(a^2 - ab + b^2)$, $a = p$ and $b = 8w$, and the factored polynomial is $(p + 8w)(p^2 - 8pw + 64w^2)$.

Review

95. What is the product of any whole number and 0?
96. What is the sum of any whole number and 0?
97. What is the quotient of 0 divided by any whole number?
98. What is the quotient of any whole number divided by 0?

SUCCESS IN COLLEGE MATHEMATICS

99. Look at your last test from this class. Find a problem for which you did not receive full credit.
100. Redo that problem correctly.

When a painter is hired to paint the interior of a new house, completing the job requires hanging protective plastic, spraying a coat of primer, spraying a coat of finish paint, and removing the plastic. In algebra, a polynomial is not factored completely until all of the factors are prime. In this section, we will use more than one factoring method to factor polynomials completely.

SECTION 6.4 Factoring Completely

After reading the text, working the practice problems, and completing assigned exercises, you should be able to:

Factor a polynomial completely.

Factoring Completely

We have used five methods for factoring polynomials.

Methods for Factoring Polynomials

- Greatest common factor
- Grouping
- Guess and check
- *ac* method
- Pattern

When a polynomial is factored completely, each factor of the polynomial is prime. To **factor a polynomial completely**, we may need to use more than one factoring method. First, factor out any common factor of all of the terms. If the remaining polynomial is not prime, choose a method to factor it.

Factoring a Polynomial Completely

1. Factor out any greatest common factor.
2. If the remaining polynomial has four terms, factor by grouping.
3. If the remaining polynomial has three terms, use guess and check, the *ac* method, or a pattern to factor.
4. If the remaining polynomial has two terms, use a pattern to factor.

When a polynomial is completely factored, each factor is prime.

EXAMPLE 1 Factor $3x^2 + 36x + 96$ completely.

SOLUTION ▶ Factor out the greatest common factor, 3.

$3x^2 + 36x + 96$

$= 3(x^2 + 12x + 32)$ Factor out the greatest common factor, 3.

The remaining polynomial, $x^2 + 12x + 32$, has three terms. Since $a = 1$, use the guess and check method to factor. Since $b = 12$ and $c = 32$, both numbers in the factors are positive. Use $(x + \_\_\_)(x + \_\_\_)$. The sum of the numbers is 12, and the product of the numbers is 32.

Guess.

$3(x^2 + 12x + 32)$ — Factor out the greatest common factor, 3.

$= 3(x + \_\_\_)(x + \_\_\_)$ — Since $x \cdot x = x^2$, the first term in both factors is x.

$= 3(x + \mathbf{4})(x + \mathbf{8})$ — Guess: The unknown numbers are 4 and 8.

Since $x + 4$ and $x + 8$ are prime, the polynomial is factored completely. To check, multiply the factors from left to right.

$3(x + 4)(x + 8)$

$= (\mathbf{3}x + \mathbf{3} \cdot 4)(x + 8)$ — Distributive property.

$= (3x + \mathbf{12})(x + 8)$ — Simplify.

$= \mathbf{3x}(x + 8) + \mathbf{12}(x + 8)$ — Distributive property.

$= \mathbf{3x}(x) + \mathbf{3x}(8) + \mathbf{12}(x) + \mathbf{12}(8)$ — Distributive property.

$= 3x^2 + 24x + 12x + 96$ — Simplify.

$= 3x^2 + \mathbf{36x} + 96$ — Simplify; the factoring is correct.

To factor a quadratic trinomial in which the lead coefficient is 1, it is often more efficient to use the guess and check method. When the lead coefficient of a quadratic trinomial is greater than 1, it is often more efficient to use the *ac* method.

EXAMPLE 2 Factor $12x^2 - 26x - 10$ completely.

SOLUTION ▶ Factor out the greatest common factor, 2.

$12x^2 - 26x - 10$

$= 2(6x^2 - 13x - 5)$ — Factor out the greatest common factor, 2.

Since $a = 6$, use the *ac* method. Rewrite $2(6x^2 - 13x - 5)$ as $2[6x^2 - 13x - 5]$ so that parentheses can be used for grouping.

$6x^2 - 13x - 5$ $a = 6$, $b = -13$, $c = -5$, $ac = -30$			
Factor pairs of −30 **One number is negative; one number is positive.**		**Sum of factor pair**	
$(-1)(30)$	$(1)(-30)$	29	−29
$(-2)(15)$	$(2)(-15)$	13	−13
Although there are more factor pairs of −30, we stop when we find a pair whose sum is −13.			

$2[6x^2 - 13x - 5]$ — Replace parentheses with brackets.

$= 2[6x^2 -15x + 2x - 5]$ — Rewrite middle term as $-15x + 2x$.

$= 2[(6x^2 - 15x) + (2x - 5)]$ — Group terms.

$= 2[3x(\_\_\_ - \_\_\_) + 1(\_\_\_ - \_\_\_)]$ — Identify a common factor of each group.

$= 2[3x(2x - 5) + 1(2x - 5)]$ — Complete the factoring of each group.

$= 2[(2x - 5)(\_\_\_ + \_\_\_)]$ — Factor out a binomial, $2x - 5$.

$= 2[(2x - 5)(3x + 1)]$ — Complete the factoring.

$= 2(2x - 5)(3x + 1)$ — Rewrite without brackets.

Since $2x - 5$ and $3x + 1$ are prime, the polynomial is factored completely.

In the next example, after we factor out the greatest common factor, the remaining polynomial has four terms. We then factor the remaining polynomial by grouping.

EXAMPLE 3 Factor $6a^2c + 3a^2d + 2acb + abd$ completely.

SOLUTION ▶ Factor out the greatest common factor, a. Since the remaining polynomial has four terms, factor by grouping.

$6a^2c + 3a^2d + 2acb + abd$	
$= a(6ac + 3ad + 2cb + bd)$	Factor out the greatest common factor, a.
$= a[6ac + 3ad + 2cb + bd]$	Replace parentheses with brackets.
$= a[(6ac + 3ad) + (2cb + bd)]$	Use parentheses to group terms.
$= a[3a(\_\_ + \_\_) + b(\_\_ + \_\_)]$	Identify a common factor of each group.
$= a[3a(2c + d) + b(2c + d)]$	Factor a common factor from each group.
$= a[(2c + d)(\_\_ + \_\_)]$	Factor out a binomial, $2c + d$.
$= a[(2c + d)(3a + b)]$	Complete the factoring.
$= a(2c + d)(3a + b)$	Rewrite without brackets.

Since all three factors, a, $2c + d$, and $3a + b$, are prime, the polynomial is factored completely.

When the remaining polynomial has four terms, it might be necessary to change the order of the terms to factor it by grouping.

EXAMPLE 4 Factor $24pw - 4x^2 - 6px + 16wx$ completely.

SOLUTION ▶ Factor out the greatest common factor, 2. Since the remaining polynomial has four terms, factor by grouping.

$24pw - 4x^2 - 6px + 16wx$	
$= 2(12pw - 2x^2 - 3px + 8wx)$	Factor out the greatest common factor, 2.
$= 2[12pw - 2x^2 - 3px + 8wx]$	Replace parentheses with brackets.
$= 2[12pw - 2x^2 \mathbf{+ -3}px + 8wx]$	Rewrite $-3px$ as $+ -3px$.
$= 2[(12pw - 2x^2) + (-3px + 8wx)]$	Group terms.
$= 2[2(\_\_ - \_\_) + x(\_\_ + \_\_)]$	Identify a common factor of each group.
$= 2[2(\mathbf{6pw - x^2}) + x(\mathbf{-3p + 8w})]$	Complete the factoring of each group.

Since the groups do not have a common factor, we cannot factor out a binomial. Change the order of the middle terms, regroup, and factor.

$2[12pw \mathbf{- 3px - 2x^2} + 8wx]$	Change the order of the terms.
$= 2[12pw - 3px \mathbf{+ -2}x^2 + 8wx]$	Rewrite $-2x^2$ as $+ -2x^2$.
$= 2[(12pw - 3px) + (-2x^2 + 8wx)]$	Group terms.
$= 2[3p(\_\_ - \_\_) + 2x(\_\_ + \_\_)]$	Identify a common factor of each group.
$= 2[3p(\mathbf{4w - x}) + 2x(\mathbf{-x + 4w})]$	Complete the factoring of each group.
$= 2[3p(4w - x) + 2x(4w - x)]$	Change the order; $-x + 4w = 4w - x$
$= 2[(4w - x)(\_\_ + \_\_)]$	Factor out a binomial, $4w - x$.
$= 2[(4w - x)(3p + 2x)]$	Complete the factoring.
$= 2(4w - x)(3p + 2x)$	Rewrite without brackets.

Since $4w - x$ and $3p + 2x$ are prime, the polynomial is factored completely.

In the next example, we factor out the greatest common factor. The remaining polynomial has four terms, so we factor by grouping. Then, since one of the factors is a difference of squares, we use the pattern $a^2 - b^2 = (a - b)(a + b)$ to factor it.

EXAMPLE 5 Factor $2x^2y + x^2z - 18y - 9y$ completely.

SOLUTION ▶ These terms have no common factor. Since $2x^2y + x^2z - 18y - 9y$ has four terms, factor by grouping.

$2x^2y + x^2z - 18y - 9z$	These terms have no common factor.
$= 2x^2y + x^2z + -18y - 9z$	Rewrite $-18y$ as $+ -18y$.
$= (2x^2y + x^2z) + (-18y - 9z)$	Group terms.
$= x^2(\_\_\_ + \_\_\_) + -9(\_\_\_ + \_\_\_)$	Identify a common factor of each group.
$= x^2(2y + z) + -9(2y + z)$	Complete the factoring of each group.
$= (2y + z)(\_\_\_ + \_\_\_)$	Factor out a binomial, $2y + z$.
$= (2y + z)(x^2 + -9)$	Complete the factoring.
$= (2y + z)(x^2 - 9)$	Rewrite $x^2 + -9$ as $x^2 - 9$.

Since $x^2 - 9$ has two terms and matches the difference of squares pattern, $a^2 - b^2 = (a - b)(a + b)$, use a pattern to factor.

$(2y + z)(x^2 - \mathbf{3^2})$	Match pattern: $a = x$; $b = 3$
$= (2y + z)\mathbf{(x + 3)(x - 3)}$	Write factors: $(a + b)(a - b)$.

Since $2y + z$, $x + 3$, and $x - 3$ are prime, the polynomial is factored completely.

In the next example, after the greatest common factor is factored out, the remaining polynomial is prime.

EXAMPLE 6 Factor $2x^2 + 6x$ completely.

SOLUTION ▶ Factor out the greatest common factor, $2x$.

$2x^2 + 6x$	
$= 2x(x + 3)$	Factor out the greatest common factor, $2x$.

Since both factors, $2x$ and $x + 3$, are prime, the polynomial is factored completely.

To provide practice in factoring, most of the polynomials in the exercises in this section are not prime. However, many polynomials are prime and cannot be factored.

EXAMPLE 7 Factor $9x^2 + 16y^2$ completely.

SOLUTION ▶ The terms have no common factor. The polynomial has two terms but is not a difference of squares, a difference of cubes, or a sum of cubes. The only factors of $9x^2 + 16y^2$ are itself and 1. It is a prime polynomial.

Practice Problems

For problems 1–19, factor completely. Check. Identify any prime polynomials.

1. $4x^2 + 20x + 24$
2. $p^3 - 3p^2 - 10p$
3. $3w^2 - 12$
4. $2a^2 + 8a$
5. $4px + 12wx + 2py + 6wy$
6. $2a^3 - 16a^2 - 18a$
7. $6n^2 + 24$
8. $60xy - 12x^2 + 160y - 32x$
9. $3x^2 - 33x - 540$
10. $18h^2 - 54k^2$
11. $6ac + 10c + 3ad + 5d$
12. $5x^2 + 32x - 23$

13. $80mp^2 - 32m^2p + 70p - 28m$	**14.** $2p^3 - p^2 - 15p$
15. $25x^2y - 16$	**16.** $12xy - 2z + 6y - 4xz$
17. $12x^2 + 22x + 10$	**18.** $x^3 - 4x + 3x^2 - 12$
19. $u^7 + 5u^6 - 24u^5$	

Factoring Other Polynomials Completely

The degree of a quadratic trinomial is 2. In the next example, after we factor out the greatest common factor, the remaining trinomial is not quadratic. However, we can use a pattern, the *ac* method, or guess and check to factor it.

EXAMPLE 8 Factor $64x^{19} + 48x^{10}y^2 + 9xy^4$ completely.

SOLUTION ▶ Factor out the greatest common factor, x. The remaining polynomial has three terms, and it matches a perfect square trinomial pattern, $a^2 + 2ab + b^2 = (a + b)(a + b)$.

$64x^{19} + 48x^{10}y^2 + 9xy^4$	
$= x(64x^{18} + 48x^9y^2 + 9y^4)$	Factor out the greatest common factor, x.
$= x[64x^{18} + 48x^9y^2 + 9y^4]$	Use brackets to separate the factors.
$= x[(8x^9)^2 + 2 \cdot 8x^9 \cdot 3y^2 + (3y^2)^2]$	Match pattern: $a = 8x^9$; $b = 3y^2$
$= x[(8x^9 + 3y^2)(8x^9 + 3y^2)]$	Write factors: $(a + b)(a + b)$.
$= x(8x^9 + 3y^2)^2$	Write in exponential notation.

Since both factors, x and $8x^9 + 3y^2$, are prime, the polynomial is factored completely.

To factor a binomial completely, look first for a greatest common factor. If the remaining polynomial is a sum or difference of perfect cubes, use a pattern to factor.

EXAMPLE 9 Factor $64x^7y^2 + x^4y^5$ completely.

SOLUTION ▶ Factor out the greatest common factor, x^4y^2. The remaining polynomial has two terms, and it matches the sum of perfect cubes pattern, $a^3 + b^3 = (a + b)(a^2 - ab + b^2)$.

$64x^7y^2 + x^4y^5$	
$= x^4y^2(64x^3 + y^3)$	Factor out the greatest common factor, x^4y^2.
$= x^4y^2[64x^3 + y^3]$	Use brackets to separate the factors.
$= x^4y^2[(4x)^3 + y^3]$	Match pattern: $a = 4x$; $b = y$
$= x^4y^2[(4x + y)((4x)^2 - 4xy + y^2)]$	Write factors: $(a + b)(a^2 - ab + b^2)$.
$= x^4y^2[(4x + y)(\mathbf{16x^2} - 4xy + y^2)]$	Simplify; $(4x)^2 = 16x^2$
$= x^4y^2(4x + y)(16x^2 - 4xy + y^2)$	Rewrite without brackets.

Although the first and last terms in $16x^2 - 4xy + y^2$ are perfect squares, the middle term does not match the perfect square trinomial pattern. Since all the factors, x^4y^2, $4x + y$, and $16x^2 - 4xy + y^2$, are prime, the polynomial is factored completely.

When factoring polynomials that include factors that are the sum or difference of cubes, write or refer to a list of perfect cubes: $2^3 = 8$, $3^3 = 27$, $4^3 = 64$, $5^3 = 125$, $6^3 = 216$, $7^3 = 343$, $8^3 = 512$, $9^3 = 729$, and $10^3 = 1000$.

EXAMPLE 10 Factor $40x^3 - 1715y^3$ completely.

SOLUTION ▶ Factor out the greatest common factor. The remaining polynomial has two terms, and it matches the difference of perfect cubes pattern, $a^3 - b^3 = (a - b)(a^2 + ab + b^2)$.

$40x^3 - 1715y^3$

$= 5(8x^3 - 343y^3)$ Factor out the greatest common factor, 5.

$= 5[8x^3 - 343y^3]$ Use brackets to separate the factors.

$= 5[(2x)^3 - (7y)^3]$ Match pattern: $a = 2x$; $b = 7y$

$= 5[(2x - 7y)((2x)^2 + 2x \cdot 7y + (7y)^2)]$ Write factors: $(a - b)(a^2 + ab + b^2)$.

$= 5[(2x - 7y)(\mathbf{4x^2 + 14xy + 49y^2})]$ Simplify; $(2x)^2 = 4x^2$; $(7y)^2 = 49y^2$

$= 5(2x - 7y)(4x^2 + 14xy + 49y^2)$ Rewrite without brackets.

Although the first and last terms in the factor $4x^2 + 14xy + 49y^2$ are perfect squares, the middle term does not match the perfect square trinomial pattern. This factor is prime. Since $2x - 7y$ and $4x^2 + 14xy + 49y^2$ are prime, the polynomial is factored completely.

Practice Problems

For problems 20–28, factor completely. Identify any prime polynomials.

20. $4p^{20} - 20p^{10}w^3 + 25w^6$
21. $18x^{10} + 12x^5y^2 + 2y^2$
22. $2x^6 - 24x^3 + 72$
23. $7x^3 - 7y^3$
24. $25p^4 + 200pw^3$
25. $4c^7d^{12} + 4c^7d^9$
26. $15mr^4z - 15mrz^4$
27. $16x^4 - 6xy^3$
28. $9n^3 + 1000p^3$

6.4 Vocabulary Practice

Match the term with its description.

1. $a^2 + 2ab + b^2 = (a + b)(a + b)$
2. $a(b + c) = ab + ac$
3. $a^3 - b^3 = (a - b)(a^2 + ab + b^2)$
4. $a^3 + b^3 = (a + b)(a^2 - ab + b^2)$
5. $a^2 - b^2 = (a + b)(a - b)$
6. A polynomial with two terms
7. $x^m \cdot x^n = x^{m+n}$
8. $(x^m)^n = x^{mn}$
9. The coefficient of the first term of a quadratic trinomial
10. The only factors of this polynomial are itself and 1.

A. binomial
B. difference of cubes pattern
C. difference of squares pattern
D. distributive property
E. lead coefficient
F. perfect square trinomial pattern
G. power rule of exponents
H. prime polynomial
I. product rule of exponents
J. sum of cubes pattern

6.4 Exercises

Follow your instructor's guidelines for showing your work.

For exercises 1–84, factor completely. Identify any prime polynomials.

1. $3x^2 - 12$
2. $2x^2 - 18$
3. $x^3 + 6x^2 + 5x$
4. $x^3 + 5x^2 + 4x$
5. $10px + 5xy + 10py + 5y^2$
6. $6ac + 3ab + 6bc + 3b^2$
7. $8n^2 + 8p^2$
8. $7c^2 + 7d^2$

9. $4x^2 + 24x + 36$
10. $5y^2 + 30y + 45$
11. $90xz + 72x + 10yz + 8y$
12. $42cf + 70c + 6df + 10d$
13. $18a^2c + 42a^2 + 45ack + 105ak$
14. $63pw^2 + 18pw + 231mw^2 + 66mw$
15. $120xy + 48xz + 20y + 8z$
16. $60hn + 80hu + 12n^2 + 16nu$
17. $24kmp + 6kp^2 + 40mp + 10p^2$
18. $216yz + 30xz^2 + 135xyz + 48z^2$
19. $12cd + 4dg - 3g - 9c$
20. $10cp - 5p + 2cw - w$
21. $7x^2 + 14x - 140$
22. $5x^2 + 10x - 150$
23. $90z^2 + 120z + 40$
24. $40k^2 + 280k + 490$
25. $6p^2 + 57p + 105$
26. $4r^2 + 26r + 30$
27. $8h^2 + 108h + 280$
28. $9k^2 + 69k + 120$
29. $x^2y - 13xy + 36y$
30. $c^2d - 15cd + 54d$
31. $100x^2 - 20x + 1$
32. $144x^2 - 24x + 1$
33. $18y^2 + 75y + 45$
34. $12z^2 + 28z + 56$
35. $c^5 + 14c^3 + 48c$
36. $m^5 + 12m^3 + 27m$
37. $3d^2 + 3d - 5$
38. $3w^2 + 5w - 7$
39. $6x^2 - 6y^2$
40. $10x^2 - 10y^2$
41. $9u^2 + 16z^2$
42. $25d^2 + 64f^2$
43. $9x^2 - 16y^2$
44. $25d^2 - 64f^2$
45. $32p^2 - 50w^2$
46. $75m^2 - 48z^2$
47. $300d^3 - 48dk^2$
48. $80p^3 - 180pv^2$
49. $4y^2 + 46y + 90$
50. $6p^2 + 57p + 72$
51. $4q^2 - 4q - 4$
52. $5n^2 - 10n - 10$
53. $3n^2x - 12n^2 + mnx - 4mn$
54. $4p^2z - 20p^2 + pwz - 5pw$
55. $3u^3 + 42u^2 + 72u$
56. $2c^3 + 28c^2 + 66c$
57. $4x^2 + 5x + 1$
58. $5z^2 + 6z + 1$
59. $2p^2 - 800$
60. $3w^2 - 2700$
61. $48mr + 32mw + 12pr + 8pw$
62. $10ac + 15ad + 10bc + 15bd$
63. $3ax^2 - 6x^2$
64. $5ky^2 - 10y^2$
65. $18x^2 + 27x^2y - 51y^2 + 24xy^2 - 210$
66. $14a^2 + 56a^2b - 42b^2 + 28ab^2 - 420$
67. $3p - 3z^2$
68. $7a - 7b^2$
69. $2c^2 + 12c + 20$
70. $3b^2 + 15b + 24$
71. $h^2 + 100k^2$
72. $v^2 + 81z^2$
73. $25w^3 - 10w^2 + w$
74. $16p^3 - 8p^2 + p$
75. $6a^2c + 3a^2 - 2abc - ab$
76. $12x^2z + 4x^2 - 3xyz - xy$
77. $7x^2 - 63y^2$
78. $2a^2 - 32b^2$
79. $2hx - 8x + 3hp - 15p$
80. $3ac - 15a + 2bc - 15b$
81. $2x^2 + 36x + 162$
82. $2z^2 + 40z + 200$
83. $6a^2b + 2a^2c - 24b - 8c$
84. $6n^2p + 3n^2w - 54p - 27w$

For exercises 85–102, factor completely. Identify any prime polynomials.

85. $2p^4 + 4p^2w^3 + 2w^6$
86. $2n^6 + 4n^3p^2 + 2p^4$
87. $3x^{14} - 12x^7y^5 + 12y^{10}$
88. $3x^{18} - 12x^9y^2 + 12y^4$
89. $x^{13} - x^5y^2$
90. $x^{17} - x^7y^2$
91. $c^3 + 8d^3$
92. $p^3 + 27z^3$
93. $n^3 - 64p^3$
94. $v^3 - 125z^3$
95. $16x^3 - 54y^3$
96. $128x^3 - 54y^3$
97. $7d^3 - 56f^3$
98. $11h^3 - 88k^3$
99. $9x^3 + 36y^3$
100. $16x^3 + 64y^3$
101. $27h^3 - k^3$
102. $125n^3 - p^3$

Problem Solving: Practice and Review

Follow your instructor's guidelines for using the five steps as outlined in Section 1.3, p. 55.

103. Find the percent decrease in the number of diagnoses of AIDS from 2008 to 2009. Round to the nearest tenth of a percent.

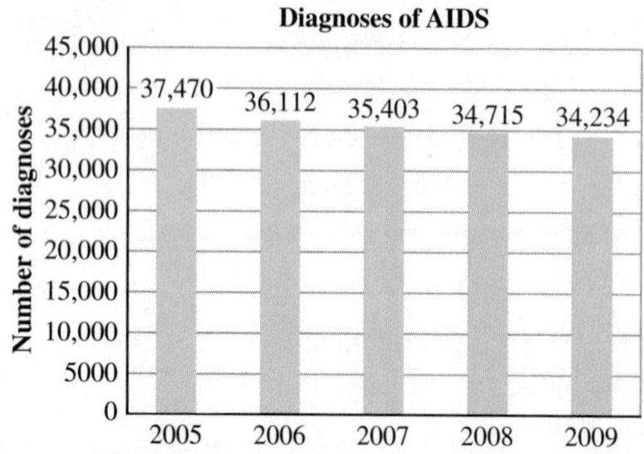

Source: gis.cdc.gov, Jan. 2012

104. A veteran drives three times a month to a VA Hospital that is 125 miles away from his home. Find the amount that the veteran will be reimbursed for his travel.

The current allowable reimbursement amount is 28.5 cents per mile with a $7.77 deductible for each one-way visit, or $15.54 for each round-trip visit. If your out-of-pocket costs exceed $46.62 in a given month, you can be reimbursed for the entire expense for any other authorized travel for that month. (*Source:* www.benefits.gov, Jan. 2012)

105. The formula for density, d, is $d = \frac{m}{v}$, where m is the mass and v is the volume. The density of a steel sphere is $7.85\ \frac{\text{g}}{\text{cm}^3}$, and its mass is 5×10^2 g. Solve the formula for v, and find the volume of this sphere. Round to the nearest whole number.

106. Find the average rate of change per year in the number of prisoners under the jurisdiction of state or federal correctional authorities. Round to the nearest hundred.

Date	Number of prisoners under the jurisdiction of state or federal correctional authorities
Dec. 31, 2000	1,391,261
Dec. 31, 2010	1,605,127

Source: www.bjs.gov, Dec. 2011

Find the Mistake

For exercises 107–110, the completed problem has one mistake.
(a) Describe the mistake in words, or copy down the whole problem and highlight or circle the mistake.
(b) Do the problem correctly.

107. Problem: Factor $12x^2 - 48y^2$ completely.

Incorrect Answer: $12x^2 - 48y^2$
$= 12(x^2 - 4y^2)$

108. Problem: Factor $6x^2 + 21x + 15$ completely.

Incorrect Answer: $6x^2 + 21x + 15$
$= 3(2x^2 + 7x + 5)$

109. Problem: Factor $5x^2 + 500y^2$ completely.

Incorrect Answer: $5x^2 + 500y^2$
$= 5(x^2 + 100y^2)$
$= 5(x + 10y)(x + 10y)$

110. Problem: Factor $45c^3 + 33c^2 - 168c$ completely.

Incorrect Answer: $45c^3 + 33c^2 - 168c$
$= 3c(15c^2 + 11c - 56)$

Review

111. What is the product of 0 and any number?

112. Solve: $9x = 0$

113. An expression is $(x - 4)(x - 3)$. Evaluate this expression when $x = 3$.

114. An expression is $(x - 4)(x - 3)$. Evaluate this expression when $x = 4$.

SUCCESS IN COLLEGE MATHEMATICS

115. Using a test from this class, count the number of points that you missed because you did not show enough work.

116. Explain why you think that your instructor wants you to show your work, not just give the correct answer, on a test.

×	0	1	2	3	4	5	6	7	8	9	10
0	0	0	0	0	0	0	0	0	0	0	0
1	0	1	2	3	4	5	6	7	8	9	10
2	0	2	4	6	8	10	12	14	16	18	20
3	0	3	6	9	12	15	18	21	24	27	30
4	0	4	8	12	16	20	24	28	32	36	40
5	0	5	10	15	20	25	30	35	40	45	50
6	0	6	12	18	24	30	36	42	48	54	60
7	0	7	14	21	28	35	42	49	56	63	70
8	0	8	16	24	32	40	48	56	64	72	80
9	0	9	18	27	36	45	54	63	72	81	90
10	0	10	20	30	40	50	60	70	80	90	100

Children often learn the multiplication facts one number at a time. They will say "I've learned my fives" or "I'm working on my sevens." They do not talk about learning their "zeros" because there is not much to learn. The product of any real number and 0 is zero. In this section, we will use the zero product property.

SECTION 6.5 The Zero Product Property

After reading the text, working the practice problems, and completing assigned exercises, you should be able to:

1. Factor a polynomial equation and use the zero product property to solve it.
2. Use a polynomial equation and the zero product property to find the dimensions of a geometric figure.

The Zero Product Property

The product of a real number and 0 is 0. If the product of two different numbers is 0, then one of the numbers must be 0. The **zero product property** summarizes what we know about factors that have a product of 0.

Zero Product Property

If a and b are real numbers and $a \cdot b = 0$, then $a = 0$ or $b = 0$. If a and b are expressions and $a \cdot b = 0$, then $a = 0$ or $b = 0$.

Mathematicians use the word "or" in a precise way. In the zero product property, "or" means that $a = 0$ is true, $b = 0$ is true, or both are true.

We use the zero product property to solve factored equations that are equal to 0. In the next example, the equation $(x + 8)(x - 2) = 0$ matches the pattern $a \cdot b = 0$. The equation is true for the values of x such that $a = 0$ or $b = 0$.

EXAMPLE 1 **(a)** Solve: $(x + 8)(x - 2) = 0$

SOLUTION ▶ To solve this equation, we find the values of x such that $(x + 8)(x - 2) = 0$ by solving $x + 8 = 0$ or $x - 2 = 0$.

$x + 8 = 0$	or	$x - 2 = 0$	Zero product property; $a = x + 8$; $b = x - 2$
$-8 \quad -8$		$+2 \quad +2$	Properties of equality.
$x + \mathbf{0} = \mathbf{-8}$		$x + \mathbf{0} = \mathbf{2}$	Simplify.
$\boldsymbol{x} = -8$		$\boldsymbol{x} = 2$	Simplify.

Solutions: $x = -8, x = 2$

(b) Check.

Check: $x = -8$

$$(x + 8)(x - 2) = 0$$
$$(\mathbf{-8} + 8)(\mathbf{-8} - 2) = 0$$
$$\mathbf{0 \cdot (-10)} = 0$$
$$\mathbf{0} = 0 \quad \text{True.}$$

Check: $x = 2$

$$(x + 8)(x - 2) = 0$$
$$(\mathbf{2} + 8)(\mathbf{2} - 2) = 0$$
$$\mathbf{10(0)} = 0$$
$$\mathbf{0} = 0 \quad \text{True.}$$

We can also use the zero product property to solve an equation with three or more factors that is equal to 0. In this situation, if a, b, and c are expressions and $a \cdot b \cdot c = 0$, then $a = 0$ or $b = 0$ or $c = 0$.

EXAMPLE 2 **(a)** Solve: $p(p + 1)(p + 5) = 0$

SOLUTION To solve this equation, we find the values of p such that $p(p + 1)(p + 5) = 0$ by solving $p = 0$ or $p + 1 = 0$ or $p + 5 = 0$.

$p = 0$ or	$p + 1 = 0$	or $p + 5 = 0$	Zero product property.
	$\underline{-1 \quad -1}$	$\underline{-5 \quad -5}$	Subtraction property of equality.
	$\mathbf{p + 0 = -1}$	$\mathbf{p + 0 = -5}$	Simplify.
	$\mathbf{p} = -1$ or	$\mathbf{p} = -5$	Simplify.

Solutions: $p = 0$, $p = -1$, $p = -5$

(b) Check.

Check: $p = 0$

$$p(p + 1)(p + 5) = 0$$
$$\mathbf{0}(\mathbf{0} + 1)(\mathbf{0} + 5) = 0$$
$$\mathbf{0 \cdot 1 \cdot 5} = 0$$
$$\mathbf{0} = 0$$

True.

Check: $p = -1$

$$p(p + 1)(p + 5) = 0$$
$$\mathbf{-1}(\mathbf{-1} + 1)(\mathbf{-1} + 5) = 0$$
$$-1 \cdot \mathbf{0} \cdot \mathbf{4} = 0$$
$$\mathbf{0} = 0$$

True.

Check: $p = -5$

$$p(p + 1)(p + 5) = 0$$
$$\mathbf{-5}(\mathbf{-5} + 1)(\mathbf{-5} + 5) = 0$$
$$(-5)(\mathbf{-4})(\mathbf{0}) = 0$$
$$\mathbf{0} = 0$$

True.

In the next example, the equation again has three factors whose product is equal to 0. Since one of the factors does not include a variable, it cannot equal 0.

EXAMPLE 3 Solve: $5(3n - 2)(n + 10) = 0$

SOLUTION To solve this equation, we find the values of n such that $5(3n - 2)(n + 10) = 0$. Since $5 \neq 0$, the only equations to solve are $3n - 2 = 0$ and $n + 10 = 0$.

$3n - 2 = 0$ or	$n + 10 = 0$	Zero product property.
$\underline{+2 \quad +2}$	$\underline{-10 \quad -10}$	Properties of equality.
$\mathbf{3n + 0 = 2}$	$\mathbf{n + 0 = -10}$	Simplify.
$\dfrac{\mathbf{3n}}{3} = \dfrac{2}{3}$	$\mathbf{n} = -10$	Division property of equality; simplify.
$n = \dfrac{2}{3}$		Simplify.

Solutions: $n = \dfrac{2}{3}$, $n = -10$

Practice Problems

For problems 1–5,
(a) solve.
(b) check.

1. $6(x - 8) = 0$ **2.** $(z + 4)(z - 9) = 0$ **3.** $x(x + 20) = 0$
4. $d(d + 5)(d - 2) = 0$ **5.** $3(x + 1)(x - 7) = 0$

Factoring and the Zero Product Property

To solve $2x + 6 = 20$, we subtract 6 from both sides and then divide both sides by 2. We are using the properties of equality to isolate the variable, x. For equations such as $x^2 - 12x = 0$, we cannot isolate x by using the properties of equality. Instead, we factor the polynomial expression completely and use the zero product property to solve.

EXAMPLE 4 **(a)** Solve: $x^2 - 12x = 0$

SOLUTION ▶ Factor $x^2 - 12x = 0$ completely.

$x^2 - 12x = 0$

$\mathbf{x(x - 12)} = 0$ Factor out the greatest common factor, x.

To solve this equation, we find the values of x such that $x(x - 12) = 0$ by solving $x = 0$ or $x - 12 = 0$.

$x = 0$ or $x - 12 = 0$ Zero product property.

$+12 \; +12$ Addition property of equality.

$\mathbf{x + 0 = 12}$ Simplify.

$\mathbf{x} = 12$ Simplify.

Solutions: $x = 0, x = 12$

(b) Check.

▶ Check: $x = 0$

$x^2 - 12x = 0$

$(\mathbf{0})^2 - 12(\mathbf{0}) = 0$

$\mathbf{0 - 0} = 0$

$\mathbf{0} = 0$ True.

Check: $x = 12$

$x^2 - 12x = 0$

$(\mathbf{12})^2 - 12(\mathbf{12}) = 0$

$\mathbf{144 - 144} = 0$

$\mathbf{0} = 0$ True.

A trinomial in the form $ax^2 + bx + c$ is a quadratic trinomial. In the next example, the equation is a **quadratic equation in standard form,** $ax^2 + bx + c = 0$.

EXAMPLE 5 **(a)** Solve: $2y^2 + 15y - 27 = 0$

SOLUTION ▶ Factor $2y^2 + 15y - 27 = 0$ completely. Since $a \neq 1$, use the ac method.

$2y^2 + 15y - 27 = 0$ $a = 2, b = 15, c = -27, ac = -54$			
Factor pairs of −54 **One number is negative; one number is positive.**		**Sum of factor pair**	
$(-1)(54)$	$(1)(-54)$	53	−53
$(-2)(27)$	$(2)(-27)$	25	−25
$(-3)(18)$	$(3)(-18)$	15	−15

$$2y^2 + 15y - 27 = 0$$

$2y^2 - 3y + 18y - 27 = 0$ Rewrite middle term as $-3y + 18y$.

$(2y^2 - 3y) + (18y - 27) = 0$ Group terms.

$y(\_\_\_ - \_\_\_) + 9(\_\_\_ - \_\_\_) = 0$ Identify a common factor in each group.

$y(2y - 3) + 9(2y - 3) = 0$ Complete the factoring of each group.

$(2y - 3)(\_\_\_ + \_\_\_) = 0$ Factor out a binomial, $2y - 3$.

$(2y - 3)(y + 9) = 0$ Complete the factoring.

Use the zero product property to solve $(2y - 3)(y + 9) = 0$.

$2y - 3 = 0$ or $y + 9 = 0$ Zero product property.

$+3 \quad +3$ $\quad -9 \quad -9$ Properties of equality.

$2y + \mathbf{0} = \mathbf{3}$ $\quad y + \mathbf{0} = \mathbf{-9}$ Simplify.

$\frac{\mathbf{2y}}{2} = \frac{3}{2}$ $\quad y = -9$ Division property of equality; simplify.

$y = \frac{3}{2}$ Simplify.

Solutions: $y = \frac{3}{2}, y = -9$

(b) Check.

Check: $y = \frac{3}{2}$

$$2y^2 + 15y - 27 = 0$$
$$2\left(\frac{\mathbf{3}}{\mathbf{2}}\right)^2 + 15\left(\frac{\mathbf{3}}{\mathbf{2}}\right) - 27 = 0$$
$$2\left(\frac{\mathbf{9}}{\mathbf{4}}\right) + \frac{\mathbf{45}}{\mathbf{2}} - 27 = 0$$
$$\frac{9}{2} + \frac{45}{2} - \frac{\mathbf{54}}{\mathbf{2}} = 0$$
$$\frac{\mathbf{54}}{\mathbf{2}} - \frac{54}{2} = 0$$
$$\mathbf{0} = 0 \quad \text{True.}$$

Check: $y = -9$

$$2y^2 + 15y - 27 = 0$$
$$2(\mathbf{-9})^2 + 15(\mathbf{-9}) - 27 = 0$$
$$2(\mathbf{81}) - \mathbf{135} - 27 = 0$$
$$\mathbf{162} - 135 - 27 = 0$$
$$\mathbf{27} - 27 = 0$$
$$\mathbf{0} = 0 \quad \text{True.}$$

The zero product property applies to polynomial equations in which $a \cdot b = 0$. In the next example, to solve $9x^2 + 30x = -25$, we first add 25 to both sides so that the polynomial expression equals 0.

EXAMPLE 6 Solve: $9x^2 + 30x = -25$

SOLUTION Rewrite $9x^2 + 30x = -25$ so that the polynomial expression equals 0.

$9x^2 + 30x = -25$

$\quad +25 \quad +25$ Addition property of equality.

$9x^2 + 30x \mathbf{+ 25} = \mathbf{0}$ Simplify.

Factor $9x^2 + 30x + 25 = 0$ completely. Since $9x^2 + 30x + 25 = 0$ is equivalent to $(3x)^2 + 2 \cdot 3x \cdot 5 + 5^2$, it matches the perfect square trinomial pattern, $a^2 + 2ab + b^2 = (a + b)(a + b)$, where $a = 3x$ and $b = 5$.

$9x^2 + 30x + 25 = 0$

$(3x)^2 + 2 \cdot 3x \cdot 5 + 5^2 = 0$ Match the pattern: $a = 3x; b = 4$

$(3x + 5)(3x + 5) = 0$ Write the factors: $(a + b)(a + b)$.

Use the zero product property to solve $(3x + 5)(3x + 5) = 0$. Since the factors are the same, we solve only one equation, $3x + 5 = 0$.

$3x + 5 = 0$	Zero product property.
$\underline{-5 \quad -5}$	Subtraction property of equality.
$3x + \mathbf{0} = \mathbf{-5}$	Simplify.
$\frac{\mathbf{3x}}{3} = \frac{-5}{3}$	Division property of equality.
$x = -\frac{5}{3}$	Simplify; there is only one solution.

Solution: $x = -\frac{5}{3}$

In the next example, before factoring and using the zero product property, we subtract $27w$ from both sides so that the polynomial expression equals 0.

EXAMPLE 7 Solve: $3w^3 = 27w$

SOLUTION ▶ Rewrite $3w^3 = 27w$ so that the equation equals 0.

$3w^3 = 27w$	
$\underline{-27w \quad -27w}$	Subtraction property of equality.
$3w^3 \mathbf{- 27w} = \mathbf{0}$	Simplify.

Factor $3w^3 - 27w = 0$ completely.

$3w^3 - 27w = 0$	
$\mathbf{3w(w^2 - 9)} = 0$	Factor out the greatest common factor, $3w$.
$3w(w^2 - \mathbf{3^2}) = 0$	Difference of squares. Match the pattern: $a = w$, $b = 3$
$3w\mathbf{(w + 3)(w - 3)} = 0$	Write the factors: $(a + b)(a - b)$.

Use the zero product property to solve $3w(w + 3)(w - 3) = 0$.

$3w = 0$	or	$w + 3 = 0$	or	$w - 3 = 0$	Zero product property.
$\frac{3w}{3} = \frac{0}{3}$		$\underline{-3 \quad -3}$		$\underline{+3 \quad +3}$	Properties of equality.
		$w + \mathbf{0} = \mathbf{-3}$		$w + \mathbf{0} = \mathbf{3}$	Simplify.
$w = 0$		$\mathbf{w} = -3$		$\mathbf{w} = 3$	Simplify.

Solutions: $w = 0, w = -3, w = 3$

Practice Problems

For problems 6–14,

(a) solve.

(b) check.

6. $c^2 + 7c + 12 = 0$ **7.** $3y^2 - y - 10 = 0$ **8.** $p^2 + 4p = 0$

9. $4u^3 - 100u = 0$ **10.** $14x - x^2 = 0$ **11.** $x^2 - 10x = -25$

12. $p^3 = 4p^2 + 12p$ **13.** $2v^2 = 32$ **14.** $60m = 2m^3 - 2m^2$

Applications

In Section 5.4, we wrote polynomial *expressions* that represented the perimeter and the area of a rectangle. Now we will find the length and width of a rectangle by solving a polynomial *equation.*

EXAMPLE 8 The length of a rectangle is 1 ft less than twice its width. The area of the rectangle is 15 ft². Use a polynomial equation in one variable to find its length and width.

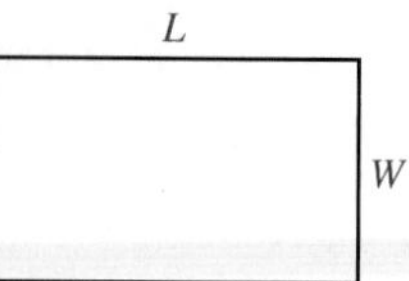

SOLUTION ▶ **Step 1 Understand the problem.**
Although there are two unknowns, the length and the width, we are to use a polynomial equation in one variable to find these unknowns. The information about the length depends on the width, so the unknown is the width of the rectangle.

$W =$ width

Step 2 Make a plan.
Since the length is 1 ft less than twice the width, the length equals $2W - 1$ ft. Use the formula for the area of a rectangle, $A = LW$, replacing A with 15 ft² and L with $2W - 1$ ft. Remove the units, rewrite the equation so that it equals 0, factor, and use the zero product property to solve. A solution of this equation is the width, W. Replace W in $2W - 1$ ft to find the length.

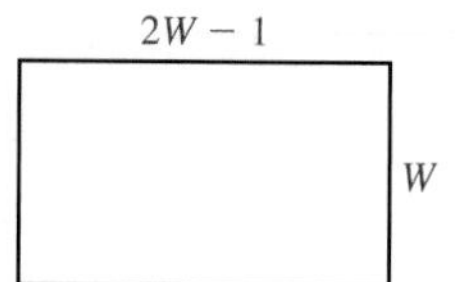

Step 3 Carry out the plan.
Write a polynomial equation in one variable.

$A = LW$	Formula for area of a rectangle.
$\mathbf{15\ ft^2} = \mathbf{(2W - 1\ ft)}W$	Replace A with 15 ft² and L with $2W - 1$ ft.
$\mathbf{15} = \mathbf{(2W - 1)}W$	Remove the units.
$15 = 2W \cdot W - 1 \cdot \mathbf{W}$	Distributive property.
$15 = \mathbf{2W^2 - W}$	Simplify.
$-15 \qquad -15$	Subtraction property of equality.
$\mathbf{0} = 2W^2 - W \mathbf{- 15}$	Simplify.

Rewriting the equation with the variables on the left side, $2W^2 - W - 15 = 0$. Factor completely. Since $a \neq 1$, use the ac method.

$2W^2 - W - 15 = 0$ $a = 2,\ b = -1,\ c = -15,\ ac = -30$			
Factor pairs of −30 **One number is negative; one number is positive.**		**Sum of factor pair**	
$(-1)(30)$	$(1)(-30)$	29	−29
$(-2)(15)$	$(2)(-15)$	13	−13
$(-3)(10)$	$(3)(-10)$	7	−7
$(-5)(6)$	$(5)(-6)$	1	−1

$2W^2 - W - 15 = 0$	
$2W^2 + 5W + -6W - 15 = 0$	Rewrite $-W$ as $5W + -6W$.
$(2W^2 + 5W) + (-6W - 15) = 0$	Group terms.
$W(\_\_\_ + \_\_\_) + -3(\_\_\_ + \_\_\_) = 0$	Identify a common factor of each group.
$W(2W + 5) + -3(2W + 5) = 0$	Complete the factoring of each group.
$(2W + 5)(\_\_\_ + \_\_\_) = 0$	Factor out a binomial, $2W + 5$.
$(2W + 5)(W + -3) = 0$	Complete the factoring.
$(2W + 5)(W \mathbf{- 3}) = 0$	Rewrite $W + -3$ as $W - 3$.

Use the zero product property to solve $(2W + 5)(W - 3) = 0$.

$2W + 5 = 0$	or	$W - 3 = 0$	Zero product property.
$\underline{-5 \quad -5}$		$\underline{+3 \quad +3}$	Properties of equality.
$2W + \mathbf{0} = \mathbf{-5}$		$W + \mathbf{0} = \mathbf{3}$	Simplify.
$\frac{\mathbf{2W}}{2} = -\frac{5}{2}$		$W = 3$	Simplify; division property of equality.
$W = -\frac{5}{2}$			Simplify.

Solutions: $W = -\frac{5}{2}$, $W = 3$

Since distance is not a negative number, $W = -\frac{5}{2}$ ft is a solution of this *equation* but is not a solution of the *problem*. The only solution to the problem is $W = 3$ ft. To find the length, replace W in $2W - 1$ ft with 3 ft.

$2W - 1$ ft	The length is 1 ft less than twice the width.
$= 2 \cdot \mathbf{3\ ft} - 1$ ft	Replace W with 3 ft.
$= \mathbf{6\ ft} - 1$ ft	Simplify.
$= 5$ ft	Simplify.

Step 4 Look back.
The length of the rectangle is 1 ft less than twice its width, and the area is 15 ft^2. Since $2(3\text{ ft}) - 1$ ft equals the length of 5 ft and since 5 ft $\cdot$ 3 ft equals the area of 15 ft^2, the answer seems reasonable.

Step 5 Report the solution.
The length of the rectangle is 5 ft, and the width of the rectangle is 3 ft.

Practice Problems

For problems 15–16, use a polynomial equation in one variable.

15. The length of a rectangle is 2 ft more than its width. The area of the rectangle is 15 ft^2. Find the length and width.

16. The length of a rectangle is 3 ft less than twice its width. The area of the rectangle is 20 ft^2. Find the length and width.

Using Technology: Solving Equations

We can estimate the solution of $x + 5 = 8$ by finding the point of intersection of the graphs of two linear equations, $y = x + 5$ and $y = 8$. The x-coordinate of this intersection point is the solution of the original equation.

EXAMPLE 9 Solve $x + 5 = 8$ by graphing.

On the Y= screen, enter two equations, $y = x + 5$ and $y = 8$. On the CALCULATE screen, choose 5: intersect. The calculator asks "First curve?" The cursor is on one of the lines. Press ENTER.

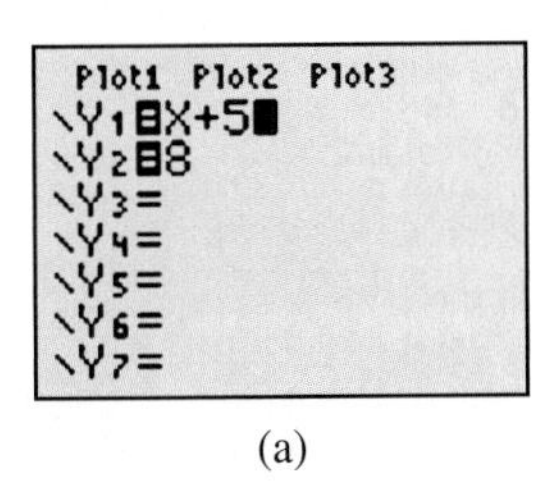

(a)

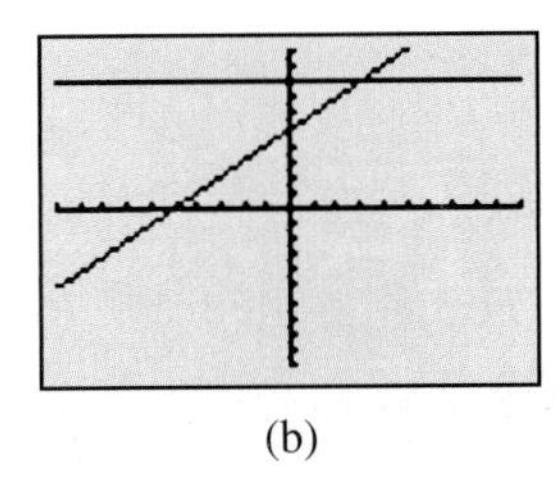
(b)

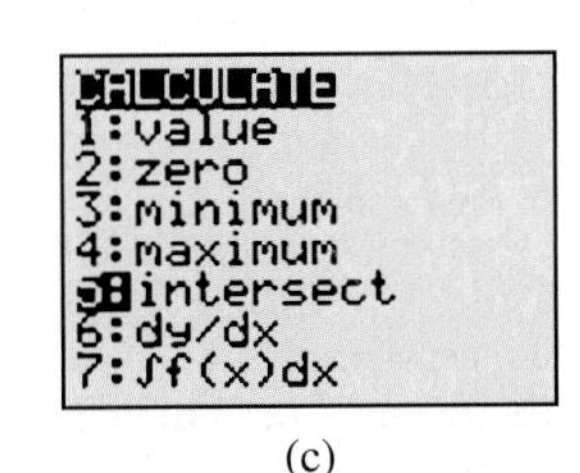

(c)

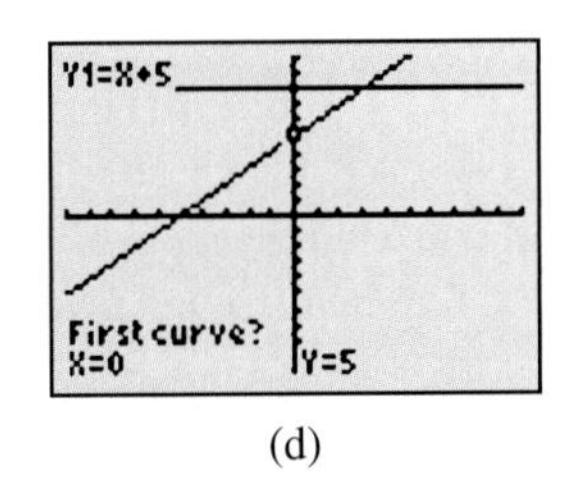

(d)

The cursor moves to the other line, and the calculator asks "Second curve?" Press ENTER. The calculator asks "Guess?" Move the cursor to the intersection point. Press ENTER. The coordinates of the intersection point appear. The solution of $x + 5 = 8$ is the x-coordinate of this point, $x = 3$.

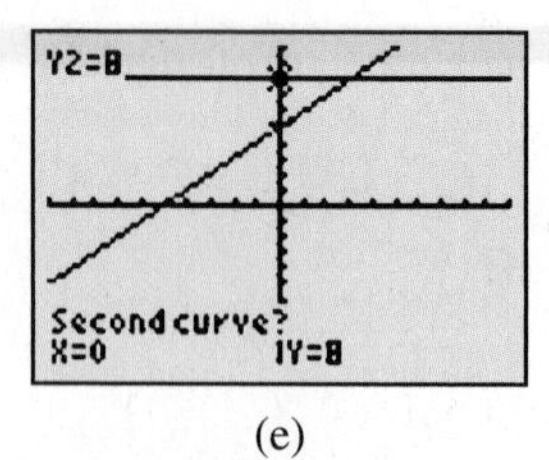

(e)

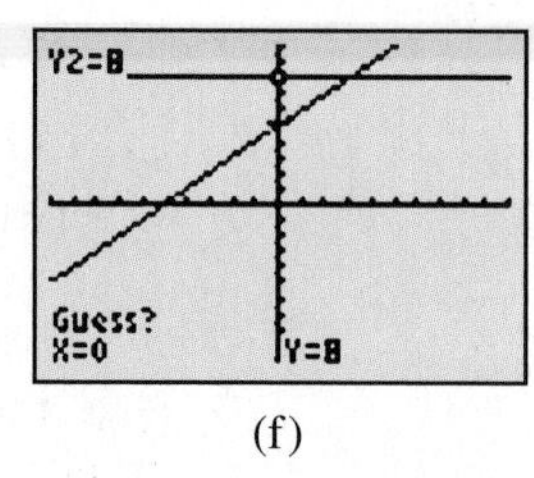

(f)

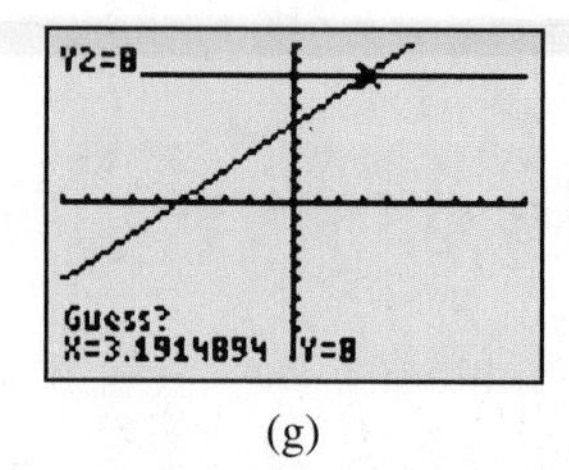

(g)

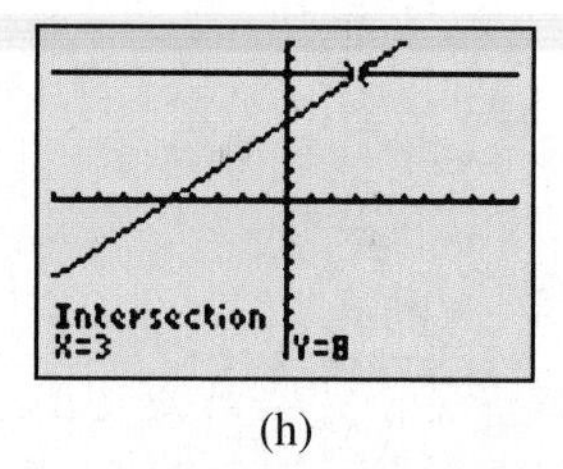

(h)

We can use this same method to estimate the solution of a quadratic equation.

EXAMPLE 10 Estimate the solutions of $x^2 - 3x - 4 = 0$.

In the Y= screen, enter $y = x^2 - 3x - 4$ and $y = 0$. Graph in a standard window. The line $y = 0$ is a horizontal line that is the same as the x-axis. The lines intersect in two points. To find the point on the left side, go to the CALCULATE screen. Choose 5: intersect.

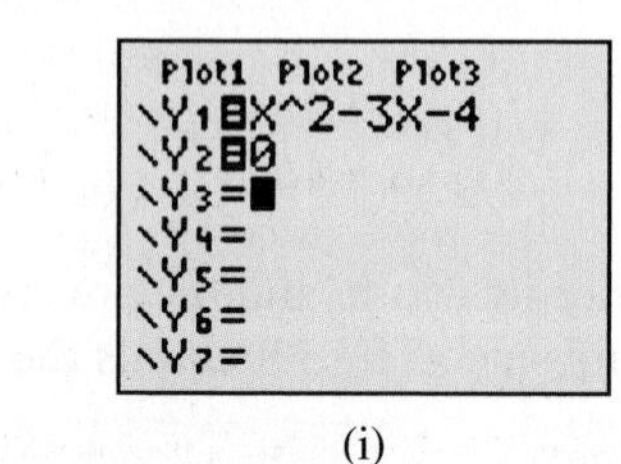

(i)

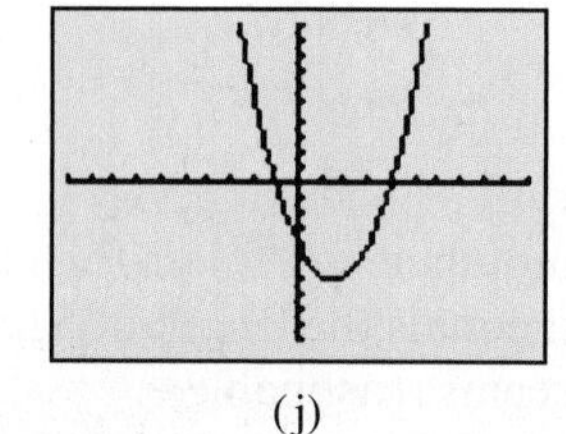
(j)

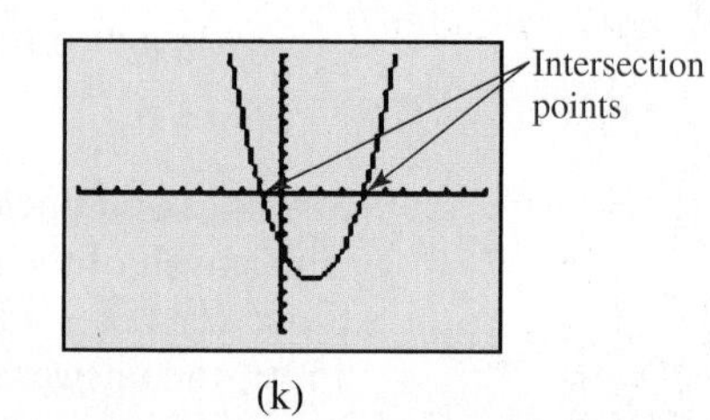

(k)

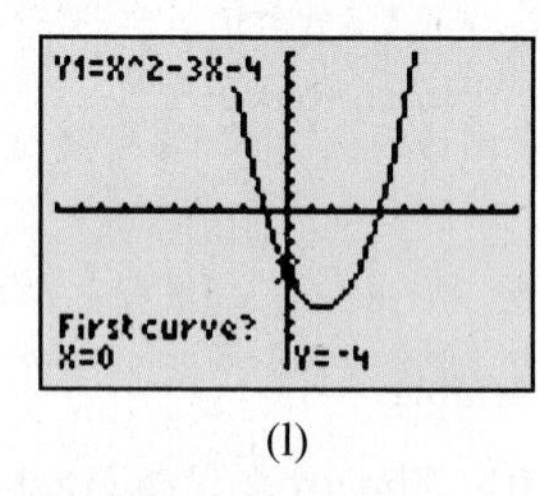

(l)

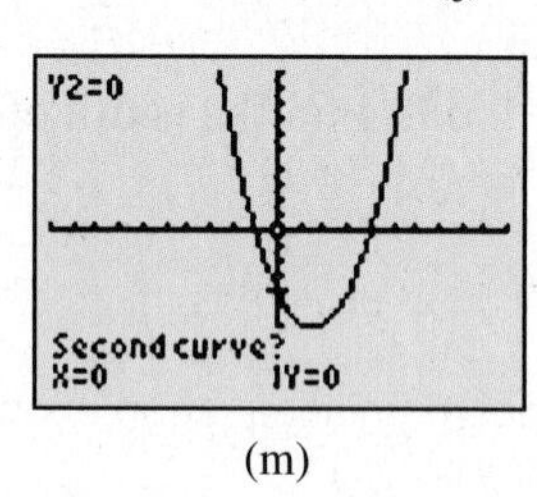

(m)

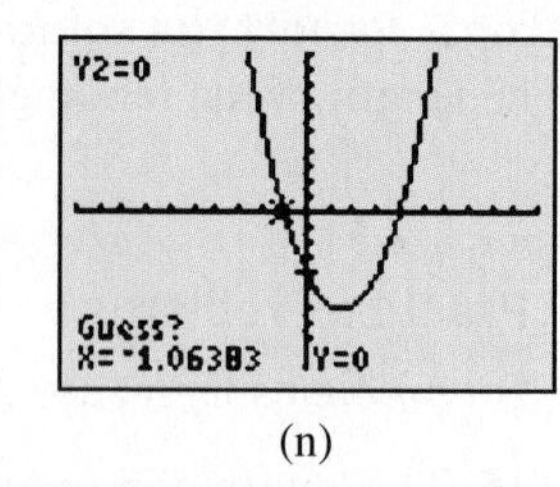

(n)

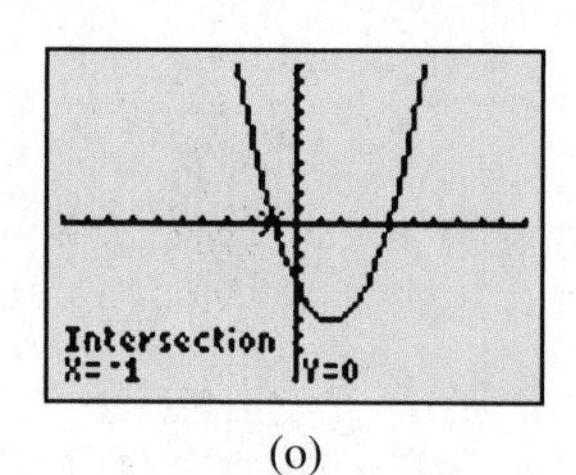

(o)

Since this intersection point is $(-1, 0)$, a solution of this equation is $x = -1$. To find the other solution, repeat the process to identify the x-coordinate of the other point of intersection, $x = 4$.

Practice Problems For problems 17–19,

(a) use the graphical method to solve each equation. Sketch the graph; describe the window.

(b) identify the intersection point(s).

(c) identify the solution of the equation.

17. $x + 5 = 3$ **18.** $x^2 - x - 6 = 0$ **19.** $x^2 + 2x - 15 = 0$

6.5 Vocabulary Practice

Match the term with its description.

1. $ax^2 + bx + c$
2. The result of multiplication
3. $ax^2 + bx + c = 0$
4. A polynomial whose only factors are itself and 1
5. $a^2 - b^2 = (a + b)(a - b)$
6. $a(b + c) = ab + ac$
7. The distance around a rectangle
8. The product of the length and the width of a rectangle
9. If $a \cdot b = 0$, then either $a = 0$ or $b = 0$.
10. A polynomial expression with three terms

A. area of a rectangle
B. difference of squares pattern
C. distributive property
D. perimeter
E. prime polynomial
F. product
G. quadratic trinomial
H. standard form of a quadratic equation
I. trinomial
J. zero product property

6.5 Exercises

Follow your instructor's guidelines for showing your work.

For problems 1–28, solve.

1. $x(x - 15) = 0$
2. $y(y - 21) = 0$
3. $(c + 3)c = 0$
4. $(k + 8)k = 0$
5. $0 = b(b + 4)$
6. $0 = d(d + 9)$
7. $(p + 4)(p - 7) = 0$
8. $(h + 2)(h - 10) = 0$
9. $(2a + 1)(a + 3) = 0$
10. $(2m + 3)(m + 8) = 0$
11. $(3p + 5)(2p - 1) = 0$
12. $(3n + 7)(4n - 1) = 0$
13. $x(x - 3)(x - 7) = 0$
14. $y(y - 5)(y - 9) = 0$
15. $5(x - 3)(x - 7) = 0$
16. $2(y - 5)(y - 9) = 0$
17. $-5(x - 3)(x - 7) = 0$
18. $-2(y - 5)(y - 9) = 0$
19. $3d(2d - 15) = 0$
20. $3w(4w - 9) = 0$
21. $4(3x + 5) = 0$
22. $5(4y + 3) = 0$
23. $(x - 4)(x - 4) = 0$
24. $(x - 6)(x - 6) = 0$
25. $h(h + 5)(h + 5) = 0$
26. $k(k + 10)(k + 10) = 0$
27. $c(4 - c) = 0$
28. $d(7 - d) = 0$
29. A student said that the solutions of $(x + 2)(x - 6) = 15$ are $x = -2$ and $x = 6$, since $-2 + 2 = 0$ and $6 - 6 = 0$. Explain what is wrong with this thinking.
30. A student said that the solutions of $3(x + 8) = 0$ are $x = 3$ and $x = -8$. Explain what is wrong with this thinking.
31. Instead of using the zero product property, use the properties of equality to solve $4(3x + 5) = 0$.
32. Explain why we cannot use the zero product property to solve $4y^2 - 15 = 10$ when it is written in this form.
33. Explain why the solutions of the equation $2(x + 5)(x - 1) = 0$ are the same as the solutions of the equation $(2x + 10)(x - 1) = 0$.
34. The equation $x(x - 9)(x - 9) = 0$ has three factors. However, it has only two solutions. Explain why.

For exercises 35–80,
(a) solve.
(b) check.

35. $x^2 + 4x - 45 = 0$
36. $x^2 + 4x - 21 = 0$
37. $y^2 + 8y + 12 = 0$
38. $y^2 + 9y + 18 = 0$
39. $n^2 - n - 56 = 0$
40. $n^2 - n - 30 = 0$
41. $p^2 + p - 12 = 0$
42. $p^2 + p - 42 = 0$
43. $x^2 - 7x + 10 = 0$
44. $x^2 - 5x + 4 = 0$
45. $m^2 + 9m = 0$
46. $a^2 + 15a = 0$
47. $k^2 - 25 = 0$
48. $p^2 - 49 = 0$
49. $k^3 - 25k = 0$
50. $p^3 - 49p = 0$
51. $w^2 + 6w + 8 = 0$
52. $m^2 + 8m + 15 = 0$
53. $w^3 + 6w^2 + 8w = 0$

54. $m^3 + 8m^2 + 15m = 0$

55. $b^2 + 10b + 25 = 0$

56. $c^2 + 8c + 16 = 0$

57. $b^2 - 10b + 25 = 0$

58. $c^2 - 8c + 16 = 0$

59. $b^2 - 25 = 0$

60. $c^2 - 16 = 0$

61. $2d^2 + 11d + 5 = 0$

62. $2w^2 + 9w + 9 = 0$

63. $6m^2 + 29m + 35 = 0$

64. $6v^2 + 31v + 35 = 0$

65. $j^2 - 17j + 72 = 0$

66. $c^2 - 15c + 54 = 0$

67. $6x - x^2 = 0$

68. $12y - y^2 = 0$

69. $x^2 + x = 90$

70. $y^2 + 3y = 40$

71. $2p^2 + 17p = -8$

72. $2u^2 + 15u = -7$

73. $z^2 = 11z - 18$

74. $b^2 = 19b - 88$

75. $v^2 = 2v$

76. $n^2 = 5n$

77. $4x^2 + 12x = -9$

78. $9x^2 + 12x = -4$

79. $2x^2 = -17x - 8$

80. $2x^2 = -15x - 7$

For exercises 81–86, use a polynomial equation to find the length and width of the rectangle.

81. A rectangle is 3 in. longer than it is wide. Its area is 108 in.^{2}.

82. A rectangle is 4 in. longer than it is wide. Its area is 117 in.^{2}.

83. The width of a rectangle is 6 ft less than its length. Its area is 112 ft^2.

84. The width of a rectangle is 7 ft less than its length. Its area is 120 ft^2.

85. The width of a rectangle is 13 ft less than twice its length. Its area is 45 ft^2.

86. The width of a rectangle is 11 ft less than twice its length. Its area is 63 ft^2.

For exercises 87–88, use the five steps and a polynomial equation to find the base b and height h of the triangle. The formula for the area A of a triangle is $A = \frac{1}{2}bh$.

87. The height h of a triangle is 2 ft more than the length of its base b. The area of the triangle is 60 ft^2.

88. The height of a triangle is 4 ft more than the length of its base. The area of the triangle is 70 ft^2.

Problem Solving: Practice and Review

Follow your instructor's guidelines for using the five steps as outlined in Section 1.3, p. 55.

89. The graph represents the relationship of the retail sales of Oxycontin in millions of grams, y, and the number of years since 1997, x.

a. Write the equation of the line of best fit.

b. Describe what the slope represents.

c. Describe what the y-coordinate of the y-intercept represents.

d. Use the equation to find the retail sales of Oxycontin in 2012. Round to the nearest million.

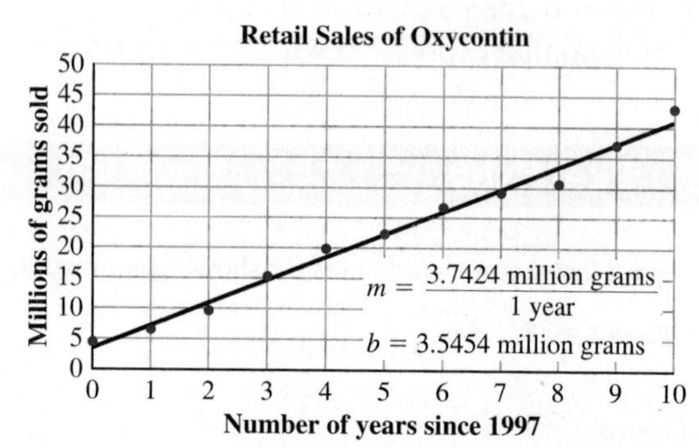

Source: Manchikanti, et al., "Therapeutic Use, Abuse, and Nonmedical Use of Opiods: a Ten Year Perspective," *Pain Physician,* 2010

90. Grease blocked a 12-in.-diameter sewer pipe in Inglewood, California, resulting in a spill of about 11,200 gal of sewage. Find the area of a cross-section of the pipe. ($\pi \approx 3.14$.) Round to the nearest tenth. (*Source:* www.latimes.com, Jan. 17, 2012)

91. The table organizes information about 2011 strawberry production in the United States.

a. Explain how the total supply was calculated.

b. Explain how the total utilization was calculated.

c. *Per capita consumption* means the consumption per person. Find the population used to calculate the per capita consumption. Round to the nearest million.

Fresh Strawberries 2011

Supply		
Utilized production	**Imports**	**Total supply**
2324.5 million lb	203.7 million lb	2528.2 million lb
Utilization		
Exports	**Consumption**	
	Total utilization	**Per capita**
274.5 million lb	2253.7 million lb	7.20 lb

Source: www.ers.usda.gov, May 27, 2011

92. Miami Dade College enrolls more minority students than any other college or university in the United States. Find the percent increase in the number of Hispanic students from 1970 to 2010. Round to the nearest percent.

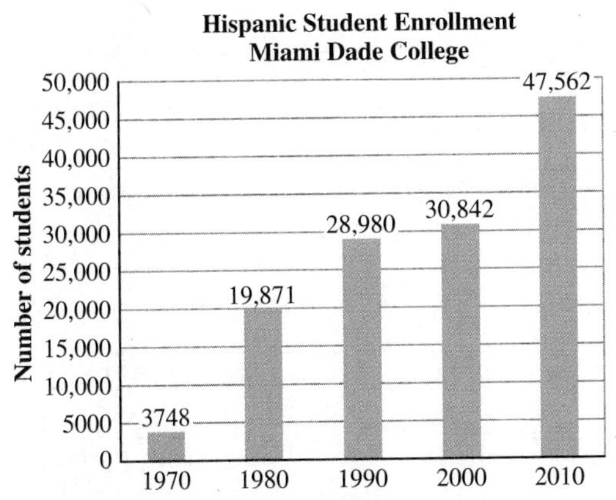

Source: www.mdc.edu, April 2011

Technology

For exercises 93–96,

(a) use the graphical method to solve each equation. Sketch the graph; describe the window.

(b) identify the intersection points.

(c) identify the solution of the equation.

93. $3x + 5 = 8$

94. $-6x + 27 = 3x - 9$

95. $x^2 - x - 2 = 0$

96. $x^2 - 5x = 6$

Find the Mistake

For exercises 97–100, the completed problem has one mistake.

(a) Describe the mistake in words, or copy down the whole problem and highlight or circle the mistake.

(b) Do the problem correctly.

97. Problem: Solve: $x^2 + 8x + 15 = -1$

Incorrect Answer:

$$x^2 + 8x + 15 = -1$$
$$(x + 5)(x + 3) = -1$$
$$x + 5 = 0 \quad \text{or} \quad x + 3 = 0$$
$$\underline{-5 \quad -5} \qquad \underline{-3 \quad -3}$$
$$x = -5 \quad \text{or} \quad x = -3$$

Solutions: $x = -5, x = -3$

98. Problem: Solve: $2x^2 - 12x - 54 = 0$

Incorrect Answer:

$$2x^2 - 12x - 54 = 0$$
$$2(x^2 - 6x - 27) = 0$$
$$2(x - 9)(x + 3) = 0$$
$$2 = 0 \quad \text{or} \quad x - 9 = 0 \quad \text{or} \quad x + 3 = 0$$
$$\underline{+9 \quad +9} \qquad \underline{-3 \quad -3}$$
$$x = 9 \quad \text{or} \quad x = -3$$

Solutions: $x = 2, x = 9, x = -3$

99. Problem: Solve: $x^2 + 12x = 0$

Incorrect Answer:

$$x^2 + 12x = 0$$
$$\frac{x^2}{x} + \frac{12x}{x} = \frac{0}{x}$$
$$x + 12 = 0$$
$$\underline{-12 \quad -12}$$
$$x = -12$$

Solution: $x = -12$

100. Problem: Solve: $x^3 + 7x^2 + 10x = 0$

Incorrect Answer:

$$x^3 + 7x^2 + 10x = 0$$
$$x(x^2 + 7x + 10) = 0$$
$$(x + 5)(x + 2) = 0$$
$$x + 5 = 0 \quad \text{or} \quad x + 2 = 0$$
$$\underline{-5 \quad -5} \qquad \underline{-2 \quad -2}$$
$$x = -5 \quad \text{or} \quad x = -2$$

Solutions: $x = -5, x = -2$

Review

101. Describe the difference between a rational number and an irrational number.

102. Simplify $\frac{108}{124}$ into lowest terms.

103. Explain how to determine if a fraction is in lowest terms.

104. Explain how to divide two fractions.

SUCCESS IN COLLEGE MATHEMATICS

105. When a student earns 75% on a test, what do you think that the student should do about the incorrect problems?

106. If you earn a C on each of your tests, do you think you will know enough t[illegible] Explain.

Study Plan for Review of Chapter 6

SECTION 6.1 Introduction to Factoring

Ask Yourself	Test Yourself	Help Yourself
Can I...	**Do 6.1 Review Exercises**	**See these Examples and Practice Problems**
find the greatest common factor of two or more terms? write a definition of *prime number*? write a definition of *prime polynomial*? write the prime factorization of a term?	1–8	Ex. 1–5, PP 1–6
factor out the greatest common factor from a polynomial expression and check the factoring?	9–14	Ex. 6–10, PP 7–12
factor a polynomial expression with four terms by grouping and check the factoring?	15–18	Ex. 11–17, PP 13–18

6.1 Review Exercises

For exercises 1–4, write the prime factorization of the number or term.

1. 100 **2.** 126

3. $36xy^2$ **4.** $20ab^2$

For exercises 5–8, find the greatest common factor of the terms.

5. 54; 36 **6.** 84; 24

7. $48x^2y$; $16xy^2$; $96xy$ **8.** $45a^3b^2c$; $18a^2b^2$; $54a^2c$

For exercises 9–11, factor out the greatest common factor.

9. $15xy^2 - 12xy + 51$ **10.** $x^5y^2z^3 - x^3y^4z^2 + x^2y^3z$

11. $45h^2km + 120hk^2 - 28hm^2$

12. Choose the greatest common factor of $60x^2y + 24xy^2 - 20x^2$.
 a. $4x$ **b.** $2xy$
 c. $2x$ **d.** $6x$
 e. $4x^2$

13. A student said that the greatest common factor of $30a^2 + 12ab + 18b$ is 2. Explain why this answer is wrong.

14. Choose the correct factoring of $45p^3w - 15pw + 30pw^2$.
 a. $5pw(9p^2 - 3 + 6w)$
 b. $15p(3p^2w - w + 2w^2)$
 c. $15pw(3p^2 - 1 + 2w)$

For exercises 15–18, factor by grouping and check the factoring. Identify any prime polynomials.

15. $5a - 20ad + 3bc - 12bcd$

16. $np - 3kp + 2n^2 - 6kn$

17. $21mw - 28m + 6pw + 8p$

18. $18wx - yz - 9xz + 2wy$

SECTION 6.2 Factoring Trinomials

Ask Yourself	Test Yourself	Help Yourself
Can I...	**Do 6.2 Review Exercises**	**See these Examples and Practice Problems**
identify a quadratic trinomial? identify the lead coefficient, middle term, or constant in a quadratic trinomial? use "guess and check" to factor a trinomial?	19–23	Ex. 1–6, PP 1–4
the *ac* factoring method to factor a trinomial and check the factoring?	24–30	Ex. 7–12, PP 5–8
ne whether a number is a perfect square? criminant of a quadratic trinomial and use it determine whether the ime?	31–35	Ex. 13, 14, PP 9–11
that is not quadratic?	36–38	Ex. 15, PP 12–15

6.2 Review Exercises

19. Choose the quadratic trinomials.

a. $x^2 + 3x + 4$ **b.** $2x^2 + 5x - 3$
c. $2x^4 + 5x^2 - 3$ **d.** $x^3 - 5x^2 + 7$

For exercises 20–23, use the guess and check method to factor.

20. $p^2 + 9p + 18$

21. $h^2 - 15h + 56$

22. $a^2 - 6a - 16$

23. $d^2 - d - 42$

For exercises 24–30, use the *ac* method to factor. Check.

24. $2x^2 + 3x - 5$

25. $2u^2 - 3u - 54$

26. $5r^2 + 8r + 3$

27. $4m^2 + 4m - 15$

28. $n^2 + 16n + 64$

29. $w^2 - 4w - 21$

30. $9c^2 + 9c - 4$

31. Choose the numbers that are perfect squares.

a. 144 **b.** 6
c. 49 **d.** 1
e. 0 **f.** 44

32. Explain why 56 is not a perfect square.

For exercises 33–35,
(a) find the discriminant.
(b) use the discriminant to determine whether the quadratic trinomial is prime.

33. $5x^2 + 6x - 8$

34. $x^2 - 20x + 96$

35. $3x^2 - 4x - 9$

For exercises 36–38, factor.

36. $x^{10} + 9x^5 + 14$

37. $n^{14} - 11n^7 + 24$

38. $y^{22} - 4y^{11} + 4$

SECTION 6.3 Patterns

Ask Yourself	Test Yourself	Help Yourself
Can I...	**Do 6.3 Review Exercises**	**See these Examples and Practice Problems**
use a pattern to factor a perfect square quadratic trinomial?	39, 40, 47	Ex. 1–3, PP 1–3
use a pattern to factor a perfect square trinomial that is not quadratic?	43, 44	Ex. 4, PP 4, 5
use a pattern to factor a binomial that is a difference of squares?	41, 42, 48	Ex. 5, 6, PP 6–9
use a pattern to factor a binomial that is a sum or difference of cubes?	45, 46, 50	Ex. 7, 8, PP 10–13

6.3 Review Exercises

For exercises 39–46,
(a) identify a pattern that matches the polynomial.
(b) use the pattern to factor the polynomial.
(c) check.

39. $x^2 + 10x + 25$

40. $x^2 - 8x + 16$

41. $x^2 - 81$

42. $100z^2 - 9y^2$

43. $9b^4 - 42b^2c + 49c^2$

44. $9x^2 + 6xy^3 + y^6$

45. $8h^3 - k^3$

46. $1000h^3 + 27k^3$

47. Choose the polynomials that match the perfect square trinomial factoring pattern, $a^2 + 2ab + b^2 = (a + b)(a + b)$.

a. $x^2 + 8x + 16$ **b.** $4x^2 + 12x + 9$
c. $9x^2 + 12x + 4$ **d.** $16x^2 + 8x + 1$

48. The difference of squares factoring pattern is $a^2 - b^2 = (a + b)(a - b)$. For the polynomial $49n^2 - 100p^2$, identify a and b.

49. A student said that the polynomial $9x^2 + 16$ is equal to $(3x + 4)^2$. Explain what is wrong with this thinking.

50. The sum of cubes factoring pattern is $a^3 + b^3 = (a + b)(a^2 - ab + b^2)$. For the polynomial $343d^3 + 512m^3$, identify a and b.

SECTION 6.4 Factoring Completely

Ask Yourself	Test Yourself	Help Yourself
Can I . . .	**Do 6.4 Review Exercises**	**See these Examples and Practice Problems**
factor a binomial completely?	51, 55, 58, 59, 64, 65, 68	Ex. 6, 7, 9, 10, PP 3, 4, 7, 10, 15, 23–28
factor a trinomial completely?	52, 54, 57, 60, 62, 66	Ex 1, 2, 8, PP 1, 2, 6, 9, 12, 14, 17, 19, 20–22
factor a polynomial with four terms completely?	53, 56, 61, 63, 67	Ex. 3, 4, 5, PP 5, 8, 11, 13, 16, 18

6.4 Review Exercises

For exercises 51–65, factor completely. Identify any prime polynomials.

51. $2d^2 - 128$

52. $n^3 + 7n^2 - 44n$

53. $3xz + xyz - 3yz - y^2z$

54. $12x^2 + 12x + 3$

55. $20u^2 + 45$

56. $2x^2 + 4x + 2xm + m$

57. $3v^2 - 12v - 288$

58. $5p^2 - 20p$

59. $72g^2 - 2h^2$

60. $6x^2 + 6x + 1$

61. $6x^2z - 2xz^2 + 3xyz - yz^2$

62. $8x^3 - 24x^2 + 18x$

63. $6x^2y - 12x^2z + xy - 2xz$

64. $5x^3 + 40y^3$

65. $27x^4 - 64xy^3$

66. A student who was asked to completely factor $3x^2 - 6x - 24$ said that the factors were $(3x + 6)(x - 4)$. Explain what is wrong with this answer.

67. A student who was asked to completely factor $x^3 - 9x + x^2 - 9$ said that the factors were $(x^2 - 9)(x + 1)$. Explain what is wrong with this answer.

68. A student who was asked to completely factor $x^2 - 11x$ said that this polynomial is prime because 11 is not a perfect square and this polynomial is not a difference of squares. Explain what is wrong with this answer.

SECTION 6.5 The Zero Product Property

Ask Yourself	Test Yourself	Help Yourself
Can I . . .	**Do 6.5 Review Exercises**	**See these Examples and Practice Problems**
write the zero product property? use the zero product property to solve a polynomial equation?	69–78	Ex. 1–7, PP 1–14
use a polynomial equation to find the dimensions of a geometric figure such as a rectangle or triangle?	78	Ex. 8, PP 15, 16

6.5 Review Exercises

For exercises 69–77,
(a) solve.
(b) check.

69. $5(2x + 9)(x - 3) = 0$

70. $x(x - 2)(x + 15) = 0$

71. $(3x - 4)(3x - 4) = 0$

72. $a^2 - a - 12 = 0$

73. $p^3 - 64p = 0$

74. $24x - x^2 = 0$

75. $n^3 = -10n^2 - 25n$

76. $2y^2 + 7y = -6$

77. $a^3 = a$

78. The length of a rectangle is 6 cm less than three times the width. The area of the rectangle is 144 cm^2. Use a polynomial equation and the zero product property to find the length and width of this rectangle.

79. The equation $3(x + 2)(x - 2) = 0$ has three factors but only two solutions. Explain why.

Chapter 6 Test

1. State the difference of squares factoring pattern.
2. State a perfect square trinomial factoring pattern.

For problems 3–16, factor completely.

3. $a^2 - 2a - 48$
4. $7x^2 - 7x$
5. $p^2 - 9$
6. $m^3 + z^3$
7. $15xz + 24x + 5yz + 8y$
8. $2x^2 + 12x - 54$
9. $8h^2 + 32h - 40$
10. $6dm - 14m - 15dp + 35p$
11. $2x^2 - 8x + 90$
12. $m^3 - 8z^3$
13. $9x^2 + 12x + 4$
14. $16a^2 + 64b^2$
15. $12ax + 27a$
16. $45x^2 + 32x + 3$
17. The discriminant is $b^2 - 4ac$.
    a. Use the discriminant to determine whether $3x^2 - 7x + 2$ is a prime polynomial.
    b. Explain how you know whether it is prime.

For problems 18–19, use the zero product property to solve.

18. $(k + 9)(k - 12) = 0$
19. $w(w - 3) = 0$

For problems 20–23,
(a) use the zero product property to solve.
(b) check.

20. $u^2 - 11u + 28 = 0$
21. $2b^3 - 4b^2 - 30b = 0$
22. $4x^2 - 100 = 0$
23. $2n^2 = 11n + 40$
24. The width of a rectangle is 5 in. less than its length. The area of the rectangle is 84 in.^{2}. Use a polynomial equation and the five steps to find the length and width of this rectangle.

Cumulative Review Chapters 4–6

1. Describe or sketch the graph of a linear system with no solutions.
2. Describe or sketch the graph of a linear system with infinitely many solutions.
3. The graph of a system of equations is two perpendicular lines. Identify the number of solutions of this system.
4. The equation of a line is $y = \frac{7}{9}x + 2$. Identify the slope of a line that is parallel to this line.

For exercises 5–8, solve by graphing.

5. $y = -4x + 2$; $3x + 2y = -6$
6. $x = 4$; $y = -3$
7. $y = \frac{3}{4}x - 5$; $y = \frac{3}{4}x + 1$
8. $3x + 8y = 30$; $5x - 2y = 4$
9. Graph $\begin{aligned} 4x + 3y &\le 24 \\ x &\ge 0 \\ y &\ge 4 \end{aligned}$. Label the solution region.

For exercises 10–13, solve by substitution.

10. $y = -5x + 3$; $7x + 4y = -14$
11. $x = 3y - 2$; $-4x + 12y = 8$
12. $y = -6x + 3$; $y = 2x - 5$
13. $y = \frac{3}{4}x - 5$; $3x - 4y = -4$

For exercises 14–17, solve by elimination.

14. $3x + 4y = 8$; $6x + 8y = 40$
15. $2x + 3y = -5$; $5x + 4y = -2$
16. $6x + y = 3$; $-2x + y = -5$
17. $2x - 3y = -17$; $8x - 5y = -26$
18. Write the distributive property.

For exercises 19–31, simplify.

19. $(7x^2 - 9x + 2) + (15x^2 - 41x - 8)$
20. $(7x^2 - 9x + 2) - (15x^2 - 41x - 8)$
21. $12x(3x^2 + 5x - 4)$
22. $(8x + 9)(5x - 3)$
23. $(3x - 9)(7x^2 - 2x + 6)$
24. $x(x - 9)(2x + 3)$
25. $(a^2 - 3a - 9) - (5a - 17)$
26. $c^2 - (5c^2 + 4c - 8)$
27. $(5x^3)(2x^9)$
28. $\frac{12a^{10}}{20a^3}$
29. $\frac{18p}{2p^8}$
30. $(5x)^0$
31. $\frac{8x^2}{16x^2}$
32. Write an example of a trinomial.
33. Identify the number of terms in a monomial.
34. Identify the number of terms in $8x^5 - 3x^4 + 2x^3 - x + 8$.

35. Identify the degree of $8x^5 - 3x^4 + 2x^3 - x + 8$.

36. Explain why $8x^2 + \frac{2}{x} - 9$ is not a polynomial.

37. Find the greatest common factor of 24 and 56.

38. Find the greatest common factor of 84 and 104.

For exercises 39–48, factor completely. Identify any prime polynomials.

39. $6x^2 - 5x - 4$

40. $7x^2 + 61x - 18$

41. $x^2 + x - 90$

42. $3n^2 - 48n + 189$

43. $5p^2 - 20p$

44. $72g^2 - 2h^2$

45. $5a^2 + 5a + 3$

46. $6ac^2 - 3acp + 2acm - amp$

47. $18p^3 - 12p + 2p$

48. $2abc - 8ac^2 + 3bc - 12c^2$

49. Write the zero product property.

For exercises 50–55, use the zero product property to solve.

50. $3(2x + 1)(x - 4) = 0$

51. $n(n - 4)(n + 6) = 0$

52. $x^2 - 8x + 16 = 0$

53. $y^2 - y = 42$

54. $2x^2 - 9x - 5 = 0$

55. $36x - x^2 = 0$

For exercises 56–60, use a system of two linear equations.

56. The perimeter of a rectangle is 84 in. Seven times the length plus the width is 192 in. Find the length and width of the rectangle.

57. The cost to make a product is \$5. The fixed monthly costs to make the product are \$1800. The company sells the product for \$8. Find the number of products that must be made and sold to break even.

58. Mixture A is 21% beetle juice. Mixture B is 35% beetle juice. Find the amount of Mixture A and Mixture B needed to make 4500 gal of a new mixture that is 29% beetle juice. Round to the nearest tenth.

59. A salt solution is 26.5% salt. Find the amount of the salt solution and pure water to mix together to make 150 gal of a solution that is 11% salt. Round to the nearest whole number.

60. A copy center has 180 reams of paper. There are nine times as many reams of white paper than reams of colored paper. Find the number of reams of white paper and the number of reams of colored paper.

Rational Expressions and Equations

7

SUCCESS IN COLLEGE MATHEMATICS

Math Anxiety

For some math students, anxiety changes their ability to think and recall information. Their worries about doing well on a test, in particular on a math test, can cause physical and emotional symptoms. Blood pressure and heart rate can increase. Other physical symptoms may include faster breathing, excess sweating, dry mouth, restlessness, and nausea. Students who have math anxiety may have trouble concentrating, sitting still, reading, or keeping their legs from bouncing. They may feel desperate and want to skip class to avoid their feelings.

Anxiety affects students of all ages, genders, ethnicities, and incomes. Negative experiences in math at school, especially feeling stupid, ashamed, or embarrassed, can result in anxiety, and family members who had such negative experiences can pass their anxieties on to their children.

Some anxiety about tests or learning math is normal and may help motivate you to study. However, if your anxiety is blocking you from showing what you know or preventing you from learning in class, then you may need help. Perhaps reading a book about dealing with math anxiety or finding an article on the Internet will give you the information you need, or the counselors at your college may be able to help you learn to cope with your anxiety.

7.1 Simplifying Rational Expressions
7.2 Multiplication and Division of Rational Expressions
7.3 Combining Rational Expressions with the Same Denominator
7.4 Combining Rational Expressions with Different Denominators
7.5 Complex Rational Expressions
7.6 Rational Equations and Proportions
7.7 Variation

$$\frac{\text{integer}}{\text{integer}} = \frac{7}{16}$$

$$\frac{\text{polynomial}}{\text{polynomial}} = \frac{x^2 + 7x + 12}{x^2 - 16}$$

We can write a *rational number* as a fraction in which the numerator and denominator are integers. We can write a *rational expression* as a fraction in which the numerator and denominator are polynomials. In this section, we begin our study of rational expressions.

SECTION 7.1

Simplifying Rational Expressions

After reading the text, working the practice problems, and completing assigned exercises, you should be able to:

Simplify a rational expression.

Simplifying Rational Expressions

To simplify a *rational number*, we find common factors in the numerator and denominator. A fraction with the same nonzero numerator and denominator is equal to 1.

Rational Number in Lowest Terms

The numerator and the denominator of a rational number in **lowest terms** have no common factors.

EXAMPLE 1 Simplify: $\frac{18}{30}$

SOLUTION ▶

$\frac{18}{30}$

$= \frac{6 \cdot 3}{6 \cdot 5}$ Find common factors.

$= \frac{6}{6} \cdot \frac{3}{5}$ Rewrite as the product of two fractions.

$= \mathbf{1} \cdot \frac{3}{5}$ A fraction with the same numerator and denominator is equal to 1.

$= \frac{3}{5}$ Simplify.

Since the product of 1 and a number is the number, we often combine some of these steps and simplify fractions that are equal to 1 by drawing a line through the numerator and denominator.

EXAMPLE 2 Simplify: $\frac{18}{30}$

SOLUTION ▶

$\frac{18}{30}$

$= \frac{\not{6} \cdot 3}{\not{6} \cdot 5}$ Find common factors.

$= \frac{3}{5}$ Simplify.

A **rational expression** is a fraction in which the numerator and denominator are polynomials. A monomial is a polynomial with one term. When the numerator and denominator of a rational expression are monomials, simplify the coefficients to lowest terms, use the quotient rule of exponents, $\frac{x^m}{x^n} = x^{m-n}$, to simplify the variables, and use the relationship $x^{-m} = \frac{1}{x^m}$ to rewrite any negative exponents as positive exponents.

EXAMPLE 3 Simplify: $\frac{40xy^5}{32x^3y}$

SOLUTION

$\frac{40xy^5}{32x^3y}$

$= \frac{\mathbf{8 \cdot 5 \cdot x^{1-3}y^{5-1}}}{\mathbf{8 \cdot 4}}$ Find common factors; quotient rule of exponents.

$= \frac{5x^{-2}y^4}{4}$ Simplify; the expression includes a negative exponent.

$= \frac{5y^4}{4x^2}$ Rewrite x^{-2} as $\frac{1}{x^2}$; all of the exponents are positive.

In Chapter 6, we learned strategies for factoring polynomials. To simplify some rational expressions, we factor the numerator and the denominator completely and then simplify common factors.

EXAMPLE 4 Simplify: $\frac{9x + 27}{27}$

SOLUTION

$\frac{9x + 27}{27}$

$= \frac{\mathbf{9(x + 3)}}{27}$ Factor the numerator; the greatest common factor is 9.

$= \frac{\cancel{9}(x + 3)}{\mathbf{\cancel{9} \cdot 3}}$ Find common factors.

$= \frac{x + 3}{3}$ Simplify.

When simplifying rational expressions, look for common factors in the numerator and denominator, and simplify fractions that are equal to 1. Factors are multiplied together, not added together. We can "cross off" or simplify only common factors.

Simplifying Fractions That Are Equal to 1

Correct: $\frac{\cancel{2} \cdot 5}{\cancel{2} \cdot 3} = \frac{5}{3}$

Correct: $\frac{\cancel{(x + 3)}(x + 1)}{\cancel{(x + 3)}(x - 2)} = \frac{x + 1}{x - 2}$

Correct: $\frac{2 + 5}{2 + 3} = \frac{7}{5}$

Wrong: $\frac{\cancel{x} + 3}{\cancel{x} + 1} = \frac{3}{1}$

Wrong: $\frac{\cancel{6}}{x + \cancel{6}} = \frac{1}{x}$

Wrong: $\frac{\cancel{2} + 5}{\cancel{2} + 3} = \frac{5}{3}$

Simplifying Rational Expressions

1. Factor the numerator and the denominator completely.
2. Find common factors in the numerator and denominator.
3. Simplify expressions that are equal to 1.
4. Multiply any remaining factors in the numerator; either multiply the factors in the denominator or leave it factored.
5. The expression is simplified if the numerator and denominator have no common factors.

EXAMPLE 5 Simplify: $\dfrac{p^2 + 9p + 20}{p^2 - 25}$

SOLUTION ▸

$\dfrac{p^2 + 9p + 20}{p^2 - 25}$

$= \dfrac{\mathbf{(p + 5)(p + 4)}}{p^2 - 25}$ Factor the numerator.

$= \dfrac{(p + 5)(p + 4)}{\mathbf{(p - 5)(p + 5)}}$ Factor the denominator; difference of squares.

$= \dfrac{\cancel{(p + 5)}(p + 4)}{(p - 5)\cancel{(p + 5)}}$ Find common factors.

$= \dfrac{p + 4}{p - 5}$ Simplify.

In the next example, the factored rational expression is $\dfrac{5}{(x + 12)(x - 3)}$. If $x = -12$, then the denominator is $(0)(-15)$, which equals 0. If $x = 3$, then the denominator is $(15)(0)$, which equals 0. Since division by 0 is undefined, x cannot equal -12 or 3.

EXAMPLE 6 A rational expression is $\dfrac{5}{x^2 + 9x - 36}$. Find any values of the variable for which this expression is undefined.

SOLUTION ▸

$\dfrac{5}{x^2 + 9x - 36}$

$= \dfrac{5}{\mathbf{(x + 12)(x - 3)}}$ Factor the denominator.

If $x + 12 = 0$ or $x - 3 = 0$, the denominator equals 0, and the expression is undefined.

$x + 12 = 0$	Find when the denominator equals 0.	$x - 3 = 0$
$-12 \quad -12$	Properties of equality.	$+3 \quad +3$
$x + \mathbf{0} = \mathbf{-12}$	Simplify.	$x + \mathbf{0} = \mathbf{3}$
$\mathbf{x} = -12$	Simplify.	$\mathbf{x} = 3$

Because $x = -12$ and $x = 3$ result in division by zero, the expression is undefined for these values of the variable. *Throughout this chapter, we will assume that all rational expressions are defined.* A variable cannot represent a number that results in a denominator of 0.

To simplify a rational expression, both the numerator and the denominator must be factored completely.

EXAMPLE 7 Simplify: $\dfrac{4x^2 + 8x}{2x^3 - 14x^2 - 36x}$

SOLUTION

$$\frac{4x^2 + 8x}{2x^3 - 14x^2 - 36x}$$

$$= \frac{\mathbf{4x(x + 2)}}{2x^3 - 14x^2 - 36x}$$ Greatest common factor in numerator is $4x$.

$$= \frac{4x(x + 2)}{\mathbf{2x(x^2 - 7x - 18)}}$$ Greatest common factor in denominator is $2x$.

$$= \frac{4x(x + 2)}{2x\mathbf{(x - 9)(x + 2)}}$$ Factor $x^2 - 7x - 18$.

$$= \frac{\mathbf{2 \cdot 2 \cdot x \cancel{(x + 2)}}}{2 \cdot x \cdot (x - 9)\cancel{(x + 2)}}$$ Find common factors.

$$= \frac{2}{x - 9}$$ Simplify.

Before simplifying the expression in the next example, we multiply the numerator by 1. After we find and simplify common factors, the numerator equals 1.

EXAMPLE 8 Simplify: $\dfrac{c - 4}{c^2 - 8c + 16}$

SOLUTION

$$\frac{c - 4}{c^2 - 8c + 16}$$

$$= \frac{c - 4}{\mathbf{(c - 4)(c - 4)}}$$ Factor the denominator.

$$= \frac{\cancel{(c - 4)} \cdot \mathbf{1}}{\cancel{(c - 4)}(c - 4)}$$ Multiply the numerator by 1; find common factors.

$$= \frac{1}{c - 4}$$ Simplify.

After simplifying common factors, multiply any remaining factors in the numerator. Ask your instructor whether you should leave the denominator factored or multiply any remaining factors in the denominator.

EXAMPLE 9 Simplify: $\dfrac{2w^2 + 14w + 24}{w^3 - 6w^2 - 27w}$

SOLUTION

$$\frac{2w^2 + 14w + 24}{w^3 - 6w^2 - 27w}$$

$$= \frac{2(w^2 + 7w + 12)}{w(w^2 - 6w - 27)}$$ Factor out the greatest common factors, 2 and w.

$$= \frac{\mathbf{2(w + 3)(w + 4)}}{\mathbf{w(w - 9)(w + 3)}}$$ Factor the trinomials.

$$= \frac{2\cancel{(w + 3)}(w + 4)}{w(w - 9)\cancel{(w + 3)}}$$ Find common factors.

$$= \frac{2(w + 4)}{w(w - 9)}$$ Simplify.

$$= \frac{\mathbf{2w + 8}}{w(w - 9)}$$ Multiply factors in the numerator.

The factors $2 - n$ and $n - 2$ are not the same. In the next example, we factor out -1 from $2 - n$ and rewrite it as $-1(n - 2)$.

EXAMPLE 10 Simplify: $\dfrac{2 - n}{n - 2}$

SOLUTION ▸

$$\frac{2 - n}{n - 2}$$

$$= \frac{\mathbf{-1(-2 + n)}}{n - 2} \quad \text{Factor out } -1 \text{ in the numerator.}$$

$$= \frac{-1\cancel{(n - 2)}}{\cancel{n - 2}} \quad \text{Change order; } -2 + n = n - 2$$

$$= -1 \quad \text{Simplify.}$$

The difference of squares factoring pattern is $a^2 - b^2 = (a - b)(a + b)$. In the next example, we use this pattern to factor $4 - n^2$.

EXAMPLE 11 Simplify: $\dfrac{4 - n^2}{n^2 - 5n + 6}$

SOLUTION ▸

$$\frac{4 - n^2}{n^2 - 5n + 6}$$

$$= \frac{\mathbf{(2 - n)(2 + n)}}{n^2 - 5n + 6} \quad \text{Factor the numerator; difference of squares pattern.}$$

$$= \frac{(2 - n)(2 + n)}{\mathbf{(n - 3)(n - 2)}} \quad \text{Factor the denominator.}$$

$$= \frac{\mathbf{-1(-2 + n)}(2 + n)}{(n - 3)(n - 2)} \quad \text{Factor out } -1;\ 2 - n = -1(-2 + n)$$

$$= \frac{-1\mathbf{(n - 2)(n + 2)}}{(n - 3)(n - 2)} \quad \text{Change order; } -2 + n = n - 2 \text{ and } 2 + n = n + 2$$

$$= \frac{-1\cancel{(n - 2)}(n + 2)}{(n - 3)\cancel{(n - 2)}} \quad \text{Find common factors.}$$

$$= \frac{-1(n + 2)}{n - 3} \quad \text{Simplify.}$$

$$= \frac{\mathbf{-1n - 1}(2)}{n - 3} \quad \text{Multiply factors in the numerator.}$$

$$= \frac{\mathbf{-n - 2}}{n - 3} \quad \text{Simplify.}$$

Practice Problems

For problems 1–8, simplify.

1. $\dfrac{2x - 14}{6}$
2. $\dfrac{x^2 - x - 12}{x^2 + 4x + 3}$
3. $\dfrac{2x^2 - 11x - 6}{x^2 - 36}$
4. $\dfrac{24w^6}{30w^2}$
5. $\dfrac{a^3 - 49a}{a^3 - 6a^2 - 7a}$
6. $\dfrac{6c^2 - 21c - 45}{24c + 36}$
7. $\dfrac{25 - x^2}{x^2 + 7x + 10}$
8. $\dfrac{7 - w}{w - 7}$

For problems 9–11, find any values of the variable for which the expression is undefined.

9. $\dfrac{3}{x^2 - 25}$
10. $\dfrac{1}{x^2 + 9x + 14}$
11. $\dfrac{7}{x^3 + 4x^2 - 21x}$

7.1 Vocabulary Practice

Match the term with its description.

1. $a^2 - b^2 = (a + b)(a - b)$
2. Division by zero
3. A polynomial with three terms
4. A polynomial with two terms
5. $ax^2 + bx + c$
6. The only factors of this polynomial are itself and 1.
7. $a(b + c) = ab + ac$
8. The value of an expression with the same numerator and denominator
9. An expression that can be written as a fraction in which the numerator and denominator are polynomials
10. $\dfrac{x^m}{x^n} = x^{m-n}$

A. binomial
B. difference of squares pattern
C. distributive property
D. 1
E. prime polynomial
F. quadratic trinomial
G. quotient rule of exponents
H. rational expression
I. trinomial
J. undefined

7.1 Exercises

Follow your instructor's guidelines for showing your work.

For exercises 1–66, simplify.

1. $\dfrac{180}{420}$
2. $\dfrac{240}{540}$
3. $\dfrac{48a^2b^3}{56ab}$
4. $\dfrac{54c^2d^5}{72cd}$
5. $\dfrac{90n^2p^8}{42n^5p^6}$
6. $\dfrac{80w^3z^7}{48w^9z^5}$
7. $\dfrac{28xy^5}{56xy}$
8. $\dfrac{27hk^4}{54hk}$
9. $\dfrac{2x - 8}{10}$
10. $\dfrac{3x - 12}{15}$
11. $\dfrac{2x - 8}{10x}$
12. $\dfrac{3x - 12}{15x}$
13. $\dfrac{10y + 20}{y^2 + 5y + 6}$
14. $\dfrac{9z + 18}{z^2 + 6z + 8}$
15. $\dfrac{4z^2 + 20z}{32z}$
16. $\dfrac{5m^2 + 30m}{75m}$
17. $\dfrac{y + 9}{y^2 + 9y}$
18. $\dfrac{z + 8}{z^2 + 8z}$
19. $\dfrac{y^2 + 9y}{y + 9}$
20. $\dfrac{z^2 + 8z}{z + 8}$
21. $\dfrac{5}{5x^2 + 10x}$
22. $\dfrac{3}{3z^2 + 12z}$
23. $\dfrac{3x - 6}{4x - 8}$
24. $\dfrac{5x - 15}{9x - 27}$
25. $\dfrac{a - b}{b - a}$
26. $\dfrac{c - d}{d - c}$
27. $\dfrac{6x + 6}{3x - 3}$
28. $\dfrac{8x + 8}{4x - 4}$
29. $\dfrac{x^2 - 1}{x + 1}$
30. $\dfrac{x^2 - 4}{x + 2}$
31. $\dfrac{h - 9}{h^2 - 81}$
32. $\dfrac{u - 8}{u^2 - 64}$
33. $\dfrac{k^2 + 5k + 6}{k^2 + 7k + 10}$
34. $\dfrac{v^2 + 9v + 20}{v^2 + 10v + 24}$
35. $\dfrac{z^2 - 3z - 40}{z^2 - 64}$
36. $\dfrac{b^2 - 7b - 18}{b^2 - 81}$
37. $\dfrac{x^2 + 6x + 9}{x^2 + 3x}$
38. $\dfrac{x^2 + 8x + 16}{x^2 + 4x}$
39. $\dfrac{m^2 - 25}{m^2 + 6m + 5}$
40. $\dfrac{n^2 - 9}{n^2 + 4n + 3}$
41. $\dfrac{u}{u^2 + 6u}$
42. $\dfrac{v}{v^2 + 7v}$
43. $\dfrac{k^2 - 3k - 40}{k^2 - 4k - 45}$
44. $\dfrac{p^2 - 6p - 27}{p^2 - 2p - 15}$
45. $\dfrac{x^2 - 4x + 3}{x^2 - 3x + 2}$
46. $\dfrac{x^2 - 5x + 4}{x^2 - 4x + 3}$
47. $\dfrac{y^2 + 11y + 28}{y^2 - 2y - 63}$
48. $\dfrac{y^2 + 11y + 30}{y^2 - 2y - 48}$
49. $\dfrac{6x^3 + 18x^2 + 12x}{9x^2 + 9x - 18}$
50. $\dfrac{6x^3 + 36x^2 + 48x}{3x^2 - 15x - 108}$
51. $\dfrac{2c^2 - 13c - 45}{2c^2 + 13c + 20}$
52. $\dfrac{2w^2 + 9w + 7}{2w^2 + 13w + 21}$
53. $\dfrac{9n^2 - 48n + 64}{9n^2 - 64}$
54. $\dfrac{9p^2 - 30p + 25}{9p^2 - 25}$
55. $\dfrac{4v^2 + 5v - 6}{2v^2 + 10v + 12}$
56. $\dfrac{10a^2 + 20a - 30}{5a^2 + 20a + 15}$
57. $\dfrac{4p^2 - 100}{6p^2 - 12p - 90}$
58. $\dfrac{3k^2 - 48}{9k^2 + 18k - 216}$
59. $\dfrac{x^3 - x^2 - 72x}{x^4 + 5x^3 - 24x^2}$
60. $\dfrac{y^3 - y^2 - 56y}{y^4 + 5y^3 - 14y^2}$
61. $\dfrac{2a^3 - 4a^2 - 6a}{4a^3 - 16a^2 - 20a}$
62. $\dfrac{2c^3 - 2c^2 - 4c}{4c^3 - 8c^2 - 12c}$

63. $\frac{9 - n^2}{n^2 + 5n - 24}$

64. $\frac{16 - k^2}{k^2 + 4k - 32}$

65. $\frac{18 - 2n^2}{4n^2 + 20n - 96}$

66. $\frac{32 - 2k^2}{4k^2 + 16k - 128}$

For exercises 67–72, simplify.

67. $\frac{x^3 + 8}{x^2 - 4}$

68. $\frac{x^3 + 27}{x^2 - 9}$

69. $\frac{a^3 - 125}{a^2 - 10a + 25}$

70. $\frac{p^3 - 8}{p^2 - 4p + 4}$

71. $\frac{h + 4}{h^3 + 64}$

72. $\frac{k + 5}{k^3 + 125}$

For exercises 73–76, write a rational expression that simplifies to the given expression.

73. $\frac{x + 3}{x - 5}$

74. $\frac{x + 2}{x - 9}$

75. $\frac{a}{a + 6}$

76. $\frac{z}{z + 2}$

For exercises 77–86, find any values of the variable for which this expression is undefined.

77. $\frac{c - 9}{c + 3}$

78. $\frac{a - 7}{a + 2}$

79. $\frac{x^2 + 3x}{x^2 - 7x - 18}$

80. $\frac{y^2 + 4y}{y^2 - 8y - 20}$

81. $\frac{5}{p^2 - 25}$

82. $\frac{8}{z^2 - 64}$

83. $\frac{3}{81 - k^2}$

84. $\frac{9}{100 - m^2}$

85. $\frac{2}{x^2 + 3}$

86. $\frac{5}{x^2 + 6}$

Problem Solving: Practice and Review

Follow your instructor's guidelines for using the five steps as outlined in Section 1.3, p. 55.

87. A truck is traveling at a speed of $\frac{65 \text{ mi}}{1 \text{ hr}}$. Find the distance it travels in 8 min. Round to the nearest tenth.

88. A group of military veterans that took a survey included 134 women and 577 men. Find the percent of the group who were women. Round to the nearest percent. (*Source:* www.pewsocialtrends.org, Dec. 22, 2011)

89. In November, 2011, North Dakota natural gas production was 15,635,813 million cubic feet. Because of a shortage of gas processing plants and other infrastructure, more than one-third of the gas is burned off or "flared" instead of being processed and sold. Find the minimum amount of natural gas that was burned off in November 2011. Round to the nearest million cubic feet. (*Sources:* www.businessweek.com, Jan. 13, 2012; www.dmr.nd.gov, 2011)

90. A fruit drink is 15% white grape juice. Use a system of two linear equations to find the amount of pure white grape juice and the amount of this fruit drink needed to make 15 gal of a new drink that is 21% white grape juice. Round to the nearest tenth.

Find the Mistake

For exercises 91–94, the completed problem has one mistake.
(a) Describe the mistake in words, or copy down the whole problem and highlight or circle the mistake.
(b) Do the problem correctly.

91. **Problem:** Simplify: $\frac{x^2 - 9}{x - 3}$

Incorrect Answer: $\frac{x^2 - 9}{x - 3}$

$= \frac{\cancel{x} \cdot x - 3 \cdot \cancel{3}}{\cancel{x} - \cancel{3}}$

$= x - 3$

92. **Problem:** Simplify: $\frac{4x + 24}{24}$

Incorrect Answer: $\frac{4x + 24}{24}$

$= \frac{4x + \cancel{24}}{\cancel{24}}$

$= 4x$

93. **Problem:** Simplify: $\frac{x^2 - x - 6}{x^2 - 9}$

Incorrect Answer: $\frac{x^2 - x - 6}{x^2 - 9}$

$= \frac{\cancel{(x + 3)}(x - 2)}{(x - 3)\cancel{(x + 3)}}$

$= \frac{x - 2}{x - 3}$

94. **Problem:** Simplify: $\frac{2x + 10}{8x^2 + 48x + 40}$

Incorrect Answer: $\frac{2x + 10}{8x^2 + 48x + 40}$

$= \frac{2(x + 5)}{8(x^2 + 6x + 5)}$

$= \frac{2\cancel{(x + 5)}}{4 \cdot 2 \cdot \cancel{(x + 5)}(x + 1)}$

$= 4(x + 1)$

Review

95. Describe how to divide two fractions.

96. Evaluate: $\frac{3}{4} \div \frac{5}{6}$

97. Evaluate: $8 \div \frac{1}{2}$

98. Evaluate: $\frac{3}{8} \div 6$

SUCCESS IN COLLEGE MATHEMATICS

99. Write a paragraph or two describing your "math history." Include any positive or negative experiences that have strongly influenced your attitudes about math.

100. Using a scale of 1 to 10, with 10 describing overwhelming anxiety that makes it difficult for you to do math, rate your own math anxiety.

$$\frac{a}{b}\cdot\frac{c}{d}=\frac{a}{b}\cdot\frac{c}{d}$$

$$\frac{a}{b}\div\frac{c}{d}=\frac{a}{b}\cdot\frac{d}{c}$$

To multiply two rational numbers, we can first simplify common factors and then multiply the numerators and multiply the denominators. To divide two rational numbers, we multiply the dividend by the reciprocal of the divisor. In this section, we will use this same process to multiply and divide rational expressions.

SECTION 7.2

Multiplication and Division of Rational Expressions

After reading the text, working the practice problems, and completing assigned exercises, you should be able to:

1. Multiply rational expressions.
2. Divide rational expressions.

Multiplying Rational Expressions

To multiply rational numbers, we can multiply the numerators, multiply the denominators, and then simplify the product, or we can find the product by simplifying before multiplying. To do this, we find common factors in the numerator and denominator, simplify fractions that are equal to 1, and then multiply the remaining factors in the numerator and in the denominator.

EXAMPLE 1 Evaluate: $\frac{9}{20}\cdot\frac{5}{12}$

SOLUTION ▶

$$\frac{9}{20}\cdot\frac{5}{12}$$

$$=\frac{\mathbf{\cancel{3}\cdot 3}}{\mathbf{4\cdot\cancel{5}}}\cdot\frac{\mathbf{\cancel{5}}}{\mathbf{\cancel{3}\cdot 4}}$$ Find common factors.

$$=\frac{3}{\mathbf{16}}$$ Simplify; multiply remaining factors.

We can use the same process to multiply rational expressions. Ask your instructor whether you should multiply the factors in the denominator of the simplified expression or leave the denominator factored.

EXAMPLE 2 Simplify: $\frac{x^2+5x}{x^2-16}\cdot\frac{x+4}{x^2-25}$

SOLUTION ▶

$$\frac{x^2+5x}{x^2-16}\cdot\frac{x+4}{x^2-25}$$

$$=\frac{\mathbf{x(x+5)}}{x^2-16}\cdot\frac{x+4}{x^2-25}$$ Factor out the greatest common factor, x.

$$=\frac{x(x+5)}{\mathbf{(x+4)(x-4)}}\cdot\frac{(x+4)}{\mathbf{(x+5)(x-5)}}$$ Factor denominators; difference of squares.

$$=\frac{x\cancel{(x+5)}}{\cancel{(x+4)}(x-4)}\cdot\frac{\cancel{(x+4)}}{\cancel{(x+5)}(x-5)}$$ Find common factors.

$$=\frac{x}{(x-4)(x-5)}$$ Simplify.

To show multiplication of rational expressions, we can either write a multiplication dot between the expressions or, as in the next example, write the expressions in parentheses.

EXAMPLE 3 Simplify: $\left(\dfrac{a^2 - 9a + 18}{10a^2 - 80a}\right)\left(\dfrac{2a}{2a^2 - a - 15}\right)$

SOLUTION ▶

$\left(\dfrac{a^2 - 9a + 18}{10a^2 - 80a}\right)\left(\dfrac{2a}{2a^2 - a - 15}\right)$

$= \dfrac{\mathbf{(a - 3)(a - 6)}}{10a^2 - 80a} \cdot \dfrac{2a}{2a^2 - a - 15}$ Factor $a^2 - 9a + 18$.

$= \dfrac{(a - 3)(a - 6)}{10a^2 - 80a} \cdot \dfrac{2a}{\mathbf{(a - 3)(2a + 5)}}$ Factor $2a^2 - a - 15$.

$= \dfrac{(a - 3)(a - 6)}{\mathbf{10a(a - 8)}} \cdot \dfrac{2a}{(a - 3)(2a + 5)}$ Factor out the greatest common factor, $10a$.

$= \dfrac{\cancel{(a - 3)}(a - 6)}{\mathbf{2 \cdot 5 \cdot \cancel{a} \cdot}(a - 8)} \cdot \dfrac{\cancel{2} \cdot \cancel{a}}{\cancel{(a - 3)}(2a + 5)}$ Find common factors.

$= \dfrac{a - 6}{5(a - 8)(2a + 5)}$ Simplify.

Multiplying Rational Expressions

1. Factor the numerator and the denominator in all expressions completely.
2. Find common factors in the numerators and denominators.
3. Simplify expressions that are equal to 1.
4. Multiply any remaining factors in the numerator; either multiply the factors in the denominator or leave it factored.

In the next example, each factor in the numerator has a common factor in the denominator. The numerator in the final product is 1.

EXAMPLE 4 Simplify: $\dfrac{c + 6}{3c^2 - 3c} \cdot \dfrac{3c}{c^2 + 8c + 12}$

SOLUTION ▶

$\dfrac{c + 6}{3c^2 - 3c} \cdot \dfrac{3c}{c^2 + 8c + 12}$

$= \dfrac{c + 6}{3c^2 - 3c} \cdot \dfrac{3c}{\mathbf{(c + 6)(c + 2)}}$ Factor $c^2 + 8c + 12$.

$= \dfrac{c + 6}{\mathbf{3c(c - 1)}} \cdot \dfrac{3c}{(c + 6)(c + 2)}$ Factor out the greatest common factor, $3c$.

$= \dfrac{c + 6}{3c(c - 1)} \cdot \dfrac{3c \cdot \mathbf{1}}{(c + 6)(c + 2)}$ Multiply the numerator by 1.

$= \dfrac{\cancel{c + 6}}{\cancel{3c}(c - 1)} \cdot \dfrac{\cancel{3c} \cdot 1}{\cancel{(c + 6)}(c + 2)}$ Find common factors.

$= \dfrac{1}{(c - 1)(c + 2)}$ Simplify.

Practice Problems

1. Evaluate: $\frac{7}{8} \cdot \frac{10}{21}$

For problems 2–4, simplify.

2. $\frac{x - 6}{3x - 9} \cdot \frac{x - 3}{x^2 - 4x - 12}$

3. $\frac{8w^7}{w^2 + 6w + 5} \cdot \frac{w^2 - 25}{4w^2 - 20w}$

4. $\left(\frac{2c^2 + 7c + 3}{c^2 - c - 12}\right)\left(\frac{c^2 - 8c + 16}{c^2 + 2c - 24}\right)$

Dividing Rational Expressions

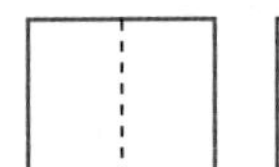
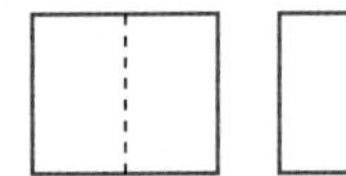
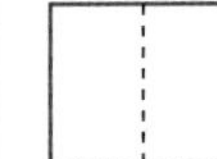

There are 6 one-halves in 3 rectangles.

Figure 1

In a division problem such as $3 \div \frac{1}{2} = 6$, the dividend is 3, the divisor is $\frac{1}{2}$, and the quotient is 6. We can visualize this problem by dividing three rectangles into halves (Figure 1). There are 6 one-halves in 3 rectangles. We can also find this quotient by multiplying the dividend by the reciprocal of the divisor. This process is sometimes described as "invert and multiply."

$3 \div \frac{1}{2}$

$= \frac{3}{1} \cdot \frac{\mathbf{2}}{\mathbf{1}}$ Rewrite division as multiplication by the reciprocal.

$= 6$ Simplify.

In the next example, $\frac{14}{15} \div \frac{2}{3}$, $\frac{14}{15}$ is the dividend, and $\frac{2}{3}$ is the divisor. We rewrite division as multiplication by the reciprocal of the divisor.

EXAMPLE 5 Evaluate: $\frac{14}{15} \div \frac{2}{3}$

SOLUTION ▶

$\frac{14}{15} \div \frac{2}{3}$

$= \frac{14}{15} \cdot \frac{\mathbf{3}}{\mathbf{2}}$ Rewrite division as multiplication by the reciprocal.

$= \frac{\mathbf{2 \cdot 7}}{\mathbf{3 \cdot 5}} \cdot \frac{3}{2}$ Find common factors.

$= \frac{7}{5}$ Simplify.

In the next example, we simplify $\frac{2x^5y^7}{3} \div \frac{4x^2y^3}{9}$. The dividend is $\frac{2x^5y^7}{3}$, and the divisor is $\frac{4x^2y^3}{9}$. After rewriting division as multiplication by the reciprocal of the divisor, we use the quotient rule of exponents, $\frac{x^m}{x^n} = x^{m-n}$, to simplify the variables.

EXAMPLE 6 Simplify: $\frac{2x^5y^7}{3} \div \frac{4x^2y^3}{9}$

SOLUTION

$\frac{2x^5y^7}{3} \div \frac{4x^2y^3}{9}$

$= \frac{2x^5y^7}{3} \cdot \frac{\mathbf{9}}{\mathbf{4x^2y^3}}$ Rewrite division as multiplication by the reciprocal.

$= \frac{2x^{5-2}y^{7-3}}{3} \cdot \frac{9}{4}$ Quotient rule of exponents.

$= \frac{2x^3y^4}{3} \cdot \frac{9}{4}$ Simplify.

$= \frac{2 \cdot x^3y^4}{3} \cdot \frac{3 \cdot 3}{2 \cdot 2}$ Find common factors.

$= \frac{3x^3y^4}{2}$ Simplify.

In the next example, we factor the polynomials to find common factors and then simplify.

EXAMPLE 7 Simplify: $\frac{n^2 + 8n + 12}{n^2 - 8n - 20} \div \frac{n^2 - 7n + 12}{n^3 - 14n^2 + 40n}$

SOLUTION

$\frac{n^2 + 8n + 12}{n^2 - 8n - 20} \div \frac{n^2 - 7n + 12}{n^3 - 14n^2 + 40n}$

$= \frac{n^2 + 8n + 12}{n^2 - 8n - 20} \cdot \frac{\mathbf{n^3 - 14n^2 + 40n}}{\mathbf{n^2 - 7n + 12}}$ Rewrite as multiplication by the reciprocal.

$= \frac{n^2 + 8n + 12}{n^2 - 8n - 20} \cdot \frac{\mathbf{n(n^2 - 14n + 40)}}{n^2 - 7n + 12}$ Factor out the greatest common factor, n.

$= \frac{\mathbf{(n + 6)(n + 2)}}{\mathbf{(n - 10)(n + 2)}} \cdot \frac{n \cdot \mathbf{(n - 10)(n - 4)}}{\mathbf{(n - 3)(n - 4)}}$ Factor the trinomials.

$= \frac{(n + 6)(n + 2)}{(n - 10)(n + 2)} \cdot \frac{n \cdot (n - 10)(n - 4)}{(n - 3)(n - 4)}$ Find common factors.

$= \frac{(n + 6)n}{n - 3}$ Simplify.

$= \frac{\mathbf{n^2 + 6n}}{n - 3}$ Multiply the factors in the numerator.

Dividing Rational Expressions

1. Rewrite division as multiplication by the reciprocal of the divisor.
2. Factor the numerator and the denominator in all expressions completely.
3. Find common factors in the numerators and denominators.
4. Simplify expressions that are equal to 1.
5. Multiply any remaining factors in the numerator; either multiply the factors in the denominator or leave it factored.

The only factors of a prime polynomial are itself and 1. In the next example, the polynomial $n^2 + 1$ is prime.

EXAMPLE 8 Simplify: $\dfrac{n^2+1}{n^2-1} \div \dfrac{n^3+3n^2}{n^2+2n-3}$

SOLUTION ▶

$$\frac{n^2+1}{n^2-1} \div \frac{n^3+3n^2}{n^2+2n-3}$$

$$= \frac{n^2+1}{n^2-1} \cdot \frac{\mathbf{n^2+2n-3}}{\mathbf{n^3+3n^2}}$$ Rewrite as multiplication by the reciprocal.

$$= \frac{n^2+1}{\mathbf{(n+1)(n-1)}} \cdot \frac{n^2+2n-3}{n^3+3n^2}$$ Factor $n^2 - 1$; difference of squares.

$$= \frac{n^2+1}{(n+1)(n-1)} \cdot \frac{\mathbf{(n+3)(n-1)}}{n^3+3n^2}$$ Factor $n^2 + 2n - 3$.

$$= \frac{n^2+1}{(n+1)(n-1)} \cdot \frac{(n+3)(n-1)}{\mathbf{n^2(n+3)}}$$ Factor out the greatest common factor, n^2.

$$= \frac{n^2+1}{(n+1)\cancel{(n-1)}} \cdot \frac{\cancel{(n+3)}\cancel{(n-1)}}{n^2\cancel{(n+3)}}$$ Find common factors.

$$= \frac{n^2+1}{n^2(n+1)}$$ Simplify.

We *multiply* factors and the result is a *product*. Do not make the mistake of thinking that n^2 in the numerator and n^2 in the denominator are common factors. Since the numerator is a *sum*, not a *product*, n^2 is not a common *factor*.

Practice Problems

5. Evaluate: $\dfrac{16}{35} \div \dfrac{4}{15}$

For problems 6–9, simplify.

6. $\dfrac{d^2-5d-24}{d^2+6d+9} \div \dfrac{2d^2-11d-40}{d^2+10d+21}$

7. $\dfrac{7k^4}{15m^8} \div \dfrac{49k^9}{3m^5}$

8. $\dfrac{x^2-4x-12}{x^2-6x} \div \dfrac{2x^2+6x+4}{8x^6}$

9. $\dfrac{y^2+9y+8}{3y^2+7y+4} \div \dfrac{y^2+10y+16}{3y^2+10y+8}$

7.2 Vocabulary Practice

Match the term with its description.

1. The only factors of this are itself and 1.
2. In $8 \div 2 = 4$, 2 is this.
3. Division by zero
4. For $\dfrac{a}{b}$, this is $\dfrac{b}{a}$.
5. $ax^2 + bx + c$
6. The result of division
7. An expression that can be written as a fraction in which the numerator and denominator are polynomials
8. $\dfrac{x^m}{x^n} = x^{m-n}$
9. The result of multiplication
10. When we rewrite a fraction as a quotient, this is the divisor.

A. denominator
B. divisor
C. prime polynomial
D. product
E. quadratic trinomial
F. quotient
G. quotient rule of exponents
H. rational expression
I. reciprocal
J. undefined

7.2 Exercises

Follow your instructor's guidelines for showing your work.

1. Describe how to divide two fractions.
2. Describe how to multiply two fractions.

For exercises 3–6, evaluate or simplify.

3. $\frac{3}{20} \cdot \frac{2}{15}$
4. $\frac{4}{15} \cdot \frac{5}{12}$
5. $\frac{2x}{15} \cdot \frac{3}{28}$
6. $\frac{3x}{10} \cdot \frac{5}{24}$

For exercises 7–32, simplify.

7. $\left(\frac{9}{4a + 20}\right)\left(\frac{4}{27}\right)$
8. $\left(\frac{8}{5w + 10}\right)\left(\frac{5}{24}\right)$
9. $\left(\frac{6z - 18}{10z + 30}\right)\left(\frac{25}{36}\right)$
10. $\left(\frac{8p - 24}{9p + 18}\right)\left(\frac{27}{32}\right)$
11. $\left(\frac{x^2 + 5x}{x^2}\right)\left(\frac{3x}{x + 5}\right)$
12. $\left(\frac{y^2 + 8y}{y^2}\right)\left(\frac{9y}{y + 8}\right)$
13. $\left(\frac{h^2}{h^2 + 3h}\right)\left(\frac{h^2 - 9}{h}\right)$
14. $\left(\frac{r^2}{r^2 + 2r}\right)\left(\frac{r^2 - 4}{r}\right)$
15. $\frac{x^2 + x}{x - 8} \cdot \frac{2x - 16}{x^2 - x}$
16. $\frac{y^2 - y}{y + 7} \cdot \frac{3y + 21}{y^2 + y}$
17. $\frac{z^2 - 7z - 18}{z^2 + 4z + 4} \cdot \frac{z^2 - 4z - 12}{z^2 - 11z + 18}$
18. $\frac{n^2 - 9n + 18}{n^2 - 6n + 9} \cdot \frac{n^2 - 2n - 3}{n^2 - 9n + 18}$
19. $\frac{p^2 + 11p + 18}{p^2 - 2p - 15} \cdot \frac{p^2 + 3p - 40}{p^2 + 10p + 16}$
20. $\frac{k^2 + 12k + 27}{k^2 + 7k - 18} \cdot \frac{k^2 + 5k - 14}{k^2 + 7k + 12}$
21. $\frac{a^2 + 16a + 64}{a^2 - 6a + 9} \cdot \frac{a^2 - 8a + 15}{a^2 + 3a - 40}$
22. $\frac{c^2 + 18c + 81}{c^2 - 4c + 4} \cdot \frac{c^2 - 5c + 6}{c^2 + 6c - 27}$
23. $\frac{2r^2 - 4r - 6}{r^2 + 5r - 24} \cdot \frac{r + 8}{2r}$
24. $\frac{3d^2 + 9d - 12}{d^2 + 10d + 24} \cdot \frac{d + 6}{3d}$
25. $\frac{m^2 - 2m - 80}{m^2 - m - 90} \cdot \frac{m^2 + 6m - 27}{m^2 + 5m - 24}$
26. $\frac{a^2 + 7a - 44}{a^2 + 9a - 22} \cdot \frac{a^2 - 9a + 14}{a^2 - 11a + 28}$
27. $\frac{2x^2 - 5x - 3}{x^2 - 12x + 27} \cdot \frac{x^2 - 15x + 54}{2x^2 + 13x + 6}$
28. $\frac{3x^2 + 14x + 8}{x^2 - 5x - 36} \cdot \frac{x^2 - 4x - 45}{3x^2 - 13x - 10}$
29. $\frac{12r^5 + 60r^4}{r^4 - r^3} \cdot \frac{r^2 - 1}{27r + 135}$
30. $\frac{28x^6 + 42x^5}{x^3 - x^2} \cdot \frac{x^2 - 1}{42x + 63}$
31. $\frac{w^2 - 3w + 5}{w^2 - 4} \cdot \frac{w^2 + 10w + 16}{w^2 - 1}$
32. $\frac{z^2 - 4z + 6}{z^2 - 9} \cdot \frac{z^2 + 7z + 12}{z^2 - 1}$
33. Fill in the numerator of $\frac{?}{x^2 + 4x - 32} \cdot \frac{x^2 + 5x - 24}{x^2 - 12x + 27}$ so that the product is $\frac{x + 2}{x - 9}$.
34. Fill in the numerator of $\frac{?}{x^2 + 4x - 32} \cdot \frac{x^2 + 5x - 24}{x^2 - 11x + 24}$ so that the product is $\frac{x + 2}{x - 8}$.

For exercises 35–38, evaluate.

35. $\frac{9}{28} \div \frac{3}{4}$
36. $\frac{16}{27} \div \frac{4}{9}$
37. $2 \div \frac{1}{2}$
38. $3 \div \frac{1}{3}$

For exercises 39–82, simplify.

39. $\frac{xy}{z^2} \div \frac{x^2}{z}$
40. $\frac{cd}{b^2} \div \frac{d^2}{b}$
41. $x \div \frac{1}{x}$
42. $z \div \frac{1}{z}$
43. $\frac{3a}{5} \div \frac{9}{10a^2}$
44. $\frac{4c}{7} \div \frac{8}{21c^2}$
45. $\frac{a}{3b} \div \frac{a}{6c}$
46. $\frac{h}{5k} \div \frac{h}{10m}$
47. $\frac{8ab}{21c^2} \div \frac{2a^2}{3c}$
48. $\frac{9hk}{40n^2} \div \frac{3h^2}{8n}$
49. $\frac{9x^2}{8y} \div \frac{x}{y}$
50. $\frac{15a^2}{14d} \div \frac{a}{d}$
51. $\frac{3b}{8d} \div \frac{3b}{20d}$
52. $\frac{7w}{9p} \div \frac{7w}{30p}$
53. $\frac{3x - 12}{4} \div \frac{3}{2}$
54. $\frac{5x - 20}{4} \div \frac{5}{2}$
55. $\frac{2a^2 + 2a}{9} \div \frac{a + 1}{3a^2}$
56. $\frac{5b^2 + 10b}{8} \div \frac{b + 2}{2b^3}$
57. $5f \div \frac{20f^3}{3}$
58. $9k \div \frac{27k^4}{4}$
59. $\frac{42u^4}{25} \div 6u$
60. $\frac{40n^5}{21} \div 8n$
61. $\frac{x - 4}{5x + 30} \div \frac{x + 2}{x + 6}$
62. $\frac{x - 4}{6x + 48} \div \frac{x + 2}{x + 8}$
63. $\frac{2z + 6}{z^2 + 3z + 2} \div \frac{z + 3}{z + 2}$
64. $\frac{3a + 6}{a^2 + 4a + 4} \div \frac{a + 4}{a + 2}$

65. $\dfrac{3p - 1}{8p} \div \dfrac{3p^2 + 14p - 5}{6p^2}$

66. $\dfrac{3y - 5}{12y} \div \dfrac{3y^2 + y - 10}{10y^2}$

67. $\dfrac{p^2 - 64}{p + 4} \div \dfrac{p + 8}{-p - 4}$

68. $\dfrac{k^2 - 36}{k + 3} \div \dfrac{k + 6}{-k - 3}$

69. $\dfrac{z^2 + 18z + 81}{z^2 + 7z - 18} \div \dfrac{z^2 - 5z + 6}{z^2 - 4z + 4}$

70. $\dfrac{a^2 + 12a + 36}{a^2 - 3a - 54} \div \dfrac{a^2 + 11a + 30}{a^2 - a - 72}$

71. $\dfrac{2k^2 + 3k}{2k^2 - 13k - 24} \div \dfrac{k^4 - k^3}{k^2 - 9k + 8}$

72. $\dfrac{2d^2 + 7d}{2d^2 + 11d + 14} \div \dfrac{d^4 - d^3}{d^2 + d - 2}$

73. $\dfrac{5b + 15}{4b + 4} \div \dfrac{2b + 6}{7b + 7}$

74. $\dfrac{2n - 16}{5n + 30} \div \dfrac{7n - 56}{3n + 18}$

75. $\dfrac{u^2 + 8u + 15}{u^2 + 2u + 1} \div \dfrac{u^2 + 7u + 10}{u^2 + 3u + 2}$

76. $\dfrac{w^2 + 10w + 16}{w^2 + 2w + 1} \div \dfrac{w^2 + 12w + 32}{w^2 + 5w + 4}$

77. $\dfrac{z^2 + 6z - 16}{z - 2} \div \dfrac{z + 8}{z + 3}$

78. $\dfrac{b^2 + 7b - 18}{b - 2} \div \dfrac{b + 9}{b + 4}$

79. $\dfrac{36a^2 + 12a + 1}{18a^2 + 15a + 2} \div \dfrac{6a^2 - 17a - 3}{3a^2 - 16a - 12}$

80. $\dfrac{16x^2 + 8x + 1}{8x^2 - 10x - 3} \div \dfrac{4x^2 + 17x + 4}{2x^2 + x - 6}$

81. $\dfrac{4x^2 + 8x - 12}{x^2 + 5x + 6} \div \dfrac{32}{8x + 16}$

82. $\dfrac{3p^2 + 15p + 12}{p^2 + 4p + 3} \div \dfrac{18}{6p + 18}$

For exercises 83–86, simplify.

83. $\dfrac{x^3 + 8}{x^2} \div \dfrac{x^2 - 4}{4x}$

84. $\dfrac{a^3 + 27}{a^2} \div \dfrac{a^2 - 9}{9a}$

85. $\dfrac{z^3 - 27}{z + 6} \div \dfrac{z^2 - 3z}{z^2 + 12z + 36}$

86. $\dfrac{w^3 - 64}{w + 9} \div \dfrac{w^2 - 4w}{w^2 + 18w + 81}$

Problem Solving: Practice and Review

Follow your instructor's guidelines for using the five steps as outlined in Section 1.3, p. 55.

87. A convenience store owner is setting the regular price for 1 gal of 2% milk. When the milk goes on sale at a discount of 30% later this week, the sale price will be \$2.99 per gallon. Find the regular price that the owner should charge for the milk. Round to the nearest hundredth.

88. If \$500 is invested for 10 years and earns simple interest, find the annual simple interest rate needed to earn \$425 in interest.

89. The cost to make a product is \$80, and the fixed overhead costs per month are \$15,000. The price of the product is \$100. Use the graph of a system of two linear equations to find the break-even point for this product.

90. The proposed Nabucco pipeline will transport oil from the eastern coast of Turkey to Austria. The pipeline travels 3900 km through Bulgaria, Romania, and Hungary and will cost 7.9 billion euros. Find the cost of the pipeline per mile in U.S. dollars. (1 mi = 1.6 km; 1 euro = 1.3145 U.S. dollars.) Round to the nearest hundredth of a million. (*Sources:* www.nabucco-pipeline.com; www.ecb.int, Jan. 27, 2012)

Find the Mistake

For exercises 91–94, the completed problem has one mistake.

(a) Describe the mistake in words, or copy down the whole problem and highlight or circle the mistake.

(b) Do the problem correctly.

91. Problem: Simplify: $\dfrac{2x + 4}{4} \div \dfrac{x^2 - 4}{6}$

Incorrect Answer: $\dfrac{2x + 4}{4} \div \dfrac{x^2 - 4}{6}$

$= \dfrac{2x + 4}{4} \cdot \dfrac{6}{x^2 - 4}$

$= \dfrac{2x + \cancel{4}}{\cancel{4}} \cdot \dfrac{6}{x^2 - 4}$

$= \dfrac{12x}{x^2 - 4}$

92. Problem: Simplify: $\dfrac{b^2 - b - 6}{b^2 - 2b - 3} \div \dfrac{b^2 - 2b - 8}{b^2 + 3b + 2}$

Incorrect Answer: $\dfrac{b^2 - b - 6}{b^2 - 2b - 3} \div \dfrac{b^2 - 2b - 8}{b^2 + 3b + 2}$

$= \dfrac{b^2 - 2b - 3}{b^2 - b - 6} \cdot \dfrac{b^2 - 2b - 8}{b^2 + 3b + 2}$

$= \dfrac{\cancel{(b - 3)}\cancel{(b + 1)}}{\cancel{(b - 3)}\cancel{(b + 2)}} \cdot \dfrac{(b - 4)\cancel{(b + 2)}}{(b + 2)\cancel{(b + 1)}}$

$= \dfrac{b - 4}{b + 2}$

93. Problem: Simplify: $\dfrac{x^2 + 3x}{x^2 - 9} \div \dfrac{x^2 - 1}{2x - 6}$

Incorrect Answer: $\dfrac{x^2 + 3x}{x^2 - 9} \div \dfrac{x^2 - 1}{2x - 6}$

$= \dfrac{x^2 + 3x}{x^2 - 9} \cdot \dfrac{2x - 6}{x^2 - 1}$

$= \dfrac{\cancel{(x + 1)}\cancel{(x + 3)}}{\cancel{(x - 3)}\cancel{(x + 3)}} \cdot \dfrac{2\cancel{(x - 3)}}{\cancel{(x + 1)}(x - 1)}$

$= \dfrac{2}{x - 1}$

94. Problem: Simplify: $\dfrac{z^2 + 4z}{z^2 - 16} \div \dfrac{2z}{z - 4}$

Incorrect Answer: $\dfrac{z^2 + 4z}{z^2 - 16} \div \dfrac{2z}{z - 4}$

$= \dfrac{z^2 + 4z}{z^2 - 16} \cdot \dfrac{z - 4}{2z}$

$= \dfrac{z^2 + 4z}{z^2 - 16} \cdot \dfrac{z - 4}{2 \cdot z}$

$= \dfrac{4z}{-16} \cdot (-2)$

$= \dfrac{-8z}{-16}$

$= \dfrac{z}{2}$

Review

For exercises 95–97, evaluate.

95. $\dfrac{5}{21} + \dfrac{2}{21}$

96. $\dfrac{8}{15} + \dfrac{7}{15}$

97. $\dfrac{16}{21} - \dfrac{2}{21}$

98. Explain why we can add only fractions that have a common denominator.

SUCCESS IN COLLEGE MATHEMATICS

99. Are there counselors at your college who can help students who have math anxiety? If so, describe where they are located and how to make an appointment to see one. If not, identify a different person or resource for learning how to cope with math anxiety.

$$\frac{a}{b} + \frac{c}{b} = \frac{a + c}{b}$$

We can add only rational numbers that have the same denominator. If we think of these rational numbers as a fraction or piece of a pie, we are adding pieces of pies that are the same size. In this section, we will add and subtract rational expressions that have the same denominator.

SECTION 7.3 Combining Rational Expressions with the Same Denominator

After reading the text, working the practice problems, and completing assigned exercises, you should be able to:

1. Add rational expressions with the same denominator.
2. Subtract rational expressions with the same denominator.
3. Find the least common multiple of two expressions.

Combining Rational Expressions

The rectangle in Figure 1 represents a cake that was originally cut into 7 pieces. Two pieces of the cake are left in the pan. This represents $\frac{2}{7}$ of the cake. The rectangle in Figure 2 represents a cake that was originally cut into 7 pieces. Three pieces of this cake are left in the pan. This represents $\frac{3}{7}$ of the cake. There are a total of 5 pieces of cake left in the pans that are of equal size. Using fractions to represent the remaining pieces, $\frac{2}{7}$ cake + $\frac{3}{7}$ cake equals $\frac{5}{7}$ cake (Figure 3). The numerator represents the number of remaining pieces of cake, and the denominator represents the number of pieces in a whole cake. To find the amount of remaining cake, we added the numerators and did not change the denominator.

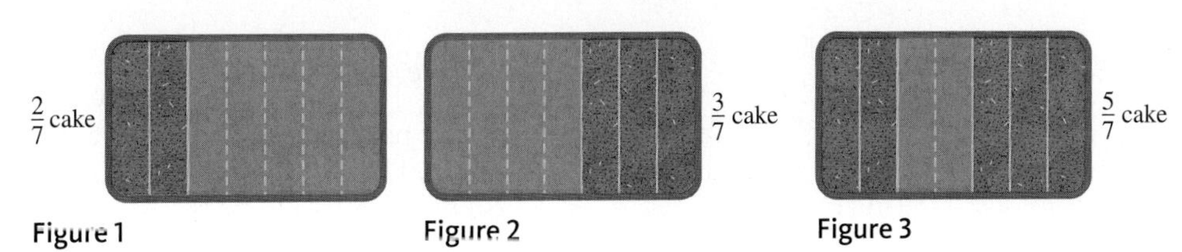

Figure 1 Figure 2 Figure 3

In the next example, we again add two fractions with the same denominator, $\frac{2}{15}$ and $\frac{7}{15}$. Since the sum, $\frac{9}{15}$, is not in lowest terms, we find common factors and simplify.

EXAMPLE 1 Evaluate: $\frac{2}{15} + \frac{7}{15}$

SOLUTION ▶

$\frac{2}{15} + \frac{7}{15}$

$= \frac{2 + 7}{15}$ Add the numerators; the denominator does not change.

$= \frac{\mathbf{9}}{15}$ Simplify.

$= \frac{3 \cdot \cancel{3}}{5 \cdot \cancel{3}}$ Find common factors.

$= \frac{3}{5}$ Simplify.

Just as with rational numbers, we can add or subtract only rational expressions that have the same denominator.

EXAMPLE 2 Simplify: $\frac{5}{x + 4} + \frac{3}{x + 4}$

SOLUTION ▶

$\frac{5}{x + 4} + \frac{3}{x + 4}$

$= \frac{5 + 3}{x + 4}$ Add the numerators; the denominator does not change.

$= \frac{\mathbf{8}}{x + 4}$ Simplify.

After adding two rational expressions, factor the numerator and/or denominator, look for common factors, and simplify. If there are no common factors, the expression is simplified.

EXAMPLE 3 Simplify: $\frac{x^2 + 6x}{x + 5} + \frac{5}{x + 5}$

SOLUTION ▶

$\frac{x^2 + 6x}{x + 5} + \frac{5}{x + 5}$

$= \frac{x^2 + 6x + 5}{x + 5}$ Add the numerators; the denominator does not change.

$= \frac{\cancel{(x + 5)}\mathbf{(x + 1)}}{\cancel{(x + 5)} \cdot \mathbf{1}}$ Factor the numerator; find common factors.

$= \frac{x + 1}{1}$ Simplify.

$= x + 1$ Simplify.

In the next example, we simplify $\frac{a^2 + 5a + 4}{2a + 6} - \frac{5a + 13}{2a + 6}$. When we combine the numerators, we write parentheses around the numerator of the second expression, $\frac{a^2 + 5a + 4 - (5a + 13)}{2a + 6}$.

EXAMPLE 4 Simplify: $\frac{a^2 + 5a + 4}{2a + 6} - \frac{5a + 13}{2a + 6}$

SOLUTION ▶

$\frac{a^2 + 5a + 4}{2a + 6} - \frac{5a + 13}{2a + 6}$

$= \frac{a^2 + 5a + 4 - (5a + 13)}{2a + 6}$ Subtract the numerators; use parentheses.

$= \frac{a^2 + 5a + 4 - \mathbf{1}(5a + 13)}{2a + 6}$ Multiply by 1; $-(5a + 13) = -1(5a + 13)$

$= \frac{a^2 + 5a + 4 - \mathbf{1}(5a) - \mathbf{1}(13)}{2a + 6}$ Distributive property.

$= \frac{a^2 + 5a + 4 - \mathbf{5a} - \mathbf{13}}{2a + 6}$ Simplify.

$= \frac{a^2 - \mathbf{9}}{2a + 6}$ Simplify.

$= \frac{\mathbf{(a + 3)(a - 3)}}{2a + 6}$ Factor $a^2 - 9$; difference of squares.

$= \frac{(a + 3)(a - 3)}{\mathbf{2(a + 3)}}$ Factor out the greatest common factor, 2.

$= \frac{(a - 3)\cancel{(a + 3)}}{2\cancel{(a + 3)}}$ Find common factors.

$= \frac{a - 3}{2}$ Simplify.

Adding or Subtracting Rational Expressions with the Same Denominator

1. Add or subtract the numerators. Do not change the denominator.
2. Factor the numerator and denominator completely.
3. Simplify common factors.

EXAMPLE 5 Simplify: $\frac{h^2 - 6h}{h^2 - 8h - 20} - \frac{7h - 30}{h^2 - 8h - 20}$

SOLUTION ▶

$\frac{h^2 - 6h}{h^2 - 8h - 20} - \frac{7h - 30}{h^2 - 8h - 20}$

$= \frac{h^2 - 6h - (7h - 30)}{h^2 - 8h - 20}$ Use parentheses.

$= \frac{h^2 - 6h - \mathbf{1}(7h - 30)}{h^2 - 8h - 20}$ Multiply by 1.

$= \frac{h^2 - 6h - \mathbf{1}(7h) - \mathbf{1}(-30)}{h^2 - 8h - 20}$ Distributive property.

$$= \frac{h^2 - 6h - \mathbf{7h + 30}}{h^2 - 8h - 20}$$ Simplify.

$$= \frac{h^2 - \mathbf{13h} + 30}{h^2 - 8h - 20}$$ Combine like terms.

$$= \frac{(h-3)\cancel{(h-10)}}{\cancel{(h-10)}(h+2)}$$ Factor; find common factors.

$$= \frac{h-3}{h+2}$$ Simplify.

Practice Problems

1. Evaluate: $\frac{9}{14} + \frac{3}{14}$

For problems 2–6, simplify.

2. $\frac{5}{x-9} + \frac{7}{x-9}$ **3.** $\frac{2}{x+1} - \frac{x+8}{x+1}$ **4.** $\frac{c^2 - 12c}{c-6} + \frac{36}{c-6}$

5. $\frac{p^2}{p+2} - \frac{p+6}{p+2}$ **6.** $\frac{k^2 + 8k}{k^2 + 7k + 12} + \frac{k + 20}{k^2 + 7k + 12}$

Prime Factorization and Least Common Multiples

In Section 6.1, we used prime factorization to find the greatest common factor of two numbers. To write the **prime factorization** of a whole number, write the number as the product of prime factors. Use exponential notation for repeated factors.

EXAMPLE 6 Write the prime factorizations of 12 and 18.

SOLUTION ▶

12		18
$= 2 \cdot 6$	Factor.	$= 2 \cdot 9$
$= 2 \cdot \mathbf{2} \cdot \mathbf{3}$	Factor again; all factors are prime.	$= 2 \cdot \mathbf{3} \cdot \mathbf{3}$
$= \mathbf{2^2} \cdot 3$	Rewrite in exponential notation.	$= 2 \cdot \mathbf{3^2}$

The **least common multiple** of two or more whole numbers is the smallest number that is a multiple of all of the numbers. To find the least common multiple using prime factorization, write the prime factorization of each number. Identify prime factors and the *greatest* exponent on the factor. The least common multiple is the product of these prime factors with these exponents.

EXAMPLE 7 Use prime factorization to find the least common multiple of 24 and 90.

SOLUTION ▶

24		90
$= 4 \cdot 6$	Factor.	$= 9 \cdot 10$
$= 2 \cdot 2 \cdot 2 \cdot 3$	Factor again; all factors are prime.	$= 3 \cdot 3 \cdot 2 \cdot 5$
$= \mathbf{2^3} \cdot 3$	Rewrite in exponential notation.	$= 2 \cdot \mathbf{3^2} \cdot 5$

The least common multiple of 24 and 90

$= 2^3 \cdot 3^2 \cdot 5^1$ Identify prime factors and the greatest exponent on each factor.

$= 8 \cdot 9 \cdot 5$ Evaluate.

$= 360$ Evaluate.

We can also use prime factorization to find the least common multiple of two or more expressions.

Using Prime Factorization to Find the Least Common Multiple

1. Write the prime factorization of each expression.
2. Identify prime factors and variables and the *greatest* exponent on these factors or variables.
3. The least common multiple is the product of the prime factors and variables with these exponents.

EXAMPLE 8 Use prime factorization to find the least common multiple of $48x^5y^2$ and $54x^3y^9$.

SOLUTION ▶

$48x^5y^2$		$54x^3y^9$
$= \mathbf{6 \cdot 8} \cdot x^5y^2$	Factor.	$= \mathbf{2 \cdot 27} \cdot x^3y^9$
$= \mathbf{2 \cdot 3 \cdot 2 \cdot 4} \cdot x^5y^2$	Factor again.	$= 2 \cdot \mathbf{3 \cdot 9} \cdot x^3y^9$
$= 2 \cdot 3 \cdot 2 \cdot \mathbf{2 \cdot 2} \cdot x^5y^2$	Factor again; all factors are prime.	$= 2 \cdot 3 \cdot \mathbf{3 \cdot 3} \cdot x^3y^9$
$= \mathbf{2^4 \cdot 3^1} \cdot x^5y^2$	Rewrite in exponential notation.	$= \mathbf{2^1 \cdot 3^3} \cdot x^3y^9$

The least common multiple of $48x^5y^2$ and $54x^3y^9$

$= 2^4 \cdot 3^3 \cdot x^5y^9$ Identify prime factors and the greatest exponent on each factor.

$= \mathbf{16 \cdot 27} \cdot x^5y^9$ Simplify.

$= \mathbf{432}x^5y^9$ Simplify.

In Example 8, the least common multiple is $16 \cdot 27 \cdot x^5y^9$. We simplified this expression by multiplying the numbers 16 and 27. In the next example, the least common multiple is $x(x + 2)(x + 3)$. Do not multiply the factors.

EXAMPLE 9 Use prime factorization to find the least common multiple of $x^2 + 5x + 6$ and $x^2 + 3x$.

SOLUTION ▶

$x^2 + 5x + 6$		$x^2 + 3x$
$= (x + 3)(x + 2)$	Factor; all factors are prime.	$= x(x + 3)$
$= (x + 3)^1(x + 2)^1$	Rewrite in exponential notation.	$= x^1(x + 3)^1$

The least common multiple of $x^2 + 5x + 6$ and $x^2 + 3x$

$= x^1(x + 2)^1(x + 3)^1$ Identify prime factors and the greatest exponent on each factor.

$= x(x + 2)(x + 3)$ Do not multiply the factors.

In the next example, the least common multiple is $(x - 6)(x + 6)(x + 6)$. In exponential notation, the least common multiple is $(x - 6)(x + 6)^2$. Do not multiply the factors.

EXAMPLE 10 Use prime factorization to find the least common multiple of $x^2 - 36$ and $x^2 + 12x + 36$.

SOLUTION ▶

$x^2 - 36$		$x^2 + 12x + 36$
$= (x - 6)(x + 6)$	Factor; all factors are prime.	$= (x + 6)(x + 6)$
$= (x - 6)^1(x + 6)^1$	Rewrite in exponential notation.	$= (x + 6)^2$

The least common multiple of $x^2 - 36$ and $x^2 + 12x + 36$

$= (x - 6)^1(x + 6)^2$ Identify prime factors and the greatest exponent on each factor.

$= (x - 6)(x + 6)^2$ Do not multiply the factors.

Practice Problems

7. Use prime factorization to find the least common multiple of 25 and 80.

For problems 8–12, use prime factorization to find the least common multiple of the expressions.

8. $24x$; $16xy$
9. $32x^2y$; $12xy^2$
10. $x^2 + 5x + 6$; $x^2 - 9$
11. $x^2 - 8x + 16$; $x^2 - 16$
12. $p^3 + 2p^2 + p$; $p^2 + 5p + 4$

7.3 Vocabulary Practice

Match the term with its description.

1. The result of multiplying these together is a product.
2. The result of addition
3. The result of subtraction
4. A polynomial whose only factors are itself and 1
5. For the expressions $9xy$ and $2y^2$, this is $18xy^2$.
6. A fraction in which the numerator and denominator have no common factors
7. In x^2, x is an example of this.
8. In x^2, 2 is an example of this.
9. A number whose only factors are itself and 1
10. For 24, this is $2^3 \cdot 3$.

A. base
B. difference
C. exponent
D. factors
E. least common multiple
F. lowest terms
G. prime factorization
H. prime number
I. prime polynomial
J. sum

7.3 Exercises

Follow your instructor's guidelines for showing your work.

For exercises 1–4, evaluate.

1. $\dfrac{9}{100} + \dfrac{3}{100}$
2. $\dfrac{9}{50} + \dfrac{3}{50}$
3. $\dfrac{12}{35} - \dfrac{2}{35}$
4. $\dfrac{16}{21} - \dfrac{4}{21}$

For exercises 5–48, simplify.

5. $\dfrac{2}{x+8} + \dfrac{8}{x+8}$
6. $\dfrac{3}{x+5} + \dfrac{5}{x+5}$
7. $\dfrac{15}{x-9} - \dfrac{6}{x-9}$
8. $\dfrac{14}{x-5} - \dfrac{9}{x-5}$
9. $\dfrac{6m}{m+9} - \dfrac{3m}{m+9}$
10. $\dfrac{8p}{p+10} - \dfrac{3p}{p+10}$
11. $\dfrac{4n}{n+3} + \dfrac{n}{n+3}$
12. $\dfrac{6w}{w+2} + \dfrac{w}{w+2}$
13. $\dfrac{a^2+6a}{a+4} + \dfrac{8}{a+4}$
14. $\dfrac{z^2+9z}{z+6} + \dfrac{18}{z+6}$
15. $\dfrac{r^2-12r}{r+2} - \dfrac{28}{r+2}$
16. $\dfrac{w^2-10w}{w+2} - \dfrac{24}{w+2}$
17. $\dfrac{n^2}{n^2+3n+2} - \dfrac{1}{n^2+3n+2}$
18. $\dfrac{u^2}{u^2+6u+8} - \dfrac{4}{u^2+6u+8}$
19. $\dfrac{2c}{c^2+3c} + \dfrac{7c}{c^2+3c}$
20. $\dfrac{9d}{d^2+2d} + \dfrac{8d}{d^2+2d}$
21. $\dfrac{w^2}{w+8} - \dfrac{64}{w+8}$
22. $\dfrac{k^2}{k+7} - \dfrac{49}{k+7}$
23. $\dfrac{w}{w^2-64} - \dfrac{8}{w^2-64}$
24. $\dfrac{k}{k^2-49} - \dfrac{7}{k^2-49}$
25. $\dfrac{w}{w^2+64} - \dfrac{8}{w^2+64}$
26. $\dfrac{k}{k^2+49} - \dfrac{7}{k^2+49}$
27. $\dfrac{x^2}{x-9} - \dfrac{7x+18}{x-9}$
28. $\dfrac{x^2}{x-7} - \dfrac{3x+28}{x-7}$
29. $\dfrac{z^2}{z+3} - \dfrac{5z+24}{z+3}$
30. $\dfrac{a^2}{a+6} - \dfrac{5a+66}{a+6}$
31. $\dfrac{c^2}{2c-16} - \dfrac{10c-16}{2c-16}$
32. $\dfrac{v^2}{2v-14} - \dfrac{9v-14}{2v-14}$
33. $\dfrac{x^2}{x^2-x-12} - \dfrac{2x+15}{x^2-x-12}$
34. $\dfrac{x^2}{x^2+3x-28} - \dfrac{10x-24}{x^2+3x-28}$
35. $\dfrac{2n^2}{2n^2-11n-21} - \dfrac{-5n-3}{2n^2-11n-21}$
36. $\dfrac{2w^2}{2w^2-11w-6} - \dfrac{-5w-2}{2w^2-11w-6}$
37. $\dfrac{x^3+7x^2}{x^3-9x} - \dfrac{30x}{x^3-9x}$
38. $\dfrac{y^3+5y^2}{y^3-16y} - \dfrac{36y}{y^3-16y}$
39. $\dfrac{4p^2+36p}{12p^2+36p+24} + \dfrac{32}{12p^2+36p+24}$

40. $\dfrac{3k^2 + 21k}{12k^2 + 36k + 24} + \dfrac{18}{12k^2 + 36k + 24}$

41. $\dfrac{2a^2}{3a^3 + 24a^2 + 45a} - \dfrac{10a + 48}{3a^3 + 24a^2 + 45a}$

42. $\dfrac{2c^2}{3c^3 + 18c^2 + 24c} - \dfrac{6c + 56}{3c^3 + 18c^2 + 24c}$

43. $\dfrac{10w^2}{5w^2 + 50w + 125} - \dfrac{250}{5w^2 + 50w + 125}$

44. $\dfrac{8p^2}{4p^2 + 32p + 64} - \dfrac{128}{4p^2 + 32p + 64}$

45. $\dfrac{3y^2}{5y^2 + 60y + 180} - \dfrac{12y + 180}{5y^2 + 60y + 180}$

46. $\dfrac{2x^2}{3x^2 + 24x + 48} - \dfrac{8x + 64}{3x^2 + 24x + 48}$

47. $\dfrac{2v^2}{2v^2 + 5v - 12} + \dfrac{13v}{2v^2 + 5v - 12} - \dfrac{24}{2v^2 + 5v - 12}$

48. $\dfrac{3r^2}{3r^2 + 17r - 6} + \dfrac{26r}{3r^2 + 17r - 6} - \dfrac{9}{3r^2 + 17r - 6}$

For exercises 49–52, simplify.

49. $\dfrac{n^3}{n^2 + n - 12} - \dfrac{27}{n^2 + n - 12}$

50. $\dfrac{z^3}{z^2 + 5z - 14} - \dfrac{8}{z^2 + 5z - 14}$

51. $\dfrac{k^3}{k^2 + 14k + 40} + \dfrac{64}{k^2 + 14k + 40}$

52. $\dfrac{m^3}{m^2 + 12m + 27} + \dfrac{27}{m^2 + 12m + 27}$

53. Explain why we cannot add or subtract fractions with different denominators.

54. A student is simplifying $\dfrac{x + 3}{x + 4}$. He thinks that the x in the numerator and the x in the denominator are common factors and that the expression will simplify to $\dfrac{3}{4}$. Explain why he cannot simplify the expression in this way.

For exercises 55–86, use prime factorization to find the least common multiple.

55. $12; 20$

56. $15; 20$

57. $x; 4x^2$

58. $c; 5c^2$

59. $6c^2d; 21cd^2$

60. $12a^2b; 18ab^2$

61. $n + 4; 3n - 27$

62. $p - 8; 6p + 12$

63. $b^2 - 7b; b^2 - 2b$

64. $k^2 - 9k; k^2 - 3k$

65. $x^2 - 36; x^2 + 12x + 36$

66. $x^2 - 25; x^2 + 10x + 25$

67. $z^2 + z - 12; z^2 - 5z + 6$

68. $a^2 + a - 30; a^2 - 7a + 10$

69. $x^3 + 6x^2; x^2 - 36$

70. $b^3 + 7b^2; b^2 - 49$

71. $3w^2 - 9w - 30; 6w^2 - 24w - 30$

72. $2c^2 - 2c - 24; 6c^2 - 18c - 24$

73. $28x^2y^5; 84xy^3$

74. $21x^3y^5; 84xy^2$

75. $60xy; 45ac$

76. $60ab; 36xy$

77. $120n^2p^2; 180n^5p^2$

78. $168n^2w^6; 252n^2w^2$

79. $315a^{10}b^2c; 117a^5b^3c^2$

80. $140a^2b^{11}c^2; 52a^3b^6c$

81. $x^3 + 6x^2 + 9x; x^3 + 7x^2 + 12x$

82. $x^3 + 4x^2 + 4x; x^3 + 5x^2 + 6x$

83. $6x^2 - 150; 4x^2 + 24x + 20$

84. $4x^2 - 36; 6x^2 + 42x + 72$

85. $10x^4 + 20x^3 + 10x^2; 8x^3 + 16x^2 + 8x$

86. $6x^4 + 24x^3 + 24x^2; 9x^3 + 36x^2 + 36x$

Problem Solving: Practice and Review

Follow your instructor's guidelines for using the five steps as outlined in Section 1.3, p. 55.

87. Find the number of 1000 8-year-old children in the United States who have cerebral palsy. Round to the nearest tenth.

Cerebral palsy is the most common motor disability in childhood, affecting approximately 1 in 303 8-year-old children in the U.S. (*Source:* www.cdc.gov, Aug. 29, 2011)

88. Nonprofit arts and culture organizations' spending on services, supplies, and employees directly and indirectly creates jobs. If 1 job = 1 FTE (full time employee), find the number of jobs created by \$350,000 in spending by an arts organization. Round to the nearest tenth.

To determine the organization's total economic impact on full-time equivalent (FTE) employment in the City of Atlanta . . .

1. Determine the amount spent by the nonprofit arts and culture organization;
2. Divide the total expenditure by 100,000; and
3. Multiply that figure by the FTE employment ratio per \$100,000 for the City of Atlanta, 3.01. (*Source: Arts and Economic Prosperity III*, Americans for the Arts, 2007)

89. Find the number of licensed drivers in the United States in 2015. Round to the nearest million.

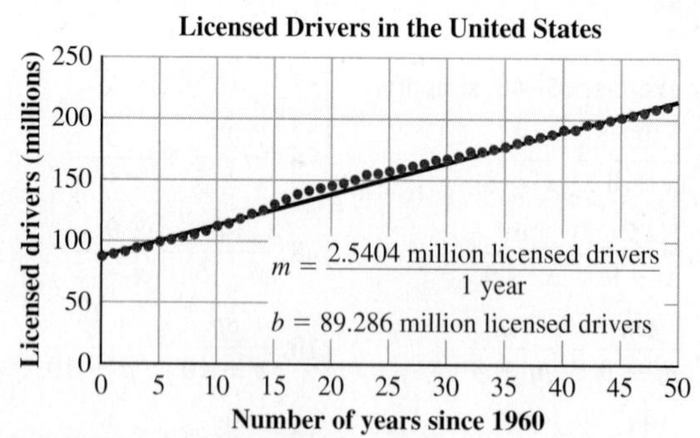

Source: www.fhwa.dot.gov, March 2011

90. An athlete and her advisor are planning her schedule for the next term. She must take at least 12 credits to keep her scholarship. Her coach does not want her to take more than 18 credits. She must take at least 6 credits in courses that meet the requirements of the general education core curriculum. Let x = credits in general education, and let y = credits outside of general education.
   a. Write four inequalities that describe the constraints on the credits the athlete can take.
   b. Graph the constraints.

Find the Mistake

For exercises 91–94, the completed problem has one mistake.
(a) Describe the mistake in words, or copy down the whole problem and highlight or circle the mistake.
(b) Do the problem correctly.

91. Problem: Evaluate: $\frac{11}{15} + \frac{2}{15}$

Incorrect Answer: $\frac{11}{15} + \frac{2}{15}$

$= \frac{13}{30}$

92. Problem: Simplify: $\frac{x^2 + 7x}{x^2 - 4x - 45} + \frac{10}{x^2 - 4x - 45}$

Incorrect Answer: $\frac{x^2 + 7x}{x^2 - 4x - 45} + \frac{10}{x^2 - 4x - 45}$

$= \frac{x^2 + 7x + 10}{x^2 - 4x - 45}$

93. Problem: Simplify: $\frac{x^2}{x^2 - 5x - 24} - \frac{3x + 40}{x^2 - 5x - 24}$

Incorrect Answer: $\frac{x^2}{x^2 - 5x - 24} - \frac{3x + 40}{x^2 - 5x - 24}$

$= \frac{x^2 - 3x + 40}{x^2 - 5x - 24}$

94. Problem: Simplify: $\frac{u^2}{u^2 - 16} - \frac{6u + 40}{u^2 - 16}$

Incorrect Answer: $\frac{u^2}{u^2 - 16} - \frac{6u + 40}{u^2 - 16}$

$= \frac{u^2 - (6u + 40)}{u^2 - 16}$

$= \frac{u^2 - 6u - 40}{u^2 - 16}$

$= \frac{(u + 10)(u - 4)}{(u + 4)(u - 4)}$

$= \frac{u + 10}{u + 4}$

Review

For exercises 95–98, evaluate.

95. $\frac{3}{4} + \frac{5}{6}$

96. $\frac{5}{7} - \frac{2}{9}$

97. $\frac{11}{12} - \frac{3}{4}$

98. $\frac{1}{3} + \frac{1}{2} + \frac{1}{4}$

SUCCESS IN COLLEGE MATHEMATICS

99. Many students have some anxiety about their classes. They worry about giving speeches, taking tests, writing papers, or simply getting all of their class work done. Worrying can be motivating, or worrying can be paralyzing. Describe how worrying affects you.

$$\frac{a}{b} + \frac{c}{b} = \frac{a + c}{b}$$

We can add or subtract only rational numbers that have the same denominator. This is also true for rational expressions. In this section, we will find least common denominators and combine rational expressions.

SECTION 7.4 Combining Rational Expressions with Different Denominators

After reading the text, working the practice problems, and completing assigned exercises, you should be able to:

1. Find the least common denominator of two or more rational expressions.
2. Build an equivalent rational expression with a given denominator.
3. Add or subtract rational expressions with different denominators.

Least Common Denominator

In Section 7.3, we used prime factorization to find the least common *multiple* of two expressions. When these expressions are denominators, we are finding the least common *denominator* (LCD) of the expressions.

EXAMPLE 1 Use prime factorization to find the least common denominator of $\frac{7}{36}$ and $\frac{1}{56}$.

SOLUTION

36		56
$= 4 \cdot 9$	Factor.	$= 8 \cdot 7$
$= 2 \cdot 2 \cdot 3 \cdot 3$	Factor again; all factors are prime.	$= \mathbf{2 \cdot 2 \cdot 2} \cdot 7$
$= 2^2 \cdot 3^2$	Rewrite in exponential notation.	$= 2^3 \cdot 7^1$

The least common denominator of $\frac{7}{36}$ and $\frac{1}{56}$

$= 2^3 \cdot 3^2 \cdot 7^1$ Identify prime factors and the greatest exponent on each factor.

$= 504$ Evaluate.

We can also use prime factorization to find a least common denominator that includes variables. In Example 2, the denominators are $24x^2y^3z$ and $40xy^4z^2$. We write the prime factorization of each denominator, identify the prime factors, and choose the greatest exponent that appears on either of the factorizations.

EXAMPLE 2 Use prime factorization to find the least common denominator of $\frac{1}{24x^2y^3z}$ and $\frac{1}{40xy^4z^2}$.

SOLUTION

$24x^2y^3z$		$40xy^4z^2$
$= \mathbf{4 \cdot 6} \cdot x^2 \cdot y^3 \cdot z$	Factor.	$= \mathbf{4 \cdot 10} \cdot x \cdot y^4 \cdot z^2$
$= \mathbf{2 \cdot 2 \cdot 2 \cdot 3} \cdot x^2 \cdot y^3 \cdot z$	Factor again; all factors are prime.	$= \mathbf{2 \cdot 2 \cdot 2 \cdot 5} \cdot x \cdot y^4 \cdot z^2$
$= \mathbf{2^3 \cdot 3^1} \cdot x^2 \cdot y^3 \cdot z$	Rewrite in exponential notation.	$= \mathbf{2^3 \cdot 5^1} \cdot x \cdot y^4 \cdot z^2$

The least common denominator of $\frac{1}{24x^2y^3z}$ and $\frac{1}{40xy^4z^2}$

$= 2^3 \cdot 3 \cdot 5 \cdot x^2 \cdot y^4 \cdot z^2$ Identify prime factors and the greatest exponent on each factor.

$= \mathbf{8 \cdot 3 \cdot 5} \cdot x^2 \cdot y^4 \cdot z^2$ Simplify.

$= \mathbf{120}x^2y^4z^2$ Simplify.

Using Prime Factorization to Find the Least Common Denominator

1. Write the prime factorization of each denominator.
2. Identify prime factors and variables and the *greatest* exponent on these factors or variables.
3. The least common denominator is the product of the prime factors and variables with these exponents.

In the next example, we factor a binomial that is a difference of squares and a trinomial.

EXAMPLE 3 Use prime factorization to find the least common denominator of $\frac{1}{x^2 - 16}$ and $\frac{3}{x^2 + 6x + 8}$.

SOLUTION

$x^2 - 16$		$x^2 + 6x + 8$
$= (x + 4)(x - 4)$	Factor; all factors are prime.	$= (x + 2)(x + 4)$
$= (x + 4)^1(x - 4)^1$	Rewrite in exponential notation.	$= (x + 2)^1(x + 4)^1$

The least common denominator of $\frac{1}{x^2 - 16}$ and $\frac{3}{x^2 + 6x + 8}$

$= (x - 4)(x + 4)(x + 2)$ Identify prime factors and the greatest exponent on each factor

We can also use prime factorization to find the least common denominator of three or more rational expressions.

EXAMPLE 4 Use prime factorization to find the least common denominator of $\frac{1}{2x - 6}$, $\frac{1}{x^2 - 6x + 9}$, and $\frac{1}{x^2 - 9}$.

SOLUTION ▶

$2x - 6$

$= 2(x - 3)$ Factor; all factors are prime.

$= 2^1(x - 3)^1$ Rewrite in exponential notation.

$x^2 - 6x + 9$

$= (x - 3)(x - 3)$

$= (x - 3)^2$

$x^2 - 9$

$= (x - 3)(x + 3)$ Factor; all factors are prime.

$= (x - 3)^1(x + 3)^1$ Rewrite in exponential notation.

The least common denominator of $\frac{1}{2x - 6}$, $\frac{1}{x^2 - 6x + 9}$, and $\frac{1}{x^2 - 9}$

$= 2(x - 3)^2(x + 3)$ Identify prime factors and the greatest exponent on each factor.

Practice Problems

For problems 1–10, use prime factorization to find the least common denominator.

1. $\frac{1}{72}; \frac{1}{48}$

2. $\frac{1}{30x}; \frac{1}{50x^2}$

3. $\frac{1}{49ab^2}; \frac{1}{25c}$

4. $\frac{1}{32a^3b^2c^5}; \frac{1}{20a^6bc}$

5. $\frac{3}{x^2 + 9x + 18}; \frac{2}{x^2 + 7x + 6}$

6. $\frac{1}{2x + 10}; \frac{3}{2x^2 - 50}$

7. $\frac{4}{x^3 - 7x^2}; \frac{5}{x^2 - 9x + 14}$

8. $\frac{1}{x - 3}; \frac{1}{x + 4}$

9. $\frac{1}{6x^2}; \frac{1}{3x^3 + 15x^2 + 6x}$

10. $\frac{1}{x^2 + 5x + 6}; \frac{1}{x^2 - 9}; \frac{1}{x^2 + 6x + 9}$

Equivalent Fractions

The rectangle in Figure 1 represents a cake cut into 3 pieces. One piece of cake is left in the pan. It is $\frac{1}{3}$ of the cake. The rectangle in Figure 2 represents a cake that was originally cut into 4 pieces. One piece of cake is left in the pan. It is $\frac{1}{4}$ of the cake. The remaining pieces of cake are not the same size. Since we can add only pieces that are the same size, we need to cut the remaining cake into equal size pieces (Figure 3). Each remaining piece of cake is now $\frac{1}{12}$ of a cake. Adding the numerators and not changing the denominators, $\frac{4}{12}$ cake + $\frac{3}{12}$ cake equals $\frac{7}{12}$ cake.

$\frac{1}{3}$ cake

Figure 1

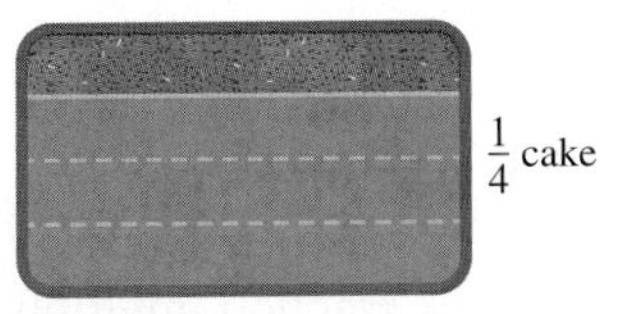

$\frac{1}{4}$ cake

Figure 2

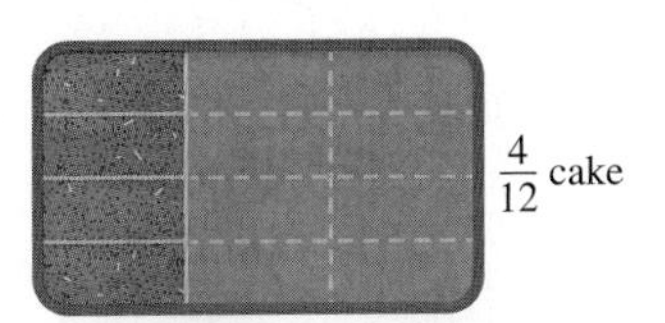

$\frac{4}{12}$ cake

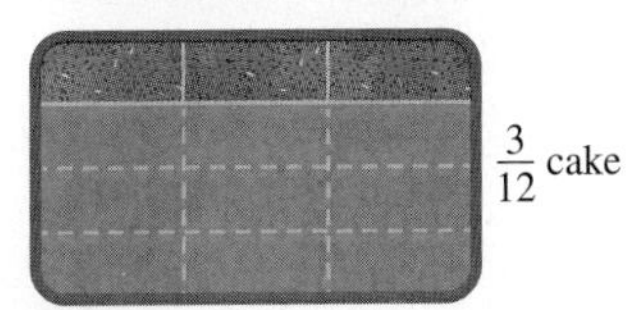

$\frac{3}{12}$ cake

Figure 3

The fractions $\frac{1}{3}$ and $\frac{4}{12}$ are equivalent fractions. To rewrite $\frac{1}{3}$ as an equivalent fraction with a denominator of 12, multiply by a fraction that is equal to 1. Choose the fraction equal to 1 so that the denominator of this fraction multiplied by the denominator of the original fraction equals the denominator of the new fraction.

EXAMPLE 5 Rewrite $\frac{1}{3}$ as an equivalent fraction with a denominator of 12.

SOLUTION ▶

$\frac{1}{3}$

$= \frac{1}{3} \cdot \frac{4}{4}$ Multiply by a fraction equal to 1; choose $\frac{4}{4}$ because $3 \cdot 4 = 12$.

$= \frac{4}{12}$ Evaluate.

In the next example, the original fraction is $\frac{11}{42}$. To build an equivalent fraction with a denominator of 210, we multiply by $\frac{5}{5}$. We choose to multiply by $\frac{5}{5}$ because $42 \cdot 5 = 210$.

EXAMPLE 6 Rewrite $\frac{11}{42}$ as an equivalent fraction with a denominator of 210.

SOLUTION ▶

$\frac{11}{42}$ $42 \cdot 5 = 210$

$= \frac{11}{42} \cdot \frac{5}{5}$ Multiply by a fraction equal to 1.

$= \frac{55}{210}$ Evaluate.

We use the same process to build equivalent fractions that include variables. In the next example, the original fraction is $\frac{7}{3a}$. To build an equivalent fraction with a denominator of $12ab$, we multiply by $\frac{4b}{4b}$. We choose to multiply by $\frac{4b}{4b}$ because $3a \cdot 4b = 12ab$.

EXAMPLE 7 Rewrite $\frac{7}{3a}$ as an equivalent expression with a denominator of $12ab$.

SOLUTION ▶

$\frac{7}{3a}$

$= \frac{7}{3a} \cdot \frac{4b}{4b}$ Multiply by a fraction equal to 1.

$= \frac{28b}{12ab}$ Simplify.

To rewrite a rational expression such as $\frac{x}{x^2 + 5x + 6}$ as an equivalent fraction, first factor the denominator completely. To simplify the new fraction, multiply the factors in the numerator. Ask your instructor whether you should leave the denominator factored or multiply any remaining factors in the denominator.

EXAMPLE 8 Rewrite $\frac{x}{x^2 + 5x + 6}$ as an equivalent expression with a denominator of $(x + 2)(x + 3)(x - 3)$.

SOLUTION ▶

$$\frac{x}{x^2 + 5x + 6}$$

$$= \frac{x}{\mathbf{(x + 2)(x + 3)}}$$ Factor the denominator.

$$= \frac{x}{(x + 2)(x + 3)} \cdot \frac{(x - 3)}{(x - 3)}$$ Multiply by a fraction equal to 1.

$$= \frac{x(x - 3)}{(x + 2)(x + 3)(x - 3)}$$ Simplify.

$$= \frac{\mathbf{x^2 - 3x}}{(x + 2)(x + 3)(x - 3)}$$ Multiply factors in the numerator.

Building an Equivalent Rational Expression with a Different Denominator

1. Factor the denominator of the expression completely.
2. Identify the factors that are needed to build the new denominator.
3. Multiply by a fraction that is equal to 1. The numerator and denominator of this fraction are the identified factors.
4. Multiply factors in the numerator; leave the denominator factored.

Practice Problems

For problems 11–17, rewrite each expression as an equivalent expression with the given denominator.

11. $\frac{3}{20}$; 80 **12.** $\frac{7}{8x}$; $24xy^2$

13. $\frac{x}{10(x + 4)}$; $30(x + 4)(x + 2)$

14. $\frac{5}{(x - 2)(x + 6)}$; $x(x - 2)(x + 6)$

15. $\frac{2}{x^2 - 49}$; $(x - 7)(x + 7)(x + 1)$

16. $\frac{3}{x^2 + 5x + 6}$; $(x + 3)(x + 2)(x - 5)$

17. $\frac{2}{5x^3 - 10x^2}$; $20x^2(x - 2)(x + 1)$

Addition and Subtraction

We can add or subtract only rational numbers that have the same denominator. To add or subtract fractions with different denominators, identify the least common denominator, rewrite both fractions as equivalent fractions with the new denominator, add the numerators, and do not change the denominator. If the sum or difference is not in lowest terms, simplify.

EXAMPLE 9 Evaluate: $\frac{5}{18} + \frac{7}{60}$

SOLUTION ▶ Find the least common denominator.

18	60	Least common denominator
$= 2 \cdot 3^2$	$= 2^2 \cdot 3 \cdot 5^1$	$= 2^2 \cdot 3^2 \cdot 5^1$
		$= 180$

$\frac{5}{18} + \frac{7}{60}$

$= \frac{5}{18} \cdot \frac{10}{10} + \frac{7}{60} \cdot \frac{3}{3}$ Multiply by fractions equal to 1.

$= \frac{50}{180} + \frac{21}{180}$ Evaluate.

$= \frac{50 + 21}{180}$ Add numerators; the denominator does not change.

$= \frac{\mathbf{71}}{\mathbf{180}}$ Evaluate.

Since the numerator and denominator of $\frac{71}{180}$ have no common factors, this expression is simplified.

We use the same process to add and subtract rational expressions.

EXAMPLE 10 Simplify: $\frac{7}{5x} + \frac{3}{4x^2}$

SOLUTION ▶ Find the least common denominator.

$5x$	$4x^2$	Least common denominator
$= 5 \cdot x$	$= 2^2 \cdot x^2$	$= 2^2 \cdot 5^1 \cdot x^2$
		$= \mathbf{20x^2}$

$\frac{7}{5x} + \frac{3}{4x^2}$

$= \frac{7}{5x} \cdot \frac{4x}{4x} + \frac{3}{4x^2} \cdot \frac{5}{5}$ Multiply by fractions equal to 1.

$= \frac{28x}{20x^2} + \frac{15}{20x^2}$ Simplify.

$= \frac{28x + 15}{20x^2}$ Add the numerators; the denominator does not change.

Since $28x + 15$ is a prime polynomial and the numerator and denominator of $\frac{28x + 15}{20x^2}$ have no common factors, this expression is simplified.

In the next example, the least common denominator is $(w - 3)(w + 3)(w + 5)$. To build equivalent fractions with this common denominator, we multiply by fractions that are equal to 1, $\frac{w + 5}{w + 5}$ and $\frac{w - 3}{w - 3}$.

EXAMPLE 11 Simplify: $\frac{w}{w^2 - 9} + \frac{6w + 1}{w^2 + 8w + 15}$

SOLUTION ▶ Find the least common denominator.

$w^2 - 9$	$w^2 + 8w + 15$	Least common denominator
$= (w - 3)(w + 3)$	$= (w + 5)(w + 3)$	$= (w - 3)(w + 3)(w + 5)$

$\frac{w}{w^2 - 9} + \frac{6w + 1}{w^2 + 8w + 15}$

$= \frac{w}{\mathbf{(w - 3)(w + 3)}} + \frac{6w + 1}{\mathbf{(w + 5)(w + 3)}}$ Factor denominators.

$= \frac{w}{(w - 3)(w + 3)} \cdot \frac{(w + 5)}{(w + 5)} + \frac{(6w + 1)}{(w + 5)(w + 3)} \cdot \frac{(w - 3)}{(w - 3)}$ Multiply by fractions equal to 1.

$= \frac{\mathbf{w \cdot} w \mathbf{+ w \cdot} 5}{(w - 3)(w + 3)(w + 5)} + \frac{\mathbf{6w}(w - 3) \mathbf{+ 1}(w - 3)}{(w + 5)(w + 3)(w - 3)}$ Distributive property.

$= \frac{\mathbf{w^2 + 5w}}{(w - 3)(w + 3)(w + 5)} + \frac{6w(w - 3) + 1(w - 3)}{(w + 5)(w + 3)(w - 3)}$ Simplify.

$= \frac{w^2 + 5w}{(w - 3)(w + 3)(w + 5)} + \frac{\mathbf{6w}(w) \mathbf{+ 6w}(-3) \mathbf{+ 1}(w) \mathbf{+ 1}(-3)}{(w + 5)(w + 3)(w - 3)}$ Distributive property.

$= \frac{w^2 + 5w}{(w - 3)(w + 3)(w + 5)} + \frac{\mathbf{6w^2 - 18w + w - 3}}{(w + 5)(w + 3)(w - 3)}$ Simplify.

$= \frac{w^2 + 5w}{(w - 3)(w + 3)(w + 5)} + \frac{6w^2 \mathbf{- 17w} - 3}{(w + 5)(w + 3)(w - 3)}$ Combine like terms.

$= \frac{w^2 + 5w + 6w^2 - 17w - 3}{(w - 3)(w + 3)(w + 5)}$ Add the numerators.

$= \frac{\mathbf{7w^2 - 12w} - 3}{(w - 3)(w + 3)(w + 5)}$ Combine like terms.

The numerator is a quadratic trinomial in the form $ax^2 + bx + c$. If the numerator is prime, the expression is simplified. If the discriminant, $b^2 - 4ac$, of this trinomial is not a perfect square or if it is less than 0, the trinomial is prime and the expression $\frac{7w^2 - 12w - 3}{(w - 3)(w + 3)(w + 5)}$ is simplified.

$b^2 - 4ac$ The discriminant; $a = 7, b = -12, c = -3$

$= (\mathbf{-12})^2 - 4(\mathbf{7})(\mathbf{-3})$ Replace a, b, and c.

$= 228$ Evaluate.

$\sqrt{228}$

$= 15.0996...$ Since 228 is not a perfect square, $7w^2 - 12w - 3$ is prime.

Since $7w^2 - 12w - 3$ is a prime polynomial and the numerator and denominator of $\frac{7w^2 - 12w - 3}{(w - 3)(w + 3)(w + 5)}$ have no common factors, this expression is simplified.

Adding or Subtracting Rational Expressions

1. Factor the denominators completely.
2. Find the least common denominator.
3. To build equivalent rational expressions with the least common denominator, multiply by a fraction that is equal to 1.
4. Combine numerators. The denominator stays the same.
5. Factor the numerator completely. Simplify common factors.
6. Multiply any factors in the numerator. Either multiply the factors in the denominator or leave the denominator factored.

When subtracting rational expressions, use parentheses around the numerator of the second expression.

EXAMPLE 12 Simplify: $\dfrac{h}{h^2 + 4h + 3} - \dfrac{3}{h^2 - 4h - 5}$

SOLUTION ▶ Find the least common denominator.

$h^2 + 4h + 3$	$h^2 - 4h - 5$	Least common denominator
$= (h + 3)^1(h + 1)^1$	$= (h - 5)^1(h + 1)^1$	$= (h + 3)(h + 1)(h - 5)$

$\dfrac{h}{h^2 + 4h + 3} - \dfrac{3}{h^2 - 4h - 5}$

$= \dfrac{h}{(h + 3)(h + 1)} \cdot \dfrac{(h - 5)}{(h - 5)} - \dfrac{3}{(h - 5)(h + 1)} \cdot \dfrac{(h + 3)}{(h + 3)}$ Multiply by fractions equal to 1.

$= \dfrac{\mathbf{h \cdot h + h(-5)}}{(h + 3)(h + 1)(h - 5)} - \dfrac{\mathbf{3 \cdot h + 3(3)}}{(h - 5)(h + 1)(h + 3)}$ Distributive property.

$= \dfrac{\mathbf{h^2 - 5h}}{(h + 3)(h + 1)(h - 5)} - \dfrac{\mathbf{3h + 9}}{(h - 5)(h + 1)(h + 3)}$ Simplify.

$= \dfrac{h^2 - 5h - (3h + 9)}{(h + 3)(h + 1)(h - 5)}$ Subtract numerators; use parentheses.

$= \dfrac{h^2 - 5h - \mathbf{1}(3h + 9)}{(h + 3)(h + 1)(h - 5)}$ Multiply by 1.

$= \dfrac{h^2 - 5h \mathbf{- 1(3h) - 1(9)}}{(h + 3)(h + 1)(h - 5)}$ Distributive property.

$= \dfrac{h^2 - 5h \mathbf{- 3h - 9}}{(h + 3)(h + 1)(h - 5)}$ Simplify.

$= \dfrac{h^2 \mathbf{- 8h} - 9}{(h + 3)(h + 1)(h - 5)}$ Combine like terms.

$= \dfrac{\mathbf{(h - 9)}\cancel{\mathbf{(h + 1)}}}{(h + 3)\cancel{(h + 1)}(h - 5)}$ Factor numerator; find common factors.

$= \dfrac{h - 9}{(h + 3)(h - 5)}$ Simplify.

Since $h - 9$ is a prime polynomial and the numerator and denominator of $\dfrac{h - 9}{(h + 3)(h - 5)}$ have no common factors, this expression is simplified.

In the next example, each denominator is one term, a monomial.

EXAMPLE 13 Simplify: $\dfrac{5}{6ab^2} - \dfrac{4}{15a^3b}$

SOLUTION ▶ Find the least common denominator.

$6ab^2$	$15a^3b$	Least common denominator
$= \mathbf{2 \cdot 3} \cdot a \cdot b^2$	$= \mathbf{3 \cdot 5} \cdot a^3 \cdot b$	$= 2 \cdot 3 \cdot 5 \cdot a^3 \cdot b^2$
		$= \mathbf{30}a^3b^2$

$$\frac{5}{6ab^2} - \frac{4}{15a^3b}$$

$$= \frac{5}{6ab^2} \cdot \frac{5a^2}{5a^2} - \frac{4}{15a^3b} \cdot \frac{2b}{2b} \qquad \text{Multiply by fractions equal to 1.}$$

$$= \frac{25a^2}{30a^3b^2} - \frac{8b}{30a^3b^2} \qquad \text{Simplify.}$$

$$= \frac{25a^2 - 8b}{30a^3b^2} \qquad \text{Add numerators.}$$

Since $25a^2 - 8b$ is a prime polynomial and the numerator and denominator of $\dfrac{25a^2 - 8b}{30a^3b^2}$ have no common factors, this expression is simplified.

In the next example, the denominators of the rational expressions do not have a common factor.

EXAMPLE 14 Simplify: $\dfrac{x}{x-9} - \dfrac{4}{x+2}$

SOLUTION ▶ Find the least common denominator.

$x - 9$	$x + 2$	Least common denominator
$= (x-9)^1$	$= (x+2)^1$	$= (x-9)(x+2)$

$$\frac{x}{x-9} - \frac{4}{x+2} \qquad \text{Least common denominator: } (x-9)(x+2).$$

$$= \frac{x}{(x-9)} \cdot \frac{(x+2)}{(x+2)} - \frac{4}{x+2} \cdot \frac{(x-9)}{(x-9)} \qquad \text{Multiply by fractions that are equal to 1.}$$

$$= \frac{\mathbf{x \cdot x + x \cdot 2}}{(x-9)(x+2)} - \frac{\mathbf{4 \cdot x + 4(-9)}}{(x+2)(x-9)} \qquad \text{Distributive property.}$$

$$= \frac{\mathbf{x^2 + 2x}}{(x-9)(x+2)} - \frac{\mathbf{4x - 36}}{(x+2)(x-9)} \qquad \text{Simplify.}$$

$$= \frac{x^2 + 2x - (4x - 36)}{(x-9)(x+2)} \qquad \text{Subtract numerators; use parentheses.}$$

$$= \frac{x^2 + 2x - \mathbf{1}(4x - 36)}{(x-9)(x+2)} \qquad \text{Multiply by 1.}$$

$$= \frac{x^2 + 2x \mathbf{- 1}(4x) \mathbf{- 1}(-36)}{(x-9)(x+2)} \qquad \text{Distributive property.}$$

$$= \frac{x^2 + 2x \mathbf{- 4x + 36}}{(x-9)(x+2)} \qquad \text{Simplify.}$$

$$= \frac{x^2 \mathbf{- 2x} + 36}{(x-9)(x+2)} \qquad \text{Combine like terms.}$$

Since $x^2 - 2x + 36$ is a prime polynomial and the numerator and denominator of $\dfrac{x^2 - 2x + 36}{(x-9)(x+2)}$ have no common factors, this expression is simplified.

Practice Problems

For problems 18–22, simplify.

18. $\frac{3}{10} + \frac{4}{15}$ **19.** $\frac{5}{108xy^2} + \frac{7}{72x^2y}$ **20.** $\frac{4x + 1}{x^2 + x - 30} + \frac{3x + 2}{x^2 - 3x - 10}$

21. $\frac{a + 5}{2a^2 - 11a - 6} - \frac{3a - 7}{a^2 - 4a - 12}$ **22.** $\frac{w}{w - 4} - \frac{9}{w + 7}$

7.4 Vocabulary Practice

Match the term with its description.

1. The "top" of a fraction
2. A polynomial whose only factors are itself and 1
3. To add two fractions, the fractions must have this.
4. A fraction in which the numerator and denominator have no common factors
5. $ax^2 + bx + c$
6. $a^2 - b^2 = (a - b)(a + b)$
7. $a(b + c) = ab + ac$
8. A polynomial with one term
9. A polynomial with two terms
10. Division by zero

A. binomial
B. common denominator
C. difference of squares pattern
D. distributive property
E. lowest terms
F. monomial
G. numerator
H. prime polynomial
I. quadratic trinomial
J. undefined

7.4 Exercises

Follow your instructor's guidelines for showing your work.

For exercises 1–12, use prime factorization to find the least common denominator.

1. $\frac{3}{4}; \frac{5}{6}$

2. $\frac{3}{8}; \frac{5}{6}$

3. $\frac{5}{18x}; \frac{1}{30x^2}$

4. $\frac{7}{15y}; \frac{1}{24y^2}$

5. $\frac{2}{a^2 - 9a + 20}; \frac{7}{a^2 - 16}$

6. $\frac{5}{c^2 - 8c + 15}; \frac{6}{a^2 - 9}$

7. $\frac{2}{x^2 - 7x}; \frac{3}{3x - 21}$

8. $\frac{1}{y^2 - 4y}; \frac{1}{3y - 12}$

9. $\frac{1}{50x^2y}; \frac{1}{35xy^3z}$

10. $\frac{1}{40xy^2z}; \frac{1}{42xyz^3}$

11. $\frac{2}{x^3 + 3x^2}; \frac{1}{x^2 - 9}$

12. $\frac{3}{y^2 - 25}; \frac{2}{y^3 + 5y^2}$

For exercises 13–24, rewrite each expression as an equivalent expression with the given denominator.

13. $\frac{3}{7}$; 56

14. $\frac{5}{7}$; 42

15. $\frac{2}{9a}$; $27a^2b$

16. $\frac{2}{9b}$; $36ab^2$

17. $\frac{7}{8x + 6}$; $10(4x + 3)$

18. $\frac{4}{15x + 6}$; $12(5x + 2)$

19. $\frac{2}{x^3 - 8x^2 - 9x}$; $5x^2(x - 9)(x + 1)$

20. $\frac{3}{x^3 - 7x^2 - 8x}$; $6x^2(x - 8)(x + 1)$

21. $\frac{2p}{p^2 - 36}$; $(p + 6)(p - 6)(p + 1)$

22. $\frac{3w}{w^2 - 49}$; $(w + 7)(w - 7)(w + 1)$

23. $\frac{3c}{c^2 - 13c + 40}$; $(c - 8)(c - 2)(c - 5)$

24. $\frac{2u}{u^2 - 11u + 28}$; $(u - 7)(u - 3)(u - 4)$

25. Explain why it is necessary to use a common denominator when adding rational expressions.

26. Explain why it is necessary to use a common denominator when subtracting rational expressions.

For exercises 27–34, evaluate.

27. $\frac{1}{12} + \frac{5}{12}$

28. $\frac{1}{14} + \frac{5}{14}$

29. $\frac{9}{50} - \frac{2}{25}$

30. $\frac{7}{16} - \frac{3}{8}$

31. $\frac{5}{9} + \frac{7}{24}$

32. $\frac{5}{8} + \frac{7}{30}$

33. $\frac{15}{19} - \frac{8}{11}$

34. $\frac{12}{13} - \frac{5}{9}$

For exercises 35–86, simplify.

35. $\frac{2}{5a} - \frac{3}{9b}$

36. $\frac{3}{8n} - \frac{2}{5p}$

37. $\frac{5}{9x} + \frac{7}{24x^2}$

38. $\frac{5}{8x} + \frac{7}{30x^2}$

39. $\frac{5r}{21} - \frac{3r}{10}$

40. $\frac{6w}{25} - \frac{3w}{10}$

41. $\frac{8}{3ab^2} + \frac{5}{6a^2b}$

42. $\frac{9}{4hk^2} + \frac{5}{8h^2k}$

43. $\frac{9}{8hk^2} - \frac{k}{2h}$

44. $\frac{11}{6ab^2} - \frac{b}{3a}$

45. $\frac{2}{p-3} + \frac{5}{p+4}$

46. $\frac{3}{w-4} + \frac{7}{w+2}$

47. $\frac{12}{a+8} - \frac{2}{a-6}$

48. $\frac{11}{m+9} - \frac{4}{m-5}$

49. $\frac{d-4}{5d+30} + \frac{2}{d+6}$

50. $\frac{c-4}{6c+48} + \frac{2}{c+8}$

51. $\frac{f+1}{9f-18} - \frac{f}{f-2}$

52. $\frac{v+1}{6v-24} - \frac{v}{v-4}$

53. $\frac{4}{z^2+3z} + \frac{5}{z^2+9z}$

54. $\frac{8}{r^2+7r} + \frac{3}{r^2+2r}$

55. $\frac{1}{x^2+5x+6} + \frac{2}{x^2+10x+16}$

56. $\frac{1}{y^2+6y+8} + \frac{3}{y^2+9y+14}$

57. $\frac{4}{n^2-9n+20} - \frac{6}{n^2-2n-15}$

58. $\frac{6}{h^2-12h+35} - \frac{2}{h^2-4h-21}$

59. $\frac{p}{p-6} - \frac{36}{p^2-6p}$

60. $\frac{z}{z-8} - \frac{64}{z^2-8z}$

61. $\frac{w-2}{2w^2-w} - \frac{2}{w}$

62. $\frac{k-4}{4k^2-k} - \frac{4}{k}$

63. $\frac{8z-5}{2z^2+5z} + \frac{1}{z}$

64. $\frac{6y-1}{3y^2+y} + \frac{1}{y}$

65. $\frac{3a}{a^2+7a+10} - \frac{2a}{a^2+6a+8}$

66. $\frac{4m}{m^2+8m+12} - \frac{3m}{m^2+7m+10}$

67. $\frac{x}{x-2} - \frac{4x}{x^2-4}$

68. $\frac{x}{x-5} - \frac{10x}{x^2-25}$

69. $\frac{p}{p+5} + \frac{8p}{p^2+2p-15}$

70. $\frac{w}{w+2} + \frac{6w}{w^2-2w-8}$

71. $\frac{4}{v^2-1} - \frac{2}{v-1}$

72. $\frac{18}{c^2-9} - \frac{3}{c-3}$

73. $\frac{-14}{a^2+a-12} + \frac{2}{a-3}$

74. $\frac{-18}{z^2+4z-5} + \frac{3}{z-1}$

75. $\frac{x}{x+1} - \frac{4}{x^2+6x+5}$

76. $\frac{x}{x+2} - \frac{1}{x^2+5x+6}$

77. $\frac{y}{y+1} - \frac{2}{y^2-1}$

78. $\frac{p}{p+2} - \frac{8}{p^2-4}$

79. $\frac{3}{n+4} - \frac{n}{n^2+3n-4}$

80. $\frac{4}{u+5} - \frac{u}{u^2+4u-5}$

81. $\frac{12}{16-x^2} + \frac{x}{x+4}$

82. $\frac{10}{9-x^2} + \frac{x}{x+3}$

83. $\frac{1}{6a} + \frac{2}{3a} - \frac{3}{4a}$

84. $\frac{1}{6n} + \frac{3}{2n} - \frac{7}{4n}$

85. $\frac{1}{x^2+5x+6} + \frac{3}{x^2-9} + \frac{2}{x^2+6x+9}$

86. $\frac{1}{x^2+5x+6} + \frac{3}{x^2-4} + \frac{2}{x^2+4x+4}$

Problem Solving: Practice and Review

Follow your instructor's guidelines for using the five steps as outlined in Section 1.3, p. 55.

87. People who live with friends or family because of economic need are considered to be living "doubled up." Find the percent increase in people living doubled up from 2008 to 2010. Round to the nearest percent.

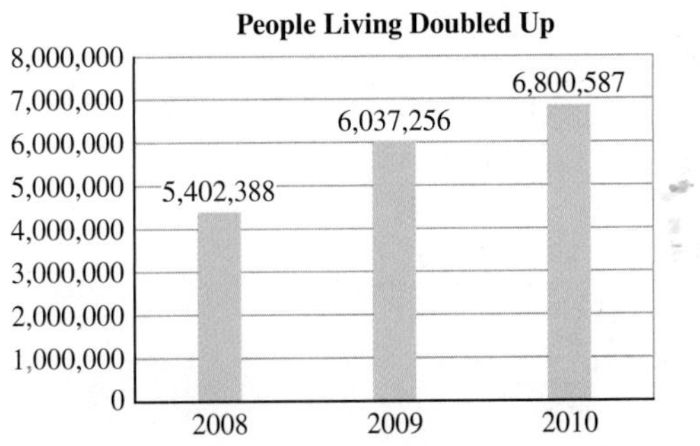

Source: www.endhomelessness.org, Jan. 17, 2012

88. Find the average pay in Austin before the jump in pay. Round to the nearest hundred.

12 of the top 20 cities for tech jobs had above average wage growth. The largest: Austin, Texas with a 13 percent jump in pay to average $89,419. (*Source:* www.diceholdings.com, Jan. 24, 2012)

89. A homeowner is comparing the price of putting a fence around the pool area in his backyard and the price of putting a fence around the entire backyard. The pool area is a rectangle that is 30 ft wide and 45 ft long. The backyard is a rectangle that is 80 ft wide and 100 ft long. The average price of fencing and gates is $10.50 per foot. Find the difference in price to fence the two areas.

90. Target Corporation paid $751,000 for 15 acres of land in Westfield, Massachusetts. Target will pay about $727,000 each year in taxes for nine years and then will pay about $1.8 million per year. Find the total amount of taxes that Target will pay in 15 years. (*Source:* St. Paul Pioneer Press, March 14, 2009)

Find the Mistake

For exercises 91–94, the completed problem has one mistake.

(a) Describe the mistake in words, or copy down the whole problem and highlight or circle the mistake.

(b) Do the problem correctly.

91. Problem: Simplify: $\dfrac{x}{x^2 - 25} + \dfrac{5}{x^2 - 10x + 25}$

Incorrect Answer: The least common denominator is $(x - 5)(x + 5)(x - 5)$.

$$\frac{x}{x^2 - 25} + \frac{5}{x^2 - 10x + 25}$$
$$= \frac{x}{(x - 5)(x + 5)} + \frac{5}{(x - 5)(x - 5)}$$
$$= \frac{x}{(x - 5)(x + 5)(x - 5)} + \frac{5}{(x - 5)(x + 5)(x - 5)}$$
$$= \frac{\cancel{x + 5}}{(x - 5)\cancel{(x + 5)}(x - 5)}$$
$$= \frac{1}{(x - 5)(x - 5)}$$

92. Problem: Simplify: $\dfrac{c}{c - 1} + \dfrac{4}{c + 2}$

Incorrect Answer: The least common denominator is $(c - 1)(c + 2)$.

$$\frac{c}{c - 1} + \frac{4}{c + 2}$$
$$= \frac{c}{c - 1} \cdot \frac{(c + 2)}{(c + 2)} + \frac{4}{c + 2} \cdot \frac{(c - 1)}{(c - 1)}$$
$$= \frac{c^2 + 2c}{(c - 1)(c + 2)} + \frac{4c - 4}{(c - 1)(c + 2)}$$
$$= \frac{c^2 + 6c - 4}{(c - 1)(c + 2)}$$
$$= \frac{\cancel{c^2} + 6c - 4}{\cancel{c^2} + c - 2}$$
$$= \frac{6c - 4}{c - 2}$$

93. Problem: Simplify: $\dfrac{6}{x + 4} - \dfrac{4}{x + 2}$

Incorrect Answer: The least common denominator is $(x + 4)(x + 2)$.

$$\frac{6}{x + 4} - \frac{4}{x + 2}$$
$$= \frac{6}{x + 4} \cdot \frac{(x + 2)}{(x + 2)} - \frac{4}{x + 2} \cdot \frac{(x + 4)}{(x + 4)}$$
$$= \frac{6(x + 2)}{(x + 4)(x + 2)} - \frac{4(x + 4)}{(x + 4)(x + 2)}$$
$$= \frac{6x + 12 - 4x - 16}{(x + 4)(x + 2)}$$
$$= \frac{2x - 4}{(x + 4)(x - 2)}$$
$$= \frac{2\cancel{(x - 2)}}{(x + 4)\cancel{(x - 2)}}$$
$$= \frac{2}{x + 4}$$

94. Problem: Simplify: $\dfrac{y + 4}{2y + 10} - \dfrac{5}{y^2 - 25}$

Incorrect Answer: The least common denominator is $2(y - 5)(y + 5)$.

$$\frac{y + 4}{2y + 10} - \frac{5}{y^2 - 25}$$
$$= \frac{y + 4}{2(y + 5)} \cdot \frac{(y - 5)}{(y - 5)} - \frac{5}{(y - 5)(y + 5)} \cdot \frac{2}{2}$$
$$= \frac{y^2 + 4y - 5y - 20}{2(y - 5)(y + 5)} - \frac{10}{2(y - 5)(y + 5)}$$
$$= \frac{y^2 - y - 30}{2(y - 5)(y + 5)}$$
$$= \frac{(y + 6)\cancel{(y - 5)}}{2\cancel{(y - 5)}(y + 5)}$$
$$= \frac{y + 6}{2(y + 5)}$$

Review

For exercises 95–98, evaluate.

95. $\dfrac{\frac{10}{3}}{\frac{2}{3}}$ **96.** $\dfrac{\frac{8}{7}}{\frac{2}{7}}$ **97.** $\dfrac{\frac{6}{1}}{\frac{1}{2}}$ **98.** $\dfrac{\frac{3}{4}}{\frac{1}{4}}$

SUCCESS IN COLLEGE MATHEMATICS

99. Some math students "negative self-talk" about math. They may speak using negative language such as "I can't do math," instead of using positive language as in "I can do some math already, and with hard work, I'll be able to do more." They may say, "I should be able to work faster" instead of "I can work fast enough to pass the course." Write a different example of negative self-talk about math. Then rewrite it as a positive statement.

$$m = \frac{\frac{1}{8} - \frac{1}{2}}{\frac{3}{8} - \frac{1}{4}}$$

A rational expression in which the numerator or denominator is itself a fraction is a **complex rational expression**. In this section, we will simplify these expressions.

SECTION 7.5 Complex Rational Expressions

After reading the text, working the practice exercises, and completing assigned exercises, you should be able to:

Simplify a complex rational expression.

Complex Rational Expressions

In Chapter 3, we used the slope formula, $m = \frac{y_2 - y_1}{x_2 - x_1}$, to find the slope of a line. In the next example, the coordinates of the points on the line, (x_1, y_1) and (x_2, y_2), are fractions.

EXAMPLE 1 Find the slope of the line that passes through $\left(\frac{1}{9}, \frac{1}{15}\right)$ and $\left(\frac{5}{9}, \frac{14}{15}\right)$.

SOLUTION ▶

$m = \frac{y_2 - y_1}{x_2 - x_1}$ The slope formula.

$m = \dfrac{\frac{\mathbf{14}}{\mathbf{15}} - \frac{\mathbf{1}}{\mathbf{15}}}{\frac{\mathbf{5}}{\mathbf{9}} - \frac{\mathbf{1}}{\mathbf{9}}}$ $(x_1, y_1) = \left(\frac{1}{9}, \frac{1}{15}\right)$; $(x_2, y_2) = \left(\frac{5}{9}, \frac{14}{15}\right)$

$m = \dfrac{\frac{\mathbf{13}}{\mathbf{15}}}{\frac{\mathbf{4}}{\mathbf{9}}}$ Evaluate the numerator and denominator.

$m = \frac{13}{15} \div \frac{\mathbf{4}}{\mathbf{9}}$ Rewrite the fraction as a quotient.

$m = \frac{13}{15} \cdot \frac{\mathbf{9}}{\mathbf{4}}$ Rewrite division as multiplication by the reciprocal of the divisor.

$m = \frac{13}{\cancel{3} \cdot 5} \cdot \frac{\cancel{3} \cdot 3}{4}$ Find common factors.

$m = \frac{\mathbf{39}}{\mathbf{20}}$ Simplify.

We can use the same process that we used in Example 1 to simplify a complex rational expression. Rewrite the complex rational expression as a quotient, rewrite division as multiplication by the reciprocal of the divisor, identify common factors, and simplify.

EXAMPLE 2 Simplify: $\dfrac{\dfrac{x^2 + 2x}{2x + 8}}{\dfrac{12x}{x + 4}}$

SOLUTION ▶

$\dfrac{\dfrac{x^2 + 2x}{2x + 8}}{\dfrac{12x}{x + 4}}$

$= \dfrac{x^2 + 2x}{2x + 8} \div \dfrac{\mathbf{12x}}{\mathbf{x + 4}}$ Rewrite the fraction as a quotient.

$= \dfrac{x^2 + 2x}{2x + 8} \cdot \dfrac{\mathbf{x + 4}}{\mathbf{12x}}$ Rewrite division as multiplication by the reciprocal.

$= \dfrac{\mathbf{x(x + 2)}}{\mathbf{2(x + 4)}} \cdot \dfrac{(x + 4)}{12x}$ Factor completely; find common factors.

$= \dfrac{x + 2}{24}$ Simplify.

In the next example, we rewrite the numerator of the complex fraction as a fraction with a denominator of 1.

EXAMPLE 3 Simplify: $\dfrac{\dfrac{c + 4}{c^2 - 16}}{2c}$

SOLUTION ▶

$\dfrac{\dfrac{c + 4}{c^2 - 16}}{2c}$

$= \dfrac{\dfrac{c + 4}{\mathbf{1}}}{\dfrac{c^2 - 16}{2c}}$ Rewrite the numerator as a fraction.

$= \dfrac{c + 4}{1} \div \dfrac{\mathbf{c^2 - 16}}{\mathbf{2c}}$ Rewrite the fraction as a quotient.

$= \dfrac{c + 4}{1} \cdot \dfrac{\mathbf{2c}}{\mathbf{c^2 - 16}}$ Rewrite division as multiplication by the reciprocal.

$= \dfrac{c + 4}{1} \cdot \dfrac{2c}{\mathbf{(c + 4)(c - 4)}}$ Factor completely; find common factors.

$= \dfrac{2c}{c - 4}$ Simplify.

Practice Problems

For problems 1–4, simplify.

1. $\dfrac{\frac{4}{21}}{\frac{2}{49}}$ 2. $\dfrac{\frac{x+3}{x-2}}{\frac{x-3}{2x-4}}$ 3. $\dfrac{\frac{x+9}{x^2+10x+9}}{x-1}$ 4. $\dfrac{\frac{a^2+a}{2a-16}}{\frac{a^2+4a}{a^2-4a-32}}$

Simplifying Complex Rational Expressions (Strategy 1)

Some complex rational expressions include addition or subtraction of fractions with different denominators. In the first strategy for simplifying these expressions, we find the least common denominator for the fractions in the numerator and for the fractions in the denominator. We then rewrite the fractions with these denominators, add or subtract the fractions, and simplify the remaining complex fraction.

EXAMPLE 4 Evaluate: $\dfrac{\frac{2}{3}+\frac{1}{6}}{\frac{1}{4}+\frac{5}{6}}$

SOLUTION ▶

$\dfrac{\frac{2}{3}+\frac{1}{6}}{\frac{1}{4}+\frac{5}{6}}$ The least common denominator for the numerator is 6. The least common denominator for the denominator is 12.

$= \dfrac{\frac{2}{3}\left(\frac{2}{2}\right)+\frac{1}{6}}{\frac{1}{4}\left(\frac{3}{3}\right)+\frac{5}{6}\left(\frac{2}{2}\right)}$ Multiply by fractions equal to 1.

$= \dfrac{\frac{\mathbf{4}}{\mathbf{6}}+\frac{1}{6}}{\frac{\mathbf{3}}{\mathbf{12}}+\frac{\mathbf{10}}{\mathbf{12}}}$ Simplify.

$= \dfrac{\frac{5}{6}}{\frac{13}{12}}$ Add numerators; the denominators do not change.

$= \frac{5}{6} \div \frac{\mathbf{13}}{\mathbf{12}}$ Rewrite the fraction as a quotient.

$= \frac{5}{6} \cdot \frac{\mathbf{12}}{\mathbf{13}}$ Rewrite division as multiplication by the reciprocal.

$= \frac{5}{\cancel{6}} \cdot \frac{\mathbf{2 \cdot \cancel{6}}}{13}$ Find common factors.

$= \frac{10}{13}$ Simplify.

We can use the same process to simplify complex rational expressions when the denominators include variables.

EXAMPLE 5 Simplify: $\dfrac{\dfrac{1}{x} - \dfrac{1}{x+1}}{\dfrac{1}{x} + \dfrac{1}{x-1}}$

SOLUTION ▶

$$\dfrac{\dfrac{1}{x} - \dfrac{1}{x+1}}{\dfrac{1}{x} + \dfrac{1}{x-1}}$$

The least common denominator in the numerator is $x(x+1)$.

The least common denominator in the denominator is $x(x-1)$.

$$= \dfrac{\dfrac{1}{x}\left(\dfrac{x+1}{x+1}\right) - \dfrac{1}{x+1}\left(\dfrac{x}{x}\right)}{\dfrac{1}{x}\left(\dfrac{x-1}{x-1}\right) + \dfrac{1}{x-1}\left(\dfrac{x}{x}\right)}$$

Multiply by fractions equal to 1.

$$= \dfrac{\dfrac{\mathbf{x+1}}{x(x+1)} - \dfrac{\mathbf{x}}{x(x+1)}}{\dfrac{\mathbf{x-1}}{x(x-1)} + \dfrac{\mathbf{x}}{x(x-1)}}$$

Simplify numerators; leave the denominators factored.

$$= \dfrac{\dfrac{x+1-x}{x(x+1)}}{\dfrac{x-1+x}{x(x-1)}}$$

Add numerators; the denominators do not change.

$$= \dfrac{\dfrac{\mathbf{1}}{x(x+1)}}{\dfrac{\mathbf{2x-1}}{x(x-1)}}$$

Simplify numerators.

$$= \dfrac{1}{x(x+1)} \div \dfrac{\mathbf{2x-1}}{\mathbf{x(x-1)}}$$

Rewrite the fraction as a quotient.

$$= \dfrac{1}{\cancel{x}(x+1)} \cdot \dfrac{\cancel{x}(x-1)}{2x-1}$$

Rewrite as multiplication by the reciprocal.

$$= \dfrac{x-1}{(x+1)(2x-1)}$$

Simplify common factors.

Simplifying a Complex Rational Expression (Strategy 1)

1. Find the least common denominator for the fractions in the numerator. Build equivalent fractions with this denominator.
2. Find the least common denominator for the fractions in the denominator. Build equivalent fractions with this denominator.
3. Add or subtract the fractions in the numerator and the denominator.
4. Rewrite the rational expression as a quotient.
5. Rewrite division as multiplication by the reciprocal of the divisor.
6. Factor; find common factors.
7. Simplify.

In the next example, there are whole numbers in the numerator and the denominator. We write these whole numbers as fractions with a denominator of 1.

EXAMPLE 6 Simplify: $\dfrac{2 + \dfrac{1}{y}}{\dfrac{1}{y+1} + 1}$

SOLUTION

$\dfrac{2 + \dfrac{1}{y}}{\dfrac{1}{y+1} + 1}$ The common denominator for the numerator is y. The common denominator for the denominator is $y + 1$.

$= \dfrac{\dfrac{2}{\mathbf{1}} + \dfrac{1}{y}}{\dfrac{1}{y+1} + \dfrac{1}{\mathbf{1}}}$ Rewrite whole numbers as fractions with denominators of 1.

$= \dfrac{\dfrac{2}{1}\left(\dfrac{y}{y}\right) + \dfrac{1}{y}}{\dfrac{1}{y+1} + \dfrac{1}{1}\left(\dfrac{y+1}{y+1}\right)}$ Multiply by fractions equal to 1.

$= \dfrac{\dfrac{\mathbf{2y}}{\mathbf{y}} + \dfrac{1}{y}}{\dfrac{1}{y+1} + \dfrac{\mathbf{y+1}}{\mathbf{y+1}}}$ Simplify.

$= \dfrac{\dfrac{2y+1}{y}}{\dfrac{1+y+1}{y+1}}$ Add numerators; the denominators do not change.

$= \dfrac{\dfrac{2y+1}{y}}{\dfrac{\mathbf{y+2}}{y+1}}$ Simplify.

$= \dfrac{2y+1}{y} \div \dfrac{\mathbf{y+2}}{\mathbf{y+1}}$ Rewrite the fraction as a quotient.

$= \dfrac{2y+1}{y} \cdot \dfrac{\mathbf{y+1}}{\mathbf{y+2}}$ Rewrite as multiplication by the reciprocal.

$= \dfrac{(2y+1)(y+1)}{y(y+2)}$ There are no common factors.

$= \dfrac{\mathbf{2y^2 + 3y + 1}}{y(y+2)}$ Multiply factors in the numerator.

Practice Problems

For problems 5–7, use Strategy 1 to evaluate or simplify.

5. $\dfrac{\dfrac{3}{5} + \dfrac{1}{10}}{\dfrac{4}{15} + \dfrac{1}{3}}$ **6.** $\dfrac{\dfrac{1}{x+2} + \dfrac{2}{x}}{\dfrac{1}{x+3} + \dfrac{1}{x}}$ **7.** $\dfrac{3 - \dfrac{1}{x+1}}{3 - \dfrac{1}{x}}$

Complex Rational Expressions (Strategy 2)

In Strategy 2 for simplifying complex rational expressions that include addition or subtraction, we find the least common denominator of all of the individual fractions and multiply the numerator and denominator of the complex rational expression by this least common denominator. This clears the fractions in the complex rational expression.

EXAMPLE 7 Evaluate: $\dfrac{\frac{2}{3}+\frac{1}{6}}{\frac{1}{4}+\frac{5}{6}}$

SOLUTION ▶

$\dfrac{\frac{2}{3}+\frac{1}{6}}{\frac{1}{4}+\frac{5}{6}}$ The least common denominator of 3, 6, and 4 is 12.

$= \dfrac{\frac{2}{3}+\frac{1}{6}}{\frac{1}{4}+\frac{5}{6}} \cdot \dfrac{12}{12}$ Multiply by a fraction equal to 1.

$= \dfrac{\left(\frac{2}{3}+\frac{1}{6}\right)\cdot\frac{\mathbf{12}}{\mathbf{1}}}{\left(\frac{1}{4}+\frac{5}{6}\right)\cdot\frac{\mathbf{12}}{\mathbf{1}}}$ $\dfrac{12}{12} = \dfrac{\frac{12}{1}}{\frac{12}{1}}$

$= \dfrac{\frac{2}{3}\cdot\frac{\mathbf{12}}{\mathbf{1}}+\frac{1}{6}\cdot\frac{\mathbf{12}}{\mathbf{1}}}{\frac{1}{4}\cdot\frac{\mathbf{12}}{\mathbf{1}}+\frac{5}{6}\cdot\frac{\mathbf{12}}{\mathbf{1}}}$ Distributive property.

$= \dfrac{\frac{2}{\cancel{3}}\cdot\frac{\cancel{\mathbf{3}}\cdot\mathbf{4}}{1}+\frac{1}{\cancel{6}}\cdot\frac{\mathbf{2}\cdot\cancel{\mathbf{6}}}{1}}{\frac{1}{\cancel{4}}\cdot\frac{\mathbf{3}\cdot\cancel{\mathbf{4}}}{1}+\frac{5}{\cancel{6}}\cdot\frac{\mathbf{2}\cdot\cancel{\mathbf{6}}}{1}}$ Find common factors.

$= \dfrac{\frac{8}{1}+\frac{2}{1}}{\frac{3}{1}+\frac{10}{1}}$ Simplify.

$= \dfrac{8+2}{3+10}$ Simplify; the fractions are cleared.

$= \dfrac{10}{13}$ Simplify.

Simplifying a Complex Rational Expression (Strategy 2)

1. Find the least common denominator of all of the individual fractions.
2. Multiply the complex rational expression by a fraction equal to 1.
3. Use the distributive property.
4. Factor; find common factors.
5. Simplify.

In the next example, the least common denominator has two factors.

EXAMPLE 8 Simplify: $\dfrac{2 + \dfrac{1}{y}}{\dfrac{1}{y+1} + 1}$

SOLUTION ▶

$$\frac{2 + \frac{1}{y}}{\frac{1}{y+1} + 1}$$ The least common denominator is $y(y + 1)$.

$$= \frac{2 + \frac{1}{y}}{\frac{1}{y+1} + 1} \cdot \frac{y(y+1)}{y(y+1)}$$ Multiply by a fraction equal to 1.

$$= \frac{\left(2 + \frac{1}{y}\right) \cdot \frac{y(y+1)}{\mathbf{1}}}{\left(\frac{1}{y+1} + 1\right) \cdot \frac{y(y+1)}{\mathbf{1}}}$$ $\dfrac{y(y+1)}{y(y+1)} = \dfrac{\frac{y(y+1)}{1}}{\frac{y(y+1)}{1}}$

$$= \frac{\frac{2}{1} \cdot \frac{\mathbf{y(y+1)}}{\mathbf{1}} + \frac{1}{y} \cdot \frac{\mathbf{y(y+1)}}{\mathbf{1}}}{\frac{1}{y+1} \cdot \frac{\mathbf{y(y+1)}}{\mathbf{1}} + \frac{1}{1} \cdot \frac{\mathbf{y(y+1)}}{\mathbf{1}}}$$ Distributive property.

$$= \frac{\frac{2y(y+1)}{1} + \frac{1}{\cancel{y}} \cdot \frac{\cancel{y}(y+1)}{1}}{\frac{1}{\cancel{(y+1)}} \cdot \frac{y\cancel{(y+1)}}{1} + \frac{1}{1} \cdot \frac{y(y+1)}{1}}$$ Find common factors.

$$= \frac{2y(y+1) + 1(y+1)}{y + y(y+1)}$$ Simplify; the fractions are cleared.

$$= \frac{\mathbf{2y^2 + 2y + y + 1}}{y + \mathbf{y^2 + y}}$$ Distributive property.

$$= \frac{2y^2 + \mathbf{3y} + 1}{y^2 + \mathbf{2y}}$$ Combine like terms.

$$= \frac{(2y+1)(y+1)}{y(y+2)}$$ Factor; there are no common factors.

$$= \frac{\mathbf{2y^2 + 3y + 1}}{y(y+2)}$$ Multiply the factors in the numerator.

In the next example, the least common denominator has three factors.

EXAMPLE 9 Simplify: $\dfrac{\dfrac{1}{x} - \dfrac{1}{x+1}}{\dfrac{1}{x} + \dfrac{1}{x-1}}$

SOLUTION ▶

$$\dfrac{\dfrac{1}{x} - \dfrac{1}{x+1}}{\dfrac{1}{x} + \dfrac{1}{x-1}}$$

The least common denominator is $x(x + 1)(x - 1)$.

$$= \dfrac{\dfrac{1}{x} - \dfrac{1}{x+1}}{\dfrac{1}{x} + \dfrac{1}{x-1}} \cdot \dfrac{x(x+1)(x-1)}{x(x+1)(x-1)}$$

Multiply by a fraction equal to 1.

$$= \dfrac{\left(\dfrac{1}{x} - \dfrac{1}{x+1}\right) \cdot \dfrac{x(x+1)(x-1)}{\mathbf{1}}}{\left(\dfrac{1}{x} + \dfrac{1}{x-1}\right) \cdot \dfrac{x(x+1)(x-1)}{\mathbf{1}}}$$

$$\dfrac{x(x+1)(x-1)}{x(x+1)(x-1)} = \dfrac{\dfrac{x(x+1)(x-1)}{1}}{\dfrac{x(x+1)(x-1)}{1}}$$

$$= \dfrac{\dfrac{1}{x} \cdot \dfrac{\mathbf{x(x+1)(x-1)}}{\mathbf{1}} - \dfrac{1}{x+1} \cdot \dfrac{\mathbf{x(x+1)(x-1)}}{\mathbf{1}}}{\dfrac{1}{x} \cdot \dfrac{\mathbf{x(x+1)(x-1)}}{\mathbf{1}} + \dfrac{1}{x-1} \cdot \dfrac{\mathbf{x(x+1)(x-1)}}{\mathbf{1}}}$$

Distributive property.

$$= \dfrac{\dfrac{\cancel{x}(x+1)(x-1)}{\cancel{x}} - \dfrac{x\cancel{(x+1)}(x-1)}{\cancel{x+1}}}{\dfrac{\cancel{x}(x+1)(x-1)}{\cancel{x}} + \dfrac{x(x+1)\cancel{(x-1)}}{\cancel{x-1}}}$$

Find common factors.

$$= \dfrac{(x+1)(x-1) - x(x-1)}{(x+1)(x-1) + x(x+1)}$$

Simplify; the fractions are cleared.

$$= \dfrac{\mathbf{x}(x-1) + \mathbf{1}(x-1) - \mathbf{x} \cdot x - \mathbf{x}(-1)}{\mathbf{x}(x-1) + \mathbf{1}(x-1) + \mathbf{x} \cdot x + \mathbf{x}(1)}$$

Distributive property.

$$= \dfrac{\mathbf{x} \cdot x + \mathbf{x}(-1) + \mathbf{1} \cdot x + \mathbf{1}(-1) - x \cdot x - x(-1)}{\mathbf{x} \cdot x + \mathbf{x}(-1) + \mathbf{1} \cdot x + \mathbf{1}(-1) + x \cdot x + x(1)}$$

Distributive property.

$$= \dfrac{x^2 - x + x - 1 - x^2 + x}{x^2 - x + x - 1 + x^2 + x}$$

Simplify.

$$= \dfrac{x - 1}{2x^2 + x - 1}$$

Combine like terms.

$$= \dfrac{x - 1}{\mathbf{(2x - 1)(x + 1)}}$$

Factor; no common factors; the expression is simplified.

Practice Problems

For problems 8–10, use Strategy 2 to evaluate or simplify.

8. $\dfrac{\dfrac{3}{5} + \dfrac{1}{10}}{\dfrac{4}{15} + \dfrac{1}{3}}$ **9.** $\dfrac{3 - \dfrac{1}{x+1}}{3 - \dfrac{1}{x}}$ **10.** $\dfrac{\dfrac{1}{x+2} + \dfrac{2}{x}}{\dfrac{1}{x+3} + \dfrac{1}{x}}$

7.5 Vocabulary Practice

Match the term with its description.

1. For the numbers 4, 6, and 9, this is 36.
2. For the fractions $\frac{1}{4}$, $\frac{1}{6}$, and $\frac{1}{9}$, this is 36.
3. The value of a fraction in which the numerator and denominator are equal
4. A fraction in which the numerator and denominator have no common factors
5. For the fraction $\frac{x}{y}$, this is $\frac{y}{x}$.
6. Division by zero
7. $a(b + c) = ab + ac$
8. An expression in which the numerator and/or denominator are fractions
9. The result of division
10. The result of multiplication

A. complex rational expression
B. distributive property
C. least common denominator
D. least common multiple
E. lowest terms
F. 1
G. product
H. quotient
I. reciprocal
J. undefined

7.5 Exercises

Follow your instructor's guidelines for showing your work.

For exercises 1–8, find the slope of the line that passes through the given points.

1. $(9, 15)\ (18, 42)$
2. $(8, 14)\ (15, 42)$
3. $\left(\frac{1}{9}, \frac{2}{15}\right)\left(\frac{5}{9}, \frac{11}{15}\right)$
4. $\left(\frac{3}{8}, \frac{4}{9}\right)\left(\frac{7}{8}, \frac{8}{9}\right)$
5. $\left(\frac{3}{8}, \frac{2}{5}\right)\left(\frac{9}{16}, \frac{3}{4}\right)$
6. $\left(\frac{2}{3}, \frac{9}{10}\right)\left(\frac{5}{8}, \frac{11}{20}\right)$
7. $\left(\frac{3}{8}, -\frac{1}{2}\right)\left(-\frac{5}{8}, -\frac{5}{2}\right)$
8. $\left(\frac{5}{9}, -\frac{1}{3}\right)\left(-\frac{7}{9}, -\frac{5}{3}\right)$

For exercises 9–24, evaluate or simplify.

9. $\dfrac{\frac{2}{3}}{\frac{5}{6}}$
10. $\dfrac{\frac{4}{5}}{\frac{3}{10}}$
11. $\dfrac{\frac{5x}{4}}{\frac{2x^2}{15}}$
12. $\dfrac{\frac{4p}{9}}{\frac{2p^2}{3}}$
13. $\dfrac{\frac{3a + 12}{2}}{\frac{9a + 36}{8}}$
14. $\dfrac{\frac{4n + 32}{3}}{\frac{8n + 64}{15}}$
15. $\dfrac{\frac{5p - 5}{4p + 12}}{\frac{10p + 10}{7p + 21}}$
16. $\dfrac{\frac{2w - 6}{4w + 16}}{\frac{12w + 36}{5w + 20}}$
17. $\dfrac{\frac{a^2 + 6a + 5}{6a}}{\frac{a^2 - 1}{24a^4}}$
18. $\dfrac{\frac{r^2 + 11r + 24}{9r}}{\frac{r^2 - 64}{27r^3}}$
19. $\dfrac{\frac{b^2 + 7b + 10}{b^2 + 8b + 15}}{\frac{b^2 + 6b + 8}{b^2 - b - 12}}$
20. $\dfrac{\frac{c^2 + 6c + 8}{c^2 + 7c + 10}}{\frac{c^2 - 3c - 4}{c^2 + 6c + 5}}$
21. $\dfrac{\frac{v^2 - 5v + 4}{v^2 - 6v + 8}}{\frac{v^2 + 2v - 3}{v^2 + v - 6}}$
22. $\dfrac{\frac{p^2 - 11p + 30}{p^2 - 2p - 24}}{\frac{p^2 - 4p - 5}{p^2 + 5p + 4}}$
23. $\dfrac{1}{\frac{x^2 + 6x + 8}{x^2 - 4}}$
24. $\dfrac{1}{\frac{z^2 + 9z + 14}{z^2 - 49}}$

For exercises 25–68, evaluate or simplify.

25. $\dfrac{\frac{1}{2} + \frac{1}{3}}{\frac{1}{3} + \frac{1}{5}}$
26. $\dfrac{\frac{1}{3} + \frac{1}{2}}{\frac{1}{2} + \frac{1}{7}}$
27. $\dfrac{\frac{1}{2}}{\frac{1}{3} + \frac{1}{5}}$
28. $\dfrac{\frac{1}{3}}{\frac{1}{2} + \frac{1}{7}}$
29. $\dfrac{\frac{2}{5} - \frac{1}{3}}{\frac{1}{5} + \frac{1}{3}}$
30. $\dfrac{\frac{2}{3} - \frac{1}{2}}{\frac{1}{3} + \frac{1}{6}}$
31. $\dfrac{\frac{4}{9} + \frac{3}{2}}{\frac{5}{9} - \frac{1}{6}}$
32. $\dfrac{\frac{5}{6} + \frac{1}{3}}{\frac{2}{3} - \frac{1}{12}}$

33. $\dfrac{\frac{1}{x}}{\frac{1}{x}+\frac{1}{x+3}}$

34. $\dfrac{\frac{1}{x}}{\frac{1}{x}+\frac{1}{x+2}}$

35. $\dfrac{\frac{1}{x}}{\frac{1}{x-3}+\frac{1}{x+3}}$

36. $\dfrac{\frac{1}{x}}{\frac{1}{x-2}+\frac{1}{x+2}}$

37. $\dfrac{\frac{1}{x+3}+\frac{1}{x}}{\frac{1}{x+3}}$

38. $\dfrac{\frac{1}{x+2}+\frac{1}{x}}{\frac{1}{x+2}}$

39. $\dfrac{\frac{1}{x+1}-\frac{1}{x}}{\frac{1}{x+1}}$

40. $\dfrac{\frac{1}{x+4}-\frac{1}{x}}{\frac{1}{x+4}}$

41. $\dfrac{\frac{5}{x+1}+\frac{3}{x-2}}{\frac{5}{x+1}+\frac{2}{x-2}}$

42. $\dfrac{\frac{4}{x-1}+\frac{2}{x+1}}{\frac{3}{x-1}+\frac{1}{x+1}}$

43. $\dfrac{\frac{2}{x+4}+\frac{5}{x-2}}{\frac{3}{x+4}+\frac{4}{x-2}}$

44. $\dfrac{\frac{2}{x-2}+\frac{3}{x+3}}{\frac{2}{x-2}+\frac{1}{x+3}}$

45. $\dfrac{\frac{3}{x-3}-\frac{4}{x+1}}{\frac{4}{x-3}+\frac{6}{x+1}}$

46. $\dfrac{\frac{3}{x-4}-\frac{1}{x+2}}{\frac{6}{x-4}+\frac{5}{x+2}}$

47. $\dfrac{\frac{1}{x+3}+\frac{2}{x-2}}{\frac{2}{x+3}-\frac{3}{x-2}}$

48. $\dfrac{\frac{3}{x+1}+\frac{1}{x-2}}{\frac{1}{x+1}-\frac{4}{x-2}}$

49. $\dfrac{\frac{1}{x-4}-2}{\frac{4}{x+5}-4}$

50. $\dfrac{\frac{3}{x-1}-4}{\frac{4}{x+2}-1}$

51. $\dfrac{5-\frac{1}{x+3}}{2+\frac{4}{x-1}}$

52. $\dfrac{5-\frac{3}{x-1}}{4+\frac{3}{x+3}}$

53. $\dfrac{\frac{x-2}{x+1}}{\frac{x}{x-1}}$

54. $\dfrac{\frac{x}{x-2}}{\frac{x+4}{x+3}}$

55. $\dfrac{\frac{x-1}{x-2}}{\frac{x+2}{x+1}}$

56. $\dfrac{\frac{x+1}{x-3}}{\frac{x+3}{x-1}}$

57. $\dfrac{x+y}{\frac{1}{x}+\frac{1}{y}}$

58. $\dfrac{x-y}{\frac{1}{y}-\frac{1}{x}}$

59. $\dfrac{\frac{1}{x}-\frac{1}{y}}{y-x}$

60. $\dfrac{\frac{1}{y}+\frac{1}{x}}{y+x}$

61. $\dfrac{\frac{a}{2}+\frac{b}{3}}{\frac{a}{3}+\frac{b}{5}}$

62. $\dfrac{\frac{c}{3}+\frac{d}{2}}{\frac{c}{2}+\frac{d}{7}}$

63. $\dfrac{\frac{2}{a}+\frac{3}{b}}{\frac{3}{a}+\frac{5}{b}}$

64. $\dfrac{\frac{3}{c}+\frac{2}{d}}{\frac{2}{c}+\frac{5}{d}}$

65. $\dfrac{3}{3+\frac{3}{3+x}}$

66. $\dfrac{2}{2+\frac{2}{2+x}}$

67. $\dfrac{x}{x+\frac{x}{x+3}}$

68. $\dfrac{x}{x+\frac{x}{x+2}}$

For exercises 69–70, resistors restrict the flow of electrons in an electric circuit. The unit of resistance is an *ohm*.

The formula $R = \dfrac{1}{\frac{1}{R_1}+\frac{1}{R_2}}$ describes the total resistance R in ohms of a parallel circuit with two resistors, R_1 and R_2.

12 V R_1 R_2

69. In a parallel circuit, Resistor R_1 has a resistance of 5 ohms, and Resistor R_2 has a resistance of 15 ohms. Find the total resistance of the circuit.

70. In a parallel circuit, Resistor R_1 has a resistance of 8 ohms, and Resistor R_2 has a resistance of 12 ohms. Find the total resistance of the circuit.

Problem Solving: Practice and Review

Follow your instructor's guidelines for using the five steps as outlined in Section 1.3, p. 55.

71. The wholesale price of a display with 24 flashlights is \$16.40. The company wants to make a profit of 45% on each flashlight. Find the retail price that the company should charge for one flashlight. Round to the nearest hundredth.

72. The pulp used to make Cascades Moka unbleached toilet paper is 80% post-consumer recycled material and 20% recovered corrugated boxes. Currently, 3.4 million tons of toilet paper are used per year in the United States. Fifty-three percent of this toilet paper is made from virgin (nonrecycled) fiber sources. A switch to 100% recycled bath tissue could save 30.6 million trees and 68 million gigajoules of energy per year. Find the amount of toilet paper used per year that is made from virgin fiber sources. Write the answer in place value notation. (*Source:* www.bradenton.com, Jan. 25, 2012)

73. An American born in February 1950 will reach full retirement age in 2015. His social security payments are based on his lifetime earnings. His first option is to retire at age 65 years and 10 months and receive \$1000 a month for the rest of his life. His second option is to retire at age 62 (46 months before full retirement age) and receive \$758 a month for the rest of his life. What age will he be in years and months when the payments he has received from either option are equal? Round up to the nearest month. (*Source:* www.ssa.gov)

74. RAND Texas is a statistical service that contains more than 80 databases. As shown in the table, the cost of a group subscription to this service depends on the number of people in the group of users. Find the total annual cost of a subscription for 75 users.

Number of users	Cost per user per year
1	\$270
2–9	\$175
10–99	\$95

Source: txrand.org

Find the Mistake

For exercises 75–78, one part of simplifying a rational expression is completed.

(a) Describe the mistake in words, or copy down this part of the problem and highlight or circle the mistake.

(b) Do this part of the problem correctly.

75. Problem: To simplify $\dfrac{\frac{4}{15x}}{\frac{8}{15}}$, rewrite the expression as $\dfrac{4}{15x} \div \dfrac{8}{15}$ and simplify.

Incorrect Answer:
$$\frac{4}{15x} \div \frac{8}{15}$$
$$= \frac{4}{15x} \cdot \frac{15}{8}$$
$$= \frac{\cancel{4}}{\cancel{15}x} \cdot \frac{\cancel{15}}{\cancel{4} \cdot 2}$$
$$= 2x$$

76. Problem: The first step in using Strategy 1 to simplify $\dfrac{2 + \frac{3}{x}}{3 + \frac{1}{x}}$ is to add the expressions in the numerator and add the expressions in the denominator.

Incorrect Answer:
$$\frac{2 + \frac{3}{x}}{3 + \frac{1}{x}}$$
$$= \frac{\frac{5}{x}}{\frac{4}{x}}$$

77. Problem: The first step in using Strategy 2 to simplify $\dfrac{\frac{1}{x+1} - \frac{1}{x-2}}{\frac{2}{x+1} + \frac{3}{x-2}} \cdot \dfrac{(x+1)(x-2)}{(x+1)(x-2)}$ is to find the least common denominator of all the individual fractions, multiply by a fraction equal to 1, use the distributive property, and simplify.

Incorrect Answer:
$$\frac{\frac{1}{x+1} - \frac{1}{x-2}}{\frac{2}{x+1} + \frac{3}{x-2}} \cdot \frac{(x+1)(x-2)}{(x+1)(x-2)}$$
$$= \frac{\frac{1}{\cancel{x+1}} \cdot (\cancel{x+1})(x-2) - \frac{1}{\cancel{x-2}}(x+1)(\cancel{x-2})}{\frac{2}{\cancel{x+1}} \cdot (\cancel{x+1})(x-2) + \frac{3}{\cancel{x-2}} \cdot (x+1)(\cancel{x-2})}$$
$$= \frac{x - 2 - x + 1}{2x - 4 + 3x + 3}$$

78. Problem: To simplify $\dfrac{\frac{x^2 - 2x - 15}{x^2 - 7x + 10}}{\frac{x^2 + 2x - 8}{x^2 + 2x - 3}}$, the first step is to factor the polynomials completely.

Incorrect Answer:
$$\frac{\frac{x^2 - 2x - 15}{x^2 - 7x + 10}}{\frac{x^2 + 2x - 8}{x^2 + 2x - 3}}$$
$$= \frac{\frac{(x-5)(x+3)}{(x-2)(x-5)}}{\frac{(x+2)(x-4)}{(x+3)(x-1)}}$$

Review

For exercises 79–82,

(a) clear the fractions and solve.

(b) check.

79. $\dfrac{5}{6} + \dfrac{2}{3}x = \dfrac{5}{2}$

80. $\dfrac{3}{2}u + \dfrac{3}{4} = \dfrac{9}{2}$

81. $1 = \dfrac{7}{6}w + \dfrac{5}{12}$

82. $\dfrac{5}{6}h + 8 = 12$

SUCCESS IN COLLEGE MATHEMATICS

83. When a student with math anxiety is given a test, feelings of anxiety and panic can make the student feel that he or she cannot do a single problem on the test. What do you think a student should do if this happens?

© Chiyacat/Shutterstock.com

The rate at which a person completes a job such as shoveling a driveway is called a *rate of work*. In this section, we will use rational equations and rates of work to solve application problems.

SECTION 7.6

Rational Equations and Proportions

After reading the text, working the practice problems, and completing assigned exercises, you should be able to:

1. Clear the denominators from a rational equation in one variable.
2. Solve a rational equation in one variable.
3. Identify any extraneous solutions of a rational equation.
4. Solve an application problem using a rate of work.
5. Solve an application problem using a proportion.

Solving a Rational Equation

In the previous sections in this chapter, we simplified rational expressions. In this section, we solve rational equations. A *rational expression* such as $\frac{2}{x+3} + \frac{5}{x+1}$ does *not* include an equals sign. We evaluate or simplify expressions. We do not know what number a variable in an expression represents. We cannot clear the fractions from an expression. A *rational equation* such as $\frac{2}{x+3} + \frac{5}{x+1} = \frac{31}{15}$ includes an equals sign. We solve equations, isolating the variable so that we can find out what number the variable represents. To clear the fractions from an equation, we multiply both sides of the equation by the least common denominator (LCD).

EXAMPLE 1 (a) Solve: $\frac{5}{6}a + \frac{1}{4} = \frac{19}{20}$

SOLUTION ▸

$\frac{5}{6}a + \frac{1}{4} = \frac{19}{20}$ Least common denominator: 60.

$60\left(\frac{5}{6}a + \frac{1}{4}\right) = 60\left(\frac{19}{20}\right)$ Multiplication property of equality.

$\mathbf{\frac{60}{1}}\left(\frac{5}{6}a\right) + \mathbf{\frac{60}{1}}\left(\frac{1}{4}\right) = \frac{60}{1}\left(\frac{19}{20}\right)$ Distributive property; $60 = \frac{60}{1}$

$\frac{\cancel{6}\cdot 10\cdot 5\cdot a}{\cancel{6}} + \frac{15\cdot\cancel{4}}{\cancel{4}} = \frac{3\cdot\cancel{20}\cdot 19}{\cancel{20}}$ Find common factors.

$50a + 15 = 57$ Simplify; the fractions are cleared.

$\underline{\quad -15 \; -15}$ Subtraction property of equality.

$50a + \mathbf{0} = \mathbf{42}$ Simplify.

$\frac{\mathbf{50a}}{50} = \frac{42}{50}$ Division property of equality.

$$a = \frac{\mathbf{2 \cdot 21}}{\mathbf{2 \cdot 25}}$$ Find common factors.

$$a = \frac{\mathbf{21}}{\mathbf{25}}$$ Simplify.

(b) Check.

$$\frac{5}{6}a + \frac{1}{4} = \frac{19}{20}$$

$$\frac{5}{6}\left(\frac{\mathbf{21}}{\mathbf{25}}\right) + \frac{1}{4} = \frac{19}{20}$$ Replace the variable with the solution.

$$\frac{\mathbf{5 \cdot 3 \cdot 7}}{\mathbf{2 \cdot 3 \cdot 5 \cdot 5}} + \frac{1}{4} = \frac{19}{20}$$ Find common factors.

$$\frac{\mathbf{7}}{\mathbf{10}} + \frac{1}{4} = \frac{19}{20}$$ Simplify.

$$\frac{7}{10} \cdot \frac{2}{2} + \frac{1}{4} \cdot \frac{5}{5} = \frac{19}{20}$$ Multiply by fractions equal to 1.

$$\frac{\mathbf{14}}{\mathbf{20}} + \frac{\mathbf{5}}{\mathbf{20}} = \frac{19}{20}$$ Evaluate.

$$\frac{\mathbf{19}}{\mathbf{20}} = \frac{19}{20}$$ True. Evaluate; the solution is correct.

In the next example, the equation $\frac{8}{x} - \frac{2}{3} = \frac{2}{x}$ includes a variable in the denominator. To clear the fractions, we multiply both sides of the equation by the least common denominator, $3x$.

EXAMPLE 2 **(a)** Solve: $\frac{8}{x} - \frac{2}{3} = \frac{2}{x}$

SOLUTION

$$\frac{8}{x} - \frac{2}{3} = \frac{2}{x}$$ Least common denominator: $3x$.

$$3x\left(\frac{8}{x} - \frac{2}{3}\right) = 3x\left(\frac{2}{x}\right)$$ Multiplication property of equality.

$$\frac{\mathbf{3x}}{\mathbf{1}}\left(\frac{8}{x}\right) + \frac{\mathbf{3x}}{\mathbf{1}}\left(-\frac{2}{3}\right) = \frac{\mathbf{3x}}{\mathbf{1}}\left(\frac{2}{x}\right)$$ Distributive property; find common factors.

$$24 - 2x = 6$$ Simplify; the fractions are cleared.

$$\underline{-24 \qquad\qquad -24}$$ Subtraction property of equality.

$$\mathbf{0} - 2x = \mathbf{-18}$$ Simplify.

$$\frac{\mathbf{-2x}}{-2} = \frac{-18}{-2}$$ Division property of equality.

$$x = 9$$ Simplify.

(b) Check.

$$\frac{8}{x} - \frac{2}{3} = \frac{2}{x}$$

$$\frac{8}{\mathbf{9}} - \frac{2}{3} = \frac{2}{\mathbf{9}}$$ Replace the variables with the solution.

$$\frac{8}{9} - \frac{2}{3} \cdot \frac{3}{3} = \frac{2}{9}$$ Multiply by a fraction equal to 1.

$$\frac{8}{9} - \frac{\mathbf{6}}{\mathbf{9}} = \frac{2}{9}$$ Evaluate.

$$\frac{\mathbf{2}}{\mathbf{9}} = \frac{2}{9}$$ True. Evaluate; the solution is correct.

In the next example, the least common denominator (LCD) of $\frac{3}{x-2} + \frac{3}{10} = \frac{3}{5}$ is $10(x - 2)$.

EXAMPLE 3 **(a)** Solve: $\frac{3}{x-2} + \frac{3}{10} = \frac{3}{5}$

SOLUTION

$\frac{3}{x-2} + \frac{3}{10} = \frac{3}{5}$	LCD: $10(x-2)$.
$10(x-2)\left(\frac{3}{x-2} + \frac{3}{10}\right) = 10(x-2)\left(\frac{3}{5}\right)$	Multiplication property of equality.
$\mathbf{10(x-2)}\left(\frac{3}{x-2}\right) + \mathbf{10(x-2)}\left(\frac{3}{10}\right) = \mathbf{10(x-2)}\left(\frac{3}{5}\right)$	Distributive property.
$\frac{10\cancel{(x-2)}}{1}\left(\frac{3}{\cancel{x-2}}\right) + \frac{\cancel{10}(x-2)}{1}\left(\frac{3}{\cancel{10}}\right) = \frac{10(x-2)}{1}\left(\frac{3}{5}\right)$	Find common factors.
$10(3) + (x-2)(3) = \frac{10(x-2)}{1}\left(\frac{3}{5}\right)$	Simplify.
$10(3) + (x-2)(3) = \frac{2\cdot\cancel{5}(x-2)}{1}\left(\frac{3}{\cancel{5}}\right)$	Find common factors.
$10(3) + (x-2)(3) = \mathbf{2}(x-2)(\mathbf{3})$	Simplify; the fractions are cleared.
$\mathbf{30} + (x-2)(3) = \mathbf{6}(x-2)$	Simplify.
$30 + \mathbf{3x} + \mathbf{3}(-2) = \mathbf{6x} + \mathbf{6}(-2)$	Distributive property.
$30 + 3x \mathbf{- 6} = 6x \mathbf{- 12}$	Simplify.
$\mathbf{24} + 3x = 6x - 12$	Combine like terms.
$-24 \qquad -24$	Subtraction property of equality.
$\mathbf{0} + 3x = 6x \mathbf{- 36}$	Simplify.
$\mathbf{3x} = 6x - 36$	Simplify.
$-6x \quad -6x$	Subtraction property of equality.
$\mathbf{-3x} = \mathbf{0} - 36$	Simplify.
$\frac{-3x}{-3} = \frac{\mathbf{-36}}{-3}$	Division property of equality.
$x = 12$	Simplify.

(b) Check.

$\frac{3}{x-2} + \frac{3}{10} = \frac{3}{5}$	
$\frac{3}{\mathbf{12} - 2} + \frac{3}{10} = \frac{3}{5}$	Replace the variable with the solution.
$\frac{3}{\mathbf{10}} + \frac{3}{10} = \frac{3}{5}$	Evaluate.
$\frac{\mathbf{6}}{\mathbf{10}} = \frac{3}{5}$	Evaluate.
$\frac{\mathbf{3 \cdot 2}}{\mathbf{5 \cdot 2}} = \frac{3}{5}$	Find common factors.
$\frac{\mathbf{3}}{\mathbf{5}} = \frac{3}{5}$ True.	Simplify; the solution is correct.

In the next example, the least common denominator is $(a + 4)(a - 8)$.

EXAMPLE 4 **(a)** Solve: $\frac{3}{a+4} = \frac{6}{a-8}$

SOLUTION

$\frac{3}{a+4} = \frac{6}{a-8}$	LCD: $(a+4)(a-8)$.
$(a+4)(a-8)\left(\frac{3}{a+4}\right) = (a+4)(a-8)\left(\frac{6}{a-8}\right)$	Muliplication property of equality.
$\frac{\cancel{(a+4)}(a-8)}{1}\left(\frac{3}{\cancel{a+4}}\right) = \frac{(a+4)\cancel{(a-8)}}{1}\left(\frac{6}{\cancel{a-8}}\right)$	Find common factors.
$3(a-8) = 6(a+4)$	Simplify; the fractions are cleared.
$3a + 3(-8) = 6a + 6(4)$	Distributive property.
$3a - 24 = 6a + 24$	Simplify.
$-6a \quad -6a$	Subtraction property of equality.
$-3a - 24 = 0 + 24$	Simplify.
$+24 \quad +24$	Addition property of equality.
$-3a + 0 = 48$	Simplify.
$\frac{-3a}{-3} = \frac{48}{-3}$	Division property of equality.
$a = -16$	Simplify.

(b) Check.

$\frac{3}{a+4} = \frac{6}{a-8}$	
$\frac{3}{-16+4} = \frac{6}{-16-8}$	Replace the variables in the original equation.
$\frac{3}{-12} = \frac{6}{-24}$	Evaluate.
$-\frac{\cancel{3}\cdot 1}{\cancel{3}\cdot 4} = -\frac{\cancel{6}\cdot 1}{\cancel{6}\cdot 4}$	Find common factors.
$-\frac{1}{4} = -\frac{1}{4}$ True.	Simplify; the solution is correct.

Practice Problems

For problems 1–4,
(a) clear the fractions and solve.
(b) check.

1. $\frac{5}{8}x + \frac{1}{6} = \frac{3}{4}$ **2.** $\frac{9}{x} + \frac{4}{7} = \frac{17}{x}$ **3.** $\frac{7}{x-4} + \frac{2}{5} = \frac{3}{4}$ **4.** $\frac{7}{x-5} = \frac{9}{x+3}$

Extraneous Solutions

When we clear the fractions from a rational equation, the result is a new equation without denominators. If a solution of the new equation causes division by zero in the original equation, it is an **extraneous solution**. Ask your instructor how you should mark extraneous solutions in your work. Extraneous solutions result from the process of clearing the denominators, not from making an algebra or arithmetic mistake.

In the next example, we clear the fractions, and the result is a quadratic equation. To solve a quadratic equation in standard form, $ax^2 + bx + c = 0$, factor and use the zero product property (Section 6.5).

EXAMPLE 5 (a) Solve: $\frac{1}{x-1} + \frac{3}{x} = \frac{x}{x-1}$

SOLUTION

$$\frac{1}{x-1} + \frac{3}{x} = \frac{x}{x-1} \quad \text{LCD: } x(x-1).$$

$$x(x-1)\left(\frac{1}{x-1} + \frac{3}{x}\right) = x(x-1)\left(\frac{x}{x-1}\right) \quad \text{Multiplication property of equality.}$$

$$x(x-1)\left(\frac{1}{x-1} + \frac{3}{x}\right) = \frac{x\cancel{(x-1)}}{\mathbf{1}}\left(\frac{x}{\cancel{x-1}}\right) \quad \text{Find common factors.}$$

$$x(x-1)\left(\frac{1}{x-1} + \frac{3}{x}\right) = \mathbf{x^2} \quad \text{Simplify.}$$

$$\frac{\mathbf{x(x-1)}}{\mathbf{1}}\left(\frac{1}{x-1}\right) + \frac{\mathbf{x(x-1)}}{\mathbf{1}}\left(\frac{3}{x}\right) = x^2 \quad \text{Distributive property.}$$

$$\frac{x\cancel{(x-1)}}{1}\left(\frac{1}{\cancel{x-1}}\right) + \frac{\cancel{x}(x-1)}{1}\left(\frac{3}{\cancel{x}}\right) = x^2 \quad \text{Find common factors.}$$

$$\mathbf{x + (x-1)(3)} = x^2 \quad \text{Simplify; the fractions are cleared.}$$

$$x + \mathbf{3x + 3(-1)} = x^2 \quad \text{Distributive property.}$$

$$x + 3x \mathbf{- 3} = x^2 \quad \text{Simplify.}$$

$$\mathbf{4x} - 3 = x^2 \quad \text{Combine like terms.}$$

$$\underline{-4x + 3 \qquad -4x + 3} \quad \text{Properties of equality.}$$

$$\mathbf{0} = x^2 \mathbf{- 4x + 3} \quad \text{Simplify.}$$

$$0 = \mathbf{(x-1)(x-3)} \quad \text{Factor.}$$

$$x - 1 = 0 \quad \text{or} \quad x - 3 = 0 \quad \text{Zero product property.}$$

$$\underline{+1 \;\; +1} \qquad \underline{+3 \;\; +3} \quad \text{Addition property of equality.}$$

$$\mathbf{x + 0 = 1} \quad \text{or} \quad \mathbf{x + 0 = 3} \quad \text{Simplify.}$$

$$\cancel{\mathbf{x = 1}} \quad \text{or} \quad \mathbf{x} = 3 \quad \text{Simplify.}$$

The equation $4x - 3 = x^2$ has two solutions: $x = 1$ or $x = 3$. However, $x = 1$ is not a solution of the original equation, $\frac{1}{x-1} + \frac{3}{x} = \frac{x}{x-1}$. As the check shows, replacing x with 1 results in division by zero. So $x = 1$ is an extraneous solution.

(b) Check.

Check: $x = 1$

$$\frac{1}{x-1} + \frac{3}{x} = \frac{x}{x-1}$$

$$\frac{1}{\mathbf{1}-1} + \frac{3}{\mathbf{1}} = \frac{\mathbf{1}}{\mathbf{1}-1} \quad \text{Replace variables.}$$

$$\frac{1}{\mathbf{0}} + \mathbf{3} = \frac{1}{\mathbf{0}} \quad \text{Undefined.} \quad \text{Simplify.}$$

Check: $x = 3$

$$\frac{1}{x-1} + \frac{3}{x} = \frac{x}{x-1}$$

$$\frac{1}{\mathbf{3}-1} + \frac{3}{\mathbf{3}} = \frac{\mathbf{3}}{\mathbf{3}-1}$$

$$\frac{1}{\mathbf{2}} + \mathbf{1} = \frac{3}{\mathbf{2}}$$

$$\frac{\mathbf{3}}{\mathbf{2}} = \frac{3}{2} \quad \text{True.}$$

Since $x = 1$ is an extraneous solution, the only solution of this equation is $x = 3$.

Solving a Rational Equation

1. Find the least common denominator (LCD).
2. Clear the fractions by multiplying both sides of the equation by the least common denominator. Simplify.
3. Solve the equation.
4. Check the solution(s). A value that results in division by zero is not a solution.

In the next example, we identify possible extraneous solutions before solving the rational equation.

EXAMPLE 6 Solve: $\dfrac{3}{x-3} = \dfrac{x}{x-3} - \dfrac{3}{2}$

SOLUTION ▶ Since $3 - 3 = 0$, the solution $x = 3$ results in division by 0. If $x = 3$ is a solution, it is an extraneous solution.

$$\frac{3}{x-3} = \frac{x}{x-3} - \frac{3}{2}$$ LCD: $2(x-3)$.

$$\frac{2(x-3)}{1}\left(\frac{3}{x-3}\right) = \frac{2(x-3)}{1}\left(\frac{x}{x-3} - \frac{3}{2}\right)$$ Multiplication property of equality.

$$\frac{\mathbf{2(x-3)}(3)}{x-3} = \frac{\mathbf{2(x-3)}(x)}{x-3} \mathbf{+} \frac{\mathbf{2(x-3)}(-3)}{2}$$ Distributive property.

$$\frac{2\cancel{(x-3)}(3)}{\cancel{x-3}} = \frac{2\cancel{(x-3)}(x)}{\cancel{x-3}} + \frac{\cancel{2}(x-3)(-3)}{\cancel{2}}$$ Find common factors.

$$6 = 2x + (x-3)(-3)$$ Simplify; the fractions are cleared.

$$6 = 2x + \mathbf{(-3)}x \mathbf{- 3(-3)}$$ Distributive property.

$$6 = 2x \mathbf{- 3x} - 3(-3)$$ Simplify.

$$6 = 2x - \mathbf{3x + 9}$$ Simplify.

$$6 = \mathbf{-x} + 9$$ Combine like terms.

$$\underline{-9 \qquad\quad -9}$$ Addition property of equality.

$$\mathbf{-3} = -x \mathbf{+ 0}$$ Simplify.

$$\frac{-3}{-1} = \frac{\mathbf{-x}}{-1}$$ Division property of equality.

$$\cancel{3 = x}$$ Simplify.

The only solution of the equation $6 = 2x - 3(x - 3)$ is $x = 3$. However, since this solution results in division by zero in the original rational equation, $x = 3$ is an extraneous solution. The equation $\dfrac{3}{x-3} = \dfrac{x}{x-3} - \dfrac{3}{2}$ has no solution.

Practice Problems

For problems 5–7,
(a) solve.
(b) check.

5. $\dfrac{6}{x-6} + \dfrac{4}{x} = \dfrac{x}{x-6}$ **6.** $\dfrac{2}{x+4} + \dfrac{4}{x} = \dfrac{x}{x+4}$ **7.** $\dfrac{5}{x} + \dfrac{1}{x-8} = \dfrac{2}{3}$

Work

A rate of work can describe the amount of work done in a given time. For example, a production line might make products at a rate of $\frac{75 \text{ products}}{1 \text{ hr}}$, or a worker might prune grapevines at a rate of $\frac{45 \text{ grapevines}}{1 \text{ hr}}$. These rates describe the amount of work done in 1 unit of time. Other rates of work describe how long it takes to complete 1 job. For example, nursing assistants give sponge baths to patients. We do not describe the area of skin that a nursing assistant can clean per minute. Instead, we use a rate such as $\frac{1 \text{ bath}}{12 \text{ min}}$. This describes how long it takes for the assistant to complete the sponge bath.

To find the fraction of a job completed in a given amount of time, multiply the rate of work and the time worked.

Fraction of a Job Completed

fraction of a job completed = (rate of work)(time worked)

EXAMPLE 7 The work rate of a nursing assistant giving a sponge bath is $\frac{1 \text{ bath}}{12 \text{ min}}$. What fraction of the job is complete after the assistant works 4 min?

SOLUTION ▶ fraction of job completed

$= \left(\frac{1 \text{ bath}}{12 \text{ min}}\right)\left(\frac{4 \text{ min}}{1}\right)$ (rate of work)(time)

$= \frac{1 \text{ bath} \cdot \cancel{4} \cancel{\text{min}}}{3 \cdot \cancel{4} \cancel{\text{min}}}$ Find common factors.

$= \frac{1}{3} \text{ bath}$ Simplify.

In some jobs, two people can work together to complete a job. If their individual rates of work do not change,

$$(\text{worker A rate})(\text{time A works}) + (\text{worker B rate})(\text{time B works}) = 1 \text{ job}$$

EXAMPLE 8 Brother and Sister are shoveling the driveway. Brother can shovel the driveway in 45 min. Sister can shovel the driveway in 38 min. Find the time for Brother and Sister to shovel the driveway working together. Round to the nearest whole number. (Assume that their rates of work do not change when they work together.)

SOLUTION ▶ **Step 1 Understand the problem.**

The unknown is the time to shovel the driveway working together.

T = time to shovel working together

Step 2 Make a plan.

The product of the rate of work and the time spent working is the fraction of the job completed. Brother's rate of work is $\left(\frac{1 \text{ job}}{45 \text{ min}}\right)$. Sister's rate of is work is $\left(\frac{1 \text{ job}}{38 \text{ min}}\right)$.

A word equation is (Brother's rate)(time) + (Sister's rate)(time) = 1 job.

Step 3 Carry out the plan.

$$\left(\frac{1\text{ job}}{45\text{ min}}\right)T + \left(\frac{1\text{ job}}{38\text{ min}}\right)T = 1\text{ job} \quad \text{Brother's work + Sister's work = 1 job}$$

$$\frac{1}{45}T + \frac{1}{38}T = 1 \quad \text{Remove the units; LCD is 1710.}$$

$$1710\left(\frac{1}{45}T + \frac{1}{38}T\right) = 1710(1) \quad \text{Multiplication property of equality.}$$

$$\frac{\mathbf{1710}}{\mathbf{1}}\left(\frac{1}{45}T\right) + \frac{\mathbf{1710}}{\mathbf{1}}\left(\frac{1}{38}T\right) = \mathbf{1710} \quad \text{Distributive property; simplify.}$$

$$\frac{38 \cdot \cancel{45} \cdot 1}{1 \cdot \cancel{45}}T + \frac{\cancel{38} \cdot 45 \cdot 1}{1 \cdot \cancel{38}}T = 1710 \quad \text{Find common factors.}$$

$$\mathbf{38T + 45T} = 1710 \quad \text{Simplify; the fractions are cleared.}$$

$$\mathbf{83T} = 1710 \quad \text{Combine like terms.}$$

$$\frac{83T}{83} = \frac{1710}{83} \quad \text{Division property of equality.}$$

$$T = 20.60\ldots \quad \text{Simplify.}$$

$$T \approx \mathbf{21\text{ min}} \quad \text{Round; include the units.}$$

Step 4 Look back.
Since Brother's rate of work is $\frac{1\text{ job}}{45\text{ min}}$, in 21 min he completes $\frac{21}{45}$ job, or about 0.47 job. Since Sister's rate of work is $\frac{1\text{ job}}{38\text{ min}}$, in 21 min she completes $\frac{21}{38}$ job, or about 0.55 job. In 21 min, the total work completed by both Brother and Sister is 0.47 job + 0.55 job, which equals 1.02 job. Since 1.02 job rounds to 1 job, the answer seems reasonable.

Step 5 Report the solution.
Working together, Brother and Sister can shovel the driveway in about 21 min.

In some situations, the work of one person undoes the work of another. In the next example, the person doing a job is working faster than the person "undoing" the job. Eventually, the job is completed.

EXAMPLE 9 A baker makes cupcakes to sell at a cupcake shop and cupcakes for special orders. She can make 8 dozen cupcakes in 60 min. If 1 dozen cupcakes are sold every 45 min in the shop and 14 dozen cupcakes are needed for a special order, find the time for the baker to complete the special order.

SOLUTION **Step 1 Understand the problem.**
The unknown is the time to make the cupcakes.

T = time to complete the special order

Step 2 Make a plan.
The fraction of the job completed equals (rate of work)(time). As the baker works to fill the special order, some of the work is undone by selling cupcakes. One job equals completing the special order of 14 dozen cupcakes. A word equation is (making cupcakes rate)(time) − (selling cupcakes rate)(time) = 14 dozen cupcakes.

Step 3 Carry out the plan.

$$\left(\frac{8 \text{ dozen}}{60 \text{ min}}\right)T - \left(\frac{1 \text{ dozen}}{45 \text{ min}}\right)T = 14 \text{ dozen}$$ making cupcakes − selling cupcakes = 1 job

$$\frac{8}{60}T - \frac{1}{45}T = 14$$ Remove units; LCD is 180.

$$180\left(\frac{8}{60}T - \frac{1}{45}T\right) = 180(14)$$ Multiplication property of equality.

$$\mathbf{180}\left(\frac{8}{60}T\right) + \mathbf{180}\left(-\frac{1}{45}T\right) = \mathbf{2520}$$ Distributive property; simplify.

$$\frac{\mathbf{3 \cdot \cancel{60} \cdot} 8}{\cancel{60}}T + \frac{\mathbf{4 \cdot \cancel{45}}(-1)}{\cancel{45}}T = 2520$$ Find common factors.

$$\mathbf{24T - 4T} = 2520$$ Simplify; the fractions are cleared.

$$\mathbf{20T} = 2520$$ Combine like terms.

$$\frac{20T}{20} = \frac{2520}{20}$$ Division property of equality.

$$T = 126 \text{ min}$$ Simplify; include the units.

Step 4 Look back.

The baker makes $\left(\frac{8 \text{ dozen}}{60 \text{ min}}\right)(126 \text{ min})$ cupcakes, or 16.8 dozen cupcakes. The shop sells $\left(\frac{1 \text{ dozen cupcakes}}{45 \text{ min}}\right)(126 \text{ min})$, or 2.8 dozen cupcakes. Since 16.8 dozen cupcakes – 2.8 dozen cupcakes equals the 14 dozen cupcakes needed for the special order, the answer seems reasonable.

Step 5 Report the solution.

The baker will fill the special order in about 126 min.

Practice Problems

For problems 8–9, use the five steps. Assume that the rate of work is the same if done individually or if done together.

8. A developmental math instructor can grade the quizzes for a class in 12 min. A different instructor can grade the quizzes in 15 min. Find the time for the instructors to do the job working together. Round to the nearest whole number.

9. Every 8 min, an employee squeezes 1 qt of orange juice and pours the juice into a container. Every 15 min, a different employee sells a total of 1 qt of juice from the container. At the beginning of their shift, the 8-qt container is empty. Find the amount of time needed to fill the container. Round to the nearest whole number.

Proportions

In Section 2.2, we used proportions to solve application problems. To solve a proportion in which the variable is in a denominator, first clear the fractions.

EXAMPLE 10 A bottle contains 750 mg of antibiotic in 15 mL of liquid. A dog with a bladder infection needs 80 mg of antibiotic per day. Use a proportion to find the amount of liquid that the dog needs per day. Write the answer as a decimal number.

SOLUTION ▶ **Step 1 Understand the problem.**

The unknown is the amount of liquid that the dog needs per day.

a = amount of liquid

Step 2 Make a plan.

The unknown ratio is $\frac{80\text{ mg}}{a}$. The known ratio is $\frac{750\text{ mg}}{15\text{ mL}}$. We can write and solve a proportion: unknown ratio = known ratio.

Step 3 Carry out the plan.

$$\frac{80\text{ mg}}{a} = \frac{750\text{ mg}}{15\text{ mL}}$$

$$\frac{80}{a} = \frac{750}{15} \qquad \text{Remove the units; LCD is } 15a.$$

$$\left(\frac{15\cancel{a}}{1}\right)\left(\frac{80}{\cancel{a}}\right) = \left(\frac{\cancel{15}a}{1}\right)\left(\frac{750}{\cancel{15}}\right) \qquad \text{Multiplication property of equality.}$$

$$1200 = 750a \qquad \text{Simplify; the fractions are cleared.}$$

$$\frac{1200}{750} = \frac{750a}{750} \qquad \text{Division property of equality.}$$

$$1.6\text{ mL} = a \qquad \text{Simplify; include the units.}$$

Step 4 Look back.

Since the unknown ratio $\frac{80}{1.6}$ equals 50 and the known ratio $\frac{750}{15}$ also equals 50, the answer seems reasonable.

Step 5 Report the solution.

The dog needs 1.6 mL of the solution per day.

Practice Problems

For problems 10–11, use the five steps and a proportion.

10. A child with Rocky Mountain spotted fever needs 43 mg of doxycycline. Five milliliters of a raspberry-apple-flavored liquid contains 50 mg of doxycycline. Find the amount of liquid that the child needs.

11. The resident population of the United States in December 2011 was about 312,603,000 people. Find the number of people who were incarcerated. Round to the nearest thousand.

The United States continues to maintain the highest rate of incarceration in the world at 731 per 100,000 population. (*Sources:* www.census.gov; sentencingproject .org, Feb. 2012)

7.6 Vocabulary Practice

Match the term with its description.

1. Undefined
2. If $ab = 0$, then $a = 0$ or $b = 0$.
3. $a(b + c) = ab + ac$
4. $ax^2 + bx + c$
5. A number that can be written as a fraction in which the numerator and denominator are integers
6. A polynomial with two terms
7. An expression that can be written as a fraction in which the numerator and denominator are polynomials
8. When we check this value, it is not a solution of the original equation.
9. A polynomial with one term
10. The value of a fraction in which the numerator and denominator are equal

A. binomial
B. distributive property
C. division by zero
D. extraneous solution
E. monomial
F. 1
G. quadratic trinomial
H. rational expression
I. rational number
J. zero product property

7.6 Exercises

Follow your instructor's guidelines for showing your work.

For exercises 1–10,
(a) solve.
(b) check.

1. $\frac{2}{3}x - \frac{1}{8} = \frac{11}{12}$

2. $\frac{3}{5}x - \frac{1}{4} = \frac{9}{10}$

3. $\frac{4}{9}p - \frac{1}{8} = \frac{25}{72}$

4. $\frac{3}{8}z + \frac{1}{16} = \frac{5}{32}$

5. $\frac{4}{15}k + \frac{3}{4} = -2$

6. $\frac{1}{6}w + \frac{23}{8} = -3$

7. $\frac{2}{3}x + \frac{3}{2} = \frac{1}{3}x + \frac{1}{6}$

8. $\frac{2}{9}x + \frac{5}{3} = \frac{5}{9}x + \frac{7}{3}$

9. $\frac{1}{2}u = \frac{3}{4}u + \frac{5}{4}$

10. $\frac{2}{3}n = \frac{5}{6}n + \frac{5}{3}$

For exercises 11–30,
(a) solve.
(b) check.

11. $\frac{1}{6} + \frac{1}{a} = \frac{2}{3}$

12. $\frac{1}{10} + \frac{1}{v} = \frac{3}{5}$

13. $\frac{11}{c} - \frac{1}{3} = \frac{2}{5}$

14. $\frac{13}{d} - \frac{5}{9} = \frac{1}{6}$

15. $\frac{4}{5} + \frac{1}{4} = \frac{21}{x}$

16. $\frac{5}{6} + \frac{1}{3} = \frac{28}{x}$

17. $\frac{3}{10} + \frac{7}{m} = \frac{14}{m} + \frac{1}{15}$

18. $\frac{1}{5} + \frac{11}{c} = \frac{6}{c} + \frac{2}{5}$

19. $\frac{15}{4z} + \frac{2}{3} = \frac{1}{24}$

20. $\frac{8}{3y} + \frac{5}{16} = -\frac{17}{48}$

21. $\frac{2}{3} + \frac{4}{5d} = \frac{34}{45}$

22. $\frac{5}{9} + \frac{3}{8b} = \frac{107}{144}$

23. $\frac{3}{2f} + \frac{1}{6} = \frac{9}{f} - \frac{1}{3}$

24. $\frac{15}{4h} + \frac{1}{9} = \frac{5}{h} + \frac{1}{24}$

25. $\frac{5}{r - 2} = \frac{10}{r - 8}$

26. $\frac{15}{a - 4} = \frac{5}{a + 12}$

27. $\frac{8}{b - 16} = \frac{40}{b}$

28. $\frac{10}{p - 7} = \frac{45}{p}$

29. $\frac{1}{x - 14} = \frac{1}{3x - 44}$

30. $\frac{4}{x - 7} = \frac{24}{2x - 6}$

For exercises 31–40,
(a) solve.
(b) check.

31. $\frac{x}{x + 3} = \frac{1}{x + 3} + \frac{6}{x}$

32. $\frac{x}{x + 5} = \frac{1}{x + 5} + \frac{6}{x}$

33. $\frac{3}{c} + 8c = 10$

34. $\frac{5}{d} + 12d = 23$

35. $\frac{3}{w - 3} + \frac{4}{w} = \frac{w}{w - 3}$

36. $\frac{6}{x - 6} + \frac{5}{x} = \frac{x}{x - 6}$

37. $\frac{n^2}{n - 9} - \frac{9n}{n - 9} = -9$

38. $\frac{p^2}{p - 8} - \frac{8p}{p - 8} = -8$

39. $\frac{q}{q + 8} = -\frac{8}{q + 8} + \frac{3}{q}$

40. $\frac{c}{c + 8} = -\frac{4}{c + 8} + \frac{6}{c}$

41. If both sides of the equation $\frac{1}{x - 1} + \frac{2}{x} = \frac{x}{x - 1}$ are multiplied by $x(x - 1)$, the simplified equation is $1x + 2(x - 1) = x^2$. Rewriting in standard form and factoring, the equation is $(x - 2)(x - 1) = 0$ and its solutions are $x = 1$ or $x = 2$. Explain why the solution $x = 1$ is extraneous.

42. Author Colin Tudge writes in his book *The Variety of Life*: "For some have argued that works of art should be self-contained and need no *extraneous* information to be appreciated: no biography, no history, no referents of any kind." Explain the meaning of extraneous in this statement. (*Source:* Word of the Day, October 27, 2000, Merriam-Webster Online)

For exercises 43–58,
(a) solve.
(b) check.

43. $\frac{2}{w} + \frac{8}{15} = \frac{2}{5}$

44. $\frac{3}{k} + \frac{7}{18} = \frac{5}{9}$

45. $\frac{2x}{x + 1} = \frac{4}{x + 1} + \frac{5}{x}$

46. $\frac{2x}{x + 1} = \frac{2}{x + 1} + \frac{3}{x}$

47. $\frac{2}{p + 6} = \frac{6}{p + 4}$

48. $\frac{4}{a + 6} = \frac{9}{a - 4}$

49. $\frac{9}{10}v + \frac{1}{3} = -\frac{22}{15}$

50. $\frac{8}{9}r + \frac{4}{5} = -\frac{28}{15}$

51. $\frac{9}{r} = 2$

52. $\frac{8}{b} = 5$

53. $\frac{2}{x} = 0$

54. $\frac{4}{y} = 0$

55. $\frac{c}{c + 9} - \frac{4}{c} = -\frac{9}{c + 9}$

56. $\frac{h}{h + 10} - \frac{3}{h} = -\frac{10}{h + 10}$

57. $\frac{d + 1}{3} = \frac{d - 3}{6}$

58. $\frac{z + 2}{4} = \frac{z - 8}{12}$

For exercises 59–66, use the five steps. Assume that the rate of work does not change if done individually or together.

59. A worker can prune one row of grapevines in 44 min. Another worker can prune one row in 33 min. Find the time for these workers to do the job together. Round to the nearest whole number.

60. A worker can clean the old straw from the horse stalls in a barn and replace it with new straw in 52 min. Another worker can do the same job in 83 min. Find the time for these workers to do the job together. Round to the nearest whole number.

61. The water from a garden hose turned on at full pressure fills a hot tub in 45 min. If the drain is open, the hot tub empties in 62 min. Find the amount of time to fill the hot tub with the drain open. Round to the nearest whole number.

62. The faucet in a bathtub fills the tub to the overflow drain in 15 min. The drain on the bathtub empties a full tub in 6 min. If a bathtub is full, the faucet is turned on, and the drain is opened, find the amount of time for the bathtub to empty.

63. An employee at a grocery store can restock the aisles in 4 hr 10 min. Another employee can do the same job in 3 hr 40 min. Find the time it will take the employees to do this job together. Round to the nearest whole number.

64. A laundry worker at a hotel can wash and fold all the towels from a Saturday night in 4 hr 50 min. Another worker can do the same job in 4 hr 35 min. Find the time it will take the employees to do this job together. Round to the nearest whole number.

65. The newsletters for a company can be printed on a small personal printer in 65 min or on the network office printer in 48 min. Find the time to print the newsletters if both printers are used. Round to the nearest whole number.

66. A mother and a grandmother are sewing the blocks for a queen-size quilt. The mother can sew all of the blocks in about 35 hr. The grandmother can sew all the blocks in about 50 hr. Find the time it will take mother and grandmother to do this job working together. Round to the nearest whole number.

For exercises 67–82, use the five steps and a proportion.

67. About five of 100 pregnant women have pre-eclampsia, a condition that results in high blood pressure. About 300,000 pregnant women per year in the United States have pre-eclampsia. Find the number of pregnant women in the United States used to create this ratio. (*Source:* www.nytimes.com, March 17, 2009)

68. Find the number of adults used to create the ratio "four out of five."

Four out of five adults now use the Internet. 184 million adults are online from their homes, offices, schools or other locations. (*Source:* www.harrisinteractive.com, Nov. 17, 2008)

69. Find the number of 725,000 women in their mid-40s with a history of normal pregnancy who would be expected to have a heart attack or stroke some 10 years later.

Of 100 women in their mid-40's with a history of normal pregnancy, about 4 would be expected to have a heart attack or stroke some 10 years later. (*Source:* www.nytimes.com, March 17, 2009)

70. A survey asked 505 companies whether they would continue to match their employees' contributions to their 401k retirement plans. Find the number of companies that will continue to match the contributions.

Three out of five employers maintain 401(k) match despite economic crisis. (*Source:* www.americanbenefitscouncil.org, March 17, 2009)

71. In 2010, there were 426.0 cases of chlamydia per 100,000 Americans with a total of 1,307,893 cases of chlamydia. Find the population of Americans used to create this ratio. Round to the nearest hundred. (*Source:* www.cdc.gov, 2011)

72. In 2010, there were 14.9 cases of syphilis per 100,000 Americans with a total of 45,834 cases of syphilis. Find the population of Americans used to create this ratio. Round to the nearest hundred. (*Source:* www.cdc.gov, 2011)

73. In 2010, 3.5 per 100,000 full-time equivalent workers were killed on the job with a total of 547 workers killed on the job. Find the number of full-time equivalent workers used to create this ratio. Round to the nearest whole number. (*Source:* www.osha.gov)

74. The "Fatal Four" causes of death on construction sites are falls, electrocution, being struck by an object, and being caught-in or -between. They were responsible for about three out of five construction worker deaths in 2010. Find the total number of construction worker deaths. Round to the nearest whole number.

Eliminating the Fatal Four would save 431 workers' lives in America every year. (*Source:* www.osha.gov)

75. In 2010, about 2,465,940 Americans died. Find the number of Americans who died without a will. Round to the nearest hundred. (*Source:* www.cdc.gov, Jan. 11, 2012)

Seven out of ten Americans die without a will. (*Source:* extension.umd.edu)

76. In 2010, about 2,465,940 Americans died. Find the number of these deaths that were from chronic diseases. Round to the nearest hundred. (*Source:* www.cdc.gov, Jan. 11, 2012)

7 out of 10 deaths among Americans each year are from chronic diseases. (*Source:* www.cdc.gov, July 7, 2010)

77. Find the number of 2,200,000 adults who binge drink about four times a month. Round to the nearest thousand.

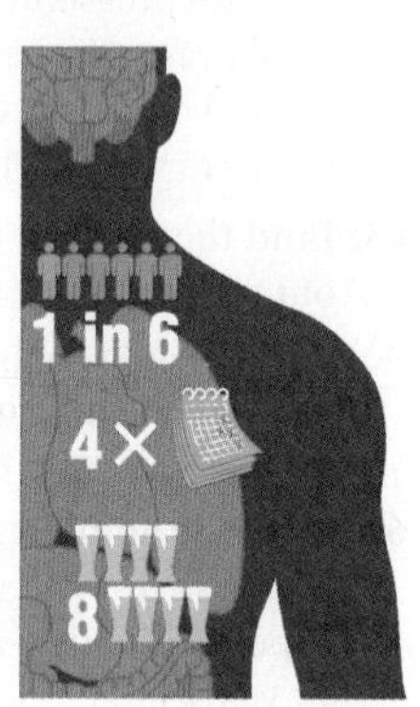

Binge drinking is a nationwide problem and bigger than previously thought. One in six adults binge drinks about four times a month. Binge drinking is defined as consuming four or more drinks for women or five or more drinks for men over a short period of time. Most binge drinkers are not alcohol-dependent. (*Source:* www.cdc.gov, Jan. 2012)

78. Find the number of 80,500 children and adolescents who are obese. Round to the nearest hundred.

Approximately one in six children and adolescents are obese. (*Source:* www.cdc.gov, Nov. 2011)

79. MRI scans of women with the BRCA1 and BRCA2 genetic mutations that were positive for cancer were wrong five out of six times. (These results are "false positives.") If 1500 women with these mutations had MRI scans that indicated cancer, predict how many of these women did not have cancer. (*Source:* www.telegraph.co.uk, March 26, 2008)

80. About two out of three smokers want to quit smoking. Find the number of people who want to quit smoking.

An estimated 45.3 million people, or 19.3% of all adults (aged 18 years or older), in the United States smoke cigarettes. (*Source:* www.cdc.gov, Nov. 2011)

81. Pentobarbital is a sedative. A vial with 50 mL of solution contains 2.5 g of pentobarbital. A patient needs 175 mg of pentobarbital. Find the volume of the solution that the patient needs.

82. Cyclosporine is an anti-rejection drug given to organ transplant patients. A bottle contains 50 mL of liquid. Each milliliter of liquid contains 100 mg of cyclosporine. A kidney transplant patient needs to take 850 mg of cyclosporine each day. Find the amount of solution that the patient should take each day.

Problem Solving: Practice and Review

Follow your instructor's guidelines for using the five steps as outlined in Section 1.3, p. 55.

83. A regulation basketball court in the NBA and the NCAA is 94 ft long and 50 ft wide. A regulation high school basketball court is 84 ft long and 50 ft wide. Find the percent increase in the area of an NCAA court compared to a high school court. Round to the nearest percent.

84. The height of a triangle is 3 ft more than the length of its base, and its area is 54 ft^2. Use a quadratic equation to find the base and height of this triangle. $\left(A = \frac{1}{2}bh.\right)$

85. The table compares the number of ads and cost of advertising in the GOP presidential primaries from Jan. 1, 2007, to Jan. 25, 2008, and from Jan. 1, 2011, to Jan. 25, 2012. Find the percent increase in the number of ads paid for by interest groups. Round to the nearest percent.

		Candidate-sponsored ads	Interest group–sponsored ads
2008	Number of ads	66,557	1763
	Cost of ads	\$48.7 million	\$1.1 million
2012	Number of ads	39,429	30,442
	Cost of ads	\$13.7 million	\$15.2 million

Source: www.mediaproject.wesleyan.edu, Jan. 30, 2012

86. The rulebook of the U.S. Lawn Mower Racing Association describes how to award points.

100 points for registration
100 points for starting a race
100 points for finishing a race
300 points for first place
250 points for second place
200 points for third place
150 points for fourth place
100 points for fifth place

© Gamma-Rapho via Getty Images

Source: www.letsmow.com

A lawn mower racer registered for a day of racing. She started and completed three races. She placed fourth in the first race, third in the second race, and first in the final race. Find the total number of points she earned.

Find the Mistake

For exercises 87–90, the completed problem has one mistake.

(a) Describe the mistake in words, or copy down the whole problem and highlight or circle the mistake.

(b) Do the problem correctly.

87. Problem: Solve: $\frac{11}{x} + \frac{13}{12} = 1$

Incorrect Answer: Least common denominator is $12x$.

$$12x\left(\frac{11}{x} + \frac{13}{12}\right) = 1$$

$$12x\left(\frac{11}{x}\right) + 12x\left(\frac{13}{12}\right) = 1$$

$$132 + 13x = 1$$

$$-132 \qquad\quad -132$$

$$0 + 13x = -131$$

$$\frac{13x}{13} = \frac{-131}{13}$$

$$x = -\frac{131}{13}$$

88. Problem: Solve: $\frac{2}{x-2} + \frac{8}{x} = \frac{x}{x-2}$

Incorrect Answer: $\frac{2}{x-2} + \frac{8}{x} = \frac{x}{x-2}$

$$x(x-2)\left(\frac{2}{x-2} + \frac{8}{x}\right) = x(x-2)\left(\frac{x}{x-2}\right)$$

$$x(x-2)\left(\frac{2}{x-2}\right) + x(x-2)\left(\frac{8}{x}\right) = x^2$$

$$2x + (x-2)8 = x^2$$

$$2x + 8x - 16 = x^2$$

$$10x - 16 = x^2$$

$$0 = x^2 - 10x + 16$$

$$0 = (x-8)(x-2)$$

$x - 8 = 0$ or $x - 2 = 0$

$x = 8$ or $x = 2$

89. Problem: Solve: $\frac{a}{a+14} = -\frac{3}{a+14} + \frac{4}{a}$

Incorrect Answer:

$$\frac{a}{a+14} = -\frac{3}{a+14} + \frac{4}{a}$$

$$a(a+14)\left(\frac{a}{a+14}\right) = a(a+14)\left(-\frac{3}{a+14} + \frac{4}{a}\right)$$

$$a^2 = a(a+14)\left(-\frac{3}{a+14}\right) + a(a+14)\left(\frac{4}{a}\right)$$

$$a^2 = -3a + (a+14)(4)$$

$$a^2 = -3a + 4a + 56$$

$$a^2 = a + 56$$

$$a^2 - a - 56 = 0$$

$$(a+8)(a-7) = 0$$

$a + 8 = 0$ or $a - 7 = 0$

$-8 \quad -8$ $\qquad$ $+7 \quad +7$

$a = -8$ or $a = 7$

90. Problem: If t = time to mow the lawn together, write an equation that represents this problem situation and solve it.

Worker A can mow the lawn in 84 min. Worker B can mow the lawn in 63 min. Find the time for the workers to mow the lawn together. Assume that their rates working together are the same as their rates working alone.

Incorrect Answer: t = time to mow the lawn together

$$t = \frac{(84 \text{ min} + 63 \text{ min})}{2}$$

$$t = 73.5 \text{ min}$$

Review

91. Identify the slope of the line represented by $y = \left(\frac{40 \text{ mi}}{1 \text{ hr}}\right)x$.

92. Is the slope of the line in the graph negative or positive?

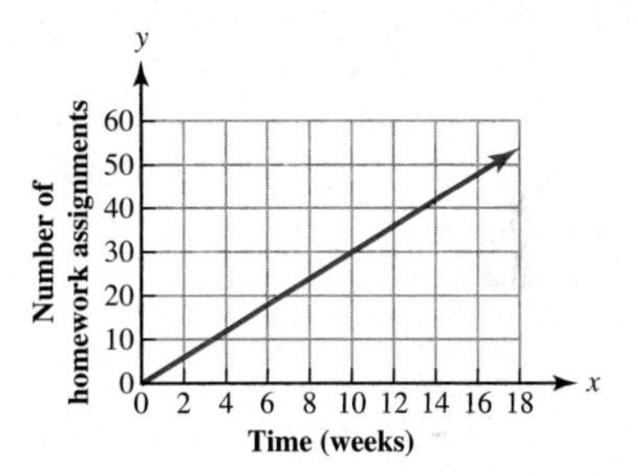

93. Solve: $800 = 5k$

94. Solve: $0.75 = \frac{k}{60}$

SUCCESS IN COLLEGE MATHEMATICS

95. When you leave your math class and do not understand the new material enough to do your homework, you may feel anxious. Outside of class, where can you get help with doing homework?

96. Describe how often you seek out help with your homework.

In many jobs, our pay depends on how many hours we work. We make more money if we work more hours. This is an example of a *direct variation*. In this section, we will study *direct variations* and *inverse variations*.

SECTION 7.7

Variation

After reading the text, working the practice problems, and completing assigned exercises, you should be able to:

1. Write and evaluate an equation that is a direct variation.
2. Write and evaluate an equation that is an inverse variation.
3. Graph an equation that is a direct variation.
4. Identify whether the relationship between two variables in a formula is direct or inverse.

Direct Variation

If a hair stylist earns \$25 an hour, an equation that represents the relationship of the number of hours worked, x, and the gross earnings, y, is $y = \left(\frac{\$25}{1\text{ hr}}\right)x$. The slope of the graph of this equation is $\frac{\$25}{1\text{ hr}}$. As the number of hours worked increases, the amount of gross earnings also increases. This equation is an example of a **direct variation**.

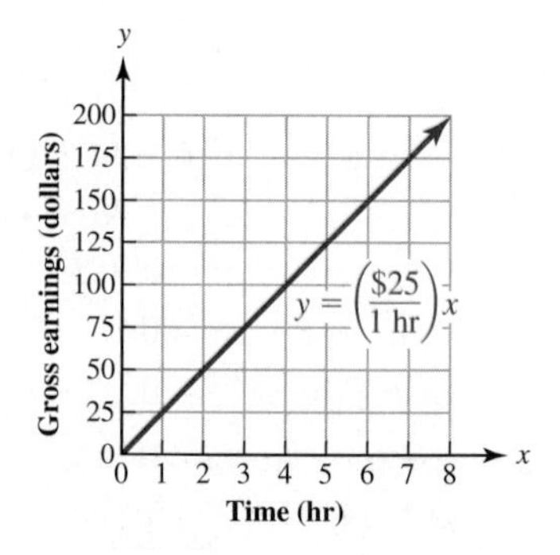

> **Direct Variation**
>
> For an equation that is a direct variation, $y = kx$, the constant of proportionality, which is also the slope of the graph of the equation, is k, where $k > 0$. As x increases, y increases. As x decreases, y decreases.

EXAMPLE 1 The relationship of x and y is a direct variation. When $x = 2$, then $y = 16$.

(a) Find the constant of proportionality, k.

SOLUTION ▶

$y = kx$ Write the equation for a direct variation.

$\mathbf{16} = k(\mathbf{2})$ Replace x and y.

$\frac{16}{2} = \frac{k(2)}{2}$ Division property of equality.

$8 = k$ Simplify.

(b) Write an equation that represents this direct variation.

$y = kx$ The equation for a direct variation.

$y = \mathbf{8}x$ Replace k.

(c) Find y when $x = 2$.

$y = 8x$

$y = 8\mathbf{(2)}$ Replace x with 2.

$y = \mathbf{16}$ Simplify.

(d) Use slope-intercept graphing to graph $y = 8x$.

The y-intercept is $(0, 0)$. The slope is $\frac{8}{1}$.

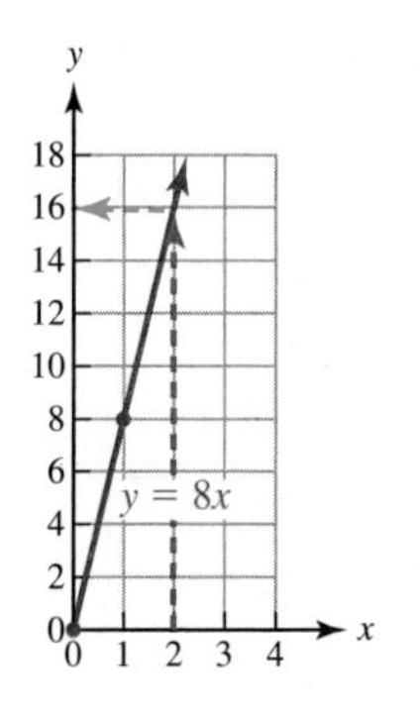

(e) Use the graph to find y when $x = 2$.

Draw a vertical dashed line up from 2 on the x-axis. When the dashed line intersects the graph, pivot and draw a horizontal dashed line to the y-axis. Since this horizontal dashed line intersects the y-axis at 16, when $x = 2$, $y = 16$.

In the next example, the constant of proportionality, k, includes units.

EXAMPLE 2 The relationship of the amount of butter, x, and the number of calories in butter, y, is a direct variation. In 6 T of butter, there are 600 cal.

(a) Find the constant of proportionality, k.

SOLUTION

$y = kx$ Write the equation for a direct variation.

$\mathbf{600\ cal} = k\mathbf{(6\ T)}$ Replace x and y.

$\frac{600\text{ cal}}{6\text{ T}} = \frac{k(6\text{ T})}{6\text{ T}}$ Division property of equality.

$\frac{100\text{ cal}}{1\text{ T}} = k$ Simplify.

(b) Write an equation that represents this direct variation.

$y = kx$ The equation for a direct variation.

$y = \left(\frac{\mathbf{100\ cal}}{\mathbf{1\ T}}\right)x$ Replace k.

(c) Find the number of calories in 16 T of butter (the amount in two sticks of butter or in one recipe of chocolate chip cookies).

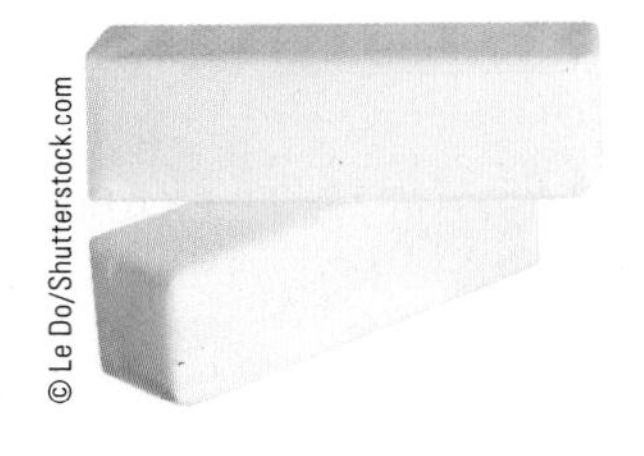
© Le Do/Shutterstock.com

$y = \left(\frac{100\text{ cal}}{1\text{ T}}\right)x$

$y = \left(\frac{100\text{ cal}}{1\cancel{\text{T}}}\right)\mathbf{(16\ \cancel{T})}$ Replace x with 16 T.

$y = \mathbf{1600\ cal}$ Simplify.

Practice Problems

1. The relationship of x and y is a direct variation. When $x = 3$, then $y = 12$.
   a. Find the constant of proportionality, k.
   b. Write an equation that represents this direct variation.
   c. Find y when $x = 5$.
   d. Use slope-intercept graphing to graph the equation.
   e. Use the graph to find y when $x = 2$.
2. The relationship of the miles driven, x, and the amount in dollars that the IRS considers deductible costs for operating an automobile, y, is a direct variation. In 2011, for driving 500 mi to move, the IRS allowed a deductible cost of $117.50. (*Source:* www.irs.gov)
   a. Find the constant of proportionality, k.
   b. Write an equation that represents this direct variation.
   c. Find y when $x = 375$ mi. Round to the nearest hundredth.

Inverse Variation

A color copier delivers finished copies into a tray. An equation that represents the relationship of the time in seconds for the delivery of a copy, x, and the number of copies per hour that the copier can make, y, is $y = \dfrac{\frac{3600 \text{ copies} \cdot \text{s}}{1 \text{ hr}}}{x}$. This is an **inverse variation**. As x increases, y decreases. The constant of proportionality, k, is the product of x and y. Since the units of k in inverse variations are often complicated, they are not always included in the equation.

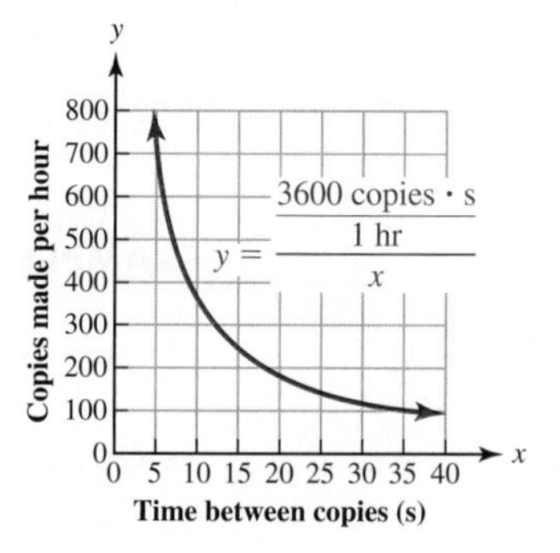

Inverse Variation

For an equation that is an inverse variation, $y = \dfrac{k}{x}$, the constant of proportionality is k, where $k > 0$. As x increases, y decreases. As x decreases, y increases.

In the next example, we find the constant of proportionality for an inverse variation.

EXAMPLE 3 The relationship of x and y is an inverse variation. When $x = 2$, $y = 10$.

(a) Find the constant of proportionality, k.

SOLUTION ▶

$y = \dfrac{k}{x}$ The equation for an inverse variation.

$\mathbf{10} = \dfrac{k}{\mathbf{2}}$ Replace x and y.

$2 \cdot 10 = 2 \cdot \dfrac{k}{2}$ Multiplication property of equality.

$20 = k$ Simplify.

(b) Write an equation that represents this inverse variation.

$$y = \frac{k}{x}$$ The equation for an inverse variation.

$$y = \frac{\mathbf{20}}{x}$$ Replace k.

(c) Find y when $x = 4$.

$$y = \frac{20}{x}$$

$$y = \frac{20}{\mathbf{4}}$$ Replace x with 4.

$$y = \mathbf{5}$$ Simplify.

In the next example, we find the constant of proportionality for an inverse variation.

EXAMPLE 4 When a car travels a fixed distance of 100 mi, the relationship of the speed of the car, x, and the time it travels, y, is an inverse variation. If the car travels at a speed of $\frac{50 \text{ mi}}{1 \text{ hr}}$, it takes 2 hr to travel the distance.

(a) Find the constant of proportionality, k.

SOLUTION

$$y = \frac{k}{x}$$ Equation for an inverse variation.

$$2 \text{ hr} = \frac{k}{\frac{\mathbf{50\ mi}}{\mathbf{1\ hr}}}$$ Replace x and y.

$$\left(\frac{50 \text{ mi}}{1 \cancel{\text{hr}}}\right)(2 \cancel{\text{hr}}) = \left(\frac{50 \cancel{\text{mi}}}{\cancel{1} \text{ hr}}\right)\left(\frac{k}{\frac{50 \cancel{\text{mi}}}{\cancel{1} \text{ hr}}}\right)$$ Multiplication property of equality.

$$100 \text{ mi} = k$$ Simplify.

(b) Write an equation that represents this inverse variation.

$$y = \frac{k}{x}$$ The equation for an inverse variation.

$$y = \frac{\mathbf{100\ mi}}{x}$$ $y = \frac{k}{x}$

(c) Find the time when the speed is $\frac{20 \text{ mi}}{1 \text{ hr}}$.

$$y = \frac{100 \text{ mi}}{x}$$

$$y = 100 \text{ mi} \div \frac{\mathbf{20\ mi}}{\mathbf{1\ hr}}$$ Replace x; rewrite the fraction as division.

$$y = (100 \cancel{\text{mi}})\left(\frac{\mathbf{1\ hr}}{\mathbf{20}\ \cancel{\mathbf{mi}}}\right)$$ Rewrite as multiplication by the reciprocal.

$$y = \mathbf{5\ hr}$$ Simplify.

Practice Problems

3. The relationship of x and y is an inverse variation. When $x = 5, y = 6$.
  a. Find the constant of proportionality, k.
  b. Write an equation that represents this inverse variation.
  c. Find y when $x = 2$.

4. For a rectangle with an area of 24 in.^{2}, the relationship of the length, x, and the width, y, is an inverse variation. When the length of the rectangle is 4 in., the width is 6 in.
  a. Find the constant of proportionality, k. Include the units.
  b. Write an equation that represents this inverse variation.
  c. Find the length of this rectangle when the width is 8 in.

Formulas and Variation

A formula represents a relationship of variables. If the formula includes only multiplication and division, the relationship of any two variables is either a direct variation or an inverse variation.

EXAMPLE 5 The formula $V = LWH$ represents the relationship of the volume of a box V and its length L, width W, and height H.

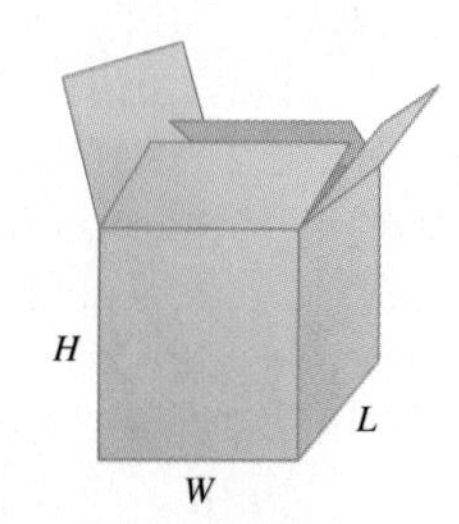

(a) Is the relationship between V and L a direct variation or an inverse variation?

SOLUTION ▶ Imagine this equation as a balanced scale. We assume that the values of W and H cannot change. If L increases, what must happen to V so that the scale remains balanced?

$V = LWH$ If L increases, the scale will tip to the right. To balance the scale, V must also increase. This is a *direct variation*.

(b) Is the relationship of L and H a direct variation or an inverse variation?

▶ Imagine this equation as a balanced scale. We assume that the values of V and W cannot change. If L increases, what must happen to H so that the scale remains balanced?

$V = LWH$ If L increases, the scale will tip to the right. The only way to balance this scale is for H to decrease. This is an *inverse variation*.

Practice Problems

For problems 5–7, the formula $R = \dfrac{336gm}{T}$ represents the relationship of tire diameter, T; gear ratio, g; speed, m; and number of revolutions of the tires per minute, R.

5. Is the relationship of T and R a direct variation or an inverse variation?

6. Is the relationship of g and m a direct variation or an inverse variation?

7. Is the relationship of T and m a direct variation or an inverse variation?

7.7 Vocabulary Practice

Match the equation with the term. A term can be used more than once.

1. $y = \frac{k}{x}$
2. $y = kx$
3. $d = \left(60\frac{\text{mi}}{\text{hr}}\right)t$
4. $t = \frac{125\text{ mi}}{r}$
5. $F = m\left(32\frac{\text{m}}{\text{s}^2}\right)$
6. $C = 2\pi r$
7. $y = 120x$
8. $P = \frac{nRT}{V}$, when n, R, and T are constant
9. $P = \frac{nRT}{V}$, when n, R, and V are constant
10. $W = \frac{A}{L}$, when A is constant

A. Direct variation
B. Inverse variation

7.7 Exercises

Follow your instructor's guidelines for showing your work.

1. Explain why the relationship of the number of apples picked, x, and the number of apple pies that can be made from the apples, y, is a direct variation.
2. Explain why the relationship of the number of square feet of carpet that need to be vacuumed, x, and the amount of time it takes to vacuum the carpet, y, is a direct variation.
3. Explain why the relationship of the number of bags of leaves per hour that are raked, x, and the hours it takes to rake a yard, y, is an inverse variation.
4. Explain why the relationship of the size of a garbage bag, x, and the number of bags that can fit in a dumpster, y, is an inverse variation.
5. The relationship of x and y is a direct variation. When $x = 2$, $y = 6$.
   a. Find the constant of proportionality, k.
   b. Write an equation that represents this direct variation.
   c. Find y when $x = 4$.
   d. Use slope-intercept graphing to graph this equation.
   e. Use the graph to find y when $x = 5$.
6. The relationship of x and y is a direct variation. When $x = 2$, $y = 8$.
   a. Find the constant of proportionality, k.
   b. Write an equation that represents this direct variation.
   c. Find y when $x = 3$.
   d. Use slope-intercept graphing to graph this equation.
   e. Use the graph to find y when $x = 5$.
7. The relationship of x and y is a direct variation. When $x = 1$, $y = 6$.
   a. Find the constant of proportionality, k.
   b. Write an equation that represents this direct variation.
   c. Find y when $x = 4$.
   d. Use slope-intercept graphing to graph this equation.
   e. Use the graph to find y when $x = 2$.
8. The relationship of x and y is a direct variation. When $x = 1$, $y = 5$.
   a. Find the constant of proportionality, k.
   b. Write an equation that represents this direct variation.
   c. Find y when $x = 2$.
   d. Use slope-intercept graphing to graph this equation.
   e. Use the graph to find y when $x = 3$.
9. The relationship of the amount of weed killer concentrate, x, and the amount of mixed weed killer spray, y, is a direct variation. A gardener uses 2 oz of concentrate to make 1 gal of weed killer spray.
   a. Find the constant of proportionality, k. Include the units of measurement.
   b. Write an equation that represents this relationship.
   c. Find the amount of mixed weed killer spray that can be made with 8 oz of concentrate.
   d. Use slope-intercept graphing to graph this equation.
   e. Use the graph to find the amount of mixed weed killer spray that can be made with 6 oz of concentrate.
10. The relationship of the amount of salad dressing, x, and the amount of sodium in the dressing, y, is a direct variation. Six servings of dressing contain 1800 mg of sodium.
    a. Find the constant of proportionality, k. Include the units of measurement.
    b. Write an equation that represents this relationship.
    c. Find the amount of sodium in a bottle that contains 16 servings of salad dressing.

d. Use slope-intercept graphing to graph this equation.
e. Use the graph to find the amount of sodium in 3 servings of salad dressing.

11. The relationship of x and y is an inverse variation. When $x = 2$, $y = 6$.
a. Find the constant of proportionality, k.
b. Write an equation that represents this inverse variation.
c. Find y when $x = 4$.

12. The relationship of x and y is an inverse variation. When $x = 3$, $y = 6$.
a. Find the constant of proportionality, k.
b. Write an equation that represents this inverse variation.
c. Find y when $x = 9$.

13. The relationship of x and y is an inverse variation. When $x = 2$, $y = 10$.
a. Find the constant of proportionality, k.
b. Write an equation that represents this inverse variation.
c. Find y when $x = 5$.

14. The relationship of x and y is an inverse variation. When $x = 4$, $y = 5$.
a. Find the constant of proportionality, k.
b. Write an equation that represents this inverse variation.
c. Find y when $x = 10$.

15. For a fixed number of windows, the number of windows washed per hour, x, and the number of hours it takes to wash the windows, y, is an inverse variation. If a person can wash 20 windows per hour, it takes 9 hr to wash the windows.
a. Find the constant of variation, k. Include the units of measurement.
b. Write an equation that represents this relationship.
c. If a person can wash 30 windows per hour, find the time needed to wash the windows.

16. For a fixed number of hotel rooms, the number of rooms cleaned per hour, x, and the number of hours it takes to clean the rooms, y, is an inverse variation. If a person can clean 8 rooms per hour, it takes 15 hr to clean the rooms.
a. Find the constant of variation, k. Include the units of measurement.
b. Write an equation that represents this relationship.
c. If a person can clean 6 rooms per hour, find the time needed to clean the rooms.

17. The relationship of the radius of a circle, x, and the circumference of the circle, y, is a direct variation. The radius of a circle is 10 cm, and the circumference is 62.8 cm.
a. Find the constant of proportionality, k.
b. Write an equation that represents this relationship.
c. Find the circumference of a circle with a radius of 20 cm.

18. The relationship of the diameter of a circle, x, and the circumference of the circle, y, is a direct variation. The diameter of a circle is 20 cm, and the circumference is 62.8 cm.
a. Find the constant of proportionality, k.
b. Write an equation that represents this relationship.
c. Find the circumference of a circle with a diameter of 40 cm.

19. If the price per share of a company's stock is constant, the relationship of the earnings per share, x, and the financial ratio *price to earnings*, y, is an inverse variation. The earnings per share of a company is \$3.50, and its price to earnings ratio is 16.
a. Find the constant of proportionality, k. Include the units of measurement.
b. Write an equation that represents this relationship.
c. Find the price to earnings ratio when the earnings per share is \$2.

20. If the annual credit sales are constant, the relationship of the accounts receivable, x, and the financial ratio *receivables turnover*, y, is an inverse variation. The accounts receivable of a company are \$150 million, and its receivables turnover ratio is 12.
a. Find the constant of proportionality. Include the units of measurement.
b. Write an equation that represents this relationship.
c. Find the receivables turnover ratio when the accounts receivable are \$200 million.

21. For a fixed length of household copper wire, the relationship of the cross-sectional area, x, and the resistance, y, is an inverse variation. When the cross-sectional area is $3.14 \times 10^{-6}\ \text{m}^2$, the resistance is 5.4×10^{-3} ohm.
a. Find the constant of proportionality, k. Use scientific notation. Include the units of measurement.
b. Write an equation that represents this relationship.
c. Find the resistance when the cross-sectional area is $2.05 \times 10^{-6}\ \text{m}^2$. Round the mantissa to the nearest tenth.

22. If the force acting on an object is constant, the relationship of the mass of the object, x, and the acceleration of the object, y, is an inverse variation. When the mass is 1000 kg, the acceleration is $\frac{4\ \text{m}}{1\ \text{s}^2}$.
a. Find the constant of proportionality, k. Include the units of measurement.
b. Write an equation that represents this relationship.
c. Find the acceleration when the mass is 1500 kg. Round to the nearest tenth.

23. The relationship of the number of tickets sold, x, and the total ticket receipts for an outdoor concert, y, is a direct variation. When 11,000 tickets are sold, the total ticket receipts are \$495,000.
a. Find the constant of proportionality, k. Include the units of measurement.
b. Write an equation that represents this relationship.
c. Find the number of tickets sold when the total ticket receipts are \$562,500.
d. Find the total ticket receipts from the sale of 7575 tickets.
e. What does k represent in this equation?

24. The relationship of the taxable value of a property, x, and the annual property tax, y, is a direct variation. When the taxable value of a property is \$250,000, the annual property tax bill is \$5375.
a. Find the constant of proportionality, k.
b. Write an equation that represents this relationship.

c. Find the taxable value of a property with an annual property tax bill of $8062.50.
d. Find the tax owed for a property with an assessed value of $185,000. Round to the nearest whole number.
e. What does k represent in this equation?

25. The relationship of the time traveled by a charter jet, x, and the cost to charter the jet, y, is a direct variation. When a charter jet flies for 3 hr, the cost is $18,867.
a. Find the constant of proportionality. Round to the nearest whole number. Include the units of measurement.
b. Write an equation that represents this relationship.
c. Find the cost to charter a jet for 12 hr.
d. What does k represent in this equation?

26. The relationship of the time a tour guide works, x, and the cost to hire the tour guide, y, is a direct variation. When a tour guide works for 15 hr, the cost is $1125.
a. Find the constant of proportionality, k. Include the units of measurement.
b. Write an equation that represents this relationship.
c. Find the cost to hire a tour guide for 8 hr.
d. What does k represent in this equation?

27. The relationship of the distance driven, x, and the cost of gasoline, y, is a direct variation. For a trip of 400 mi, the cost is $60.
a. Find the constant of proportionality. Include the units of measurement.
b. Write an equation that represents this relationship.
c. Find the cost of gasoline to drive 225 mi.
d. What does k represent in this equation?

28. The relationship of the distance driven, x, and the cost of gasoline, y, is a direct variation. For a trip of 250 mi, the cost is $90.
a. Find the constant of proportionality. Include the units of measurement.
b. Write an equation that represents this relationship.
c. Find the cost of gasoline to drive 225 mi.
d. What does k represent in this equation?

29. When a car travels a fixed distance, the relationship of the speed of the car, x, and the time it travels, y, is an inverse variation. When the speed is $\frac{35 \text{ mi}}{1 \text{ hr}}$, the time is 0.5 hr.
a. Find the constant of proportionality. Include the units of measurement.
b. Write an equation that represents this relationship.
c. Find the time in hours to travel this distance at a speed of $\frac{50 \text{ mi}}{1 \text{ hr}}$.
d. Change the time in part c to minutes.

30. When a car travels a fixed distance, the relationship between the speed of the car, x, and the time it travels, y, is an inverse variation. When the speed is $\frac{48 \text{ mi}}{1 \text{ hr}}$, the time is 0.75 hr.
a. Find the constant of proportionality. Include the units of measurement.
b. Write an equation that represents this relationship.
c. Find the time in hours to travel this distance at a speed of $\frac{80 \text{ mi}}{1 \text{ hr}}$.
d. Change the time in part c to minutes.

31. In radiography, a grid reduces the effect of X-ray scattering. The relationship of the interspace distance on the grid, x, and the grid ratio, y, is an inverse variation. When the interspace distance on a grid is 300 micrometers, the grid ratio is 8. Write an equation that represents this variation. Include the units.

32. When the radiation is constant, the relationship of the current in an X-ray tube, x, and the exposure time, y, is an inverse variation. When the current is 600 milliamp, the exposure time is 0.2 s. Write an equation that represents this variation. Include the units.

33. The formula $f = \frac{c}{L}$ represents the relationship of the frequency of light f, the wavelength of light L, and the speed of light c. Assume that c is constant. Is the relationship of f and L a direct variation or an inverse variation?

34. The formula $R = \frac{V}{I}$ represents the relationship of the resistance R, voltage V, and current I in an electric circuit. Assume that V is constant. Is the relationship of R and I a direct variation or an inverse variation?

For exercises 35–36, $T = \frac{336gm}{R}$ represents the relationship of tire diameter, T; gear ratio, g; speed, m; and revolutions of the tire per minute, R. Is the relationship of the given variables a direct variation or an inverse variation?

35. g and m are constant; the relationship of T and R.

36. g and R are constant; the relationship of T and m.

For exercises 37–38, $T = \frac{R}{A}$ represents the relationship of the asset turnover ratio, T; the sales revenue of a company, R; and the total revenues of a company, A. Is the relationship of the given variables a direct variation or an inverse variation?

37. R is constant; the relationship of A and T.

38. A is constant; the relationship of R and T.

For exercises 39–40, $C = \frac{A}{L}$ represents the relationship of the current ratio, C; the current assets of a company, A; and the current liabilities of the company, L. Is the relationship of the given variables a direct variation or an inverse variation?

39. L is constant; the relationship of C and A.

40. A is constant; the relationship of C and L.

For exercises 41–44, the formula $R = \frac{VC}{T}$ describes the flow rate of fluid R through an intravenous drip. Is the relationship of the given variables a direct variation or an inverse variation?

41. V and T are constant; the relationship of R and C.

42. C and T are constant; the relationship of R and V.

43. R and T are constant; the relationship of V and C.

44. V and C are constant; the relationship of R and T.

For exercises **45–48**, the formula $R = \frac{UF}{P}$ describes the glomular filtration rate by a kidney R. Is the relationship of the given variables a direct variation or an inverse variation?

45. F and P are constant; the relationship of R and U.

46. U and P are constant; the relationship of R and F.

47. U and F are constant; the relationship of R and P.

48. R and P are constant; the relationship of U and F.

For exercises **49–52**, the formula $C = \frac{P_m P_i}{TF}$ describes the cost of insurance, C. Is the relationship of the given variables a direct variation or an inverse variation?

49. P_i, T, and F are constant; the relationship of C and P_m.

50. P_m, T, and F are constant; the relationship of C and P_i.

51. C, T, and F are constant; the relationship of P_i and P_m.

52. C, P_i, and P_m are constant; the relationship of T and F.

For exercises **53–56**, the formula $F = \frac{100 S_u C_p}{S_p C_u}$ describes the fractional excretion of sodium, F. Is the relationship of the given variables a direct variation or an inverse variation?

53. S_u, S_p, and C_u are constant; the relationship of F and C_p.

54. C_p, S_p, and C_u are constant; the relationship of F and S_u.

55. F, S_u, and C_p are constant; the relationship of S_p and C_u.

56. F, S_p, and C_u are constant; the relationship of C_p and S_u.

Problem Solving: Practice and Review

Follow your instructor's guidelines for using the five steps as outlined in Section 1.3, p. 55.

57. When the top of a cone is removed, the formula for the volume of the remaining cone (the frustrum) is $V = \frac{1}{3}\pi(R^2 + Rr + r^2)h$, where r is the radius of the circle at the top of the frustrum and R is the radius of the circle at the bottom of the frustrum. In 1856, an American army officer, Henry Hopkins Sibley, invented and received a patent for the design of a conical tent that could sleep 12 soldiers. (The apex is the diameter of the top of the frustrum.) Find the volume of the tent in cubic feet. Use $\pi \approx 3.14$. Round to the nearest whole number.

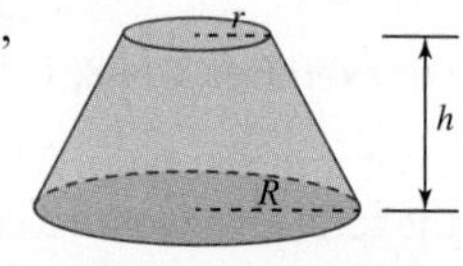

Be it known that I, H.H. Sibley, United States Army, have invented a new and improved Conical Tent . . . the tent is in shape the frustrum of a cone; the base 18 feet; the height 12 feet; the apex 1 foot 6 inches [1.5 ft]. (*Source:* patimg1.uspto.gov)

58. A student overdraws a bank account about five times each month. Predict the total overdraft fees the student will pay in 1 year.

Chase's overdraft fees are \$25 for the first fee each year, \$32 for the next four and \$35 after that. (*Source:* www.nytimes.com, Sept. 23, 2009)

59. Medical researchers collected data on 272 patients who were hospitalized for at least 24 hours with the 2009 H1N1 influenza in the United States from April 2009 to mid-June 2009. One out of four of these patients were admitted to an intensive care unit. About 9 out of 20 patients were children under the age of 18 years. Find the number of patients who were children. Round to the nearest whole number. (*Source:* www.nejm.org, Nov. 12, 2009)

60. In 2011, the total property tax millage rate for Fort Lauderdale, Florida, was 20.1705. (For every \$1000 in taxable property, an owner owes a tax of \$20.1705.) If a property owner pays the tax in four installments, a discount is applied to the first three installments. Find the total amount of tax paid by installments on taxable property of \$175,000. Round to the nearest hundredth.

Installment due date	Discount on the payment
June 30	6%
September 30	4.5%
December 31	3%
March 31	None

Sources: www.broward.org; www.bcpa.net.millage.asp

Find the Mistake

For exercises **61–64**, the completed problem has one mistake.
(a) Describe the mistake in words or copy down the whole problem and highlight or circle the mistake.
(b) Do the problem correctly.

61. Problem: The relationship of the number of gallons of gas, x, and the total cost of the gas, y, is a direct variation. If 8 gallons of gas costs \$24, find the constant of proportionality.

Incorrect Answer: $k = xy$
$k = (8 \text{ gal})(\$24)$
$k = \$192 \text{ gal}$

62. Problem: The relationship of the number of weeks a box of garbage bags is used, x, and the number of bags left in the box, y, is an inverse variation. When x is 8 weeks, y is 168 bags. Find the constant of proportionality, k.

Incorrect Answer: $k = \frac{y}{x}$
$k = \frac{168 \text{ bags}}{8 \text{ weeks}}$

63. Problem: Use slope-intercept graphing to graph $y = \frac{3}{4}x$.

Incorrect Answer:

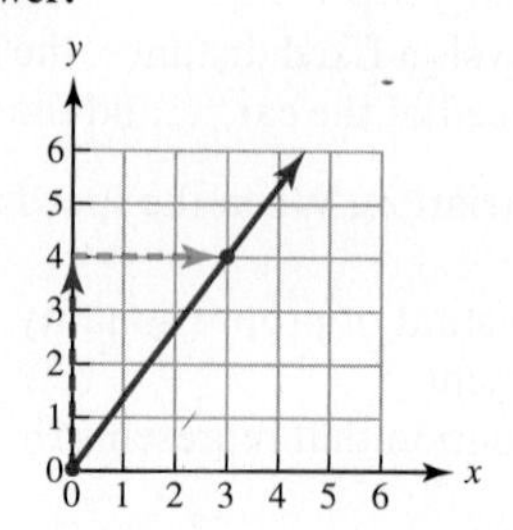

64. Problem: In the formula $A = \frac{10}{B}$, is the relationship between A and B a direct variation or an inverse variation?

Incorrect Answer: Since as B increases, A also increases, this is a direct variation.

Review

For exercises 65–68, evaluate.

65. $\sqrt{16}$

66. $\sqrt[3]{64}$

67. $\sqrt{-100}$

68. $\sqrt[3]{-125}$

SUCCESS IN COLLEGE MATHEMATICS

69. Some students "go blank" on tests because they are too anxious to think. Other students "go blank" on tests because they do not have effective study habits. They did not prepare successfully for the test. One way to check whether you are prepared for a test is to answer the Study Plan Questions in the review for this chapter. Complete the Study Plan Questions in the Study Plan for Review of Section 7.1 by answering "yes" or "no."

70. Now complete the review exercises for this section, and check your answers with the answers in the back of the book. Notice that each review exercise is paired to a study plan question.

71. Were your answers to the Study Plan Questions correct? (If you answered "yes" to a Study Plan Question, you should have been able to complete the review exercises for that question correctly.)

Study Plan for Review of Chapter 7

SECTION 7.1 Simplifying Rational Expressions

Ask Yourself	Test Yourself	Help Yourself
Can I . . .	**Do 7.1 Review Exercises**	**See these Examples and Practice Problems**
simplify a rational number and use the quotient rule of exponents to simplify a rational expression?	2, 4, 6	Ex. 3, PP 4
simplify a rational number or expression by factoring and simplifying common factors?	1, 3, 5, 7–9	Ex. 1–11, PP 1–8
identify any value(s) of a variable in a rational expression that result in division by 0?	10–13	Ex. 6, PP 9–11

7.1 Review Exercises

For exercises 1–4, explain why the expression is not simplified.

1. $\frac{x(x-1)(x+4)}{(x+4)(x+3)}$

2. $\frac{10a^3b^7}{24}$

3. $\frac{x^2-4x}{10x}$

4. $\frac{5x^{-5}y^4}{8}$

5. An expression is $\frac{(4-c)(c+6)}{(c-4)(c+5)}$. Describe the next step in simplifying this expression.

For exercises 6–9, simplify.

6. $\frac{16a^2bc^9}{44ab^4c^2}$

7. $\frac{4x-36}{24}$

8. $\frac{x^2-9}{x^2+8x+15}$

9. $\frac{a^2+8a}{a^3-a^2-72a}$

For exercises 10–12, find any values of the variable for which this expression is undefined.

10. $\frac{5}{x^2-8x-9}$

11. $\frac{3}{u^2-81}$

12. $\frac{2}{n^3-12n^2+32n}$

13. The denominator of the expression $\frac{4}{x^2+1}$ can never equal 0. Explain why.

SECTION 7.2 Multiplication and Division of Rational Expressions

Ask Yourself	Test Yourself	Help Yourself
Can I...	**Do 7.2 Review Exercises**	**See these Examples and Practice Problems**
multiply rational expressions?	14–17	Ex. 1–4, PP 1–4
divide rational expressions?	18–22	Ex. 5–8, PP 5–9

7.2 Review Exercises

For exercises 14–17, simplify.

14. $\left(\dfrac{5}{8x}\right)\left(\dfrac{2x^3}{15}\right)$

15. $\dfrac{6x + 10}{x^2 - 36} \cdot \dfrac{x - 6}{12x + 20}$

16. $\dfrac{x^2 + 7x - 18}{x^2 + 17x + 72} \cdot \dfrac{x^2 - x - 20}{x^2 - 7x + 10}$

17. $\dfrac{6p^4 - 12p^3}{p^2 - 6p + 8} \cdot \dfrac{p^2 - 4p}{6p^3 + 48p^2}$

For exercises 18–22, simplify.

18. $\dfrac{9ab^2}{5} \div \dfrac{3a^2b^4}{20}$

19. $\dfrac{8x^5y}{15z} \div \dfrac{20x^4}{3z}$

20. $\dfrac{12w - 72}{w^2 - 9w + 18} \div \dfrac{18w + 72}{w^2 + w - 12}$

21. $\dfrac{2z^2 - 13z - 45}{3z^2 + 10z + 8} \div \dfrac{4z^2 + 20z + 25}{3z^2 + 7z + 2}$

22. $\dfrac{c^2 - 9}{8c} \div \dfrac{c + 3}{2c^3 + 2c^2 - 4c}$

SECTION 7.3 Combining Rational Expressions with the Same Denominator

Ask Yourself	Test Yourself	Help Yourself
Can I...	**Do 7.3 Review Exercises**	**See these Examples and Practice Problems**
add or subtract rational expressions with the same denominator?	23–28	Ex. 1–5, PP 1–6
write the prime factorization of a number or expression?	29–31	Ex. 6, PP 7–12
use prime factorization to find the least common multiple of two or more expressions?	32–34	Ex. 7–10, PP 8–12

7.3 Review Exercises

For exercises 23–28, simplify.

23. $\dfrac{5x}{9} - \dfrac{2x}{9}$

24. $\dfrac{2x^2}{x + 4} + \dfrac{8x}{x + 4}$

25. $\dfrac{2p^2}{2p + 3} + \dfrac{27p + 36}{2p + 3}$

26. $\dfrac{u^2 + 10u}{u^2 - 81} + \dfrac{9}{u^2 - 81}$

27. $\dfrac{5h^2}{25h^2 + 20h + 4} - \dfrac{38h + 16}{25h^2 + 20h + 4}$

28. $\dfrac{x^2 + 5x + 4}{x^2 - 2x - 24} - \dfrac{x^2 + 4x + 10}{x^2 - 2x - 24}$

For problems 29–31, write the prime factorization of the number.

29. 75

30. 144

31. 23

For problems 32–34, find the least common multiple of the expressions.

32. $18xy^2$; $30x^2y$

33. $u - 6$; $u^2 - u - 30$

34. $7x$; $x^3 - 9x^2 + 14x$

SECTION 7.4 Combining Rational Expressions with Different Denominators

Ask Yourself	Test Yourself	Help Yourself
Can I...	**Do 7.4 Review Exercises**	**See these Examples and Practice Problems**
use prime factorization to find the least common denominator of two or more rational expressions?	35–37, 47	Ex. 1–4, PP 1–10
rewrite a rational expression as an equivalent expression with a different denominator?	38–47	Ex. 5–8, PP 11–17
add or subtract rational expressions with different denominators?	40–46	Ex. 9–14, PP 18–22

7.4 Review Exercises

For problems 35–37, find the least common denominator.

35. $\dfrac{1}{12h^2p}; \dfrac{1}{30hp}$

36. $\dfrac{1}{x^2+8x}; \dfrac{1}{x^2-64}$

37. $\dfrac{1}{x^2+10x+9}; \dfrac{1}{x^2+2x+1}$

38. Rewrite $\dfrac{2}{3(x+9)}$ as an equivalent expression with the denominator $6(x+9)(x-2)$.

39. Rewrite $\dfrac{x}{x^2-6x-7}$ as an equivalent expression with the denominator $(x-7)(x+1)(x+8)$.

For exercises 40–42, simplify.

40. $\dfrac{4}{3x^2}+\dfrac{2}{5xy}$

41. $\dfrac{3q+1}{q^2-1}+\dfrac{q}{q+1}$

42. $\dfrac{8}{4a^2-4a-3}+\dfrac{6}{2a-3}$

For exercises 43–46, simplify.

43. $\dfrac{1}{7np}-\dfrac{3}{14pw}$

44. $\dfrac{d}{d+2}-\dfrac{d}{d-2}$

45. $\dfrac{8}{v^2-8v+12}-\dfrac{4}{v^2-3v-18}$

46. $\dfrac{3}{2x}-\dfrac{x}{12}$

47. A student is adding $\dfrac{3}{14x}+\dfrac{2}{21x}$. The student multiplied 14 and 21 and rewrote the expressions with the common denominator $294x$. Explain why the student may want to instead use the common denominator, $42x$.

SECTION 7.5 Complex Rational Expressions

Ask Yourself	Test Yourself	Help Yourself
Can I...	**Do 7.5 Review Exercises**	**See these Examples and Practice Problems**
simplify a complex rational number or expression by rewriting the expression as division of the numerator by the denominator?	48	Ex. 1–3, PP 1–4
simplify a complex expression using Strategy 1?	49–51	Ex. 4–6, PP 5–7
simplify a complex rational expression using Strategy 2?	49–50	Ex. 7–9, PP 8–10

7.5 Review Exercises

48. Simplify: $\dfrac{\dfrac{x^2+3x}{x^2-5x-24}}{\dfrac{x^2+2x}{x^2-4x-32}}$

For exercises 49–50,
(a) use Strategy 1 to simplify.
(b) use Strategy 2 to simplify.

49. $\dfrac{\dfrac{2}{c}+\dfrac{3}{d}}{\dfrac{5}{d}-\dfrac{6}{c^2}}$

50. $\dfrac{3+\dfrac{1}{x}}{4+\dfrac{2}{x}}$

51. A capacitor stores electric charge. The unit of capacitance is a *farad.* Two capacitors in a series circuit, C_1 and C_2, have a total capacitance C given by the formula $C=\dfrac{1}{\dfrac{1}{C_1}+\dfrac{1}{C_2}}$. If Capacitor C_1 has a capacitance of 4 microfarads and Capacitor C_2 has a capacitance of 6 microfarads, find the total capacitance of the circuit in microfarads. Write the answer as a decimal number.

SECTION 7.6 Rational Equations and Proportions

Ask Yourself	Test Yourself	Help Yourself
Can I...	**Do 7.6 Review Exercises**	**See these Examples and Practice Problems**
clear the fractions from a rational equation, solve, and check the solution?	53–58	Ex. 1–6, PP 1–7
identify an extraneous solution of a rational equation?	55, 57, 59–61	Ex. 5–6, PP 5–7
use a rational equation to solve an application problem with rates of work?	62	Ex. 7–9, PP 8–9
use a proportion to solve an application problem?	63–64	Ex. 10, PP 10–11

7.6 Review Exercises

52. Describe the difference between an expression and an equation.

For exercises 53–57,
(a) clear the fractions from the equation and solve.
(b) check.

53. $\frac{5}{n} + \frac{2}{21} = \frac{1}{3}$

54. $\frac{7}{c + 15} = \frac{1}{c - 15}$

55. $\frac{x}{x + 6} = -\frac{6}{x + 6} + \frac{9}{x}$

56. $\frac{2f}{f + 1} = \frac{2}{f + 1} + \frac{3}{f}$

57. $\frac{3}{p - 3} + \frac{3}{2} = \frac{p}{p - 3}$

58. A student is solving $\frac{k}{k + 4} + \frac{3}{k - 2} = \frac{18}{k^2 + 2k - 8}$. After factoring, the equation is $\frac{k}{k + 4} + \frac{3}{k - 2} = \frac{18}{(k + 4)(k - 2)}$. Describe what the student should do next.

59. Explain why some rational equations have an extraneous solution.

60. To solve $\frac{x}{x - 9} = \frac{x}{x - 9} + 81$, a student cleared the fractions and correctly solved the equation $x = x + 81(x - 9)$. The student's solution was $x = 9$. Explain why this is not a solution of the equation.

61. A student solving $\frac{1}{a + 2} + \frac{1}{a + 3} = \frac{1}{a^2 + 5a + 6}$ says that the solution is $a = -2$. Is the student correct? Explain.

For exercise 62, use the five steps.

62. A motel housekeeper can clean all the rooms on a floor in 145 min. Another housekeeper can do the same job in 160 min. Find the amount of time it would take them to do the job together. Assume that they work at the same rates working together as they do alone. Round to the nearest whole number.

For exercises 63–64, use a proportion and the five steps.

63. According to the Alzheimer's Association, one in eight people age 65 and older has Alzheimer's disease. Predict the number of 2500 people age 65 and older who have Alzheimer's disease. Round to the nearest whole number. (*Source:* www.alz.org)

64. Predict the number of 6500 people with adult children who say that they have paid off debts for their offspring.

It's not just big banks that are getting bailouts these days. Two out of five people with adult children say they have paid off debts for their offspring—most notably, car loans and medical bills, according to a new CreditCards.com scientific poll. (*Source:* www.creditcards.com)

SECTION 7.7 Variation

Ask Yourself	Test Yourself	Help Yourself
Can I...	**Do 7.7 Review Exercises**	**See these Examples and Practice Problems**
find the constant of proportionality for a direct variation and write an equation that represents the direct variation?	65	Ex. 1–2, PP 1–2
find the constant of proportionality for an inverse variation and write an equation that represents an inverse variation?	66	Ex. 3–4, PP 3–4
determine whether the relationship between two variables in a formula is a direct variation or an inverse variation?	67–69	Ex. 5, PP 5–7

7.7 Review Exercises

65. The relationship between the number of city blocks walked, x, and the distance walked in miles, y, is a *direct variation*. The average city block in the eastern United States is about 0.05 mi.
   a. Find the constant of proportionality. Include the units of measurement.
   b. Write an equation that describes this relationship.
   c. Find the number of miles walked by a person who walks 45 city blocks.
   d. Graph this equation.
   e. Use the graph to find the number of miles walked by a person who walks 60 city blocks.

66. A wheel of cheese weighs 320 oz. The relationship of the size of a serving of cheese, x, and the number of servings in the wheel of cheese, y, is represented by $y = \frac{320 \text{ oz}}{x}$.
   a. Is this a direct variation or an inverse variation?
   b. What is the constant of proportionality?
   c. If the size of each serving is 6 oz, find the number of servings in the wheel of cheese. Round to the nearest whole number.

67. A formula is $A = \frac{100P}{T}$.
   a. Is the relationship between A and P a direct variation or an inverse variation?

b. Is the relationship between A and T a direct variation or an inverse variation?
c. Is the relationship between P and T a direct variation or an inverse variation?
d. If P stays constant and T increases, will A increase or decrease?
e. If A stays constant and T increases, will P increase or decrease?
f. If T stays constant and P decreases, will A increase or decrease?

68. When one buys apples at a grocery store, the charge depends on the number of pounds of apples. If the number of pounds of apples increases, the price also increases. Is this an example of a direct variation or an inverse variation?

69. As the length of a board increases, the force needed to break the board decreases. Is this an example of a direct variation or an inverse variation?

Chapter 7 Test

1. Explain why $\frac{(x+2)(x-7)}{(x-4)(x+2)}$ is equivalent to $\frac{x-7}{x-4}$.

For problem 2, use the five steps.

2. A caterer can prepare 18 dozen appetizers in 1 hr and 30 min. Another caterer can do the same job in 1 hr and 50 min. Find the time in minutes that it will take them to prepare the appetizers working together. Assume that they work at the same rates together and alone. Round to the nearest minute.

3. a. Describe the mistake made in doing this problem.
   b. Redo the problem correctly.

Problem: Simplify: $\frac{x^2+5x+6}{x^2-4}$

Incorrect Answer: $\frac{x^2+5x+6}{x^2-4}$

$$= \frac{x^2+5x+6}{(x-2)(x+2)}$$

$$= \cancel{(x-2)(x+2)}\frac{x^2+5x+6}{\cancel{(x-2)(x+2)}}$$

$$= x^2+5x+6$$

For problems 4–11, simplify.

4. $\frac{n^2+10n+25}{n^2+8n+15} \cdot \frac{n^2+2n-8}{n^2+3n-10}$

5. $\frac{24x^3-30x^2}{18x^7}$

6. $\frac{u+2}{u-4} \div \frac{u+8}{u-4}$

7. $\frac{2p^2-6p}{4p^2-24p+36} \div \frac{p^2+5p}{p^2-6p-55}$

8. $\frac{3k}{2k+16} + \frac{k+18}{2k+16}$

9. $\frac{w+3}{w-8} - \frac{w+1}{w+6}$

10. $\frac{x^2}{x^2-16x+63} - \frac{3x+28}{x^2-16x+63}$

11. $\frac{x}{x^2+12x+36} + \frac{x+8}{x^2+6x}$

For problems 12–15,
(a) solve.
(b) check.

12. $\frac{c}{c+5} = -\frac{5}{c+5} + \frac{10}{c}$

13. $\frac{3}{5} + \frac{7}{y} = \frac{5}{6}$

14. $\frac{40}{z+13} = \frac{8}{z-7}$

15. $\frac{7}{10}x - \frac{3}{2} = \frac{5}{6}$

For problem 16, use a proportion and the five steps.

16. Of 30,000 two-year-olds, find the number who will vist an emergency department for an unintentional medication overdose.

Each year, one of every 150 two-year-olds visits an emergency department in the United States for an unintentional medication overdose, most often after finding and eating or drinking medicines without adult supervision. (*Source:* www.cdc.gov, Dec. 13, 2011)

17. a. Is $y = \left(\frac{\$1.25}{1 \text{ day}}\right)x$ a direct variation or an inverse variation?
    b. If x is time in days, what are the units for y?
    c. Draw a graph of this equation.
    d. Use your graph to find y when x is 5 days.

18. a. Is $y = \frac{\$750}{x}$ a direct variation or an inverse variation?
    b. If x increases, does y increase or decrease?
    c. If $x = \frac{\$25}{1 \text{ ft}^2}$, find y.

19. A formula is $AB = CDE$. Is the relationship between A and B a direct variation or an inverse variation? Explain.

Radical Expressions and Equations

8

SUCCESS IN COLLEGE MATHEMATICS

Learning from Tutors

Your college may have a tutoring center, or you may be enrolled in a TRIO program in which you can receive free math tutoring. Perhaps you can hire a tutor for yourself or together with a group of students. Or you might have a friend, family member, or high school teacher who is willing to help. To get the most benefit from a tutor, you need to continue to use good study habits and time management. Visiting a tutor does not replace reading your book, taking and rewriting notes, and asking questions in class. A tutor's job is not to be your instructor. A tutor's job is not to just help you finish your homework. Instead, a tutor helps you identify and learn what you do not understand.

Tutors often do not show you how to do something. A tutor uses questions, asking you what you think should be done next, until the misunderstanding is revealed. If you say, "I don't even know where to start," the tutor will help you look at the problem until you do find the place to start. The tutor is showing you the questions you should be asking yourself, helping you learn the way to find your own misunderstandings.

In very busy drop-in tutoring centers, a tutor might be able to help you for only a few minutes and then must move on to another student. In this situation, it can help to bring another student with you. As you wait for your turn for help, you can work on the problems together.

8.1 Square Roots
8.2 Adding and Subtracting Square Roots
8.3 Multiplying Square Roots
8.4 Dividing Square Roots and Rationalizing Denominators
8.5 Radical Equations
8.6 Higher Index Radicals and Rational Exponents

© Tom Biegalski/Shutterstock.com

In slang, *radical* means something that is cool or excellent. In politics, a radical is someone who holds an extreme point of view. The Latin word *radicalis* means "having roots." In mathematics, a **radical** is another word for a *root* of a number. In this section, we will simplify radicals.

SECTION 8.1 Square Roots

After reading the text, working the practice problems, and completing assigned exercises, you should be able to:

1. Evaluate a square root.
2. Use the product rule of radicals to simplify a square root.

Square Roots and Radicals

A square has four sides of equal length, s. To find its area, we multiply the length of a side by itself. To *square* a number, we also multiply it by itself: $6^2 = 36$. To find the **square root** of a number, we undo the process of squaring. For example, since $6 \cdot 6 = 36$ and $(-6)(-6) = 36$, the square roots of 36 are 6 and -6. The positive square root of a number is its **principal square root**. The notation for principal square root is $\sqrt{}$. For example, $\sqrt{36} = 6$.

The square root of a negative number is not a real number. For example, $\sqrt{-36}$ is not a real number. The product of two positive numbers or two negative numbers is a positive number. Since the product of any real number except 0 and itself is a positive number, the square root of a negative number cannot be a real number.

EXAMPLE 1 **(a)** What are the square roots of 49?

SOLUTION ▸ The square roots of 49 are 7 and -7.

(b) Evaluate: $\sqrt{49}$

▸ $\sqrt{49}$ What is the principal square root of 49?

$= 7$ $(7)(7) = 49$

(c) Evaluate: $\sqrt{-49}$

▸ $\sqrt{-49}$ is not a real number.

The expression $\sqrt{49}$ is a square root. It is also a **radical**. The **radicand** is 49. The **index** of a square root is 2. Although it usually is not shown, the square root $\sqrt{49}$ can also be written with its index as $\sqrt[2]{49}$.

Square Roots

Each real number greater than 0 has two real square roots. The positive square root is the principal square root. The notation for a principal square root is $\sqrt{}$. The square root of a negative number is not a real number. For the radical $\sqrt{x}$, where x is a real number greater than 0, x is the radicand.

In the next example, the radical is an irrational number that is a nonterminating, nonrepeating decimal number. We use a calculator to find its approximate value and round to a given place value.

EXAMPLE 2 Evaluate $\sqrt{5}$. Round to the nearest hundredth.

SOLUTION ▶

$\sqrt{5}$	Use the square root key on a calculator to evaluate.
$= 2.23606\ldots$	This irrational number does not repeat or terminate.
≈ 2.24	Round to the nearest hundredth.

Practice Problems

For problems 1–5, evaluate. If the number is irrational, round to the nearest hundredth.

1. $\sqrt{4}$ **2.** $\sqrt{81}$ **3.** $\sqrt{-64}$ **4.** $\sqrt{2}$ **5.** $\sqrt{15}$

Perfect Squares and the Product Rule of Radicals

To simplify square roots, we identify factors of the radicand that are **perfect squares**.

Perfect Squares

If a whole number greater than 1 is a perfect square, its principal square root is a rational number. If a whole number greater than 1 is not a perfect square, its principal square root is an irrational number.

List of perfect squares: 4, 9, 16, 25, 36, 49, 64, 81, 100, 121, 144, 169, 196, 225, . . .

EXAMPLE 3 Find the greatest whole number factor of the given number that is a perfect square.

SOLUTION **(a)** 32

▶ Perfect squares less than 32: **4**, 9, 16, 25

Factors of 32: 1, 2, **4**, 8, 16, 32

The greatest whole number factor of 32 that is a perfect square is 16.

(b) 108

▶ Perfect squares less than 108: **4**, **9**, 16, 25, 36, 49, 64, 81, 100

Factors of 108: 1, 2, 3, **4**, 6, **9**, 12, 18, 27, 36, 54, 108

The greatest whole number factor of 108 that is a perfect square is 36.

The notation for multiplication of radicals is either a multiplication dot or writing the radicals next to each other: $\sqrt{x} \cdot \sqrt{y} = \sqrt{x}\sqrt{y}$. We use the **product rule of radicals** to simplify radicals.

Product Rule of Radicals

If a and b are real numbers greater than or equal to 0 and n is an even whole number greater than 0, then $\sqrt[n]{ab} = \sqrt[n]{a} \cdot \sqrt[n]{b}$ and $\sqrt[n]{a} \cdot \sqrt[n]{b} = \sqrt[n]{ab}$.

To simplify a square root, find the greatest perfect square factor of the radicand. Use the product rule of radicals to rewrite the square root as a product of two square roots. The greatest perfect square factor will be one of the radicands. Simplify. Do not evaluate irrational numbers such as $\sqrt{2}$.

EXAMPLE 4 Simplify: $\sqrt{32}$

SOLUTION ▶ $\sqrt{32}$

$= \sqrt{16}\sqrt{2}$ Greatest perfect square factor is 16; product rule of radicals.

$= \mathbf{4\sqrt{2}}$ Simplify.

The perfect square factors of 72 are 4, 9, and 36. To simplify $\sqrt{72}$ in the next example, we use the greatest perfect square factor, 36.

EXAMPLE 5 Simplify: $\sqrt{72}$

SOLUTION ▶ $\sqrt{72}$

$= \sqrt{36}\sqrt{2}$ Greatest perfect square factor is 36; product rule of radicals.

$= \mathbf{6\sqrt{2}}$ Simplify.

Practice Problems

For problems 6–9, find the greatest perfect square factor of the number.

6. 54 **7.** 80 **8.** 200 **9.** 60

For problems 10–13, simplify.

10. $\sqrt{18}$ **11.** $\sqrt{50}$ **12.** $\sqrt{75}$ **13.** $\sqrt{700}$

Simplifying Radicals with Variables

The notation $\sqrt{x}$ represents the principal square root of x. A principal square root is always greater than or equal to 0. In this chapter, we will assume that all factors in the radicands are nonnegative real numbers so that $\sqrt{x^2} = x$. In Chapter 9, we will change this assumption so that we can solve quadratic equations with solutions that are not real numbers.

If a perfect square has a base that is a variable, the exponent is a multiple of the index, 2. For example, $\sqrt{x^6} = x^3$ because $x^3 \cdot x^3 = x^6$. The exponent, 6, is a multiple of the index, 2. The exponent in the simplified expression, 3, equals the original exponent, 6, divided by the index, 2.

EXAMPLE 6 Simplify: $\sqrt{x^{10}}$

SOLUTION ▶ $\sqrt[2]{x^{10}}$ The index of a square root is 2.

$= \mathbf{x^5}$ Simplify; $10 \div 2 = 5$

Exponential Expressions with Variables That Are Perfect Squares

A variable raised to a power is a perfect square if its exponent is a multiple of 2.

A square root is simplified when the radicand has no factors that are perfect squares.

EXAMPLE 7 Simplify: $\sqrt{45x^3y}$

SOLUTION ▶ $\sqrt{45x^3y}$

$= \sqrt{9}\sqrt{5}\sqrt{x^2}\sqrt{x}\sqrt{y}$ Find perfect square factors; product rule of radicals.

$= 3 \cdot \sqrt{5} \cdot x \cdot \sqrt{x} \cdot \sqrt{y}$ Simplify.

$= \mathbf{3x\sqrt{5xy}}$ Product rule of radicals; the radicand has no perfect square factors.

Simplifying a Square Root

Assume that all factors in radicands represent nonnegative real numbers.

1. Use the product rule of radicals to rewrite the square root as a product of square roots. Choose radicands that are perfect squares.
2. Simplify. When a square root is simplified, the radicand has no factors that are perfect squares.
3. Use the product rule of radicals to rewrite any remaining square roots as a single square root.

Practice Problems

For problems 14–17, simplify.

14. $\sqrt{x^8}$ **15.** $\sqrt{p^{20}}$ **16.** $\sqrt{x^4y^6}$ **17.** $\sqrt{a^2b^8c}$

For problems 18–20, simplify.

18. $\sqrt{48p^7w^5}$ **19.** $\sqrt{25h^4k}$ **20.** $\sqrt{108a^9b^{10}}$

8.1 Vocabulary Practice

Match the term with its description.

1. A number that can be written as a fraction in which the numerator and denominator are integers
2. A nonterminating, nonrepeating decimal number
3. In $\sqrt[n]{x}$, this is n.
4. In $\sqrt[n]{x}$, this is x.
5. $\sqrt{ab} = \sqrt{a}\sqrt{b}$
6. Another name for a root
7. If x^n is a perfect square, then n is divisible by this.
8. The rational numbers and the irrational numbers are elements of this set.
9. The square root of this number is a rational number.
10. The result of multiplication

A. index
B. irrational number
C. perfect square
D. product
E. product rule of radicals
F. radical
G. radicand
H. rational number
I. real numbers
J. 2

8.1 Exercises

Follow your instructor's guidelines for showing your work. Assume that all factors in radicands represent real numbers that are greater than or equal to 0.

For exercises 1–12, evaluate. If the number is irrational, round to the nearest hundredth.

1. $\sqrt{225}$ **2.** $\sqrt{169}$
3. $\sqrt{64}$ **4.** $\sqrt{10{,}000}$
5. $\sqrt{7}$ **6.** $\sqrt{3}$
7. $\sqrt{101}$ **8.** $\sqrt{102}$
9. $\sqrt{5}$ **10.** $\sqrt{11}$
11. $\sqrt{-4}$ **12.** $\sqrt{-9}$

13. Explain why we know that $\sqrt{10}$ is between 3 and 4.

14. Explain why we know that $\sqrt{19}$ is between 4 and 5.

For exercises 15–30, simplify.

15. $\sqrt{18}$ **16.** $\sqrt{27}$
17. $\sqrt{75}$ **18.** $\sqrt{50}$
19. $\sqrt{98}$ **20.** $\sqrt{162}$
21. $\sqrt{125}$ **22.** $\sqrt{80}$
23. $\sqrt{20}$ **24.** $\sqrt{32}$
25. $\sqrt{405}$ **26.** $\sqrt{448}$
27. $\sqrt{800}$ **28.** $\sqrt{600}$
29. $\sqrt{72}$ **30.** $\sqrt{128}$

For exercises 31–48, simplify.

31. $\sqrt{z^{12}}$
32. $\sqrt{p^{22}}$
33. $\sqrt{a^{20}}$
34. $\sqrt{b^{60}}$
35. $\sqrt{a^{21}}$
36. $\sqrt{b^{61}}$
37. $\sqrt{a^{22}}$
38. $\sqrt{b^{62}}$
39. $\sqrt{a^{23}}$
40. $\sqrt{b^{63}}$
41. $\sqrt{h^{20}k^{30}}$
42. $\sqrt{n^{40}p^{50}}$
43. $\sqrt{r^4z^7}$
44. $\sqrt{u^6z^9}$
45. $\sqrt{x^{15}y^{19}}$
46. $\sqrt{x^{13}y^{17}}$
47. $\sqrt{x^7y^4z}$
48. $\sqrt{x^9y^6z}$

For exercises 49–72, simplify.

49. $\sqrt{25a^2b^{10}}$
50. $\sqrt{36c^2d^{12}}$
51. $\sqrt{50a^3b^{10}}$
52. $\sqrt{72c^3d^{12}}$
53. $\sqrt{21n^5p}$
54. $\sqrt{15w^7z}$
55. $\sqrt{24np}$
56. $\sqrt{28yz}$
57. $\sqrt{288a^6b^3}$
58. $\sqrt{242f^6g^5}$
59. $\sqrt{27a^5b^9}$
60. $\sqrt{32c^7d^9}$
61. $\sqrt{180r}$
62. $\sqrt{125j}$
63. $\sqrt{36x^2y}$
64. $\sqrt{49a^2b}$
65. $\sqrt{3xy^2}$
66. $\sqrt{5cd^2}$
67. $\sqrt{32x^2y^3}$
68. $\sqrt{40x^3y^2}$
69. $\sqrt{45a^4b^6c^8}$
70. $\sqrt{54a^6b^8c^{10}}$
71. $\sqrt{18xy^5}$
72. $\sqrt{8n^5p}$
73. Explain why the square root of a negative number cannot be a real number.
74. Explain why $\sqrt{x^2}$ does not equal x when x is a negative number.

For exercises 75–82, write a square root that when simplified equals the given expression.

75. $8y^2z^3$
76. $9a^2b^9$
77. $3\sqrt{2y}$
78. $5\sqrt{3y}$
79. $2a\sqrt{5b}$
80. $3c\sqrt{7d}$
81. $5n^3\sqrt{p}$
82. $4m^5\sqrt{u}$
83. Is π greater than or less than $\sqrt{6}$?
84. Is π greater than or less than $\sqrt{5}$?

Problem Solving: Practice and Review

Follow your instructor's guidelines for using the five steps as outlined in Section 1.3, p. 55.

85. Find the percent decrease in the amount of interest payments per year. Round to the nearest percent.

During the past 10 years, state agencies and public universities have paid out \$9.4 million worth of interest on late payments to vendors and contractors, an American-Statesman analysis has found. . . . The numbers compiled by the state comptroller's office show an encouraging trend: Interest payments have dropped dramatically in recent years, from \$1.25 million in 2009 to about \$467,000 last year. (*Source:* www.statesman.com, Feb. 11, 2012)

86. The average rate of increase in the amount of strawberries grown in the United States is $\frac{0.107 \text{ billion lb}}{1 \text{ year}}$. In 2001, 1.26 billion lb of strawberries were grown in the United States. (*Source:* www.ers.usda.gov, May 5, 2011)
   a. Write a linear equation that represents the relationship of the number of pounds of strawberries grown in the United States, y, and the number of years since 2001, x.
   b. Find the number of pounds of strawberries grown in the United States in 2013. Round to the nearest tenth of a billion.
87. Since the 2007–2008 school year, the overall capital budget for the Miami-Dade school district has dropped 43% to \$614 million in the 2010–2011 school year. Find the overall capital budget in 2007–2008. Write the answer in millions of dollars, rounding to the nearest million. (*Source:* www.miamiherald.com, Feb. 11, 2012)
88. For every 3500 calories eaten but not used by the body, a child will gain about a pound of weight. There are 97 calories in 8 oz (one serving) of Coca-Cola Classic®. The volume of 1 can of Coca-Cola Classic is 12 oz. If a child drinks 2 cans per day and the child's body does not use the calories, find how many pounds the child will gain per year. (1 year = 365 days.) Round to the nearest whole number.

Find the Mistake

For exercises 89–92, the completed problem has one mistake.
(a) Describe the mistake in words, or copy down the whole problem and highlight or circle the mistake.
(b) Do the problem correctly.

89. **Problem:** Simplify: $\sqrt{50}$
Incorrect Answer: $\sqrt{50}$
$= \sqrt{25} + \sqrt{25}$
$= 5 + 5$
$= 10$

90. **Problem:** Simplify: $\sqrt{36x^{16}}$
Incorrect Answer: $\sqrt{36x^{16}}$
$= \sqrt{36}\sqrt{x^{16}}$
$= 6x^4$

91. **Problem:** Simplify: $\sqrt{200x^3}$
Incorrect Answer: $\sqrt{200x^3}$
$= \sqrt{100}\sqrt{2}\sqrt{x^3}$
$= 10\sqrt{2x^3}$

92. **Problem:** Simplify: $\sqrt{192x^8y^7}$
Incorrect Answer: $\sqrt{192x^8y^7}$
$= \sqrt{64}\sqrt{3}\sqrt{x^8}\sqrt{y^4}\sqrt{y^3}$
$= 8x^4y^2\sqrt{3y^3}$

Review

For exercises 93–96, simplify.

93. $8x^2 + 5x^2$

94. $9a + 15c - 12a - 21c$

95. $(9a)(15a)$

96. Explain how to determine whether two terms are *like terms*. Include an example in your explanation.

SUCCESS IN COLLEGE MATHEMATICS

97. List the place(s) and hours of free math tutoring available to you on your campus. If you are taking this course online, list any resources for tutoring that are provided by your college or the publisher of your textbook.

© Carey Bates/Shutterstock.com

Although carrots, turnips, and radishes are roots, they are not "like roots." In this section, we will add and subtract radicals that are square roots. Only radicals that are "like radicals" can be combined.

SECTION 8.2 Adding and Subtracting Square Roots

After reading the text, working the practice problems, and completing assigned exercises, you should be able to:

1. Identify and combine like radicals.
2. Simplify an expression before combining like radicals.

Combining Like Radicals

To simplify the expression $3x + 5x + 2y - 9y$, we combine like terms. The simplified expression is $8x - 7y$. To simplify radical expressions, we combine **like radicals**. Like radicals have the same radicand and the same index. In the expression $4\sqrt{5} + 3\sqrt{5}$, the index of the like radicals is 2, the radicand is 5, and the coefficients are 4 and 3. To simplify, add the coefficients. The radical does not change.

EXAMPLE 1 Simplify: $4\sqrt{5} + 3\sqrt{5}$

SOLUTION

$4\sqrt{5} + 3\sqrt{5}$

$= 7\sqrt{5}$ Combine like radicals.

When combining like radicals, work from left to right.

EXAMPLE 2 Simplify: $9\sqrt{2} - 3\sqrt{2} - 15\sqrt{2}$

SOLUTION

$9\sqrt{2} - 3\sqrt{2} - 15\sqrt{2}$

$= \mathbf{6}\sqrt{2} - 15\sqrt{2}$ Combine like radicals.

$= -9\sqrt{2}$ Combine like radicals.

Like radicals have exactly the same radicand. In the next example, we cannot combine $10\sqrt{a}$ and $4\sqrt{ac}$ because the radicands are not the same.

EXAMPLE 3 Simplify: $7\sqrt{a} + 9\sqrt{ac} + 3\sqrt{a} - 5\sqrt{ac}$

SOLUTION ▶

$7\sqrt{a} + 9\sqrt{ac} + 3\sqrt{a} - 5\sqrt{ac}$

$= \mathbf{10\sqrt{a} + 4\sqrt{ac}}$ Combine like radicals.

In the next example, after combining like radicals, we simplify the radicand.

EXAMPLE 4 Simplify: $6\sqrt{50} - 13\sqrt{50}$

SOLUTION ▶

$6\sqrt{50} - 13\sqrt{50}$

$= -7\sqrt{50}$ Combine like radicals.

$= -7\sqrt{25}\sqrt{2}$ Find perfect square factors; product rule of radicals.

$= -7 \cdot 5\sqrt{2}$ Simplify.

$= \mathbf{-35\sqrt{2}}$

Practice Problems

For problems 1–7, simplify.

1. $5\sqrt{w} + 3\sqrt{w}$ **2.** $8\sqrt{m} - 1\sqrt{m}$ **3.** $\sqrt{x} - 8\sqrt{x} + 20\sqrt{x}$

4. $5\sqrt{c} + 9\sqrt{c} + 15\sqrt{d} - 8\sqrt{d}$ **5.** $5\sqrt{h} + 3\sqrt{k} - 11\sqrt{h}$

6. $9\sqrt{xy} - 3\sqrt{y} - \sqrt{y} + \sqrt{xy}$ **7.** $4\sqrt{18} - 3\sqrt{18}$

Simplifying Before Combining Like Radicals

To simplify $\sqrt{28} + \sqrt{7}$, first look for perfect square factors of 28. After simplifying $\sqrt{28}$, combine like radicals.

EXAMPLE 5 Simplify: $\sqrt{28} + \sqrt{7}$

SOLUTION ▶

$\sqrt{28} + \sqrt{7}$

$= \sqrt{4}\sqrt{7} + \sqrt{7}$ Find perfect square factors; product rule of radicals.

$= \mathbf{2}\sqrt{7} + \mathbf{1}\sqrt{7}$ Simplify; multiply by 1.

$= 3\sqrt{7}$ Combine like radicals.

To simplify $\sqrt{28} - \sqrt{175}$ in the next example, look for perfect square factors of both radicands. Since $28 = 4 \cdot 7$, check to see whether 175 is divisible by 7. If so, the simplified radicals might be like radicals.

EXAMPLE 6 Simplify: $\sqrt{28} - \sqrt{175}$

SOLUTION ▶

$\sqrt{28} - \sqrt{175}$

$= \sqrt{4}\sqrt{7} - \sqrt{25}\sqrt{7}$ Find perfect square factors; product rule of radicals.

$= \mathbf{2}\sqrt{7} - \mathbf{5}\sqrt{7}$ Simplify.

$= -3\sqrt{7}$ Combine like radicals.

To find a perfect square factor of a large number like 833, divide by each number in the list of perfect squares until a factor appears. For 833, first divide by 4; since $833 \div 4 = 208.25$, 4 is not a factor. Continue the process with 9, then 16, then 25, then 36, and 49. Since $833 \div 49 = 17$, 49 is a perfect square factor of 833.

EXAMPLE 7 Simplify: $\sqrt{153} + \sqrt{833}$

SOLUTION ▶

$\sqrt{153} + \sqrt{833}$

$= \sqrt{9}\sqrt{17} + \sqrt{49}\sqrt{17}$ Find perfect square factors; product rule of radicals.

$= 3\sqrt{17} + 7\sqrt{17}$ Simplify.

$= 10\sqrt{17}$ Combine like radicals.

A radical such as $\sqrt{x^n}$ is a perfect square if n is divisible by 2. For example, $\sqrt{b^6} = b^3$.

EXAMPLE 8 Simplify: $\sqrt{98b^7} + \sqrt{18b^7}$

SOLUTION ▶

$\sqrt{98b^7} + \sqrt{18b^7}$

$= \sqrt{49}\sqrt{2}\sqrt{b^6}\sqrt{b} + \sqrt{9}\sqrt{2}\sqrt{b^6}\sqrt{b}$ Find perfect square factors; product rule of radicals.

$= 7\sqrt{2}\cdot b^3\sqrt{b} + 3\sqrt{2}\cdot b^3\sqrt{b}$ Simplify.

$= 7b^3\sqrt{2b} + 3b^3\sqrt{2b}$ Product rule of radicals.

$= 10b^3\sqrt{2b}$ Combine like radicals; $7b^3 + 3b^3 = 10b^3$

Practice Problems

For problems 8–15, simplify.

8. $\sqrt{75} + \sqrt{48}$
9. $\sqrt{76} + \sqrt{304}$
10. $\sqrt{20} - \sqrt{125} + \sqrt{80}$
11. $\sqrt{100p} + \sqrt{25p}$
12. $\sqrt{18u^2} + \sqrt{50u^2}$
13. $\sqrt{72a} - \sqrt{162a}$
14. $\sqrt{175h^5} + \sqrt{343h^5}$
15. $\sqrt{54p^2w} + \sqrt{24p^2w}$

8.2 Vocabulary Practice

Match the term with its description.

1. $\sqrt{\ }$
2. In $\sqrt[n]{x}$, this is n.
3. $\sqrt{ab} = \sqrt{a}\sqrt{b}$
4. The square root of this number is a rational number.
5. Radicals that have the same radicand and index
6. A variable with an exponent that is divisible by 2
7. A nonterminating, nonrepeating decimal number
8. In $\sqrt[n]{x}$, this is x.
9. Another name for a root
10. A number that can be written as a fraction in which the numerator and denominator are integers

A. index
B. irrational number
C. like radicals
D. perfect square number
E. perfect square variable
F. principal square root
G. product rule of radicals
H. radical
I. radicand
J. rational number

8.2 Exercises

Follow your instructor's guidelines for showing your work. Assume that variables in the radicand are greater than or equal to 0.

For exercises 1–28, simplify.

1. $2\sqrt{x} + 9\sqrt{x}$
2. $3\sqrt{h} + 7\sqrt{h}$
3. $1\sqrt{w} + 1\sqrt{w}$
4. $1\sqrt{v} + 1\sqrt{v}$
5. $7\sqrt{a} - \sqrt{a}$
6. $9\sqrt{b} - \sqrt{b}$
7. $2\sqrt{x} - 9\sqrt{x}$
8. $3\sqrt{h} - 7\sqrt{h}$
9. $3\sqrt{c} + 8\sqrt{c} + 10\sqrt{d} - 5\sqrt{d}$
10. $5\sqrt{x} + 3\sqrt{x} + 9\sqrt{y} - 4\sqrt{y}$
11. $8\sqrt{x} + 9\sqrt{3y} - 6\sqrt{x} - 21\sqrt{3y}$
12. $12\sqrt{a} + 8\sqrt{5b} - 7\sqrt{a} - 17\sqrt{5b}$
13. $3\sqrt{w} - \sqrt{p} - 9\sqrt{w} - \sqrt{p}$

14. $4\sqrt{n} - \sqrt{u} - 13\sqrt{n} - \sqrt{u}$

15. $2\sqrt{b} + 3\sqrt{5c} - 4\sqrt{b} + 8\sqrt{3c}$

16. $7\sqrt{x} + 4\sqrt{2y} - 9\sqrt{x} + 6\sqrt{3y}$

17. $\sqrt{8} + \sqrt{8}$

18. $\sqrt{12} + \sqrt{12}$

19. $\sqrt{72} + \sqrt{72}$

20. $\sqrt{32} + \sqrt{32}$

21. $\sqrt{18} + \sqrt{18} + \sqrt{18}$

22. $\sqrt{12} + \sqrt{12} + \sqrt{12}$

23. $8\sqrt{63} - 4\sqrt{63}$

24. $7\sqrt{54} - 2\sqrt{54}$

25. $7\sqrt{20} - 10\sqrt{20}$

26. $5\sqrt{18} - 11\sqrt{18}$

27. $\sqrt{27} - 3\sqrt{27}$

28. $\sqrt{8} - 7\sqrt{8}$

For exercises **29–72**, simplify.

29. $\sqrt{75} + \sqrt{12}$

30. $\sqrt{50} + \sqrt{18}$

31. $2\sqrt{75} + 8\sqrt{12}$

32. $3\sqrt{50} + 5\sqrt{18}$

33. $\sqrt{24} - \sqrt{54}$

34. $\sqrt{28} - \sqrt{63}$

35. $7\sqrt{24} - 8\sqrt{54}$

36. $4\sqrt{28} - 5\sqrt{63}$

37. $\sqrt{28} + \sqrt{63}$

38. $\sqrt{20} + \sqrt{45}$

39. $\sqrt{2} + \sqrt{8}$

40. $\sqrt{3} + \sqrt{12}$

41. $\sqrt{300} - \sqrt{200}$

42. $\sqrt{75} - \sqrt{50}$

43. $\sqrt{54} - \sqrt{24}$

44. $\sqrt{50} - \sqrt{32}$

45. $\sqrt{75} - \sqrt{12} + \sqrt{3}$

46. $\sqrt{48} - \sqrt{12} + \sqrt{3}$

47. $-\sqrt{200} + \sqrt{50} - \sqrt{32}$

48. $-\sqrt{300} + \sqrt{27} - \sqrt{75}$

49. $\sqrt{100} + \sqrt{36}$

50. $\sqrt{100} + \sqrt{49}$

51. $\sqrt{8h} + \sqrt{32h}$

52. $\sqrt{12x} + \sqrt{48x}$

53. $\sqrt{180z} - \sqrt{500z}$

54. $\sqrt{80m} - \sqrt{125m}$

55. $\sqrt{7x^2} + \sqrt{700x^2}$

56. $\sqrt{11v^2} + \sqrt{44v^2}$

57. $\sqrt{8x^2} + \sqrt{32x^2}$

58. $\sqrt{18y^2} + \sqrt{8y^2}$

59. $\sqrt{8x^6} + \sqrt{32x^6}$

60. $\sqrt{18y^6} + \sqrt{32y^6}$

61. $\sqrt{25a^3} + \sqrt{64a^3}$

62. $\sqrt{16p^3} + \sqrt{49p^3}$

63. $\sqrt{16x^2y} + \sqrt{9x^2y}$

64. $\sqrt{81n^2p} + \sqrt{4n^2p}$

65. $\sqrt{a^3b} - 4\sqrt{a^3b}$

66. $\sqrt{x^3y} - 5\sqrt{x^3y}$

67. $\sqrt{243u^2z} + \sqrt{48u^2z}$

68. $\sqrt{50a^2b} + \sqrt{288a^2b}$

69. $\sqrt{252n^5z^2} + \sqrt{28n^5z^2}$

70. $\sqrt{343a^6b^3} + \sqrt{28a^6b^3}$

71. $\sqrt{36u^3} + \sqrt{49u^3} - \sqrt{25u^3}$

72. $\sqrt{16z^3} + \sqrt{9z^3} - \sqrt{4z^3}$

For exercises **73–82**, evaluate $\sqrt{b^2 - 4ac}$ for the given values of a, b, and c.

73. $a = 1, b = 11, c = 24$

74. $a = 1, b = 14, c = 45$

75. $a = 1, b = 9, c = 18$

76. $a = 1, b = 8, c = 15$

77. $a = 1, b = 4, c = -12$

78. $a = 1, b = 2, c = -15$

79. $a = 1, b = 0, c = -81$

80. $a = 1, b = 0, c = -64$

81. $a = 2, b = -11, c = -21$

82. $a = 2, b = -9, c = -18$

Problem Solving: Practice and Review

Follow your instructor's guidelines for using the five steps as outlined in Section 1.3, p. 55.

83. According to the 2000 Census, *Garcia* is the eighth most common surname in the United States. For every 100,000 Americans, 318 are named *Garcia*. The 2000 Census reported that the population of the United States was 281,421,906. Predict the number of these Americans who had a surname of Garcia. Round to the nearest thousand. (*Source:* www.census.gov)

84. A TEU is the volume of a standard shipping container that is 20 ft long. If the container is 8 ft wide and 8.5 ft high, find the volume of shipping in January 2011. Write the answer in *cubic feet*. Round to the nearest million.

© Aleksey Kondratyuk/ Shutterstock.com

January 2012 is forecast at 1.1 million TEUs, down 8.7 percent from January 2011. (*Source:* www.neworleanscitybusiness.com, Nov. 8, 2011)

85. Ordinary corn syrup contains dextrose sugar. Enzyme conversion changes it into a syrup that is 42% fructose. This new syrup can be used to make high-fructose corn syrup that is 95% fructose. Manufacturers blend 95% fructose syrup and 42% fructose syrup to produce a 55% fructose syrup. Use a system of two linear equations to find the amount of 42% fructose syrup and the amount of 95% fructose syrup needed to produce 5000 gal of 55% fructose syrup. Round to the nearest whole number.

86. An employee has \$500 in her 403B retirement account at work. She is going to use automatic withdrawal from her checking account to add \$35 to this retirement account each month.

a. Write a linear equation that represents the relationship of the total amount in her retirement account, y, and the number of months that she has used automatic withdrawal, x.

b. Find the balance in her retirement account after 2 years of using automatic withdrawal.

Find the Mistake

For exercises **87–90**, the completed problem has one mistake.

(a) Describe the mistake in words, or copy down the whole problem and highlight or circle the mistake.

(b) Do the problem correctly.

87. Problem: Simplify: $\sqrt{20} + \sqrt{20}$

Incorrect Answer: $\sqrt{20} + \sqrt{20}$

$= 2\sqrt{20}$

88. Problem: Simplify: $\sqrt{4x} + \sqrt{4x}$

Incorrect Answer: $\sqrt{4x} + \sqrt{4x}$

$= \sqrt{8x}$

$= \sqrt{4}\sqrt{2}\sqrt{x}$

$= 2\sqrt{2x}$

89. Problem: Simplify: $\sqrt{48a^2b} + \sqrt{27a^2b}$

Incorrect Answer: $\sqrt{48a^2b} + \sqrt{27a^2b}$
$= \sqrt{16}\sqrt{3}\sqrt{a^2b} + \sqrt{9}\sqrt{3}\sqrt{a^2b}$
$= 4\sqrt{3a^2b} + 3\sqrt{3a^2b}$
$= 7\sqrt{3a^2b}$

90. Problem: Simplify: $\sqrt{50x} + \sqrt{8y}$

Incorrect Answer: $\sqrt{50x} + \sqrt{8y}$
$= \sqrt{25}\sqrt{2}\sqrt{x} + \sqrt{4}\sqrt{2}\sqrt{x}$
$= 5\sqrt{2x} + 2\sqrt{2x}$
$= 7\sqrt{2x}$

Review

91. Simplify: $\frac{3}{5} \cdot \frac{7}{7}$

92. Simplify: $\frac{2}{9} \cdot \frac{4}{4}$

93. In the set of real numbers, 1 is the *multiplicative identity.* Explain why you think it is called the multiplicative identity.

94. In the set of real numbers, 0 is the *additive identity.* Explain why you think it is called the additive identity.

SUCCESS IN COLLEGE MATHEMATICS

95. When you seek help from a tutor, it is important to bring both your textbook and your class notes. Explain why you think it is important to bring your class notes.

$a(b + c) = ab + ac$

The distributive property, $a(b + c) = ab + ac$, is true for all real numbers a, b, and c. When radicals represent real numbers, we can use the distributive property to multiply expressions that include radicals. Throughout this section, we assume that all factors in radicands represent real numbers that are greater than or equal to 0 and that all variables that are bases of exponential expressions represent nonnegative real numbers.

SECTION 8.3 Multiplying Square Roots

After reading the text, working the practice problems, and completing assigned exercises, you should be able to:

1. Multiply square roots and simplify the product.
2. Use the distributive property to multiply radical expressions.

Multiplying Square Roots

To multiply square roots, use the product rule of radicals to write the product as a single square root. Factor each number and write perfect squares in exponential notation. Rewrite as a product of square roots and simplify.

EXAMPLE 1 Simplify: $\sqrt{6}\sqrt{15}$

SOLUTION

$\sqrt{6}\sqrt{15}$

$= \sqrt{6 \cdot 15}$ Product rule of radicals.

$= \sqrt{2 \cdot 3 \cdot 3 \cdot 5}$ Factor each number.

$= \sqrt{3^2}\sqrt{10}$ Exponential notation; product rule of radicals.

$= \mathbf{3\sqrt{10}}$ Simplify.

In the next example, the radicands include numbers and variables.

EXAMPLE 2 Simplify: $\sqrt{10x}\sqrt{35xy}$

SOLUTION ▶

$\sqrt{10x}\sqrt{35xy}$

$= \sqrt{10 \cdot x \cdot 35 \cdot x \cdot y}$ Product rule of radicals.

$= \sqrt{\mathbf{2 \cdot 5} \cdot x \cdot \mathbf{5 \cdot 7} \cdot x \cdot y}$ Factor each number.

$= \sqrt{5^2}\sqrt{x^2}\sqrt{2 \cdot 7 \cdot y}$ Exponential notation; product rule of radicals.

$= \mathbf{5x\sqrt{14y}}$ Simplify.

Multiplying Square Roots

1. If the radicands include perfect square factors, simplify before multiplying.
2. Use the product rule of radicals to rewrite as a single radical. Factor each number.
3. Write perfect squares in exponential notation. Rewrite as a product of radicals.
4. Simplify.

In the next example, since it is often easier to find perfect square factors of smaller numbers, we choose to first simplify each radical and then multiply the remaining radicals.

EXAMPLE 3 Simplify: $\sqrt{72ab}\sqrt{32b}$

SOLUTION ▶

$\sqrt{72ab}\sqrt{32b}$

$= \sqrt{36}\sqrt{2ab}\sqrt{16}\sqrt{2b}$ Find perfect square factors; product rule of radicals.

$= \mathbf{6} \cdot \sqrt{2ab} \cdot \mathbf{4} \cdot \sqrt{2b}$ Simplify.

$= 24\sqrt{2ab \cdot 2b}$ Simplify; product rule of radicals.

$= 24\sqrt{2^2}\sqrt{b^2}\sqrt{a}$ Exponential notation; product rule of radicals.

$= 24 \cdot \mathbf{2} \cdot \mathbf{b}\sqrt{a}$ Simplify.

$= \mathbf{48b\sqrt{a}}$ Simplify.

Practice Problems

For problems 1–4, simplify.

1. $\sqrt{15p}\sqrt{10p}$ **2.** $\sqrt{21pw}\sqrt{70w}$ **3.** $\sqrt{45x^2y}\sqrt{35y}$ **4.** $\sqrt{18ab}\sqrt{48a}$

The Distributive Property

The distributive property, $a(b + c) = ab + ac$, is true for all real numbers a, b, and c. Since irrational numbers such as $\sqrt{3}$ and $\sqrt{2}$ are real numbers, we can use the distributive property to multiply expressions that include radicals.

EXAMPLE 4 Simplify: $\sqrt{3}(\sqrt{2x} + 5)$

SOLUTION ▶

$\sqrt{3}(\sqrt{2x} + 5)$

$= \sqrt{\mathbf{3}}\sqrt{2x} + \sqrt{\mathbf{3}} \cdot 5$ Distributive property.

$= \sqrt{6x} + 5\sqrt{3}$ Product rule of radicals; change order.

In the next example, we use the distributive property and then simplify the product. If a square root is simplified, the radicand has no perfect square factors.

EXAMPLE 5 Simplify: $\sqrt{6x}(\sqrt{3x} + \sqrt{2})$

SOLUTION ▶

$\sqrt{6x}(\sqrt{3x} + \sqrt{2})$

$= \sqrt{\mathbf{6x}}\sqrt{3x} + \sqrt{\mathbf{6x}}\sqrt{2}$ Distributive property.

$= \sqrt{2 \cdot 3 \cdot x \cdot 3 \cdot x} + \sqrt{2 \cdot 3 \cdot x \cdot 2}$ Product rule of radicals; factor each number.

$= \sqrt{3^2}\sqrt{x^2}\sqrt{2} + \sqrt{2^2}\sqrt{3x}$ Use exponential notation; product rule of radicals.

$= \mathbf{3x}\sqrt{2} + \mathbf{2}\sqrt{3x}$ Simplify.

In Chapter 5, we used the distributive property to multiply $(a - b)(a + b)$. In the next example, we use the distributive property to multiply $(\sqrt{a} - \sqrt{b})(\sqrt{a} + \sqrt{b})$.

EXAMPLE 6 Simplify: $(\sqrt{a} - \sqrt{b})(\sqrt{a} + \sqrt{b})$

SOLUTION ▶

$(\sqrt{a} - \sqrt{b})(\sqrt{a} + \sqrt{b})$

$= \sqrt{\mathbf{a}}(\sqrt{a} + \sqrt{b}) - \sqrt{\mathbf{b}}(\sqrt{a} + \sqrt{b})$ Distributive property.

$= \sqrt{\mathbf{a}}\sqrt{a} + \sqrt{\mathbf{a}}\sqrt{b} - \sqrt{\mathbf{b}}\sqrt{a} - \sqrt{\mathbf{b}}\sqrt{b}$ Distributive property.

$= a + \sqrt{ab} - \sqrt{ba} - b$ Simplify; product rule of radicals.

$= a + \sqrt{ab} - \sqrt{\mathbf{ab}} - b$ Change order; commutative property.

$= a + \mathbf{0} - b$ Simplify.

$= a - b$ Simplify.

In Example 6, we see that $(\sqrt{a} - \sqrt{b})(\sqrt{a} + \sqrt{b})$ equals $a - b$, an expression that does not include any radicals. The expressions $\sqrt{a} - \sqrt{b}$ and $\sqrt{a} + \sqrt{b}$ are **conjugates**. In the next example, $\sqrt{3x} - \sqrt{5}$ and $\sqrt{3x} + \sqrt{5}$ are conjugates.

EXAMPLE 7 Simplify: $(\sqrt{3x} - \sqrt{5})(\sqrt{3x} + \sqrt{5})$

SOLUTION ▶

$(\sqrt{3x} - \sqrt{5})(\sqrt{3x} + \sqrt{5})$

$= \sqrt{\mathbf{3x}}(\sqrt{3x} + \sqrt{5}) - \sqrt{\mathbf{5}}(\sqrt{3x} + \sqrt{5})$ Distributive property.

$= \sqrt{\mathbf{3x}}\sqrt{3x} + \sqrt{\mathbf{3x}}\sqrt{5} - \sqrt{\mathbf{5}}\sqrt{3x} - \sqrt{\mathbf{5}}\sqrt{5}$ Distributive property.

$= 3x + \sqrt{15x} - \sqrt{15x} - 5$ Simplify; product rule of radicals.

$= 3x + \mathbf{0} - 5$ Simplify.

$= 3x - 5$ Simplify.

Practice Problems

For problems 5–8, simplify.

5. $\sqrt{2}(\sqrt{7x} + \sqrt{5})$

6. $\sqrt{5a}(\sqrt{10a} + 6)$

7. $(\sqrt{2x} - \sqrt{7})(\sqrt{2x} + \sqrt{7})$

8. $(\sqrt{5x} - 1)(\sqrt{5x} - 1)$

8.3 Vocabulary Practice

Match the term with its description.

1. Radicals that have the same radicand and index
2. $\sqrt{a} - \sqrt{b}$ and $\sqrt{a} + \sqrt{b}$
3. $\sqrt{ab} = \sqrt{a}\sqrt{b}$
4. The square root of this number is a rational number.
5. $\sqrt{}$
6. A variable with an exponent that is divisible by 2
7. A number that can be written as a fraction in which the numerator and denominator are integers
8. In $\sqrt[n]{x}$, this is x.
9. A nonrepeating, nonterminating decimal number
10. $a(b + c) = ab + ac$

A. conjugates
B. distributive property
C. irrational number
D. like radicals
E. perfect square number
F. perfect square variable
G. principal square root
H. product rule of radicals
I. radicand
J. rational number

8.3 Exercises

Follow your instructor's guidelines for showing your work. Assume that all factors in radicands represent real numbers that are greater than or equal to 0. In simplified expressions, denominators are rational, and rational numbers are in lowest terms.

For exercises 1–20, simplify.

1. $\sqrt{3}\sqrt{3}$
2. $\sqrt{5}\sqrt{5}$
3. $\sqrt{a}\sqrt{a}$
4. $\sqrt{c}\sqrt{c}$
5. $\sqrt{7m}\sqrt{7m}$
6. $\sqrt{3n}\sqrt{3n}$
7. $\sqrt{2ab}\sqrt{2ab}$
8. $\sqrt{3xy}\sqrt{3xy}$
9. $\sqrt{6}\sqrt{21}$
10. $\sqrt{10}\sqrt{14}$
11. $\sqrt{12}\sqrt{20}$
12. $\sqrt{28}\sqrt{8}$
13. $\sqrt{42}\sqrt{33}$
14. $\sqrt{35}\sqrt{30}$
15. $\sqrt{30}\sqrt{60}$
16. $\sqrt{40}\sqrt{50}$
17. $\sqrt{10}\sqrt{15}$
18. $\sqrt{12}\sqrt{18}$
19. $\sqrt{8}\sqrt{18}$
20. $\sqrt{8}\sqrt{32}$

For exercises 21–40, simplify.

21. $\sqrt{14ab}\sqrt{21bc}$
22. $\sqrt{10hk}\sqrt{14kw}$
23. $\sqrt{5a}\sqrt{15b}$
24. $\sqrt{3c}\sqrt{15d}$
25. $\sqrt{10n}\sqrt{2m}$
26. $\sqrt{21a}\sqrt{3p}$
27. $\sqrt{11x}\sqrt{11y}$
28. $\sqrt{13r}\sqrt{13v}$
29. $\sqrt{12xy}\sqrt{3y}$
30. $\sqrt{18ab}\sqrt{2b}$
31. $\sqrt{10a}\sqrt{10b}$
32. $\sqrt{20x}\sqrt{20y}$
33. $\sqrt{7abc}\sqrt{21ac}$
34. $\sqrt{3pwz}\sqrt{21pz}$
35. $\sqrt{18a}\sqrt{2a}$
36. $\sqrt{8p}\sqrt{2p}$
37. $\sqrt{200p}\sqrt{32p}$
38. $\sqrt{300n}\sqrt{12n}$
39. $\sqrt{6}\sqrt{10}\sqrt{15}$
40. $\sqrt{6}\sqrt{14}\sqrt{21}$
41. The product of $\sqrt{6}$ and another radical is $3\sqrt{22}$. Find the other radical.
42. The product of $\sqrt{10}$ and another radical is $5\sqrt{6}$. Find the other radical.
43. The product of $\sqrt{x^5}$ and another radical is x^4. Find the other radical.
44. The product of $\sqrt{z^7}$ and another radical is z^6. Find the other radical.

For exercises 45–78, simplify.

45. $3(2x + 5)$
46. $5(3x + 2)$
47. $\sqrt{3}(\sqrt{2x} + 5)$
48. $(\sqrt{5})(\sqrt{3x} + 2)$
49. $\sqrt{3}(\sqrt{2x} + \sqrt{5})$
50. $(\sqrt{5})(\sqrt{3x} + \sqrt{2})$
51. $(x + \sqrt{2})(x + \sqrt{3})$
52. $(x + \sqrt{2})(x + \sqrt{5})$
53. $(x - \sqrt{2})(x - \sqrt{3})$
54. $(x - \sqrt{2})(x - \sqrt{5})$
55. $(x - \sqrt{2})(x + \sqrt{3})$
56. $(x - \sqrt{2})(x + \sqrt{5})$
57. $\sqrt{7n}(\sqrt{14n} + 5)$
58. $\sqrt{2n}(\sqrt{6n} + 3)$
59. $\sqrt{3p}(\sqrt{6p} - 2)$
60. $\sqrt{5a}(\sqrt{10a} - 7)$
61. $\sqrt{3p}(\sqrt{p} - \sqrt{2})$
62. $\sqrt{5a}(\sqrt{a} - \sqrt{3})$
63. $(\sqrt{x} + \sqrt{y})(\sqrt{x} + \sqrt{y})$
64. $(\sqrt{n} + \sqrt{p})(\sqrt{n} + \sqrt{p})$
65. $(\sqrt{x} + y)(\sqrt{x} + y)$

66. $(\sqrt{n} + p)(\sqrt{n} + p)$
67. $(\sqrt{x} - \sqrt{y})(\sqrt{x} - \sqrt{y})$
68. $(\sqrt{n} - \sqrt{p})(\sqrt{n} - \sqrt{p})$
69. $(\sqrt{x} - y)(\sqrt{x} - y)$
70. $(\sqrt{n} - p)(\sqrt{n} - p)$
71. $(a - \sqrt{6})(a - \sqrt{15})$
72. $(u - \sqrt{10})(u - \sqrt{6})$
73. $(a + \sqrt{6})(a - \sqrt{15})$
74. $(u + \sqrt{10})(u - \sqrt{6})$
75. $(\sqrt{3x} + 1)(\sqrt{3x} + 1)$
76. $(\sqrt{5x} + 1)(\sqrt{5x} + 1)$
77. $(\sqrt{x} + \sqrt{15})(\sqrt{x} + \sqrt{21})$
78. $(\sqrt{x} + \sqrt{22})(\sqrt{x} + \sqrt{6})$

For exercises 79–84, write the conjugate of the expression.

79. $\sqrt{n} - \sqrt{6}$
80. $\sqrt{w} - \sqrt{3}$
81. $\sqrt{a} + \sqrt{5}$
82. $\sqrt{v} + \sqrt{7}$
83. $\sqrt{x} - 2$
84. $\sqrt{y} - 5$

Problem Solving: Practice and Review

Follow your instructor's guidelines for using the five steps as outlined in Section 1.3, p. 55.

85. Find the percent increase in the price for a gallon of gas. Round to the nearest tenth of a percent.

    The average price for a gallon of gas was \$3.51 on Monday, up 12 cents from a month earlier. (*Source:* www.pnj.com, Feb. 14, 2012)

86. Find the total number of adults used to find the percent of adults who lived in households with only wireless telephones. Round to the nearest million.

    In the first 6 months of 2011, more than 3 of every 10 households (31.6%) did not have a landline telephone but did have at least one wireless telephone. Approximately 30.2% of all adults (approximately 69 million adults) lived in households with only wireless telephones; 36.4% of all children (approximately 27 million children) lived in households with only wireless telephones. (*Source:* www.cdc.gov, Dec. 21, 2011)

87. The neon Heinz 57 sign at the Heinz History Center in Pittsburgh is 42.5 ft high and 32 ft wide. It "refills" the Heinz logo with ketchup every 30 s. Find the number of times the Heinz Ketchup bottle will "refill" the logo per year. (1 year = 365 days.) (*Source:* www.heinzhistorycenter.org)

Courtesy Heinz History Center

88. A truck owner is buying four studded snow tires. If she also buys wheels for the tires for \$110 per wheel, the shop mounts and balances the tires at no cost. If she does not buy wheels, she will pay for mounting and balancing the tires two times a year when she changes tires. The cost of mounting and balancing is \$20 per tire. Find the number of years at which the cost of buying tires without wheels (paying for mounting and balancing twice a year) is the same as the cost of buying tires with wheels. Round up to the nearest whole number.

Find the Mistake

For exercises 89–92, the completed problem has one mistake.
(a) Describe the mistake in words, or copy down the whole problem and highlight or circle the mistake.
(b) Do the problem correctly.

89. **Problem:** Simplify: $\sqrt{50}\sqrt{6}$

    Incorrect Answer: $\sqrt{50}\sqrt{6}$
    $= \sqrt{2}\sqrt{25}\sqrt{6}$
    $= 5\sqrt{12}$

90. **Problem:** Simplify: $\sqrt{3}(\sqrt{x} - 5)$

    Incorrect Answer: $\sqrt{3}(\sqrt{x} - 5)$
    $= \sqrt{3x} - \sqrt{15}$

91. **Problem:** Simplify: $\sqrt{18x}\sqrt{15x}$

    Incorrect Answer: $\sqrt{18x}\sqrt{15x}$
    $= \sqrt{2 \cdot 3 \cdot 3 \cdot x \cdot 3 \cdot 5 \cdot x}$
    $= \sqrt{3^2}\sqrt{2 \cdot 3 \cdot 5 \cdot x}$
    $= 3\sqrt{30x}$

92. **Problem:** Simplify: $\sqrt{2a}\sqrt{6ab}$

    Incorrect Answer: $\sqrt{2a}\sqrt{6ab}$
    $= \sqrt{2 \cdot a \cdot 2 \cdot 3 \cdot a \cdot b}$
    $= \sqrt{2^2}\sqrt{a^2}\sqrt{b^2}\sqrt{3}$
    $= 2ab\sqrt{3}$

Review

For exercises 93–96, use a pattern to factor. Identify any prime polynomials.

93. $x^2 - y^2$
94. $9a^2 - b^2$
95. $16n^2 - 25p^2$
96. $25x^2 + 4y^2$

SUCCESS IN COLLEGE MATHEMATICS

97. In this section, the product rule of radicals is used to simplify radical expressions. The steps that show the use of the product rule of radicals are included in the examples. If you seek help from a tutor on your homework assignment, the tutor might demonstrate a "shortcut" for finding the right answer that does not include showing work with the product rule of radicals. Describe any disadvantages of using this shortcut.

$$1.41421\overline{)1.00000}$$?

$$2\overline{)1.41421}$$?

The division problem on the top is more difficult than the one on the bottom. This is one reason that we rationalize denominators. In this section, we will learn how to rationalize a denominator.

SECTION 8.4

Dividing Square Roots and Rationalizing Denominators

After reading the text, working the practice problems, and completing assigned exercises, you should be able to:

1. Use the quotient rule of radicals to evaluate or simplify a square root.
2. Rationalize the denominator of a radical expression.
3. Rationalize the denominator of a radical expression that includes a sum or difference.

The Quotient Rule of Radicals

We use the **quotient rule of radicals** to rewrite a radical expression as the quotient of two radicals.

Quotient Rule of Radicals

If a, b are real numbers, $a \geq 0$, $b > 0$, and n is a whole number greater than 1,

$$\sqrt[n]{\frac{a}{b}} = \frac{\sqrt[n]{a}}{\sqrt[n]{b}} \text{ and } \frac{\sqrt[n]{a}}{\sqrt[n]{b}} = \sqrt[n]{\frac{a}{b}}.$$

In Example 1, the numerator of the radicand is a perfect square, 16, and the denominator of the radicand is a perfect square, 25. The simplified expression is a rational number.

EXAMPLE 1 Evaluate: $\sqrt{\frac{16}{25}}$

SOLUTION ▶ $\sqrt{\frac{16}{25}}$

$= \frac{\sqrt{16}}{\sqrt{25}}$ Quotient rule of radicals.

$= \frac{4}{5}$ Simplify.

In the next example, the numerator of the simplified expression includes a square root.

EXAMPLE 2 Simplify: $\sqrt{\frac{50}{9}}$

SOLUTION ▶

$$\sqrt{\frac{50}{9}}$$

$$= \frac{\sqrt{50}}{\sqrt{9}}$$ Quotient rule of radicals.

$$= \frac{\sqrt{25}\sqrt{2}}{\sqrt{9}}$$ Find perfect square factors; product rule of radicals.

$$= \frac{5\sqrt{2}}{3}$$ Simplify.

Many traditions for simplifying expressions make arithmetic easier and more efficient. By tradition, simplified expressions have no common factors in the numerator and denominator and have only positive exponents. Also by tradition, *simplified expressions have no radicals in the denominator.* To simplify an expression such as $\frac{1}{\sqrt{2}}$, multiply by a fraction that is equal to 1, $\frac{\sqrt{2}}{\sqrt{2}}$. Although the numerator of the product now includes a radical, the denominator does not include a radical.

EXAMPLE 3 Simplify: $\frac{1}{\sqrt{2}}$

SOLUTION ▶

$$\frac{1}{\sqrt{2}}$$

$$= \frac{1}{\sqrt{2}} \cdot \frac{\sqrt{2}}{\sqrt{2}}$$ Multiply by a fraction equal to 1.

$$= \frac{\sqrt{2}}{\sqrt{4}}$$ Product rule of radicals.

$$= \frac{\sqrt{2}}{2}$$ Simplify; there are no radicals in the denominator.

$$\begin{array}{r} .70710... \\ 1.41421\overline{)1.0000000000} \\ -989947 \\ \hline 100530 \\ -0 \\ \hline 1005300 \\ -989947 \\ \hline 153530 \\ -141421 \\ \hline 121090 \end{array}$$

Figure 1

Since the denominator in the simplified expression is a rational number, we say that we have **rationalized the denominator**. We rationalize denominators because it is easier to add fractions with rational denominators than it is to add fractions with irrational denominators. It is also easier to divide by a rational number divisor than by an irrational number divisor. For example, since $\sqrt{2} \approx 1.41421$, $\frac{1}{\sqrt{2}}$ is approximately equal to $1 \div 1.41421$ (Figure 1), and $\frac{\sqrt{2}}{2}$ is approximately equal to $1.41421 \div 2$ (Figure 2). The arithmetic in Figure 2 is much less complicated than the arithmetic in Figure 1.

$$\begin{array}{r} .70710... \\ 2\overline{)1.41421} \\ -14 \\ \hline 01 \\ -0 \\ \hline 14 \\ -14 \\ \hline 02 \\ -2 \\ \hline 01 \end{array}$$

Figure 2

Dividing Square Roots

1. Use the quotient rule of radicals to rewrite the expression with a radical in the numerator and a radical in the denominator.
2. If the radicands include perfect square factors, simplify.
3. If the denominator is irrational, multiply the expression by a fraction that is equal to 1 to rationalize the denominator.
4. Simplify.

In the next example, we first use the quotient rule of radicals so that we can simplify the numerator and simplify the denominator. Since the denominator includes a radical, we then multiply by 1 to rationalize the denominator.

EXAMPLE 4 Simplify: $\sqrt{\frac{72}{5y}}$

SOLUTION ▶ $\sqrt{\frac{72}{5y}}$

$= \frac{\sqrt{72}}{\sqrt{5y}}$ Quotient rule of radicals.

$= \frac{\sqrt{36}\sqrt{2}}{\sqrt{5y}}$ Find perfect square factors; product rule of radicals.

$= \frac{6\sqrt{2}}{\sqrt{5y}}$ Simplify.

$= \frac{6\sqrt{2}}{\sqrt{5y}} \cdot \frac{\sqrt{5y}}{\sqrt{5y}}$ To rationalize the denominator, multiply by a fraction equal to 1.

$= \frac{6\sqrt{10y}}{5y}$ Product rule of radicals; simplify.

Practice Problems

For problems 1–4, simplify.

1. $\sqrt{\frac{25}{64}}$ 2. $\sqrt{\frac{12x}{y}}$ 3. $\sqrt{\frac{2}{5u}}$ 4. $\sqrt{\frac{20a}{7c}}$

Rationalizing Denominators That Are a Sum or a Difference

The conjugate of $\sqrt{x} - \sqrt{5}$ is $\sqrt{x} + \sqrt{5}$. When we multiply $\sqrt{x} - \sqrt{5}$ by its conjugate, the product does not include any radicals. In the next example, we rationalize the denominator of $\frac{3}{\sqrt{x} - \sqrt{5}}$ by multiplying by a fraction that is equal to 1, $\frac{\sqrt{x} + \sqrt{5}}{\sqrt{x} + \sqrt{5}}$.

EXAMPLE 5 Simplify: $\dfrac{3}{\sqrt{x} - \sqrt{5}}$

SOLUTION ▶

$$\frac{3}{\sqrt{x} - \sqrt{5}}$$

$$= \frac{3}{\sqrt{x} - \sqrt{5}} \cdot \frac{\sqrt{x} + \sqrt{5}}{\sqrt{x} + \sqrt{5}}$$ Multiply by a fraction equal to 1.

$$= \frac{3\sqrt{x} + 3\sqrt{5}}{\sqrt{x}\sqrt{x} + \sqrt{x}\sqrt{5} - \sqrt{5}\sqrt{x} - \sqrt{5}\sqrt{5}}$$ Distributive property.

$$= \frac{3\sqrt{x} + 3\sqrt{5}}{x + \sqrt{5x} - \sqrt{5x} - 5}$$ Simplify; product rule of radicals.

$$= \frac{3\sqrt{x} + 3\sqrt{5}}{x + 0 - 5}$$ Simplify.

$$= \frac{3\sqrt{x} + 3\sqrt{5}}{x - 5}$$ Simplify; there are no radicals in the denominator.

In Example 5, we multiplied the expression by $\dfrac{\sqrt{x} + \sqrt{5}}{\sqrt{x} + \sqrt{5}}$ because $\sqrt{x} + \sqrt{5}$ is the conjugate of $\sqrt{x} - \sqrt{5}$. In the next example, we rationalize the denominator of $\dfrac{7}{\sqrt{a} + 4}$ by multiplying by $\dfrac{\sqrt{a} - 4}{\sqrt{a} - 4}$ because $\sqrt{a} + 4$ and $\sqrt{a} - 4$ are conjugates.

EXAMPLE 6 Simplify: $\dfrac{7}{\sqrt{a} + 4}$

SOLUTION ▶

$$\frac{7}{\sqrt{a} + 4}$$

$$= \frac{7}{\sqrt{a} + 4} \cdot \frac{\sqrt{a} - 4}{\sqrt{a} - 4}$$ Multiply by a fraction equal to 1.

$$= \frac{7\sqrt{a} + 7(-4)}{\sqrt{a}\sqrt{a} + \sqrt{a}(-4) + 4\sqrt{a} + 4(-4)}$$ Distributive property.

$$= \frac{7\sqrt{a} - 28}{a - 4\sqrt{a} + 4\sqrt{a} - 16}$$ Simplify.

$$= \frac{7\sqrt{a} - 28}{a + 0 - 16}$$ Simplify.

$$= \frac{7\sqrt{a} - 28}{a - 16}$$ Simplify; there are no radicals in the denominator.

Practice Problems

For problems 5–7, simplify.

5. $\dfrac{2}{\sqrt{c} - \sqrt{3}}$ **6.** $\dfrac{2}{\sqrt{c} + \sqrt{3}}$ **7.** $\dfrac{5}{\sqrt{x} + 8}$

8.4 VOCABULARY PRACTICE

Match the term with its description.

1. Radicals that have the same radicand and index
2. In $\sqrt[n]{x}$, this is n.
3. $\sqrt{ab} = \sqrt{a}\sqrt{b}$
4. The square root of this number is a rational number.
5. $\sqrt{}$
6. A variable with an exponent that is divisible by 2
7. A number that can be written as a fraction in which the numerator and denominator are integers
8. In $\sqrt[n]{x}$, this is x.
9. A nonrepeating, nonterminating decimal number
10. $\sqrt{\dfrac{a}{b}} = \dfrac{\sqrt{a}}{\sqrt{b}}$

A. index
B. irrational number
C. like radicals
D. perfect square number
E. perfect square variable
F. principal square root
G. product rule of radicals
H. quotient rule of radicals
I. radicand
J. rational number

8.4 Exercises

Follow your instructor's guidelines for showing your work. Assume that all factors in radicands represent real numbers that are greater than or equal to 0. A denominator cannot equal 0. In simplified expressions, denominators are rational and rational numbers are in lowest terms.

For exercises 1–20, simplify.

1. $\sqrt{\dfrac{9}{25}}$
2. $\sqrt{\dfrac{16}{49}}$
3. $\sqrt{\dfrac{x^2}{y^2}}$
4. $\sqrt{\dfrac{a^2}{b^2}}$
5. $\sqrt{\dfrac{64w^2}{9}}$
6. $\sqrt{\dfrac{81p^2}{4}}$
7. $\sqrt{\dfrac{a^6}{b^{10}}}$
8. $\sqrt{\dfrac{x^8}{y^6}}$
9. $\sqrt{\dfrac{32}{49}}$
10. $\sqrt{\dfrac{20}{81}}$
11. $\sqrt{\dfrac{125}{36}}$
12. $\sqrt{\dfrac{128}{9}}$
13. $\sqrt{\dfrac{8x^2}{9}}$
14. $\sqrt{\dfrac{27x^2}{16}}$
15. $\sqrt{\dfrac{16x^3}{25}}$
16. $\sqrt{\dfrac{9x^3}{16}}$
17. $\sqrt{\dfrac{3p}{64}}$
18. $\sqrt{\dfrac{5w}{16}}$
19. $\sqrt{\dfrac{15x}{4}}$
20. $\sqrt{\dfrac{17a}{25}}$

For exercises 21–40, simplify.

21. $\sqrt{\dfrac{4}{u}}$
22. $\sqrt{\dfrac{9}{r}}$
23. $\sqrt{\dfrac{25}{n}}$
24. $\sqrt{\dfrac{16}{h}}$
25. $\sqrt{\dfrac{3}{x}}$
26. $\sqrt{\dfrac{7}{y}}$
27. $\sqrt{\dfrac{3}{7}}$
28. $\sqrt{\dfrac{7}{5}}$
29. $\sqrt{\dfrac{4}{7}}$
30. $\sqrt{\dfrac{9}{10}}$
31. $\sqrt{\dfrac{80}{27}}$
32. $\sqrt{\dfrac{125}{18}}$
33. $\sqrt{\dfrac{15x}{4y}}$
34. $\sqrt{\dfrac{21a}{16b}}$
35. $\sqrt{\dfrac{36}{5y}}$
36. $\sqrt{\dfrac{25}{3y}}$
37. $\sqrt{\dfrac{75x}{7y}}$
38. $\sqrt{\dfrac{125a}{11b}}$
39. $\sqrt{\dfrac{6p}{5w}}$
40. $\sqrt{\dfrac{10n}{3p}}$

For exercises 41–64, simplify.

41. $\dfrac{4}{\sqrt{x} + 3}$
42. $\dfrac{3}{\sqrt{x} + 2}$
43. $\dfrac{4}{\sqrt{x} - 3}$
44. $\dfrac{3}{\sqrt{x} - 2}$
45. $\dfrac{x}{\sqrt{x} + 3}$
46. $\dfrac{x}{\sqrt{x} + 2}$
47. $\dfrac{\sqrt{x}}{\sqrt{x} + 3}$
48. $\dfrac{\sqrt{x}}{\sqrt{x} + 2}$

49. $\dfrac{4}{\sqrt{x} + \sqrt{3}}$

50. $\dfrac{3}{\sqrt{x} + \sqrt{2}}$

51. $\dfrac{\sqrt{u}}{\sqrt{u} + 5}$

52. $\dfrac{\sqrt{v}}{\sqrt{v} + 7}$

53. $\dfrac{\sqrt{y}}{\sqrt{y} - 6}$

54. $\dfrac{\sqrt{n}}{\sqrt{n} - 4}$

55. $\dfrac{\sqrt{3}}{\sqrt{x} + 8}$

56. $\dfrac{\sqrt{5}}{\sqrt{x} + 3}$

57. $\dfrac{\sqrt{2}}{\sqrt{m} - \sqrt{5}}$

58. $\dfrac{\sqrt{3}}{\sqrt{w} - \sqrt{2}}$

59. $\dfrac{\sqrt{10}}{\sqrt{x} - \sqrt{5}}$

60. $\dfrac{\sqrt{6}}{\sqrt{x} - \sqrt{3}}$

61. $\dfrac{\sqrt{10}}{\sqrt{x} + \sqrt{2}}$

62. $\dfrac{\sqrt{6}}{\sqrt{x} + \sqrt{3}}$

63. $\dfrac{\sqrt{10}}{\sqrt{x} + 2}$

64. $\dfrac{\sqrt{6}}{\sqrt{x} + 3}$

For exercises 65–80, simplify.

65. $\sqrt{30}\sqrt{10}$

66. $\sqrt{20}\sqrt{10}$

67. $\sqrt{\dfrac{20}{pw}}$

68. $\sqrt{\dfrac{54}{hk}}$

69. $\sqrt{50} + \sqrt{8}$

70. $\sqrt{48} + \sqrt{27}$

71. $\sqrt{50}\sqrt{8}$

72. $\sqrt{48}\sqrt{27}$

73. $\sqrt{\dfrac{50}{3}}$

74. $\sqrt{\dfrac{27}{2}}$

75. $\sqrt{200u} - \sqrt{50u}$

76. $\sqrt{300n} - \sqrt{75n}$

77. $\dfrac{9}{\sqrt{3}}$

78. $\dfrac{4}{\sqrt{2}}$

79. $\sqrt{72a^2b} + \sqrt{98a^2b}$

80. $\sqrt{20xy^2} + \sqrt{125xy^2}$

Problem Solving: Practice and Review

Follow your instructor's guidelines for using the five steps as outlined in Section 1.3, p. 55.

81. The length of a rectangle is 3 ft more than twice its width. The area of the rectangle is 230 ft^2. Use a quadratic equation in one variable to find the length and width.

82. The diameter of a bolt sized with the SAE system is represented by the relationship $D = (0.13)(\text{SAE number}) + 0.06$, where D is in inches. Find the diameter of a bolt in millimeters with an SAE number of 4. (1 in. = 25.4 mm.) (*Source:* engineering.ucdavis.edu)

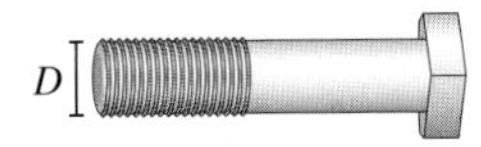

83. A queen-size mattress is marked down 70% to a sale price of $399. Find the regular price of the mattress.

84. Write a linear equation that describes the total cost of storing cord blood, y, for x years at the Utah Cord Bank. Use the equation to find the cost of storing cord blood for 25 years.

The service (cord blood banking) is offered in Utah through a for-profit private company called Utah Cord Bank . . . at a one-time cost of $940 plus $85 a year for storage. (*Source:* www.sltrib.com, March 2, 2009)

Find the Mistake

For exercises 85–88, the completed problem has one mistake.

(a) Describe the mistake in words, or copy down the whole problem and highlight or circle the mistake.

(b) Do the problem correctly.

85. Problem: Simplify: $\sqrt{\dfrac{100}{c}}$

Incorrect Answer: $\sqrt{\dfrac{100}{c}}$

$= \dfrac{\sqrt{100}}{\sqrt{c}}$

$= \dfrac{\sqrt{100}}{\sqrt{c}} \cdot \dfrac{\sqrt{c}}{\sqrt{c}}$

$= \dfrac{10}{c}$

86. Problem: Simplify: $\sqrt{\dfrac{16}{7w}}$

Incorrect Answer: $\sqrt{\dfrac{16}{7w}}$

$= \dfrac{\sqrt{16}}{\sqrt{7w}}$

$= \dfrac{4}{\sqrt{7w}}$

87. Problem: Simplify: $\dfrac{\sqrt{6c}}{\sqrt{3}}$

Incorrect Answer: $\dfrac{\sqrt{6c}}{\sqrt{3}}$

$= \dfrac{\sqrt{6c}}{\sqrt{3}} \cdot \dfrac{\sqrt{3}}{\sqrt{3}}$

$= \dfrac{\sqrt{18c}}{3}$

88. Problem: Simplify: $\sqrt{\dfrac{64x}{w}}$

Incorrect Answer: $\sqrt{\dfrac{64x}{w}}$

$= \dfrac{\sqrt{64x}}{\sqrt{w}}$

$= \dfrac{\sqrt{64x}}{\sqrt{w}} \cdot \dfrac{\sqrt{w}}{\sqrt{w}}$

$= \dfrac{\sqrt{64wx}}{w}$

Review

For exercises 89–92,
(a) solve.
(b) check.

89. $3(2x - 5) + 8 = -6x$

90. $\frac{3}{4}x - 9 = 15$

91. $x^2 - 8x + 12 = 0$

92. $3x^2 - 6x = 0$

SUCCESS IN COLLEGE MATHEMATICS

93. Suppose you have a homework problem that you do not understand and you ask a friend for help. Your friend says, "This is the way to do it," and does the problem for you. For the rest of the assignment, you follow your friend's method even though you do not understand why it works. Describe any disadvantages of using your friend's method.

© trekandshoot/Shutterstock.com

We can use a radical equation in two variables to represent the relationship of the speed of a tsunami wave and the depth of an ocean. In this section, we will solve radical equations in one variable and use radical equations in two variables to represent application problems.

SECTION 8.5

Radical Equations

After reading the text, working the practice problems, and completing assigned exercises, you should be able to:

1. Solve a radical equation that includes one square root.
2. Identify an extraneous solution of a radical equation.
3. Use a radical equation in two variables to solve an application problem.

Radical Equations

To solve an equation in one variable, we isolate the variable on one side of the equation. To isolate the variable in a radical equation that includes one square root, first isolate the square root that includes the variable. Then raise each side of the equation to the second power and simplify. This is called "squaring both sides."

EXAMPLE 1 Solve: $\sqrt{p} = 13$

SOLUTION

$\sqrt{p} = 13$

$(\sqrt{p})^2 = (13)^2$ The square root is isolated; square both sides.

$p = 169$ Simplify.

In the next example, we square both sides and then use the properties of equality to isolate the variable.

EXAMPLE 2 **(a)** Solve: $\sqrt{x + 4} = 6$

SOLUTION

$$\sqrt{x + 4} = 6$$

$$(\sqrt{x + 4})^2 = 6^2$$ The square root is isolated; square both sides.

$$x + 4 = 36$$ Simplify.

$$-4 \quad -4$$ Subtraction property of equality.

$$x + \mathbf{0} = 32$$ Simplify.

$$\boldsymbol{x} = 32$$ Simplify.

(b) Check.

$$\sqrt{x + 4} = 6$$ Use the original equation.

$$\sqrt{\mathbf{32} + 4} = 6$$ Replace the variable with the solution, 32.

$$\sqrt{\mathbf{36}} = 6$$ Evaluate.

$$\mathbf{6} = 6 \quad \text{True.}$$ Evaluate; the solution is correct.

In the next example, before squaring both sides, we use the properties of equality to isolate the square root.

EXAMPLE 3 **(a)** Solve: $\sqrt{2x + 1} - 13 = 20$

SOLUTION

$$\sqrt{2x + 1} - 13 = 20$$

$$+13 \quad +13$$ Addition property of equality.

$$\sqrt{2x + 1} + \mathbf{0} = \mathbf{33}$$ Simplify.

$$(\sqrt{2x + 1})^2 = (33)^2$$ Square both sides.

$$2x + 1 = 1089$$ Simplify.

$$-1 \quad -1$$ Subtraction property of equality.

$$2x + \mathbf{0} = \mathbf{1088}$$ Simplify.

$$\frac{\mathbf{2x}}{2} = \frac{1088}{2}$$ Division property of equality.

$$x = 544$$ Simplify.

(b) Check.

$$\sqrt{2x + 1} - 13 = 20$$ Use the original equation.

$$\sqrt{2(\mathbf{544}) + 1} - 13 = 20$$ Replace the variable with the solution, 544.

$$\sqrt{\mathbf{1089}} - 13 = 20$$ Simplify the radicand.

$$\mathbf{33} - 13 = 20$$ Evaluate the square root.

$$\mathbf{20} = 20 \quad \text{True.}$$ Simplify; the solution is correct.

Solving a Radical Equation with a Square Root

1. Isolate the square root.
2. Square both sides.
3. Isolate the variable.
4. Check the solution.

When we square both sides of a radical equation, the result may be a quadratic equation. If this equation can be factored, we can use the zero product property to solve it.

EXAMPLE 4 **(a)** Solve: $4z = \sqrt{16z - 3}$

SOLUTION ▶

$$4z = \sqrt{16z - 3}$$

$$(4z)^2 = (\sqrt{16z - 3})^2 \quad \text{Square both sides.}$$

$$16z^2 = 16z - 3 \quad \text{Simplify.}$$

$$-16z + 3 \quad -16z + 3 \quad \text{Properties of equality.}$$

$$\mathbf{16z^2 - 16z + 3 = 0} \quad a = 16, b = -16, c = 3, ac = 48$$

$$16z^2 - \mathbf{12z} + \mathbf{-4z} + 3 = 0 \quad \text{Rewrite } -16z \text{ as } -12z + -4z.$$

$$(16z^2 - 12z) + (-4z + 3) = 0 \quad \text{Group terms.}$$

$$\mathbf{4z}(4z - 3) + \mathbf{(-1)}(4z - 3) = 0 \quad \text{Factor a common factor from each group.}$$

$$\mathbf{(4z - 3)}(4z - 1) = 0 \quad \text{Complete the factoring.}$$

$$4z - 3 = 0 \quad \text{or} \quad 4z - 1 = 0 \quad \text{Zero product property.}$$

$$+3 \; +3 \qquad +1 \; +1 \quad \text{Addition property of equality.}$$

$$4z + \mathbf{0} = 3 \quad \text{or} \quad 4z + \mathbf{0} = \mathbf{1} \quad \text{Simplify.}$$

$$\frac{\mathbf{4z}}{4} = \frac{3}{4} \qquad \frac{\mathbf{4z}}{4} = \frac{1}{4} \quad \text{Division property of equality.}$$

$$z = \frac{3}{4} \quad \text{or} \quad z = \frac{1}{4} \quad \text{Simplify.}$$

(b) Check.

▶ Check: $z = \frac{3}{4}$

$$4z = \sqrt{16z - 3}$$

$$4\left(\frac{\mathbf{3}}{\mathbf{4}}\right) = \sqrt{16\left(\frac{\mathbf{3}}{\mathbf{4}}\right) - 3}$$

$$\mathbf{3} = \sqrt{\mathbf{12} - 3}$$

$$3 = \sqrt{\mathbf{9}}$$

$$3 = \mathbf{3} \quad \text{True.}$$

Check: $z = \frac{1}{4}$

$$4z = \sqrt{16z - 3}$$

$$4\left(\frac{\mathbf{1}}{\mathbf{4}}\right) = \sqrt{16\left(\frac{\mathbf{1}}{\mathbf{4}}\right) - 3}$$

$$1 = \sqrt{\mathbf{4} - 3}$$

$$1 = \sqrt{\mathbf{1}}$$

$$1 = \mathbf{1} \quad \text{True.}$$

Practice Problems

For problems 1–5,
(a) solve.
(b) check.

1. $\sqrt{x + 5} = 6$ **2.** $\sqrt{x + 5} - 3 = 6$ **3.** $\sqrt{x + 1} - 8 = 5$
4. $x = \sqrt{9x - 20}$ **5.** $2x = \sqrt{7x - 3}$

Extraneous Solutions

To solve a radical equation with a square root, we square both sides, creating an equation that does not have a square root, and solve this new equation. If a solution of the new equation is not a solution of the original equation, it is an **extraneous solution**.

EXAMPLE 5 (a) Solve: $\sqrt{x+5} = -15$

SOLUTION ▶ The principal square root of a real number cannot be negative. If we square both sides and isolate the variable, the result is an extraneous solution. The check shows that the solution is false. This equation has no solution.

$\sqrt{x+5} = -15$	
$(\sqrt{x+5})^2 = (-15)^2$	Square both sides.
$x + 5 = 225$	Simplify.
$-5 \quad -5$	Subtraction property of equality.
$x + \mathbf{0} = \mathbf{220}$	Simplify.
$x = 220$	Simplify.

(b) Check.

$\sqrt{x+5} = -15$		Use the original equation.
$\sqrt{\mathbf{220} + 5} = -15$		Replace the variable with the solution, 220.
$\sqrt{\mathbf{225}} = -15$		Evaluate.
$\mathbf{15} = -15$	False.	Evaluate.

The solution $x = 220$ is an extraneous solution. It is a solution of $x + 5 = 225$, but it is *not* a solution of the original equation, $\sqrt{x+5} = -15$. The original equation has no solution. Ask your instructor how you should mark extraneous solutions.

An extraneous solution does not indicate that we have made an algebraic or arithmetic mistake. Instead, it happens because the process for solving the equation by squaring both sides eliminates the square root.

EXAMPLE 6 (a) Solve: $x = \sqrt{3x+4}$

SOLUTION ▶

$x = \sqrt{3x+4}$	
$x^2 = (\sqrt{3x+4})^2$	Square both sides.
$x^2 = 3x + 4$	Simplify.
$-3x - 4 \quad -3x - 4$	Subtraction property of equality.
$x^2 - \mathbf{3x} - \mathbf{4} = \mathbf{0}$	Simplify.
$(\mathbf{x} - \mathbf{4})(\mathbf{x} + \mathbf{1}) = 0$	Factor.
$x - 4 = 0$ or $x + 1 = 0$	Zero product property.
$+4 \quad +4$ $\quad -1 \quad -1$	Properties of equality.
$x + \mathbf{0} = \mathbf{4}$ or $x + \mathbf{0} = \mathbf{-1}$	Simplify.
$\mathbf{x} = 4$ or $x = -1$	Simplify.

(b) Check.

Check: $x = 4$			Check: $x = -1$	
$x = \sqrt{3x+4}$		Original equation.	$x = \sqrt{3x+4}$	
$\mathbf{4} = \sqrt{3(\mathbf{4}) + 4}$		Replace variables.	$\mathbf{-1} = \sqrt{3(\mathbf{-1}) + 4}$	
$4 = \sqrt{\mathbf{16}}$		Simplify.	$-1 = \sqrt{\mathbf{1}}$	
$4 = \mathbf{4}$	True.	Simplify.	$-1 = \mathbf{1}$	False.

Since $x = -1$ is not a solution of the original equation, the only solution of $x = \sqrt{3x+4}$ is $x = 4$.

In the next example, both solutions are extraneous. The equation has no solution.

EXAMPLE 7 **(a)** Solve: $x = \sqrt{-8x - 15}$

SOLUTION

$x = \sqrt{-8x - 15}$

$x^2 = (\sqrt{-8x - 15})^2$ Square both sides.

$x^2 = -8x - 15$ Simplify.

$+8x + 15 \quad +8x + 15$ Addition property of equality.

$\mathbf{x^2 + 8x + 15 = 0}$ Simplify.

$(x + 3)(x + 5) = 0$ Factor.

$x + 3 = 0$ or $x + 5 = 0$ Zero product property.

$-3 \quad -3 \qquad -5 \quad -5$ Subtraction property of equality.

$x + 0 = -3$ or $x + 0 = -5$ Simplify.

$x = -3$ or $x = -5$ Simplify.

(b) Check.

Check: $x = -3$

$x = \sqrt{-8x - 15}$

$-3 = \sqrt{-8(-3) - 15}$

$-3 = \sqrt{9}$

$-3 = 3$ False.

Check: $x = -5$

$x = \sqrt{-8x - 15}$

$-5 = \sqrt{-8(-5) - 15}$

$-5 = \sqrt{25}$

$-5 = 5$ False.

Both solutions are extraneous. The equation $x = \sqrt{-8x - 15}$ has no solution.

Practice Problems

For problems 6–8,
(a) solve.
(b) check.

6. $x = \sqrt{10x - 21}$ **7.** $h = \sqrt{4h + 45}$ **8.** $2x = \sqrt{-4x + 3}$

Solving Radical Equations with Two Square Roots

To solve equations with two square roots, we need to square both sides of the equation two times.

Solving a Radical Equation with Two Square Roots

1. Isolate one of the square roots.
2. Square both sides. Simplify.
3. Isolate the remaining square root. Simplify.
4. Square both sides. Simplify.
5. Isolate the variable.
6. Check the solution.

EXAMPLE 8 **(a)** Solve: $\sqrt{2x + 6} = 1 + \sqrt{x + 4}$

SOLUTION ▶

$\sqrt{2x + 6} = 1 + \sqrt{x + 4}$ — $\sqrt{2x + 6}$ is isolated.

$(\sqrt{2x + 6})^2 = (1 + \sqrt{x + 4})^2$ — Square both sides.

$2x + 6 = (1 + \sqrt{x + 4})(1 + \sqrt{x + 4})$ — Simplify.

$2x + 6 = \mathbf{1 \cdot 1 + 1\sqrt{x + 4} + \sqrt{x + 4}(1) + \sqrt{x + 4}\sqrt{x + 4}}$ — Distributive property.

$2x + 6 = \mathbf{1} + 1\sqrt{x + 4} + \mathbf{1\sqrt{x + 4} + x + 4}$ — Simplify.

$2x + 6 = \mathbf{2\sqrt{x + 4}} + x + \mathbf{5}$ — Simplify.

$\underline{-x \quad -5 \qquad\qquad -x \quad -5}$ — Subtraction property of equality.

$\mathbf{x + 1} = 2\sqrt{x + 4} + \mathbf{0 + 0}$ — Simplify.

$(x + 1)^2 = (2\sqrt{x + 4})^2$ — Square both sides.

$(x + 1)(x + 1) = (2\sqrt{x + 4})(2\sqrt{x + 4})$ — Simplify.

$\mathbf{x \cdot x + x \cdot 1 + 1x + 1 \cdot 1} = 4(x + 4)$ — Distributive property.

$\mathbf{x^2 + 2x + 1} = \mathbf{4x + 4(4)}$ — Simplify; distributive property.

$x^2 + 2x + 1 = 4x + \mathbf{16}$ — Simplify.

$\underline{\quad -4x \quad -16 \quad -4x \quad -16}$ — Subtraction property of equality.

$x^2 \mathbf{- 2x - 15 = 0}$ — Simplify.

$\mathbf{(x - 5)(x + 3)} = 0$ — Factor.

$x - 5 = 0$ or $x + 3 = 0$ — Zero product property.

$\underline{\quad +5 \quad +5}$ $\qquad \underline{\quad -3 \quad -3}$ — Properties of equality.

$x + \mathbf{0 = 5}$ or $x + \mathbf{0 = -3}$

$\mathbf{x} = 5$ or $x = -3$ — Simplify.

(b) Check.

Check: $x = 5$	Check: $x = -3$
$\sqrt{2x + 6} = 1 + \sqrt{x + 4}$	$\sqrt{2x + 6} = 1 + \sqrt{x + 4}$
$\sqrt{2(\mathbf{5}) + 6} = 1 + \sqrt{\mathbf{5} + 4}$	$\sqrt{2(\mathbf{-3}) + 6} = 1 + \sqrt{\mathbf{-3} + 4}$
$\sqrt{\mathbf{16}} = 1 + \sqrt{\mathbf{9}}$	$\sqrt{\mathbf{0}} = 1 + \sqrt{\mathbf{1}}$
$4 = 4$ True.	$0 = 2$ False.

Since $x = -3$ is not a solution of the original equation, the only solution of $\sqrt{2x + 6} = 1 + \sqrt{x + 4}$ is $x = 5$.

Practice Problems

For problems 9–10,

(a) solve.

(b) check.

9. $\sqrt{2x + 3} = \sqrt{x + 2} + 2$ **10.** $\sqrt{4 - x} + \sqrt{x + 6} = 4$

Applications

A radical equation in two variables represents the relationship of the two variables. In the next example, the equation describes the relationship of the depth of the ocean at some location and the wave speed of a tsunami (tidal wave) at that location.

EXAMPLE 9 The equation $y = \sqrt{\left(\frac{9.8 \text{ m}}{1 \text{ s}^2}\right)x}$ represents the relationship of the wave speed of a tsunami in meters per second, y, and the depth of the ocean in meters, x. Find the speed of the wave when the ocean is 4000 m deep. Round to the nearest whole number.

SOLUTION ▶

$$y = \sqrt{\left(\frac{9.8 \text{ m}}{1 \text{ s}^2}\right)x}$$

$$y = \sqrt{\left(\frac{9.8 \text{ m}}{1 \text{ s}^2}\right)(\mathbf{4000 \text{ m}})} \quad \text{Replace } x \text{ with 4000 m.}$$

$$y = \sqrt{\frac{\mathbf{39{,}200 \text{ m}^2}}{\mathbf{1 \text{ s}^2}}} \quad \text{Simplify the radicand.}$$

$$y = \frac{\mathbf{197.989\ldots \text{ m}}}{\mathbf{1 \text{ s}}} \quad \text{Evaluate the square root; } \sqrt{\frac{\text{m}^2}{\text{s}^2}} = \frac{\text{m}}{\text{s}}$$

$$y \approx \frac{\mathbf{198} \text{ m}}{1 \text{ s}} \quad \text{Round to the nearest whole number.}$$

Practice Problems

11. A Circular Area Profile uses census data to estimate the population within a circular area. The software requires the researcher to enter the radius of the circle. The equation $y = \sqrt{\frac{x}{\pi}}$ represents the relationship of the area of a circle, x, and its radius, y. Find the radius of a circle with an area of 100 mi^2. ($\pi \approx 3.14$.) Round to the nearest tenth.

Using Technology: Solving an Equation with a Graphing Calculator

In Section 6.5, we used a graphical method to solve polynomial equations. We can also use this strategy to solve a radical equation such as $x = \sqrt{2x + 3}$. We enter the equations $y = x$ and $y = \sqrt{2x + 3}$ on the Y= screen, graph the equations in a standard window, and use the **intersect** command in the CALCULATE screen to find the intersection point of the graphs. The x-coordinate of the intersection point is the solution of the original equation.

EXAMPLE 10 Solve: $x = \sqrt{2x + 3}$

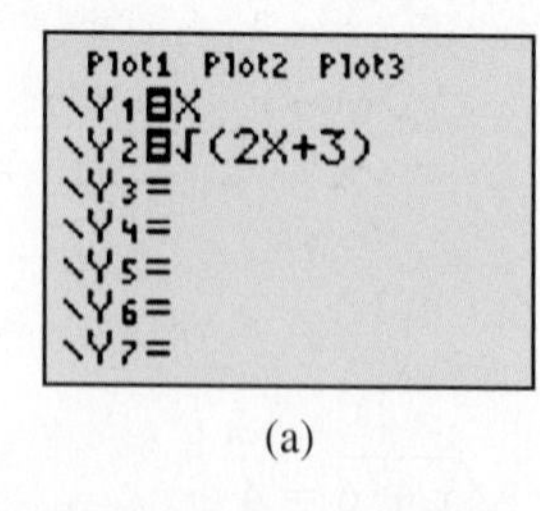

(a)

(b)

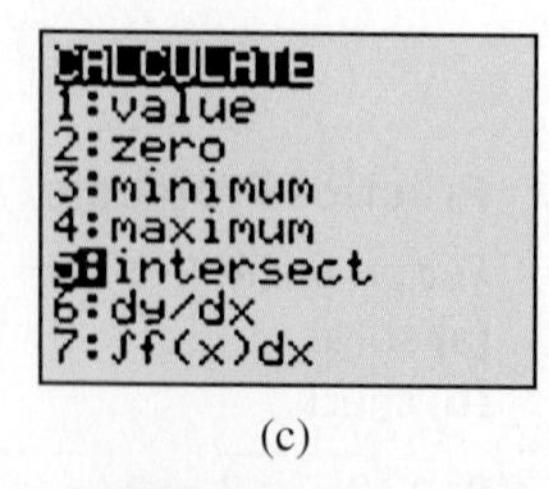

(c)

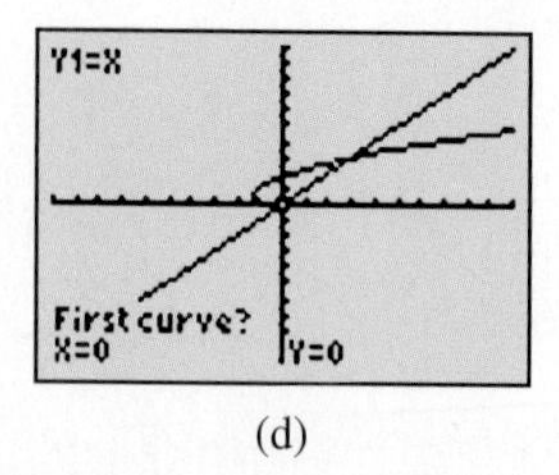

(d)

Type the equation in the Y= screen. Graph in a standard window. Go to the CALCULATE screen; choose the **intersect** command. The calculator asks "First curve?" Press ENTER.

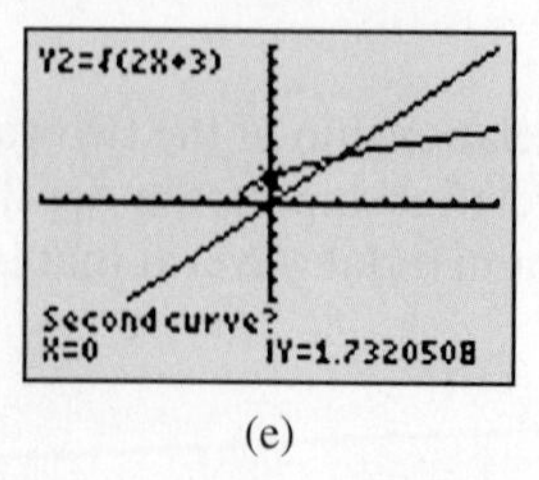

(e)

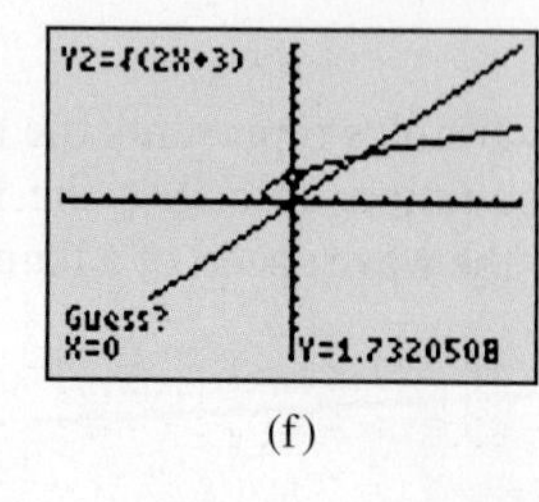

(f)

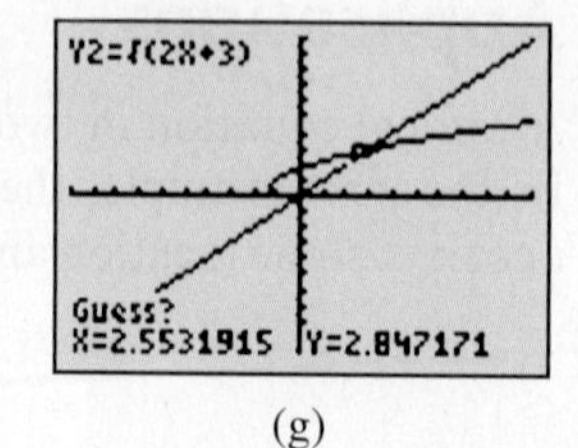

(g)

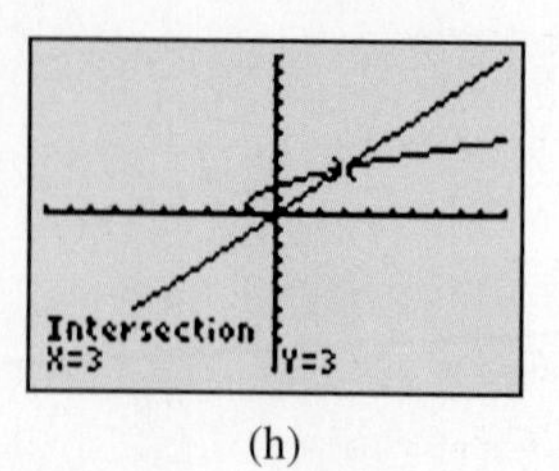

(h)

The calculator asks "Second curve?" Press ENTER. The calculator asks "Guess?" Move the cursor to the approximate intersection point of the graphs. Press ENTER. The calculator shows the point of intersection, (3, 3). The solution is the x-coordinate of the intersection point, $x = 3$.

Practice Problems For problems 12–15,
(a) solve by graphing. Sketch the graph; describe the window.
(b) identify the solution.

12. $2x - 8 = -3x + 2$ **13.** $x^2 - 2x = 8$ **14.** $\sqrt{x + 9} = 2$ **15.** $x = \sqrt{5x}$

8.5 Vocabulary Practice

Match the term with its description.

1. Squaring a radical equation removes the square root. It can also result in this kind of solution.
2. To solve a radical equation with a square root, isolate the radical and then raise each side to this power.
3. $\{0, 1, 2, 3, \ldots\}$
4. Another way of saying "raising both sides to the second power"
5. $\sqrt{}$
6. If $ab = 0$, then $a = 0$ or $b = 0$.
7. In $\sqrt[n]{x}$, this is x.
8. In $\sqrt[n]{x}$, this is n.
9. Another word for *power*
10. $ax^2 + bx + c = 0$

A. exponent
B. extraneous solution
C. index
D. principal square root
E. radicand
F. second
G. set of whole numbers
H. squaring both sides
I. standard form of a quadratic equation
J. zero product property

8.5 Exercises

Follow your instructor's guidelines for showing your work. Assume that all factors in radicands represent real numbers that are greater than or equal to 0.

For exercises 1–30,
(a) solve.
(b) check.

1. $\sqrt{x} = 4$ **2.** $\sqrt{x} = 5$
3. $\sqrt{x} = -4$ **4.** $\sqrt{x} = -5$
5. $\sqrt{x + 1} = 4$ **6.** $\sqrt{x + 1} = 5$
7. $\sqrt{x - 1} = 4$ **8.** $\sqrt{x - 1} = 5$
9. $\sqrt{x} + 1 = 4$ **10.** $\sqrt{x} + 1 = 5$
11. $\sqrt{x} - 1 = 4$ **12.** $\sqrt{x} - 1 = 5$
13. $\sqrt{x} + 6 = 4$ **14.** $\sqrt{x} + 8 = 5$
15. $\sqrt{5x} = 30$ **16.** $\sqrt{4x} = 20$
17. $\sqrt{-5x} = 30$ **18.** $\sqrt{-4x} = 20$
19. $\sqrt{-x} = 20$ **20.** $\sqrt{-x} = 30$
21. $\sqrt{3x + 6} = 15$ **22.** $\sqrt{2x + 8} = 12$
23. $\sqrt{3x + 6} - 9 = 15$
24. $\sqrt{2x + 8} - 6 = 12$
25. $\sqrt{3x + 6} + 21 = 15$
26. $\sqrt{2x + 8} + 18 = 12$
27. $5 = \sqrt{5u + 10}$
28. $6 = \sqrt{6w + 12}$
29. $5 + \sqrt{5u + 10} = 5$
30. $6 + \sqrt{6w + 12} = 6$

For exercises 31–40, use factoring and the zero product property to solve.

31. $r^2 + 7r + 10 = 0$ **32.** $u^2 + 9u + 18 = 0$
33. $m^2 - m - 72 = 0$ **34.** $n^2 - n - 30 = 0$
35. $4z^2 - 12z - 7 = 0$ **36.** $4w^2 - 4w - 15 = 0$
37. $6k^2 + 7k - 5 = 0$
38. $6h^2 + 11h - 7 = 0$
39. $9p^2 + 24p + 16 = 0$
40. $4d^2 + 12d + 9 = 0$

For exercises 41–54,
(a) solve.
(b) check.

41. $x = \sqrt{7x - 10}$
42. $x = \sqrt{11x - 18}$
43. $2n = \sqrt{8n - 3}$
44. $2w = \sqrt{12w - 5}$
45. $x + 3 = \sqrt{16x - 12}$
46. $z - 3 = \sqrt{4z - 15}$

47. $n = \sqrt{7n - 12}$
48. $p = \sqrt{9p - 14}$
49. $w = \sqrt{5w}$
50. $x = \sqrt{7x}$
51. $-w = \sqrt{5w}$
52. $-x = \sqrt{7x}$
53. $d = \sqrt{6d - 9}$
54. $z = \sqrt{4z - 4}$

For exercises 55–60,
(a) solve.
(b) check.

55. $2n = \sqrt{-8n - 3}$
56. $2x = \sqrt{-12x - 5}$
57. $p - 2 = \sqrt{4p + 13}$
58. $f - 1 = \sqrt{3f + 15}$
59. $y + 1 = \sqrt{-3y - 5}$
60. $w + 4 = \sqrt{-9w - 56}$

For exercises 61–66,
(a) solve.
(b) check.

61. $\sqrt{x + 5} = \sqrt{x} + 1$
62. $\sqrt{x + 12} = \sqrt{x} + 2$
63. $\sqrt{7 - x} - 6 = \sqrt{x + 11}$
64. $\sqrt{m + 3} + 1 = \sqrt{m - 8}$
65. $\sqrt{a - 5} + \sqrt{a + 6} = 11$
66. $\sqrt{y - 4} + \sqrt{y + 7} = 11$

For exercises 67–68, if an object travels away from the earth at a fast enough speed, the *escape velocity*, the force of gravity cannot bring the object back to the earth. For any planet with the radius of the earth and a mass in kilograms of x, the escape velocity y in meters per second is $y = (4.56 \times 10^{-4})\sqrt{x}$.

67. The mass of the earth is about 5.98×10^{24} kg. Find the escape velocity from the earth. Write the answer in scientific notation. Round the mantissa to the nearest hundredth.

68. A planet with the same radius as the earth has a mass of 1.6×10^{25} kg. Find the escape velocity from this planet. Write the answer in scientific notation. Round the mantissa to the nearest hundredth.

For exercises 69–70, a plane with emergency supplies is traveling at $\frac{144 \text{ km}}{1 \text{ hr}}$. The equation $y = \sqrt{\frac{2x}{9.8}}$ represents the relationship of the time it takes for a rescue package to hit the ground in seconds, y, and the height of the plane in meters, x. Find the time it takes for a package to hit the ground for the given height of the plane. Round to the nearest whole number.

69. 150 m
70. 100 m

For exercises 71–72, the SMOG (Simple Measure of Gobbledygook) Readability Formula $y = \sqrt{x} + 3$ represents the relationship of the grade reading level of a book or article, y, and the total number of polysyllabic words in a group of ten sentences near the beginning of the book or article, a group of ten sentences in the middle, and a group of ten sentences near the end, x. (A polysyllabic word has three or more syllables. Include repetitions of the same word.) Find the SMOG reading level. Round to the nearest tenth.

71. An article has a total of 80 polysyllabic words in the selected sentences.

72. Select a book or article assigned in one of your classes. Include the title of the book or article in your answer.

For exercises 73–74, a town uses the lineal frontage of a business on the street in feet, x, to determine the allowable area of a sign for the business in square feet, y. (*Source:* www.ci.amesbury.ma.us)

73. For a hanging sign that is the primary sign of the business, the allowable area is represented by $y = 2.5\sqrt{x}$. Find the allowable area of this sign for a business with lineal frontage of 25 ft.

74. For a hanging banner that is a secondary sign of the business, the allowable area is represented by $y = 1.5\sqrt{x}$. Find the allowable area of this banner for a business with lineal frontage of 36 ft.

Problem Solving: Practice and Review

Follow your instructor's guidelines for using the five steps as outlined in Section 1.3, p. 55.

75. The length of a rectangle is 3 ft more than twice its width. The perimeter of the rectangle is 66 ft. Use a system of linear equations to find its length and width.

76. The size of a bicycle tire is often reported as two measurements. The first measurement is the outside diameter of the tire. The second measurement is the width of the tire. Find the outer circumference of this tire. The formula for the circumference of a circle is $C = 2\pi r$, where r is the radius. ($\pi \approx 3.14$.) ($1'' = 1$ in.) Round to the nearest tenth.

Schwalbe HS159 Puncture Protection 27 × 1 1/4
Schwalbe is a German tire company known for its durable puncture-resistant tires. This model is a robust, rugged tire, perfect for bad weather or commuting. 27″ × 1 1/4″. 85PSI Max, SBC tread compound, black with off-white sidewall, wire bead, 550g. (*Source:* www.biketiresdirect.com)

77. The formula for density is $D = \frac{M}{V}$, where M is the mass, V is the volume, and D is the density. The density of 18-carat gold is $\frac{15 \text{ g}}{1 \text{ cm}^3}$. The mass of an 18-carat gold ring is 0.25 oz. Find its volume in cubic centimeters. (1 oz ≈ 28.35 g.) Round to the nearest tenth.

78. The joining fee for the Health Zone athletic club is $175. The monthly dues are $54. The joining fee for the North Little Rock Athletic Club is $150. The monthly dues are $56. Find the number of months of membership at which the cost of belonging to each club is equal. Round to the nearest whole number.

Technology

For exercises 79–82,
(a) solve. Choose a window that shows the point of intersection. Sketch the graph; describe the window.
(b) identify any solutions.

79. $\sqrt{2x + 18} = 4$
80. $\sqrt{3x + 24} = 6$
81. $\sqrt{3x} - 6 = 3$
82. $\sqrt{3x - 4} = -6$

Find the Mistake

For exercises 83–86, the completed problem has one mistake.
(a) Describe the mistake in words, or copy down the whole problem and highlight or circle the mistake.
(b) Do the problem correctly.

83. Problem: Solve: $\sqrt{3x} = -15$

Incorrect Answer: $\sqrt{3x} = -15$

$$(\sqrt{3x})^2 = (-15)^2$$
$$3x = 225$$
$$x = 75$$

84. Problem: Solve: $\sqrt{2x} - 48 = -28$

Incorrect Answer: Since a principal square root cannot equal a negative number, this equation has no solution.

85. Problem: Solve: $6 = \sqrt{6k + 18}$

Incorrect Answer: $6 = \sqrt{6k + 18}$

$$6^2 = (\sqrt{6k + 18})^2$$
$$12 = 6k + 18$$
$$-18 \quad\quad -18$$
$$-6 = 6k + 0$$
$$\frac{-6}{6} = \frac{6k}{6}$$
$$-1 = k$$

86. Problem: Solve: $x = \sqrt{9x + 22}$

Incorrect Answer: $x = \sqrt{9x + 22}$

$$x^2 = (\sqrt{9x + 22})^2$$
$$x^2 = 9x + 22$$
$$-9x \; -22 \quad -9x \; -22$$
$$x^2 - 9x - 22 = 0$$
$$(x - 11)(x + 2) = 0$$
$$x - 11 = 0 \quad \text{or} \quad x + 2 = 0$$
$$+11 \; +11 \quad\quad -2 \; -2$$
$$x + 0 = 11 \quad\quad x + 0 = -2$$
$$x = 11 \quad \text{or} \quad x = -2$$

Review

For exercises 87–90, evaluate.

87. $\sqrt{64}$
88. $\sqrt{-64}$
89. $\sqrt[3]{-64}$
90. $\sqrt[3]{64}$

SUCCESS IN COLLEGE MATHEMATICS

91. Suppose you miss class and decide to go to the tutoring center for help on your homework. What should you do before you go to the tutoring center so that you can make the best use of your time with the tutor?

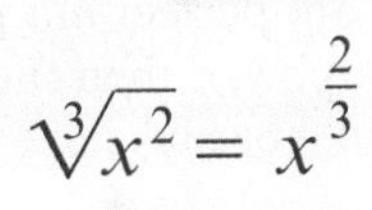

The expression $\sqrt[3]{x^2}$ is in radical notation. In exponential notation, this expression is $x^{\frac{2}{3}}$. In this section, we will use both notations. Throughout this section, we assume that factors including variables in radicands and bases including variables of exponential expressions represent nonnegative real numbers.

SECTION 8.6 Higher Index Radicals and Rational Exponents

After reading the text, working the practice problems, and completing assigned exercises, you should be able to:

1. Evaluate or simplify an expression that includes radicals with an index greater than 2.
2. Rewrite a radical expression in exponential notation and simplify.
3. Rewrite an exponential notation in radical notation and simplify.
4. Simplify an expression that includes rational exponents.

Higher Order Roots

The expression $\sqrt[3]{64}$ is a **cube root**. The index of this cube root is 3. Since $4 \cdot 4 \cdot 4 = 64$, $\sqrt[3]{64} = 4$. Unlike square roots, the cube root of a negative number is a real number. For example, since $(-5)(-5)(-5) = -125$, $\sqrt[3]{-125} = -5$.

The square root of a perfect square is a rational number. A perfect square is an example of a **perfect power**. For example, since $2^4 = 16$, 16 is a perfect fourth power. The table includes perfect squares, perfect cubes, and other perfect powers.

Perfect Squares, Perfect Cubes, and Other Perfect Powers

Base, n	2	3	4	5	6	7	8	9	10	11	12	13	14	15
Perfect square n^2	4	9	16	25	36	49	64	81	100	121	144	169	196	225
Perfect cube n^3	8	27	64	125	216	343	512	729	1000					
Perfect fourth power n^4	16	81	256	625										
Perfect fifth power n^5	32	243	1024	3125										
Perfect sixth power n^6	64	729	4096											

EXAMPLE 1 Evaluate.

(a) $\sqrt[4]{256}$

SOLUTION ▶ $= 4$ $\quad 4 \cdot 4 \cdot 4 \cdot 4 = 256$

(b) $\sqrt{144}$

▶ $= 12$ $\quad 12 \cdot 12 = 144$

(c) $\sqrt[3]{-64}$

▶ $= -4$ $\quad (-4)(-4)(-4) = -64$

(d) $\sqrt[5]{243}$

▶ $= 3$ $\quad 3 \cdot 3 \cdot 3 \cdot 3 \cdot 3 = 243$

(e) $\sqrt[4]{625}$

▶ $= 5$ $\quad 5 \cdot 5 \cdot 5 \cdot 5 = 625$

To simplify a root with an index n, we find factors that are perfect *nth* powers and use the product rule of radicals. If a variable is a perfect power, then the exponent is divisible by the index. For example, x^6 is a perfect cube because $6 \div 3 = 2$. Simplifying, $\sqrt[3]{x^6} = x^2$.

Product Rule of Radicals for a Radical with an Index, n

If a and b are nonnegative real numbers and n is an even whole number greater than 0, $\sqrt[n]{ab} = \sqrt[n]{a}\sqrt[n]{b}$.

If a and b are real numbers and n is an odd whole number, $\sqrt[n]{ab} = \sqrt[n]{a}\sqrt[n]{b}$.

EXAMPLE 2 Simplify.

(a) $\sqrt[3]{432a^3}$

SOLUTION ▶ $= \sqrt[3]{216}\sqrt[3]{2}\sqrt[3]{a^3}$ Find perfect cube factors.

$= \mathbf{6a\sqrt[3]{2}}$ Simplify; product rule of radicals.

(b) $\sqrt[3]{-432n^5}$

▶ $= \sqrt[3]{-216}\sqrt[3]{2}\sqrt[3]{n^3}\sqrt[3]{n^2}$ Find perfect cube factors.

$= -6n\sqrt[3]{2n^2}$ Simplify; product rule of radicals.

(c) $\sqrt[4]{162x}$

▶ $= \sqrt[4]{81}\sqrt[4]{2}\sqrt[4]{x}$ Find perfect fourth power factors.

$= 3\sqrt[4]{2x}$ Simplify; product rule of radicals.

Practice Problems

For problems 1–4, evaluate.

1. $\sqrt[3]{125}$ **2.** $\sqrt[3]{-125}$ **3.** $\sqrt[6]{64}$ **4.** $\sqrt[6]{-64}$

For problems 5–6, simplify.

5. $\sqrt[4]{32x}$ **6.** $\sqrt[3]{-343x^6}$

Simplifying Radical Expressions

When simplifying radical expressions, we can combine only like radicals. Like radicals have the same index and the same radicand.

EXAMPLE 3 Simplify: $2\sqrt[4]{x} + 5\sqrt[4]{x}$

SOLUTION ▶

$2\sqrt[4]{x} + 5\sqrt[4]{x}$

$= 7\sqrt[4]{x}$ Combine like radicals.

In the next example, we simplify each radical before combining like radicals.

EXAMPLE 4 Simplify: $\sqrt[3]{54} - \sqrt[3]{16}$

SOLUTION ▶

$\sqrt[3]{54} - \sqrt[3]{16}$

$= \sqrt[3]{27}\sqrt[3]{2} - \sqrt[3]{8}\sqrt[3]{2}$ Product rule of radicals.

$= \mathbf{3}\sqrt[3]{2} - \mathbf{2}\sqrt[3]{2}$ Simplify.

$= 1\sqrt[3]{2}$ Combine like radicals.

$= \sqrt[3]{2}$ Simplify.

In the next example, we multiply cube roots and look for perfect cube factors.

EXAMPLE 5 Simplify: $\sqrt[3]{4}\sqrt[3]{10}$

SOLUTION ▶

$\sqrt[3]{4}\sqrt[3]{10}$

$= \sqrt[3]{4 \cdot 10}$ Product rule of radicals.

$= \sqrt[3]{2 \cdot 2 \cdot 2 \cdot 5}$ Factor the numbers.

$= \sqrt[3]{2^3}\sqrt[3]{5}$ Use exponential notation; product rule of radicals.

$= \mathbf{2}\sqrt[3]{5}$ Simplify; $\sqrt[3]{8} = 2$

In the next example, we use the quotient rule of radicals to rewrite the expression as a single radical. Both radicals have the same index.

EXAMPLE 6 Simplify: $\dfrac{\sqrt[4]{96x}}{\sqrt[4]{3}}$

SOLUTION ▶

$\dfrac{\sqrt[4]{96x}}{\sqrt[4]{3}}$

$= \sqrt[4]{\dfrac{96x}{3}}$ Quotient rule of radicals.

$= \sqrt[4]{\mathbf{32}x}$ Simplify.

$= \sqrt[4]{\mathbf{2 \cdot 2 \cdot 2 \cdot 2 \cdot 2} \cdot x}$ Factor the number.

$= \sqrt[4]{2^4}\sqrt[4]{2x}$ Use exponential notation; product rule of radicals.

$= \mathbf{2}\sqrt[4]{2x}$ Simplify.

Practice Problems

For problems 7–13, simplify.

7. $10\sqrt[5]{y} - 4\sqrt[5]{y}$ **8.** $\sqrt[3]{128} + \sqrt[3]{-54}$ **9.** $\sqrt[3]{128} - \sqrt[3]{54}$

10. $\sqrt[4]{12}\sqrt[4]{8}$ **11.** $\sqrt[3]{6x^2}\sqrt[3]{9x^2}$ **12.** $\dfrac{\sqrt[3]{48}}{\sqrt[3]{2}}$ **13.** $\dfrac{\sqrt[3]{400x^2}}{\sqrt[3]{5x}}$

Radicals and Rational Exponents

Since $\sqrt[3]{2^3} = 2$ and $2^{\frac{3}{3}} = 2^1$ or 2, then $\sqrt[3]{2^3} = 2^{\frac{3}{3}}$. This example shows the relationship of a radical and exponential notation, $\sqrt[n]{x^m} = x^{\frac{m}{n}}$.

Radicals and Exponential Notation

If m and n are whole numbers and $n > 1$, then $\sqrt[n]{x^m} = x^{\frac{m}{n}}$.

In some situations, it is easier to work with radicals. In other situations, it is easier to work in exponential notation. It is important to be able to change from one notation into the other. To rewrite a radical in exponential notation, identify m and n and use the relationship $\sqrt[n]{x^m} = x^{\frac{m}{n}}$.

EXAMPLE 7 Rewrite the radical in exponential notation.

(a) $\sqrt[3]{x^2}$ $\quad m = 2, n = 3$

SOLUTION ▶ $= x^{\frac{2}{3}}$ $\quad \sqrt[n]{x^m} = x^{\frac{m}{n}}$

(b) $\sqrt[5]{p}$

▶ $= \sqrt[5]{p^1}$ $\quad m = 1, n = 5$

$= p^{\frac{1}{5}}$ $\quad \sqrt[n]{x^m} = x^{\frac{m}{n}}$

(c) $\sqrt{c}$

▶ $= \sqrt[2]{c^1}$ $\quad m = 1, n = 2$

$= c^{\frac{1}{2}}$ $\quad \sqrt[n]{x^m} = x^{\frac{m}{n}}$

In rewriting a radical in exponential notation, exponents may be improper fractions but must be in lowest terms. Since rational numbers can be written as fractions, exponents that are fractions are often called **rational exponents**.

EXAMPLE 8 Rewrite the radical in exponential notation and simplify.

(a) $\sqrt{x^3y^{10}}$

SOLUTION ▶ $= \sqrt{x^3}\sqrt{y^{10}}$ $\quad$ Product rule of radicals.

$= x^{\frac{3}{2}}y^{\frac{10}{2}}$ $\quad \sqrt[n]{x^m} = x^{\frac{m}{n}}$

$= x^{\frac{3}{2}}y^5$ $\quad$ Simplify.

(b) $\sqrt[3]{x^6y^{10}}$

$= \sqrt[3]{x^6}\sqrt[3]{y^{10}}$ Product rule of radicals.

$= x^{\frac{6}{3}}y^{\frac{10}{3}}$ $\sqrt[n]{x^m} = x^{\frac{m}{n}}$

$= x^2y^{\frac{10}{3}}$ Simplify.

We can also use the relationship $x^{\frac{m}{n}} = \sqrt[n]{x^m}$ to rewrite exponential expressions with rational exponents in radical notation.

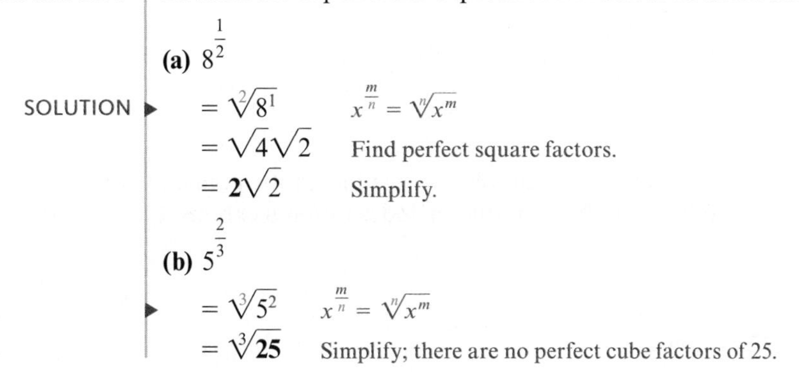

EXAMPLE 9 Rewrite the exponential expression in radical notation and simplify.

(a) $8^{\frac{1}{2}}$

SOLUTION ▸ $= \sqrt[2]{8^1}$ $x^{\frac{m}{n}} = \sqrt[n]{x^m}$

$= \sqrt{4}\sqrt{2}$ Find perfect square factors.

$= \mathbf{2\sqrt{2}}$ Simplify.

(b) $5^{\frac{2}{3}}$

▸ $= \sqrt[3]{5^2}$ $x^{\frac{m}{n}} = \sqrt[n]{x^m}$

$= \mathbf{\sqrt[3]{25}}$ Simplify; there are no perfect cube factors of 25.

Practice Problems

For problems 14–15, rewrite the exponential expression in radical notation and simplify.

14. $32^{\frac{1}{2}}$ **15.** $6^{\frac{2}{3}}$

For problems 16–18, rewrite the radical in exponential notation and simplify.

16. $\sqrt{x^4y^9}$ **17.** $\sqrt[3]{x^4y^9}$ **18.** $\sqrt[4]{x^4y^{12}}$

The Exponent Rules and Rational Exponents

We studied the rules of exponents in Section 5.1. We can use the product rule of exponents, $x^m \cdot x^n = x^{m+n}$, to simplify expressions that include rational exponents. In a simplified expression, all fractions are in lowest terms.

EXAMPLE 10 Simplify: $x^{\frac{2}{9}}y^{\frac{1}{4}}x^{\frac{1}{9}}$

SOLUTION ▸ $x^{\frac{2}{9}}y^{\frac{1}{4}}x^{\frac{1}{9}}$

$= x^{\left(\frac{2}{9}+\frac{1}{9}\right)}y^{\frac{1}{4}}$ Product rule of exponents; add the exponents.

$= x^{\frac{3}{9}}y^{\frac{1}{4}}$ Simplify.

$= x^{\left(\frac{3 \cdot 1}{3 \cdot 3}\right)}y^{\frac{1}{4}}$ Find common factors.

$= \mathbf{x^{\frac{1}{3}}y^{\frac{1}{4}}}$ Simplify.

When an exponent is raised to another power, use the power rule of exponents, $(x^m)^n = x^{m \cdot n}$, and multiply the exponents.

EXAMPLE 11 Simplify $\left(a^{\frac{3}{4}}\right)^{\frac{2}{9}}$ and rewrite in radical notation.

SOLUTION ▶ $\left(a^{\frac{3}{4}}\right)^{\frac{2}{9}}$

$= a^{\left(\frac{3}{4}\right)\left(\frac{2}{9}\right)}$ Power rule of exponents; multiply the exponents.

$= a^{\left(\frac{3 \cdot 1}{2 \cdot 2}\right)\left(\frac{2}{3 \cdot 3}\right)}$ Find common factors.

$= a^{\frac{1}{6}}$ Simplify.

$= \sqrt[6]{a}$ Rewrite in radical notation: $x^{\frac{m}{n}} = \sqrt[n]{x^m}$

In the next example, we use the quotient rule of exponents, $\frac{x^m}{x^n} = x^{m-n}$, to simplify an expression that includes rational exponents.

EXAMPLE 12 Simplify: $\dfrac{x^{\frac{11}{12}}}{x^{\frac{5}{12}}}$

SOLUTION ▶ $\dfrac{x^{\frac{11}{12}}}{x^{\frac{5}{12}}}$

$= x^{\left(\frac{11}{12} - \frac{5}{12}\right)}$ Quotient rule of exponents; subtract the exponents.

$= x^{\frac{6}{12}}$ Simplify.

$= x^{\frac{1 \cdot 6}{2 \cdot 6}}$ Find common factors.

$= x^{\frac{1}{2}}$ Simplify.

Because a simplified expression has only positive exponents, in the next example we rewrite $x^{\frac{-2}{11}}$ as $\dfrac{1}{x^{\frac{2}{11}}}$.

Negative Exponents

If x and m are real numbers and x is not equal to 0, $x^{-m} = \dfrac{1}{x^m}$ and $\dfrac{1}{x^{-m}} = x^m$.

EXAMPLE 13 Simplify: $\dfrac{5x^{\frac{1}{11}}}{9x^{\frac{3}{11}}}$

SOLUTION ▶

$$\frac{5x^{\frac{1}{11}}}{9x^{\frac{3}{11}}}$$

$= \dfrac{5x^{\left(\frac{1}{11}-\frac{3}{11}\right)}}{9}$ Quotient rule of exponents; subtract exponents.

$= \dfrac{5x^{-\frac{2}{11}}}{9}$ Simplify.

$= \dfrac{5}{9x^{\frac{2}{11}}}$ Rewrite with only positive exponents.

Practice Problems

For problems 19–21, simplify.

19. $z^{\frac{2}{5}}z^{\frac{1}{5}}$ **20.** $x^{\frac{5}{18}}x^{\frac{7}{24}}$ **21.** $x^{\frac{1}{8}}y^{\frac{1}{9}}x^{\frac{3}{20}}y^{\frac{2}{15}}$

For problems 22–24, simplify and rewrite in radical notation.

22. $\left(h^{\frac{2}{9}}\right)^{\frac{3}{10}}$ **23.** $\left(x^{\frac{3}{8}}\right)^{\frac{8}{3}}$ **24.** $\left(p^{\frac{1}{2}}\right)^{2}$

For problems 25–28, simplify.

25. $\dfrac{a^{\frac{5}{7}}}{a^{\frac{3}{7}}}$ **26.** $\dfrac{n^{\frac{5}{9}}}{n^{\frac{2}{9}}}$ **27.** $\dfrac{x^{\frac{5}{7}}}{x^{\frac{3}{8}}}$ **28.** $\dfrac{3x^{\frac{1}{9}}}{7x^{\frac{5}{9}}}$

8.6 Vocabulary Practice

Match the term with its description.

1. $\sqrt[n]{a}\cdot\sqrt[n]{b} = \sqrt[n]{ab}$
2. $a(b + c) = ab + ac$
3. $x^m \cdot x^n = x^{m+n}$
4. The positive square root of a positive number
5. $\dfrac{x^m}{x^n} = x^{m-n}$
6. The expression under a radical sign
7. The 3 in $\sqrt[3]{x}$ is an example of this.
8. $(x^m)^n = x^{mn}$
9. $\sqrt[n]{\dfrac{a}{b}} = \dfrac{\sqrt[n]{a}}{\sqrt[n]{b}}$
10. A number that is an element of the set $\{\ldots, -2, -1, 0, 1, 2, 3, \ldots\}$

A. distributive property
B. index
C. integer
D. power rule of exponents
E. product rule of exponents
F. product rule of radicals
G. principle square root
H. quotient rule of exponents
I. quotient rule of radicals
J. radicand

8.6 Exercises

Follow your instructor's guidelines for showing your work. Assume that all factors including variables in radicands represent real numbers that are greater than or equal to 0 and that all variables that are bases of exponential expressions represent nonnegative real numbers.

For exercises 1–8, evaluate.

1. $\sqrt[3]{27}$
2. $\sqrt[3]{8}$
3. $\sqrt[3]{-27}$
4. $\sqrt[3]{-8}$
5. $\sqrt[4]{81}$
6. $\sqrt[4]{16}$
7. $\sqrt[4]{-81}$
8. $\sqrt[4]{-16}$

For exercises 9–40, simplify.

9. $\sqrt[3]{54}$
10. $\sqrt[3]{40}$
11. $\sqrt[5]{96}$
12. $\sqrt[5]{64}$
13. $\sqrt[4]{32}$
14. $\sqrt[4]{48}$
15. $\sqrt[3]{4}$
16. $\sqrt[3]{9}$
17. $\sqrt[4]{-64}$
18. $\sqrt[4]{-48}$
19. $\sqrt[3]{-54}$
20. $\sqrt[3]{-40}$
21. $\sqrt{18}$
22. $\sqrt{12}$
23. $\sqrt[5]{-1}$
24. $\sqrt[3]{-1}$
25. $8\sqrt[4]{z} + 3\sqrt[4]{z}$
26. $9\sqrt[4]{x} + 2\sqrt[4]{x}$
27. $8\sqrt[4]{z} - 3\sqrt[4]{z}$
28. $9\sqrt[4]{x} - 2\sqrt[4]{x}$
29. $\sqrt[3]{54} + \sqrt[3]{16}$
30. $\sqrt[3]{24} + \sqrt[3]{81}$
31. $\sqrt[3]{192} + 5\sqrt[3]{3}$
32. $\sqrt[3]{128} + 4\sqrt[3]{2}$
33. $\sqrt[3]{54} - \sqrt[3]{16}$
34. $\sqrt[3]{24} - \sqrt[3]{81}$
35. $\sqrt[3]{54} + \sqrt[3]{-16}$
36. $\sqrt[3]{24} + \sqrt[3]{-81}$
37. $\sqrt[3]{9}\sqrt[3]{18}$
38. $\sqrt[3]{4}\sqrt[3]{12}$
39. $\sqrt[4]{8}\sqrt[4]{6}$
40. $\sqrt[4]{8}\sqrt[4]{14}$

For exercises 41–54, rewrite the radical expression in exponential notation.

41. $\sqrt{x}$
42. $\sqrt{y}$
43. $\sqrt{x^3}$
44. $\sqrt{y^5}$
45. $\sqrt[4]{x^3}$
46. $\sqrt[4]{y^3}$
47. $\sqrt[3]{x^2y}$
48. $\sqrt[3]{xy^2}$
49. $\sqrt[5]{x}$
50. $\sqrt[5]{y}$
51. $\sqrt[99]{x}$
52. $\sqrt[97]{y}$
53. $\sqrt[3]{h^4}$
54. $\sqrt[3]{k^4}$

For exercises 55–62, rewrite the radical expression in exponential notation and simplify.

55. $\sqrt[4]{x^2}$
56. $\sqrt[4]{y^2}$
57. $\sqrt[20]{a^4}$
58. $\sqrt[20]{a^5}$
59. $\sqrt[24]{d^3}$
60. $\sqrt[24]{d^4}$
61. $\sqrt[4]{a^6b^8}$
62. $\sqrt[4]{a^8b^{10}}$

For exercises 63–66, rewrite the exponential expression in radical notation and simplify.

63. $m^{\frac{3}{2}}$
64. $p^{\frac{3}{2}}$
65. $b^{\frac{4}{3}}$
66. $a^{\frac{5}{3}}$

For exercises 67–86, simplify.

67. $a^{\frac{1}{3}}a^{\frac{1}{3}}$
68. $u^{\frac{1}{5}}u^{\frac{2}{5}}$
69. $x^{\frac{1}{6}}x^{\frac{1}{6}}$
70. $y^{\frac{1}{8}}y^{\frac{1}{8}}$
71. $x^{\frac{2}{9}}x^{\frac{1}{5}}$
72. $y^{\frac{1}{8}}y^{\frac{3}{5}}$
73. $\left(x^{\frac{1}{2}}\right)^{\frac{1}{4}}$
74. $\left(y^{\frac{1}{6}}\right)^{\frac{1}{3}}$
75. $\left(x^{\frac{3}{8}}\right)^{2}$
76. $\left(y^{\frac{5}{9}}\right)^{3}$
77. $\left(a^{\frac{1}{8}}\right)^{\frac{2}{3}}$
78. $\left(c^{\frac{1}{6}}\right)^{\frac{2}{5}}$
79. $x^{\frac{2}{9}}x^{\frac{3}{5}}x^{\frac{1}{6}}$
80. $y^{\frac{3}{10}}y^{\frac{1}{4}}y^{\frac{2}{15}}$
81. $\dfrac{x^{\frac{4}{5}}}{x^{\frac{1}{5}}}$
82. $\dfrac{x^{\frac{7}{9}}}{x^{\frac{2}{9}}}$
83. $\dfrac{a^{\frac{3}{4}}}{a^{\frac{1}{4}}}$
84. $\dfrac{a^{\frac{5}{8}}}{a^{\frac{1}{8}}}$
85. $\dfrac{2a^{\frac{3}{7}}}{3a^{\frac{5}{7}}}$
86. $\dfrac{4c^{\frac{3}{5}}}{7c^{\frac{4}{5}}}$

Problem Solving: Practice and Review

Follow your instructor's guidelines for using the five steps as outlined in Section 1.3, p. 55.

87. Find the percent increase in the value of Boeing stock per share. Round to the nearest hundredth of a percent.
Boeing shares rose 17 cents to $75.02 in morning trading.
(*Source:* seattletimes.nwsource.com, Feb. 14, 2012)

88. Anesthesiologists use a cuffed endotracheal tube to intubate children. For children older than 1 year, a formula for finding the correct size of the internal diameter of the tube in millimeters, D, is $D = \dfrac{A}{4} + 3.5$, where A is the age of the child in years. Predict the correct internal diameter of this tube for a 6-year-old child.
(*Source:* Duracher et al., *Paediatric Anaethesia*, Feb. 2008)

For exercises 89–90, in a study of broiler chickens, the daily water consumption per chicken, *y*, depended on the age of the chicken in days, *x*. A linear model of this relationship is

$$y = \left(\frac{5.28 \dfrac{\text{mL}}{\text{chicken}}}{\text{age in days}} \right) x.$$

(*Source:* Pesti et al., *Poultry Science*, May 1985)

89. Predict the amount of water needed per day by a chicken that is 20 days old.

90. Find the percent increase in the amount of water needed by a chicken that is 20 days old compared to a chicken that is 12 days old. Round to the nearest percent.

Find the Mistake

For exercises 91–94, the completed problem has one mistake.

(a) Describe the mistake in words, or copy down the whole problem and highlight or circle the mistake.

(b) Do the problem correctly.

91. Problem: Rewrite $x^{\frac{4}{7}}$ in radical notation.

Incorrect Answer: $x^{\frac{4}{7}}$

$= \sqrt[4]{x^7}$

92. Problem: Simplify: $\sqrt[3]{48}$

Incorrect Answer: $\sqrt[3]{48}$

$= \sqrt[3]{16}\sqrt[3]{3}$

$= 4\sqrt[3]{3}$

93. Problem: Simplify: $x^{\frac{2}{7}}x^{\frac{3}{7}}$

Incorrect Answer: $x^{\frac{2}{7}}x^{\frac{3}{7}}$

$= x^{\left(\frac{2}{7}\right)\left(\frac{3}{7}\right)}$

$= x^{\frac{6}{49}}$

94. Problem: Simplify: $\dfrac{x^{\frac{9}{10}}}{x^{\frac{3}{10}}}$

Incorrect Answer: $\dfrac{x^{\frac{9}{10}}}{x^{\frac{3}{10}}}$

$= x^{\frac{6}{10}}$

Review

For exercises 95–98, factor completely.

95. $4x^2 - 4x - 15$

96. $2x^3 - 7x^2 - 4x$

97. $8x^2 - 32$

98. $4x^2 - 9$

SUCCESS IN COLLEGE MATHEMATICS

99. Describe what you think is the difference between *teaching* and *tutoring*.

Study Plan for Review of Chapter 8

SECTION 8.1 Square Roots

Ask Yourself	Test Yourself	Help Yourself
Can I . . .	**Do 8.1 Review Exercises**	**See these Examples and Practice Problems**
evaluate a square root, rounding irrational numbers to a given place value?	1–3	Ex. 1–2, PP 1–5
use the product rule of radicals to simplify a square root?	5–10	Ex. 4–7, PP 10–20

8.1 Review Exercises

For exercises 1–3, evaluate. If the number is irrational, use a calculator and round to the nearest hundredth.

1. $\sqrt{121}$

2. $\sqrt{52}$

3. $\sqrt{-100}$

4. Explain why the square root of a negative number is not a real number.

For exercises 5–10, simplify.

5. $\sqrt{800}$

6. $\sqrt{128}$

7. $\sqrt{245}$

8. $\sqrt{45}$

9. $\sqrt{288x^3}$

10. $\sqrt{98x^3y^2}$

SECTION 8.2 Adding and Subtracting Square Roots

Ask Yourself	Test Yourself	Help Yourself
Can I...	**Do 8.2 Review Exercises**	**See these Examples and Practice Problems**
identify and combine like radicals?	11–17	Ex. 1–4, PP 1–7
simplify an expression before combining like radicals?	18–20	Ex. 5–8, PP 8–15

8.2 Review Exercises

11. A radical is $\sqrt{2x}$. Identify the like radicals.

a. $2\sqrt{x}$ **b.** $-\sqrt{2x}$

c. $\sqrt{x^2}$ **d.** $2\sqrt{2x}$

e. $\sqrt{-2x}$

12. Explain why $\sqrt{5x}$ and $\sqrt{5y}$ are not like radicals.

For exercises 13–17, simplify.

13. $13\sqrt{2} - 8\sqrt{2}$

14. $\sqrt{x} - 5\sqrt{x}$

15. $-\sqrt{p} + 9\sqrt{p} - 15\sqrt{p}$

16. $3\sqrt{x} + 7\sqrt{y} + 10\sqrt{x} - 9\sqrt{y}$

17. $5\sqrt{20} - 3\sqrt{20}$

For exercises 18–20, simplify.

18. $4\sqrt{72} + 9\sqrt{200}$

19. $\sqrt{18} - \sqrt{50}$

20. $\sqrt{32x} + \sqrt{50x}$

SECTION 8.3 Multiplying Square Roots

Ask Yourself	Test Yourself	Help Yourself
Can I...	**Do 8.3 Review Exercises**	**See these Examples and Practice Problems**
multiply square roots and simplify the product?	21–24	Ex. 1–3, PP 1–4
use the distributive property to multiply radical expressions?	25–27	Ex. 4–7, PP 5–8

8.3 Review Exercises

For exercises 21–24, simplify.

21. $\sqrt{15}\sqrt{35}$

22. $\sqrt{30xy}\sqrt{3x}$

23. $\sqrt{21a}\sqrt{9a}$

24. $\sqrt{75p^2q}\sqrt{6pq}$

For exercises 25–27, simplify.

25. $\sqrt{2}(\sqrt{6x} - 14)$

26. $\sqrt{2}(\sqrt{6x} - \sqrt{14})$

27. $(x - \sqrt{3})(x + \sqrt{15})$

SECTION 8.4 Dividing Square Roots and Rationalizing Denominators

Ask Yourself	Test Yourself	Help Yourself
Can I...	**Do 8.4 Review Exercises**	**See these Examples and Practice Problems**
use the quotient rule of radicals to evaluate or simplify a square root?	28–32	Ex. 1, 2, 4, PP 1–4
rationalize the denominator of a radical expression?	30–33	Ex. 3–4, PP 2–4
rationalize the denominator of a radical expression that includes a sum or difference?	34–35	Ex. 5–6, PP 5–7

8.4 Review Exercises

28. Evaluate: $\sqrt{\dfrac{64}{49}}$

For exercises 29–32, simplify.

29. $\sqrt{\dfrac{9x}{16}}$

30. $\sqrt{\dfrac{11x}{3}}$

31. $\sqrt{\dfrac{27}{20}}$

32. $\sqrt{\dfrac{40c}{63d}}$

33. Choose the expression(s) in which the denominator is rational.

a. $\dfrac{3}{\sqrt{x}}$ b. $\dfrac{\sqrt{x}}{\sqrt{49}}$

c. $\dfrac{\sqrt{x}+3}{\sqrt{3}}$ d. $\dfrac{\sqrt{x}+3}{5}$

For exercises 34–35, simplify.

34. $\dfrac{3}{\sqrt{x}-\sqrt{6}}$

35. $\dfrac{\sqrt{3}}{\sqrt{x}+\sqrt{6}}$

SECTION 8.5 Radical Equations

Ask Yourself	Test Yourself	Help Yourself
Can I . . .	**Do 8.5 Review Exercises**	**See these Examples and Practice Problems**
solve a radical equation that includes one square root?	36–41	Ex. 1–4, PP 1–5
identify an extraneous solution of a radical equation?	37, 39, 41, 43–44	Ex. 5–8, PP 7–9
solve a radical equation that includes two square roots?	42	Ex. 8, PP 9–10
use a radical equation in two variables to solve an application problem?	45	Ex. 9, PP 11

8.5 Review Exercises

For exercises 36–42,

(a) solve.
(b) check.

36. $\sqrt{6x-12}=24$

37. $\sqrt{w}+15=4$

38. $\sqrt{c}-9=4$

39. $p=\sqrt{6p+72}$

40. $\sqrt{15a-54}=a$

41. $\sqrt{-15a-54}=a$

42. $\sqrt{n}=\sqrt{n-5}+1$

43. A student solved $x+4=\sqrt{38-x}$ and said that the solutions were $x=2$ and $x=-11$. Explain why $x=-11$ is an extraneous solution.

44. Explain why the equation $\sqrt{5a-2}=-1$ has no solution.

45. The formula for the volume of a cylinder is $V=\pi r^2 h$, where V is the area, r is the radius, and h is the height. Solved for r, this formula is $r=\sqrt{\dfrac{V}{\pi h}}$. If the volume of a circular swimming pool is 3052 ft^3 and its depth is 3 ft, find its radius. ($\pi \approx 3.14$.) Round to the nearest whole number.

SECTION 8.6 Higher Index Radicals and Rational Exponents

Ask Yourself	Test Yourself	Help Yourself
Can I . . .	**Do 8.6 Review Exercises**	**See these Examples and Practice Problems**
evaluate or simplify a radical with an index greater than 2?	46–51	Ex. 1–6, PP 1–13
rewrite a radical expression in exponential notation and simplify?	52–53	Ex. 7–8, PP 16–18
rewrite an exponential expression in radical notation and simplify?	54	Ex. 9, PP 14–15
simplify an expression that includes rational exponents?	55–62	Ex. 10–13, PP 19–28

8.6 Review Exercises

46. Evaluate: $\sqrt[4]{625}$

For exercises 47–51, simplify.

47. $\sqrt[3]{48}$

48. $\sqrt[4]{48}$

49. $\sqrt[3]{10}\sqrt[3]{50}$

50. $\sqrt[3]{40}+2\sqrt[3]{40}$

51. $\sqrt[5]{96x^4y^6}$

52. Rewrite $\sqrt[3]{d^5}$ in exponential notation.

53. Rewrite $\sqrt[3]{x^9y^2}$ in exponential notation and simplify.

54. Rewrite $a^{\frac{12}{5}}$ in radical notation and simplify.

55. Explain why the expression $a^{\frac{12}{10}}$ is not simplified.

56. Explain why the expression $a^{\frac{3}{4}}b^{-\frac{1}{8}}$ is not simplified.

57. Fractions can be added, subtracted, multiplied, and divided. For which of these operations do you need to find a common denominator?

For exercises 58–62, simplify.

58. $a^{\frac{1}{8}}a^{\frac{1}{8}}a^{\frac{5}{8}}$

59. $x^{\frac{1}{4}}x^{\frac{2}{5}}$

60. $\left(a^{\frac{2}{3}}\right)^{\frac{3}{5}}$

61. $\dfrac{c^{\frac{3}{17}}}{c^{\frac{2}{17}}}$

62. $\dfrac{4x^{\frac{5}{7}}}{9x^{\frac{6}{7}}}$

Chapter 8 Test

Assume that all factors in radicands represent real numbers that are greater than or equal to 0 and that all variables that are bases of exponential expressions represent nonnegative real numbers.

For problems 1–14, simplify.

1. $\sqrt{12}$
2. $\sqrt{192x^5}$
3. $\sqrt{15a^4b^3}$
4. $\sqrt{98}$
5. $\sqrt{75a}$
6. $\sqrt{60n^{10}}$
7. $8\sqrt{3} + 5\sqrt{3}$
8. $7\sqrt[3]{x} - 12\sqrt[3]{x}$
9. $5\sqrt{a} - \sqrt{b} - 3\sqrt{b} + 2\sqrt{a}$
10. $\sqrt{32} + \sqrt{50}$
11. $\sqrt{\dfrac{5p}{64}}$
12. $\sqrt{\dfrac{5p}{3z}}$
13. $\dfrac{3}{\sqrt{5u}}$
14. $\dfrac{7}{\sqrt{x} - 4}$

15. A radical is $\sqrt[3]{x^2}$.
   a. Identify the radicand.
   b. Identify the index.
   c. Rewrite the radical in exponential notation.

16. Explain how to identify like radicals.

17. Explain why $\dfrac{\sqrt{3}}{\sqrt{5}}$ is equivalent to $\dfrac{\sqrt{15}}{5}$.

For problems 18–20,
(a) solve.
(b) check.

18. $\sqrt{2x - 20} - 8 = 14$

19. $w = \sqrt{-4w + 45}$

20. $\sqrt{x + 9} + 21 = 15$

21. In a car engine, the *bore*, *B*, is the diameter of an engine cylinder. The *stroke*, *S*, is the distance that the piston travels up or down in the cylinder. The *displacement* of a cylinder is the volume swept by the piston as it makes one full stroke. For a displacement of 366 in.[2], the relationship of the bore and the stroke is described by the equation $B = \sqrt{\dfrac{366}{2\pi S}}$. A mechanic is modifying a Ford engine with eight cylinders. The new displacement will be 366 in.[2]. The stroke is 3.5 in. Find the new bore. ($\pi \approx 3.14$.) Round *down* to the nearest hundredth of an inch.

For problems 22–27, simplify.

22. $\sqrt[4]{32x^5}$

23. $\sqrt[3]{16} - 5\sqrt[3]{128}$

24. $x^{\frac{3}{4}}x^{\frac{1}{5}}$

25. $\left(a^{\frac{2}{3}}\right)^{\frac{5}{12}}$

26. $\dfrac{a^{\frac{5}{9}}}{a^{\frac{1}{9}}}$

27. $\dfrac{2a^{\frac{3}{25}}}{3a^{\frac{9}{25}}}$

Quadratic Equations

9

SUCCESS IN COLLEGE MATHEMATICS

Studying for a Final Exam

The questions on a comprehensive final exam assess what you have learned from the beginning of the term. Your score on the final exam also is part of your final grade. The process of reviewing for a final exam may help you better understand the connections between concepts learned in different chapters. Reviewing may also help transfer your knowledge more permanently into your long-term memory. The Success in College Mathematics exercises in each section of this chapter suggest some ways to review for the final exam.

By the end of the term, you are probably tired. You might have big projects or papers due in other classes that make it difficult to find time to do your math homework as well as to study for the final exam. However, it is important to find that time and make this studying part of your schedule. Studying for a final exam should not be limited to the last few hours before the test.

9.1 Quadratic Equations in One Variable
9.2 Completing the Square
9.3 The Quadratic Formula
9.4 Quadratic Equations with Nonreal Solutions
9.5 Quadratic Equations in the Form $y = ax^2 + bx + c$

DOE/NREL/Jessica Raker

To find the length of a guy wire that stabilizes a wind turbine tower, we can write and solve a quadratic equation. In this section, we will write and solve quadratic equations in the form $x^2 = k$.

SECTION 9.1 Quadratic Equations in One Variable

After reading the text, working the practice problems, and completing assigned exercises, you should be able to:

1. Solve a quadratic equation in the form $x^2 = k$.
2. Rearrange a quadratic equation into the form $x^2 = k$.
3. Use the Pythagorean theorem to solve an application problem.
4. Use a quadratic equation to solve an application problem.

Quadratic Equations in the Form $x^2 = k$

A quadratic equation is a polynomial equation with a degree of 2. The quadratic equation $x^2 = 25$ has two solutions, $x = 5$ or $x = -5$. Since $\sqrt{25} = 5$, these solutions can be written as $x = \sqrt{25}$ or $x = -\sqrt{25}$. We can extend this example to solve other quadratic equations. The solutions of quadratic equations that can be written in the form $x^2 = k$, where k is a real number greater than or equal to 0, are $x = \sqrt{k}$ or $x = -\sqrt{k}$. In **plus-minus notation**, the solutions are $x = \pm\sqrt{k}$.

EXAMPLE 1 (a) Solve: $p^2 = 100$

SOLUTION

$p^2 = 100$ — The equation is in the form $x^2 = k$.

$p = \sqrt{100}$ or $p = -\sqrt{100}$ — The solutions are $x = \sqrt{k}$ or $x = -\sqrt{k}$.

$p = \mathbf{10}$ or $p = \mathbf{-10}$ — Simplify.

$p = \pm 10$ — $\pm$ notation.

(b) Check.

Check: $p = 10$

$p^2 = 100$

$(\mathbf{10})^2 = 100$

$(\mathbf{10})(\mathbf{10}) = 100$

$\mathbf{100} = 100$ True.

Check: $p = -10$

$p^2 = 100$

$(\mathbf{-10})^2 = 100$

$(\mathbf{-10})(\mathbf{-10}) = 100$

$\mathbf{100} = 100$ True.

Solutions of a Quadratic Equation in the Form $x^2 = k$

The solutions of $x^2 = k$ are $x = \sqrt{k}$ or $x = -\sqrt{k}$, where k is a real number greater than or equal to 0. In $\pm$ notation, the solutions are $x = \pm\sqrt{k}$.

In the next example, the solutions of $z^2 = 77$ are irrational numbers.

EXAMPLE 2 (a) Solve: $z^2 = 77$

SOLUTION

$z^2 = 77$ — The equation is in the form $x^2 = k$.

$z = \sqrt{77}$ or $z = -\sqrt{77}$ — The solutions are $x = \sqrt{k}$ or $x = -\sqrt{k}$.

$z = \pm\sqrt{77}$ — $\pm$ notation.

(b) Check.

Check: $z = \sqrt{77}$

$$z^2 = 77$$
$$(\sqrt{77})^2 = 77$$
$$77 = 77 \quad \text{True.}$$

Check: $z = -\sqrt{77}$

$$z^2 = 77$$
$$(-\sqrt{77})^2 = 77$$
$$(-1\sqrt{77})^2 = 77$$
$$(-1\sqrt{77})(-1\sqrt{77}) = 77$$
$$77 = 77 \quad \text{True.}$$

In the next example, we use the product rule of radicals, $\sqrt{ab} = \sqrt{a}\sqrt{b}$, to simplify the solutions.

EXAMPLE 3 (a) Solve: $a^2 = 200$

SOLUTION

$a^2 = 200$ — The equation is in the form $x^2 = k$.

$a = \sqrt{200}$ or $a = -\sqrt{200}$ — The solutions are $x = \sqrt{k}$ or $x = -\sqrt{k}$.

$a = \pm\sqrt{200}$ — $\pm$ notation.

$a = \pm\sqrt{100}\sqrt{2}$ — Product rule of radicals.

$a = \pm 10\sqrt{2}$ — Simplify.

(b) Check.

Check: $a = 10\sqrt{2}$

$$a^2 = 200$$
$$(10\sqrt{2})^2 = 200$$
$$(10\sqrt{2})(10\sqrt{2}) = 200$$
$$100\sqrt{4} = 200$$
$$100 \cdot 2 = 200$$
$$200 = 200 \quad \text{True.}$$

Check: $a = -10\sqrt{2}$

$$a^2 = 200$$
$$(-10\sqrt{2})^2 = 200$$
$$(-10\sqrt{2})(-10\sqrt{2}) = 200$$
$$100\sqrt{4} = 200$$
$$100 \cdot 2 = 200$$
$$200 = 200 \quad \text{True.}$$

Practice Problems

For problems 1–4,
(a) solve.
(b) check.

1. $x^2 = 49$ **2.** $h^2 = 12$ **3.** $z^2 = 40$ **4.** $c^2 = 75$

Rewriting Quadratic Equations in the Form $x^2 = k$

The solutions of $x^2 = k$ are $x = \pm\sqrt{k}$. In the next example, $15x^2 = 90$ is not in the form $x^2 = k$. To rewrite it in this form, we divide both sides by the lead coefficient, 15.

EXAMPLE 4 Solve: $15x^2 = 90$

SOLUTION ▶

$15x^2 = 90$	The equation is not in the form $x^2 = k$.
$\frac{15x^2}{15} = \frac{90}{15}$	Division property of equality.
$x^2 = 6$	Simplify; the equation is in the form $x^2 = k$.
$x = \sqrt{6}$ or $x = -\sqrt{6}$	The solutions are $x = \sqrt{k}$ or $x = -\sqrt{k}$.
$x = \pm\sqrt{6}$	$\pm$ notation.

In the next example, to rewrite $x^2 - 7 = 104$ in the form $x^2 = k$, we add 7 to both sides.

EXAMPLE 5 **(a)** Solve: $x^2 - 7 = 104$

SOLUTION ▶

$x^2 - 7 = 104$	The equation is not in the form $x^2 = k$.
$+7 \quad +7$	Addition property of equality.
$x^2 + \mathbf{0} = \mathbf{111}$	Simplify.
$x^2 = 111$	Simplify; the equation is in the form $x^2 = k$.
$x = \sqrt{111}$ or $x = -\sqrt{111}$	The solutions are $x = \sqrt{k}$ or $x = -\sqrt{k}$.
$x = \pm\sqrt{111}$	$\pm$ notation.

(b) Check.

▶

Check: $x = \sqrt{111}$	Check: $x = -\sqrt{111}$
$x^2 - 7 = 104$	$x^2 - 7 = 104$
$(\sqrt{\mathbf{111}})^2 - 7 = 104$	$(-\sqrt{\mathbf{111}})^2 - 7 = 104$
$\mathbf{111} - 7 = 104$	$\mathbf{111} - 7 = 104$
$\mathbf{104} = 104$ True.	$\mathbf{104} = 104$ True.

In the next example, we use two properties of equality to rewrite the equation in the form $x^2 = k$.

EXAMPLE 6 Solve: $-5w^2 - 20 = -145$

SOLUTION ▶

$-5w^2 - 20 = -145$	The equation is not in the form $x^2 = k$.
$+20 \quad +20$	Addition property of equality.
$-5w^2 + \mathbf{0} = \mathbf{-125}$	Simplify.
$\frac{\mathbf{-5w^2}}{-5} = \frac{-125}{-5}$	Division property of equality.
$w^2 = 25$	Simplify; the equation is in the form $x^2 = k$.
$w = \sqrt{25}$ or $w = -\sqrt{25}$	The solutions are $x = \sqrt{k}$ or $x = -\sqrt{k}$.
$w = \mathbf{5}$ or $w = \mathbf{-5}$	Simplify.
$w = \pm 5$	$\pm$ notation.

If $x = p + 2$, then the equation $(p + 2)^2 = 16$ is in the form $x^2 = k$, and the solutions of the equation are $p + 2 = \sqrt{16}$ or $p + 2 = -\sqrt{16}$.

EXAMPLE 7 **(a)** Solve: $(p + 2)^2 = 16$

SOLUTION ▶

$(p + 2)^2 = 16$	The equation is in the form $x^2 = k$.
$p + 2 = \sqrt{16}$ or $p + 2 = -\sqrt{16}$	The solutions are $x = \sqrt{k}$ or $x = -\sqrt{k}$.

$p + 2 = \mathbf{4}$	or	$p + 2 = \mathbf{-4}$	Simplify.
$\underline{-2 \quad -2}$		$\underline{-2 \quad -2}$	Subtraction property of equality.
$p + \mathbf{0} - \mathbf{2}$	or	$p + \mathbf{0} = \mathbf{-6}$	Simplify.
$\mathbf{p} = 2$	or	$\mathbf{p} = -6$	Simplify.

(b) Check.

Check: $p = 2$	Check: $p = -6$
$(p + 2)^2 = 16$	$(p + 2)^2 = 16$
$(\mathbf{2} + 2)^2 = 16$	$(\mathbf{-6} + 2)^2 = 16$
$\mathbf{4}^2 = 16$	$(\mathbf{-4})^2 = 16$
$\mathbf{16} = 16$ True.	$\mathbf{16} = 16$ True.

Practice Problems

For problems 5–10,
(a) solve.
(b) check.

5. $x^2 - 18 = 82$ **6.** $8w^2 - 15 = 785$ **7.** $-3u^2 + 12 = -132$
8. $(a + 3)^2 = 25$ **9.** $(a - 3)^2 = 25$ **10.** $(c + 7)^2 = 9$

Applications

If we drop an object such as a pumpkin from a height, the force of gravity causes it to fall. Ignoring air resistance, the formula $d = \left(\frac{16 \text{ ft}}{1 \text{ s}^2}\right)t^2$ represents the relationship of the distance the object falls in feet, d, and the time it falls in seconds, t.

EXAMPLE 8 A pumpkin is dropped from a height of 66 ft. Find the time for the pumpkin to hit the ground. Round to the nearest tenth.

SOLUTION **Step 1 Understand the problem.**
The unknown is the time needed for the pumpkin to fall 66 ft.

$t =$ time for the pumpkin to fall 66 ft

Step 2 Make a plan.
Use the formula $d = \left(\frac{16 \text{ ft}}{1 \text{ s}^2}\right)t^2$, replacing d with 66 ft.

Step 3 Carry out the plan.

$d = \left(\frac{16 \text{ ft}}{1 \text{ s}^2}\right)t^2$	Formula for distance during free fall.
$\mathbf{66 \text{ ft}} = \left(\frac{16 \text{ ft}}{1 \text{ s}^2}\right)t^2$	Replace d with 66 ft.
$66 = 16t^2$	Remove units; the equation is not in the form $x^2 = k$.
$\frac{66}{16} = \frac{16t^2}{16}$	Division property of equality.
$4.125 = t^2$	Simplify.
$t^2 = 4.125$	Reverse sides; the equation is in the form $x^2 = k$.

$t = \sqrt{4.125}$	or	$t = -\sqrt{4.125}$	The solutions are $x = \sqrt{k}$ or $x = -\sqrt{k}$.
$t = \mathbf{2.03...}$	or	$t = \mathbf{-2.03...}$	Simplify.
$t \approx \mathbf{2.0 \text{ s}}$	or	$t \approx \mathbf{-2.0 \text{ s}}$	Round; replace the units.

Step 4 Look back.
Only 2.0 s can be a solution of the problem because elapsed time cannot be negative. Working backwards, $\left(\frac{16\text{ ft}}{1\text{ s}^2}\right)\left(\frac{2.0\text{ s}}{1}\right)^2$ equals 64 ft, which is close to the exact distance of 66 ft. Since the time is rounded down to the nearest tenth, we expect the distance to be a little less. The answer seems reasonable.

Step 5 Report the solution.
The pumpkin hits the ground in about 2 s.

Triangles have three sides and three angles. The sum of the measure of the angles in every triangle is 180 degrees. A **right triangle** has one 90-degree angle. The length of the side opposite the right angle is c. This side is the **hypotenuse**. The lengths of the other two sides are a and b. These sides are the **legs**. The Pythagorean theorem for right triangles states that $a^2 + b^2 = c^2$.

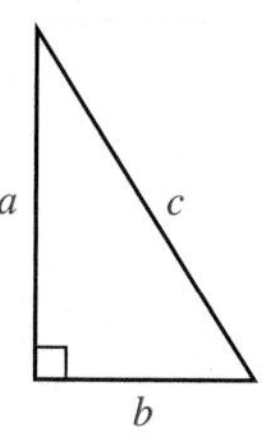

EXAMPLE 9 A small home wind turbine is mounted on top of a 70-ft tower. As shown in the diagram, the tower is stabilized by two upper guy wires that are attached 7 ft below the top of the tower and two lower guy wires that are attached to the tower 35 ft above the ground. Each guy wire is anchored to the ground 35 ft from the base of the tower. Find the length of the upper guy wire labeled in the diagram. Round to the nearest whole number.

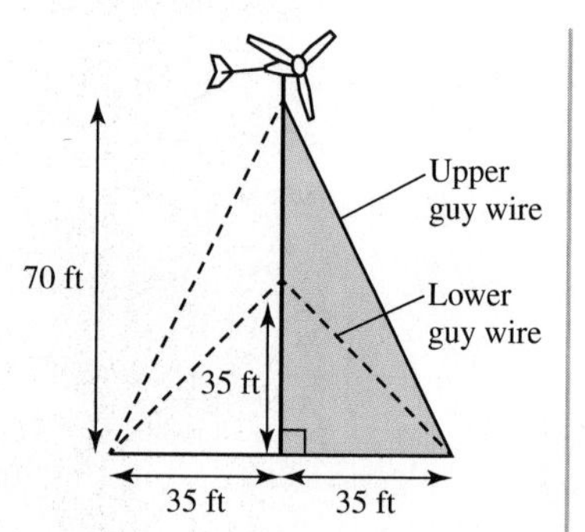

SOLUTION ▶ **Step 1 Understand the problem.**
The unknown is the length of the upper guy wire. The distance from the base of the tower to the point where the lower guy wires are attached to the tower is extraneous information.

c = length of an upper guy wire

Step 2 Make a plan.
The distance from the ground to the point where an upper guy wire connects to the tower is 70 ft – 7 ft, which equals 63 ft. The tower, an upper guy wire, and a line drawn from the base of the tower to the point where the upper guy wire is anchored to the ground form a right triangle. To find the length of an upper guy wire, we can use the Pythagorean theorem, $a^2 + b^2 = c^2$.

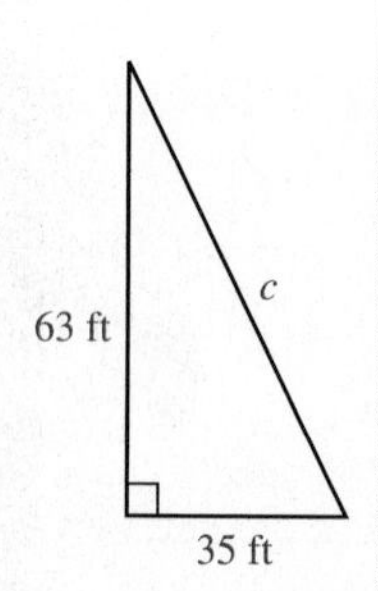

Step 3 Carry out the plan.

$\mathbf{(63\text{ ft})^2 + (35\text{ ft})^2} = c^2$ Replace variables; $a = 63$ ft and $b = 35$ ft.

$\mathbf{3969\text{ ft}^2 + 1225\text{ ft}^2} = c^2$ Simplify.

$\mathbf{5194\text{ ft}^2} = c^2$ Simplify.

$c^2 = 5194\text{ ft}^2$ Reverse sides; the equation is in the form $x^2 = k$.

$c = \sqrt{5194\text{ ft}^2}$ or $c = -\sqrt{5194\text{ ft}^2}$ The solutions are $x = \sqrt{k}$ or $x = -\sqrt{k}$.

$c = \mathbf{72.069...}$ ft or $c = \mathbf{-72.069...}$ ft Simplify.

$c \approx \mathbf{72\text{ ft}}$ or $c \approx \mathbf{-72\text{ ft}}$ Round.

Step 4 Look back.
A distance is a positive measurement. Although −72 ft is a solution of the equation, it is not a solution of this problem. Since $\sqrt{(72\text{ ft})^2}$ equals 72 ft and $\sqrt{(63\text{ ft})^2 + (35\text{ ft}^2)}$ also equals about 72 ft, the answer seems reasonable.

Step 5 Report the solution.
The length of one of the upper guy wires is about 72 ft.

Practice Problems

For problems 11–12, use the five steps.

11. A rock is dropped from a cliff that is 40 ft above the water. Ignoring air resistance, use the formula $d = \left(\frac{16\text{ ft}}{1\text{ s}^2}\right)t^2$ to find the time for the rock to hit the water. Round to the nearest tenth.

12. Using the information from the diagram in Example 9, find the length of the lower guy wire. Round to the nearest tenth.

9.1 Vocabulary Practice

Match the term with its description.

1. For a right triangle, $a^2 + b^2 = c^2$.
2. We can add a term to or subtract a term from both sides of an equality. The solution is not changed.
3. We can multiply or divide both sides of an equation by a real number that is not equal to 0. The solution is not changed.
4. The degree of $x^2 = 15$
5. In the term $5x^2$, 5 is an example of this.
6. $\pm$
7. The side opposite the 90-degree angle in a right triangle
8. A number that cannot be written as a fraction in which the numerator and denominator are integers
9. $\sqrt{\ }$
10. A triangle with a 90-degree angle

A. addition or subtraction property of equality
B. coefficient
C. hypotenuse
D. irrational number
E. multiplication or division property of equality
F. plus-minus notation
G. principal square root
H. Pythagorean theorem
I. right triangle
J. two

9.1 Exercises

Follow your instructor's guidelines for showing your work.

For exercises 1–20, solve.

1. $x^2 = 81$ **2.** $y^2 = 16$
3. $c^2 = 4$ **4.** $w^2 = 81$
5. $64 = m^2$ **6.** $100 = u^2$
7. $d^2 = 144$ **8.** $x^2 = 169$
9. $w^2 = 484$ **10.** $p^2 = 576$
11. $4x^2 = 100$ **12.** $3x^2 = 243$
13. $\frac{1}{2}x^2 = 50$ **14.** $\frac{1}{3}x^2 = 27$
15. $-x^2 = -169$ **16.** $-x^2 = -196$
17. $-5p^2 = -20$ **18.** $-6r^2 = -216$
19. $-\frac{1}{4}x^2 = -4$ **20.** $-\frac{1}{2}x^2 = -18$

For exercises 21–30, solve.

21. $z^2 = 128$ **22.** $a^2 = 162$
23. $x^2 = 150$ **24.** $d^2 = 20$
25. $p^2 = 147$ **26.** $v^2 = 192$
27. $500 = k^2$ **28.** $700 = h^2$
29. $y^2 = 47$ **30.** $z^2 = 17$

For exercises 31–38,
(a) solve.
(b) check.

31. $5x^2 = 60$ **32.** $7x^2 = 84$
33. $10x^2 = 80$ **34.** $10y^2 = 180$
35. $-5z^2 = -45$ **36.** $-8p^2 = -32$
37. $-2w^2 = -180$ **38.** $-3m^2 = -60$

For exercises 39–50, solve.

39. $v^2 + 18 = 99$
40. $w^2 + 35 = 99$
41. $v^2 - 18 = 7$
42. $w^2 - 35 = 14$
43. $3n^2 + 5 = 368$
44. $5a^2 + 7 = 852$
45. $d^2 + 15 = 71$
46. $f^2 + 18 = 94$
47. $2d^2 + 15 = 71$
48. $2f^2 + 14 = 94$
49. $4x^2 - 7 = 37$
50. $6x^2 - 5 = 73$

For exercises 51–56, solve.

51. $(x - 10)^2 = 49$
52. $(x - 1)^2 = 64$
53. $(x + 10)^2 = 49$
54. $(x + 1)^2 = 64$
55. $(x - 9)^2 = 100$
56. $(x - 3)^2 = 144$

For exercises 57–62, the lengths of the legs of a right triangle are given. Use the Pythagorean theorem to find the exact length of the hypotenuse.

57. $a = 30$ in.; $b = 40$ in.
58. $a = 6$ in.; $b = 8$ in.
59. $a = 24$ ft; $b = 32$ ft
60. $a = 27$ ft; $b = 36$ ft
61. $a = 10$ in.; $b = 10$ in.
62. $a = 20$ in.; $b = 20$ in.

For exercises 63–68, use the five steps and the Pythagorean theorem.

For exercises 63–64, the infield of a baseball field is a diamond. The distance between adjacent bases is 90 ft. We can divide the infield into right triangles.

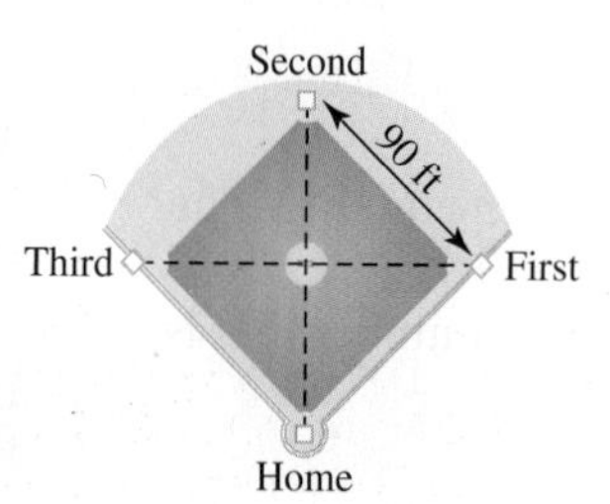

63. Find the distance between home plate and second base. Round to the nearest tenth.

64. Find the distance between first base and third base. Round to the nearest tenth.

65. Find the length of the dog-loading ramp in the photograph. Round to the nearest tenth.

© Laura Bracken

66. A standard doorway is an obstacle for a person in a wheelchair. A portable ramp makes the doorway accessible. The height of a doorway is 75 mm. The bottom of the ramp is 598 mm from the doorway. Find the length of the ramp. Round to the nearest hundredth. (*Source:* www.thorworld.co.uk)

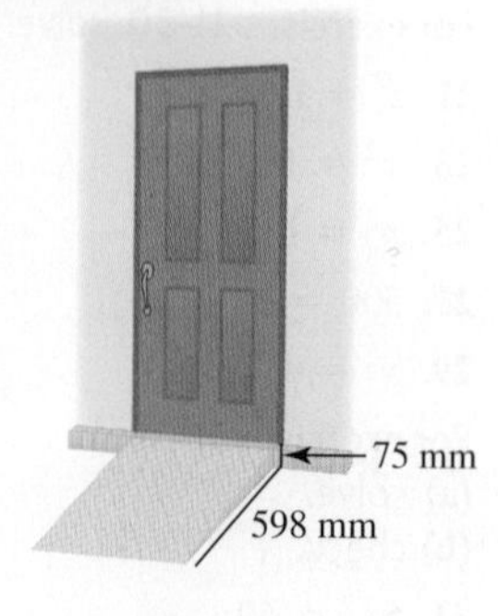

67. The height of a tower is 80 ft. A guy wire is anchored at a distance from the base that is equal to 60% of the height of the tower. The other end of the guy wire is connected to the top of the tower. Find the length of this guy wire. Round to the nearest whole number.

68. The height of a tower is 80 ft. A guy wire is anchored at a distance from the base that is equal to 80% of the height of the tower. The other end of this guy wire is connected to the tower 20 ft below its top. Find the length of this guy wire. Round to the nearest whole number.

69. The height of the IDS Center in Minneapolis is 792 ft. Ignoring air resistance, use the formula $d = \left(\frac{16\text{ ft}}{1\text{ s}^2}\right)t^2$ to find the time for an object to fall this distance. Round to the nearest hundredth. (*Source:* www.glasssteeland.com)

© Laura Bracken

70. The height of the building at 225 South Sixth in Minneapolis is 776 ft. Ignoring air resistance, use the formula $d = \left(\frac{16\text{ ft}}{1\text{ s}^2}\right)t^2$ to find the time for an object to fall this distance. Round to the nearest hundredth. (*Source:* www.emporis.com)

Problem Solving: Practice and Review

Follow your instructor's guidelines for using the five steps as outlined in Section 1.3, p. 55.

71. The diagram shows the dimensions of a small moving truck.
    a. Change the measurements into inches.
    b. Find the volume of the storage area in the truck in cubic inches.
    c. Change the volume into cubic feet. Round to the nearest tenth.

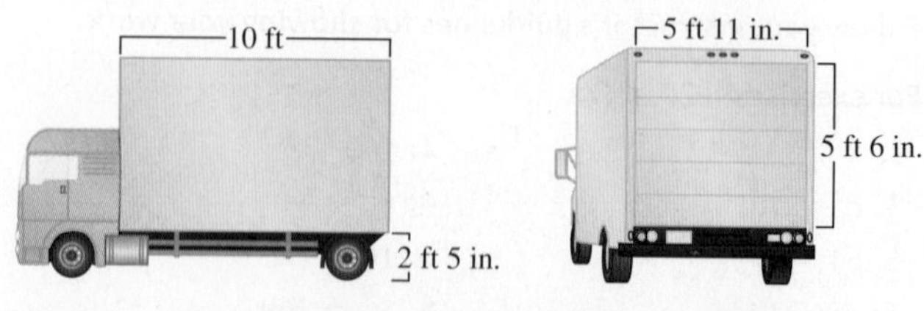

72. A travel agent is booking an extended stay of 8 weeks at a hotel. The hotel rate is \$79.50 a night. According to the hotel website, "A travel agent commission of 10% is paid on (or up to) the first seven nights' stay for all commissionable rates." Find the commission the travel agent will earn for this booking.

73. According to the NFL Digest of Rules, "The field is 360 feet long and 160 feet wide. The end zones are 30 feet deep. The line used in try-for-point plays is two yards out from the goal line." Find the percent that the combined area of the end zones is of the total area of the field. Round to the nearest percent. (*Source:* www.nfl.com)

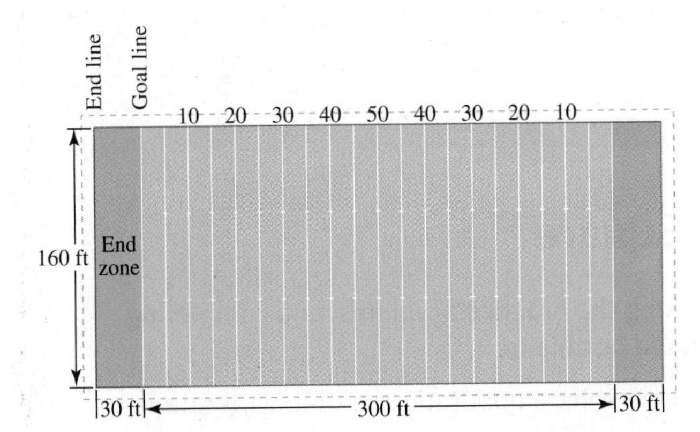

74. A small business owner creates trivets from used wine corks and sells them over the Internet. The cost to make each trivet is \$3. She wants to pay herself a salary of \$400 a week. She sells each trivet for \$18 plus shipping. Find the number of trivets she needs to make and sell each week to break even. Round up to the nearest whole number.

Find the Mistake

For exercises 75–78, the completed problem has one mistake.

(a) Describe the mistake in words, or copy down the whole problem and highlight or circle the mistake.

(b) Do the problem correctly.

75. Problem: Solve: $x^2 + 2 = 36$

Incorrect Answer: $x^2 + 2 = 36$

$$\sqrt{x^2} + 2 = \sqrt{36} \quad \text{or} \quad \sqrt{x^2} + 2 = -\sqrt{36}$$
$$x + 2 = 6 \quad \text{or} \quad x + 2 = -6$$
$$\underline{-2 \;\; -2} \qquad \underline{-2 \;\; -2}$$
$$x + 0 = 4 \quad \text{or} \quad x + 0 = -8$$
$$x = 4 \quad \text{or} \quad x = -8$$

76. Problem: Solve: $x^2 = 324$

Incorrect Answer: $x^2 = 324$

$$x = \sqrt{324}$$
$$x = 18$$

77. Problem: Solve: $m^2 = 125$

Incorrect Answer: $m^2 = 125$

$$m = \sqrt{125} \quad \text{or} \quad m = -\sqrt{125}$$
$$m = \pm\sqrt{125}$$

78. Problem: Solve: $3m^2 = 36$

Incorrect Answer: $3m^2 = 36$

$$\frac{3m}{3} = \frac{36}{3} \quad \text{or} \quad \frac{3m}{3} = \frac{-36}{3}$$
$$m = 12 \quad \text{or} \quad m = -12$$

Review

For exercises 79–82, factor completely.

79. $x^2 + 6x + 9$

80. $x^2 + 8x + 16$

81. $x^2 - 10x + 25$

82. $x^2 - 14x + 49$

SUCCESS IN COLLEGE MATHEMATICS

83. This textbook includes Cumulative Reviews for Chapters 1–3, 4–6, 7–9, and 1–9. Select one of the reviews. You can select one that you have already done and do it again. As you complete and check the answers for the exercises in these reviews, classify each question into three types: A, B, and C. Mark any exercise that you can answer correctly without looking anything up as an "A." Mark any exercise that you can answer correctly after you refresh your memory by looking at an example as a "B." Mark any exercise that you cannot do at all or that you do incorrectly with a "C."

© Alexandr Vlassyuk/Shutterstock.com

In working on a jigsaw puzzle, the last piece has to be an exact match for the last "hole" in the puzzle. In this section, we find a term that matches a "hole" in a quadratic expression. This term "completes the square." We then solve quadratic equations by "completing the square."

SECTION 9.2 Completing the Square

After reading the text, working the practice problems, and completing assigned exercises, you should be able to:

1. Use completing the square to solve a quadratic equation with rational solutions.
2. Use completing the square to solve a quadratic equation with irrational solutions.

Completing the Square

The product of a binomial and itself is a perfect square trinomial.

EXAMPLE 1 Simplify: $(x + 3)(x + 3)$

SOLUTION ▶

$(x + 3)(x + 3)$	The product of a binomial and itself.
$= \mathbf{x}(x + 3) + \mathbf{3}(x + 3)$	Distributive property.
$= x^2 + 3x + 3x + 9$	Distributive property.
$= x^2 + \mathbf{6x} + 9$	Simplify; this is a perfect square trinomial.

We can build a visual model that represents $x^2 + 6x + 9$.

Since area = (length)(width), the area of this square is x^2.

The area of this rectangle is $6x$. Divide it into two rectangles with equal areas of $3x$.

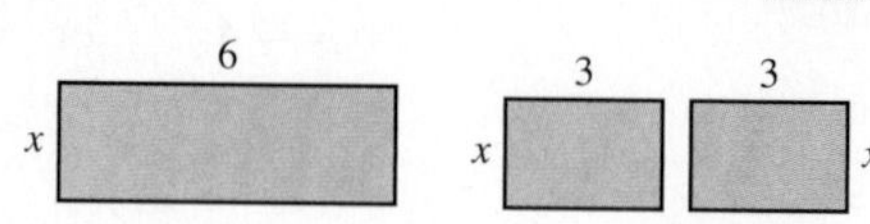

The area of this square is 9.

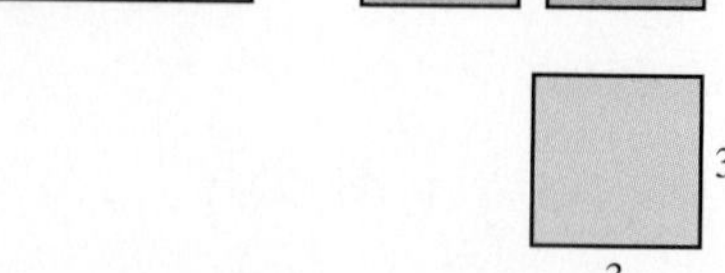

The area of the complete square is $x^2 + 3x + 3x + 9$. Combining like terms, the area is $x^2 + 6x + 9$.

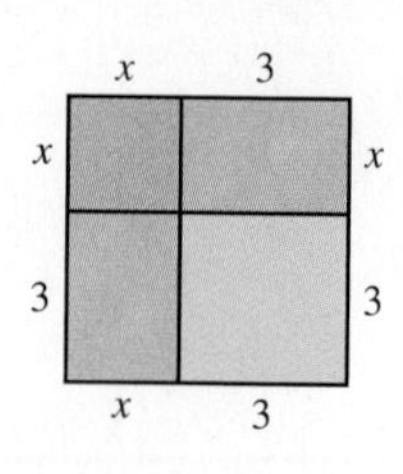

We use the same kind of visual model in the next example.

EXAMPLE 2 A binomial is $x^2 + 8x$. Build a visual model to discover the term that is needed to change this binomial into a perfect square trinomial.

SOLUTION ▶ We can build a visual model that represents $x^2 + 8x$. The area of this square is x^2.

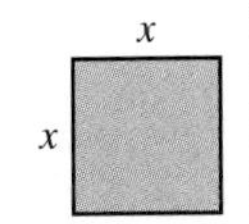

The area of this rectangle is $8x$. Divide it into two rectangles with equal areas of $4x$.

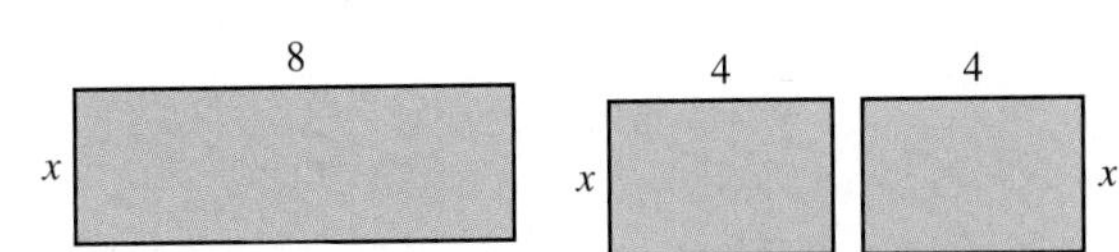

The shaded areas represent $x^2 + 4x + 4x$, which equals $x^2 + 8x$. Since each side of the missing square is 4, its area is 4^2, which equals 16.

To change $x^2 + 8x$ into a perfect square trinomial, add 16; $x^2 + 8x + 16$ is a perfect square trinomial.

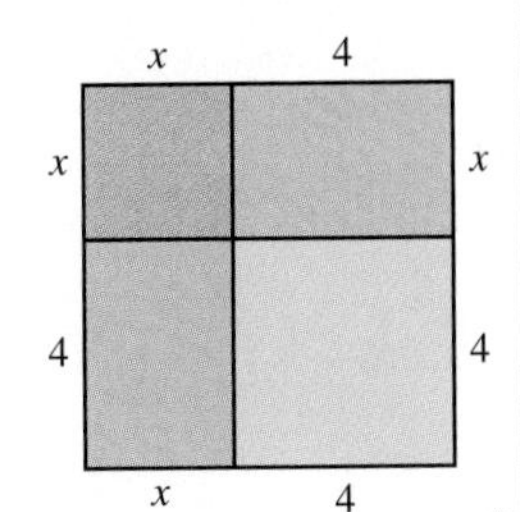

In Example 2, we completed the square by adding a constant and used geometric reasoning to find the value of the constant. We can also use algebraic reasoning to complete the square.

EXAMPLE 3 For $x^2 + bx$, show that the term needed to complete the square is $\left(\frac{1}{2}b\right)^2$.

SOLUTION ▶ If the term needed to complete the square is $\left(\frac{1}{2}b\right)^2$, then $x^2 + bx + \left(\frac{1}{2}b\right)^2$ is a perfect square trinomial, and its factors are identical binomials. We will factor this expression and see whether the factors are identical.

$x^2 + bx + \left(\frac{1}{2}b\right)^2$	Is this expression a perfect square trinomial?
$= x^2 + bx + \left(\frac{1}{2}b\right)\left(\frac{1}{2}b\right)$	Rewrite exponential notation as multiplication.
$= x^2 + bx + \frac{1}{4}b^2$	Simplify.
$= x^2 + \frac{1}{2}bx + \frac{1}{2}bx + \frac{1}{4}b^2$	*ac* factoring; rewrite the middle term.
$= \left(x^2 + \frac{1}{2}bx\right) + \left(\frac{1}{2}bx + \frac{1}{4}b^2\right)$	Group.
$= x(\_\_ + \_\_) + \frac{1}{2}b(\_\_ + \_\_)$	Factor a common factor from each group.
$= x\left(x + \frac{1}{2}b\right) + \frac{1}{2}b\left(x + \frac{1}{2}b\right)$	Complete the factoring of each group.
$= \left(x + \frac{1}{2}b\right)(\_\_ + \_\_)$	Factor out a binomial, $x + \frac{1}{2}b$.
$= \left(x + \frac{1}{2}b\right)\left(x + \frac{1}{2}b\right)$	Complete the factoring.

Since the factors are identical binomials, $\left(\frac{1}{2}b\right)^2$ is the correct term to add to $x^2 + bx$ to complete the square.

To add a term to a binomial to create a perfect square trinomial, identify b and add $\left(\frac{1}{2}b\right)^2$.

EXAMPLE 4 Add a term to the binomial to create a perfect square trinomial.

(a) $x^2 + 20x$

SOLUTION ▶

$\left(\frac{1}{2}b\right)^2$ The term needed to complete the square.

$= \left(\frac{1}{2} \cdot \mathbf{20}\right)^2$ Replace b with 20.

$= (\mathbf{10})^2$ Simplify.

$= 100$ Simplify.

The expression $x^2 + 20x + 100$ is a perfect square trinomial.

(b) $x^2 - 16x$

▶

$\left(\frac{1}{2}b\right)^2$ The term needed to complete the square.

$= \left(\frac{1}{2}(\mathbf{-16})\right)^2$ Replace b with -16.

$= (\mathbf{-8})^2$ Simplify.

$= 64$ Simplify.

The expression $x^2 - 16x + 64$ is a perfect square trinomial.

Practice Problems

For problems 1–4, add a term to the binomial to create a perfect square trinomial.

1. $z^2 + 10z$ **2.** $z^2 - 10z$ **3.** $p^2 + 18p$ **4.** $n^2 - 14n$

Solving Quadratic Equations by Completing the Square

A quadratic equation in standard form is $ax^2 + bx + c = 0$. To solve a quadratic equation in which $a = 1$ by completing the square, add $\left(\frac{1}{2}b\right)^2$ to both sides and factor. Since the factored equation is in the form $x^2 = k$, the solutions are $x = \sqrt{k}$ or $x = -\sqrt{k}$.

EXAMPLE 5 **(a)** Solve $x^2 + 8x = 0$ by completing the square.

SOLUTION ▶ Find the term needed to complete the square.

$\left(\frac{1}{2}b\right)^2$ $a = 1$; the term needed to complete the square.

$= \left(\frac{1}{2} \cdot \mathbf{8}\right)^2$ Replace b with 8.

$= (\mathbf{4})^2$ Simplify.

$= 16$ Simplify.

Since $\left(\frac{1}{2}b\right)^2 = 16$, add 16 to both sides of the equation.

$x^2 + 8x = 0$		
$+16 \quad +16$		Addition property of equality.
$x^2 + 8x + \mathbf{16} = \mathbf{16}$		Simplify.
$\mathbf{(x + 4)(x + 4)} = 16$		Factor.
$(x + 4)^2 = 16$		Write in the form $x^2 = k$.
$x + 4 = \sqrt{16}$ or	$x + 4 = -\sqrt{16}$	The solutions are $x = \sqrt{k}$ or $x = -\sqrt{k}$.
$x + 4 = \mathbf{4}$ or	$x + 4 = \mathbf{-4}$	Simplify.
$-4 \quad -4$	$-4 \quad -4$	Subtraction property of equality.
$x + \mathbf{0} = \mathbf{0}$	$x + \mathbf{0} = \mathbf{-8}$	Simplify.
$x = 0$ or	$x = -8$	Simplify.

(b) Check.

Check: $x = 0$	Check: $x = -8$
$x^2 + 8x = 0$	$x^2 + 8x = 0$
$(\mathbf{0})^2 + 8(\mathbf{0}) = 0$	$(\mathbf{-8})^2 + 8(\mathbf{-8}) = 0$
$\mathbf{0 + 0} = 0$	$\mathbf{64 - 64} = 0$
$\mathbf{0} = 0$ True.	$\mathbf{0} = 0$ True.

The variable in a quadratic equation does not have to be x. In the next example, the quadratic equation is $u^2 - 12u - 13 = 0$, and the variable is u. We use the addition property of equality to isolate $u^2 - 12u$ before completing the square.

EXAMPLE 6 **(a)** Solve $u^2 - 12u - 13 = 0$ by completing the square.

SOLUTION Isolate $u^2 - 12u$.

$u^2 - 12u - 13 = 0$	
$+13 \quad +13$	Addition property of equality.
$u^2 - 12u + \mathbf{0} = \mathbf{13}$	Simplify.

Find the term needed to complete the square.

$\left(\frac{1}{2}b\right)^2$	
$= \left(\frac{1}{2}(\mathbf{-12})\right)^2$	Replace b with -12.
$= (\mathbf{-6})^2$	Simplify.
$= 36$	Simplify.

Since $\left(\frac{1}{2}b\right)^2 = 36$, add 36 to both sides of the equation.

$u^2 - 12u + 0 = 13$		
$+36 \quad +36$		Addition property of equality.
$u^2 - 12u + \mathbf{36} = \mathbf{49}$		Simplify.
$\mathbf{(u - 6)(u - 6)} = 49$		Factor.
$(u - 6)^2 = 49$		Write in the form $x^2 = k$.
$u - 6 = \sqrt{49}$ or	$u - 6 = -\sqrt{49}$	The solutions are $x = \sqrt{k}$ or $x = -\sqrt{k}$.
$u - 6 = \mathbf{7}$ or	$u - 6 = \mathbf{-7}$	Simplify.
$+6 \quad +6$	$+6 \quad +6$	Subtraction property of equality.
$u + \mathbf{0} = \mathbf{13}$ or	$u + \mathbf{0} = \mathbf{-1}$	Simplify.
$u = 13$ or	$u = -1$	Simplify.

(b) Check in the original equation.

Check: $u = 13$

$$u^2 - 12u - 13 = 0$$
$$(\mathbf{13})^2 - 12(\mathbf{13}) - 13 = 0$$
$$\mathbf{169 - 156} - 13 = 0$$
$$\mathbf{0} = 0 \quad \text{True.}$$

Check: $u = -1$

$$u^2 - 12u - 13 = 0$$
$$(\mathbf{-1})^2 - 12(\mathbf{-1}) - 13 = 0$$
$$\mathbf{1 + 12} - 13 = 0$$
$$\mathbf{0} = 0 \quad \text{True.}$$

Solving a Quadratic Equation, $x^2 + bx + c = 0$, by Completing the Square

1. Isolate $x^2 + bx$.
2. Complete the square by adding $\left(\frac{1}{2}b\right)^2$ to both sides.
3. Factor the perfect square trinomial, and solve the equation.
4. Check in the original equation.

Practice Problems

For problems 5–7,
(a) solve by completing the square.
(b) check.

5. $z^2 + 10z + 21 = 0$ **6.** $z^2 - 10z - 75 = 0$ **7.** $p^2 + 18p - 40 = 0$

Irrational Solutions

In the next example, the solutions of the quadratic equation are irrational numbers. An irrational number cannot be written as a fraction in which the numerator and denominator are integers.

EXAMPLE 7 Solve $k^2 + 6k - 2 = 0$ by completing the square.

SOLUTION ▶ Isolate $k^2 + 6k$.

$$k^2 + 6k - 2 = 0$$
$$\underline{ +2 \;\; +2} \quad \text{Isolate } k^2 + 6k\text{; addition property of equality.}$$
$$k^2 + 6k + \mathbf{0} = \mathbf{2} \quad \text{Simplify.}$$

Find the term needed to complete the square.

$$\left(\frac{1}{2}b\right)^2$$
$$= \left(\frac{1}{2}\cdot\mathbf{6}\right)^2 \quad \text{Replace } b \text{ with 6.}$$
$$= (\mathbf{3})^2 \quad \text{Simplify.}$$
$$= 9 \quad \text{Simplify.}$$

Since $\left(\frac{1}{2}b\right)^2 = 9$, add 9 to both sides of the equation.

$k^2 + 6k + 0 = 2$	
$\quad +9 \quad +9$	Addition property of equality.
$k^2 + 6k + \mathbf{9} = \mathbf{11}$	Simplify.
$(k + 3)(k + 3) = 11$	Factor.
$(k + 3)^2 = 11$	Write in the form $x^2 = k$.
$k + 3 = \sqrt{11}$ or $k + 3 = -\sqrt{11}$	The solutions are $x = \sqrt{k}$ or $x = -\sqrt{k}$.
$-3 \quad -3$ or $-3 \quad -3$	Subtraction property of equality.
$k + \mathbf{0} = \mathbf{\sqrt{11} - 3}$ or $k + \mathbf{0} = \mathbf{-\sqrt{11} - 3}$	Simplify; $\sqrt{11}$ and 3 are not like terms.
$k = -3 + \sqrt{11}$ or $k = -3 - \sqrt{11}$	Simplify; change order.
$k = -3 \pm \sqrt{11}$	$\pm$ notation.

Practice Problems

For problems 8–10, solve by completing the square.

8. $z^2 + 4z - 7 = 0$ **9.** $h^2 - 14h - 12 = 0$ **10.** $k^2 + 18k + 2 = 0$

9.2 Vocabulary Practice

Match the term with its description.

1. A rectangle with four equal sides
2. $\sqrt{\ }$
3. $\pm$
4. A polynomial with three terms
5. A polynomial with two terms
6. $a^2 + 2ab + b^2$
7. The product of the length and the width of a rectangle
8. A number that cannot be written as a fraction in which the numerator and denominator are integers
9. $ax^2 + bx + c = 0$
10. A number that can be written as a fraction in which the numerator and denominator are integers

A. area of a rectangle
B. binomial
C. irrational number
D. perfect square trinomial
E. plus-minus notation
F. principal square root
G. rational number
H. square
I. standard form of a quadratic equation
J. trinomial

9.2 Exercises

Follow your instructor's guidelines for showing your work.

For exercises 1–2,
(a) Represent the binomial visually as an incomplete square (see Examples 1 and 2).
(b) Write a term that represents the area of the missing part of the square.

1. $x^2 + 10x$ **2.** $x^2 + 4x$

For exercises 3–10, add a term to the binomial to create a perfect square trinomial.

3. $p^2 + 30p$ **4.** $h^2 + 40h$
5. $p^2 - 30p$ **6.** $h^2 - 40h$
7. $x^2 - 16x$ **8.** $x^2 - 12x$
9. $m^2 + 24m$ **10.** $n^2 + 22n$

For exercises 11–32,
(a) solve by completing the square.
(b) check.

11. $w^2 + 14w - 32 = 0$
12. $k^2 + 16k - 17 = 0$
13. $w^2 - 14w - 32 = 0$
14. $k^2 - 16k - 17 = 0$
15. $z^2 + 12z + 11 = 0$
16. $p^2 + 10p + 9 = 0$
17. $z^2 - 12z + 11 = 0$
18. $p^2 - 10p + 9 = 0$

19. $x^2 + 40x + 76 = 0$

20. $r^2 + 50r + 225 = 0$

21. $w^2 - 2w - 35 = 0$

22. $y^2 - 2y - 15 = 0$

23. $m^2 + 8m - 84 = 0$

24. $p^2 + 8p - 33 = 0$

25. $k^2 + 8k + 12 = 0$

26. $z^2 + 6z + 8 = 0$

27. $u^2 - 26u + 25 = 0$

28. $v^2 - 16v + 15 = 0$

29. $a^2 - 4a - 12 = 0$

30. $d^2 - 6d - 16 = 0$

31. $x^2 - 2x - 8 = 0$

32. $x^2 - 3x - 10 = 0$

For exercises 33–64, solve by completing the square.

33. $x^2 + 4x - 9 = 0$

34. $x^2 + 4x - 11 = 0$

35. $p^2 + 6p + 2 = 0$

36. $p^2 + 6p + 3 = 0$

37. $k^2 - 8k - 2 = 0$

38. $k^2 - 8k - 4 = 0$

39. $n^2 + 10n + 4 = 0$

40. $n^2 + 10n + 2 = 0$

41. $z^2 + 2z + 1 = 0$

42. $z^2 - 2z + 1 = 0$

43. $a^2 - 12a - 4 = 0$

44. $a^2 - 12a - 8 = 0$

45. $r^2 - 16r + 1 = 0$

46. $r^2 - 16r - 1 = 0$

47. $m^2 - 18m + 79 = 0$

48. $m^2 - 18m + 78 = 0$

49. $u^2 + 20u - 3 = 0$

50. $u^2 + 20u + 3 = 0$

51. $a^2 + 4a - 10 = 0$

52. $b^2 + 6b - 10 = 0$

53. $c^2 + 14c + 48 = 0$

54. $d^2 + 10d + 16 = 0$

55. $b^2 - 6b - 16 = 0$

56. $c^2 - 12c - 28 = 0$

57. $a^2 - 12a + 36 = 0$

58. $p^2 - 14p + 49 = 0$

59. $w^2 + 4w + 3 = 0$

60. $k^2 + 6k + 5 = 0$

61. $m^2 = 20m + 6$

62. $n^2 = 10n + 8$

63. $b^2 = 2b + 4$

64. $c^2 = 2c + 5$

Problem Solving: Practice and Review

Follow your instructor's guidelines for using the five steps as outlined in Section 1.3, p. 55.

65. At many colleges and universities, students pay tuition and fees, which do not include the costs of room, board, or books. Find the percent of the total amount of tuition and fees that is from fees. Round to the nearest percent.

The university is charging a \$51 "energy surcharge" for rising electricity costs. A \$270 "technology fee" for computer service . . . the \$371.25 fee for the campus health center, a \$135 fee to maintain buildings and grounds and a \$624 "incidental fee" for student activities. . . . all told, fees add up to \$1,542 . . . on top of tuition of \$3,984. (*Source:* www.nytimes.com, Sept. 7, 2007)

66. Find the difference in the price per magnet for an order of 8 magnets and the price per magnet for an order of 40 magnets. Round to the nearest hundredth.

Number of custom magnets	Total price
1	\$9.99
2	\$17.99
4	\$29.99
6	\$40.99
8	\$50.99
10	\$59.99
12	\$68.99
16	\$84.99
20	\$99.99
40	\$179.99

67. A student buys health insurance through his college. The student pays the first \$250 in medical bills each year (the deductible). The insurance covers 80% of the cost of X-rays, MRIs, other diagnostic imaging, and doctor bills. When the student visits a doctor because of sharp pains radiating down his right leg, his doctor orders an MRI of the student's spine. The bill for the MRI is \$1500, and the bill for the doctor visit is \$95. If the student has not yet paid any of his deductible, find the amount of these bills that the insurance will pay.

68. The population of Galveston in 2000 was 57,247 people (Census 2000). Of these people, 6.6% were children under 5 years of age. Predict how many of these children test positive for lead poisoning. Round to the nearest whole number.

Nearly one in five children in Galveston has enough lead in their blood to cause learning disabilities and behavioral problems. (*Source:* www.chron.com, Jan. 27, 2008)

Technology

For exercises 69–72,

(a) use the method from Section 6.5 to solve the equation. Choose a window that shows both intersection points. Sketch the graph; describe the window.

(b) identify the solution(s). If the solution is irrational, round to the nearest tenth.

69. $x^2 + 4x = 4$

70. $x^2 + 6x = 9$

71. $x^2 + 6x = -2$

72. $x^2 - 5x = -4$

Find the Mistake

For exercises **73–76**, the completed problem has one mistake.

(a) Describe the mistake in words, or copy down the whole problem and highlight or circle the mistake.

(b) Do the problem correctly.

73. Problem: Solve $x^2 + 8x - 9 = 0$ by completing the square.

Incorrect Answer: $b = 8; \left(\frac{1}{2} \cdot 8\right)^2 = 16$

$$x^2 + 8x - 9 = 0$$
$$+9 \quad +9$$
$$x^2 + 8x + 0 = 9$$
$$x^2 + 8x + 16 = 9$$
$$(x + 4)(x + 4) = 9$$
$$(x + 4)^2 = 9$$
$$x + 4 = \sqrt{9} \quad \text{or} \quad x + 4 = -\sqrt{9}$$
$$x + 4 = 3 \quad \text{or} \quad x + 4 = -3$$
$$-4 \quad -4 \qquad -4 \quad -4$$
$$x + 0 = -1 \quad \text{or} \quad x + 0 = -7$$
$$x = -1 \quad \text{or} \quad x = -7$$

74. Problem: Solve $x^2 + 10x - 11 = 0$ by completing the square.

Incorrect Answer: $b = 10; \left(\frac{1}{2} \cdot 10\right)^2 = 25$

$$x^2 + 10x - 11 = 0$$
$$+11 \quad +11$$
$$x^2 + 10x = 11$$
$$+25 \quad +25$$
$$x^2 + 10x + 25 = 36$$
$$(x + 5)(x + 5) = 36$$
$$(x + 5)^2 = 36$$
$$x + 5 = \sqrt{36}$$
$$x + 5 = 6$$
$$-5 \quad -5$$
$$x + 0 = 1$$
$$x = 1$$

75. Problem: Solve $x^2 + 12x - 64 = 0$ by completing the square.

Incorrect Answer: $b = 12; \left(\frac{1}{2} \cdot 12\right)^2 = 36$

$$x^2 + 12x - 64 = 0$$
$$+64 \quad +64$$
$$x^2 + 12x + 0 = 64$$
$$+36 \quad +36$$
$$x^2 + 12x + 36 = 100$$
$$(x + 6)(x + 6) = 100$$
$$x + 6 = 100 \quad \text{or} \quad x + 6 = -100$$
$$-6 \quad -6 \qquad -6 \quad -6$$
$$x + 0 = 94 \quad \text{or} \quad x + 0 = -106$$
$$x = 94 \quad \text{or} \quad x = -106$$

76. Problem: Solve: $x^2 - 2x - 35 = 0$

Incorrect Answer: $b = -2; \left(\frac{1}{2}(-2)\right)^2 = 1$

$$x^2 - 2x - 35 = 0$$
$$+35 \quad +35$$
$$x^2 - 2x + 0 = 35$$
$$+1 \quad +1$$
$$x^2 - 2x + 1 = 36$$
$$(x + 1)(x + 1) = 36$$
$$(x + 1)^2 = 36$$
$$x + 1 = \sqrt{36} \quad \text{or} \quad x + 1 = -\sqrt{36}$$
$$x + 1 = 6 \quad \text{or} \quad x + 1 = -6$$
$$-1 \quad -1 \qquad -1 \quad -1$$
$$x + 0 = 5 \quad \text{or} \quad x + 0 = -7$$
$$x = 5 \quad \text{or} \quad x = -7$$

Review

For exercises **77–80**, evaluate.

77. $5^2 - 4 \cdot 1 \cdot 2$

78. $\sqrt{10^2 - 4 \cdot 2 \cdot 8}$

79. $\sqrt{2^2 - 4(-2)(4)}$

80. $\sqrt{(-6)^2 - 4(1)(9)}$

SUCCESS IN COLLEGE MATHEMATICS

81. In completing exercise 83 in Section 9.1, you might have marked some of the Cumulative Exercises with a "C." For each of these problems, identify the section in the book in which this material was first taught, and create a practice test for these exercises by copying a similar practice problem from that section. Write the section and practice problem number next to each problem.

© Ken Hurst/Shutterstock.com

If a light-rail train is accelerating at a constant rate, we can use a quadratic equation in one variable to find the time it will take the train to travel a given distance. An efficient way to solve this equation is to use the quadratic formula. In this section, we will use the quadratic formula to solve quadratic equations.

SECTION 9.3 The Quadratic Formula

After reading the text, working the practice problems, and completing assigned exercises, you should be able to:

1. Use the quadratic formula to solve a quadratic equation with rational solutions.
2. Use the quadratic formula to solve a quadratic equation with irrational solutions.
3. Use the discriminant to describe the solution(s) of a quadratic equation.
4. Solve an application problem using a quadratic equation and the quadratic formula.

The Quadratic Formula

The standard form of a quadratic equation is $ax^2 + bx + c = 0$. If we solve this equation by completing the square (Appendix 2), the solutions are $x = \dfrac{-b + \sqrt{b^2 - 4ac}}{2a}$ or $x = \dfrac{-b - \sqrt{b^2 - 4ac}}{2a}$. In plus-minus notation, the solutions are $x = \dfrac{-b \pm \sqrt{b^2 - 4ac}}{2a}$. This is the quadratic formula.

> **The Quadratic Formula**
> For the equation $ax^2 + bx + c = 0$, where a, b, and c are real numbers and $a \neq 0$, $x = \dfrac{-b \pm \sqrt{b^2 - 4ac}}{2a}$.

To use the quadratic formula to solve an equation written in the form $ax^2 + bx + c = 0$, identify the values of a, b, and c. Replace a, b, and c in the formula with these values, and simplify to find the solutions of the equation.

EXAMPLE 1 (a) Use the quadratic formula to solve $x^2 + 2x - 168 = 0$.

SOLUTION ▶

$\mathbf{1}x^2 + \mathbf{2}x - \mathbf{168} = 0$ $a = 1, b = 2, c = -168$

$x = \dfrac{\mathbf{-2} \pm \sqrt{(\mathbf{2})^2 - 4(\mathbf{1})(\mathbf{-168})}}{2(\mathbf{1})}$ $x = \dfrac{-b \pm \sqrt{b^2 - 4ac}}{2a}$

$x = \dfrac{-2 \pm \sqrt{\mathbf{4 + 672}}}{\mathbf{2}}$ Simplify.

$x = \dfrac{-2 \pm \sqrt{\mathbf{676}}}{2}$ Simplify.

$x = \dfrac{-2 \pm \mathbf{26}}{2}$ Simplify.

$x = \dfrac{-2 + 26}{2}$ or $x = \dfrac{-2 - 26}{2}$ Write the solutions using "or."

$x = \dfrac{\mathbf{24}}{2}$ or $x = \dfrac{\mathbf{-28}}{2}$ Simplify.

$x = \mathbf{12}$ or $x = \mathbf{-14}$ Simplify.

(b) Check.

Check: $x = 12$

$$x^2 + 2x - 168 = 0$$
$$(\mathbf{12})^2 + 2(\mathbf{12}) - 168 = 0$$
$$\mathbf{144 + 24} - 168 = 0$$
$$\mathbf{168} - 168 = 0$$
$$\mathbf{0} = 0 \quad \text{True.}$$

Check: $x = -14$

$$x^2 + 2x - 168 = 0$$
$$(\mathbf{-14})^2 + 2(\mathbf{-14}) - 168 = 0$$
$$\mathbf{196 - 28} - 168 = 0$$
$$\mathbf{168} - 168 = 0$$
$$\mathbf{0} = 0 \quad \text{True.}$$

In the next example, b is a negative number, -3.

EXAMPLE 2 Use the quadratic formula to solve $2x^2 - 3x - 35 = 0$.

SOLUTION

$\mathbf{2}x^2 - \mathbf{3}x - \mathbf{35} = 0$ $a = 2, b = -3, c = -35$

$x = \dfrac{-(\mathbf{-3}) \pm \sqrt{(\mathbf{-3})^2 - 4(\mathbf{2})(\mathbf{-35})}}{2(\mathbf{2})}$ $x = \dfrac{-b \pm \sqrt{b^2 - 4ac}}{2a}$

$x = \dfrac{\mathbf{3} \pm \sqrt{\mathbf{9 + 280}}}{\mathbf{4}}$ Simplify.

$x = \dfrac{3 \pm \sqrt{\mathbf{289}}}{4}$ Simplify.

$x = \dfrac{3 \pm \mathbf{17}}{4}$ Simplify.

$x = \dfrac{3 + 17}{4}$ or $x = \dfrac{3 - 17}{4}$ Write the solutions using "or."

$x = \dfrac{\mathbf{20}}{4}$ or $x = \dfrac{\mathbf{-14}}{4}$ Simplify.

$x = \mathbf{5}$ or $x = -\dfrac{\mathbf{7}}{\mathbf{2}}$ Simplify.

If the radicand in the quadratic formula, $b^2 - 4ac$, equals 0, the equation has only one solution. In the next example, since the variable in the equation is u, the quadratic formula is $u = \dfrac{-b \pm \sqrt{b^2 - 4ac}}{2a}$.

EXAMPLE 3 Use the quadratic formula to solve $9u^2 - 6u + 1 = 0$.

SOLUTION ▶

$9u^2 - 6u + 1 = 0$ — $a = 9, b = -6, c = 1$

$u = \dfrac{-(-6) \pm \sqrt{(-6)^2 - 4(9)(1)}}{2(9)}$ — $u = \dfrac{-b \pm \sqrt{b^2 - 4ac}}{2a}$

$u = \dfrac{6 \pm \sqrt{36 - 36}}{18}$ — Simplify.

$u = \dfrac{6 \pm \sqrt{0}}{18}$ — Simplify.

$u = \dfrac{6 \pm 0}{18}$ — Simplify.

$u = \dfrac{6}{18}$ — Simplify; $6 + 0 = 6$ and $6 - 0 = 6$

$u = \dfrac{1}{3}$ — Simplify; one solution.

Practice Problems

For problems 1–4, use the quadratic formula to solve.

1. $x^2 + 15x + 54 = 0$ **2.** $w^2 + 2w - 63 = 0$

3. $2d^2 - 11d - 21 = 0$ **4.** $p^2 + 6p + 9 = 0$

Irrational Solutions

When the radicand in the quadratic formula, $b^2 - 4ac$, is greater than 0 and is not a perfect square, the solutions of the equation are irrational numbers.

EXAMPLE 4 Use the quadratic formula to solve $2x^2 - x - 9 = 0$.

SOLUTION ▶

$2x^2 - 1x - 9 = 0$ — $a = 2, b = -1, c = -9$

$x = \dfrac{-(-1) \pm \sqrt{(-1)^2 - 4(2)(-9)}}{2(2)}$ — $x = \dfrac{-b \pm \sqrt{b^2 - 4ac}}{2a}$

$x = \dfrac{1 \pm \sqrt{1 + 72}}{4}$ — Simplify.

$x = \dfrac{1 \pm \sqrt{73}}{4}$ — Simplify.

Practice Problems

For problems 5–7, use the quadratic formula to solve.

5. $x^2 + 7x + 4 = 0$ **6.** $2w^2 + 3w - 7 = 0$ **7.** $2d^2 - 6d - 5 = 0$

The Discriminant

The discriminant, $b^2 - 4ac$, is the expression under the radical in the quadratic formula: $x = \dfrac{-b \pm \sqrt{b^2 - 4ac}}{2a}$. The value of the discriminant reveals how many solutions a quadratic equation has and whether the solutions are real or nonreal numbers. If the discriminant equals 0, $x = \dfrac{-b \pm 0}{2a}$, which simplifies to $x = -\dfrac{b}{2a}$. The equation has only one solution that is a real number. If the discriminant is greater than 0, the equation has two solutions that are real numbers. If the discriminant is less than 0, the equation has two solutions that are nonreal numbers. We will solve equations with nonreal solutions in Section 9.4.

Real numbers are either rational numbers or irrational numbers. We can use the discriminant to identify whether real solutions of a quadratic equation are rational or irrational numbers. If the discriminant is greater than 0 and it is a perfect square, the simplified square root is an integer, and both solutions are rational numbers. If the discriminant is not a perfect square, the square root is not a rational number, and both solutions are irrational numbers.

Using the Discriminant, $b^2 - 4ac$, to Describe the Solutions of $ax^2 + bx + c = 0$

If the discriminant $= 0$, the equation has one solution that is a rational number.

If the discriminant > 0 and it is a perfect square, the equation has two solutions that are rational numbers.

If the discriminant > 0 and it is not a perfect square, the equation has two solutions that are irrational numbers.

If the discriminant < 0, the equation has two solutions that are nonreal numbers.

The square root of a perfect square is a rational number. In the next example, the discriminant, 105, is not a perfect square. The solutions of the equation are irrational numbers.

EXAMPLE 5 Use the discriminant to describe the solutions of $2x^2 + 9x - 3 = 0$.

SOLUTION ▶

$b^2 - 4ac$

$= \mathbf{9}^2 - 4(\mathbf{2})(\mathbf{-3})$ $a = 2, b = 9, c = -3$

$= 81 + 24$ Evaluate.

$= 105$ Evaluate.

Since $\sqrt{105} = 10.246\ldots$, a nonrepeating nonterminating decimal number, 105 is not a perfect square. Since the discriminant, 105, is greater than 0 and is not a perfect square, $2x^2 + 9x - 3 = 0$ has two solutions that are irrational numbers.

Practice Problems

For problems 8–11, use the discriminant to describe the solutions.

8. $10x^2 - 37x - 36 = 0$

9. $2h^2 - 3h - 7 = 0$

10. $n^2 + 8n + 16 = 0$

11. $p^2 - 7p + 30 = 0$

Applications

When we use a quadratic equation to solve an application problem, the solutions are often irrational and are rounded to a given place value. In the next example, the variable in the quadratic equation is t. The quadratic formula to solve this equation is $t = \dfrac{-b \pm \sqrt{b^2 - 4ac}}{2a}$.

EXAMPLE 6 The speed of a freight train is $\frac{15\text{ ft}}{1\text{ s}}$ as it begins a constant acceleration of $\frac{0.14\text{ ft}}{1\text{ s}^2}$. The time, t, for the train to travel 1000 ft is the solution of the equation $1000\text{ ft} = \left(\frac{15\text{ ft}}{1\text{ s}}\right)t + (0.5)\left(\frac{0.14\text{ ft}}{1\text{ s}^2}\right)t^2$. Find this time. Round to the nearest tenth.

SOLUTION ▶ **Step 1 Understand the problem.**
The unknown is the time needed to travel 1000 ft.

t = time to travel 1000 ft

Step 2 Make a plan.
Since this is a quadratic equation, use the quadratic formula to solve. Rewrite the equation in standard form, $ax^2 + bx + c = 0$.

Step 3 Carry out the plan.

$$1000 = 15t + (0.5)(0.14)t^2 \qquad \text{Remove the units.}$$

$$1000 = 15t + \mathbf{0.07}t^2 \qquad \text{Simplify.}$$

$$\underline{-1000 \qquad\qquad -1000} \qquad \text{Subtraction property of equality.}$$

$$\mathbf{0} = 0.07t^2 + 15t - \mathbf{1000} \qquad \text{Simplify.}$$

$$t = \frac{-\mathbf{15} \pm \sqrt{(\mathbf{15})^2 - 4(\mathbf{0.07})(-\mathbf{1000})}}{2(\mathbf{0.07})} \qquad t = \frac{-b \pm \sqrt{b^2 - 4ac}}{2a}$$

$$t = \frac{-15 \pm \sqrt{\mathbf{505}}}{\mathbf{0.14}} \qquad \text{Simplify.}$$

$$t = \frac{-15 \pm \mathbf{22.4...}}{0.14} \qquad \text{Simplify.}$$

$$t = \frac{-15 + 22.4...}{0.14} \quad \text{or} \quad t = \frac{-15 - 22.4...}{0.14} \qquad \text{Rewrite the solutions using “or.”}$$

$$t = \frac{\mathbf{7.47...}}{0.14} \quad \text{or} \quad t = \frac{\mathbf{-37.4...}}{0.14} \qquad \text{Simplify the numerators.}$$

$$t \approx \mathbf{53.4\text{ s}} \quad \text{or} \quad t \approx \mathbf{-267.7\text{ s}} \qquad \text{Round; include the units.}$$

Step 4 Look back.
Only the positive solution can be a solution because elapsed time cannot be negative. Using the original equation and a time of 53.4 s, the distance traveled is $\left(\frac{15\text{ ft}}{1\text{ s}}\right)(\mathbf{53.4\text{ s}}) + (0.5)\left(\frac{0.14}{1\text{ s}^2}\right)(\mathbf{53.4\text{ s}})^2$, which equals about 1000 ft. Since that is the distance that the train traveled, the answer seems reasonable.

Step 5 Report the solution.
It will take the train about 53.4 s to travel 1000 ft.

We have used quadratic equations and the zero product property to find the length and width of a rectangle. We can instead use the quadratic formula to solve these equations.

EXAMPLE 7 The area of a rectangle is 182 ft^2, and its perimeter is 56.4 ft. Use a quadratic equation in one variable to find the length and width of this rectangle.

SOLUTION ▶ **Step 1 Understand the problem.**
Although there are two unknowns, the length and the width, we need to use a quadratic equation in *one* variable to find these unknowns. We can choose to

write the equation to find either unknown. We will choose to find the length of the rectangle.

L = length

Step 2 Make a plan.

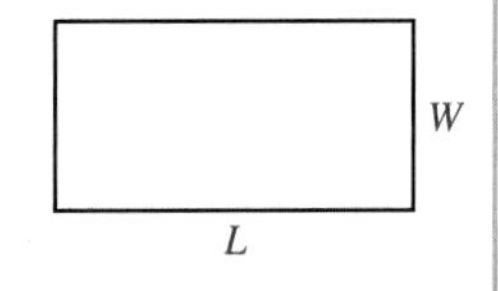

The formula for the perimeter of a rectangle is $P = 2W + 2L$, and the formula for the area of a rectangle is $A = LW$. Replace P in the formula for perimeter with 56.4 ft, and solve for W. In the area formula, replace W with this expression, and replace A with 182 ft^2. This is a quadratic equation in L. Rewrite this equation in standard form, and use the quadratic formula to solve.

Step 3 Carry out the plan.

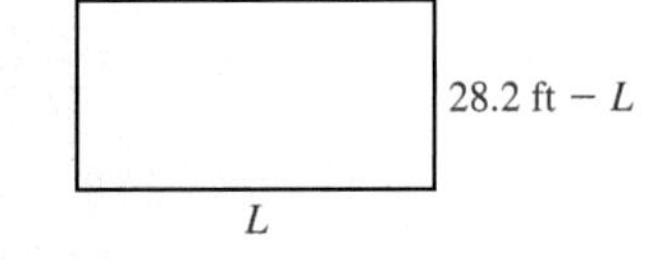

In the formula for perimeter $P = 2W + 2L$, replace P with 56.4 ft and solve for W.

$P = 2W + 2L$	Formula for perimeter.
$\mathbf{56.4\ ft} = 2W + 2L$	Replace P with 56.4 ft.
$-2L \qquad -2L$	Subtraction property of equality.
$56.4\text{ ft} \mathbf{- 2L} = 2W \mathbf{+ 0}$	Simplify.
$\dfrac{56.4\text{ ft}}{2} - \dfrac{2L}{2} = \dfrac{2W}{2}$	Division property of equality.
$28.2\text{ ft} - L = W$	Simplify.

In the formula for area, $A = LW$, replace A with 182 ft^2, replace W with 28.2 ft $- L$, and solve for L.

$A = LW$	Formula for area.
$\mathbf{182\ ft^2} = L(\mathbf{28.2\ ft - L})$	Replace A and W.
$182 = L(28.2 - L)$	Remove the units.
$182 = \mathbf{28.2L - L^2}$	Distributive property.
$+L^2 - 28.2L \qquad -28.2L + L^2$	Properties of equality.
$\mathbf{1L^2 - 28.2L} + 182 = \mathbf{0}$	Simplify.
$L = \dfrac{-(\mathbf{-28.2}) \pm \sqrt{(\mathbf{-28.2})^2 - 4(\mathbf{1})(\mathbf{182})}}{2(\mathbf{1})}$	$x = \dfrac{-b \pm \sqrt{b^2 - 4ac}}{2a}$
$L = \dfrac{\mathbf{28.2 \pm \sqrt{67.24}}}{\mathbf{2}}$	Simplify.
$L = \dfrac{28.2 \pm \mathbf{8.2}}{2}$	Simplify.
$L = \dfrac{28.2 + 8.2}{2}$ or $L = \dfrac{28.2 - 8.2}{2}$	Rewrite using "or."
$L = \mathbf{18.2\ ft}$ or $L = \mathbf{10\ ft}$	Simplify; include the units.

Since $W = 28.2$ ft $- L$, if $L = 18.2$ ft, then $W = 10$ ft. If $L = 10$ ft, then $W = 18.2$ ft. So both solutions of the equations are solutions of the problem. If we want the length to be greater than the width, we choose the solution $L = 18.2$ ft and $W = 10$ ft.

Step 4 Look back.

Since (18.2 ft)(10 ft) equals the area of 182 ft^2 and 2(18.2 ft) + 2(10 ft) equals 56.4 ft, the answer seems reasonable.

Step 5 Report the solution.

The length of the rectangle is 18.2 ft, and the width of the rectangle is 10 ft.

Practice Problems

For problems 12–13, use the five steps and the quadratic formula.

12. A pry bar slides off a roof that is 28 ft above the ground. The time for the pry bar to hit the ground is a solution of the equation $28\text{ ft} = \left(\frac{8\text{ ft}}{1\text{ s}}\right)t + \left(\frac{16\text{ ft}}{1\text{ s}^2}\right)t^2$. Find the time for the pry bar to hit the ground t. Round to the nearest tenth.

13. The area of a rectangle is 168 ft^2, and the perimeter is 58 ft. Find the length and width of the rectangle.

Methods for Solving Quadratic Equations

Which method should we use to solve a quadratic equation in the form $ax^2 + bx + c = 0$? This depends on the values of a, b, and c and often on which method seems most efficient. We can always use completing the square or the quadratic formula. If a, b, or c are decimal numbers or when solving application problems in which the quadratic equation cannot be factored or $b \neq 0$, the quadratic formula is the most efficient choice. If $b = 0$, we can rewrite the equation in the form $x^2 = k$. The solutions are $x = \sqrt{k}$ or $x = -\sqrt{k}$. If $ax^2 + bx + c = 0$ can be factored, the most efficient method might be to use the zero product property (Section 6.5). If the discriminant $b^2 - 4ac$ is a perfect square, then $ax^2 + bx + c$ can be factored.

Methods for Solving a Quadratic Equation in the Form $ax^2 + bx + c = 0$

- Completing the square
- Using the quadratic formula
- Rewriting the equation in the form $x^2 = k$
- Factoring and using the zero product property

EXAMPLE 8 Solve: $3x^2 + 5x - 7 = 0$

SOLUTION ▶ For this equation, $a = 3, b = 5, c = -7$, and the discriminant, $b^2 - 4ac$, equals 109. Since the discriminant is not a perfect square, $3x^2 + 5x - 7$ is prime. We cannot use the zero product property to solve. Since $b \neq 0$, we cannot write the equation in the form $x^2 = k$. Although we could solve by completing the square, since $\left(\frac{1}{2}b\right)^2$ is a fraction, $\frac{25}{4}$, it is probably quicker to use the quadratic formula.

$$3x^2 + 5x - 7 = 0 \qquad a = 3, b = 5, c = -7$$

$$x = \frac{-5 \pm \sqrt{5^2 - 4(3)(-7)}}{2(3)} \qquad x = \frac{-b \pm \sqrt{b^2 - 4ac}}{2a}$$

$$x = \frac{-5 \pm \sqrt{25 + 84}}{6} \qquad \text{Simplify.}$$

$$x = \frac{-5 \pm \sqrt{109}}{6} \qquad \text{Simplify.}$$

Practice Problems

For problems 14–20, solve.

14. $a^2 - 9a - 12 = 0$ **15.** $k^2 - 6k = 0$ **16.** $c^2 + 4c - 9 = 0$
17. $x^2 - 11x - 12 = 0$ **18.** $25w^2 + 40w + 16 = 0$
19. $49h^2 - 81 = 0$ **20.** $x^2 = 20$

9.3 Vocabulary Practice

Match the term with its description.

1. A number that cannot be written as a fraction in which the numerator and denominator are integers
2. $\sqrt{\ }$
3. $\pm$
4. The product of length and width of a rectangle
5. The distance around the outside of a rectangle
6. $b^2 - 4ac$
7. $a(b + c) = ab + ac$
8. The square root of this is a rational number.
9. $ax^2 + bx + c = 0$
10. $x = \dfrac{-b \pm \sqrt{b^2 - 4ac}}{2a}$

A. area of a rectangle
B. discriminant
C. distributive property
D. irrational number
E. perfect square
F. perimeter of a rectangle
G. plus-minus notation
H. principal square root
I. quadratic formula
J. standard form of a quadratic equation

9.3 Exercises

Follow your instructor's guidelines for showing your work.

For exercises 1–10,
(a) use the quadratic formula to solve.
(b) check.

1. $1x^2 + 15x + 44 = 0$
2. $1x^2 + 13x + 40 = 0$
3. $x^2 + 2x - 35 = 0$
4. $x^2 + 6x - 27 = 0$
5. $p^2 - 27p + 180 = 0$
6. $w^2 - 20p + 91 = 0$
7. $2d^2 - 11d - 40 = 0$
8. $2z^2 - 11z - 63 = 0$
9. $9h^2 - 9h - 40 = 0$
10. $9k^2 - 9k - 28 = 0$

For exercises 11–30, use the quadratic formula to solve.

11. $25m^2 + 10m + 1 = 0$ **12.** $36n^2 + 12n + 1 = 0$
13. $16u^2 + 24u + 9 = 0$ **14.** $36u^2 + 60u + 25 = 0$
15. $x^2 + 9x - 11 = 0$ **16.** $x^2 + 3x - 7 = 0$
17. $x^2 - 9x - 11 = 0$ **18.** $x^2 - 3x - 7 = 0$
19. $3p^2 - 5p - 7 = 0$ **20.** $3z^2 - 5z - 3 = 0$
21. $5u^2 + u - 1 = 0$ **22.** $7v^2 + v - 1 = 0$
23. $5u^2 - u - 1 = 0$ **24.** $7v^2 - v - 1 = 0$
25. $h^2 + h - 5 = 0$ **26.** $k^2 + k - 7 = 0$
27. $2u^2 + u - 7 = 0$ **28.** $2p^2 + p - 5 = 0$
29. $10z^2 + 11z - 15 = 0$ **30.** $10h^2 + 11h - 13 = 0$

For exercises 31–44,
(a) find the discriminant.
(b) use the discriminant to describe the solutions of the equation.

31. $1x^2 + 16x + 64 = 0$
32. $1x^2 + 18x + 81 = 0$
33. $d^2 + 8d + 15 = 0$
34. $r^2 + 9r + 14 = 0$
35. $n^2 - n - 56 = 0$
36. $p^2 - p - 72 = 0$
37. $x^2 + 5x - 9 = 0$
38. $x^2 + 7x - 9 = 0$
39. $4z^2 + 5z - 8 = 0$
40. $4v^2 + 5v - 12 = 0$
41. $m^2 - 2m + 1 = 0$
42. $n^2 + 2n + 1 = 0$

43. $3u^2 - u - 1 = 0$

44. $5k^2 - k - 8 = 0$

For exercises 45–48,
(a) choose values for a, b, and c to match the given discriminant.
(b) write a quadratic equation using these values.
(c) describe the solutions. Do not solve the equation.

45. Discriminant > 0, and it is a perfect square.

46. Discriminant > 0, and it is not a perfect square.

47. Discriminant > 0, and it is a perfect square greater than 25.

48. Discriminant $= 0$.

For exercises 49–56, use the five steps and the quadratic formula.

49. The area of a rectangle is 221 ft^2, and the perimeter is 60 ft. Find the length and width of this rectangle.

50. The area of a rectangle is 234 ft^2, and the perimeter is 62 ft. Find the length and width of this rectangle.

51. A light-rail train traveling at $\frac{14.67 \text{ ft}}{1 \text{ s}}$ is accelerating at a constant rate of $\frac{1.27 \text{ ft}}{1 \text{ s}^2}$. Find the time for the train to travel 5280 ft by solving the equation $5280 \text{ ft} = \left(\frac{14.67 \text{ ft}}{1 \text{ s}}\right)t + \frac{1}{2}\left(\frac{1.27 \text{ ft}}{1 \text{ s}^2}\right)t^2$. Round to the nearest whole number.

52. A light-rail train traveling at $\frac{29.33 \text{ ft}}{1 \text{ s}}$ is accelerating at a constant rate of $\frac{1.11 \text{ ft}}{1 \text{ s}^2}$. Find the time for the train to travel 5280 ft by solving the equation $5280 \text{ ft} = \left(\frac{29.33 \text{ ft}}{1 \text{ s}}\right)t + \frac{1}{2}\left(\frac{1.11 \text{ ft}}{1 \text{ s}^2}\right)t^2$. Round to the nearest whole number.

53. To find the concentration in moles per liter of formic acid in a solution, a chemistry student solves the equation $1x^2 + (1.77 \times 10^{-4})x - (1.77 \times 10^{-5}) = 0$. Find the concentration. Write the answer in scientific notation. Round the mantissa to the nearest hundredth.

54. To find the concentration in moles per liter of chlorous acid in a solution, a chemistry student solves the equation $1x^2 + (1.1 \times 10^{-2})x - (1.1 \times 10^{-2}) = 0$. Find the concentration. Write the answer in scientific notation. Round the mantissa to the nearest hundredth.

55. In an experiment, an electronics student finds that the total resistance of a series circuit with two resistors is 78 ohms. The total resistance of a parallel circuit with the same resistors is 8.7 ohms. The resistance of each resistor in ohms is a solution of the equation $x^2 - 78x + 678.6 = 0$. Find the resistance of each resistor. Round to the nearest whole number.

56. In an experiment, an electronics student finds that the total resistance of a series circuit with two resistors is 110 ohms. The total resistance of a parallel circuit with the same resistors is 9.09 ohms. The resistance of each resistor in ohms is a solution of the equation $x^2 - 110x + 1000 = 0$. Find the resistance of each resistor.

For exercises 57–58, use the Pythagorean theorem, the formula $d = rt$, and the five steps.

57. Two cars meet at the intersection of Interstate 29 and Highway 36 in Missouri. Then one car travels east on Highway 36 at a rate of $\frac{60 \text{ mi}}{1 \text{ hr}}$, and the other car travels south on I-29 at $\frac{75 \text{ mi}}{1 \text{ hr}}$. Find the time in *minutes* when the cars are 30 mi apart. Round to the nearest minute.

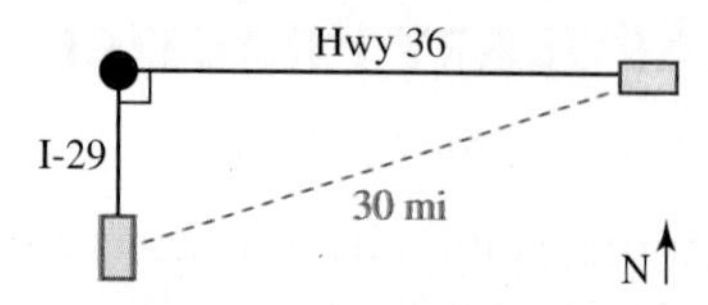

58. Two cars meet at the intersection of Interstate 80 and Highway 15 in Nebraska. Then one car travels north on Highway 15 at a rate of $\frac{60 \text{ mi}}{1 \text{ hr}}$, and the other car travels west on I-80 at $\frac{75 \text{ mi}}{1 \text{ hr}}$. Find the time in *minutes* when the cars are 40 mi apart. Round to the nearest minute.

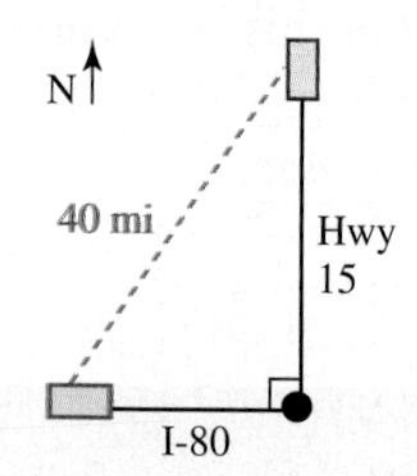

For exercises 59–60, use the five steps and a quadratic equation in one variable.

59. The length of a rectangle is 4 ft less than twice its width. The area of the rectangle is 576 ft^2. Find the length and width.

60. The length of a rectangle is 6 ft more than three times its width. The area of the rectangle is 297 ft^2. Find the length and width.

For exercises 61–68, solve.

61. $a^2 = 75$

62. $w^2 - 6w + 8 = 0$

63. $h^2 + 11h + 2 = 0$

64. $21n^2 + 32n - 5 = 0$

65. $v^2 + 6v + 3 = 0$

66. $c^2 + 7c = 0$

67. $2k^2 + k - 8 = 0$

68. $d^2 - 15 = 0$

Problem Solving: Practice and Review

Follow your instructor's guidelines for using the five steps as outlined in Section 1.3, p. 55.

69. Wal-Mart advertised a Thanksgiving dinner for a family of eight for under \$20, not including tax. The dinner included a 12-pound turkey at \$0.40 per pound, three cans of vegetables at \$0.50 each, 2 lb of sweet potatoes at \$0.25 per pound, two cans of cranberries at \$0.88 each, one 12-count package of brown and serve rolls for \$0.75, two packages of stuffing mix at \$1.50 each, and one pumpkin roll cake at \$5.50. In Idaho, the sales tax on food is 6.5%. Find the total cost of the dinner, including tax. Round to the nearest hundredth.

70. CellCept is an anti-rejection drug given to transplant patients. The concentration of a solution of CellCept is $\frac{33.3 \text{ mg}}{1 \text{ mL}}$. Find the amount of this solution and the amount of water needed to make 85 mL of a new solution with a concentration of $\frac{6 \text{ mg}}{1 \text{ mL}}$. Round to the nearest tenth.

71. The *diameter at basal height* of a tree is the diameter about 3 ft above the ground. The approximate shape of the trunk of a tree above basal height is a cone. The formula for the volume of a cone is $V = \frac{\pi r^2 h}{3}$, where r is the radius and h is the height. Measured from its basal height, a tree is 50 ft tall. The diameter of its trunk is 2.5 ft. Find the approximate volume of lumber in cubic feet in this trunk. ($\pi \approx 3.14$.) Round to the nearest whole number. (*Source:* G. John Smith; www.math.bcit.ca, 1997)

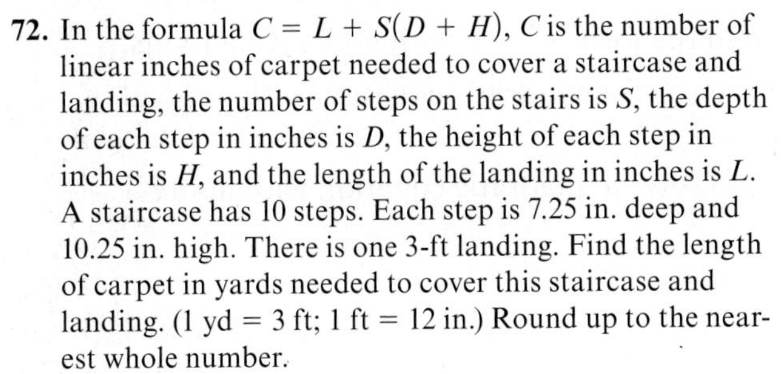

72. In the formula $C = L + S(D + H)$, C is the number of linear inches of carpet needed to cover a staircase and landing, the number of steps on the stairs is S, the depth of each step in inches is D, the height of each step in inches is H, and the length of the landing in inches is L. A staircase has 10 steps. Each step is 7.25 in. deep and 10.25 in. high. There is one 3-ft landing. Find the length of carpet in yards needed to cover this staircase and landing. (1 yd = 3 ft; 1 ft = 12 in.) Round up to the nearest whole number.

Find the Mistake

For exercises 73–76, the completed problem has one mistake.

(a) Describe the mistake in words, or copy down the whole problem and highlight or circle the mistake.

(b) Do the problem correctly.

73. Problem: Use the quadratic formula to solve $2x^2 - 51x + 270 = 0$.

Incorrect Answer: $x = \frac{-51 \pm \sqrt{(-51)^2 - 4(2)(270)}}{2(2)}$

$$x = \frac{-51 \pm \sqrt{2601 - 2160}}{4}$$

$$x = \frac{-51 \pm \sqrt{441}}{4}$$

$$x = \frac{-51 \pm 21}{4}$$

$$x = \frac{-51 + 21}{4} \quad \text{or} \quad x = \frac{-51 - 21}{4}$$

$$x = \frac{-30}{4} \quad \text{or} \quad x = \frac{-72}{4}$$

$$x = -\frac{15}{2} \quad \text{or} \quad x = -18$$

74. Problem: Use the quadratic formula to solve $2x^2 - 7x - 1 = 0$.

Incorrect Answer: $x = \frac{-(-7) \pm \sqrt{(-7)^2 - 4(2)(-1)}}{2(2)}$

$$x = \frac{7 \pm \sqrt{49 - 8}}{4}$$

$$x = \frac{7 \pm \sqrt{41}}{4}$$

75. Problem: Use the quadratic formula to solve $x^2 - 30 = 0$.

Incorrect Answer: $x = \frac{-(-30) \pm \sqrt{(-30)^2 - 4(1)(0)}}{2(1)}$

$$x = \frac{30 \pm \sqrt{900 - 0}}{2}$$

$$x = \frac{30 \pm \sqrt{900}}{2}$$

$$x = \frac{30 \pm 30}{2}$$

$$x = \frac{30 + 30}{2} \quad \text{or} \quad x = \frac{30 - 30}{2}$$

$$x = \frac{60}{2} \quad \text{or} \quad x = \frac{0}{2}$$

$$x = 30 \quad \text{or} \quad x = 0$$

76. Problem: Use the quadratic formula to solve $x^2 - 3x - 7 = 0$.

Incorrect Answer: $x = \frac{-(-3) \pm \sqrt{-3^2 - 4(1)(-7)}}{2(1)}$

$$x = \frac{3 \pm \sqrt{-9 + 28}}{2}$$

$$x = \frac{3 \pm \sqrt{19}}{2}$$

Review

77. Use roster notation to represent the set of whole numbers.

78. Use roster notation to represent the set of integers.

79. Describe the difference between a rational number and an irrational number.

80. Explain why $\sqrt{-2}$ is not a real number.

SUCCESS IN COLLEGE MATHEMATICS

81. For exercise 81 in Section 9.2, you wrote a practice test. Using the answers for the Practice Problems that are provided at the back of this textbook, prepare an answer key for your practice test.

© Peter Dean/Shutterstock.com

A turbine at a hydroelectric dam produces alternating current (AC) electricity. To describe alternating current, we use the nonreal number i. The solutions of some quadratic equations also include the number i. In this section, we will learn how to solve these equations.

SECTION 9.4 Quadratic Equations with Nonreal Solutions

After reading the text, working the practice problems, and completing assigned exercises, you should be able to:

1. Simplify a radical in which the radicand is less than 0.
2. Determine whether a number is an element of a given set.
3. Add and subtract complex numbers.
4. Solve a quadratic equation in the form $x^2 = k$ with nonreal solutions.
5. Use completing the square to solve a quadratic equation with nonreal solutions.
6. Use the quadratic formula to solve a quadratic equation with nonreal solutions.

Imaginary Numbers

The number 3.14159. . . is an irrational number that equals the quotient of the circumference of a circle and its diameter. We usually name this number with the Greek letter π. Another number that is named with a letter is the number i. By definition, $i^2 = -1$. So $\sqrt{-1} = i$. Since the square root of -1 is not a real number, i is not a real number. However, it is an *imaginary* number. The set of **imaginary numbers** includes i and multiples of i such as $3i$. In daily life, *imaginary* often means "make-believe" or "pretend." But imaginary numbers are not make-believe. For example, in physics classes, students analyze alternating current (AC). They use imaginary numbers to describe the phase shift of the voltage of the current.

To simplify a square root in which the radicand is negative, rewrite it as a product of a square root and $\sqrt{-1}$. Then $\sqrt{-1}$ simplifies to i.

EXAMPLE 1 Simplify: $\sqrt{-3}$

SOLUTION ▶

$\sqrt{-3}$

$= \sqrt{3}\sqrt{-1}$ Rewrite as a product of a square root and $\sqrt{-1}$.

$= \sqrt{3}\,i$ Simplify; $\sqrt{-1} = i$

Ask your instructor whether you should write $\sqrt{3}i$ or $i\sqrt{3}$. If you write $\sqrt{3}i$, be sure that the i is *not* written under the radical sign.

EXAMPLE 2 Simplify: $\sqrt{-12}$

SOLUTION

$\sqrt{-12}$

$= \sqrt{12}\sqrt{-1}$ Rewrite as a product of a square root and $\sqrt{-1}$.

$= \sqrt{4}\sqrt{3}\sqrt{-1}$ Perfect square factor: 4.

$= 2\sqrt{3}i$ Simplify; $\sqrt{-1} = i$

Practice Problems

For problems 1–4, simplify.

1. $\sqrt{-5}$ **2.** $\sqrt{-25}$ **3.** $\sqrt{-50}$ **4.** $\sqrt{-48}$

Complex Numbers

We organize the real numbers in sets including the whole numbers, integers, rational numbers, and irrational numbers. Numbers such as $7i$ are elements of the set of imaginary numbers. This set includes the number i and its multiples. Since $\sqrt{-49}$ simplifies to $7i$, it is an imaginary number. However, $6 + 2i$ is not an imaginary number because it is a sum of a real number and an imaginary number.

The sum of a real number and an imaginary number is a **complex number**. The elements in the set of **complex numbers** are numbers that can be written in the form $a + bi$, where a and b are real numbers. For example, the number $2 + 3i$ is a complex number. Every real number and every imaginary number is a complex number. In $a + bi$ form, the real number 12 is $12 + 0i$. The imaginary number $7i$ is $0 + 7i$.

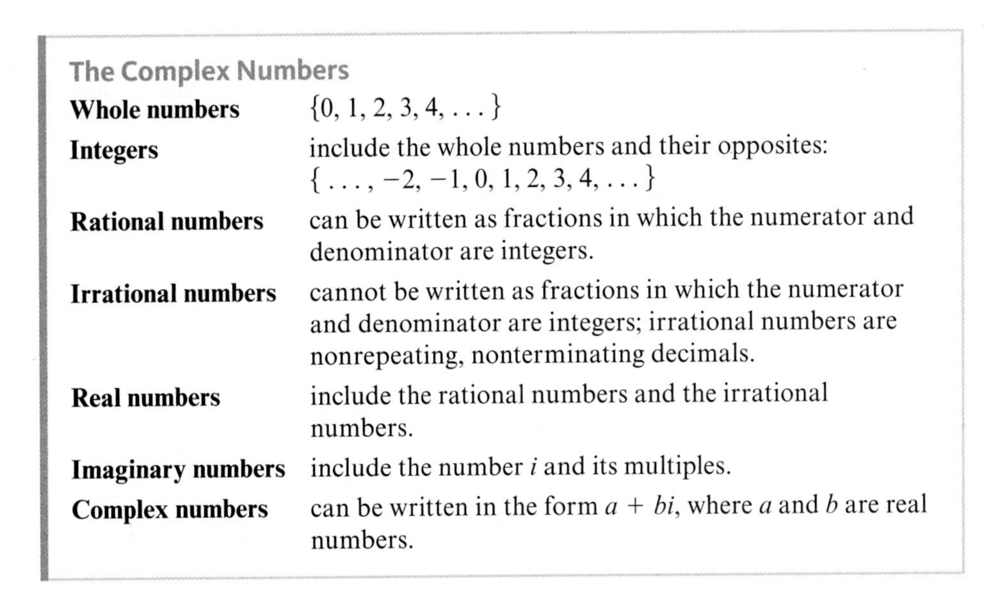

The Complex Numbers

Whole numbers	$\{0, 1, 2, 3, 4, \ldots\}$
Integers	include the whole numbers and their opposites: $\{\ldots, -2, -1, 0, 1, 2, 3, 4, \ldots\}$
Rational numbers	can be written as fractions in which the numerator and denominator are integers.
Irrational numbers	cannot be written as fractions in which the numerator and denominator are integers; irrational numbers are nonrepeating, nonterminating decimals.
Real numbers	include the rational numbers and the irrational numbers.
Imaginary numbers	include the number i and its multiples.
Complex numbers	can be written in the form $a + bi$, where a and b are real numbers.

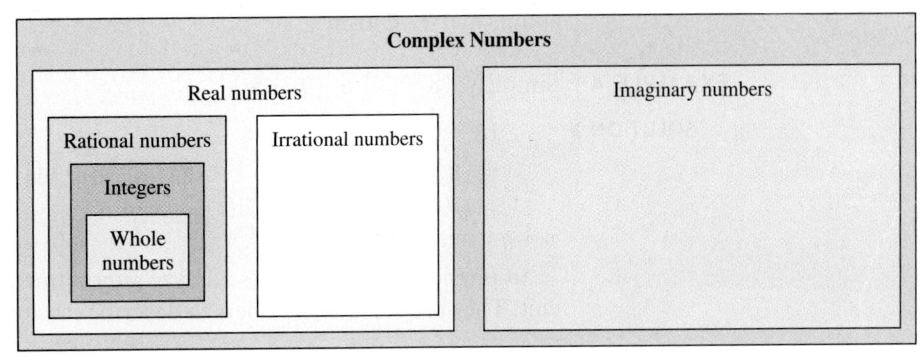

We use subsets to describe the relationship of sets. Since all of the imaginary numbers are complex numbers, the set of imaginary numbers is a subset of the complex numbers. The set of real numbers is also a subset of the complex numbers. The set of whole numbers is a subset of the set of integers, the set of rational numbers, the set of real numbers, and the set of complex numbers. The set of irrational numbers is a subset of the real numbers and the complex numbers, but it is *not* a subset of the rational numbers or of the imaginary numbers.

EXAMPLE 3 Put an X in the box if the number is an element of the set at the top of the column.

SOLUTION ▶

	Complex numbers	Imaginary numbers	Real numbers	Rational numbers	Irrational numbers	Integers	Whole numbers
2	X		X	X		X	X
-2	X		X	X		X	
$\frac{1}{2}$	X		X	X			
$\sqrt{2}$	X		X		X		
$2i$	X	X					
$\sqrt{-2}$	X	X					
$3 + 2i$	X						

Practice Problems

5. Copy the chart. Put an X in the box if the number is an element of the set at the top of the column.

	Complex numbers	Imaginary numbers	Real numbers	Rational numbers	Irrational numbers	Integers	Whole numbers
$5 + 8i$							
5							
$\sqrt{5}$							
$5i$							
-5							
0							
$\sqrt{25}$							
$\sqrt{-25}$							

Simplifying Complex Numbers

To add two complex numbers, add the real parts and add the imaginary parts. The sum is in $a + bi$ form.

EXAMPLE 4 Simplify: $(8 + 3i) + (9 - 5i)$

SOLUTION ▶

$(8 + 3i) + (9 - 5i)$

$= (8 + 9) + (3i - 5i)$ Add real parts; add imaginary parts.

$= 17 - 2i$ Simplify.

In physics classes, students analyze alternating current (AC) flowing through a circuit. They use complex numbers to describe the voltage of the current. The magnitude of the voltage is the real part of the complex number. The phase shift of the voltage is the imaginary part of the complex number.

EXAMPLE 5 Simplify: $(9.8 + 6.9i)$ volts $+ (9.6 - 8.7i)$ volts

SOLUTION ▶

$(9.8 + 6.9i)$ volts $+ (9.6 - 8.7i)$ volts

$= [(9.8 + 9.6) + (6.9 - 8.7)i]$ volts — Add real numbers; add imaginary numbers.

$= \mathbf{(19.4 - 1.8i)}$ volts — Simplify.

When we subtract polynomials with two or more terms, we put parentheses around the polynomial being subtracted. When subtracting complex numbers, we also use parentheses.

EXAMPLE 6 Simplify: $(6 + 2i) - (15 - 10i)$

SOLUTION ▶

$(6 + 2i) - (15 - 10i)$

$= (6 + 2i) - \mathbf{1}(15 - 10i)$ — Multiply by 1.

$= 6 + 2i - \mathbf{1}(15) - \mathbf{1}(-10i)$ — Distributive property.

$= 6 + 2i \mathbf{- 15 + 10i}$ — Simplify.

$= (6 - 15) + (2i + 10i)$ — Add real numbers; add imaginary numbers.

$= -9 + 12i$ — Simplify.

Practice Problems

For problems 6–8, simplify.

6. $(8 + 3i) + (5 + 6i)$ **7.** $(8 + 3i) - (5 + 6i)$

8. $(8 + 3i) - (5 - 6i)$

Quadratic Equations with Nonreal Solutions

In the next example, since the quadratic equation is in the form $x^2 = k$, the solutions of the equation are in the form $x = \sqrt{k}$ and $x = -\sqrt{k}$.

EXAMPLE 7 **(a)** Solve: $p^2 = -36$

SOLUTION ▶

$p^2 = -36$ — The equation is in the form $x^2 = k$.

$p = \sqrt{-36}$ or $p = -\sqrt{-36}$ — The solutions are $x = \sqrt{k}$ or $x = -\sqrt{k}$.

$p = \sqrt{36}\sqrt{-1}$ or $p = -\sqrt{36}\sqrt{-1}$ — Rewrite as a product of a square root and $\sqrt{-1}$.

$p = \mathbf{6i}$ or $p = \mathbf{-6i}$ — Simplify; $\sqrt{-1} = i$

$p = \pm 6i$ — $\pm$ notation.

(b) Check.

▶

Check: $p = 6i$

$p^2 = -36$

$\mathbf{(6i)}^2 = -36$

$\mathbf{6i \cdot 6i} = -36$

$\mathbf{36 \cdot i^2} = -36$

$36(\mathbf{-1}) = -36$

$\mathbf{-36} = -36$ True.

Check: $p = -6i$

$p^2 = -36$

$\mathbf{(-6i)}^2 = -36$

$\mathbf{(-6i)(-6i)} = -36$

$\mathbf{36 \cdot i^2} = -36$

$36(\mathbf{-1}) = -36$

$\mathbf{-36} = -36$ True.

Practice Problems

For problems 9–11,
(a) solve.
(b) check.

9. $x^2 = -64$ **10.** $m^2 = -200$ **11.** $v^2 + 20 = -16$

Completing the Square

We can use completing the square to solve a quadratic equation with nonreal solutions.

EXAMPLE 8 Solve $d^2 - 2d + 3 = 0$ by completing the square.

SOLUTION

$d^2 - 2d + 3 = 0$	Isolate $d^2 - 2d$.
$-3 \quad -3$	Subtraction property of equality.
$d^2 - 2d + \mathbf{0} = \mathbf{-3}$	$b = -2;\ \left(\frac{1}{2}(-2)\right)^2 = 1$
$+1 \quad +1$	Addition property of equality.
$d^2 - 2d + \mathbf{1} = \mathbf{-2}$	Simplify.
$(d - 1)(d - 1) = -2$	Factor.
$(d - 1)^2 = -2$	Write in the form $x^2 = k$.
$d - 1 = \sqrt{-2}$ or $d - 1 = -\sqrt{-2}$	The solutions are $x = \sqrt{k}$ or $x = -\sqrt{k}$.
$d - 1 = \sqrt{2}\sqrt{-1}$ or $d - 1 = -\sqrt{2}\sqrt{-1}$	Rewrite as a product of a square root and $\sqrt{-1}$.
$d - 1 = \sqrt{2}i$ or $d - 1 = -\sqrt{2}i$	Simplify; $\sqrt{-1} = i$
$+1 \quad +1$ $\quad$ $+1 \quad +1$	Addition property of equality.
$d + \mathbf{0} = \sqrt{2}i + \mathbf{1}$ or $d + \mathbf{0} = -\sqrt{2}i + \mathbf{1}$	Simplify.
$d = 1 \pm \sqrt{2}i$	Simplify; $\pm$ notation.

Practice Problems

For problems 12–14, solve by completing the square.

12. $z^2 + 4z + 20 = 0$ **13.** $h^2 - 14h + 51 = 0$ **14.** $k^2 + 18k + 100 = 0$

The Quadratic Formula

We can use the quadratic formula to solve quadratic equations with nonreal solutions.

EXAMPLE 9 Use the quadratic formula to solve $3x^2 - 5x + 9 = 0$.

SOLUTION

$3x^2 - 5x + 9 = 0$	$a = 3, b = -5, c = 9$
$x = \dfrac{-(-5) \pm \sqrt{(-5)^2 - 4(3)(9)}}{2(3)}$	$x = \dfrac{-b \pm \sqrt{b^2 - 4ac}}{2a}$
$x = \dfrac{5 \pm \sqrt{25 - 108}}{6}$	Simplify.

$x = \dfrac{5 \pm \sqrt{-83}}{6}$ Simplify.

$x = \dfrac{5 \pm \sqrt{83} \cdot \sqrt{-1}}{6}$ Rewrite as a product of a square root and $\sqrt{-1}$.

$x = \dfrac{5 \pm \sqrt{83}i}{6}$ Simplify; $\sqrt{-1} = i$

Since there are no perfect square factors of 83, $\sqrt{83}$ cannot be simplified. We usually write nonreal numbers in $a + bi$ form as a sum or difference of a real number and an imaginary number: $x = \dfrac{5}{6} \pm \dfrac{\sqrt{83}}{6}i$.

Practice Problems

For problems 15–17, use the quadratic formula to solve.

15. $z^2 + 3z + 13 = 0$ **16.** $2h^2 - 3h + 5 = 0$ **17.** $2k^2 + k + 3 = 0$

9.4 Vocabulary Practice

Match the term with its description.

1. This set of numbers is the union of the rational numbers and the irrational numbers.
2. All numbers in this set can be written as a fraction in which the numerator and denominator are integers.
3. All numbers in this set are nonrepeating, nonterminating decimal numbers.
4. The numbers in the set $\{0, 1, 2, 3, \ldots\}$
5. The numbers in the set $\{\ldots, -3, -2, -1, 0, 1, 2, \ldots\}$
6. The set of all numbers that can be written in the form $a + bi$, where a and b are real numbers
7. $x = \dfrac{-b \pm \sqrt{b^2 - 4ac}}{2a}$
8. The expression under a radical
9. $a(b + c) = ab + ac$
10. This set of numbers includes only i and real number multiples of i.

A. complex numbers
B. distributive property
C. imaginary numbers
D. integers
E. irrational numbers
F. quadratic formula
G. radicand
H. rational numbers
I. real numbers
J. whole numbers

9.4 Exercises

Follow your instructor's guidelines for showing your work.

For exercises 1–12, simplify.

1. $\sqrt{-11}$ **2.** $\sqrt{-15}$

3. $\sqrt{-44}$ **4.** $\sqrt{-60}$

5. $\sqrt{-144}$ **6.** $\sqrt{-100}$

7. $\sqrt{3^2 - 4(2)(8)}$ **8.** $\sqrt{3^2 - 4(8)(1)}$

9. $\sqrt{3^2 - 4(2)(-8)}$ **10.** $\sqrt{3^2 - 4(8)(-1)}$

11. $\sqrt{6^2 - 4(2)(9)}$ **12.** $\sqrt{8^2 - 4(4)(8)}$

13. Explain why the square root of a negative number cannot be a real number.

14. The square root of a negative number is not a real number. However, the cube root of a negative number is a real number. Explain why.

15. Copy the chart below. Put an X in the box if the number is an element of the set at the top of the column.

16. Copy the chart below. Put an X in the box if the number is an element of the set at the top of the column.

For exercises 17–24, simplify.

17. $(5 + 3i) + (8 + 7i)$

18. $(15 + 8i) + (2 + 4i)$

19. $(5 - 3i) + (8 - 7i)$

20. $(15 - 8i) + (2 - 4i)$

21. $(2 - 3i) - (6 - 9i)$

22. $(3 - 4i) - (10 - 9i)$

23. $(2 - 3i) - (6 + 9i)$

24. $(3 - 4i) - (10 + 9i)$

25. The voltage drops in an AC circuit are $3.1460 + 17.175i$ volts, $-18.697 + 3.0820i$ volts, and $135.61 - 20.252i$ volts. Find the sum of these voltages.

26. The voltage drops in an AC circuit are $15 - 26.6491i$ volts, $-9.7294 - 6.8813i$ volts, and $9.7452 - 19.7729i$ volts. Find the sum of these voltages.

For exercises 27–30,
(a) solve.
(b) check.

27. $x^2 = -100$

28. $x^2 = -225$

29. $m^2 = -4$

30. $n^2 = -9$

For exercises 31–44, solve.

31. $p^2 = -5$

32. $w^2 = -3$

33. $v^2 = -50$

34. $k^2 = -72$

35. $4h^2 = -196$

36. $7a^2 = -252$

37. $c^2 + 8 = -15$

38. $d^2 + 7 = -12$

39. $a^2 - 3 = -11$

40. $z^2 - 5 = -17$

41. $k^2 + 8 = -41$

42. $p^2 + 7 = -57$

43. $2w^2 + 4 = -18$

44. $2n^2 + 6 = -20$

For exercises 45–58, solve the equation by completing the square.

45. $x^2 + 4x + 8 = 0$

46. $x^2 + 4x + 20 = 0$

47. $p^2 + 14p + 74 = 0$

48. $p^2 + 14p + 53 = 0$

49. $z^2 - 20z + 164 = 0$

50. $z^2 - 20z + 200 = 0$

51. $a^2 + 12a + 157 = 0$

52. $a^2 + 12a + 45 = 0$

53. $v^2 + 2v + 8 = 0$

54. $v^2 + 2v + 14 = 0$

55. $w^2 - 18w + 86 = 0$

56. $w^2 - 18w + 83 = 0$

57. $m^2 + 24m + 167 = 0$

58. $m^2 + 24m + 150 = 0$

For exercises 59–68, use the quadratic formula to solve the equation.

59. $x^2 - 3x + 3 = 0$

60. $x^2 - 3x + 5 = 0$

61. $p^2 + 3p + 12 = 0$

62. $z^2 + 3z + 14 = 0$

Chart for exercise 15

	Complex numbers	Imaginary numbers	Real numbers	Rational numbers	Irrational numbers	Integers	Whole numbers
$\sqrt{-5}$							
$\sqrt{81}$							
$\frac{2}{5}$							
$-6i$							
$3 + 2i$							
$\sqrt{3}$							

Chart for exercise 16

	Complex numbers	Imaginary numbers	Real numbers	Rational numbers	Irrational numbers	Integers	Whole numbers
$9 + i$							
$\sqrt{7}$							
$\sqrt{-7}$							
$\sqrt{16}$							
$\frac{1}{4}$							
$-8i$							

63. $3u^2 + 2u + 9 = 0$

64. $3v^2 + 2v + 10 = 0$

65. $5r^2 - 3r + 3 = 0$

66. $5d^2 - 3d + 2 = 0$

67. $2n^2 + 5n + 21 = 0$

68. $2k^2 + 5k + 23 = 0$

For exercises 69–76, solve.

69. $3x^2 = -9$

70. $5x^2 = -35$

71. $3b^2 - 11b + 1 = 0$

72. $3a^2 - 13a + 1 = 0$

73. $n^2 + 5n + 6 = 0$

74. $p^2 + 9p + 14 = 0$

75. $3x^2 - 12x = 0$

76. $4x^2 - 12x = 0$

Problem Solving: Practice and Review

Follow your instructor's guidelines for using the five steps as outlined in Section 1.3, p. 55.

77. In 2012, there were 360,000 Florida home loans in foreclosure. Find the number of Florida home loans. Round to the nearest hundred.

With 14 percent of Florida home loans in foreclosure, abandoned houses are scattered across the state. (*Source:* www.nytimes.com, Feb. 23, 2012)

78. Find the amount of the exhibition and sales space devoted to exhibits. Round to the nearest hundred.

Rolex opened a new. . . 8,600 square-foot exhibition and sales space, the Rolex Experience, along the Bund, the riverfront avenue in Shanghai. About two-thirds of the surface area is devoted to exhibits, tracing the history of Oyster and other watches that have established Rolex's popularity. (*Source:* www.nytimes.com, Feb. 21, 2012)

79. Find the amount of yen equal to 1 U.S. dollar (the exchange rate). Round to the nearest tenth.

The Insight first went on sale in Japan last month, where it sells for 1.89 million yen ($19,260). (*Source:* http://latimesblogs.latimes.com, March 2009)

© jpskenn/Shutterstock.com

80. In an isosceles triangle, the measures of two of the angles are equal. The third angle of an isosceles triangle measures 42°. The total measure of the angles in the triangle are 180°. Find the measure of each of the equal angles.

Find the Mistake

For exercises 81–84, the completed problem has one mistake.

(a) Describe the mistake in words, or copy down the whole problem and highlight or circle the mistake.

(b) Do the problem correctly.

81. Problem: Simplify: $\sqrt{-18}$

Incorrect Answer: $\sqrt{-18}$

$= \sqrt{18}\sqrt{-1}$

$= \sqrt{18}\,i$

82. Problem: Solve: $x^2 = -11$

Incorrect Answer: $x^2 = -11$

$x = \sqrt{-11}$ or $x = -\sqrt{-11}$

$x = \pm\sqrt{-11}$

83. Problem: Use the quadratic formula to solve $x^2 - 3x + 9 = 0$.

Incorrect Answer: $x = \dfrac{-(-3) \pm \sqrt{(-3)^2 - 4(1)(9)}}{2(1)}$

$x = \dfrac{3 \pm \sqrt{9 - 36}}{2}$

$x = \dfrac{3 \pm \sqrt{-27}}{2}$

$x = \dfrac{3 \pm \sqrt{9}\sqrt{3}}{2}$

$x = \dfrac{3 \pm 3\sqrt{3}}{2}$

84. Problem: Use the quadratic formula to solve $x^2 - 3x - 7 = 0$.

Incorrect Answer: $x = \dfrac{-3 \pm \sqrt{(-3)^2 - 4(1)(7)}}{2(1)}$

$x = \dfrac{3 \pm \sqrt{9 - 28}}{2}$

$x = \dfrac{3 \pm \sqrt{-19}}{2}$

$x = \dfrac{3 \pm \sqrt{19}\,i}{2}$

Review

For exercises 85–88, graph the equation on an *xy*-coordinate system.

85. $y = \dfrac{3}{4}x - 2$

86. $5x - 3y = 21$

87. $y - 3 = 2(x + 1)$

88. $\dfrac{2}{3}x + \dfrac{1}{8}y = 19$

SUCCESS IN COLLEGE MATHEMATICS

89. Use the Vocabulary Practice in each section of Chapter 3 to write a list of the terms in this chapter. Circle the terms that you cannot define without looking them up.

90. Describe how you plan to review the terms that you do not know.

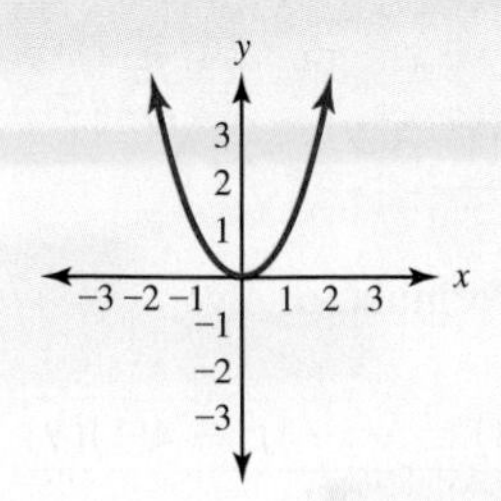

The graph of a quadratic equation in the form $y = ax^2 + bx + c$ is a parabola. In this section, we will find the vertex, axis of symmetry, and x-intercepts of these graphs.

SECTION 9.5

Quadratic Equations in the Form $y = ax^2 + bx + c$

After reading the text, working the practice problems, and completing assigned exercises, you should be able to:

1. Complete a table of solutions and graph a quadratic equation in the form $y = ax^2 + bx + c$.
2. Find the x-intercepts of the graph of a quadratic equation in the form $y = ax^2 + bx + c$.
3. Use $x = -\frac{b}{2a}$ to find the vertex of the graph of a quadratic equation in the form $y = ax^2 + bx + c$.
4. Use the intercepts, vertex, and axis of symmetry to sketch the graph of a quadratic equation in the form $y = ax^2 + bx + c$.

The Graph of a Quadratic Equation

The standard form of a quadratic equation in two variables is $y = ax^2 + bx + c$. The graph of $y = ax^2 + bx + c$ (Figure 1) is a U shape that is called a **parabola**. The graph can "open up" or "open down." The "bottom" or "top" point of the parabola is its **vertex**. The left "arm" and right "arm" of a parabola are mirror images. If a vertical line is drawn through the vertex, it divides the parabola into two halves. If we could fold the graph on the line, the two halves would exactly match. This vertical line is the **axis of symmetry**. To graph a parabola, graph enough ordered pairs to see the "U" shape.

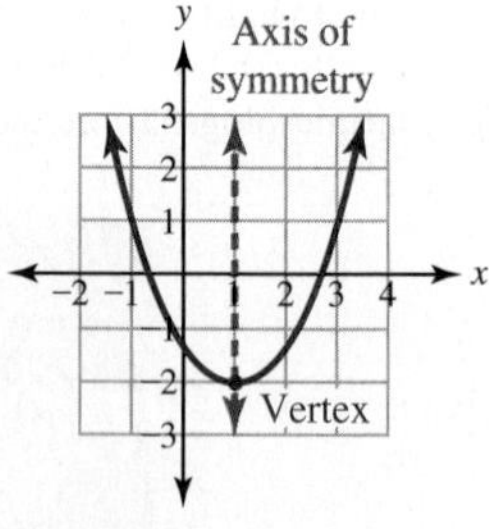

Figure 1

EXAMPLE 1 A quadratic equation is $y = x^2 + 3$.

(a) Identify a, b, and c.

SOLUTION ▸ $y = \mathbf{1}x^2 + \mathbf{0}x + \mathbf{3}$ $\quad a = 1, b = 0, c = 3$

(b) Use a table of solutions to graph the equation. Graph the axis of symmetry.

▸ To graph a *linear* equation, we find three solutions of the equation, graph the solutions, and draw a straight line through the points. To graph a *quadratic* equation in the form $y = ax^2 + bx + c$, we need to find and graph enough solutions to see the U shape of the graph. Choose a value for x, replace x in the equation with this value, solve for y, and graph the ordered pair. Continue this process until the graphed points show the shape of the parabola.

$y = \mathbf{0}^2 + 3$ $y = \mathbf{3}$
$y = \mathbf{1}^2 + 3$ $y = \mathbf{4}$
$y = \mathbf{2}^2 + 3$ $y = \mathbf{7}$

x	y
0	3
1	4
2	7

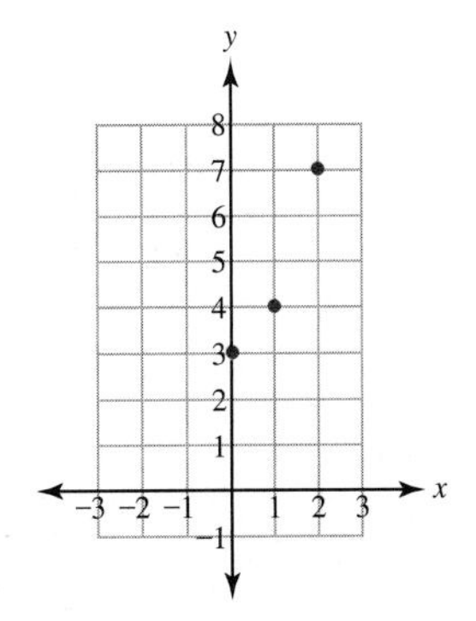

These three points appear to be on the "right arm" of the parabola. To see the "left arm," we need to find more points to the left of the y-axis.

$y = (\mathbf{-1})^2 + 3$ $y = \mathbf{4}$
$y = (\mathbf{-2})^2 + 3$ $y = \mathbf{7}$

x	y
0	3
1	4
2	7
−1	4
−2	7

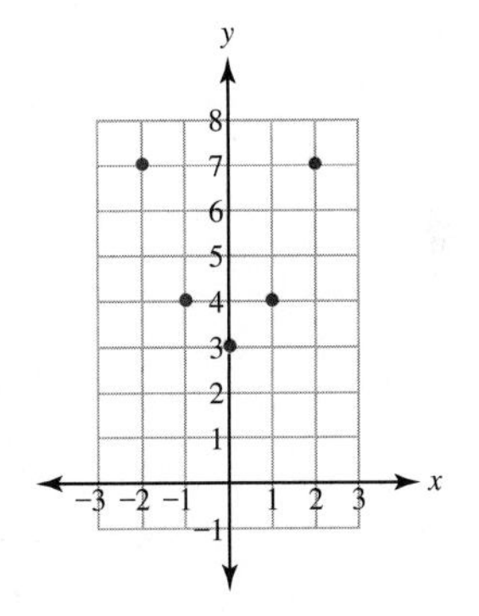

Draw a smooth curve through the points to graph the parabola. Since the vertical line that is the axis of symmetry is not part of the graph of the equation, draw it as a dashed line.

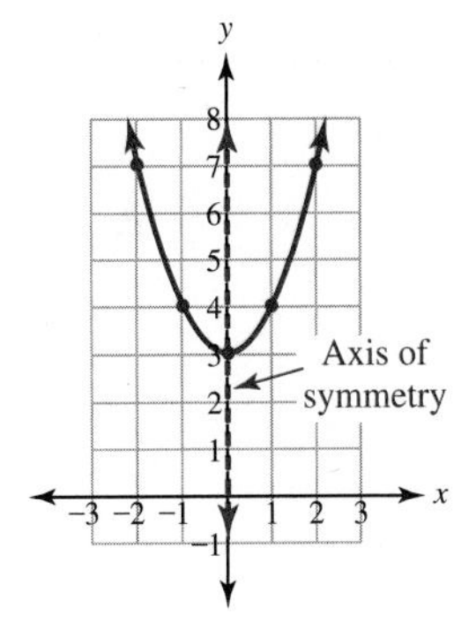

(c) Identify the vertex.

The vertex is $(0, 3)$.

(d) Identify whether the parabola opens up or opens down.

The parabola opens up.

(e) Write the equation of the axis of symmetry.

Since the axis of symmetry is the vertical line that passes through the vertex, $(0, 3)$, its equation is $x = 0$.

In Example 1, the value of a is 1, and the parabola opens up. The lead coefficient, a, is 1, so $a > 0$. In the next example, $a < 0$. The graph opens down.

EXAMPLE 2 A quadratic equation is $y = -x^2 + 6x - 5$.

(a) Identify a, b, and c.

SOLUTION $\mathbf{y = -1x^2 + 6x - 5}$ $\quad a = -1, b = 6, c = -5$

(b) Use a table of solutions to graph the equation. Graph the axis of symmetry.

Choose a value for x, replace x in the equation with this value, solve for y, and graph the ordered pair. Continue this process until the graphed points show the shape of the parabola.

$y = -(\mathbf{0}^2) + 6(\mathbf{0}) - 5$ $y = \mathbf{-5}$
$y = -(\mathbf{1}^2) + 6(\mathbf{1}) - 5$ $y = \mathbf{0}$
$y = -(\mathbf{2}^2) + 6(\mathbf{2}) - 5$ $y = \mathbf{3}$
$y = -(\mathbf{3}^2) + 6(\mathbf{3}) - 5$ $y = \mathbf{4}$
$y = -(\mathbf{4}^2) + 6(\mathbf{4}) - 5$ $y = \mathbf{3}$
$y = -(\mathbf{5}^2) + 6(\mathbf{5}) - 5$ $y = \mathbf{0}$

x	y
0	−5
1	0
2	3
3	4
4	3
5	0

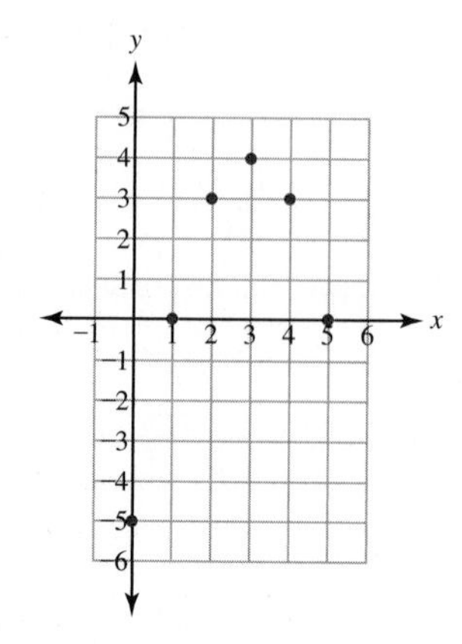

When the input value is greater than 3, the output values begin to decrease. The point (4, 3) is on the right arm of the parabola. We can finish the graph by finding more ordered pairs. Or we can draw the right arm of the parabola as a mirror image of the already graphed left arm. Since the vertical line that is the axis of symmetry is not part of the graph of the equation, draw it as a dashed line.

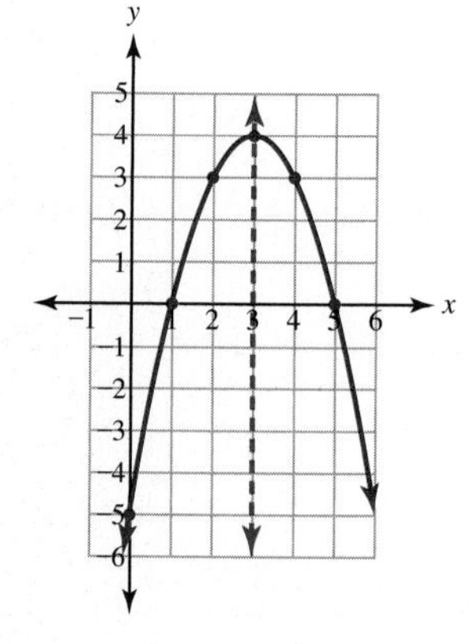

(c) Identify the vertex.

The vertex is $(3, 4)$.

(d) Identify whether the parabola opens up or opens down.

The parabola opens down (the lead coefficient, -1, is less than 0).

(e) Write the equation of the axis of symmetry.

Since the axis of symmetry is the vertical line that passes through the vertex, $(3, 4)$, its equation is $x = 3$.

The **lead coefficient** of the quadratic equation $y = ax^2 + bx + c$ is a. For example, the lead coefficient of $y = 3x^2 + 1$ is 3. As we have already seen, if the lead coefficient is greater than 0, the graph of the equation opens up. If the lead coefficient is less than 0, the graph of the equation opens down. The lead coefficient also affects the shape of the graph in another way. In the next example, we compare the graphs of two quadratic equations with the same values for b and c but different values for a, $y = 3x^2 + 1$, and $y = 1x^2 + 1$. The graph of $y = 3x^2 + 1$ is "taller" than the graph of $y = 1x^2 + 1$. This happens because for a given value of x, the y-value of $y = 3x^2 + 1$ is greater than the y-value of $y = 1x^2 + 1$.

To predict how tall the graphs will be, identify the absolute value of each lead coefficient, $|a|$. The absolute value of the lead coefficient of the taller graph is greater than the absolute value of the lead coefficient of the shorter graph.

EXAMPLE 3 The graphs represent $y = 3x^2 + 1$ and $y = 1x^2 + 1$.

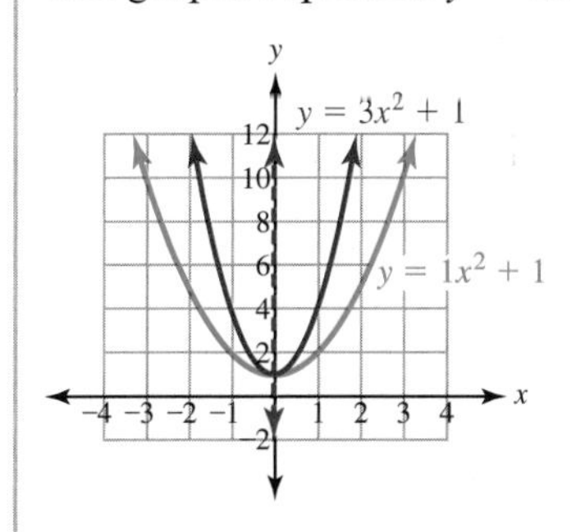

(a) Identify the "taller" graph.

SOLUTION ▶ We describe the graph of $y = 3x^2 + 1$ as "taller" than the graph of $y = 1x^2 + 1$ because at a given value of x, the y-value of $y = 3x^2 + 1$ is greater than the y-value of $y = 1x^2 + 1$. Notice that $|3| > |1|$.

(b) Identify the vertex of each graph.

▶ The vertex of each graph is $(0, 1)$.

(c) Write the equation of the axis of symmetry of each graph.

▶ The equation of the axis of symmetry of each graph is $x = 0$.

Practice Problems

For problems 1–4,
(a) identify *a*, *b*, and *c*.
(b) identify whether the graph opens up or down.
(c) use a table of solutions to graph the equation. Graph the axis of symmetry.
(d) identify the vertex.
(e) write the equation of the axis of symmetry.

1. $y = -x^2$ **2.** $y = x^2 + 1$ **3.** $y = -\frac{1}{3}x^2$ **4.** $y = x^2 - 6x + 10$

5. Is the graph of $y = 2x^2 - 1$ taller or shorter than the graph of $y = 6x^2 - 1$? Explain.

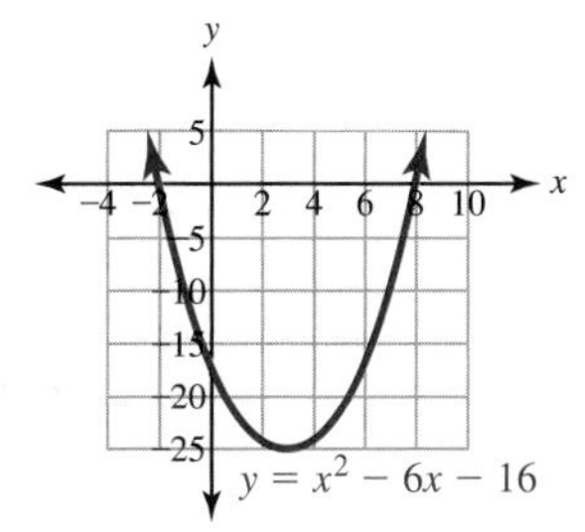

Figure 2

The *x*-Intercepts and Vertex of $y = ax^2 + bx + c$

The x-intercepts of the graph of $y = x^2 - 6x - 16$ (Figure 2) are $(-2, 0)$ and $(8, 0)$. The y-coordinate of an x-intercept is 0. To algebraically find the x-intercepts of a quadratic equation in the form $y = ax^2 + bx + c$, replace y with 0 and solve for x.

EXAMPLE 4 Find the x-intercepts of the graph of $y = x^2 - 6x - 16$.

SOLUTION ▶

$y = x^2 - 6x - 16$

$\mathbf{0} = x^2 - 6x - 16$ Replace y with 0.

$0 = \mathbf{(x - 8)(x + 2)}$ Factor.

$x - 8 = 0$ or $x + 2 = 0$ Zero product property.

$+8\ +8$ $\quad -2\ -2$ Properties of equality.

$\mathbf{x + 0 = 8}$ $\quad \mathbf{x + 0 = -2}$ Simplify.

$\mathbf{x} = 8$ $\quad \mathbf{x} = -2$ Simplify.

These solutions are the x-coordinates of the x-intercepts of the graph. The x-intercepts are $(8, 0)$ and $(-2, 0)$.

When a quadratic equation is in the form $y = ax^2 + bx + c$, the x-coordinate of the vertex of its graph equals $-\frac{b}{2a}$.

EXAMPLE 5 Find the vertex of the graph of $y = x^2 - 6x - 16$.

SOLUTION ▶ Find the x-coordinate of the vertex.

$x = -\frac{b}{2a}$

$x = -\frac{(\mathbf{-6})}{2(\mathbf{1})}$ Replace a and b.

$x = \mathbf{3}$ Simplify.

Find the y-coordinate of the vertex.

$y = x^2 - 6x - 16$

$y = (\mathbf{3})^2 - 6(\mathbf{3}) - 16$ Replace x with 3.

$y = \mathbf{9 - 18} - 16$ Simplify.

$y = \mathbf{-25}$ Simplify.

The vertex of the graph is $(3, -25)$.

The graph of a quadratic equation in the form $y = ax^2 + bx + c$ is a parabola with a vertical axis of symmetry that passes through the vertex. If $a > 0$, the graph opens up; if $a < 0$, the graph opens down. We can use the x-intercepts, the vertex, and the axis of symmetry to sketch the graph of a quadratic equation without completing a table of solutions.

EXAMPLE 6 A quadratic equation is $y = x^2 - 2x - 3$.

(a) Find the x-intercepts.

SOLUTION ▶

$y = x^2 - 2x - 3$

$\mathbf{0} = x^2 - 2x - 3$ Replace y with 0.

$0 = \mathbf{(x - 3)(x + 1)}$ Factor.

$x - 3 = 0$ or $x + 1 = 0$ Zero product property.

$\mathbf{+3\ +3}$ $\qquad$ $\mathbf{-1\ -1}$ Properties of equality.

$x + \mathbf{0} = \mathbf{3}$ $\qquad$ $x + \mathbf{0} = \mathbf{-1}$ Simplify.

$\boldsymbol{x} = 3$ $\qquad$ $\boldsymbol{x} = -1$ Simplify.

The x-intercepts are $(3, 0)$ and $(-1, 0)$.

(b) Use $x = -\frac{b}{2a}$ to find the vertex.

▶ Find the x-coordinate of the vertex.

$x = -\frac{b}{2a}$

$x = -\frac{(\mathbf{-2})}{2(\mathbf{1})}$ Replace a and b.

$x = \mathbf{1}$ Simplify.

Find the y-coordinate of the vertex.

$y = x^2 - 2x - 3$	
$y = \mathbf{1}^2 - 2(\mathbf{1}) - 3$	Replace x with 1.
$y = 1 - \mathbf{2} - 3$	Simplify.
$y = \mathbf{-4}$	Simplify.

The vertex of the graph is $(1, -4)$.

(c) Write the equation of the axis of symmetry.

The axis of symmetry is a vertical line that includes the vertex. Since the vertex is $(1, -4)$, the equation of the axis of symmetry is $x = 1$.

(d) Graph the x-intercepts, the vertex, and the axis of symmetry. Sketch the graph of the parabola.

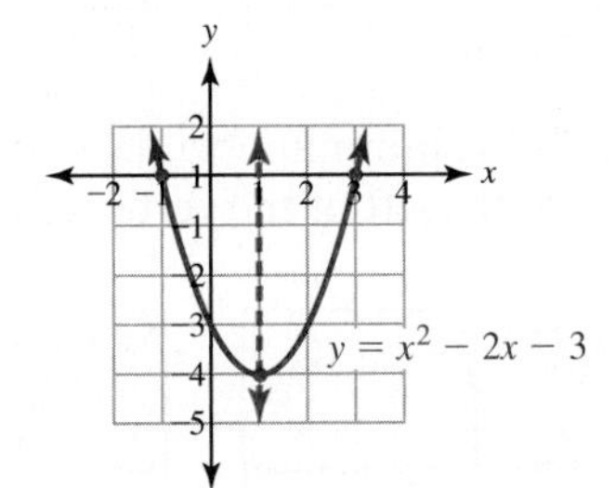

Graphing a Quadratic Equation in Two Variables

To graph a quadratic equation in standard form, $y = ax^2 + bx + c$,

1. Find the x-intercepts of the parabola, by replacing y in the equation with 0 and solving for x.
2. Find the vertex of the parabola. The x-coordinate of the vertex is $x = -\dfrac{b}{2a}$.

   To find the y-coordinate of the vertex, replace x in the equation with the x-coordinate of the vertex and solve for y.
3. Graph the x-intercepts and the vertex. Draw a dashed vertical line through the vertex to represent the axis of symmetry. Draw a parabola that passes through the x-intercepts and the vertex that is symmetric to this dashed vertical line.

Practice Problems

For problems 6–9,

(a) find the x-intercepts.

(b) use $x = -\dfrac{b}{2a}$ to find the vertex.

(c) write the equation of the axis of symmetry.

(d) graph the x-intercepts, the vertex, and the axis of symmetry. Sketch the graph of the parabola.

6. $y = x^2 - 2x - 24$ **7.** $y = x^2 - 4x - 12$ **8.** $y = x^2 - 4$

9. $y = x^2 - 6x + 9$

Using Technology: Finding the Vertex of the Graph of a Quadratic Equation

The CALCULATE screen includes choices called "minimum" and "maximum." For a parabola that opens up, the "minimum" is the y-coordinate of the vertex. For a parabola that opens down, the "maximum" is the y-coordinate of the vertex. These commands allow us to find the vertex of the graph of a quadratic equation.

EXAMPLE 7 Graph $y = x^2 + 5x + 6$, and find its vertex.

Type the equation in the Y= screen. Press GRAPH. Go to the CALCULATE screen.

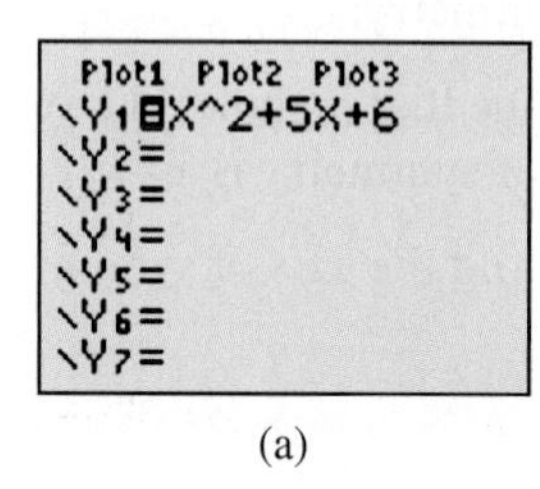

(a)

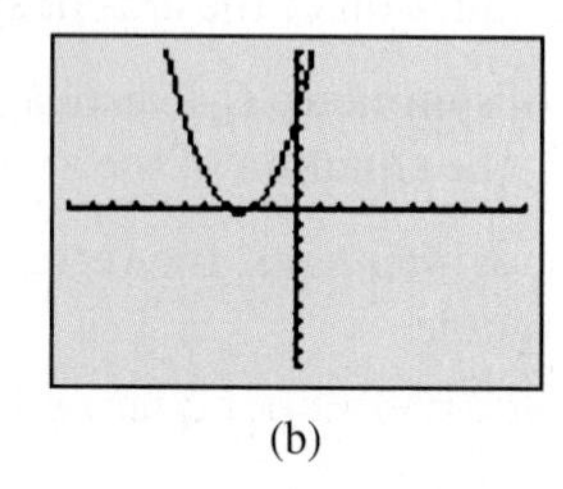
(b)

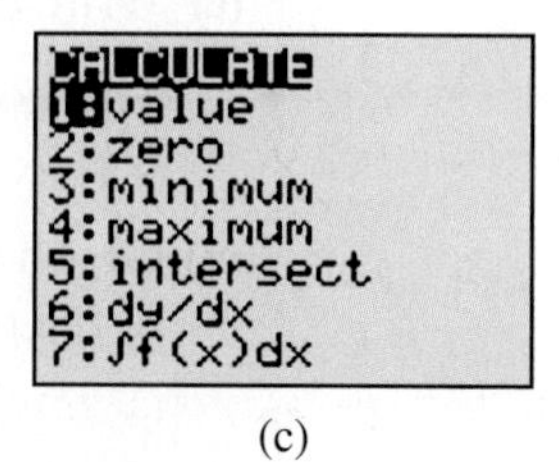

(c)

The graph opens up. The y-coordinate of the vertex is the minimum value. Choose 3: minimum. Press ENTER. The calculator asks "Left Bound?" Move the cursor to the left of the vertex of the parabola.

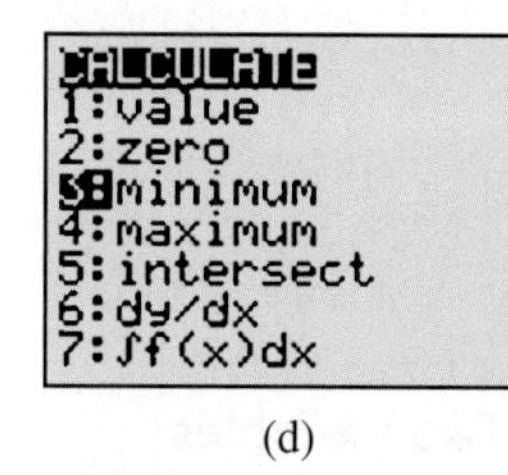

(d)

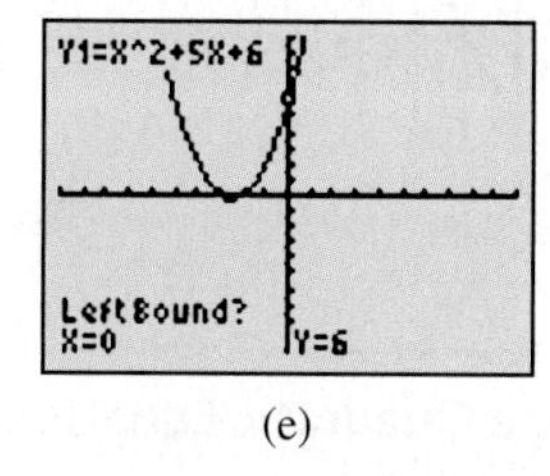

(e)

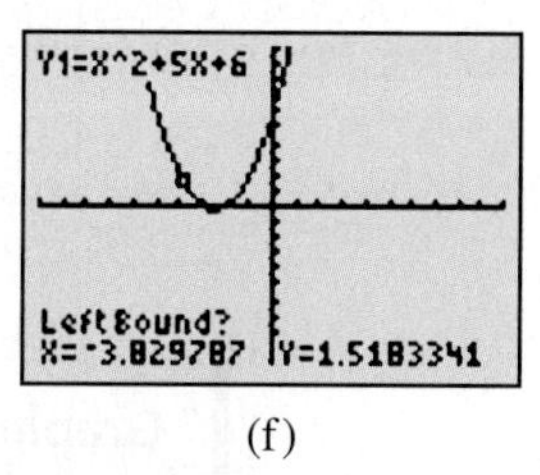

(f)

Press ENTER. The calculator asks "Right Bound?" Move the cursor to the right of the vertex. Press ENTER. The calculator asks "Guess?"

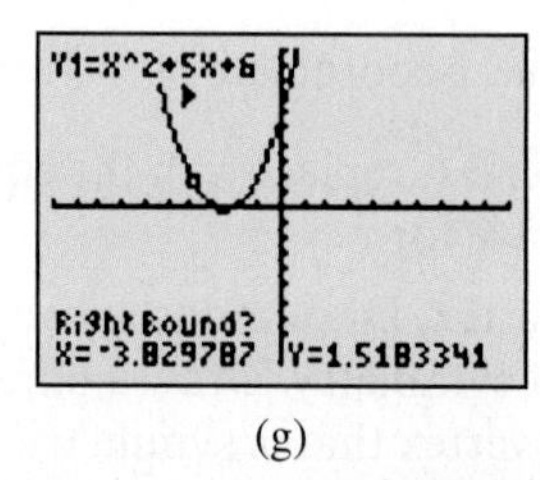

(g)

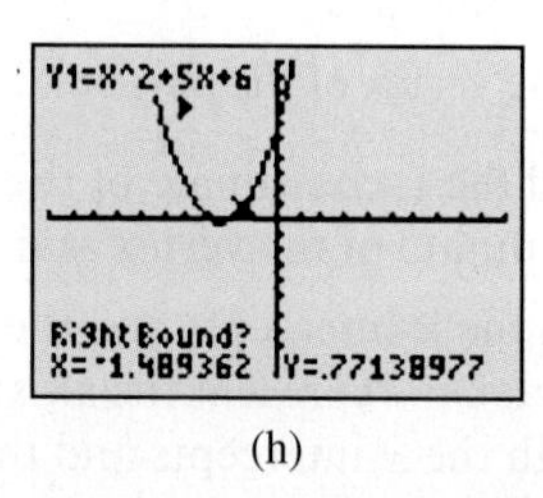

(h)

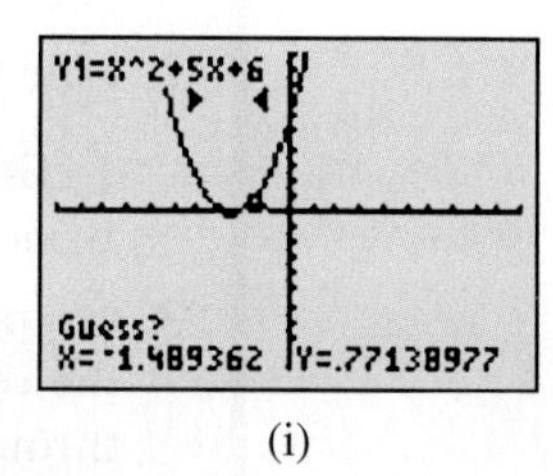

(i)

Move the cursor to the vertex. Press ENTER. The vertex of the parabola is $(-2.5, -0.25)$.

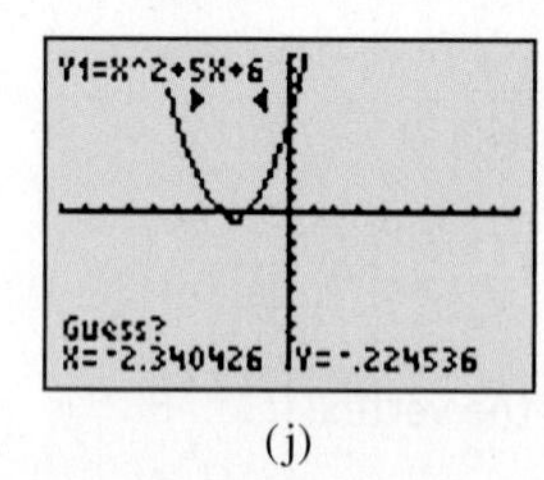

(j)

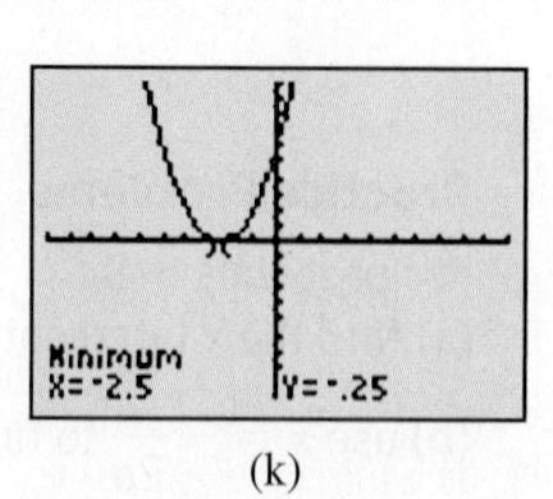

(k)

Practice Problems For problems 10–12,

(a) graph. Sketch the graph; describe the window.

(b) find the vertex.

10. $y = x^2 - 18x + 74$ **11.** $y = -x^2 - 8x - 10$ **12.** $y = x^2 - 9$

9.5 Vocabulary Practice

Match the term with its description.

1. $y = ax^2 + bx + c$
2. $x = -\dfrac{b}{2a}$
3. In $y = ax^2 + bx + c$, a is this.
4. The shape of the graph of a quadratic equation in two variables
5. If $a > 0$, the graph of a quadratic equation in two variables does this.
6. If $a < 0$, the graph of a quadratic equation in two variables does this.
7. The x-coordinate of this is 0.
8. The y-coordinate of this is 0.
9. The vertical line that passes through the vertex of the graph of a quadratic equation
10. The "bottom" or "top" point of the graph of a quadratic equation in two variables

A. axis of symmetry
B. lead coefficient
C. opens down
D. opens up
E. parabola
F. standard form of a quadratic equation in two variables
G. vertex
H. x-coordinate of the vertex
I. x-intercept
J. y-intercept

9.5 Exercises

Follow your instructor's guidelines for showing your work.

1. If the lead coefficient of $y = ax^2 + bx + c$ is a positive number, does the graph of the equation open up or open down?
2. If the lead coefficient of $y = ax^2 + bx + c$ is a negative number, does the graph of the equation open up or open down?

For exercises 3–24,
(a) identify a, b, and c.
(b) identify whether the graph opens up or down.
(c) use a table of solutions to graph the equation. Graph the axis of symmetry.
(d) identify the vertex.
(e) write the equation of the axis of symmetry.

3. $y = x^2 + 2$
4. $y = x^2 + 3$
5. $y = -x^2 + 2$
6. $y = -x^2 + 3$
7. $y = 2x^2$
8. $y = 4x^2$
9. $y = \dfrac{1}{4}x^2$
10. $y = \dfrac{1}{3}x^2$
11. $y = x^2 + 4x - 6$
12. $y = x^2 + 4x - 5$
13. $y = x^2 - 6x + 9$
14. $y = x^2 - 4x + 4$
15. $y = x^2 + 6x + 9$
16. $y = x^2 + 4x + 4$
17. $y = x^2 - 7$
18. $y = x^2 - 6$
19. $y = -x^2 + 2x + 4$
20. $y = -x^2 + 2x + 5$
21. $y = 3x^2 + 2$
22. $y = 3x^2 - 5$
23. $y = x^2 - 2x + 4$
24. $y = x^2 - 2x + 2$
25. Is the graph of $y = 5x^2$ shorter or taller than the graph of $y = 3x^2$? Explain.
26. Is the graph of $y = 6x^2$ shorter or taller than the graph of $y = 3x^2$? Explain.

For exercises 27–34, find the x-intercepts.

27. $y = x^2 - 10x + 21$
28. $y = x^2 - 9x + 20$
29. $y = 2x^2 + 15x - 8$
30. $y = 2x^2 + 3x - 5$
31. $y = 9x^2 - 16$
32. $y = 4x^2 - 25$
33. $y = x^2 + 4x$
34. $y = x^2 + 6x$

For exercises 35–50, use $x = -\dfrac{b}{2a}$ to find the vertex.

35. $y = x^2 - 10x + 15$
36. $y = x^2 - 8x + 10$
37. $y = x^2 + 10x + 15$
38. $y = x^2 + 8x + 10$
39. $y = x^2 - 7x$

40. $y = x^2 - 9x$

41. $y = x^2 - 6x + 9$

42. $y = x^2 - 4x + 4$

43. $y = 2x^2 + 12x + 6$

44. $y = 2x^2 + 16x + 9$

45. $y = 2x^2 - 12x + 6$

46. $y = 2x^2 - 16x + 9$

47. $y = 3x^2 - 12x + 5$

48. $y = 3x^2 - 6x + 4$

49. $y = -x^2 + 8x + 1$

50. $y = -x^2 + 6x + 3$

For exercises 51–58,
(a) find the x-intercepts.
(b) find the vertex.
(c) write the equation of the axis of symmetry.
(d) graph the x-intercepts, the vertex, and the axis of symmetry. Sketch the graph of the parabola.

51. $y = x^2 - 6x - 16$

52. $y = x^2 + 2x - 24$

53. $y = x^2 - 9$

54. $y = x^2 - 16$

55. $y = x^2 - 8x + 16$

56. $y = x^2 - 4x + 4$

57. $y = 4x^2 + 8x - 5$

58. $y = 4x^2 - 8x - 5$

Problem Solving: Practice And Review

Follow your instructor's guidelines for using the five steps as outlined in Section 1.3, p. 55.

59. The NCAA rules for women's basketball state that the rim of the hoop shall be 10 ft above the floor. The three-point line on the floor is 19 ft 9 in. away from a spot on the floor directly below the center of the hoop.
   a. Rewrite the measurements in inches.
   b. Use the Pythagorean theorem to find the diagonal distance in inches from the three-point line on the floor to the center of the hoop. Round to the nearest whole number.
   c. Change the distance in part b into feet. Round to the nearest tenth.

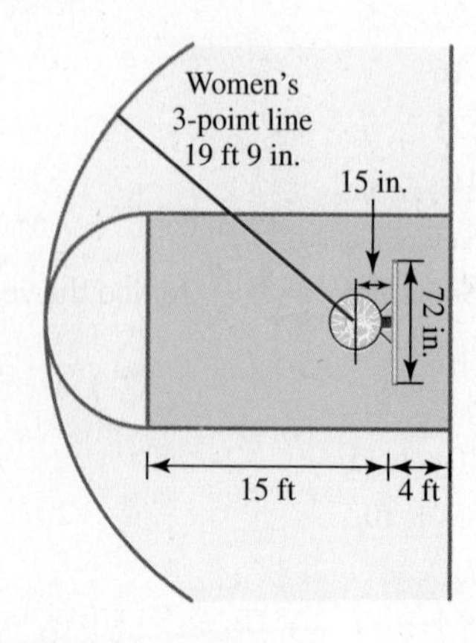

60. Find the average rate of change in new Alzheimer's disease cases per year between 2000 and 2010.

In 2000, there were an estimated 411,000 new cases of Alzheimer's disease. By 2010, that number is expected to increase to 454,000 new cases per year; by 2029, to 615,000; and by 2050, to 959,000. (*Source:* www.alz.org)

61. Scale is removed from steel before galvanizing with an 18% hydrochloric acid solution. Hydrochloric acid solution is delivered by rail in a tanker. This solution is 31.45% hydrochloric acid. Find the amount of water and the amount of acid solution needed to make 30,000 gallons of a new solution that is 18% hydrochloric acid. Round to the nearest whole number.

62. The NFL quarterback passing rating formula is

$$R = 100\left(\frac{\left[\dfrac{100\left(\frac{C}{A}\right) - 30}{20}\right] + \left[(0.25)\left(\frac{Y}{A} - 3\right)\right] + \left[20\left(\frac{T}{A}\right)\right] + \left[2.375 - \left(25\frac{I}{A}\right)\right]}{6}\right)$$

The number of attempted passes is A, the number of completed passes is C, the number of touchdown passes is T, the number of interceptions is I, and the total yards from passes is Y. In 1994, Steve Young completed 324 of 461 passes for 3969 yards, 35 touchdowns, and 10 interceptions. Find his passing rating. Round to the nearest tenth.

Technology

For exercises 63–66, graph each equation on a graphing calculator. Choose a window that shows the vertex.
(a) Sketch the graph; describe the window.
(b) Use the calculator to identify the vertex.

63. $y = x^2 - 10x + 29$

64. $y = -x^2 - 12x - 38$

65. $y = -3x^2 - 18x - 34$

66. $y = 2x^2 + 4x + 5$

Find The Mistake

For exercises 67–70, the completed problem has one mistake.
(a) Describe the mistake in words, or copy down the whole problem and highlight or circle the mistake.
(b) Do the problem correctly.

67. Problem: Use a table of solutions to graph $y = -x^2 + 3$.

Incorrect Answer:

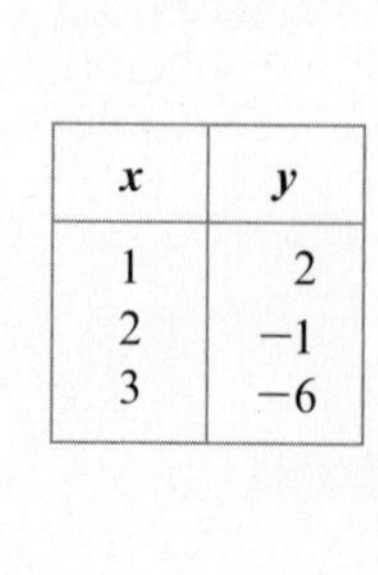

x	y
1	2
2	−1
3	−6

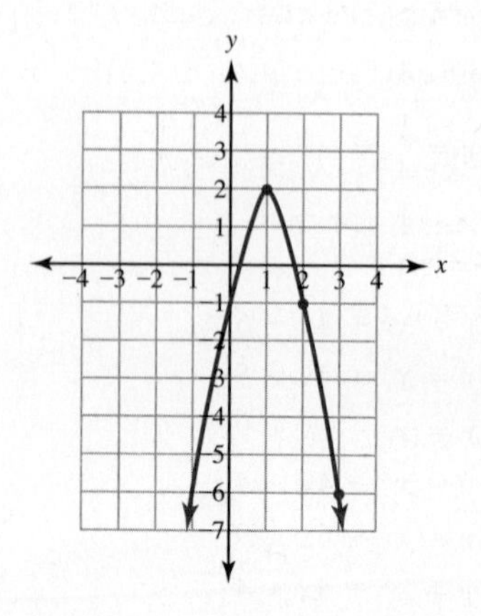

68. **Problem:** Use $x = -\frac{b}{2a}$ to find the vertex of $y = x^2 + 6x - 10$.

Incorrect Answer: $x = -\frac{b}{2a}$

$x = -\frac{6}{2(1)}$

$x = -3$

The vertex is $x = -3$.

69. **Problem:** Find the x-intercepts of the graph of $y = x^2 + 2x - 24$.

Incorrect Answer: $0 = (x - 6)(x + 4)$

$x - 6 = 0$ or $x + 4 = 0$

$\underline{+6 \quad +6}$ $\quad$ $\underline{-4 \quad -4}$

$x + 0 = 6$ $\quad$ $x + 0 = -4$

$x = 6$ $\quad$ $x = -4$

The x-intercepts are $(6, 0)$ and $(-4, 0)$.

70. **Problem:** The vertex of a parabola that opens up is $(4, -5)$. Write the equation of the axis of symmetry.

Incorrect Answer: $y = -5$

Review

71. Graph: $2x - y > 18$

72. Evaluate: $\frac{7}{0}$

73. Write the equation of the graphed line.

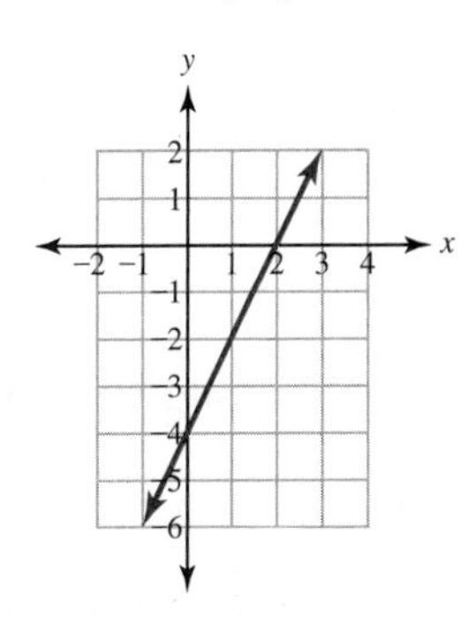

74. Find the slope of the line that passes through $(-8, 5)$ and $(-1, -3)$.

SUCCESS IN COLLEGE MATHEMATICS

75. Using a test from this semester, copy the questions for which you did not receive full credit on a fresh piece of paper. Complete this practice test. You might need to go to a tutoring center or visit your instructor during office hours to make sure that you have done them correctly.

Study Plan for Review of Chapter 9

SECTION 9.1 Quadratic Equations in One Variable

Ask Yourself	Test Yourself	Help Yourself
Can I . . .	**Do 9.1 Review Exercises**	**See these Examples and Practice Problems**
solve a quadratic equation in the form $x^2 = k$ and check the solution(s)?	1–6	Ex. 1–7, PP 1–10
rewrite a quadratic equation in the form $x^2 = k$?	3–6	Ex. 4–7, PP 5–10
use a quadratic equation in the form $x^2 = k$ to solve an application problem?	7–8	Ex. 8–9, PP 11–12

9.1 Review Exercises

For exercises 1–6,
(a) solve.
(b) check.

1. $x^2 = 36$
2. $z^2 = 45$
3. $8v^2 = 56$
4. $54 = \frac{1}{2}k^2$
5. $u^2 + 15 = 35$
6. $21 - x^2 = -28$

For exercises 7–8, use the five steps.

7. Use the Pythagorean theorem to find the height of the top of the ladder above the ground in *inches*. Round to the nearest whole number.

© Laura Bracken

8. Passengers on the Giant Drop Ride at Dreamworld in Australia sit in open air eight-seat gondolas. It takes 90 s for the gondolas to travel from the ground to the top. The gondola then free-falls for about 400 ft. A magnetic braking system stops the gondolas before they hit the ground. Use the formula $d = \left(\frac{16\text{ ft}}{1\text{ s}^2}\right)t^2$ to find the time to free fall 400 ft. (*Source:* www.dreamworld.com.au)

9. Explain why the diagonal drawn between opposite corners of a door is the hypotenuse of a right triangle.

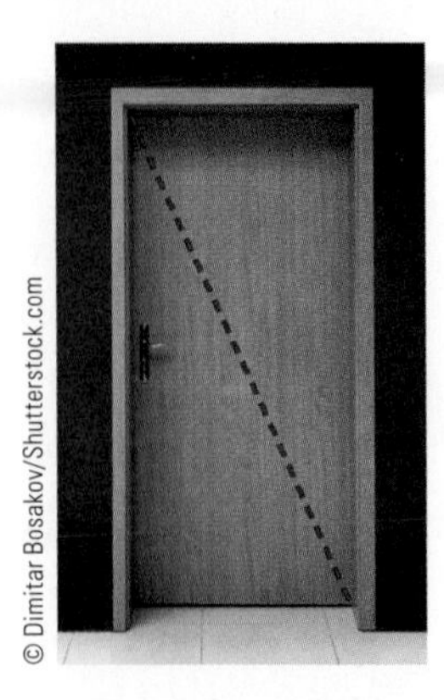
© Dimitar Bosakov/Shutterstock.com

SECTION 9.2 Completing the Square

Ask Yourself	Test Yourself	Help Yourself
Can I...	**Do 9.2 Review Exercises**	**See these Examples and Practice Problems**
add a term to a binomial to create a perfect square trinomial?	10–11	Ex. 2–4, PP 1–4
solve a quadratic equation with rational solutions by completing the square?	12–13	Ex. 5, 6, PP 5–7
solve a quadratic equation with irrational solutions by completing the square?	14–16	Ex. 7, PP 8–10

9.2 Review Exercises

For exercises 10–11, identify the constant needed to complete the square.

10. $x^2 + 18x$

11. $z^2 - 24z$

For exercises 12–16, solve by completing the square.

12. $x^2 + 4x + 3 = 0$

13. $z^2 - 4z - 45 = 0$

14. $p^2 + 6p - 37 = 0$

15. $n^2 - 12n - 54 = 0$

16. $x^2 - 10x = 8$

SECTION 9.3 The Quadratic Formula

Ask Yourself	Test Yourself	Help Yourself
Can I...	**Do 9.3 Review Exercises**	**See these Examples and Practice Problems**
use the quadratic formula to solve a quadratic equation?	17–20, 27	Ex. 1–4, 8, PP 1–7
find the discriminant and use it to describe the solutions of a quadratic equation?	21–26, 28, 50	Ex. 5, PP 8–11
use a quadratic equation to solve an application problem?	36	Ex. 6–7, PP 12–13

9.3 Review Exercises

For exercises 17–20, use the quadratic formula to solve the equation.

17. $x^2 + 10x + 16 = 0$

18. $2x^2 + 2x - 21 = 0$

19. $5x^2 + 7x - 3 = 0$

20. $x^2 - 15x - 2 = 0$

For exercises 21–24, use the discriminant to describe the solutions of each equation.

21. $x^2 + 22x + 121 = 0$

22. $x^2 + 19x + 88 = 0$

23. $2x^2 + 15x + 21 = 0$

24. $3x^2 + 7x + 21 = 0$

25. Explain why a quadratic equation with a discriminant that equals 0 has only one solution.

26. Explain why a quadratic equation with a discriminant that is less than 0 has nonreal solutions.

27. Identify the solutions of the quadratic equation $ax^2 + bx + c = 0$.

28. Identify the discriminants that are from quadratic equations with two irrational solutions.

a. 12 **b.** 9
c. -3 **d.** 0
e. 21 **f.** -100

For exercises 29–35, use any method from this chapter to solve.

29. $a^2 + 9a = 0$

30. $x^2 = 48$

31. $x^2 + 6x + 7 = 0$

32. $2p^2 + 5p - 7 = 0$

33. $3x^2 = 33$

34. $3n^2 - 9 = 0$

35. $c^2 - 12c + 20 = 0$

36. A drinking fountain spouts water approximately in the shape of a parabola. The times at which the water reaches a height of 0.3 ft are the solutions of the equation $0.3 = 5.65t - 16t^2$, where t is in seconds. Find these values for t. Round to the nearest hundredth.

© Kris Schmidt/Shutterstock.com

SECTION 9.4 Quadratic Equations with Nonreal Solutions

Ask Yourself	Test Yourself	Help Yourself
Can I...	**Do 9.4 Review Exercises**	**See these Examples and Practice Problems**
simplify a radical in which the radicand is less than 0?	37–38	Ex. 1–2, PP 1–4
determine whether a number is an element of a given set?	41–42, 51	Ex. 3, PP 5
add and subtract complex numbers?	39–40	Ex. 4–6, PP 6–8
solve a quadratic equation with nonreal solutions?	43–49	Ex. 7–9, PP 9–17

9.4 Review Exercises

For exercises 37–40, simplify.

37. $\sqrt{-16}$

38. $\sqrt{-48}$

39. $(5 + 3i) + (2 - 9i)$

40. $(8 + 2i) - (17 - 3i)$

41. Identify the numbers that are elements of the set of imaginary numbers.
- **a.** 13
- **b.** $\sqrt[3]{-8}$
- **c.** $2i$
- **d.** $3 - 4i$
- **e.** $\sqrt{-8}$

42. **a.** Is every real number a complex number?
b. Is every imaginary number a complex number?
c. Is every complex number a real number?

For exercises 43–45,
(a) solve.
(b) check.

43. $n^2 = -36$

44. $h^2 - 15 = -96$

45. $-2x^2 = 36$

For exercises 46–47, solve by completing the square.

46. $w^2 + 10w + 30 = 0$

47. $x^2 + 6x + 21 = 0$

For exercises 48–49, use the quadratic formula to solve the equation.

48. $x^2 + 3x + 5 = 0$

49. $5x^2 + x + 12 = 0$

50. Choose the discriminants for quadratic equations with two nonreal solutions.
- **a.** 12
- **b.** 9
- **c.** -3
- **d.** 0
- **e.** 21
- **f.** -100

51. Copy the chart below. Put an X in the box if the number is an element of the set at the top of the column.

Chart for exercise 51

	Complex numbers	Imaginary numbers	Real numbers	Rational numbers	Irrational numbers	Integers	Whole numbers
8							
-10							
$-\frac{1}{3}$							
0.21							
$\sqrt{5}$							
0							
$3i$							
$\sqrt{-7}$							
$9 - 2i$							

SECTION 9.5 Quadratic Equations in the Form $y = ax^2 + bx + c$

Ask Yourself	Test Yourself	Help Yourself
Can I...	**Do 9.5 Review Exercises**	**See these Examples and Practice Problems**
graph a quadratic equation in the form $y = ax^2 + bx + c$, identify the vertex, and write the equation of the axis of symmetry?	52–54	Ex. 1–3, PP 1–4
describe the effect of a on the graph of quadratic equation?	55, 56	Ex 3, PP 5
given a quadratic equation, find the x-intercepts, use $x = -\frac{b}{2a}$ to find its vertex, and sketch its graph?	57–59	Ex. 4–6, PP 6–9

9.5 Review Exercises

For exercises 52–54,
(a) identify *a*, *b*, and *c*.
(b) use a table of solutions to graph the equation. Graph the axis of symmetry.
(c) identify the vertex.
(d) write the equation of the axis of symmetry.

52. $y = x^2 - 10$

53. $y = -x^2 - 4$

54. $y = x^2 - 4x - 5$

55. Identify the quadratic equations with a graph that opens down.
a. $y = x^2 + 5x + 6$ **b.** $y = -x^2 + 5x + 6$
c. $y = -x^2 + 5$ **d.** $y = x^2$

56. Identify the quadratic equation with the "tallest" graph.
a. $y = \frac{1}{2}x^2$ **b.** $y = x^2$
c. $y = 2x^2$ **d.** $y = 3x^2$

57. Find the x-intercepts of the graph of $y = x^2 - x - 56$.

58. Use $x = -\frac{b}{2a}$ to find the vertex of the graph of $y = 2x^2 - 10x + 4$.

59. A quadratic equation is $y = x^2 + 8x + 7$.
a. Find the x-intercepts.
b. Find the vertex.
c. Write the equation of the axis of symmetry.
d. Graph the x-intercepts, the vertex, and the axis of symmetry. Sketch the graph of the parabola.

Chapter 9 Test

For problems 1–3,
(a) rewrite the equation in the form $x^2 = k$.
(b) solve.
(c) check.

1. $3x^2 - 15 = 12$

2. $5z^2 = 15$

3. $u^2 + 40 = 33$

For problems 4–5, solve by completing the square.

4. $x^2 + 8x - 13 = 0$

5. $n^2 + 2n - 15 = 0$

For problems 6–7, use the quadratic formula to solve.

6. $x^2 - 7x + 6 = 0$

7. $6x^2 + x - 8 = 0$

For problems 8–10, use the discriminant to describe the solutions of the equation.

8. $x^2 + 9x - 22 = 0$

9. $3x^2 - x - 9 = 0$

10. $x^2 + 15 = 0$

11. A quadratic equation is $y = x^2 + 8x + 12$.
a. Identify *a*, *b*, and *c*.
b. Use a table of solutions to graph the equation. Graph the axis of symmetry.
c. Identify the vertex.
d. Write the equation of the axis of symmetry.

12. A quadratic equation is $y = x^2 + 10x + 16$.
a. Find the x-intercepts.
b. Find the vertex.
c. Write the equation of the axis of symmetry.
d. Graph the x-intercepts, the vertex, and the axis of symmetry. Sketch the graph of the parabola.

13. A utility company needs to replace the guy wire on a utility pole. Find the length of the guy wire in feet. Round to the nearest tenth.

Cumulative Review Chapters 7–9

For exercises 1–17, simplify.

1. $\sqrt{12x^7y^3}$
2. $\dfrac{x^2 + 9x + 20}{x^2 - 16}$
3. $\dfrac{3p}{p^2 + 3p - 10} - \dfrac{6}{p^2 + 3p - 10}$
4. $\dfrac{a^2 - 1}{3a + 3} \div \dfrac{a - 1}{6}$
5. $\sqrt{18} - \sqrt{72} + \sqrt{50}$
6. $\dfrac{x}{x + 1} - \dfrac{4}{x^2 + 6x + 5}$
7. $\dfrac{8c}{c - 3} + \dfrac{5}{9 - c^2}$
8. $\sqrt{15xy}\sqrt{30xz}$
9. $\dfrac{x^2 - 6x}{x^2 - 14x + 48} \cdot \dfrac{x^2 + x - 72}{x^2 + 7x}$
10. $\dfrac{\dfrac{3}{w^2} - \dfrac{4}{w}}{\dfrac{w}{6}}$
11. $\sqrt{\dfrac{18y^3}{25x^2}}$
12. $\dfrac{x + 5}{\sqrt{6x}}$
13. $\dfrac{k}{\sqrt{k} + 2}$
14. $\dfrac{28a^5b^2c}{40a^2bc^8}$
15. $\sqrt[3]{40x^6y^7}$
16. $\sqrt{-16}$
17. $\sqrt{-98}$

For exercises 18–29,
(a) solve.
(b) check.

18. $x^2 - 11x + 24 = 0$
19. $\sqrt{x} + 9 = 15$
20. $\sqrt{x + 9} = 15$
21. $\sqrt{p - 14} = 5$
22. $a^2 - 8 = 28$
23. $\dfrac{6}{n + 3} + \dfrac{20}{n^2 + n - 6} = \dfrac{5}{n - 2}$
24. $\dfrac{1}{x - 1} + \dfrac{2}{x} = \dfrac{x}{x - 1}$
25. $x^2 + 3x - 8 = 0$
26. $\dfrac{5}{x} = \dfrac{3}{8}$
27. $x = \sqrt{7x + 18}$
28. $3p^2 - 12p = 0$
29. $\dfrac{2}{x} + \dfrac{x}{8} = \dfrac{x}{4}$
30. Use the quadratic formula to solve $x^2 - 9x + 7 = 0$.
31. Solve $x^2 - 12x - 15 = 0$ by completing the square.
32. A quadratic equation is $4x^2 + 9x + 15 = 0$.
    a. Find the discriminant.
    b. Use the discriminant to describe the solutions.
33. A formula for a quick estimation of the body surface area of a human adult is $B = \dfrac{1}{6}\sqrt{WH}$, where B is the body surface area in square meters, H is body height in meters, and W is body weight in kilograms. Find the body surface area of an adult who is 6 ft tall and weighs 210 lb. (1 ft ≈ 0.305 m; 1 lb. ≈ 0.454 kg.) Round to the nearest tenth. (*Source:* Reading et al., *Clinical Anatomy* 18: 126–130, 2005)
34. The length of one leg of a right triangle is 8 in. The length of the other leg is 15 in. Use the Pythagorean theorem to find the length of the hypotenuse.
35. The perimeter of a rectangle is 46 in. Its area is 120 in.^{2}. Use a quadratic equation in one variable to find the length and width of this rectangle.
36. The length of a rectangle is 1 in. less than four times its width. The area of the rectangle is 95 in.^{2}. Use a quadratic equation to find the length and width of this rectangle.
37. Copy the chart below. Put an X in the box if the number is an element of the set at the top of the column.

Chart for exercise 37

	Complex numbers	Imaginary numbers	Real numbers	Rational numbers	Irrational numbers	Integers	Whole numbers
$\sqrt{5}$							
5							
$\sqrt{-5}$							
0.5							
$5i$							
$\dfrac{1}{5}$							

38. a. Use a table of solutions to graph $y = x^2 - 5$.
b. Use the graph to identify the vertex.

39. Use an algebraic method to find the x-intercepts of the graph of $y = x^2 + 5x - 14$.

For exercises 40–41, use the five steps.

40. A meat grinder can grind the meat for one batch of sausage in 5 hr 10 min. Another kind of meat grinder can grind the same amount of meat in 6 hr 48 min. If both grinders are used, find the time needed to grind the meat. Round to the nearest minute.

41. Find the number of 1850 college students who would report binge drinking.

The pressure to use alcohol, drugs and cigarettes can be huge for some college students, especially when trying to make friends and become part of a group. Drinking on some college campuses is more pervasive and destructive than many people realize. Studies show that four out of five college students drink alcohol. Two out of five report binge drinking (defined as five or more drinks for men and four or more for women in one sitting). One in five students reports three or more binge episodes in the prior two weeks. (*Source:* www.cdc.gov, Jan. 15, 2009)

Cumulative Review Chapters 1–9

1. Describe the order of operations.
2. Evaluate: $(27 \div 3^2) + 4 - 18 \div 2 \cdot 3$
3. Evaluate: $\frac{3}{4} + \frac{5}{9} \div \frac{1}{3}$
4. Simplify: $6(3x - 5) - (20x - 4)$
5. Change 8 ft into inches.
6. Describe the difference between a rational number and an irrational number.
7. Find the area of a circle with a diameter of 8 cm. ($\pi \approx 3.14$.)
8. The width of a rectangle is 8 m, and the length is 12 m. Find the area of the rectangle.
9. Explain why the square root of a negative number is not a real number.
10. Clear the fractions from $\frac{2}{3}x - \frac{4}{5} = \frac{1}{6}x + 2$. Do not solve.
11. Solve $5x - 9y = 15$ for y.

For exercises 12–15, solve.

12. $8x - 14 = 3x - (x + 2)$
13. $\frac{3}{5}x - 9 = 21$
14. $2(n + 3) + 5n = 7(n - 4)$
15. $3p = -7p$

For exercises 16–18,
(a) solve.
(b) use a number line graph to represent the solution.

16. $-3x + 7 < 19$
17. $5(x - 1) \geq 7(x + 3)$
18. Find the y-intercept and the x-intercept of the line represented by $9x + 4y = -27$.

For exercises 19–24, graph.

19. $3x - 5y = 21$
20. $x = 4$
21. $y = -2$
22. $y = -\frac{5}{6}x + 4$
23. $9x - 2y < 18$
24. $y \geq 4x - 6$
25. Find the slope of the line that passes through $(-5, -9)$ and $(4, 12)$.
26. Find the slope of the line represented by $7x - 3y = 18$.
27. Write the equation in slope-intercept form that passes through the points $(1, 5)$ and $(3, 13)$.
28. Identify the slope of a line that is parallel to the line $y = 7x - 4$.
29. Write the equation in slope-intercept form of the graphed line.

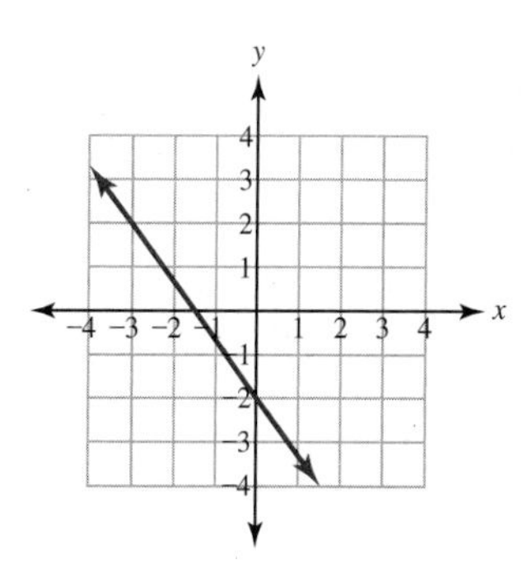

30. The graph represents the relationship of the number of years since 1980, x, and the number of end-stage renal disease patients in the United States. Write a linear equation that represents this relationship, and use it to find the number of end-stage renal disease patients in 2015.

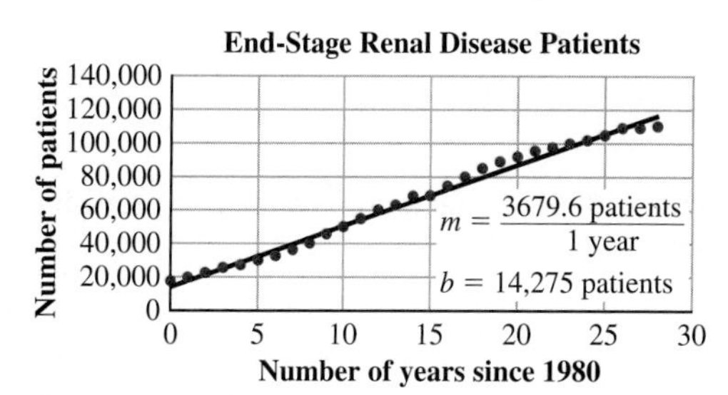

Source: www.cdc.gov, 2011

31. Explain why not every relation is a function.
32. If $f(x) = x^2 - 6x + 2$, evaluate $f(-4)$.
33. Use graphing to solve the system of equations $\begin{array}{l} y = 2x - 3 \\ y = -x + 9 \end{array}$.
34. Solve $\begin{array}{l} y = -3x + 1 \\ 4x + 5y = 27 \end{array}$ by substitution.
35. Solve $\begin{array}{l} 3x + 4y = 11 \\ 7x - 6y = 18 \end{array}$ by elimination.
36. Solve $\begin{array}{l} y = \frac{3}{4}x - 5 \\ 3x - 4y = 18 \end{array}$ by elimination.
37. Graph the system $\begin{array}{l} y \leq -\frac{2}{3}x + 6 \\ x \geq 0 \\ y \geq x + 1 \end{array}$.

For exercises 38–41, simplify.

38. $x^5y^3x^7$
39. $(a^3b^4)^2$
40. $\frac{u^9}{u^4}$
41. $\frac{3n^2}{8n^{11}}$

For exercises 42–43, rewrite the measurement in scientific notation.

42. 7,000,000 kg
43. 0.00061 m

For exercises 44–46, evaluate. Write the answer in scientific notation.

44. $(8 \times 10^5)(4 \times 10^6)$

45. $\dfrac{8 \times 10^6}{2 \times 10^{-3}}$

46. $9 \times 10^5 \text{ kg} + 3 \times 10^4 \text{ kg}$

For exercises 47–51, simplify.

47. $(3x^2 + 8x - 9) - (5x^2 - 6x - 1)$

48. $(6x - 5)(3x + 9)$

49. $(2x + 1)(5x^2 - 3x - 4)$

50. $(14x^3 + 12x^2 - 8x) \div 2x$

51. $(x^2 - 7x - 18) \div (x - 2)$

For exercises 52–58, factor completely.

52. $x^2 - 9$

53. $a^2 - 9a + 18$

54. $n^3 - 4n^2 - 5n$

55. $4x^2 + 12x + 9$

56. $15ac - 21a + 25c - 35$

57. $8n^2 + 2p - 6w + 10$

58. $6x^2 + 13x - 5$

For exercises 59–61, use factoring and the zero product property to solve.

59. $x^2 + 8x - 9 = 0$

60. $c^3 - 16c = 0$

61. $12a^2 - 3a = 0$

For exercises 62–65, simplify.

62. $\dfrac{a^2 - 2a - 15}{a^2 - 4a - 5}$

63. $\dfrac{2d + 4}{4} \div \dfrac{d^2 - 4}{6}$

64. $\dfrac{1}{x - 1} + \dfrac{x}{x^2 - 1}$

65. $\dfrac{n}{n^2 + 4n + 3} - \dfrac{3}{n^2 - 4n - 5}$

66. Solve: $\dfrac{4}{x - 4} = \dfrac{x}{x - 4} - \dfrac{4}{3}$

For exercises 67–72, simplify.

67. $\sqrt{240n^3xy^6}$

68. $\sqrt{8x} + \sqrt{18x}$

69. $\sqrt{\dfrac{50a^3}{9b^2}}$

70. $(\sqrt{2x} + \sqrt{5})(\sqrt{2x} - \sqrt{5})$

71. $\sqrt[3]{54a^7b^6}$

72. $\left(x^{\frac{3}{4}}\right)^{\frac{5}{6}}$

73. Rewrite $\sqrt[5]{a^4}$ in exponential notation.

74. Rewrite $n^{\frac{2}{3}}$ in radical notation.

75. Solve: $x^2 + 6 = 31$

76. Solve $x^2 + 8x - 5 = 0$ by completing the square.

77. Use the quadratic formula to solve $2x^2 + 5x - 1 = 0$.

78. Simplify: $\sqrt{-15}$

79. Simplify: $\sqrt{-20}$

80. Solve: $x^2 = -17$

81. Use a table of solutions to graph $y = x^2 - 4x + 6$.

82. Use $x = -\dfrac{b}{2a}$ to find the vertex of the graph of $y = x^2 - 6x + 13$.

For exercises 83–99, use the five steps.

83. Find the number of 35,500 young adults who believe that graduates today have more debt than they can manage.

"Three in four Americans now say that college is too expensive for most people to afford," Mr. Duncan said. "That belief is even stronger among young adults—three-fourths of whom believe that graduates today have more debt than they can manage." (*Source:* www.nytimes.com, Nov. 29, 2011)

84. A computer is marked down 18% to a sale price of \$650. Find the regular price. Round to the nearest whole number.

85. A college pool is 25 yd long. Find the number of lengths of the pool that a student must swim in order to swim 1 mi. (1 yd = 3 ft; 1 mi = 5280 ft.) Round to the nearest whole number.

86. A home business accepts PayPal payments from buyers. The business pays PayPal a 2.9% transaction fee on the total sale amount plus a fee of \$0.30 per transaction. Find the total amount paid by the business for 18 transactions with a total sale amount of \$1500.

87. A student works 32 hours per month at a job paying \$7.25 per hour after withholding. The student's bills per month are \$800. Find the number of hours he needs to work per month at a second job that pays \$9.50 per hour after withholding so that he can pay his bills. Round to the nearest whole number.

88. The equation $y = \left(\dfrac{\$307.60}{1 \text{ year}}\right)x + \$4006.20$ represents the relationship of the number of years since 1994, x, and the average fee for service payment per enrollee in Medicare, y, Find the average fee for service payment in 2013. Round to the nearest ten. (*Source:* www.cdc.gov)

89. A manufacturer has some steel that is 12% nickel and some steel that is 15% nickel. Use a system of linear equations to find the amount of each kind of steel needed to make 2500 lb of steel that is 14% nickel. Round to the nearest whole number.

90. The cost to make a product is \$25. The fixed overhead costs per month are \$60,000. The price of the product is \$65. Use a system of linear equations to find the break-even point.

91. The perimeter of a rectangle is 24 in. The length is 3 in. less than twice the width. Use a system of linear equations to find the length and width of the rectangle.

92. If the side supports on a swing set are 10 ft long and the distance between the bases of the side supports is 5 ft, find the height of the swing set. Round to the nearest tenth.

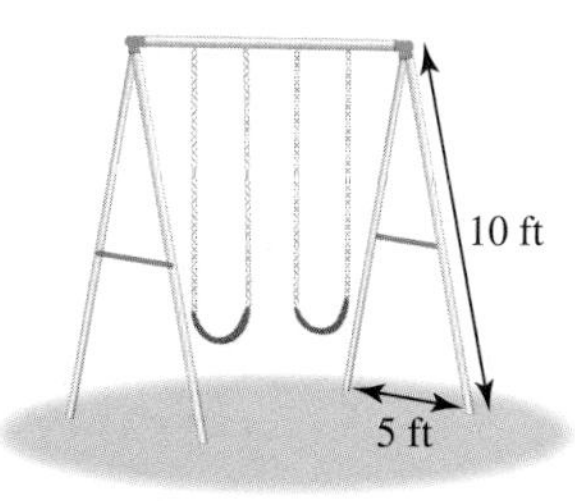

93. In 2009–2010, fiscal support for higher education by the state of Arizona was $1,088,561,900. In 2011–2012, this support decreased to $814,457,600. Find the percent decrease. Round to the nearest percent. (*Source:* grapevine .illinoisstate.edu, Jan. 2012)

94. A take-out order cost $15.50. A bicycle delivery worker in New York City received a tip of $2 for delivering the order. Find the percent of the amount of the order that the worker received as a tip. Round to the nearest tenth of a percent. (*Source:* www.nytimes.com, March 2, 2011)

95. Find the number of intravenous drug users interviewed who had a positive HIV test result and who were unaware of their infection. Round to the nearest whole number.

This report summarizes data from 10,073 IDUs [intravenous drug users] interviewed and tested in 20 MSAs [metropolitan statistical areas] in 2009. Of IDUs tested, 9% had a positive HIV test result, and 45% of those testing positive were unaware of their infection. (*Source:* www.cdc .gov, March 2, 2012)

96. In Microsoft Word 2010, the standard page size is 8.5 in. wide and 11 in. long. In the *moderate* page layout, the top and bottom margins are each 1 in. high, and the left and right margins are each 0.75 in. wide. Find the area of the margins.

97. A stainless steel cylindrical roller is used to make chimney cakes. The dough is wrapped in a spiral around the roller and baked in an oven. The roller is 50 cm long, and its diameter is 5 cm. Find the surface area of the cylinder. ($A = 2\pi rh$; $\pi \approx 3.14$.)

© foodfolio/Alamy

98. Find the maximum volume of a basketball that meets NCAA regulations. $\left(V = \frac{4\pi r^3}{3}; \pi \approx 3.14.\right)$ Round to the nearest tenth.

The N.C.A.A. rule book mandates that basketballs must meet certain seemingly self-evident requirements. . . Section 15, Article 1 says, "the ball shall be spherical." . . . The finer points state that basketballs used in men's games can be a maximum 30 inches and a minimum 29 ½ inches in circumference. (*Source:* www.nytimes.com, March 1, 2012)

99. In a study of foodborne disease outbreaks from 1993–2006 caused by contaminated dairy products, 1571 illnesses resulted from nonpasteurized dairy products; these made up 36% of the total illnesses. Find the number of total illnesses. Round to the nearest whole number. (*Source:* Langer et al., *Emerging Infectious Diseases,* March 2012)

APPENDIX 1

Reasonability and Problem Solving

To find the amount of an antibiotic solution to inject into the muscle of a patient, a nursing student wrote and solved the equation $A = \left(\frac{500}{1.8}\right)(250\text{ mL})$. She looked at her answer of 69,444 mL and thought, "That can't be right!" Injecting 69,444 mL (about 35 large bottles of soda) into someone's muscle is not reasonable. *Looking back* and judging the reasonability of a proposed answer make up an essential part of solving application problems.

Reasonability

Part of looking back is deciding whether an answer is reasonable; a reasonable answer makes sense and seems possible. We can use our experience to tell us that an answer such as a car traveling a distance of 600 mi in a time of 5 min is *not* reasonable. But how do we convince someone else that our answer *is* reasonable?

In Example 1, we use *estimation* and *working backwards* to explain why the answer seems reasonable. This and the following examples do not give explanations of why the original thinking and/or the equations used to solve the problems are correct. Instead, they give explanations of why the proposed answers are reasonable solutions to the problems.

EXAMPLE 1 **Problem:** The suggested retail price of a Ford Focus sedan is $17,505. The dealer will sell the car for $16,499. Find the change in price.

Equation: $c = \$17{,}505 - \$16{,}499$

Answer: The change in price is $1006.

Explain why this answer is reasonable. To *estimate*, round the numbers in the problem so the arithmetic can be redone quickly, usually without a calculator. Rounding, $17,505 is about $17,500, and $16,499 is about $16,500. Since the difference between $17,500 and $16,500 is $1000 and $1000 is very close to the answer of $1006, the answer seems reasonable.

Working backwards from the answer to the original numbers, the sum of the change in price and the sale price should be the original price. Since $\$1006 + \$16{,}499$ equals $17,505, a change in price of $1006 seems reasonable.

In some situations, *estimation* is not a good method for checking. In the next example, we show that the answer seems reasonable by *working backwards* and by *doing the problem another way.*

EXAMPLE 2 **Problem:** At a warehouse store, a student can buy hamburger meat in 10-lb packages. How many 0.25-lb hamburgers can be made from this package?

Equation: $p = 10\text{ lb} \div 0.25\text{ lb}$

Answer: The student can make 40 hamburgers from this package.

Explain why this answer is reasonable. *Estimation* does not work well here. If we round 0.25 lb to 0.2 lb, the arithmetic is not any easier. If we round it to 0 lb, it does not work at all.

Working backwards, since $(0.25 \text{ lb})(40) = 10 \text{ lb}$, the answer of 40 hamburgers seems reasonable.

Doing the problem another way, if each hamburger is 0.25 lb (a quarter pounder), then there are four hamburgers in each pound. Since $\left(\frac{4 \text{ hamburgers}}{1 \text{ lb}}\right)(10 \text{ lb})$ equals 40 hamburgers, the answer seems reasonable.

In baseball, a ball in play can be anywhere *in the ballpark:* in the outfield, the infield, or in the catcher's mitt. A reasonable answer is also *in the ballpark.* We are checking whether the answer is within an acceptable or expected range of answers.

EXAMPLE 3 **Problem:** The price of a textbook is \$114. The sales tax rate is 6.5%. Find the total cost to buy the textbook.

Equation: $p = \$114 + (0.065)(\$114)$

Answer: The total cost is \$121.41.

Explain why this answer seems reasonable. Since the original price of the book, without tax, is \$114, the bottom end of the *ballpark* is \$114.

The total cost of the book is \$114 plus a tax of 6.5%. Since 6.5% is less than 10%, the total cost of the book must be less than \$114 plus a tax of 10%. (Why choose 10%? To find 10% of a number, just divide it by 10.) Since 10% of \$114 is \$11.40. and $\$114 + \$11.40 = \$125.40$, the top end of the *ballpark* is \$125.40.

Since the answer of \$121.41 is between \$114 and \$125.40, it is *in the ballpark,* and it seems reasonable.

For work with formulas, changing the value of just one of the variables in the formula can establish an expected range of reasonable answers.

EXAMPLE 4 **Problem:** A rectangular room is 6 ft wide and 12 ft long. Find its area.

Formula and Equation: area = (width)(length); $A = (6 \text{ ft})(12 \text{ ft})$

Answer: The area of the room is 72 ft^2.

Explain why this answer seems reasonable. To show that the answer is *in the ballpark,* choose a value for the length that is less than 12 ft for the bottom of the ballpark, 10 ft. Choose a value for the length that is greater than 12 ft for the top of the ballpark, 15 ft. Do not change the width.

If the length is 10 ft, the area of the room is 60 ft^2. If the length is 15 ft, the area is 90 ft^2. Since 72 ft^2 is between 60 ft^2 and 90 ft^2, the answer seems reasonable.

To *do the problem another way,* divide the rectangle into twelve strips that are each 6 ft wide and 1 ft long. The area of each strip is $(6 \text{ ft})(1 \text{ ft})$, which equals 6 ft^2, and the total area is $(12)(6 \text{ ft}^2)$, which equals 72 ft^2. Since this is the same as the original answer, it seems reasonable.

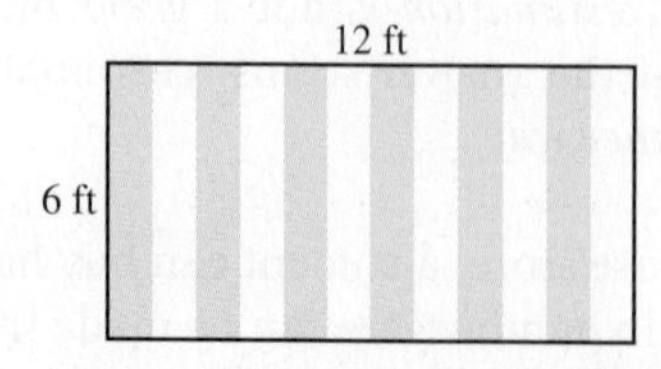

When a proportion is used to solve a problem, replace the variable with the solution and write each fraction as a decimal number. If the decimal numbers are very close to equal (any difference caused by rounding), the two ratios are the same, and the answer therefore seems reasonable.

EXAMPLE 5 **Problem:** In August 2009, officials expected that as many as two out of five Oregonians would have swine flu in the fall and winter flu season. If the population was about 3,800,000, find the number of Oregonians that were expected to come down with swine flu. (*Source:* lmtribune.com, Aug. 22, 2009)

Equation: $\frac{x}{3{,}800{,}000} = \frac{2}{5}$

Answer: In the fall and winter flu season, 1,520,000 Oregonians were expected to have swine flu.

Explain why this answer is reasonable. Replace x with the answer. Since $\frac{1{,}520{,}000}{3{,}800{,}000} = 0.4$ and $\frac{2}{5} = 0.4$, the answer of 1,520,000 people seems reasonable.

When *looking back* at an answer and deciding whether it is reasonable, first think about whether the answer makes sense. Use your experience and knowledge, and evaluate whether it is a possible answer to the problem situation. Then use one of the approaches in this section to explain why the solution is reasonable.

Practice Problems

For problems 1–6, explain why the answer seems reasonable.

1. **Problem:** Find the change in the population of New York.

   Since 1960, New York has lost 7.3 million residents to the rest of the country. This was partially offset by an influx of 4.8 million foreign immigrants. (*Source:* www.empirecenter.org, Aug. 2011)

   Equation: $c = 4.8 \text{ million} - 7.3 \text{ million}$

   Answer: The population of New York has decreased by 2.5 million people.

2. **Problem:** A student works 13 hours per week at a job that pays $8.25 per hour and 8 hours per week at a job that pays $10.15 an hour. Find the total amount earned at both jobs.

   Equation: $A = (13 \text{ hr})\left(\frac{\$8.25}{1 \text{ hr}}\right) + (8 \text{ hr})\left(\frac{\$10.15}{1 \text{ hr}}\right)$

   Answer: The total amount earned at both jobs per week is $188.45.

3. **Problem:** Find the number of people enrolled in community colleges who earn a degree or certificate within eight years. Round to the nearest tenth of a million.

   Nationally, 11.7 million people are enrolled in community colleges, but only about 39 percent of students earn a degree or certificate within eight years. (*Source:* www.nytimes.com, Aug. 14, 2009)

   Equation: $p = (0.39)(11.7 \text{ million people})$.

   Answer: About 4.6 million community college students earn a degree or certificate within eight years.

4. **Problem:** A room is 14 ft long, 12 ft wide, and 10 ft high. Find the volume of this room.

   Formula and equation: volume = (length)(width)(height); $V = (14 \text{ ft})(12 \text{ ft})(10 \text{ ft})$

   Answer: The volume of the room is 1680 ft^3.

5. **Problem:** Four out of five personal computers in the world run on Intel microchips. Find the number out of 16,000 PCs that run on Intel microchips. (*Source:* www.nytimes.com, Oct. 27, 2009)

   Equation: $\frac{4}{5} = \frac{x}{16{,}000}$

   Answer: Of the 16,000 computers, 12,800 of them run on Intel microchips.

6. **Problem:** The price of a ticket to the Free to Be Music and Arts Festival in Los Angeles is marked down 85% to $10. Find the original price of the ticket. Round to the nearest whole number. (*Source:* www.latimes.com, Aug. 2, 2011)

   Equation: $x - 0.85x = \$10$

   Answer: The original price of the ticket was about $67.

Appendix 1 Exercises

For exercises 1–24, explain why the answer seems reasonable.

1. **Problem:** At the University of North Carolina–Chapel Hill, the in-state tuition is $3865 per year. Out-of-state students pay $21,753 per year. Find the difference for 4 years of tuition for an in-state student and an out-of-state student. (*Source:* www.newsobserver.com, Aug. 25, 2009)

   Equation: $x = 4(\$21{,}753) - 4(\$3865)$

   Answer: The difference in tuition is $71,552.

2. **Problem:** Students at Florida International University had 557 people yo-yoing at the same time in a 2-minute period, a new world record. The previous record was 493 people. Find the difference in the number of people. (*Source:* www.miamiherald.com, Aug. 25, 2009)

   Equation: $n = 557 \text{ people} - 493 \text{ people}$

   Answer: The difference in the number of people yo-yoing is 64 people.

3. **Problem:** The scoreboard at TCF Bank Stadium at the University of Minnesota is a rectangle that is 48 ft high by 108 ft wide. Find the area of the scoreboard. (*Source:* www.twincities.com, Aug. 16, 2009)

   Equation: $A = (48 \text{ ft})(108 \text{ ft})$

   Answer: The area of the scoreboard is 5184 ft^2.

4. **Problem:** The scoreboard at Yankee Stadium is 59 ft high and 101 ft wide. Find the perimeter of the scoreboard. (*Source:* www.signindustry.com)

   Equation: $P = 2(59 \text{ ft}) + 2(101 \text{ ft})$

   Answer: The perimeter of the scoreboard is 320 ft.

5. **Problem:** The city of Richardson, Texas, pays monthly car allowances to employees. The cost per year for 141 employees is $842,000. Find the average monthly car allowance per employee. Round to the nearest whole number. (*Source:* www.dallasnews.com, Aug. 26, 2009)

   Equation: $A = \frac{\$842{,}000 \div 12}{141 \text{ employees}}$

   Answer: The average monthly car allowance is about $498.

6. **Problem:** A light rail line in Seattle is 14 mi long, extending from Westlake Center to Tukwila. The cost of construction was $2.3 billion. Find the average cost per mile. Round to the nearest *million* dollars. (*Source:* seattletimes.nwsource.com, July 26, 2009)

   Equation: $C = \frac{\$2.3 \text{ billion}}{14 \text{ mi}}$

   Answer: The average cost per mile is about $164 million.

7. **Problem:** The regular price of a computer is $999. It is on sale at 15% off. Find the sale price.

   Equation: $P = \$999 - (0.15)(\$999)$

   Answer: The sale price is $849.15.

8. **Problem:** The number of Washington students completing the Free Application for Student Aid (FAFSA) for the 2009–2010 school year rose 20% compared to the 2008–2009 school year when 201,500 students completed the financial aid forms. Find the number of students who completed the FAFSA for 2009–2010.

   Equation: $N = 201{,}500 \text{ students} + (0.20)(201{,}500 \text{ students})$

   Answer: For 2009–2010, 241,800 students completed the FAFSA.

9. **Problem:** A costumer has 25 yd of fabric. She needs $1\frac{1}{4}$ yd of this fabric for each costume. Find the number of costumes she can make from this fabric.

   Equation: $N = 25 \text{ yd} \div 1\frac{1}{4} \text{ yd}$

   Answer: The costumer can make 20 costumes from this fabric.

10. **Problem:** A recipe requires $\frac{3}{4}$ cup of flour. A canister contains 9 cups of flour. Find how many recipes can be made with this flour.

    Equation: $N = 9 \text{ cups} \div \frac{3}{4} \text{ cup}$

    Answer: Twelve recipes can be made with this flour.

11. **Problem:** The cost of a DVD is $35.99. Shipping is $1.99 per order plus $0.99 per item. Find the total cost (without tax) for an order of 12 DVDs.

    Equation: $C = (12 \text{ DVD})\left(\frac{\$35.99}{1 \text{ DVD}}\right) + \$1.99 + (12 \text{ items})\left(\frac{\$0.99}{1 \text{ item}}\right)$

    Answer: The total cost for an order of 12 DVDs is $445.75.

12. **Problem:** A floor seat ticket to see *Smackdown and ECW Live* at the Canton Civic Center costs $60. The convenience charge per ticket is $9.70, and the building facility charge per ticket is $1. Find the total cost for five tickets.

    Equation: $C = (5 \text{ tickets})\left(\frac{\$60}{1 \text{ ticket}}\right) + (5 \text{ tickets})\left(\frac{\$9.70}{1 \text{ ticket}}\right) + (5 \text{ tickets})\left(\frac{\$1}{1 \text{ ticket}}\right)$

    Answer: The total cost for five tickets is $353.50.

13. **Problem:** From 2006 to 2008, speed was a factor in 58% of crashes in Ohio caused by juvenile drivers. These drivers caused 61,784 traffic crashes. Find the number of these crashes in which speed was a factor. Round to the nearest ten. (*Source:* www.cleveland.com, Aug. 26, 2009)

    Equation: $N = (0.58)(61{,}784)$

    Answer: Speed was a factor in about 35,830 crashes caused by juvenile drivers.

14. **Problem:** Find the number of people in the U.S. population predicted for 2030. Round to the nearest tenth of a million.

    By 2030, Census Bureau data show, the 72 million people expected to be ages 65 and older will represent 19 percent of the U.S. population—up from 13 percent in 2010. (*Source:* bls.gov, Spring 2011)

    Equation: $0.19N = 72$ million people

    Answer: The population is predicted to be about 378.9 million people.

15. **Problem:** South Salt Lake and Salt Lake City agreed to each pay $2.5 million of the $46 million cost to build two miles of streetcar line. The federal government will pay $35 million. Find the percent that the two cities are paying of the total cost of the line. Round to the nearest percent. (*Source:* www.sltrib.com, Aug. 27, 2009)

    Equation: $P = \frac{2(\$2.5 \text{ million})}{\$46 \text{ million}} \cdot 100\%$

    Answer: South Salt Lake and Salt Lake City will pay about 11% of the total cost of the line.

16. **Problem:** On August 2, 2011, the closing price of Harley-Davidson stock was $42.57 per share. This price was 22.8% higher than it was on January 2, 2011. Find the price on January 2, 2011. Round to the nearest hundredth. (*Source:* investor.harley-davidson.com)

    Equation: $P + 0.228P = \$42.57$

    Answer: The price on January 2, 2011 was about $34.67.

17. **Problem:** According to the Centers for Disease Control and Prevention, one out of five children in the United States is obese or overweight. Use a proportion to predict how many of 12,500 children are obese or overweight (*Source:* www.cdc.gov)

    Equation: $\frac{1}{5} = \frac{n}{12{,}500}$

    Answer: Of 12,500 children, 2500 are overweight.

18. **Problem:** According to the National Retail Federation, four out of five Americans say that the recession is affecting their back-to-school spending and college plans. Use a proportion to predict how many of 130,000 Americans say that the recession is affecting their back-to-school spending and college plans. (*Source:* www.nrf.com, July 14, 2009)

    Equation: $\frac{4}{5} = \frac{N}{130{,}000}$

    Answer: Of 130,000 Americans, about 104,000 say that the recession is affecting their back-to-school spending and college plans.

19. **Problem:** The Spokane Valley City Council raised the motel tax by 50 cents to $2 a night. Find the percent increase in the motel tax per night. Round to the nearest percent. (*Source:* seattletimes.nwsource.com, Aug. 27, 2009)

    Equation: $P = \frac{\$0.50}{\$1.50} \cdot 100\%$

    Answer: The percent increase in the motel tax per night is about 33%.

20. **Problem:** The average electricity bill for a home in Austin was $88 in April 2009. During a record heat wave in July 2009, it was $235. Find the percent increase. Round to the nearest percent. (*Source:* www.statesman.com, Aug. 25, 2009)

    Equation: $P = \frac{\$235 - \$88}{\$88} \cdot 100\%$

    Answer: The electricity bills increased by about 167%.

21. **Problem:** As the H1N1 flu vaccine was produced in August 2009, the National Center for Immunization and Respiratory Diseases was concerned that supplies might not meet the demand. If rationing were necessary, it recommended that pregnant women, persons who live with or provide care for infants less than 6 months of age, health-care and emergency medical personnel, children between 6 months and 4 years, and children and adolescents ages 5–18 years with certain medical conditions should have priority. These groups included about 159 million people. The population of the United States in August 2009 was about 307 million people. Find the percent that the population of these groups is of the entire population of the United States. Round to the nearest percent. (*Source:* www.cdc.gov, Aug. 28, 2009)

    Equation: $P = \frac{159 \text{ million people}}{307 \text{ million people}} \cdot 100\%$

    Answer: These groups are about 52% of the population of the United States.

22. **Problem:** On average, each year in California, there will be approximately 46,100 job openings in occupations that require a degree in science, technology, engineering, or mathematics. About 24,000 of the jobs will require at least a bachelor's degree. Find the percent of the jobs that will require at least a bachelor's degree. Round to the nearest percent. (*Source:* Institute for Higher Education Leadership & Policy, www.csus.edu/ihe, June 2009)

    Equation: $P = \dfrac{24{,}000 \text{ jobs}}{46{,}100 \text{ jobs}} \cdot 100\%$

    Answer: About 52% of the jobs will require at least a bachelor's degree.

23. **Problem:** Find the operating budget before the reductions. Round to the nearest tenth of a million.

    [The Texas Education Agency] is reducing its operating budget by $48 million, or 36 percent . . . layoffs will account for the majority of the savings. (*Source:* www.statesman.com, July 12, 2011)

    Equation: $0.36B = \$48$ million

    Answer: The operating budget before the reductions was about $133.3 million.

24. **Problem:** An Austin Energy spokesperson said that about $14.1 million in electricity bills due in July had not been paid. This was 7.8% of the total amount of bills. Find the total amount of bills. Round to the nearest tenth of a million. (*Source:* www.statesman.com, Aug. 25, 2009)

    Equation: $0.078B = \$14.1$ million

    Answer: The total amount of bills was about $180.8 million.

APPENDIX 2

Deriving the Quadratic Formula

In Section 9.2, we solved quadratic equations by completing the square. If we use completeing the square to solve the quadratic equation $ax^2 + bx + c = 0$, the solutions are $x = \dfrac{-b \pm \sqrt{b^2 - 4ac}}{2a}$. This is the quadratic formula. In this section, we will confirm that the quadratic formula represents the solutions of $ax^2 + bx + c = 0$. To avoid confusion when finding $\frac{1}{2}b^2$ to complete the square, we will rewrite the coefficients with capital letters as $Ax^2 + Bx + C = 0$.

EXAMPLE 1 Solve $Ax^2 + Bx + C = 0$ by completing the square.

The first step to solve this equation by completing the square is to isolate $Ax^2 + Bx$.

$$
\begin{array}{rll}
Ax^2 + Bx + C &= 0 & \\
-C & \;\;-C & \text{Subtraction property of equality.} \\
\hline
Ax^2 + Bx + \mathbf{0} &= \mathbf{-C} & \text{Simplify.}
\end{array}
$$

Now divide both sides of the equation by A.

$$\frac{A}{A}x^2 + \frac{B}{A}x = \frac{-C}{A} \qquad \text{Division property of equality.}$$

$$x^2 + \frac{B}{A}x = \frac{-C}{A} \qquad \text{Simplify.}$$

For $x^2 + bx$, the term needed to complete the square is $\left(\frac{1}{2}b\right)^2$. In this equation, $b = \frac{B}{A}$. To find the term needed to complete the square, we simplify $\left(\frac{1}{2} \cdot \frac{B}{A}\right)^2$.

$$\left(\frac{1}{2}b\right)^2$$

$$= \left(\frac{1}{2} \cdot \frac{\mathbf{B}}{\mathbf{A}}\right)^2 \qquad \text{For this equation, } b = \frac{B}{A}.$$

$$= \left(\frac{\mathbf{B}}{\mathbf{2A}}\right)^2 \qquad \text{Simplify.}$$

$$= \frac{B^2}{4A^2} \qquad \text{Simplify.}$$

We now complete the square by adding $\frac{B^2}{4A^2}$ to both sides of the equation. We then factor the equation and rewrite it in exponential notation.

$$x^2 + \frac{B}{A}x + \frac{B^2}{4A^2} = \frac{-C}{A} + \frac{B^2}{4A^2} \qquad \text{Addition property of equality.}$$

$$\left(\mathbf{x} + \frac{\mathbf{B}}{\mathbf{2A}}\right)\left(\mathbf{x} + \frac{\mathbf{B}}{\mathbf{2A}}\right) = \frac{-C}{A} + \frac{B^2}{4A^2} \qquad \text{Factor the perfect square trinomial.}$$

$$\left(\mathbf{x} + \frac{\mathbf{B}}{\mathbf{2A}}\right)^{\mathbf{2}} = \frac{-C}{A} + \frac{B^2}{4A^2} \qquad \text{Use exponential notation.}$$

This equation is in the form $x^2 = k$, and the solutions of this equation are $x = \pm\sqrt{k}$. We write the solutions, change the order of the terms in the radicand, and then combine the terms.

$$x + \frac{B}{2A} = \pm\sqrt{-\frac{C}{A} + \frac{B^2}{4A^2}}$$ The solutions of $x^2 = k$ are $x = \pm\sqrt{k}$.

$$x + \frac{B}{2A} = \pm\sqrt{\frac{B^2}{4A^2} - \frac{C}{A}}$$ Change the order of terms in the radicand.

$$x + \frac{B}{2A} = \pm\sqrt{\frac{B^2}{4A^2} - \frac{C}{A}\cdot\frac{4A}{4A}}$$ LCD is $4A^2$; multiply by a fraction equal to 1.

$$x + \frac{B}{2A} = \pm\sqrt{\frac{B^2}{4A^2} - \frac{4AC}{4A^2}}$$ Simplify.

$$x + \frac{B}{2A} = \pm\sqrt{\frac{B^2 - 4AC}{4A^2}}$$ Subtract the fractions in the radicand.

To simplify the denominator, we rewrite the radical as a quotient of two radicals. We then use the subtraction property of equality to isolate x. Finally, we combine the two fractions.

$$x + \frac{B}{2A} = \pm\frac{\sqrt{B^2 - 4AC}}{\sqrt{4A^2}}$$ Quotient rule of radicals.

$$x + \frac{B}{2A} = \pm\frac{\sqrt{B^2 - 4AC}}{2A}$$ Simplify.

$$-\frac{B}{2A} \qquad -\frac{B}{2A}$$ Subtraction property of equality.

$$x + 0 = -\frac{B}{2A} \pm \frac{\sqrt{B^2 - 4AC}}{2A}$$ Simplify.

$$x = -\frac{B}{2A} \pm \frac{\sqrt{B^2 - 4AC}}{2A}$$ Simplify; both fractions have the same denominator, $2A$.

$$x = \frac{-B \pm \sqrt{B^2 - 4AC}}{2A}$$ Combine fractions.

Changed back to lowercase letters, the solutions of $ax^2 + bx + c = 0$ are $x = \frac{-b \pm \sqrt{b^2 - 4ac}}{2a}$, the quadratic formula. To find the solutions of a quadratic equation, we replace a, b, and c in the formula with the values from a quadratic equation and simplify.

The Quadratic Formula

For the equation $ax^2 + bx + c = 0$, where a, b, and c are real numbers and $a \neq 0$, $x = \frac{-b \pm \sqrt{b^2 - 4ac}}{2a}$.

Answers to Practice Problems

CHAPTER R

Section R.1

1. 8 **2.** 3 **3.** 15 **4.** 9 **5.** 100 **6.** 16 **7.** 20 **8.** 35 **9.** 26 **10.** 36 **11.** 2 **12.** 123 **13.** 6 **14.** 0 **15.** undefined **16.** 0 **17.** 6 **18.** 6 **19.** 6 **20.** 6 **21.** 0 **22.** undefined **23.** 81 **24.** 56 **25.** 59 **26.** 26

Section R.2

1. 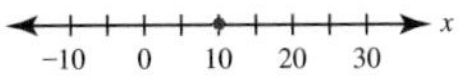

2. 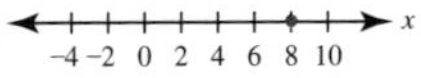

3. 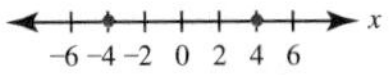

4. 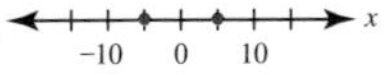

5. -4 **6.** -8 **7.** -8 **8.** 4 **9.** -4 **10.** 5 **11.** -5 **12.** 5 **13.** -11 **14.** -11 **15.** 11 **16.** 5 **17.** 7 **18.** 7 **19.** -15 **20.** 24 **21.** -24 **22.** 24 **23.** -24 **24.** 14 **25.** -10 **26.** -14 **27.** 10 **28.** 49 **29.** 49 **30.** -49 **31.** 66 **32.** -51 **33.** 16 **34.** 38 **35.** -1 **36.** -2 **37.** -13 **38.** -7 **39.** 3 **40.** 32

Section R.3

1. $\frac{1}{3}$ **2.** $\frac{5}{6}$ **3.** $\frac{1}{16}$ **4.** 1 **5.** $\frac{3}{5}$ **6.** $\frac{21}{25}$ **7.** $\frac{5}{3}$ **8.** $\frac{18}{5}$ **9.** $\frac{15}{8}$ **10.** $6\frac{1}{5}$ **11.** $4\frac{4}{5}$ **12.** $26\frac{2}{3}$ **13.** $\frac{21}{40}$ **14.** $\frac{1}{4}$ **15.** 1 **16.** $\frac{2}{15}$ **17.** $\frac{2}{3}$ **18.** $-\frac{7}{4}$ **19.** $\frac{3}{32}$ **20.** -27 **21.** $\frac{1}{2}$ **22.** $\frac{3}{20}$ **23.** $\frac{11}{45}$ **24.** $\frac{34}{75}$ **25.** $-\frac{3}{5}$ **26.** $-\frac{11}{24}$ **27.** $-\frac{31}{6}$ **28.** $-\frac{7}{36}$ **29.** $\frac{25}{144}$ **30.** $-\frac{25}{2}$ **31.** $\frac{8}{63}$ **32.** $\frac{1972}{1407}$ **33.** $\frac{91}{160}$

Section R.4

1. $\frac{81}{100}$ **2.** $\frac{3999}{1000}$ **3.** $\frac{3}{5}$ **4.** $\frac{1}{2500}$ **5.** 0.75 **6.** 4.4 **7.** $0.\overline{5}$ **8.** $0.\overline{27}$ **9.** 220 **10.** 200 **11.** 31.7 **12.** 31.75 **13.** 0.060 **14.** 0.1 **15.** 35,000 **16.** 30,000 **17.** $\frac{3}{100}$ **18.** $\frac{17}{100}$ **19.** $\frac{3}{5}$ **20.** $\frac{2}{25}$ **21.** 0.18 **22.** 0.02 **23.** 0.065 **24.** 1.37 **25.** 60% **26.** 62.5% **27.** 6 **28.** 6 **29.** 1.5 **30.** $<$ **31.** $>$ **32.** $>$ **33.** $<$ **34.** $<$ **35.** $<$ **36.** $<$ **37.** $<$ **38.** $>$ **39.** $<$ **40.** $<$

CHAPTER 1

Section 1.1

1. a. $\{0, 1, 2, 3, 4, \ldots\}$ **b.** $\{\ldots, -3, -2, -1, 0, 1, 2, 3, \ldots\}$ **2.** Answers may vary. 2 and 3 are factors of 6. Prime numbers have only 1 and themselves as factors.

3. 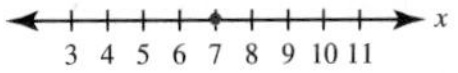

4. -7 **5.** 7 **6.** 7 **7.** 25 **8.** 25 **9.** -1 **10.** 8 **11.** 8 **12.** $\frac{3}{7}$ **13.** $\frac{11}{17}$ **14.** $\frac{1}{3}$ **15.** $\frac{3}{10}$ **16.** yes **17.** no **18.** see table below **19.** 15 **20.** 19 **21.** 7 **22.** 25 **23.** 108 **24.** -34 **25.** -405 **26.** 36 **27.** 3 **28.** -4

Section 1.2

1. a. 2 **b.** $\textcircled{3}\underline{x} + \boxed{8}$ **2. a.** 3 **b.** $\textcircled{\frac{3}{4}}\underline{x} + \textcircled{6}\underline{y} + \boxed{5}$ **3.** $22b$ **4.** $-12w$ **5.** $-8m + 18p + 4$ **6.** $-\frac{1}{2}p + \frac{1}{2}$ **7.** $\frac{5}{3}p$ **8.** $-\frac{25}{24}a + 19$ **9.** $15xy$ **10.** $18ab$ **11.** $\frac{5}{24}x^2$ **12.** $9b$ **13.** C **14.** B

Table for Section 1.1 practice problem 18

	Real numbers	Rational numbers	Irrational numbers	Integers	Whole numbers
6	X	X		X	X
$-\frac{5}{9}$	X	X			
π	X		X		
0.8	X	X			
-43	X	X		X	
$\frac{4}{5}$	X	X			

15. A **16.** A **17.** C **18.** B **19.** $x - \frac{5}{2}$
20. $12a + 17b$ **21.** $-12z + 37$ **22.** $-40x - 20$
23. $8x + 48y - 20$ **24.** $13x + 2$ **25.** $2x^2 + 11x + 7$
26. $6a^2 - 7a - 16$

Section 1.3

For these application problems, the given answer is not in a complete sentence. However, students should report all solutions using complete sentences. Follow your instructor's guidelines for showing your work.

1. \$302.94 **2.** 8.2 billion metric tons **3.** 720 false alarms **4.** \$427 **5.** 16 pieces **6.** 36 students
7. 255 fuel assemblies

Section 1.4

For all application problems, the given answer is not in a complete sentence. However, students should report all solutions using complete sentences. Follow your instructor's guidelines for showing your work.

1. 1709 adults **2.** 74% **3.** 14% **4.** 52%
5. 21% decrease **6.** 2.4% decrease **7.** 3.1% increase
8. \$24.38 **9.** \$15,299 **10.** \$28.14 **11.** 8.2%

Section 1.5

1. 11 in. **2.** 3 in. **3.** 28 in.^{2} **4.** 2 **5.** 30 in.^{3}
6. 20.32 cm **7.** 11.2 km **8.** 0.065 km **9.** 4.7 in.
10. $\frac{3.18 \text{ qt}}{1 \text{ hr}}$ **11.** 11,880 ft **12.** 24.2 lb
13. 6.008 km **14.** 6008 m **15.** 227.5 mi **16.** 384 mi
17. \$45 **18.** \$126 **19.** 18 ft **20.** 18 ft^2 **21.** 30.4 cm
22. 120 in.^{2} **23.** 360 in.^{2} **24.** 7.8 ft **25.** 384 cm^3
26. 30,240 in.^{3} **27.** 30 ft **28.** 7.5 in.^{2}

Section 1.6

1. $0.\overline{6}$ **2.** $0.\overline{45}$ **3.** $0.\overline{857142}$ **4.** 56.5 ft **5.** 803.8 in.^{2}
6. 201.0 in.^{2} **7.** 1695.6 cm^3 **8.** 113.0 ft^3 **9.** 937.8 in.^{3}
10. 9748.6 in.^{3} **11.** 38 ft^2 **12.** 42.4 in.^{3} **13.** 33,493 cm^3
14. 3 and -3 **15.** 9 **16.** m **17.** 9 m **18.** 0
19. 1 **20.** 4.5 **21.** 2.2 **22.** not a real number **23.** 2
24. -2 **25.** 10 ft **26.** 0 **27.** 1 **28.** -1 **29.** 5 ft
30. 102.5 in. **31.** 6 knots

CHAPTER 2

Section 2.1

1. a. $x = -27$ **2. a.** $k = 39$ **3. a.** $c = -12$
4. a. $d = 108$ **5. a.** $y = -24$ **6. a.** $x = -9$
7. a. $p = 6$ **8. a.** $h = \frac{4}{5}$ **9. a.** $x = \frac{4}{5}$
10. a. $x = \frac{5}{4}$ **11. a.** $d = \frac{7}{30}$ **12. a.** $p = -\frac{17}{2}$
13. a. $c = 11.5$ **14. a.** $m = 0.3$ **15. a.** $h = 75$
16. a. $p = 400$ **17. a.** $p = 240$ **18.** \$920
19. 13,368 tickets **20.** 33,802 people
21. 1000 businesses **22.** 997 births
23. 8.25 mL **24.** 125 mi

Section 2.2

1. a. $c = 3$ **2. a.** $k = -10$ **3. a.** $x = -165$
4. a. $x = 12$ **5. a.** $w = -5$ **6. a.** $x = -10$
7. a. $x = -6$ **8. a.** $w = -23$ **9. a.** $x = \frac{3}{10}$
10. a. $x = 6.2$ **11. a.** $z = \frac{5}{34}$ **12.** 18,240 mi
13. 1800 products

Section 2.3

1. a. $w = -25$ **2. a.** $w = 2$ **3. a.** $k = -4$
4. a. $x = -\frac{7}{4}$ **5. a.** $x = -57$ **6. a.** $x = 112.5$
7. a. no solution **8. a.** the set of real numbers
9. a. $x = 0$ **10.** 15 products **11.** 10 quilts
12. 2.5 hr

Section 2.4

1. $x = -\frac{4}{3}y + 4$ **2.** $y = \frac{5}{2}x + \frac{15}{2}$
3. $y = -\frac{8}{3}x + 240$ **4.** $t = \frac{x}{v_1}$ **5.** $V = \frac{nRT}{P}$
6. a. $W = \frac{A}{L}$ **b.** 81 ft **7. a.** $L = \frac{P - 2W}{2}$ **b.** 493 m

Section 2.5

1. 1 2 3 4 5 6 7 8 x
2. −9 −8 −7 −6 −5 −4 −3 −2 x
3. −3 −2 −1 0 1 2 3 4 x
4. −5 −4 −3 −2 −1 0 1 2 x
5. a. $x > -2$ **b.** −4 −3 −2 −1 0 1 2 3 x
6. a. $x \geq -8$ **b.** −10 −9 −8 −7 −6 −5 −4 −3 x
7. a. $x > 2$ **b.** 0 1 2 3 4 5 6 7 x
8. a. $x < 8$ **b.** 3 4 5 6 7 8 9 10 x
9. a. $x > 12$ **b.** 10 11 12 13 14 15 16 17 x
10. a. $x \geq 0$ **b.** −2 −1 0 1 2 3 4 5 x
11. a. $x < \frac{33}{10}$ **b.** −1 0 1 2 3 4 5 x
12. a. $x \leq -\frac{1}{6}$ **b.** −1 0 x
13. a. $x \leq -13$ **b.** −17 −15 −13 −11 x
14. a. $x < -12$ **b.** −16 −14 −12 −10 x
15. a. $x \geq \frac{10}{3}$ **b.** 3 4 5 x
16. no more than 1470 ft^2 **17.** no fewer than 1000 sales
18. up to and including 400 min

CHAPTER 3

Section 3.1

1. a. (1, 4) **b.** 1st quadrant **2. a.** (−2, −4)
b. 3rd quadrant **3. a.** (−3, 70) **b.** 2nd quadrant
4. a. (20, −300) **b.** 4th quadrant

5.

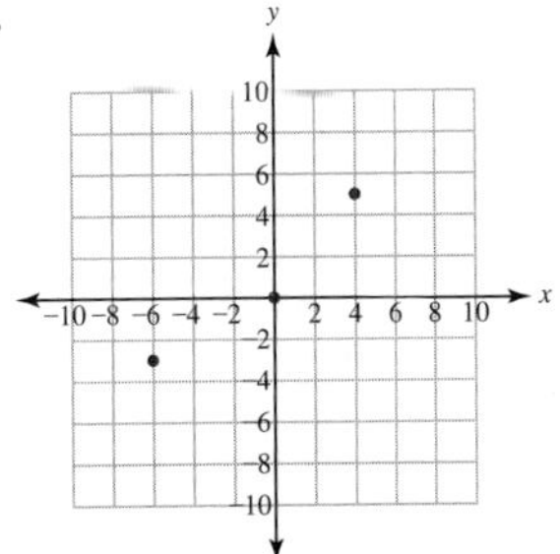

6.

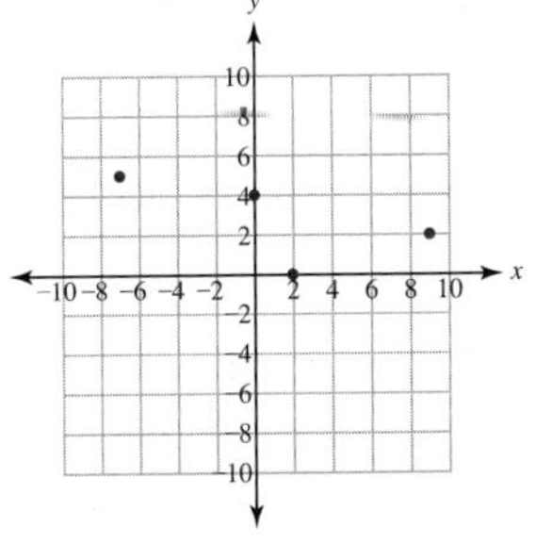

14. a.

x	y
0	−3
−6	0
2	−4

b.

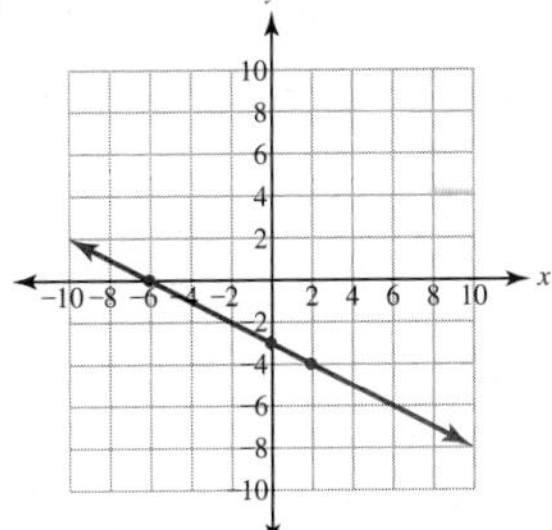

7.

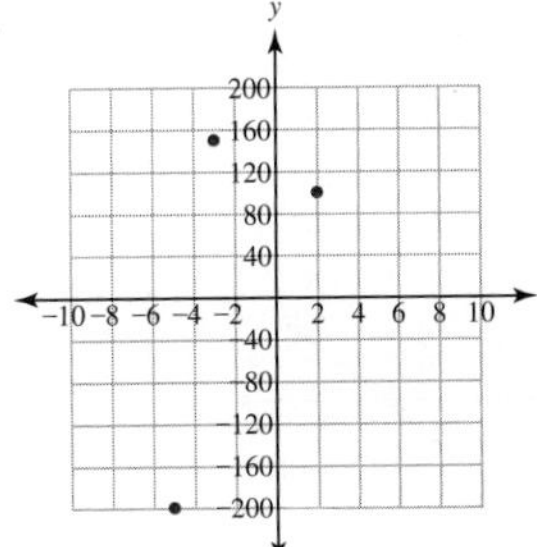

8.

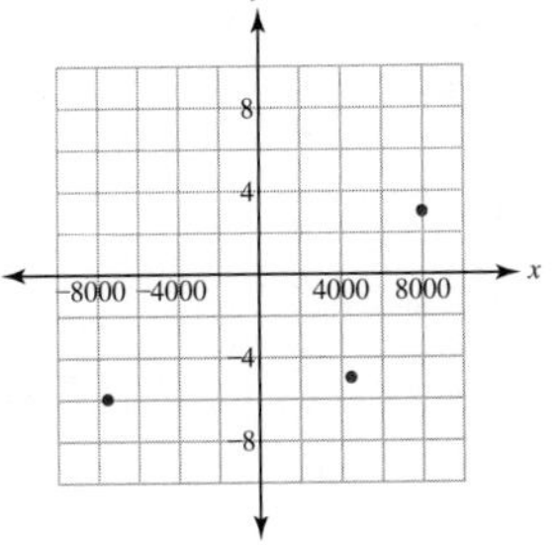

15. a.

x	y
0	6
9	0
6	2

b.

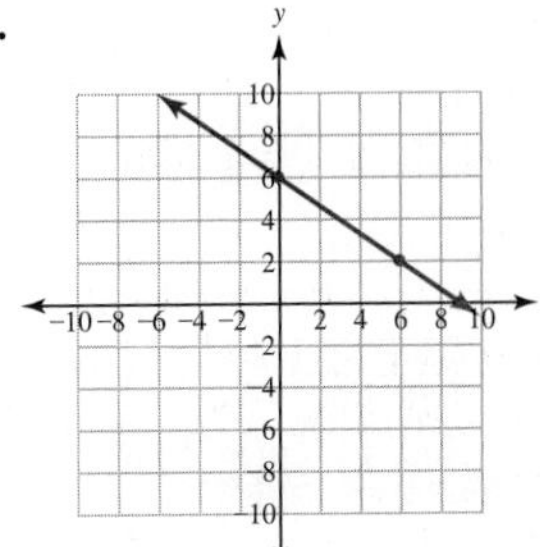

9.

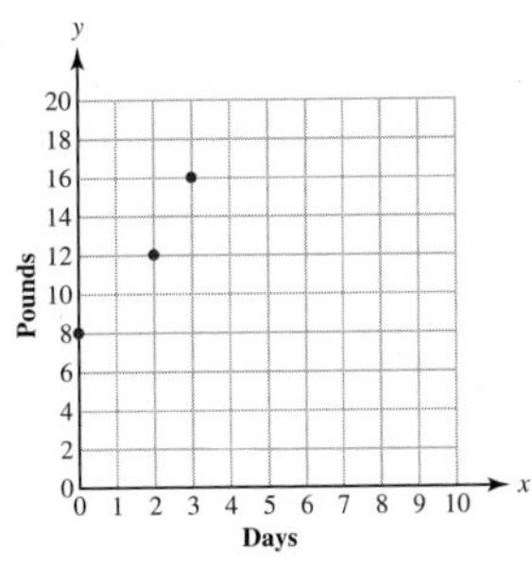

10.

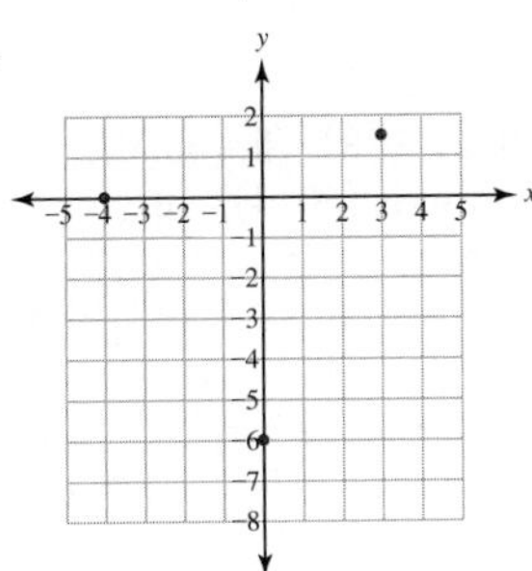

16. a.

x	y
0	−15
5	0
3	−6

b.

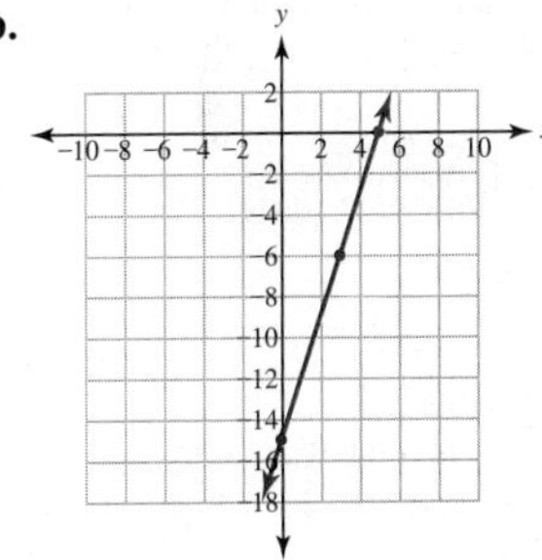

11.

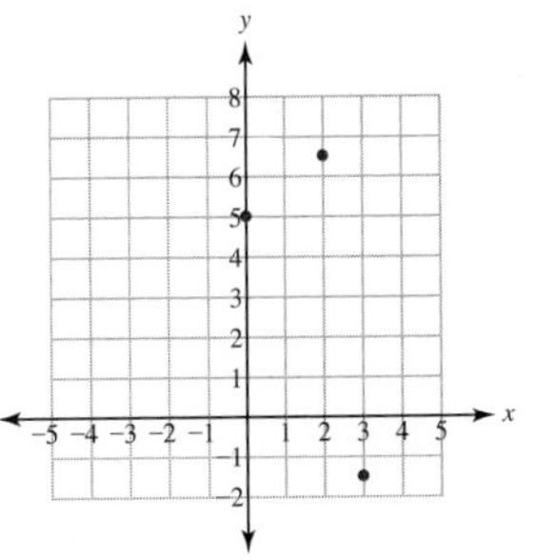

17. a.

x	y
0	−3
$-\frac{9}{2}$	0
−3	−1

b.

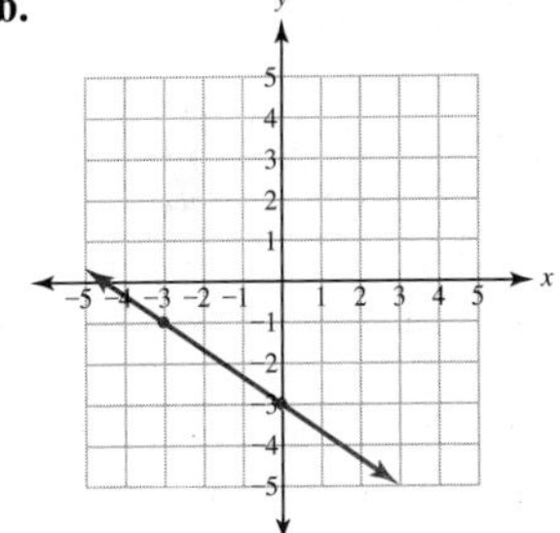

12 a.

x	y
0	6
6	0
−3	9

b.

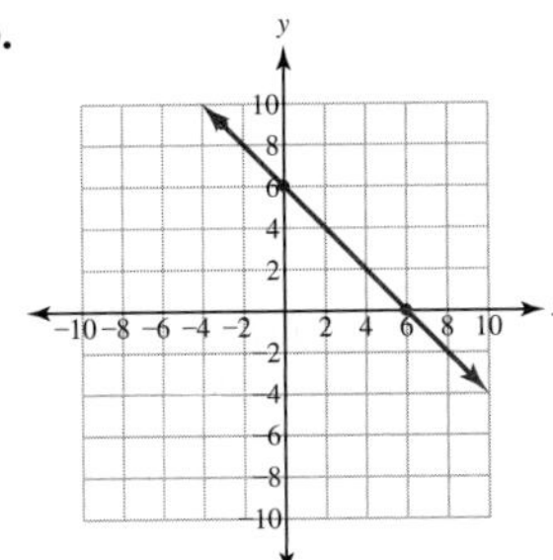

18. a.

x	y
3	−2
0	−4
−3	−6

b.

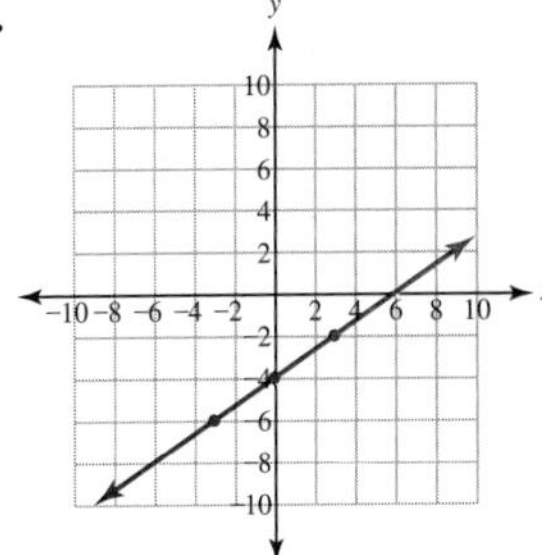

13. a.

x	y
0	−6
6	0
−3	−9

b.

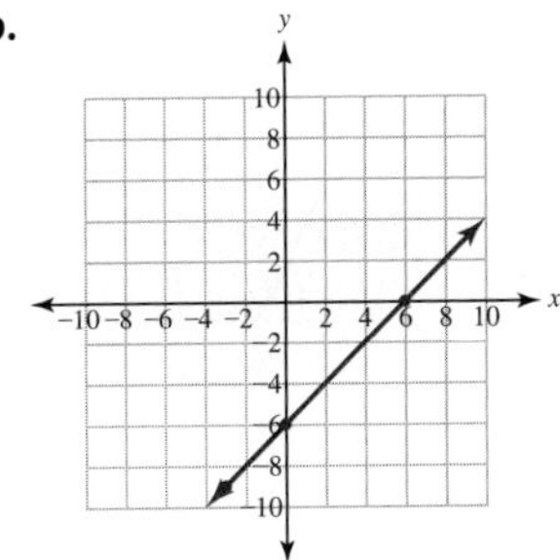

19. a.

x (lb)	y ($)
0	3
2	7
4	11

b.

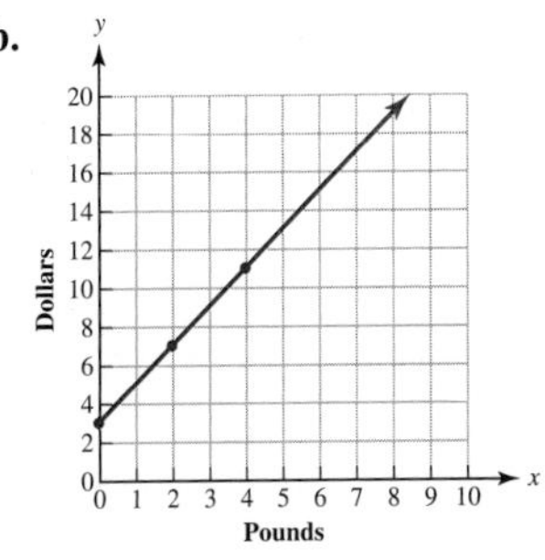

20. a.

x	y
6	−3
0	2
−6	7

b.

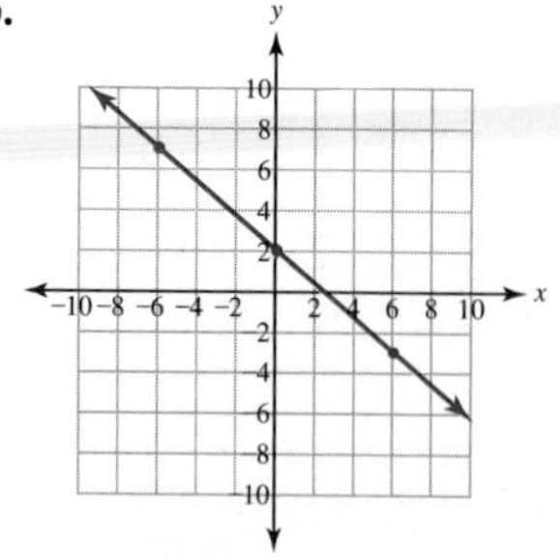

21.

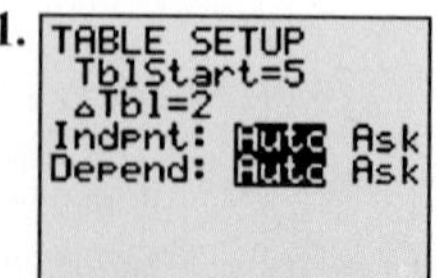

X	Y₁
5	10
7	30
9	50
11	70
13	90
15	110
17	130

X=5

22.

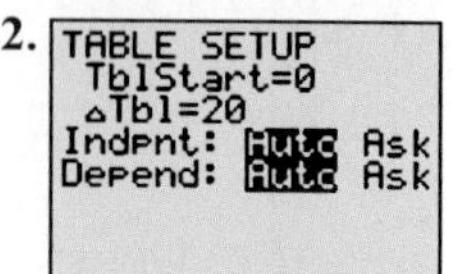

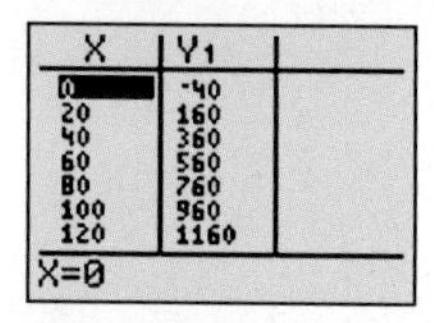

23.

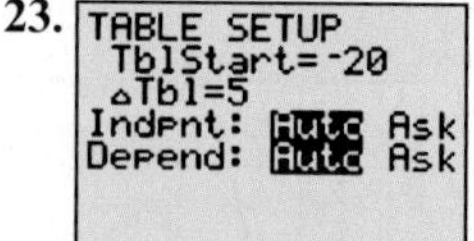

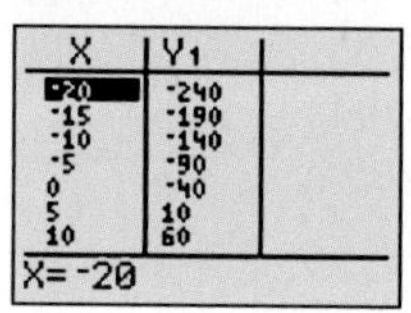

Section 3.2

1. $7x - y = 5$ or $-7x + y = -5$ **2.** $2x - y = 2$
3. $3x + y = 54$ **4.** $35x - 42y = 23$
5. a. $(0, 5)$ **b.** $(5, 0)$ **c.** Answers may vary. One example point is $(2, 3)$. **d.**

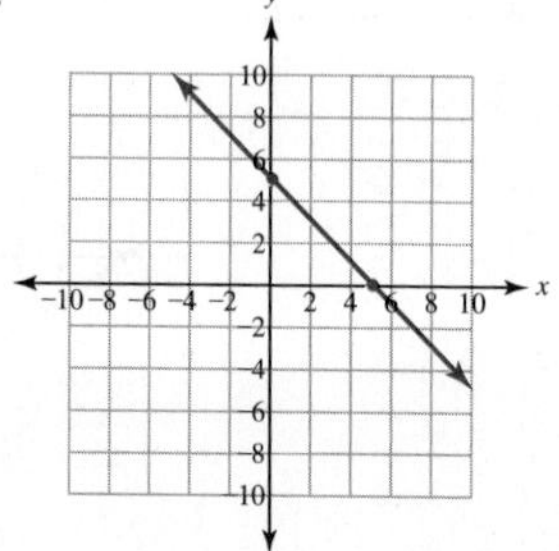

6. a. $(0, 6)$ **b.** $(3, 0)$ **c.** Answers may vary. One example point is $(1, 4)$. **d.**

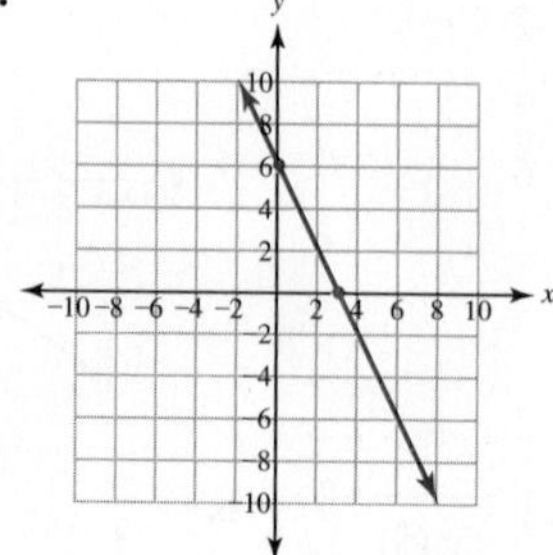

7. a. $(0, -3)$ **b.** $(9, 0)$ **c.** Answers may vary. One example point is $(6, -1)$. **d.**

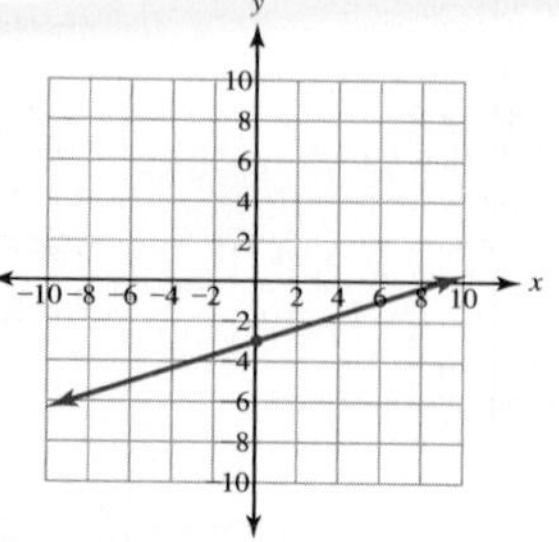

8. a. $(0, -4)$ **b.** $(5, 0)$ **c.** Answers may vary. One example point is $(-5, -8)$. **d.**

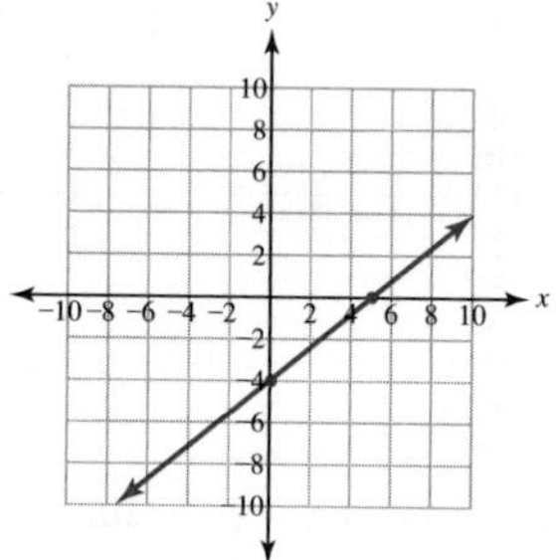

9. a. $(0, 8)$ **b.** $(40, 0)$ **c.** Answers may vary. One example point is $(20, 4)$. **d.**

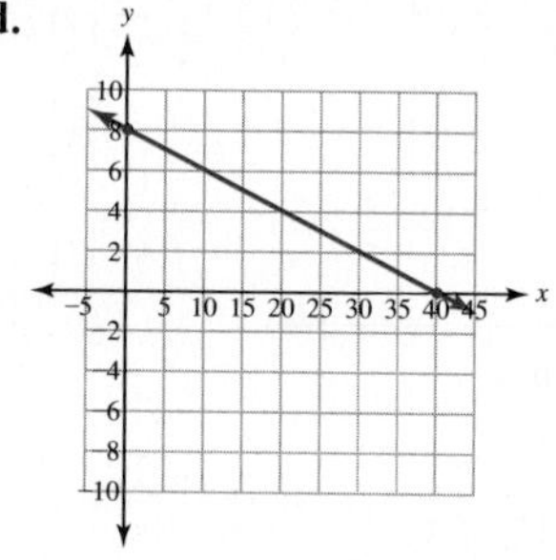

10. a. $\left(0, \frac{9}{2}\right)$ **b.** $(-3, 0)$ **c.** Answers may vary. One example point is $(-1, 3)$. **d.**

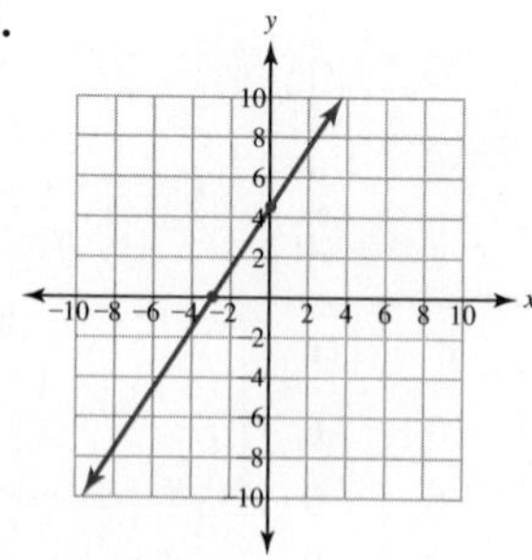

11. a. ($0, $800) **b.** ($400, $0) **c.** Answers may vary. One example point is ($200, $400).
d.

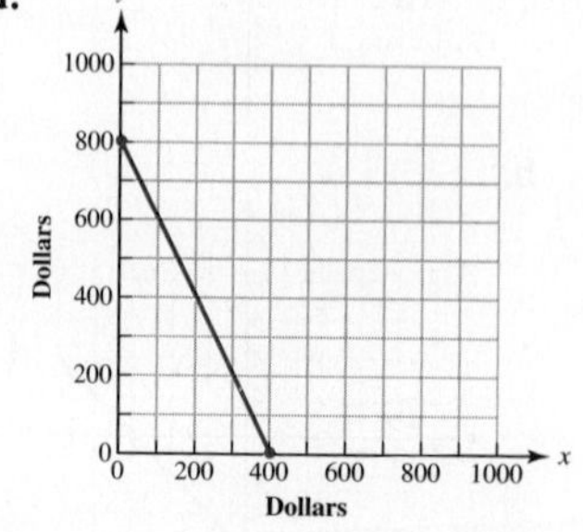

12. a. Answers may vary. Three example points are:

x	y
−2	−5
0	1
2	7

b.

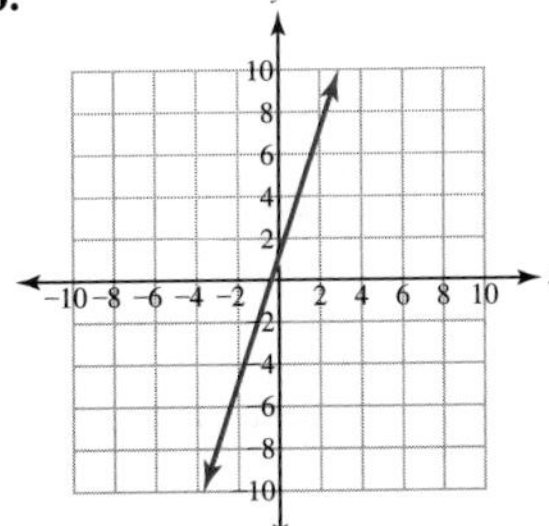

13. a. Answers may vary. Three example points are:

x	y
−4	25
0	9
4	−7

b.

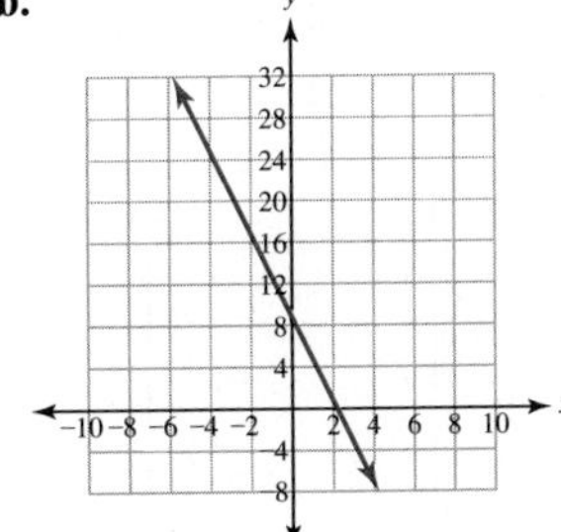

14. a. Answers may vary. Three example points are:

x	y
0	−5
6	−1
3	−3

b.

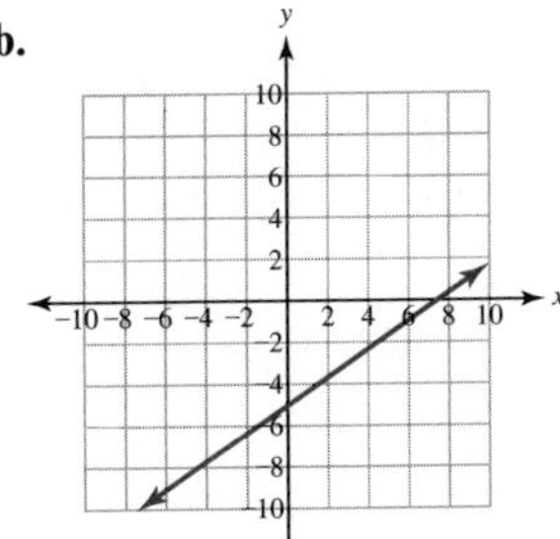

15. a. $y = -5$ **b.**

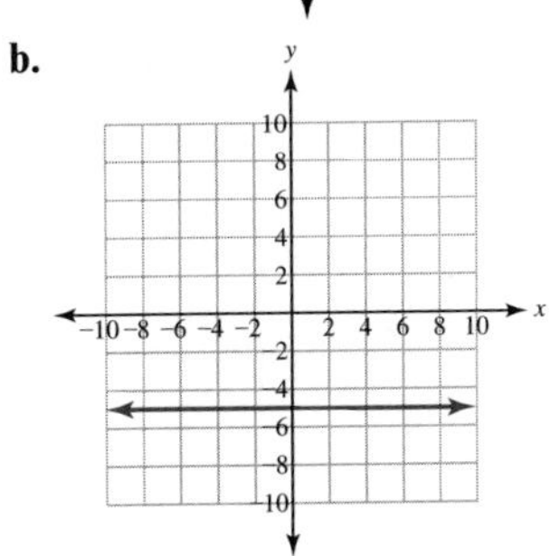

16. a. $x = 4$ **b.**

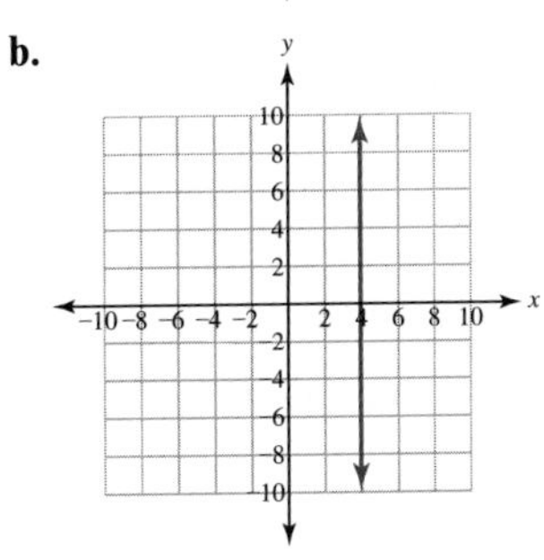

17.

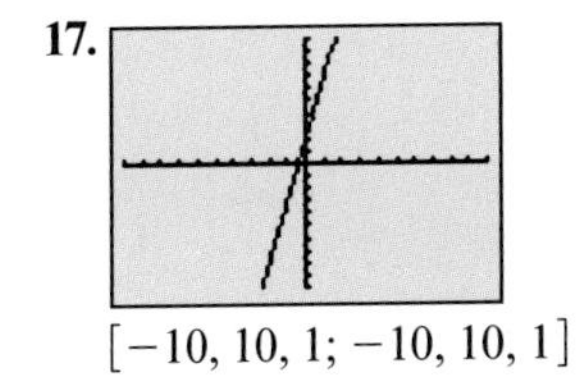

$[-10, 10, 1; -10, 10, 1]$

18.

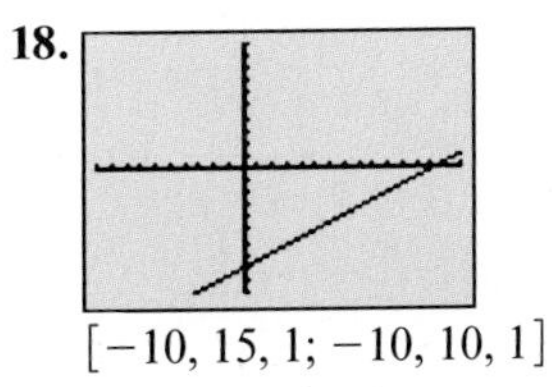

$[-10, 15, 1; -10, 10, 1]$

19.

$[-10, 10, 1; -10, 10, 1]$

20.

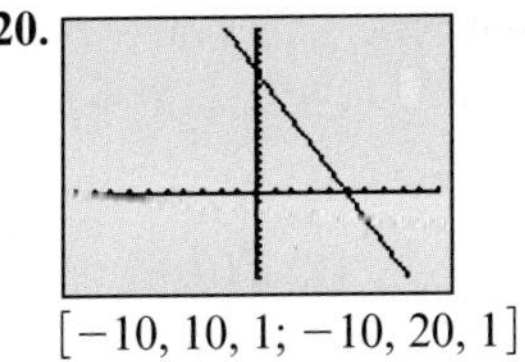

$[-10, 10, 1; -10, 20, 1]$

Section 3.3

1. a. (2004, 8.87 billion lb), (2010, 10.1 billion lb)

b. $\frac{0.21 \text{ billion lb}}{1 \text{ year}}$

2. a. (1998, 9.5 lb), (2011, 7.6 lb) **b.** $-\frac{0.15 \text{ lb}}{1 \text{ year}}$

3. a. (2007, 69.5 births per 1000 women), (2010, 64.7 births per 1000 women)

b. $-\frac{1.6 \text{ births per 1000 women}}{1 \text{ year}}$

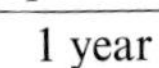

4. a.

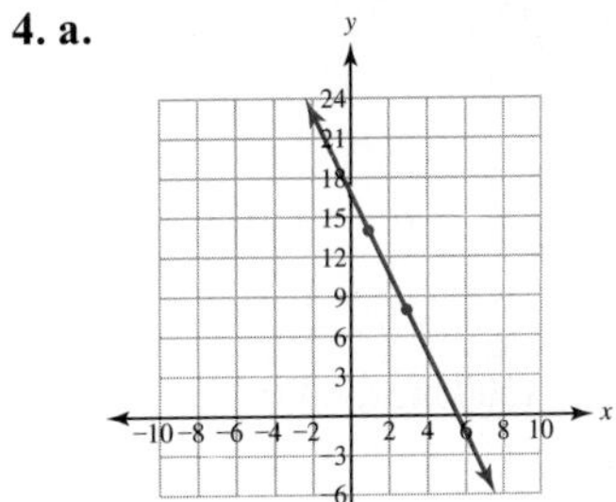

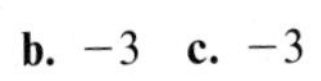

b. −3 **c.** −3

5. a.

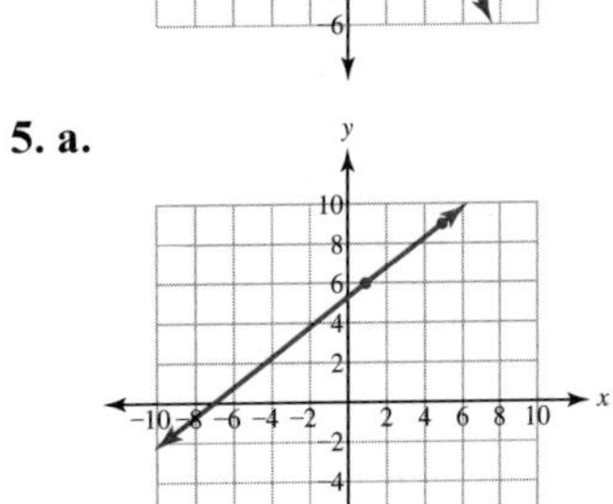

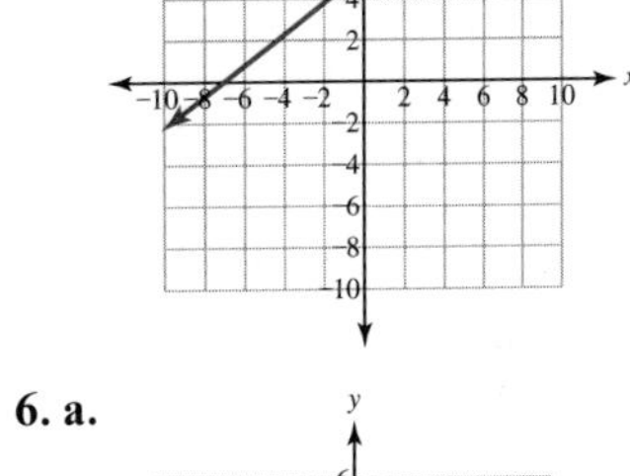

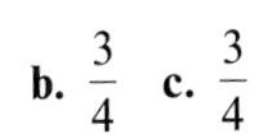

b. $\frac{3}{4}$ **c.** $\frac{3}{4}$

6. a.

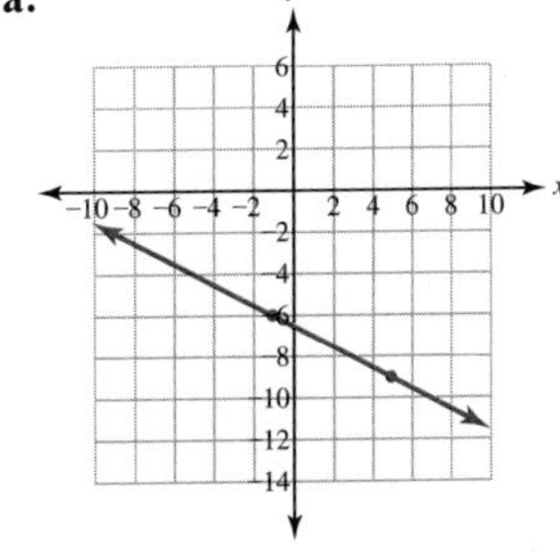

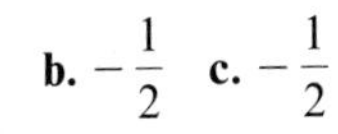

b. $-\frac{1}{2}$ **c.** $-\frac{1}{2}$

7. a.

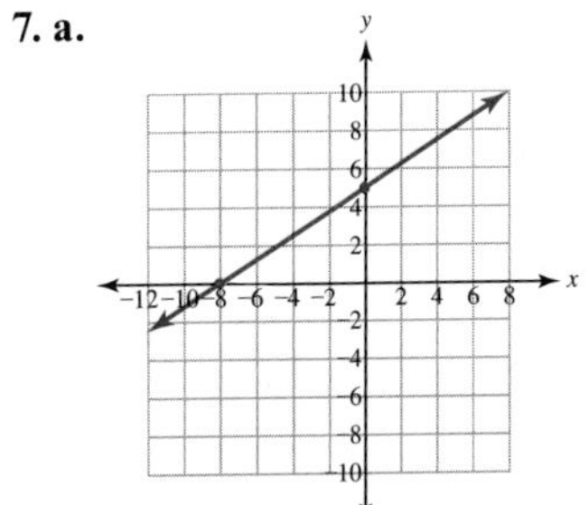

b. $\frac{5}{8}$ **c.** $\frac{5}{8}$

8. a. Answers may vary. Two possible points are $(-1, -1)$ and $(0, 1)$. **b.** 2 **9. a.** Answers may vary. Two possible points are (0 s, 90 cm) and (5 s, 40 cm). **b.** $-\dfrac{10 \text{ cm}}{1 \text{ s}}$

10. $-\dfrac{1}{6}$ **11. a.** $(0, 6)$ **b.** $(8, 0)$ **c.** $-\dfrac{3}{4}$ **12. a.** 0 **b.** horizontal **13. a.** undefined **b.** vertical

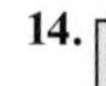

14.

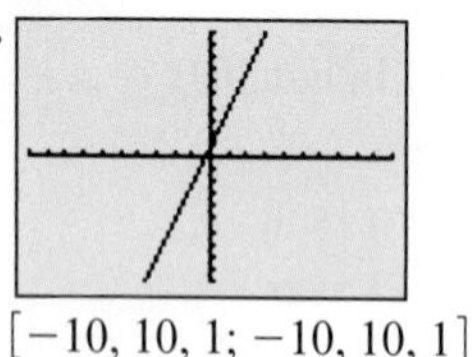

$[-10, 10, 1; -10, 10, 1]$

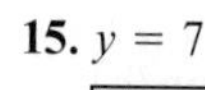

15. $y = 7$

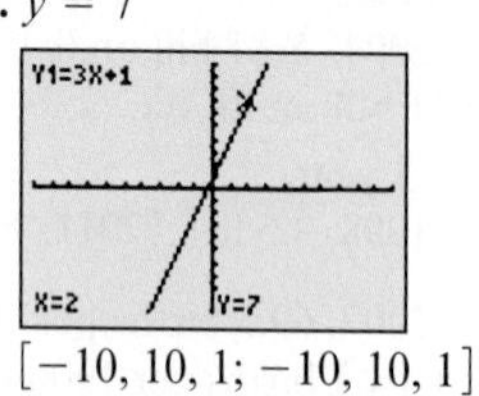

$[-10, 10, 1; -10, 10, 1]$

16. Answers may vary. The calculator displays an error.

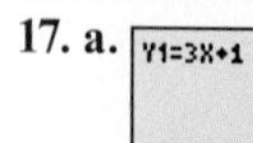

17. a.

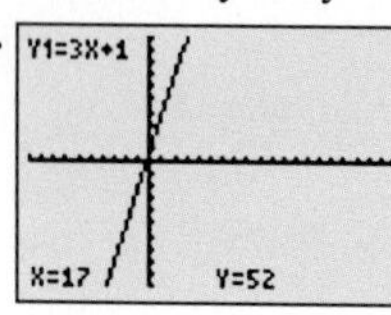

$[-10, 20, 1; -10, 10, 1]$

b. $y = 52$

Section 3.4

1. a. (0 s, 2100 megabits), (200 s, 1500 megabits)

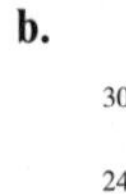

b.

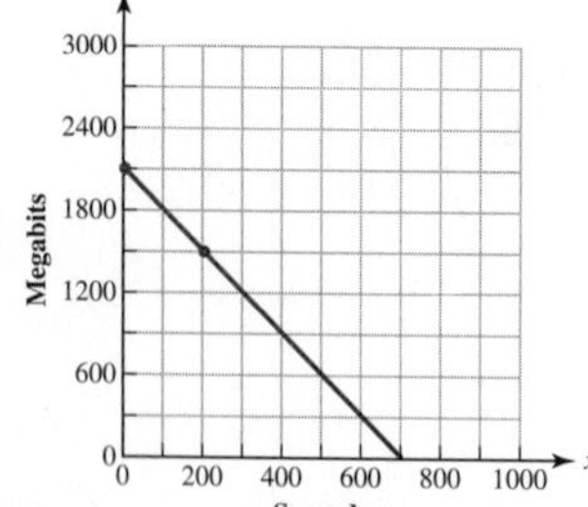

c. (0 s, 2100 megabits); answers may vary. The y-coordinate represents the amount left to download when the download starts. **d.** $-\dfrac{3 \text{ megabits}}{1 \text{ s}}$; answers may vary. The slope represents the speed of the download.

e. $y = \left(-\dfrac{3 \text{ megabits}}{1 \text{ s}}\right)x + 2100 \text{ megabits}$ **f.** 300 megabits

g. (700 s, 0 megabits); answers may vary. The x-coordinate represents the time needed to complete the download.

2. a. $y = \left(\dfrac{60 \text{ bricks}}{1 \text{ hr}}\right)x + 2400 \text{ bricks}$ **b.** 9600 bricks

3. $y = 5x + 9$ **4.** $y = -8x - 5$ **5.** $y = \dfrac{4}{3}x - 9$

6. $y = 7x + \dfrac{1}{2}$ **7. a.** $y = \dfrac{9}{2}x - 18$ **b.** $\dfrac{9}{2}$ **c.** $(0, -18)$

d. $(4, 0)$ **8. a.** $y = -\dfrac{3}{5}x - \dfrac{12}{5}$ **b.** $-\dfrac{3}{5}$ **c.** $\left(0, -\dfrac{12}{5}\right)$

d. $(-4, 0)$ **9. a.** $y = \dfrac{1}{6}x - \dfrac{3}{8}$ **b.** $\dfrac{1}{6}$ **c.** $\left(0, -\dfrac{3}{8}\right)$ **d.** $\left(\dfrac{9}{4}, 0\right)$

10. a. $y = \dfrac{3}{7}x + \dfrac{50}{7}$ **b.** $\dfrac{3}{7}$ **c.** $\left(0, \dfrac{50}{7}\right)$ **d.** $\left(-\dfrac{50}{3}, 0\right)$

11. a. $(0, -5)$ **b.** 4 **c.** $y = 4x - 5$

12. a. $(0, 2)$ **b.** -6 **c.** $y = -6x + 2$

13. a. $(0, -4)$ **b.** $\dfrac{2}{5}$ **c.** $y = \dfrac{2}{5}x - 4$

14. a.

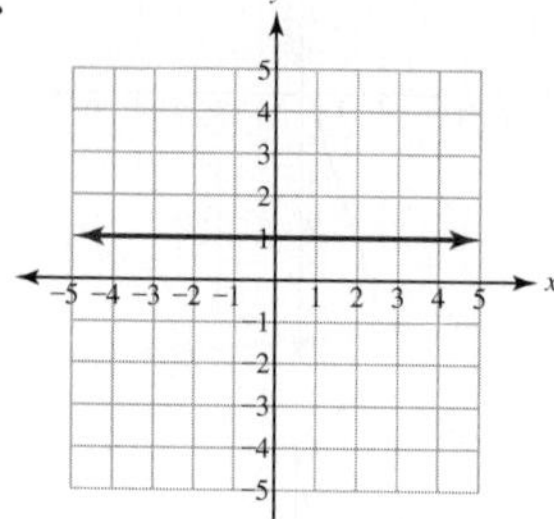

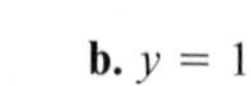

b. $y = 1$

15. a.

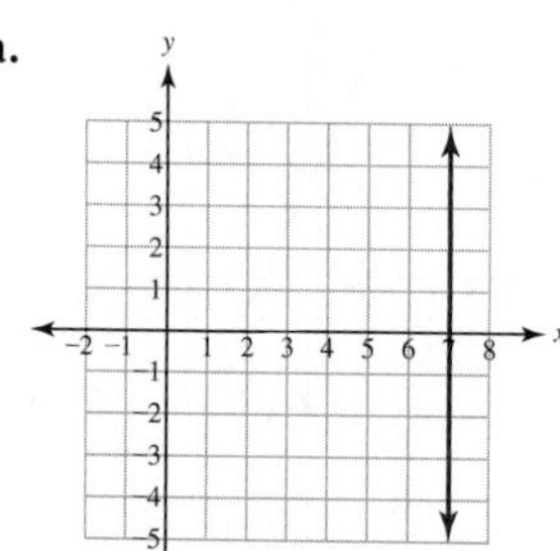

b. $x = 7$

16.

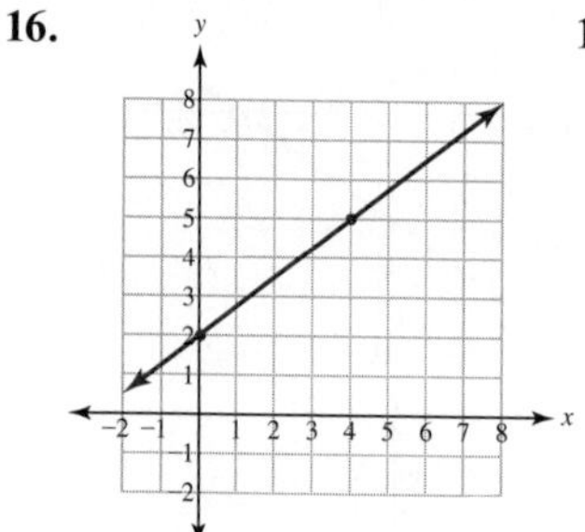

17.

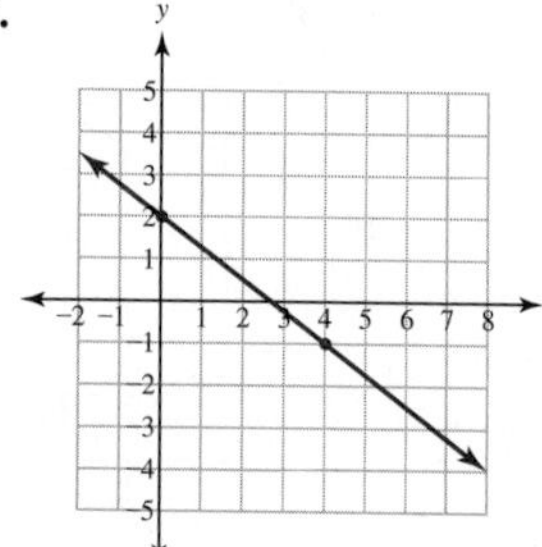

18.

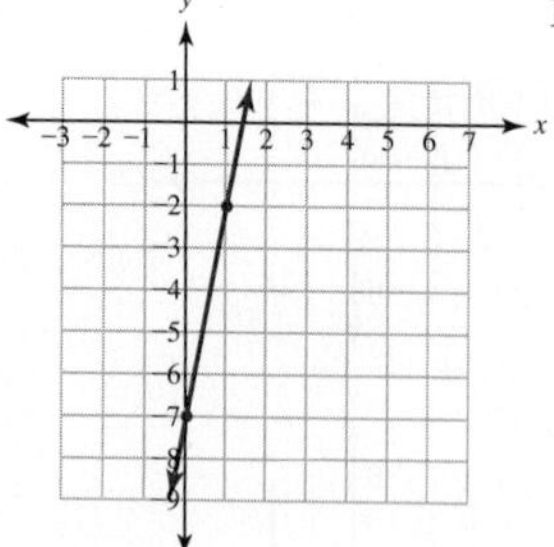

19.

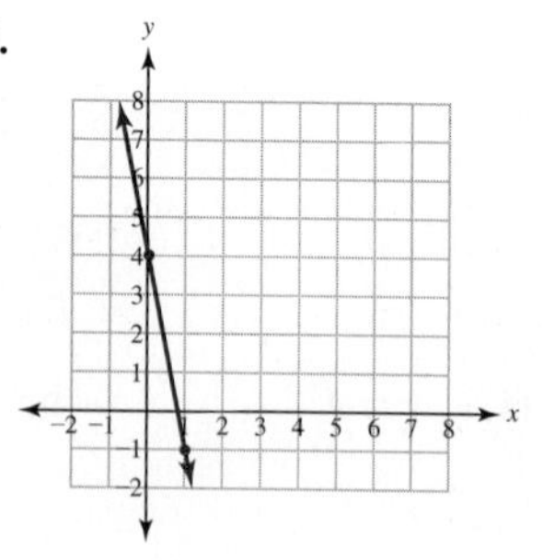

20. a. $y = \dfrac{2}{3}x - 4$ **b.**

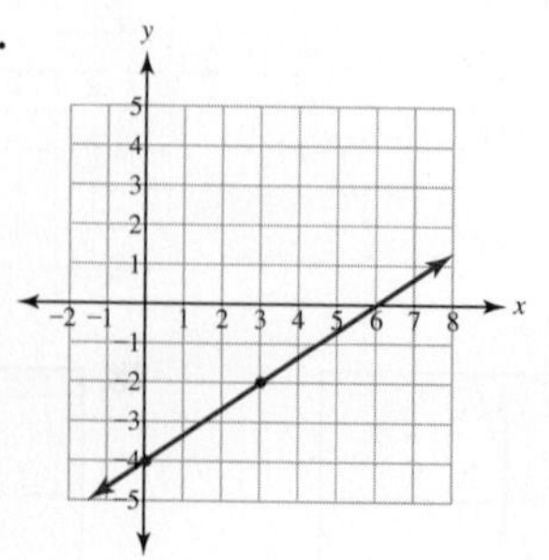

21. a. $y = -\frac{6}{5}x + 8$ **b.**

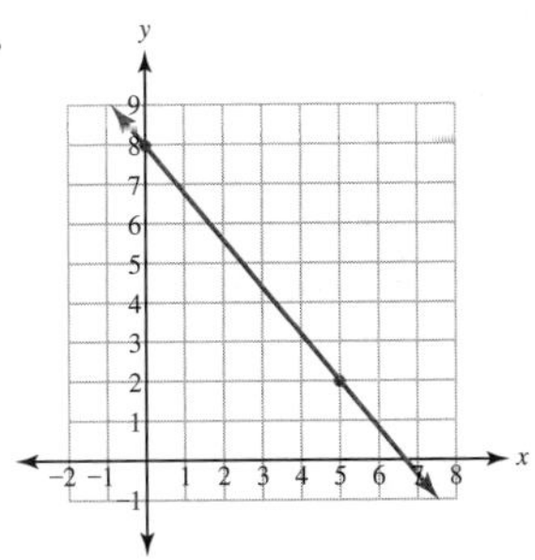

Section 3.5

1. a. $y = 3x - 3$ **b.** $(0, -3)$

2. $y = -\frac{5}{8}x + 11$ **3.** $y = 7x - 41$

4. $y = -\frac{3}{5}x + \frac{19}{5}$ **5.** $y = \frac{3}{7}x + \frac{5}{7}$

6. a. $y = \left(\frac{\$0.549 \text{ trillion}}{1 \text{ year}}\right) + \$8.717 \text{ trillion}$ **b.** Answers may vary. The slope is the annual increase in gross domestic product. **c.** Answers may vary. The y-coordinate is the gross domestic product in 1998. **d.** $16.403 trillion
e. Answers may vary. The U.S. economy was shrinking.

7. a. 7 **b.** 7 **c.** $y = 7x + 43$ **8. a.** 3 **b.** 3
c. $y = 3x - 19$ **9.** $y = -4x + 7$

10. a. 2 **b.** $-\frac{1}{2}$ **c.** $y = -\frac{1}{2}x - 2$

11. a. 3 **b.** $-\frac{1}{3}$ **c.** $y = -\frac{1}{3}x + \frac{17}{3}$

12. Answers may vary. These lines are not perpendicular because the product of $\frac{5}{6}$ and $\frac{6}{5}$ is not -1.

13. a.

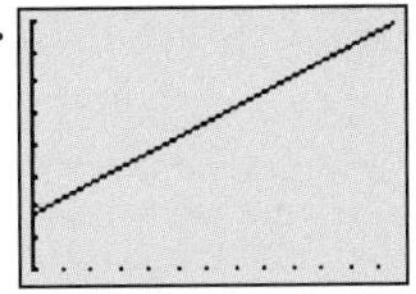

[0, 25, 2; 10,000, 90,000, 10,000]

b. Answers may vary. 54,010 medical emergencies

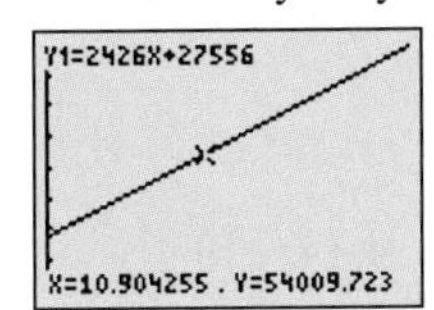

[0, 25, 2; 10,000, 90,000, 10,000]

c. 54,240 medical emergencies

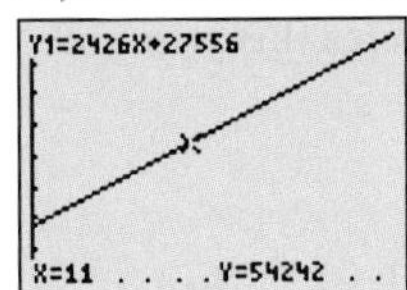

[0, 25, 2; 10,000, 90,000, 10,000]

Section 3.6

1.

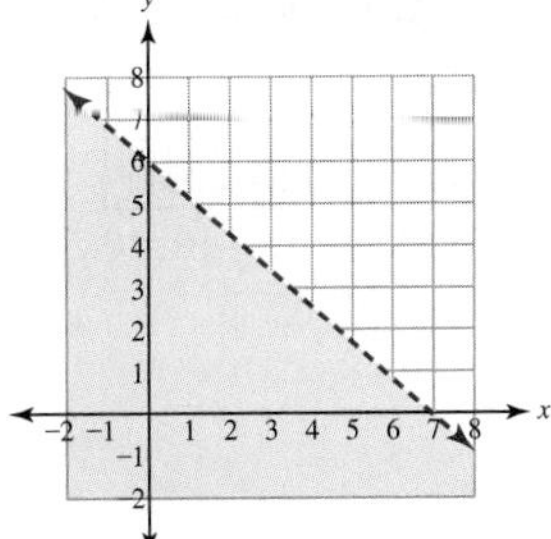

2.

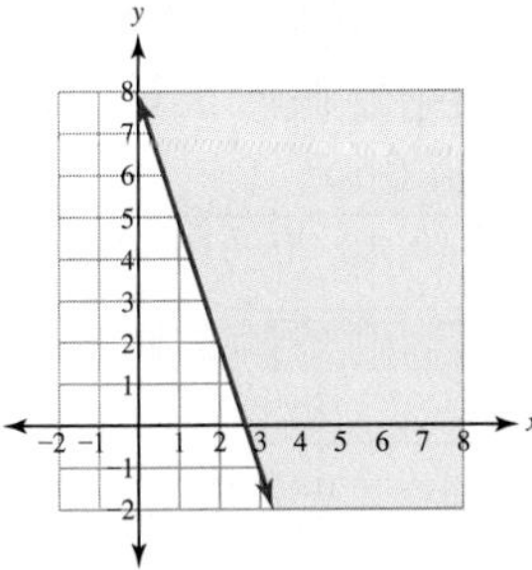

3.

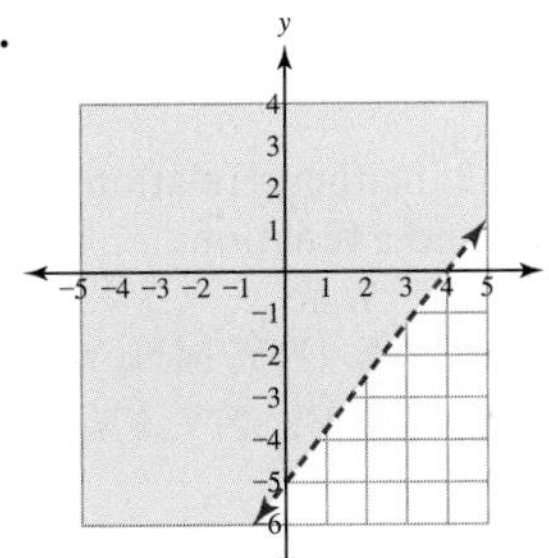

4.

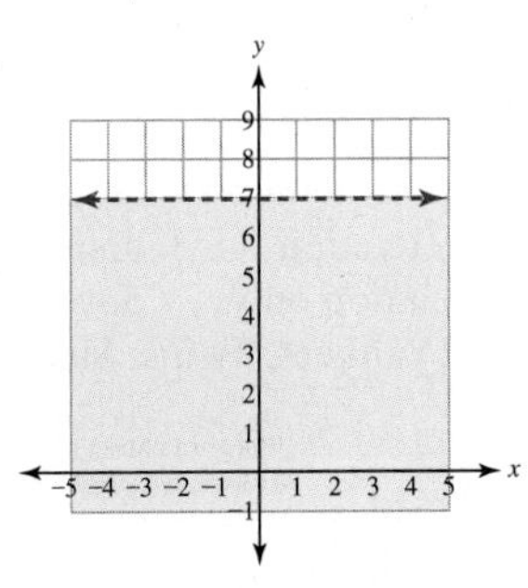

5. $y < -4x + 4$ **6.** $y \geq -\frac{7}{5}x - \frac{8}{5}$ **7.** $x < -3$

8. x = amount of clear opening; $x \geq 5.7 \text{ ft}^2$
9. x = amount of rent per month; $x \leq \$550$
10. x = amount of Guatemala Antigua coffee; y = amount of Colombia Nariño coffee; $x + y \leq 7000$ lb
11. x = employee's total salary; y = rise in employee's pension contribution; $y \geq 0.01x$
12. x = volume of garbage per household; $x \leq 90$ lbs
13. a. x = number of digital photo files; y = number of digital video files; $x + y \leq 100$

b.

14. a. x = number of USB drives; y = number of coffee mugs; $\left(\frac{\$5}{1 \text{ USB drive}}\right)x + \left(\frac{\$8}{1 \text{ coffee mug}}\right)y \leq \$4000$

b.

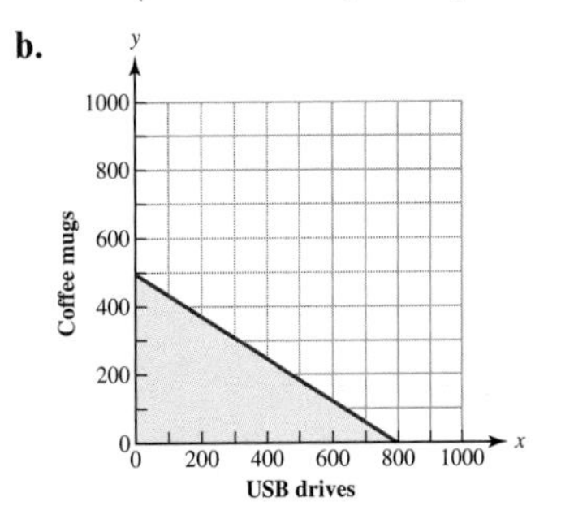

15. 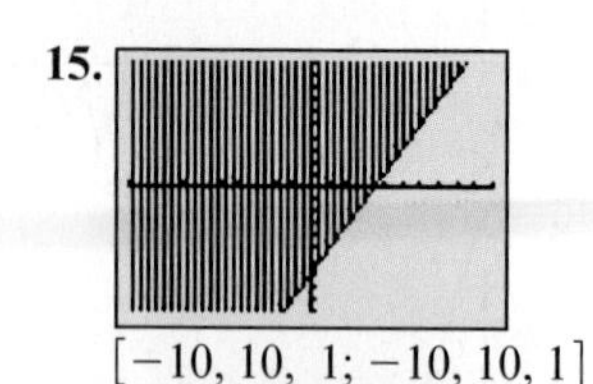

[−10, 10, 1; −10, 10, 1]

16. 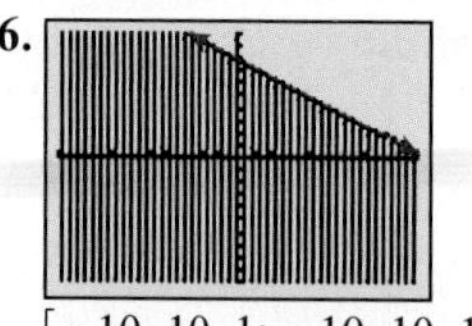

[−10, 10, 1; −10, 10, 1]

17. 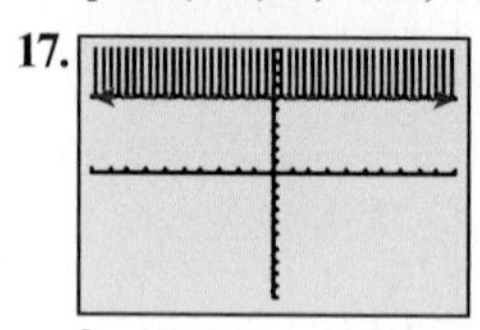

[−10, 10, 1; −10, 10, 1]

Section 3.7

1. both a relation and a function **2.** neither a relation nor a function **3.** both a relation and a function **4.** a relation only **5.** yes **6.** yes **7.** no **8. a.** {Yankees, Twins, Mets, Mariners} **b.** {NY, MN, WA} **9. a.**

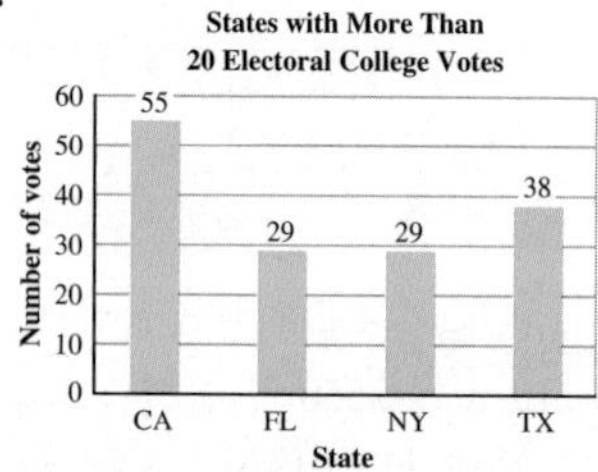

b. {CA, FL, NY, TX} **c.** {55 votes, 29 votes, 38 votes} **10. a.** x **b.** y **c.** the set of real numbers **d.** the set of real numbers **11.** −17 **12.** 53 **13.** Pitt **14.** 2

15. 

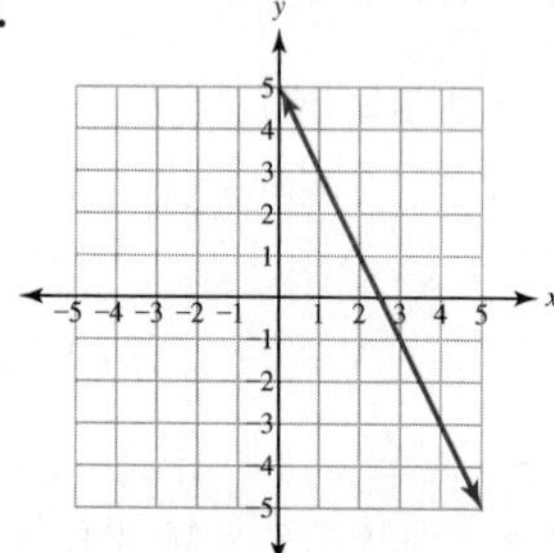

16.

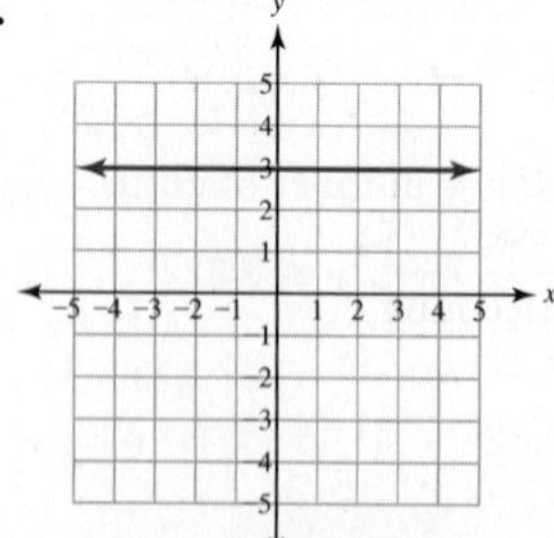

17. a. 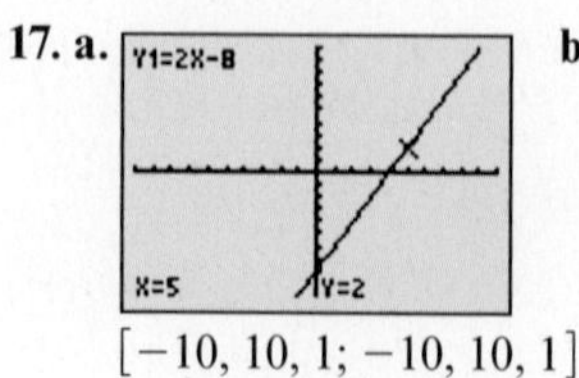

[−10, 10, 1; −10, 10, 1]

b. 2

18. a. 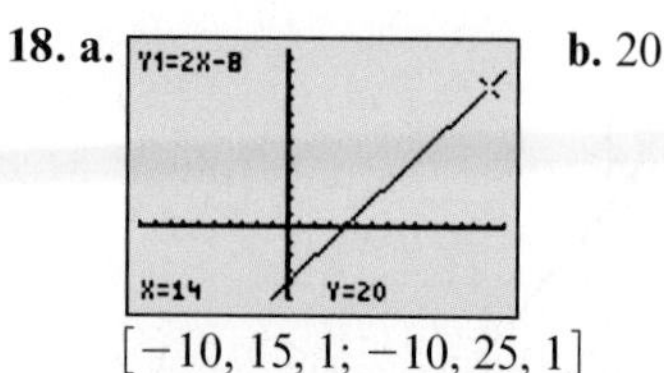

[−10, 15, 1; −10, 25, 1]

b. 20

19. a. Y1=2X−10, X=−3, Y=−16

[−10, 10, 1; −20, 10, 1]

b. −16

CHAPTER 4

Section 4.1

1. no solution **2.** infinitely many solutions **3.** one solution **4.** one solution **5.** (1, −1) **6.** (3, 5) **7.** (5, 3) **8.** (0, 4) **9.** (3, −2) **10.** infinitely many solutions; $\left\{(x, y) \mid y = -\frac{2}{3}x + 2\right\}$ **11.** (2000 products, \$4000) **12. a.** $y = \left(\frac{\$3.50}{1\text{ product}}\right)x$ **b.** (1000 products, \$3500)

13.

[−10, 10, 1; −10, 10, 1]; (1, 3)

14. 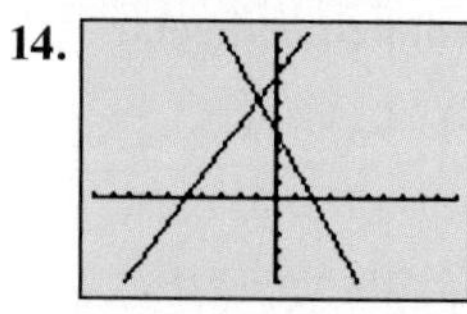

[−10, 10, 1; −10, 20, 2]; (−1, 12)

15. 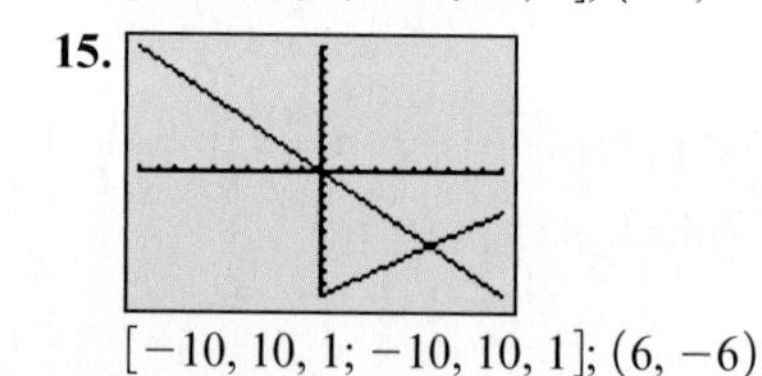

[−10, 10, 1; −10, 10, 1]; (6, −6)

Section 4.2

1. (2, 5) **2.** (1, 4) **3.** (−9, −2) **4.** (4, −5) **5.** $\left(\frac{63}{16}, \frac{15}{8}\right)$ **6.** no solution **7.** (0, 1) **8.** (0, 0) **9.** infinitely many solutions; $\{(x, y) \mid y = 4x + 2\}$ **10.** 250 products, \$1,075,000 **11.** 60 acres this year and 15 acres next year

Section 4.3

1. $\frac{1}{3}$ **2.** $\frac{5}{9}$ **3.** $\frac{1}{9}$ **4.** $\frac{13}{99}$ **5. a.** (1, 2) **6. a.** (−2, 3) **7. a.** (2, −3) **8. a.** $\left(-2, -\frac{1}{2}\right)$ **9. a.** (10, −3) **10. a.** (6, −5) **11. a.** (1, −4) **12.** no solution **13.** infinitely many solutions; $\{(x, y) \mid 6x - y = 9\}$ **14.** no solution **15.** no solution **16.** Volume of large cup is 10 oz, and volume of small cup is 4 oz. **17.** 5 movies for \$9.99 and 3 movies for \$14.99

Section 4.4

1. 0.8 L Mixture A and 1.2 L Mixture B
2. 40 oz of insecticide concentrate and 216 oz of water
3. 4.4 L 45% antifreeze mixture and 1.6 L pure antifreeze
4. 224 L of water; the total volume is 424 L
5. 4 reams white paper and 5 reams colored paper
6. 158 lb ham and 92 lb turkey
7. length 10 in. and width 6 in.
8. length 25 in. and width 17 in. **9.** 180 min, 210 mi

Section 4.5

1.

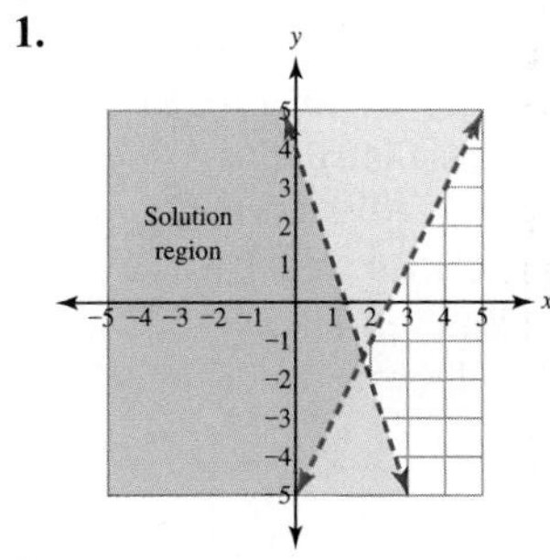

2.

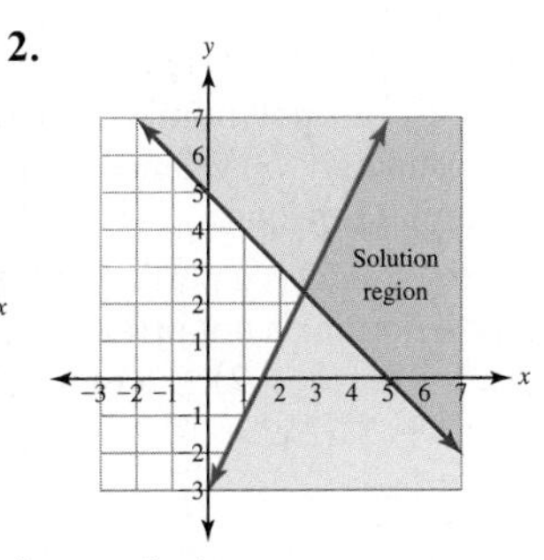

3.

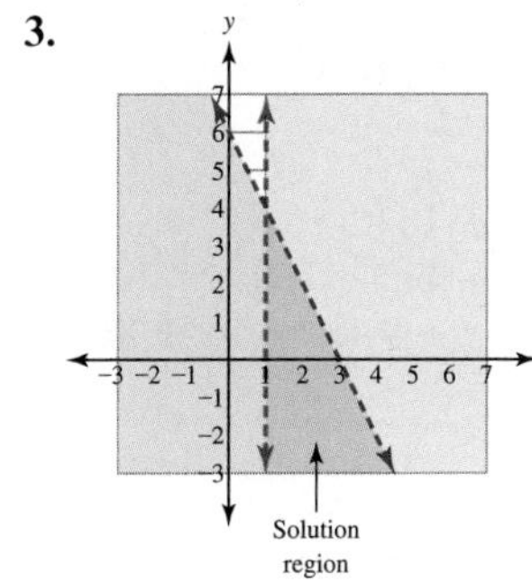

4. no solution;

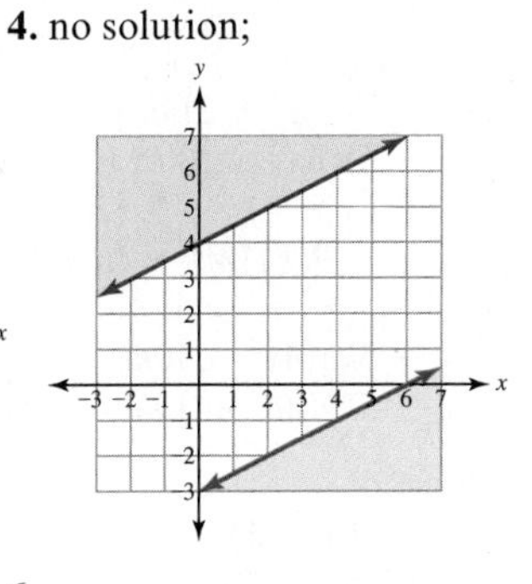

5.

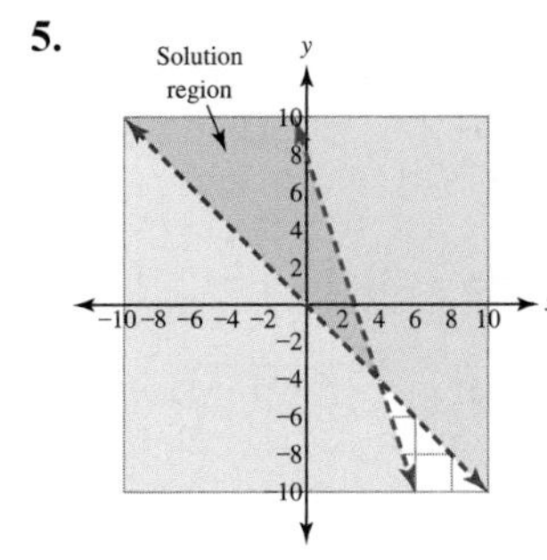

6.

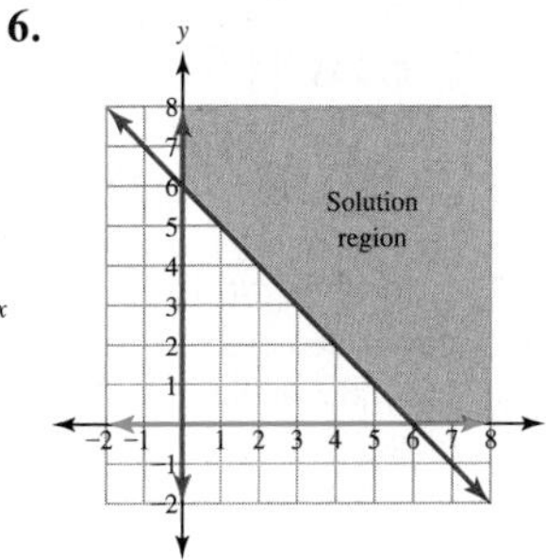

7.

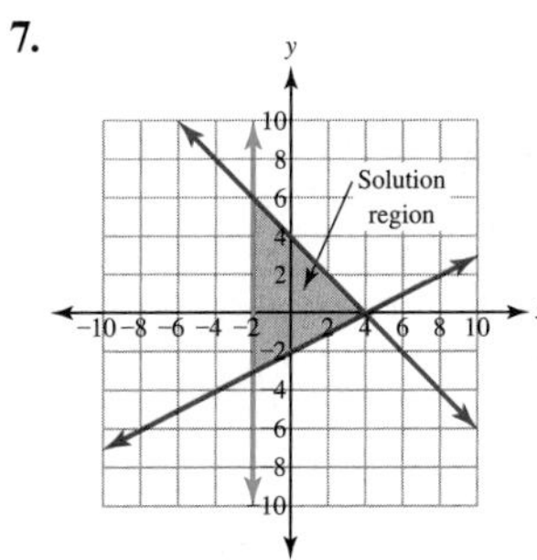

8. yes **9.** no **10.** $x \geq 400$ lb
11. $x \geq 0$; $y \geq 0$; $x + y \leq 30$
12. a. $\left(\dfrac{10\text{ min}}{\text{windows washed}}\right)x + \left(\dfrac{15\text{ min}}{\text{windows caulked}}\right)y \leq 180$ min; $y \geq 8$ windows caulked; $x \geq 4$ windows washed

b.

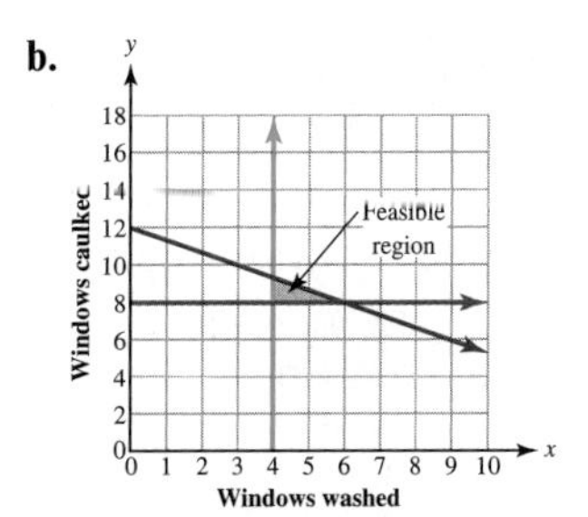

13.

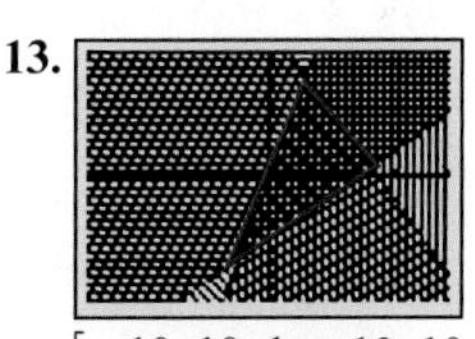

$[-10, 10, 1; -10, 10, 1]$

CHAPTER 5

Section 5.1

1. n^4 **2.** $7n^4$ **3.** $(7n)^4$ **4.** $-8z^3$ **5.** $5x^2 - 4x - 9$
6. $-10x$ **7.** $15x^9$ **8.** p^9 **9.** $-2x^4$ **10.** $\dfrac{5a^5}{3}$
11. m^{45} **12.** h^3k^3 **13.** $\dfrac{w^8}{p^8}$ **14.** $25x^6$ **15.** $\dfrac{9a^2}{b^8}$
16. $\dfrac{16a^2}{25b^8}$ **17.** 1 **18.** 1 **19.** 1 **20.** $\dfrac{1}{x^5}$ **21.** $\dfrac{1}{y^{15}}$
22. 1 **23.** $\dfrac{z^4}{p^6}$ **24.** $\dfrac{5}{f^4}$ **25.** $\dfrac{3m^2}{4}$ **26.** $144p^{14}$
27. $\dfrac{32w^8}{9}$

Section 5.2

1. 8.5×10^{-8} m **2.** $3 \times 10^{10}\ \dfrac{\text{cm}}{\text{s}}$ **3.** 9.8×10^3 kg
4. 6.3×10^{-5} L **5.** $\$2 \times 10^9$ **6.** $\$3.5 \times 10^{12}$
7. $6.8 \times 10^{15}\text{ km}^2$ **8.** $1.7 \times 10^{16}\text{ km}^2$ **9.** $1.7 \times 10^{-14}\text{ km}^2$
10. $3 \times 10^{-3}\ \dfrac{\text{g}}{\text{mL}}$ **11.** 7×10^2 m **12.** $6.4 \times 10^7\text{ m}^3$
13. 8×10^5 km **14.** 1.2×10^6 m **15.** 3.9×10^6 km
16. 4.5×10^{-8} m **17.** 3.8×10^{-4} km **18.** 9.05×10^{-2} s
19. $E = 5.44 \times 10^2\ \dfrac{\text{kg}\cdot\text{m}^2}{\text{s}^2}$ **20.** $v = -1.2 \times 10^{-1}\ \dfrac{\text{m}}{\text{s}}$
21. $8.5 \times 10^{15}\text{ km}^2$ **22.** $2.2 \times 10^{-3}\ \dfrac{\text{g}}{\text{mL}}$
23. $1.5625 \times 10^7\text{ m}^3$

Section 5.3

1. a. 3 **b.** x **c.** 1 **2.** a, c, d, e **3.** Answers may vary. A monomial such as $4x^2$ has one term, a binomial such as $3x + 2$ has two terms, and a trinomial such as $x^2 + x + 1$ has three terms. **4.** $9x^4 - x^3 - 14x^2 + 8x + 12$; 9
5. 3 **6.** 0 **7.** $-4d^3 - 3d^2 + d + 9$
8. $9x^2 + 16x - 13$ **9.** $x^2 + 14x - 5$ **10.** $2x^2 + 5x - 9$
11. $12h^3 + 3h^2k - 7hk^2 - 26$ **12.** $\dfrac{21}{40}x^2 + \dfrac{5}{12}x - 7$
13. $18x - 60$ **14.** $18x^3 - 60x^2$ **15.** $28a^5 + 35a^4 - 14a^3$
16. $18x^3 - 60x^2y$ **17.** $\dfrac{1}{6}c^2 - \dfrac{4}{5}c$ **18.** $\dfrac{1}{4}n^4 + \dfrac{5}{9}n^3$
19. $\dfrac{9}{4}x + 9$

Section 5.4
1. $10k^2 + 8k$ **2.** $10k^2 + 48k + 32$
3. $3a^3 + 22a^2 + 30a + 8$ **4.** $40c^2 - 35c$
5. $2d^2 + 7d - 30$ **6.** $6w^3 - 22w^2 - 40w$
7. $x^2 + 10x + 25$ **8.** $9x^2 + 30x + 25$ **9.** $x^2 - 10x + 25$
10. $9x^2 - 30x + 25$ **11.** $x^2 - 25$ **12.** $9x^2 - 25$
13. a. $W + 4$ **b.** $4W + 8$ **c.** $W^2 + 4W$

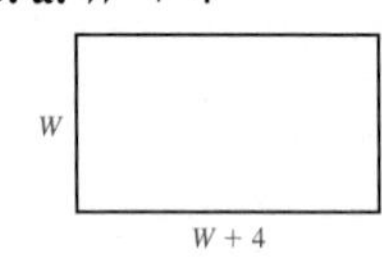

14. a. $a + 6$; $a - 2$ **b.** $3a + 4$

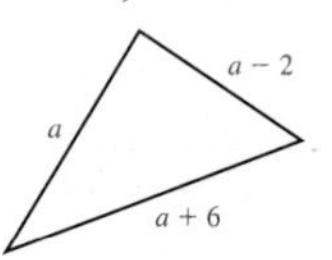

15. a. $2b - 10$ **b.** $b^2 - 5b$

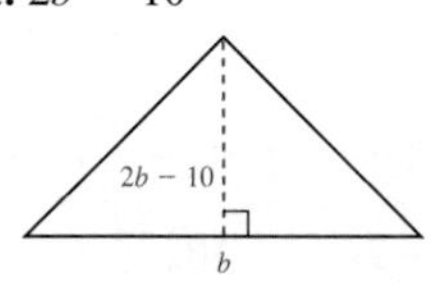

Section 5.5
1. a. dividend: $28a^6b^{11}c$; divisor: $7a^2bc$ **b.** $4a^4b^{10}$
2. a. dividend: $21x^3 - 27x^2 + 6x - 9$; divisor: $3x$
b. $7x^2 - 9x + 2 - \dfrac{3}{x}$
3. a. dividend: $28a^6b^{11}c - 28a^5bc + 35$; divisor: $7a^2bc$
b. $4a^4b^{10} - 4a^3 + \dfrac{5}{a^2bc}$ **4. a.** $x + 10$ **5. a.** $x - 3$
6. a. $3x + 4$ **7. a.** $x + 8$ **8. a.** $(x + 1)$ R(17)
9. a. $(x + 1)$ R(6) **10. a.** $(x + 3)$ R(57)
11. a. $(x - 4)$ R(32)
12. a. $x - 9$ **b.**

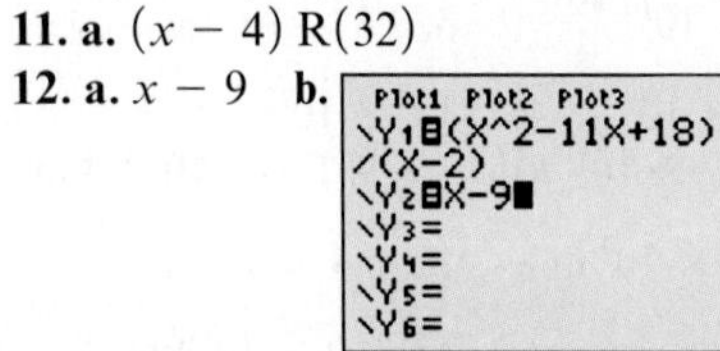

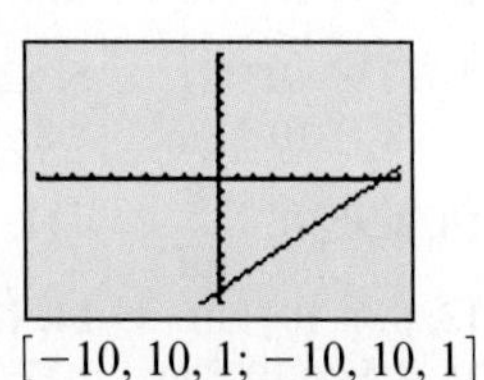

$[-10, 10, 1; -10, 10, 1]$

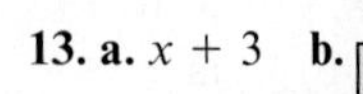

13. a. $x + 3$ **b.**

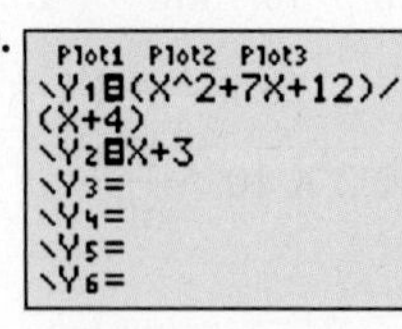

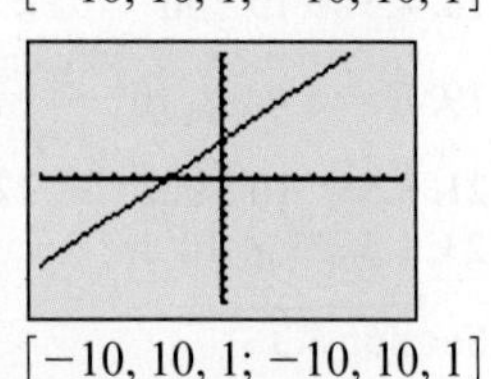

$[-10, 10, 1; -10, 10, 1]$

14. a. $x + 5$ **b.**

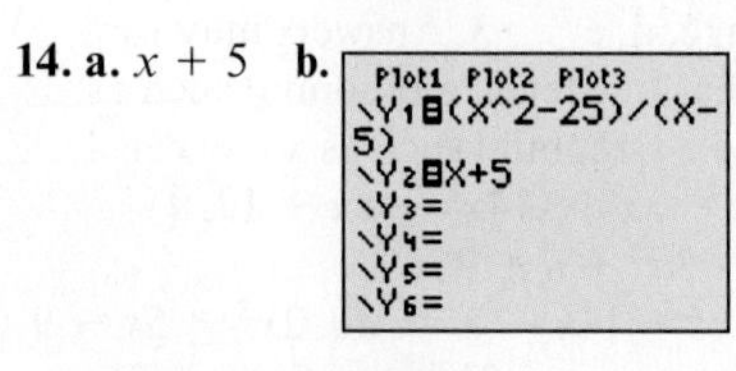

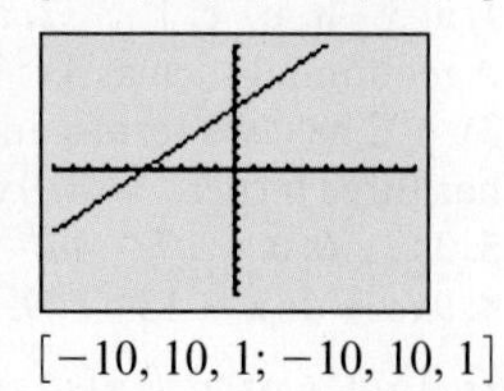

$[-10, 10, 1; -10, 10, 1]$

CHAPTER 6

Section 6.1
1. $4xy$ **2.** $4x^2y^2$ **3.** $15x$ **4.** 18
5. $3xy$ **6.** a^2d **7. a.** $6cd(4d^3 + 3c^2)$
8. a. $2ac(4ac^4 + 15)$ **9. a.** $12ab(3a^2b^3 - 1)$
10. a. $m^2p^2(7m + 3p^2 + 5p)$ **11. a.** prime
12. a. $x(x^2 + 3x - 5)$ **13. a.** $(3x + 2y)(4x + z)$
14. a. $(2a - 3b)(5z + 3c)$ **15. a.** $(7h + 8)(5k - 3)$
16. a. $(x + 5y^2)(7x - 3y)$ **17. a.** $(3y - z)(4x - 5)$
18. a. $(p + 6)(p + 6)$

Section 6.2
1. $(x + 4)(x + 9)$ **2.** $(z - 3)(z - 6)$
3. $(w + 9)(w - 6)$ **4.** $(x + 8)(3x + 5)$
5. $(x + 6)(2x + 5)$ **6.** $(x - 6)(2x - 5)$
7. $(3x + 5)(2x + 1)$ **8.** $(3x - 2)(4x + 1)$
9. a. 1 **b.** Answers may vary. Since 1 is a perfect square ($\sqrt{1} = 1$), this polynomial is not prime. **10. a.** -287 **b.** Answers may vary. Since -287 is less than 0, this polynomial is prime. **11. a.** 1369 **b.** Answers may vary. Since 1369 is a perfect square ($\sqrt{1369} = 37$), this polynomial is not prime. **12.** $(x^3 + 4)(x^3 + 5)$
13. $(a^4 - 8)(a^4 + 2)$ **14.** $(2x^2 + 1)(x^2 - 5)$
15. $(2x^3 - 3)(3x^3 + 4)$

Section 6.3
1. $(x + 6)(x + 6)$ or $(x + 6)^2$
2. $(2h + 3)(2h + 3)$ or $(2h + 3)^2$ **3.** prime
4. $(7h^3 - 2)(7h^3 - 2)$ or $(7h^3 - 2)^2$
5. $(6x^5 + 5)(6x^5 + 5)$ or $(6x^5 + 5)^2$
6. $(x + 6)(x - 6)$ **7.** $(5d + f)(5d - f)$
8. prime **9.** $(11m^4 + 4n^2)(11m^4 - 4n^2)$
10. $(x - 9)(x^2 + 9x + 81)$ **11.** $(x + 9)(x^2 - 9x + 81)$
12. $(2h + 5k)(4h^2 - 10hk + 25k^2)$
13. $(2h - 5k)(4h^2 + 10hk + 25k^2)$

Section 6.4
1. $4(x + 2)(x + 3)$ **2.** $p(p - 5)(p + 2)$
3. $3(w - 2)(w + 2)$ **4.** $2a(a + 4)$
5. $2(p + 3w)(2x + y)$ **6.** $2a(a - 9)(a + 1)$
7. $6(n^2 + 4)$ **8.** $4(5y - x)(3x + 8)$
9. $3(x - 20)(x + 9)$ **10.** $18(h^2 - 3k^2)$
11. $(3a + 5)(2c + d)$ **12.** prime
13. $2(5p - 2m)(8mp + 7)$ **14.** $p(p - 3)(2p + 5)$
15. prime **16.** $2(2x + 1)(3y - z)$
17. $2(x + 1)(6x + 5)$ **18.** $(x + 2)(x - 2)(x + 3)$
19. $u^5(u + 8)(u - 3)$ **20.** $(2p^{10} - 5w^3)^2$
21. $2(9x^{10} + 6x^5y^2 + y^2)$ **22.** $2(x^3 - 6)^2$
23. $7(x - y)(x^2 + xy + y^2)$
24. $25p(p + 2w)(p^2 - 2pw + 4w^2)$
25. $4c^7d^9(d + 1)(d^2 - d + 1)$
26. $15mrz(r - z)(r^2 + rz + z^2)$ **27.** $2x(8x^3 - 3y^3)$
28. prime

Section 6.5
1. a. $x = 8$ **2. a.** $z = -4$ or $z = 9$
3. a. $x = 0$ or $x = -20$ **4. a.** $d = 0$ or $d = -5$ or $d = 2$
5. a. $x = -1$ or $x = 7$ **6. a.** $c = -3$ or $c = -4$
7. a. $y = -\dfrac{5}{3}$ or $y = 2$ **8. a.** $p = 0$ or $p = -4$
9. a. $u = 0$ or $u = -5$ or $u = 5$ **10. a.** $x = 0$ or $x = 14$
11. a. $x = 5$ **12. a.** $p = 0$ or $p = 6$ or $p = -2$
13. a. $v = 4$ or $v = -4$ **14. a.** $m = 0$ or $m = 6$ or $m = -5$
15. The width is 3 ft, and the length is 5 ft.
16. The width is 4 ft, and the length is 5 ft.

17. a. 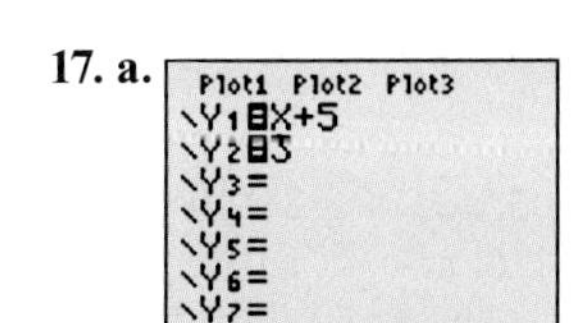

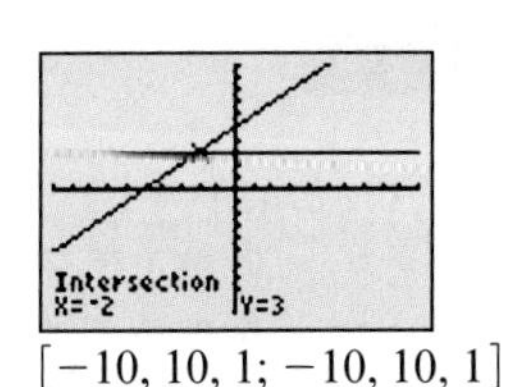

$[-10, 10, 1; -10, 10, 1]$

b. $(-2, 3)$ **c.** $x = -2$

18. a. 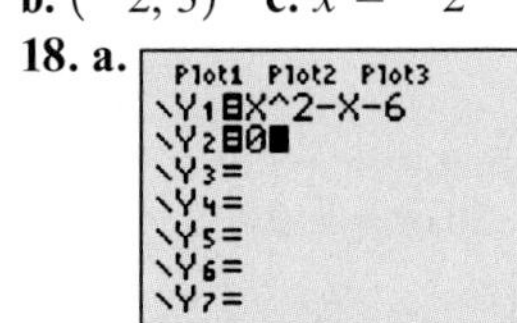

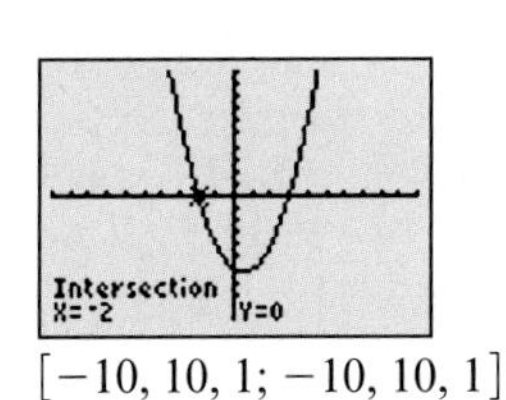

$[-10, 10, 1; -10, 10, 1]$

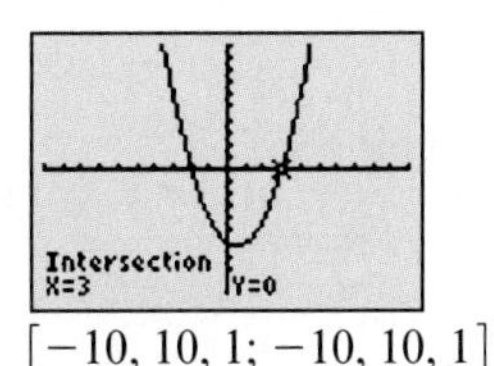

$[-10, 10, 1; -10, 10, 1]$

b. $(-2, 0), (3, 0)$ **c.** $x = -2$ or $x = 3$

19. a. 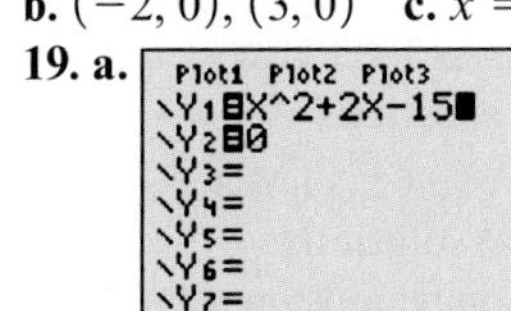

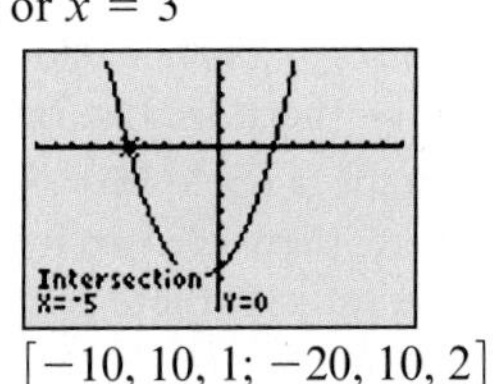

$[-10, 10, 1; -20, 10, 2]$

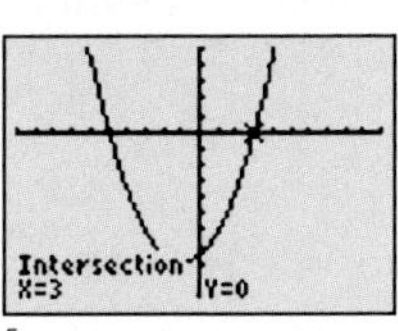

$[-10, 10, 1; -20, 10, 2]$

b. $(-5, 0), (3, 0)$ **c.** $x = -5$ or $x = 3$

CHAPTER 7

Section 7.1

1. $\frac{x-7}{3}$ **2.** $\frac{x-4}{x+1}$ **3.** $\frac{2x+1}{x+6}$ **4.** $\frac{4w^4}{5}$ **5.** $\frac{a+7}{a+1}$
6. $\frac{c-5}{4}$ **7.** $\frac{-x+5}{x+2}$ **8.** -1 **9.** $x = 5, x = -5$
10. $x = -2, x = -7$ **11.** $x = 3, x = -7, x = 0$

Section 7.2

1. $\frac{5}{12}$ **2.** $\frac{1}{3(x+2)}$ **3.** $\frac{2w^6}{w+1}$ **4.** $\frac{2c+1}{c+6}$ **5.** $\frac{12}{7}$
6. $\frac{d+7}{2d+5}$ **7.** $\frac{1}{35k^5m^3}$ **8.** $\frac{4x^5}{x+1}$ **9.** 1

Section 7.3

1. $\frac{6}{7}$ **2.** $\frac{12}{x-9}$ **3.** $\frac{-x-6}{x+1}$ **4.** $c - 6$ **5.** $p - 3$
6. $\frac{k+5}{k+3}$ **7.** 400 **8.** $48xy$ **9.** $96x^2y^2$
10. $(x+2)(x-3)(x+3)$ **11.** $(x-4)^2(x+4)$
12. $p(p+1)^2(p+4)$

Section 7.4

1. 144 **2.** $150x^2$ **3.** $1225ab^2c$ **4.** $160a^6b^2c^5$
5. $(x+6)(x+1)(x+3)$ **6.** $2(x-5)(x+5)$
7. $x^2(x-7)(x-2)$ **8.** $(x-3)(x+4)$
9. $6x^2(x^2+5x+2)$ **10.** $(x+2)(x-3)(x+3)^2$
11. $\frac{12}{80}$ **12.** $\frac{21y^2}{24xy^2}$ **13.** $\frac{3x^2+6x}{30(x+4)(x+2)}$
14. $\frac{5x}{x(x-2)(x+6)}$ **15.** $\frac{2x+2}{(x-7)(x+7)(x+1)}$
16. $\frac{3x-15}{(x+3)(x+2)(x-5)}$ **17.** $\frac{8x+8}{20x^2(x-2)(x+1)}$
18. $\frac{17}{30}$ **19.** $\frac{10x+21y}{216x^2y^2}$ **20.** $\frac{7x^2+29x+14}{(x-5)(x+2)(x+6)}$
21. $\frac{-5a^2+18a+17}{(a-6)(a+2)(2a+1)}$ **22.** $\frac{w^2-2w+36}{(w-4)(w+7)}$

Section 7.5

1. $\frac{14}{3}$ **2.** $\frac{2x+6}{x-3}$ **3.** $\frac{x-1}{x+1}$ **4.** $\frac{a+1}{2}$ **5.** $\frac{7}{6}$
6. $\frac{3x^2+13x+12}{(x+2)(2x+3)}$ **7.** $\frac{3x^2+2x}{(x+1)(3x-1)}$ **8.** $\frac{7}{6}$
9. $\frac{3x^2+2x}{(x+1)(3x-1)}$ **10.** $\frac{3x^2+13x+12}{(x+2)(2x+3)}$

Section 7.6

1. a. $x = \frac{14}{15}$ **2. a.** $x = 14$ **3. a.** $x = 24$ **4. a.** $x = 33$
5. a. $x = 4$ **6. a.** $x = -2$ or $x = 8$ **7. a.** $x = 5$ or $x = 12$
8. 7 min **9.** 137 min **10.** 4.3 mL **11.** 2,285,000 people

Section 7.7

1. a. $k = 4$ **b.** $y = 4x$ **c.** $y = 20$
d. 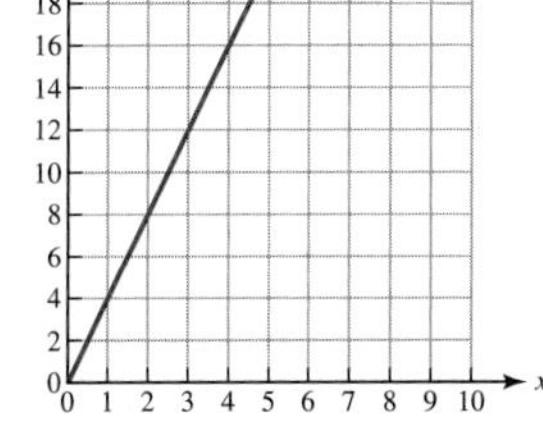

e. $y = 8$

2. a. $k = \frac{\$0.235}{1 \text{ mi}}$ **b.** $y = \left(\frac{\$0.235}{1 \text{ mi}}\right)x$ **c.** \$88.13

3. a. $k = 30$ **b.** $y = \frac{30}{x}$ **c.** $y = 15$

4. a. $k = 24 \text{ in.}^2$ **b.** $y = \frac{24 \text{ in.}^2}{x}$ **c.** 3 in.

5. inverse **6.** inverse **7.** direct

CHAPTER 8

Section 8.1

1. 2 **2.** 9 **3.** not a real number **4.** 1.41 **5.** 3.87
6. 9 **7.** 16 **8.** 100 **9.** 4 **10.** $3\sqrt{2}$ **11.** $5\sqrt{2}$
12. $5\sqrt{3}$ **13.** $10\sqrt{7}$ **14.** x^4 **15.** p^{10} **16.** x^2y^3
17. $ab^4\sqrt{c}$ **18.** $4p^3w^2\sqrt{3pw}$ **19.** $5h^2\sqrt{k}$ **20.** $6a^4b^5\sqrt{3a}$

Section 8.2

1. $8\sqrt{w}$ **2.** $7\sqrt{m}$ **3.** $13\sqrt{x}$ **4.** $14\sqrt{c} + 7\sqrt{d}$
5. $-6\sqrt{h} + 3\sqrt{k}$ **6.** $10\sqrt{xy} - 4\sqrt{y}$ **7.** $3\sqrt{2}$
8. $9\sqrt{3}$ **9.** $6\sqrt{19}$ **10.** $\sqrt{5}$ **11.** $15\sqrt{p}$ **12.** $8u\sqrt{2}$
13. $-3\sqrt{2a}$ **14.** $12h^2\sqrt{7h}$ **15.** $5p\sqrt{6w}$

Section 8.3

1. $5p\sqrt{6}$ **2.** $7w\sqrt{30p}$ **3.** $15xy\sqrt{7}$ **4.** $12a\sqrt{6b}$
5. $\sqrt{14x} + \sqrt{10}$ **6.** $5a\sqrt{2} + 6\sqrt{5a}$ **7.** $2x - 7$
8. $5x - 2\sqrt{5x} + 1$

Section 8.4

1. $\frac{5}{8}$ **2.** $\frac{2\sqrt{3xy}}{y}$ **3.** $\frac{\sqrt{10u}}{5u}$ **4.** $\frac{2\sqrt{35ac}}{7c}$
5. $\frac{2\sqrt{c} + 2\sqrt{3}}{c - 3}$ **6.** $\frac{2\sqrt{c} - 2\sqrt{3}}{c - 3}$ **7.** $\frac{5\sqrt{x} - 40}{x - 64}$

Section 8.5

1. a. $x = 31$ **2. a.** $x = 76$ **3. a.** $x = 168$
4. a. $x = 4$ or $x = 5$ **5. a.** $x = \frac{3}{4}$ or $x = 1$
6. a. $x = 3$ or $x = 7$ **7. a.** $h = 9$ **8. a.** $x = \frac{1}{2}$
9. a. $x = 23$ **10. a.** $x = -5$ or $x = 3$ **11.** 5.6 mi
12. a.

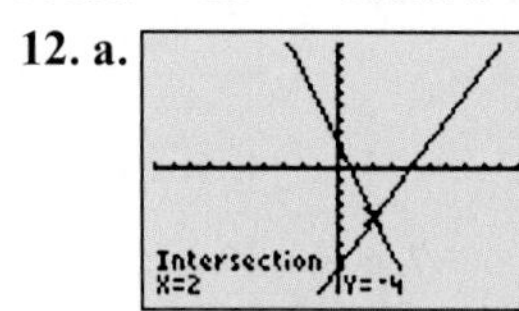

[−10, 10, 1; −10, 10, 1]

b. $x = 2$

13. a.

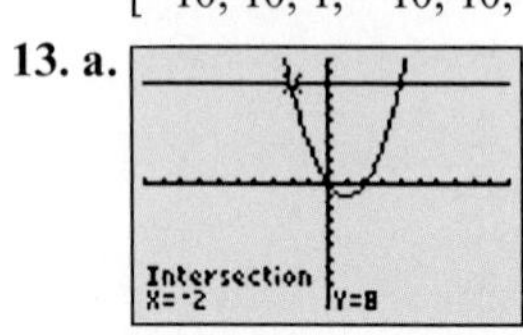

[−10, 10, 1; −10, 10, 1] [−10, 10, 1; −10, 10, 1]

b. $x = -2$; $x = 4$

14. a. [−10, 10, 1, −10, 10, 1]

b. $x = -5$

15. a.

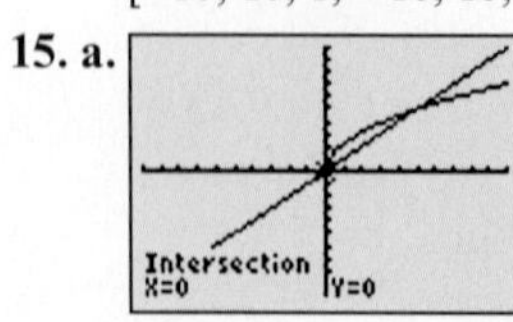

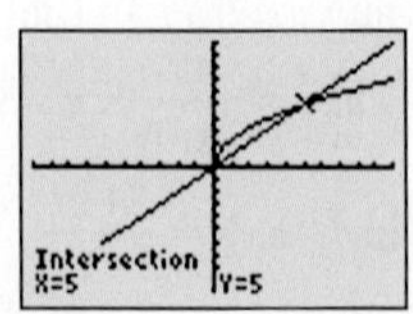

[−10, 10, 1; −10, 10, 1] [−10, 10, 1; −10, 10, 1]

b. $x = 0$; $x = 5$

Section 8.6

1. 5 **2.** −5 **3.** 2 **4.** not a real number **5.** $2\sqrt[4]{2x}$
6. $-7x^2$ **7.** $6\sqrt[5]{y}$ **8.** $\sqrt[3]{2}$ **9.** $\sqrt[3]{2}$ **10.** $2\sqrt[4]{6}$
11. $3x\sqrt[3]{2x}$ **12.** $2\sqrt[3]{3}$ **13.** $2\sqrt[3]{10x}$ **14.** $4\sqrt{2}$
15. $\sqrt[3]{36}$ **16.** $x^2y^{\frac{9}{2}}$ **17.** $x^{\frac{4}{3}}y^3$ **18.** xy^3 **19.** $z^{\frac{3}{5}}$
20. $x^{\frac{41}{72}}$ **21.** $x^{\frac{11}{40}}y^{\frac{11}{45}}$ **22.** $h^{\frac{1}{15}}$ **23.** x **24.** p **25.** $a^{\frac{2}{7}}$
26. $n^{\frac{1}{3}}$ **27.** $x^{\frac{19}{56}}$ **28.** $\frac{3}{7x^{\frac{4}{9}}}$

CHAPTER 9

Section 9.1

1. a. $x = \pm 7$ **2. a.** $h = \pm 2\sqrt{3}$ **3. a.** $z = \pm 2\sqrt{10}$
4. a. $c = \pm 5\sqrt{3}$ **5. a.** $x = \pm 10$ **6. a.** $w = \pm 10$
7. a. $u = \pm 4\sqrt{3}$ **8. a.** $a = -8$ or $a = 2$
9. a. $a = -2$ or $a = 8$ **10. a.** $c = -10$ or $c = -4$
11. $t = 1.6$ s **12.** 49.5 ft

Section 9.2

1. 25 **2.** 25 **3.** 81 **4.** 49 **5. a.** $z = -7$ or $z = -3$
6. a. $z = -5$ or $z = 15$ **7. a.** $p = -20$ or $p = 2$
8. $z = -2 \pm \sqrt{11}$ **9.** $h = 7 \pm \sqrt{61}$
10. $k = -9 \pm \sqrt{79}$

Section 9.3

1. $x = -9$ or $x = -6$ **2.** $w = -9$ or $w = 7$
3. $d = -\frac{3}{2}$ or $d = 7$ **4.** $p = -3$ **5.** $x = \frac{-7 \pm \sqrt{33}}{2}$
6. $w = \frac{-3 \pm \sqrt{65}}{4}$ **7.** $d = \frac{3 \pm \sqrt{19}}{2}$
8. discriminant = 2809; $\sqrt{2809} = 53$, so 2809 is a perfect square; there are two rational solutions
9. discriminant = 65; $\sqrt{65} = 8.06\ldots$, so 65 is not a perfect square; there are two irrational solutions
10. discriminant = 0; there is one rational solution
11. discriminant = −71; there are two nonreal solutions
12. 1.1 s **13.** width = 8 ft; length = 21 ft
14. $a = \frac{9 \pm \sqrt{129}}{2}$ **15.** $k = 0$ or $k = 6$
16. $c = -2 \pm \sqrt{13}$ **17.** $x = 12$ or $x = -1$ **18.** $w = -\frac{4}{5}$
19. $h = \pm\frac{9}{7}$ **20.** $x = \pm 2\sqrt{5}$

Section 9.4

1. $\sqrt{5}i$ **2.** $5i$ **3.** $5\sqrt{2}i$ **4.** $4\sqrt{3}i$
5.

	Complex numbers	Imaginary numbers	Real numbers	Rational numbers	Irrational numbers	Integers	Whole numbers
$5 + 8i$	X						
5	X		X	X		X	X
$\sqrt{5}$	X		X		X		
$5i$	X	X					
−5	X		X	X		X	
0	X		X	X		X	X
$\sqrt{25}$	X		X	X		X	X
$\sqrt{-25}$	X	X					

6. $13 + 9i$ **7.** $3 - 3i$ **8.** $3 + 9i$ **9. a.** $x = \pm 8i$
10. a. $m = \pm 10\sqrt{2}i$ **11. a.** $v = \pm 6i$ **12.** $z = -2 \pm 4i$
13. $h = 7 \pm \sqrt{2}i$ **14.** $k = -9 \pm \sqrt{19}i$
15. $z = -\frac{3}{2} \pm \frac{\sqrt{43}i}{2}$ **16.** $h = \frac{3}{4} \pm \frac{\sqrt{31}i}{4}$
17. $k = -\frac{1}{4} \pm \frac{\sqrt{23}i}{4}$

Section 9.5

1. a. $a = -1, b = 0, c = 0$ **b.** down

c. 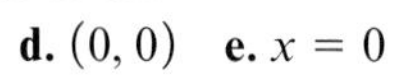 **d.** $(0, 0)$ **e.** $x = 0$

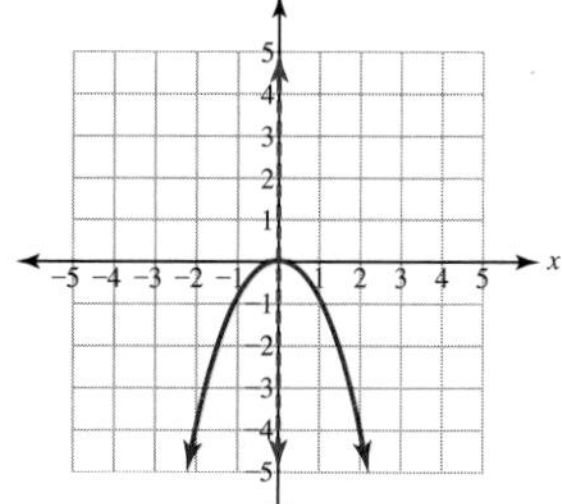

2. a. $a = 1, b = 0, c = 1$ **b.** up

c. **d.** $(0, 1)$ **e.** $x = 0$

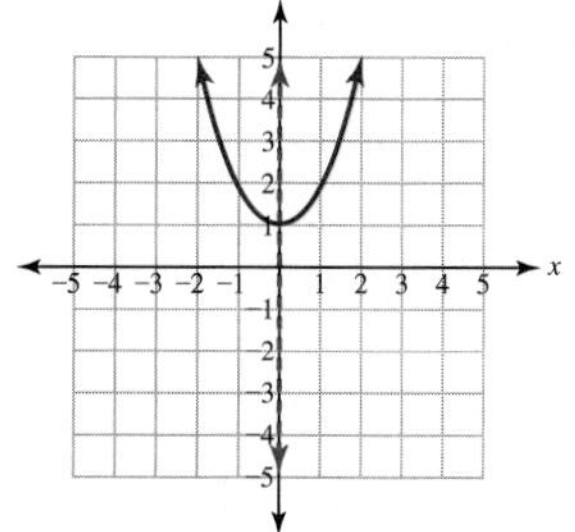

3. a. $a = -\frac{1}{3}, b = 0, c = 0$ **b.** down

c. **d.** $(0, 0)$ **e.** $x = 0$

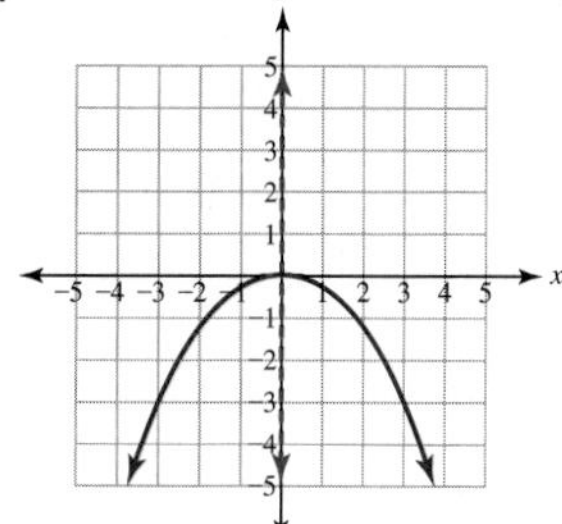

4. a. $a = 1, b = -6, c = 10$ **b.** up

c. 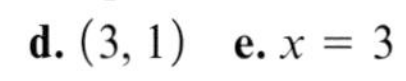 **d.** $(3, 1)$ **e.** $x = 3$

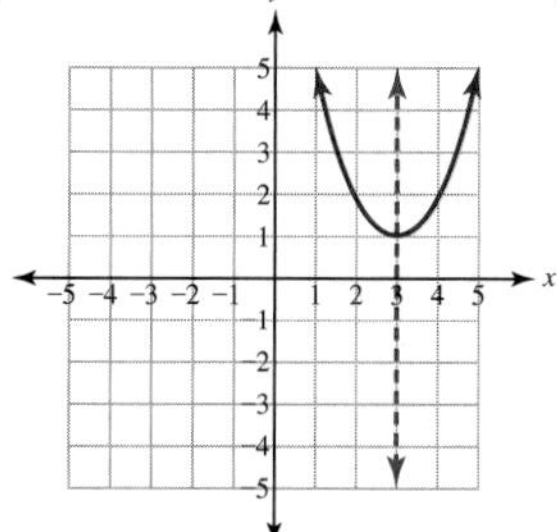

5. shorter; 2 is less than 6

6. a. $(6, 0), (-4, 0)$ **b.** $(1, -25)$ **c.** $x = 1$

d.

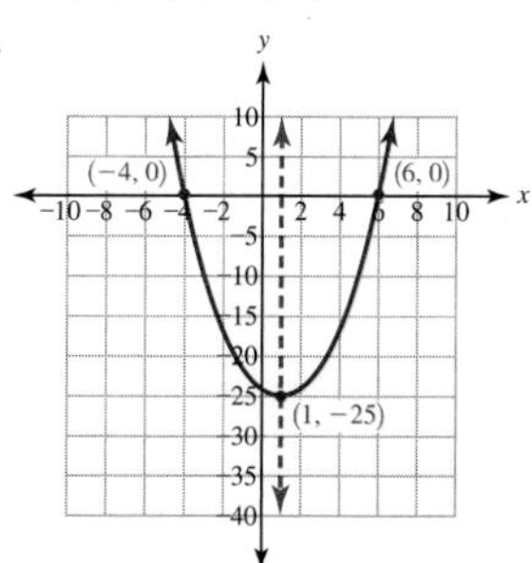

7. a. $(6, 0), (-2, 0)$ **b.** $(2, -16)$ **c.** $x = 2$

d.

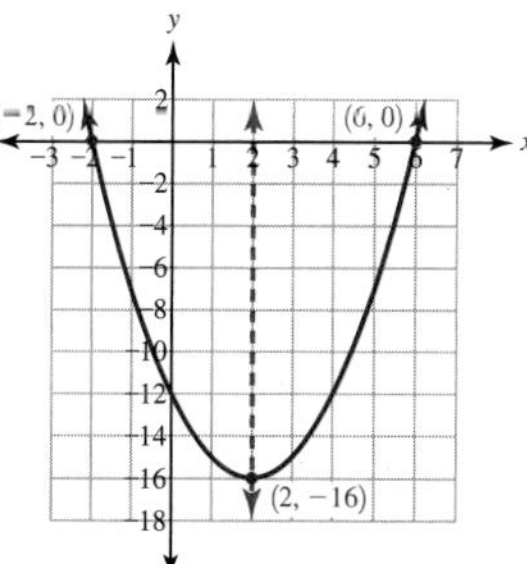

8. a. $(2, 0), (-2, 0)$ **b.** $(0, -4)$ **c.** $x = 0$

d.

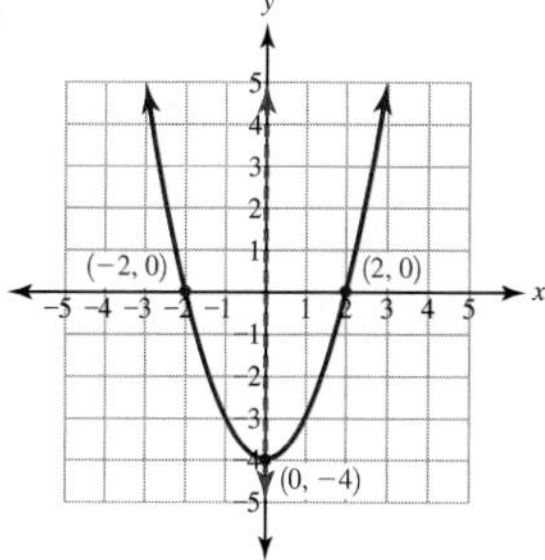

9. a. $(3, 0)$ **b.** $(3, 0)$ **c.** $x = 3$

d.

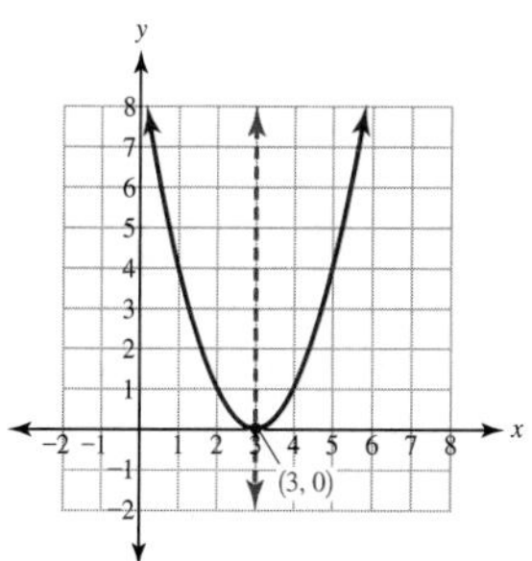

10. a. 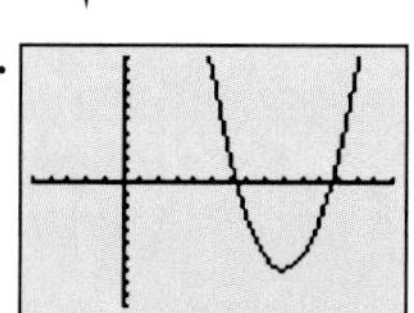 **b.** $(9, -7)$

[−5, 15, 1; −10, 10, 1]

11. a. 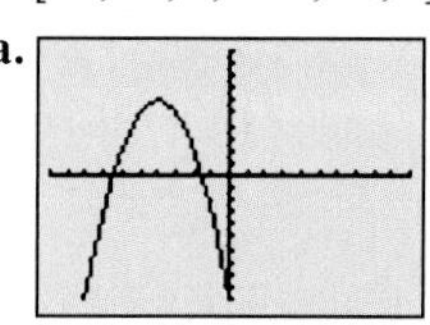 **b.** $(-4, 6)$

[−10, 10, 1; −10, 10, 1]

12. a. 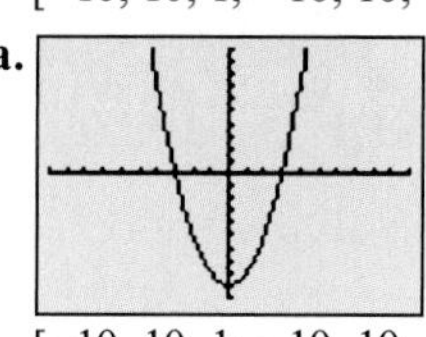 **b.** $(0, -9)$

[−10, 10, 1; −10, 10, 1]

Answers to Selected Exercises

CHAPTER R

Chapter R Pretest

1. -9 **2.** -3 **3.** 0 **4.** undefined **5.** 14 **6.** -15 **7.** 37 **8.** -6 **9.** 35 **10.** $\frac{2}{5}$ **11.** $\frac{3}{4}$ **12.** $\frac{5}{6}$ **13.** $\frac{115}{72}$ **14.** $\frac{13}{12}$ **15.** $\frac{10}{7}$ **16.** -18 **17.** 9 **18.** -9 **19.** 0 **20.** $\frac{1}{3}$ **21.** 3.8 **22.** 189,500 **23.** 20% **24.** 30% **25.** 36 **26.** 17.5 **27.** $-8 < 5$ **28.** $-8 < -5$ **29.** $\frac{3}{4} > \frac{2}{3}$ **30.** $0.18 > -0.3$

Section R.1 Exercises

1. 36 **3.** 64 **5.** 64 **7.** 76 **9.** 48 **11.** 180 **13.** 20 **15.** 48 **17.** 8 **19.** 18 **21.** 18 **23.** 8 **25.** 64 **27.** 20 **29.** 28 **31.** 12 **33.** 6 **35.** 6 **37.** 8 **39.** undefined **41.** 22 **43.** 32 **45.** 42 **47.** 23 **49.** 5 **51.** 14 **53.** 70 **55.** 94 **57.** 11 **59.** 36 **61.** 3 **63.** undefined **65.** 1 **67.** 0 **69.** 26 **71.** 48 **73.** 26 **75.** 1 **77.** undefined **79.** 4 **81. b.** 1 **85.** 18 **87.** 68

Section R.2 Exercises

1. 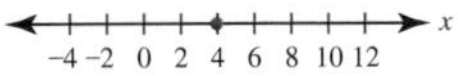

3. 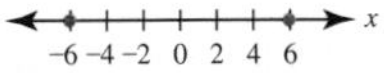

5. 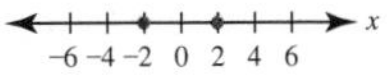

7. 7 **9.** -13 **11.** -7 **13.** -13 **15.** 5 **17.** 1 **19.** -5 **21.** -1 **23.** -5 **25.** 1 **27.** 6 **29.** -6 **31.** -6 **33.** 6 **35.** 3 **37.** -3 **39.** -3 **41.** 3 **43.** -3 **45.** 3 **47.** 64 **49.** 64 **51.** -64 **53.** -54 **55.** 12 **57.** 24 **59.** -12 **61.** -25 **63.** -25 **65.** -1 **67.** 11 **69.** 0 **71.** -3 **73.** 8 **75.** 28 **77.** undefined **79.** 0 **81.** -11 **83.** -25 **85.** -7 **87.** -25 **89.** -148 **91.** 140 **93.** 34 **95.** -18 **97.** -23 **99.** 115 **101.** 184 **103.** 65

Section R.3 Exercises

1. $\frac{7}{16}$ **3.** $\frac{8}{11}$ **5.** $\frac{27}{14}$ **7.** $\frac{7}{18}$ **9.** $\frac{1}{73}$ **11.** $\frac{2}{3}$ **13.** $1\frac{3}{7}$ **15.** $10\frac{1}{2}$ **17.** $3\frac{4}{9}$ **19.** $1\frac{50}{51}$ **21.** $6\frac{5}{6}$ **23.** $\frac{4}{27}$ **25.** $\frac{4}{7}$ **27.** $\frac{1}{10}$ **29.** $\frac{15}{28}$ **31.** $\frac{6}{5}$ **33.** 6 **35.** $\frac{5}{24}$ **37.** $-\frac{5}{6}$ **39.** $-\frac{3}{10}$ **41.** $-\frac{5}{16}$ **43.** $\frac{8}{3}$ **45.** $\frac{3}{8}$ **47.** $\frac{9}{2}$ **49.** $\frac{11}{8}$ **51.** 12 **53.** $\frac{1}{16}$ **55.** $-\frac{54}{5}$ **57.** $-\frac{16}{9}$ **59.** $\frac{5}{9}$ **61.** $\frac{4}{5}$ **63.** $\frac{6}{11}$ **65.** $\frac{3}{5}$ **67.** $-\frac{1}{3}$ **69.** $\frac{2}{3}$ **71.** $-\frac{3}{4}$ **73.** -1 **75.** $\frac{80}{100}$ **77.** $\frac{9}{24}$ **79.** $\frac{12}{63}$ **81.** $\frac{71}{36}$ **83.** $\frac{52}{15}$ **85.** $\frac{11}{60}$ **87.** $\frac{41}{100}$ **89.** $-\frac{13}{12}$ **91.** $-\frac{2}{9}$ **93.** $\frac{5}{16}$ **95.** $-\frac{15}{88}$ **97.** $\frac{7}{12}$ **99.** $\frac{3}{4}$ **101.** 1 **103.** $\frac{34}{15}$ **105.** $\frac{64}{25}$ **107.** $\frac{1}{10}$ **109.** $-\frac{1}{4}$ **111.** -30 **113.** $-\frac{15}{4}$ **115.** $\frac{377}{714}$ **117.** $\frac{20}{357}$

Section R.4 Exercises

1. $\frac{7}{10}$ **3.** $\frac{63}{100}$ **5.** $\frac{6}{5}$ **7.** $\frac{7}{50}$ **9.** $\frac{19}{20}$ **11.** $\frac{1}{500}$ **13.** 0.8 **15.** 1.6 **17.** 0.375 **19.** $0.1\overline{6}$ **21.** 0.1 **23.** $0.\overline{428571}$ **25.** 350 **27.** 0.9 **29.** 0.24 **31.** 65,000 **33.** 5.50 **35.** 0.008 **37.** 8.0 **39.** $\frac{7}{100}$ **41.** $\frac{3}{25}$ **43.** $\frac{17}{20}$ **45.** $\frac{113}{100}$ **47.** 0.02 **49.** 0.18 **51.** 0.084 **53.** 0.0125 **55.** 2 **57.** 0.006

59.

Fraction notation	Percent notation	Decimal number notation
$\frac{4}{5}$	80%	0.8
$\frac{61}{100}$	61%	0.61
$\frac{91}{100}$	91%	0.91
$\frac{7}{100}$	7%	0.07
$\frac{311}{100}$	311%	3.11
$\frac{1}{100}$	1%	0.01
$\frac{57}{100}$	57%	0.57

61. 25% **63.** 4% **65.** 18.75% **67.** 1% **69.** 0.1% **71.** 3 **73.** 12 **75.** 16 **77.** 0.75 **79.** 56.25 **81.** 3 **83.** 1.26 **85.** $12 > 5$ **87.** $-12 < 5$ **89.** $-12 < -5$ **91.** $0.4 < 0.9$ **93.** $-0.4 > -0.9$ **95.** $0.4 > -0.9$ **97.** $-\frac{1}{4} > -\frac{3}{4}$ **99.** $\frac{5}{8} < \frac{2}{3}$ **101.** $\frac{3}{4} > \frac{5}{9}$ **103.** $\frac{2}{11} < \frac{3}{16}$ **105.** $0 > -\frac{1}{4}$ **107.** $\frac{35}{3} > \frac{49}{5}$

Chapter R Posttest

1. 8 **2.** 0 **3.** -16 **4.** 16 **5.** -15 **6.** $\frac{23}{24}$ **7.** -39 **8.** undefined **9.** 17 **10.** -24 **11.** 49 **12.** $\frac{1}{6}$ **13.** 56 **14.** $\frac{1}{6}$ **15.** $-\frac{20}{3}$ **16.** -32 **17.** $\frac{19}{8}$

18. $\frac{2}{7}$ **19.** $\frac{5}{12}$ **20.** 0 **21.** 3.47 **22.** 140 **23.** 35
24. 7.6 **25.** 12.5% **26.** 20% **27.** $-11 < -2$
28. $7 > -4$ **29.** $\frac{5}{6} > \frac{7}{9}$ **30.** $0.03 > -0.11$

CHAPTER 1

Section 1.1 Vocabulary Practice
1. D **2.** J **3.** I **4.** A **5.** F **6.** H **7.** C **8.** E **9.** G **10.** B

Section 1.1 Exercises
1. $\{0, 1, 2, 3, 4, \ldots\}$ **3.** 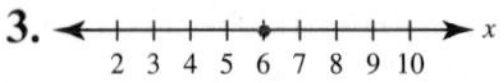

5. 9
7. 9 **9.** -9 **11.** 9 **13.** 9 **15.** 0 **17.** 0
19. undefined **21.** undefined
23. $\{\ldots, -3, -2, -1, 0, 1, 2, 3, 4, \ldots\}$ **25.** 6 **27.** 100
29. 100 **31.** -100 **33.** 8 **35.** -8 **37.** 25
39. 25 **43. b.** yes **45.** $\frac{2}{1}$ **47.** $\frac{-9}{1}$ **49.** $\frac{3}{10}$
51. $\frac{47}{100}$ **53.** $\frac{809}{1000}$ **55.** see chart below **57.** 11
59. 4 **61.** 1 **63.** 18 **65.** -18 **67.** -2 **69.** -4
71. 8 **73.** $\frac{2}{3}$ **75.** $\frac{5}{36}$ **77.** $\frac{49}{36}$ **79.** $-\frac{1}{2}$ **81.** 4
83. -10 **85.** 4 **87.** 1 **89.** 40 **91.** 51 **93.** -18
95. 121 **97.** 1 **99.** undefined **101.** $\frac{1}{20}$ **103.** $\frac{7}{4}$
105. $-\frac{4}{7}$ **107.** $\frac{16}{7}$ **109.** 0 **111.** 9 **113.** 42
119. $-\frac{87}{5}$ **121.** 15 **123. b.** 95 **125. b.** 4 **127.** 1508
129. 14,094

Section 1.2 Vocabulary Practice
1. I **2.** H **3.** E **4.** B **5.** D **6.** G **7.** C **8.** A **9.** F **10.** J

Section 1.2 Exercises
1. a. 3 **b.** 8, 2 **c.** 5 **d.** x, y **3. a.** 2 **b.** 8, -6
c. no constant **d.** a, b **5.** $22k$ **7.** $-15b$ **9.** $-\frac{1}{2}m + \frac{3}{8}$
11. $\frac{47}{45}u - \frac{8}{11}$ **13.** $10a - 14b + 1$ **15.** $2c - 12w - 5$
17. $-\frac{4}{5}p + \frac{1}{2}$ **19.** $\frac{17}{20}c + \frac{1}{10}d$ **21.** $\frac{3}{2}x + \frac{6}{5}$
23. $-\frac{28}{5}u + \frac{17}{3}y$ **25.** $10x^2 - 3x - 1$ **27.** $2c^2 + 9c + 19$
29. $3k^2 - 8k - 2$ **31.** $5x^2 - x + 6$ **33.** B **35.** A
37. C **39.** C **41.** $40yz$ **43.** $15x^2$ **45.** $-21ab$
47. $-72m$ **49.** $-5x + 5y$ **51.** $350a + 24b - 47$
53. $6x - 5$ **55.** $-6a - 20$ **57.** $-a + 5$ **59.** $-x + 1$
61. $-22x + 66$ **63.** $2x + 12$ **65.** $4w - 1$
67. $-15w + 9$ **69.** $-\frac{39}{2}a + \frac{3}{8}$ **71.** $\frac{3}{20}c + \frac{13}{24}d$
73. $4x + \frac{8}{3}$ **75.** $\frac{2}{15}p + \frac{7}{25}w$ **77.** $8z - 9$
79. $-3c + \frac{51}{4}$ **81.** $\frac{142}{35}z + \frac{141}{35}$ **91. b.** $-6x + 28$
93. b. $26x + 7$ **97.** undefined

Section 1.3 Vocabulary Practice
1. B **2.** A **3.** B **4.** A **5.** B **6.** A **7.** B **8.** A **9.** B **10.** A

Section 1.3 Exercises
1. \$19.75 **3.** \$225,700 **5.** \$7650 **7.** 13 hours
9. 100 text messages **11.** \$62.40 **13.** \$2.2 billion
15. \$36,250 **17.** 41.4 tons **19.** \$12,238
21. 1266.1 million people **23.** 9.855 million pounds
25. 24 billion singles **27.** \$650,750,000
29. \$1073 per square foot **31.** \$31.50 **33.** 47,018 acres
35. \$80 **37.** 720 nonfatal injuries **39.** 45 additional hours
41. \$1,421,700,000 **43.** 18% **45.** \$55,119,200,000
47. \$1372 **49.** 32 weeks **51.** 13°F **53.** \$615.03
55. \$22,288 **57.** 1880 policyholders **59.** $\frac{3}{8}$ inch
61. $\frac{19}{48}$ of the estate **63.** 128 strips **65.** 25 districts
67. $\frac{5}{8}$ pound **69.** 1192 freshmen **71.** \$161.25 million
73. 814 fatalities **75.** 32 sheets of cardboard
79. b. 41% **81.** 0.9 **83.** 6.25%

Section 1.4 Vocabulary Practice
1. F **2.** J **3.** H **4.** G **5.** A **6.** D **7.** E **8.** I **9.** C **10.** B

Section 1.4 Exercises
1. 0.05 **3.** 0.83 **5.** 0.091 **7.** 1.42 **9.** 0.0215
11. 2.56 **13.** 121.5 **15.** 56.25 **17.** 195 **19.** 35%
21. 3.5% **23.** 200% **25.** 2.5% **27.** 25,160 visits
29. 0.9 acre **31.** 14% **33.** 15% **35.** 79% **37.** 27%
39. 14% **41.** 38% **43.** 69% **45.** 14% **47.** 29%
49. 33% increase **51.** 24% decrease **53.** 20% increase
55. 42% increase **57.** 41% increase **59.** 16% increase
61. 54% decrease **63.** \$45 **65.** \$55.20 **67.** \$692.23
69. 15% discount **71.** \$6.73 **73.** \$14,070.38
75. \$503.96 **77.** 22% discount **79.** 233% markup
81. \$68.34 **83. b.** 6% decrease **85. b.** 25% decrease
87. 25 **89.** -8

Chart for Section 1.1 exercise 55

	Real numbers	Rational numbers	Irrational numbers	Integers	Whole numbers
-6	X	X		X	
0	X	X		x	X
π	X		X		
-0.75	X	X			
18	X	X		X	X
$\frac{1}{3}$	X	X			

Section 1.5 Vocabulary Practice

1. J **2.** B **3.** D **4.** G **5.** C **6.** F **7.** E **8.** A **9.** I **10.** H

Section 1.5 Exercises

1. 32 mm^2 **3.** 128 mm^3 **5.** 128 mm^3 **7.** 6 **9.** 26,400 ft **11.** 24 qt **13.** 3000 m **15.** 12,000 g **17.** 0.5 gal **19.** 0.75 ft **21.** 8 km **23.** $\frac{96 \text{ km}}{1 \text{ hr}}$ **25.** 11.3 mi **27.** 8.2 kg **29.** $\frac{37.5 \text{ mi}}{1 \text{ hr}}$ **31.** 8004 m **33.** 8.004 km **35.** \$1800 **37.** \$156.25 **39.** \$150 **41.** 260 mi **43.** 22 mi **45.** 26 mi **47.** 69 mi **49.** 24 in. **51.** 58 in. **53.** 52 in.^2 **55.** 270 in.^2 **57.** 24 cm **59.** $\frac{13}{12}$ ft **61.** 90 ft^2 **63.** $\frac{1}{64}$ in.^2 **65.** 9 in.^2 **67.** 21.8 yd^2 **69.** 9 ft **71.** $\frac{0.8 \text{ g}}{1 \text{ cm}^3}$ **73.** 2431 calories **75.** 1343 calories **77.** 13% **79.** 14°F **81.** −9°C **83.** 21.70 **87.** 333.3 board feet **89.** 76 **91.** 60.4 **93.** 60,500 ft^2 **95.** 8.9 yd **97. b.** \$67.50 **99. b.** 24 cm^2 **101.** $\frac{33}{10}x - \frac{4}{3}$ **103.** $-20z + 41$

Section 1.6 Vocabulary Practice

1. H **2.** I **3.** A **4.** E **5.** J **6.** F **7.** B **8.** D **9.** C **10.** G

Section 1.6 Exercises

3. $0.8\overline{3}$ **5.** $0.\overline{428571}$ **7.** 18.8 cm **9.** 3.1 in. **11.** 113.0 cm^2 **13.** 113.0 m^2 **15.** 1695.6 in.^3 **17.** 24,416.6 in.^3 **19.** 4239.0 in.^3 **21.** 8177.1 in.^3 **23.** 414.5 in.^3 **25.** 49.46 in. **27.** 194.7 in.^2 **29.** 1100.7 in.^3 **31.** 8,021,826 mm^3 **33.** yes **35.** yes **37.** no **39.** no **41.** 2600 **43.** 7 **45.** 3.873 **47.** 7 km **49.** not a real number **55. a.** $A^2 = 9$; $A = 3$; $b = 3$ **b.** 3.4643 **c.** 3.4641 **57.** 216 **59.** 6 **61.** −6 **63.** 10 **65.** see chart below **67.** 10 ft **69.** 9.49 ft **71.** 11.31 m **73.** 9079 votes **75.** 4.1 nautical mi **77.** 4.2 knots **79.** $\frac{8.4 \text{ mi}}{1 \text{ hr}}$ **81.** 125 knots **83. a.** 6 ft^2 **b.** 6 ft^2 **85.** $\frac{950 \text{ gal}}{1 \text{ min}}$ **87.** 11.4 nautical mi **89. a.** 706.858 m^2 **b.** 706.5 m^2 **91. a.** 628.319 in.^3 **b.** 628 in.^3 **93. b.** 78.5 in.^2 **95. b.** −6 **97.** {0, 1, 2, 3, 4, . . . }

Chapter 1 Review Exercises

1. a. {0, 1, 2, 3, 4, . . . } **b.** { . . . , −3, −2, −1, 0, 1, 2, 3, . . . } **3.** b, c, f, h **4. a.** 35 **b.** 7, 5 **5.**

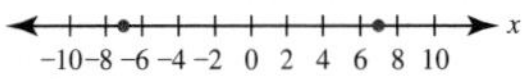

6. 64 **7.** 64 **8.** −64 **9.** 41 **10.** 73 **11.** 4 **12.** 0 **13.** 0 **14.** 4 **15.** −4 **16.** undefined **17.** 0 **18.** undefined **19.** 12 **20.** $\frac{2}{5}$ **21.** $\frac{1}{9}$ **22.** $\frac{1}{3}$ **23.** $\frac{8}{5}$ **24.** $\frac{24}{61}$ **25.** $-\frac{27}{4}$ **27.** see chart below **29.** Divide 12 by 2. **30.** Divide 12 by 2. **31.** 21 **32.** 6 **33.** undefined **34.** $\frac{47}{20}$ **35.** −23 **36.** −17 **37.** 313 **38.** −20 **39.** 0 **40.** −1 **41.** −10 **42.** −40 **43.** 0 **44. a.** 8, 2, 7 **b.** 9 **c.** x, y, z **d.** 4 **45.** $-5p$ **46.** $-7u + 3y + 8$ **47.** $4x^2 - 6x - 10$ **48.** $-30wz$ **49.** $\frac{1}{6}a^2b$ **50. a.** the commutative property **51.** C **52.** B **53.** B **54.** A **55.** A **56.** C **57.** $6a + 7$ **58.** $-23x - 4$ **59.** $4x - 10y - 38$ **60.** $-p - 13$

Chart for Section 1.6 exercise 65

	Real numbers	Rational numbers	Irrational numbers	Integers	Whole numbers
$\frac{3}{5}$	X	X			
π	X		X		
−8	X	X		X	
$\sqrt{7}$	X		X		
$\sqrt{100}$	X	X		X	X
−0.4	X	X			
0	X	X		X	X
$0.\overline{3}$	X	X			

Chart for Chapter 1 Review exercise 27

	Real numbers	Rational numbers	Irrational numbers	Integers	Whole numbers
0.3	X	X			
$\frac{1}{8}$	X	X			
−9	X	X		X	
$\sqrt{11}$	X		X		
0	X	X		X	X
π	X		X		

61. $-\frac{2}{15}x+\frac{19}{15}z$ **62.** $26z+3$ **67.** \$242
68. \$70 million **69.** 82 ft **70.** 100,000 students
71. 72 strips **72.** \$187,500 **73.** $\frac{19}{24}$ cup **74.** 0.07
75. 0.083 **76.** a, b, d, e **77.** 2.8 billion gal **78.** 219%
79. 30 institutions **80.** 1.9 million people **81.** 45%
82. \$2.94 **83.** 20 m **84.** 8 **85.** 42 in.^{3} **86.** 99 kg
87. $\frac{1\text{ kg}}{1000\text{ g}};\ \frac{1000\text{ g}}{1\text{ kg}}$ **88.** 2.8 L **89.** 18.4 lb **90.** \$1710
91. 26.25 mi **92.** 1232 in.^{3} **93.** 410 cm **94.** 9 in.^{2}
95. a, c, d **96.** 9.4 **97.** $\frac{693}{1000}$ **98.** $0.\overline{7}$ **100.** 44 in.
101. 267.9 cm^3 **102.** 14.9 in.^{2} **103.** 276 gal
104. -10 and 10 **105.** 9 **106.** 0 **107.** not a real number
108. 3.46 **109.** 7.14 **110.** -3 **111.** 2
112. rational numbers **114.** 10 ft **115.** 1.1 m^2

Chapter 1 Test

1. see chart below **2.** 34 **3.** undefined **4.** 0
5. 17 **6.** -2 **7.** $\frac{3}{20}$ **8.** 22 **9.** -76 **10.** 14
11. $\frac{91}{12}h-60$ **12.** $-8w$ **13.** $-6p-15$ **16.** C
17. A **18.** B **19.** 8820 ft^2 **20.** 19 points
21. 1.4 m^3 **22.** 176.625 in.^{2} **23.** 4.1% increase
24. 305 adults ages 18–29 **25.** 8% decrease
26. \$34.47

CHAPTER 2

Section 2.1 Vocabulary Practice

1. J **2.** G **3.** A **4.** I **5.** B **6.** C **7.** E **8.** F **9.** H **10.** D

Section 2.1 Exercises

1. $x=16$ **3.** $x=32$ **5.** $x=192$ **7.** $x=192$
9. $x=3$ **11. a.** $p=-15$ **13. a.** $c=5$ **15. a.** $h=20$
17. a. $w=-1.1$ **19. a.** $a=4.5$ **21. a.** $x=\frac{4}{3}$
23. a. $m=-\frac{11}{36}$ **25. a.** $z=0$ **27. a.** $x=-40$
29. a. $p=1050$ **31. a.** $x=-1.6$ **33. a.** $k=\frac{29}{24}$
35. a. $a=12$ **37. a.** $v=40$ **39. a.** $n=675$
41. a. $p=-0.5$ **43. a.** $p=-50$ **45. a.** $x=1.75$
49. \$160 **51.** \$188,474 **53.** 49,500 people **55.** \$49.99
57. 116 adults **59.** \$3298 **61.** 22,430 new cases
63. 619 people **65.** 400,000 children **67.** 691 drivers
69. 335 adults **71.** 3.6 lb **73.** 0.35 mL **75.** 128 mi
77. 3789 crashes **79.** 180 stitches **81.** 60 cal
83. 2100 adult Americans **85.** 16% increase **87.** 47%
89. $p=114{,}055$ **91.** $d=51.02$ **93. b.** $x=648$
95. b. $x=\frac{22}{3}$ **97.** $6x-25$ **99.** $\frac{22}{15}x-\frac{62}{15}$

Section 2.2 Vocabulary Practice

1. C **2.** G **3.** A **4.** J **5.** H **6.** D **7.** B **8.** F **9.** E **10.** I

Section 2.2 Exercises

1. a. $x=7$ **3. a.** $x=17$ **5. a.** $x=5$ **7. a.** $x=\frac{1}{2}$
9. a. $x=-6$ **11. a.** $x=-3$ **13. a.** $x=-3$
15. a. $x=-9$ **17. a.** $x=68$ **19. a.** $x=-28$
21. a. $x=9$ **23. a.** $p=20$ **25. a.** $x=9$
27. a. $x=-23$ **29. a.** $w=24$ **31. a.** $h=81$
33. a. $k=-26$ **35. a.** $x=\frac{1}{3}$ **37. a.** $u=\frac{7}{8}$
39. a. $z=-29$ **41. a.** $x=0$ **43. a.** $b=12.1$
45. a. $x=17.9$ **47. a.** $x=32$ **49. a.** $x=-11$
51. a. $z=3$ **53. a.** $k=-3$ **55. a.** $x=4$
57. a. $a=5.5$ **59. a.** $h=-\frac{1}{2}$ **61. a.** $p=\frac{17}{24}$
63. a. $x=3$ **65. a.** $w=-9$ **67. a.** $y=9$
69. a. $m=-12$ **71. a.** $k=12$ **73.** 22,500 mi
75. 1750 products **77.** \$360,000 of additional sales
79. \$14.88 **81.** \$21.91 **83.** \$17.94 **85.** 2.5 hr
87. 6 additional hr **89.** 6 additional mi **91.** 1031 mi
93. 9.1 mi **95.** 1219 babies **97.** 6,164,717 gal
99. $p=\frac{5933}{350}$ **101.** $d=\frac{14{,}381}{1630}$ **103. b.** $x=6$
105. b. $x=21$ **107.** 40 **109.** 168

Section 2.3 Vocabulary Practice

1. E **2.** J **3.** I **4.** G **5.** D **6.** C **7.** B **8.** F **9.** H **10.** A

Section 2.3 Exercises

1. a. $x=4$ **3. a.** $w=-10$ **5. a.** $y=5$
7. a. $h=-7$ **9. a.** $k=-12$ **11. a.** $x=-\frac{13}{2}$
13. a. $p=420$ **15. a.** $x=111$ **17. a.** $a=\frac{15}{2}$
19. a. $z=72$ **21. a.** $v=0$ **23. a.** $x=18$
25. a. $x=44$ **27. a.** $u=48$ **29. a.** $c=-\frac{3}{2}$
31. a. $a=-\frac{7}{3}$ **33. a.** $n=\frac{5}{8}$ **35. a.** $n=\frac{5}{16}$
37. a. $p=\frac{3}{40}$ **39. a.** $y=76$ **41. a.** $b=-48$
43. a. $y=\frac{3}{2}$ **45. a.** $p=6$ **47. a.** $x=-\frac{8}{5}$

Chart for Chapter 1 Test problem 1

	Real numbers	Rational numbers	Irrational numbers	Integers	Whole numbers
-4	X	X		X	
$\sqrt{7}$	X		X		
$\frac{2}{9}$	X	X			
18	X	X		X	X
π	X		X		
-0.13	X	X			

49. a. $x = -\frac{12}{5}$ **51. a.** $x = 5$ **53. a.** $a = 240$
55. a. $x = 45$ **57. a.** $x = -7.5$ **59. a.** $w = 277.5$
61. a. $n = 1$ **63. a.** $k = -2.25$ **65. a.** no solution
67. a. the set of real numbers **69. a.** $x = 0$
71. a. the set of real numbers **73. a.** $p = 0$
75. a. no solution **77. a.** no solution **79. a.** $c = 0$
81. a. the set of real numbers **83. a.** no solution
85. a. $r = 2$ **91.** 300 products **93.** 127 products
95. 0.6 mi **97.** 6 hr 40 min **99.** 92% increase
101. \$350.4 million **103.** $c = 6$ **105.** the set of real numbers **107. b.** $x = 16$ **109. b.** $x = -\frac{15}{11}$
111. 0.75 hr **113.** $3x + 8y$

Section 2.4 Vocabulary Practice

1. C **2.** I **3.** D **4.** E **5.** H **6.** J **7.** B **8.** G **9.** F **10.** A

Section 2.4 Exercises

1. $y = -3x + 20$ **3.** $y = 3x - 20$ **5.** $y = \frac{1}{2}x - \frac{10}{3}$
7. $y = -\frac{7}{2}x + \frac{15}{2}$ **9.** $y = -\frac{1}{12}x - 2$
11. $y = -\frac{5}{6}x + 15$ **13.** $y = 3x - 21$ **15.** $y = \frac{3}{4}x$
17. $y = -12x$ **19.** $y = \frac{3}{4}x - \frac{11}{4}$ **21.** $y = \frac{40}{51}x + \frac{4}{17}$
23. $y = \frac{9}{2}x$ **25.** $y = -\frac{50}{27}x + \frac{200}{3}$ **27.** $y = \frac{5}{6}x + \frac{5}{12}$
29. $x = \frac{1}{3}y + \frac{20}{3}$ **31.** $x = 2y + \frac{20}{3}$ **33.** $x = -\frac{6}{5}y + 18$
35. $x = \frac{4}{3}y$ **37.** $A = -\frac{5}{2}B + \frac{25}{2}$ **39.** $P = \frac{7}{2}R + 7$
41. $G = -F - 300$ **43.** $R = -\frac{2}{5}W + 42$ **45.** $L = \frac{A}{W}$
47. $n = \frac{PV}{RT}$ **49.** $R = \frac{A - P}{PT}$ **51.** $H = \frac{V}{LW}$
53. $b = \frac{2A}{h}$ **55.** $C = \frac{5}{9}F - \frac{160}{9}$ **57.** $T = \frac{144V}{LW}$
59. $x_1 = 3A - x_2 - x_3$ **61.** $r = \frac{C}{2\pi}$ **63.** $b = \frac{3P}{h}$
65. $H = C - L - \frac{1}{5}T$ **67.** $C = LR - E$ **69.** $L = \frac{P}{3}$
71. a. $r = \frac{d}{t}$ **b.** $72\ \frac{\text{mi}}{\text{hr}}$ **73. a.** $R = \frac{A - P}{PT}$ **b.** 6%
75. a. $V = \frac{M}{D}$ **b.** 67.8 mL **77. a.** $L = \frac{(144\text{ in.}^3)V}{TW}$
b. 4608 in. **c.** 384 ft **79. a.** $a = \frac{V_f - V_i}{t}$ **b.** $16.6\ \frac{\text{ft}}{\text{s}^2}$
81. a. $b_1 = \frac{2A}{h} - b_2$ **b.** 25 in.

83. a. $E = C + F - 2$ **b.** 12 edges **85.** \$108.76
87. 5896 tools **89.** 160 ft **91.** \$4,230,431.25
93. b. $y = \frac{12}{19}x - 2$ **95.** $x = \frac{1}{4}y + \frac{27}{4}$ **97.** > **99.** >

Section 2.5 Vocabulary Practice

1. J **2.** F **3.** B **4.** E **5.** G **6.** D **7.** H **8.** I **9.** C **10.** A

Section 2.5 Exercises

1. 3 4 5 6 7 8 9 10 x **3.** −6 −5 −4 −3 −2 −1 0 1 x
5. −2 −1 0 1 x **7.** 3 4 5 6 7 8 9 10 x
9. a. $p > -9$ **11. a.** $p \geq -15$ **13. a.** $p < 9$
15. a. $p > -15$ **17. a.** $c < 23$ **19. a.** $c \geq -23$
21. a. $c < -17$ **23. a.** $z \geq 5$ **25. a.** $x \geq 8$
27. a. $w \leq -\frac{1}{3}$ **29. a.** $a > 5$ **31. a.** $x < -15$
33. a. $z < 6$ **35. a.** $y \geq -\frac{23}{2}$
37. a. $x < 8$ **b.** 3 4 5 6 7 8 9 10 x
39. a. $x \leq -3$ **b.** −8 −7 −6 −5 −4 −3 −2 −1 x
41. a. $x \geq 25$ **b.** 15 20 25 30 35 40 45 50 x
43. a. $x \geq \frac{3}{2}$ **b.** 1 2 3 4 x
45. a. $x > 0$ **b.** −2 −1 0 1 2 3 4 5 x
47. a. $x \leq -7$ **b.** −11 −9 −7 −5 x
49. a. $x \geq 9$ **b.** 7 8 9 10 11 12 13 14 x
51. a. $x < 5$ **b.** 0 1 2 3 4 5 6 7 x
53. a. $m > \frac{10}{3}$ **55. a.** $k \geq \frac{15}{2}$
57. a. $w < 18$ **59. a.** $p \leq -5$ **61. a.** $u > 16$
65. n = the number of passengers; $n \leq 12$ passengers
67. c = the number of customers without electricity; $c > 2.3$ million customers **69.** p = the amount of protein; $p \leq 84$ g **71.** h = height; $h \geq 54$ in. **73.** 392 mi
75. 143 hr **77.** 9.725 **79.** \$3.59 per day
81. 34 million tons **83.** 1.05 million people **85.** 91%
87. \$8.50 **89.** $h > \frac{167}{375}$ **91.** $x < \frac{11}{10}$ **93. b.** $x < -12$
95. b. 0 1 2 3 4 5 6 7 x **97.** see chart below
99. no solution

Chapter 2 Review Exercises

1. a. $x = -18$ **2. a.** $x = 75$ **3. a.** $a = 3$
4. a. $b = -32$ **5. a.** $x = \frac{13}{12}$ **6. a.** $x = -\frac{6}{5}$

Chart for Section 2.5 exercise 97

	Real numbers	Rational numbers	Irrational numbers	Integers	Whole numbers
$\sqrt{12}$	X		X		
$\frac{3}{11}$	X	X			
-25	X	X		X	
0.68	X	X			

7. a. $h = 60$ **8. a.** $k = 0$ **9. a.** $p = 180$
10. a. $z = \frac{1}{4}$ **11. a.** $x = 0$ **12. a.** $z = 9$ **13. a.** $a = 40$
14. a. $x = 1.6$ **15. a.** $a = -13.5$ **16.** \$893 per month
17. \$680 **18.** \$25,347,000,000 **19.** 3960 announcements
20. 15 fatal work injuries **21.** 64,600 psychologists
22. a. $x = 9$ **23. a.** $m = 24$ **24. a.** $b = 0$
25. a. $z = -92$ **26. a.** $x = \frac{29}{2}$ **27. a.** $x = 3$
28. 14 referrals who sign a contract **29.** 960,000 cell phones
30. \$1043.88 **31. a.** $a = -14$ **32. a.** $x = -\frac{1}{2}$
33. a. the set of real numbers **34. a.** $x = 32$
35. a. no solution **36. a.** $x = 1$ **37. a.** $x = 0$
38. a. $z = 24$ **39. a.** $d = -200$ **40. a.** $k = \frac{9}{8}$
41. a. no solution **42.** a, c **43.** b, c, e
44. 13 months **45.** 300 products
46. $y = \frac{3}{2}x - \frac{9}{2}$ **47.** $x = \frac{2}{3}y + 3$ **48.** $S = \frac{AD}{N}$
49. $a = \frac{24d - D}{D}$ **50. a.** $A = \frac{NS}{D}$ **b.** $\frac{63 \text{ mi}}{1 \text{ hr}}$
51. a. $R = \frac{C}{P - W}$ **b.** 2% commission
52. 3 4 5 6 7 8 9 10 **53.** −1 0 1 2
54. a. $x < 7$ **b.** 2 3 4 5 6 7 8 9
55. a. $x \leq -3$ **b.** −8 −7 −6 −5 −4 −3 −2 −1
56. a. $x > 6$ **b.** 4 5 6 7 8 9 10 11
57. a. $x < 0$ **b.** −5 −4 −3 −2 −1 0 1 2
58. a. $x < -5$ **b.** −10 −9 −8 −7 −6 −5 −4 −3
59. a. $x \geq -1$ **b.** −3 −2 −1 0 1 2 3 4
60. a. $x < 26$ **b.** 21 22 23 24 25 26 27 28
61. a, b, e, f **62.** b, c, d **63.** a, b **64.** d
65. a, b **66.** n = number of pieces of art; $n > 340{,}000$ pieces of art **67.** I = income level; $I \leq \$126{,}600$
68. 1.425 gal **69.** 77 strokes

Chapter 2 Test

1. a. $x = -4$ **2. a.** $p = 24$ **3. a.** no solution
4. a. $c = -3$ **5. a.** $w = -10$ **6. a.** $x = 15$
7. a. $d = 9$ **8. a.** $a = \frac{27}{2}$ **9. a.** the set of real numbers
10. a. $p = 6.25$ **11. a.** $n = 0$
12. a. $x \geq -5$ **b.** −7 −6 −5 −4 −3 −2 −1 0
13. a. $x < 4$ **b.** −1 0 1 2 3 4 5 6
14. $x = -\frac{1}{3}$ **15.** $h = -8$
17. $y = \frac{12}{5}x - 8$ **18.** $C = \frac{D - A + 2B}{2}$
19. $W = \frac{9A}{1.06L}$ **20.** P = the amount of a Pell Grant award; $P \leq \$5550$ **21.** 11.7 million men **22.** 40.9 million people **23.** 634 products **24.** \$695 **25.** 15.2 oz

CHAPTER 3

Section 3.1 Vocabulary Practice

1. G **2.** I **3.** C **4.** A **5.** E **6.** F **7.** H **8.** J **9.** D **10.** B

Section 3.1 Exercises

1. (0, 0) **3.** (6, 7) **5.** (−4, 2) **7.** (−6, −2)
9. (0, −3) **11.** 1st quadrant **13.** 2nd quadrant
15. 3rd quadrant **17.** 4th quadrant
19–29. odd

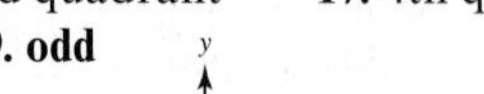

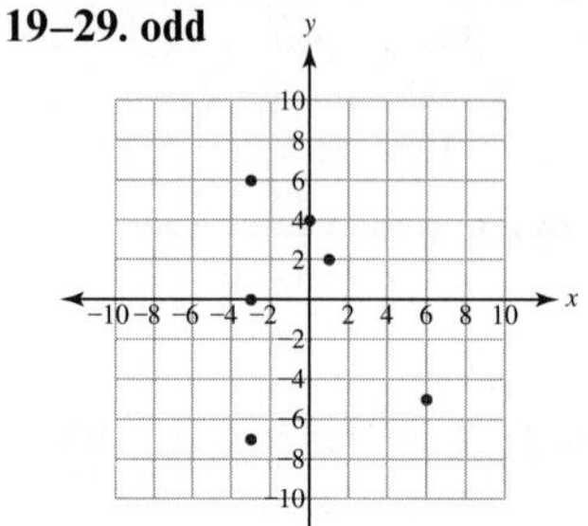

31.

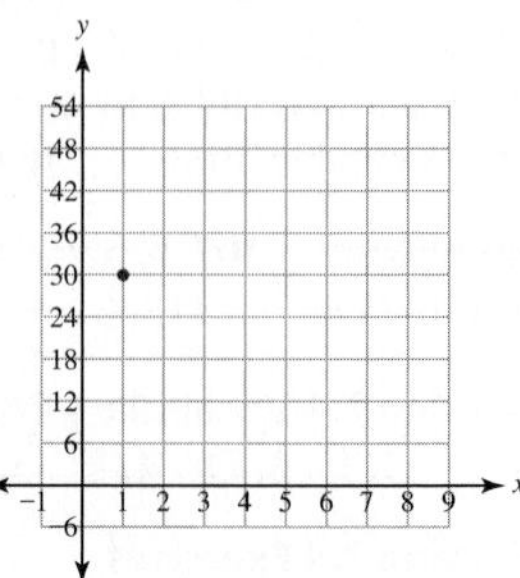

33.

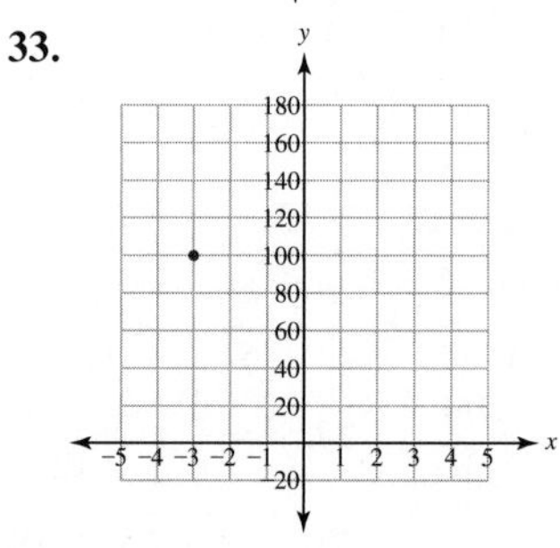

35.

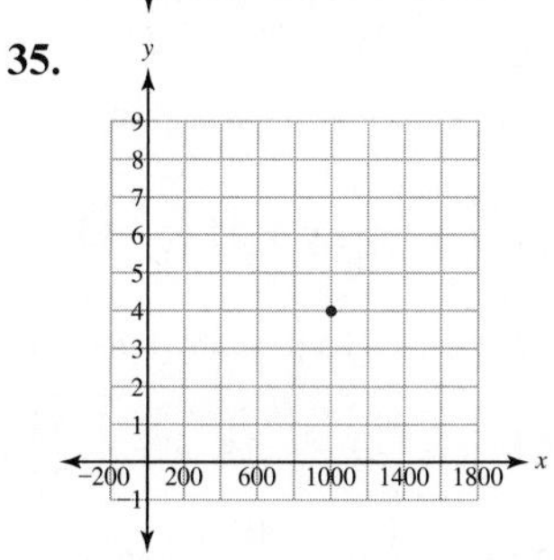

37.

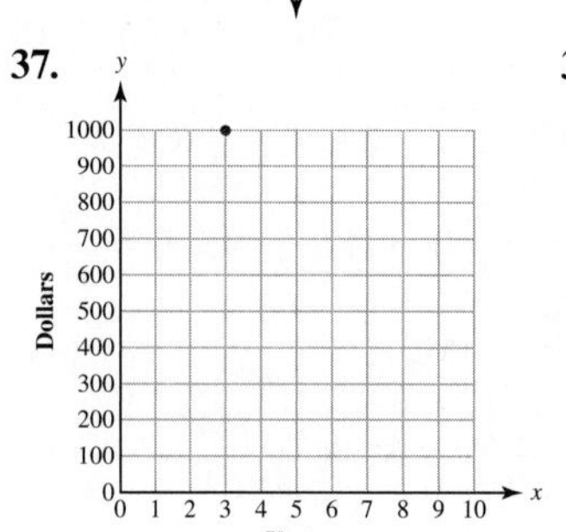

39.

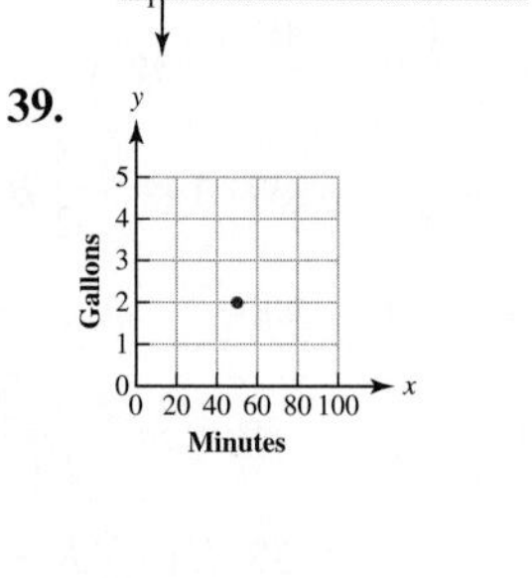

41.

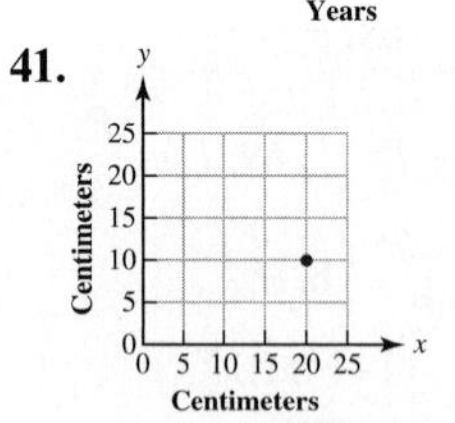

43.

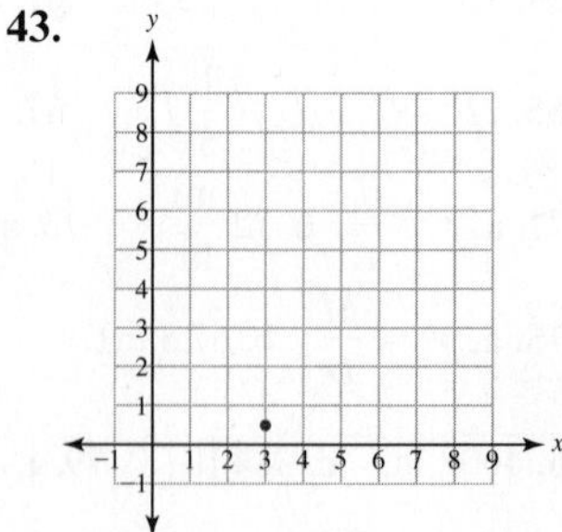

45.

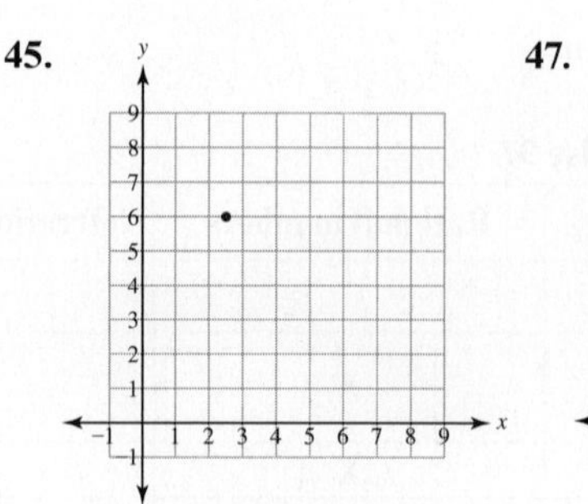

47.

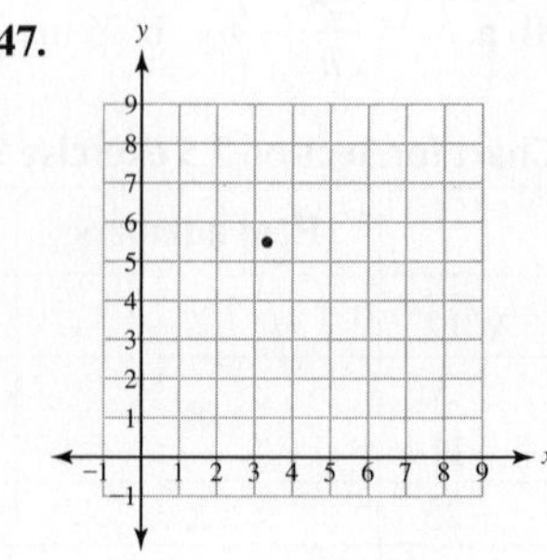

49. a.

x (\$)	y (\$)
0	10
10	0
3	7

b.

51. a.

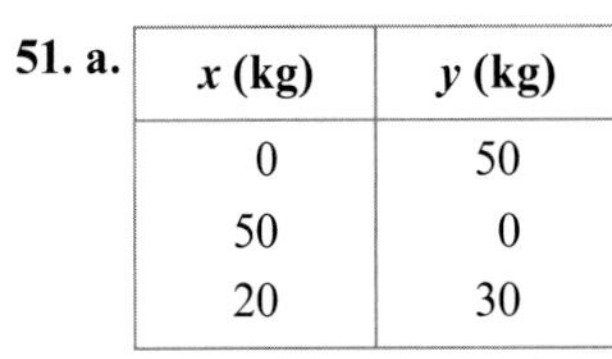

x (kg)	y (kg)
0	50
50	0
20	30

b.

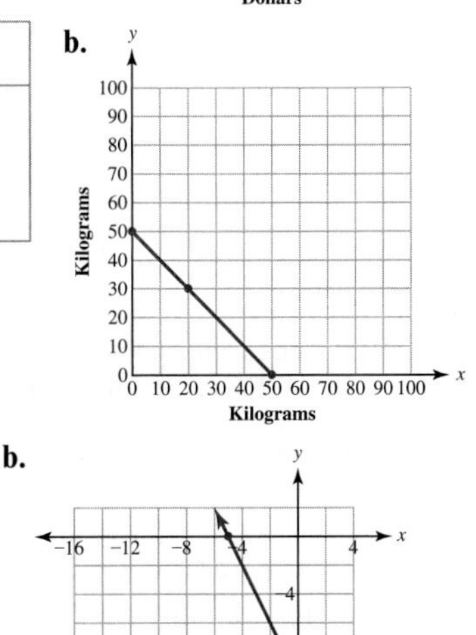

53. a.

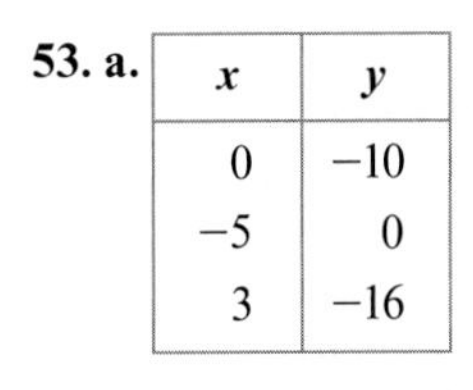

x	y
0	−10
−5	0
3	−16

b.

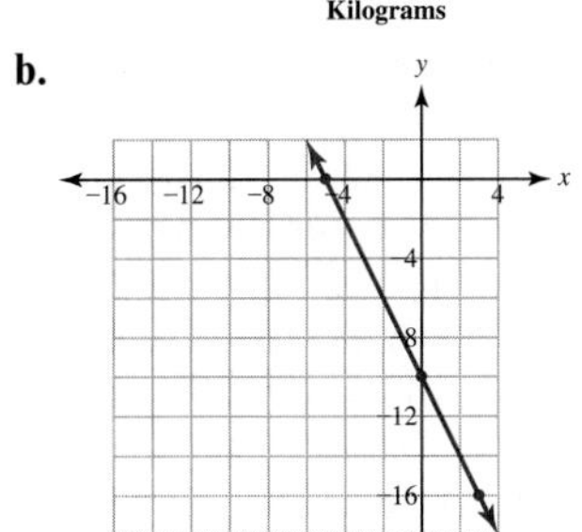

55. a.

x	y
0	12
−4	0
3	21

b.

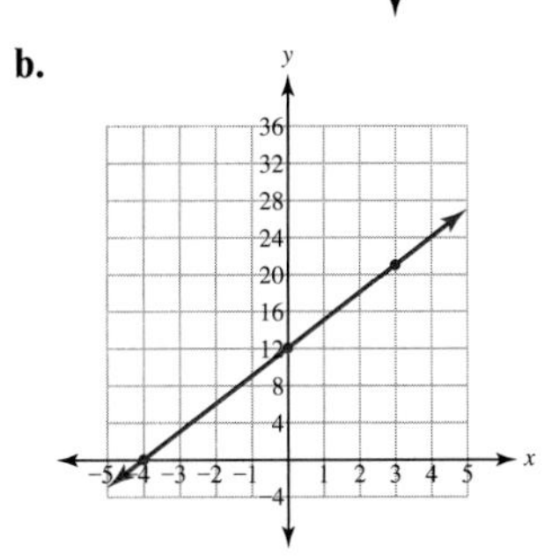

57. a.

x	y
0	3
5	0
−3	$\frac{24}{5}$

b.

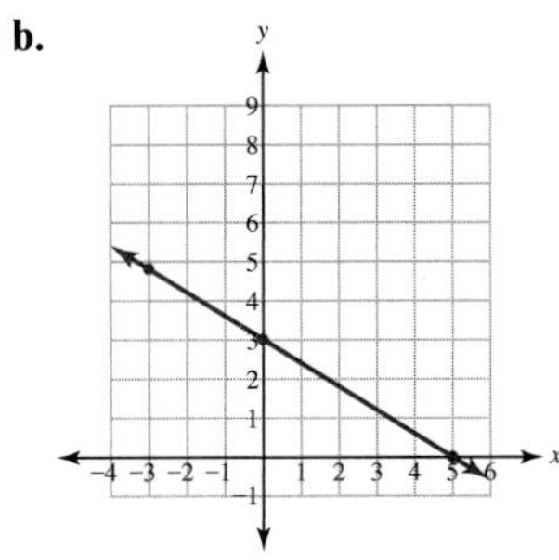

59. a.

x	y
0	−3
5	0
−5	−6

b.

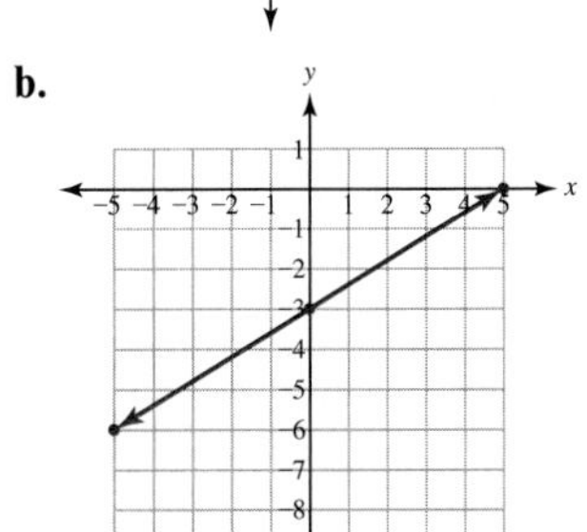

61. a.

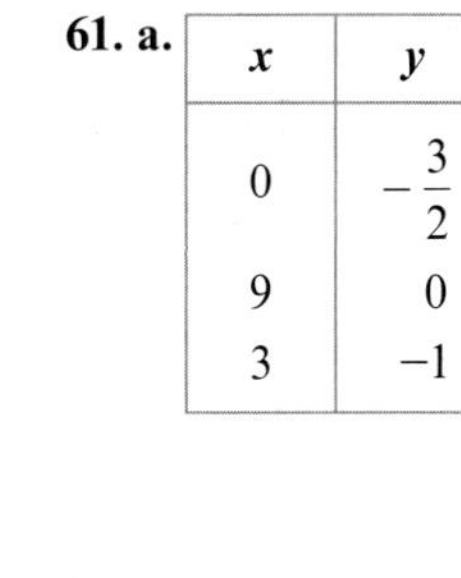

x	y
0	$-\frac{3}{2}$
9	0
3	−1

b.

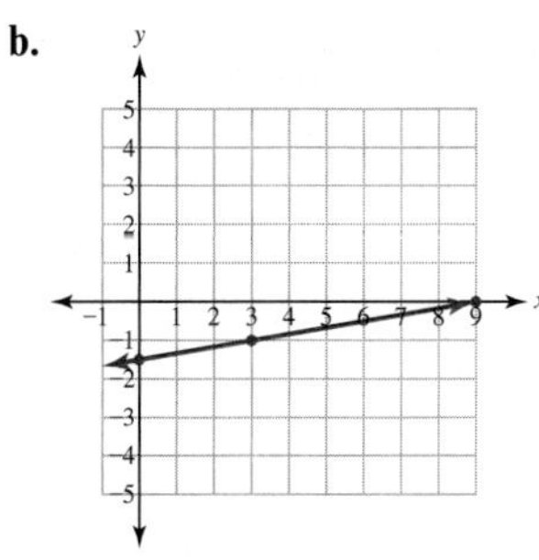

63. a.

x	y
0	−9
$\frac{3}{2}$	0
−3	−27

b.

65. a.

x	y
0	$\frac{21}{4}$
3	0
−3	$\frac{21}{2}$

b.

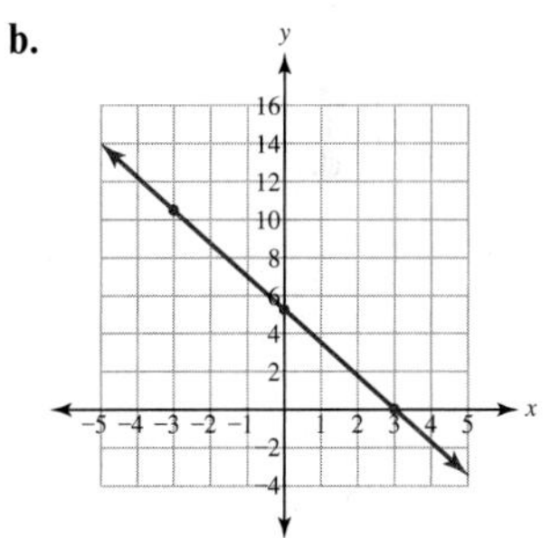

67. a.

x	y
0	30
30	0
3	27

b.

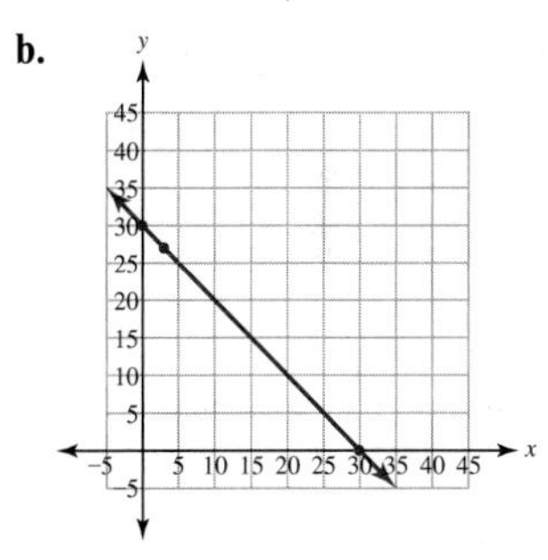

69. a.

x	y
0	2
120	0
−60	3

b.

71. a.

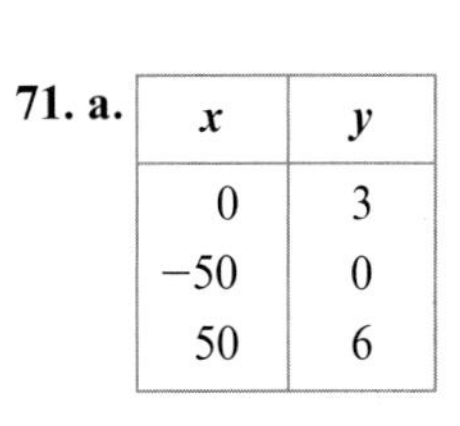

x	y
0	3
−50	0
50	6

b.

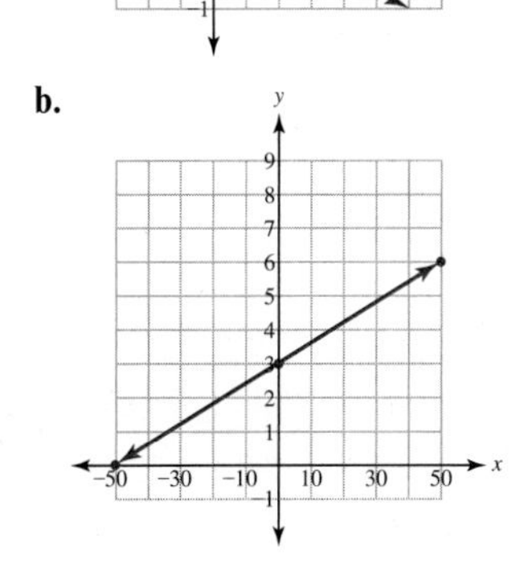

73. a.

x	y
0	2
2	4
4	6

b.

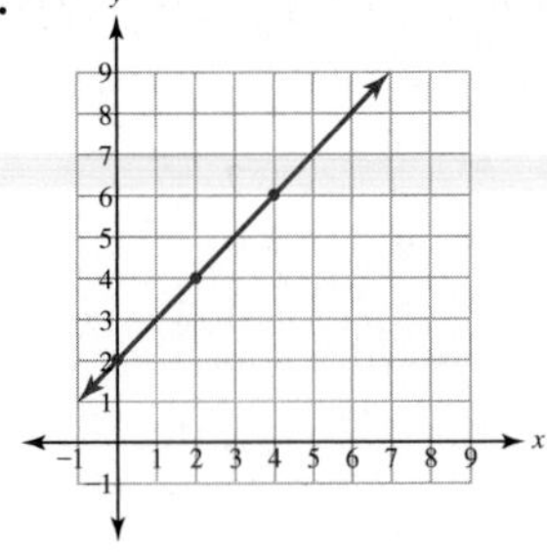

75. a.

x (ft)	y (\$)
0	10
5	20
10	30

b.

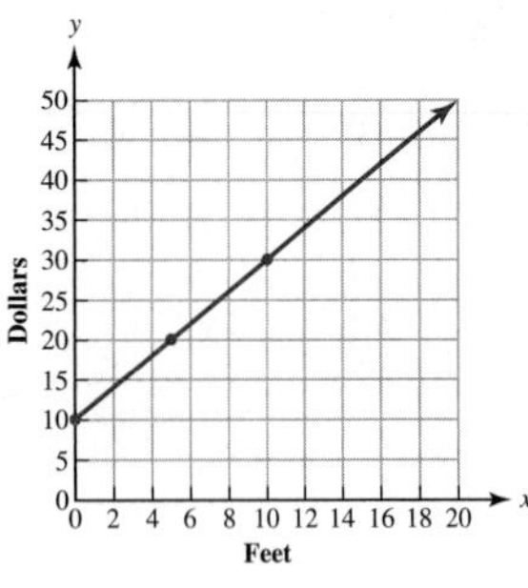

77. a.

x	y
−2	5
0	1
2	−3

b.

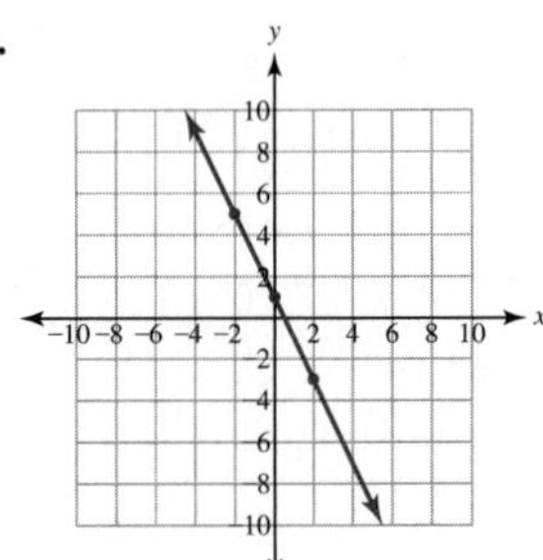

79. a.

x	y
−4	−16
0	−4
4	8

b.

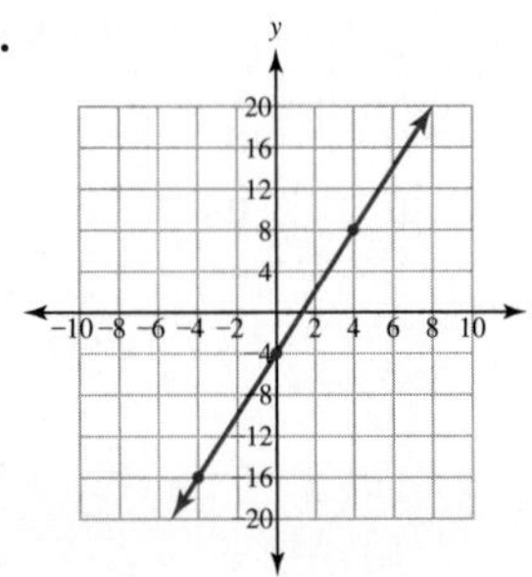

81. a.

x	y
−3	−8
0	−6
3	−4

b.

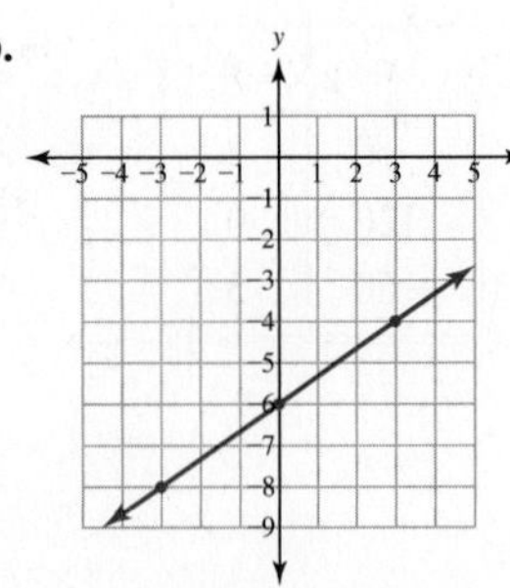

83. a.

x	y
−6	13
0	8
6	3

b.

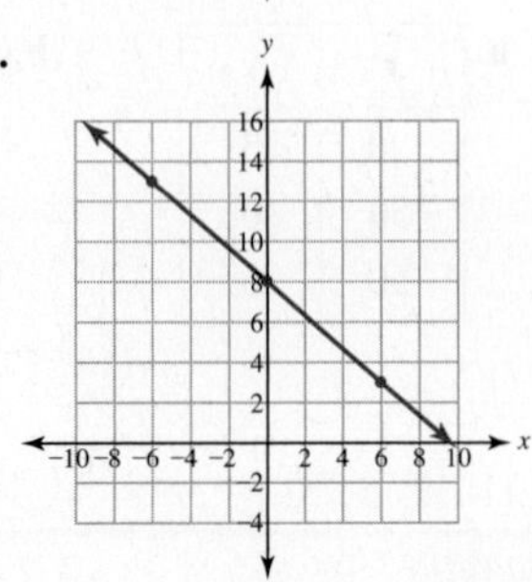

85. \$1635 **87.** \$2280.88

89.

X	Y1
1	-2
6	18
11	38
16	58
21	78
26	98
31	118

X=1

91.

X	Y1
-8	-38
-5	-26
-2	-14
1	-2
4	10
7	22
10	34

X=-8

93. b. $(-4, 3)$ **95. b.**

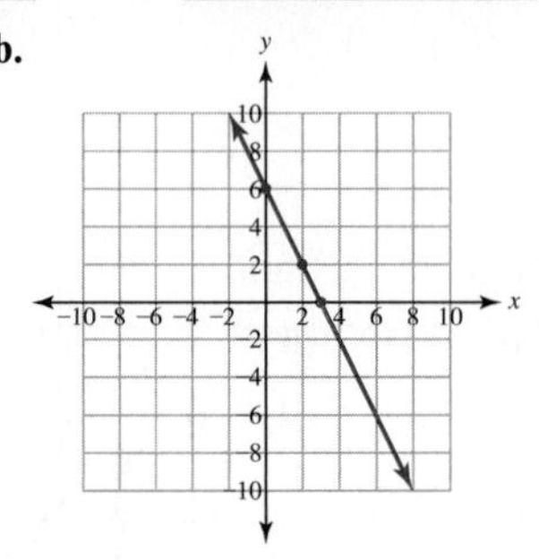

97. 6 **99.** −7

Section 3.2 Vocabulary Practice

1. G **2.** I **3.** D **4.** B **5.** E **6.** A **7.** F **8.** J **9.** H **10.** C

Section 3.2 Exercises

1. $-6x + y = -2$ or $6x - y = 2$ **3.** $16x + 15y = 800$
5. $2x - y = -11$ or $-2x + y = 11$
7. $-3x + 4y = -44$ or $3x - 4y = 44$
9. $-5x + 8y = 29$ or $5x - 8y = -29$
11. a. (0, 2) **b.** (3, 0) **13. a.** (0, 3) **b.** (−3, 0)
15. a. (0, −4) **b.** (−7, 0) **17. a.** (0, 140) **b.** (160, 0)
19. a. (0 s, 3 gal) **b.** (6 s, 0 gal)
21. a. (0, 6) **b.** (6, 0) **d.**

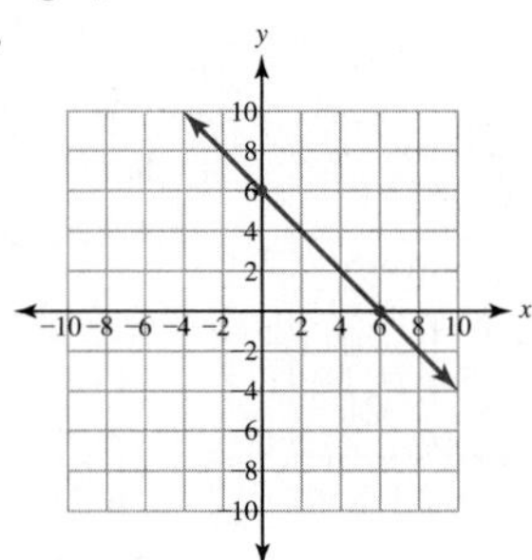

23. a. (0, −5) **b.** (−5, 0) **d.**

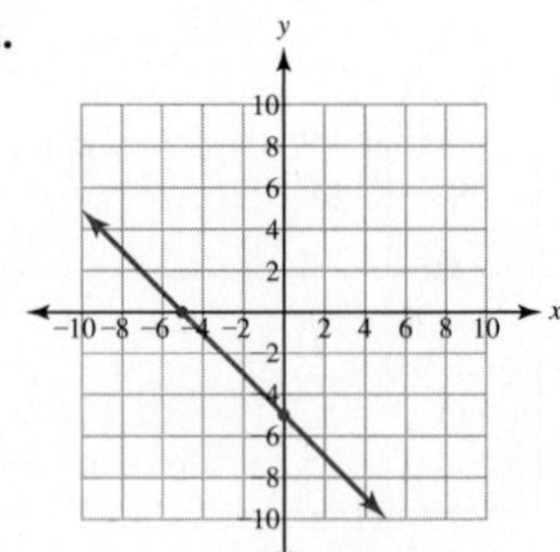

25. a. (0, −6) **b.** (6, 0) **d.**

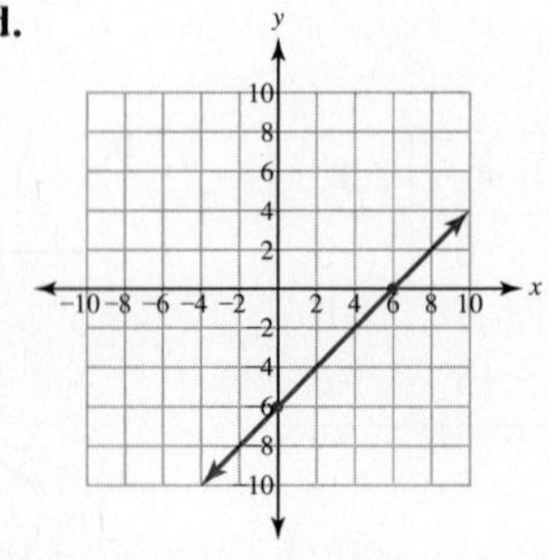

27. a. $(0, 5)$ **b.** $(-5, 0)$ **d.**

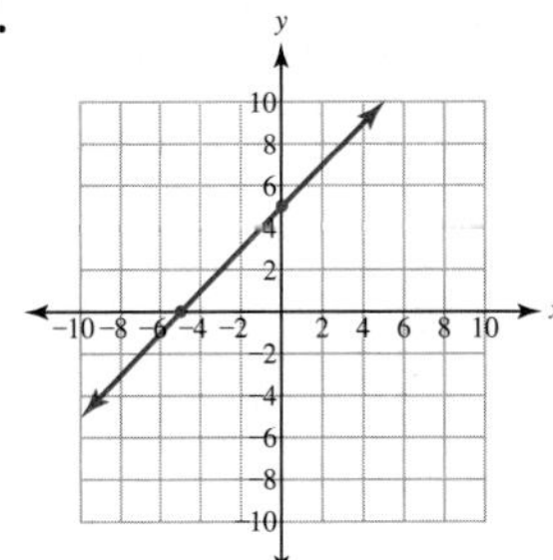

29. a. $(0, -9)$ **b.** $(-9, 0)$ **d.**

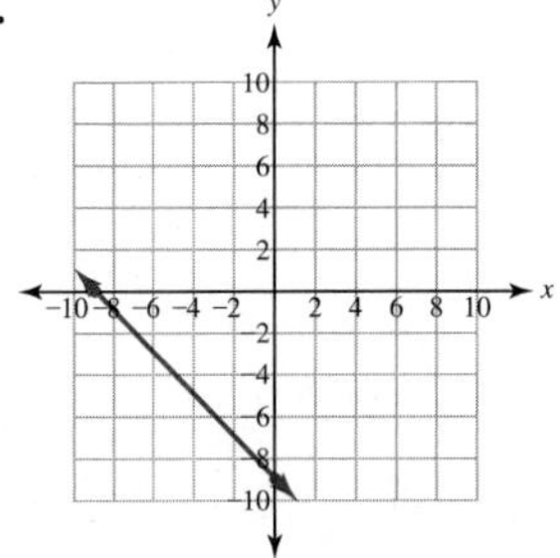

31. a. $(0, -1)$ **b.** $(1, 0)$ **d.**

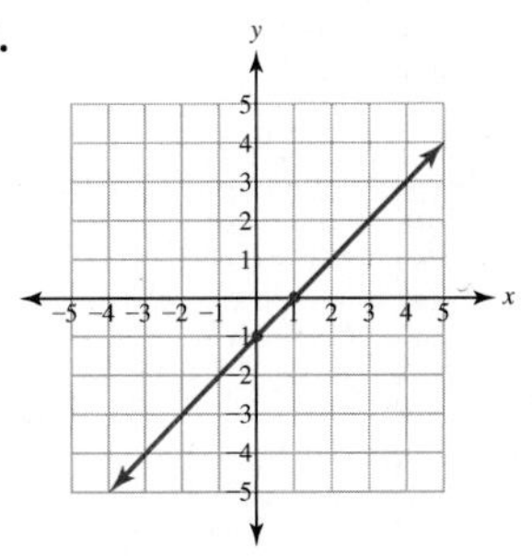

33. a. $(0, 12)$ **b.** $(4, 0)$ **d.**

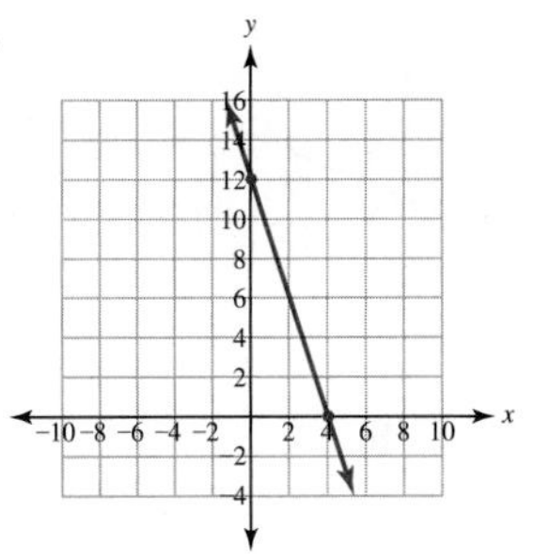

35. a. $(0, 4)$ **b.** $(12, 0)$ **d.**

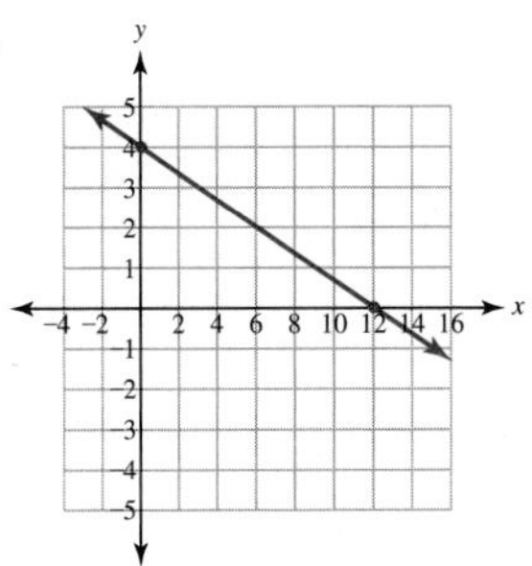

37. a. $(0, 14)$ **b.** $(4, 0)$ **d.**

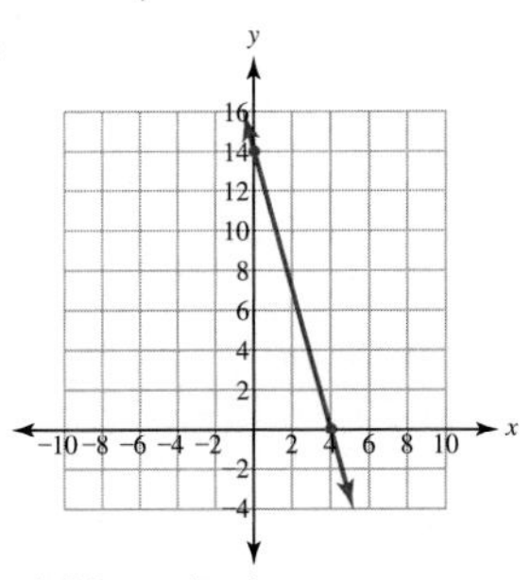

39. a. $(0, -4)$ **b.** $(10, 0)$ **d.**

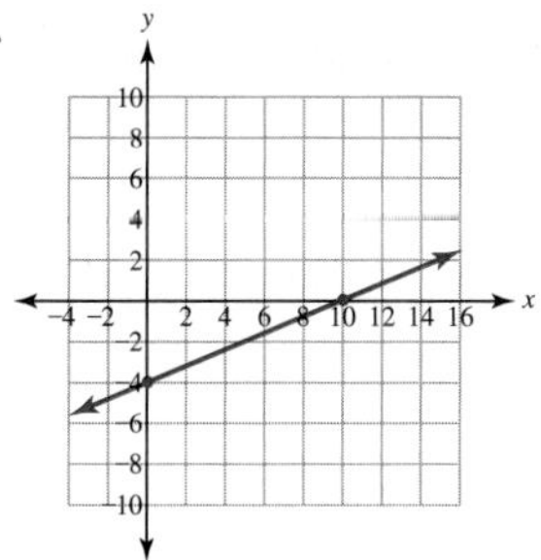

41. a. $(0, 12)$ **b.** $(-10, 0)$ **d.**

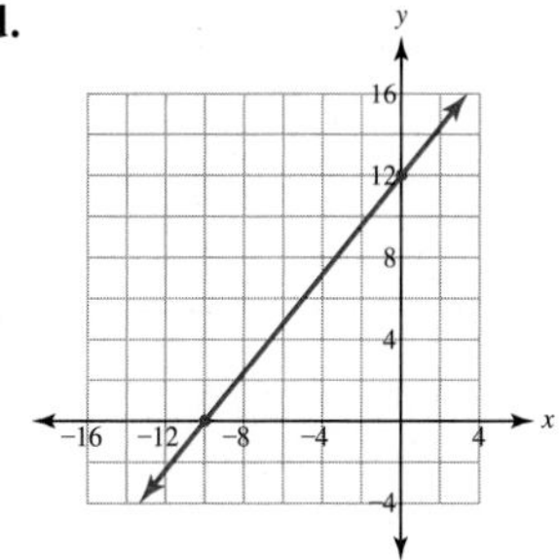

43. a. $(0, 3)$ **b.** $\left(-\frac{27}{2}, 0\right)$ **d.**

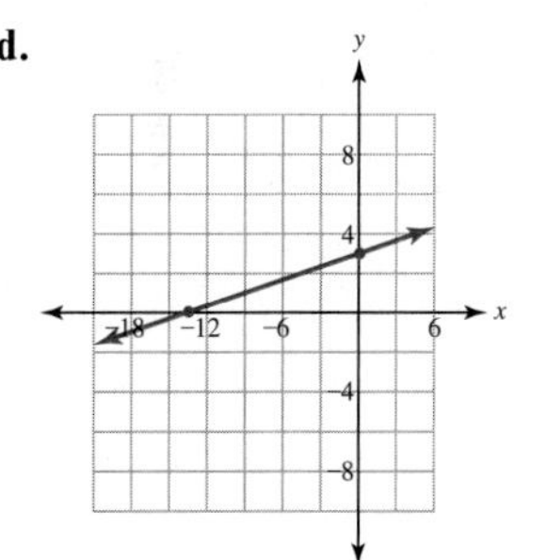

45. a. $(0, 8)$ **b.** $(3, 0)$ **d.**

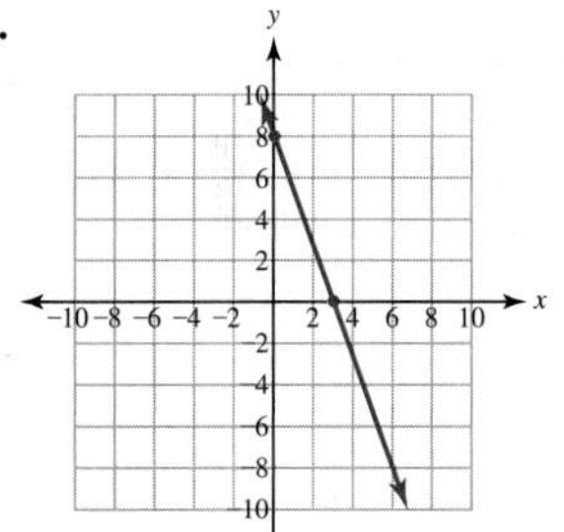

47. a. $(0, 8)$ **b.** $(-3, 0)$ **d.**

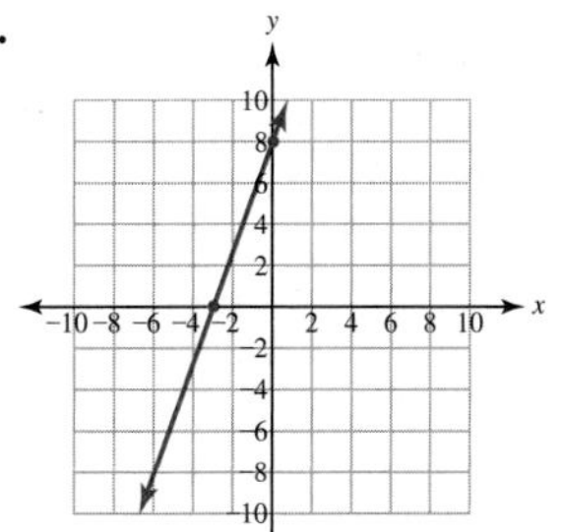

49. a. $\left(0, \frac{20}{3}\right)$ **b.** $(5, 0)$ **d.**

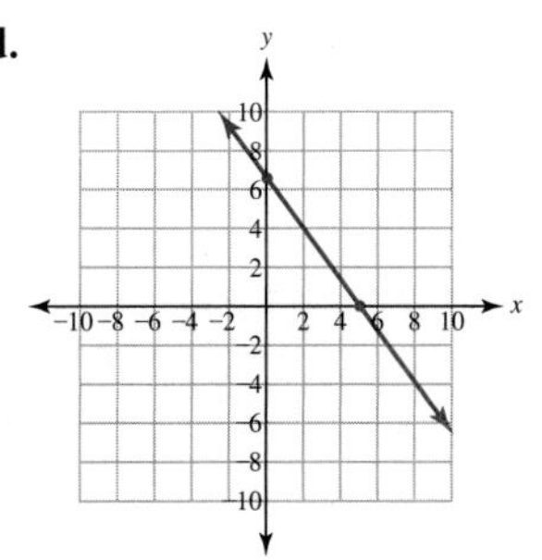

51. a. $\left(0, -\dfrac{15}{2}\right)$ **b.** $(-5, 0)$ **d.**

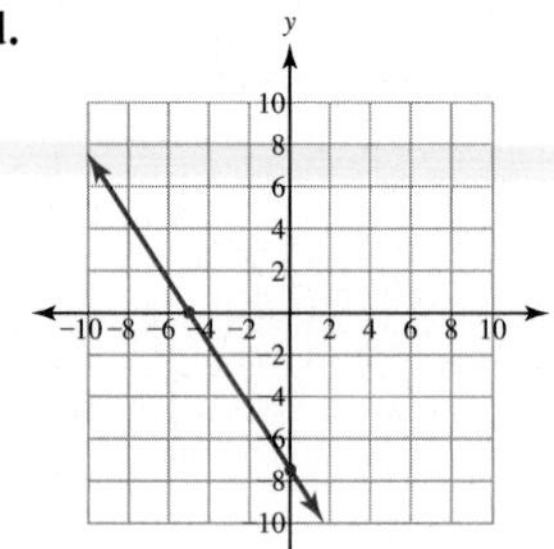

53. a. $(0, 400)$ **b.** $(400, 0)$ **d.**

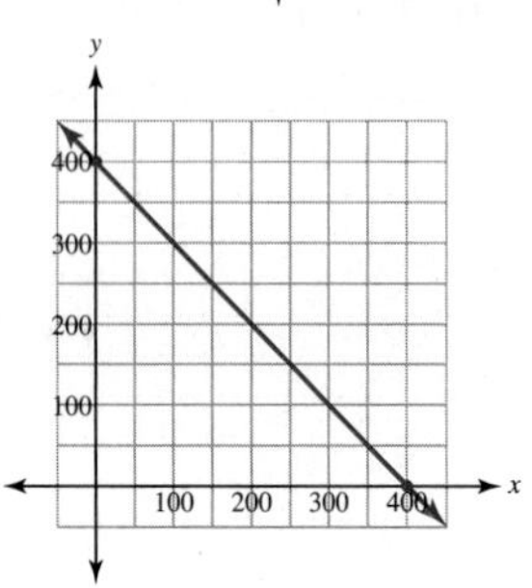

55. a. $(0, -1000)$ **b.** $(-1000, 0)$ **d.**

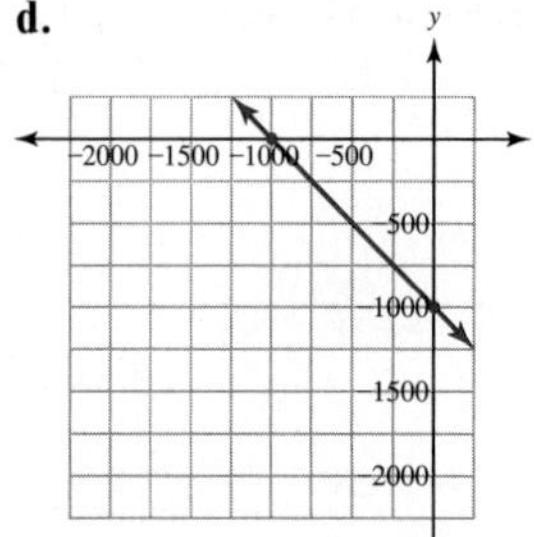

57. a. $(0, 500)$ **b.** $(25, 0)$ **d.**

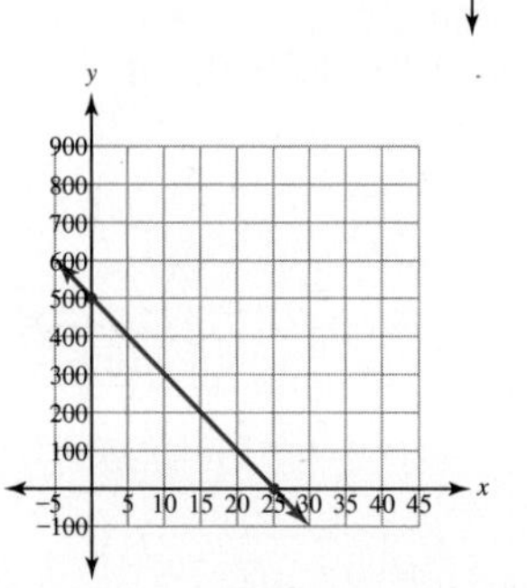

59. a. $(0, 500)$ **b.** $(-5, 0)$ **d.**

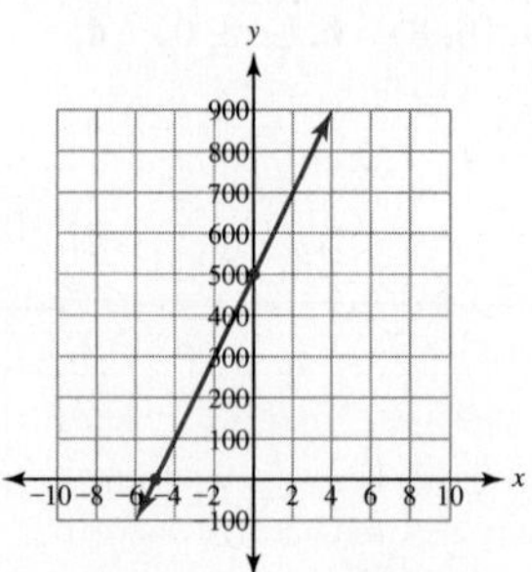

61. a. $(0, 100)$ **b.** $(-30, 0)$ **d.**

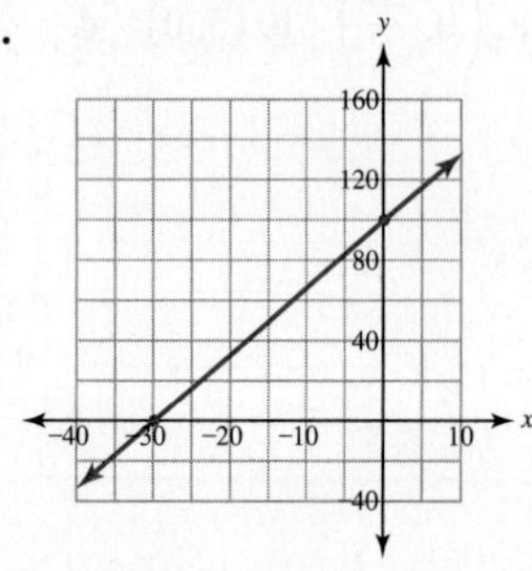

63. a. $\left(0, \dfrac{4}{3}\right)$ **b.** $(2, 0)$ **d.**

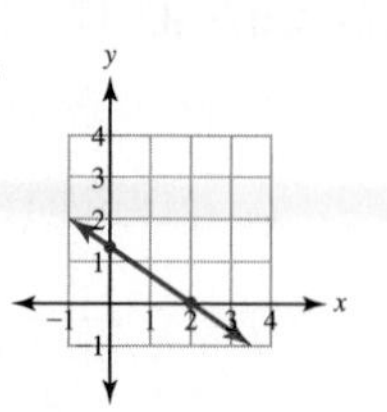

65. b.

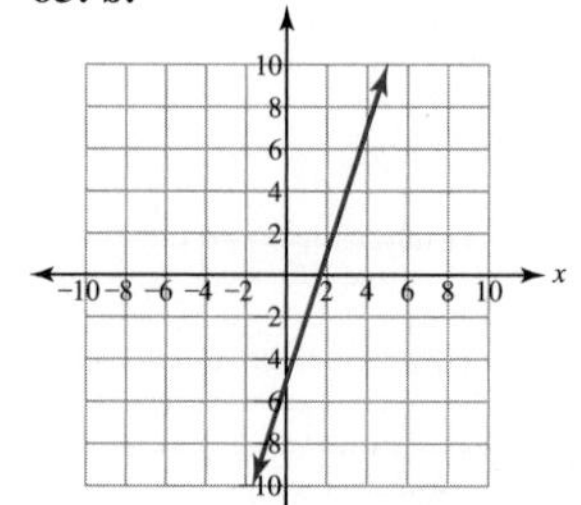

67. b.

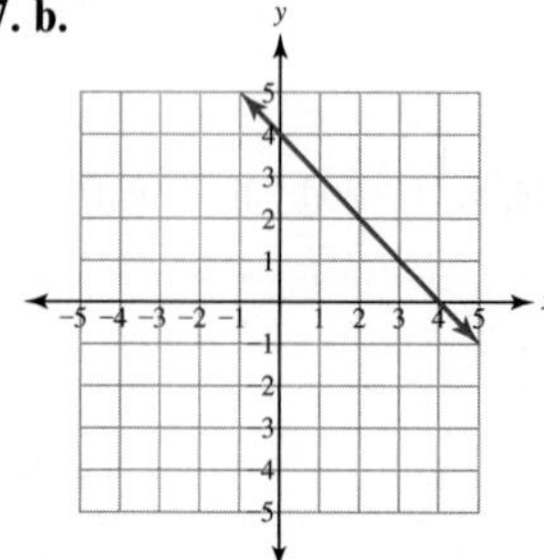

69. b.

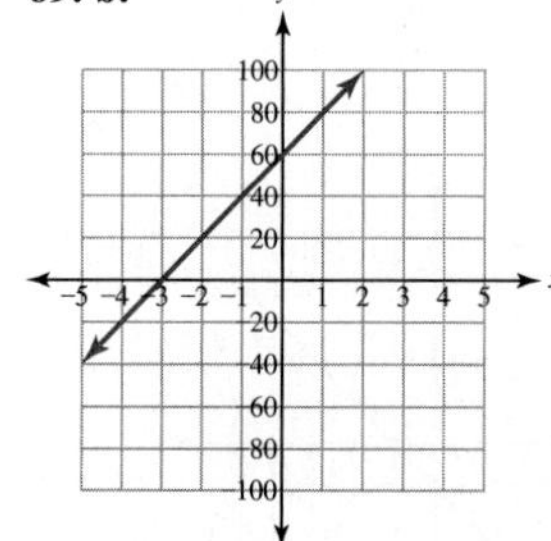

71. b.

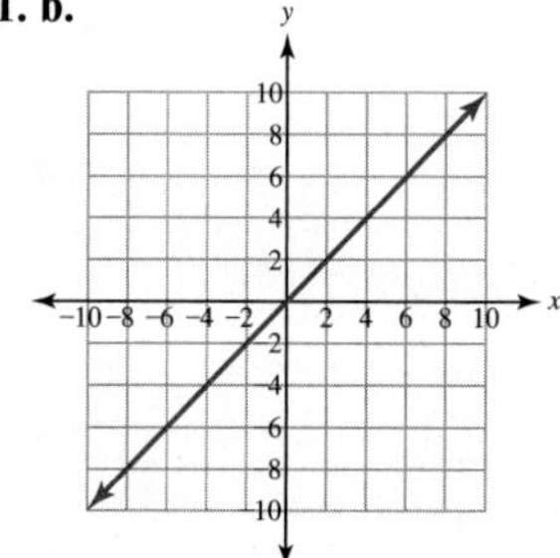

73. b.

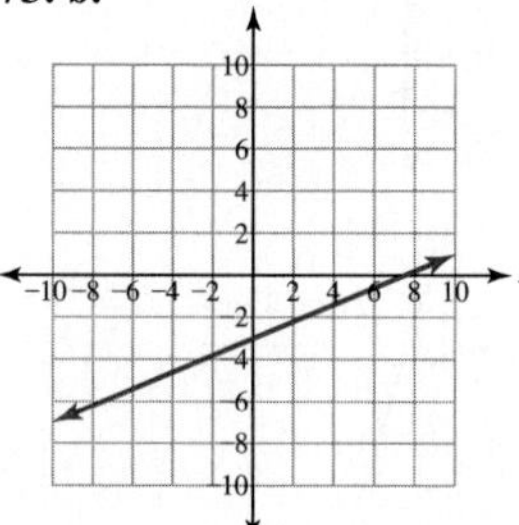

75. b.

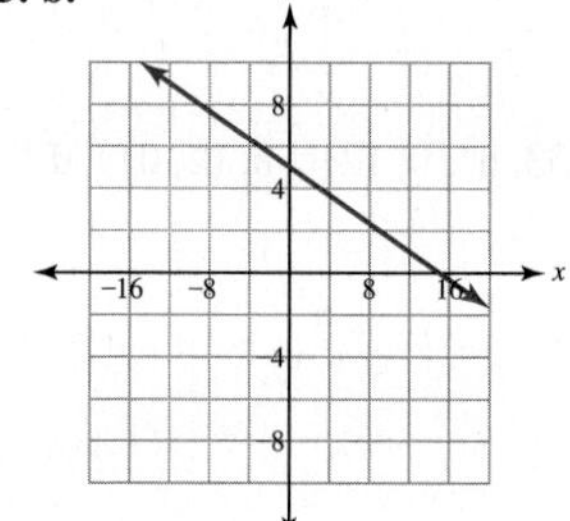

79. a. $x = 4$ **b.**

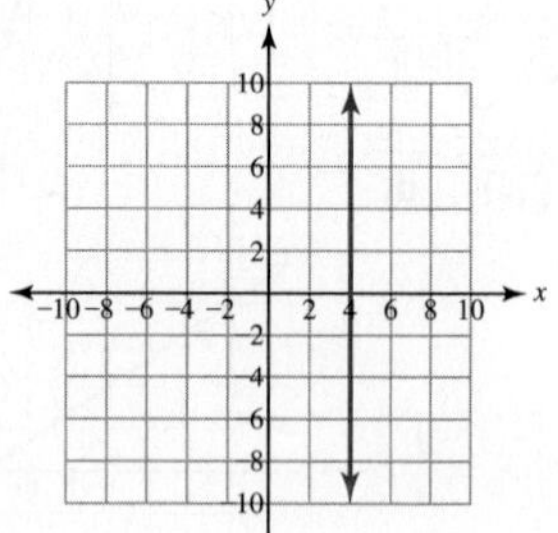

81. a. $x = -2$ **b.**

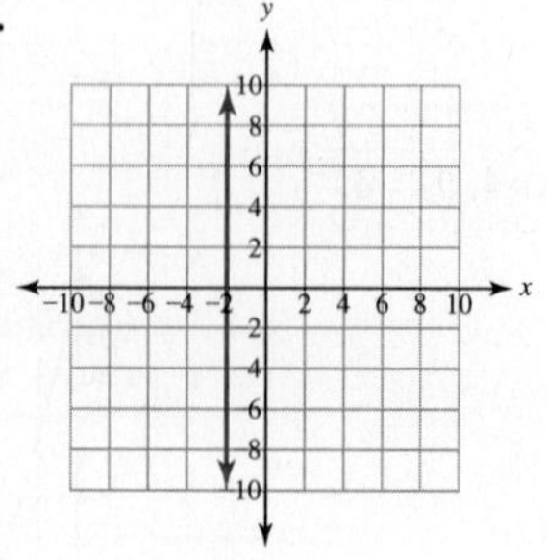

83. a. $y = 5$ **b.**

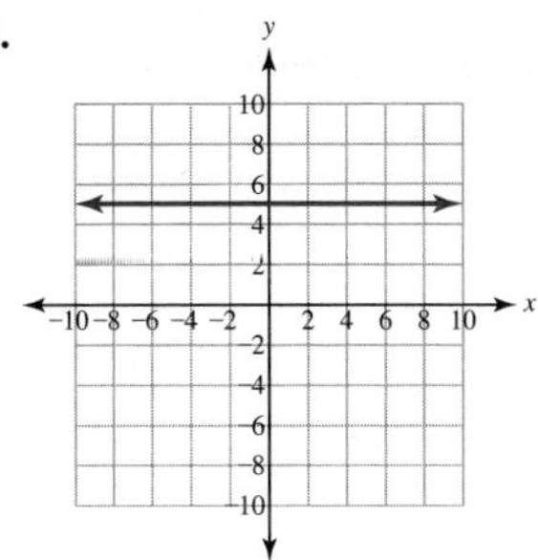

85. a. $y = -1$ **b.**

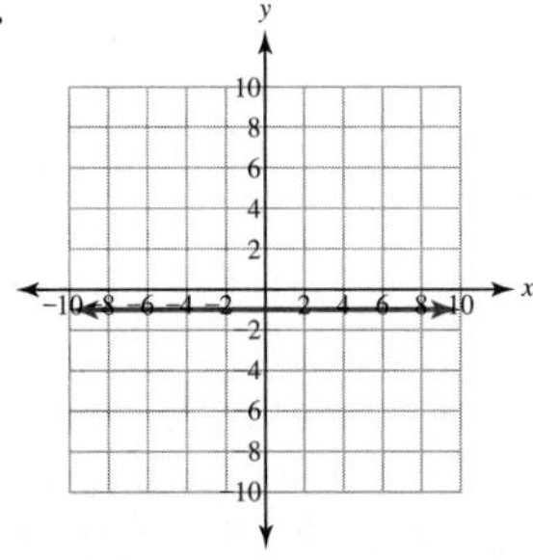

87. 768 ft^2 **89.** 33.3 million people

91.

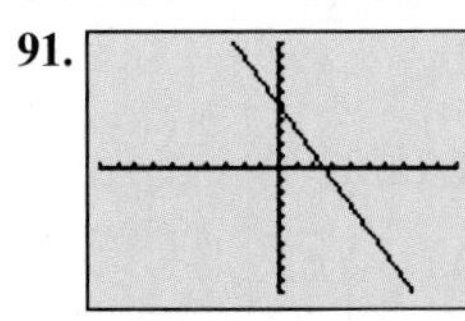

$[-10, 10, 1; -10, 10, 1]$

93.

$[-10, 10, 1; -10, 10, 1]$

95. b. The y-intercept is $(0, 6)$. **97. b.** $x = 6$

99. $m = 3$ **101.** $m = 0$

Section 3.3 Vocabulary Practice

1. I **2.** H **3.** B **4.** G **5.** J **6.** A **7.** C **8.** D **9.** E **10.** F

Section 3.3 Exercises

1. a. (2007, 28 acres), (2010, 252 acres) **b.** $\dfrac{75 \text{ acres}}{1 \text{ year}}$

3. a. (2006, 7,575,000 men), (2009, 8,770,000 men)
b. $\dfrac{398{,}000 \text{ men}}{1 \text{ year}}$

5. a. (2005, 985 deaths), (2010, 760 deaths) **b.** $-\dfrac{45 \text{ deaths}}{1 \text{ year}}$

7. a. (2005, 207,896,198 connections), (2010, 302,859,674 connections)
b. $\dfrac{18{,}993{,}000 \text{ connections}}{1 \text{ year}}$

9. a.

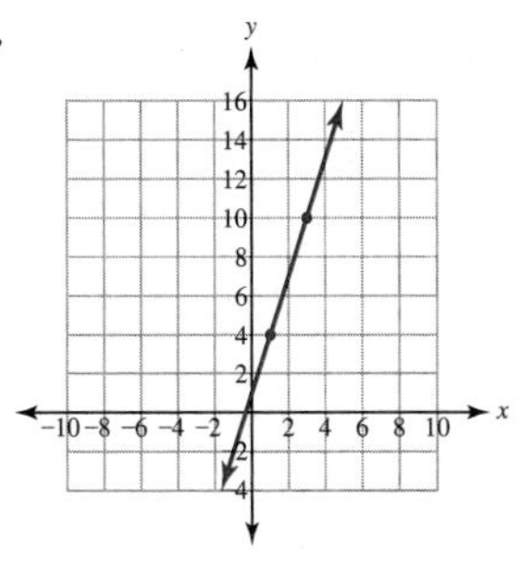

b. 3 **c.** 3

11. a.

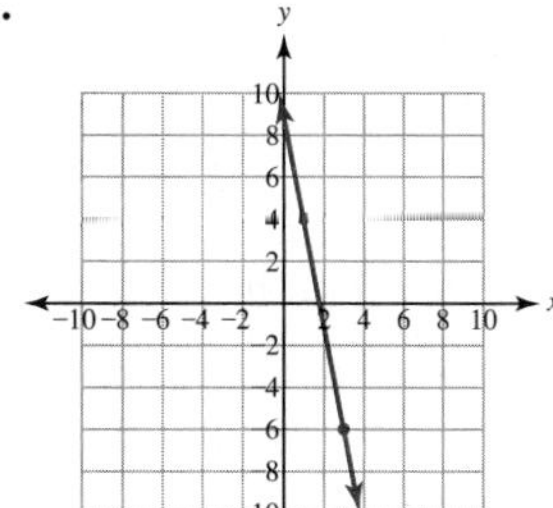

b. -5 **c.** -5

13. a.

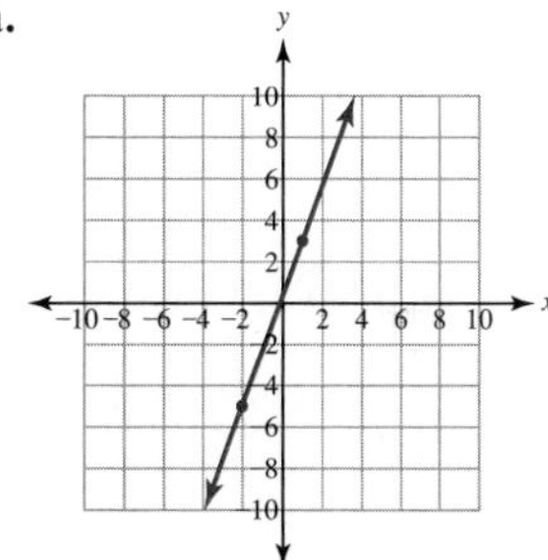

b. $\dfrac{8}{3}$ **c.** $\dfrac{8}{3}$

15. a.

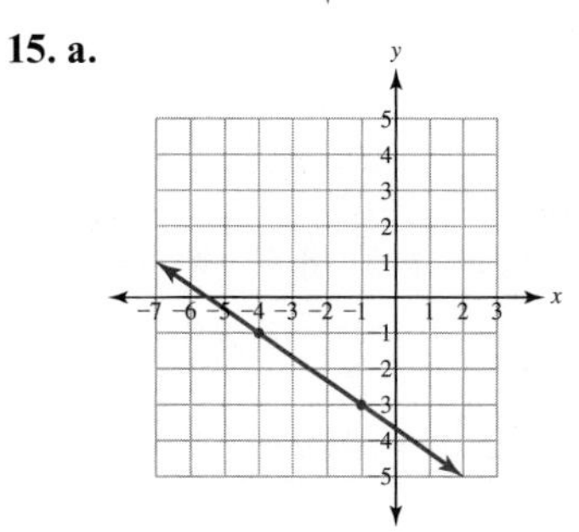

b. $-\dfrac{2}{3}$ **c.** $-\dfrac{2}{3}$

17. a.

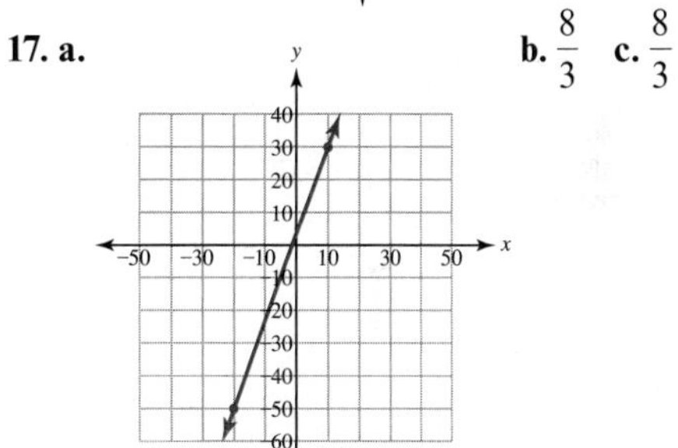

b. $\dfrac{8}{3}$ **c.** $\dfrac{8}{3}$

19. a.

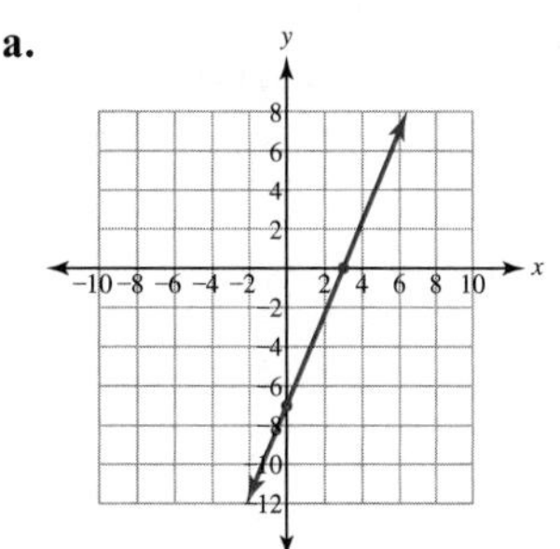

b. $\dfrac{7}{3}$ **c.** $\dfrac{7}{3}$

21. b. 1 **23. b.** -1 **25. b.** undefined **27. b.** 0

29. b. -1 **31. b.** $-\dfrac{4}{3}$ **33. b.** $\dfrac{1}{6}$ **35. b.** $\dfrac{\$2000}{1 \text{ year}}$

37. $\dfrac{5}{2}$ **39.** 3 **41.** $\dfrac{15}{14}$ **43.** $\dfrac{31}{8}$ **45.** $-\dfrac{4}{3}$ **47.** 9

49. $\dfrac{1}{3}$ **51.** $\dfrac{5}{3}$ **53.** $-\dfrac{1}{3}$ **55.** -3 **57.** $-\dfrac{3}{10}$ **59.** $\dfrac{8}{3}$

61. $\frac{1}{64}$ **63.** $-\frac{9}{4}$ **65.** $\frac{\$9}{1 \text{ hr}}$ **67. a.** $(0, 8)$ **b.** $(18, 0)$ **c.** $-\frac{4}{9}$ **69. a.** $(0, -9)$ **b.** $(27, 0)$ **c.** $\frac{1}{3}$ **71. a.** $(0, 20)$ **b.** $(-8, 0)$ **c.** $\frac{5}{2}$ **73. a.** $(0, 3)$ **b.** $\left(\frac{15}{2}, 0\right)$ **c.** $-\frac{2}{5}$ **75. a.** $(0, -8)$ **b.** $(8, 0)$ **c.** 1 **77. a.** $(0, 3)$ **b.** $(3, 0)$ **c.** -1 **81.** \$23 more per week **83.** 1.75 billion gal

85. a.

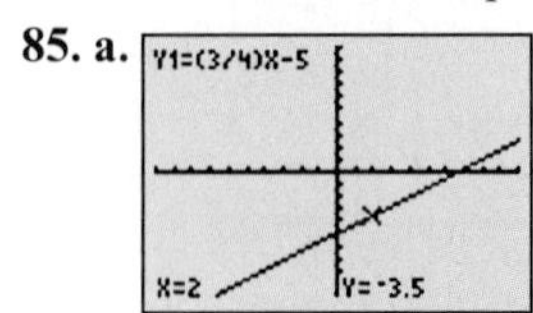

$[-10, 10, 1; -10, 10, 1]$

b. $y = -3.5$

87. a.

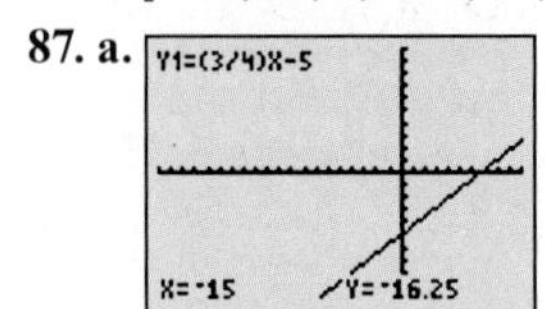

$[-20, 10, 1; -10, 10, 1]$

b. $y = -16.25$

89. b. $m = \frac{3}{2}$ **91. b.** $\frac{2}{7}$ **93.** $(0, 1)$ **95.** $(0, -3)$

Section 3.4 Vocabulary Practice

1. F **2.** G **3.** H **4.** I **5.** J **6.** B **7.** A **8.** C **9.** D **10.** E

Section 3.4 Exercises

1. a. (0 hr, 24 gal), (2 hr, 18 gal)

b.

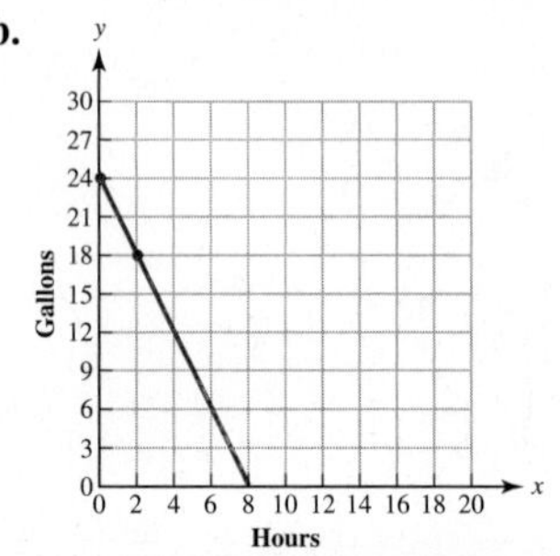

c. (0 hr, 24 gal) **d.** $-\frac{3 \text{ gal}}{1 \text{ hr}}$ **e.** $y = \left(-\frac{3 \text{ gal}}{1 \text{ hr}}\right)x + 24 \text{ gal}$ **f.** 15 gal **g.** (8 hr, 0 gal)

3. a. (3 hr, 6000 lb), (5 hr, 10,000 lb)

b.

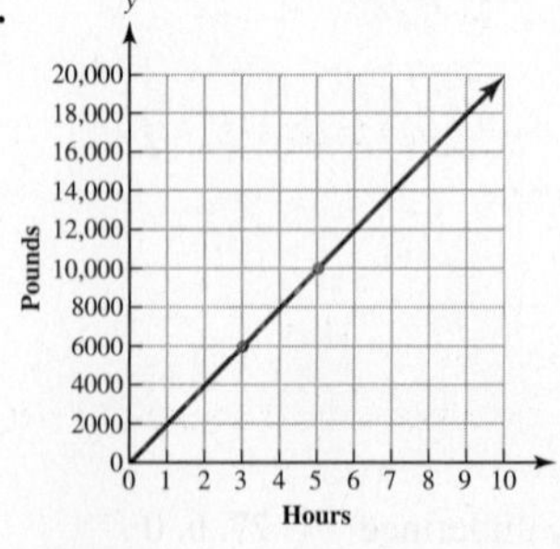

c. (0 hr, 0 lb) **d.** $\frac{2000 \text{ lb}}{1 \text{ hr}}$ **e.** $y = \left(\frac{2000 \text{ lb}}{1 \text{ hr}}\right)x$ **f.** 14,000 lb of potato chips **g.** (0 hr, 0 lb)

5. a. $y = \left(\frac{\$0.99}{1 \text{ book}}\right)x + \$3.00$ **b.** \$27.75

7. a. $y = \left(-\frac{0.25 \text{ g}}{1 \text{ hr}}\right)x + 6 \text{ g}$ **b.** 4 g **c.** (24 hr, 0 g)

9. a. $y = \left(\frac{\$8.90}{1 \text{ lb}}\right)x + \$9.95$ **b.** \$54.45

11. a. $y = \left(\frac{\$800}{1 \text{ month}}\right)x + \$1200$ **b.** \$5200 **13.** 15,200 ft **15.** 28.5 miles **17.** $y = -4x + 3$ **19.** $y = 5x$ **21.** $y = \frac{2}{3}x + 4$ **23.** $y = \left(\frac{30 \text{ mi}}{1 \text{ hr}}\right)x + 28 \text{ mi}$

25. a. 3 **b.** $(0, 2)$ **c.** $\left(-\frac{2}{3}, 0\right)$ **27. a.** -5 **b.** $(0, 10)$ **c.** $(2, 0)$ **29. a.** $y = -\frac{7}{9}x - 7$ **b.** $-\frac{7}{9}$ **c.** $(0, -7)$ **d.** $(-9, 0)$ **31. a.** $y = \frac{7}{9}x + 7$ **b.** $\frac{7}{9}$ **c.** $(0, 7)$ **d.** $(-9, 0)$

33. a. $y = -x + 4$ **b.** -1 **c.** $(0, 4)$ **d.** $(4, 0)$

35. a. $y = x - 6$ **b.** 1 **c.** $(0, -6)$ **d.** $(6, 0)$

37. a. $y = -x$ **b.** -1 **c.** $(0, 0)$ **d.** $(0, 0)$

39. a. $y = \frac{4}{3}x + 8$ **b.** $\frac{4}{3}$ **c.** $(0, 8)$ **d.** $(-6, 0)$

41. a. $y = \frac{2}{9}x - 3$ **b.** $\frac{2}{9}$ **c.** $(0, -3)$ **d.** $\left(\frac{27}{2}, 0\right)$

43. a. $y = \frac{9}{2}x - \frac{27}{2}$ **b.** $\frac{9}{2}$ **c.** $\left(0, -\frac{27}{2}\right)$ **d.** $(3, 0)$

45. a. $y = \frac{9}{8}x + 12$ **b.** $\frac{9}{8}$ **c.** $(0, 12)$ **d.** $\left(-\frac{32}{3}, 0\right)$

47. a. $y = \frac{1}{2}x + 1$ **b.** $\frac{1}{2}$ **c.** $(0, 1)$ **d.** $(-2, 0)$

49. a. $y = \frac{4}{9}x - \frac{67}{9}$ **b.** $\frac{4}{9}$ **c.** $\left(0, -\frac{67}{9}\right)$ **d.** $\left(\frac{67}{4}, 0\right)$

51. a. $y = 2x + \frac{1}{12}$ **b.** 2 **c.** $\left(0, \frac{1}{12}\right)$ **d.** $\left(-\frac{1}{24}, 0\right)$

53. a. $(0, -3)$ **b.** 1 **c.** $y = x - 3$ **55. a.** $(0, -3)$ **b.** $\frac{3}{4}$ **c.** $y = \frac{3}{4}x - 3$ **57. a.** $(0, 1)$ **b.** -2 **c.** $y = -2x + 1$ **59. a.** $(0, 2)$ **b.** $-\frac{1}{5}$ **c.** $y = -\frac{1}{5}x + 2$ **61. a.** $(0, 0)$ **b.** -1 **c.** $y = -x$ **63. a.** $(0, -4)$ **b.** 3 **c.** $y = 3x - 4$

65. a. (0 s, 10 cm) **b.** $\frac{5 \text{ cm}}{1 \text{ s}}$ **c.** $y = \left(\frac{5 \text{ cm}}{1 \text{ s}}\right)x + 10 \text{ cm}$

67. a. (0 s, 25 cm) **b.** $-\frac{5 \text{ cm}}{1 \text{ s}}$ **c.** $y = \left(-\frac{5 \text{ cm}}{1 \text{ s}}\right)x + 25 \text{ cm}$

69. a. $(0, -3)$ **b.** 0 **c.** $y = -3$

71. a. $y = 5$ **b.**

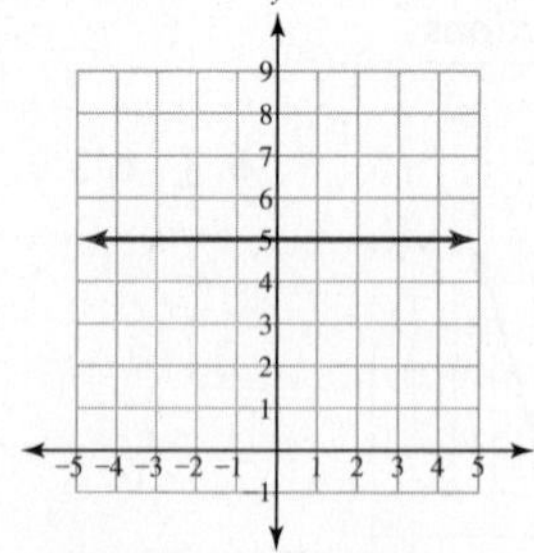

c. $x = 4$

d.

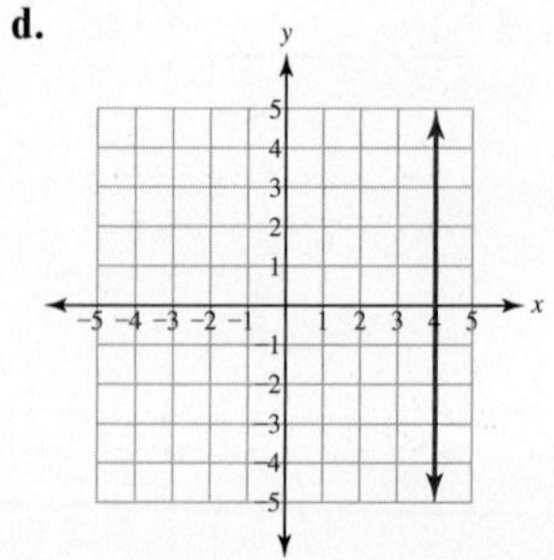

73. a. $y = -1$ **b.**

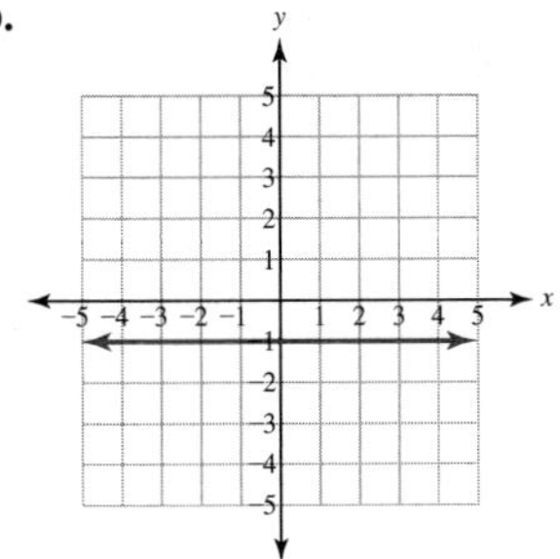

c. $x = -4$ **d.**

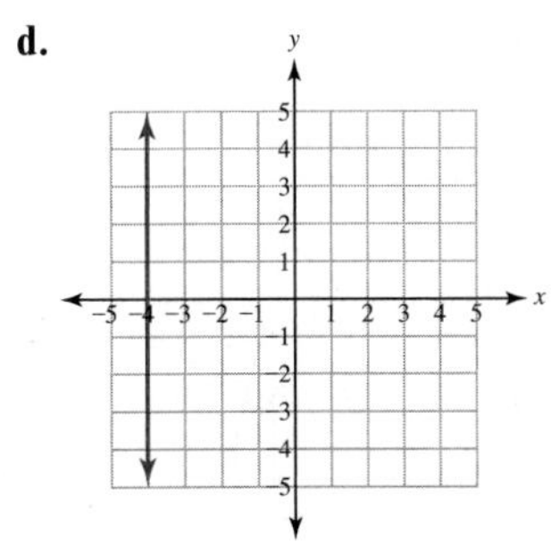

75. a. 0 **b.** $y = 4$ **77. a.** undefined **b.** $x = -3$

79.

Equation	Slope	y-intercept
$y = -\frac{5}{6}x + 2$	$-\frac{5}{6}$	(0, 2)
$y = 8$	0	(0, 8)
$y = 8x$	8	(0, 0)
$x = -12$	undefined	none

81.

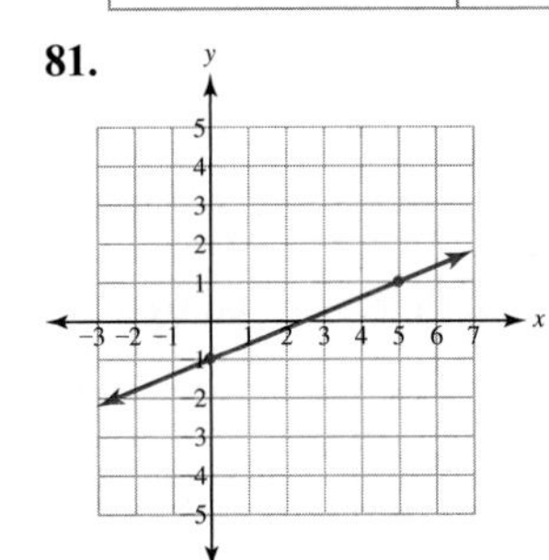

83.

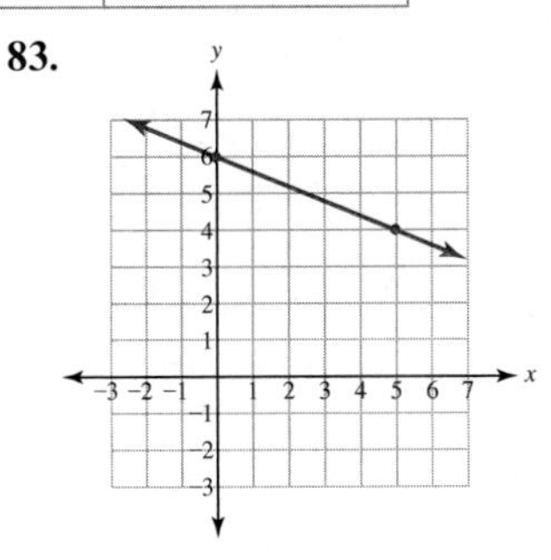

85.

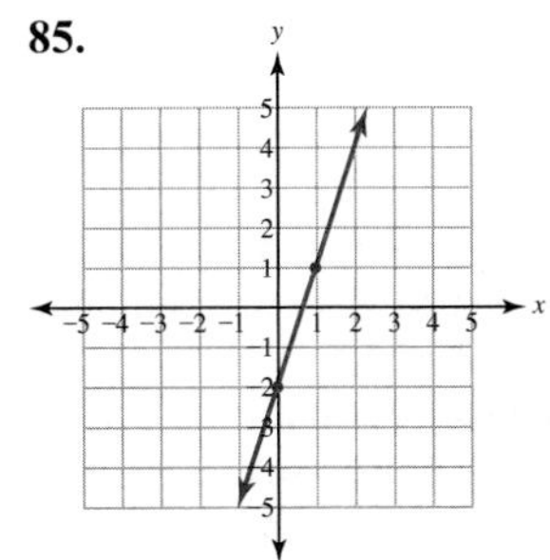

87.

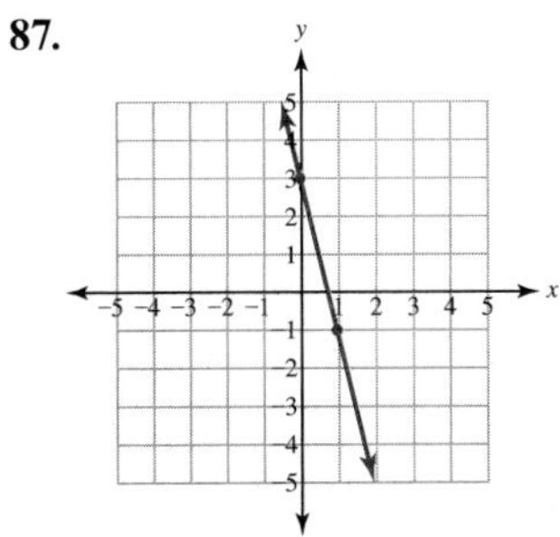

89.

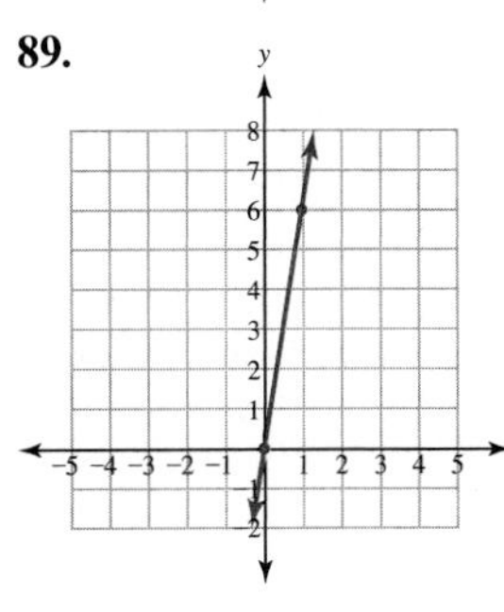

91.

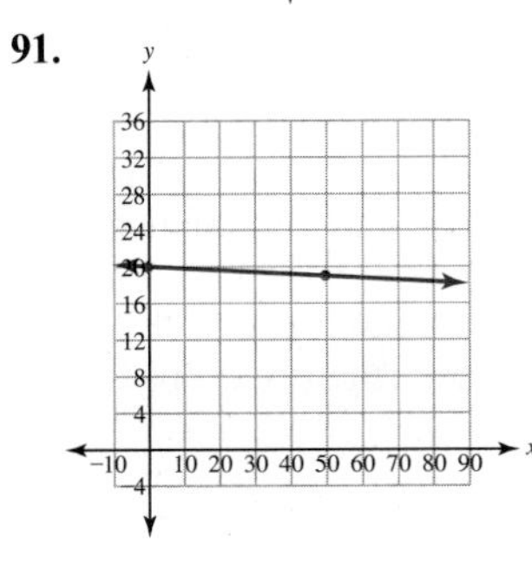

93. 18,300 lb **95.** 3620 workers **97. b.** $y = \frac{6}{5}x - 6$

99. b.

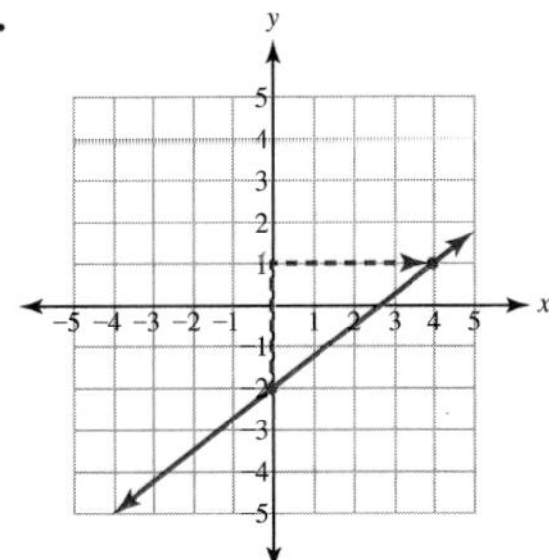

101. parallel **103.** neither

Section 3.5 Vocabulary Practice

1. B **2.** C **3.** F **4.** J **5.** E **6.** I **7.** D **8.** A **9.** G **10.** H

Section 3.5 Exercises

1. $y = 8x - 25$ **3.** $y = -4x + 13$ **5.** $y = -2x - 23$
7. $y = \frac{1}{4}x + 10$ **9.** $y = -\frac{2}{3}x - 4$ **11.** $y = \frac{2}{3}x + 4$
13. $y = -\frac{3}{4}x + \frac{19}{4}$ **15.** $y = 8x + 11$ **17.** $y = -3$
19. $y = 0.2x + 6.2$ **21. a.** 5 **b.** $y = 5x - 29$
23. a. -4 **b.** $y = -4x + 48$ **25. a.** 5 **b.** $y = 5x - 8$
27. a. $\frac{3}{4}$ **b.** $y = \frac{3}{4}x + \frac{69}{4}$ **29. a.** 0 **b.** $y = -9$
31. a. 1 **b.** $y = x$ **33. a.** -1 **b.** $y = -x$
35. a. 1 **b.** $y = x + \frac{1}{10}$ **37. a.** $\frac{1}{2}$ **b.** $y = \frac{1}{2}x + \frac{5}{36}$
39. a. $\frac{21}{80}$ **b.** $y = \frac{21}{80}x - \frac{27}{40}$
41. a. $(-4, -1), (2, 3)$ **b.** $\frac{2}{3}$ **c.** $y = \frac{2}{3}x + \frac{5}{3}$
43. a. $(-2, 4), (3, 1)$ **b.** $-\frac{3}{5}$ **c.** $y = -\frac{3}{5}x + \frac{14}{5}$
45. a. $y = \left(\frac{1.195 \text{ million people}}{1 \text{ year}}\right)x + 2.51$ million people
d. 62.3 million people
47. a. $y = \left(\frac{1096.7 \text{ registrations}}{1 \text{ year}}\right)x + 42{,}489$ registrations
d. 61,100 registrations
49. a. $y = \left(\frac{\$0.24}{1 \text{ year}}\right)x + \$5.53$ **d.** \$8.65
51. a. $y = \left(\frac{5264.1 \text{ teachers}}{1 \text{ year}}\right)x + 241{,}421$ teachers
d. 399,300 teachers **53.** 8 **55.** $-\frac{1}{8}$ **57.** -1
59. $y = 9x - 58$ **61.** $y = \frac{3}{4}x + \frac{1}{2}$ **63.** $y = -\frac{8}{5}x + 5$
65. $y = \frac{3}{2}x + 12$ **69.** There are an infinite number of lines.
71. 1 line **73. a.** 5 **b.** $-\frac{1}{5}$ **c.** $y = -\frac{1}{5}x - \frac{26}{5}$
75. a. 5 **b.** 5 **c.** $y = 5x - 26$ **77.** $y = -\frac{3}{4}x + \frac{17}{4}$
79. $y = -\frac{2}{3}x + \frac{1}{3}$ **81. a.** $y = \frac{5}{4}x - \frac{3}{2}$ **b.** $y = \frac{5}{4}x + \frac{27}{4}$
83. a. $y = -\frac{5}{7}x + \frac{13}{7}$ **b.** $y = -\frac{5}{7}x + \frac{31}{7}$

85. a. $y = -3$ **b.** $y = 1$ **87. a.** $y = 3x$ **b.** $y = 3x + 8$
89. 14.1 lb **91.** \$0.30 per minute
93. 70,800 people

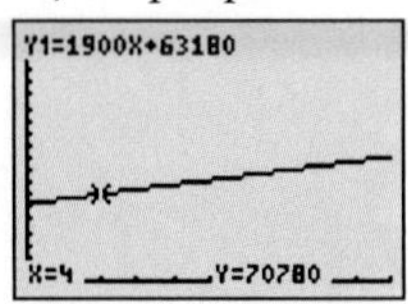

[0, 20, 2; 0, 200,000, 10,000]

95. b. $y = 6x - 3$ **97. b.** $y = -\frac{3}{2}x + 21$

99. $\{3, 5, 9, 15\}$ **101.** $y = -3$

Section 3.6 Vocabulary Practice

1. F **2.** D **3.** G **4.** E **5.** J **6.** A **7.** I **8.** C **9.** B **10.** H

Section 3.6 Exercises

1.

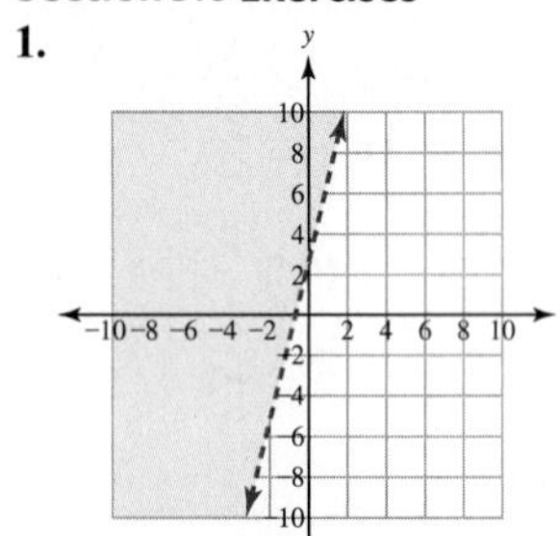

3.

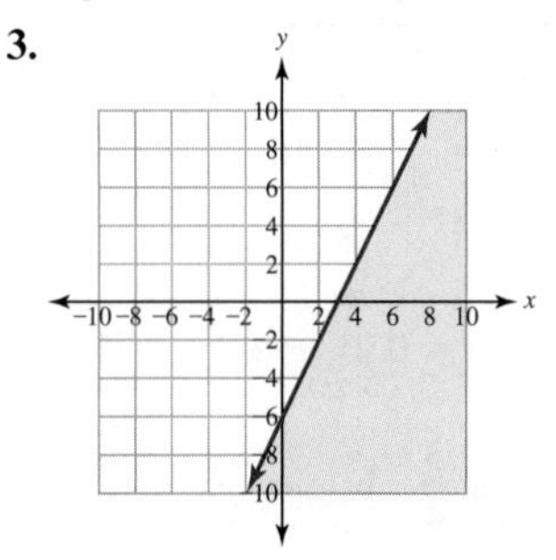

5.

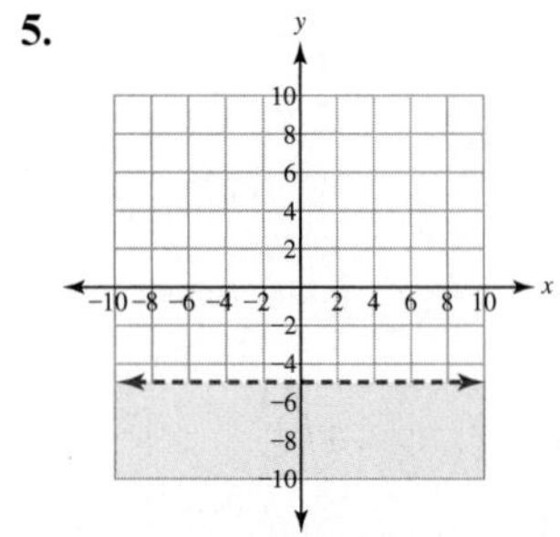

7.

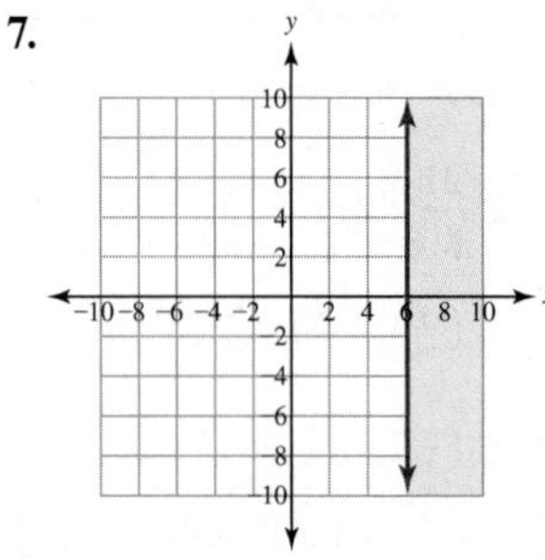

9.

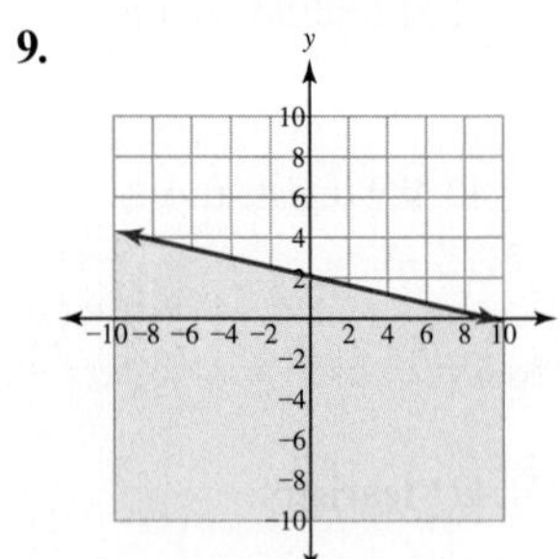

11.

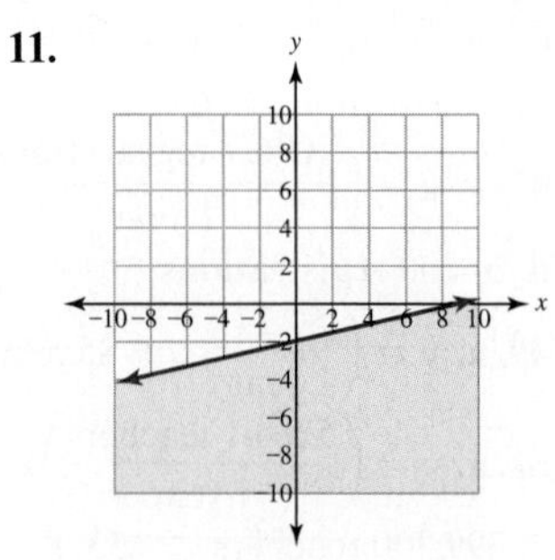

13.

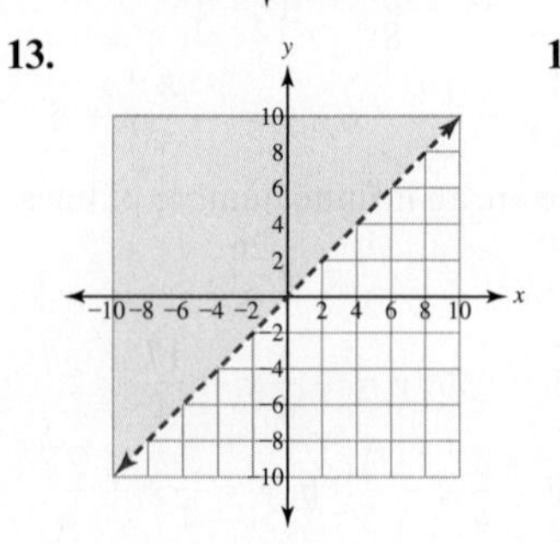

15.

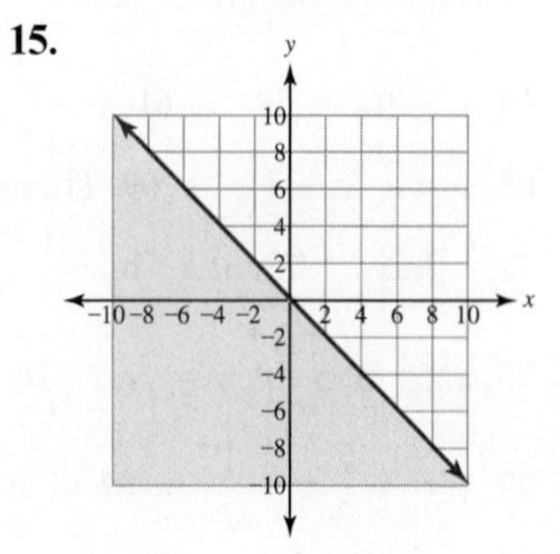

17.

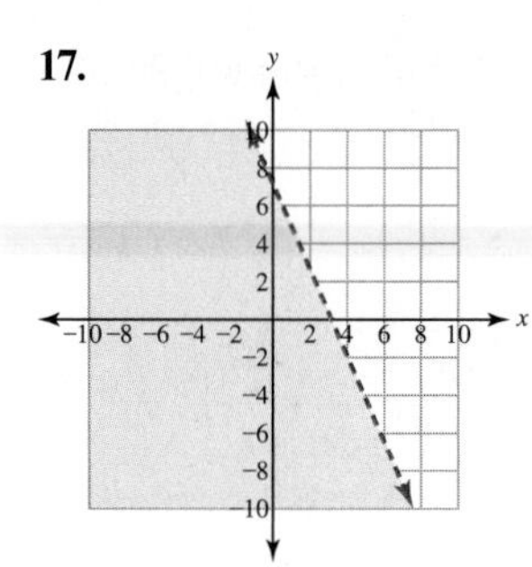

19.

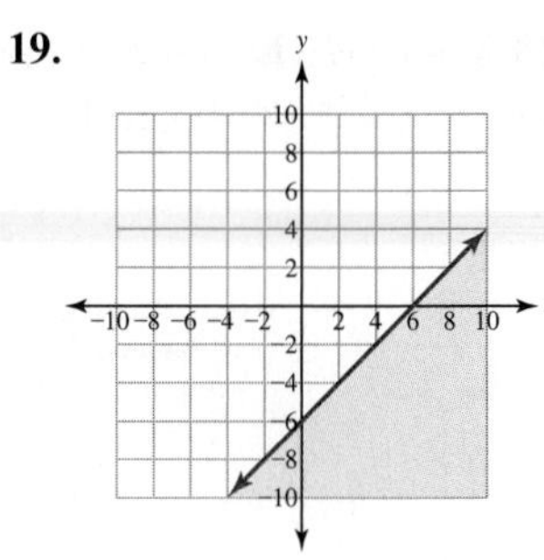

21.

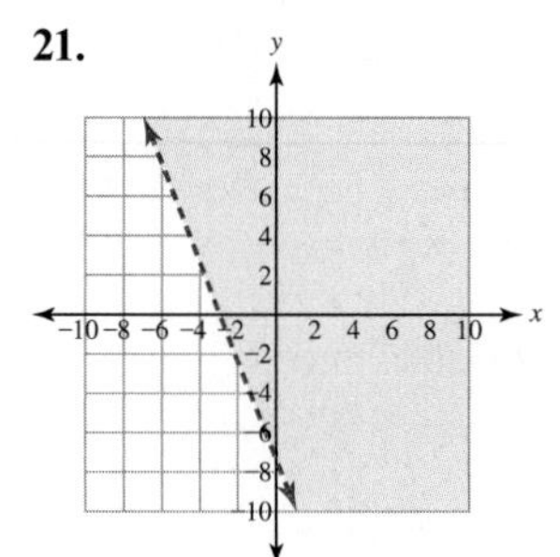

23.

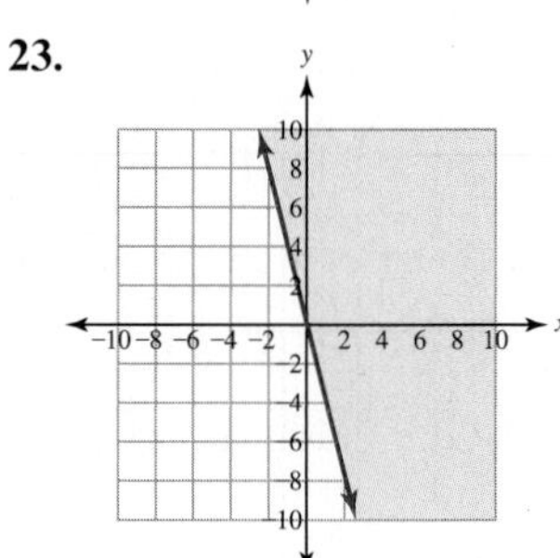

25.

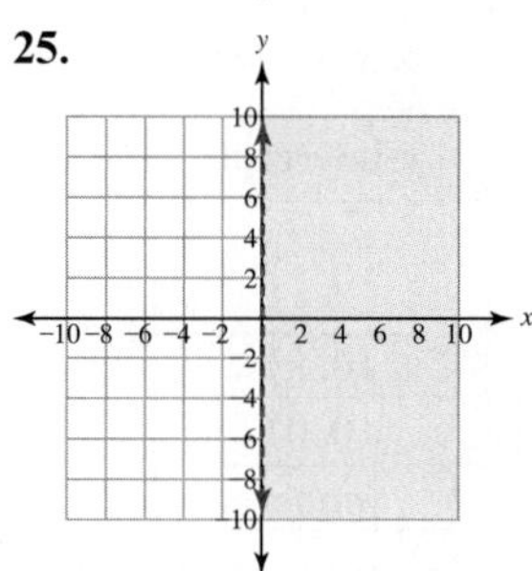

27.

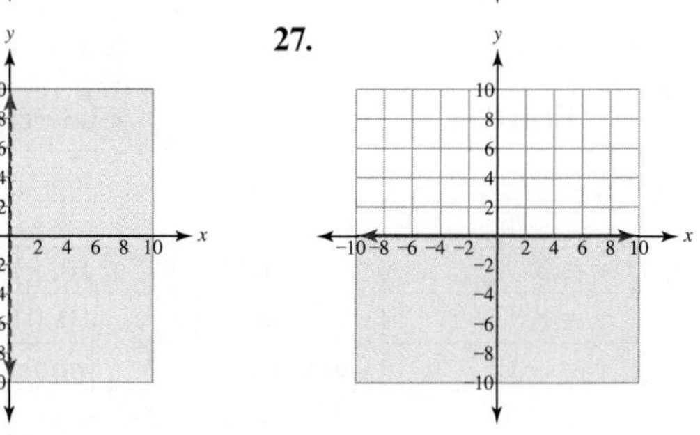

29. $y \geq -2x + 1$ **31.** $y \geq -3$ **33.** $y < x$ **35.** $x < 1$
37. $y > -\frac{2}{3}x + \frac{8}{3}$ **39.** $y < \frac{4}{3}x + 3$ **47.** x = area of land; $x \geq \frac{1}{2}$ acre **49.** x = number of shares owned by a director on the board; $x \geq 2500$ shares **51.** x = exposure to flour dust; $x \leq \frac{10\text{ mg}}{1\text{ m}^3}$ **53.** x = amount of folate per day; $x \geq 400$ micrograms **55.** x = balance of deposits; y = balance of loans; $x + y \geq \$1000$ **57.** x = number of single music downloads; y = number of television show downloads; $\$0.99x + \$1.99y \leq \$35$ **59.** x = amount of first contribution; y = amount of second contribution; $x + y \leq \$2500$ **61.** x = number of years of employment; y = age of the employee; $x + y \geq 90$ years **63.** x = amount of final average salary; y = benefit level; $y \leq 8x$
65. x = amount of final average salary; y = benefit level; $y \leq 8x + \$500{,}000$ **67.** x = amount invested in the stock fund; y = amount invested in the bond fund; $x + y \leq \$15{,}500$ **69.** x = credits required for a program; y = credits that can be paid for by financial aid; $y \leq 1.25x$ **71.** x = amount of loan; y = amount of loan guaranteed by the federal government; $y \leq 0.80x$
73. x = undergraduate GPA; y = score on the GMAT; $200x + y \geq 1120$ **75.** x = amount of jewelry; y = amount of cadmium; $y \leq 0.0003x$ **77.** x = distance from planes in normal flight; $x \geq 500$ ft

79. a. x = weight of the first piece of luggage; y = weight of the second piece of luggage; $x + y \leq 20$ kg

b.

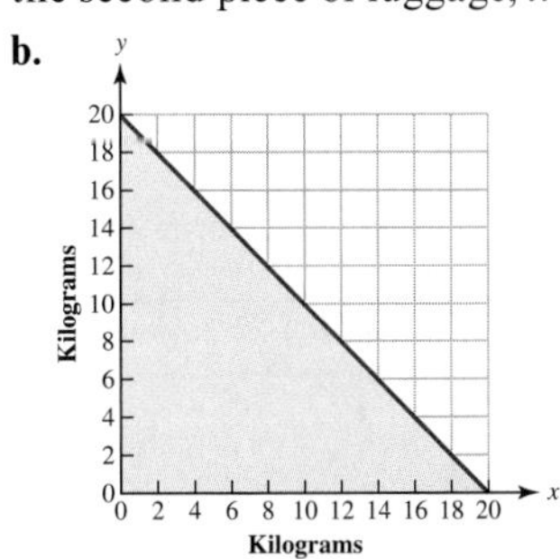

81. a. x = amount of ground beef; y = amount of ground lamb; $\left(\frac{\$3}{1\text{ lb}}\right)x + \left(\frac{\$4}{1\text{ lb}}\right)y \leq \$90$

b.

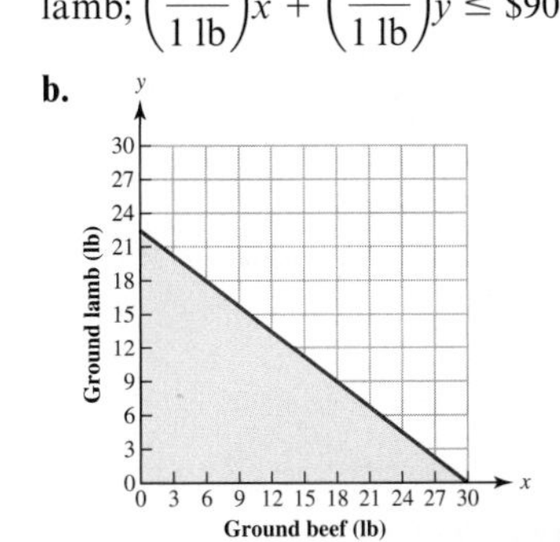

83. 2094 in.^{3}

85. \$0.9 billion **87.**

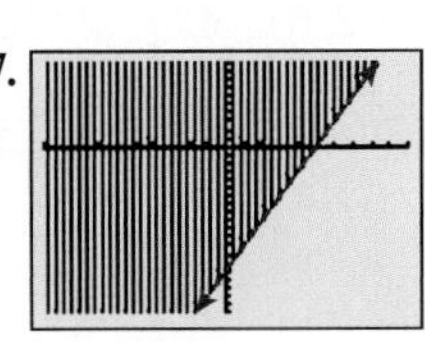

$[-10, 10, 1; -20, 10, 1]$

89.

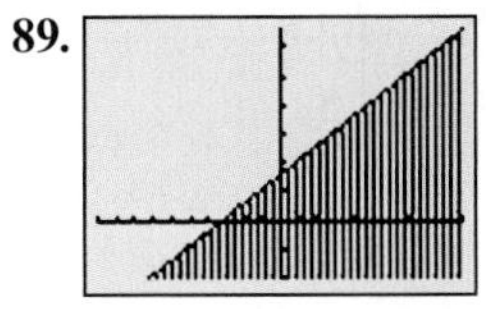

$[-10, 10, 1; -10000, 33000, 5000]$

91. b.

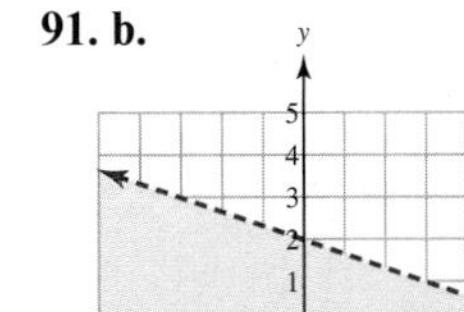

93. b.

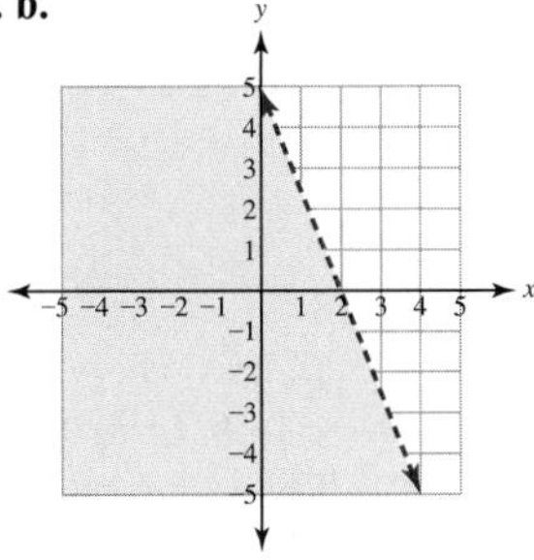

95. 1875 products **97.** $-\frac{1}{8}$

Section 3.7 Vocabulary Practice

1. C **2.** A **3.** F **4.** G **5.** B **6.** D **7.** E **8.** H **9.** I **10.** J

Section 3.7 Exercises

1. yes **3.** no **5.** yes **7.** both a relation and a function **9.** both a relation and a function **11.** both a relation and a function **13.** a relation only **15.** both a relation and a function **17.** both a relation and a function **19.** a relation only **25. a.** $\{(-1, 1), (1, 2), (2, -2), (3, 1)\}$ **b.** This relation is a function. **27.** yes **29.** no

31. a. **b.** This relation is a function.

33. a. {Buddy, Joey, Skunkface, Oscar} **b.** {dog, goat, cat}
35. a. {June 6, June 7, June 8, June 9}
b. {cloudy, partly cloudy, clear}
37. a. {20 buckets, 21 buckets, . . . , 119 buckets, 120 buckets}
b. {\$96.00, \$100.80, . . . , \$571.20, \$576.00}
39. a. {Monday, Tuesday, Wednesday, Thursday, Friday, Saturday, Sunday} **b.** {6, 7, 8, 9}

41. a.

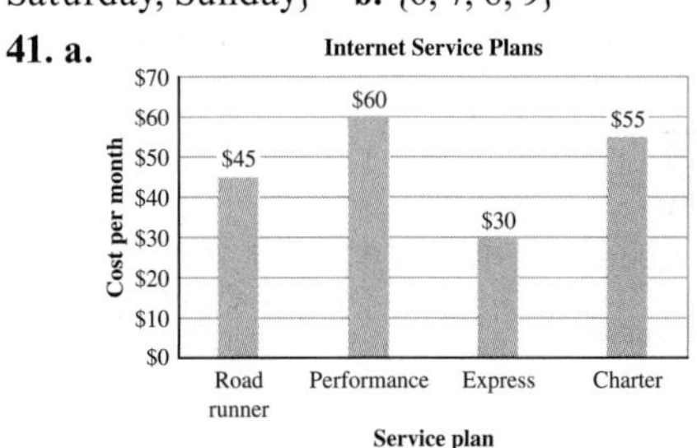

b. {Road Runner, Performance, Express, Charter}
c. {\$45, \$60, \$30, \$55} **43. a.** x **b.** y **c.** the set of real numbers **d.** the set of real numbers
45. a. the set of real numbers **b.** the set of real numbers
47. $y = -13$ **49.** $y = -8$ **51.** $y = 20$ **53.** 15
55. 400 **57.** 5 **59.** 50 **61.** $-\frac{11}{3}$ **63.** 3 **65.** 0
67. Ocala **69.** 5280 ft **71.** -1 **73.** 252.3 million devices **75.** 50% **77.** 14

79.

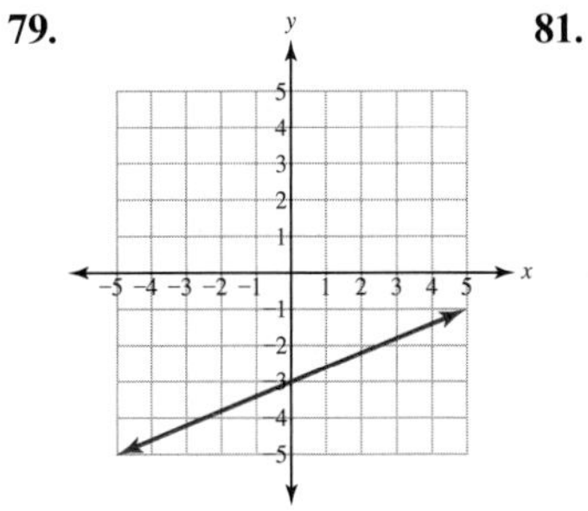

81.

83.

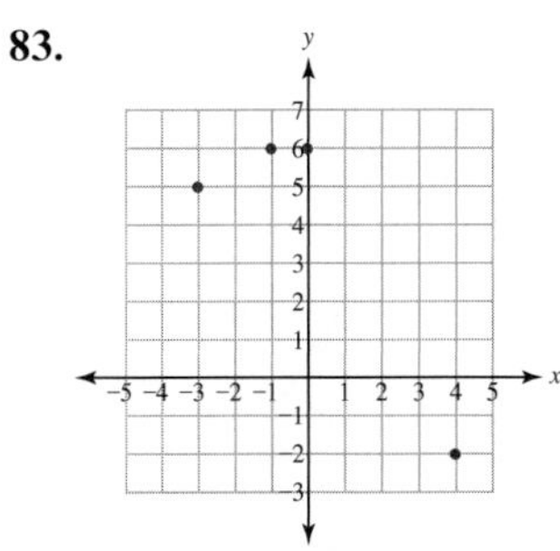

85. 57% **87.** 2.6 mL

89. a. 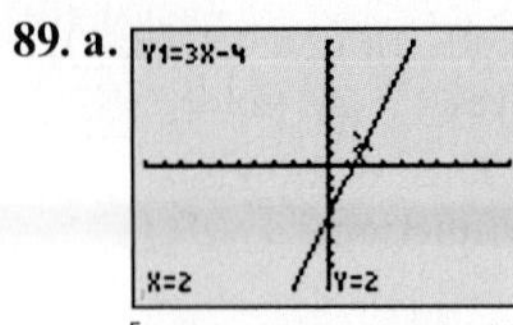

$[-10, 10, 1; -10, 10, 1]$ **b.** 2

91. a. 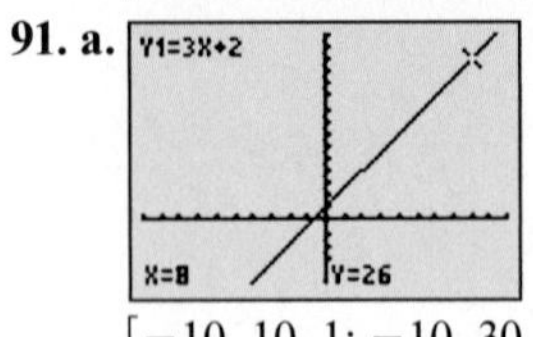

$[-10, 10, 1; -10, 30, 2]$ **b.** 26

93. b. The domain is $\{8, 2, 0, 5\}$.

95. b.

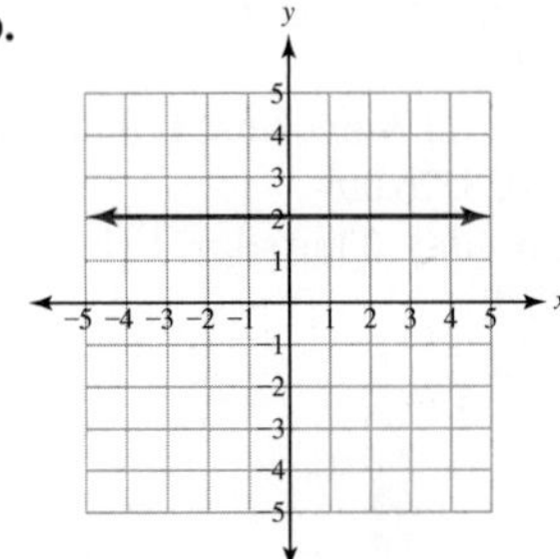

97. the set of real numbers **99.** $x = 0$

Chapter 3 Review Exercises

1.

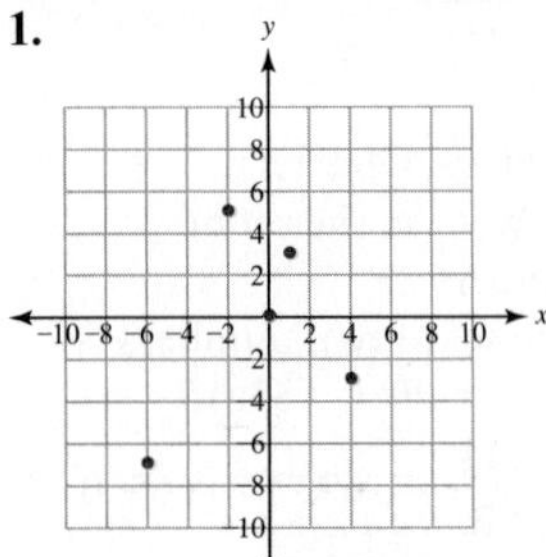

2. a.

x	y
0	-2
5	0
-5	-4

b.

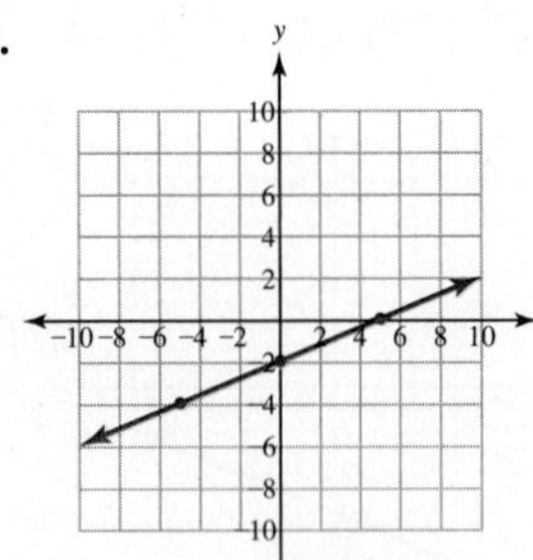

3. a.

x	y
0	$-\frac{9}{2}$
-3	0
3	-9

b.

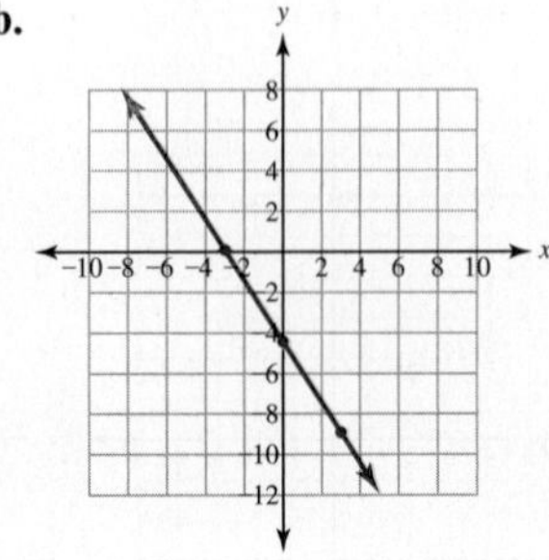

4. a.

x	y
0	6
-1	0
2	18

b.

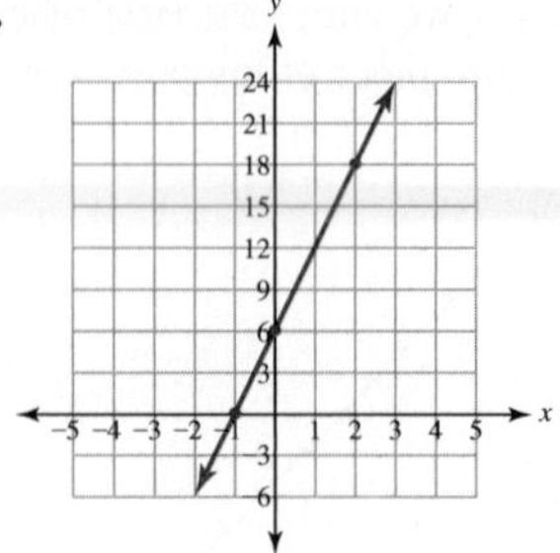

5. a.

x (gigabytes)	y (gigabytes)
0	8
16	0
8	4

b.

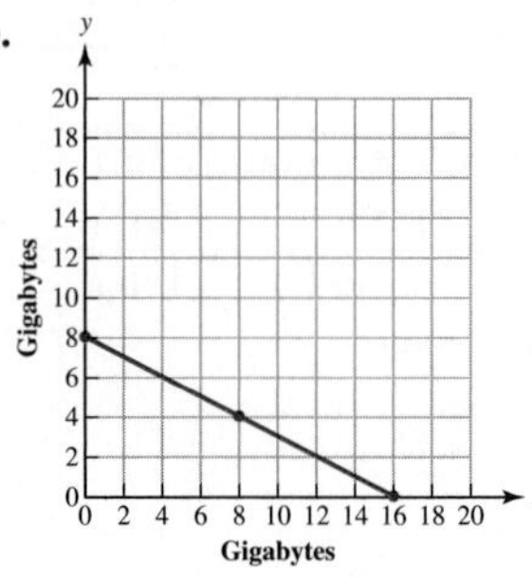

6. a. $9x + y = 15$ **b.** $80x - 5y = 32$

7. a. $(0, 6)$ **b.** $(2, 0)$ **8.** $(0, 10)$ **b.** $(2, 0)$

9. a. $(0, 3)$ **b.** no x-intercept

10. a. no y-intercept **b.** $(2, 0)$

11. a. $(0, 0)$ **b.** $(0, 0)$

12. a. $(0, -12)$ **b.** $(-2, 0)$ **d.**

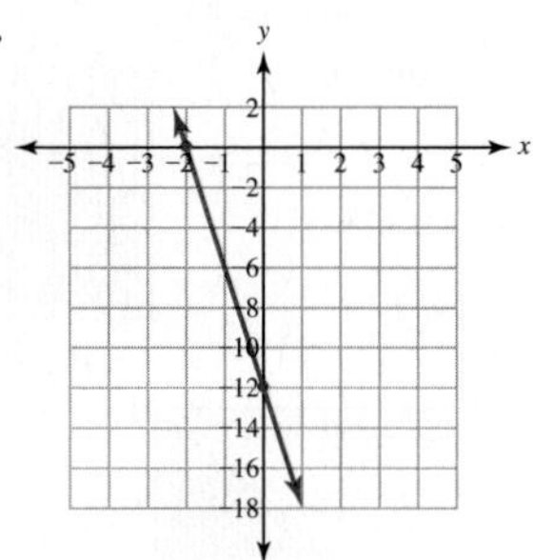

13. a. $(0, -36)$ **b.** $(30, 0)$ **d.**

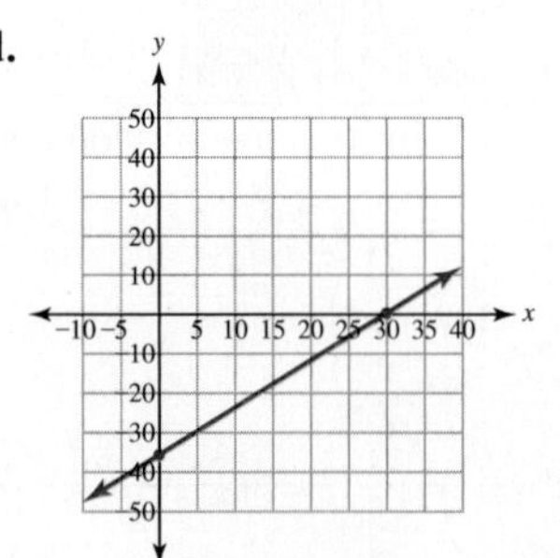

14. a. $\left(0, \frac{15}{2}\right)$ **b.** $(-3, 0)$ **d.**

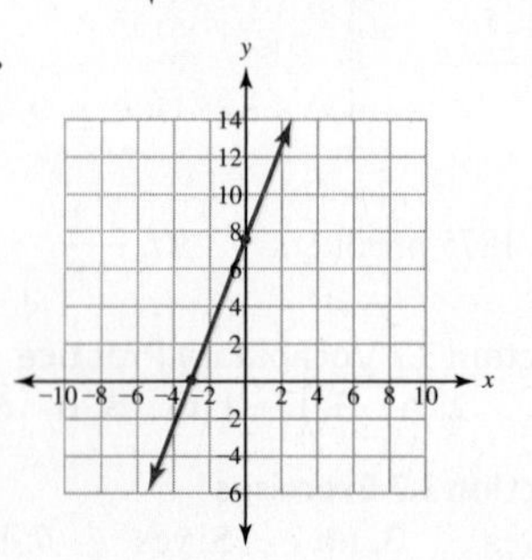

15. b.

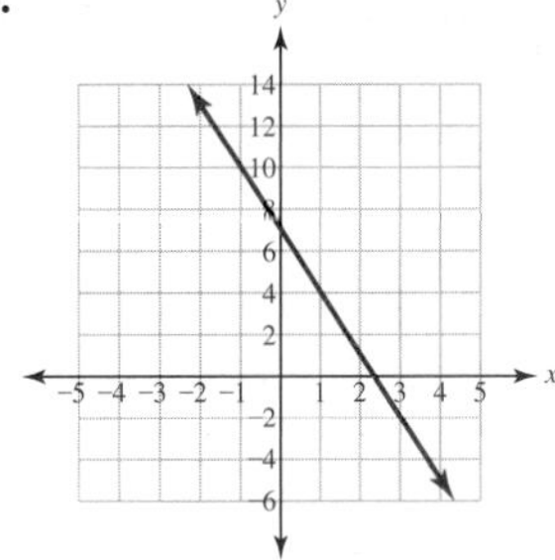

16. b.

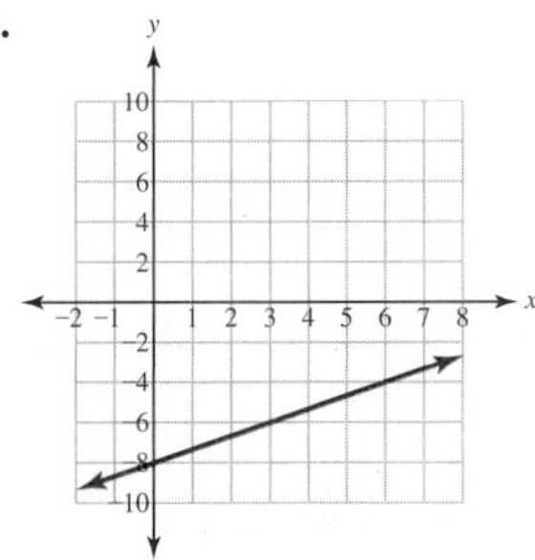

18. b, c, d **19.** infinitely many solutions

20. a. $x = 8$ **b.**

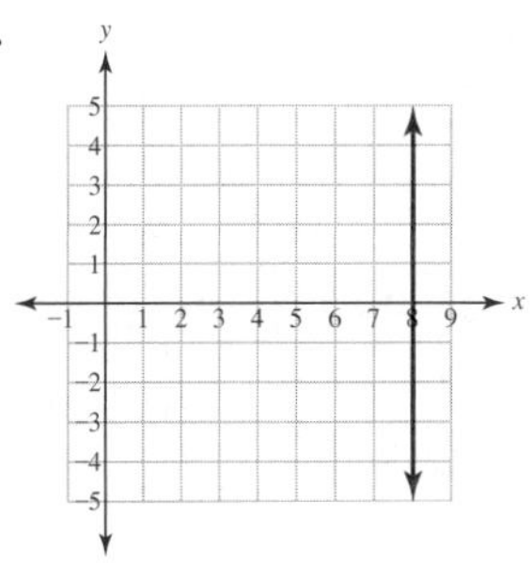

21. a. $y = 2$ **b.**

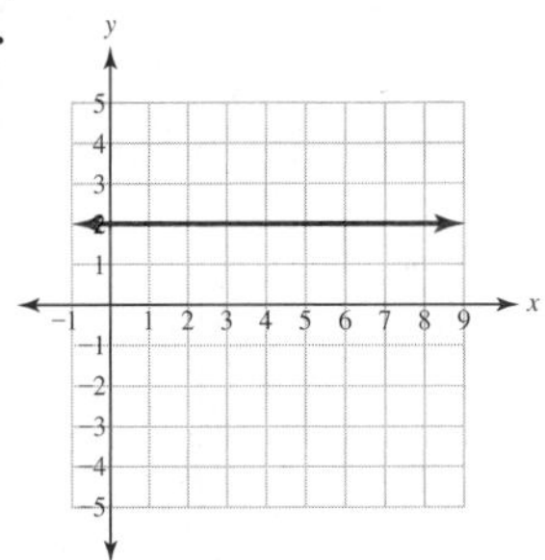

22. a. (0 hr, 55 cups), (2 hr, 30 cups) **b.** $\dfrac{-12.5\text{ cups}}{1\text{ hr}}$

23. a. (15 min, 800 mL), (20 min, 750 mL) **b.** $\dfrac{-10\text{ mL}}{1\text{ min}}$

24. The average rate of change will be a negative number.

25. $-\dfrac{19}{3}$ **26.** 7 **27. b.** $-\dfrac{2}{7}$ **28. b.** $\dfrac{3}{2}$ **29. b.** 0

30. b. undefined **31.** $\dfrac{2}{15}$ **32.** undefined **33.** 0

34. a. 9759 students **b.** (0 years, 4059.2 students) **c.** $\dfrac{407.13\text{ students}}{1\text{ year}}$

35. a. $y = \left(-\dfrac{4177\text{ tons}}{1\text{ day}}\right)x + 2{,}000{,}000\text{ tons}$ **b.** 1,624,000 tons **c.** (479 days, 0 tons) **d.** (0 days, 2,000,000 tons)

36. $y = -6x + 8$ **37. a.** $y = \dfrac{3}{8}x - \dfrac{21}{8}$ **b.** $\dfrac{3}{8}$ **c.** $\left(0, -\dfrac{21}{8}\right)$ **d.** $(7, 0)$ **38. a.** $y = -\dfrac{9}{8}x + 9$ **b.** $-\dfrac{9}{8}$ **c.** $(0, 9)$ **d.** $(8, 0)$

39. a. $y = -7$ **b.** 0 **c.** $(0, -7)$ **d.** no x-intercept

40. a. $y = x$ **b.** 1 **c.** $(0, 0)$ **d.** $(0, 0)$ **41. a.** $(0, 4)$ **b.** 3 **c.** $y = 3x + 4$ **42. a.** $(0, -2)$ **b.** $-\dfrac{5}{3}$ **c.** $y = -\dfrac{5}{3}x - 2$ **43. a.** $(0, 1)$ **b.** 0 **c.** $y = 1$

44. a.

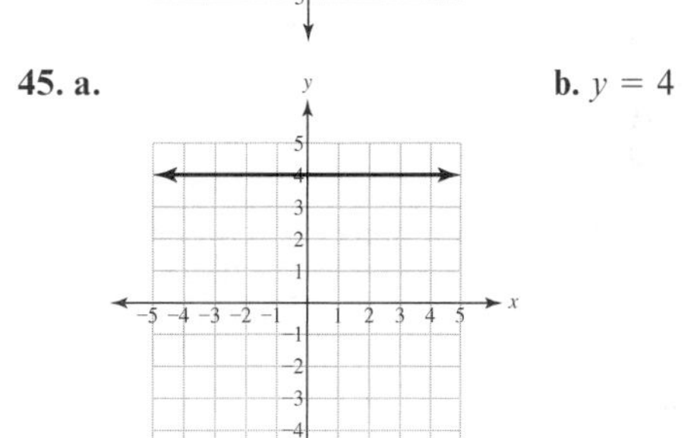

b. $x = -1$

45. a. **b.** $y = 4$

46. a.

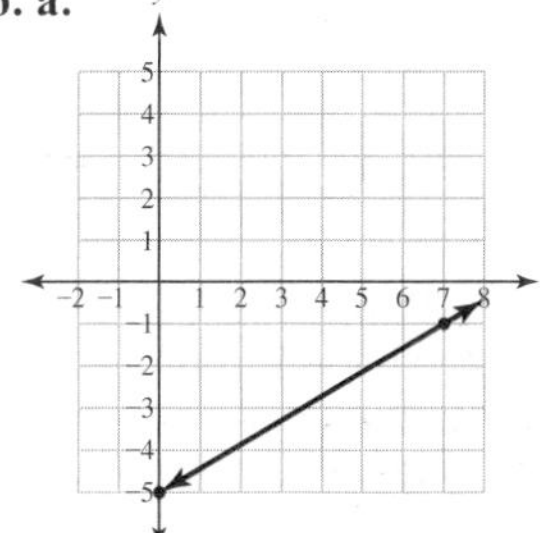

b.

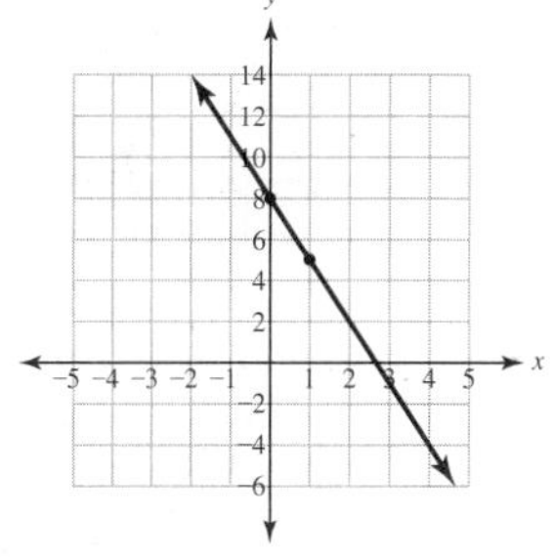

47. a, c **48.** $y = 8x - 149$ **49.** $y = \dfrac{11}{2}x + 36$

50. $y = -9x + 2$ **51.** $y = -5$ **52.** $y = -2x + 51$

53. $y = \dfrac{1}{4}x + \dfrac{17}{2}$ **54.** $y = -2$ **55.** $y = x$

56. $y = -\dfrac{2}{7}x + 4$ **57.** $y = 1$ **58.** $y = -\dfrac{7}{5}x + \dfrac{16}{5}$

59. a. 157,560 violent crimes **b.** $-\dfrac{4655\text{ crimes}}{1\text{ year}}$ **c.** (0 years, 213,420 crimes) **60. a.** $-\dfrac{5}{6}$ **b.** $\dfrac{6}{5}$

61. e **62.** a, d **63. a.** $y = 2x - 9$ **b.** $y = -\dfrac{1}{2}x + \dfrac{7}{2}$

64. a. $y = -\dfrac{2}{3}x + 10$ **b.** $y = \dfrac{3}{2}x - \dfrac{19}{2}$

65. a. $y = 6x - 41$ **b.** $y = -\dfrac{1}{6}x + 20$

66. a. $-\dfrac{5}{3}$ **b.** $\dfrac{3}{4}$ **67. a.** dashed **b.** dashed **c.** solid **d.** dashed **e.** solid **68.** b

70.

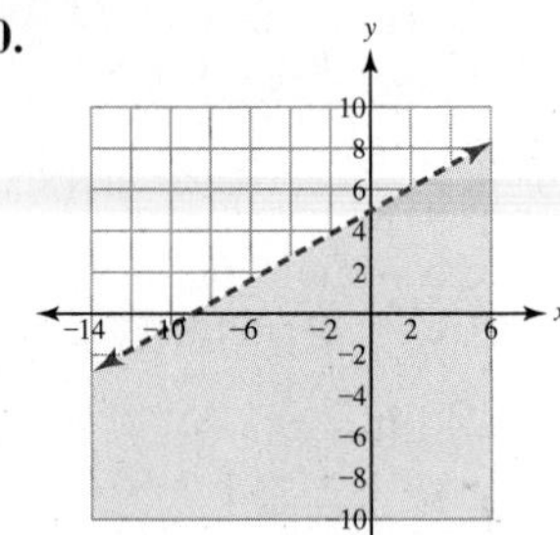

71.

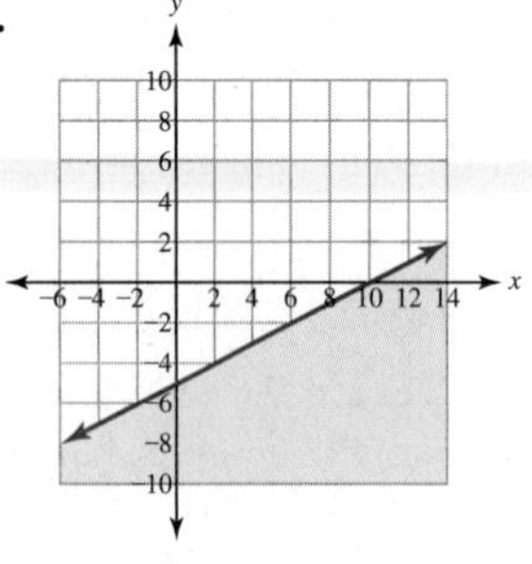

72.

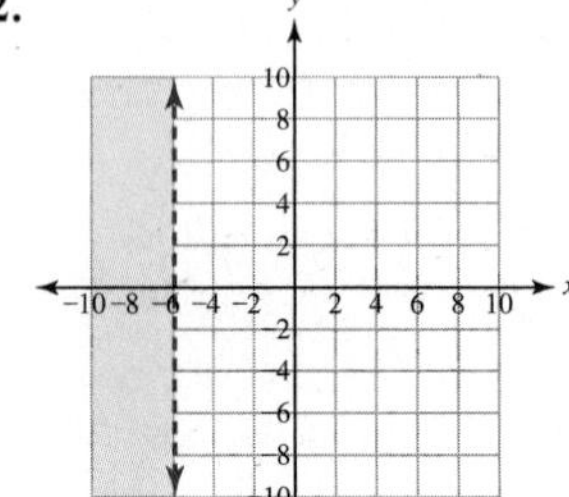

73.

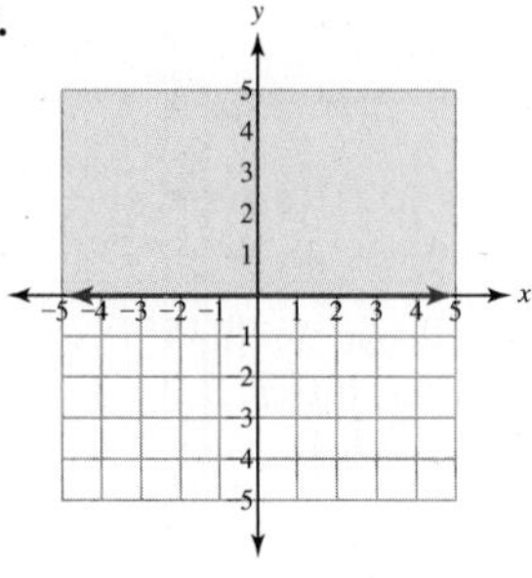

74. $y > 2x + 3$ **75.** $y \leq \frac{2}{3}x - \frac{5}{3}$ **76.** $x > -1$

77. x = number of peach trees per acre; $x \leq 201$ trees

78. a. x = arithmetic reasoning score; y = word knowledge and paragraph comprehension score; $x + y \geq 100$

b.

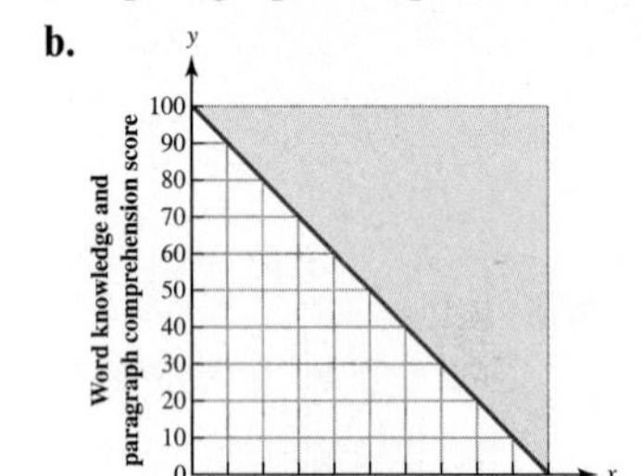

79. x = minimum wage; y = worker's pay; $y \geq 0.60x$

81. This relation is a function. **82.** This relation is a function. **83.** This relation is not a function.

84. This relation is not a function. **85.** no **86.** yes

87. -15 **88.** -3 **89.** 4 **90.** $7\frac{1}{4}$ **91. a.** 2 **b.** -2

92. a. x **b.** y **c.** the set of real numbers **d.** the set of real numbers **93.**

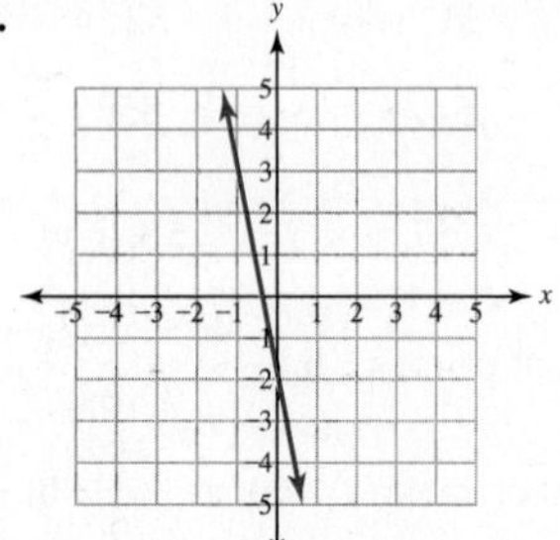

94.

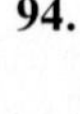

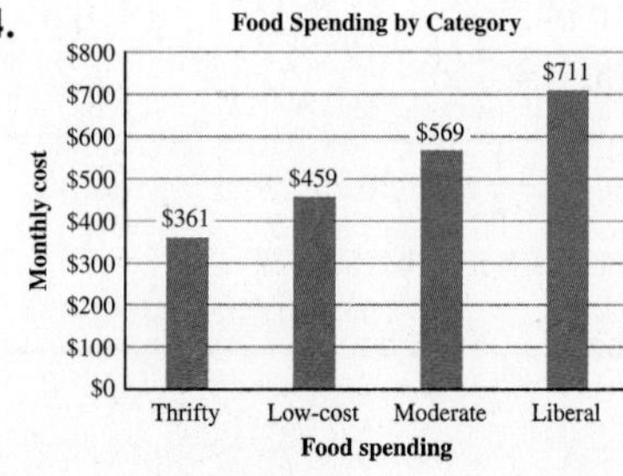

95. a. {Laurie, Judy, Jennifer, Jane}
b. {21 years, 23 years, 24 years}

Chapter 3 Test

1. a. $(0, 2)$ **b.** $(3, 0)$ **2. a.** $(0, -9)$ **b.** $(4, 0)$

d.

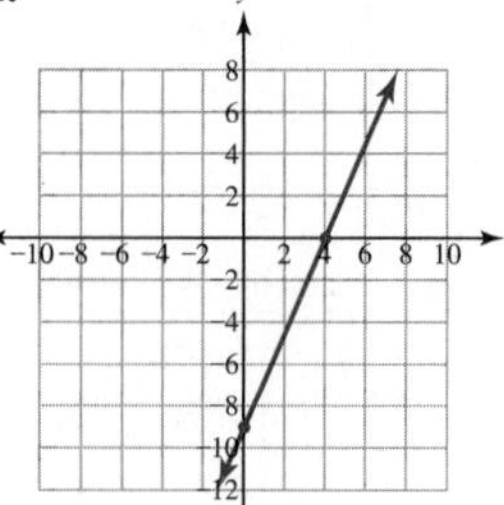

3.

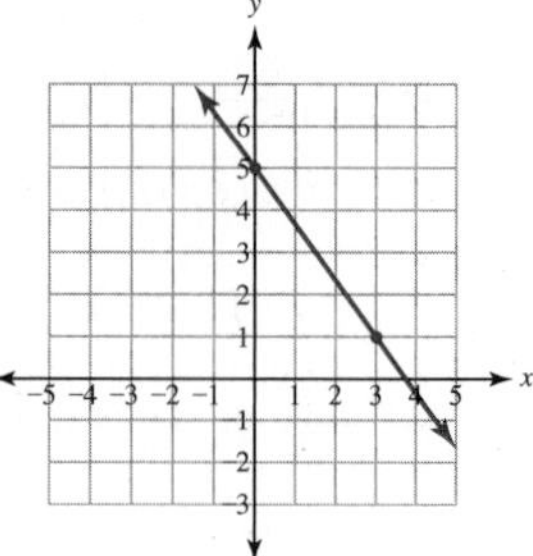

4.

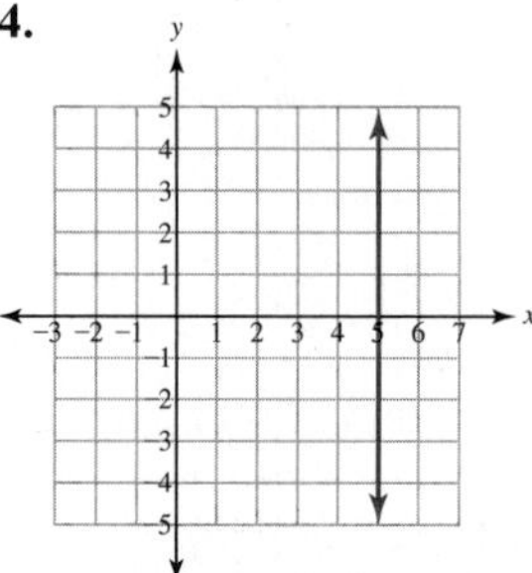

5.

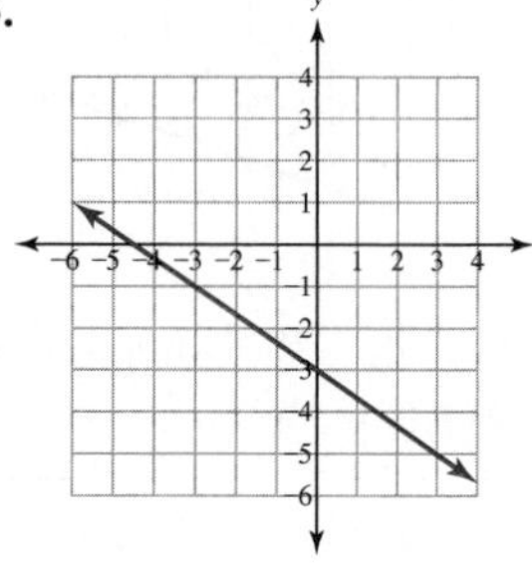

6.

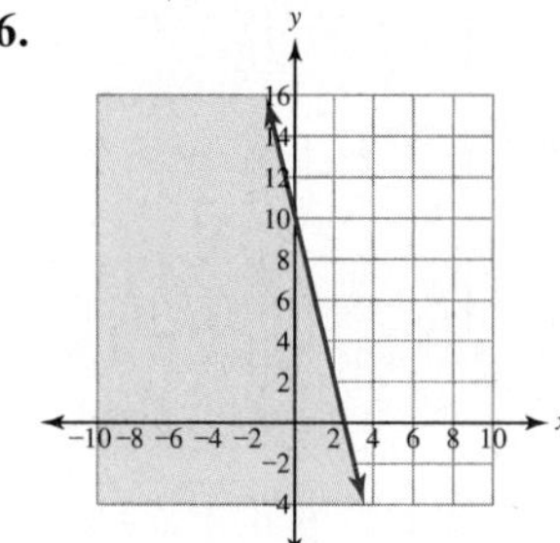

7.

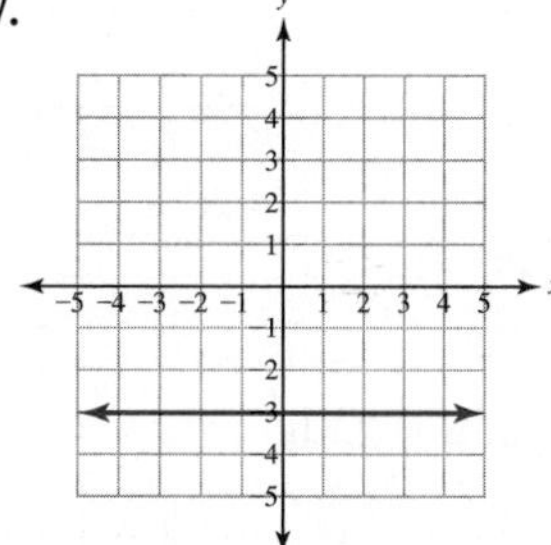

8.

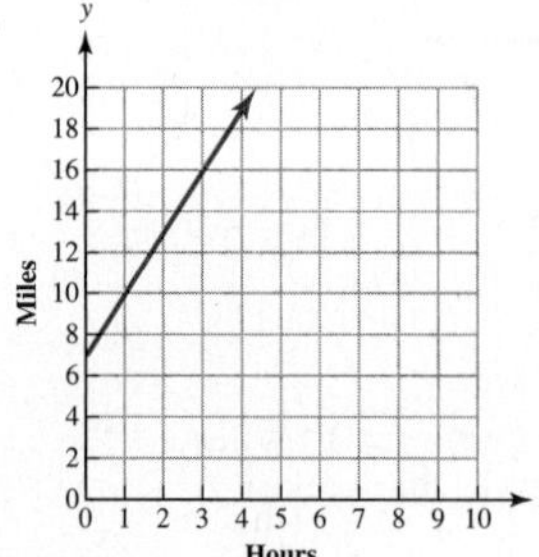

9.

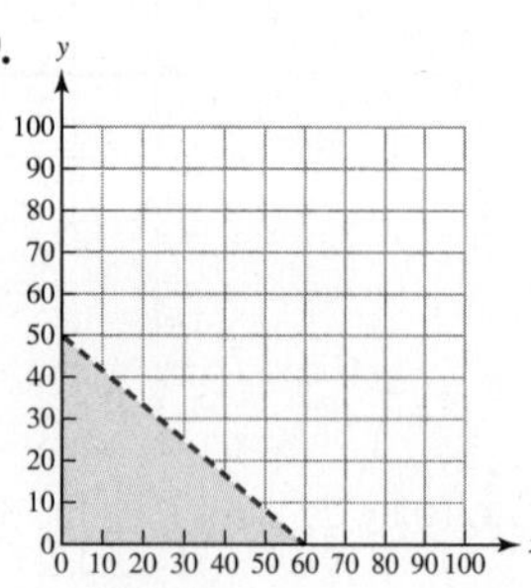

10. $\frac{5}{4}$ **11.** $(-4, 0)$ **12.** $y = \frac{1}{2}x + 7$ **13.** $y = 800$

14. $y = -3x + 5$ **15.** $y \geq -\frac{7}{4}x + 6$ **17.** 24

18. -3 **19.** $-5x + y = 51$ or $5x - y = -51$

20. a. 326,000,000 people **b.** $\frac{2{,}970{,}814 \text{ people}}{1 \text{ year}}$

Chapters 1–3 Cumulative Review

1. $\{\ldots, -3, -2, -1, 0, 1, 2, 3, \ldots\}$ **2.** -32 **3.** $5x - \frac{61}{6}$

4. $-8x^2 + 23x - 12$ **5.** the commutative property of addition **6.** $a(b + c) = a \cdot b + a \cdot c$ **7.** 17 in. **8.** \$24

9. 30.2 in.^{2} **10.** 27,648 in.^{3} **11.** 11 **12.** 5 in.

13. a. $p = \frac{7}{6}$ **14. a.** $x = 21$ **15. a.** no solution
16. a. $a = 0$ **17. a.** $w = 315$ **18. a.** $n = 104$
19. a. $x < -28$ **20. a.** $v \leq 22$ **21. a.** $a > 5$
22. **23.**
24. $y = \frac{8}{9}x - \frac{5}{3}$ **25.** $A = 4D - 12$
26. x-intercept $(8, 0)$; y-intercept $(0, -14)$

27.

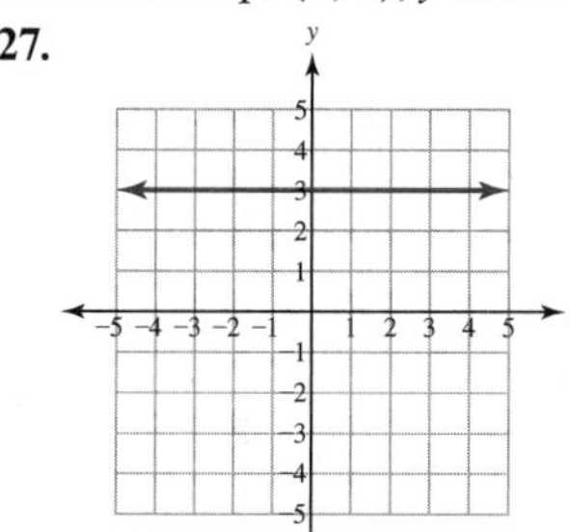

28.

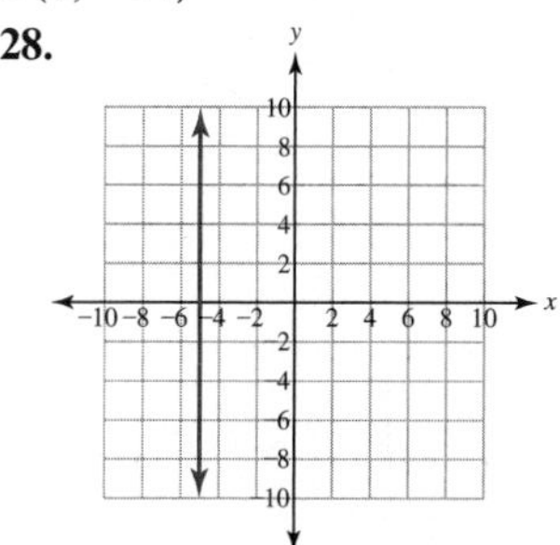

29.

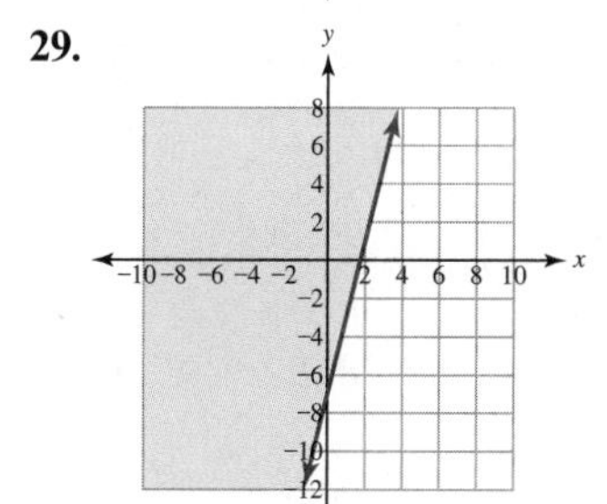

30.

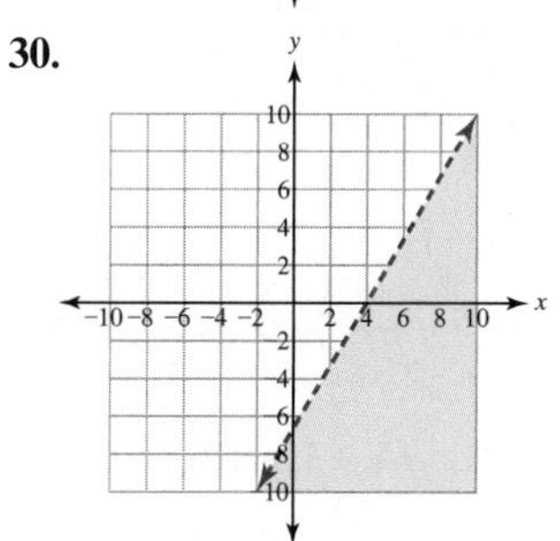

31. b.

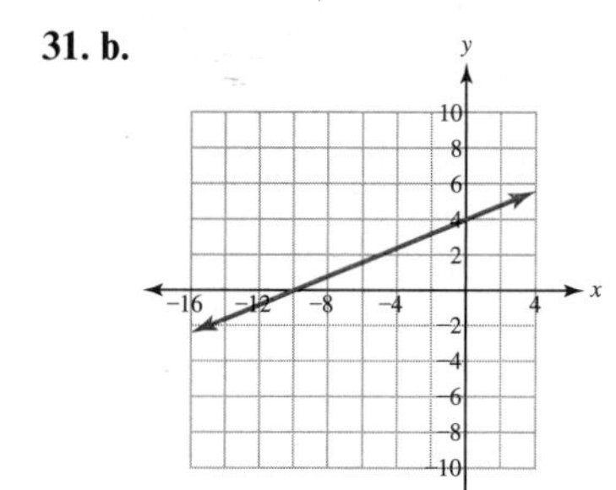

32.

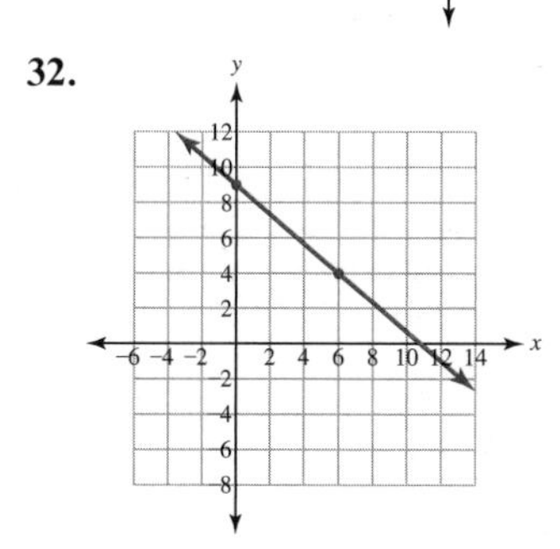

33. $y = \frac{3}{2}x - 3$ **34.** undefined **35.** $-\frac{16}{3}$ **36.** 0
37. 11 **38.** $y = 7$ **39.** $y = \frac{1}{11}x + \frac{43}{11}$
40. $y = 3x + 17$ **41.** 3.5 in. **42. a.** yes
43. 42 **44. a.** yes **45.** 4%
46. 158,000 newspapers **47.** 37,671 complaints
48. 42 Americans **49.** 5200 air travelers **50.** 1518 adults
51. 13.4 mi **52.** 0.9% **53.** 3215 gal

CHAPTER 4

Section 4.1 Vocabulary Practice

1. E **2.** B **3.** C **4.** H **5.** G **6.** D **7.** F **8.** J **9.** I **10.** A

Section 4.1 Exercises

1. one solution **3.** infinitely many solutions **5.** no solution **7.** $(-1, 2)$ **9.** no solution **11.** $(2, -3)$ **13.** $(2, 1)$ **15.** $(-20, 75)$ **17.** $(-6, 4)$ **19.** infinitely many solutions; $\{(x, y) \mid y = -x + 2\}$ **21.** (300, 300) **23.** (5 pies, 10 bags of cookies) **25.** (600 products, \$6000) **27.** $(1, 4)$ **29.** $(2, -3)$ **31.** $(0, 0)$ **33.** $(0, 5)$ **35.** $(4, -1)$ **37.** $(4, -4)$ **39.** no solution **41.** $(6, 2)$ **43.** $(3, -2)$ **45.** $(-2, 2)$ **47.** $(2, 2)$ **49.** $(9, -4)$ **51.** $(9, -2)$ **53.** no solution **55.** $(-2, -2)$
57. infinitely many solutions; $\left\{(x, y) \mid y = \frac{1}{9}x - 2\right\}$
59. $(0, 0)$ **61.** $(10, 1)$ **63.** $(4, 0)$ **65.** $(4, 1)$
67. no solution
69. a. $x + 2y = 7$

x	y
-1	4
0	$\frac{7}{2}$
1	3
2	$\frac{5}{2}$
3	2
4	$\frac{3}{2}$
5	1
6	$\frac{1}{2}$

$y = 2x - 9$

x	y
-1	-11
0	-9
1	-7
2	-5
3	-3
4	-1
5	1
6	3

b. $(5, 1)$

73. (600 products, \$6000) **77. a.** $y = \left(\frac{\$300}{1 \text{ product}}\right)x$ **b.** (60 products, \$18,000) **79.** \$12,000 in short-term investments and \$6000 in long-term investments **81.** 600 large bags and 2400 small bags **83.** 109 s **85.** 10%

87.

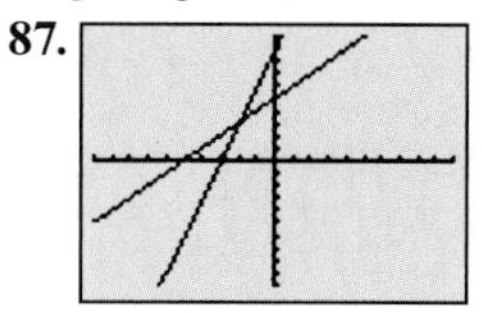

$[-10, 10, 1; -10, 10, 1]$; $(-2, 3)$

89.

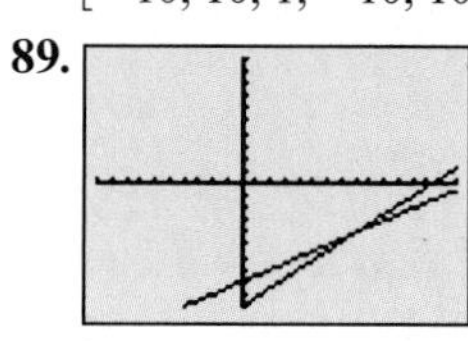

$[-10, 15, 1; -10, 10, 1]$; $(8, -4)$

91. b. infinitely many solutions; $\left\{(x, y) \mid y = -\frac{1}{3}x + 2\right\}$
93. b. no solution **95. a.** $x = 5$
97. a. the set of real numbers

Section 4.2 Vocabulary Practice

1. F **2.** E **3.** C **4.** D **5.** J **6.** I **7.** G **8.** A **9.** H **10.** B

Section 4.2 Exercises

1. a. $x = 2$ **b.** $y = 8$ **c.** (2, 8) **3. a.** $2x + 2$ **b.** $x = 1$ **c.** $y = 4$ **d.** (1, 4) **5. a.** (3, 4) **7. a.** (3, −3) **9. a.** no solution **11. a.** (2, 2) **13. a.** (9, −4) **15. a.** no solution **17. a.** (−2, −2) **19. a.** infinitely many solutions; $\left\{(x, y) \mid y = \frac{1}{9}x - 2\right\}$ **21. a.** (0, 0) **23. a.** (20, 10) **25. a.** (4, 1) **27. a.** no solution **29. a.** $\left(6, \frac{1}{2}\right)$ **31. a.** $\left(-\frac{1}{4}, 2\right)$ **33. a.** $\left(\frac{1}{2}, 0\right)$ **35. a.** $\left(\frac{1}{4}, \frac{3}{4}\right)$ **37. a.** infinitely many solutions; $\left\{(x, y) \mid y = \frac{3}{4}x + 2\right\}$ **39. a.** $\left(\frac{1}{2}, \frac{1}{3}\right)$ **41.** (5, 10) **43.** $\left(\frac{1}{2}, -\frac{1}{2}\right)$ **45.** (1, 2) **47.** no solution **49.** (400 products, \$24,000) **51.** (800 products, \$40,000) **53.** (200 lb, 250 lb) **55.** (1000 acres, 4000 acres) **57.** (16 hr, 1040 mi) **59.** (40 adult tickets, 110 youth tickets) **61. a.** (−2, 8) **63. a.** no solution **65. a.** (4, −6) **67. a.** (0, 2) **69. a.** infinitely many solutions; $\left\{(x, y) \mid y = -\frac{1}{2}x + 2\right\}$ **71. a.** (0, 8) **73.** (2400 products, \$24,000) **75.** (120 products, \$300,000) **77.** 21 lb of potatoes and 7 lb of carrots **79.** \$24,000 in investments and \$8000 for paying off debts **81.** \$9000 for print journals and \$36,000 for online journals **83.** \$1213 **85.** (−1, 3)

87.

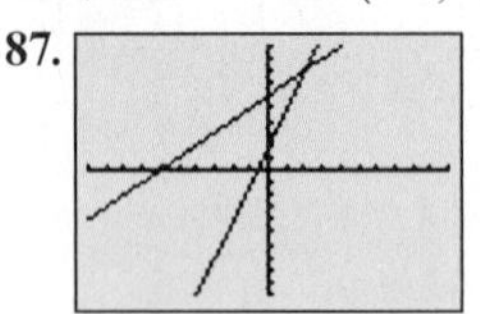

[−10, 10, 1; −10, 10, 1]; (2, 8)

89.

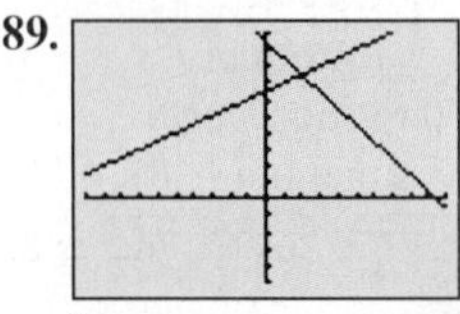

[−10, 10, 1; −10, 20, 2]; (2, 15)

91. b. (5, −6) **93. b.** infinitely many solutions; $\{(x, y) \mid y = 2x + 3\}$ **95.** $3x + 4y = 36$

Section 4.3 Vocabulary Practice

1. C **2.** I **3.** H **4.** G **5.** E **6.** B **7.** F **8.** D **9.** A **10.** J

Section 4.3 Exercises

1. $\frac{4}{9}$ **3.** $\frac{38}{99}$ **5. a.** (−1, 4) **7. a.** (2, −3) **9. a.** (1, 3) **11. a.** (5, 3) **13. a.** (−1, −6) **15. a.** (0, 5) **17. a.** $\left(\frac{1}{2}, 6\right)$ **19. a.** no solution **21. a.** (8, 2) **23. a.** (6, −1) **25. a.** (−3, 2) **27. a.** (4, 1) **29. a.** no solution **31. a.** (0, 2) **33. a.** infinitely many solutions; $\{(x, y) \mid -2x + y = -3\}$ **35. a.** infinitely many solutions; $\{(x, y) \mid x - 5y = 8\}$ **37. a.** (−1, −1) **39. a.** no solution **41. a.** (2, −1) **43. a.** (3, 0) **49.** (1, 8) **51.** (1000, 4000) **53.** no solution **55.** infinitely many solutions; $\{(x, y) \mid -3x + y = 9\}$ **59.** (6, 5) **61.** $\left(\frac{1}{2}, \frac{3}{2}\right)$ **63.** $\left(\frac{1}{4}, \frac{1}{4}\right)$ **65.** (6, 3) **67.** (12, 21) **69.** (6, −2) **71. a.** (7, 5) **73. a.** (−2, 3) **75.** A burger costs \$1.79. An order of fries costs \$0.89. **77.** 290 student tickets and 30 other tickets **79.** 75 chocolate bars and 25 boxes of thin mints **81.** 278 gal **83.** 2205 teenagers

85.

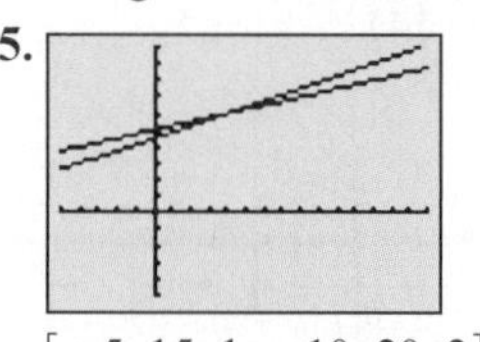

[−5, 15, 1; −10, 20, 2]; (4, 12)

87.

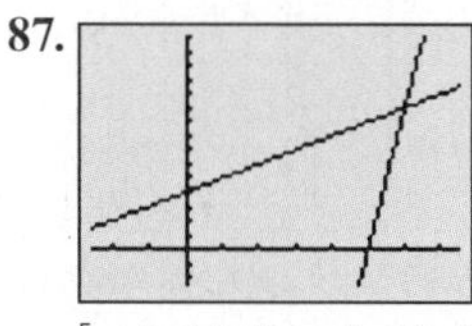

[−5, 15, 2; −5, 30, 2]; (12, 20)

89. b. (2, 2) **91. b.** infinitely many solutions; $\{(x, y) \mid -3x + 4y = 18\}$ **93.** $t = \frac{d}{r}$ **95.** 0%

Section 4.4 Vocabulary Practice

1. D **2.** G **3.** C **4.** J **5.** F **6.** B **7.** I **8.** E **9.** A **10.** H

Section 4.4 Exercises

1. 218 gal Drink A; 582 gal Drink B **3.** 249 L Mixture A; 71 L water **5.** 38 gal Mixture A; 21 gal pure merlot grape juice **7.** 925 gal regular; 275 gal premium **9.** 13 lb apples; 37 lb oranges **11.** length 16 in. and width 7 in. **13.** length 9 cm and width 4 cm **15.** 220 min; 220 mi **17.** 150 min; 11.25 mi **19.** 10 lb Snack Mix A; 30 lb Snack Mix B **21.** 16 L Mixture A; 4 L water **23.** 129 tickets for children; 171 tickets for adults **25.** 467 gal of Drink A; 933 gal of Drink B **27.** 142 L of the salt solution and 158 L of water **29.** 2.66 oz of 14-karat gold and 5.34 oz of 20-karat gold **31.** 6.26 oz of 24-karat gold and 8.75 oz of copper **33.** 3.0 g of 14-karat gold; 9.0 g of 18-karat gold **35.** 3550 gal of the leftover winter fuel and 3550 gal of pure No. 2 Diesel **37.** The second group will hike 6 hr and 30 mi. **39.** 12.8 lb of almonds; 27.2 lb of Brazil nuts. **41.** 1 can of chili costs \$0.89, and 1 can of soup costs \$2.00. **43.** 250 reams white paper and 50 reams colored paper **45.** 8496 gal bottled and 2832 gal stored in barrels **47.** 384 products, \$6048 **49.** The length is 20 in.; the width is 5 in. **51.** The length is 24 in.; the width is 15 in. **53.** 0.2 L concentrated solution and 2.8 L water **55.** 8 students; \$1200 **57.** 780 lb of distillers grain and 720 lb of cornstalks **59.** 60 minutes; 5 mi **61.** 5 timber bolts and 15 hex bolts **63.** 13.3 lb Pigeon Feed A and 36.7 lb Pigeon Feed B **65.** \$6750 in Mutual Fund A and \$10,250 in Mutual Fund B **67.** boat $\frac{25 \text{ mi}}{1 \text{ hr}}$; current $\frac{5 \text{ mi}}{1 \text{ hr}}$ **69.** 14.4 lb Golden German Finch Millet and 3.6 lb Red Proso Millet **71.** 3.75 lb jelly beans and 36.25 lb pomegranate seeds **73.** 455 nonsmoking rooms and 65 smoking rooms **75.** dog food \$35 and sweet feed \$12 **77.** 72 million children **79.** 2074 drivers

81. 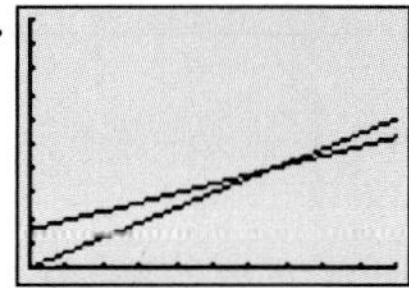

$[0, 150, 15; 0, 10{,}000, 1000]$; (100 products, \$4000)

83. 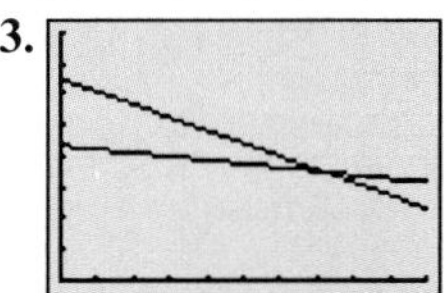

$[0, 30, 3; 0, 400, 50]$; (21.7 years, 174.8 deaths)

85. b. $x + y = 80$ gal; $0.06x + 0.15y = (0.11)(80 \text{ gal})$

87. b. $2x + 2y = 54$ in.; $y = x + 15$ in.

89. 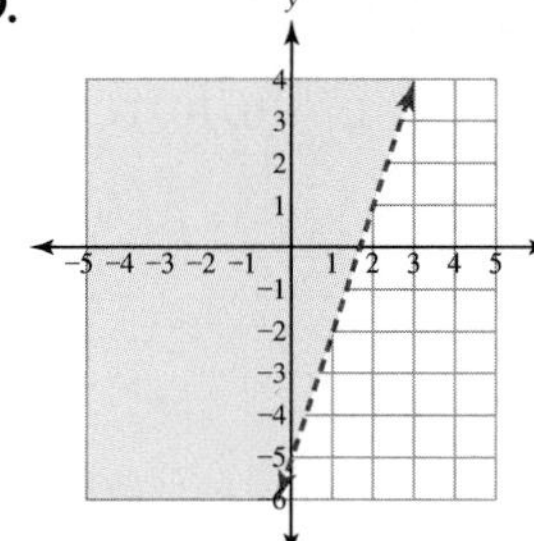

91. 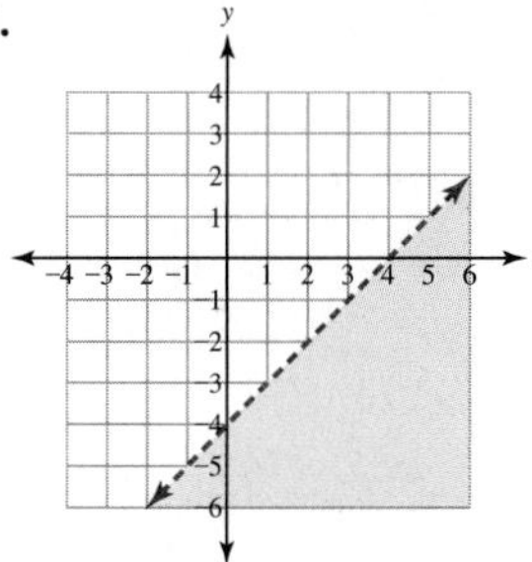

Section 4.5 Vocabulary Practice

1. F **2.** D **3.** G **4.** E **5.** J **6.** A **7.** I **8.** C **9.** B **10.** H

Section 4.5 Exercises

3. 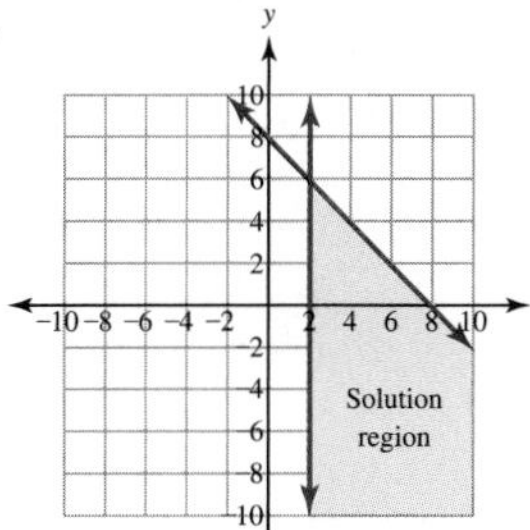

5. 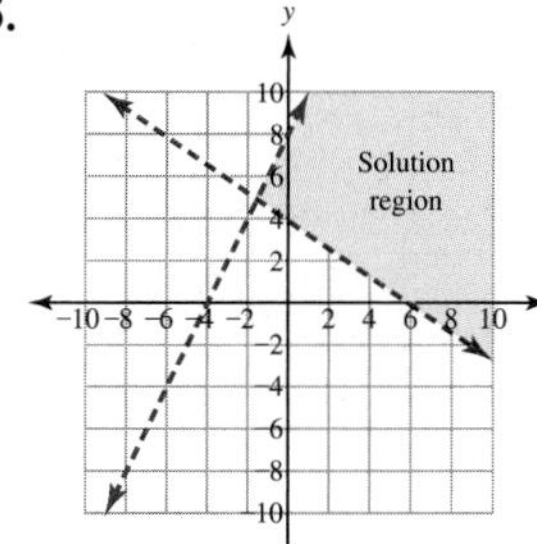

7. 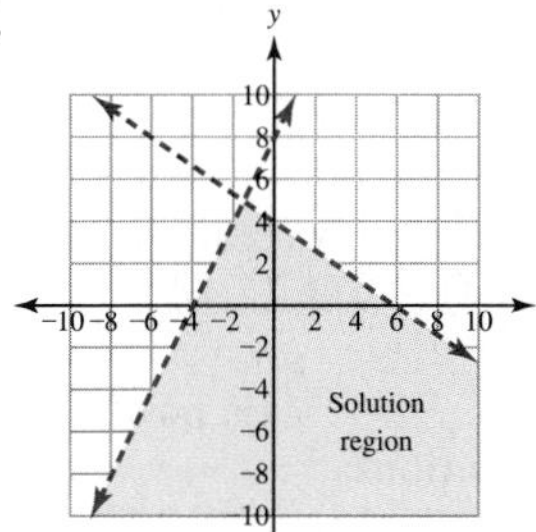

9. 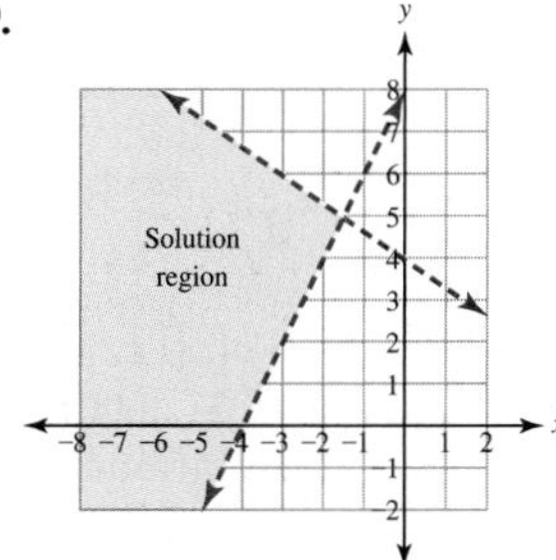

11. 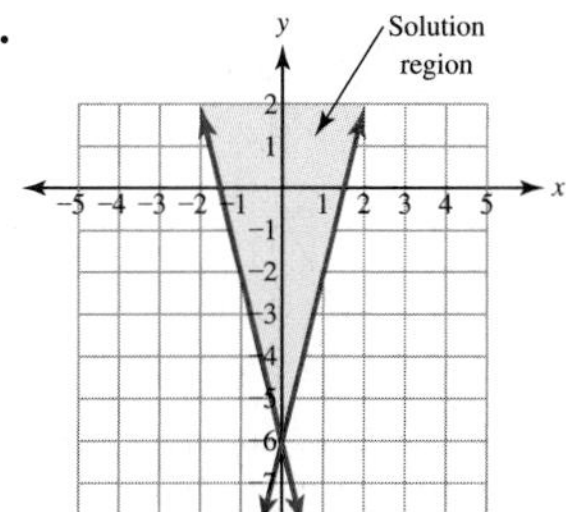

13. 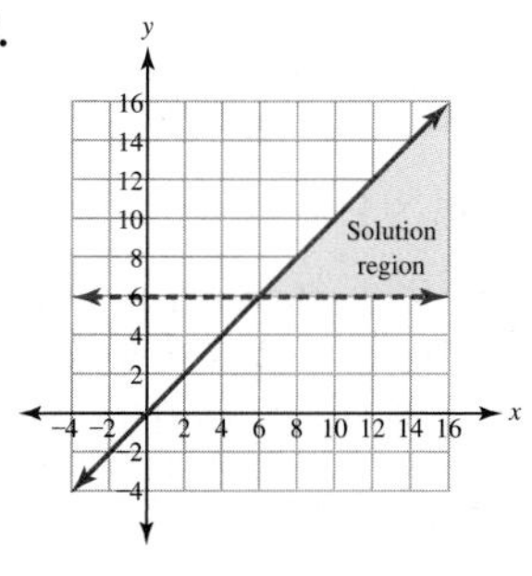

15. 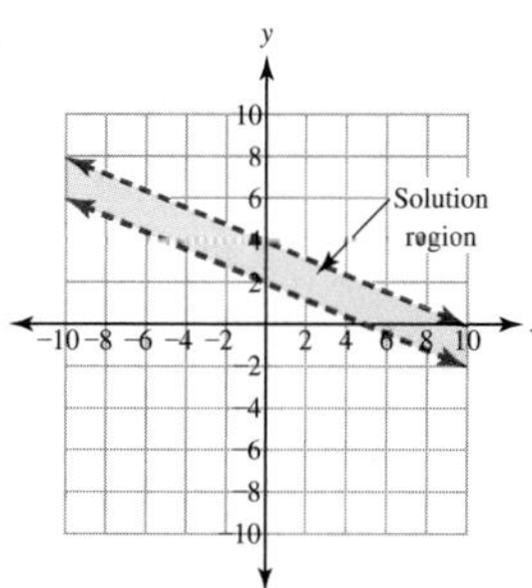

17. 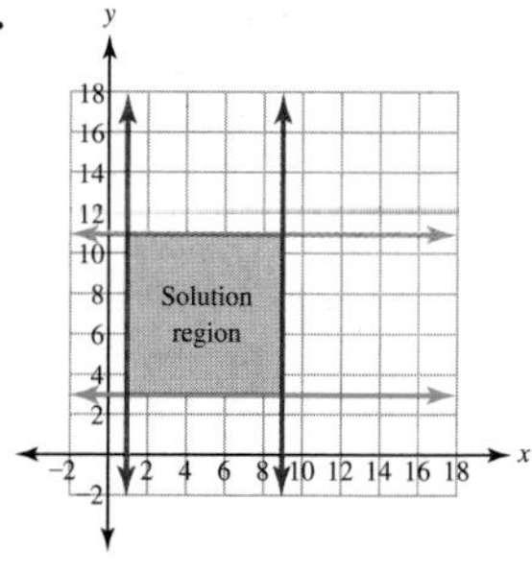

19. 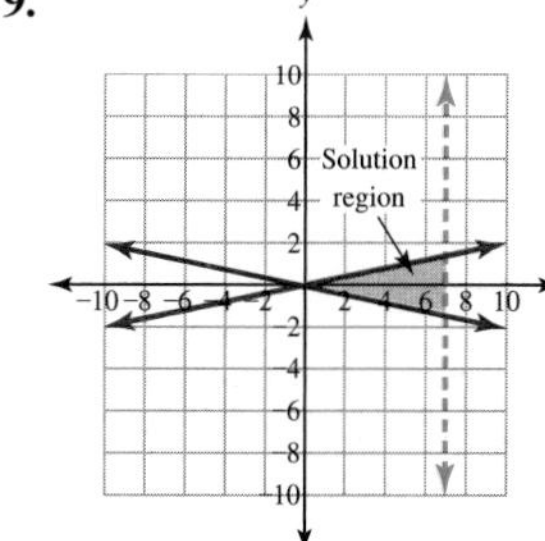

21. 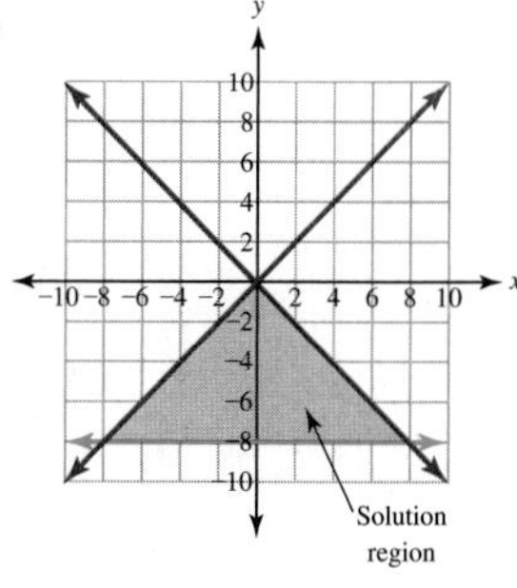

23. 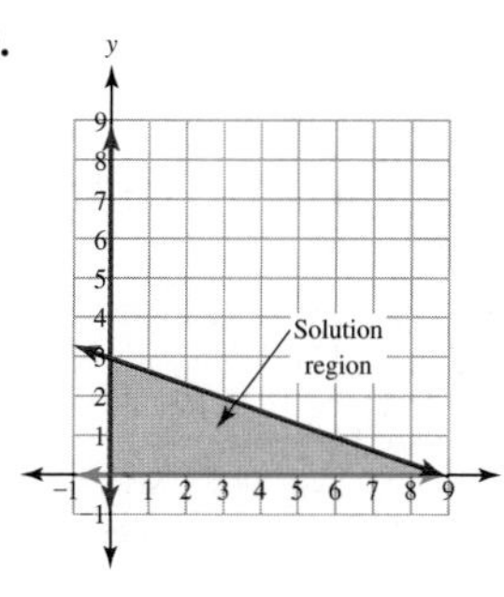

25. 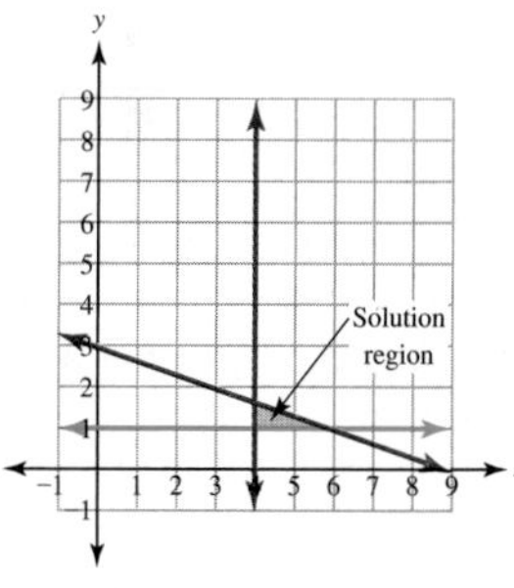

27. 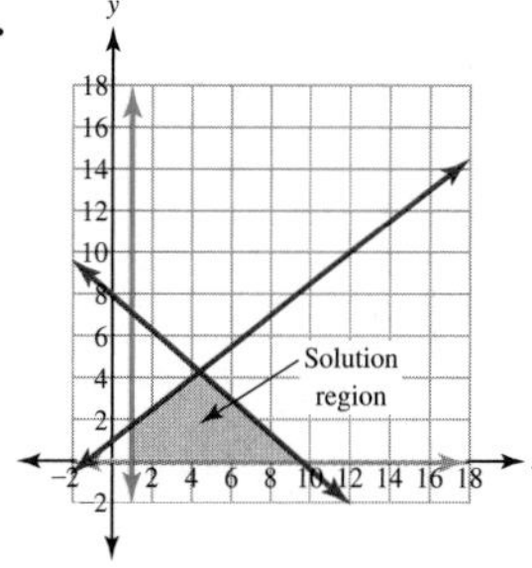

29. 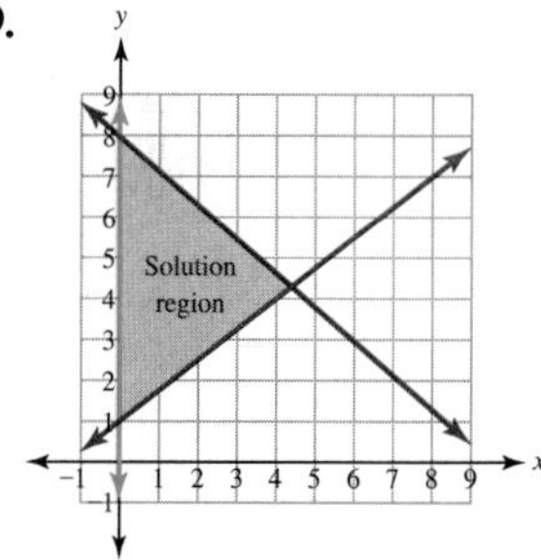

31. 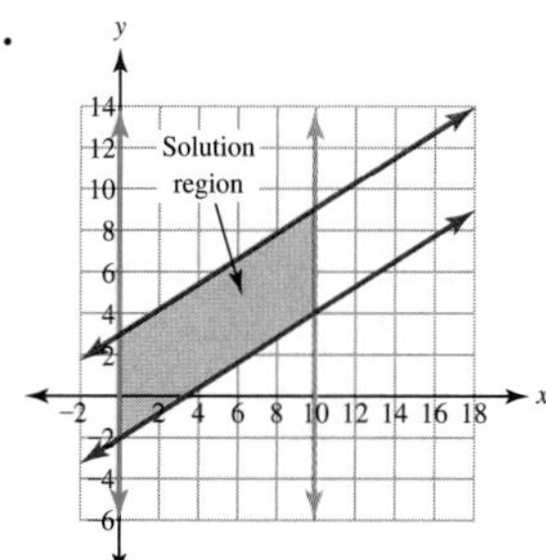

33. 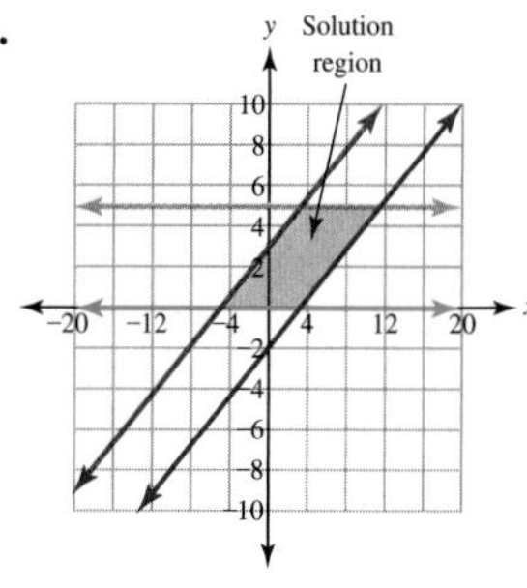

35. 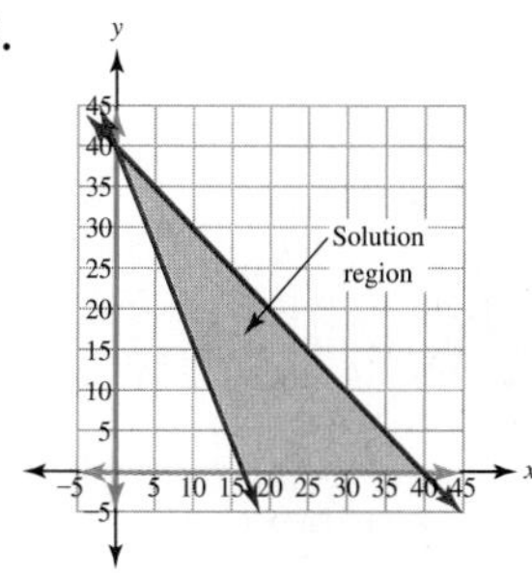

37. 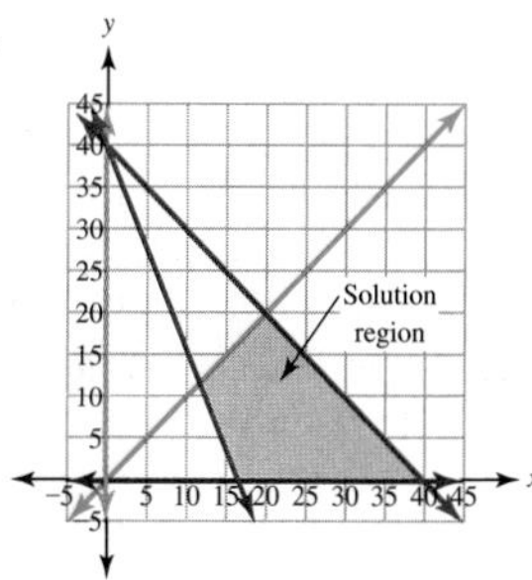

39.

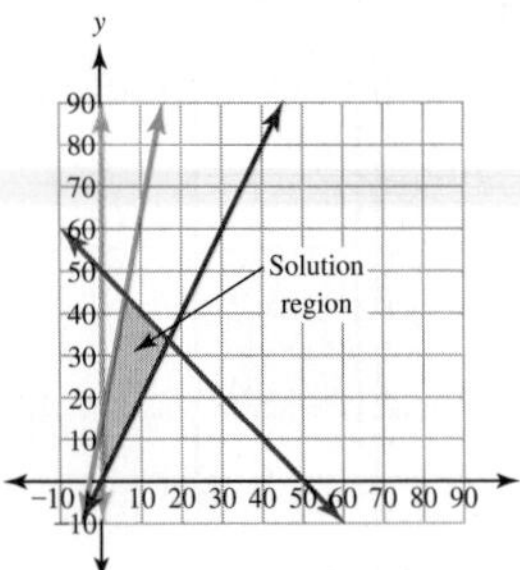

41.

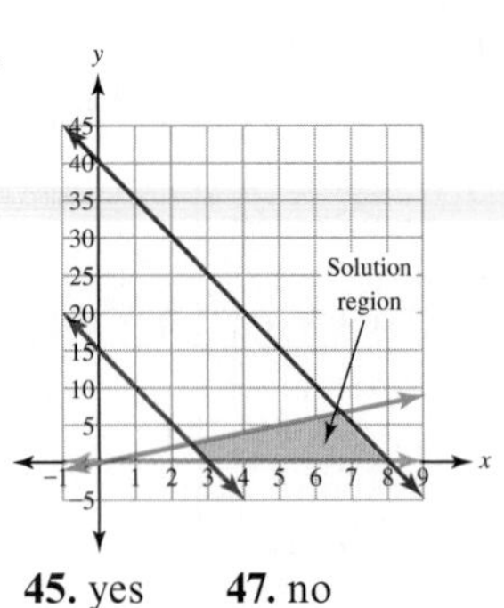

43.

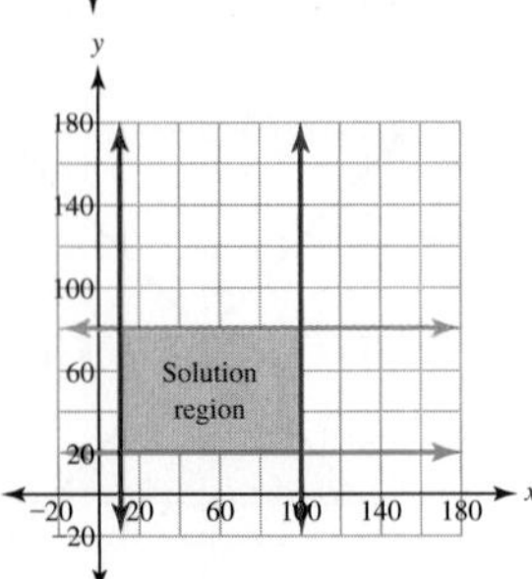

45. yes **47.** no

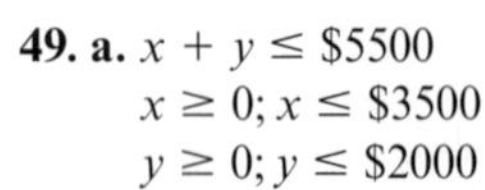

49. a. $x + y \leq \$5500$
$x \geq 0; x \leq \$3500$
$y \geq 0; y \leq \$2000$

b.

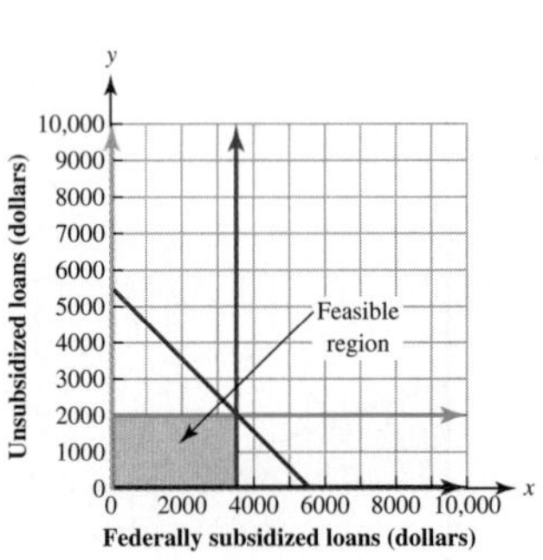

51. a. $x + y \leq 100$
$x \geq 60$
$x \leq 70$
$y \geq 30$
$y \leq 40$

b.

53. a. $x + y \leq 50$
$x \geq 35$
$y \geq 0$

b.

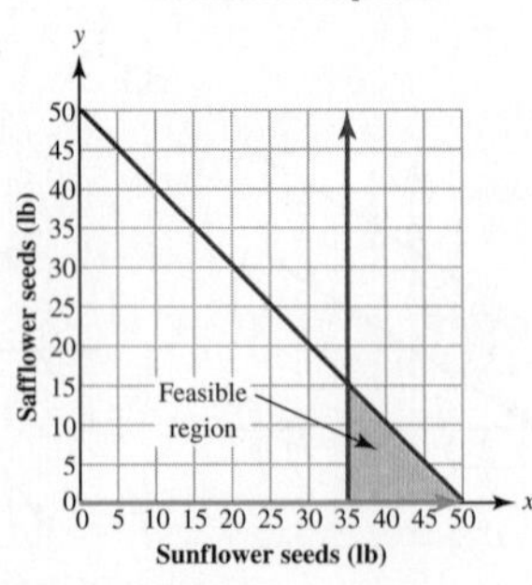

55. a. $x + y \leq 25{,}500$
$y \geq 4x$
$x \geq 2000$
$y \geq 0$

b.

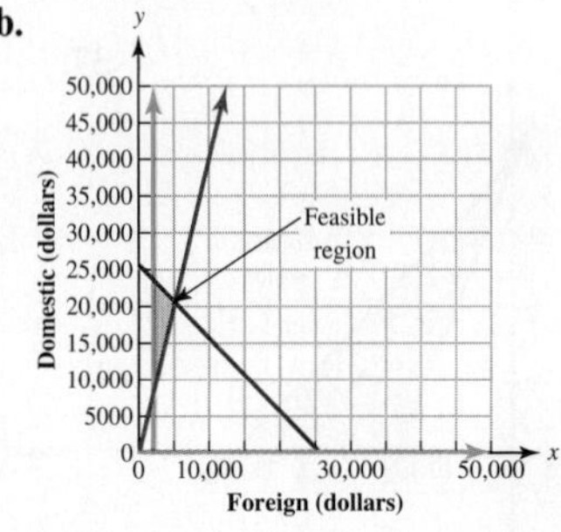

57. a. $2.50x + 2.00y \leq 120$
$x \geq 3y$
$y \geq 5$
$x \geq 0$

b.

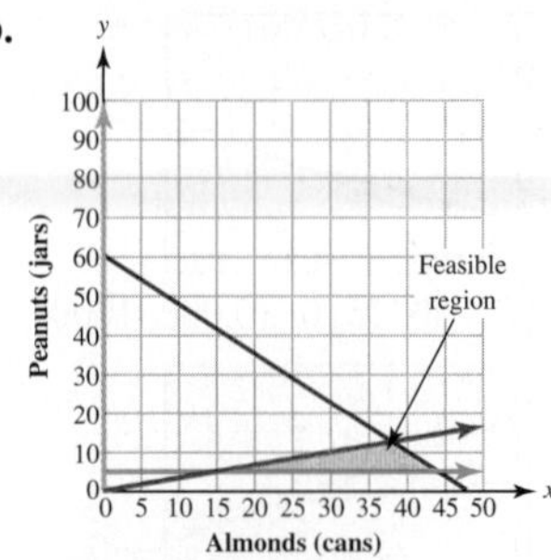

59. 8140 ft **61.** 24% increase

63.

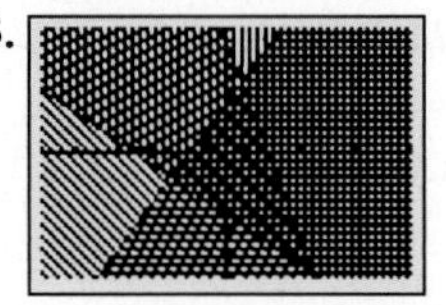

$[-10, 10, 1; -10, 10, 1]$

65.

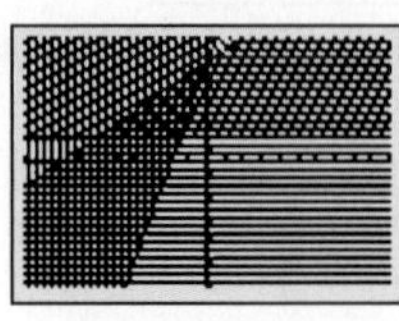

$[-10, 10, 1; -10, 10, 1]$

67. b.

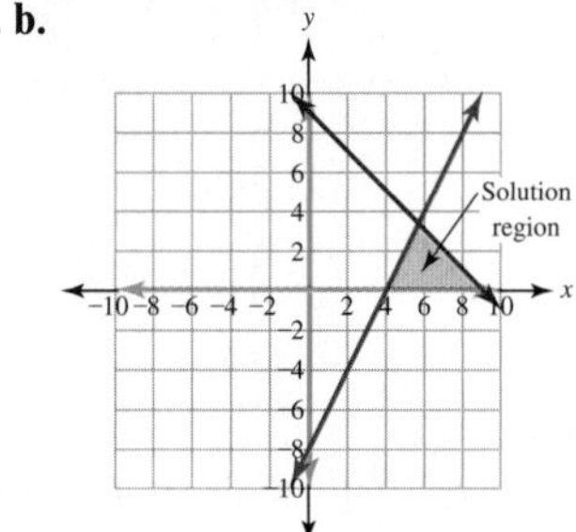

69. b.

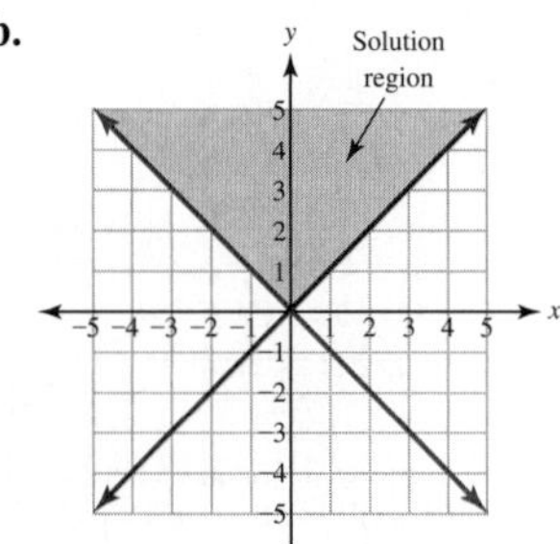

71. 5 **73.** $8x - 11$

Chapter 4 Review Exercises

1. no solution **2.** infinitely many solutions **3.** one solution **4.** $(1, 1)$ **5.** no solution **6.** $(-2, 2)$ **7.** $(3, -5)$ **8.** $(1, 2)$ **9.** infinitely many solutions; $\{(x, y) \mid y = -2x - 3\}$ **10.** $(-5, 4)$ **11.** $(2, 6)$ **12.** $(6, -2)$ **14.** $\{(x, y) \mid y = 3x + 2\}$ **16.** (40 products, \$12,000) **18.** $(6, -1)$ **19.** no solution **20.** $(1, 3)$ **21.** infinitely many solutions; $\left\{(x, y) \mid y = \frac{1}{2}x - \frac{3}{4}\right\}$ **22.** $(-6, -3)$ **23.** $\left(6, \frac{1}{4}\right)$ **24.** c **25.** a, b, c, d, e **26.** 320 bottles of carbonated water and 160 bottles of noncarbonated water **27.** (50 products, \$16,250) **28.** $\frac{5}{9}$ **29.** $\frac{4}{33}$ **30.** $2x + 3y = 14$ **31. a.** $(-7, 11)$ **32. a.** no solution **33. a.** $(0, 8)$ **34. a.** infinitely many solutions; $\{(x, y) \mid y = x - 2\}$ **35. a.** $\left(5, -\frac{1}{2}\right)$ **36. a.** $(9, -10)$ **37.** y **40.** paperback: \$8.37; hardback: \$16.29

41. 20 mochas and 80 lattes **42.** width: 7 in.; length: 17 in. **43.** 1.7 gal of Mixture A; 0.3 gal of Mixture B **44.** 160 gal of water; 660 gal of the 20% salt solution **45.** 400 kg of Brass A; 100 kg of Brass B **46.** \$1560 on VISA; \$390 on Mastercard **47.** lemon drops: \$0.50 per pound; chocolates: \$2.50 per pound **48.** width: 8 cm; length: 14 cm

49.

50.

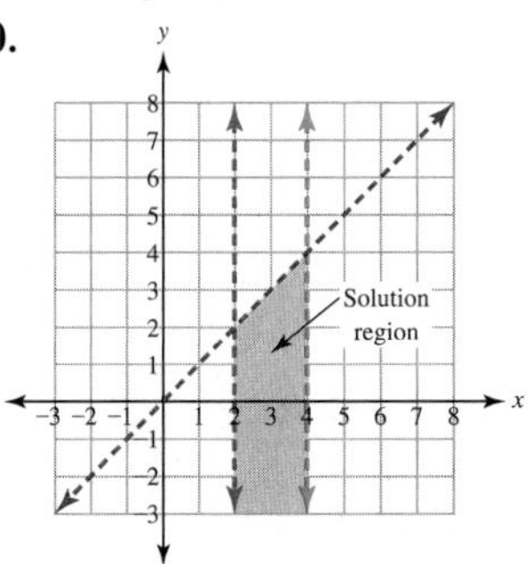

51.

52.

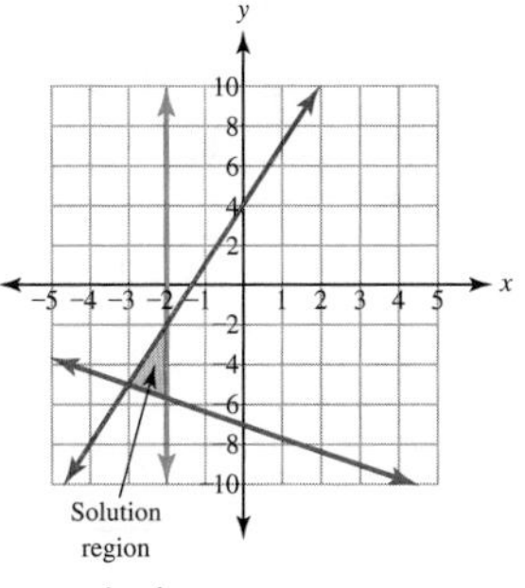

53.

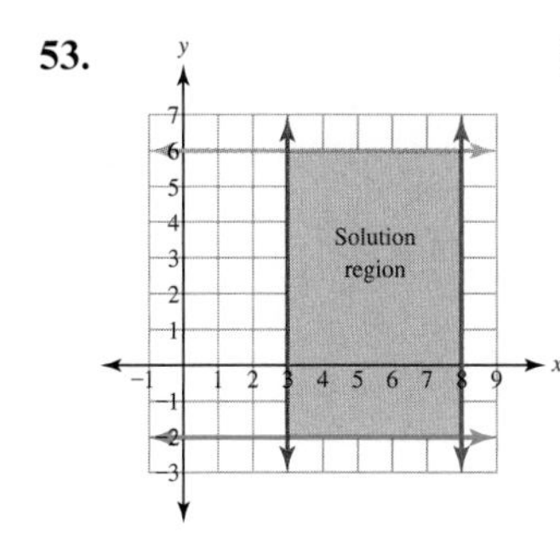

54. no solution;

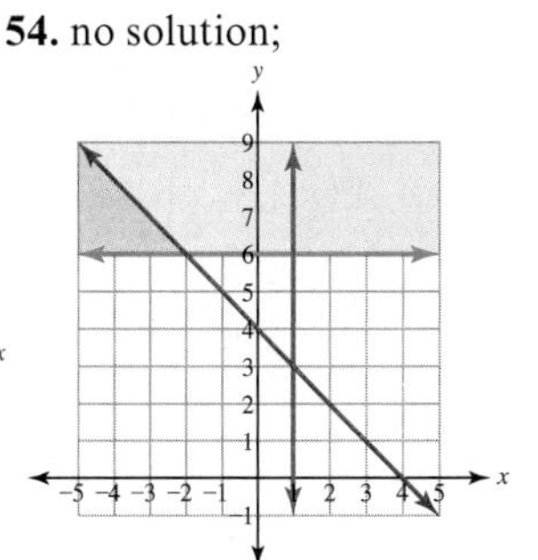

55. a.

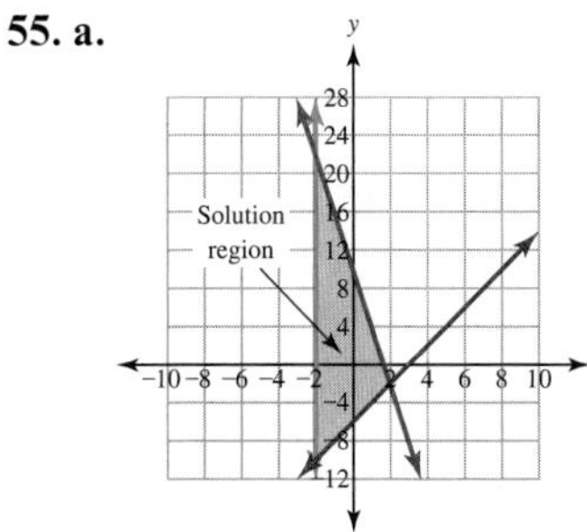

b. (1, −2) is a solution of the system.

56. a. $x + y \le 2000$ **b.**
$x \ge 400$
$x \le 700$
$y \ge 900$
$y \le 1300$

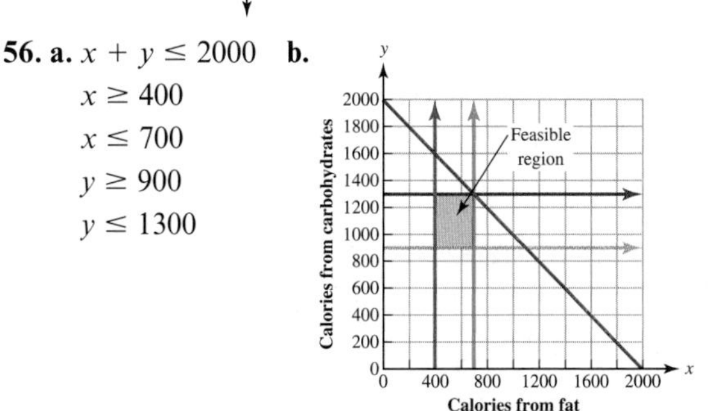

Chapter 4 Test

1. (2, 3) **2.** (−4, 30) **3.** (1, 8) **4.** no solution **5.** $\left(\frac{1}{4}, 17\right)$ **6.** infinitely many solutions; $\left\{(x, y) \mid y = \frac{5}{6}x - \frac{3}{4}\right\}$ **7. a.** a graph with coinciding lines **b.** a graph with parallel lines **c.** a graph with lines that intersect in only one point **9.** 175 slices pizza and 125 soft drinks **10.** 0.6 gal diaper pail bleach solution and 1.4 gal water **11.** 5000 products with \$400,000 in costs and \$400,000 in revenue **12.** 15 sheep and 3 goats

13.

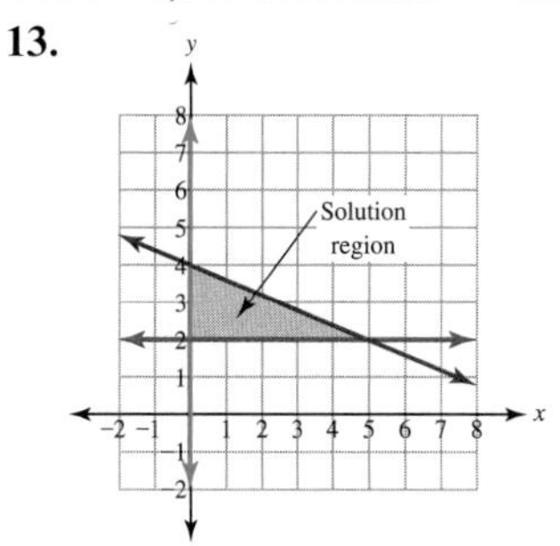

14. a. $4x + 3y \le 600$
$x \ge 20$
$y \ge 40$

b.

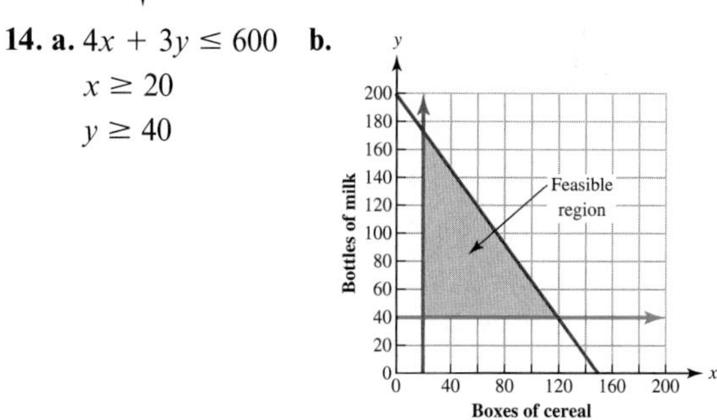

CHAPTER 5

Section 5.1 Vocabulary Practice

1. B **2.** A **3.** F **4.** H **5.** D **6.** G **7.** C **8.** E **9.** I **10.** J

Section 5.1 Exercises

1. $-2x^2 - 6$ **3.** $11x^4 + 16y^2$ **5.** $-\frac{3}{4}x - \frac{1}{3}$ **7.** $\frac{4}{3}h^2 - \frac{5}{12}h$ **9.** $k^3 - 5k^2 - k + 6$ **11.** b^{15} **13.** $36x^8$ **15.** x^7 **17.** $4h^5$ **19.** $\frac{30x^{10}}{17z^3}$ **21.** $-\frac{7}{15}z^5$ **23.** u^{14} **25.** c^3d^3 **27.** a^6b^2 **29.** $9a^2$ **31.** $-27a^3$ **33.** $36n^6$ **35.** $\frac{h^7}{k^7}$ **37.** $\frac{x^4}{y^{10}}$ **39.** $\frac{9x^4}{16y^{10}}$ **41.** $32z^{10}$ **43.** 1 **45.** 1 **47.** 1 **49.** 1 **51.** $\frac{1}{y^6}$ **53.** $\frac{1}{64}$ **55.** $\frac{1}{10^4}$ **57.** $\frac{9x^5}{y^{11}}$ **59.** $\frac{15}{2p}$ **61.** $288x^5$ **63.** $-x^6$ **65.** $-x^{14}$ **67.** $\frac{1}{3x}$ **69.** $\frac{1}{9x^2}$ **71.** $-\frac{x^{11}}{6y^{15}}$ **73.** $\frac{3a^2b^4}{5}$ **75.** 1 **77.** $\frac{w^5}{2}$ **79.** $\frac{64p^4}{9q^2}$ **81.** $6z^3 + 48z^2 + 24z$ **83. a.** x^{12} **b.** x^{12} **c.** The two expressions are the same. **85.** $y = \left(\frac{\$75}{1\text{ hr}}\right)x + \$50$ **87.** 288 mm^3 **89.** 7776 **91.** −48,828,125 **93. b.** $36n^6$ **95. b.** $5n^8$ **97.** $a(b + c) = ab + ac$ **99.** $17x - 39$

Section 5.2 Vocabulary Practice

1. G **2.** I **3.** B **4.** H **5.** J **6.** F **7.** D **8.** E **9.** A **10.** C

Section 5.2 Exercises

1. 4×10^{-8} m **3.** 1.5×10^4 nucleotides **5.** 6.02×10^{23}
7. $6.626 \times 10^{-34} \dfrac{\text{m}^2 \cdot \text{kg}}{\text{s}}$ **9.** 8.88×10^4 mi
11. greater than 12 **13.** less than 7
15. $\$1.1341 \times 10^{12}$ **17.** 1.83×10^8 acres **19.** 8×10^7
21. 1.2×10^8 **23.** 1.2×10^{-6} **25.** 3×10^5
27. 3×10^{11} **29.** 3×10^{-11} **31.** 8.2×10^3
33. 8.2×10^{13} **35.** 8.2×10^{-11} **37.** 8.1×10^7
39. 8.1×10^{-5} **41.** 9×10^3 **43.** 5×10^3
45. 1.4×10^3 **47.** 8.3×10^5 **49.** 3.8×10^{-4}
51. $7.2 \times 10^{13}\ \text{cm}^2$ **53.** $3.2 \times 10^{12}\ \text{in.}^3$
55. $1.80864 \times 10^{-15}\ \text{km}^2$ **57.** $1.1304 \times 10^{-1}\ \text{cm}^3$
59. 4.5×10^3 cm **61.** 4.5×10^{-3} cm **63.** $8 \times 10^{-4} \dfrac{\text{g}}{\text{mL}}$
65. 1.6×10^5 in. **67.** $2.16 \times 10^{-10}\ \text{m}^3$
69. $2.16 \times 10^{14}\ \text{m}^3$ **71.** 2.5×10^{-9} **73.** 6.5×10^{-8} g
75. 1.82×10^{-8} g **77.** 1.41×10^4 kg
79. 5.73×10^4 kg **81.** $C = 1.54 \times 10^{-7}$
83. $n \approx 8.21 \times 10^{-6}$ **85.** $M = 2.4 \times 10^{-4}$
87. $m \approx 5.98 \times 10^{24}$ **89.** $v \approx 1.56 \times 10^1$
91. $h \approx 2.48 \times 10^{-1}$ **93.** $v \approx 3.71$ **95.** $C = 1 \times 10^{-1}$
97. $P = 1.4 \times 10^3 \dfrac{\text{joule}}{\text{s}}$ **99.** $P = 2.2 \dfrac{\text{kg} \cdot \text{m}}{\text{s}}$
101. $E \approx 3.75 \times 10^3 \dfrac{\text{kg} \cdot \text{m}^2}{\text{s}^2}$ **103.** $P = 2.64 \times 10^4 \dfrac{\text{kg} \cdot \text{m}}{\text{s}}$
105. $D \approx 5.55 \times 10^8$ mi **107.** 28,800,000 gal **109.** 8.6 in.
111. $8 \times 10^{-4} \dfrac{\text{g}}{\text{mL}}$ **113.** $K_a \approx 1.82 \times 10^{-5}$
115. b. 7.9×10^{-7} **117. b.** 8×10^8 **121.** $-4x$

Section 5.3 Vocabulary Practice

1. H **2.** B **3.** C **4.** J **5.** E **6.** F **7.** D **8.** G **9.** I **10.** A

Section 5.3 Exercises

1. a. 1 **b.** binomial **3. a.** 3 **b.** trinomial **5. a.** 1 **b.** monomial **7. a.** 0 **b.** monomial **9. a.** 2 **b.** binomial **11. a.** 2 **b.** trinomial **17. a.** yes **19. a.** yes **21. a.** yes **23. a.** no **b.** There is a variable in the denominator. **25. a.** yes **27. a.** no **b.** A variable is an exponent. **29. a.** no **b.** An exponent is not a whole number. **31. a.** $-4x^3 + 3x^2 - 2x + 7$ **b.** -4
33. a. $a^3 + 7a - 216$ **b.** 1 **37.** $7x + 6$ **39.** $-x + 24$
41. $-2x + 3$ **43.** $3x^2 + 4x - 24$ **45.** $3x^2 - 19x + 9$
47. $13y^2 + y - 15$ **49.** $-5y^2 + 11y - 11$ **51.** $-x^2 + x$
53. $\dfrac{19}{24}p + \dfrac{5}{9}$ **55.** $2b^2 - \dfrac{3}{5}b + 2$ **57.** $-8c - 9$
59. $-10r^2 - 9r + 8$ **61.** $15m^{11}$ **63.** $4a^8$ **65.** $6x - \dfrac{27}{4}$
67. $12x^2 - 27x$ **69.** $-8p^2 + p$ **71.** $-27c^3 + 36c^2$
73. $\dfrac{2}{3}x^3 - 4x^2 + 6x$ **75.** $-6a^2 - \dfrac{3}{10}a$
77. $16k^3 - 8k^2 + 40k$ **79.** $6x^2 + 5xy$
81. $\dfrac{5}{6}x^2 - \dfrac{13}{2}x - 18$ **83.** $4x^2 + 2x + 18$
85. $11c^3 - 90c^2 + 12c$ **87.** 0.541
89. 6.96 million people **91. b.** $3x^2 + 3x - 3$
93. b. $5x^2 - 3x + 1$ **95.** x^4 **97.** $\dfrac{1}{x^4}$

Section 5.4 Vocabulary Practice

1. G **2.** D **3.** E **4.** A **5.** H **6.** J **7.** B **8.** I **9.** F **10.** C

Section 5.4 Exercises

1. $h^2 + 2h$ **3.** $h^2 + 6h + 8$ **5.** $h^2 - 2h - 8$
7. $h^2 - 6h + 8$ **9.** $n^2 - n - 12$ **11.** $q^2 - 2q + 1$
13. $q^2 - 1$ **15.** $h^2 + 2hk + k^2$ **17.** $3x^2 + 17x + 20$
19. $4a^2 - 2a - 2$ **21.** $4x^2 - 20x + 21$
23. $7z^2 + 34z - 48$ **25.** $15h^2 - 22h + 8$
27. $m^2 + 2mn - 3n^2$ **29.** $6k^2 + 13k - 5$
31. $7b^2 - 60b - 27$ **33.** $80u^3 - 8u^2 + 24u$
35. $-c^3 + c^2 - 6c$ **37.** $6a^3 + 29a^2 + 17a - 45$
39. $x^4 + 9x^3 + 28x^2 + 31x - 4$ **41.** $k^3 + 6k^2 + 8k + 3$
43. $p^3 + p^2 - 21p + 4$ **45.** $b^3 - 7b^2 + 17b - 20$
47. $2r^3 - r^2 - 15r + 18$ **49.** $3x^3 + 26x^2 + 41x + 10$
51. a. $x^2 + 13x + 40$ **b.** $x^2 - 13x + 40$
53. a. $x^2 + 3x - 40$ **b.** $x^2 - 3x - 40$ **55.** $2x^2 + 8x + 6$
57. $3a^3 - 6a^2 - 24a$ **59.** $12w^3 - 92w^2 + 160w$
61. $x^2 + 10x + 25$ **63.** $x^2 - 8x + 16$ **65.** $x^2 - 81$
67. $25x^2 - 36$ **69. a.** $a^2 + 2ab + b^2$ **b.** yes
71. a. $a^2 - b^2$ **b.** yes
75. a. $W + 8$ **b.** $4W + 16$ **c.** $W^2 + 8W$

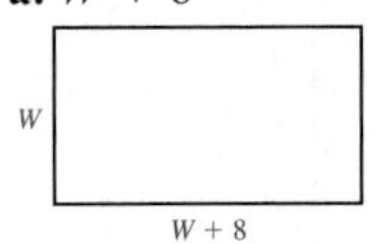

77. a. $3L - 8$ **b.** $8L - 16$ **c.** $3L^2 - 8L$

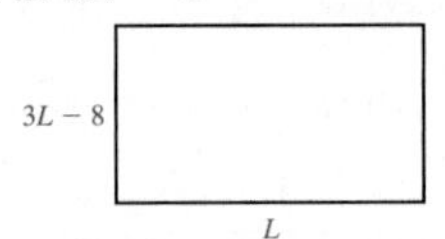

79. a. $a + 4$; $a - 3$ **b.** $3a + 1$

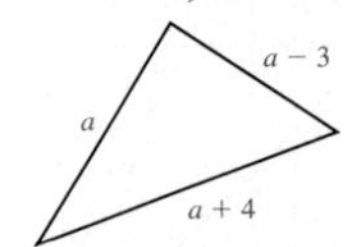

81. a. $b - 6$ **b.** $\dfrac{1}{2}b^2 - 3b$

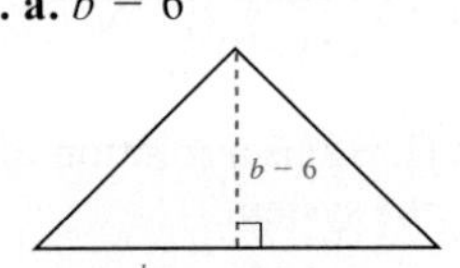

83. a. $2h + 4$ **b.** $h^2 + 2h$ **85.** $4s$

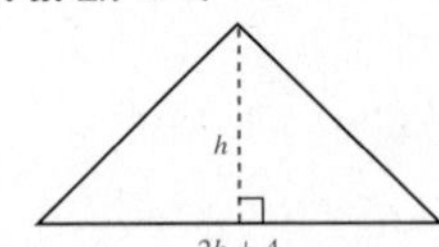

87. a. $\dfrac{5}{4}W$ **b.** $\dfrac{9}{2}W$ **c.** $\dfrac{5}{4}W^2$

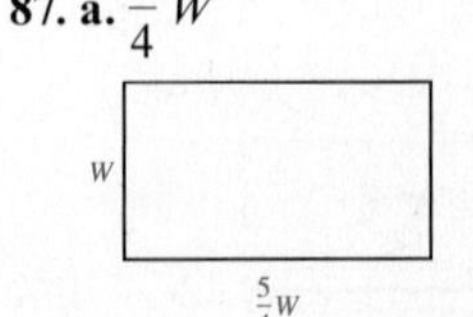

89. a. $3L$ **b.** $8L$ **c.** $3L^2$

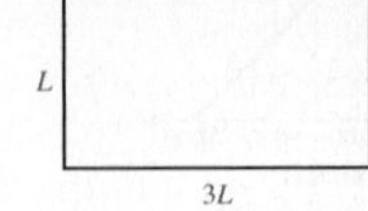

91. 1.7 L concentrated sulfuric acid; 3.3 L water

93. a. $C = \left(\dfrac{\$41.50}{1 \text{ X-ray}}\right)n + \$51$ **b.** \$341.50

95. b. $5x^2 - 17x + 6$ **97. b.** $24x^2 - 2x - 1$

99. 5 **101.** 20

Section 5.5 Vocabulary Practice

1. I **2.** E **3.** B **4.** C **5.** G **6.** A **7.** J **8.** D **9.** F **10.** H

Section 5.5 Exercises

1. a. $x^2 + 5x$ **b.** x **3. a.** $x^2 - 25$ **b.** $x + 5$

5. $4x^2 + 7x + 3$ **7.** $4a^4 + 2a^3 - a^2$ **9.** $3a^2 - 1$

11. $7d^3 - 4d^2 - d + 2$ **13.** $7u^3 - 8u + 9$

15. $-7h^2 + 3h - 6$ **17.** $w^3 - 3w + 12 + \dfrac{4}{w}$

19. $-w^3 + 3w - 12 - \dfrac{4}{w}$ **21.** $\dfrac{1}{4}h^8 - h^2 + 25h + 5$

23. $\dfrac{1}{4}h^6 - 1 + \dfrac{25}{h} + \dfrac{5}{h^2}$ **25.** $\dfrac{1}{2}m - 2 + \dfrac{4}{m}$

27. $2x + 1 - \dfrac{3}{x}$ **29.** $\dfrac{12}{5}y^2 - 2y + \dfrac{6}{5}$ **31. a.** 204

33. a. 354 **35. a.** 312 **37. a.** $x + 2$ **39. a.** $x - 2$

41. a. $2x + 3$ **43. a.** $x + 7$ **45. a.** $2x + 3$ **47. a.** $y + 9$

49. a. $a - 12$ **51. a.** $3z + 2$ **53. a.** $3x - 1$

55. a. $(x + 6)\ \text{R}(-2)$ **57. a.** $(x - 10)\ \text{R}(30)$

59. a. $(x + 8)\ \text{R}(-9)$ **61. a.** $(x + 14)\ \text{R}(57)$

63. a. $(2x + 1)\ \text{R}(16)$ **65. a.** $(3x + 17)\ \text{R}(-13)$

67. a. $(a - 5)\ \text{R}(4)$ **69. a.** $(c - 6)\ \text{R}(44)$

71. \$237.50 **73.** \$5763

75. a. $x - 3$ **b.**

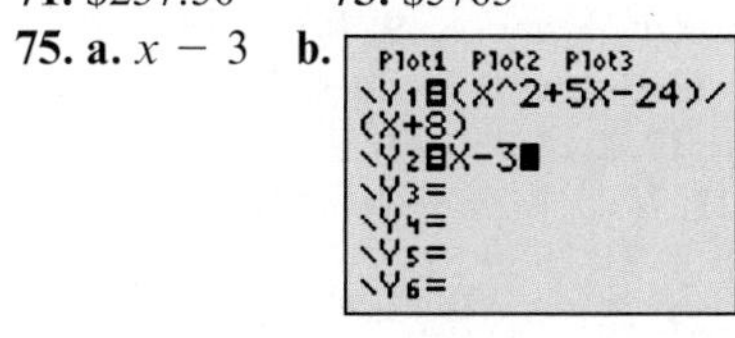

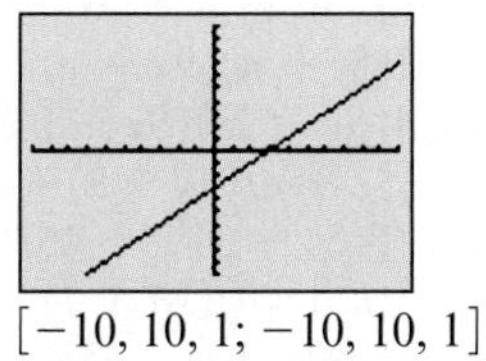

[−10, 10, 1; −10, 10, 1]

77. a. $x + 15$ **b.**

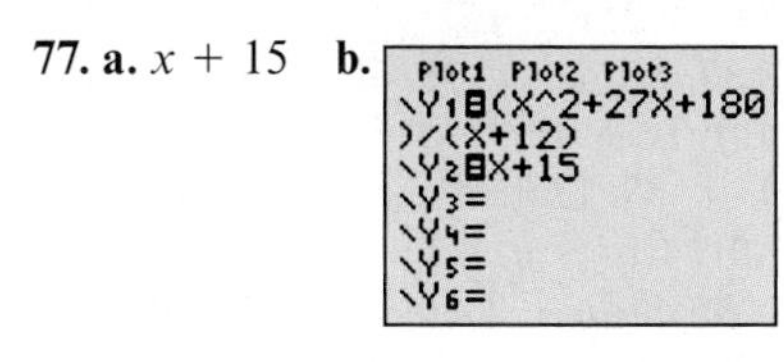

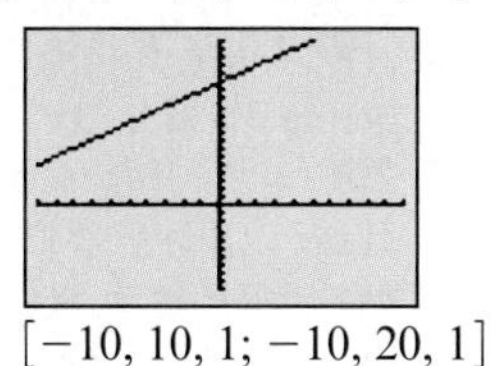

[−10, 10, 1; −10, 20, 1]

79. b. $6x^2 + 2x - \dfrac{1}{2}$ **81. b.** $(2x + 21)\ \text{R}(171)$

Chapter 5 Review Exercises

1. $-6w + 23p - 4$ **2.** $-8x^2 + 5x + 6$ **3.** $48x^{10}$

4. $\dfrac{5a^5b^2}{z^4}$ **5.** $\dfrac{7p^5}{13q^7}$ **6.** $-18x^7y^2$ **7.** 1 **8.** a^{63}

9. $9x^2$ **10.** $16x^6y^8$ **11.** $\dfrac{25a^6}{16b^{10}}$ **12.** $\dfrac{81}{w^6}$ **13.** $\dfrac{1}{10d^2f}$

14. $2x^5y$ **15.** $180x^{18}y^{19}$ **16.** a, b, c, d, e **18.** a, b, e

19. 3×10^{-12} L **20.** 4.22×10^{27} atoms

21. \$$1.5125 \times 10^{13}$ **22.** 3.01×10^{27} atoms

23. $2.81 \times 10^{-7}\ \text{m}^2$ **24.** $9.09 \times 10^{-1}\ \dfrac{\text{kg}}{\text{L}}$ **25.** 8.5×10^3 L

26. 5.7×10^{-3} g **27.** $7.5 \times 10^7\ \dfrac{\text{kg} \cdot \text{m}^2}{\text{s}^2}$

28. $E \approx 2.13 \times 10^{-2}$ **31. a.** yes **32. a.** yes **33. a.** no

34. a. no **35. a.** yes **36. a.** yes **37. a.** no

41. a. $-x^5 + 6x^3 + 21x^2 + 9x + 15$ **b.** 5 **c.** −1 **d.** 15 **e.** 5

42. a. $\dfrac{3}{4}x^3 + \dfrac{2}{3}x^2 - x + \dfrac{5}{6}$ **b.** 3 **c.** $\dfrac{3}{4}$ **d.** $\dfrac{5}{6}$ **e.** 4

43. $22x^2 - 50x - 6$ **44.** $-8x^2 + 32x + 10$

45. $3x^2 - 7x - 28$ **46.** $\dfrac{31}{24}y^2 + 8y - \dfrac{1}{2}$ **47.** $45p^3$

48. $\dfrac{7}{2}x^3 + 3x^2 - 2x$ **49.** $-6x^4 + x^3 - 9x^2 + 8x$

50. $-12x^4 + 2x^3 - 18x^2 + 16x$ **51.** the product rule of exponents **52.** b, d **53.** $x^2 + 5x - 36$

54. $x^2 - 16x + 60$ **55.** $7x^2 - 61x - 18$

56. $6a^2 + 43a + 77$ **57.** $81x^2 - 25$

58. $x^3 - 9x^2 + 29x - 36$ **59.** $x^2 - 4xy - 21y^2$

60. $14w^3 - 61w^2 - 94w + 33$ **61.** $6p^3 + 34p^2 - 12p$

62. $x^2 + 16x + 64$ **63.** $a^2 - 100$ **64.** $16y^2 - 40y + 25$

65. $100z^2 - y^2$ **66. a.** $a = 5$; $b = 2$ **b.** $a = 3h$; $b = 4$

67. a, c, d

68. a. $3W + 6$ **b.** $8W + 12$ **c.** $3W^2 + 6W$

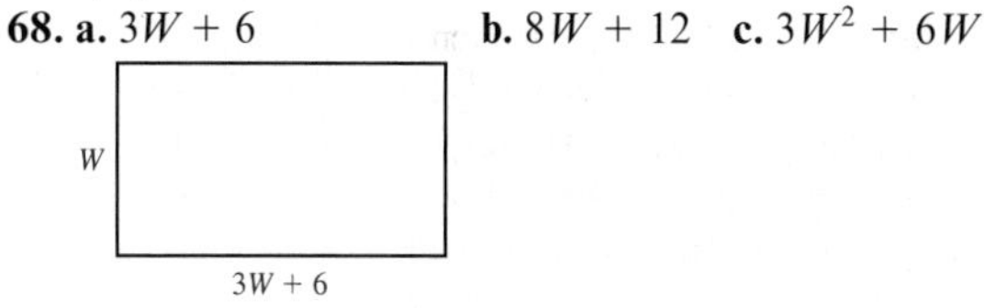

69. a. $\dfrac{1}{2}L$ **b.** $3L$ **c.** $\dfrac{1}{2}L^2$

70. a. $3b + 5$ **b.** $\dfrac{3}{2}b^2 + \dfrac{5}{2}b$

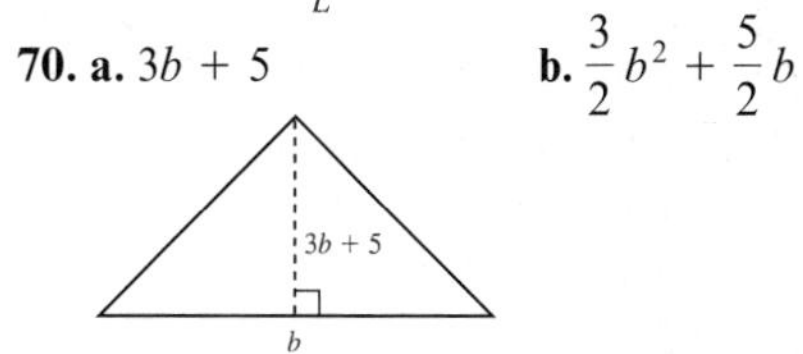

71. a. $10x^3 + 6x^2 + \dfrac{3}{2}x + 4$ **72. a.** $\dfrac{9}{2}x^9 - 3x^6 + x^2 + \dfrac{4}{x}$

73. a. $x + 8$ **74. a.** $x + 8$ **75. a.** $(x + 4)\ \text{R}(18)$

76. a. $(x + 3)\ \text{R}(-3)$ **77.** $3a^2 + 11a - 20$

78. $n^2 + 5n + 7$ **79.** $w + 7$

Chapter 5 Test

1. $100x^7y$ **2.** $\dfrac{19y^6}{2x^3}$ **3.** $81a^{10}b^4c^2$ **4.** 1 **5.** $\dfrac{25}{p^8}$

6. $\dfrac{49w^2}{36c^2}$ **7.** $\dfrac{1}{25x^2y^6}$ **8.** $9 \times 10^{-6}\ \dfrac{\text{g}}{\text{mL}}$ **9.** 4.37×10^{20} kg

10. $2.16 \times 10^{-25}\ \text{m}^3$ **11.** $4h^5 - 36h^4 + 20h^3$

12. $8x^2 + 3x - 5$ **13.** $x^3 - 3x^2 - 31x + 21$

14. $3x^3 + 9x^2 - 12x$ **15.** $8x^2 + 3x - 20$

16. $-2p^2 + 4p - 13$ **17.** $1.04 \times 10^{18}\ \text{m}^2$ **18.** $9x^2 - 49$

19. a, d **21. a.** $x^3 - 9x^2 + 4x + 12$ **b.** 3 **c.** four

22. a. the difference of squares pattern **b.** $49x^2 - 4$

23. a. $5W + 6$ **b.** $12W + 12$ **c.** $5W^2 + 6W$

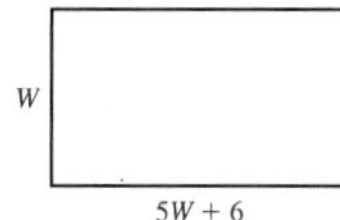

24. 2×10^{-6} g per kilogram per minute

25. 1.8986×10^{27} kg **26.** $E = 6.75 \times 10^3\ \dfrac{\text{kg} \cdot \text{m}^2}{\text{s}^2}$

27. $3x^3 - 6x^2 + 1 - \frac{9}{x}$ **28. a.** $x - 5$
29. a. $(x + 6)$ R(2)

CHAPTER 6

Section 6.1 Vocabulary Practice

1. I **2.** H **3.** J **4.** F **5.** B **6.** E **7.** C **8.** D **9.** G **10.** A

Section 6.1 Exercises

1. 6 **3.** $12xy^2$ **5.** $12x^2y$ **7.** 3 **9.** x
11. a. $7(2p + 3q)$ **13. a.** prime **15. a.** $4x(x + 5)$
17. a. $7(8x - 5z)$ **19. a.** $20x(3x - 5)$ **21. a.** $60x(x - 1)$
23. a. $5ab^5(3a^2b - 18b^2 + 7a)$
25. a. $2x(-10x - 3y + 7z + 1)$
27. a. $10(5m^2n^2 + 2m^2n + 3mn^2 + 10)$
29. a. $x^2y^5(x^2y^2 + x^3 + xy + 1)$
31. a. $13(2mn + 6m - 3n)$ **33. a.** prime
35. a. $6u(5u^2 + 3u + 9)$ **37. a.** $2w(5w^2 - 16w + 4)$
39. a. $p(p^3 - p^2 - p - 1)$ **41. a.** $(p - 5)(2p + 3w)$
43. a. $(3f + 4)(10f - g)$ **45. a.** $(6c + 7)(5d - 4)$
47. a. prime **49. a.** $(m + v)(3m + 1)$
51. a. $(m + v)(3m - 1)$ **53. a.** prime
55. a. $(f + g)(2j + h)$ **57. a.** $(x - z)(x^2 + 1)$
59. a. $(7w - 2k)(3u - 5h)$ **61. a.** $(5c + 4d)(3m + 2a)$
63. a. $(c - d)(3a - b)$ **65. a.** prime
67. $(x + 7)(x + 4)$ **69.** $(p - 9)(p + 6)$
71. $(c - 8)(c - 3)$ **73.** $(2a + 3)(5a - 2)$
75. $(4x - 1)(9x - 5)$ **77.** $(a + b)(a - b)$
79. $(a + b)(a + b)$ **81.** $(5x + 3)(5x + 3)$
83. $2(xy + 2ab + 3cd)$ **85.** $(x + y)(a + 2)$
87. $a^3(1 + a + a^2 + a^4)$ **89.** 160 death sentences
91. 5.4 million albums **93. b.** $12x^3y^2(5x^2y^2 + 3)$
95. b. $-2x(3x - 2z)$ **97.** 8, 9 **99.** -12, 8

Section 6.2 Vocabulary Practice

1. I **2.** G **3.** H **4.** J **5.** E **6.** F **7.** D **8.** B **9.** A **10.** C

Section 6.2 Exercises

1. $(x + 7)(x + 9)$ **3.** $(p - 5)(p - 4)$ **5.** prime
7. $(z - 9)(z + 2)$ **9.** $(z + 9)(z - 2)$
11. $(x - 7)(x - 7)$ **13.** $(x + 7)(x + 7)$ **15.** prime
17. $(c - 9)(2c + 1)$ **19.** $(v + 2)(3v + 2)$
21. $(m - 1)(5m + 3)$ **23.** $(2c - 7)(c + 3)$
25. $(j - 3)(2j + 13)$ **27.** $(j + 3)(2j - 13)$
29. prime **31.** $(v + 2)(3v + 2)$ **33.** $(5y + 1)(y - 4)$
35. $(b + 8)(3b + 4)$ **37.** $(3w - 5)(3w - 1)$
39. $(2x + 5)(2x + 5)$ **41.** $(f - 8)(2f + 11)$
43. $(x + 9)(x + 11)$ **45.** $(p - 6)(p - 4)$
47. $(z - 9)(z + 3)$ **49.** $(z + 8)(z - 6)$
53. $(y^2 - 8)(y^2 + 4)$ **55.** $(x^3 + 9)(x^3 + 2)$
57. $(z^4 - 5)(z^4 - 6)$ **59.** $(b + 8)(3b + 4)$
61. $(2y + 9)(3y - 8)$ **63.** $(2d + 3)(4d + 7)$
65. $(2k - 9)(2k - 9)$ or $(2k - 9)^2$ **67.** prime
69. $2(x^2 + 3x + 1)$ **71.** $(2m + w)(3p - 4)$
73. prime **75.** $(x - 6)(x + 6)$ **77. a.** 1
79. a. 2544 **81. a.** 900 **83. a.** 729 **85.** 5.5 acres
87. 15,876,000 ft^3 **89. b.** $(x - 6)(x + 1)$
91. b. $(2c + 1)(3c - 4)$ **93.** $16x^2 + 40x + 25$
95. $36x^2 - 49$

Section 6.3 Vocabulary Practice

1. E **2.** H **3.** B **4.** J **5.** C **6.** F **7.** D **8.** A **9.** I **10.** G

Section 6.3 Exercises

1. $(x + 4)^2$ **3.** $(y - 6)^2$ **5.** $(u + 10)^2$ **7.** $(y - 3)^2$
9. $(k + 8)^2$ **11.** $(3k + 8)^2$ **13.** $(3k + 8m)^2$
15. $(5c - 6)^2$ **17.** $(y - 3z)^2$ **19.** $(y^{10} - 3z^{10})^2$
21. $x^{20} = (x^{10})^2$ **23.** $(f + 5)(f - 5)$
25. $(w + 10)(w - 10)$ **27.** $(2f + 5)(2f - 5)$
29. $(10f + 3h)(10f - 3h)$ **31.** $(6 + y)(6 - y)$
33. $(x^9 + y^{25})(x^9 - y^{25})$ **37.** $(x + 5)(x^2 - 5x + 25)$
39. $(h - 3)(h^2 + 3h + 9)$ **41.** $(p - 10)(p^2 + 10p + 100)$
43. $(2h - 3k)(4h^2 + 6hk + 9k^2)$
45. $(4m + 1)(16m^2 - 4m + 1)$ **47.** $(x + 1)(x - 1)$
49. $(c + 1)^2$ **51.** $(n + 10)(n^2 - 10n + 100)$
53. $(9w + 2)^2$ **55.** $(3p + w^4)(3p - w^4)$
57. $(n - 12)(n + 5)$ **59.** prime **61.** $6(a^2 + 8a + 10)$
63. $(2x - 1)(9x + 7)$ **65.** $(v - 9)(p + 6)$
67. $(5p - 3r)(3x + 2y)$ **69.** $(5u^2 - 9z^3)(5u^2 + 9z^3)$
71. $(v + 9)^2$ **73.** $7p(6p^4 - 4p^3 + 8p^2 - 10p + 3)$
75. $(3w - 5)(3w - 1)$ **77.** $9(m^2 + 9n^2)$
79. $(b + 8)(3b + 4)$ **81.** $7(2x^2 + x - 7)$
83. $(2c - 1)(2c + 5)$ **85.** commutative property of multiplication **87.** garden cart: 5 ft^3; wheelbarrow: 3 ft^3
89. $1,170,845,000 **91. b.** $(4x + 5)(4x + 5)$ or $(4x + 5)^2$
93. b. prime **95.** 0 **97.** 0

Section 6.4 Vocabulary Practice

1. F **2.** D **3.** B **4.** J **5.** C **6.** A **7.** I **8.** G **9.** E **10.** H

Section 6.4 Exercises

1. $3(x - 2)(x + 2)$ **3.** $x(x + 5)(x + 1)$
5. $5(2p + y)(x + y)$ **7.** $8(n^2 + p^2)$ **9.** $4(x + 3)^2$
11. $2(5z + 4)(9x + y)$ **13.** $3a(3c + 7)(2a + 5k)$
15. $4(5y + 2z)(6x + 1)$ **17.** $2p(4m + p)(3k + 5)$
19. $(3c + g)(4d - 3)$ **21.** $7(x^2 + 2x - 20)$
23. $10(3z + 2)^2$ **25.** $3(p + 7)(2p + 5)$
27. $4(h + 10)(2h + 7)$ **29.** $y(x - 9)(x - 4)$
31. $(10x - 1)^2$ **33.** $3(6y^2 + 25y + 15)$
35. $c(c^2 + 6)(c^2 + 8)$ **37.** prime **39.** $6(x - y)(x + y)$
41. prime **43.** $(3x - 4y)(3x + 4y)$
45. $2(4p - 5w)(4p + 5w)$ **47.** $12d(5d - 2k)(5d + 2k)$
49. $2(y + 9)(2y + 5)$ **51.** $4(q^2 - q - 1)$
53. $n(x - 4)(3n + m)$ **55.** $3u(u + 2)(u + 12)$
57. $(x + 1)(4x + 1)$ **59.** $2(p - 20)(p + 20)$
61. $4(3r + 2w)(4m + p)$ **63.** $3x^2(a - 2)$
65. $3(6x^2 + 9x^2y - 17y^2 + 8xy^2 - 70)$ **67.** $3(p - z^2)$
69. $2(c^2 + 6c + 10)$ **71.** prime **73.** $w(5w - 1)^2$
75. $a(2c + 1)(3a - b)$ **77.** $7(x - 3y)(x + 3y)$
79. prime **81.** $2(x + 9)^2$ **83.** $2(3b + c)(a - 2)(a + 2)$
85. $2(p^2 + w^3)^2$ **87.** $3(x^7 - 2y^5)^2$
89. $x^5(x^4 - y)(x^4 + y)$ **91.** $(c + 2d)(c^2 - 2cd + 4d^2)$
93. $(n - 4p)(n^2 + 4np + 16p^2)$
95. $2(2x - 3y)(4x^2 + 6xy + 9y^2)$
97. $7(d - 2f)(d^2 + 2df + 4f^2)$ **99.** $9(x^3 + 4y^3)$
101. $(3h - k)(9h^2 + 3hk + k^2)$ **103.** 1.4%
105. $v = \frac{m}{d}$; 64 cm^3 **107. b.** $12(x - 2y)(x + 2y)$
109. b. $5(x^2 + 100y^2)$ **111.** 0 **113.** 0

Section 6.5 Vocabulary Practice

1. G **2.** F **3.** H **4.** E **5.** B **6.** C **7.** D **8.** A **9.** J **10.** I

Section 6.5 Exercises

1. $x = 0$ or $x = 15$ **3.** $c = -3$ or $c = 0$
5. $b = 0$ or $b = -4$ **7.** $p = -4$ or $p = 7$

9. $a = -\frac{1}{2}$ or $a = -3$ **11.** $p = -\frac{5}{3}$ or $p = \frac{1}{2}$
13. $x = 0$ or $x = 3$ or $x = 7$ **15.** $x = 3$ or $x = 7$
17. $x = 3$ or $x = 7$ **19.** $d = 0$ or $d = \frac{15}{2}$
21. $x = -\frac{5}{3}$ **23.** $x = 4$ **25.** $h = 0$ or $h = -5$
27. $c = 0$ or $c = 4$ **31.** $x = -\frac{5}{3}$ **35. a.** $x = -9$ or $x = 5$
37. a. $y = -6$ or $y = -2$ **39. a.** $n = -7$ or $n = 8$
41. a. $p = -4$ or $p = 3$ **43. a.** $x = 2$ or $x = 5$
45. a. $m = 0$ or $m = -9$ **47. a.** $k = 5$ or $k = -5$
49. a. $k = 0$ or $k = 5$ or $k = -5$ **51. a.** $w = -4$ or $w = -2$
53. a. $w = 0$ or $w = -4$ or $w = -2$ **55. a.** $b = -5$
57. a. $b = 5$ **59. a.** $b = 5$ or $b = -5$
61. a. $d = -5$ or $d = -\frac{1}{2}$ **63. a.** $m = -\frac{5}{2}$ or $m = -\frac{7}{3}$
65. a. $j = 8$ or $j = 9$ **67. a.** $x = 0$ or $x = 6$
69. a. $x = -10$ or $x = 9$ **71. a.** $p = -8$ or $p = -\frac{1}{2}$
73. a. $z = 2$ or $z = 9$ **75. a.** $v = 0$ or $v = 2$
77. a. $x = -\frac{3}{2}$ **79. a.** $x = -8$ or $x = -\frac{1}{2}$
81. width = 9 in.; length = 12 in. **83.** width = 8 ft; length = 14 ft **85.** width = 5 ft; length = 9 ft
87. base = 10 ft; height = 12 ft
89. a. $y = \left(\frac{3.7424 \text{ million g}}{1 \text{ year}}\right)x + 3.5454$ million g
d. 60 million g **91. c.** 313 million people
93. a.

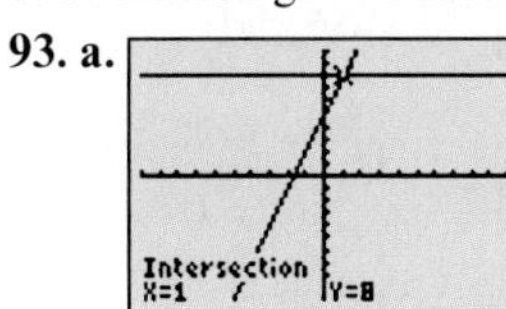
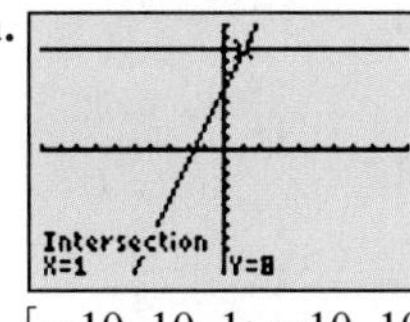

$[-10, 10, 1; -10, 10, 1]$

b. $(1, 8)$ **c.** $x = 1$

95. a.

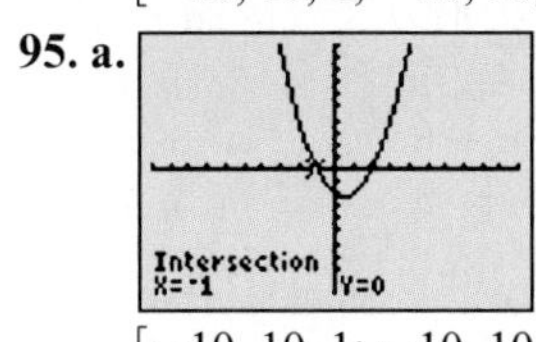
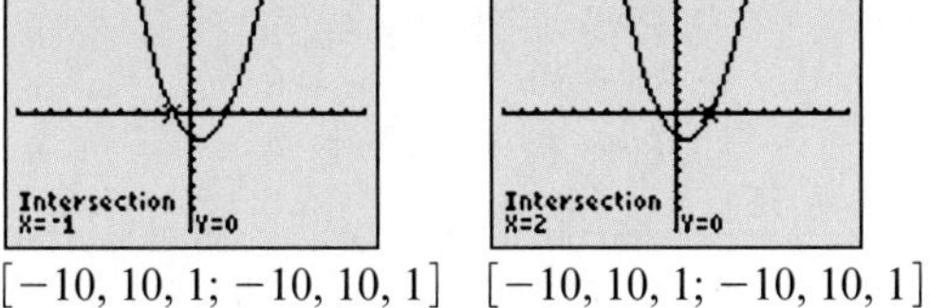

$[-10, 10, 1; -10, 10, 1]$ $[-10, 10, 1; -10, 10, 1]$

b. $(-1, 0), (2, 0)$ **c.** $x = -1$ or $x = 2$
97. b. $x = -4$ **99. b.** $x = -12$ or $x = 0$

Chapter 6 Review Exercises

1. $2^2 \cdot 5^2$ **2.** $2 \cdot 3^2 \cdot 7$ **3.** $6^2 \cdot x \cdot y^2$ **4.** $2^2 \cdot 5 \cdot a \cdot b^2$
5. 18 **6.** 12 **7.** $16xy$ **8.** $9a^2$
9. $3(5xy^2 - 4xy + 17)$ **10.** $x^2y^2z(x^3z^2 - xy^2z + y)$
11. $h(45hkm + 120k^2 - 28m^2)$ **12.** a **14.** c
15. $(1 - 4d)(5a + 3bc)$ **16.** $(n - 3k)(p + 2n)$
17. prime **18.** $(2w - z)(9x + y)$ **19.** a, b
20. $(p + 6)(p + 3)$ **21.** $(h - 7)(h - 8)$
22. $(a - 8)(a + 2)$ **23.** $(d - 7)(d + 6)$
24. $(x - 1)(2x + 5)$ **25.** $(u - 6)(2u + 9)$
26. $(r + 1)(5r + 3)$ **27.** $(2m - 3)(2m + 5)$
28. $(n + 8)^2$ **29.** $(w - 7)(w + 3)$
30. $(3c - 1)(3c + 4)$ **31.** a, c, d, e
33. a. 196 **b.** not prime **34. a.** 16 **b.** not prime
35. a. 124 **b.** prime **36.** $(x^5 + 2)(x^5 + 7)$
37. $(n^7 - 8)(n^7 - 3)$ **38.** $(y^{11} - 2)^2$
39. a. $a^2 + 2ab + b^2 = (a + b)(a + b)$ **b.** $(x + 5)^2$
40. a. $a^2 - 2ab + b^2 = (a - b)(a - b)$ **b.** $(x - 4)^2$
41. a. $a^2 - b^2 = (a + b)(a - b)$ **b.** $(x - 9)(x + 9)$
42. a. $a^2 - b^2 = (a + b)(a - b)$ **b.** $(10z + 3y)(10z - 3y)$
43. a. $a^2 - 2ab + b^2 = (a - b)(a - b)$ **b.** $(3b^2 - 7c)^2$
44. a. $a^2 + 2ab + b^2 = (a + b)(a + b)$ **b.** $(3x + y^3)^2$
45. a. $a^3 - b^3 = (a - b)(a^2 + ab + b^2)$
b. $(2h - k)(4h^2 + 2hk + k^2)$
46. a. $a^3 + b^3 = (a + b)(a^2 - ab + b^2)$
b. $(10h + 3k)(100h^2 - 30hk + 9k^2)$ **47.** a, b, c, d
48. $a = 7n$; $b = 10p$ **50.** $a = 7d$; $b = 8m$
51. $2(d + 8)(d - 8)$ **52.** $n(n - 4)(n + 11)$
53. $z(3 + y)(x - y)$ **54.** $3(2x + 1)^2$
55. $5(4u^2 + 9)$ **56.** prime **57.** $3(v - 12)(v + 8)$
58. $5p(p - 4)$ **59.** $2(6g - h)(6g + h)$ **60.** prime
61. $z(3x - z)(2x + y)$ **62.** $2x(2x - 3)^2$
63. $x(6x + 1)(y - 2z)$ **64.** $5(x + 2y)(x^2 - 2xy + 4y^2)$
65. $x(3x - 4y)(9x^2 + 12xy + 16y^2)$
69. a. $x = -\frac{9}{2}$ or $x = 3$ **70. a.** $x = 0$ or $x = 2$ or $x = -15$
71. a. $x = \frac{4}{3}$ **72. a.** $a = -3$ or $a = 4$
73. a. $p = 0$ or $p = 8$ or $p = -8$ **74. a.** $x = 0$ or $x = 24$
75. a. $n = 0$ or $n = -5$ **76. a.** $y = -\frac{3}{2}$ or $y = -2$
77. a. $a = 0$ or $a = 1$ or $a = -1$
78. width = 8 cm; length = 18 cm

Chapter 6 Test

1. $a^2 - b^2 = (a + b)(a - b)$ **2.** $a^2 + 2ab + b^2 = (a + b)^2$ or $a^2 - 2ab + b^2 = (a - b)^2$ **3.** $(a - 8)(a + 6)$
4. $7x(x - 1)$ **5.** $(p - 3)(p + 3)$
6. $(m + z)(m^2 - mz + z^2)$ **7.** $(5z + 8)(3x + y)$
8. $2(x - 3)(x + 9)$ **9.** $8(h - 1)(h + 5)$
10. $(3d - 7)(2m - 5p)$ **11.** $2(x^2 - 4x + 45)$
12. $(m - 2z)(m^2 + 2mz + 4z^2)$ **13.** $(3x + 2)^2$
14. $16(a^2 + 4b^2)$ **15.** $3a(4x + 9)$
16. $(5x + 3)(9x + 1)$ **17. a.** not prime
18. $k = -9$ or $k = 12$ **19.** $w = 0$ or $w = 3$
20. a. $u = 7$ or $u = 4$ **21. a.** $b = -3$ or $b = 5$ or $b = 0$
22. a. $x = 5$ or $x = -5$ **23. a.** $n = 8$ or $n = -\frac{5}{2}$
24. width = 7 in.; length = 12 in.

Chapters 4–6 Cumulative Review

1. parallel lines **2.** coinciding lines **3.** one solution
4. $\frac{7}{9}$ **5.** $(2, -6)$ **6.** $(4, -3)$ **7.** no solution **8.** $(2, 3)$
9.

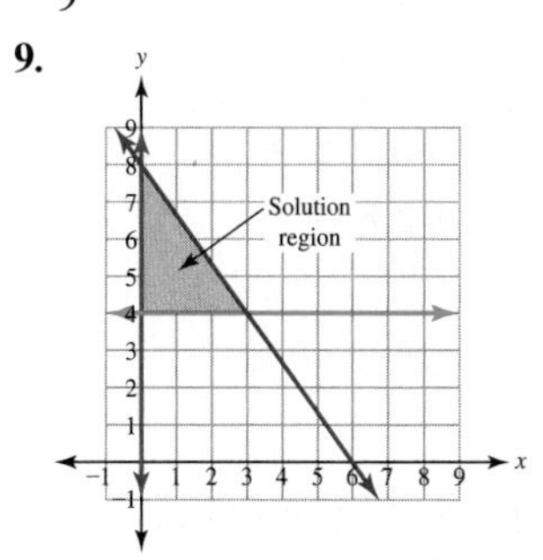

10. $(2, -7)$
11. infinitely many solutions; $\{(x, y) \mid x = 3y - 2\}$
12. $(1, -3)$ **13.** no solution **14.** no solution
15. $(2, -3)$ **16.** $(1, -3)$ **17.** $\left(\frac{1}{2}, 6\right)$
18. $a(b + c) = ab + ac$ **19.** $22x^2 - 50x - 6$
20. $-8x^2 + 32x + 10$ **21.** $36x^3 + 60x^2 - 48x$

22. $40x^2 + 21x - 27$ **23.** $21x^3 - 69x^2 + 36x - 54$
24. $2x^3 - 15x^2 - 27x$ **25.** $a^2 - 8a + 8$
26. $-4c^2 - 4c + 8$ **27.** $10x^{12}$ **28.** $\frac{3a^7}{5}$ **29.** $\frac{9}{p^7}$
30. 1 **31.** $\frac{1}{2}$ **33.** 1 **34.** 5 **35.** 5 **37.** 8
38. 4 **39.** $(2x + 1)(3x - 4)$ **40.** $(x + 9)(7x - 2)$
41. $(x - 9)(x + 10)$ **42.** $3(n - 9)(n - 7)$
43. $5p(p - 4)$ **44.** $2(6g + h)(6g - h)$ **45.** prime
46. $a(3c + m)(2c - p)$ **47.** $2p(3p - 1)^2$
48. $c(2a + 3)(b - 4c)$ **49.** If a and b are expressions and $a \cdot b = 0$, then either $a = 0$ or $b = 0$.
50. $x = -\frac{1}{2}$ or $x = 4$ **51.** $n = 0$ or $n = 4$ or $n = -6$
52. $x = 4$ **53.** $y = 7$ or $y = -6$
54. $x = -\frac{1}{2}$ or $x = 5$ **55.** $x = 0$ or $x = 36$
56. length 25 in., width 17 in. **57.** 600 products
58. 1928.6 gallons Mixture A, 2571.4 gallons Mixture B
59. 62 gallons salt solution, 88 gallons pure water
60. 162 reams white paper, 18 reams colored paper

CHAPTER 7

Section 7.1 Vocabulary Practice

1. B **2.** J **3.** I **4.** A **5.** F **6.** E **7.** C **8.** D **9.** H **10.** G

Section 7.1 Exercises

1. $\frac{3}{7}$ **3.** $\frac{6ab^2}{7}$ **5.** $\frac{15p^2}{7n^3}$ **7.** $\frac{y^4}{2}$ **9.** $\frac{x - 4}{5}$
11. $\frac{x - 4}{5x}$ **13.** $\frac{10}{y + 3}$ **15.** $\frac{z + 5}{8}$ **17.** $\frac{1}{y}$ **19.** y
21. $\frac{1}{x(x + 2)}$ **23.** $\frac{3}{4}$ **25.** -1 **27.** $\frac{2x + 2}{x - 1}$
29. $x - 1$ **31.** $\frac{1}{h + 9}$ **33.** $\frac{k + 3}{k + 5}$ **35.** $\frac{z + 5}{z + 8}$
37. $\frac{x + 3}{x}$ **39.** $\frac{m - 5}{m + 1}$ **41.** $\frac{1}{u + 6}$ **43.** $\frac{k - 8}{k - 9}$
45. $\frac{x - 3}{x - 2}$ **47.** $\frac{y + 4}{y - 9}$ **49.** $\frac{2x^2 + 2x}{3(x - 1)}$ **51.** $\frac{c - 9}{c + 4}$
53. $\frac{3n - 8}{3n + 8}$ **55.** $\frac{4v - 3}{2(v + 3)}$ **57.** $\frac{2p + 10}{3(p + 3)}$ **59.** $\frac{x - 9}{x(x - 3)}$
61. $\frac{a - 3}{2(a - 5)}$ **63.** $\frac{-n - 3}{n + 8}$ **65.** $\frac{-n - 3}{2(n + 8)}$
67. $\frac{x^2 - 2x + 4}{x - 2}$ **69.** $\frac{a^2 + 5a + 25}{a - 5}$ **71.** $\frac{1}{h^2 - 4h + 16}$
77. $c = -3$ **79.** $x = -2, x = 9$ **81.** $p = 5, p = -5$
83. $k = -9, k = 9$ **85.** none **87.** 8.7 mi
89. 5,211,938 million ft³ **91. b.** $x + 3$
93. b. $\frac{x + 2}{x + 3}$ **97.** 16

Section 7.2 Vocabulary Practice

1. C **2.** B **3.** J **4.** I **5.** E **6.** F **7.** H **8.** G **9.** D **10.** A

Section 7.2 Exercises

3. $\frac{1}{50}$ **5.** $\frac{x}{70}$ **7.** $\frac{1}{3(a + 5)}$ **9.** $\frac{5z - 15}{12(z + 3)}$ **11.** 3
13. $h - 3$ **15.** $\frac{2x + 2}{x - 1}$ **17.** $\frac{z - 6}{z - 2}$ **19.** $\frac{p + 9}{p + 3}$
21. $\frac{a + 8}{a - 3}$ **23.** $\frac{r + 1}{r}$ **25.** 1 **27.** $\frac{x - 6}{x + 6}$
29. $\frac{4r^2 + 4r}{9}$ **31.** $\frac{w^3 + 5w^2 - 19w + 40}{(w - 2)(w + 1)(w - 1)}$
33. $x^2 - 2x - 8$ **35.** $\frac{3}{7}$ **37.** 4 **39.** $\frac{y}{xz}$ **41.** x^2
43. $\frac{2a^3}{3}$ **45.** $\frac{2c}{b}$ **47.** $\frac{4b}{7ac}$ **49.** $\frac{9x}{8}$ **51.** $\frac{5}{2}$
53. $\frac{x - 4}{2}$ **55.** $\frac{2a^3}{3}$ **57.** $\frac{3}{4f^2}$ **59.** $\frac{7u^3}{25}$ **61.** $\frac{x - 4}{5(x + 2)}$
63. $\frac{2}{z + 1}$ **65.** $\frac{3p}{4(p + 5)}$ **67.** $-p + 8$ **69.** $\frac{z + 9}{z - 3}$
71. $\frac{1}{k^2}$ **73.** $\frac{35}{8}$ **75.** $\frac{u + 3}{u + 1}$ **77.** $z + 3$ **79.** $\frac{a - 6}{a - 3}$
81. $x - 1$ **83.** $\frac{4x^2 - 8x + 16}{x(x - 2)}$ **85.** $\frac{z^3 + 9z^2 + 27z + 54}{z}$
87. \$4.27 **89.** (750 products, \$75,000) **91. b.** $\frac{3}{x - 2}$
93. b. $\frac{2x}{(x + 1)(x - 1)}$ **95.** $\frac{1}{3}$ **97.** $\frac{2}{3}$

Section 7.3 Vocabulary Practice

1. D **2.** J **3.** B **4.** I **5.** E **6.** F **7.** A **8.** C **9.** H **10.** G

Section 7.3 Exercises

1. $\frac{3}{25}$ **3.** $\frac{2}{7}$ **5.** $\frac{10}{x + 8}$ **7.** $\frac{9}{x - 9}$ **9.** $\frac{3m}{m + 9}$
11. $\frac{5n}{n + 3}$ **13.** $a + 2$ **15.** $r - 14$ **17.** $\frac{n - 1}{n + 2}$
19. $\frac{9}{c + 3}$ **21.** $w - 8$ **23.** $\frac{1}{w + 8}$ **25.** $\frac{w - 8}{w^2 + 64}$
27. $x + 2$ **29.** $z - 8$ **31.** $\frac{c - 2}{2}$ **33.** $\frac{x - 5}{x - 4}$
35. $\frac{n + 1}{n - 7}$ **37.** $\frac{x + 10}{x + 3}$ **39.** $\frac{p + 8}{3(p + 2)}$ **41.** $\frac{2a - 16}{3a(a + 5)}$
43. $\frac{2w - 10}{w + 5}$ **45.** $\frac{3y - 30}{5(y + 6)}$ **47.** $\frac{v + 8}{v + 4}$ **49.** $\frac{n^2 + 3n + 9}{n + 4}$
51. $\frac{k^2 - 4k + 16}{k + 10}$ **55.** 60 **57.** $4x^2$ **59.** $42c^2d^2$
61. $3(n - 9)(n + 4)$ **63.** $b(b - 7)(b - 2)$
65. $(x - 6)(x + 6)^2$ **67.** $(z + 4)(z - 2)(z - 3)$
69. $x^2(x + 6)(x - 6)$ **71.** $6(w - 5)(w + 2)(w + 1)$
73. $84x^2y^5$ **75.** $180acxy$ **77.** $360n^5p^2$ **79.** $4095a^{10}b^3c^2$
81. $x(x + 3)^2(x + 4)$ **83.** $12(x + 1)(x - 5)(x + 5)$
85. $40x^2(x + 1)^2$ **87.** 3.3 8-year-old children
89. 229 million licensed drivers **91. b.** $\frac{13}{15}$ **93. b.** $\frac{x + 5}{x + 3}$
95. $\frac{19}{12}$ **97.** $\frac{1}{6}$

Section 7.4 Vocabulary Practice

1. G **2.** H **3.** B **4.** E **5.** I **6.** C **7.** D **8.** F **9.** A **10.** J

Section 7.4 Exercises

1. 12 **3.** $90x^2$ **5.** $(a - 4)(a + 4)(a - 5)$ **7.** $3x(x - 7)$
9. $350x^2y^3z$ **11.** $x^2(x - 3)(x + 3)$ **13.** $\frac{24}{56}$
15. $\frac{6ab}{27a^2b}$ **17.** $\frac{35}{10(4x + 3)}$ **19.** $\frac{10x}{5x^2(x - 9)(x + 1)}$
21. $\frac{2p^2 + 2p}{(p + 6)(p - 6)(p + 1)}$ **23.** $\frac{3c^2 - 6c}{(c - 8)(c - 2)(c - 5)}$

27. $\frac{1}{2}$ **29.** $\frac{1}{10}$ **31.** $\frac{61}{72}$ **33.** $\frac{13}{209}$ **35.** $\frac{-5a + 6b}{15ab}$
37. $\frac{40x + 21}{72x^2}$ **39.** $\frac{-13r}{210}$ **41.** $\frac{16a + 5b}{6a^2b^2}$ **43.** $\frac{-4k^3 + 9}{8hk^2}$
45. $\frac{7p - 7}{(p - 3)(p + 4)}$ **47.** $\frac{10a - 88}{(a - 6)(a + 8)}$ **49.** $\frac{1}{5}$
51. $\frac{-8f + 1}{9(f - 2)}$ **53.** $\frac{9z + 51}{z(z + 3)(z + 9)}$
55. $\frac{3x + 14}{(x + 2)(x + 3)(x + 8)}$ **57.** $\frac{-2n + 36}{(n - 5)(n - 4)(n + 3)}$
59. $\frac{p + 6}{p}$ **61.** $\frac{-3}{2w - 1}$ **63.** $\frac{10}{2z + 5}$
65. $\frac{a}{(a + 4)(a + 5)}$ **67.** $\frac{x}{x + 2}$ **69.** $\frac{p}{p - 3}$ **71.** $\frac{-2}{v + 1}$
73. $\frac{2}{a + 4}$ **75.** $\frac{x^2 + 5x - 4}{(x + 1)(x + 5)}$ **77.** $\frac{y - 2}{y - 1}$
79. $\frac{2n - 3}{(n - 1)(n + 4)}$ **81.** $\frac{x^2 - 4x - 12}{(x - 4)(x + 4)}$ **83.** $\frac{1}{12a}$
85. $\frac{6x^2 + 13x - 3}{(x - 3)(x + 2)(x + 3)^2}$ **87.** 26% **89.** $2205
91. b. $\frac{x^2 + 25}{(x - 5)(x - 5)(x + 5)}$ **93. b.** $\frac{2x - 4}{(x + 2)(x + 4)}$
95. 5 **97.** 12

Section 7.5 Vocabulary Practice

1. D **2.** C **3.** F **4.** E **5.** I **6.** J **7.** B **8.** A **9.** H **10.** G

Section 7.5 Exercises

1. 3 **3.** $\frac{27}{20}$ **5.** $\frac{28}{15}$ **7.** 2 **9.** $\frac{4}{5}$ **11.** $\frac{75}{8x}$ **13.** $\frac{4}{3}$
15. $\frac{7p - 7}{8(p + 1)}$ **17.** $\frac{4a^4 + 20a^3}{a - 1}$ **19.** $\frac{b - 4}{b + 4}$ **21.** 1
23. $\frac{x - 2}{x + 4}$ **25.** $\frac{25}{16}$ **27.** $\frac{15}{16}$ **29.** $\frac{1}{8}$ **31.** 5
33. $\frac{x + 3}{2x + 3}$ **35.** $\frac{x^2 - 9}{2x^2}$ **37.** $\frac{2x + 3}{x}$ **39.** $-\frac{1}{x}$
41. $\frac{8x - 7}{7x - 8}$ **43.** $\frac{7x + 16}{7x + 10}$ **45.** $\frac{-x + 15}{2(5x - 7)}$
47. $\frac{-3x - 4}{x + 13}$ **49.** $\frac{2x^2 + x - 45}{4(x - 4)(x + 4)}$
51. $\frac{5x^2 + 9x - 14}{2(x + 1)(x + 3)}$ **53.** $\frac{x^2 - 3x + 2}{x(x + 1)}$
55. $\frac{x^2 - 1}{(x - 2)(x + 2)}$ **57.** xy **59.** $\frac{1}{xy}$ **61.** $\frac{15a + 10b}{2(5a + 3b)}$
63. $\frac{3a + 2b}{5a + 3b}$ **65.** $\frac{x + 3}{x + 4}$ **67.** $\frac{x + 3}{x + 4}$ **69.** 3.75 ohms
71. $0.99 **73.** 77 years 11 months **75. b.** $\frac{1}{2x}$
77. b. $\frac{x - 2 - x - 1}{2x - 4 + 3x + 3}$ **79. a.** $x = \frac{5}{2}$ **81. a.** $w = \frac{1}{2}$

Section 7.6 Vocabulary Practice

1. C **2.** J **3.** B **4.** G **5.** I **6.** A **7.** H **8.** D **9.** E **10.** F

Section 7.6 Exercises

1. a. $x = \frac{25}{16}$ **3. a.** $p = \frac{17}{16}$ **5. a.** $k = -\frac{165}{16}$
7. a. $x = -4$ **9. a.** $u = -5$ **11. a.** $a = 2$ **13. a.** $c = 15$
15. a. $x = 20$ **17. a.** $m = 30$ **19. a.** $z = -6$
21. a. $d = 9$ **23. a.** $f = 15$ **25. a.** $r = -4$
27. a. $b = 20$ **29. a.** $x = 15$ **31. a.** $x = -2$ or $x = 9$
33. a. $c = \frac{1}{2}$ or $c = \frac{3}{4}$ **35. a.** $w = 4$ **37. a.** $n = -9$
39. a. $q = 3$ **43. a.** $w = -15$ **45. a.** $x = -\frac{1}{2}$ or $x = 5$
47. a. $p = -7$ **49. a.** $v = -2$ **51. a.** $r = \frac{9}{2}$
53. a. no solution **55. a.** $c = 4$ **57. a.** $d = -5$
59. 19 min **61.** 164 min **63.** 117 min (1 hr 57 min)
65. 28 min **67.** 6,000,000 pregnant women
69. 29,000 women **71.** 307,017,100 Americans
73. 15,628,571 full-time equivalent workers **75.** 1,726,200 Americans **77.** 367,000 adults **79.** 1250 women
81. 3.5 mL **83.** 12% **85.** 1627% increase
87. b. $x = -132$ **89. b.** $a = -7$ or $a = 8$ **91.** $\frac{40 \text{ mi}}{1 \text{ hr}}$
93. $k = 160$

Section 7.7 Vocabulary Practice

1. B **2.** A **3.** A **4.** B **5.** A **6.** A **7.** A **8.** B **9.** A **10.** B

Section 7.7 Exercises

5. a. $k = 3$ **b.** $y = 3x$ **c.** $y = 12$
d.

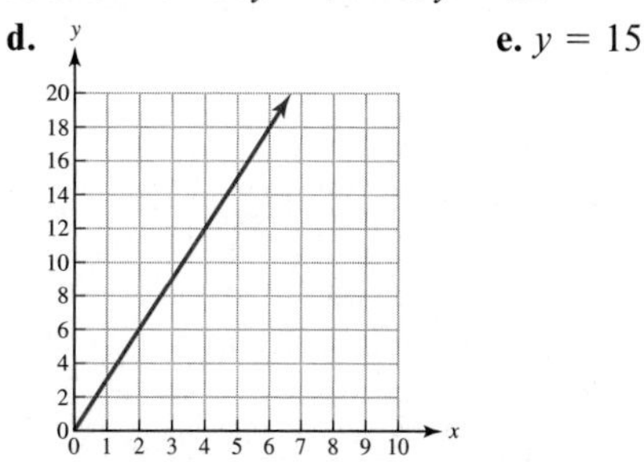

e. $y = 15$

7. a. $k = 6$ **b.** $y = 6x$ **c.** $y = 24$
d.

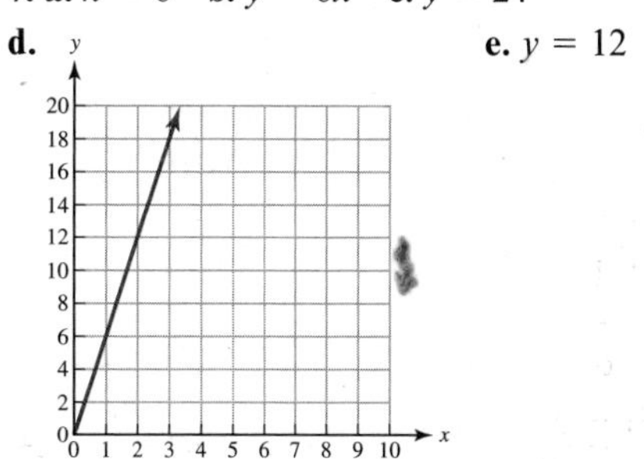

e. $y = 12$

9. a. $k = \frac{0.5 \text{ gal}}{1 \text{ oz}}$ **b.** $y = \left(\frac{0.5 \text{ gal}}{1 \text{ oz}}\right)x$ **c.** 4 gal
d.

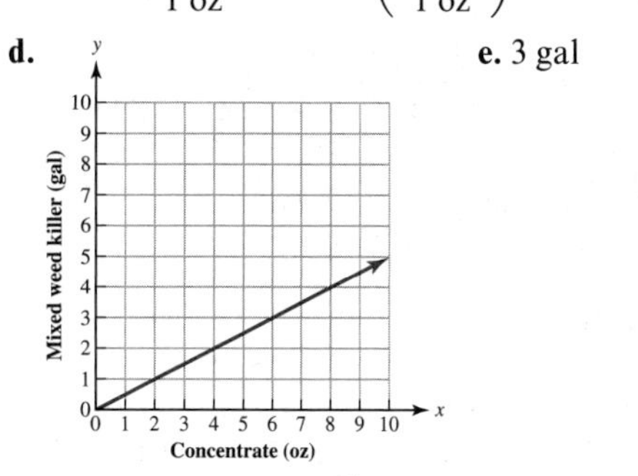

e. 3 gal

11. a. $k = 12$ **b.** $y = \frac{12}{x}$ **c.** $y = 3$
13. a. $k = 20$ **b.** $y = \frac{20}{x}$ **c.** $y = 4$
15. a. $k = 180$ windows **b.** $y = \frac{180 \text{ windows}}{x}$ **c.** 6 hr

17. a. $k = 6.28$ **b.** $y = 6.28x$ **c.** 125.6 cm **19. a.** $k = \$56$ **b.** $y = \frac{\$56}{x}$ **c.** 28 **21. a.** $k = 1.6956 \times 10^{-8}$ ohm · m² **b.** $y = \frac{1.6956 \times 10^{-8}\ \text{ohm} \cdot \text{m}^2}{x}$ **c.** 8.3×10^{-3} ohm

23. a. $k = \frac{\$45}{1\ \text{ticket}}$ **b.** $y = \left(\frac{\$45}{1\ \text{ticket}}\right)x$ **c.** 12,500 tickets **d.** \$340,875 **e.** the price of a ticket **25. a.** $k = \frac{\$6289}{1\ \text{hr}}$ **b.** $y = \left(\frac{\$6289}{1\ \text{hr}}\right)x$ **c.** \$75,468 **d.** the cost per hour to rent the jet **27. a.** $k = \frac{\$0.15}{1\ \text{mi}}$ **b.** $y = \left(\frac{\$0.15}{1\ \text{mi}}\right)x$ **c.** \$33.75 **d.** the cost per mile for gasoline **29. a.** $k = 17.5$ mi **b.** $y = \frac{17.5\ \text{mi}}{x}$ **c.** 0.35 hr **d.** 21 min

31. $y = \frac{2400\ \text{micrometers}}{x}$ **33.** inverse **35.** inverse **37.** inverse **39.** direct **41.** direct **43.** inverse **45.** direct **47.** inverse **49.** direct **51.** inverse **53.** direct **55.** inverse **57.** 4437 ft³ **59.** 122 children

61. b. $k = \frac{\$3}{1\ \text{gal}}$ **63. b.**

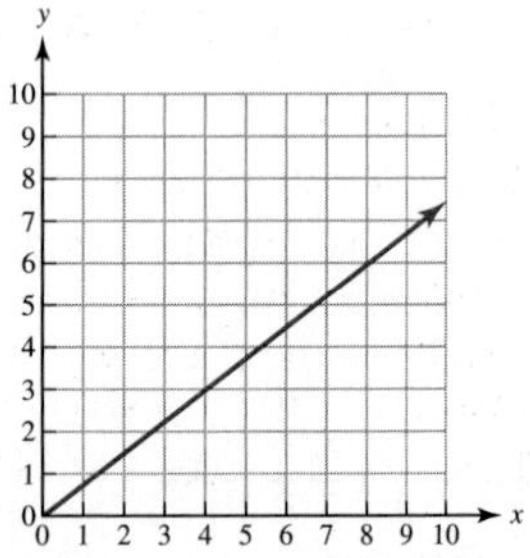

65. 4 **67.** not a real number

Chapter 7 Review Exercises

6. $\frac{4ac^7}{11b^3}$ **7.** $\frac{x-9}{6}$ **8.** $\frac{x-3}{x+5}$ **9.** $\frac{1}{a-9}$
10. $x = 9; x = -1$ **11.** $u = 9; u = -9$
12. $n = 0; n = 8; n = 4$ **14.** $\frac{x^2}{12}$ **15.** $\frac{1}{2(x+6)}$
16. $\frac{x+4}{x+8}$ **17.** $\frac{p^2}{p+8}$ **18.** $\frac{12}{ab^2}$ **19.** $\frac{2xy}{25}$ **20.** $\frac{2}{3}$
21. $\frac{3z^2 - 26z - 9}{(2z+5)(3z+4)}$ **22.** $\frac{c^3 - 2c^2 - 5c + 6}{4}$ **23.** $\frac{x}{3}$
24. $2x$ **25.** $p + 12$ **26.** $\frac{u+1}{u-9}$ **27.** $\frac{h-8}{5h+2}$
28. $\frac{1}{x+4}$ **29.** $3 \cdot 5^2$ **30.** $2^4 \cdot 3^2$ **31.** 23 **32.** $90x^2y^2$
33. $(u-6)(u+5)$ **34.** $7x(x-7)(x-2)$ **35.** $60h^2p$
36. $x(x+8)(x-8)$ **37.** $(x+9)(x+1)^2$
38. $\frac{4x-8}{6(x+9)(x-2)}$ **39.** $\frac{x^2+8x}{(x-7)(x+1)(x+8)}$
40. $\frac{6x+20y}{15x^2y}$ **41.** $\frac{q+1}{q-1}$ **42.** $\frac{12a+14}{(2a-3)(2a+1)}$
43. $\frac{-3n+2w}{14npw}$ **44.** $\frac{-4d}{(d-2)(d+2)}$
45. $\frac{4v+32}{(v-6)(v-2)(v+3)}$ **46.** $\frac{-x^2+18}{12x}$
48. $\frac{x+4}{x+2}$ **49. a.** $\frac{3c^2+2cd}{5c^2-6d}$ **b.** $\frac{3c^2+2cd}{5c^2-6d}$

50. a. $\frac{3x+1}{2(2x+1)}$ **b.** $\frac{3x+1}{2(2x+1)}$ **51.** 2.4 microfarads
53. a. $n = 21$ **54. a.** $c = 20$ **55. a.** $x = 9$
56. a. $f = -\frac{1}{2}$ or $f = 3$ **57. a.** no solution **61.** No
62. 76 min **63.** 313 people **64.** 2600 people
65. a. $k = \frac{0.05\ \text{mi}}{1\ \text{city block}}$ **b.** $y = \left(\frac{0.05\ \text{mi}}{1\ \text{city block}}\right)x$ **c.** 2.25 mi
d.

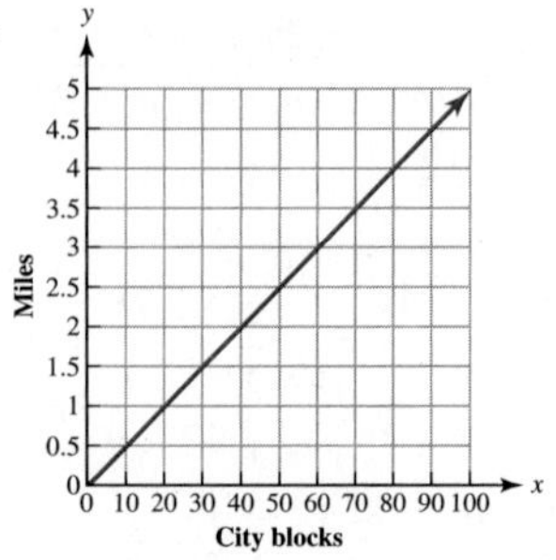

e. 3 mi

66. a. inverse variation **b.** 320 oz **c.** 53 servings
67. a. direct variation **b.** inverse variation **c.** direct variation **d.** decrease **e.** increase **f.** decrease
68. direct variation **69.** inverse variation

Chapter 7 Test

2. 50 min **3. b.** $\frac{x+3}{x-2}$ **4.** $\frac{n+4}{n+3}$ **5.** $\frac{4x-5}{3x^5}$
6. $\frac{u+2}{u+8}$ **7.** $\frac{p-11}{2(p-3)}$ **8.** $\frac{2k+9}{k+8}$ **9.** $\frac{16w+26}{(w-8)(w+6)}$
10. $\frac{x+4}{x-9}$ **11.** $\frac{2x^2+14x+48}{x(x+6)(x+6)}$ **12.** $c = 10$
13. $y = 30$ **14.** $z = 12$ **15.** $x = \frac{10}{3}$
16. 200 two-year-olds
17. a. direct variation **b.** dollars
c.

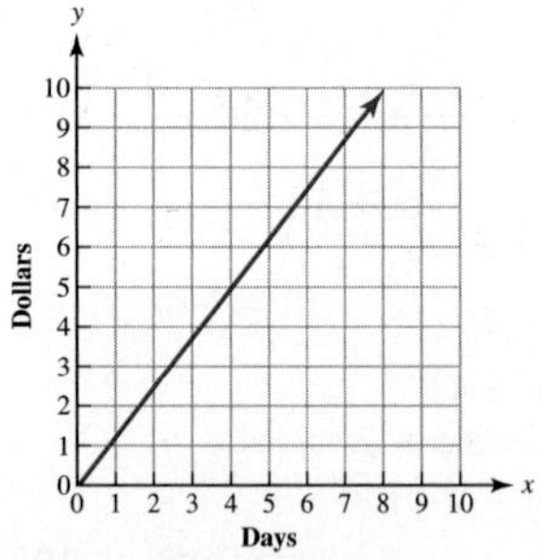

d. \$6.25

18. a. inverse variation **b.** decrease **c.** 30 ft²

CHAPTER 8

Section 8.1 Vocabulary Practice

1. H **2.** B **3.** A **4.** G **5.** E **6.** F **7.** J **8.** I **9.** C **10.** D

Section 8.1 Exercises

1. 15 **3.** 8 **5.** 2.65 **7.** 10.05 **9.** 2.24
11. not a real number **15.** $3\sqrt{2}$ **17.** $5\sqrt{3}$ **19.** $7\sqrt{2}$
21. $5\sqrt{5}$ **23.** $2\sqrt{5}$ **25.** $9\sqrt{5}$ **27.** $20\sqrt{2}$ **29.** $6\sqrt{2}$
31. z^6 **33.** a^{10} **35.** $a^{10}\sqrt{a}$ **37.** a^{11} **39.** $a^{11}\sqrt{a}$
41. $h^{10}k^{15}$ **43.** $r^2z^3\sqrt{z}$ **45.** $x^7y^9\sqrt{xy}$ **47.** $x^3y^2\sqrt{xz}$
49. $5ab^5$ **51.** $5ab^5\sqrt{2a}$ **53.** $n^2\sqrt{21np}$ **55.** $2\sqrt{6np}$
57. $12a^3b\sqrt{2b}$ **59.** $3a^2b^4\sqrt{3ab}$ **61.** $6\sqrt{5r}$ **63.** $6x\sqrt{y}$
65. $y\sqrt{3x}$ **67.** $4xy\sqrt{2y}$ **69.** $3a^2b^3c^4\sqrt{5}$

71. $3y^2\sqrt{2xy}$ **75.** $\sqrt{64y^4z^6}$ **77.** $\sqrt{18y}$ **79.** $\sqrt{20a^2b}$
81. $\sqrt{25n^6p}$ **83.** greater than **85.** 63% decrease
87. \$1077 million **89. b.** $5\sqrt{2}$ **91. b.** $10x\sqrt{2x}$
93. $13x^2$ **95.** $135a^2$

Section 8.2 Vocabulary Practice
1. F **2.** A **3.** G **4.** D **5.** C **6.** E **7.** B **8.** I **9.** H **10.** J

Section 8.2 Exercises
1. $11\sqrt{x}$ **3.** $2\sqrt{w}$ **5.** $6\sqrt{a}$ **7.** $-7\sqrt{x}$
9. $11\sqrt{c} + 5\sqrt{d}$ **11.** $2\sqrt{x} - 12\sqrt{3y}$
13. $-6\sqrt{w} - 2\sqrt{p}$ **15.** $-2\sqrt{b} + 8\sqrt{3c} + 3\sqrt{5c}$
17. $4\sqrt{2}$ **19.** $12\sqrt{2}$ **21.** $9\sqrt{2}$ **23.** $12\sqrt{7}$
25. $-6\sqrt{5}$ **27.** $-6\sqrt{3}$ **29.** $7\sqrt{3}$ **31.** $26\sqrt{3}$
33. $-\sqrt{6}$ **35.** $-10\sqrt{6}$ **37.** $5\sqrt{7}$ **39.** $3\sqrt{2}$
41. $10\sqrt{3} - 10\sqrt{2}$ **43.** $\sqrt{6}$ **45.** $4\sqrt{3}$ **47.** $-9\sqrt{2}$
49. 16 **51.** $6\sqrt{2h}$ **53.** $-4\sqrt{5z}$ **55.** $11x\sqrt{7}$
57. $6x\sqrt{2}$ **59.** $6x^3\sqrt{2}$ **61.** $13a\sqrt{a}$ **63.** $7x\sqrt{y}$
65. $-3a\sqrt{ab}$ **67.** $13u\sqrt{3z}$ **69.** $8n^2z\sqrt{7n}$ **71.** $8u\sqrt{u}$
73. 5 **75.** 3 **77.** 8 **79.** 18 **81.** 17
83. 895,000 Americans **85.** 3774 gal of 42% fructose syrup, 1226 gal of 95% fructose syrup **87. b.** $4\sqrt{5}$
89. b. $7a\sqrt{3b}$ **91.** $\frac{3}{5}$

Section 8.3 Vocabulary Practice
1. D **2.** A **3.** H **4.** E **5.** G **6.** F **7.** J **8.** I **9.** C **10.** B

Section 8.3 Exercises
1. 3 **3.** a **5.** $7m$ **7.** $2ab$ **9.** $3\sqrt{14}$ **11.** $4\sqrt{15}$
13. $3\sqrt{154}$ **15.** $30\sqrt{2}$ **17.** $5\sqrt{6}$ **19.** 12 **21.** $7b\sqrt{6ac}$
23. $5\sqrt{3ab}$ **25.** $2\sqrt{5mn}$ **27.** $11\sqrt{xy}$ **29.** $6y\sqrt{x}$
31. $10\sqrt{ab}$ **33.** $7ac\sqrt{3b}$ **35.** $6a$ **37.** $80p$ **39.** 30
41. $\sqrt{33}$ **43.** $\sqrt{x^3}$ **45.** $6x + 15$ **47.** $\sqrt{6x} + 5\sqrt{3}$
49. $\sqrt{6x} + \sqrt{15}$ **51.** $x^2 + x\sqrt{3} + x\sqrt{2} + \sqrt{6}$
53. $x^2 - x\sqrt{3} - x\sqrt{2} + \sqrt{6}$
55. $x^2 + x\sqrt{3} - x\sqrt{2} - \sqrt{6}$ **57.** $7n\sqrt{2} + 5\sqrt{7n}$
59. $3p\sqrt{2} - 2\sqrt{3p}$ **61.** $p\sqrt{3} - \sqrt{6p}$
63. $x + 2\sqrt{xy} + y$ **65.** $x + 2y\sqrt{x} + y^2$
67. $x - 2\sqrt{xy} + y$ **69.** $x - 2y\sqrt{x} + y^2$
71. $a^2 - a\sqrt{15} - a\sqrt{6} + 3\sqrt{10}$
73. $a^2 - a\sqrt{15} + a\sqrt{6} - 3\sqrt{10}$ **75.** $3x + 2\sqrt{3x} + 1$
77. $x + \sqrt{21x} + \sqrt{15x} + 3\sqrt{35}$ **79.** $\sqrt{n} + \sqrt{6}$
81. $\sqrt{a} - \sqrt{5}$ **83.** $\sqrt{x} + 2$ **85.** 3.5% increase
87. 1,051,200 times **89. b.** $10\sqrt{3}$ **91. b.** $3x\sqrt{30}$
93. $(x + y)(x - y)$ **95.** $(4n + 5p)(4n - 5p)$

Section 8.4 Vocabulary Practice
1. C **2.** A **3.** G **4.** D **5.** F **6.** E **7.** J **8.** I **9.** B **10.** H

Section 8.4 Exercises
1. $\frac{3}{4}$ **3.** $\frac{x}{y}$ **5.** $\frac{8w}{3}$ **7.** $\frac{a^3}{b^5}$ **9.** $\frac{4\sqrt{2}}{7}$ **11.** $\frac{5\sqrt{5}}{6}$
13. $\frac{2x\sqrt{2}}{3}$ **15.** $\frac{4x\sqrt{x}}{5}$ **17.** $\frac{\sqrt{3p}}{8}$ **19.** $\frac{\sqrt{15x}}{2}$
21. $\frac{2\sqrt{u}}{u}$ **23.** $\frac{5\sqrt{n}}{n}$ **25.** $\frac{\sqrt{3x}}{x}$ **27.** $\frac{\sqrt{21}}{7}$ **29.** $\frac{2\sqrt{7}}{7}$
31. $\frac{4\sqrt{15}}{9}$ **33.** $\frac{\sqrt{15xy}}{2y}$ **35.** $\frac{6\sqrt{5y}}{5y}$ **37.** $\frac{5\sqrt{21xy}}{7y}$
39. $\frac{\sqrt{30pw}}{5w}$ **41.** $\frac{4\sqrt{x} - 12}{x - 9}$ **43.** $\frac{4\sqrt{x} + 12}{x - 9}$
45. $\frac{x\sqrt{x} - 3x}{x - 9}$ **47.** $\frac{x - 3\sqrt{x}}{x - 9}$ **49.** $\frac{4\sqrt{x} - 4\sqrt{3}}{x - 3}$
51. $\frac{u - 5\sqrt{u}}{u - 25}$ **53.** $\frac{y + 6\sqrt{y}}{y - 36}$ **55.** $\frac{\sqrt{3x} - 8\sqrt{3}}{x - 64}$
57. $\frac{\sqrt{2m} + \sqrt{10}}{m - 5}$ **59.** $\frac{\sqrt{10x} + 5\sqrt{2}}{x - 5}$ **61.** $\frac{\sqrt{10x} - 2\sqrt{5}}{x - 2}$
63. $\frac{\sqrt{10x} - 2\sqrt{10}}{x - 4}$ **65.** $10\sqrt{3}$ **67.** $\frac{2\sqrt{5pw}}{pw}$ **69.** $7\sqrt{2}$
71. 20 **73.** $\frac{5\sqrt{6}}{3}$ **75.** $5\sqrt{2u}$ **77.** $3\sqrt{3}$ **79.** $13a\sqrt{2b}$
81. 10 ft wide, 23 ft long **83.** \$1330 **85. b.** $\frac{10\sqrt{c}}{c}$
87. b. $\sqrt{2c}$ **89. a.** $x = \frac{7}{12}$ **91. a.** $x = 2$ or $x = 6$

Section 8.5 Vocabulary Practice
1. B **2.** F **3.** G **4.** H **5.** D **6.** J **7.** E **8.** C **9.** A **10.** I

Section 8.5 Exercises
1. a. $x = 16$ **3. a.** no solution **5. a.** $x = 15$
7. a. $x = 17$ **9. a.** $x = 9$ **11. a.** $x = 25$
13. a. no solution **15. a.** $x = 180$ **17. a.** $x = -180$
19. a. $x = -400$ **21. a.** $x = 73$ **23. a.** $x = 190$
25. a. no solution **27. a.** $u = 3$ **29. a.** $u = -2$
31. $r = -5$ or $r = -2$ **33.** $m = -8$ or $m = 9$
35. $z = -\frac{1}{2}$ or $z = \frac{7}{2}$ **37.** $k = -\frac{5}{3}$ or $k = \frac{1}{2}$
39. $p = -\frac{4}{3}$ **41. a.** $x = 2$ or $x = 5$
43. a. $n = \frac{1}{2}$ or $n = \frac{3}{2}$ **45. a.** $x = 3$ or $x = 7$
47. a. $n = 3$ or $n = 4$ **49. a.** $w = 0$ or $w = 5$
51. a. $w = 0$ **53. a.** $d = 3$ **55. a.** no solution
57. a. $p = 9$ **59. a.** no solution **61. a.** $x = 4$
63. a. no solution **65. a.** $a = 30$ **67.** $1.12 \times 10^9 \frac{m}{s}$
69. 6 s **71.** 11.9 grade level **73.** 12.5 ft^2
75. width = 10 ft, length = 23 ft **77.** 0.47 cm^3
79. a. **b.** $x = -1$

$[-10, 10, 1; -10, 10, 1]$

81. a. **b.** $x = 5$

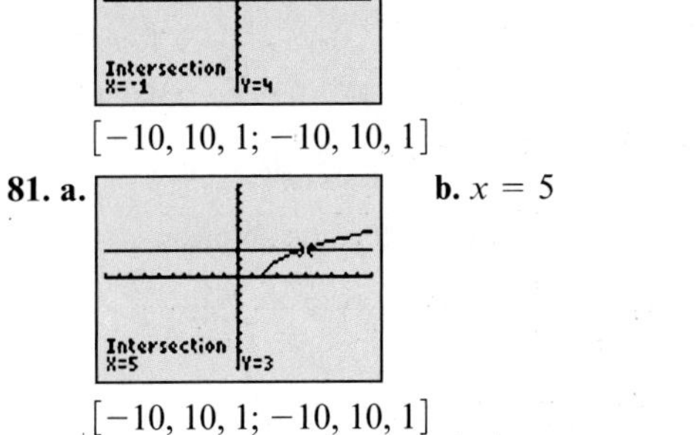

$[-10, 10, 1; -10, 10, 1]$

83. b. no solution **85. b.** $k = 3$ **87.** 8 **89.** -4

Section 8.6 Vocabulary Practice
1. F **2.** A **3.** E **4.** G **5.** H **6.** J **7.** B **8.** D **9.** I **10.** C

Section 8.6 Exercises

1. 3 **3.** −3 **5.** 3 **7.** not a real number **9.** $3\sqrt[3]{2}$
11. $2\sqrt[5]{3}$ **13.** $2\sqrt[4]{2}$ **15.** $\sqrt[3]{4}$ **17.** not a real number
19. $-3\sqrt[3]{2}$ **21.** $3\sqrt{2}$ **23.** −1 **25.** $11\sqrt[4]{z}$ **27.** $5\sqrt[4]{z}$
29. $5\sqrt[3]{2}$ **31.** $9\sqrt[3]{3}$ **33.** $\sqrt[3]{2}$ **35.** $\sqrt[3]{2}$ **37.** $3\sqrt[3]{6}$
39. $2\sqrt[4]{3}$ **41.** $x^{\frac{1}{2}}$ **43.** $x^{\frac{3}{2}}$ **45.** $x^{\frac{3}{4}}$ **47.** $x^{\frac{2}{3}}y^{\frac{1}{3}}$
49. $x^{\frac{1}{5}}$ **51.** $x^{\frac{1}{99}}$ **53.** $h^{\frac{4}{3}}$ **55.** $x^{\frac{1}{2}}$ **57.** $a^{\frac{1}{5}}$ **59.** $d^{\frac{1}{8}}$
61. $a^{\frac{3}{2}}b^2$ **63.** $\sqrt{m^3}$; $m\sqrt{m}$ **65.** $\sqrt[3]{b^4}$; $b\sqrt[3]{b}$ **67.** $a^{\frac{2}{3}}$
69. $x^{\frac{1}{3}}$ **71.** $x^{\frac{19}{45}}$ **73.** $x^{\frac{1}{8}}$ **75.** $x^{\frac{3}{4}}$ **77.** $a^{\frac{1}{12}}$ **79.** $x^{\frac{89}{90}}$
81. $x^{\frac{3}{5}}$ **83.** $a^{\frac{1}{2}}$ **85.** $\dfrac{2}{3a^{\frac{2}{7}}}$ **87.** 0.23% **89.** 105.6 mL
91. b. $\sqrt[7]{x^4}$ **93. b.** $x^{\frac{5}{7}}$ **95.** $(2x-5)(2x+3)$
97. $8(x-2)(x+2)$

Chapter 8 Review Exercises

1. 11 **2.** 7.21 **3.** not a real number **5.** $20\sqrt{2}$
6. $8\sqrt{2}$ **7.** $7\sqrt{5}$ **8.** $3\sqrt{5}$ **9.** $12x\sqrt{2x}$
10. $7xy\sqrt{2x}$ **11.** b, d **13.** $5\sqrt{2}$ **14.** $-4\sqrt{x}$
15. $-7\sqrt{p}$ **16.** $13\sqrt{x} - 2\sqrt{y}$ **17.** $4\sqrt{5}$ **18.** $114\sqrt{2}$
19. $-2\sqrt{2}$ **20.** $9\sqrt{2x}$ **21.** $5\sqrt{21}$ **22.** $3x\sqrt{10y}$
23. $3a\sqrt{21}$ **24.** $15pq\sqrt{2p}$ **25.** $2\sqrt{3x} - 14\sqrt{2}$
26. $2\sqrt{3x} - 2\sqrt{7}$ **27.** $x^2 + x\sqrt{15} - x\sqrt{3} - 3\sqrt{5}$
28. $\dfrac{8}{7}$ **29.** $\dfrac{3\sqrt{x}}{4}$ **30.** $\dfrac{\sqrt{33x}}{3}$ **31.** $\dfrac{3\sqrt{15}}{10}$
32. $\dfrac{2\sqrt{70cd}}{21d}$ **33.** b, d **34.** $\dfrac{3\sqrt{x} + 3\sqrt{6}}{x-6}$
35. $\dfrac{\sqrt{3x} - 3\sqrt{2}}{x-6}$ **36. a.** $x = 98$ **37. a.** no solution
38. a. $c = 169$ **39. a.** $p = 12$ **40. a.** $a = 6$ or $a = 9$
41. a. no solution **42. a.** $n = 9$ **45.** 18 ft **46.** 5
47. $2\sqrt[3]{6}$ **48.** $2\sqrt[4]{3}$ **49.** $5\sqrt[3]{4}$ **50.** $6\sqrt[3]{5}$
51. $2y\sqrt[5]{3x^4y}$ **52.** $d^{\frac{5}{3}}$ **53.** $x^3y^{\frac{2}{3}}$ **54.** $a^2\sqrt[5]{a^2}$
58. $a^{\frac{7}{8}}$ **59.** $x^{\frac{13}{20}}$ **60.** $a^{\frac{2}{5}}$ **61.** $c^{\frac{1}{17}}$ **62.** $\dfrac{4}{9x^{\frac{1}{7}}}$

Chapter 8 Test

1. $2\sqrt{3}$ **2.** $8x^2\sqrt{3x}$ **3.** $a^2b\sqrt{15b}$ **4.** $7\sqrt{2}$
5. $5\sqrt{3a}$ **6.** $2n^5\sqrt{15}$ **7.** $13\sqrt{3}$ **8.** $-5\sqrt[3]{x}$
9. $7\sqrt{a} - 4\sqrt{b}$ **10.** $9\sqrt{2}$ **11.** $\dfrac{\sqrt{5p}}{8}$ **12.** $\dfrac{\sqrt{15pz}}{3z}$
13. $\dfrac{3\sqrt{5u}}{5u}$ **14.** $\dfrac{7\sqrt{x} + 28}{x-16}$ **15. a.** x^2 **b.** 3 **c.** $x^{\frac{2}{3}}$
18. a. $x = 252$ **19. a.** $w = 5$ **20. a.** no solution
21. 4.08 in. **22.** $2x\sqrt[4]{2x}$ **23.** $-18\sqrt[3]{2}$ **24.** $x^{\frac{19}{20}}$
25. $a^{\frac{5}{18}}$ **26** $a^{\frac{4}{9}}$ **27.** $\dfrac{2}{3a^{\frac{6}{25}}}$

CHAPTER 9

Section 9.1 Vocabulary Practice

1. H **2.** A **3.** E **4.** J **5.** B **6.** F **7.** C **8.** D **9.** G **10.** I

Section 9.1 Exercises

1. $x = \pm 9$ **3.** $c = \pm 2$ **5.** $m = \pm 8$ **7.** $d = \pm 12$
9. $w = \pm 22$ **11.** $x = \pm 5$ **13.** $x = \pm 10$ **15.** $x = \pm 13$
17. $p = \pm 2$ **19.** $x = \pm 4$ **21.** $z = \pm 8\sqrt{2}$
23. $x = \pm 5\sqrt{6}$ **25.** $p = \pm 7\sqrt{3}$ **27.** $k = \pm 10\sqrt{5}$
29. $y = \pm\sqrt{47}$ **31. a.** $x = \pm 2\sqrt{3}$ **33. a.** $x = \pm 2\sqrt{2}$
35. a. $z = \pm 3$ **37. a.** $w = \pm 3\sqrt{10}$ **39.** $v = \pm 9$
41. $v = \pm 5$ **43.** $n = \pm 11$ **45.** $d = \pm 2\sqrt{14}$
47. $d = \pm 2\sqrt{7}$ **49.** $x = \pm\sqrt{11}$ **51.** $x = 3$ or $x = 17$
53. $x = -17$ or $x = -3$ **55.** $x = -1$ or $x = 19$
57. 50 in. **59.** 40 ft **61.** $10\sqrt{2}$ in. **63.** 127.3 ft
65. 64.9 in. **67.** 93 ft **69.** 7.04 s
71. a. 5 ft 11 in. = 71 in.; 5 ft 6 in. = 66 in.; 10 ft = 120 in.; 2 ft 5 in. = 29 in. **b.** 562,320 $in.^3$ **c.** 325.4 ft^3
73. 17% **75. b.** $x = \pm\sqrt{34}$ **77. b.** $m = \pm 5\sqrt{5}$
79. $(x+3)^2$ **81.** $(x-5)^2$

Section 9.2 Vocabulary Practice

1. H **2.** F **3.** E **4.** J **5.** B **6.** D **7.** A **8.** C **9.** I **10.** G

Section 9.2 Exercises

1. a.

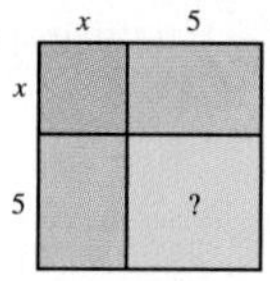

b. 25 **3.** 225 **5.** 225 **7.** 64

9. 144 **11. a.** $w = -16$ or $w = 2$ **13. a.** $w = -2$ or $w = 16$
15. a. $z = -11$ or $z = -1$ **17. a.** $z = 11$ or $z = 1$
19. a. $x = -38$ or $x = -2$ **21. a.** $w = 7$ or $w = -5$
23. a. $m = -14$ or $m = 6$ **25. a.** $k = -6$ or $k = -2$
27. a. $u = 1$ or $u = 25$ **29. a.** $a = 6$ or $a = -2$
31. a. $x = -2$ or $x = 4$ **33.** $x = -2 \pm \sqrt{13}$
35. $p = -3 \pm \sqrt{7}$ **37.** $k = 4 \pm 3\sqrt{2}$ **39.** $n = -5 \pm \sqrt{21}$
41. $z = -1$ **43.** $a = 6 \pm 2\sqrt{10}$ **45.** $r = 8 \pm 3\sqrt{7}$
47. $m = 9 \pm \sqrt{2}$ **49.** $u = -10 \pm \sqrt{103}$
51. $a = -2 \pm \sqrt{14}$ **53.** $c = -8$ or $c = -6$
55. $b = -2$ or $b = 8$ **57.** $a = 6$ **59.** $w = -3$ or $w = -1$
61. $m = 10 \pm \sqrt{106}$ **63.** $b = 1 \pm \sqrt{5}$ **65.** 28%
67. $1076

69. a.

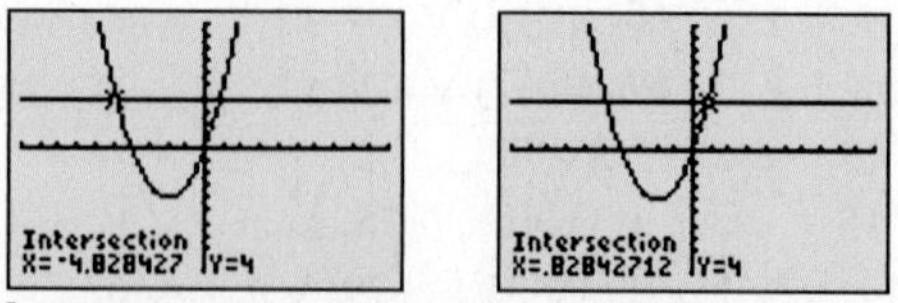

[−10, 10, 1; −10, 10, 1] [−10, 10, 1; −10, 10, 1]

b. $x = -4.8$ or $x = 0.8$

71. a.

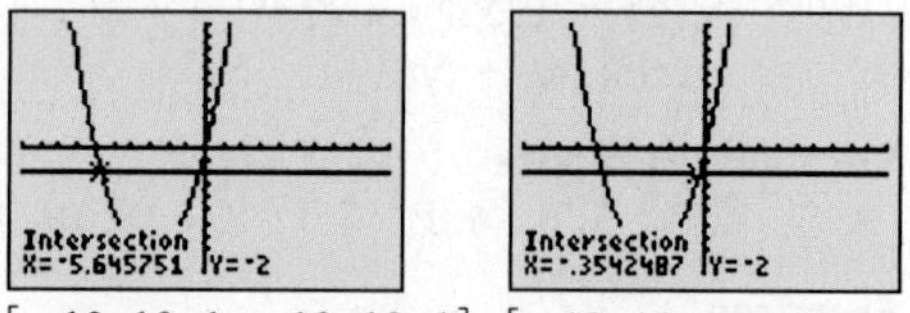

[−10, 10, 1; −10, 10, 1] [−10, 10, 1; −10, 10, 1]

b. $x = -5.6$ or $x = -0.4$
73. b. $x = -9$ or $x = 1$ **75. b.** $x = -16$ or $x = 4$
77. 17 **79.** 6

Section 9.3 Vocabulary Practice

1. D **2.** H **3.** G **4.** A **5.** F **6.** B **7.** C **8.** E **9.** J **10.** I

Section 9.3 Exercises

1. a. $x = -11$ or $x = -4$ **3. a.** $x = -7$ or $x = 5$

5. a. $p = 12$ or $p = 15$ **7. a.** $d = -\frac{5}{2}$ or $d = 8$

9. a. $h = -\frac{5}{3}$ or $h = \frac{8}{3}$ **11.** $m = -\frac{1}{5}$ **13.** $u = -\frac{3}{4}$

15. $x = \frac{-9 \pm 5\sqrt{5}}{2}$ **17.** $x = \frac{9 \pm 5\sqrt{5}}{2}$

19. $p = \frac{5 \pm \sqrt{109}}{6}$ **21.** $u = \frac{-1 \pm \sqrt{21}}{10}$

23. $u = \frac{1 \pm \sqrt{21}}{10}$ **25.** $h = \frac{-1 \pm \sqrt{21}}{2}$

27. $u = \frac{-1 \pm \sqrt{57}}{4}$ **29.** $z = \frac{-11 \pm \sqrt{721}}{20}$

31. a. 0 **b.** one rational solution **33. a.** 4 **b.** two rational solutions **35. a.** 225 **b.** two rational solutions **37. a.** 61 **b.** two irrational solutions **39. a.** 153 **b.** two irrational solutions **41. a.** 0 **b.** one rational solution **43. a.** 13 **b.** two irrational solutions **45. c.** two rational solutions **47. c.** two rational solutions **49.** length = 17 ft; width = 13 ft **51.** 80 s **53.** 4.12×10^{-3} moles per liter **55.** 10 ohms and 68 ohms **57.** 19 min **59.** width = 18 ft; length = 32 ft

61. $a = \pm 5\sqrt{3}$ **63.** $h = \frac{-11 \pm \sqrt{113}}{2}$

65. $v = -3 \pm \sqrt{6}$ **67.** $k = \frac{-1 \pm \sqrt{65}}{4}$ **69.** \$18.97

71. 82 ft^3 **73. b.** $x = \frac{15}{2}$ or $x = 18$ **75. b.** $x = \pm\sqrt{30}$

77. $\{0, 1, 2, 3, \ldots\}$

Section 9.4 Vocabulary Practice

1. I **2.** H **3.** E **4.** J **5.** D **6.** A **7.** F **8.** G **9.** B **10.** C

Section 9.4 Exercises

1. $\sqrt{11}i$ **3.** $2\sqrt{11}i$ **5.** $12i$ **7.** $\sqrt{55}i$ **9.** $\sqrt{73}$

11. $6i$ **15.** see chart below **17.** $13 + 10i$

19. $13 - 10i$ **21.** $-4 + 6i$ **23.** $-4 - 12i$

25. $120.059 + 0.005i$ volts **27. a.** $x = \pm 10i$

29. a. $m = \pm 2i$ **31.** $p = \pm\sqrt{5}i$ **33.** $v = \pm 5\sqrt{2}i$

35. $h = \pm 7i$ **37.** $c = \pm\sqrt{23}i$ **39.** $a = \pm 2\sqrt{2}i$

41. $k = \pm 7i$ **43.** $w = \pm\sqrt{11}i$ **45.** $x = -2 \pm 2i$

47. $p = -7 \pm 5i$ **49.** $z = 10 \pm 8i$ **51.** $a = -6 \pm 11i$

53. $v = -1 \pm \sqrt{7}i$ **55.** $w = 9 \pm \sqrt{5}i$

57. $m = -12 \pm \sqrt{23}i$ **59.** $x = \frac{3}{2} \pm \frac{\sqrt{3}}{2}i$

61. $p = -\frac{3}{2} \pm \frac{\sqrt{39}}{2}i$ **63.** $u = -\frac{1}{3} \pm \frac{\sqrt{26}}{3}i$

65. $r = \frac{3}{10} \pm \frac{\sqrt{51}}{10}i$ **67.** $n = -\frac{5}{4} \pm \frac{\sqrt{143}}{4}i$

69. $x = \pm\sqrt{3}i$ **71.** $b = \frac{11 \pm \sqrt{109}}{6}$

73. $n = -3$ or $n = -2$ **75.** $x = 0$ or $x = 4$

77. 2,571,400 home loans **79.** 98.1 yen **81. b.** $3\sqrt{2}i$

83. b. $x = \frac{3}{2} \pm \frac{3\sqrt{3}}{2}i$

85. **87.**

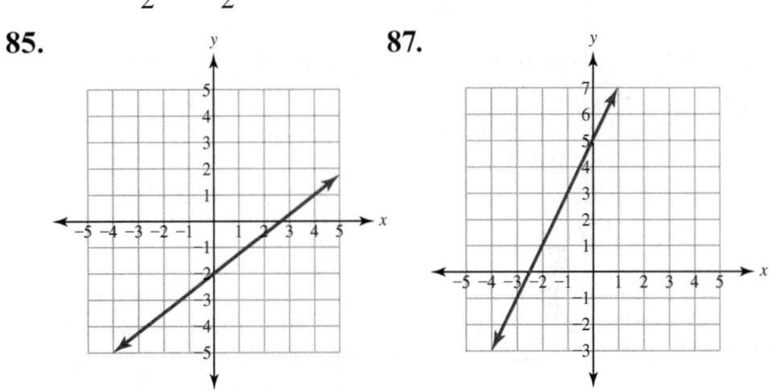

Section 9.5 Vocabulary Practice

1. F **2.** H **3.** B **4.** E **5.** D **6.** C **7.** J **8.** I **9.** A **10.** G

Section 9.5 Exercises

1. up **3. a.** $a = 1, b = 0, c = 2$ **b.** up

c. **d.** $(0, 2)$ **e.** $x = 0$

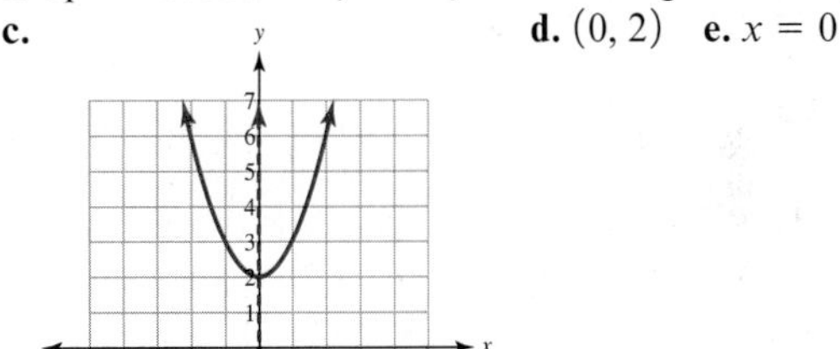

Chart for Section 9.4 exercise 15

	Complex numbers	Imaginary numbers	Real numbers	Rational numbers	Irrational numbers	Integers	Whole numbers
$\sqrt{-5}$	X	X					
$\sqrt{81}$	X		X	X		X	X
$\frac{2}{5}$	X		X	X			
$-6i$	X	X					
$3 + 2i$	X						
$\sqrt{3}$	X		X		X		

5. a. $a = -1, b = 0, c = 2$ **b.** down

c.

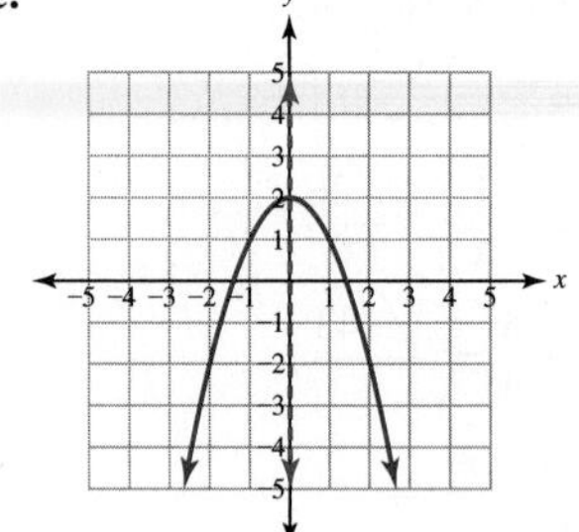

d. $(0, 2)$ **e.** $x = 0$

7. a. $a = 2, b = 0, c = 0$ **b.** up

c.

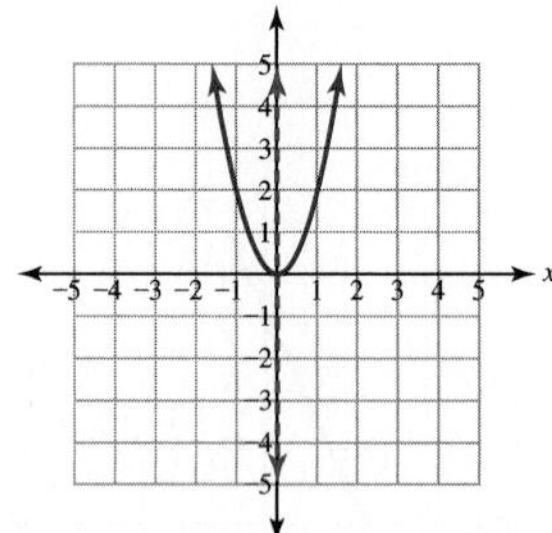

d. $(0, 0)$ **e.** $x = 0$

9. a. $a = \frac{1}{4}, b = 0, c = 0$ **b.** up

c.

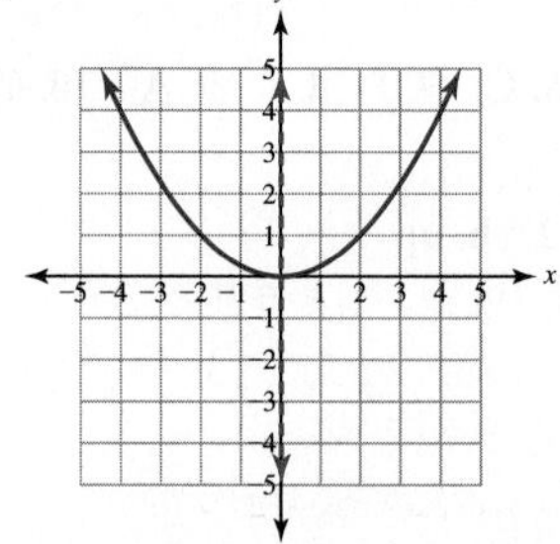

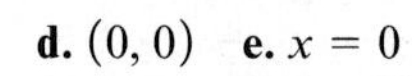

d. $(0, 0)$ **e.** $x = 0$

11. a. $a = 1, b = 4, c = -6$ **b.** up

c.

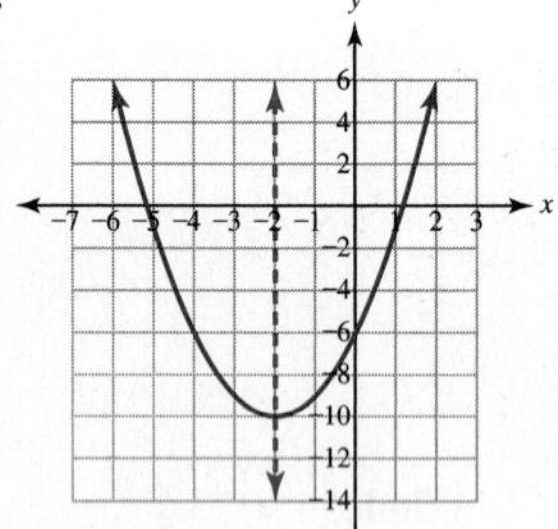

d. $(-2, -10)$ **e.** $x = -2$

13. a. $a = 1, b = -6, c = 9$ **b.** up

c.

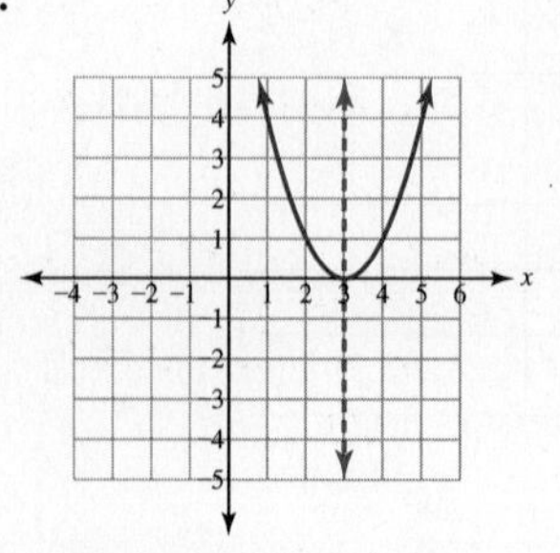

d. $(3, 0)$ **e.** $x = 3$

15. a. $a = 1, b = 6, c = 9$ **b.** up

c.

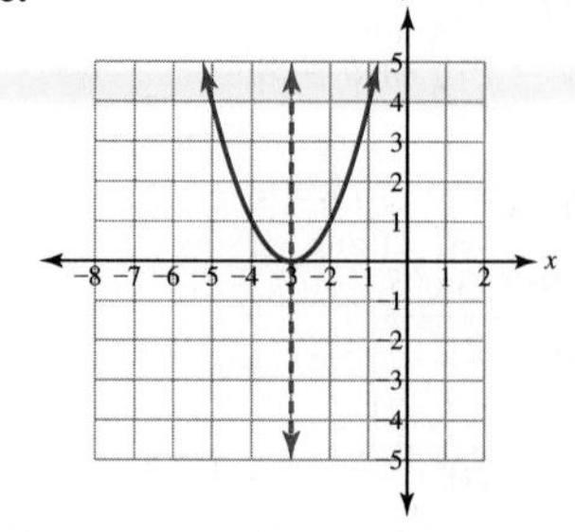

d. $(-3, 0)$ **e.** $x = -3$

17. a. $a = 1, b = 0, c = -7$ **b.** up

c.

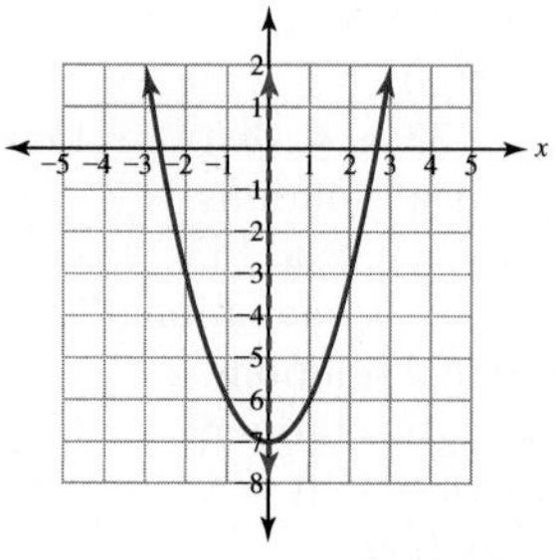

d. $(0, -7)$ **e.** $x = 0$

19. a. $a = -1, b = 2, c = 4$ **b.** down

c.

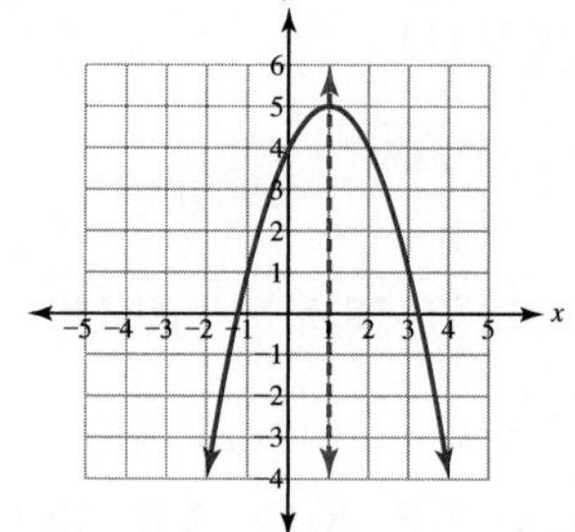

d. $(1, 5)$ **e.** $x = 1$

21. a. $a = 3, b = 0, c = 2$ **b.** up

c.

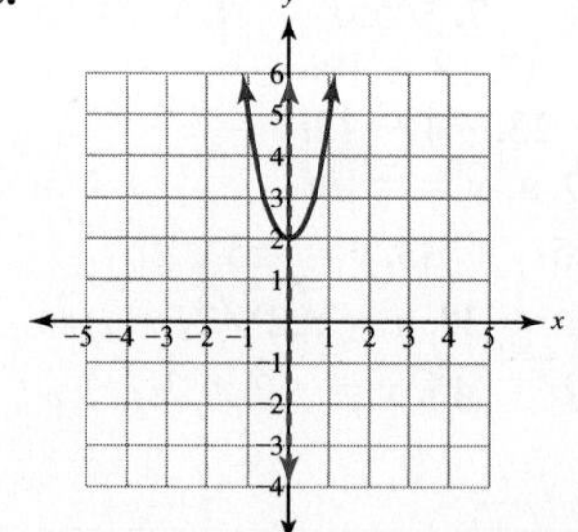

d. $(0, 2)$ **e.** $x = 0$

23. a. $a = 1, b = -2, c = 4$ **b.** up

c.

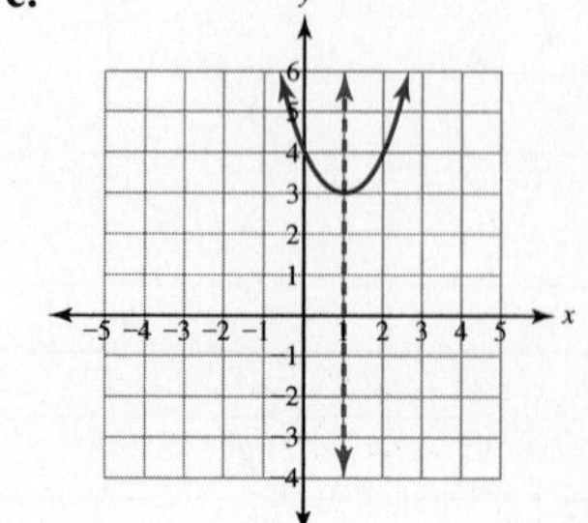

d. $(1, 3)$ **e.** $x = 1$

25. taller **27.** $(3, 0), (7, 0)$ **29.** $(-8, 0), \left(\frac{1}{2}, 0\right)$
31. $\left(-\frac{4}{3}, 0\right), \left(\frac{4}{3}, 0\right)$ **33.** $(-4, 0), (0, 0)$
35. $(5, 10)$ **37.** $(5, -10)$
39. $\left(\frac{7}{2}, -\frac{49}{4}\right)$ **41.** $(3, 0)$ **43.** $(-3, -12)$
45. $(3, -12)$ **47.** $(2, -7)$ **49.** $(4, 17)$
51. a. $(-2, 0), (8, 0)$ **b.** $(3, -25)$ **c.** $x = 3$
d.

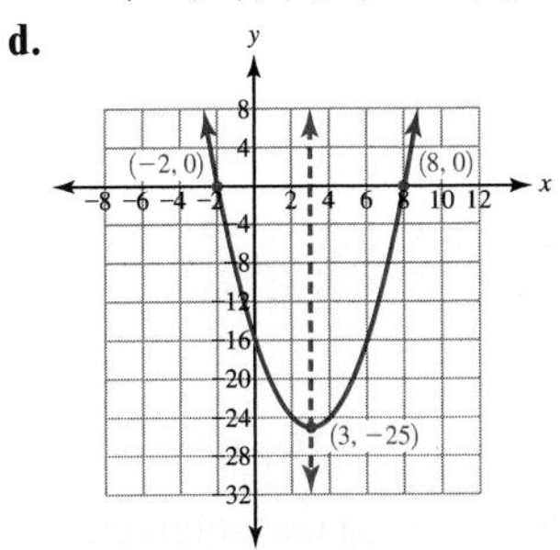

53. a. $(-3, 0), (3, 0)$ **b.** $(0, -9)$ **c.** $x = 0$
d.

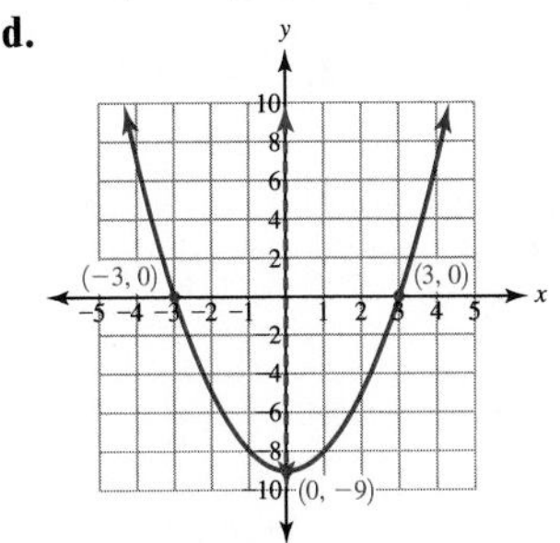

55. a. $(4, 0)$ **b.** $(4, 0)$ **c.** $x = 4$
d.

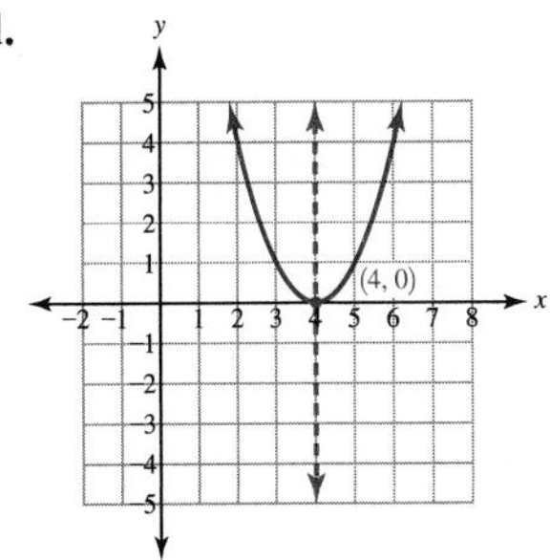

57. a. $\left(-\frac{5}{2}, 0\right), \left(\frac{1}{2}, 0\right)$ **b.** $(-1, -9)$ **c.** $x = -1$
d.

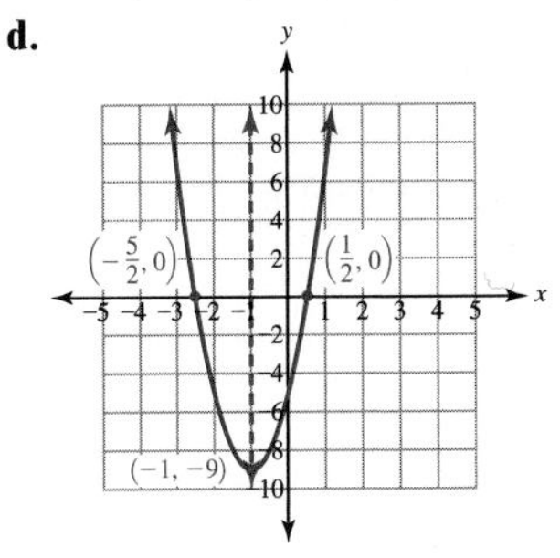

59. a. rim 120 in. above the floor; three-point line 237 in. from the center of the hoop **b.** 266 in. **c.** 22.2 ft
61. 17,170 gal acid solution; 12,830 gal of water

63. a.

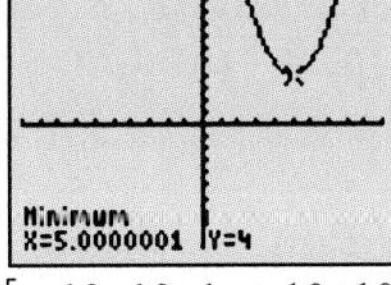

$[-10, 10, 1; -10, 10, 1]$

b. $(5, 4)$

65. a.

$[-10, 10, 1; -20, 10, 1]$

b. $(-3, -7)$

67. b.

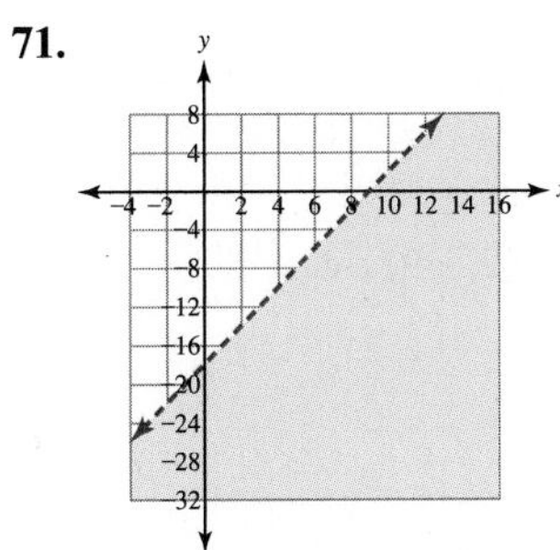

69. b. $(-6, 0); (4, 0)$

71.

73. $y = 2x - 4$

Chapter 9 Review Exercises

1. a. $x = \pm 6$ **2. a.** $z = \pm 3\sqrt{5}$ **3. a.** $v = \pm\sqrt{7}$
4. a. $k = \pm 6\sqrt{3}$ **5. a.** $u = \pm 2\sqrt{5}$ **6. a.** $x = \pm 7$
7. 108 in. **8.** 5 s **10.** 81 **11.** 144
12. $x = -3$ or $x = -1$ **13.** $z = -5$ or $z = 9$
14. $p = -3 \pm \sqrt{46}$ **15.** $n = 6 \pm 3\sqrt{10}$
16. $x = 5 \pm \sqrt{33}$ **17.** $x = -8$ or $x = -2$
18. $x = \dfrac{-1 \pm \sqrt{43}}{2}$ **19.** $x = \dfrac{-7 \pm \sqrt{109}}{10}$
20. $x = \dfrac{15 \pm \sqrt{233}}{2}$ **21.** discriminant 0, one rational solution **22.** discriminant 9, two rational solutions
23. discriminant 57, two irrational solutions
24. discriminant -203, two nonreal solutions
27. $x = \dfrac{-b \pm \sqrt{b^2 - 4ac}}{2a}$ **28.** a, e **29.** $a = 0$ or $a = -9$
30. $x = \pm 4\sqrt{3}$ **31.** $x = -3 \pm \sqrt{2}$ **32.** $p = -\frac{7}{2}$ or $p = 1$
33. $x = \pm\sqrt{11}$ **34.** $n = \pm\sqrt{3}$ **35.** $c = 2$ or $c = 10$
36. 0.07 s and 0.29 s **37.** $4i$ **38.** $4\sqrt{3}i$ **39.** $7 - 6i$
40. $-9 + 5i$ **41.** c, e **42. a.** yes **b.** yes **c.** no
43. a. $n = \pm 6i$ **44. a.** $h = \pm 9i$ **45. a.** $x = \pm 3\sqrt{2}i$
46. $w = -5 \pm \sqrt{5}i$ **47.** $x = -3 \pm 2\sqrt{3}i$

48. $x = -\frac{3}{2} \pm \frac{\sqrt{11}}{2}i$ **49.** $x = -\frac{1}{10} \pm \frac{\sqrt{239}}{10}i$

50. c, f **51.** see chart below

52. a. $a = 1, b = 0, c = -10$

b.

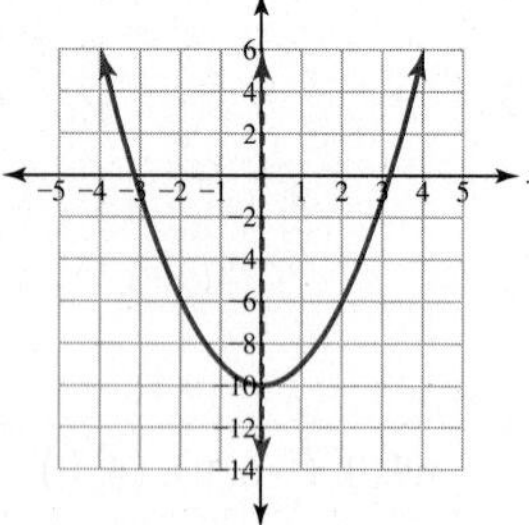

c. $(0, -10)$ **d.** $x = 0$

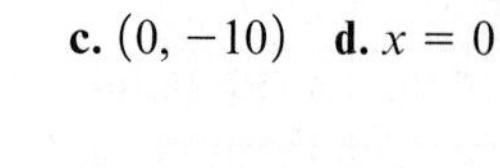

53. a. $a = -1, b = 0, c = -4$

b.

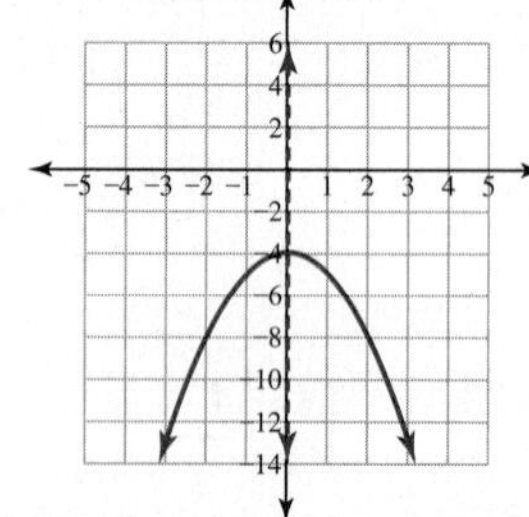

c. $(0, -4)$ **d.** $x = 0$

54. a. $a = 1, b = -4, c = -5$

b.

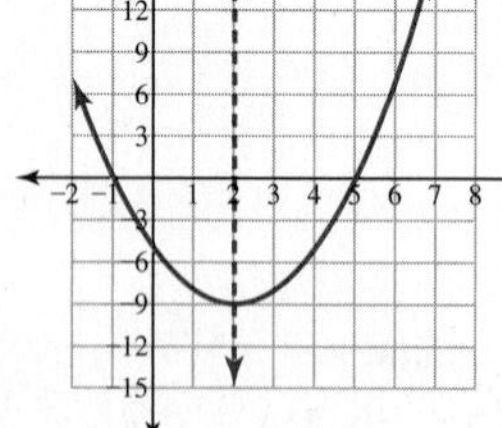

c. $(2, -9)$ **d.** $x = 2$

55. b, c **56.** d **57.** $(-7, 0), (8, 0)$ **58.** $\left(\frac{5}{2}, -\frac{17}{2}\right)$

59. a. $(-7, 0), (-1, 0)$ **b.** $(-4, -9)$ **c.** $x = -4$

d.

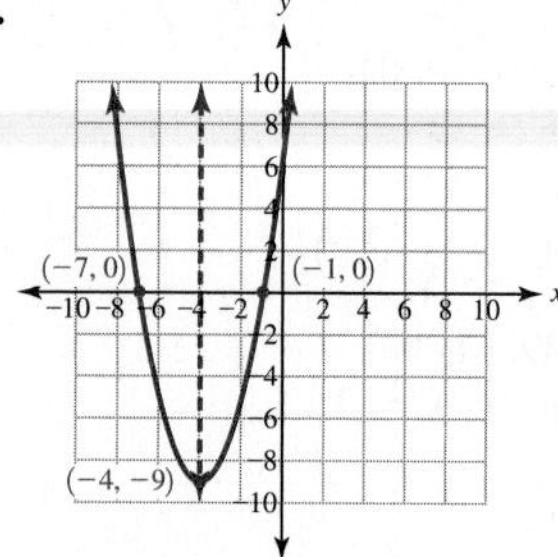

Chapter 9 Test

1. a. $x^2 = 9$ **b.** $x = \pm 3$ **2. a.** $z^2 = 3$ **b.** $z = \pm\sqrt{3}$

3. a. $u^2 = -7$ **b.** $u = \pm\sqrt{7}i$ **4.** $x = -4 \pm \sqrt{29}$

5. $n = -5$ or $n = 3$ **6.** $x = 1$ or $x = 6$

7. $x = \frac{-1 \pm \sqrt{193}}{12}$ **8.** discriminant 169, two rational solutions **9.** discriminant 109, two irrational solutions

10. discriminant -60, two nonreal solutions

11. a. $a = 1, b = 8, c = 12$

b.

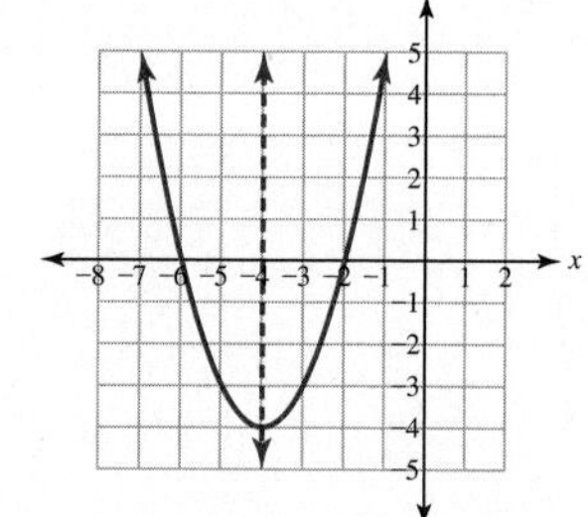

c. $(-4, -4)$ **d.** $x = -4$

12. a. $(-2, 0); (-8, 0)$ **b.** $(-5, -9)$ **c.** $x = -5$

d.

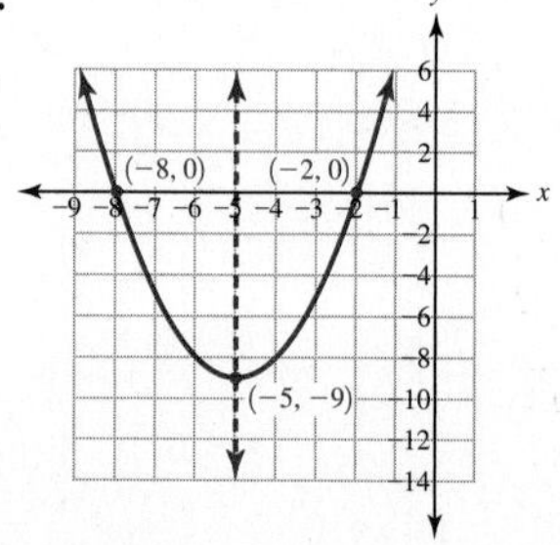

13. 15.6 ft

Chart for Chapter 9 Review exercise 51

	Complex numbers	Imaginary numbers	Real numbers	Rational numbers	Irrational numbers	Integers	Whole numbers
8	X		X	X		X	X
-10	X		X	X		X	
$-\frac{1}{3}$	X		X	X			
0.21	X		X	X			
$\sqrt{5}$	X		X		X		
0	X		X	X		X	X
$3i$	X	X					
$\sqrt{-7}$	X	X					
$9 - 2i$	X						

Chapters 7–9 Cumulative Review

1. $2x^3y\sqrt{3xy}$ **2.** $\dfrac{x+5}{x-4}$ **3.** $\dfrac{3}{p+5}$ **4.** 2 **5.** $2\sqrt{2}$

6. $\dfrac{x^2+5x-4}{(x+1)(x+5)}$ **7.** $\dfrac{8c^2+24c-5}{(c-3)(c+3)}$ **8.** $15x\sqrt{2yz}$

9. $\dfrac{x+9}{x+7}$ **10.** $\dfrac{-24w+18}{w^3}$ **11.** $\dfrac{3y\sqrt{2y}}{5x}$

12. $\dfrac{x\sqrt{6x}+5\sqrt{6x}}{6x}$ **13.** $\dfrac{k\sqrt{k}-2k}{k-4}$ **14.** $\dfrac{7a^3b}{10c^7}$

15. $2x^2y^2\sqrt[3]{5y}$ **16.** $4i$ **17.** $7\sqrt{2}i$

18. a. $x = 3$ or $x = 8$ **19. a.** $x = 36$ **20. a.** $x = 216$

21. a. $p = 361$ **22. a.** $a = \pm 6$ **23. a.** $n = 7$

24. a. $x = 2$ **25. a.** $x = \dfrac{-3 \pm \sqrt{41}}{2}$ **26. a.** $x = \dfrac{40}{3}$

27. a. $x = 9$ **28. a.** $p = 0$ or $p = 4$

29. a. $x = -4$ or $x = 4$ **30.** $x = \dfrac{9 \pm \sqrt{53}}{2}$

31. $x = 6 \pm \sqrt{51}$ **32. a.** -159 **b.** two nonreal solutions

33. 2.2 m^2 **34.** 17 in. **35.** length 15 in.; width 8 in.

36. length 19 in.; width 5 in. **37.** see chart below

38. a.

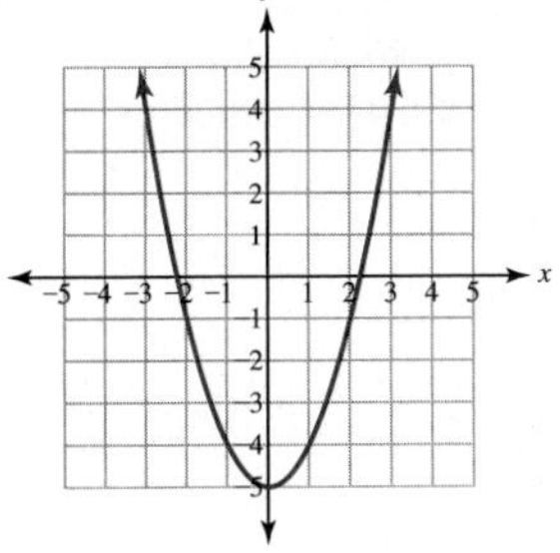

b. $(0, -5)$

39. $(-7, 0)$, $(2, 0)$ **40.** 2 hr 56 min **41.** 740 students

Chapters 1–9 Cumulative Review

2. -20 **3.** $\dfrac{29}{12}$ **4.** $-2x - 26$ **5.** 96 in. **7.** 50.24 cm^2

8. 96 m^2 **10.** $20x - 24 = 5x + 60$ **11.** $y = \dfrac{5}{9}x - \dfrac{5}{3}$

12. $x = 2$ **13.** $x = 50$ **14.** no solution **15.** $p = 0$

16. a. $x > -4$ **b.** 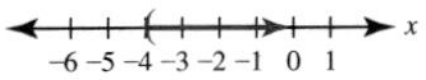

17. a. $x \le -13$ **b.** 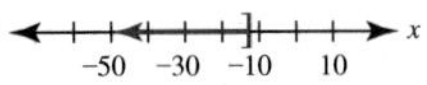

18. x-intercept $(-3, 0)$; y-intercept $\left(0, -\dfrac{27}{4}\right)$

19.

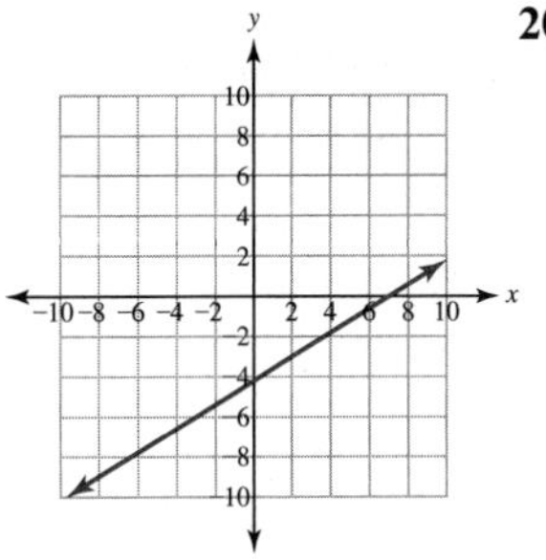

20.

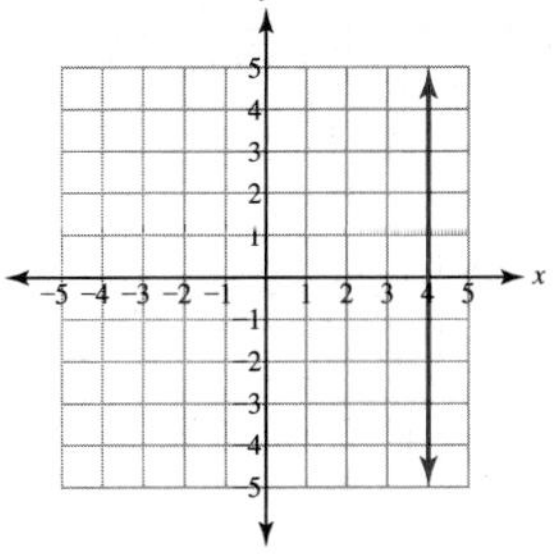

21.

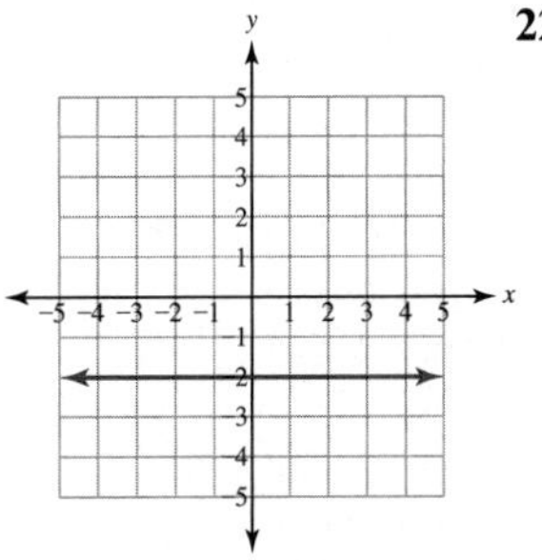

22.

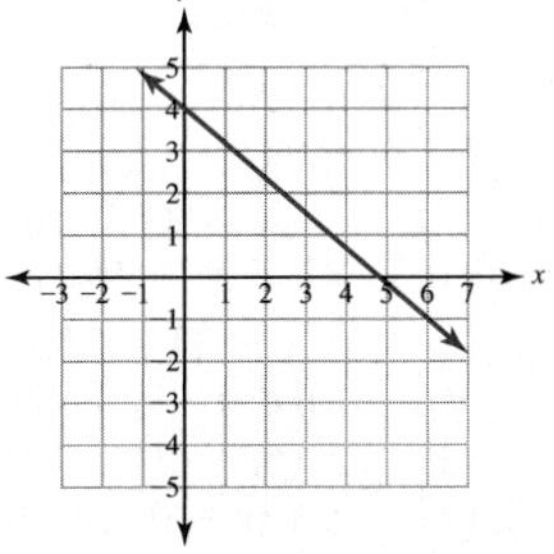

23.

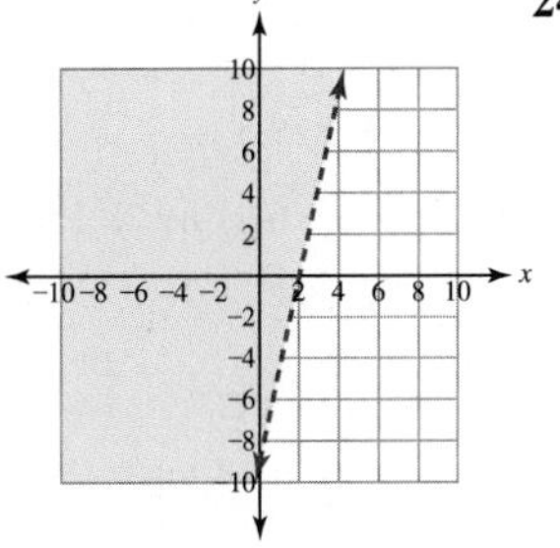

24.

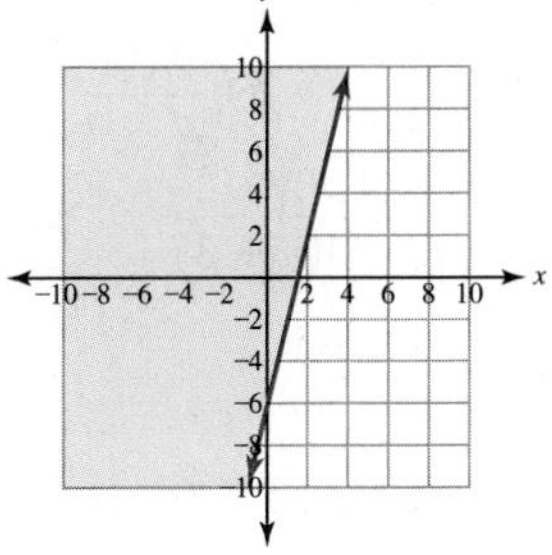

25. $\dfrac{7}{3}$ **26.** $\dfrac{7}{3}$ **27.** $y = 4x + 1$

28. 7 **29.** $y = -\dfrac{4}{3}x - 2$

30. $y = \left(\dfrac{3679.6 \text{ patients}}{1 \text{ year}}\right)x + 14{,}275$ patients; 143,061 patients

32. 42 **33.** $(4, 5)$ **34.** $(-2, 7)$

35. $\left(3, \dfrac{1}{2}\right)$ **36.** no solution

Chart for Chapters 7–9 Cumulative Review exercise 37

	Complex numbers	Imaginary numbers	Real numbers	Rational numbers	Irrational numbers	Integers	Whole numbers
$\sqrt{5}$	X		X		X		
5	X		X	X		X	X
$\sqrt{-5}$	X	X					
0.5	X		X	X			
$5i$	X	X					
$\dfrac{1}{5}$	X		X	X			

37. 

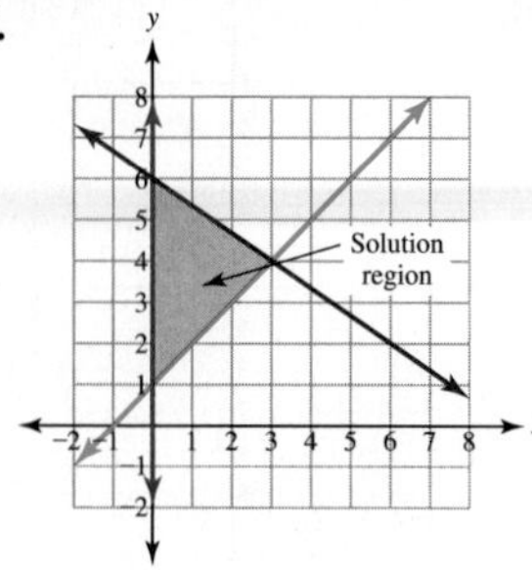

38. $x^{12}y^3$ **39.** a^6b^8 **40.** u^5 **41.** $\dfrac{3}{8n^9}$ **42.** 7×10^6 kg
43. 6.1×10^{-4} m **44.** 3.2×10^{12} **45.** 4×10^9
46. 9.3×10^5 kg **47.** $-2x^2 + 14x - 8$
48. $18x^2 + 39x - 45$ **49.** $10x^3 - x^2 - 11x - 4$
50. $7x^2 + 6x - 4$ **51.** $(x - 5)$ R(-28)
52. $(x - 3)(x + 3)$ **53.** $(a - 6)(a - 3)$
54. $n(n - 5)(n + 1)$ **55.** $(2x + 3)^2$
56. $(3a + 5)(5c - 7)$ **57.** $2(4n + p - 3w + 5)$
58. $(2x + 5)(3x - 1)$ **59.** $x = -9$ or $x = 1$
60. $c = 0$ or $c = 4$ or $c = -4$ **61.** $a = 0$ or $a = \dfrac{1}{4}$
62. $\dfrac{a + 3}{a + 1}$ **63.** $\dfrac{3}{d - 2}$ **64.** $\dfrac{2x + 1}{(x - 1)(x + 1)}$
65. $\dfrac{n - 9}{(n - 5)(n + 3)}$ **66.** no solution **67.** $4ny^3\sqrt{15nx}$
68. $5\sqrt{2x}$ **69.** $\dfrac{5a\sqrt{2a}}{3b}$ **70.** $2x - 5$ **71.** $3a^2b^2\sqrt[3]{2a}$
72. $x^{\frac{5}{8}}$ **73.** $a^{\frac{4}{5}}$ **74.** $\sqrt[3]{n^2}$ **75.** $x = 5$ or $x = -5$
76. $x = -4 \pm \sqrt{21}$ **77.** $x = \dfrac{-5 \pm \sqrt{33}}{4}$ **78.** $\sqrt{15}i$
79. $2\sqrt{5}i$ **80.** $x = \pm\sqrt{17}i$
81.

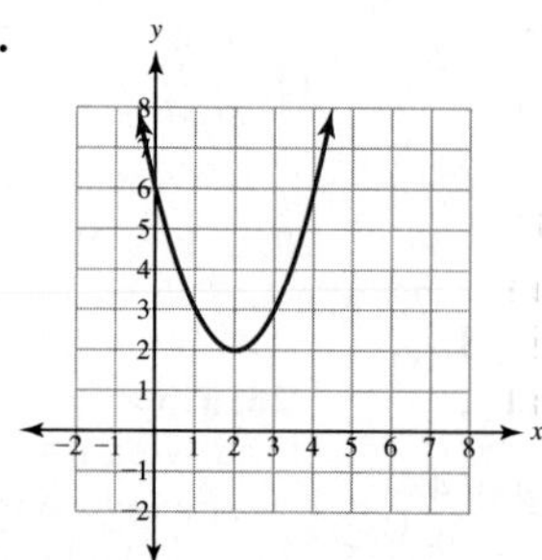

82. $(3, 4)$ **83.** 26,625 young adults **84.** \$793
85. 70 lengths **86.** \$48.90 **87.** 60 hr **88.** \$9850
89. 1667 lb 15% nickel steel, 833 lb 12% nickel steel
90. (1500 items, \$97,500) **91.** length 7 in., width 5 in.
92. 9.7 ft **93.** 25% decrease **94.** 12.9%
95. 408 intravenous drug users **96.** 30.5 in.^{2}
97. 785 cm^2 **98.** 456.4 in.^{3} **99.** 4364 illnesses

Index

A
Abbreviations of units, 79
Absolute value of number, 7, 37
ac method of factoring trinomials
quadratic trinomials, 435–438, **437**
trinomials that are not quadratic, 439–440
Adding
complex numbers, 618–619
fractions with same denominator, **19**
fractions with unlike denominators, **20**–21
linear equations, 313–314
measurements, 81–82
measurements in scientific notation, **380**
polynomial expressions, 388–390
rational expressions with different denominators, 500–**502**
rational expressions with same denominator, 489, **490**
rational numbers, 488–489
signed numbers, 7–9
square roots, 553–554, 555
Addition method. *See* Elimination method
Addition property of equality, **110**–111, 135
Addition property of inequality, **154**
Addition rule of exponents, **362**
Additive identity, 36
Algebra, derivation of term, 46
Algebraic expressions, 46–54
simplifying, 46, 47, 50–52
al-Khwārizmī, Muhammad ibn Mūsā, 46
Angle, right, 96
Anxiety, math, 473
Application problems. *See also* Problem solving
cost problems, 333–336
five steps for solving, **55**
involving fractions, 60
involving radical equations in two variables, 573–574
involving scientific notation, 381–382
linear models describing, 212–215
mixture problems, 327–333
rate of work, 524–526
retail, 72–73
rewriting numbers including word names in place value notation, **57**
solving using elimination, 321–323
solving using equation in one variable, 116–120, 129–130
solving using equation in one variable with variables on both sides, 140–142
solving using formula, 147–149
solving using inequality in one variable, 159–160
solving using proportions, 118–120, 526–527
solving using quadratic equations, 593–594
solving using quadratic formula, 609–612
solving using substitution, 307–309
solving using systems of linear equations, 289–291, 307–309, 321–323, 327–344
solving using zero product property, 461–463
using Pythagorean theorem to solve, 594–595
using roots, 97
using slope, 199, 202–203
Areas
of circle, **92**
formulas, 85–86, **87**, **92**
of rectangle, 85, **87**, 398, 611
of square, 95
of trapezoid, 90
of triangle, 85–86, **87**, 100, 466
writing polynomial expressions in one variable representing, 398–400
Arithmetic, fundamental theorem of, 423
Arithmetic mean, 90
Arithmetic review
decimal numbers, 25–28
evaluating expressions, 2–7, 11–12
fractions, 14–25
order, of numbers on number line, 30
order of operations, **2**–4, 11–12
percent, 28–30, **29**
pretest, 1
signed numbers, 7–13
Associative property
of addition, **50**
of multiplication, **50**
Auditory learner, 195, 211
Average rate of change, 196–198
slope as, 198–205, **199**, 213
Axes of Cartesian coordinate system, 170–172
Axis of symmetry of parabola, 624, 625, 629

B
Bakhshali Manuscript, 99
Base, 2, 37
of exponential expressions, 362
of exponential expressions, in power of ten, **375**–376
of exponential expressions, rules for, 366–367
many bases rule for product, **366**, 367
many bases rule for quotient, **366**–367
of triangle, 86
Binomials
multiplying, 390, 394–396
FOIL, 396
squaring, 397
Boundary line of graph of linear inequality
dashed versus solid, 249
finding equation of, 252–255
horizontal, 252
vertical, 251–252
Boundary value, 153
Bounded solution region, 347
Brackets, set, 36
Break even, **140**–142
Break-even point, writing system of linear equations for, **290**–291, **307**–308

C
Calculators. *See* Using technology
Cartesian coordinate system, 170
graphing ordered pair on, 170–172
Checking solution of equation, 111
for reasonability, 55, A-1–A-6
Circle, formulas for, **92**–93
Circumference of circle, **92**
Clearing equations of fractions or decimals, 136–**138**
in inequalities in one variable, 157–158
in rational equations, 518, 519, 520
Coefficients
defined, 46, 386
lead, 387, **388**, 431, 450, 626–627

Page numbers in bold indicate location of terms when appearing in Definition, Summary, or Procedure boxes.

Coinciding lines, 286, **287**
  recognizing using elimination, 320
  recognizing using substitution, 306
Combined units, 381–382
Combining like radicals, 553–557
  simplifying before, 554–555
Combining like terms, 47, 115–116, 388–390, 391
Common factors, 422
  greatest (GCF), 422–425
Commutative property
  of addition, **50**
  of multiplication, **50**
Completing the square, 598–600
  solving quadratic equations by, 600–603, **602**, 620
  solving quadratic equations with irrational solutions by, 602–603
  visual models for, 598–599
Complex numbers, **617**–619
  adding, 618–619
  set of, 617–618
  simplifying, 618–619
  subtracting, 619
Complex rational expressions, 507–517
  simplifying, 508–514, **510**, **512**
Composite number, 36
Concept map, 248
Conditional equations, 138
Cone, volume of, **92**, 615
  frustrum, 540
Conjugates, 559
  rationalizing denominators that are a sum or difference using, 564–565
Constant of proportionality
  for direct variation, 532–533
  for inverse variation, 534–535
Constants, 46, 386
Constraints, 159
  inequalities representing, 255–258
  systems of linear inequalities representing, 349–351
Contradiction (linear equation in one variable), 138
Conversion factors, 80–81
Coordinate
  x, 171
  y, 171
Coordinate system. *See* Cartesian coordinate system
Correspondence, 263
Cube of a number, 96
Cube roots, 96, 577
Cubes
  difference of cubes pattern, **446**
  perfect, **445**, 578
  sum of cubes pattern, **446**
Cylinder, volume of, **92**, 93–94

D

Decimal numbers
  clearing equations of, 137–**138**
  clearing inequalities of, 158
  in expanded notation, 25
  in place value notation, 25
  repeating, 26–27, 91–92
  repeating, rewriting as fraction, 313–314
  rewriting fractions as, 26, 60
  rounding to given place value, **27**–28
  solving equations in one variable with, 115–120, **116**
  terminating, 91–92
  writing percent as, 28, 67–68
Degree of polynomial expression, 387, **388**
Denominators, 38
  adding fractions with same, **19**
  adding fractions with unlike, **20**–21
  adding rational expressions with different, 500–**502**
  adding rational expressions with same, 489, **490**
  building equivalent rational expression with different, 498–**499**
  least common, 20–21, 495–497
  least common, using prime factorization to find, **496**–497
  proportions with variable in, 526–527
  rationalizing, 563–564
  rationalizing, that are a sum or difference, 564–565
  rewriting fraction as equivalent fraction with different, **19**
  subtracting fractions with same, **19**
  subtracting fractions with unlike, **20**–21
  subtracting rational expressions with different, **502**–503
  subtracting rational expressions with same, **490**–491
Dependent variable, 267
Descartes, René, 170
Descending order, polynomial in one variable in, 387
Diameter of circle, **92**
Difference
  defined, 2
  rationalizing denominators that are a, 564–565
  of squares, 396
Difference of cubes pattern, **446**
Difference of squares pattern, **398**, 444–**445**
Direct variation, **532**–534, 536
  graph of equation representing, 532, 533
Discount, 72–73
Discriminant
  in quadratic formula, 608
  of quadratic trinomials, 438
  using to describe solutions of quadratic equations, 608–**609**
Discriminant test for quadratic trinomials, **438**–439
Distance formula, 82–83, 147, 337
Distributive property, 49, **50**
  multiplying polynomial expressions using, 390–391, 395–396
  multiplying square roots using, 558–559
  using in solving equation in one variable, 127–128
Dividend, 18, 405
Dividing
  exponential expressions, 364–365
  fractions, **18**
  measurements in scientific notation, 378–**379**
  notation for, 36
  polynomials, 404–414
  polynomials, remainders after, 409–411
  rational expressions, 483–485, **484**
  signed numbers, 9–11
  sign rules, **10**
  square roots, **564**
Division
  long, 405–411
  rewriting as multiplication by reciprocal of divisor, 483
  by zero, 5, 36
Division property of equality, **110**, 111, 113
Division property of inequality, **154**
Divisor, 18, 405
Domain
  of function, 266–267
  of relation, 266–267

E

Elements of set, 36
Elimination method
  applications of, 321–323
  recognizing coinciding lines using, 320
  recognizing parallel lines using, 319–320
  solving systems of linear equations using, 315–327, **316**–**317**
  solving systems with infinitely many solutions using, 320
  solving systems with no solution using, 319–320
Ellipsis, 36
Equality, properties of, **110**–115
  solving equation for a variable using, 145–146

solving equation in one variable using, 110–115, **113**, 124–126
Equations, 49. *See also* Linear equations in two variables; Quadratic equations; Systems of linear equations
clearing fractions or decimals, 136–**138**
conditional, 138
with decimal numbers, solving, 115–120
of lines, 215–220, 230–241
of lines, finding slope using, 204, 205
of lines of best fit in slope-intercept form, writing, **234**–236
in one variable, simplifying, 126–129
proportions in one variable, solving, 118–120
radical, 568–574
solutions of, 110
solving, in one variable, 110–133, **113**, **116**, **127**, **135**
solving, in one variable with variables on both sides, 133–145, **135**
solving for a variable, 145–146
solving polynomial, using zero product property, **457**–463
solving rational, 518–**523**
solving using properties of equality, **110**–115, 124–126
Equivalent fractions, 497–499
building equivalent rational expression with different denominator, 498–**499**
rewriting fraction as, with different denominator, **19**
Escape velocity, 576
Estimation, 55, A-1
Evaluating
exponential expressions, 37
expressions, 2–7, 11–12
expressions, including fractions, 22
expressions, order of operations in, **2**–4, **41**–43
functions, 267–269
order of operations and, **2**–4, 11–12
perfect powers, 578
Expanded notation, 25
Exponential expressions, 2
evaluating, 37
multiplying, 363–364
power of ten, **375**–376
simplifying using exponent rules, 362–368, 371–372
with variables that are perfect squares, **550**
writing in radical notation, 581
Exponential notation, 362
radicals in, **580**–581
Exponent rules, 362–368
Exponents, 2, 37
addition rule of, **362**
negative, 369–371, **370**, **582**
perfect powers, 578
on power of ten, **375**–376
power rule of, **365**
power rule of, simplifying rational exponents using, 582
product rule of, **363**–364
quotient rule of, **364**–365
quotient rule of, simplifying rational exponents using, 582, 583
raising measurement in scientific notation to a power, **379**
rational, 580–581
rational, simplifying using exponent rules, 581–583
subtraction rule of, **362**
zero as, 368–**369**
Expressions. *See also* Polynomial expressions; Radical expressions; Rational expressions
evaluating, 2–7, 11–12
evaluating, order of operations in, **2**–4, **41**–43
exponential, 362–363
including fractions, evaluating, 22
numerical, 46
substitution, 299
Extraneous information, 55
Extraneous solutions
to radical equations with square root, 570–572
to rational equations, 521–523
Extrapolation, 235

F

Factoring. *See also* Prime factorization
ac method, 435–440, **437**
completely, **449**–454
defined, 424
factoring out greatest common factor, 424–425
guess and check method, 431–434, **432**
polynomials, 424–430
polynomials, by grouping, 425–428, **427**
polynomials, methods for, **446**, **449**
quadratic trinomials, 431–**438**
to simplify rational expressions, 475–478, **476**
trinomials that are not quadratic, 439–440, 444
using difference of cubes pattern, **446**
using difference of squares pattern, 444–**445**
using perfect square trinomial patterns, **443**–444
using sum of cubes pattern, **446**
zero product property and, 459–461
Factors, 36
common, 422
greatest common (GCF), 422–425
perfect square factor of large number, finding, 554
Farad (unit of capacitance), 543
Feasible region
for linear inequalities in two variables, 256–258
for systems of linear inequalities, 350–351
Five steps to problem solving, **55**
using, 56–61, 68–73
FOIL, 396
Formulas, 82–91
for approximating square root of nonperfect square, 99
area of circle, **92**
area of rectangle, 85, **87**, 168, 398, 611
area of trapezoid, 90
area of triangle, 85–86, **87**, 100, 399–400, 466
arithmetic mean or average, 90
baseball statistics, 393
body surface area of human adult, 637
capacitance, 543
carpet needed to cover staircase and landing, 615
Celsius-Fahrenheit temperature conversion, 90
circumference of circle, **92**
Clark's rule for converting adult dose to child dose of medication, 90
cost of insurance, 540
daily water intake for dairy cow, 284
density, 89, 147, 456, 576
diameter of circle, **92**
direct variation in, 536
distance, 82–83, 147, 337
distance, for falling object, 593, 596, 634
distance in nautical miles from person in boat to horizon, 100
final velocity, 147
finding range of reasonable answers when using, A-2
Flesch-Kincaid Grade Level Readability, 106
flow rate of fluid through intravenous drip, 539
fractional excretion of sodium, 540
frequency of light, 539
Fried's rule for converting adult dose to child dose of medication, 90
future value, 148
glomular filtration rate by kidney, 540
Harris-Benedict, 89
Heron's formula, 100
hydroplaning, 100
including π, **92**–94

Formulas (*continued*)
intensity of X-ray radiation, 90
internal diameter of cuffed endotracheal tube, 584
inverse variation in, 536
Khine, 283
Mosteller's formula, 107
NFL quarterback passing rating, 632
P/E ratio, 90
percent markup, 77
percent of X-rays absorbed by antiscatter grid, 90
perimeter of rectangle, 84, **87**, 336, 398, 611
perimeter of triangle, 84, **87**, 399
quadratic, **606**–612, 620–621, **A-8**
quadratic, deriving, A-7–A-8
radius of circle, **92**
range in nautical miles of VHF marine antenna, 101
resistance, 516, 539
sales commission, 166
simple interest, 83–84
slope, **199**
SMOG (Simple Measure of Gobbledygook) Readability, 576
solving for a variable, 147–149
solving systems of linear equations involving, 336–339
speed in knots of water wave in deep water, 100
speed of car before braking, 97
speed relative to tire diameter, 166
square root formula to allocate votes per country, 100
surface area of cylinder, 641
volume of cone, **92**
volume of cylinder, **92**, 356
volume of frustrum of cone, 540
volume of lumber in board feet, 90, 151
volume of rectangular solid, **87**, 536
volume of sphere, **92**, 641
volume of water in gallons per minute delivered by fire hose, 101
weight-loss program, 107
Fraction bar, 4
Fraction of job completed, **524**–526
Fractions, 14–25. *See also* Rational expressions
adding, with same denominator, **19**
adding, with unlike denominators, **20**–21
adding polynomial expressions with, 389–390
application problems with, 60
clearing from equations, 136–**138**
clearing from inequalities, 157–158
clearing from rational equations, 518, 519, 520
dividing, **18**
equal to 1, simplifying, **475**
equivalent, 497–499
equivalent rational expression with different denominator, building, 498–**499**
improper, 16, 26, 48
mixed numbers, 16
multiplying, **17**–18
ordering, 30
order of operations, 22
problem solving with, 60–61
proper, 16, 48
rewriting as decimal numbers, 26, 60
rewriting as equivalent fraction with different denominator, **19**
rewriting repeating decimal as, 313–314
simplifying into lowest terms, 14–**15**, 38–39
subtracting, with same denominator, **19**
subtracting, with unlike denominators, **20**–21
writing percent as, 67–68
Function rule, 268
Functions
defined, 263
domain of, 266–267
evaluating, 267–269
graphing, 269–270
notation, 268
range of, 266–267
as relations, 263–264
vertical line test of, 264–**265**
Fundamental theorem of arithmetic, 423
Future value, 148

G

GCF. *See* Greatest common factor (GCF)
Grams, 78
Graphing. *See also* Number line graph
functions, 269–270
horizontal lines, 190
intercept graphing of linear equations, **185**–187
linear equations in two variables, 174–178
linear equations with isolated variable, 187–**189**
linear inequalities, 249–252, **251**, **345**
ordered pair, 170–172
quadratic equation in two variables, 624–**629**
slope-intercept method, 220–224, **221**, 287–289
solving system of linear equations by, 287–289, 290–291
systems of linear inequalities, 345–351
vertical lines, 190
Graphing calculator, using. *See* Using technology
Graphs
of equation representing direct variation, 532, 533
of equation representing inverse variation, 534
evaluating function using, 269
of feasible region, 256–258, 350–351
finding slope of line using, 200, 201, 203, 205
of functions, vertical line test of, 264–**265**
of linear model, extrapolation using, 235
of parabola, 624–629
of system of two linear equations, 286–**287**
writing linear inequality represented by, 252–**255**
Greater than symbol (>), 30, 249
Greatest common factor (GCF)
factoring out, 424–425
finding, 422–424
using prime factorization to find, **423**–424
Grouping, factoring four-term polynomials by, 425–428, **427**
Grouping symbols, 2–4
order of operations and, 42–43
Guess and check method of factoring trinomials, 431–434, **432**

H

Height of triangle, 86
Horizontal lines, 190–191
as boundary line in graphs of linear inequalities, 252
equation in slope-intercept form of, 219–220
graphing, 190
slope of, 206
Hypotenuse, 96, 594

I

Identifying
contradictions, 138
identities, 138–139
intercepts of line, 185
solution region of linear inequality, 250–251, 252, 256, 257
solutions of systems of linear inequalities, 348–349
systems of linear equations with infinitely many solutions, 286, 287, 306, 320
systems of linear equations with no solution, 286, 287, 305–306, 319–320
Identities
additive, 36
multiplicative, 36
Imaginary numbers, 616–**617**

Improper fractions, 16, 48
rewriting as decimal number, 26
Independent variable, 267
Index
for cube root, 96
product rule of radicals for radical with index, *n*, **578**
of square root, 548
Input values, 263
domain of relation or function as set of, 266–267
Inspection, solving by, 112
Integers, 37, **40**, **617**, 618
set of, 37
Intercepts. *See also* Slope-intercept form
intercept graphing of linear equations, **185**–187
x-intercepts, 185–187, 627–629
y-intercepts, 185–187, 212–213, 215
Interest, simple, 83–84
International System of Units (S.I.), 78–79
abbreviations, 79
prefixes, 79
relationship to U.S. customary units, 80
Inverse variation, **534**–536
graph of equation representing, 534
Irrational numbers, **40**, 92–94, **617**
as subset of real numbers and complex numbers, 617, 618
Irrational solutions to quadratic equations, 602–603, 608, 609
Isolating the variable, 110–113
using properties of inequality, **154**

K

Kinesthetic learner, 195, 211
Kitab al-Jabr wa-l-Muqabala (al-Khwārizmī), 46

L

LCD. *See* Least common denominator (LCD)
Lead coefficient, 387, **388**
of perfect square trinomial, 443
of quadratic equations in two variables, 626–627
of quadratic trinomial, 431, 450
Leading zero, 25
Learning preferences, 169, 195, 211, 229, 248
Least common denominator (LCD), 20
finding, 20–21
of rational expressions, 495–497
using prime factorization to find, **496**–497
Least common multiple
defined, 491
using prime factorization to find, 491–493, **492**
Legs of triangle, 594
right triangle, 96
Length, units of, 79
Less than symbol (<), 30, 249
Like radicals
combining, 553–557
defined, 553
Like terms, 47
combining, 47, 115–116, 388–390, 391
Linear equations in two variables, 173–179. *See also* Equations; Systems of linear equations
finding solution of, **174**–179
forms of, **215**, **230**–**231**
graphing, 174–178
graphing, with isolated variable, 187–**189**
intercept graphing of, **185**–187
modeling using, 212–215
point-slope form of, 230, **231**–234
slope-intercept form of, **215**–219, **230**
slope-intercept form of, methods for writing equation in, **218**, **231**, **233**
slope-intercept graphing of, 220–224, **221**
standard form of, 183–**184**, **215**, **230**
Linear inequalities. *See also* Linear inequalities in two variables
constraints represented by, 255–258
in one variable, 153–160
in one variable, application problems using, 159–160
in one variable, on number line graph, 153
in one variable, solving, 154–157, **156**
in one variable, with fractions or decimals, 157–158
properties of, **154**–158
strict and not strict, 249
Linear inequalities in two variables, 249–262. *See also* Systems of linear inequalities
constraints represented by, 255–258
defined, 249
graphing, 249–252, **251**, **345**
graphing with vertical or horizontal boundary lines, 251–252
solution region of, 249, 256–258
strict versus not strict inequalities, 249
writing inequality represented by graph, 252–**255**
Linear models
extrapolation using equation or graph of, 235
solving application problems using, 212–215
using lines of best fit to write, **234**–236
Line of best fit, 230
writing equation of, in slope-intercept form, **234**–236
Lines
boundary, 249, 251–255
coinciding, 286, **287**, 306, 320
equations of, 215–220, 230–241
horizontal, 190–191, 206, 219–220, 252
parallel, 236–238, **240**, 286, 287, 305–306, 319–320
perpendicular, 238–241, **240**
point-slope form, 230, **231**–234, 253–254
slope-intercept form, **215**–219, **230**
slope of, 198–205, **199**
vertical, 190–191, 206, 220, 251–252
Long division, dividing polynomials using, 405–411
remainders, 409–411
Loss and profit, 140, 289
Lowest terms
rational number in, **474**
simplifying fractions into, 14–**15**, 38–39

M

Mantissa, **376**
Many bases rule for product, **366**, 367
Many bases rule for quotient, **366**–367
Market share, 99
Mass/weight, units of, 79
Math anxiety, 473
Maximum values, using inequality in one variable to solve applications including, 159
Mean, arithmetic, 90
Measurement(s)
adding, 81–82, **380**
combined units of, 381–382
International System of Units (S.I.), 78–79, 80
in scientific notation, 377–382
in scientific notation, multiplying or dividing, 377–378, **379**
subtracting, 81–82, **380**
unit conversion, 80–81
U.S. Customary System, 78–79, 80
units of, 78–80
Minimum values, using inequality in one variable to solve applications including, 159
Mixed numbers, 16
Mixture problems, 327–333
volume of ingredient in mixture, **328**–333
Models, linear, 212–215
using lines of best fit to write, **234**–236
Monomials
dividing polynomials by, 404–405
multiplying, 390–391
Multiplication property of equality, **110**, 112, 113
Multiplication property of inequality, **154**

Multiplicative identity, 36
Multiplying
  binomials, 390, 394–396
  distributive property of multiplication over addition, **50**
  exponential expressions, 363–364
  fractions, **17**–18
  measurements in scientific notation, 377–378, **379**
  monomials, 390–391
  notation for, 36
  polynomial by monomial, 390–391
  polynomials, patterns for, 397–**398**
  rational expressions, 481–483, **482**
  rational numbers, 481
  rewriting division as multiplication by reciprocal of divisor, 483
  signed numbers, 9–11
  sign rules, **10**
  square roots, 557–559, **558**
  terms, 49

N

Negative exponents, 369–371, **370**, **582**
Negative numbers
  cube root of, 577
  problem solving with, 59–60
  square root of, 95, 548
Notation, 2, 36
  boundary value on number line graph, 153
  for division, 36
  expanded, 25
  exponential, 362
  exponential, radicals in, **580**–581
  function, 268
  for multiplication, 36
  place value, 25, **57**
  plus-minus, 590
  for principal square root, 548
  radical, exponential expressions in, 581
  roster, 36
  scientific, **376**–386
  set-builder, 289
  square root, 95
Number line graph, 7–9
  adding/subtracting on, 8–9
  boundary value on, 153
  inequalities in one variable on, 153
  order of numbers on, 30
  point on, 7
  scale of, 7, 37, 170
  tick marks on, 7, 37, 170
Number of parts, **29**
Numbers. *See also* Decimal numbers; Negative numbers; Rational numbers; Real numbers; Whole numbers
  absolute value, 7, 37
  complex, **617**–619
  composite, 36
  imaginary, 616–**617**
  including word names, rewriting to place value notation, **57**
  irrational, **40**, 92–94, **617**, 618
  mixed, 16
  order, on number line, 30
  prime, 36, 423
  signed, 7–13
Numerator, 38
Numerical expression, 46

O

One, evaluating expressions with, 4
Operations. *See also* Adding; Dividing; Multiplying; Subtracting
  defined, 2
  order of, **2**–4, 11–12, 22, **41**–43
Opposite reciprocals, 239, 240
Opposites, 37
  on number line graph, 7
Optimization, 351
"Or," use of term in zero product property, **457**
Ordered pairs, 170–173
  graphing, 170–172
  in relation, 263
  as solutions, 173–178
Order of numbers, 30
Order of operations
  evaluating expressions, **2**–4, **41**–43
  in expressions with fractions, 22
  signed numbers and, 11–12
Origin, 170
Output values, 263
  range of relation or function as set of, 266–267
Overhead, 140

P

Parabola
  axis of symmetry, 624, 625, 629
  graph of, 624–629
  vertex of, 624, 625, 628–629
  x-intercepts of, 627–629
Parallel lines, **240**, 286, 287
  recognizing using elimination, 319–320
  recognizing using substitution, 305–306
  slopes of, 236–238, 240
Parentheses, 2, 11
Patterns
  difference of cubes, **446**
  difference of squares, **398**, **444**–**445**
  factoring using, **443**–446, **445**, **446**
  for multiplying polynomials, 397–**398**
  perfect square trinomial, 397, **398**, **443**–444
  sum of cubes, **446**
Percent, **68**, **327**
  decrease and increase, 71–72, 329–330
  finding and using, **29**–30
  mixture problems involving, 327–333
  number of parts, **29**
  problem solving with, 67–78
  retail applications, 72–73
  writing as decimal number, 28, 67–68
  writing as fraction, 67–68
  writing ratio as, 69–70
Perfect cubes, **445**, 578
Perfect powers, 578
Perfect squares, 95, **442**, **549**, 578
  exponential expressions with variables that are, **550**
  finding perfect square factor of large number, 554
Perfect square trinomial, 397
Perfect square trinomial patterns, 397, **398**, **443**–444
Perimeter
  of rectangle, formula for, 84, **87**, 336, 398, 611
  of triangle, formula for, 84, **87**, 399
  writing polynomial expressions in one variable representing, 398–399
Perpendicular lines, **240**
  slopes of, 238–241
π (pi), 40, 616
  formulas including, **92**–94
  rounded value, 93
Placeholders in polynomial division, 408
Place value notation, 25
  for powers of ten, 376
  writing numbers including word names in, **57**
Place values
  rounding decimal number to given, **27**–28
  writing numbers in scientific notation and counting, **377**
Plus-minus notation, 590
Point
  on Cartesian coordinate system, 170
  on number line graph, 7
Point-slope form
  using to write equation of boundary line, 253–254
  writing equations of lines using, 230, **231**–234
Polya, George, 55
Polynomial equations, using zero product property to solve, **457**–463
Polynomial expressions, 386–**388**. *See also* Trinomials
  adding, 388–390
  degree of, 387, **388**
  in descending order, 387
  dividing, 404–414
  dividing by monomial, 404–405
  dividing by polynomial, 405–411
  factoring, 424–430
  factoring completely, **449**–454

factoring four-term polynomial by grouping, 425–428, **427**
factoring methods, **446**, **449**
factoring out greatest common factor, 424–425
multiplying, patterns for, 397–**398**
multiplying, using distributive property, 390–391, 395–396
multiplying binomials, 390, 394–396
multiplying polynomial by monomial, 390–391
in one variable, 398–400
prime polynomials, 424, 428, 438–439, 444, 445, 452
simplifying by combining like terms, 388–390, 391
subtracting, 389
writing, 398–400
Power. *See* Exponents
Power of ten, **375**–376
scientific notation and, **376**–377
word names for, 376
Power rule of exponents, **365**
simplifying rational exponents using, 582
Prefixes
for polynomials, 387
S.I., 79
Price, sale, 72–73
Prime factorization
finding greatest common factor using, **423**–424
finding least common denominator using, **496**–497
finding least common multiple using, 491–493, **492**
writing, of whole number, **423**
Prime numbers, 36, 423
Prime polynomials, 424, 428, 444, 445, 452
discriminant test for quadratic trinomials, **438**–439
Principal, 148
Principal cube root, 96
Principal square root, 95, 548
notation for, 548
Problem solving. *See also* Application problems; Formulas; Solving
in the ballpark, 55, 56, A-2
cost problems, 333–336
doing problems another way, 55, 58, A-2
estimation in, 55, A-1
five steps for, **55**
five steps for, using, 56–61, 68–73
with fractions, 60–61
involving rate of work, 524–526
mixture problems, 327–333
with negative numbers, 59–60
with percent, 67–78
reasonability and, 55–58, A-1–A-6
using proportions, 118–120, 526–527
using Pythagorean theorem, 594–595
using quadratic equations, 593–594, 609–612
using systems of linear equations, 289–291, 307–309, 321–323, 327–344
using zero product property, 461–463
with whole numbers, 54–59
working backwards in, 55, 57, A-1–A-2
Product, 2, 36
many bases rule for, **366**, 367
zero product property, **457**–467
Product rule of exponents, **363**–364
Product rule of radicals, **549**
for radical with index, *n*, **578**
simplifying square roots using, 549–550
Profit and loss, 140, 289
Proper fractions, 16, 48
Properties
associative, 49, **50**
commutative, 49, **50**
distributive, 49, **50**
of equality, **110**–115
of equality, solving equation for a variable using, 145–146
of equality, solving equation in one variable using, 110–115, **113**, 124–126
of inequality, **154**–158
of real numbers, **50**
zero product, **457**–467
Proportions
defined, 118
inequality in one variable as, 159–160
in one variable, solving, 118–120
solving applications problems using, 118–120, 526–527
with variable in denominator, 526–527
Pythagorean theorem, 96–97
problem solving using, 594–595

Q

Quadrants, 170
Quadratic equations
with irrational solutions, 602–603, 608, 609
lead coefficient of, 626–627
methods for solving, **612**
with nonreal solutions, solving, 619–621
rewriting, in form $x^2 = k$, 591–593
solutions of, A-7
solutions of, in form $x^2 = k$, **590**–591
solutions to, using discriminant to describe, 608–**609**
solving, by completing the square, 600–603, **602**, 620
solving applications using, 593–594, 609–612
solving using quadratic formula, **606**–608, 620–621
in standard form, 459, 624
in two variables, graphing, 624–**629**
in two variables, graphing using x-intercepts and vertex, 627–**629**
in two variables, table of solutions for graphing, 624–625, 626
Quadratic formula, **606**–612, **A-8**
deriving, A-7–A-8
solving application problems using, 609–612
solving quadratic equations using, 606–608, 620–621
Quadratic trinomials
discriminant test for, **438**–439
factoring, 431–**438**
factoring using *ac* method, 435–438, **437**
factoring using guess and check method, 431–434, **432**
lead coefficient of, 431, 450
Quotient, 2, 18, 405
many bases rule for, **366**–367
Quotient rule
of exponents, **364**–365
of exponents, simplifying rational exponents using, 582, 583
of radicals, **562**–564

R

Radical equations
applications of, 573–574
with one square root, extraneous solutions to, 570–572
solving, with one square root, 568–570, **569**
solving, with two square roots, **572**–573
Radical expressions
rationalizing denominator of, 563–564
simplifying, 579–580
Radical notation, writing exponential expressions in, 581
Radicals
combining like, 553–557
in exponential notation, **580**–581
higher order roots, 577–579
product rule of, **549**–550
product rule of, for radical with index, *n*, **578**
quotient rule of, **562**–564
with variables, simplifying, 550–551
Radicand, 548
Radius of circle, **92**
Range
of function, 266–267
of relation, 266–267
Rate of work, 524–526
Rates of change, average, 196–198
slope as, 198–205, **199**, 213

Ratio, writing as percent, 69–70
Rational equations
  clearing fractions from, 518, 519, 520
  extraneous solutions to, 521–523
  solving, 518–**523**
  solving rate of work problems using, 524–526
Rational exponents, 580–581
  simplifying, using exponent rules, 581–583
Rational expressions
  adding, with different denominators, 500–**502**
  adding, with same denominator, 489, **490**
  building equivalent, with different denominator, 498–**499**
  complex, 507–517
  complex, simplifying, 508–514, **510**, **512**
  defined, 475
  dividing, 483–485, **484**
  least common denominator of, 495–497
  least common denominator of, using prime factorization to find, **496**–497
  multiplying, 481–483, **482**
  simplifying, 475–478, **476**
  subtracting, with different denominators, **502**–503
  subtracting, with same denominator, **490**–491
  undefined, 476
  using prime factorization to find least common denominator of, **496**–497
  using prime factorization to find least common multiple of, 491–493, **492**
Rationalizing denominators, 563–564
  that are a sum or difference, 564–565
Rational numbers, 38–39, **40**, 91–92, **617**, 618
  adding, 488–489
  in lowest terms, 38, **474**
  multiplying, 481
Real numbers, 36–46, **617**
  absolute value of, 37
  classifying, 36
  properties of, **50**
  sets of, **40**
  as subset of complex numbers, 617, 618
Reasonability, problem solving and, 55–58, A-1–A-6
Reciprocals
  defined, 17
  opposite, 239, 240
  using, to divide fractions, 18
Rectangle
  area, 85, **87**, 168, 398, 611
  perimeter, 84, **87**, 336, 398, 611
Rectangular (Cartesian) coordinate system, 170–172
Rectangular solid, volume of, **87**, 536
Regression methods, 234
Relations
  defined, 263
  domain of, 266–267
  functions as, 263–264
  range of, 266–267
Remainders in polynomial division, 409–411
Repeating decimal numbers, 26–27, 91–92
  rewriting as fraction, 313–314
Retail applications, 72–73
Right angle, 96
Right triangle, 96–97, 238, 594
Rise, 198, 199
Roots. *See also* Square roots
  applications using, 97
  cube, 96, 577
  higher order, 577–579
Roster notation, 36
Rounding
  decimal number to given place value, **27**–28
  to nearest whole number, 60–61
  using approximately equal sign, 60
Rounding digit, 27
Rules
  exponent, 362–368
  exponent, simplifying rational exponents using, 581–583
  function, 268
  product rule of radicals, **549**–550
  product rule of radicals for radical with index *n*, **578**
  quotient rule of radicals, **562**–564
  sign, for division and multiplication, **10**
Run, 198, 199

S

S.I. (International System of Units), 78–79
  abbreviations, 79
  prefixes, 79
  relationship to U.S. customary units, 80
Sale price, 72–73
Scale
  of Cartesian coordinate system, 170–172
  on graph, slope and, 202
  of number line graph, 7, 37, 170
Scientific calculators. *See* Using technology
Scientific notation, **376**–386
  adding measurements in, **380**
  application problems using, 381–382
  dividing measurements in, 378–**379**
  multiplying measurements in, 377–378, **379**
  raising measurement in, to a power, **379**
  subtracting measurements in, **380**
  writing numbers in, **377**
Set brackets, 36
Set-builder notation, 289
Sets, 36
  of complex numbers, 617–618
  of integers, 37
  of real numbers, **40**
  of whole numbers, 36
Signed numbers, 7–13
  adding, 7–9
  dividing, 9–11
  multiplying, 9–11
  order of operations and, 11–12
  subtracting, 7–9
Sign rules
  for division, **10**
  for multiplication, **10**
Simple interest formula, 83–84
Simplifying
  algebraic expression, 46, 47, 50–52
  before combining like radicals, 554–555
  complex numbers, 618–619
  complex rational expressions, 508–514, **510**, **512**
  equations in one variable, 126–129
  exponential expressions, using exponent rules, 362–368, 371–372
  fractions equal to 1, **475**
  fractions into lowest terms, 14–**15**, 38–39
  higher order roots, 578
  polynomial expression by combining like terms, 388–390, 391
  radical expressions, 579–580
  radicals with variables, 550–551
  rational exponents, using exponent rules, 581–583
  rational expressions, 475–478, **476**
  rational number, 38, 474
  square roots, 549–550, **551**
  square roots with negative radicands, 616–617
  traditions for simplifying expressions, 563
Slope(s)
  application problems using, 199, 202–203
  as average rate of change, 198–205, **199**, 213
  finding from equation of line, 204, 205
  finding from formula, **199**, 200–204, 205, 206, 213, 507
  finding from graph, 200, 201, 203, 205
  of horizontal lines, 206
  of a line, 198–205, **199**
  of linear model, 213
  methods for finding, **205**

of parallel lines, 236–238, 240
of perpendicular lines, 238–241
of vertical lines, 206
Slope-intercept form
equation of horizontal line in, 219–220
of linear equation in two variables, **215**–219, **230**
using to write equation of boundary line, 253
writing equation of line in, given slope and point on line, **231**–232
writing equation of line in, given two points or graph of line, **233**–234
writing equation of line in, methods for, **218**
Slope-intercept method, 220–224, **221**
solving systems of equations by, 287–289
Solution region
bounded, 347
feasible regions, 256–258, 350–351
of linear inequality, 249, 256–258
of systems of linear inequalities, 345, 350–351
test point to identify, 250–251, 252, 256, 257
Solutions
of equations, 110
of equations, checking, 111
extraneous, 521–523
extraneous, to radical equations with square root, 570–572
infinitely many, 138–139, 286, 287, 306, 320
irrational, to quadratic equations, 602–603, 608, 609
of linear equation in two variables, finding, **174**–179
ordered pair as, 173–178
of quadratic equations, A-7
of quadratic equations, using discriminant to describe, 608–**609**
of quadratic equations in form $x^2 = k$, **590**–591
quadratic equations with nonreal, solving, 619–621
reasonability of, 55
reasonability of, checking, 55–58, A-1–A-6
reporting, 55
of systems of linear equations, identifying, 286–287, 289, 305–306, 319–320
of systems of linear inequalities, identifying, 348–349
Solving. *See also* Problem solving
application problems using equations in one variable, 116–120, 129–130
application problems using quadratic equations and quadratic formula, 609–612
application problems using systems of linear equations, 289–291, 307–309, 321–323, 327–344
equation for a variable, 145–146
formula for a variable, 147–149
by inspection, 112
linear equations in one variable, 110–133, **113**, **116**, **127**, **135**
linear equations in one variable using properties of equality, **110**–115, 124–126
linear equations in one variable with decimal numbers, 115–120, **116**
linear equations in one variable with variables on both sides, 133–145, **135**
linear inequalities in one variable, 154–157, **156**
polynomial equations using zero product property, **457**–463
quadratic equations, methods for, **612**
quadratic equations by completing the square, 600–603, **602**, 620
quadratic equations using quadratic formula, **606**–608, 620–621
quadratic equations with nonreal solutions, 619–621
radical equations with one square root, 568–570, **569**
radical equations with two square roots, **572**–573
rational equations, 518–**523**
systems of linear equations by elimination, 315–327, **316–317**
systems of linear equations by graphing, 287–289, 290–291
systems of linear equations by substitution, 298–312, **304**
systems of linear equations involving formulas, 336–339
Sphere, volume of, **92**, 94, 641
Square brackets, 2
Square of a number, 95
Square roots, 95–96, **548**–549
adding, 553–554, 555
dividing, **564**
index of, 548
multiplying, 557–559, **558**
of negative numbers, 95, 548
with negative radicands, simplifying, 616–617
of nonperfect square, formula for approximating, 99
notation, 95
principal, 95, 548
simplifying, 549–550, **551**
solving radical equations with one, 568–570, **569**
solving radical equations with two, **572**–573
subtracting, 554
Squares
area of, 95
completing, 598–600
completing, solving quadratic equations by, 600–603, **602**, 620
difference of, 396
difference of squares pattern, **398**, 444–**445**
perfect, 95, **442**, **549**, 578
perfect, exponential expression with variables that are, **550**
perfect square trinomial patterns, 397, **398**, **443**–444
Squaring both sides, 568
Standard form
of linear equation in two variables, 183–**184**, **215**, **230**
of quadratic equation, 459, 624
Subscripts, 147
Subsets, 617–618
Substitution expression, 299
Substitution method
recognizing coinciding lines using, 306
recognizing parallel lines using, 305–306
solving application problems using, 307–309
solving systems of linear equations using, 298–312, **304**
Subtracting
complex numbers, 619
fractions with same denominators, **19**
fractions with unlike denominators, **20**–21
linear equations, 313–314
measurements, 81–82
measurements in scientific notation, **380**
polynomial expressions, 389
rational expressions with different denominators, **502**–503
rational expressions with same denominator, **490**–491
signed numbers, 7–9
square roots, 554
Subtraction property of equality, **110**, 111, 135
Subtraction property of inequality, **154**
Subtraction rule of exponents, **362**
Sum, 2
rationalizing denominators that are a, 565
Sum of cubes pattern, **446**
Symbolic representation of relationship, 173. *See also* Equations
Symbols. *See also* Notation
less than (<), 30
approximately equal sign, 60
delta (Δ), 179
equal sign, 49
greater than (>), 30, 249

Symbols (*continued*)
grouping, 2–4
grouping, order of operations and, 42–43
inequality, 249
for multiplication, 11
Symmetry, axis of, 624, 625, 629
Systems of linear equations
defined, 286
graph of system of two linear equations, 286–**287**
identifying systems with infinitely many solutions, 286, 287, 306, 320
identifying systems with no solution, 286, 287, 305–306, 319–320
solutions of, 286–287
solutions of, in set-builder notation, 289, 320
solving application problems using, 289–291, 307–309, 321–323, 327–344
solving by graphing, 287–289, 290–291
solving using elimination, 315–327, **316–317**
solving using substitution, 298–312, **304**
writing, for break-even point, **290**–291, **307**–308
Systems of linear inequalities, 344–356
constraints represented by, 349–351
defined, 345
graphing, 345–351
identifying solutions of, 348–349
with no solution, 347–348
solution region of, 345, 350–351

T
Table of solutions of linear equations in two variables, 174–178
Terminating decimal numbers, 91–92
Terms
combining like, 47, 115–116, 388–390, 391
defined, 46, 386
like, 47
multiplying, 49
simplifying fractions into lowest, 14–**15**, 38–39
Test point to identify solution region of linear inequality, 250–251, 252, 256, 257
Tick marks
on number line graph, 7, 37, 170
on *x*- and *y*-axes of Cartesian coordinate system, 170–172
Time, units of, 79
Trapezoid, 90
Triangle
base of, 86
formula for area, 85–86, **87**, 100, 399–400, 466
formula for perimeter of, 84, **87**, 399
height of, 86
legs of, 594
right, 96–97, 238, 594
Trinomials
discriminant test for quadratic, **438**–439
factoring quadratic, 431–**438**
factoring trinomials that are not quadratic, 439–440, 444
factoring using *ac* method, 435–440, **437**
factoring using guess and check method, 431–434, **432**
perfect square, 397
quadratic, 431–439, 450

U
U.S. Customary System, 78–79
abbreviations, 79
relationship to S.I. units, 80
Units of measurement (units), 78–80
abbreviations, 79
combined units, 381–382
conversion of, 80–81
relationships of, 80
slope of line including, 201–202
Unknowns, variables assigned to, 55, 386
Using technology
building table of solutions, 179
checking division, 411–412
finding outputs and Zoom Out, 270–271
finding vertex of graph of quadratic equation, 630
fraction arithmetic, 22–23
graphing inequalities, 259
graphing linear equation on graphing calculator, 191–192
graphing systems of linear inequalities, 351–353
order of operations on calculator, 5–6
scientific notation, 382
signed numbers on calculator, 12
solving equations, 463–464
solving equation with graphing calculator, 574–575
systems of linear equations, 292
Trace, Value, and ZoomFit, 241–242
using graph to find ordered pairs, 207

V
Variables
assigning, 55, 386
on both sides of equation in one variable, 133–145
defined, 46, 386
dependent, 267
independent, 267
isolating, 110–113
Variation, 532–541
direct, **532**–534, 536
formulas and, 536
inverse, **534**–536
Velocity
escape, 576
final, 147
Verbal representation, 173
Vertex of parabola, 624, 625, 628–629
Vertical lines, 190–191, 220
as boundary line in graphs of linear inequalities, 251–252
graphing, 190
slope of, 206
Vertical line test, 264–**265**
Visual learner, 195, 211
Volume
of cone, **92**, 615
of cylinder, **92**, 93–94, 356
of ingredient in mixture, **328**–333
of rectangular solid, **87**, 536
of sphere, **92**, 94, 641
units of, 79

W
Whole numbers, **40**, **617**, 618
graphing on number line, 170
problem solving with, 54–59
set of, 36
writing prime factorization of, **423**
Work, rate of, 524–526
Working backwards, 55, 57, A-1–A-2

X
x-axis, 170
x-coordinate, 171
x-intercepts, 185–187, 627–629

Y
y-axis, 170
y-coordinate, 171
y-intercepts, 185–187
in linear models, 212–213
in slope-intercept form of linear equation, 215

Z
Zero
as additive identity, 36
division by, 5, 36
evaluating expressions with, 4–5
as exponent, 368–**369**
leading, 25
Zero product property, **457**–467
applications of, 461–463
factoring and, 459–461

BUFFALO TRACE

BUFFALO TRACE

CARVING THE TRAIL TO GREAT BOURBON

PHOTOGRAPHS AND TEXT BY

DAVID TOCZKO

FOREWORD BY

RICHARD TAYLOR

Acclaim Press
MORLEY, MISSOURI

P.O. Box 238
Morley, MO 63767
(573) 472-9800
www.acclaimpress.com

Book Design: M. Frene Melton
Cover Design: M. Frene Melton

Library of Congress Control Number: 2014910407

ISBN-13: 978-1-938905-67-4
ISBN-10: 1-938905-67-9

First Printing 2014
Printed in the United States of America
10 9 8 7 6 5 4 3 2 1

This publication was produced using available information.
The publisher regrets it cannot assume responsibility for errors or omissions.

Contents

Dedication 6
Acknowledgments 7
Foreword 8
The Legends 10
Welcome to Buffalo Trace 12
Stony Point Mansion 18
The Grounds 20
The Mash 46
The Fermenters 54
The Still 64
The Micro Still 66
Spent Mash 68
The Barrels 72
The Branding Shed 80
The Cistern Building (Barrel Filling) 82
The Warehouses 94
Warehouse "X" 110
Regauge (Barrel Dumping) 112
Warehouse "H" 128
Quality Assurance 130
Specialty Bottling 134
The Bottling Process 136
The Hands 148
Elmer T. Lee 156
The River 160
The Stream 162
Colonel Blanton 169
The Big Picture 172
The Water Tower 174
The Lesson 176
Egg Hunt 178
National Historic Landmark 182
"White Dog" Day 184
Ghost Tour 188
6,000,000th Barrel 192
The Lighting of the Trace 194
The Master Distiller 198
The Firehouse 204
The Past 206
Index 208

Dedication

To Beth, for all that you do and all that you are. You are the rock that keeps me grounded and the star by which I steer.

Acknowledgments

This book is a result of the help, support, encouragement and input from a long list of people who I am proud to call "friends." Without them, this book would not have been possible. Those who know nothing of bourbon and how it is made provided me an outsider's view to help explain the history and process behind the scenes and make sense to the "novice" bourbon lover. Their gentle persuasion and constructive criticism helped make order of my chaos.

Mark Brown, President and CEO of Buffalo Trace Distillery, threw his support behind this project from the very beginning. Amy Preske, Public Relations and Events Manager, shared my vision and arranged the access and opportunities for me to capture what makes Buffalo Trace such a special place in Kentucky as well as bourbon history.

On one of my visits to the distillery, I had the privilege of meeting Elmer T. Lee, former Master Distiller at Buffalo Trace and one of Kentucky Bourbon's legends. It was on one of his regular Tuesday visits to the distillery to "check on things" and was certainly the highlight of this endeavor. His kind and gentle way, along with his passion for bourbon, gave me the inspiration to do my best as he had always done. To sit with him and savor the smell of bourbon wafting in the air of the bottling building was an experience I will never forget.

A special thanks to my sister Eileen who has encouraged me, cheered me on and cheered me up. Her quiet way and words of wisdom have had an impact greater than she will ever know.

To everyone at Acclaim Press, my heartfelt thanks. I feel privileged to have worked with such a group of talented professionals. Their experience and insights helped make this vision a reality.

Foreword

Because of Kentucky's early isolation west of the mountains and its skein of waterways along which much of the region's history has been written, it is blessed with an abundance of magical places, places instilling a sense that the past is still with us. In northern Kentucky there is Big Bone Lick where Pleistocene mammals came to forage for the salt so necessary to their diets, where North America's first-comers were drawn to hunt wooly mammoths and other salt-craving creatures, and where the earliest explorers marveled at the enormous ossuary and other remnants of the past they found there. In southeastern Clark County there is the mosaic of cornfields that mark the site of the last permanent Shawnee settlement, Eskippakithiki ("place of blue licks") in what was later to be called Indian Old Fields. Morning mist there is reminiscent of the smoke from its camp fires and the ubiquitous trader John Finley, who reputedly first told Boone of the great hunting ground that was Kentucky.

Buffalo Trace Distillery, occupying the former site of Leestown in what is now Frankfort, is another such place. Beneath the aroma of working mash and barrels of bourbon sleeping in their ricks is a rich history, starting with a now-vanished ford across the Kentucky over which buffalo made their way south from Big Bone and northern Kentucky along the west side of the river. Crossing at one of the river's only natural fords and passing through a notch in the surrounding bluffs, they wore a trail, or trace, eastward into Central Kentucky to Lexington along what is known then and now as Leestown Road. Another trail split off to Georgetown and Stamping Ground (named for the buffalo congregating at its salt deposits) to still other licks and the rich grazing lands of the Inner Bluegrass. The history of this area parallels the settlement of Kentucky, the birth of Frankfort as Kentucky's capital, and the rise of commerce along the Kentucky River. This history also includes the formation of the one of the first modern distilling operations at what is now the oldest continuously operated distillery in the country. At Buffalo Trace, winner of numerous international distilling awards and recently designated as a National Historic Landmark, that history continues.

With a little imagination we can summon from history the now-vanished fort at Leestown, the depredations of marauding Indians, a bustling river trade, and the area's first modern distillery, using corn from local farmers and its own fertile fields while drawing its water from the pure stream of Cove Spring Branch whose waters still flow through the property. The location was ideal for capitalizing on the opening of waterways that took Kentucky farm products, including its bourbon, to markets downriver as far as New Orleans. The property is invested with the spirits of the Shawnee hunters and Euro- and African Americans who settled here as well as generations, men and women, of coopers, distillers, bottlers, warehousemen, and boatmen who spent much of their working lives in the river bottom producing distilled spirits invented by Kentuckians to be consumed in every corner of the world. Its leaders are memorialized on the labels of its brands—George T. Stagg, E.H. Taylor, Jr., Albert Blanton, and, most recently, Master Distiller Elmer T. Lee. The distillery that now occupies 130 acres has structures that span four centuries, and it proudly values its history and traditions.

For those of us whose imaginations profit from a little guidance we have the photographs of David Toczko, who has skillfully managed to recreate a sense of the magic of one of Kentucky's special places, one whose history is evolving, unlike Big Bone, which is preserved as a park, and Indian Old Fields, an expanse of cornfields not so different now from when Native Americans subsisted in part on the corn grown there. There is something more than the smell of spirits that populates the air at Buffalo Trace, perhaps most palpable at early morning or dusk when the lowering sun filters through trees along the river, and Toczko captures as much of the distillery's visible and invisible past as he does its present bustle—an ambience conferred by age and history, a place of harmony in which even the trees seem aware of their relation to history. The lens of his camera focuses on the particulars of this active distillery, capturing in light a sense of its special qualities, its people, its textures, its natural and built environments, its smells and subtle beauties.

Richard Taylor
Kentucky Poet Laureate
1999-2000

George T. Stagg (1835-1893) Known for his prowess as a whiskey salesman, Stagg teamed up with E. H. Taylor to distribute Taylor's fine bourbon. In 1878, he purchased OFC Distillery from Taylor and set about an aggressive expansion project. In 1904, the Distillery was renamed to bear Stagg's name, a title that was maintained for nearly a century. Photo courtesy of Buffalo Trace Distillery Archives.

Colonel Albert B. Blanton (1881-1959) Blanton began his career in bourbon as an office boy for the George T. Stagg Distillery at the age of 16. Over the next several years, Blanton worked in every department in the Distillery. By the time he was 20 years old, Colonel Blanton, as he was to become known, was appointed Superintendent of the Distillery, its warehouses and bottling shop. In 1921, Colonel Blanton was named President of the George T. Stagg Bourbon Distillery. Photo courtesy of Buffalo Trace Distillery Archives.

E. H. Taylor, Jr. (1830-1923) A descendant of two American presidents, Taylor is best known as the "Father of the Modern Bourbon Industry" for his many innovations in distilling and work on behalf of the bourbon industry, most notably the Bottled-in-Bond Act of 1897. Taylor purchased the Distillery in 1869 and renamed it "OFC" for Old Fire Copper. Among his innovations were copper fermentation tanks, state-of-the-art grain equipment, columnar stills and a steam heating system for the warehouses still in use today. In 1878, Taylor sold the Distillery to his associate, George T. Stagg. Photo courtesy of Buffalo Trace Distillery Archives.

The entrance to the Distillery is a welcome sight to the more than 100,000 visitors hosted each year.

RACE

Various views of the grounds as you make your way down the driveway to the Distillery.

Views during various seasons of the year of "Stony Point Mansion." The Mansion, built by Colonel Blanton for himself and his new bride in 1934, was where he lived until his death in 1959. The mansion now serves as administrative offices for the Distillery's staff.

The Distillery's grounds are ablaze with color during the fall of the year.

Various exterior views of Warehouse "C." Built in 1885, it is one of the oldest aging warehouses at Buffalo Trace.

The Barrel Crossing is a popular photo opportunity for visitors. The "tracks" throughout the Distillery are used to roll barrels from one location to another.

BUFFALO TRACE
DISTILLERY
BT
BARREL
CROSSING
STOP
BTD

Sunrise gives a golden glow to a barrel outside the gift shop as well as stalks of wheat on display inside.

19

A mural along the tour route employs the painting technique Trompe l'Oeil, a French term meaning "to trick the eye." The perspective of the mural changes based on the position from which it is viewed.

This life size sculpture of a buffalo outside the gift shop is another popular photo opportunity for visitors. The polished horns give an idea to the many folks who pose with him during their visit.

BUFFALO TRACE
DISTILLERY
THIS LIFE SIZE SCULPTURE PAYS HOMAGE TO THE
MILLIONS OF BUFFALO THAT ONCE LIVED IN KENTUCKY.
OUR DISTILLERY SITS ALONG THE GREAT BUFFALO TRACE; A
GREAT TRAIL THAT MIGRATING BUFFALO AND EARLY PIONEERS
ONCE TROD ON THEIR JOURNEY THROUGH KENTUCKY.
WHILE FEW BUFFALO SURVIVE TO THIS DAY, THE RUGGED
INDEPENDENCE AND SPIRIT OF ADVENTURE THEY EMBODY
CAN BE FOUND IN THE AMERICAN PEOPLE TODAY.

An exterior detail of the Fermenter Building, built in 1936, is an example of the masonry found adorning many of the buildings at Buffalo Trace.

Warehouse "V," the world's only single-barrel bonded warehouse, currently holds the 6,000,000th barrel filled at Buffalo Trace Distillery.

An early morning exterior view of the Chill Room shows steam rising into the air from the start of another batch of bourbon in the background.

Below and right: Exterior views of the Still House.

A window of the Chill Room reflecting the adjacent grounds.

Construction of the Clubhouse was commissioned by Colonel Blanton in 1935 and was built by the employees, for the employees. The building is constructed of four log cabins that were located across the state. These cabins were disassembled and then reassembled on-site. The Clubhouse was dedicated to Elmer T. Lee upon his retirement in 1986 and is now called the Elmer T. Lee Clubhouse. The lower level is still used as the employee break room.

THUNDER
CARVED FROM A FALLEN 300-YEAR OLD SYCAMORE TREE
TO COMMEMORATE THE MAJESTY OF MIGRATING BUFFALO
IN A WILDERNESS PAST
CREATED BY KENTUCKY SCULPTOR STAN SCHU
JUNE 1 1999

THUNDER
CARVED FROM A FALLEN 300-YEAR OLD SYCAMORE TREE
TO COMMEMORATE THE MAJESTY OF MIGRATING BUFFALO
IN A WILDERNESS PAST
CREATED BY KENTUCKY SCULPTOR STAN SCHU
JUNE 1 1999

The "Wishing Well" is one of the many features that adorn the grounds of the Distillery.

This log cabin, located adjacent to the Clubhouse, was originally built as a guest quarters, but now serves as the offices of Buffalo Trace President and CEO Mark Brown.

The Distillery's mascot is incorporated into the railings and fences across the grounds.

A detail of the Distillery's antique fire engine on display outside the Firehouse Café. In previous years, each distillery had its own volunteer fire department on-site due to the high risk of fires.

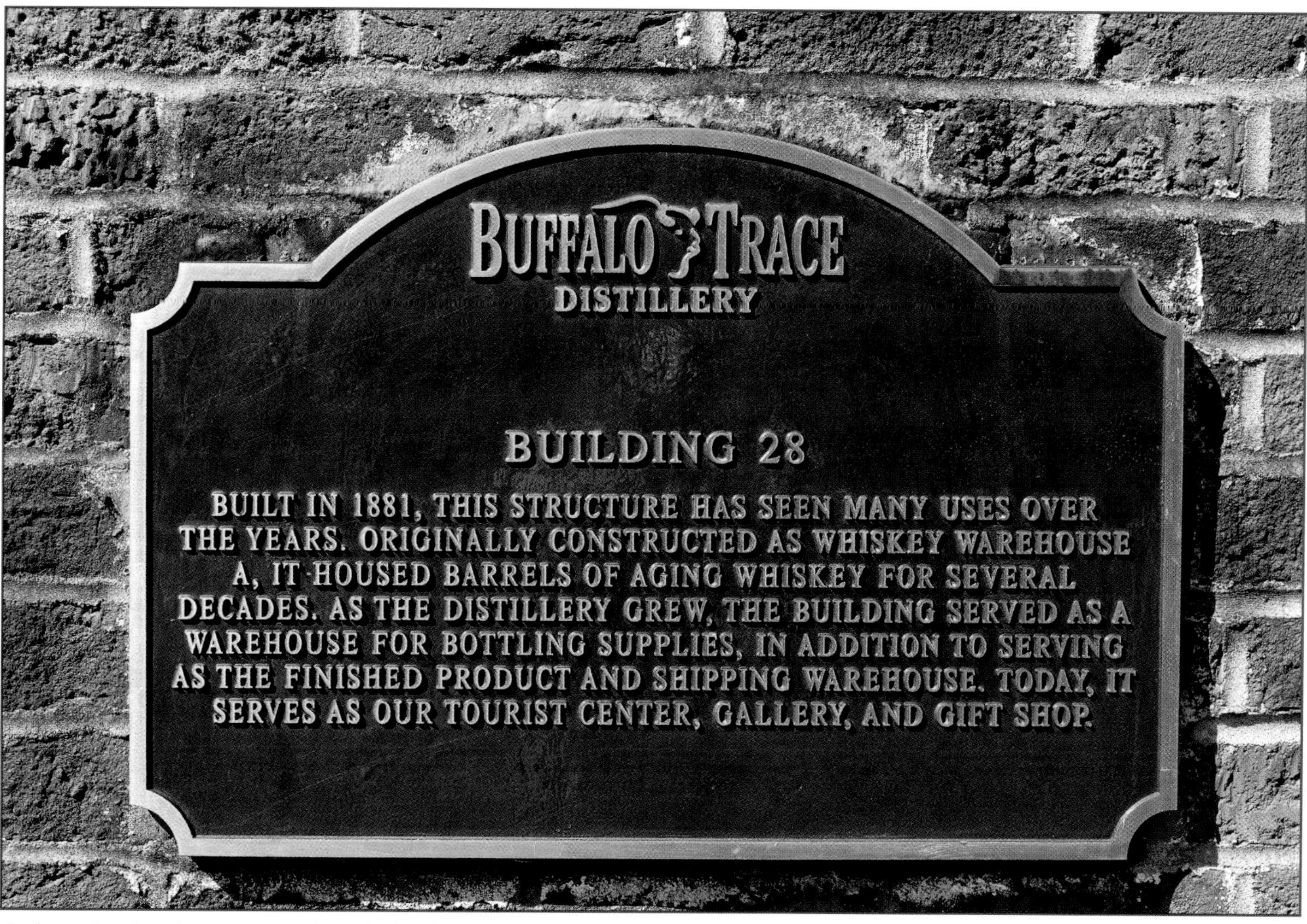

A plaque detailing the various uses the Gift Shop building has seen over the years.

A sign currently stored in the soon-to-be restored Dickel Building.

Grains used by Buffalo Trace in its bourbon production are delivered by tractor-trailer trucks and then conveyed to silos where they are stored prior to use.

TIMPTE

EXIT

The Traveling Scale Hopper, with a capacity of more than 200 bushels, is used to weigh the grains in the process. The hopper moves along rails in the floor and is easily operated by a hand wheel as demonstrated by employee John Hood.

MALL GRAIN
COOKER
#1
CAP. 1287 GALS.
SMALL GRA
COOKER
CAP. 1287
Rye

BUFFALO TRACE
DISTILLERY
ATMOSPHERIC COOKING
PROCESS
YEAST MASH
No 3
CAP. 2385

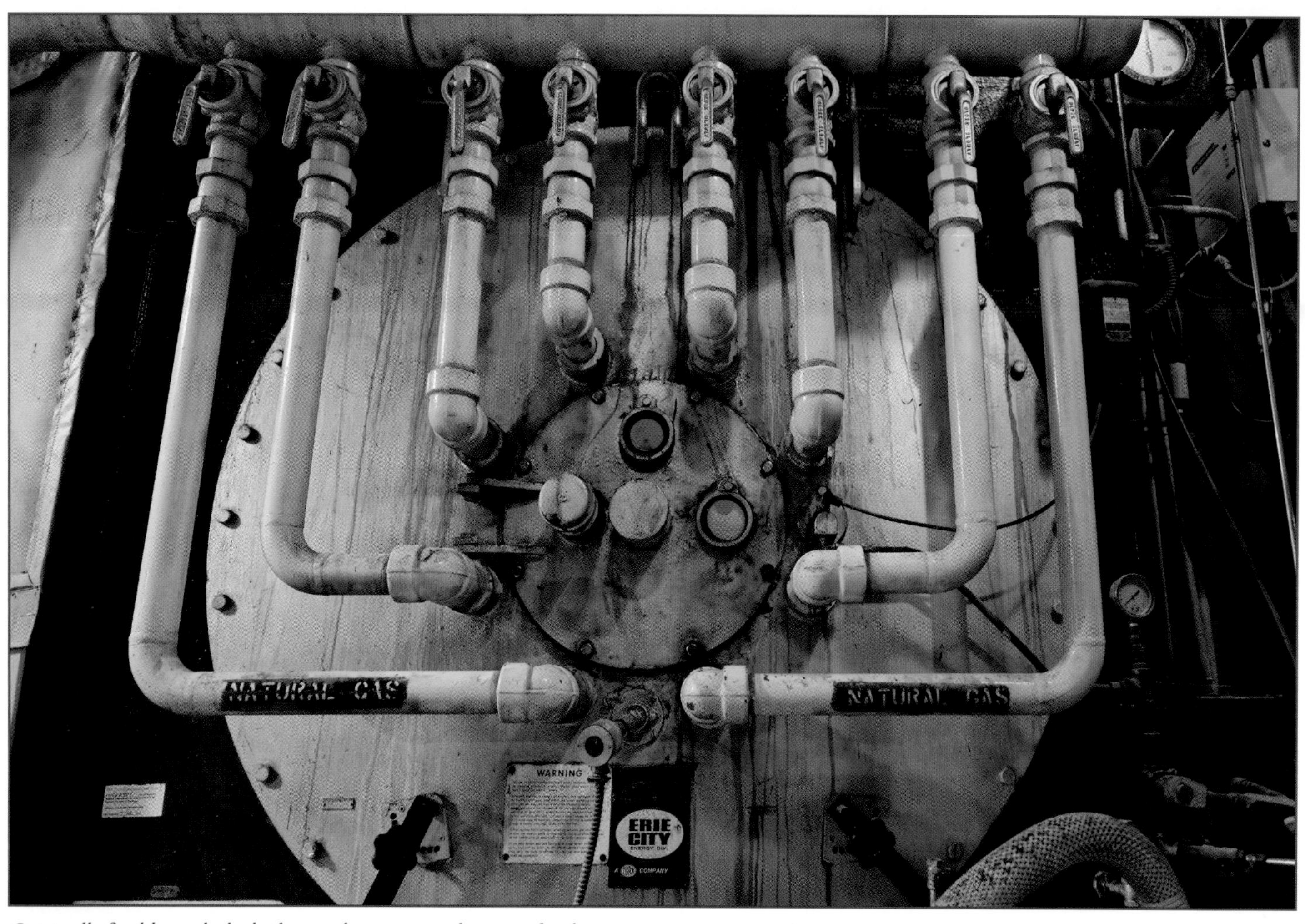

Originally fired by coal, the boilers used to generate the steam for the process now use natural gas.

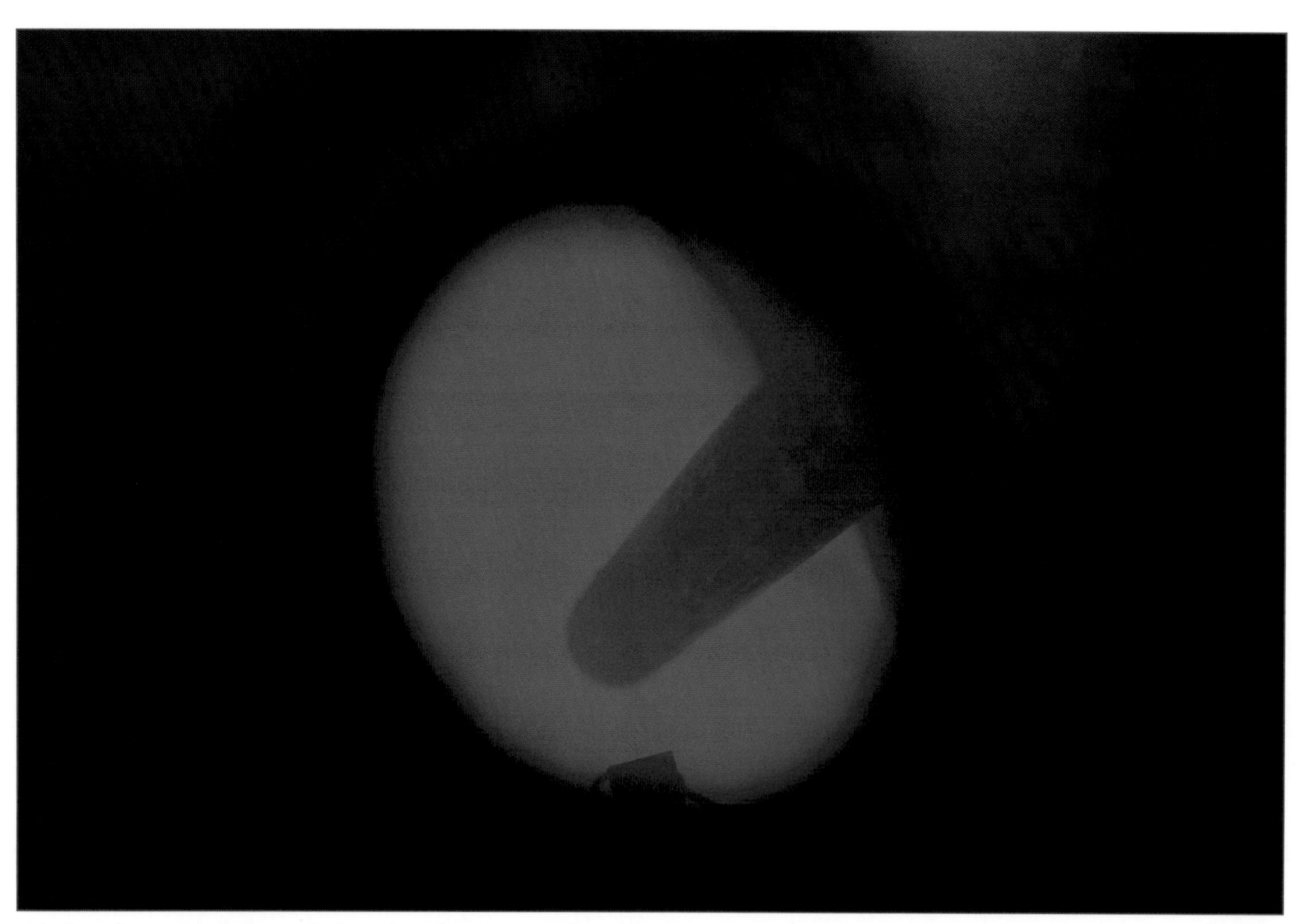

A look inside one of the boilers as the flames heat the water to produce steam.

YEAST MASH COOKER
No 3
CAP. 2385 GALS.

BUFFALO TRACE
DISTILLERY
ATMOSPHERIC COOKING
PROCESS

The first batch of the season flows into an empty fermenting tank. This is a sight rarely seen at Buffalo Trace. Each of the 12 fermenting tanks has a capacity of 92,000 gallons, the largest in the industry. More views on the following two pages.

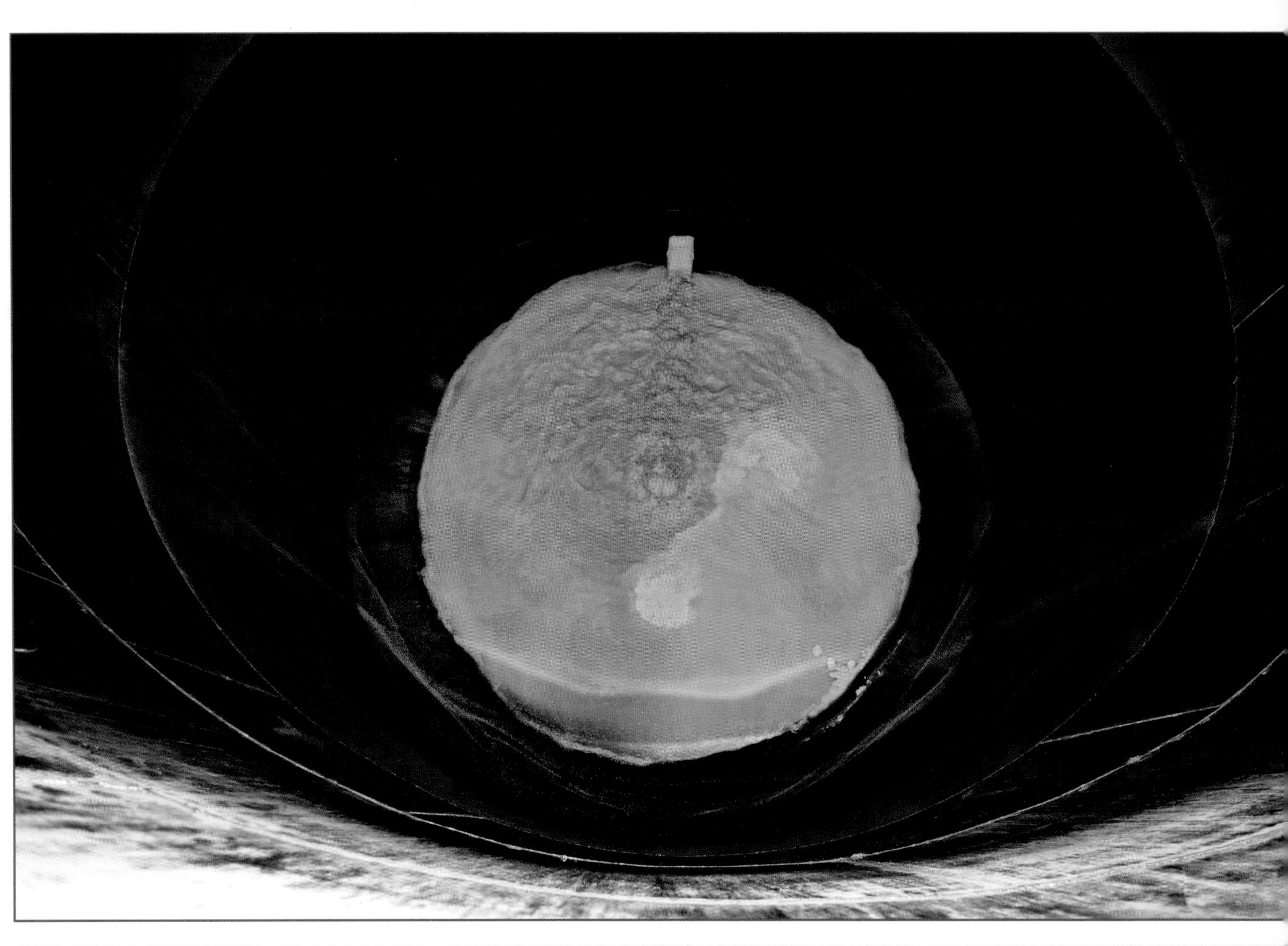

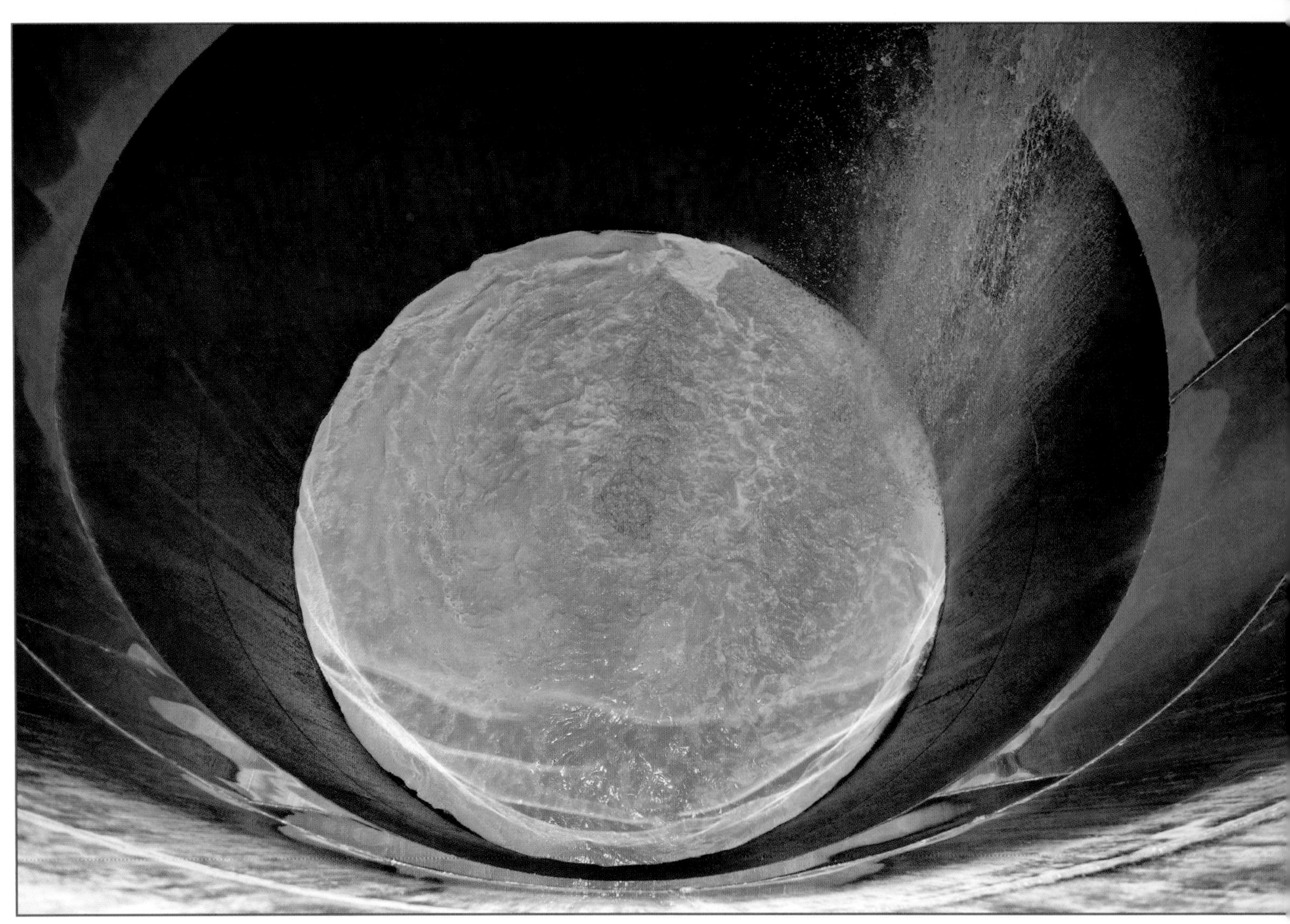

Corn oil rises to the surface as part of the fermentation process. The hood, visible in the upper left portion of the photograph, is used to vent the CO2 that is given off during fermentation.

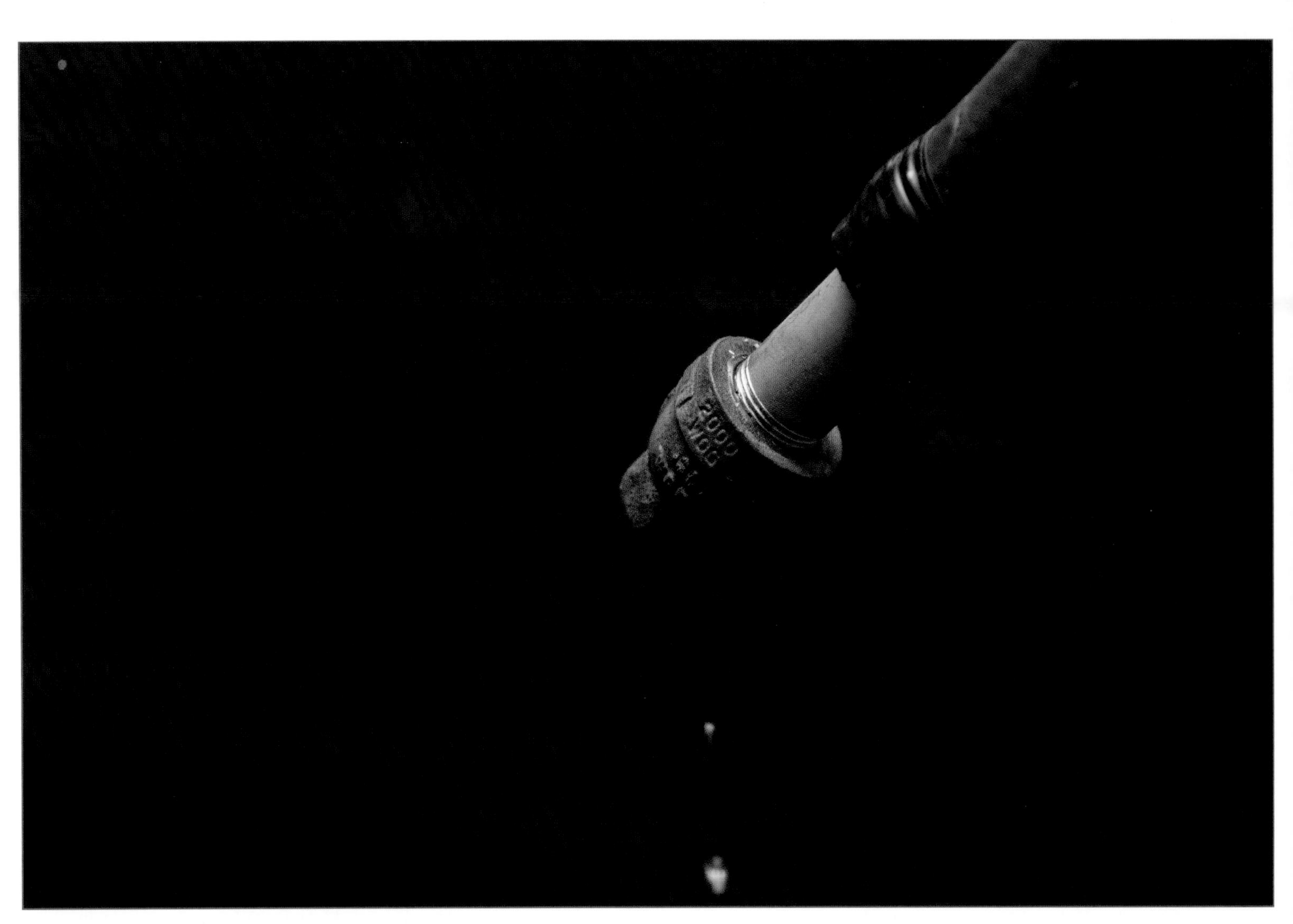

A sign posted next to one of the 92,000 gallon fermentation tanks is always a crowd favorite for pictures.

LACE
MING

Views of the Try Boxes where samples of the White Dog (un-aged bourbon) are drawn off for sampling.

Opposite: The sight glass in the side of Buffalo Trace's 60,000 gallon beer still allows visitors to take a sneak peek into the distillation process.

BUFFALO TRACE
DISTILLERY
Beer Still
Capacity
60,000 gallons

The "Micro Still," a combination of a Pot Still and a Column Still, serves as the heart of the Distillery's research and development efforts in the E. H. Taylor Room. The mural of Colonel Taylor on the wall serves as a reminder of the heritage and standards to which the company's products must uphold.

Employee Johnny Estes keeps a watchful eye on the spent mash drying process. Once dried, the mash is loaded into tractor trailers and sold to local farmers for use as feed for livestock.

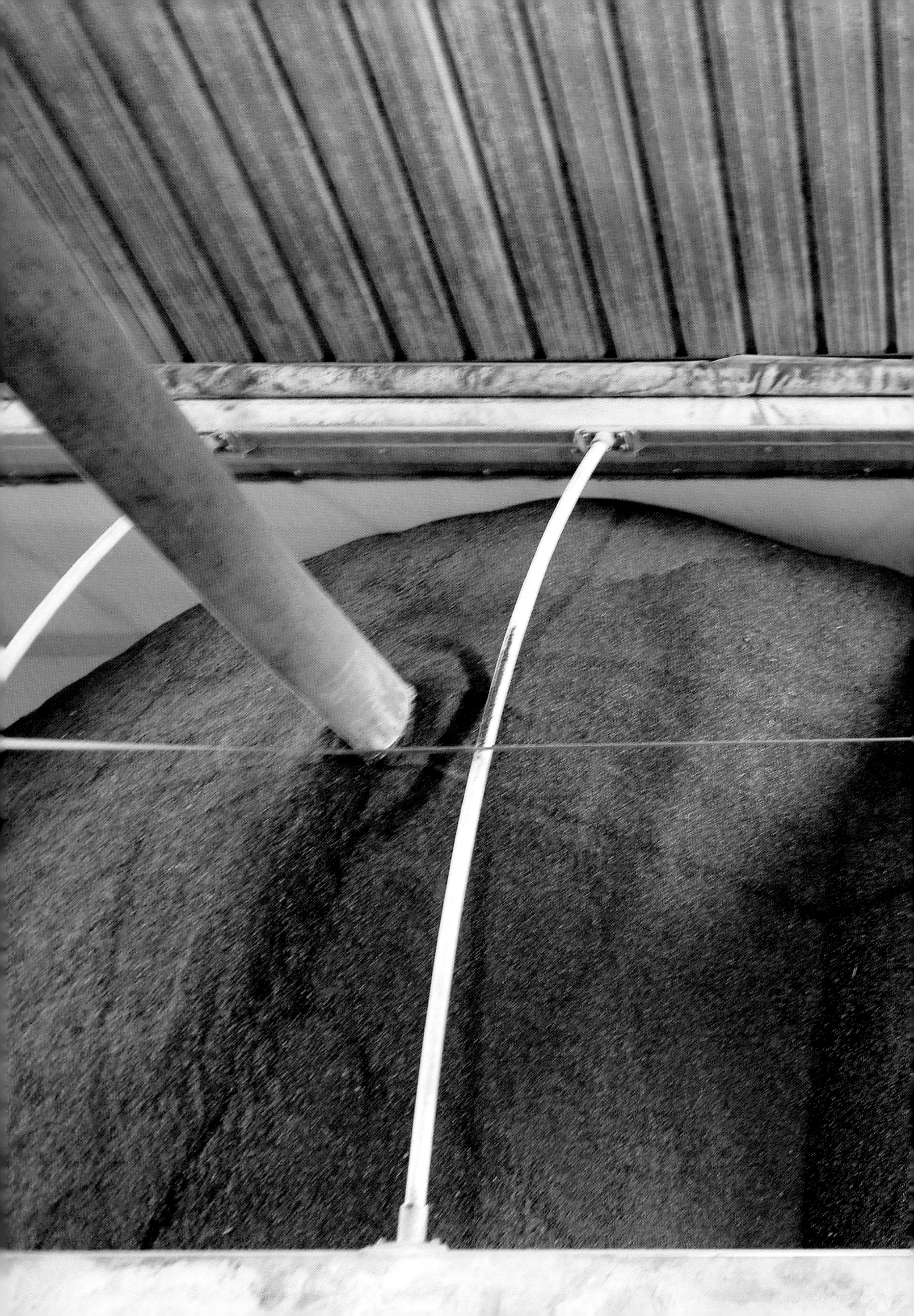

Barrel lids are charred on one side at Independent Stave Company, which supplies Buffalo Trace with its barrels.

Newly "raised" barrels make their way down the line to the next step in the process.

Barrels pass through a steam tunnel to expose the wood to heat and moisture, making the wood pliable prior to bending.

Right: The inside of a barrel after being charred.

After being exposed to steam, the barrels pass through a dry fires tunnel in order to maintain the temperature needed for bending the wood.

By law, the barrels used to age bourbon must be made of charred white oak.

Another view of the barrel charring process.

A new shipment of barrels arrives at the Distillery.

New barrels are staged in the Branding Shed prior to being filled.

An area of the Branding Shed that is used as part of the "Bourbon Barrel Tour" which explains the barrel-making process.

Barrels used in the Single Oak Project shown here with the "lily pads" of the actual trees from which the barrels were made.

Another view of new barrels in the Branding Shed.

Barrels staged for filling in the Cistern Building.

Each barrel is filled with 52 gallons of un-aged bourbon, referred to as "White Dog." Up to 20% can be lost during the aging process due to evaporation and absorption into the barrel. This loss is affectionately called the "Angel's Share."

The last few drops of White Dog enter a barrel in the filling process.

A filled barrel with the bung in place is ready to make its trip to the warehouse for aging.

A filled barrel is rolled from the filling station to be loaded on a truck and taken to a warehouse for aging.

This full barrel hoist lifts the barrels from the filling area to a loading ramp.

Gravity is used to move barrels whenever possible. Here, the filled barrels roll downhill from the Cistern Building to a waiting truck.

A truck with freshly filled barrels makes ready to transfer them to one of 13 warehouses for on-site aging.

A view of the barrels loaded and ready to go to the warehouse.

An old fashioned barrel wagon on the grounds of the Distillery harkens back to the "good old days."

Barrels are raised by an elevator system to the designated floor in the warehouse for aging.

LOT NO.
BUFFALO

Exterior views of the warehouses at Buffalo Trace Distillery. Depending on the weather, the windows of the warehouses will be opened or closed to help regulate the temperature inside.

Exterior view of Warehouse "C."

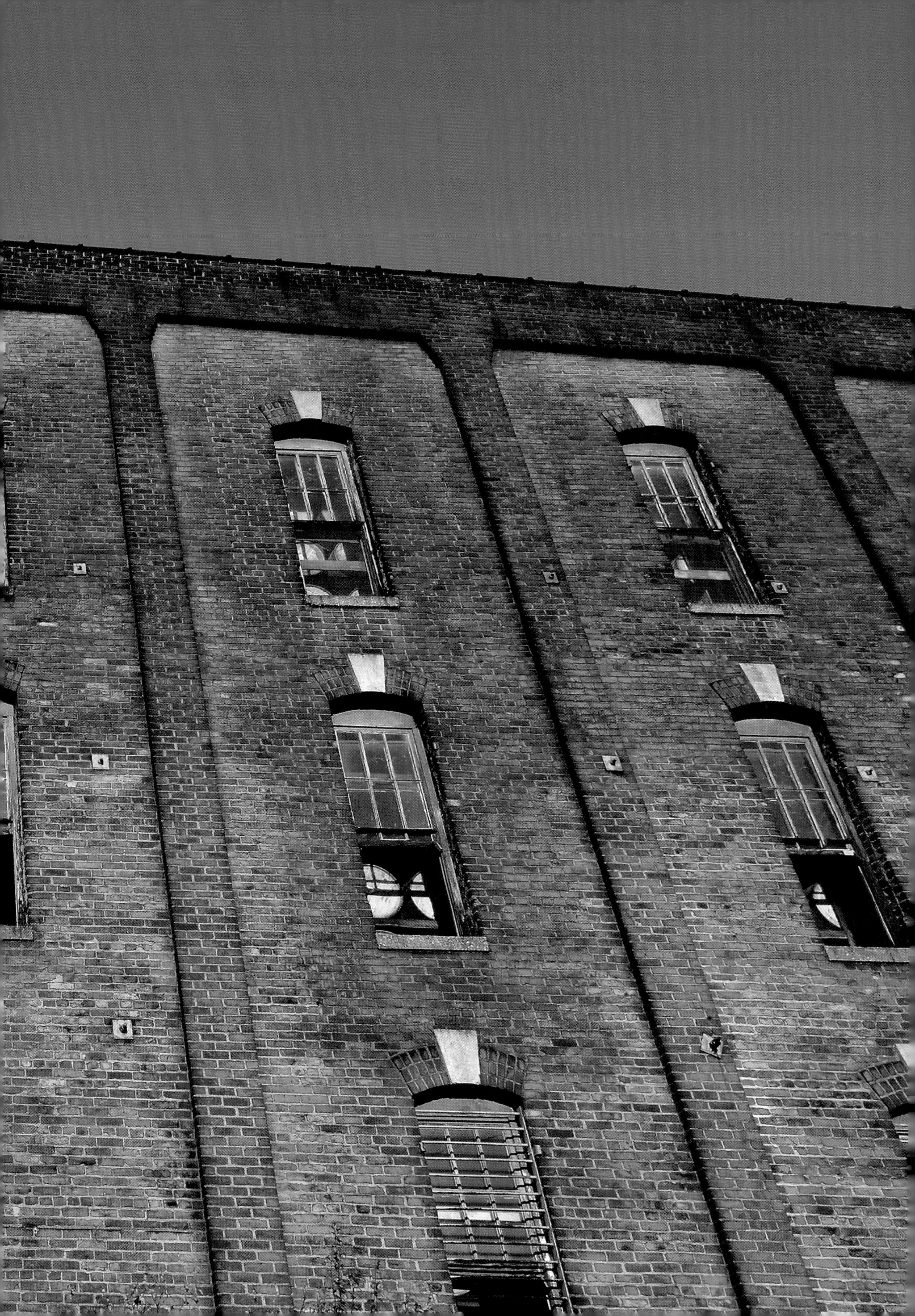

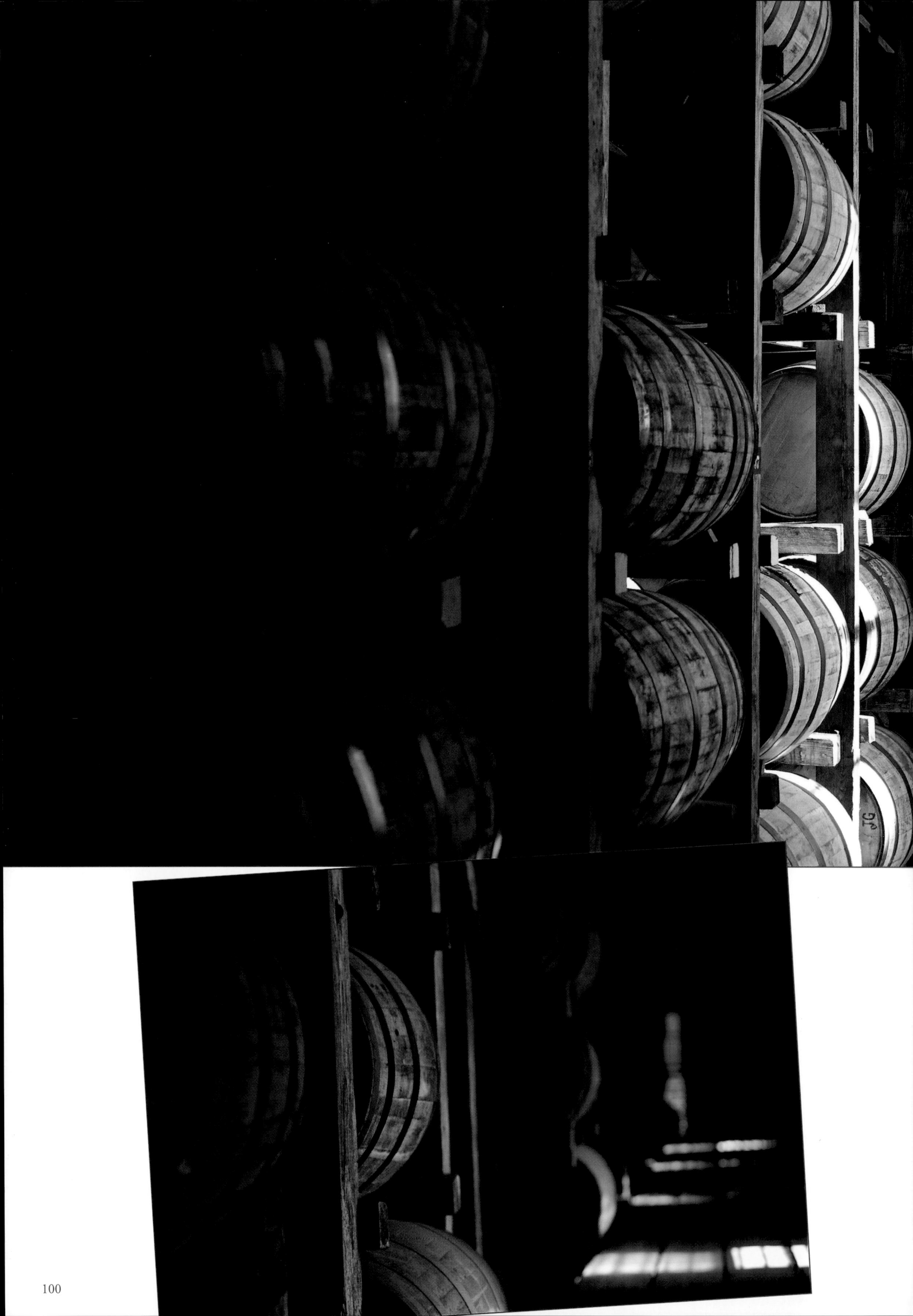

Barrels aging inside Warehouse "C."

Exterior views of Warehouse "B."

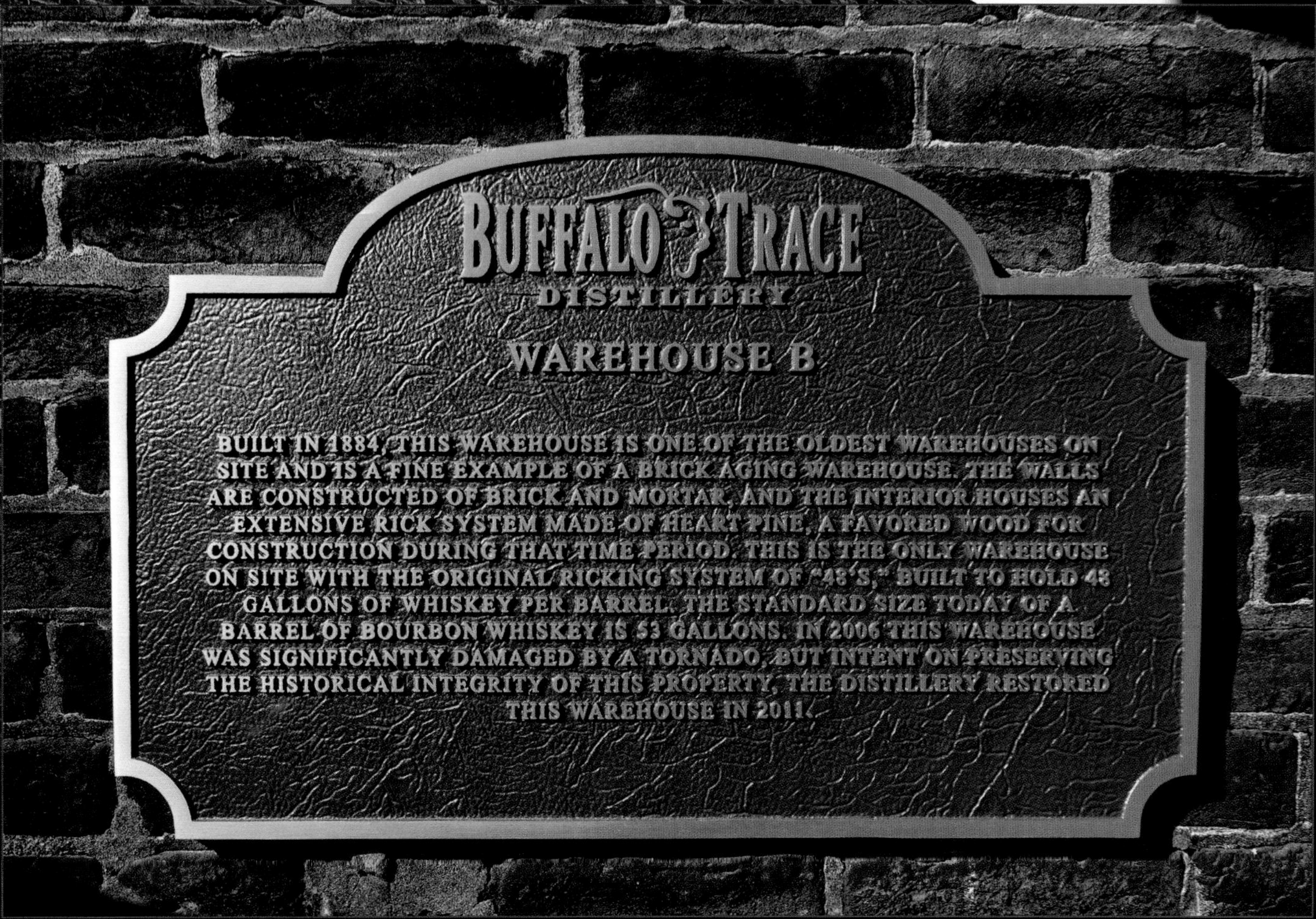

BUFFALO TRACE
DISTILLERY
WAREHOUSE B
BUILT IN 1884, THIS WAREHOUSE IS ONE OF THE OLDEST WAREHOUSES ON SITE AND IS A FINE EXAMPLE OF A BRICK AGING WAREHOUSE. THE WALLS ARE CONSTRUCTED OF BRICK AND MORTAR, AND THE INTERIOR HOUSES AN EXTENSIVE RICK SYSTEM MADE OF HEART PINE, A FAVORED WOOD FOR CONSTRUCTION DURING THAT TIME PERIOD. THIS IS THE ONLY WAREHOUSE ON SITE WITH THE ORIGINAL RICKING SYSTEM OF "48'S," BUILT TO HOLD 48 GALLONS OF WHISKEY PER BARREL. THE STANDARD SIZE TODAY OF A BARREL OF BOURBON WHISKEY IS 53 GALLONS. IN 2006 THIS WAREHOUSE WAS SIGNIFICANTLY DAMAGED BY A TORNADO, BUT INTENT ON PRESERVING THE HISTORICAL INTEGRITY OF THIS PROPERTY, THE DISTILLERY RESTORED THIS WAREHOUSE IN 2011.

A view through the empty barrel racks of Warehouse "B."

Tools of the trade used for replacing damaged or leaking barrel heads.

An interior view of one of the windows in Warehouse "B."

The process of replacing a damaged barrel head.

A seldom seen view of bourbon aging in the barrel. The lid had been damaged and was being replaced. A beautiful sight and a wonderful aroma!

The ribbon-cutting celebration for Warehouse "X." This state-of-the-art facility represents an investment in the ongoing research on how the environment influences the quality of the whiskey.

Pictured left to right are: Frankfort Mayor Bill May; Kentucky State Auditor Adam Edelin; Buffalo Trace President and CEO Mark Brown; former Governor of Kentucky Julian Carroll; Kentucky State Representative Derrick Graham, and Kentucky State Senator Damon Thayer.

Examples of how gravity and the "tracks" embedded in the pavement are used to move the barrels from the dock to the barrel dumping process. Another view on the following spread.

Another view of the "tracks" embedded in the pavement.

Left: Shutters and bars on one of the windows of the Regauge Building, a reminder of the days when tax revenuers were on site and windows had bars and doors had padlocks. The revenuer had one key and the warehouse manager had the other. Each could not open the door without the other one being present.

After years of maturing in the warehouse, this bourbon is ready to make its appearance.

Barrels lined up in the Regauge Building awaiting their turn to be emptied.

Employee Lloyd Riley prepares barrels to be emptied by removing the bungs from them just prior to dumping.

These silver hooks are called "breathers." Occasionally, after the bung is drilled, pieces of the bung may remain and block the bung hole. The breather is inserted, allowing air to enter the barrel in order to completely empty it.

Opposite: A close up view of a bung being drilled from the barrel.

CAUTION
MEN
WORKING

Warehouse "H," the only metal clad warehouse on-site, was built in 1935 by Colonel Blanton. The majority of the Blanton's Bourbon is aged in this warehouse because it was his favorite aging warehouse. It is also used as a tasting warehouse for people wanting to purchase an entire barrel of bourbon.

While many modern devices are used to ensure the quality of the products, there is none more sophisticated than that of the human nose and pallet.

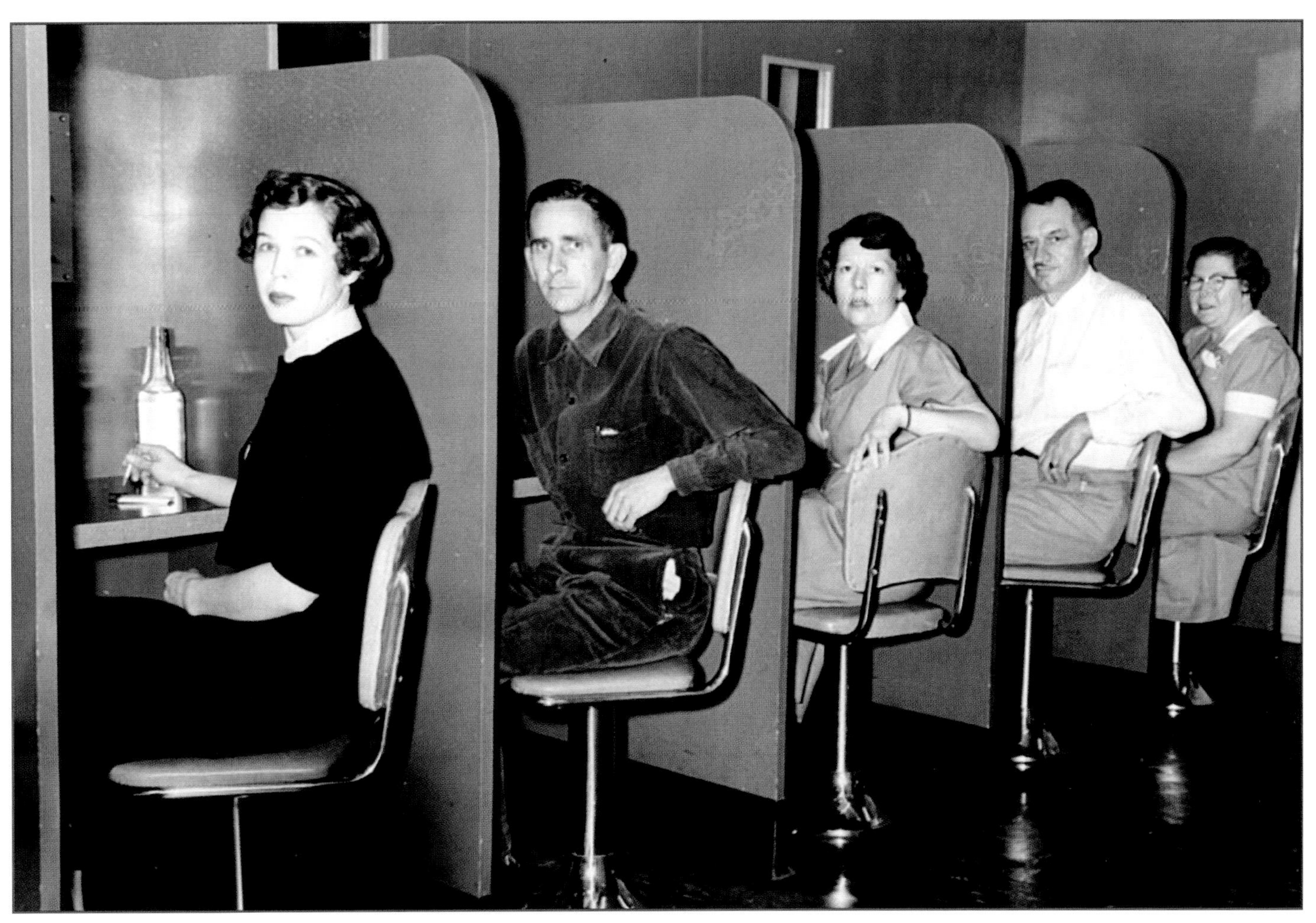

The Distillery uses a panel of employees as taste testers to ensure each profile of its bourbons remains the same. This 1955 photo shows the panel "hard" at work. Photo courtesy of Buffalo Trace Distillery Archive.

This "Tasting Table" has a top that turns so that the tester can sit in one spot and spin the samples to them. Visible in the background are the same Tasting Booths pictured above.

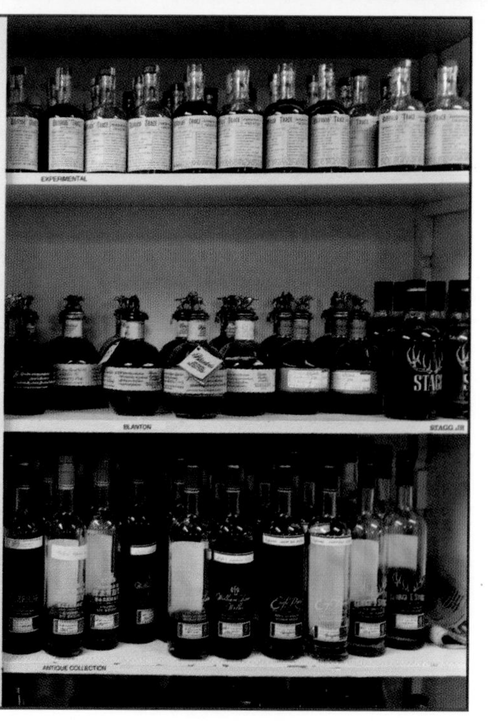

Drew Mayville, Master Blender and Director of Quality, takes time out of his busy schedule to give us a lesson on the fine art of taste testing.

The "Percolator," located in the Reguage Building, is one of several steps in the filtering process.

The Weller Bottling Hall is where individual barrels that have been purchased are bottled.

Bottles from individually purchased barrels bearing their personalized label.

Bottles of Buffalo Trace Bourbon in various stages of being filled.

750ml
RRR
750ml
STAND STRONG
BUFFALO TRACE
BUFFALO TRACE
STAND STRONG

BT 1.75
BUFFALO TRACE
KENTUCKY
STRAIGHT BOURBON
WHISKEY

BUFFALO TRACE
KENTUCKY
STRAIGHT BOURBON
WHISKEY

BUFFALO
KENT
STRAIGHT
WHIS

BUFFALO TRACE
KENTUCKY
STRAIGHT BOURBON
WHISKEY

Bottles of Eagle Rare Bourbon during the filling process.

Eagle Rare

Various ages of Pappy Van Winkle filled and on their way.

Pappy Van Winkle's
Family Reserve
Kentucky Straight Bourbon Whiskey
15 years old
750 ml. • Alc. 53.5% / Vol. (107 proof)
Bottled by Old Rip Van Winkle Distillery • Frankfort, Kentucky

Bottles of Blanton's, the world's first single-barrel bourbon, developed in 1984 by then Master Distiller Elmer T. Lee.

Blanton's
THE ORIGINAL
SINGLE BARREL
BOURBON WHISKEY
dumped on 9-23-13
from Barrel No. 72
Warehouse H
on Rick No. 40
filtered and bottled by hand at 93 Proof.
70cl e

Examples of the care and attention each bottle is given during the process.

Blanton's
THE ORIGINAL
SINGLE BARREL
BOURBON WHISKEY

Photographs showing the hand-crafting that goes into each Buffalo Trace brand.

BUFFALO TRACE
KENTUCKY
STRAIGHT BOURBON
WHISKEY

Blanton's
THE ORIGINAL
SINGLE BARREL
BOURBON WHISKEY

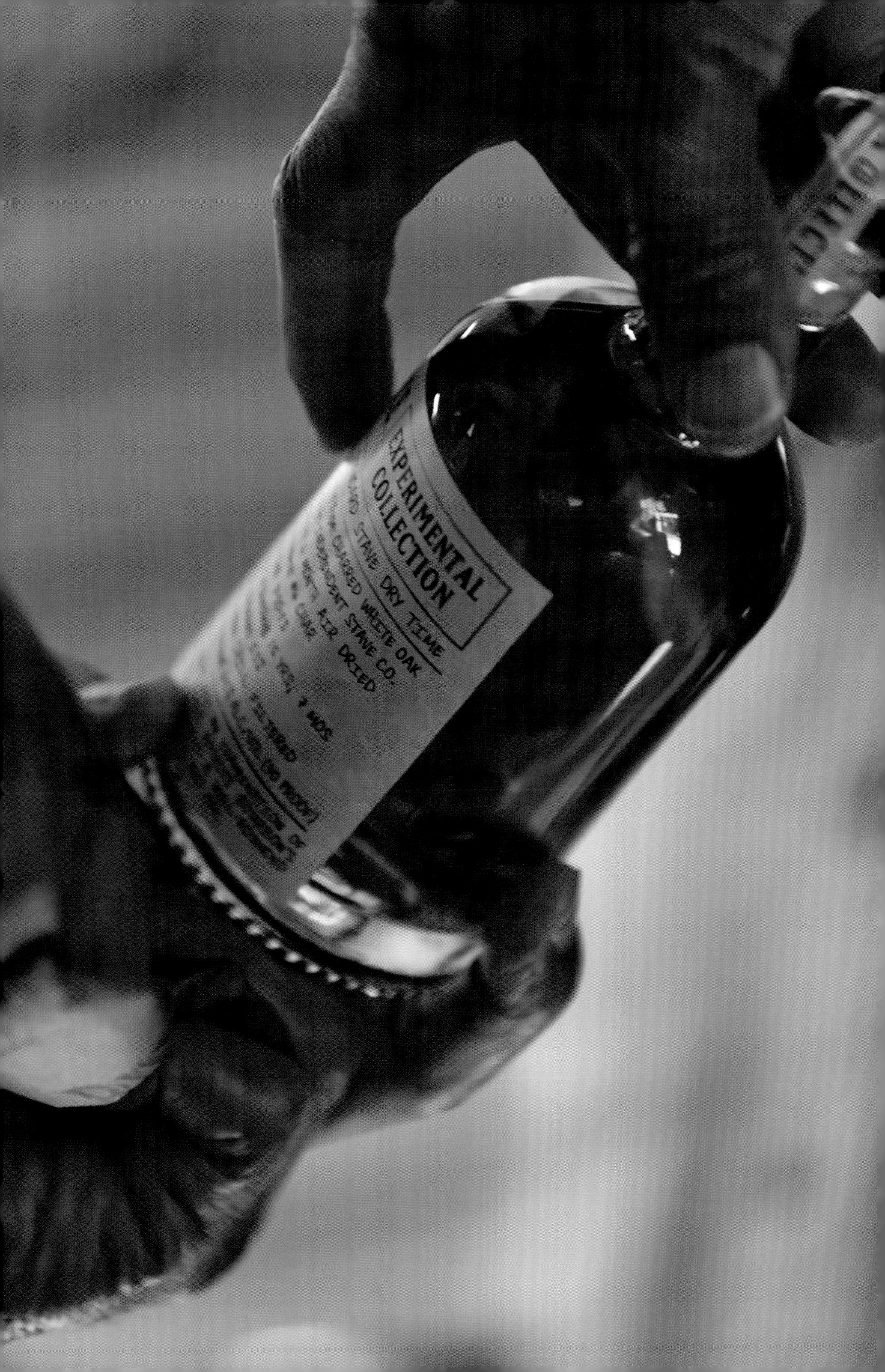
EXPERIMENTAL
COLLECTION
STAVE DRY TIME
CHARRED WHITE OAK
STAVE CO.
AIR DRIED
YRS, 7 MOS
FILTERED
(90 PROOF)

RRR 750
R 750ml
RRR 7
750ml
STAND STRONG
BUFFALO TRACE
KENTUCKY STRAIGHT BOURBON WHISKEY
45% Alc/Vol
TRACE
TRACE
BUFFALO TRACE
KENTUCKY
BUFFALO

In 1986, Buffalo Trace honored Master Distiller Emeritus Elmer T. Lee with his very own brand of single barrel bourbon bearing his name.

When Elmer T. Lee retired in 1985, he donated his barrel hammer to Buffalo Trace. After graduating with honors from the University of Kentucky with an engineering degree in 1949, he went to work in September in the engineering department of the George T. Stagg Distillery. In 1966 Elmer was promoted to Plant Superintendent and in 1969 became Plant Manager and Master Distiller.

Master Distiller Emeritus and bourbon icon Elmer T. Lee (1919-2013) While Elmer made many contributions and innovations to the bourbon industry, his most notable came in 1984 with his introduction of Blanton's, the world's first single barrel bourbon. Ironically, he honored the man, Colonel Albert B. Blanton, who was so skeptical of hiring him initially. Elmer continued to serve as an ambassador for the industry after his "retirement" in 1985. His work and legacy have been honored by his induction into the Kentucky Bourbon Hall of Fame in 2001, his receipt of the "Lifetime Achievement Award" from both Whisky Advocate in 2002 and Whisky Magazine in 2012 and his induction into Whisky Magazine's Hall of Fame.

A collection of items on display in the Gift Shop honoring Elmer T. Lee include his signature hat, his barrel hammer and a special edition of his namesake bourbon commemorating his 90th birthday.

DISTILLER

The Distillery was originally located on the banks of the Kentucky River for its means of transporting both raw materials and finished goods. The water used in producing its bourbon comes from Cove Spring which the Distillery purchased in 1892.

Being located on a river does have its dangers. This 1962 photo shows workers continuing to do their job with the river at a height of 40.5 feet.

Right: Plaques on the wall of the Dry House mark the levels of the numerous floods the Distillery has endured over the years.

December 1978, 48.5'

April 1937, 47.6'

The stream flowing through the grounds, fed by the Kentucky River, adds to both the peace and beauty of the grounds.

The Chill House is reflected in the stream flowing adjacent to the Elmer T. Lee Clubhouse.

More views of the stream flowing through the grounds.

A statue of Colonel Albert B. Blanton keeps a watchful eye over the operations. Blanton was President from 1921 to 1952. He led the company through the Prohibition days when the Distillery was one of a handful permitted by the government to produce "medicinal whiskey." There were only six active distilleries remaining at the end of Prohibition in 1933.

A panoramic view of the Distillery grounds from Stony Point Mansion on the left to Warehouse "I" on the right.

BUFFALO TRACE
DISTILLERY

Views of the landmark Buffalo Trace Water Tower and famous Barrel Crossing.

Sunset gives a golden glow to the iconic Water Tower at the Distillery.

At the conclusion of a tour, visitors are treated to a lesson in bourbon tasting. Pictured are third generation employee Freddie Johnson (Left), Jimmy Carey (Top Right) and Jim True (Bottom Right).

WHISKY

Buffalo Trace Distillery, in conjunction with the Frankfort Parks and Recreation Department holds an Easter egg hunt that is free and open to the public. 2014 was the 15th annual event and was attended by 1,937 "hunters."

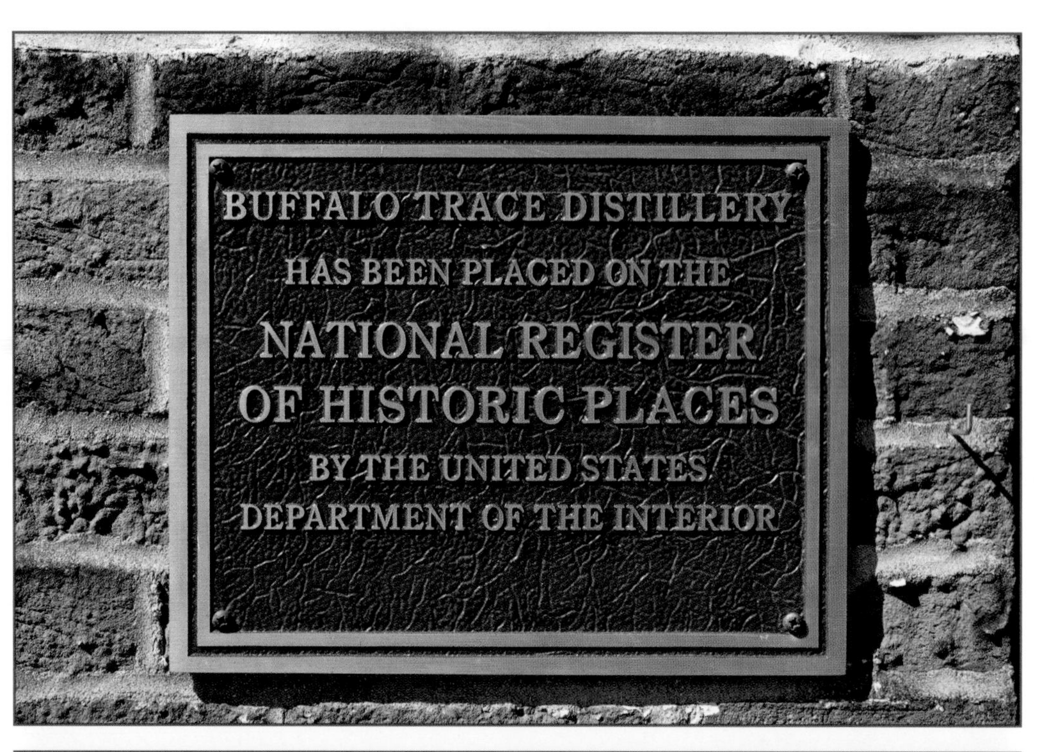

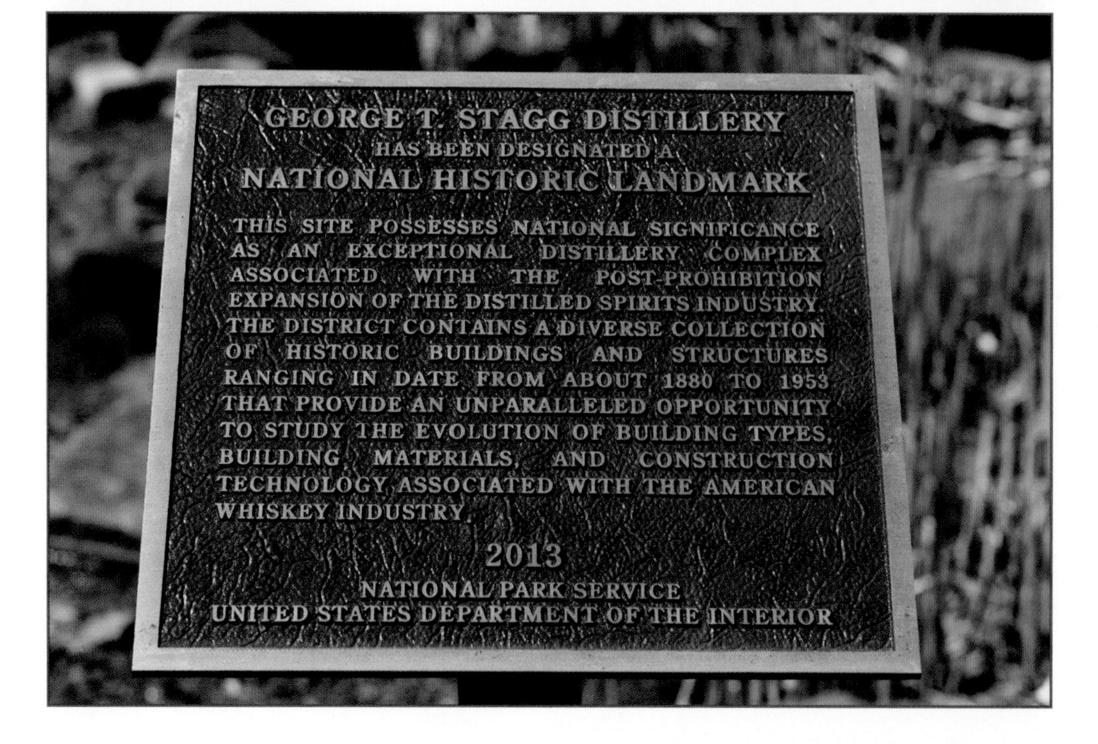

After a 15 year effort, Buffalo Trace Distillery was designated a National Historic Landmark in 2013. Pictured center left is Mark Brown, Buffalo Trace President and CEO (Left) and NHL Consultant Carolyn Brookes (Right).

BUFFALO TRACE DISTILLERY
Est.
1787
NATIONAL HISTORIC LANDMARK

A296916

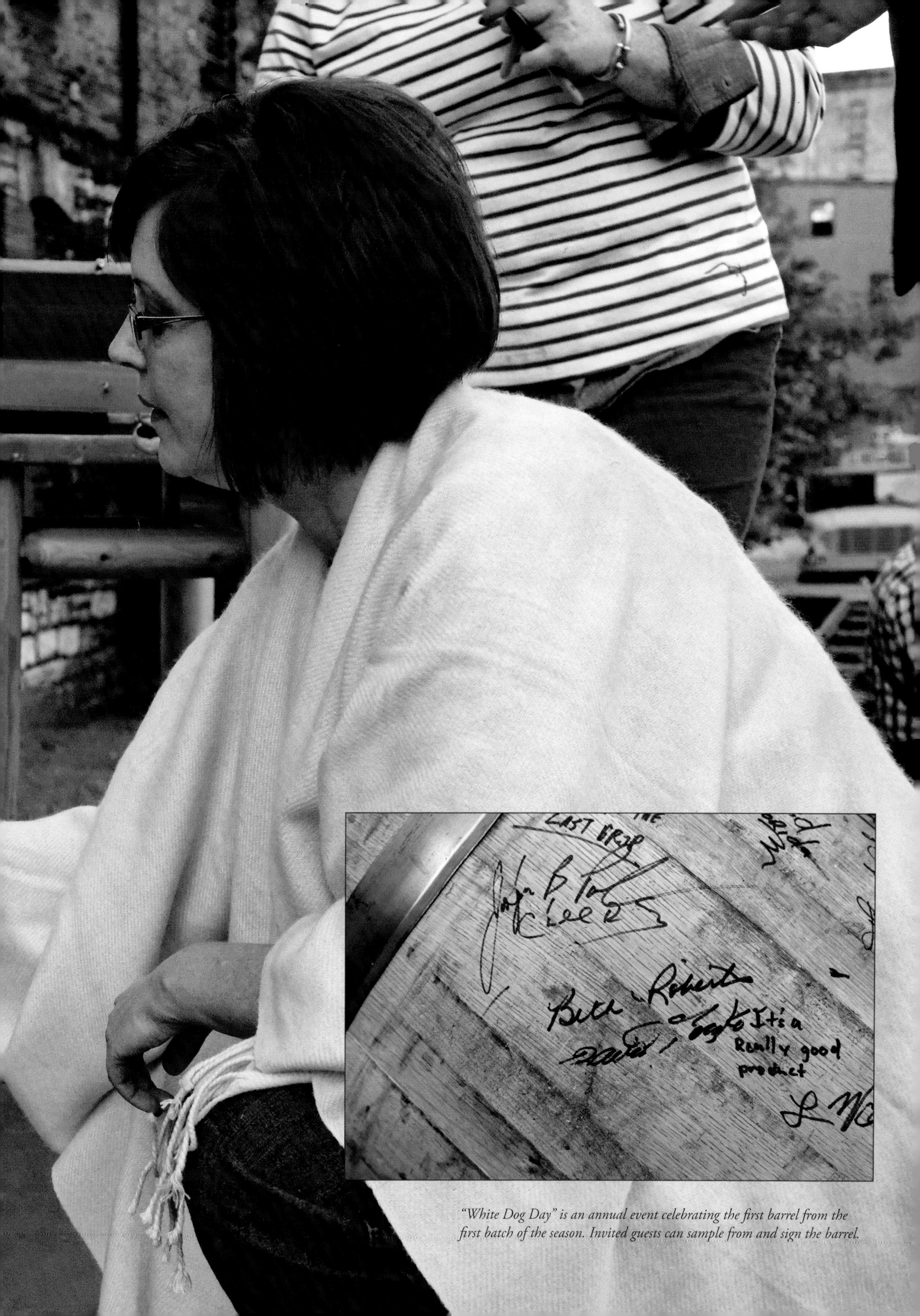

"White Dog Day" is an annual event celebrating the first barrel from the first batch of the season. Invited guests can sample from and sign the barrel.

A "Thief" is used to draw off the White Dog for sampling.

A Barrel Hammer is used to pound on alternating sides of the bung to work it from the barrel.

Featured on the television show Ghost Hunters®, Buffalo Trace has long been rumored to be haunted, most notably by Colonel Blanton himself. Here, the exterior of Warehouse "C" takes on an eerie atmosphere at night.

Visitors on the Ghost Tour are lead through the warehouses, buildings and Stony Point Mansion.

THE
6,000,000th
OF
BUFFALO
Kentucky Straight
Filled May 14,

May 14, 2008 marked the filling of the 6 millionth barrel at Buffalo Trace Distillery since the end of Prohibition. The ceremony to roll it into Warehouse "V," the only single barrel bonded warehouse in the world, featured notables, left to right, Elmer T. Lee, Master Distiller Emeritus, Gary Gayheart, former Master Distiller, Harlen Wheatley, current Master Distiller, Jimmy Johnson, former Warehouse Supervisor, and Mark Brown, President and CEO. Johnson, an employee from 1936 to 1978, rolled-in each of the millionth barrels until his passing in 2011. Inset photo courtesy of Buffalo Trace Distillery Archives.

"The Lighting of the Trace" is an annual Christmas event at Buffalo Trace. The Distillery is decorated with more than 110,000 lights and the public is invited to drive through the Distillery grounds. Last year's 15th annual event saw 4,000 carloads of visitors.

HAPPY
HOLIDAYS

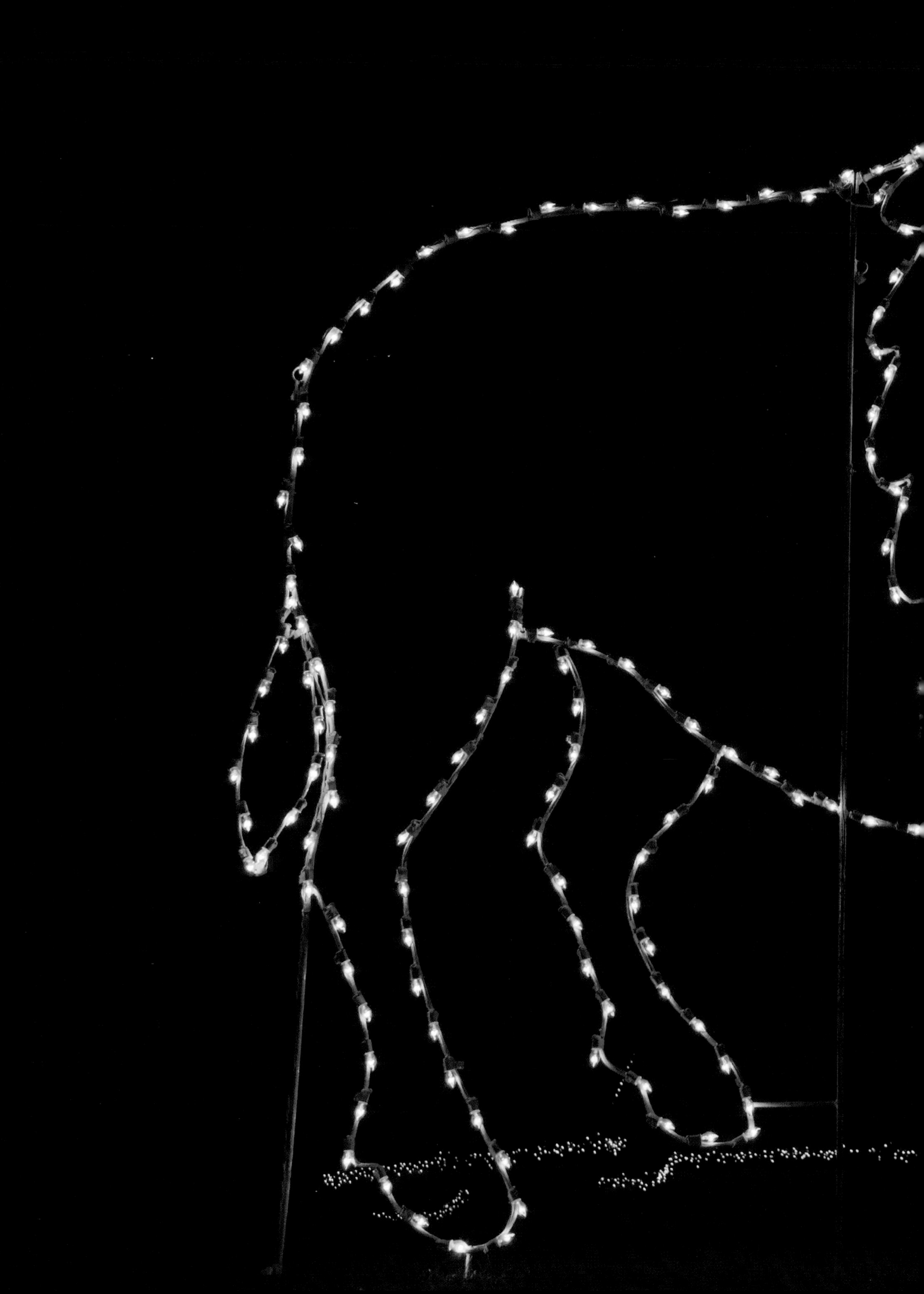

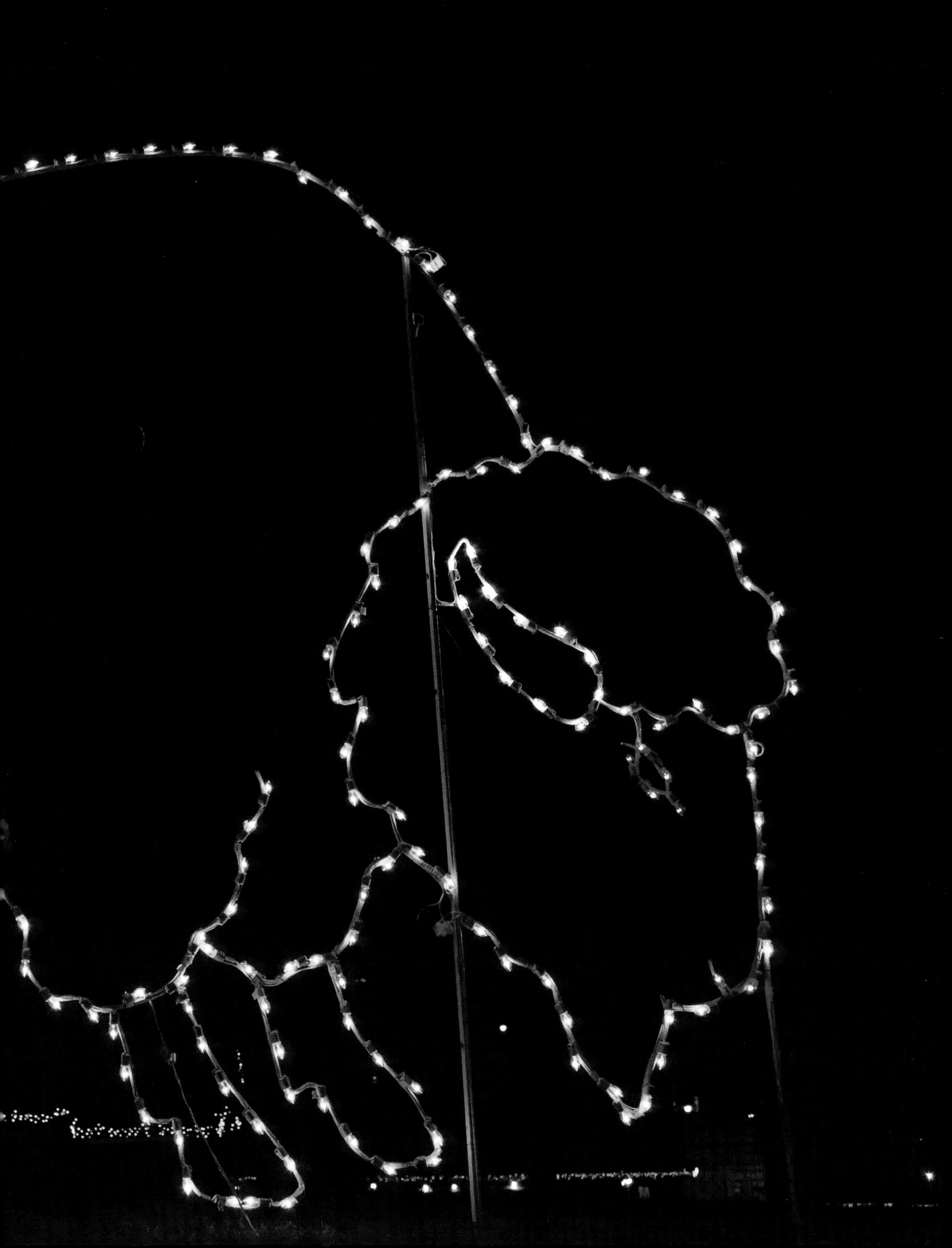

2005 saw the arrival of Buffalo Trace's current Master Distiller Harlen Wheatley. While some of his responsibilities include being an ambassador for the bourbon industry, the majority of his time is spent ensuring the quality of the products produced from start to finish.

Above and above right: Wheatley's "Baby," the Micro Still, is used for experimental batches and is where the famed Experimental Collection whiskies are produced.

Wheatley pulls a sample from the Try Box for evaluation.

FIRE HOUSE
DO NOT
BLOCK THIS DRIVE
FIRE DEPT.
BUFFALO TRACE
DISTILLERY
Firehouse Café

You can work up an appetite touring the Distillery. The original on-site firehouse, built in the 1930's, was "re-purposed" in 2010 as a café and is open seasonally from April through October.

Pictured here are the offices of the E. H. Taylor, Jr. Co. and the George T. Stagg Co. Both companies operated side by side with one making whiskey and the other distributing whiskey. George T. Stagg Co. eventually took ownership of the distillery that was owned by E. H. Taylor, Jr. Stagg changed the name of the distillery from O.F.C. to George T. Stagg Distillery. The distillery was renamed again in 1999 to the Buffalo Trace Distillery. c. Late 1800's Photo courtesy of Buffalo Trace Distillery Archives.

A group of distillery workers at the E. H. Taylor Distillery pose in front of the house known as "Riverside." This structure is located on the distillery grounds and is the oldest standing structure in Franklin County. c. 1870's Photo courtesy of Buffalo Trace Distillery Archives.

Barrels being moved from the old O.F.C. Cistern Room (today's gift shop on Left) to Warehouse "C," built in 1885, for aging. c. 1910 Photo courtesy of Buffalo Trace Distillery Archives.

The Frankfort and Cincinnati Railroad began serving the distillery in 1933. The train is running alongside Building 3 with Warehouse "C" in the background. c. 1930's Photo courtesy of Buffalo Trace Distillery Archives.

Index

Numerical

6,000,000th barrel 33, 193

A

Angel's Share 83

B

barrel/barrels 73, 74, 75, 76, 77, 79, 80, 81, 82, 83, 84, 85, 88, 89, 90, 91, 94, 101, 109, 112, 117, 118, 134, 135, 185, 207
Barrel Crossing 24, 174
barrel dumping process 112
barrel hammer 156, 158, 187
barrel head 106, 108
barrel hoist 89
Barrel lids 73
barrel wagon 93
Big Bone Lick 8, 9
Blanton, Albert B. 9, 10, 19, 38, 129, 157, 169, 188
Blanton's 146, 157
boiler 52
Bourbon Barrel Tour 80
Branding Shed 80, 81
breathers 118
Brookes, Carolyn 182
Brown, Mark 7, 42, 111, 182, 193
Buffalo Trace Water Tower 174
Building 3 207
bung/bung hole 85, 118, 187

C

Carey, Jimmy 176
Carroll, Julian 111
charred/charring (barrel) 73, 76, 78
charred white oak 77
Chill House 164
Chill Room 34, 36
Cistern Building 82, 90
Clark County, Kentucky 8
Column Still 66
Corn oil 59
Cove Spring Branch 9, 160

D

Dickel Building 45
Dry House 160

E

Eagle Rare Bourbon 142
Easter egg hunt 179
Edclin, Adam 111
E.H. Taylor Distillery 206
E.H. Taylor Room 66
elevator system 94
Elmer T. Lee Clubhouse 38, 42, 164
Eskippakithiki (Shawnee settlement) 8
Estes, Johnny 68
Experimental Collection 200

F

fermentation 59
fermentation tank 54, 62
Fermenter Building 32
Finley, John 8
firehouse 205
Firehouse Café 44, 205
Frankfort and Cincinnati Railroad 207
Frankfort, Kentucky 8
Frankfort Parks and Recreation Department 179
Franklin County, Kentucky 206

G

Gayheart, Gary 193
Georgetown, Kentucky 8
George T. Stagg Distillery 10, 156, 206
Ghost Hunters 188
Ghost Tour 190
Gift Shop 45, 158, 207
Graham, Derrick 111
grain 46, 49

H

hand-crafting 150
Hood, John 49

I

Independent Stave Company 73
Indian Old Fields 8, 9

J

Johnson, Freddie 176
Johnson, Jimmy 193

K

Kentucky Bourbon Hall of Fame 157
Kentucky River 8, 160, 162

L

Lee, Elmer T. 7, 9, 38, 146, 155, 156, 157, 158, 193
Leestown, Kentucky 8, 9
Lexington, Kentucky 8
Lighting of the Trace 194

M

mash 68
May, Bill 111
Mayville, Drew 132
Micro Still 66, 200

O

O.F.C. Cistern Room 207
OFC Distillery 10, 11

P

Pappy Van Winkle 144
Percolator 134
Pot Still 66
Preske, Amy 7
Prohibition 169, 193

R

Regauge Building 116, 117, 134
Riley, Lloyd 118
Riverside 206

S

Single Oak Project 80
Stagg, George T. 9, 10, 11
Stamping Ground, Kentucky 8
Still House 36
Stony Point Mansion 19, 172, 190

T

taste testers 131
Tasting Booths 131
Tasting Table 131
Taylor, Jr., E.H. 9, 10, 11, 32, 66, 206
Taylor, Richard 9
Thayer, Damon 111
Thief 186
tracks 24, 112, 115
Traveling Scale Hopper 49
True, Jim 176
Try Box 64, 200

W

Warehouse "B" 104, 106, 107
Warehouse "C" 22, 98, 101, 188, 207
Warehouse "H" 129
Warehouse "I" 172
Warehouse "V" 33, 193
Warehouse "X" 110
Weller Bottling Hall 134
Wheatley, Harlen 193, 198, 200
White Dog 64, 83, 84, 186
White Dog Day 185
Wishing Well 41